12th International Conference on Damage Assessment of Structures (DAMAS 2017)

Journal of Physics: Conference Series Volume 842

Kitakyushu, Japan
10 – 12 July 2017

ISBN: 978-1-5108-4186-4
ISSN: 1742-6588

Printed from e-media with permission by:

Curran Associates, Inc.
57 Morehouse Lane
Red Hook, NY 12571

Some format issues inherent in the e-media version may also appear in this print version.

Printed by Curran Associates, Inc. (2017)

For permission requests, please contact the Institute of Physics
at the address below.

Institute of Physics
Dirac House, Temple Back
Bristol BS1 6BE UK

Phone: 44 1 17 929 7481
Fax: 44 1 17 920 0979

techtracking@iop.org

Additional copies of this publication are available from:

Curran Associates, Inc.
57 Morehouse Lane
Red Hook, NY 12571 USA
Phone: 845-758-0400
Fax: 845-758-2633
Email: curran@proceedings.com
Web: www.proceedings.com

Table of contents

Volume 842

12th International Conference on Damage Assessment of Structures

10–12 July 2017, Kitakyushu, Japan

Accepted papers received: 9 May 2017
Published online: 2 June 2017

Preface

011001
OPEN ACCESS
12th International Conference on Damage Assessment of Structures

011002
OPEN ACCESS
Peer review statement

Structural Health and Condition Monitoring

012001
OPEN ACCESS
A Robust Poincare Maps Method for Damage Detection based on Single Type of Measurement

Zhi-Bo Yang, Ya-Nan Wang, Hao Zuo, Xing-Wu Zhang, Yong Xie and Xue-Feng Chen.....1

012002
OPEN ACCESS
A new fault detection strategy using the enhancement ensemble empirical mode decomposition

Jiawei Xiang and Yongteng Zhong.....9

012003
OPEN ACCESS
A study on the inclusion sizing using immersion ultrasonic C-scan imaging

D Chen, H F Xiao, M Li and J W Xu.....18

012004
OPEN ACCESS
The modal surface interpolation method for damage localization

Maria Pina Limongelli.....25

012005
OPEN ACCESS
Effect of vitro preservation on mechanical properties of brain tissue

Wei Zhang, Yi-fan Liu, Li-fu Liu, Ying Niu, Jian-li Ma and Cheng-wei Wu.....33

012006
OPEN ACCESS
Is it worth changing pattern recognition methods for structural health monitoring?

L A Bull, K Worden, E J Cross and N Dervilis.....39

012007
OPEN ACCESS
Damage localization and quantification in simply supported beams using static test data

Q Ma and M Solís.....50

012008
OPEN ACCESS
Effect of test frequency on fatigue strength of AZ31 alloy

G J Lu, Y Koyanagi, M Tsushida, H Kitahara and S Ando.....58

012009
OPEN ACCESS
A vibration-based health monitoring program for a large and seismically vulnerable masonry dome

M L Pecorelli, R Ceravolo, G De Lucia and R Epicoco.....64

012010
OPEN ACCESS
Study on the evaluation of erosion damage by using laser ultrasonic integrated with a wavelet analysis technique

T Wan, T Naoe, T Wakui, M Futakawa, H Obayashi and T Sasa.....75

012011
OPEN ACCESS
Application of vision measurements for modal analysis of wires for the purpose of overhead transmission lines monitoring

Krzysztof Mendrok, Ziemowit Dworakowski, Krzysztof Holak and Piotr Kohut.....85

012012
OPEN ACCESS
Data driven innovations in structural health monitoring

M J Rosales and R Liyanapathirana.....93

012013
OPEN ACCESS
A new frequency matching technique for FRF-based model updating

Xiuming Yang, Xinglin Guo, Huajiang Ouyang and Dongsheng Li.....104

012014
OPEN ACCESS
Application of proper orthogonal decomposition and radial basis functions for crack size estimation using particle swarm optimization

B Benaissa, M Köppen, M Abdel Wahab and S Khatir.....114

012015
OPEN ACCESS
Influence of Torsional Excitation on Dynamic Responses of Rotors with a Breathing Slant Crack

Zhiwen Lu, Chunrong Hua, Dawei Dong, Bing Yan and Kang Fan.....125

012016
OPEN ACCESS
Review on structural damage assessment via transmissibility with vibration based measurements

Yun-Lai Zhou, Cao Hongyou, Ni Zhen and Magd Abdel Wahab.....136

012017
OPEN ACCESS
Damages detection in cylindrical metallic specimens by means of statistical baseline models and updated daily temperature profiles

Rodolfo Villamizar-Mejia, Luis-Eduardo Mujica-Delgado, Magda-Liliana Ruiz-Ordóñez, Jhonatan Camacho-Navarro and Gustavo Moreno-Beltrán.....142

012018
OPEN ACCESS
Structural damage continuous monitoring by using a data driven approach based on principal component analysis and cross-correlation analysis

Jhonatan Camacho-Navarro, Magda Ruiz, Rodolfo Villamizar, Luis Mujica, Gustavo Moreno-Beltrán and Jabid Quiroga.....151

012019
OPEN ACCESS
Ensemble learning as approach for pipeline condition assessment

Jhonatan Camacho-Navarro, Magda Ruiz, Rodolfo Villamizar, Luis Mujica and Gustavo Moreno-Beltrán.....160

012020
OPEN ACCESS
Structural system identification by utilizing transmissibility coherence from limited measured data

Yun-Lai Zhou, Cao Hongyou, Ni Zhen, Li Zhang and Magd Abdel Wahab.....170

012021
OPEN ACCESS
Identification of Bouc-Wen hysteretic parameters based on enhanced response sensitivity approach

Li Wang and Zhong-Rong Lu.....177

012022
OPEN ACCESS
Multi-damage identification based on joint approximate diagonalisation and robust distance measure

S Cao and H Ouyang.....185

012023
OPEN ACCESS
Structure damage identification under ambient excitation based on wavelet packet analysis

Qingzhou Li and Dongsheng Li.....193

012024
OPEN ACCESS
Research on the spatial analysis method of seismic hazard for island

Jing Jia, Jitong Jiang, Qiuhong Zheng and Huiying Gao.....205

012025
OPEN ACCESS
Assessment of vocal cord nodules: a case study in speech processing by using Hilbert-Huang Transform

M Civera, C M Filosi, N M Pugno, M Silvestrini, C Surace and K Worden.....217

012026
OPEN ACCESS
Minimum constitutive relation error based static identification of beams using force method

Jia Guo and Izuru Takewaki.....234

012027
OPEN ACCESS
A proposal application based on strain energy for damage detection and quantification of beam composite structure using vibration data

S Tiachacht, A Bouazzouni, S Khatir, M A behtani, Y-L Zhou and M Abdel Wahab.....244

012028
OPEN ACCESS
Damage localization and quantification of composite stratified beam Structures using residual force method

A Behtani, A Bouazzouni, S Khatir, S Tiachacht, Y-L Zhou and M Abdel Wahab.....256

012029
OPEN ACCESS
Sub-surface defects detection of by using active thermography and advanced image edge detection

Peter W. Tse and Gaochao Wang.....266

012030
OPEN ACCESS
On the physical significance of the Effective Independence method for sensor placement

Yaoguang Jiang, Dongsheng Li and Gangbing Song.....275

Damage in Civil Engineering

012031
OPEN ACCESS
Condition Assessment of Corrosion-damaged Bridge Girders Strengthened with Post-tensioned Composite Strips

Yail J. Kim, Jae-Yoon Kang, Jong-Sup Park and Woo-Tai Jung.....286

012032
OPEN ACCESS
Damage Detection Sensitivity of a Vehicle-based Bridge Health Monitoring System

Ayaho Miyamoto, Akito Yabe and Válter J. G. Lúcio.....293

012033
OPEN ACCESS
Damage detection in multi-span beams based on the analysis of frequency changes

G R Gillich, J L Ntakpe, M Abdel Wahab, Z I Praisach and M C Mimis.....305

012034
OPEN ACCESS
Damage severity estimation from the global stiffness decrease

C Nitescu, G R Gillich, M Abdel Wahab, T Manescu and Z I Korka.....317

012035
OPEN ACCESS
Research of Ancient Architectures in Jin-Fen Area Based on GIS&BIM Technology

Jing Jia, Qiuhong Zheng, Huiying Gao and Hai Sun.....328

012036
OPEN ACCESS
Identification of Delamination in Concrete Slabs by SIBIE Procedure

M. Yamada, Y. Yagi and M. Ohtsu.....337

012037
OPEN ACCESS
Fatigue damage assessment of electric roads based on probabilistic load models

R Ceravolo, G Miraglia and C Surace.....343

012038
OPEN ACCESS
Energy dissipation and absorption capacity influence on experimental modal parameters of a PC girder

O S Luna Vera, C W Kim and Y Oshima.....354

012039
OPEN ACCESS
Developments in damage assessment by Marie Skłodowska-Curie TRUSS ITN project

A González.....363

012040
OPEN ACCESS
An anisotropic damage model for concrete structures under cyclic loading-uniaxial modeling

Yuchuan Long and Yuming He.....380

Damage in Machineries

012041
OPEN ACCESS
Fatigue Analysis of Proton Exchange Membrane Fuel Cell Stacks Based on Structural Stress Distribution

C.W. Wu, B. Liu, M.Y. Wei and L.F. Liu.....394

012042
OPEN ACCESS
Novel tool wear monitoring method in milling difficult-to-machine materials using cutting chip formation

P P Zhang, Y Guo and B Wang.....401

012043
OPEN ACCESS
Tuned Liquid Column Damper based Reduction of Dynamic Responses of Scaled Offshore Platforms in Different Ocean Wave Basins

Deirdre O'Donnell, Jimmy Murphy, Cian Desmond, Vesna Jaksic and Vikram Pakrashi.....409

012044
OPEN ACCESS
Potentials of Optical Damage Assessment Techniques in Automotive Crash-Concepts composed of FRP-Steel Hybrid Material Systems

M Dlugosch, B Spiegelhalter, T Soot, D Lukaszewicz, J Fritsch and S Hiermaier.....417

012045
OPEN ACCESS
Topic Correlation Analysis for Bearing Fault Diagnosis Under Variable Operating Conditions

Chao Chen, Fei Shen and Ruqiang Yan.....429

012046
OPEN ACCESS
Non-negative Matrix Factorization and Co-clustering: A Promising Tool for Multi-tasks Bearing Fault Diagnosis

Fei Shen, Chao Chen and Ruqiang Yan.....438

012047
OPEN ACCESS
A new damage diagnosis approach for NC machine tools based on hybrid Stationary subspace analysis

Chen Gao, Yuqing Zhou and Yan Ren.....448

012048
OPEN ACCESS
Study on quantitative risk assessment model of the third party damage for natural gas pipelines based on fuzzy comprehensive assessment

Zeyang Qiu, Wei Liang, Xue Wang, Yang Lin and Meng Zhang.....455

012049
OPEN ACCESS
Dependency of tensile strength of ductile cast iron on strain rate and temperature

Tomohiro Ikeda, Takuo Umetani, Nobuhiro Kai, Keisaku Ogi, Nao-Aki Noda and Yoshikazu Sano.....466

012050
OPEN ACCESS
Stress and deformation of ceramic rolls to produce high quality zinc coated steel sheet

Hiromasa Sakai, Nao-Aki Noda, Yoshikazu Sano, Yasushi Takase and Guowei Zhang.....473

012051
OPEN ACCESS
Coming out prevention by stopper for the shrink fitted sandwiched shaft from the ceramic sleeve

Guowei Zhang, Nao-Aki Noda, Yoshikazu Sano, Hiromasa Sakai and Kazuhiro Oda.....481

012052
OPEN ACCESS
Integrated condition monitoring of a fleet of offshore wind turbines with focus on acceleration streaming processing

Jan Helsen, Nicoletta Gioia, Cédric Peeters and Pieter-Jan Jordaens.....487

012053
OPEN ACCESS
Vibration-based angular speed estimation for multi-stage wind turbine gearboxes

Cédric Peeters, Quentin Leclère, Jérôme Antoni, Patrick Guillaume and Jan Helsen.....494

012054
OPEN ACCESS
Experimental investigations of contact-type damage nonlinearity

A. Klepka, K. Dziedziech, Ł. Pieczonka, M. Adamczyk and W.J. Staszewski.....502

012055
OPEN ACCESS
Fault Features Extraction and Identification based Rolling Bearing Fault Diagnosis

B Qin, G D SUN, L Y ZHANG, J G WANG and J HU.....511

012056
OPEN ACCESS
Study on observation condition and on simple evaluation method of TypeIV creep damage of Mod.9Cr-1Mo steel

F Kawashima, T Masumitsu, K Negi, K Watanabe and K Fujiwara.....524

012057
OPEN ACCESS
Fault Diagnosis for Centre Wear Fault of Roll Grinder Based on a Resonance Demodulation Scheme

Liming Wang, Yimin Shao, Lei Yin, Yilin Yuan and Jing Liu.....538

012058
OPEN ACCESS
An illustration of new methods in machine condition monitoring, Part I: stochastic resonance

K. Worden, I. Antoniadou, S. Marchesiello, C. Mba and L. Garibaldi.....545

012059
OPEN ACCESS
An Illustration of New Methods in Machine Condition Monitoring, Part II: Adaptive outlier detection

I. Antoniadou, K. Worden, S. Marchesiello, C. Mba and L. Garibaldi.....555

012060
OPEN ACCESS
A new state evaluation method of oil pump unit based on AHP and FCE

Yang Lin, Wei Liang, Zeyang Qiu, Meng Zhang and Wenqing Lu.....562

012061
OPEN ACCESS
Numerical Simulation of cracked orthotropic materials using extended isogeometric analysis

S H Habib, I Belaidi, S Khatir and M Abdel Wahab.....568

012062
OPEN ACCESS
Development of four-point bending fatigue test method using continuously hydrogen-charging pipe specimen

T Yoshimoto and T Matsuo.....579

012063
OPEN ACCESS
Failure analysis of blots for diesel engine intercooler

Ping Ren, Zongquan Li, Jiangfei Wu, Yibin Guo and Wanyou Li.....586

012064
OPEN ACCESS
Application of Risk-Based Inspection method for gas compressor station

Meng Zhang, Wei Liang, Zeyang Qiu and Yang Lin.....594

012065
OPEN ACCESS
A new testing method for investigating the shear-mode fatigue crack growth behavior in hydrogen environment

Y Akaki, T Matsuo, Y Nishimura, S Miyakawa and M Endo.....602

012066
OPEN ACCESS
Improvement of torsional fatigue limit and rendering surface defect harmless by shot peening for spring steel

K Takahashi, M Nakagawa, H Koike and H Okada.....608

012067
OPEN ACCESS
Fatigue limit prediction of ferritic-pearlitic ductile cast iron considering stress ratio and notch size

T Deguchi, H J Kim and T Ikeda.....617

012068
OPEN ACCESS
Development of testing machine for tunnel inspection using multi-rotor UAV

Tatsuya Iwamoto, Tomoya Enaka and Keijirou Tada.....625

012069
OPEN ACCESS
Estimation of dispersion curves by combining Effective Elastic Constants and SAFE Method: A case study in a plate under stress

J E Quiroga, L Mujica, R Villamizar, M Ruiz and J Camacho.....632

012070
OPEN ACCESS
Impacting load control of floating supported friction plate and its experimental verification

Keyan Ning, Yu Wang, Dingchuan Huang and Lei Yin.....642

012071
OPEN ACCESS
Research on quality evaluation and forecast control in assembly process for complex products

Jihong Pang, Xiaobo Xue, Yini Jin and Yongchao Li.....649

012072
OPEN ACCESS
Damage evolution in plates subjected to fatigue loading

J Toribio, J C Matos and B González.....655

012073
OPEN ACCESS
Adhesive properties and adhesive joints strength of graphite/epoxy composites

Anna Rudawska, Dana Stančeková, Nadezda Cubonova, Tetiana Vitenko, Miroslav Müller and Petr Valášek.....661

Damage in Composite Materials

012074
OPEN ACCESS
Experimental research and statistic analysis of polymer composite adhesive joints strength

Anna Rudawska, Izabela Miturska, Jakub Szabelski, Agnieszka Skoczylas, Paweł Droździel, Elżbieta Bociąga, Radovan Madleňák and Dariusz Kasperek.....670

012075
OPEN ACCESS
Impact localization on composite structures using time difference and MUSIC approach

Yongteng Zhong and Jiawei Xiang.....678

012076
OPEN ACCESS
Analysis of cracked RC beams under vibration

R Capozucca and E Magagnini.....684

012077
OPEN ACCESS
Micro CT Analysis of Dynamic Damage in Laminates: Impact vs. blast loading

Laurence A Coles, Anish Roy and Vadim V Silberschmidt.....692

012078
OPEN ACCESS
Evaluation of debonding strength of single lap joint by the intensity of singular stress field

Tatsujiro Miyazaki and Nao-Aki Noda.....700

012079
OPEN ACCESS
Detection of Fatigue Crack in Basalt FRP Laminate Composite Pipe using Electrical Potential Change Method

Wael A. Altabey and Mohammed Noori.....708

012080
OPEN ACCESS
A symplectic analytical singular element for the inclined crack terminating at the material interface of composite structures

XF Hu, QS Shen and WA Yao.....720

012081
OPEN ACCESS
Acoustic emission based damage localization in composites structures using Bayesian identification

A Kundu, M J Eaton, S Al-Jumali, S Sikdar and R Pullin.....726

012082
OPEN ACCESS
A Two-step Approach for Damage Identification in Axially Functionally Graded Beams

Z.R. Lu, R.Z. Yao, J.K. Liu and M. Huang.....741

012083
OPEN ACCESS
New damage-sensitive feature for structures with bolted joints

W F Luo and L Yu.....750

012084
OPEN ACCESS
Delamination detection in laminated composite using Virtual crack closure technique (VCCT) and modal flexibility based on dynamic analysis

S Khatir, A Behtani, S Tiachacht, A Bouazouni, M Abdel Wahab and Y-L Zhou.....759

012085
OPEN ACCESS
An isogeometric approach for size-dependent buckling analysis of FGM nanoplates

P Phung-Van, M Abdel-Wahab and H Nguyen-Xuan.....768

IOP Conf. Series: Journal of Physics: Conf. Series **842** (2017) 011001 doi:10.1088/1742-6596/842/1/011001

Proceedings of 12th International Conference on Damage Assessment of Structures DAMAS 2017 Kitakyushu, Japan, 10-12 July 2017

Chairman
Prof. Magd Abdel Wahab
Ghent University, Belgium

Co-chairman
Prof. Nao-Aki Noda
Kyushu Institute of Technology, Japan

International organisation committee
Prof. Huajiang Ouyang, University of Liverpool, UK
Prof. Wieslaw Ostachowicz, Polish Academy of Sciences, Poland
Prof. Spilios Fassois, Patras University, Greece

Local Organizing Committee

Prof. N-A Noda, Kyushu Institute of Technology, Japan
Prof. H Sakamoto, Doshisha University, Japan
Prof. M Endo, Fukuoka University, Japan
Prof. M Goto, Oita University, Japan
Prof. C Makabe, University of Ryukyu, Japan
Vice-Pres. E Yamaguchi, Kyushu Institute of Technology, Japan
Prof. K Kosa, Kyushu Institute of Technology, Japan
Pres. N Sugano, FUJI P.S Corporation, Japan
Prof. K Oda, Oita University, Japan
Prof. G Hotta, National Institute of Technology, Ariake College, Japan
Prof. T Yakushiji, National Institute of Technology, Oita College, Japan
Prof. T Takase, Nagasaki University, Japan
Prof. M Hibino, Kyushu Institute of Technology, Japan
Prof. T Miyazaki, University of Ryukyu, Japan
Prof. T-C Chiu, National Cheng Kung University, Taiwan
Prof. K Masuda, University of Toyama, Japan
Prof. T Hattori, Shizuoka Institute of Science and Technology, Japan
Prof. S Nishida, Saga University, Japan
Dr. Y Sano, Kyushu Institute of Technology, Japan
Dr. Y Takase, Kyushu Institute of Technology, Japan

International Scientific Committee

F Aymerich, University of Cagliari, Italy
S K Bhalla, Indian Institute of Techonology Delhi, India
F Casciati, University of Pavia, Italy
R Ceravolo, Politechnico di Torino, Italy
F Chu, Tsinghua University, China
G De Roeck, KU Leuven, Belgium
L Faravelli, University of Pavia, Italy
S Fassois, Patras University, Greece
M Friswell, University of Swansea, UK
G-R Gillich, Eftimie Murgu Univ Resita, Romania
J-C Golinval, University of Liege, Belgium
A Gonzalez, University College Dublin, Ireland
K Holford, University of Cardiff, UK
MP Limongelli, Politecnico di Milano, Italy
G Manson, University of Sheffield, UK
T McCarthy, University of Wollongong, Australia
N-A Noda, Kyushu Institute of Technology, Japan
K Oda, Oita University, Japan
P Omenzetter, University of Aberdeen, UK
W Ostachowicz, Polish Academy of Science, Poland
H Ouyang, University of Liverpool, UK
B Peeters, LMS International, Belgium
R Pullin, Cardiff University, UK
A Rudawska, Lublin University of Technology, Poland
V Silberschmidt, University of Loughborough, UK
J J Sinou, Ecole Centrale de Lyon, France
C Surace, Politecnico di Torino, Italy
K Worden, University of Sheffield, UK
Y-L Zhou, National University of Singapore, Singapore
W Zhu, University of Maryland, Baltimore, USA
X Zhu, University of Western Sidney, Australia

IOP Conf. Series: Journal of Physics: Conf. Series **842** (2017) 011001 doi:10.1088/1742-6596/842/1/011001

Editorial

This volume contains the proceedings of the 12th International Conference on Damage Assessment of Structures, DAMAS 2017, Kitakyushu, Japan, 10-12, July 2017. DAMAS has a long history of almost 22 years. The first DAMAS conference took place in 1995 (Pescara, Italy), followed by a biannual meeting in 1997 (Sheffield, UK), 1999 (Dublin, Ireland), 2001 (Cardiff, UK), 2003 (Southampton, UK), 2005 (Gdansk, Poland), 2007 (Torino, Italy), 2009 (Beijing, China), 2011 (Oxford, UK), 2013 (Dublin, Ireland) and 2015 (Ghent, Belgium). The twelfth edition of DAMAS conference series, DAMAS 2017, is hosted by Kyushu Institute of Technology, Japan, and is held at in the West Japan Industry and Trade Convention Association, located in the town centre of Kitakyushu city.

The conference is established as a major international forum for research topics relevant to damage assessment of engineering structures and systems including numerical simulations, signal processing of sensor measurements and theoretical techniques as well as experimental case studies. The presentations of DAMAS 2017 are divided into 4 main sessions, namely 1) Structural Health and Condition Monitoring, 2) Damage in Civil Engineering, 3) Damage in Machineries and 4) Damage in Composite Materials.

The organising committee is grateful to keynote speakers; Prof Kunitomo Sugiura, Department of Civil and Earth Resources Engineering, Kyoto University, Japan, for his keynote lecture entitled ‘Damage Assessment of Orthotropic Steel Bridge Decks‘, Prof Kikuo Kishimoto, Department of Transdisciplinary Science and Engineering, School of Environment and Society, Tokyo Institute of Technology, Japan, for his keynote lecture entitled ‘Interfacial Mechanics for Multi-Material Structures’ and Prof Stéphane Bordas, Faculté des Sciences, de la Technologie et de la Communication, Université du Luxembourg, for his keynote lecture entitled ‘Multi-scale fracture simulations: model and discretisation error control’.

Special thanks go to members of the Scientific Committee of DAMAS 2017 for reviewing the articles published in this volume and for judging their scientific merits. Based on the comments of reviewers and the scientific merits of the submitted manuscripts, the articles were accepted for publication in the conference proceedings and for presentation at the conference venue. The accepted papers are of a very high scientific quality and contribute to advancement of knowledge in all research topics relevant to DAMAS conference.

Finally, the organising committee would like to thank all authors, who have contributed to this volume and presented their research work at DAMAS 2017.

Professor Magd Abdel Wahab

Chairman of DAMAS 2017

Peer review statement

All papers published in this volume of *Journal of Physics: Conference Series* have been peer reviewed through processes administered by the proceedings Editors. Reviews were conducted by expert referees to the professional and scientific standards expected of a proceedings journal published by IOP Publishing.

A Robust Poincare Maps Method for Damage Detection based on Single Type of Measurement

Zhi-Bo Yang[1], Ya-Nan Wang[1], Hao Zuo[1], Xing-Wu Zhang[1], Yong Xie[2,3], Xue-Feng Chen[1]*

1 School of Mechanical Engineering, Xi'an Jiaotong University, 710049, Xi'an, P.R. China

2 School of Aerospace, Xi'an Jiaotong University, Xi'an 710049, P.R. China

3 The State Key Laboratory of Strength and Vibration for Mechanical Structures, P.R. China

E-mail: phdapple@mail.xjtu.edu.cn, wangyanan@stu.xjtu.edu.cn, zuo-hao123@sina.com, xwzhang@mail.xjtu.edu.cn yxie@mail.xjtu.edu.cn, chenxf@mail.xjtu.edu.cn

*chenxf@mail.xjtu.edu.cn (Xue-Feng Chen)

Abstract. The classical vibration-based structural health monitoring evaluates damage via the analysis of single type measurement. The relationships between different types of measurements, however, are usually neglected. To address this problem, vibration data is to be analyzed in state space in this paper based on the the Poincare map method. The Poincare map method is an effective approach in damage detection. This method, however, requires the dependent measurements for displacement, velocity and acceleration, respectively. Based on the Fourier transform properties, a robust Poincare map method for damage detection is proposed based on the single type of measurement. Numerical and experimental verifications are employed to verify the effectiveness of the developed method. Compared with the classical numerical differential method, the present method is more accurate and robust to noise.

1. Introduction

Vibration-based damage assessment is one of the most crucial issues in structural health monitoring. Due to the on-line property of vibration measurement, the corresponding technologies are widely used in many engineering fields [1-4]. During the past decade, many efficient damage detection or assessment methods are proposed based on different measurements or evaluations. Zhou et al. [5] proposed a novel damage detection method based on the transmissibility. Based on their approach, the damage in early stage can be identified rapidly. To reveal the damage effect on structure, Khatir et al. [6, 7] investigated the dynamic responses of structures with single- or multiple-damage. Yang et al. [8, 9] investigated the entropy and fractural dimension methods for damage detection. These methods paved some novel damage detection paths in different aspects. However, only limited types of measurements are employed. The relationships between measurements are usually neglected.

Different from the common approaches, the Poincare map method provides a novel view in damage detection, that is, damage are evaluated in state space rather than based on only one measurement. In the vibration-based system, the state space can be spanned by displacement, velocity and acceleration.

IOP Conf. Series: Journal of Physics: Conf. Series **842** (2017) 012001 doi:10.1088/1742-6596/842/1/012001

It has been proved that the the difference before and after the damage appearance in the state space is an effective damage index. However, the displacement, velocity and acceleration in these works are measured independently in applications. Thus, the compensation for measurement may be expensive. The natural solution for this problem is to derive the velocity and acceleration from displacement, but the poor performance of numerical derivative make this way unsubstantial. Focusing on this problem, a robust Poincare map method based on single type of measurement is proposed in this paper to overcome the problems in numerical derivative-based methods. It will be proved in the following parts that the proposed method is effective in noisy conditions.

2. Damage Detection Method and Simple Verification

2.1. Model and state space

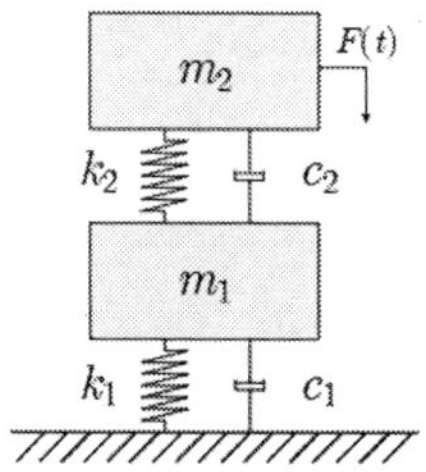

Figure 1. The 2-DOFs model.

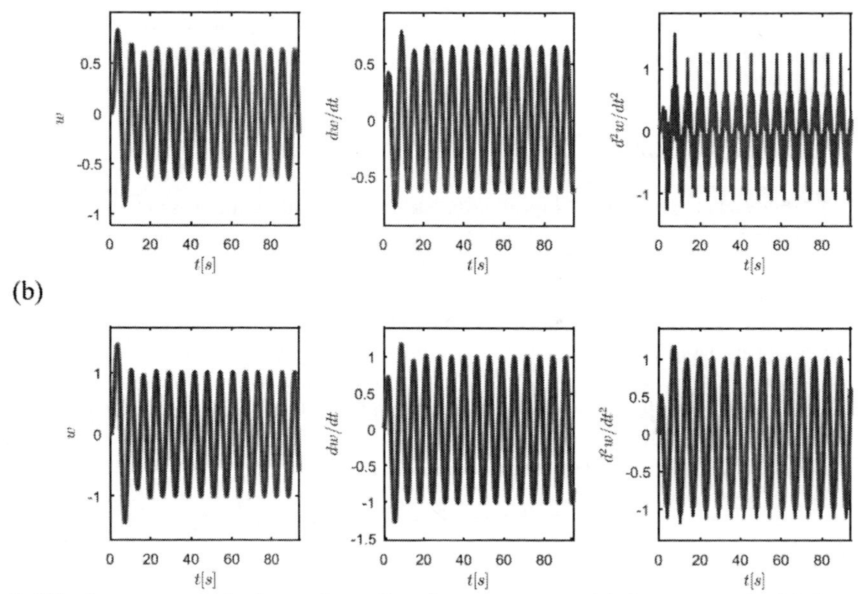

Figure 2. Displacement, velocity and acceleration responses: (a) from mass 1, (b) from mass 2. Red lines are for intact structures, gray lines are for damaged structures.

The theoretical development of this work is interpreted according to the analysis of a simple model, namely the 2-DOFs model as shown in Fig. 1. The dynamic equation of the 2-DOFs model is written by:

$$\begin{bmatrix} k_1+k_2 & -k_2 \\ -k_2 & k_2 \end{bmatrix}\begin{bmatrix} w_1 \\ w_2 \end{bmatrix}+\begin{bmatrix} c_1+c_2 & -c_2 \\ -c_2 & c_2 \end{bmatrix}\begin{bmatrix} \frac{dw_1}{dt} \\ \frac{dw_2}{dt} \end{bmatrix}+\begin{bmatrix} m_1 & 0 \\ 0 & m_2 \end{bmatrix}\begin{bmatrix} \frac{d^2w_1}{dt^2} \\ \frac{d^2w_2}{dt^2} \end{bmatrix}=\begin{bmatrix} 0 \\ \sin(2\pi f_1 t) \end{bmatrix} \tag{1}$$

where k is the stiffness, c is the damping, m is the mass, w is the displacement and t is the time. The subscript means the number of DOF for all symbols. In order to model the nonlinear defect, the stiffness of spring 1 (k_1) is assigned as the variable yields:

$$k_1 = 1 + \cos(2\pi f_2 t) \tag{2}$$

The other parameters are: $k_2 = 1$, $m_1 = m_2 = 1$, $c_1 = c_2 = 1$, and $2\pi f_1 = 1$, $2\pi f_2 = 5$. The natural frequencies of the intact system are 0.618 and 1.618. Obviously, the configurations of parameter avoid the resonance of system. By aid of the Newmark method, the calculated displacement w, velocity dw/dt and acceleration d^2w/dt^2 responses before and after the occurrence of damage are obtained and shown in Fig. 2. The red and gray lines depict the responses of the intact structure and damaged structure, respectively. The acceleration response changes sharply when damage appears and the corresponding differences of displacement and velocity responses are not evident. Since the defect k_1 is connected with mass 1 directly, the change of node 2 is smaller than that of node 1.

To enhance the nonlinear damage feature, Manoach and Trendafilova [10-12] proposed to express these responses in state space and identify damage based on the Poincare map method. In their works, the state space $\mathbf{P} = [w, dw/dt]^T$ is employed and the two basic components of state space are provided by independent measurements. In this work, we extend the state space as $\mathbf{P} = [w, dw/dt, d^2w/dt^2]^T$, and the three terms of state space will be provided by single measurement considering the differential relationships of them.

The state space projections of the responses shown in Fig. 2 are presented in Fig. 3. In order to interpret the effectiveness of Poincare map method, velocity and acceleration responses in these figures are calculated by Newmark method directly instead of the estimation by a single measurement. Compared with the Poincare maps of node 2, the Poincare maps correspond to node 1 show a structural change before and after the occurrence of damage. This is the foundation of damage detection based on Poincare maps. For clarity, these Poincare maps are also called as the 1st (w, dw/dt), the 2nd (w, d^2w/dt^2), and the 3rd (dw/dt, d^2w/dt^2) projections in this paper.

(a)

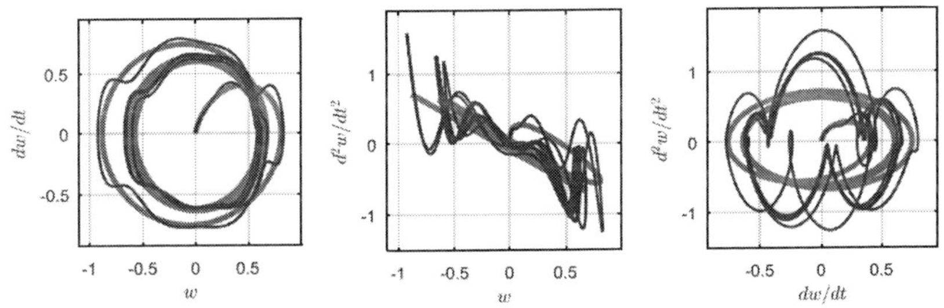

(b)

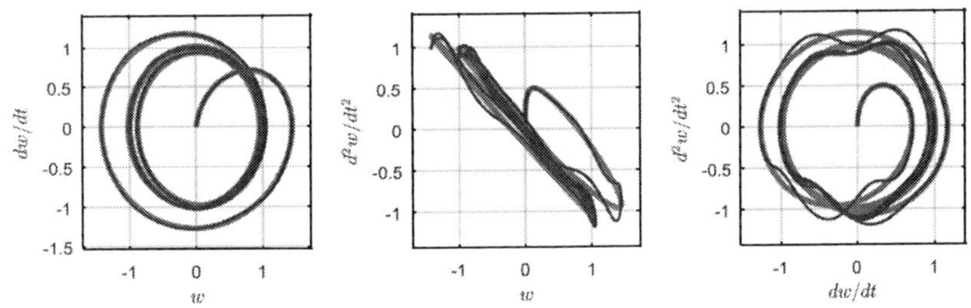

Figure 3. Poincare maps of 2-DOFs system before and after damage, the projections of node 1 (a), node 2 (b). Red lines are for damaged conditions, gray lines are for intact conditions.

2.2. Poincare Maps obtained by single type of measurement

As mentioned above, Poincare maps can be generated by a single measurement referring to the differential relationships among displacement, velocity and acceleration. It is well-know that the numerical integration is stable, thus deriving displacement and velocity from acceleration will not be discussed in this paper. Instead, we will investigate the procedure of deriving velocity and acceleration from displacement via numerical differential, which is ill-conditioned.

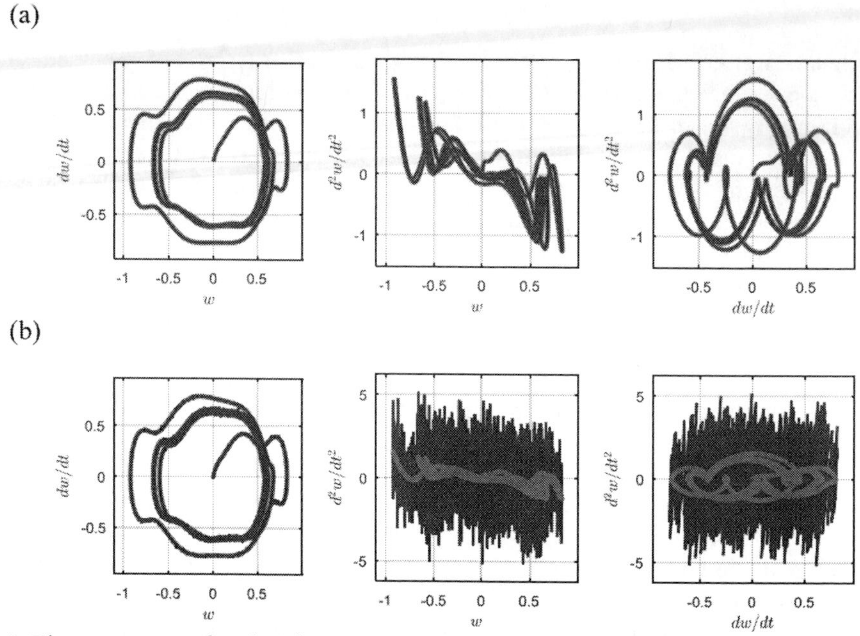

Figure 4. The responses of node 1 in state space, velocity and acceleration responses are estimated by numerical derivative: (a) noise-free, (b) 70 dB. Red lines are for calculated signals, gray lines are for reference signals.

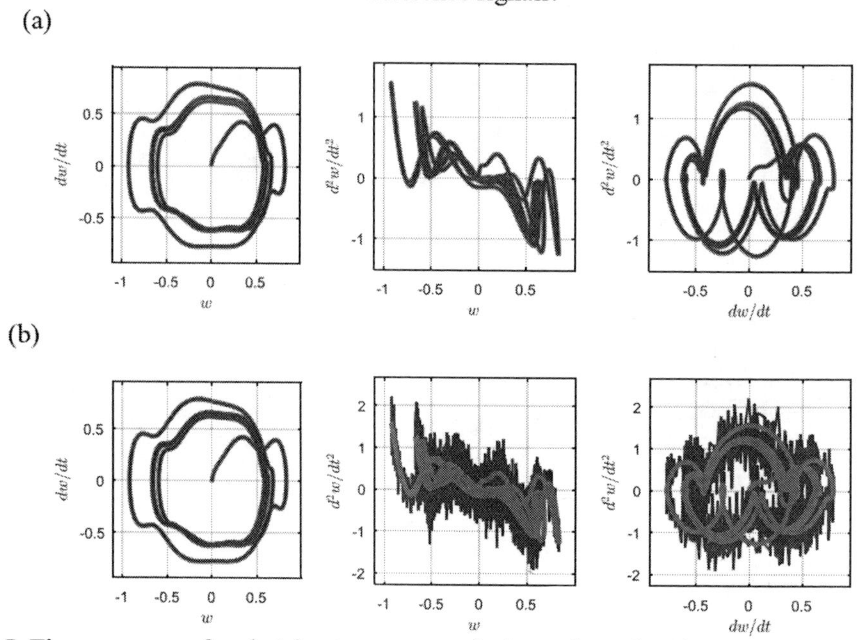

Figure 5. The responses of node 1 in state space, velocity and acceleration responses are estimated by DFT: (a) noise-free, (b) 70 dB. Red lines are for calculated signals, gray lines are for reference signals.

The central difference method, which is usually used in numerical differential, is employed to accomplish this procedure firstly. The related Poincare maps (red lines) of node 1 with different signal to noise ratios (SNR) are presented in Fig. 4, where the noisy signal is generated by invoking Matlab function "awgn" and the gray lines provide the reference signals calculated by model directly without noise. In the noise-free condition, the central difference method performances well. However, only the 1st projection agrees with the reference signal when noise exists in displacement w. This verifies the unstable problem in velocity and acceleration estimations by central difference method.

To address the similar problem, the derivative estimation based on Fourier spectrum was proposed by authors in [10-12] for modal curvature. Then the method is extended to Poincare maps estimation by single measurement in this paper. For completeness, the technological process of the method is briefly introduced in this section. The Fourier transform (FT) and the inverse Fourier transform (IFT) of displacement response *w*(*t*) is given as:

$$W(\omega)=\int_{-\infty}^{+\infty} e^{-j\omega t}w(t)dt\text{ , } w(t)=\frac{1}{2\pi}\int_{-\infty}^{+\infty} e^{j\omega t}W(\omega)d\omega \tag{3}$$

where $t\in\mathbb{R}$ and $\omega\in\mathbb{R}$. In the Fourier frame, velocity and acceleration can be obtained by:

$$\frac{dw}{dt}=\frac{1}{2\pi}j\int_{-\infty}^{+\infty} e^{j\omega t}\omega\hat{w}(\omega)d\omega\text{ , }\frac{d^2w}{dt^2}=-\frac{1}{2\pi}\int_{-\infty}^{+\infty} e^{j\omega t}\omega^2\hat{w}(\omega)d\omega \tag{4}$$

For the discrete signal, the discrete Fourier transform (DFT) is the natural choice for velocity and acceleration calculations. Without the loss of generality, we assume *w*(*t*) is defined on the interval [0, 2π], and the interval is divided by *N* grid points as *N*-1 terms with the length Δ. By the DFT synthesis, Eq. (4) is rewritten as:

$$\frac{dw}{dt}=\frac{\Delta}{2\pi}j\sum_{\omega}\omega e^{j\omega t}\Big(\sum_{t}e^{-j\omega t}w(t)\Big)\text{ , }\frac{d^2w}{dt^2}=-\frac{\Delta}{2\pi}\sum_{\omega}\omega^2 e^{j\omega t}\Big(\sum_{t}e^{-j\omega t}w(t)\Big) \tag{5}$$

Based on Eq. (5), velocity and acceleration can be calculated from displacement directly. The corresponding results are plotted in Fig. 5. Compared with the classical central difference method, the Fourier spectrum based estimation is more robust to noise.

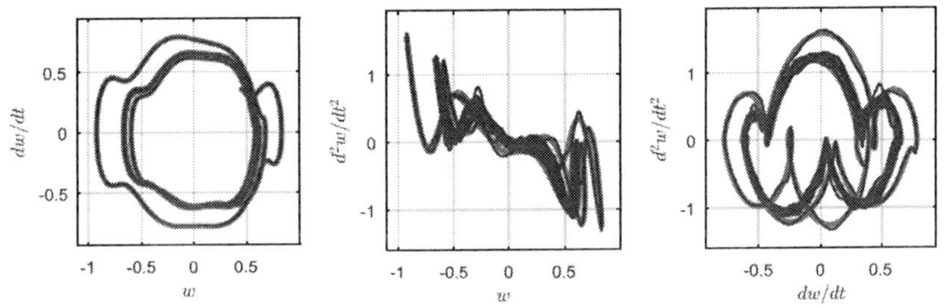

Figure 6. The responses of node 1 in state space after EEMD filtering (SNR = 70 dB), velocity and acceleration responses are estimated by DFT. Red lines are for calculated signals, gray lines are for reference signals.

In order to improve the present result further, the EEMD (Ensemble Empirical Mode Decomposition) is introduced as the filter. The filtered projections (SNR = 70 dB) are presented in Fig. 6. An obvious improvement validates the effectiveness of EEMD in this methodology. To remove the boundary effect induced by EEMD filtering, several sampling points locating at the beginning and end have been deleted.

2.3. Damage index

Some damage indices have been presented and validated in references [10-12]. They are employed in this paper with some necessary extensions to fulfil the requirements of the extended state space. The damage index I_i is defined as:

$$I_i=\frac{1}{3}\sum_{k=1}^{3}\left|\frac{S_{i,k}^d-S_{i,k}^u}{S_{i,k}^u}\right| \tag{6}$$

$$S_{i,k}^u=\sum_{j=1}^{N}\left\|\mathbf{P}_{i,j+1,k}^u-\mathbf{P}_{i,j,k}^u\right\|_2\text{ , }S_{i,k}^d=\sum_{j=1}^{N}\left\|\mathbf{P}_{i,j+1,k}^d-\mathbf{P}_{i,j,k}^d\right\|_2 \tag{7}$$

The above damage index is based on the comparison between the intact and inspected Poincare maps. The spikes presented in damage indices will be considered as damage feature. In applications, threshold or outlier analysis is required to determine the presence of damage and its location. To evaluate the effectiveness of the combination of proposed method and damage index, a noise immunity test containing 100 samples is conducted and presented in terms of box plots (as shown in Fig. 7). The meanings of the symbols in box plots could be found in the "Help" of Matlab. The increase of noise level expands the sample distribution of each node. The distributions of node 1 and 2 overlap with each other when SNR = 60 dB, and the value of I_1 is also not large enough to be deemed as the damage feature. When SNR is larger than 65 dB, the sample distributions of nodes separate from each other, and the value of I_1 also increases with SNR. I_2 changes slightly comparing with I_1. These tendencies and phenomenon verify that the combination of the proposed method and damage index provides a robust damage evaluation for the inspected system.

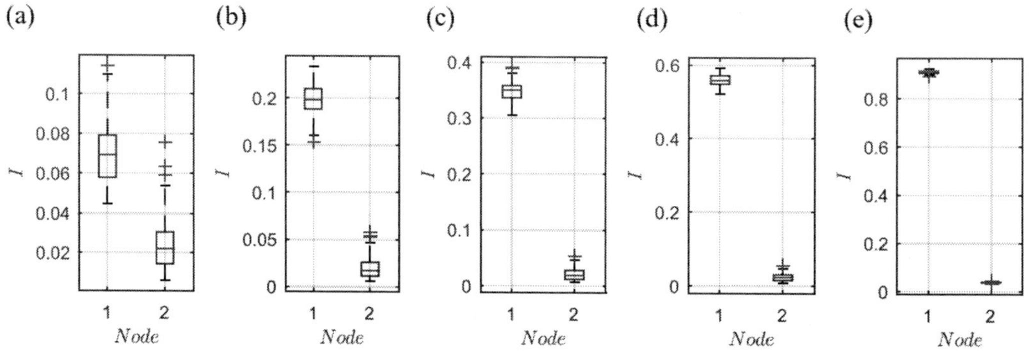

Figure 7. Box plots of damage index with different SNRs: (a) 60 dB, (b) 65 dB,(c) 70 dB, (d) 80 dB, (e) 90 dB.

3. Experimental Verification

A supplementary experimental validation is conducted on a cracked steel cantilever beam with the following dimensions: length 600 mm, crack location 350 mm from the fixed edge, cross-section 20 mm × 20 mm. Another intact beam with the same dimensions serves as the baseline. A shaker attached on the tip of the cantilever beam serves as the actuator, and the voltage acts on the shaker is assigned high enough to generate the nonlinear effect (the crash/rub within the crack). To register the response signal simultaneously, eight acceleration sensors (PCB) are implemented at 90 mm, 170 mm, 254 mm, 333 mm, 372 mm, 425 mm, 485 mm and 545 mm from the clamped end. For clarity, they are denoted as node 1-8 simply. Obviously, nodes 4 and 5 locate near the crack location. Damage characteristic should be featured at these two nodes if any. Restricted by the equipment, acceleration sensors are adopted to work as the displacement sensors in this experiment, namely, the derivatives are conducted on acceleration responses. To evaluate the algorithm in a wide range of frequency domain, four single-frequency excitations are selected, 50 Hz, 100 Hz, 150 Hz and 200 Hz.

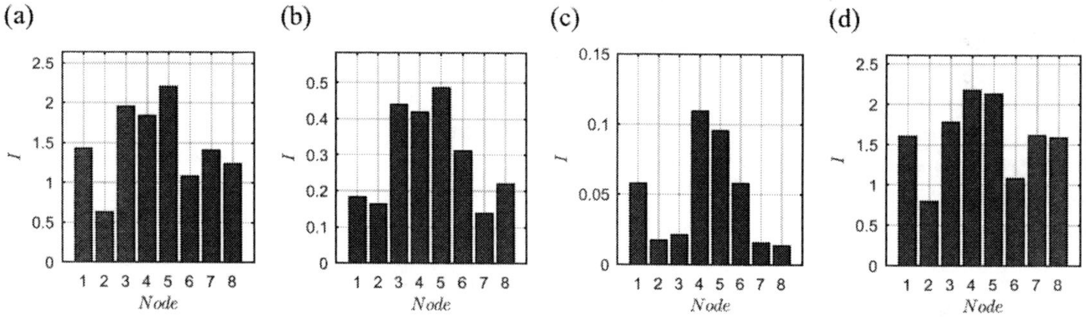

Figure 8. Damage indices for: (a) 50 Hz, (b) 100 Hz, (c) 150 Hz and (d) 200 Hz excitations.

The damage indices obtained from different excitations are presented in Fig. 8. Compared with the other damage indices, the damage index obtained from the 150 Hz excitation feature the crack clearly. The related Poincare maps are given in Fig. 9. Compared with the 100 Hz response, it is observed that the nonlinear effect become more evident under the 150 Hz excitation, which is close to the natural frequency of the inspected structure. The resonance enhances the nonlinear effect, and hence makes the damage feature distinct. Besides the 150 Hz result, damage indices given by the other excitations present the highest peak at node 4 or node 5. This is helpful to pinpoint the damage location to some extent.

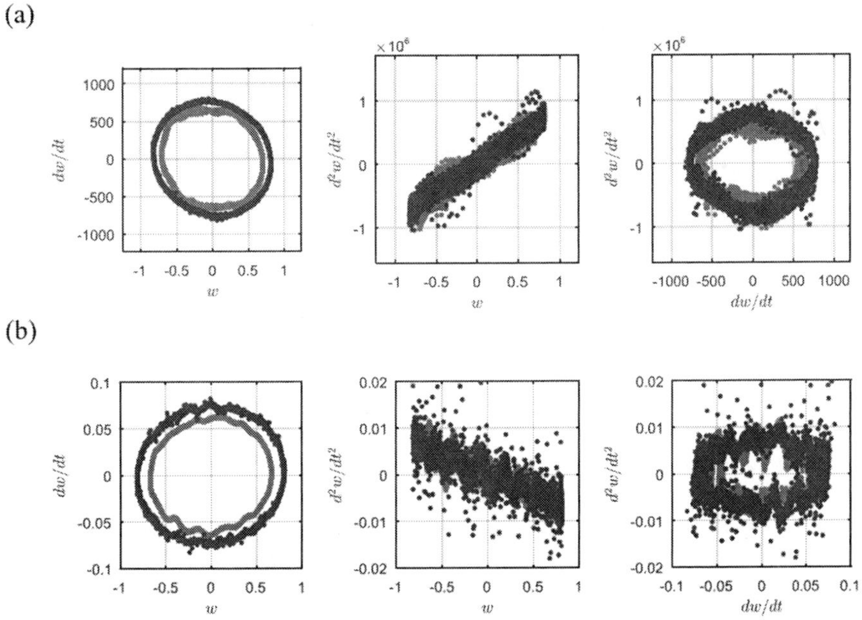

Figure 9. The Poincare maps of node 4 excited by (a) the 100 Hz and (b) 150 Hz input before (gray) and after (red) the occurrence of damage.

4. Conclusions

As validated by simulations, the developed method can construct the Poincare maps by single measurement, thus it is helpful to decrease the possible cost in measurement. However, it must be mentioned that the accuracy and effectiveness of the present method will never exceed the method based on the full measurement, as the algorithm proposed in this method only provide the more accurate approximations rather than the actual values [13-15].

Acknowledgments

The authors thank the supports given by the National Natural Science Foundation of China (Nos. 51405369 & 51335006), the National Key Basic Research Program of China (No. 2015CB057400), the National Natural Science Foundation of Shaanxi Province (No. 2016JQ5049), the China Postdoctoral Science Foundation (No. 2014M560766), the Postdoctoral Science Foundation of Shaanxi Province, and the Fundamental Research Funds for the Central Universities (No. xjj2014107).

References

[1] Khatir A, Tehami M, Khatir S and Abdel Wahab M, Multiple damage detection and localization in beam- like and complex structures using co-ordinate modal assurance criterion combined with firefly and genetic algorithms Journal of Vibroengineering 2016; 18(8): 5063 - 5073.
[2] Zhou Y L, Maia N M M, Sampaio R P C, Wahab MA. Structural damage detection using transmissibility together with hierarchical clustering analysis and similarity measure. Structural Health Monitoring, 2016; doi: 10.1177/1475921716680849.
[3] Gillich G-R, Praisach Z-I, Abdel Wahab M, Gillich N, Mituletu IC and Nitescu C, Free vibration of a perfectly clamped-free beam with stepwise eccentric distributed masses, Shock and Vibration, 2016; http://dx.doi.org/10.1155/2016/2086274, 10 pages.
[4] Zhou Y L, Maia N M M, Wahab M A, Damage detection using transmissibility compressed by principal component analysis enhanced with distance measure. Journal of Vibration and Control, 2016; doi: 10.1177/1077546316674544.

[5] Zhou Y-L and Abdel Wahab M, Rapid early damage detection using transmissibility with distance measure analysis under unknown excitation in long-term health monitoring Journal of Vibroengineering, 2016; 18(7) 4491-4499.
[6] Khatir S, Belaidi I, Serra R, Abdel Wahab M and Khatir T, Numerical study for single and multiple damage detection and localization in beam-like structures using BAT algorithm, Journal of Vibroengineering, 2016; 18(1) 202-213.
[7] Khatir S, Belaidi I, Serra R, Abdel Wahab M and Khatir T, Damage detection and localization in composite beam structures based on vibration analysis, Mechanika, 2015; 21(6) 472-479.
[8] Yang ZB, Chen XF, Xie Y, Zhang XW. The hybrid multivariate analysis method for damage detection. Structural Control and Health Monitoring. 2016;23:123-43.
[9] Yang ZB, Chen XF, Xie Y, Miao HH, Gao JJ, Qi KZ. Hybrid two-step method of damage detection for plate-like structures. Structural Control and Health Monitoring. 2016;23:267-85.
[10] Manoach E, Samborski S, Mitura A, Warminski J. Vibration based damage detection in composite beams under temperature variations using Poincare maps. International Journal of Mechanical Sciences. 2012;62:120-32.
[11] Trendafilova I, Manoach E. Vibration-based damage detection in plates by using time series analysis. Mechanical Systems and Signal Processing. 2008;22:1092-106.
[12] Manoach E, Trendafilova I. Large amplitude vibrations and damage detection of rectangular plates. Journal of sound and vibration. 2008;315:591-606.
[13] Yang Z-B, Radzienski M, Kudela P, Ostachowicz W. Two-dimensional modal curvature estimation via Fourier spectral method for damage detection. Composite Structures. 2016;148:155-67.
[14] Yang Z-B, Radzienski M, Kudela P, Ostachowicz W. Scale-Wavenumber Domain Filtering Method for Curvature Modal Damage Detection. Composite Structures. 2016;154:396-409.
[15] Yang Z-B, Radzienski M, Kudela P, Ostachowicz W. Fourier spectral-based modal curvature analysis and its application to damage detection in beams. Mechanical Systems and Signal Processing. 2017;84:763–81.

A new fault detection strategy using the enhancement ensemble empirical mode decomposition

Jiawei Xiang *, Yongteng Zhong

College of Mechanical and Electrical Engineering, Wenzhou University, Wenzhou, P.R.China

E-mail: wxw8627@163.com; zhongyongteng@wzu.edu.cn

* wxw8627@163.com. (Jiawei Xiang)

Abstract. The vibration signals of mechanical components are non-linear and non-stationary and the feature frequencies of faulty bearings will be difficult extracted. This paper presents a new approach that combines the ensemble empirical mode decomposition (EEMD), the random decrement technique (RDT), and envelope spectrum for the fast detection of faults in bearings. The proposed approach uses the optimized and fast EEMD algorithm to extract intrinsic mode functions (IMFs) from vibration signals able to tack the feature frequency of bearings. If the Impulse response signal of the first IMF is unclear, it is further extracted by the RDT, and the feature frequencies are determined by analyzing the signals using envelope spectrum. The advantages of this method are its computational efficiency, and the strong non-stationary vibration signal decomposition and impulse signal extraction abilities. Numerical simulations and experimental data collected from faulty bearings are used to validate the proposed approach. The results show that the use of the EEMD, the RDT, and the envelope spectrum is a suitable and fast on-line strategy to detect faults of mechanical components.

1. Introduction

Damage detection using vibration signals and the corresponding structural health monitoring techniques are the important topic in scientific world [1-7]. Rolling element bearings are the two key components in mechanical systems. The failure of these components (e.g., bearings) will result in the deterioration of machine operating conditions. Therefore, it is important to timely detect faults in bearings. Owing to vibration signals carry a great deal of information representing mechanical equipment health conditions, the use of vibration signals is quite common in the field of condition monitoring and diagnostics of mechanical systems [8-16]. Fault features (the impact pulse amplitude modulation is caused by the defects moving in and out of the loading zone) can be extracted from the vibration signals with signal processing techniques, such as the combinations of wavelet transform and independent component analysis (ICA) [8], the customized lifting multiwavelet packet information entropy [9], the development of an adaptive ensemble empirical mode decomposition EEMD method [10], spectral kurtosis (SK) in associate with the fast computation of the kurtogram technique [11,12], the joint integral and wavelet transform approach [13], the hybrid of PPCA and spectral kurtosis (SK) technique [14], the manifold subspace distance method [15], and the Teager energy operator (TEO) based parameter-free and broadband approach [16], etc.

For the fault detection in bearings, it is expected that a desired time–frequency analysis method has good computational efficiency and resolution in both time and frequency domains. Recently, EMD has been developed and widely applied in fault diagnosis of mechanical systems. EMD is a way to decompose a signal into IMFs along with a trend, and obtain instantaneous frequency signals (approximate stationary signals). Therefore, it is designed to work well for a signal that is non-stationary and nonlinear. Therefore, EMD is a time–frequency analysis technique, which can decompose the complicated signal (non-stationary signal) into a set of complete and almost orthogonal components (stationary signals) named intrinsic mode functions (IMFs) [17]. To reduce mode mixing, Wu and Huang proposed a new ensemble EMD (EEMD) method [18]. EEMD is a noise-assisted data analysis method by adding finite white noise to excite the original signals, which can eliminate the mode mixing phenomena. In practical applications, the key issues are how to properly select the added noise amplitude (affects EEMD decomposition results). If the added noise amplitude is inappropriate, it will cause a large amount of residual noise. To overcome this problem, Yeh et al.[19] presented the complementary of the use of EEMD method and applied it to detect simulation signals with a good performance. Recently, Wang et al. [20] proposed an optimized EMD/ EEMD algorithm to speed up the computational efficiency to one thousand times. They also proved that the computational complexity of the EMD is equivalent to fast Fourier transform (FFT). Hence, the optimized EEMD method might be a good choice to be applied to on-line detect faults in mechanical systems.

The fundamental concept of the RDT [21] is that the random response of a structure (stationary signal) can be separated into two parts, i.e., a deterministic part and a random part. By averaging enough sample responses, the random part will be enormously decreased. Ibrahim verified that the deterministic part that remains is the free-decay response (impulse response signal) associated with the initial condition in time domain [22-23]. Hence the main advantage of RDT is the strong ability to extract impulse response signals from noisy conditions. In summary, EMD/EEMD can decompose complicated non-stationary signal into a collection of stationary IMFs, whereas RDT is suitable to process stationary signals. Therefore, it might be agreeable to combine them into a hybrid one.

2. A brief review of EEMD and RDT

In 1998, Huang et al. introduced the EMD, which is able to decompose complicated signal into a collection of stationary IMFs [17]. Therefore, it has been often used in non-stationary signal processing [20]. In the EMD method, the data $x(t)$ is decomposed in terms of IMFs c_j as [17]

$$x(t) = \sum_{j=1}^{n} c_j + r_n \tag{1}$$

where r_n is the residual of data $x(t)$, after n number of IMFs are extracted.

However, mode mixing appears to be the most significant drawback of EMD [19]. Therefore, in 2009, a new artificial noise-excited EMD method was proposed by Wu and Huang, which called EEMD [20]. The procedures are similar to EMD except only one groups of white noise with finite amplitude are added into the original signals and the procedures are summarized as follows:

(1) Add a white noise $n_i(t)$ series (noise level is N_i) to the targeted data and decompose the data with added white noise into IMFs as

$$x(t) + n_i(t) = \sum_{j=1}^{n} c_j^i(t) + r_n^i(t) \tag{2}$$

where $i = 1,2,\cdots,q$ and q is the average times (ensemble number).

(2) Repeat step1 q times with different white noise series $n_i(t)$.

(3) Obtain the (ensemble) means of corresponding IMFs of the decompositions as the final result, that is

$$x(t)+n_i(t)=\frac{1}{q}\sum_{i=1}^{q}\sum_{j=1}^{n}c_j^i(t)+\frac{1}{q}\sum_{i=1}^{q}r_n^i(t)=\sum_{j=1}^{n}d_j(t)+r_n(t) \quad (3)$$

where $d_j(t)$ is the *j*th IMFs of EEMD decomposition as

$$d_j(t)=\frac{1}{q}\sum_{i=1}^{q}c_j^i(t) \quad (4)$$

and $r_n(t)$ is the final residual of EEMD decomposition as

$$r_n(t)=\frac{1}{q}\sum_{i=1}^{q}r_n^i(t) \quad (5)$$

The fast algorithm of EMD/EEMD is proposed by Wang et al. in Ref. [20], which will speed up the computational efficiency to one thousand times than that of traditional EMD/EEMD. The computational complexity of the EMD/EEMD is proved to be equivalent to FFT. Therefore, the EEMD method can be employed to quickly detect faults in mechanical components.

RDT can be used to describe the free decay response of the system (in the present, it refers to the impulse response signal). The advantage of this technique is that one obtains the free response from the stationary random response of the system [22-23]. In this regard, the response signal is divided into a number of segments *L*, each of length τ. All of these segments should have the same initial condition (triggering value), i.e., $x_i(t_i)=x_s=const.$, $i=1,2,\cdots,L$. A threshold level should be selected to obtain *L* segments. The ensemble average of the *L* segments yields the random decrement, which can be mathematically expressed by

$$x(\tau)=\frac{1}{L}\sum_{i=1}^{L}x_i(t_i+\tau) \quad (6)$$

where $x_i(t_i)=x_s$ for $i=1,2,\cdots,L$.

It points that the RDT requires no knowledge of the excitation as long as it is stationary, zero-mean Gaussian random process [22-23].

3. The diagnosis method using EEMD, RDT and Hilbert envelope spectrum

Fig.1 shows the flowchart of the present diagnosis method using EEMD, RDT and Hilbert envelope spectrum. Three steps in the proposed approach are:

(1) Obtain the first IMF of raw signal using EEMD

In order to overcome the drawback of careful selection of an IMF using EEMD method, we fix the first IMF in the present approach, and the performance will be verified using numerical simulations and experimental investigations. Moreover, for the mechanical systems, especially the bearing or gear with faults, the high frequency modulation signals is always located in the IMF1.

(2) Perform Hilbert envelope spectrum analysis and obtain the detect result

For the IMF1 or the impulse response signal, Hilbert envelope spectrum analysis is applied and the demodulation frequency is finally obtained. Compared the demodulation frequency with the theoretical fault feature frequency, the type of bearing and gear fault can be determined. If the impulse response signal of the first IMF (IMF1) is clearly revealed, we will obtain the detect result using only Hilbert envelope spectrum analysis. Otherwise, RDT is further applied to extract the purified impulse response signal.

(3) Extract impulse signals from the IMF1 using RDT

In the present, RDT is applied to extract the impulse response signal of faulty bearings and gears from the stationary signal in IMF1. Then we go back to step 2 and repeat perform Hilbert envelope spectrum analysis and finally obtain the detect result.

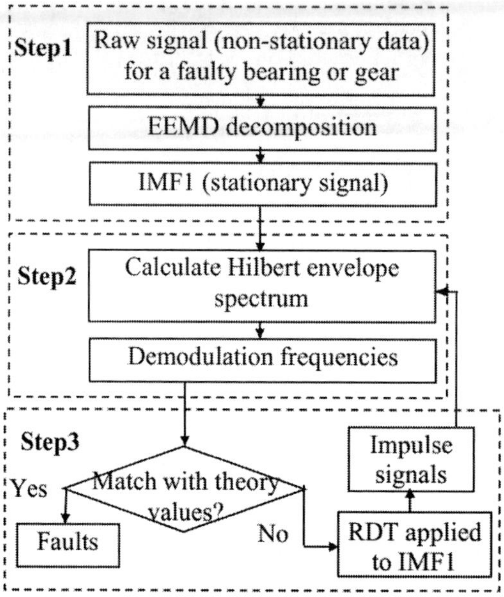

Fig. 1. The flowchart of fault detection approach

4. Numerical simulation

To simulate the bearing faults, we use an artificial simulation signal in this section. The low frequency steady state signal, high frequency transient signal and the noise are included, and the simulation signal is defined as

$$x(t) = s(t + T) + c(t) + e(t) \tag{7}$$

where $s(t)$ is the impulse response signal (simulate fault induced vibration of bearing), T is the impulse period (feature frequency), $c(t)$ is the low frequency steady state signal relative with the rotating frequency and its harmonic components, and $e(t)$ is the noise (normally distributed pseudorandom numbers).

In Eq.(7), $c(t)$ and $s(t)$ are given by

$$c(t) = 0.5\sin(2\pi f_0 t) + 0.15\sin(4\pi f_0 t) + 0.05\sin(6\pi f_0 t) \tag{8}$$

and

$$s(t) = e^{-Bt}\cos(2\pi f_n t) \tag{9}$$

in which f_0 is the fundamental frequency of the low frequency steady state signal, f_n is the natural frequency. $B = 2\pi f_n \xi$ is the attenuation coefficient depended on damping ratio ξ and natural frequency f_n. Suppose the parameters are: $T = 0.01s$ (the corresponding faulty frequency is $100Hz$), $f_0 = 30Hz$, $f_n = 4000Hz$, $\xi = 0.019894$ and the corresponding $B = 500$, $e(t)$ is a standard normal distribution with standard deviation 1.5. The sampling frequency $f_s = 20480Hz$, the sampling points are N=4096.

The simulation signal and its frequency spectrum are shown in Fig.2. The harmonic wave and impulsive components are mixed with the noise as well as the corresponding frequency spectrum.

It worth to point out here that if the noise level is higher than a certain ratio, for example, if the natural frequency $f_n = 4000Hz$ in Fig. 2 (b) is submerged by noise, it would be difficult to detect the faulty frequency for almost every signal processing method.

The parameters to run the EEMD are carefully chosen and suggested by Ref. [25], the noise level N_l =0.3, ensemble number q=100, and the number of prescribed IMFs is 8.

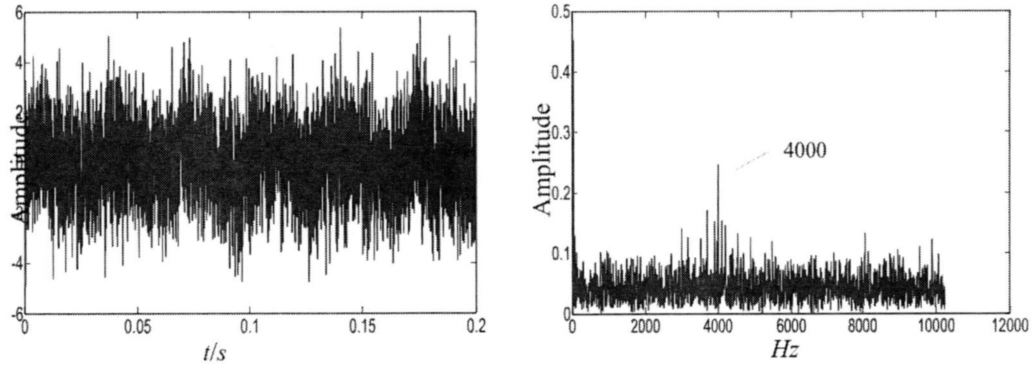

(a) The raw signal (b) The frequency spectrum of the raw signal

Fig. 2. The raw signal and the corresponding frequency spectrum

All the computations are conducted using Matlab2010a on a laptop computer with a 2.5G*Hz* CPU (Intel(R) Core (TM) I5-2450M) and 2.95GB memory. According to the TIC and TOC commands of Matlab2010a, the computing times required to finish the EEMD decomposition, RDT extraction and Hilbert envelope spectrum analysis are less than 1.76 seconds for several tests. This should be the quick enough for on-line detection of faults.

The EEMD decomposition results are shown in Fig.3. From IMF1 to IMF4, the impulse signal cannot be seen clearly. The frequency spectrum and the Hilbert envelope spectrum for the IMF1 are shown in Fig.4. In Fig4.(a), only the two frequencies 4000*Hz* and 8000*Hz* are clearly seen. However, they do not reflect the feature frequency of the fault (the modulation frequency 100*Hz* and its harmonics). The feature frequency of the fault (100 *Hz*) and the second harmonic (200*Hz*), the shaft rotating frequency (30*Hz*), and other unknown frequencies, such as 80*Hz*, 235*Hz*, 290*Hz*, 425*Hz* and 490*Hz* are shown in Fig.4 (b). Therefore, the modulation frequency (100*Hz*) and the first harmonic 200*Hz* cannot be directly employed as the indicators to determine the fault. Furthermore, the frequency spectrum and the Hilbert envelope spectrum for the IMF2 to IMF7 are also obtained (not give in the paper) and all of them are not the sensitive indicators to reveal the fault.

To verify the performance of the present scheme for the extraction of impulse signal form the IMF1, IMF2 all of the other IMFs are employed to perform RDT and Hilbert envelope spectrum analysis. In the present, the data length is $\tau = 2048$. The triggering value x_s is suggested to $1.5\sigma_s$ [22-23], where σ_s is the standard deviation of the IMF. The impulse signal extracted by RDT and its Hilbert envelope spectrum for the IMF1 are shown in Fig. 5. From Fig.5(a), we can see clearly that the impulse response signal is shown up. As shown in Fig.5(b), the feature frequency (100 *Hz*) and the 2 to 5 harmonics (200*Hz*, 300*Hz*, 400*Hz*, 500*Hz*), the shaft rotating frequency (30*Hz*) will be robustly extracted.

Based on the above investigation, we conclude that the present approach using IMF1, RDT and Hilbert envelope spectrum can be employed to determine the periodical impulses (may represent faults of bearings and gears).

5. Experimental investigations

In this section, three laboratory experiments on rolling element bearings with inner race is conducted. The experimental setup (The machinery fault simulator-magnum, MFS-MG) is shown in Fig.6 as well as shown in Ref.[24]. The test system consisted of speed monitor, manual speed governor, acceleration sensors, speed sensors, motors, spindles and computer with DAQ software [24].

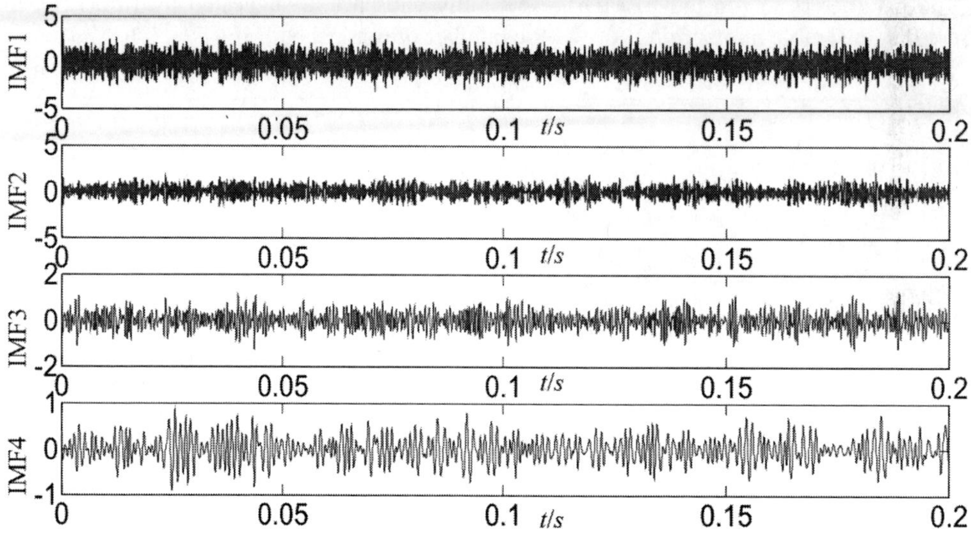

Fig. 3. The first four IMFs using the EEMD

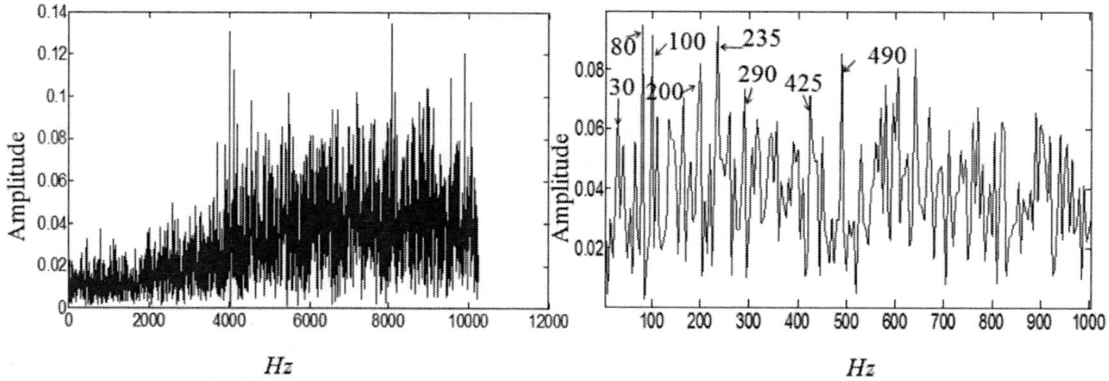

(a) The frequency spectrum for IMF1 (b) The Hilbert envelope spectrum for IMF1

Fig. 4. The frequency spectrum and the Hilbert envelope spectrum for the IMF1

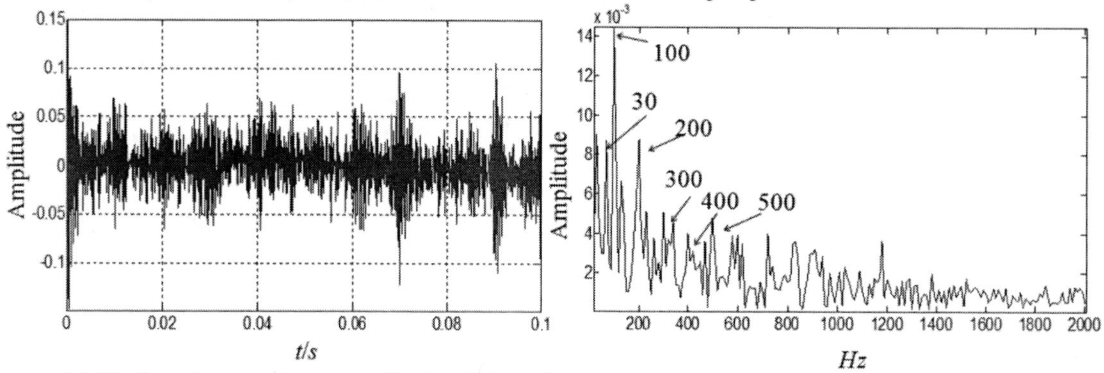

(a) The impulse signal extracted by RDT from IMF1 (b) The Hilbert envelope spectrum

Fig. 5. The impulse signal extracted by RDT and its Hilbert envelope spectrum for the IMF1

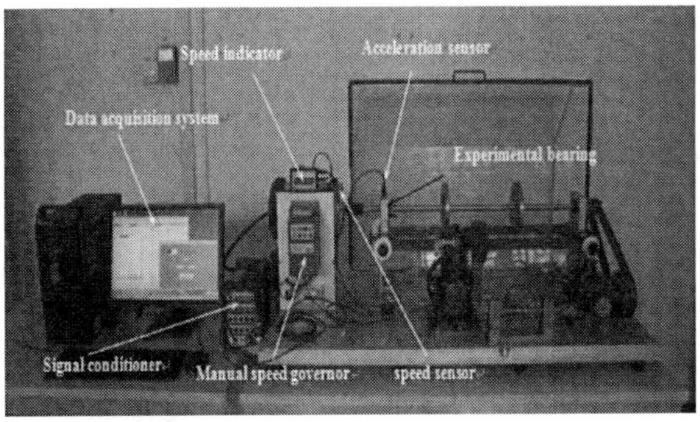

(a) The experimental setup

(b) The fault on the inner race of the bearing

Fig.6 The experimental setup and the fault on the inner race of the bearing

A deep groove bearing (ER-12K) was used in our experiment and the bearing parameters were: rolling element number $N_b = 8$, ball diameter $B_d = 0.3125$ *inch*, pitch diameter $P_d = 1.318$ *inch*, the contact angle $\alpha = 0^\circ$. In the experimental processing, the sampling frequency f_s was set to 25.6 *kHz*. For the bearing with inner race fault, the total collected data was 13824 points, the shaft rotating frequency $f_{shaft} = 30.12Hz$. Hence, the ball pass frequency of the inner race (BPFI) was [24]

$$BPFI = \frac{N_b}{2} f_{shaft} (1 + \frac{B_d}{P_d}) \cos\alpha = 4.9484 f_{shaft} = 149.05Hz \tag{15}$$

The raw signal is shown in Fig. 7(a). The present method was performed to detect the impulse response signal and further demodulate the feature frequency. The parameters to run the EEMD were similar to the numerical simulation example in Section 4, and the data length for the RDT analysis $\tau = 6800$. All the computations were conducted using the same laptop as shown in Section 4, and the computing times were less than 5.89 seconds for several tests.

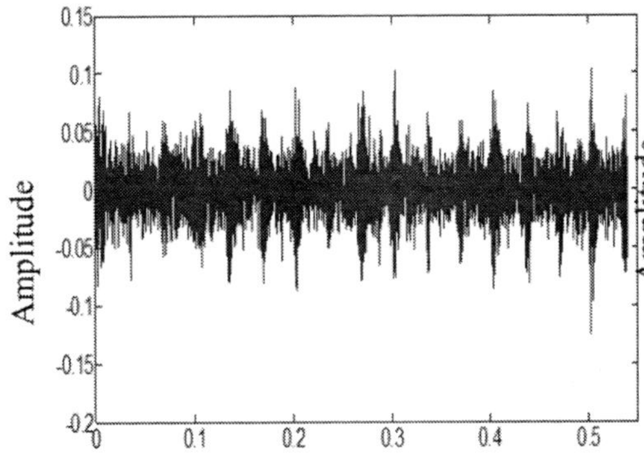

(a) The raw signal

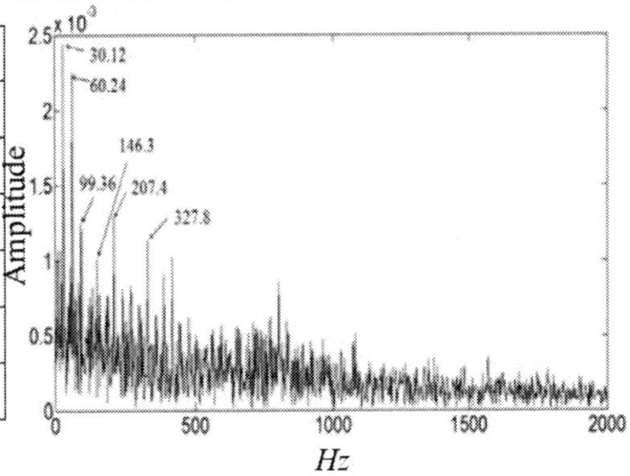

(b) The Hilbert envelope spectrum

Fig. 7. The raw signal and the corresponding Hilbert envelope spectrum for the IMF1

The Hilbert envelope spectrum of IMF1 is shown in Fig.7 (b), and we find that shaft rotating frequency and its harmonics are equal to 30.12 *Hz*, 60.24*Hz* and 99.36*Hz*. The frequency 146.3 *Hz* might be the demodulated feature frequency, which is closed to the theoretical calculation value 149.05*Hz* (as shown in Eq.(15)). However, it is submerged in other frequency components, such as 207.4 *Hz* and 327.8*Hz*.

RDT analysis is conducted to IMF1 and the results are shown in Fig.8. As shown in Fig.8(b), the shaft rotating frequency $f_{shaft} = 30.12Hz$ and the second harmonic 60.24*Hz*, the feature frequency 148.7*Hz* and other frequency components, such as 178.8 *Hz* and 297.4 *Hz* are found. It is worth pointing out here that 297.4 *Hz* is exactly the harmonic of feature frequency; 178.8 *Hz* is exactly the sum of the shafting rotating frequency (30.12 *Hz*) and the feature frequency (148.7 *Hz*). Moreover, the feature frequency 148.7*Hz* was matching with the theoretical calculation value 149.05 *Hz*. Therefore, the bearing with inner race fault was clearly detected.

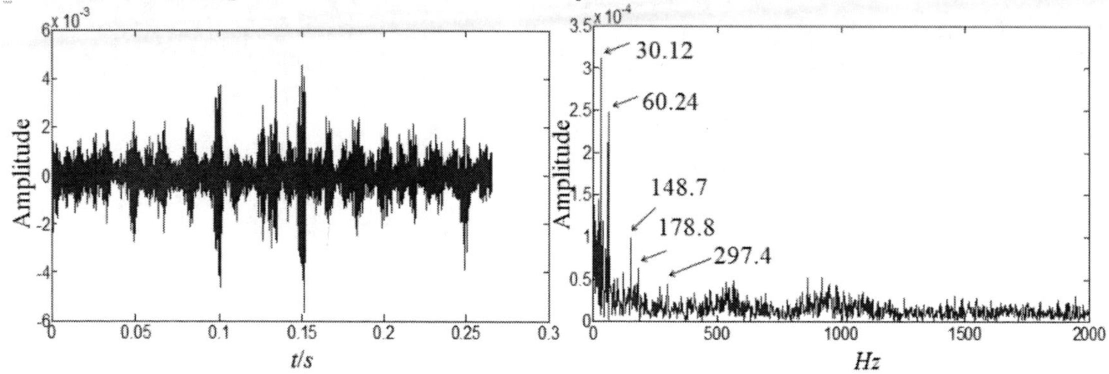

(a) The impulse signal extracted by RDT (b) The Hilbert envelope spectrum

Fig. 8. The impulse signal extracted by RDT and its Hilbert envelope spectrum for the IMF1

6. Conclusion

In this paper, the good decomposition capability of the EEMD has been explored to decompose the non-stationary vibration signal. As we known, the vibrations of a bearing with faults are commonly the amplitude modulation signals, which are located at the high frequency band. More specifically, they often locate in IMF1 with heavy noises (mostly is random noise).Then, RDT is applied to analyze IMF1 to extract the fault induced impulse signals. Finally, the feature frequency of the faulty bearing or gear is obtained using the Hilbert envelope spectrum analysis. The main advantage of this method is its computational efficiency. This is reflected by the fact that it requires much less time, in comparison to the popular EEMD method, to achieve the same data analysis. The second advantage of this method is the well combination of the strong non-stationary vibration signal decomposition and impulse signal extraction abilities of the EEMD and the RDT, respectively. More specifically, the RDT can extract impulse signal from the stationary data only. Fortunately, IMF1 is exactly the stationary data. The third advantage is the fixed IMF1 is selected to perform Hilbert envelope spectrum analysis, which will overcome the difficult for EEMD analysis on how to select the optimal IMF. From the results of the numerical simulation and experiment investigation, it can be seen that the present method is suitable to detect impulse-dominated faults in mechanical systems.

Acknowledgments

The authors are grateful to the support from the National Science Foundation of China (Nos. 51575400, 51505339), the Zhejiang Provincial Natural Science Foundation of China (LQ16105005).

References

[1] Khatir A, Tehami M, Khatir S and Abdel Wahab M 2016 *J. Vibroeng.* **18**(8) 5063- 5073.
[2] Zhou Y-L, Maia NMM, Sampaio R and Wahab MA 2016 Struct. Health Monit. doi: 10.1177/1475921716680849.
[3] Zhou Y-L, Maia N and Abdel Wahab M 2016 *J. Vib. Control* doi: 10.1177/1077546316674544.
[4] Zhou Y-L and Abdel Wahab M 2016 *J. Vibroeng.* **18**(7) 4491-4499.
[5] Khatir S, Belaidi I, Serra R, Abdel Wahab M and Khatir T 2016 *J. Vibroeng.* **18**(1) 202-213.
[6] Khatir S, Belaidi I, Serra R, Abdel Wahab M and Khatir T 2015 *Mechanika* **21**(6) 472-479.
[7] Gillich G-R, Praisach Z-I, Abdel Wahab M, Gillich N, Mituletu IC and Nitescu C 2016 *Shock Vib.* **2016**(Article ID 2086274) 10.
[8] Lin J and Zhang AM 2005 *NDT&E Int.* **38** 421-427.
[9] Chen JL, Zuo Ming J and Zi YY 2013 Smart Mater. Struct. **22** 095022.

[10] Lei YG, Li NP, Lin J and Wang SZ 2013 Sensors **13** 16950-16964.
[11] Antonia J and Randall RB 2006 Mech. Syst. Signal Pro. **20** 282-307.
[12] Antonia J and Randall RB 2006 Mech. Syst. Signal Pro. **20** 308-331.
[13] Li C and Liang M 2011 Smart Mater. Struct. **20** 045016.
[14] Xiang JW, Zhong YT and Gao HF 2015 Measurement **75** 180-191.
[15] Sun C, Zhang Z.S, Cheng W, He ZJ, Shen ZJ, Chen BQ and Zhang L 2013 Smart Mater. Struct. **22** 085012.
[16] Bozchalooi IS and Liang M 2010 Smart Mater. Struct. **19** 075008.
[17] Huang NE, Shen Z and Long SR 1998 *P. Roy. Soc. A-Math. Phy.* **454** 903-995.
[18] Wu Z and Huang NE 2009 *Adv. Data Anal. Classi.* **1** 1-41.
[19] Yeh JR, Shieh JS and Huang NE 2010 *Adv. Data Anal. Classi.* **2** 135-156.
[20] Wang YH, Yeh CH and Young HWV 2014 *Phys. A.* **400** 159-167.
[21] Cole HA 1986 *AIAA Paper.* **68** 288-319.
[22] Ibrahim SR 1977 *J. Spacecraft Rockets.* **14** 696-700.
[23] Ibrahim SR 1977 *AIAA Paper* **77** 1-9.
[24] Li PF, Jiang YY, Xiang JW 2014 *Sci. World J.* **2014** 145807.

IOP Conf. Series: Journal of Physics: Conf. Series **842** (2017) 012003 doi:10.1088/1742-6596/842/1/012003

A study on the inclusion sizing using immersion ultrasonic C-scan imaging

D Chen[1], H F Xiao[1,*], M Li[1], J W Xu[1]

[1] Collaborative Innovation Center of Steel Technology, University of Science and Technology Beijing, Beijing 100083, PR China

E-mail: huifangxiao@ustb.edu.cn

Abstract: Inclusion sizing, especially for large inclusions greater than 30μm provides important reference for metallurgical process control and fatigue life assessment of steel. Ultrasonic non-destructive testing (NDT) shows great advantages in detecting infrequently occurred large inclusions than eddy current, magnetic particle, microscopic or macroscopic examination procedures. In this paper, the performance of inclusion sizing by immersion ultrasonic C-scan imaging is studied numerically. A two-dimensional model that consists of spherically focused transducer, water couplant and steel with embedded inclusion is established and solved numerically by the finite element method. The signal intensity distributions of inclusion with different sizes are acquired and the effects of inclusion type, shape, orientation on signal intensity distribution are analysed. The results show that the 6dB-drop threshold has the smallest relative error compared with the 12dB-drop threshold and the full-drop threshold, which is better for determining inclusion size larger than 100μm. Experiment is also performed to validate the simulated results.
Key words: Ultrasonic testing; inclusion size; signal intensity distribution; numerical solution

1. Introduction

Non-metallic inclusion has a serious impact on the strength, fracture toughness, plasticity and fatigue life of steel. Analyzing and evaluating inclusion's type, shape and size will be helpful to improve steel cleanness[1]. Inclusions size is one of the most important index for evaluating steel cleanliness, which is closely related to the metallurgy technology[2].

The most commonly used methods for inclusion testing are Metallographical Microscope Observation (MMO), Scanning Electron Microscopy (SEM) and Electrolysis. These methods, however, have their own drawbacks on material destruction, sample preparation and time consumption that limit its application in industrial site[3]. Ultrasonic testing has strong penetration, high resolution and detection efficiency, which is very suitable for inclusion detection, especially for the large inclusions that occurred infrequently[4]. Currently, inclusion sizing mainly depends on the signal intensity at a particular position or the signal intensity distribution of C-scan image. The signal intensity is not only associated with the inclusion size, but also related to the inclusion type, shape and orientation[5]. The C-scan image can be used to depict inclusion size visually by signal intensity

* Corresponding author Tel: +86 010 62334255; fax: +86 010 62334255

distribution, but which resolution is limited for the inclusion size of tens of microns, even if the ultrasonic frequency is up to 100MHz.

In this paper, a two-dimensional model that consists of spherically focused transducer, water couplant and steel with embedded inclusion is established, which aims to acquire the signal intensity distribution of inclusions with sizes ranging from 30μm to 200μm. The influence of inclusion type, shape and orientation on signal intensity distribution is analyzed and the relative error of inclusion size under 6dB-drop, 12dB-drop and full-drop thresholds are discussed, which provide guidance for the inclusion sizing by ultrasonic C-scan imaging.

2. Model description

The two-dimensional model that was used to acquire signal intensity distribution is illustrated in figure 1, where *R* is the radius of piezoelectric element, *F* is the focal length in water, *h* is water thickness, *d* is the inclusion depth, *H* and *D* is the thickness and width of steel plate.

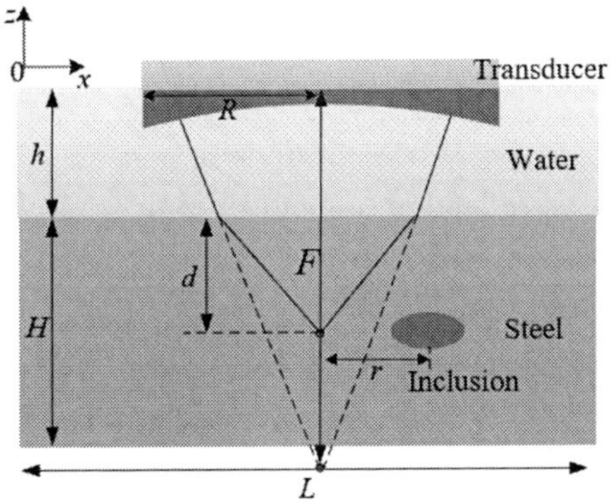

Figure 1. Model diagram.

Assuming inclusion and steel are homogeneous and isotropic, wave equation and equilibrium differential equation are expressed as:

$$\frac{1}{\rho c^2}\frac{\partial^2 p_t}{\partial t^2}+\nabla\cdot-\frac{1}{\rho}\nabla p_t=0 \quad (1)$$

$$p_t = p + p_b$$

and

$$\rho\frac{\partial^2 u}{\partial t^2}-\nabla\cdot\sigma=F_b\cdot v \quad (2)$$

where P_t is the total pressure, F_b is the body force on object, ρ and c are the material density and longitudinal wave velocity, u is the particle vibrational displacement and v is the Poisson ratio. According to the O'Neil principle, the sound wave radiated from transducer surface has identical normal vector. The pressure on the surface of acoustic lens is expressed as:

$$\begin{cases} P\ t = P_0\sin\ 2\pi ft \times 0.5-0.5\cos\ 2\pi ft/3 & ,0<t<3T \\ 0 & t\geq 3T \end{cases} \quad (3)$$

where f and T are the frequency and cycle of pulse wave whose duration is three cycles. Figure 2(a) and figure 2(b) show the transducer responses in time domain and frequency domain, respectively. The pulse duration is 30ns and the -6dB bandwidth is 60MHz.

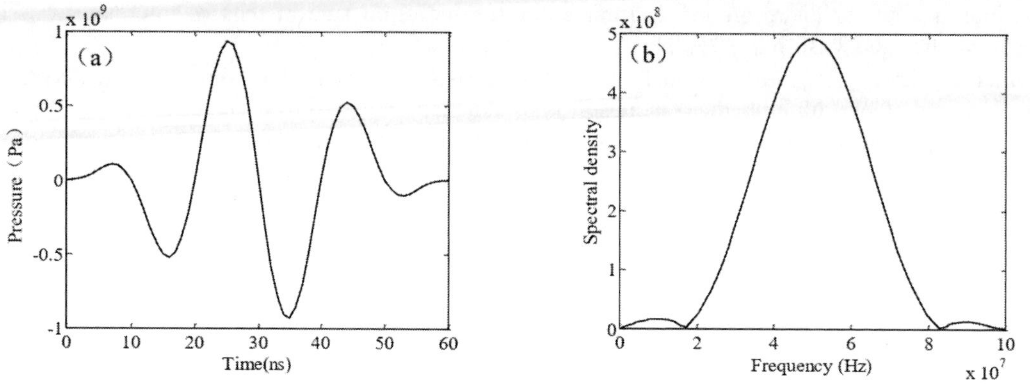

Figure 2. Characteristics of pressure (a) time domain (b) frequency domain.

In the water-steel interface the boundary function is controlled by:

$$n\cdot\left(-\frac{1}{\rho}\Delta p_{\mathrm{t}}\right)=-n\cdot u_{\mathrm{tt}}$$
$$F_{\mathrm{A}}=p_{\mathrm{t}}\cdot n \tag{4}$$

In order to eliminate sound wave reflection on both sides of steel plate, the low-reflection boundary is applied and controlled by the function of:

$$\sigma\cdot n=-\rho c_{\mathrm{p}}\left(\frac{\partial u}{\partial t}\cdot n\right)n-\rho c_{\mathrm{s}}\left(\frac{\partial u}{\partial t}\cdot\tau\right)\tau \tag{5}$$

where σ is stress, n and τ are the normal and tangential unit vector, c_p and c_s are the velocity of longitudinal wave and transverse wave.

3. The signal intensity distribution of inclusions with different sizes

The signal intensity distribution of inclusion with sizes ranged from 30μm to 200μm is acquired. The center frequency and focal length of the transducer are 50MHz and 8mm, respectively. Figure 3 shows the signal intensity distribution of inclusions whose shape is ellipse and their type is Al_2O_3. The horizontal axis is the relative position r between transducer and inclusion as shown in figure 1, and the vertical axis is the maximum echo amplitude that represents the signal intensity. With the increase of inclusion size, the signal intensity distribution becomes wider gradually. Figure 4 shows the relative errors of inclusion size by the 6dB-drop, 12dB-drop and full-drop thresholds. Comparing with the other two cases, the 6dB-drop threshold has the smallest relative error for each size of inclusion. Meanwhile, the relative error of inclusion size larger than 100μm is less than 15% by the 6dB-drop threshold.

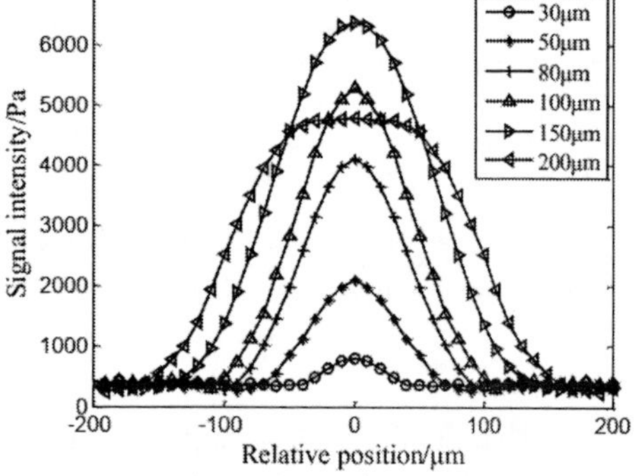

Figure 3. Signal intensity distribution of of different inclusion sizes.

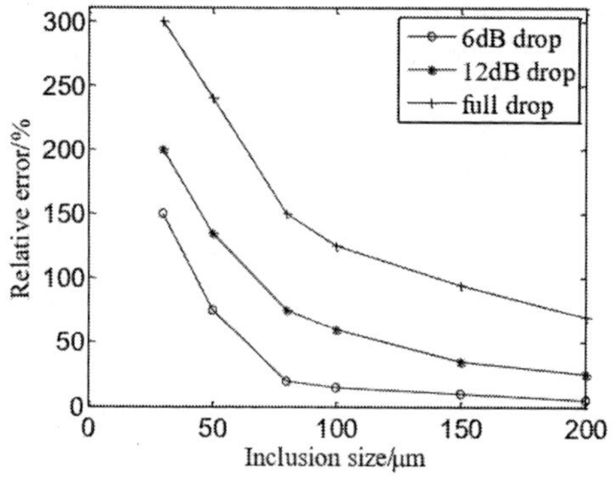

Figure 4. Relative error variation of different inclusion sizes.

4. Other factors on signal intensity distribution

For the inclusion sizing by ultrasonic C-scan imaging, the inclusion type, shape and orientation can affect signal intensity distribution, which are discussed in the following sections.

4.1. Inclusion type

Two typical types of inclusions Al_2O_3 and MnS are selected to analyse the influence of inclusion type on signal intensity distribution. The material properties of steel and inclusions are listed in table 1.

Table 1. Material properties.

Material	Density(kg/m^3)	Poisson ratio	Young's modulus(GPa)	Sound velocity(m/s)	Acoustic impedance (MRayl)
Steel	7890	0.27	209	5753	45.39
Al_2O_3	3960	0.23	375	10478	41.49
MnS	4057	0.30	138	6767	27.45

The inclusion shape is ellipse and its macro axis and minor axis are 50μm and 20μm, respectively. As shown in figure 5, the signal intensity of MnS is larger than that of Al_2O_3 at the same position. According to the Snell's law, the signal intensity is proportional to the difference of acoustic impedance. As shown in table 1, the difference of acoustic impedance between MnS and steel is 17.86 MRayl, which is larger than that of Al_2O_3 and steel (3.9 MRayl). It means that the image contrast of MnS is higher than that of Al_2O_3, which make it possible to distinguish inclusion type by the contrast of C-scan image under the same test parameters.

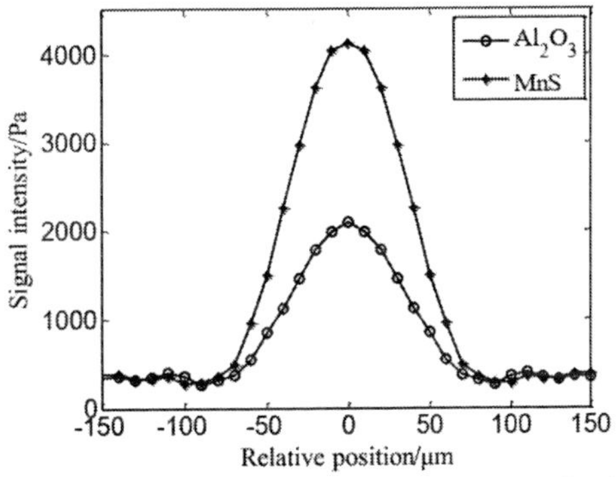

Figure 5. Signal intensity distribution of different inclusion types.

4.2. Inclusion shape

The MnS inclusion with ovality of 1:1, 5:2 and 25:3 are selected to analyze the influence of inclusion shape on signal intensity distribution. As shown in figure 6, the inclusion with ovality of 2:5 has the highest signal intensity in each position and the inclusion with ovality of 25:3 has the lowest signal intensity. When the inclusion's ovality is 25:3, its minor axis is 12μm, which is much less than the wavelength (120μm). According to the reflection law in thin-layer media[6], the sound pressure reflectivity is proportional to the inclusion's thickness. This result, from a certain extent, explains the reason why the MnS inclusion is difficult to be detected in the rolled steel.

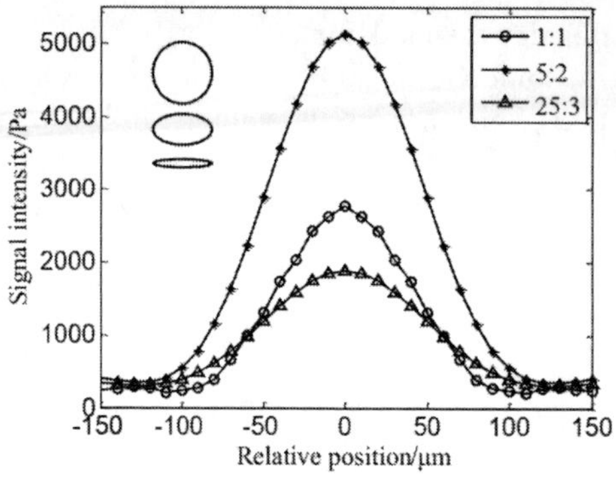

Figure 6. Signal intensity distribution of different inclusion shapes.

4.3. Inclusion orientation

The inclusion orientation is defined as the angle between macro axis of inclusion and incident direction of ultrasonic wave. Figure 7 shows the signal intensity distribution of Al_2O_3 inclusion with orientation of 45° and 90°. The inclusion's ovality is 5:2 and its macro axis is 100μm. As shown in figure 7, the signal intensity of inclusion with 45° orientation is much lower than that of inclusion with 90° orientation. The signal intensity distribution of inclusion with 45° orientation has two peaks, i.e. position 1 and 2 in figure 7. Figure 8 illustrates the relative position between transducer and inclusion. Along the moving direction, when the transducer is located in position 1, the scattering cross section of incident wave reached to the maximum and the top-surface of inclusion is located in the focal region, which leads to the signal intensity maximized. While, with transducer further moving the scattering cross section is invariant but deviate from the focal region, which leads to the signal intensity decreased. In the position 2 the top surface of inclusion is vertical to the incident wave, which leads to the signal intensity increased again.

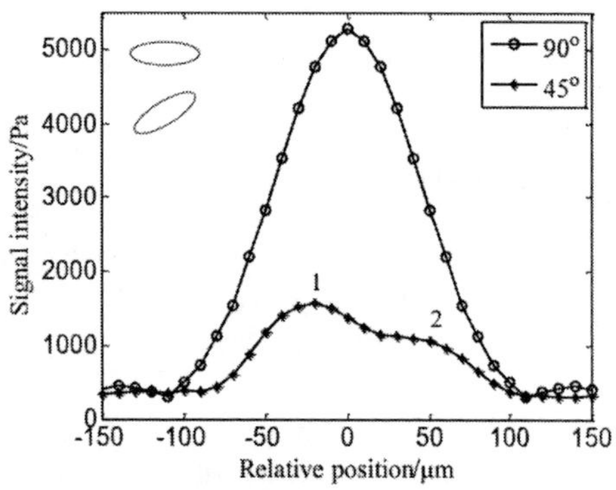

Figure 7. Signal intensity distribution of different inclusion orientations.

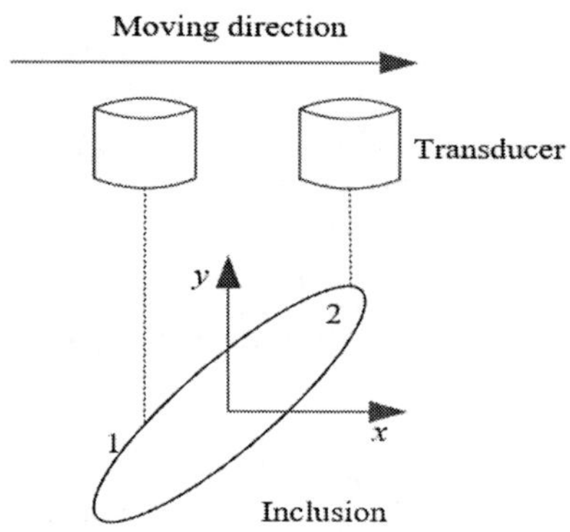

Figure 8. Schematic of transducer position

5. Experiment validation

The Acoustic Scanning Microscope SAM 300 is used to verify the accuracy of simulated results. The tested sample is interstitial steel (IF steel) that passed through cold-rolling. Table 2 lists the experimental parameters of C-scan imaging, which result is showed in figure 9. By cutting, grinding and polishing, the maximum cross-section of inclusion that exposed to the specimen's surface is 54μm in figure 10. The signal intensity distribution of this inclusion is compared with the grey value curve signed in figure 9 with dot-dash line. As shown in figure 11, the simulated result is consistent well with the experimental result, which proves the correctness of our model.

Table 2. Experimental parameters.

Center frequency(MHz)	Focus length(mm)	Numerical aperture	Step size(μm)	Focus depth(mm)	Gain(dB)	Sample frequency(GHz)
50	10	0.15	3	0.5	6	5

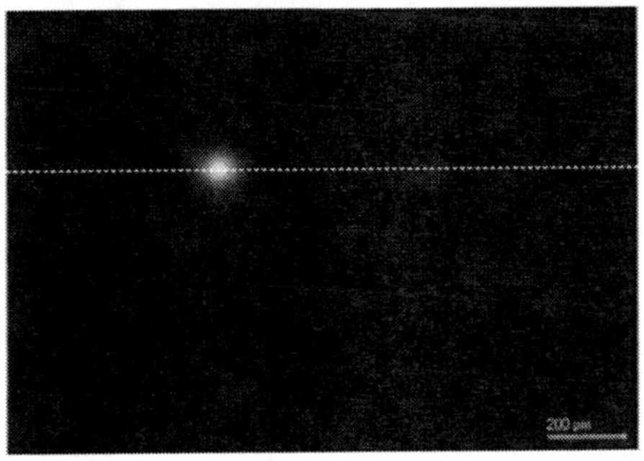

Figure 9. C-scan image of inclusion.

Figure 10. SEM photo of inclusion.

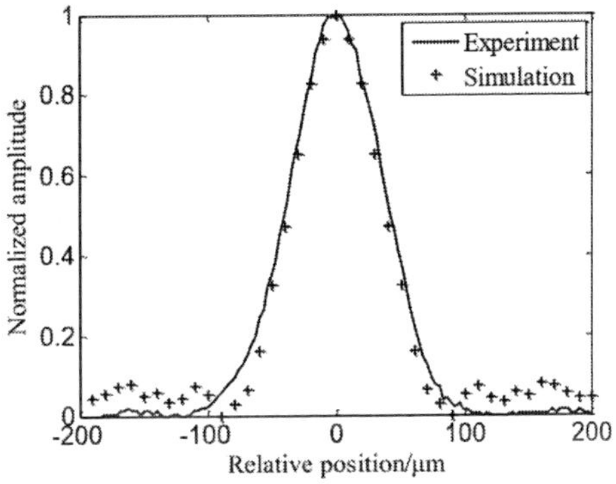

Figure 11. Comparison of Simulated and experimental results.

Conclusions

In this paper, the signal intensity distributions of inclusions with sizes ranged from 30μm to 200μm are acquired by the two-dimensional model that consists of spherically focused transducer, water couplant and steel with embedded inclusion. By comparing with the 12dB-drop threshold and the full-drop threshold the 6dB-drop threshold has the smallest relative error on inclusion sizing. For the same size and shape, the signal intensity of MnS is higher than that of Al_2O_3, which makes it possible to determine inclusion type by the contrast of C-scan image under same test parameters. The inclusion shape and orientation have serious impacts on signal intensity distribution, which make it more difficult to evaluate inclusion size.

Acknowledgement

This work was supported by the National Science and Technology Support Program of China [grant number 2015BAF30B01].

References

[1] Sahai Y 2016 Tundish technology for casting clean steel: a review. *Metallurgical and Materials Transactions B*, vol. **47**(4), pp 2095-2106

[2] Lei Z, Hong Y, Xie J, Sun C and Zhao A 2012 Effects of inclusion size and location on very-high-cycle fatigue behavior for high strength steels. *Materials Science and Engineering: A*, vol. **558**(15), pp 234-241

IOP Conf. Series: Journal of Physics: Conf. Series **842** (2017) 012003 doi:10.1088/1742-6596/842/1/012003

[3] Bartosiaki B G, Pereira J A M, Bielefeldt W V and Vilela A C F 2015 Assessment of inclusion analysis via manual and automated SEM and total oxygen content of steel. *Journal of Materials Research and Technology*, vol. **4**(3), pp 235-240
[4] Takada H, Tomura Y, Aratani M, Yamasaki T and Sasaki T 2011 On-line detection system for internal flaws in as-hot-rolled steel strip using ultrasonic probe array. *Materials Transactions*, vol. **52**(3), pp 531-538
[5] Mihaljević M, Markučič D and Runje B 2015 Analysis of influential factors for ultrasonic disc size evaluation. *Materials Testing*, vol. **57**(7-8), pp 674-679
[6] Brouard B, Lafarge D and Allard J-F 1995 A general method of modelling sound propagation in layered media. *Journal of Sound and Vibration*, vol. **183**(1), pp 129-142

IOP Conf. Series: Journal of Physics: Conf. Series **842** (2017) 012004 doi:10.1088/1742-6596/842/1/012004

The modal surface interpolation method for damage localization

Maria Pina Limongelli

Department of Architecture, Built environment and Construction Engineering, Politecnico of Milan
Piazza Leonardo da Vinci,32. 20133Milan. ITALY.

mariagiuseppina.limongelli@polimi.it

Abstract. The Interpolation Method (IM) has been previously proposed and successfully applied for damage localization in plate like structures. The method is based on the detection of localized reductions of smoothness in the Operational Deformed Shapes (ODSs) of the structure. The IM can be applied to any type of structure provided the ODSs are estimated accurately in the original and in the damaged configurations. If the latter circumstance fails to occur, for example when the structure is subjected to an unknown input(s) or if the structural responses are strongly corrupted by noise, both false and missing alarms occur when the IM is applied to localize a concentrated damage. In order to overcome these drawbacks a modification of the method is herein investigated. An ODS is the deformed shape of a structure subjected to a harmonic excitation: at resonances the ODS are dominated by the relevant mode shapes. The effect of noise at resonance is usually lower with respect to other frequency values hence the relevant ODS are estimated with higher reliability. Several methods have been proposed to reliably estimate modal shapes in case of unknown input. These two circumstances can be exploited to improve the reliability of the IM. In order to reduce or eliminate the drawbacks related to the estimation of the ODSs in case of noisy signals, in this paper is investigated a modified version of the method based on a damage feature calculated considering the interpolation error relevant only to the modal shapes and not to all the operational shapes in the significant frequency range. Herein will be reported the comparison between the results of the IM in its actual version (with the interpolation error calculated summing up the contributions of all the operational shapes) and in the new proposed version (with the estimation of the interpolation error limited to the modal shapes).

1. Introduction

Monitoring the health condition of plates is an important aspect for the assessment of the global structural condition for several application of both civil, mechanical and aeronautical engineering. The prompt identification of damage allows interventions that can increase service life with a reduction of life-cycle costs or, in emergency situations, reduce the downtime or increase the safety with respect to collapse. Damage detection techniques for plates, as well as for beams, traditionally consist in visual inspection and/or non-destructive testing (acoustic, ultrasonic, magnetic field, thermal field methods). A different approach consists in vibration based methods detecting changes of feature related to damage. Several features have been proposed in literature based on modal parameters such as frequencies, modal shapes or their derivatives. In reference [1] a comprehensive state of the art of modal methods is reported. One drawback of modal methods is related to the need of estimating modal parameters introducing error related to experimental noise that sometime hampers a correct identification of damage. This makes more feasible, approaches based on the use of operational deformed shapes [2]-[4]. The Interpolation Method (IM) is a damage localization method based on the knowledge of the Operational Shapes and that has been successfully applied to beam-like structures and, in its 2D formulation (SIM: Surface Interpolation method) for plate-like structures [5]-[6]. Recently [7] the IM has been proposed in its modal version MIM (Modal Interpolation Method). The difference between the IM and the MIM is that only the modal shapes (instead of all the operational shapes on the assigned frequency range) are considered in the computation of the error function that measures the distance between the undamaged and the damaged configurations of the structure. The aim of this paper is to present the modal version of the SIM that will be called MSIM (Modal Surface Interpolation Method) that takes into account only the modal shapes in a given frequency range for the estimation of the error function. In the following, after the presentation of the method, a numerical example is proposed to compare results that can be obtained through the modal (MSIM) and the non modal (SIM) versions of the algorithms based on the accuracy of results. The outline of the paper is as follows. In section 2 after a brief recall of the surface interpolation method (SIM), the extension to its modal version (MSIM) is presented. In section 3 the results of a numerical analysis carried out by applying the method in its two versions are reported and a discussion follows in section 4.

2. The modal surface interpolation method

The Surface Interpolation Method was previously presented in references [5] and [6]. Herein a brief recall is reported to facilitate the presentation of its 'modal version' that is based on the use of a limited number of operational shapes corresponding to resonances instead of the total family of operational shapes over the frequency range of interest. The deformed shape of a plate like structure can be described through a function of two variables x and y. If the deformed shape does not present sharp variations of slope and/or curvature and it is known at a certain number of points, a good candidate as interpolating function can be a smooth bi-cubic spline interpolation that guarantees continuity of first and second derivatives. If this is not the case, the interpolation through a smooth function gives rise to an interpolation error whose magnitude depends on the amplitude of the discontinuity in the first or second derivative. The Surface Interpolation Method (SIM) is a method of damage localization in plates that uses the variation of the interpolation error between two different configurations as damage detecting feature. Specifically, the increase of the interpolation error at a given location allows detecting a reduction of stiffness (increase of curvature) at that location. A bi-cubic spline can be defined as follows. Given the values defined on a rectangular grid of $n_{l+2} \times n_{p+2}$ nodes, consider the following equation:

$$s(x,y):\; s_p(x,y)=\sum_{j=0}^{3} c_{jp}(x)(y-y_p)^j \quad y\in[y_p, y_{p+1}] \qquad \text{with } p=0, 2..., n_p \qquad (1)$$

For each value of x, equation (1) gives a spline function along y (see black curves in Figure 1). When x varies, the coefficients $c_{jp}(x)$ also vary and a family of cubic spline functions $s(x,y)$ is obtained.

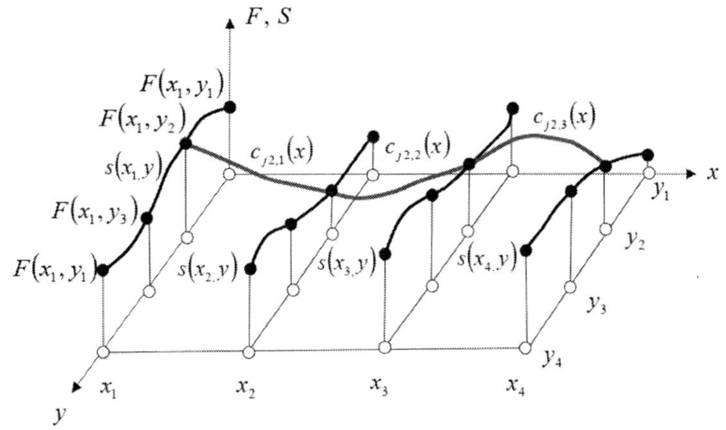

Figure 1. Bi-cubic spline interpolation.

If also the coefficients $c_{jp}(x)$ are defined by cubic spline functions (see red line in Figure 1) in each sub-area defined by $x \in [x_l, x_{l+1}]$; $y \in [y_p, y_{p+1}]$ they can be written as:

$$c_{jp}(x): \qquad c_{jp,l}(x) = \sum_{i=0}^{3} a_{jp,il}(x - x_l)^i \qquad x \in [x_l, x_{l+1}];\ y \in [y_p, y_{p+1}]; \tag{2}$$

p=0, 2..., n_p *and* l=0, 2..., n_l

The function $S(x,y)$ is a bi-cubic spline function defined through its restrictions to sub-areas as:

$$S(x,y): \qquad S_{l,p}(x,y) = \sum_{j=0}^{3}\sum_{i=0}^{3} a_{jp,il}(x - x_l)^i (y - y_p)^j \quad x \in [x_l, x_{l+1}];\ y \in [y_p, y_{p+1}] \tag{3}$$

The coefficients $a_{jp,il}$ of the bi-cubic spline function are calculated knowing the values at the nodes (x,l) and, once the coefficients $a_{jp,ol}$ are known, the bi-cubic interpolation can be finally calculated using equation (3). In the frequency domain the bi-cubic spline can be used to interpolate the deformed shape of the plate at a given frequency (usually defined as operational shape OS) and a different interpolating cubic spline can be defined at each frequency values within a given frequency range of interest. This approach has been followed in the previous applications of the SIM. For each operational shape a value of the interpolation error can be computed and, at each node, a value of the error over the considered frequency range, can be calculated by summing up the interpolation error over the frequency range of interest. For each configuration of the plate and for each node a value of the interpolation error can be defined in this way. This error quantifies the smoothness of the deformed shape of the plate at the node: the higher the error, the lower the smoothness. The comparison between the values of the nodal interpolation error in two different configurations allows to detect if a variation of smoothness is occurred thus to detect a damage. This formulation is effective if the OSs can be reliably estimated from recorded responses or from Frequency Response Functions (FRFs) or from Transmissibility Functions in case of unknown input. This is usually the case if measurement noise in recorded responses is not too high. The computation of the interpolation error at one node can be carried out with reference to a subset of Operational Shapes in a given frequency range. At resonance the effect of noise is usually lower with respect to other frequency values hence the modal shapes are usually estimated with a higher reliability with respect to operational shapes at other frequencies. In real world situations recorded signals are always affected by noise thus it is interesting to understand if better results could be obtained by defining the nodal interpolation error in a given frequency range as a function of the modal shapes in that range or as a function of the all operational shapes in the same range. The definition of the nodal interpolation error $E_i(z_i)$ as a function of the only modal shapes Assume that the beam is instrumented with n_s sensors allowing to identify n_s components of N modal shapes. The spline interpolation $\hat{\varphi}_l^{(i)}$ of the i-th modal shape at the l-th location z_l (l=1,...n_s) can be calculated using the following relationship:

$$\hat{\varphi}_l^{(i)} = \sum_{j=0}^{3} c_{j,l}^{(i)} \left(z_l - z_{l-1}\right)^j \tag{4}$$

Where the coefficients $c_{j,l}^{(i)}$ are calculated from the n_s-1 values of the modal shape $\varphi^{(i)}$ at locations z_k ($k \neq l$) imposing continuity of the spline function and of its first and second derivative at all the n_s locations:

$$c_{j,l}^{(i)} = g\left(\varphi_k^{(i)}\right) \qquad k \neq l \tag{5}$$

The interpolation error of the i-th mode at location z_l is defined as the magnitude of the difference between the real and the interpolated l-th components of the i-th modal shape:

$$E_i\left(z_l\right) = \left|\varphi_l^{(i)} - \hat{\varphi}_l^{(i)}\right| \tag{6}$$

If N modes are identified for the considered beam, the total interpolation error at location z_l is given by the norm:

$$E\left(z_l\right) = \sqrt{\sum_{i=1}^{N} E^2\left(z_l\right)} = \sqrt{\sum_{i=1}^{N} \left|\varphi_l^{(i)} - \hat{\varphi}_l^{(i)}\right|^2} \tag{7}$$

The values of $E\left(z_l\right)$ at all the locations z_l of the structure where sensors are available (that is at all locations where the components of the modal shapes are identified), characterize the current status of the structure that is are the 'signature' of the beam in a given configuration. A variation of this function at a certain location z_l, reflects a change in the structural condition (variation of curvature) : the higher the change, the higher the variation of the error $E(z_l)$. The interpolation error is thus assumed as a damage parameter and its variations between an inspection (I) phase and a reference (R) phase allow to estimate the damage feature at each instrumented location z_l :

$$\Delta E\left(z_l\right) = E_I\left(z_l\right) - E_R\left(z_l\right) \tag{8}$$

In real world conditions, due to several sources of variability influencing recorded responses, the modal shapes, hence the damage parameter $E\left(z_l\right)$ can change even if no damage occurs or, viceversa, it can exhibit no changes when damage exist, leading to false or missing detection of damage. For short term monitoring, that is if only data retrieved from one test in the undamaged and one test in the inspection phases are available, a statistical analysis of the damage feature is not possible. In this case a threshold can be defined in terms of a given percentile of the distribution of the damage feature at all the instrumented locations. Namely the threshold E_T can be defined as:

$$\Delta E_T = \mu_{\Delta E} + \nu \cdot \sigma_{\Delta E} \tag{9}$$

being μ_E and σ_E mean and standard deviation of the distribution in the inspection phase and ν the value $Z_{1-\alpha}$ of the standard Normal distribution, corresponding to the percentile α. This corresponds to assume as damaged locations the ones where the variations of the interpolation error are among the highest of all the instrumented locations. The classification of a certain z_l as a "damaged location" is thus carried out basing on the comparison of the current value of the damage feature $\Delta E\left(z_l\right)$ with the threshold E_T :

if $IDI(z_l) = \Delta E\left(z_l\right) - \Delta E_T > 0$ damage at z_l

if $IDI(z_l) = \Delta E\left(z_l\right) - \Delta E_T < 0$ no damage at z_l

3. Numerical example

In order to compare the results given by the SIM with those obtained with its modal version MSIM, the proposed method has been applied to the numerical model of the plate in Figure 2.

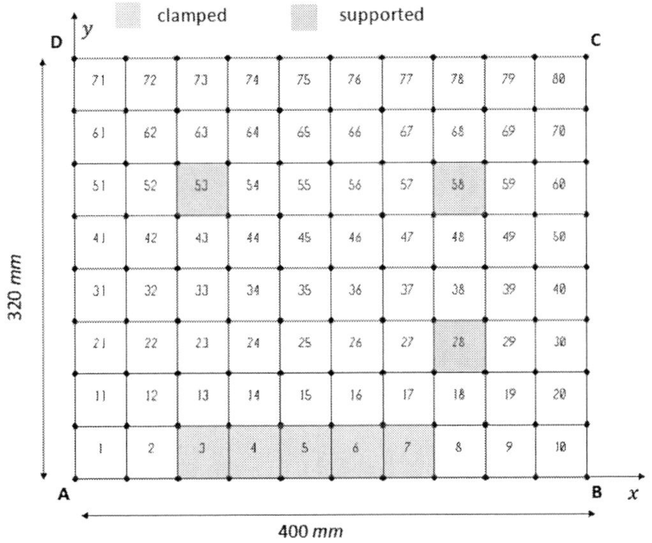

Figure 2. Numerical model of the plate and damage scenarios 1 and 2

The same numerical example has been proposed in references [8] and [9] using as a damage feature the modal deflection under a uniform unit load. The size of the plates is 420mm x 320mm with a thickness of 20mm. A finite element model of 7x10 plate elements has been built in SAP2000 [10] and reported in Figure 2. The two plates differ for the boundary conditions: the first one is clamped along the edge AB; the second one is supported on the four corners A, B, C, D and with free boundary conditions at all the other boundary joints. A total of 88 four nodes "area" elements 1m x 0.40mm x 0.20cm are used to model the plate. The material has Young's modulus in the undamaged configuration equal to 25000MPa, Poisson's ratio ν=0.3 and mass density ρ=2800kg/m^3.

Table 1. Damage scenarios

Scenario		Element n.	Damage
1	Clamped plate	3 and 7	15%
		4 and 6	25%
		5	50%
2	Supported plate	28	15%
		58	25%
		53	50%
3	Clamped plate	7	25%
		8 and 9	50%
		10	75%
4	Supported plate	36 and 44	50%

Damage has been modeled by reducing the Young's modulus in the elements located where the development of damage is more likely to occur. Namely for the clamped plate damage has been assumed located along the clamped edge in Damage scenario 1 and at one corner in Damage scenario 3; for the supported plate damage has been considered located around the middle span, in 3 separate location for Damage Scenario 2 and at two close locations for Damage Scenario 4.

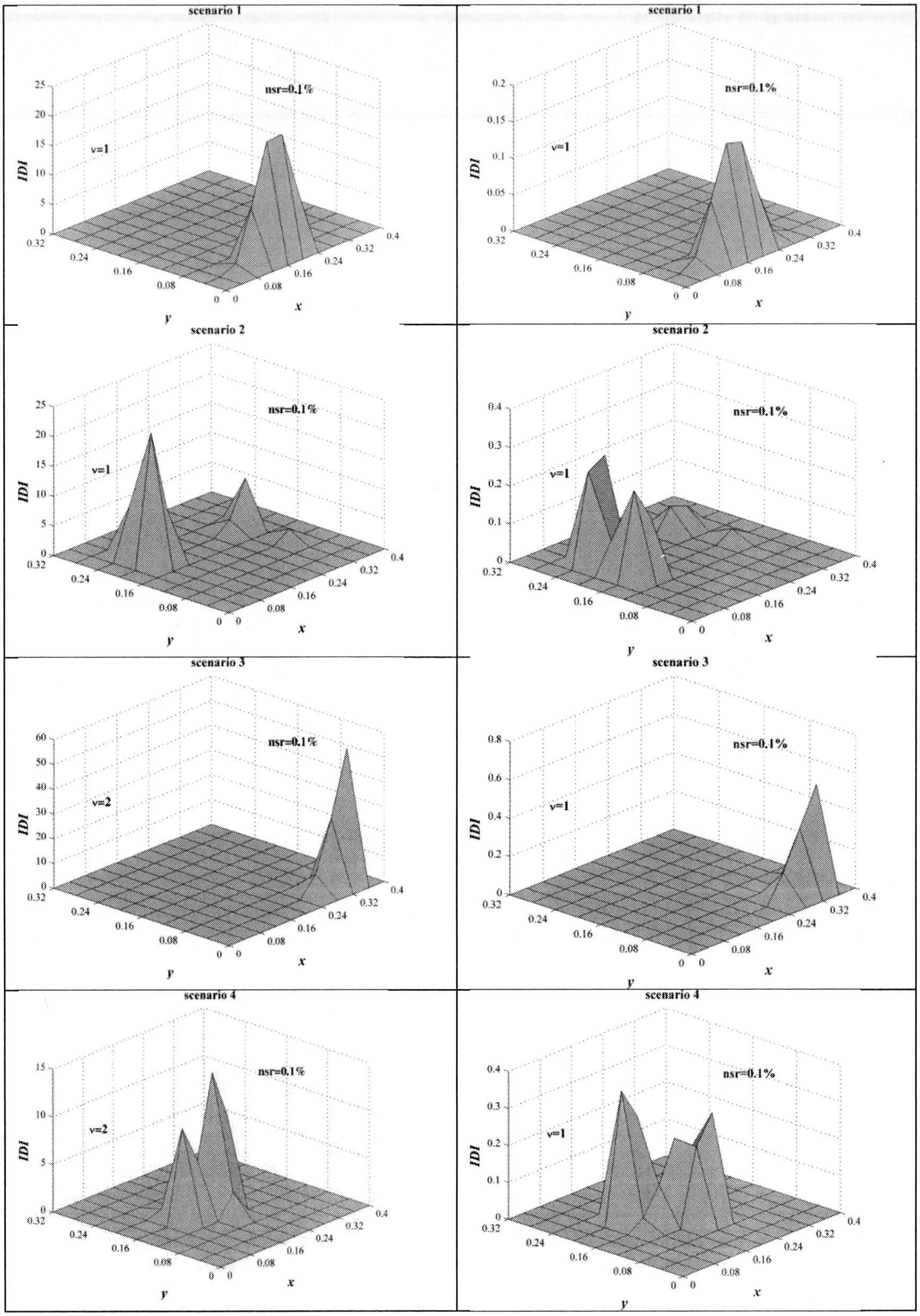

Figure 3. Comparison of results SIM (left) and MSIM (right) for 5 modes

The labels of the damaged elements and the relevant reductions of Young's modulus are reported in Table *1*. The time history response of the two plates to a random acceleration applied at supports, have been calculated both in the undamaged and in the damaged configurations, assuming a linearly elastic behavior and a sampling frequency of 5000Hz. The responses in terms of absolute accelerations have been used, together with the random excitation at supports, to calculate the frequency response functions (FRFs) at all the nodes of the plates and, following the procedure described in section 2, the interpolation error E in the undamaged and in the damaged configurations has been calculated at each node. The interpolation error has been calculated extending the summation of equation (7) to the range of frequencies 0-2500Hz that is up to the Nyquist frequency. Modal shapes retrieved form modal analysis of the numerical model have been used for the application of the MSIM. In this first application of the method noise due to sensors or environmental parameters has been neglected.

4. Discussion of results

Results for scenarios 1 to 4 are reported in Figure 3 for the SIM (left) and for the MSIM (right). These results have been obtained assuming that responses are not affected by noise: the noise to signal ratio is nsr=0.1%. In the application of equation (8) the parameter ν has been assumed equal to 1 (corresponding to a tolerable probability of false alarms P_f=15%).

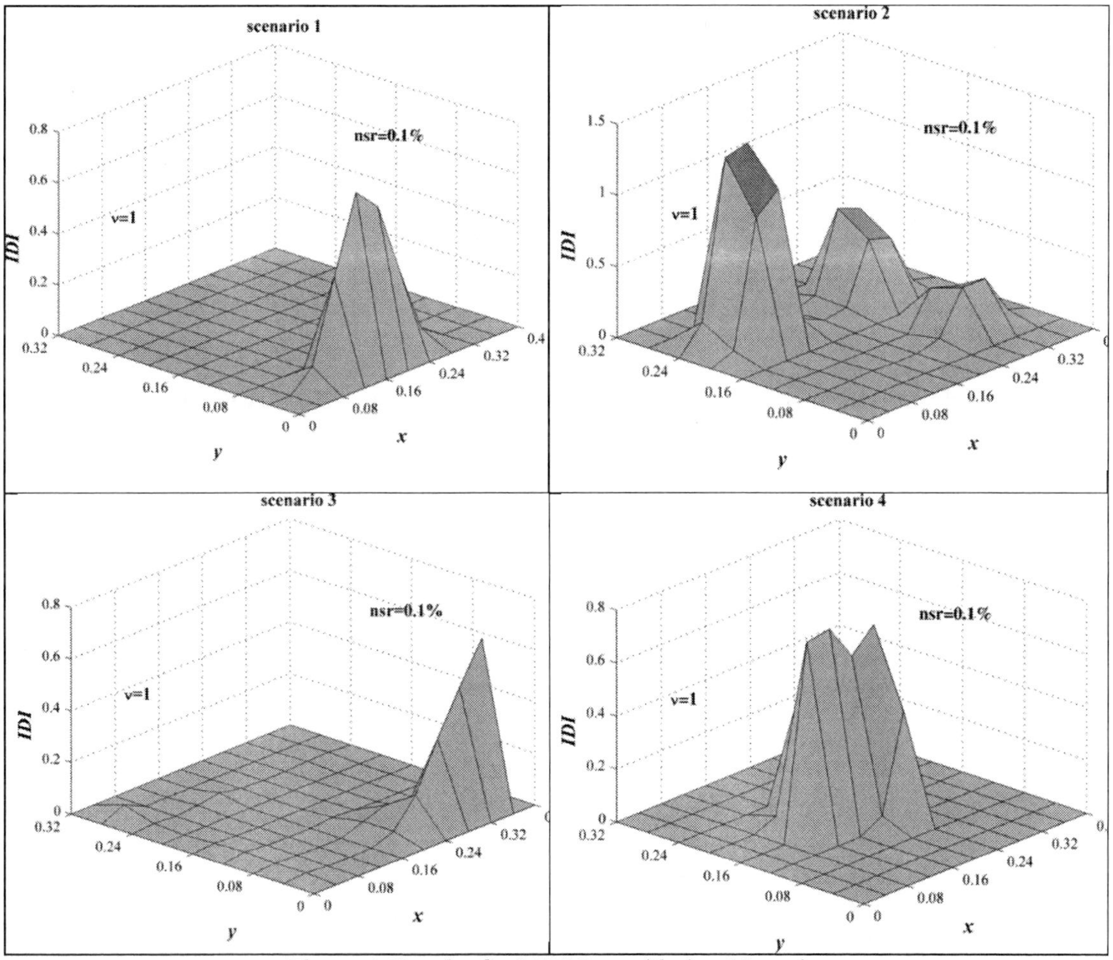

Figure 4. Results for MSIM considering 10 modes

The SIM was applied considering the frequency range between 0 and 2500Hz (Nyquist frequency) and the MSIM considering only the first 5 modes that is the one which, in a real application, would have been the only ones identifiable, given the value of the Nyquist frequency. For both the SIM and the MSIM the peaks of function IDI are correctly located at the nodes of the damaged elements clearly indicating the correct location of damage. The value of the damage index IDI is of course much lower when the MSIM is applied, due to the lower number of terms (5 modes) in the sum in equation (7). Increasing the number of modes furtherly improves the quality of the detection as shown by Figure 4 that reports results obtained considering 10 modes. The values of the damage index are higher due to the higher number of terms in the sum (see equation 7) thus the damaged regions of the plates are mode evidently singled out. However in all cases obtained with the SIM, that is considering the operational shapes in the frequency range of the fundamental modes, are always more clear with respect to those relevant to the application of MSIM. As already remarked, in all cases considered herein the effect of noise from different sources (measurement, environmental factors, etc.) has not been simulated neither for the SIM nor for the MIM. Noise in experimental data is one of the major problems for almost all damage detection algorithms since it is one of the principal sources of false or missing detections of damage thus reducing the reliability of results. Further analysis will be performed in the future in order to compare the SIM and the MSIM taking into account the effect of noise. It is expected that, due to the larger number of deformed shapes that are taken into account, the effect of noise could be better filtered in the SIM with respect to the MSIM.

References

[1] Fan W, Qiao P (2010) Vibration-based Damage Identification Methods: A Review and Comparative Study. Struct Heal Monit 10:83–111.

[2] Ratcliffe CP 2000 A frequency and curvature based experimental method for locating damage in structures. *Transactions of the American Society of Mechanical Engineers. Journal of Vibration and Acoustics*, 122: 324–329

[3] Domaneschi M., Limongelli M.P., Martinelli L. (2013). "Multi-Site Damage Localization in a Suspension Bridge via Aftershock Monitoring", *International Journal of Earthquake Engineering*, 3, pp. 56-72.

[4] Domaneschi M., Limongelli M.P., Martinelli L., (2015) "Damage detection and localization on a benchmark cable-stayed bridge", *Earthquake and Structures*, vol. 8(5), p.1113-1126.

[5] Limongelli M.P. (2013). "Two dimensional damage localization using the interpolation method". DAMAS 2013. *10th International conference on damage assessment of structures. DAMAS 2013.* Dublin Ireland. July 8-10 2013.*Key Engineering Materials*. Vol. 569-570, 860-867

[6] Limongelli M.P., Carvelli V., (2015). Damage localization in a glass fiber reinforced composite plate via the surface interpolation method". *DAMAS 2015.- 11th International conference on damage assessment of structures.* Ghent University, Belgium. 24-26 August 2015. *Key Engineering Materials.*

[7] Limongelli M.P. (2016). Structural damage localization using the modal interpolation method. *8th European Workshop On Structural Health Monitoring (EWSHM 2016)*, 5-8 July 2016, Spain, Bilbao.

[8] Wu D. Law S.S. (2004). Damage localization in plate structures from uniform load surface in the SIM

[9] Wu D. S.S. Law (2005). Sensitivity of Uniform Load Surface Curvature for Damage Identification in Plate Structures. Transactions of the American Society of Mechanical Engineers. *Journal of Vibration and Acoustics*, 127: 84–92.

[10] SAP2000. 2017 Computers and Structures. https://www.csiamerica.com/products/sap2000.

[11] Zhan Z, Martin C F 1997 Convergence and Gibbs' phenomenon in cubic spline interpolation of discontinuous functions *Journal of Computational and Applied Mathematics* **87** p. 359-371.

IOP Conf. Series: Journal of Physics: Conf. Series **842** (2017) 012005 doi:10.1088/1742-6596/842/1/012005

Effect of vitro preservation on mechanical properties of brain tissue

Wei Zhang, Yi-fan Liu, Li-fu Liu, Ying Niu, Jian-li Ma, Cheng-wei Wu

State Key Laboratory of Structural Analysis for Industrial Equipment, Department of Engineering Mechanics, Dalian University of Technology, Dalian 116024, China

Email:cwwu@dlut.edu.cn

Abstract. To develop the protective devices for preventing traumatic brain injuries, it requires the accurate characterization of the mechanical properties of brain tissue. For this, it necessary to elucidate the effect of vitro preservation on the mechanical performance of brain tissue as usually the measurements are carried out in vitro. In this paper, the thermal behavior of brain tissue preserved for various period of time was first investigated and the mechanical properties were also measured. Both reveals the deterioration with prolonged preservation duration. The observations of brain tissue slices indicates the brain tissue experiences karyorrhexis and karyorrhexis in sequence, which accounts for the deterioration phenomena.

1. Introduction

The ever rising concerns on traumatic brain injuries caused by such impact loading conditions as car accidents, blasts, sports or even daily falls have stimulated the intensive mechanical modeling of brain tissue. The simulations may also find applications in computer-integrated and robot aided surgery, surgeon training and development of protective facilities[1,2]. There is no doubt that accurate mechanical property data of brain tissue is of critical importance to validate the constitutive models and therefore to ensure the reliability of simulation results. One question facing mechanical property characterization is how to preserve the brain tissue after the dissection. The common method used is to keep the brain tissue at 0-6°C[3-5]. Nonetheless, how this vitro preservation will affect the mechanical property has not been fully investigated. In this paper, the mechanical performance of brain tissue (both white and grey matter) preserved at 1°C in vitro for various duration was studied and Differential Scanning Calorimetry (DSC) measurements were also conducted to explore the thermal behavior of brain tissue. Finally, the brain tissue slices were observed using optical microscope with the attempt to study the mechanisms of variation of mechanical property and thermal behavior with prolonged preservation duration.

IOP Conf. Series: Journal of Physics: Conf. Series **842** (2017) 012005 doi:10.1088/1742-6596/842/1/012005

2. Experiments

2.1 Preparation of samples

Small tailed han sheep (kindly provided by Dalian Shun Feng Farm) were used. These sheep are at the same age (1.5 y) and are raised in the same manner. After the euthanasia, the sheep were dissected and the brain was kept for further measurements. All these were done with the permission and monitoring of Biological and Medical Ethics Committee of Dalian University of Technology.

2.2 DSC measurement

Differential Scanning Calorimetry (DSC) measurement were conducted on a Netzsch DSC-204, for each brain, one white and one grey matter specimens were taken from the corona radiate and thalamus, as illustrated in Figure 1. Brian tissue samples (5-10mg) were sealed in standard aluminum pans (40μl) and an empty pan was used as reference.

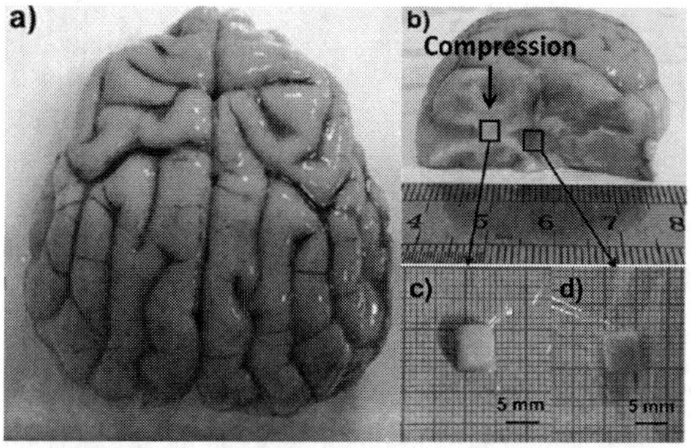

Figure 1. The schematic of brain tissue

2.3 Mechanical property measurement

The brain tissue were preserved at 1℃ for 0 h, 2 h and 14h. Using the method wedeveloped previously[6], pure white and grey samples with the size of 5 mm×5 mm×5 mm (±0.1 mm) of were obtained, as illustrated in Figure 1.

2.4 Preparation of brain tissue slice

The slices of brain tissue were prepared in reference to Cheng's method[7]. For the wax slices, steps includes selecting the materials then fixing, washing, dehydrating, transparent, dipping in wax, wrapping in wax, slicing, sticking, dewaxing and dying.

3.Results and discussion

3.1 DSC measurement

To characterize the effect of preservation condition on the thermal behavior, brain white matter of 5-10 mg were sealed in aluminum pan and then were kept at 1℃for 0 min and 5 min. Then the DSC curves were recorded, given in Figure 2. Likewise, the DSC curves of grey matter were also recorded as given in Figure 3.

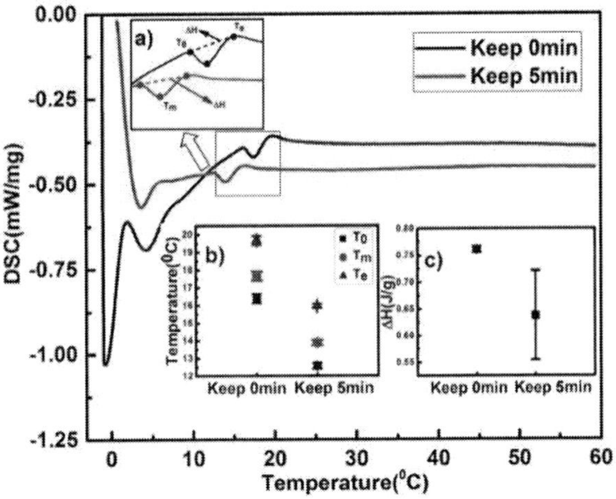

Figure 2. DSC curve of white matter

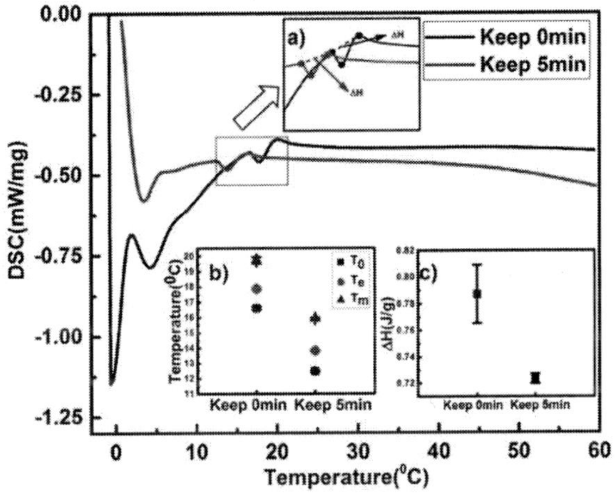

Figure 3. DSC curve of grey matter

Figure 2 indicated that for as-obtained white matter the phase transition takes places at 17.67℃and the corresponding entropy is 0.76J/g, whereas after kept for 5, the phase transition temperature and entropy decreases to 13.87℃ and 0.64J/g， respectively. Analogous trend can be observed for grey matter, see Figure 3. After preserved at 1℃ for 5 min, the phase transition temperature drops from 17.87℃ to 13.83℃ and entropy reduces from 0.79J/g to 0.72J/g.

Skrzyński et al [8] concluded that the high transition temperature can be related with the cross links of bio-molecules and consequently the mechanical properties of tissue. The higher the transition temperature, the higher density of cross linking and the better mechanical strength. As reflected by Figure 2 and Figure 3, the transition temperature drops with the prolongation of preservation time, indicating the reduction of mechanical strength of brain tissue. In terms of entropy, it also decreases with prolonged preservation, see Figure 2 and Figure 3. As entropy is indicative of capacity of absorption of energy, the decrease of entropy may suggest the deterioration of mechanical strength.

3.2 Mechanical performance in compression

To measure the low temperature preservation on the mechanical properties, white matter were preserved at 1℃for 0 h, 2 h and 4 h and then the engineering stress-strain curved in compression with a strain rate of are given 0.01/s in Figure 4.

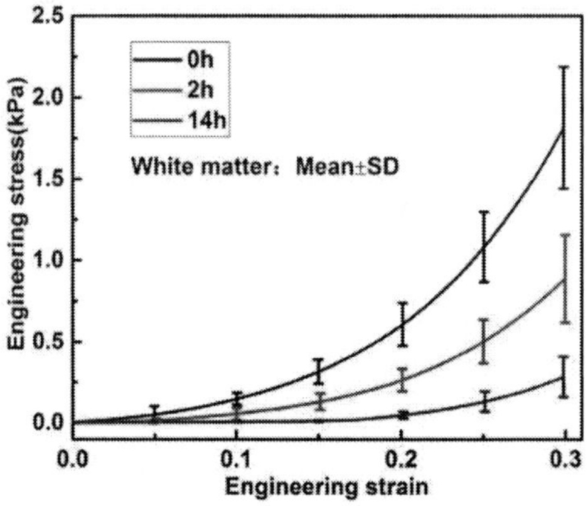

Figure 4. Engineering stress- strain curves of white matter in compression. 6 repetitions with ±standard deviations

Figure 4 demonstrates that at the strain of 0.3, the engineering stress decreases from 1.81 kPa of as obtained white matter to 0.89 kPa after preserved for 2h and then to 0.28 kPa after preservation of 14 h, indicating the pronounced decrease in mechanical strength with the prolongation of preservation time. Similar trend can be observed for grey matter as exhibited in Figure 5. At strain of 0.3, the engineering stress of grey matter of as-obtained tissue, preservation of 2 h and preservation of 14 h are 1.11 kPa, 0.66 kPa and 0.27 kPa, respectively.

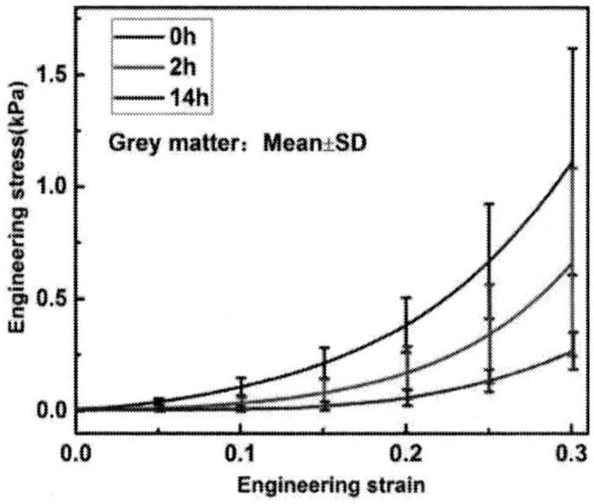

Figure 5. Engineering stress-strain curves of grey matter in compression. 6 repetitions with ±standard deviations

3.3 Observation of brain tissue slice

To investigate the mechanism of deterioration of mechanical strength with prolonged preservation time at low temperature. The brain white and grey matters were made into slice and observed on optical microscope. The images were taken using the ISCapturesoftware and are shown in Figure 6 and Figure 7.

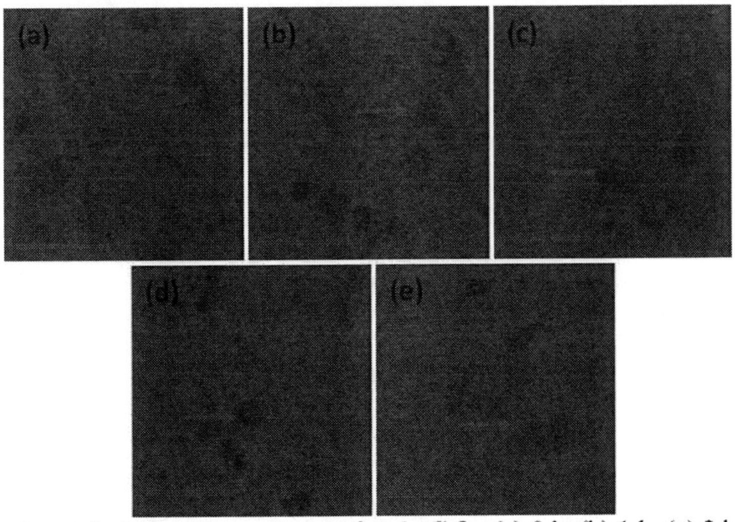

Figure 6. Cell nucleus of white matter: preserved at 1 °C for (a) 0 h, (b) 1 h, (c) 2 h, (d) 4 h, (e) 14 h. Scale bar represents 15 μm.

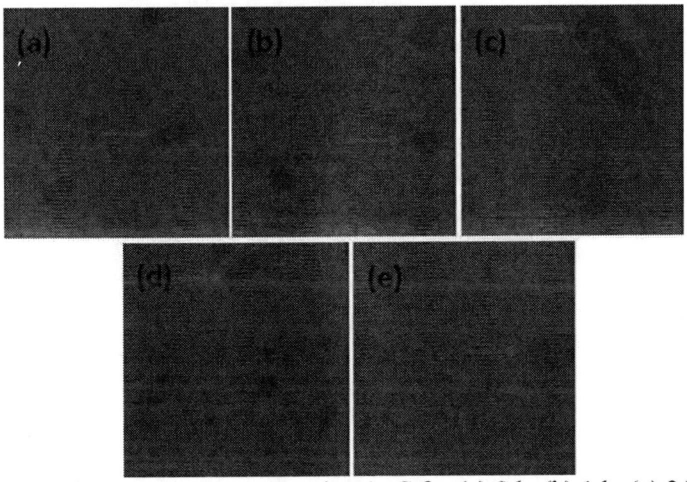

Figure 7. Cell nucleus of grey matter: preserved at 1 °C for (a) 0 h, (b) 1 h, (c) 2 h, (d) 4 h, (e) 14 h. Scale bar represents 15 μm.

Figure 6 exhibits that for the as-obtained white matter, the nuclei are large and the color is light. After preserved at 1°C for 1 h, the nuclei becomes smaller and the color becomes dark, indicating the pyconsis of cells. After preserved for 2 h, the nuclei turns out to be larger, the shape approaches to a circle and the color is light, implying the transition of nuclei from karyopycnosis to karyorrhexis. After preserved for 4h, the shape of nuclei turns from a circle to an irregular one. After 14 h, the nuclei becomes larger and the shape becomes indistinct, suggesting the karyolysis.

For the as-obtained grey matter, see Figure 7, the nuclei is a ellipse and its color is light. After preserved at 1℃up to 2h, the shape of nuclei turn to be a circle and no significant change can be observed in terms of color strength. After 4 h, the nuclei becomes smaller and the color turns to be much darker, indicating the karyopyknosis of nuclei. After preservation of 14 h, the nuclei becomes larger and thecolor becomes light, indicative of karyorrhexis.

Figure 6 and Figure 7 demonstrate that with the prolonged preservation at 1 °C the nuclei of both white and grey matters experiences karyopycnosis and then karyorrhexis, which can explain the corresponding deterioration of mechanical strength with increasing preservation duration. This may be further ascribed to effect of enzyme and the further investigation is ongoing.

4. Conclusion

The mechanical strength of both white and grey matters deteriorates with the preservation duration at 1°C. This is in agreement of the DSC measurements, which indicates the decrease of both the phase transition temperature and entropy of white and grey matters. Slice observation implies cell experiences karyopycnosis and karyorrhexis in sequence, which may takes responsibility for the deterioration of mechanical strength.

Acknowledgments

This work was supported by NSFC(11572080) and National Basic Research Program of China (2015CB057306).

References

[1] Miller K 2001 J. Biomech. **34** 651-657
[2] Taylor Z and Miller K 2005 J. Biomech. Eng. **37** 1263-1269
[3] Garo A, Hrapko M, Van Dommelen J A W and Peters G W M 2007 Biorheology **44**(1) 51-58
[4] Nicolle S, Lounis M and Willinger R 2004 Stapp Car Crash Journal **48**(48) 239-258
[5] Darvish K and Crandall J 2001 Med. Eng. Phys. **23**(9) 633-645
[6] Zhang, W, Feng, L L, Wu, F, Zhang, R R and Wu, C W (2016) Nanotechnology **27**(35) 355708
[7] Cheng X, Kang T, and Zhao Z 2007 J. NAT. MED. **61**(1) 51-55
[8] Skrzyński S, Sionkowska A, Marciniak A 2009 Journal of Biophysics 819635

IOP Conf. Series: Journal of Physics: Conf. Series **842** (2017) 012006 doi:10.1088/1742-6596/842/1/012006

Is it worth changing pattern recognition methods for structural health monitoring?

L A Bull, K Worden, E J Cross, N Dervilis

Department of Mechanical of Engineering, University of Sheffield, Sheffield, UK

Email: lbull1@sheffield.ac.uk

Abstract. The key element of this work is to demonstrate alternative strategies for using pattern recognition algorithms whilst investigating structural health monitoring. This paper looks to determine if it makes any difference in choosing from a range of established classification techniques: from decision trees and support vector machines, to Gaussian processes. Classification algorithms are tested on adjustable synthetic data to establish performance metrics, then all techniques are applied to real SHM data. To aid the selection of training data, an informative chain of artificial intelligence tools is used to explore an active learning interaction between meaningful clusters of data.

1. Introduction and problem statement

The main aim of structural health monitoring (SHM) is the development of diagnostic systems capable of detecting and classifying structural degradation. Machine learning techniques have become increasingly relevant to the field of SHM as advanced pattern recognition algorithms can learn complicated relationships from system data whilst offering a variety of techniques for novelty detection [1].

When following a data-based approach to SHM, a lack of information representing the damaged structure can result in a significant undersupply of labelled data. This makes machine learning tasks – such as classification – very difficult, as the information required for wholly supervised methods is not available. Consequently, a need for adaptive semi-supervised learning techniques, such as active learning, is very important for SHM.

The intention of this work is to demonstrate alternative strategies for using pattern recognition algorithms, and investigate their performances for SHM purposes. The main objective is to provide performance comparison metrics for a range of classification techniques, whilst illustrating an active learning methodology to select the training sets that carry the most information. Three datasets are introduced, two synthetic sets and an acoustic-emissions (AE) data set. The synthetic data is generated as a toy example in a two-dimensional feature space for SHM demonstration, and they will be used to quantify performance metrics for each classification algorithm. Each technique will then be applied to acoustic emission data, obtained from experiments concerning the box-girder of a bridge.

2. Background

2.1. *Classification algorithms*

The techniques presented in this paper are selected to provide performance metrics for alternative methods to the conventional choices in SHM. Some benchmark algorithms, such as multi-layer perceptrons, have been excluded from this work as their correct application makes these experiments highly computationally expensive and they have already been discussed extensively in an SHM context. Comprehensive comparisons including a wider range of algorithms have been suggested (see Section 6.1) and MLPs will be included in this wider remit.

Basic descriptions of the techniques tested in this work are provided, for more details the reader is directed to reference [2]. All algorithms are applied in Python, using the machine learning package *scikit-learn* [3]. During initial experiments, certain algorithm hyperparameters were defined through engineering judgement – relevant values are provided in braces. Of course, this is not good practice in practical engineering applications, as any hyperparameters must be optimised with the use of a validation subset (or otherwise); however, the elected values serve well to establish a general comparison for the purposes of this paper and will be optimised in future work.

2.1.1. Linear support vector machine (SVM). A sparse technique in which the learning task maximises the 'margin' around a linear decision boundary [2]. The margin is defined by the perpendicular distance between the decision boundary and the closest points either side (support vectors). A soft margin parameter C is introduced to allow for misclassification and prevent over fitting. $\{C = 1\}$ {stopping criteria = 1e-4} {multiclass strategy: one vs. rest classifiers}

2.1.2. Kernel support vector machine (KSVM). Applies the 'kernel trick' to SVMs, transforming data into some new space to make clusters more separable with a linear decision boundary [2]. This is achieved through the use of a specific kernel function in place of any inner products in the original space. $\{C = 1\}$ {kernels: polynomial 3rd degree γ=0.5 and Gaussian γ=0.5} {stopping criterion = 1e-4} {multiclass strategy: one vs. rest classifiers}

2.1.3. Relevance vector machine (RVM). A sparse kernel technique based on a Bayesian formulation that shares many characteristics with the SVM, but produces probabilistic outputs [4]. An Automatic Relevance Determination (ARD) prior distribution is placed over the SVM weight vector. This introduces precision hyper-parameters (α_i), which are optimised during training through likelihood maximisation. The use of an ARD prior over the parameter vector results in a number of the parameter weights concentrating at zero (large α_i); these weights have a small contribution to the classification, so if a given threshold is broken they are 'pruned' out. Relevance vectors are not available in *scikit-learn*, so an alternative implementation using the *scikit-learn* API is used here [5].
{kernel: linear} {threshold α_i = 1e09} {stopping criterion: 1e-03}

2.1.4. Decision tree (DT): A simple yet effective algorithm, shown in the literature to give excellent parametric performance, even when compared to more complex methods [6]. During classification, the classification and regression tree (CART) algorithm partitions the input/feature space into cuboidal regions according to threshold values that minimise an impurity measure [2]. This measure is assessed with the *Gini index*, which quantifies the 'mixture' of classes in each split; minimising this value encourages the formation of regions that contain a high proportion of data assigned to one specific class [2]. The tree will continue to grow, splitting at each node, until a predefined number of samples remain, or the Gini index falls below a given value. Given the greedy learning strategy of DTs, pruning techniques may be applied to reduce their size by removing the nodes that contribute least to classification, no pruning methods are applied here. {min. samples to split = 2} {threshold Gini = 1e-7}

Ensemble methods can be applied to any classifier with the aim to reduce the risk of overfitting [2]. As DTs are computationally inexpensive and ensemble methods have been shown to work well with the algorithm in the existing literature [6], the following ensemble methods are applied:

- Bootstrap aggregating (Bagging): 'Bootstrap' sample N random subsets with replacement from the training set. Train classifiers on each sample to create a committee of models and 'aggregate' by combining whilst minimising the unweighted average error of the base classifiers [2]. $\{N = 10\}$
- Boosting (*AdaBoost*): Combines multiple base classifiers; however, unlike bagging, the models are trained on the complete training set, with weights associated to each sample. Weights are updated during the learning process to prioritise misclassified samples from preceding models [2]. $\{N = 10\}$ {learning rate = 1}

2.1.5. Random forests (RF). Similar in concept to bagged DTs, however the algorithm is modified to introduce extra variability into the learner. Rather than using all the available features, the RF classifier selects a random subset of features to determine the best split at each node. $\{N = 10\}$
{min. samples to split = 2} {threshold Gini = 1e-7} {no. features in subset = 1}

2.1.6. Gaussian process classifiers (GPC). A probabilistic method that shares many characteristics with the RVM; however, the parametric models are abandoned and a Gaussian prior is placed over the basis functions directly [7]. Samples are then taken form this Gaussian process (GP) to create a technically infinite range of functions to describe the classification [7]. In the case of GPs, the kernel function is used to describe elements of the covariance matrix. Often a spherical covariance is assumed, but it is also possible to define an elliptical covariance using the Gaussian (rbf) kernel [7]; this introduces the hyper parameter γ_i, which is optimised during training through likelihood maximisation.
{kernel: Gaussian, initial $\gamma_i = 1$} {multi class method: one vs. rest classifier}

2.2. Active learning

Active-learning, or query-learning, is a semi-supervised machine learning approach; its main premise is that a learner can perform with increased efficiency given less training data, if it can select the data from which it learns. This eliminates the time-consuming process of labelling large volumes of trivial data, but more importantly for SHM, active learning significantly reduces the need for extensive training sets.

A wide range of active learning algorithms exist within the remit of machine learning. Some examples include the 'Active Support Vector Machine' (SVM_{active}) [8] and 'Manifold Adaptive Experimental Design' (MAED) [9]. Each employ various 'querying' methods to determine which unlabeled data carry the most information; measures include the distance from the decision boundary (SVM_{active}) or the average squared predicted error (MAED). The details of these algorithms are out of scope for this specific paper, however the key concept behind active learning tools is summarised in Figure 1.

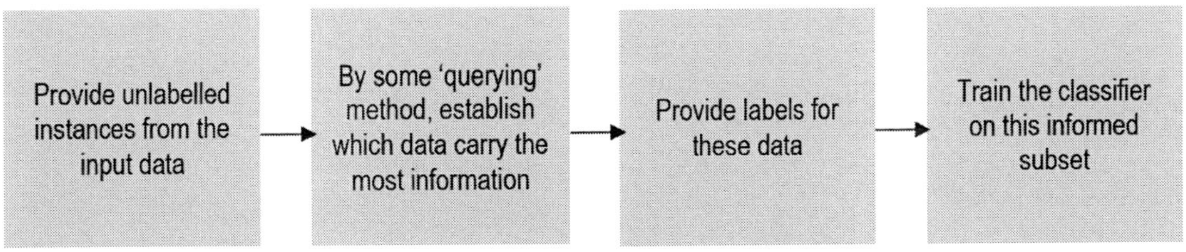

Figure 1. Active learning methodology.

2.3. Metric measures

To assess classification performance, accuracy (ACC) and F1-score ($F1$) metrics [10] will be used:

- Accuracy: percentage of correctly classified data.

$$ACC = \frac{TP + TN}{TP + TN + FP + FN} \times 100 \tag{1}$$

Where: true positives = TP, true negatives = TN, false positives = FP, false negatives = FN

- F1-score: the balanced average of precision (PRE) and recall (RE). A weighted average is used to combine PRE and RE for each class and account for class imbalance [10].

$$PRE = \frac{TP}{TP + FP} \tag{2}$$

$$REC = \frac{TP}{TP + FN} \tag{3}$$

$$F1 = 2 \times \frac{PRE \times REC}{PRE + REC} \tag{4}$$

2.4. Clustering – unsupervised

Various algorithms are available to find clusters in unlabelled data: examples include K-means [2], Affinity Propagation [11] and Gaussian Mixture Models (GMM) [2]. Initial experiments found that GMMs were particularly effective when applied to the data analysed in this work. K-means clustering is also referenced in Section 3.2, so brief descriptions are provided.

2.4.1. Gaussian mixture model (GMM). A linear superposition of K Gaussian components. After defining K number of components – and therefore clusters – a GMM is fitted to the data by updating parameters whilst using the Expectation-Maximisation algorithm [2]. {K = 3} {covariance = full} {stopping criterion = 1e-03}

2.4.2. K-means: Data is separated into some predefined number of clusters K. A cluster can be considered as a group of data whose inter-point Euclidean distances are small compared to points outside the cluster [2]. The algorithm aims to choose cluster centroids that minimise the cluster inertia or 'separation'. {K = 3} {stopping criterion = 1e-04}

3. Methodology

3.1. Data: synthetic

Synthetic 'toy' data have been generated using three 2D Gaussian distributions. These data aim to represent 3 classes of structural degradation in two-dimensional feature space, following dimension reduction. The classes are imbalanced and the total number of samples N_S = 1050 (class1=500, class2=250, class3=300). Two versions of these data were generated, a challenging set with significant cluster overlap (Figure 2) and a forgiving set with more separable clusters (Figure 3).

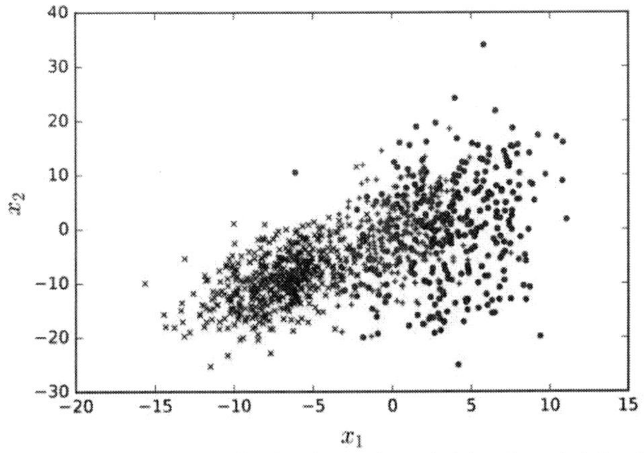

Figure 2. Challenging synthetic data: class 1 (×), class 2 (+), class 3 (○).

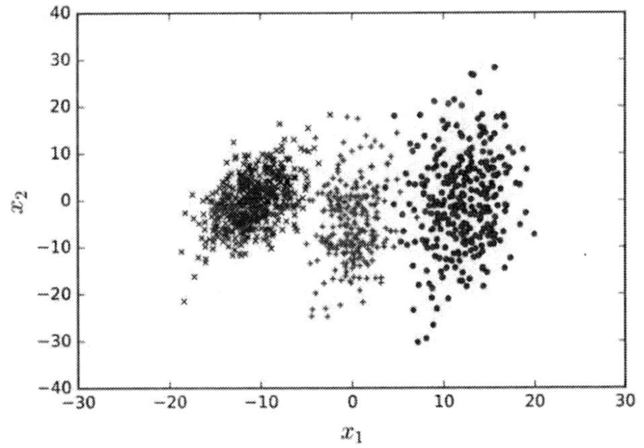

Figure 3. Easy synthetic data: class 1 (×), class 2 (+), class 3 (○).

3.2. Data: experimental acoustic emissions (AE).

Experimental data were obtained from tests at Cardiff University, concerned with collecting AE data from the box girder of a bridge [12]. Each sample represents a typical AE burst (Figure 4) and the total number of samples N_{AE} = 91. For further pre-processing and experimental details, see Reference [13]. Four traditional AE features – rise time, peak amplitude, duration and ring down count – are extracted from each burst, normalised, then projected though a linear transformation on to two dimensions using Principal Component Analysis (PCA) – Figure 5. The new co-ordinates, or principal component scores (x_1, x_2), are a linear combination of the information retained within the original features and account for the maximum variance within the data [2]. As definite labels are not available, K-means clustering is used to predict 3 class labels, which are then assumed as the definitive classes (class1=22, class2=59, class3=10).

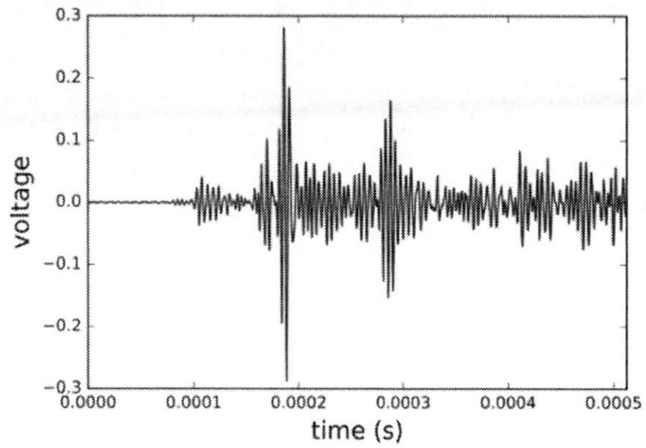

Figure 4. Time signal of a typical AE burst.

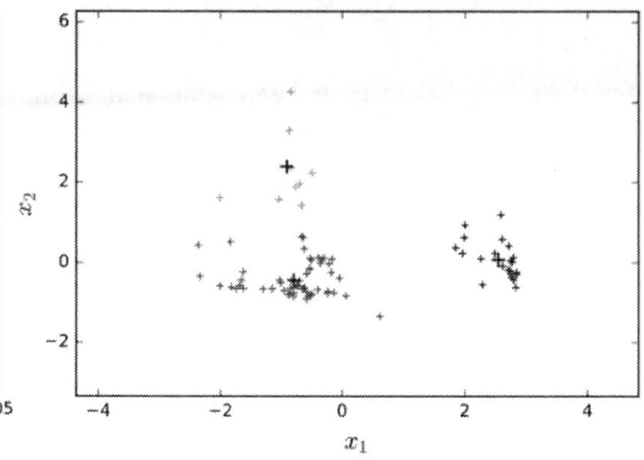

Figure 5. PCA reduction applied to AE features, labelled following K-means clustering (K = 3).

3.3. *Method*
Each data set is initially split into two subsets: a training set and test set. Basic active learning principles are demonstrated by selecting a given fraction of the most informative points for training – each classifier is then trained on these data. All the remaining data will be used as a test set to establish the accuracy and F1-score.

3.3.1. Selecting the training set. A simplified active learning regime is used to select 15% - 45% of the most informative synthetic data in 10% increments. Steps are listed below and illustrated in Figure 6:

I. Start with unlabelled data.
II. Fit a GMM model to find clusters within the data, without knowledge of labels.
III. Randomly sample a given fraction of data from each cluster to select representative examples.
IV. Provide class labels to this informed training-set.

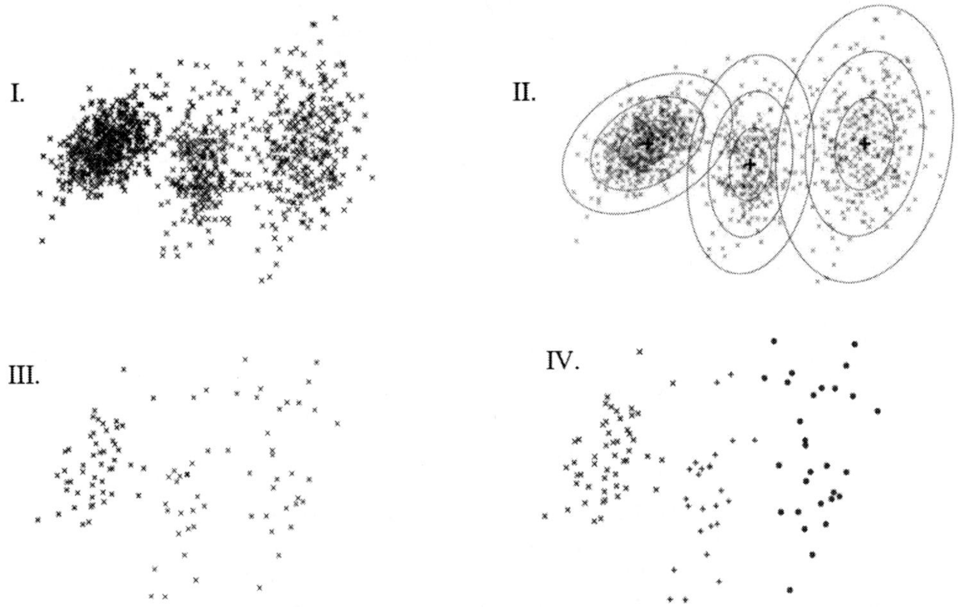

Figure 6. Visual representation of the active learning steps: 25% sample, easy synthetic data.

The querying regime used here hopes to provide a smaller, more informed training set than the alternative of one random sample across the whole data. Applying more sophisticated active learning algorithms to SHM is out of the scope of this paper but will be explored in future projects.

3.3.2. Data applications. For each training set size of the synthetic data, all classification algorithms will be trained and then tested 1000 times. The mean F1 and ACC, along with one standard deviation will be provided. Following analysis of the performance metrics, algorithms will be applied to the AE data to illustrate the relevance of this work to the remit of SHM. A 15% sample training set is used with the AE data, as this provided an array of performance metrics.

4. Performance metrics & discussion

The 'No Free Lunch Theorem' tells us that any two optimisation algorithms become equivalent when their performance is averaged across all possible problems [14]. In other words, classification results are highly data-dependant and there exists no single, optimal algorithm. The metrics presented in Figure 7 and Figure 8 are intended to illustrate algorithm behaviour across a range of scenarios to aid algorithm selection when investigating SHM.

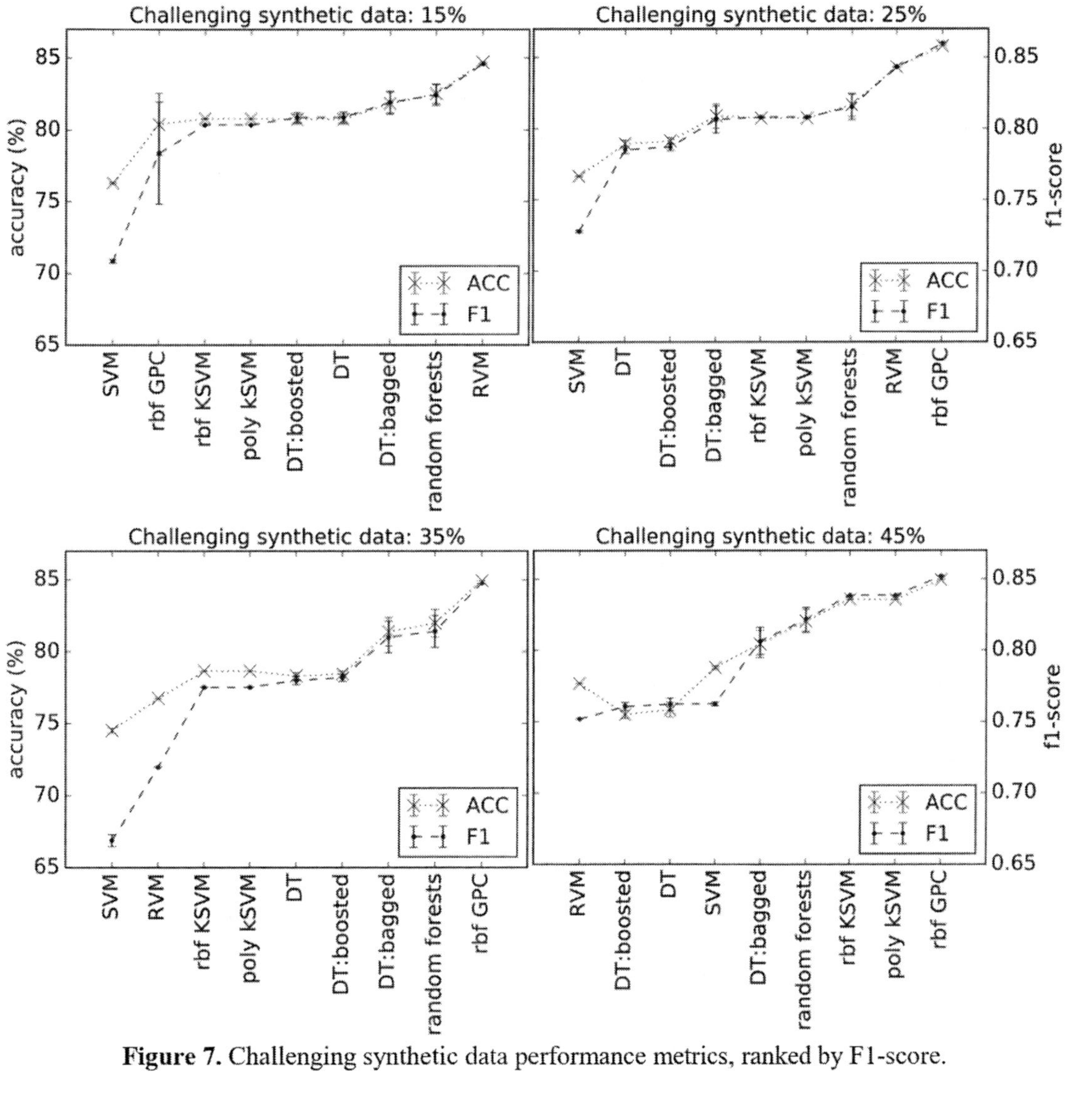

Figure 7. Challenging synthetic data performance metrics, ranked by F1-score.

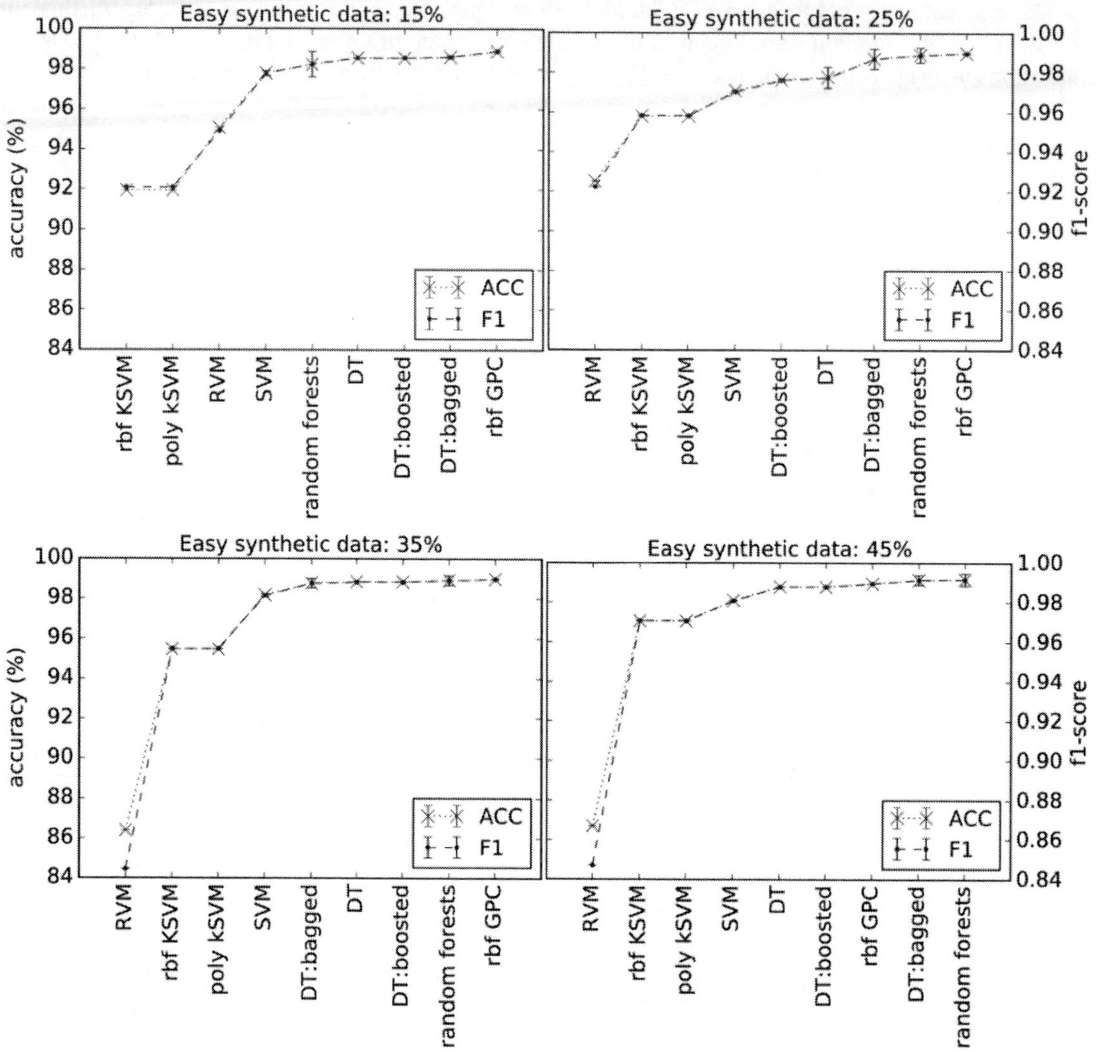

Figure 8. Easy synthetic data performance metrics, ranked by F1-score.

4.1. Linear support vector machine

Linear SVMs poorly predict class labels when applied to challenging data (Figure 7). Considering the linear decision boundary this is not surprising – it can be seen from Figure 2 that the samples are far from linearly separable. Intuitively, it then makes sense that the labels predicted for the easy data (Figure 8) are more accurate, where SVMs show good overall performance. A low variance is produced when compared to alternative methods, which is most likely due to the sparse nature of the algorithm, as every classifier is defined by similar support vectors.

4.2. Kernel support vector machine

For challenging datasets, the kernel trick succeeds in improving SVM performance – particularly with a polynomial kernel. However, when applied to the easy data, kernel techniques become a disadvantage. This is reasonable, as the data is already relatively separable in the original space (Figure 3), therefore transforming the data into some alternative space is unlikely to improve separation.

4.3. Relevance vector machine
RVMs perform very well when used with smaller training sets of the challenging data. The Bayesian framework of the classifier appears to reduce the risk of overfitting, enabling it to perform especially well with smaller training samples; however, their relative performance drops as more training data become available. When applied to the easy data, the RVMs rank lower, again their performance deteriorates as the size of the training set increases. The pruning regime could hinder competitiveness as more data becomes available and the complexity reduces, so optimisation of the precision threshold $\{\alpha_i = 1e09\}$ must still be considered. The RVMs tested in this work are shown to be optimal when applied to complicated data sets with limited training resources.

4.4. Decision tree methods
In agreement with the findings in the existing literature [6], DT methods appear to provide the best overall performance across various training sets and data complexities. Despite their simplicity, respectable ACC and F1-scores are produced throughout, even with smaller training sets. As expected, metric variance increases with the use of ensemble methods – particularly RFs. However, this increase in variance comes with increased performance, showing that averaging techniques are successfully reducing the susceptibility of DTs to overtraining.

4.5. Gaussian process classification
GPCs provide very accurate predications in most experiments, indicating the resilience of the probabilistic algorithm to overfitting. However, metric rankings fall (with increased variance) when applied to the challenging data with a 15% training set. In this case, the mapping of the Gaussian (rbf) kernel might not be suited to the specific manifold of the smallest training sample.

5. Acoustic emission application
The AE data can be described as relatively 'easy' as it has linearly separable clusters. A small training set (15%) is used to challenge each classifier. Each classification method is then trained 100 times, and the mean along with one standard deviation is provided – see Figure 9.

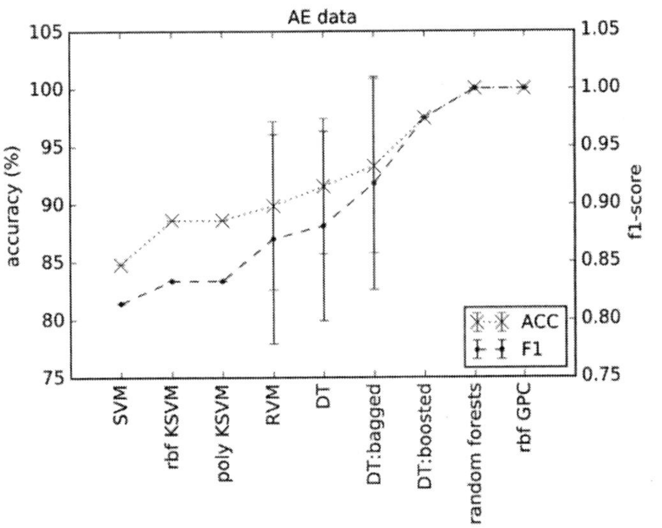

Figure 9. Classification methods applied to AE data.

In agreement with the findings from synthetic experiments, linear SVMs work well to classify these data as they are linearly separable in the original space; consequently, transformations with kernel methods show no advantage (KSVM). With this in mind, GPCs should be tested on these data with alternative kernel functions, particularly a linear kernel. RVMs provide equivalent performance to linear SVMs, which is intuitive given model similarities and the simplicity of the data. DT algorithms provide reasonable classification performance, but with a lot of variation. The use of ensemble methods offers little advantage for this application as the data set is particularly small; for this reason, introducing extra variability fails to improve the already weak learners.

6. Conclusions

This work has illustrated the necessity for a wide variety of algorithms whilst investigating SHM. The performance of any given classification is highly dependent on the properties of the data themselves, as well as the choice of any relevant parameters. Decision Tree methods provide excellent overall performance across a range of scenarios despite their simplicity, even when applied to complex data and a small training set. Probabilistic algorithms – including GPCs and RVMs – show potential, particularly when applied to challenging data. With these methods, a variety of kernel functions and pruning parameters should be explored through cross-validation for optimal performance. Traditional SHM techniques, such as SVMs, continue to provide competitive metrics; especially when the data is 'easy'. Active learning interactions have been successfully demonstrated in this work; the data-dependence of classification performance highlights the potential advantages of this tool in selecting an informative data set in the context of SHM.

6.1. Future work

For comprehensive metric comparisons, any predefined parameters must be optimised through cross validation (or otherwise); particularly those relating to non-probabilistic methods. It would be interesting to compare a wider variety of kernel functions for the computationally expensive algorithms, whilst including multi-layer perceptron metrics. Ensemble methods could be applied to more techniques in an SHM context; however, the reader is directed to the work done in reference [6] as it provides a comprehensive analysis for non-probabilistic methods. Alternative clustering algorithms should be tested to provide an array of unsupervised techniques whilst selecting data for training, along with the use of more sophisticated active learning algorithms.

Acknowledgments

The support of the UK Engineering and Physical Sciences Research Council (EPSRC) through grant reference numbers EP/J016942/1 and EP/K003836/2 is gratefully acknowledged. Further thanks are extended to Karen Holford and Rhys Pullin at Cardiff University for providing the AE data.

References

[1] Farrar C R and Worden K 2012 *Structural Health Monitoring: A Machine Learning Perspective* vol 1 (Chichester: John Wiley & Sons)

[2] Bishop C M 2006 *Pattern Recognition and Machine Learning* vol 1, ed M Jordan, J Kleinberg, et al. (New York: Springer)

[3] Pedregosa F, Varoquaux G, Gramfort A, Michel V, Thirion B, Grisel O, Blondel M, Prettenhofer P, Weiss R, Dubourg V and Vanderplas J 2011 Scikit-learn: Machine learning in python. *J. Mach. Learn. Res.* **12** 2825-30

[4] Tipping ME 2001 Sparse Bayesian learning and the relevance vector machine. *J. Mach. Learn. Res.* **1** 211-44

[5] Ritchie J and Feinberg J 2016 scikit-rvm *GitHub repository*

[6] Caruana R and Niculescu-Mizil A 2006 An empirical comparison of supervised learning algorithms. *Proc. 23rd Int. Conf. Mach. Learn.* 161-168

[7] Rasmussen C E and Williams C K I 2006 *Gaussian Processes for Machine Learning* vol 1, ed T Dietterich, D Heckerman, et al (Massachusetts: The Massachusetts Institute of Technology Press)

[8] Tong S and Koller D 2001 Support vector machine active learning with applications to text classification. *J. Mach. Learn. Res.* **2** 45-66
[9] Cai D and He X 2012 Manifold adaptive experimental design for text categorization *IEEE Trans. Know. Data Eng.* **4** 707-19
[10] Raschka S 2015 *Python Machine Learning* vol 1, ed A Hussain, R Youe, et al (Birmingham: Packt Publishing Ltd)
[11] Givoni I E and Frey B J 2009 A binary variable model for affinity propagation *Neur. Comp.* **21** 1589-1600
[12] Rippengill S, Worden K, Holford K M and Pullin R 2003 Automatic classification of acoustic emissions data *Strain* **39** 31-41
[13] Manson G, Worden K, Holford K and Pullin R 2001 Visualisation and dimension reduction of acoustic emission data for damage detection *J. Int. Mat. Sys. Struct.* **12** 529-36
[14] Wolpert D H and Macready W G 1997 No free lunch theorems for optimization *IEEE Trans. Evol. Comp.* **1** 67-82

IOP Conf. Series: Journal of Physics: Conf. Series **842** (2017) 012007 doi:10.1088/1742-6596/842/1/012007

Damage localization and quantification in simply supported beams using static test data

Q Ma and M Solís

Escuela Técnica Superior de Ingeniería, Universidad de Sevilla, Camino de los Descubrimientos, 41092 Sevilla, Spain

E-mail: msolis@us.es

Abstract. A novel simple method using static test data for damage detection, localization, and quantification in beams is presented in this paper. The method is based on the change of the deflections of the beam between a reference and a damaged state. For simply supported beams with a single damage, the maximum value of the change of deflections indicates the location of damage. Once the damage is located, one could estimate the rotational stiffness at the damaged cross section by applying a superposition scheme to isolate the effect of damage and by using basic structural analysis equilibrium equations. Afterwards, damage extent is evaluated through an existing relation between rotational stiffness and damage severity. Several static tests of a simply supported steel beam with a point load at different locations were conducted to exam the performance of the strategy. The damage is artificially introduced as an opened crack located at the bottom of the beam. The deflections of the beam were measured by using a Digital Image Correlation system. The results show that the method can accurately detect and quantify the damage. The method is non-model based and can be easily conducted. No specific loading positions are required and damage identification objective can be achieved from just one single static test.

1. Introduction

The fundamental objective of damage detection is to identify the change of properies in a structure caused by damage, which includes natural frequencies, dampings, stiffness or flexibility matrix, mode shapes, and etc [1]. Many methodologies and techniques proposed by researchers are based on system identification or parameter estimation through the dynamic response of the structure. Various numerical methods are applied for damage detection and localization, such as the transimissibility function [2–4], the BAT algorithm [5], etc.

Some researchers also applied similar ideas using static response data of the structure. Caddemi and Morassi [6] identified a single crack in beams with different boundary conditions using the static displacements. Lee and Eun [7] presented a method for locating damage through the change of curvature of static deflections. Bakhtiari-Nejad, Rahai and Esfandiari [8] developed an algorithm based on the change of stored strain energy in the elements using static noisy data for damage detection. Seyedpoor and Yazdanpanah [9] also illustrated a method through the change of strain energy using static noisy data.

In this paper, a novel non-model based simple method for single crack damage detection and localization of beams is presented. Firstly, the theory of the method is presented to illustrate how the change of pre- and post-damaged static displacements of the beam under external

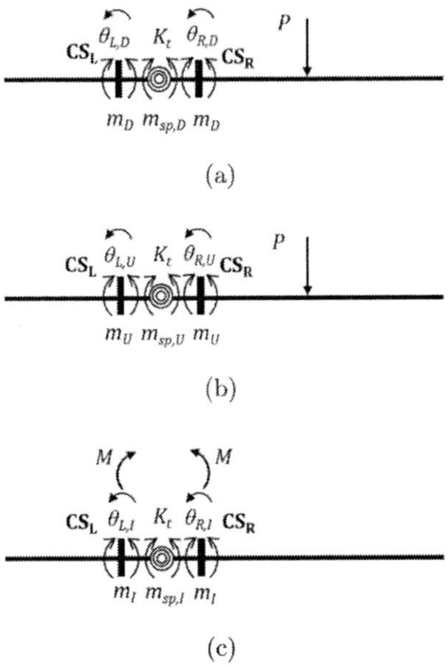

Figure 1. States of the superposition scheme: (a) Damaged State (D) , (b) reference or Undamaged State (U) and (c) Incremental State (I).

loads can be used for damage localization. From the information about the damage location, a damage quantification method based on structural analysis is introduced. Next, a series of experimental tests of a cracked simply supported beam were conducted to exam the performance of the method.

2. Theory background

It is known that the prescence of a crack will cause a reduction in the local stiffness at the cracked cross section. Hence a single damaged beam could be modeled as a rotational spring at the cracked location that connects two undamaged parts of the beam [10]. The problem of a cracked beam under some external forces (P) (*Damaged State, D*) can be decomposed into an *Undamaged State* (U) plus an *Incremental State* (I) (Fig.1). Thus, the deformation (U) and internal forces (f) can be written as

$$U_D = U_U + U_I \quad \text{and} \quad f_D = f_U + f_I \tag{1}$$

In figures Fig. 1 (a), (b) and (c), CS_L and CS_R are the left and right sides of the damage cross section respectively, K_t is the rotational stiffness of the spring that models the cracked cross section, m is the internal bending moment at CS_L and CS_R (they are equal to each other), m_{sp} is the internal torsional moment of the spring, θ_L and θ_R are the rotations at CS_L and CS_R, respectively, and the footnotes U, D, and I stand for the *Undamaged* (or reference), *Damaged*, and *Incremental States* respectively. In the *Undamaged State*, the rotations at CS_L and CS_R are set to be equal ($\theta_{L,U} = \theta_{R,U}$), which indicates the spring is not present in the undamaged beam. It is found that this superposition is valid when the applied moment (M) in the *Incremental State* is equal to the internal bending moment at damage location (m_U) in the *Undamaged State*.

The damage locations are revealed in the overall deformed shape of U_I since the external forces will introduce slope (rotation) discontinuities at damaged cross sections (Fig. 1 (c)). For a single cracked simply-supported beam, the shape of U_I is piecewise linear and it is independent from the loading position. Its magnitude depends on the magnitude of external loads, the severity of damage and the relative position of the load and the damage.

A finite element model of a simply-supported Timoshenko beam with a 1200mm length (L) and a 100 × 20mm rectangle cross section was built in ANSYS (mesh size 120mm). A spring with a rotational stiffness (K_t) of $1.8e^5$N/m^2 was used to model the crack at $0.4L$ from the left end. A concentrated load, 1kN, was applied at $0.6L$ from the left end. The deflections of the beams for the *Damaged* and *Undamaged States* are shown in Fig 2 (a). The deflection under a self-equilibrated bending moment m_U corresponding to the *Incremental State* (U_I) and the difference between the displacements of the *Undamaged* and *Damaged States* (ΔU) are shown in Fig. 2 (b). It is shown that U_I is equal to ΔU. The discontinuity in the slope indicates the damage location precisely. The slight difference between ΔU and U_I is due to numerical errors.

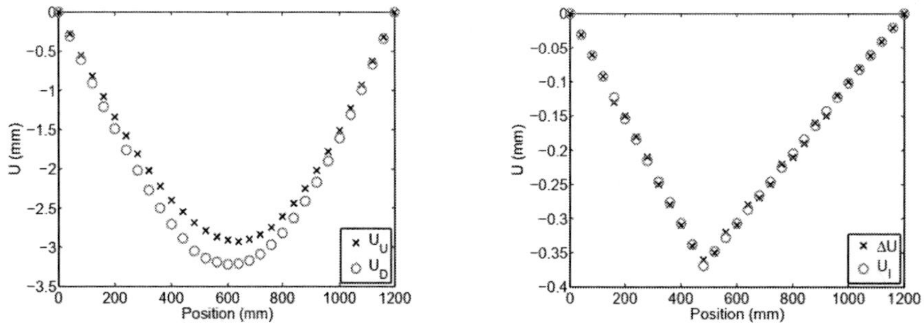

(a) (b)

Figure 2. Deflection results of the finite element model: (a) U_D and U_U of simply-supported beam; (b) ΔU and U_I of simply-supported beam

Once the damage is localized, the rotational stiffness of the cracked beam could be estimated from the *Incremental State* using the following expression:

$$K_t = \frac{m_{sp,I}}{(\theta_{L,I} - \theta_{R,I})} = \frac{m_{sp,I}}{\Delta\theta_I} \tag{2}$$

The rotation discontinuity ($\Delta\theta_I$) can be directly computed from ΔU (U_I) and the moment absorbed by the spring ($m_{sp,I}$) can be automatically calculated for a statically determinate beam since the reactions of the beam for the *Incremental State* are null and therefore $m_{sp,I}$ equals M (and m_U, as indicated previously). Once the rotational stiffness of the damaged cross-section is determined, the extent of damage can be estimated by comparing it with an existing correlation between damage size and rotational stiffness.

3. Experimental Test of A Simply-Supported Beam

3.1. Test Setup

An experimental test of a simply-supported steel beam was conducted to test the performance of the methodology. The dimension of the beam was 1200 × 100 × 20mm. A notch was cut at the bottom of the beam at $0.35L$ (425mm) from the left end. The depth of the notch was set to be 7mm (35% of the beam height). A Digital Imagine Correlation(DIC) system (Fig. 3 (a), (b) and (c)) was used for measuring the deflection of the beam under loading. A total number of 241 measuring points (damage at the 86th) were marked along the beam with an equal spacing of 5mm. A concentrated force was applied on the beam vertically through hanging a 120kg mass on it. 21 tests were performed by putting the mass at 21 equally distributed positions along the beam. The scheme of the test is shown in Fig. 3 (d).

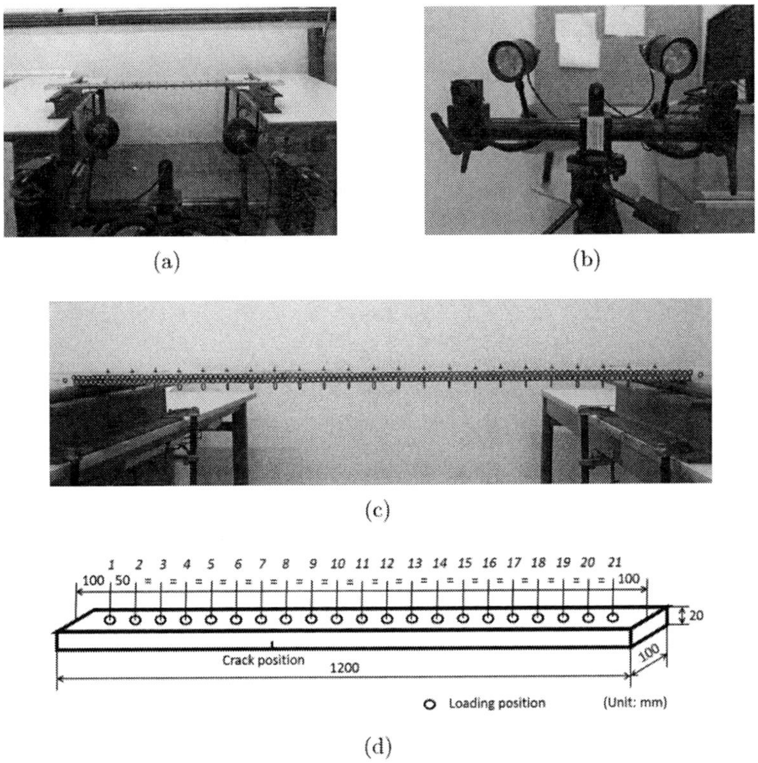

Figure 3. (a) Experimental setup, (b) DIC measuring system, (c) Tested beam and (d) Test scheme

3.2. Implementation of the methodology

Due to the effect of noise on the measured data, a trend estimate function named $l1$ Trending Filter is used to estimate the overall shape of ΔU. The $l1$ Trending Filter produces trend estimation that is piecewise linear through minimize the objective function in Eqn. (3), where λ is a nonnegative parameter. x_t is the estimated trend and y_t is the signal [11]. This trending filter automatically identifies the turning point along the piecewise shape data.

$$(1/2)\sum_{t=1}^{n}(y_t - x_t)^2 + \lambda\sum_{t=2}^{n-1}|x_{t-1} - 2x_t + x_{t+1}|^2 \tag{3}$$

The results of ΔU and the application of $l1$ Filter to ΔU (ΔU_{l1}) are displayed in Fig. 4 and 5. For all 21 loading positions, the shape of ΔU was estimated correctly. The effect of noise only takes a relative high influence when the loading positions are close to supports of the beam (at positions 1, 20 and 21). The position where the maximum value of ΔU_{l1} takes place is considered as the damage location. The predicted results are listed in table 1. All the predicted damage locations fall into a small range from the correct location. The furthest predition is at point 93 (for loading position 20), which is 35mm to the right of the real damage. Therefore, it is shown that the methodology can successfully localize the damage for this damage scenario.

For a notch type opened cracked on an elastic beam with rectangular cross-section, the equivalent rotational stiffness of the damaged cros-section (K_t) proposed by Rizos, Aspragathos, and Dimarogonas [12] (Eqn. (4) and (5)) is used in this paper for damage extent estimation:

IOP Conf. Series: Journal of Physics: Conf. Series **842** (2017) 012007 doi:10.1088/1742-6596/842/1/012007

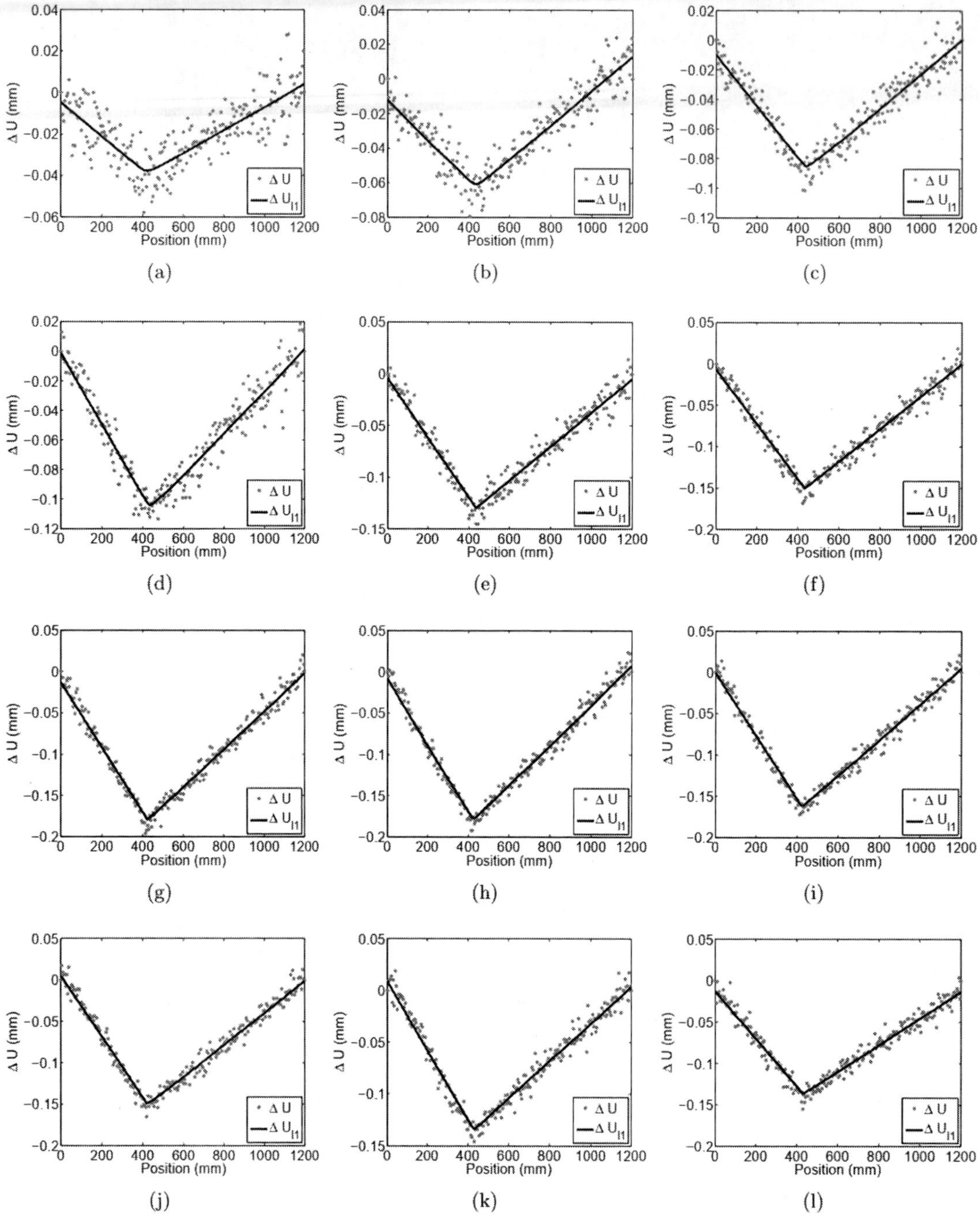

Figure 4. ΔU and ΔU_{l1} with loading at positions 1 (a) to 12 (l).

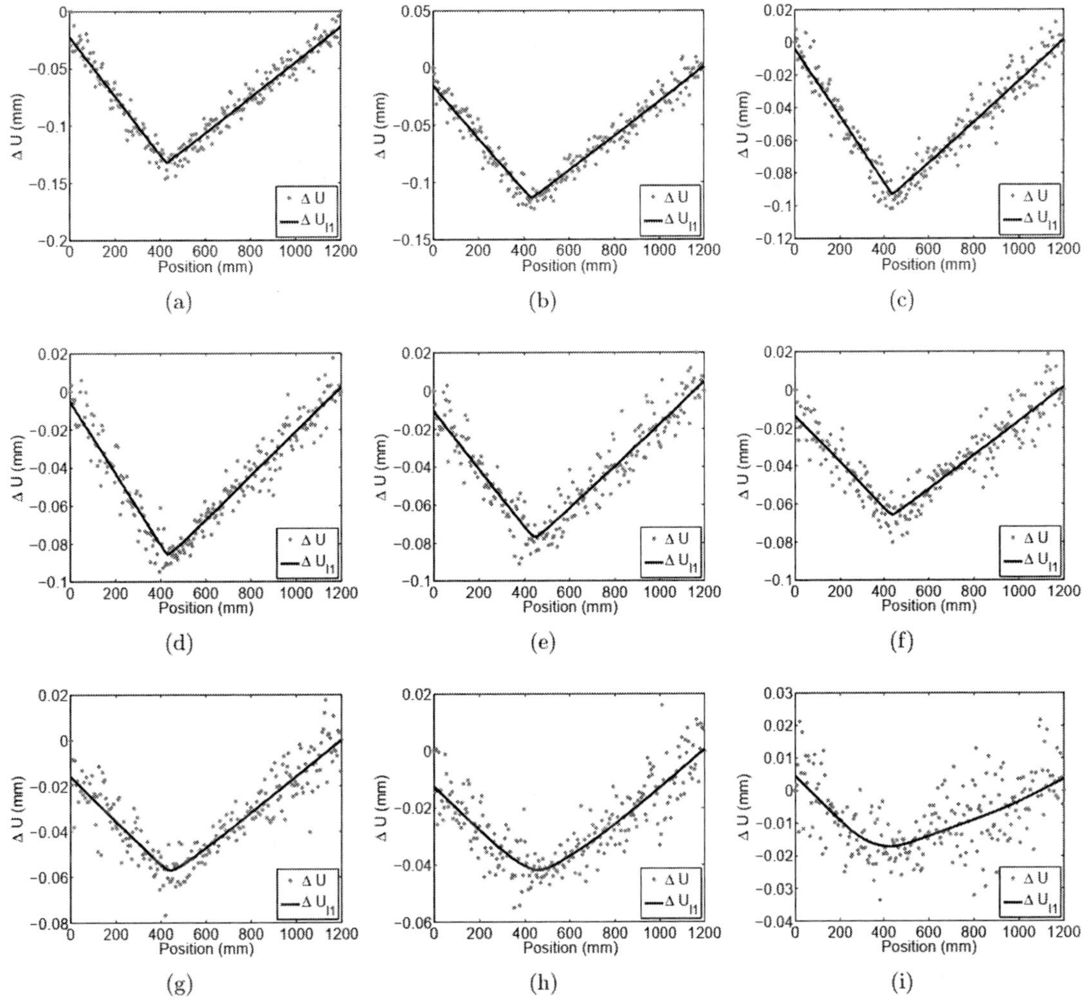

Figure 5. ΔU and ΔU_{l1} with loading at positions 13 (a) to 21 (i).

$$K_t = 1/c \qquad c = 6\pi h J(\xi)/EI \tag{4}$$

where h is the height of the beam, E the elastic modulus of the material of the beam, I the inertia of the cross-section and J is the following function of the ratio (ξ) between the notch depth and the height of the beam.

$$\begin{aligned} J(\xi) = {} & 1.86(\xi)^2 - 3.95(\xi)^3 + 16.375(\xi)^4 - 37.226(\xi)^5 \\ & + 76.81(\xi)^6 - 126.9(\xi)^7 + 172.5(\xi)^8 - 143.97(\xi)^9 + 66.56(\xi)^{10} \end{aligned} \tag{5}$$

By using this empirical relationship, damage severity can be estimated from the experimentally identified rotational stiffness (K_t) in Eqn.(2). The rotation discontinuity ($\Delta\theta_I$) can be evaluated using a piecewise linear regression function of U_I, whereas $m_{sp,I}$ (equivalent to m_U) can be easily computed from equilibrium equations of the undamaged beam. The estimated damage extent for each test are also listed in table 1. All of the estimated damage severities are bigger than the real value. This may indicate some discrepancy between the analytical model of

Table 1. Estimated damage location and severity and their errors

Localization				Quantification		
Loading Position	Measuring Point	Location (mm)	Deviation (mm)	Severity (mm)	Error (mm)	Error (%)
1	85	415	5	7.5	0.5	7
2	89	440	15	7.7	0.7	10
3	89	440	15	7.7	0.7	10
4	88	435	10	7.9	0.9	13
5	88	435	10	7.8	0.8	11
6	88	435	10	7.8	0.8	11
7	86	425	0	7.8	0.8	11
8	86	425	0	7.8	0.8	11
9	87	430	5	7.8	0.8	11
10	85	415	5	7.9	0.9	13
11	87	430	5	7.8	0.8	11
12	87	430	5	7.7	0.7	10
13	87	430	5	7.7	0.7	10
14	87	430	5	7.8	0.8	11
15	89	440	15	7.7	0.7	10
16	87	430	5	8.0	1.0	14
17	91	450	25	7.8	0.8	11
18	89	440	15	7.8	0.8	11
19	90	445	20	7.9	0.9	13
20	93	460	35	7.7	0.7	10
21	87	430	5	7.7	0.7	10

the crack as a rotational spring and the actual behavior of the damaged cross section. However, for all the loading positions, the method provides predictions with high accuracy even for those with low signal to noise ratio.

4. Conclusion

A non-model based damage detection and localization methodology based on the static displacements is presented in this paper. No specific loading positions are needed for the experimental test and structural identification is not required. Experimental results of a simply-supported steel beam with a single crack are provided. The methodology successfully predicts the crack location with a very high accuracy for all loading positions. From the predicted damage locations, the damage extent can be estimated using an existing analytical correlations between damage extent and rotational stiffness of the damaged cross section. The method provides results with high accuracy as well. In summary, the paper proves the efficiency and simplicity of the method for practical purpose.

Acknowledgments
This work was supported by the Consejería de Economía, Innovación, Ciencia y Empleo of Andalucía (Spain) under project P12-TEP-2546. The financial support is gratefully acknowledged.

Reference

[1] Hajela P and Soeiro F J 1990 *Structural Optimization* **2** 1
[2] Zhou Y L, Maia N M and Wahab M A 2016 *Journal of Vibration and Control* 1
[3] Zhou Y L and Wahab M A 2016 *Journal of Vibroengineering* **18**(7) 4491
[4] Zhou Y L, Maia N M, Sampaio R P and Wahab M A 2016 *Structural Health Monitoring* 1
[5] Khatir S, Belaidi I, Serra R, Wahab M A and Khatir T 2016 *Journal of Vibroengineering* **18**(1) 202
[6] Caddemi S and Morassi A 2007 *International Journal of Solids and Structures* **44** 5301
[7] Lee E T and Eun H C 2008 *Journal of Mechanical Science and Technology* **22** 1111
[8] Bakhtiari-Nejad F, Rahai A and Esfandiari A 2005 *Engineering Structures* **27** 1784
[9] Seyedpoor S M and Yazdanpanah O 2014 *Applied Mathematical Modelling* **38** 2661
[10] Dimarogonas A D 1996 *Engineering Fracture Mechanics* **55** 831
[11] Kim S J, Koh K, Boyd S and Gorinevsky D 2009 *Society for Industrial and Applied Mathematics* **51** 339
[12] Rizos P F, Aspragathos N and Dimarogonas A D 1990 *Journal of Sound and Vibration* **138** 381

IOP Conf. Series: Journal of Physics: Conf. Series **842** (2017) 012008 doi:10.1088/1742-6596/842/1/012008

Effect of test frequency on fatigue strength of AZ31 alloy

G J Lu[1], Y Koyanagi[1], M Tsushida[2], H Kitahara[3], and S Ando[4]

[1] Graduate School of Science and Technology, Kumamoto University, 860-8555, JAPAN

[2] Faculty of Engineering, Kumamoto University, 860-8555, JAPAN

[3] Institute of Pulsed Power Science, Kumamoto University, 860-8555, JAPAN

[4] Magnesium Research Centre, Kumamoto University, 860-8555, JAPAN

E-mail: 158d9212@st.kumamoto-u.ac.jp

Abstract. Investigation of fatigue strength of extruded AZ31 magnesium alloy was performed by ultrasonic fatigue tests (test frequency about 19500Hz) and taken to compare with conventional tension-compression fatigue tests (20Hz). The fatigue life under ultrasonic loading exhibits longer than the conventional one. Effects of ultrasonic frequency on the S-N curve, crack initiation and growth behaviour of AZ31 alloy is discussed. This work would be useful for improving the ultrasonic fatigue test method to observe the fatigue behaviour in very high cycle regime.

1. Introduction

In recent years, especially since the beginning of the new century, magnesium alloys so called "the 21th century's engineering materials" which are the lightest commercial structural alloys attract many interests. With high specific strength, high stiffness ratio, good heat dissipation, good damping capability and recycling ability, these alloys have a good prospect application for automobile and aerospace structural parts as we know the engine, power transmission and car body.

During their lifetime, some of the components have to survive more than 10^8 cycles. Therefore, fatigue strength and fracture mechanism of the alloys in very high cycle regime are of tremendous interest [1, 2]. On the other hand, significant improvements have been achieved for pursuing accelerated fatigue tests under ultrasonic resonance. For this reason, ultrasonic method is increasingly applied in order to investigate fatigue lives at very long cycle regime and explore the crack propagation behaviour [3, 4], where experiments by conventional machines are time-consuming and expensive. However, the influence of ultrasonic frequency on small crack behaviour in magnesium alloys has not been adequately studied for the prediction of fatigue damage. It is dangerous especially at giga-cycle fatigue life region [5]. Nowadays the continuum damage mechanics approach with numerical simulation is widely applied for studying the fatigue behavior [6-11]. The comparison between the estimated crack initiation by numerical models developed with the Finite Element Analysis (FEA) and the total experimental lifetime can be used to evaluate the effect of the frequency between ultrasonic and conventional fatigue test. This valuable method is considered a feasible and effective for the future work of this study.

The objective of the present study is, therefore, to investigate the fatigue strength and the small crack behaviour of extruded AZ31 magnesium alloy in the very high cycle domain under ultrasonic fatigue. The effects of frequency on the small crack behaviour and fracture mechanism are evaluated by comparing the results with those under conventional fatigue.

2. Materials and Methods

2.1. Ultrasonic fatigue test method

Schematic figure of principle of our test system is shown in Fig.1. Firstly, the power generator converts 50 or 60 Hz voltage signal into 19.5 kHz ultrasonic electrical signal. The electrical signal turns to mechanical vibration with longitudinal ultrasonic waves in the same frequency by piezoelectric converter inside the power generator. The mechanical displacement is induced to generate the stress amplitude. In this case, the specimen is designed with first mode frequency of 19.5 kHz, which is related to the specimen length. With applying this resonant vibration, the specimen gets a maximum displacement at the end and maximum stress in the central section. Finally, an ultrasonic horn and cone that amplifies the mechanical vibration generating from the converter in order to obtain the required stress amplitude in the middle section of the specimen. The stress amplitude is pre-selected and kept constant during the test.

Schematic diagram of our test system is shown in Fig. 2. There are also other devices composing the ultrasonic fatigue test machine. They are mainly divided into two parts: the recording system which includes control unit and cycle counter; the measuring system which includes the displacement sensor and thermometer. This test machine allows testing load ratio of R= -1.

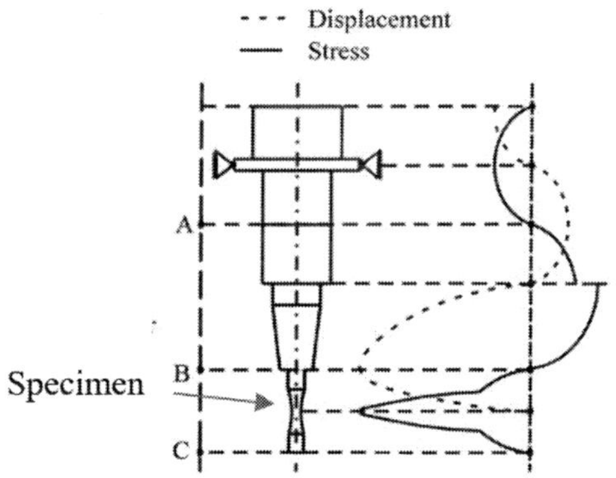

Figure 1. Schematic illustration of ultrasonic fatigue test.

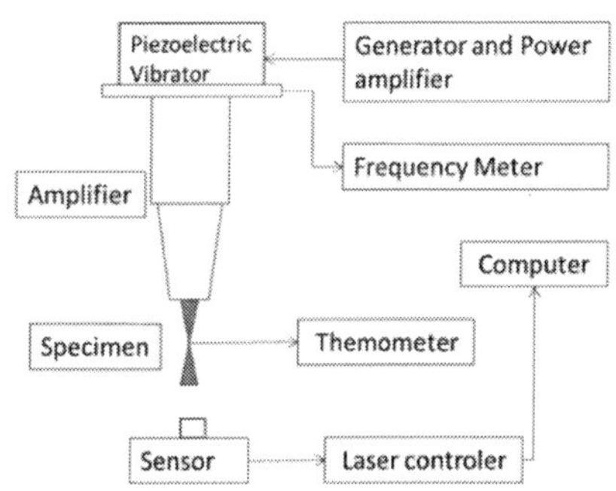

Figure 2. Schematic diagram of ultrasonic fatigue test system.[12]

2.2. Material and experimental

The material used was a extruded magnesium alloy AZ31 with a chemical composition (in wt. %) of 3.05Al, 1.04Zn, 0.39Mn, and balance Mg. Yield stress (0.2% proof stress), σ_b and tensile strengths at room temperature were 197 MPa and 286 MPa, respectively, and the elongation was 5.07%. The mean grain size of the alloy was 16.1 μm.

The ultrasonic fatigue system was tuned to resonate at a frequency of ~19.5 kHz. A longitudinal standing wave was excited in a circular taper dumbbell specimen, introducing a sinusoidal, cyclic loading along the specimen with the maximum displacement (the minimum stress amplitude) at either end of the specimen and a displacement node (the maximum stress amplitude) in between, as shown in Fig. 1. The maximum stress amplitude is obtained at the displacement node of the specimen, which can be calculated from the maximum displacement at the free end of the specimen, and the dynamic Young's modulus of the alloy. The displacement is measured by LK-G5000 laser sensor by Keyence Corporation. The dimensions of the fatigue test specimen are shown in Fig. 3.

Prior to fatigue testing, the curved surface of the specimens are polished by ~20 μm in depth after emery paper grinding in order to remove worked layer and to better observe surface damage. The observation of fatigue damage and the measurement of surface crack length are conducted directly under an optical microscope or under a scanning electron microscope (SEM). The crack length was defined vertical to the loading direction on the specimen surface.

To minimize internal friction induced temperature rise, ultrasonic fatigue tests were performed in a pulse-pause manner with pulse length varied in a range of 100 ms and pauses 3000 ms. The surface temperature at the most stressed node section of the specimen was monitored during the tests using an infrared thermal tracer. The maximum temperature rise is controlled below 3 °C during the testing. The influence of pulse-pause loading on the fatigue strength of the alloy was ignorable. Fatigue tests were carried out at ambient temperature under a load ratio of $R=-1$.

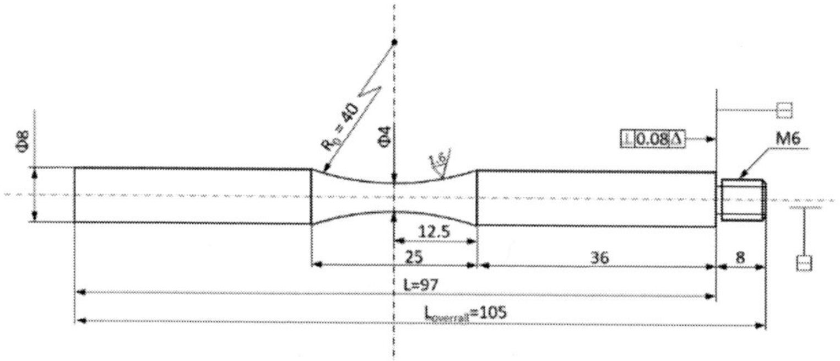

Figure 3. The geometries of the fatigue test specimen. (mm).

3. Results and Discussion

3.1. *Relationship between stress and fracture life*

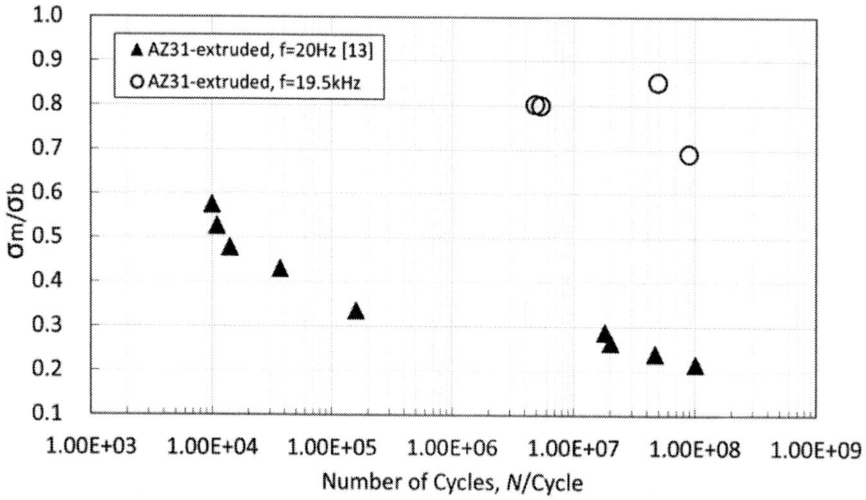

Figure 4. S-N plot of extruded AZ31 in ultrasonic fatigue tests. Results of conventional axial tension compression fatigue test of 20 Hz [13] are also plotted for comparison.

Fig. 4 shows the fatigue life of extruded AZ31 alloy under ultrasonic loading, the circular one represents data of present work. The triangle one represents data of Shiozawa [13] with extruded AZ31 alloy under conventional axial tension compression loading with 20 Hz and a load ratio of $R=-1$. For the material used in current research and Shiozawa's work, the yield stress σ_b are 197 MPa and 209 MPa, respectively [13]. Strength of the materials in these two work has a little bit different. In order to get a better comparison, the ratio of σ_m/σ_b is used to present the fatigue strength, where σ_m is the stress amplitude. Fatigue strength of current work is greater than the conventional one. It shows the existence of the effect of ultrasonic frequency. Shiozawa' data appears to have a continuous decreasing trend in the very high cycle regime and it exhibits a step-wise behaviour [13]. In case of present work, the plots shows a continuous decreasing trend insufficiently.

In the cyclic loading of 10^6 to 10^7 cycles, there is no fatigue limit which different from the traditional fatigue test data, parts of the samples cracks after 10^8 cycles in Shiozawa's work. For present work, one sample which under 136 MPa ultrasonic loading, undergone near10^8 cycles.

The fatigue strength exhibited in present work is larger than that of conventional data. The effect of loading frequency on the fatigue behaviour is considered to be obvious when the yield stress ratio (σ_m/σ_b) is smaller than 1.2 [14]. The reason is due to the behaviour of cyclic hardening and softening of the metallic materials [14]. The yield stress ratio of magnesium alloy always smaller than 1, and the ratio of the two materials compared in this paper is 0.69 (material used for ultrasonic test in present work) and 0.9 (conventional test alloy), respectively. These materials have a behaviour of cyclic softening and the frequency effect is remarkable. Besides, another major frequency effect is the temperature increasing caused by the plastic deformation under high frequent loading. In the present study, the specimens are cooled by the pulse-pause manner of the test machine. The influence of temperature on fatigue strength is considered to be limited. Other factors of the frequency effect need further investigation.

3.2. *Crack growth behaviour*

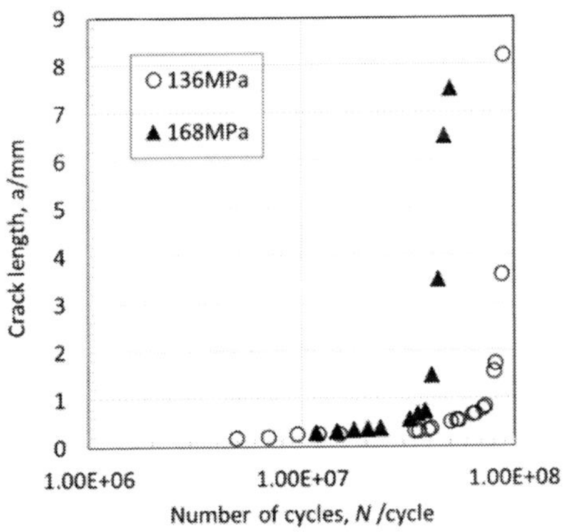

Figure 5. Crack length a vs. number of cycles *N*.

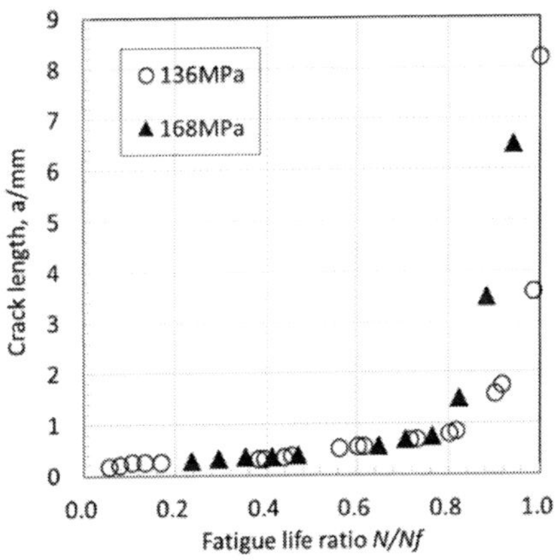

Figure 6. Crack length a vs. fatigue ratio N/N_f

Fig. 5 shows the relationship of the crack length and number of cycles with different load stress of 136 and 168 MPa. The crack grows much faster under the higher stress. After the crack grow to 1 mm, the crack growth rate gets a remarkable increasing, irrespective to the loading stress. For both higher and lower stress, the crack initiated at the early stage of fatigue cycle. The relationship of the crack length and the ratio of number of cycles to whole lifetime cycles (Fig. 6) shows crack propagation period (crack length $\leqslant$ 1 mm) takes up more than 80% of the total fatigue life, and then in the rest of less than 20% life, the crack grows from under 1 mm to near 10mm rapidly. The fracture observation shows no notable difference in different load stress (136 MPa and 168 MPa).

3.3. *Fracture surface*

After testing, the fractured sample takes to observation by scanning electronic microscope (SEM). Fig.7 shows the scanning electronic micrographs of the fracture surface of specimen at 124 MPa cyclic loading which survived more than 10^7 cycles. An interior crack initiation site was observed, and it is different from the lower cycle fatigue crack behaviour which is rarely observed of interior crack initiation. It turned out that it would be important to investigate the very high cycle fatigue behaviour of AZ31 alloy, which may surfer up to 10^8 or 10^9 stress cycling in the service life of the structural components. The crack initiation mechanism may change from the surface initiation to interior initiation.

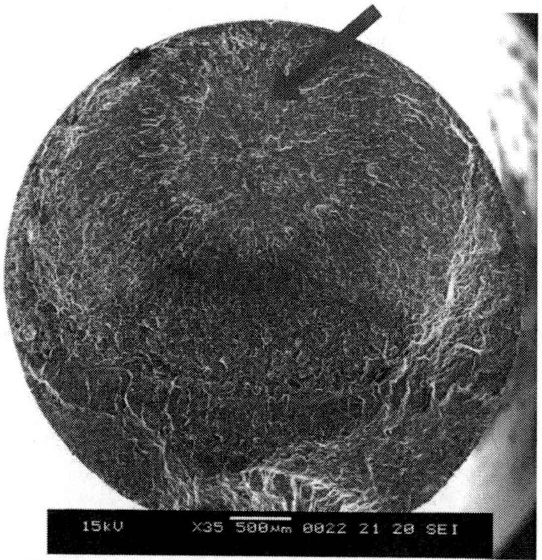

Figure 7. Scanning electronic micrographs of fracture surface of specimen with interior crack. σ_m=124MPa, N_f=9.32*10^7 cycles.

4. Conclusion

The fatigue test of an AZ31 extruded magnesium alloy in the long lifetime regime conducts at ultrasonic frequency at room temperature. The fatigue behaviour and crack propagation are examined and compared. For this study, the following conclusions can be drawn.

1. Loading frequency affects the fatigue strength of this extruded magnesium alloy.
2. Interior crack was recognized in the specimen which did not fail after cycling near 10^8 cycles, implying that the importance to investigate interior crack initiation in the very high cycle fatigue region.
3. Cracks initiated at the very beginning of fatigue cycling and the suppression of early crack propagation becomes significant under low stress levels. It is fund that more than 80% of total fatigue life is consumed in nucleating and growing cracks smaller than 1 mm.

References

[1] J K Tien and R P Gamble 1971 The Room Temperature Fatigue. Behavior of Nickel-Base Superalloy Crystals at Ultrasonic. Frequency *Metall. Trans. B.* **2** 1933-38

[2] S Stanzl 1981 A new experimental method for measuring life time and crack growth of materials under multi-stage and random loadings *Ultrasonics* **19** 269-72

[3] B Holper, H Mayer, A K Vasudevan and S E Stanzl-Tschegg 2003 Near threshold fatigue crack growth in aluminium alloys at low and ultrasonic frequency: Influences of specimen thickness, strain rate, slip behaviour and air humidity *Int. J. Fatigue* **25** 397-411

[4] Q Y Wang, C Bathias, N Kawagoishi and Q Chen 2002 Effect of inclusion on subsurface crack initiation and gigacycle fatigue strength *Int. J. Fatigue* **24** 1269-74

[5] K Tanaka and Y Akiniwa 2002 Fatigue crack propagation behaviour derived from S-N data in very high cycle regime *Fatigue Fract. Eng. Mater. Struct.* **25** 775-84

[6] Ferjaoui A, Yue T, Abdel Wahab M and Hojjati-Talemi R 2015 Prediction of fretting fatigue

crack initiation in double lap bolted joint *International Journal of Fatigue* **73** 66-76
[7] Hu P, Meng Q, Hu W, Shen F, Zhan Z and Sun L 2016 A continuum damage mechanics approach coupled with an improved pit evolution model for the corrosion fatigue of aluminum alloy *Corrosion Science* **113** 78-90
[8] Mei J and Dong P 2017 Modeling of path-dependent multi-axial fatigue damage in aluminum alloys *International Journal of Fatigue* **95** 252-263
[9] Kumar D, Biswas R, Poh LH and Abdel Wahab M 2017 Fretting fatigue stress analysis in heterogeneous material using direct numerical simulations in solid mechanics *Tribology International* **109** 124-132
[10] Bhatti NA and Abdel Wahab M 2017 Finite element analysis of fretting fatigue under out of phase loading conditions *Tribology International* **in press** http://dx.doi.org/10.1016/j.triboint.2017.1001.1022
[11] Resende Pereira KdF, Bordas S, Tomar S, Trobec R, Depolli M, Kosec G and Abdel Wahab M 2016 On the Convergence of Stresses in Fretting Fatigue *Materials* **9** 639; doi:610.3390/ma9080639
[12] D Wagner, F J Cavalieri, C Bathias, and N Ranc 2012 Ultrasonic fatigue tests at high temperature on an austenitic stainless steel *Propul. Power Res.* **1** 29-35
[13] K Shiozawa, A Ikeda and T Fukumori 2013 Effect of stress ratio and loading mode on high cycle fatigue performances of extruded magnesium alloys *Trans. Jpn. Soc. Mechanical Eng.* **79** 1366-81
[14] Y S Hong, A G Zhao and G A Qian 2009 Essential characteristics and influential factors for very-high cycle fatigue behavior of metallic materials *Acta Metall. Sin.* **45** 769-80

IOP Conf. Series: Journal of Physics: Conf. Series **842** (2017) 012009 doi:10.1088/1742-6596/842/1/012009

A vibration-based health monitoring program for a large and seismically vulnerable masonry dome

M L Pecorelli[1], R Ceravolo[1], G De Lucia[1] and R Epicoco[1]

[1]*Politecnico di Torino, Department of Structural, Building and Geotechnical Engineering, c. Duca degli Abruzzi, 24-10129 Torino, Italy*

Abstract. Vibration-based health monitoring of monumental structures must rely on efficient and, as far as possible, automatic modal analysis procedures. Relatively low excitation energy provided by traffic, wind and other sources is usually sufficient to detect structural changes, as those produced by earthquakes and extreme events. Above all, in-operation modal analysis is a non-invasive diagnostic technique that can support optimal strategies for the preservation of architectural heritage, especially if complemented by model-driven procedures. In this paper, the preliminary steps towards a fully automated vibration-based monitoring of the world's largest masonry oval dome (internal axes of 37.23 by 24.89 m) are presented. More specifically, the paper reports on signal treatment operations conducted to set up the permanent dynamic monitoring system of the dome and to realise a robust automatic identification procedure. Preliminary considerations on the effects of temperature on dynamic parameters are finally reported.

1. Introduction

Historic masonry monuments accumulate damage during their lifetime due to materials deterioration, repeated loads and exceptional events such as earthquakes. Therefore, repairs and strengthening interventions are necessary over time to preserve these monuments. Structural Health Monitoring (SHM) can help in planning maintenance activities and in assessing the effectiveness of interventions and the overall safety levels of historical buildings [1][2][3]. Among these, SHM procedures based on global vibration measurements are particularly advantageous because the dynamic response of a structure reflects its global state. Specifically, vibration-based SHM based on Operational Modal Analysis (OMA) procedures are optimal monitoring tools for heritage buildings because of their non-invasiveness. The dependence of the dynamic parameters, i.e. frequencies, modal shapes and damping, on the health state of a building has been investigated and demonstrated in several studies [4][5]. However, the use of these parameters as a damage indicator is still deemed a complex operation because of their variability under environmental conditions. The environmental factors that typically influence the values assumed by the dynamic parameters are temperature, humidity and rain [6][7][8][9], excluding the variability from excitation characteristics. Therefore, the effects of the environmental factors must be investigated. More specifically, any custom-designed SHM procedure requires the preliminary definition of the environmental variations of the dynamic parameters. Changes in the state of a building, indeed, should be related to anomalous deviations of the dynamic parameters rather than to daily or seasonal variations. To this aim, statistical recurrence of the characteristics of the parameters extracted from data recorded during several environmental conditions must be determined.

This paper proposes a protocol to automatically perform OMA on monumental buildings. The protocol is developed using the data acquired by the dynamic monitoring system installed on the dome of the Regina Montis Regalis Basilica (Piedmont, Italy), the world's largest masonry oval dome (internal axes of 37.23 by 24.89 m). As a first step, some aspects connected with the data acquisition are discussed. Secondly, a procedure to discriminate the main modes of the structure is established. Finally, some conclusions about the effect of temperature on the dynamic parameters are drawn, using the data acquired from November 21, 2016 to January 21, 2017.

2. The Regina Montis Regalis Basilica

Regina Montis Regalis Basilica is a monumental building and represents a significant historical architecture of the Italian Cultural Heritage owing its magnificent and impressive masonry dome that, with the internal axes of 37.23 by 24.89 meters, is the world's largest oval dome [10], 'figures 1(a,b)'. The construction began in 1596 based on a project of the architect Ascanio Vitozzi, and then the construction was ultimate by architect Francesco Gallo in 1735. Since the first years, several structural problems affected the building because of foundation were built on a clay layer [11]. Cracking phenomena are located especially in the dome-drum area with a total width of up to 416 mm. In 1983, concerns over the structural health state of the building prompted the decision to undertake analysis, monitoring activities and strengthening interventions.

In 1987, an hooping system was put in place, consisting of four closed rings of 56 active steel tie-bars located at the drum level 'figure 1(c)'and, in the same years, a static monitoring system was installed on the Basilica. It consists of 133 sensors subdivided in two groups: instruments to measure strains, stresses and cracks and instruments to measure the boundary conditions.

Figure 1. (a) Regina Montis Regalis Basilica; (b) The Basilica dome; (c) The strengthening system installed in 1987.

During the last decades, several non-destructive survey campaigns were conducted to better understand the structural behaviour of the Basilica [12]. In 2008, a dynamic identification campaign taken place and data were analysed by the Politecnico di Torino [13], then, preliminary analysis were conducted on the FE model of the structure [14]. The more recent studies consist in the analysis of the static monitoring data acquired in the decade 2004-2014. The analysis prove the stability of the cracks opening and, therefore, the effectiveness of the strengthening system put in place in 1987 [15]. However, the static monitoring system only provides local information about the health state of the Basilica. Consequently, in December 2015, a permanent dynamic monitoring system was installed, in order to investigate global phenomena affecting the structure.

The portion of the building interested by dynamic monitoring activities is the lantern-dome-drum system. Sensor positions were defined through Optimal Sensor Placement (OSP) techniques [16]. OSP procedures used a genetic algorithm to find the best set of instrument positions, referring to the Modal Assurance Criterion (MAC) matrix applied to a dynamically calibrated F.E. model [17]. The permanent dynamic system is composed of 12 mono-axial piezoelectric accelerometers (1000 mV/g; measure range: ± 5 g pk; broadband resolution: 0,00001 g rms; frequency range: (± 5%) $0,5<f<200$Hz, (±10%) $0,3<f<4000$Hz; resonance frequency: ≥ 10 kHz; temperature range: $-65<T<121$°C). As shows in 'figure 2', three orthogonal accelerometers are located at the base of the crypt to record the ground accelerations. A set of 9 accelerometers are located at different levels in the lantern-dome-drum area, along the longitudinal and transverse directions. In more details: (i) two accelerometers are at the base of the dome at 30 meters height; (ii) two sensors are on the dome at 45 meters height; (iii) one vertical accelerometer is located at the base of the lantern at 50 meters height; (iv) the last three accelerometers are located at the top of the lantern. The monitoring system was designed as a Master-Slave acquisition scheme (see 'figure 2'). More specifically, the data acquired by the sensors on the lantern-dome-drum system converge to the Master central acquisition unit, instead the data recorded by the accelerometers in the crypt are transmitted to the Slave element, and then, to the Master central. This configuration can reduce distortion effects in the signals by limiting the cable length to about 50 meters.

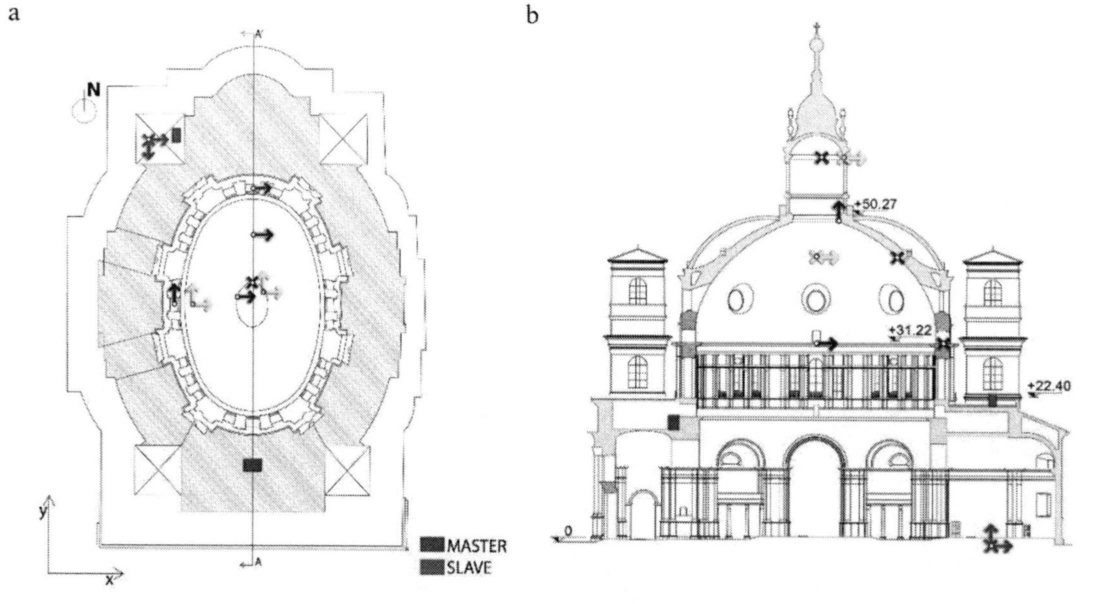

Figure 2. (a) Sensors position in plane; (b) Sensor position in section AA'.

Recorded data are transmitted, through File Transfer Protocol application, both to the Politecnico di Torino and to the Civil Protection Department, since the Basilica has been recently included in the Seismic Observatory of the Structures network owing to its cultural and architectural relevance.

3. Automated identification protocol

The Earthquake Engineering & Dynamics (EED) lab at the Politecnico di Torino is working on a fully automatic experimental modal analysis procedure to assess, in real time, the health state of the Basilica, and to evaluate the possible effects of ground motions. At present, the system identification technique, incorporated in the automatic procedure, consists in a time domain Stochastic Subspace Identification (SSI). More specifically, the third algorithm considered by the unifying theorem of Van Overschee and De Moor [18], often referred to as "canonical variate analysis" (CVA), was used.

The CVA, due to its algebraic nature, the good computational efficacy and its accuracy is adequate to be automatized. Nevertheless, manual preliminary identifications must be performed to set model orders, to perform pre-processing operations of the acquired data, as well as to set the modality of data acquisition. The sampling frequency, the length of the data acquired and its frequency content, respectively, can affect significantly the identification results. As a consequence, the main aim of the manual identification carried out using few signals are to set: (i) the order of the system; (ii) the length of the periodic records; (iii) the pre-processing operations and (iv) the time intervals of data acquisition.

The automatic identification procedure was implemented in Matlab® and its efficiency was evaluated by comparing the results of automatic and manual identification sessions. More specifically, the Matlab® code was updated in order to give the same results as the manual identification carried by the operator.

The identification procedure entails a series of data cleansing and pre-processing operations on the signals, such as mean removal, de-trending, decimation, filtering, etc. The dynamic parameters extracted by the SSI, performed on each processed signals, are analysed in terms of stability and regularity. The daily candidate modes are obtained by a clustering analysis performed on the hourly candidate mode and by averaging operations. 'Figure 3' shows the working scheme of the automatic identification procedure and the management of the acquired data. Several database with raw data, post-processed signals, hourly dynamic parameters, daily dynamic parameters and a database of the environmental conditions are created during the automatic identifications. In the following subsections, the different steps of the automatic identification procedure are described in details.

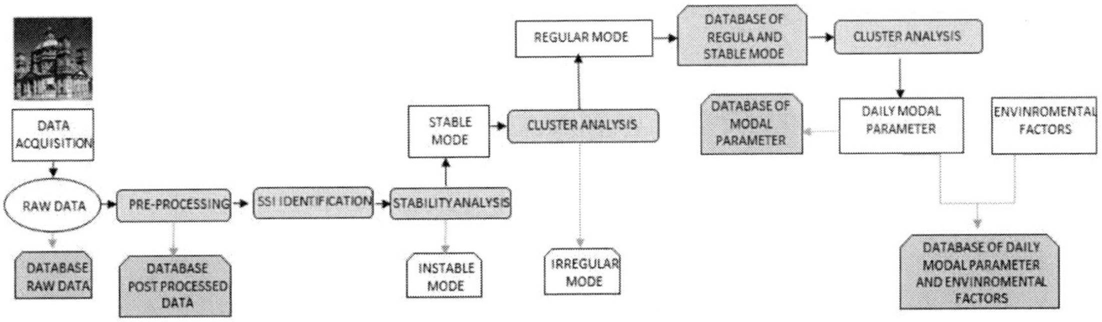

Figure 3. Flowchart of the automatic identification procedure

3.1 Pre-processing

Factors such as sampling frequency, temporal window and frequency content can significantly affect the result of the identification, independently from the model order. As a consequence they must be appropriately set.

For both manual and automated identification of the Basilica, the sampling frequency was set at 100 Hz and kept constant. Conversely, different lengths were used for the identification: 5, 10 and 20 minutes, respectively. The signals of 5 and 10 minutes were obtained by cutting the signal of 20 minutes.

The manual identifications performed evidenced that it is convenient to filter the recorded data in the band 0.5-5 Hz, independently from the length of the signal used. This filtering operation was requested to identify the first two modes, as already detected during the campaign carried out in 2008 [14]. To stress the importance of this filter, 'figure 4' reports the stabilisation diagrams related to the same recorded signal but filtered in the band 0.5-5 Hz and 0.5-15 Hz, respectively.

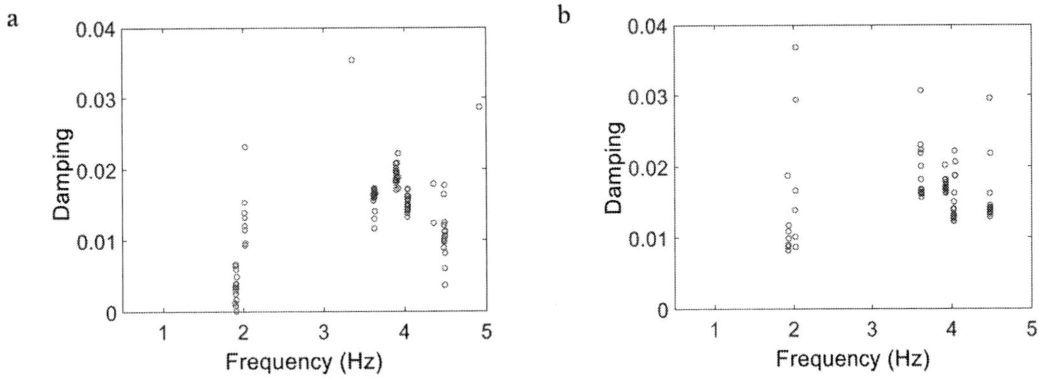

Figure 4. Stabilisation diagram in the range 0.5-5Hz obtained using records of 20 minutes and filtered in the range (a) 0.5-5Hz and (b) 0.5-15 Hz.

IOP Conf. Series: Journal of Physics: Conf. Series **842** (2017) 012009 doi:10.1088/1742-6596/842/1/012009

At the end of several manual identification sessions, a 20 minutes long signal was chosen for automatic identifications. 'Figure 5' shows the stabilisation diagrams obtained using signals of different lengths, obtained by cutting the same signal of 20 minute. The dotted lines in the diagrams indicate the identified frequencies of the Basilica (see 'Table 1'). The comparison of the stabilisation diagrams shows the difference between the parameters identified using signal of 5 and 10 minute, and some frequencies that are not captured. Furthermore, damping estimates confirms its dependence on the signal segment used in the identification [13].

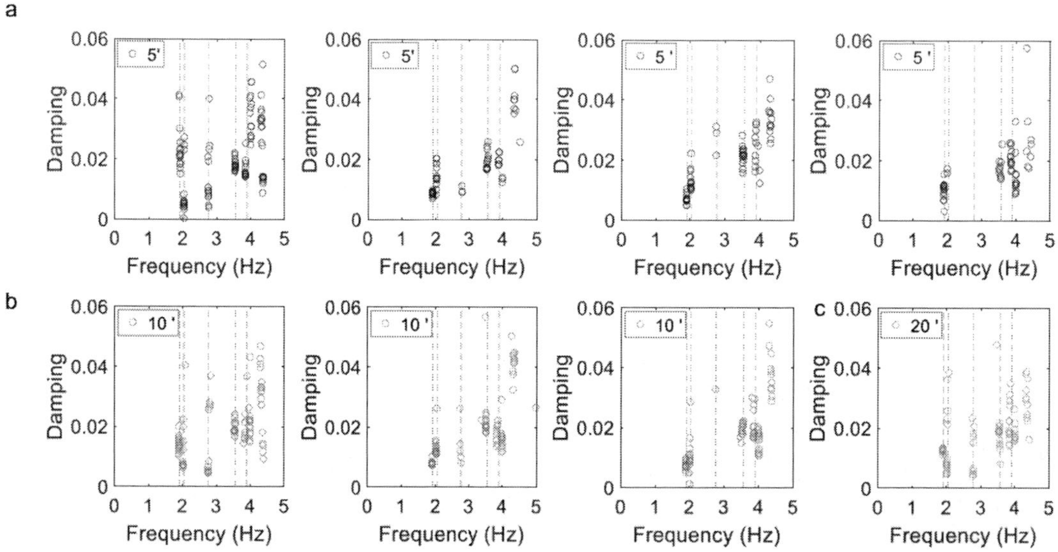

Figure 5. Stabilisation diagram in the range 0.5-5Hz obtained using records of different length, obtained copping the same signal of 20 minute (a) 5 minute signals; (b) 10 minute signal; (c) 20 minute signal.

3.2 *SSI identification and stability analysis*

The number of the parameters identified will depend on the choice about the SSI model order, whose wrong choice increases the risk to embark spurious modes characterized by a large variability. Despite being eliminated through the stabilisation criteria, spurious modes imply a worthless time-consuming computation. Consequently, a characteristic model order range needs to be chosen to identify the modes of Basilica avoiding worthless computations. To fix this interval, the model order was varied from 30 to 100 during manual identification.

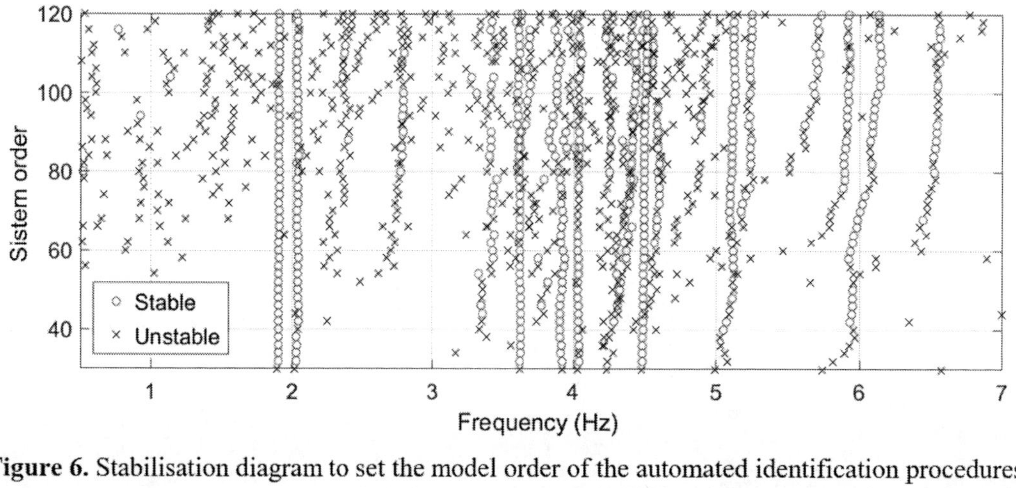

Figure 6. Stabilisation diagram to set the model order of the automated identification procedures.

The stabilisation diagram reported in 'figure 6' shows the uselessness of model orders higher than 60. Indeed, spurious parameters identified for greater model orders, because of their instability, will be removed by the introduction of the following stabilisation criteria: (i) damping in the range 0% - 10%; (ii) damping variation < 2.5 %; (iv) MAC >95%; and (iii) frequency variation < 0.03Hz. The frequency variation was assumed equal to one-third of the variance between the first two frequencies identified in 2008 [14]. Based on these results the model order of the identification algorithm was set to vary in the range 30-60 during the automatic identification

3.3 Clustering analysis and hourly candidate mode

The modes cleared with the stabilisation criteria are investigated in terms of regularity. To do this a clustering analysis is performed and the stable modes are grouped in separate sets belonging to the same structural mode. The number of the cluster is fixed through an iterative procedure considering the mutual distance in terms of frequency and MAC. The reliability of the clusters is assumed as coinciding with their element number and therefore defined through an occurrence index, I_{OCC}, defined as:

$$I_{OCC} = \frac{n}{\left(\frac{N_{max} - N_{min}}{\Delta N}\right)} \tag{1}$$

where n is the numbers of the cluster elements; N_{max} and N_{min} are maximum and minimum order of the SSI model and ΔN is the step model.
The clusters characterised by a value of I smaller than 0.3 are considered irregular and are eliminated. The candidate modes deriving from each signal are obtained as the average of the elements belonging to the same cluster. The weight w_i of i-th element of the cluster is derived by the MAC values through the formula:

$$w_i = \sum_{K=1}^{n} \frac{|\{\Psi_i\}^T\{\Psi_k{}^*\}|^2}{\{\Psi_i\}^T\{\Psi_i{}^*\}\{\Psi_k\}^T\{\Psi_k{}^*\}} \tag{2}$$

where $\{\Psi_i\}$ is the modal shape of the i-th element of the cluster; $\{\Psi_k\}$ are the modal shapes of all the n elements of the cluster.
'Figure 7' overlaps the stable and regular modes identified and the candidate mode defined by Eq. 2, using the data acquired 2 December 2016. The figure shows that the stable identified modes vary over a day. The random variability of frequencies throughout a day, as shown in 'figure 8' for the first frequency, suggests that the daily fluctuations are not related to thermal factors.

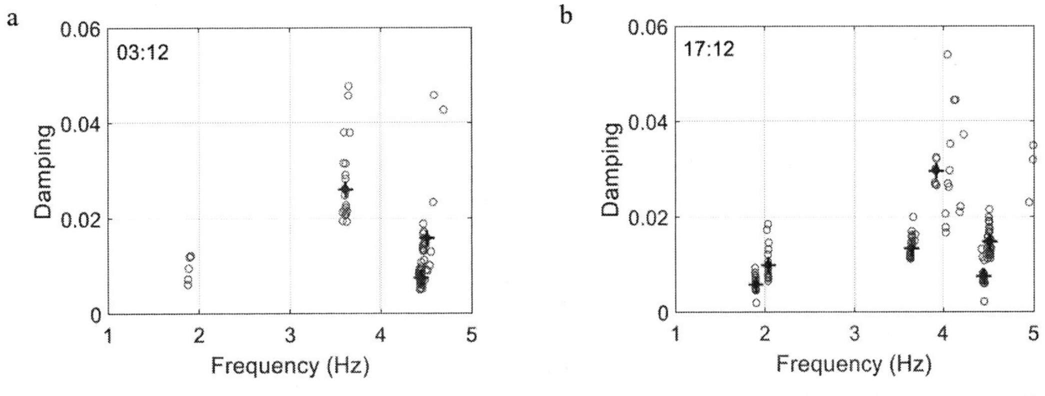

Figure 7. Frequencies identified using data acquired 02/12/2016 at different hours (a) hour 3:12 (b) hour 17:12.

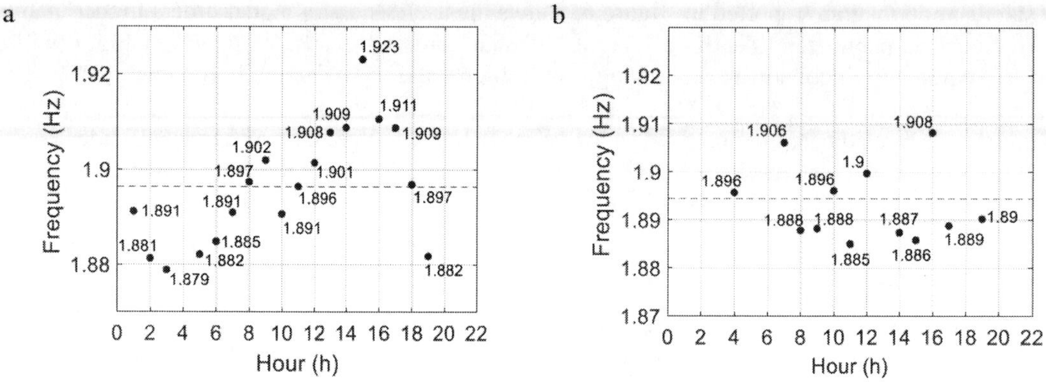

Figure 8. Variability of the first frequency during a day (a) 02/12/2016 (b) 03/12/2016.

'Figure 9' shows that the frequencies identified and their values at the same hours of different days are not constant. Consequently, monitoring should not be based on a single record acquired in the same hour because this would lead to inaccurate conclusions about the state of the Basilica and about the variability of the frequencies with temperature. Based on these results the acquisition system was set to acquire one signal hourly, and a procedure to obtain the Basilica daily candidate modes from the hourly modes was defined.

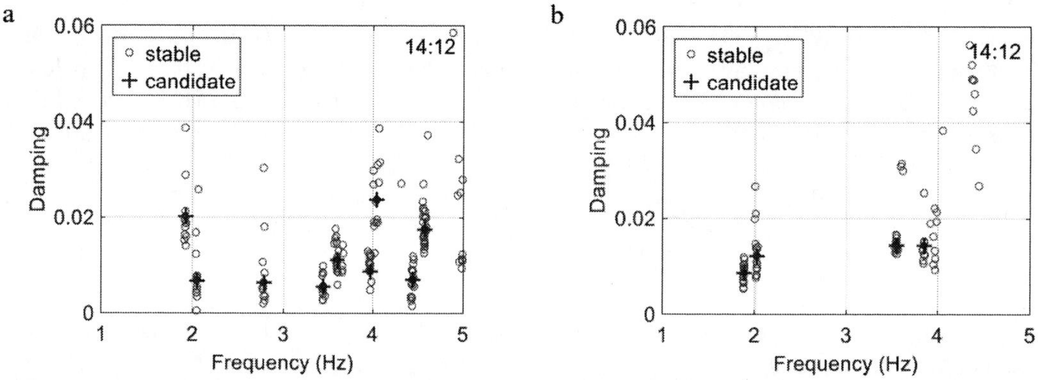

Figure 9. Frequencies identified using data acquired at hour 14:12 on (a) 02/12/2016; (b) 03/12/2016.

3.4 *Daily modes from hourly candidate modes*

The daily modes are obtained from the hourly candidate modes, through an iterative clustering analysis considering the mutual distance in terms of frequency and MAC. Once the clusters are defined, frequencies and modal shapes of the Basilica are obtained from weighted average operations. The weight of each element depends on its regularity with respect to the cluster and on its occurrence. More specifically the weight of the i-th element of the cluster is defined by:

$$w_i = \left(\sum_{k=1}^{n} \frac{|\{\Psi_i\}^T \{\Psi_k{}^*\}|^2}{\{\Psi_i\}^T \{\Psi_i{}^*\} \{\Psi_k\}^T \{\Psi_k{}^*\}} \right) \cdot \left(\frac{I_{OCC,i}}{\sum_{k=1}^{n} I_{OCC,k}} \right) \tag{3}$$

where $\{\Psi_i\}$ is the modal shape of the i-th hourly candidate mode of the cluster; $\{\Psi_k\}$ are the modal shapes of all the n hourly candidate modes of the cluster; $I_{OCC,i}$ is the occurrence index of the hourly candidate mode, while the $I_{OCC,k}$ is the occurrence index of all the n hourly candidate modes of the cluster. The reliability of the dynamic parameters at the end of the iterative clustering analysis is associated with the sum of the occurrence indices of the hourly candidate mode of the cluster. The dynamic parameters with a high value of I_{OCC} can reasonably be interpreted as structural modes of the

Basilica. Similarly, the parameters with low occurrence can be neglected. The nature of the other dynamic parameters must be investigated with the help of the FE model of the Basilica. 'Figure 10' shows for instance as the number of the daily candidate mode decreases assuming a higher occurrence index.

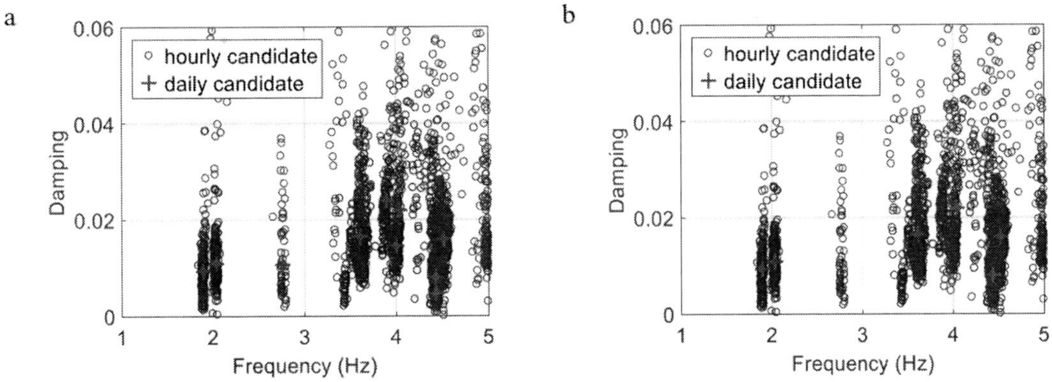

Figure 10. Daily candidate modes in the range 0-5 Hz (02/12/2016) for a minimum occurrence of (a) 50 and (b) 100.

4. Identification results

The above described procedure was used to determine the Basilica dynamic parameters over the period November 2016-January 2017. Environmental factors slightly affect the daily candidate modes that present a reduced variability. Table 1 reports the frequencies of the main modes identified over the aforementioned period. The comparison of these new results with the previous modal identification campaign shows that: (i) the frequencies of the first bending modes of the Basilica are approximately the same. The plots of the modal shapes, as reported in Fig. 11.a and Fig. 11.b, show that the first bending mode at 1.91 Hz has its main component in the Y direction, while the one at 2,04 Hz is a purely X direction bending mode; (ii) the suspect mode detected in 2008 at about 2.86 Hz in the 2008 identification is systematically confirmed in the new identification sessions. This experimental mode, depicted in Fig. 11.d, is associated to the first torsional mode of the whole Basilica's model, being governed by the four bell-towers.

With regards to the plot of the modal shapes, at the drum level there are not enough sensors to capture the torsion of the dome, see Fig. 2; (iii) the second bending modes, along the X and Y direction, respectively, are found at 3.56 Hz and 3.90 Hz, instead of 3.08 and 3.77 Hz.

Table 1. Frequencies identified in 2017.

EXPERIMENTAL 2017 - Freq.(Hz)	Mode classification
1,91	1st bending Y
2,04	1st bending X
2,77	1st torsional Basilica
3,56	2nd bending Y
3,90	2nd bending X
5,13	1 st vertical dome

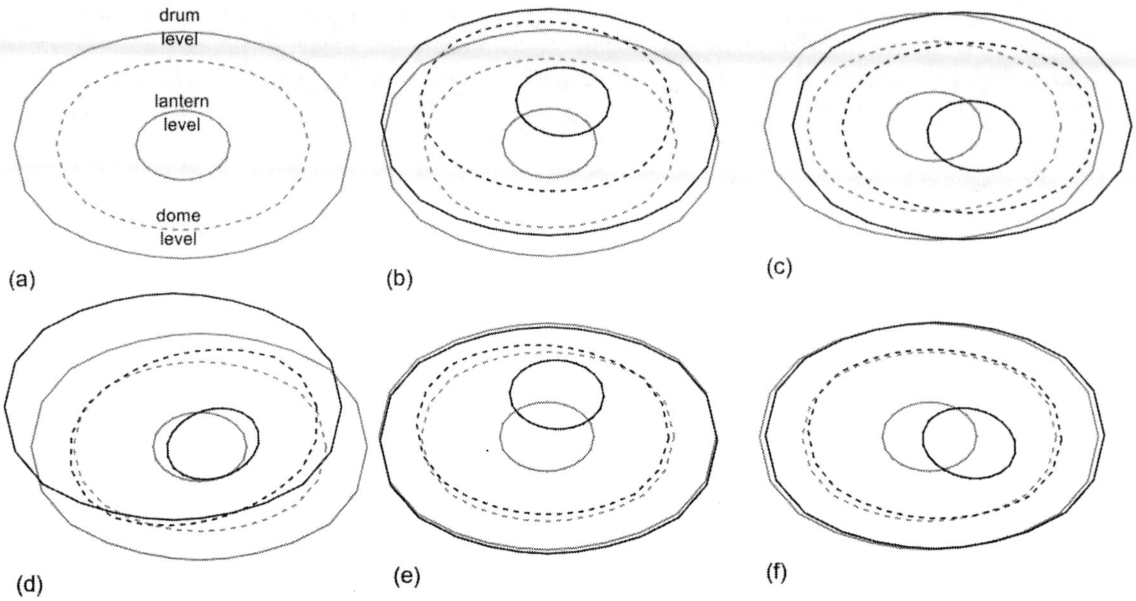

Figure 11. Illustration of the identified modal shapes of: (a) monitored levels of the Basilica (b) the 1st bending mode along Y direction; (c) the 1st bending mode along X direction; (d) the 1st torsional; (e) the 2nd bending mode along Y direction; (f) the 2nd bending mode along X direction.

5. Seasonal variation of frequencies

Assessing the dependence of the dynamic parameters of the Basilica on environmental factor is crucial to activate a robust structural health monitoring. At this stage, the dependence of the first two frequencies on the temperature is investigated. 'Figure 12' combines the frequencies automatically identified with the procedure described above, from 21 November 2016 to 21 January, and the average temperature measured in Vicoforte. Despite the importance of the temperature variation over the investigated period, the duration of this period is too short to draw a general conclusion about the dependence of frequencies on temperature. In any case, the graph seems to suggest the lack of correlation between temperature and the first two frequencies of the Basilica dome.

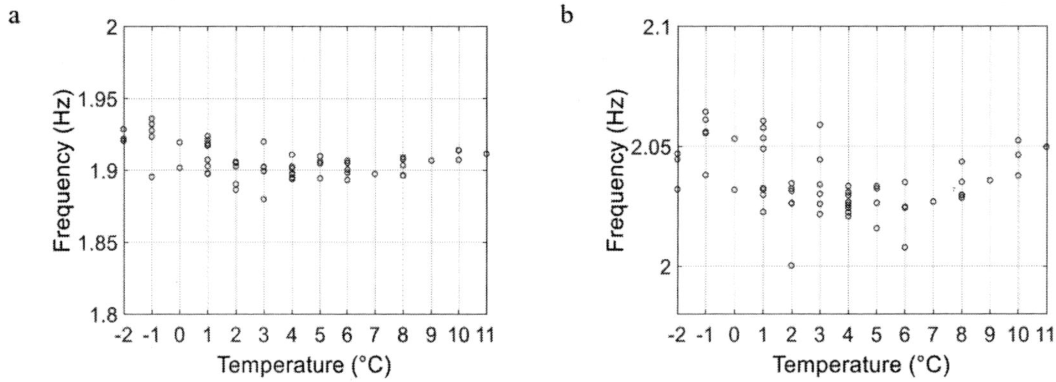

Figure 12. (a) 1st bending mode along Y direction vs. temperature; (b) 1st bending mode along X direction vs. temperature.

6. Conclusions

This paper has described the studies conducted to establish a protocol for the SHM activities to be conducted in the forthcoming years on the Regina Montis Regalis oval dome. The procedure automatically determines the hourly and daily candidate modes, using the signals recorded at each hour.

The identification protocol was defined by comparing different strategies applied to the data acquired by the permanent dynamic monitoring system installed at the beginning of 2016. The next step of this study will correlate the results of the automated identification with the environmental parameters considering a longer period. Quantifying the dependence of the dynamic parameters on environmental factors will be crucial in order to activate a reliable structural heath monitoring and to use dynamic parameters in either data-driven or model-driven SHM procedures.

References

[1] Lorenzoni F, Casarin F, Modena C, Caldon M, Islami K, Da Porto F 2013 Structural health monitoring of the Roman Arena of Verona *Journal of Civil Structural Health* Monitoring **3** 227–246

[2] Zonta D, Ceravolo R, Bursi O, Erlicher S, Zanon P and De Stefano A 2002 Issues on vibration-based identification of complex monumental structures: the Dome of S. Gaudenzio Church in Novara *In: Proc. 3rd World Conf. on Structural Control* (Como, Italy)

[3] Modena C 2015 Structural Health Monitoring: a tool for managing risks in sub-standard conditions *In Proc. 7th Int. Conf. on Structural Health Monitoring of Intelligent Infrastructures* (Turin)

[4] Sohn H 2007 Effects of environmental and operational variability on structural health monitoring *Philosophical Transactions of the Royal Society of London A: Mathematical, Physical and Engineering Sciences* **365**(1851) 539-560

[5] Salawu O S1997 Detection of structural damage through changes in frequency: a review *Engineering structures* **19**(9) 718-723

[6] Ramos L F, Marques L, Lourenço P B, De Roeck G, Campos-Costa A and Roque J 2010 Monitoring historical masonry structures with operational modal analysis: two case studies *Mechanical Systems and Signal Processing* **24**(5) 1291-1305

[7] Lorenzoni F, Casarin F, Caldon M, Islami K and Modena C 2016 Uncertainty quantification in structural health monitoring: Applications on cultural heritage buildings *Mechanical Systems and Signal Processing* **66** 268-281

[8] Ubertini F, Comanducci G, Cavalagli N, Pisello A L, Materazzi A L and Cotana F 2016 Environmental effects on natural frequencies of the San Pietro bell tower in Perugia, Italy, and their removal for structural performance assessment *Mechanical Systems and Signal Processing.*

[9] Van Buren K, Reilly J, Neal K, Edwards H and Hemez F 2017 Guaranteeing robustness of structural condition monitoring to environmental variability *Journal of Sound and Vibration* **386** 134-148

[10] Chiorino M A, Spadafora A, Calderini C and Lagomarsino S 2008 Monitoring and modelling strategies for the world's largest elliptical dome at Vicoforte, *International Journal of Architectural Heritage* **2** 274–303

[11] Casalegno C, Ceravolo R, Chiorino M A, Pecorelli M L and Zanotti Fragonara L 2014 Soil-structure modeling and updating of the "Regina Montis Regalis" basilica at Vicoforte, Italy *In Proc. 9th International Conference on Structural Analysis of Historical Constructions* (F. Peña & M. Chavez)

[12] Aoki T, Komiyama T, Tanigawa Y, Hatanaka S, Yuasa N, Hamasaki H, Chiorino M A and Roccati R 2004 Non-destructive testing of the Sanctuary of Vicoforte *In Proceedings of 13th international brick and block masonry conference (Amsterdam)*1109-1118

[13] Ceravolo R and Abbiati G 2013 Time Domain Identification of Structures: a Comparative Analysis of Output-Only Methods. Journal of Engineering Mechanics (ASCE), **139**(4) 537-544.

[14] Chiorino M A, Ceravolo R, Spadafora A, Zanotti Fragonara L and Abbiati G 2011 Dynamic Characterization of Complex Masonry Structures: The Sanctuary of Vicoforte *International Journal of Architectural Heritage* **5** 296–314

[15] Ceravolo R, De Marinis A, Pecorelli M L and Zanotti Fragonara L 2016 Monitoring of masonry historical constructions: 10 years of static monitoring of the world's largest oval dome *Structrural Control Health Monitoring* DOI 10.1002

[16] Li B, Li D, Zhao X and Ou J 2012 Optimal sensor placement in health monitoring of suspension bridge *Science China Technological Sciences* **55**(7) 2039–2047
[17] Ceravolo R, De Lucia G, Pecorelli M L and Zanotti Fragonara L 2015 Monitoring of historical buildings: Project of a dynamic monitoring system for the world's largest elliptical dome *In Environmental, Energy and Structural Monitoring Systems* (EESMS) 113-118
[18] Van Overschee P and De Moor B 1996 *Subspace identification for linear systems: Theory and implementation - applications* (Dordrecht, Netherlands Kluwer Academic Press)

Study on the evaluation of erosion damage by using laser ultrasonic integrated with a wavelet analysis technique

T Wan[1,*], T Naoe[1], T Wakui[1], M Futakawa[1], H Obayashi[1] and T Sasa[1]

[1] J-PARC Center, Japan Atomic Energy Agency, 2-4 Shirakata, Tokai-mura, Ibaraki 319-1195, Japan

* E-mail: wan.tao@jaea.go.jp

Abstract. Spallation targets are the key components of accelerator driven systems (ADSs) that are being developed in the world. Erosion damages on the target vessels are anticipated. To prevent accidents occurrence due to erosion of spallation target vessel, the damage evaluation technique is desirable. The excited vibration of LBE target vessel will be monitored remotely to establish the technique. In this study, the basic researches were carried out through experiments and numerical simulations to investigate the interaction between ultrasonic waves and damage to understand the correlation between structural vibration and damage degree. Specimens with distributed erosion damage was irradiated by laser shots, and the vibration was detected by a laser vibrometer subsequently. A technique, Wavelet Differential Analysis (WDA), was developed to quantitatively and clearly indicate the differences caused by damage in the vibration signals. The results illustrated that the developed technique is sensitive to erosion damage with small size and is capable of quantitatively evaluating erosion damage. It is expected that the developed techniques can be applied to monitor the real spallation targets in the future.

1. Introduction

The Accelerator Driven Systems (ADSs) are considered to be promising technique approaches to transmute the minor actinides contained in the high-level radioactive wastes effectively and efficiently. Japan Atomic Energy Agency (JAEA) also proposed an ADS adopting Lead-Bismuth Eutectic (LBE) as the spallation material and the coolant [1, 2]. A Beam Window (BW) is a boundary that separates the accelerator and the spallation target. Figure 1 shows the conceptual spallation target BW diagram of LBE-cooled ADS proposed by JAEA. Continuous proton beams with 1.5GeV energy will be injected into LBE through the BW for spallation reaction [3]. Flowing LBE with a speed of 2 m/s will remove the generated heat in LBE and on the BW. Obviously the integrity of BW is critical to the safe operation of ADS because it should withstand severe service conditions. To investigate the following issues for the BW: a) thermal-hydraulic behavior, b) design feasibility & lifetime estimation, and c) effects of beam trips, a LBE spallation target has the same BW structure as the ADS concept will be installed at the ADS Target Test Facility (TEF-T), which will be constructed under the frame work of Japan Proton Accelerator Research Complex (J-PARC) project. The BW will be made of type 316 SS when the operation temperature below 500 °C [4]. To take full advantage of the current J-PARC accelerator, pulsed proton beams (250 kW, 25 Hz, and 0.5 ms pulse duration) will be utilized to bombard the TEF-T LBE spallation target [5, 6].

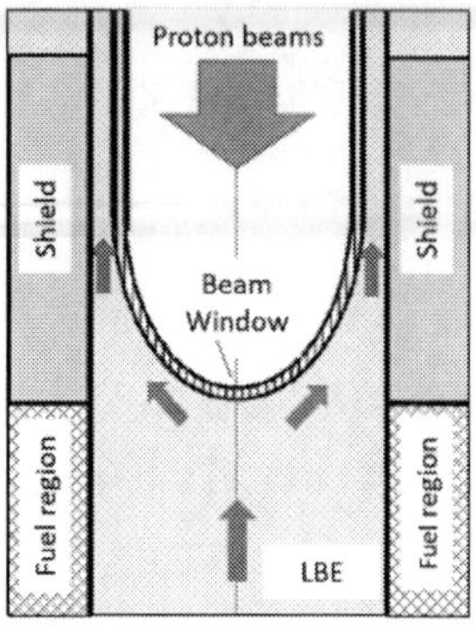

Figure 1. Conceptual diagram of spallation target beam window of LBE-cooled ADS proposed by JAEA.

The design lifetime of future ADS spallation target is several tens of years, while that of the TET-T LBE spallation target is 5000 hours/year on operation. During the operations, several kinds of damages, e.g., fatigue damage, irradiation damage, and erosion damage will be imposed on the BW. Among those sorts of damages, cavitation-induced erosion damage might be a key factor that determines the lifetime of BW. Cavitation might be generated due to negative pressure that attributes to the pressure change in heavy liquid metal (HLM). Several reasons could induce the pressure change, firstly, thermal shocks in HLM due to the pulsed proton beam injections and proton beam trips, and secondly, considerable turbulent flow of HLM. Owing to the high density of HLM, the impact force released on the BW by the collapses of cavitation bubbles close to the BW is very strong and can give rise to the occurrence of cavitation damage on the BW [7, 8]. The lifetime of BW might be significantly shorten by the cavitation damage [9].

Considering the high-radioactive environment around the target, it is expected to develop a remote and noncontact in-situ structural integrity evaluation technique to monitor the service status of target vessel. For this purpose, the laser ultrasonic technique is a reasonable alternative, which adopts a pulsed laser beam remotely and noncontact heating up the objective surface to generate ultrasonic waves and another laser beam to remotely detect the propagation of ultrasonic waves in a certain direction. By employing the laser ultrasonic technique, the dynamic responses of the excited target vessel vibration can be monitored by laser vibrometer or interferometer. It has been reported that the vibration response of target vessel can be affected by the occurrence of cavitation damage for the mercury target in Japan Spallation Neutron Source (JSNS) [10]. On the basis of the detected vibration signals, the structural integrity evaluation technique for the target vessel is expected to be established.

In the present study, the basic researches involving the interactions between ultrasonic waves and cavitation damage were carried out through experiments and numerical simulations in order to investigate correlation between structural vibration and damage degree. In the experiments, laser generation and detection of ultrasonic waves were used to remotely and noncontact evaluate the cavitation damage. A pulsed Nd-YAG laser was employed to impact the specimen, and subsequently the specimen vibration was detected by a Laser Doppler Vibrometer (LDV). In the numerical simulations, a Finite Difference Method (FDM) method was employed to systematically investigate the dependency of vibration signals on damage. Furthermore, a technique, Wavelet Differential Analysis (WDA), was developed to clearly indicate damage dependency in the vibration signals.

In our previous study [11], the vibrational signals obtained from the plate specimens were studied from the viewpoint of propagation behavior of ultrasonic waves and time-of-flight of each wave mode. In this continuous work, new developed method was focused on aims at evaluating the cavitation damage clearly and quantitatively. It is expected the developed technique can be applied to various structures that adopting HLM as spallation/coolant material.

2. Experiments

2.1. Laser ultrasonic system

Figure 2 shows the schematic of the LDV system employed in the experiments. A Q-switched Nd-YAG laser (QUANTEL USA, Ultra 100-55-20) with a wavelength of 532 nm was adopted to impact the specimen without imposing any damage on the surface. The laser beam is 3 mm in diameter with a Gaussian profile. The output energy of the pulsed laser (1 Hz) is 2.5 mJ in 6.5 ns. The bottom surface of the specimen was damaged, which was to simulate the damage on the inner surface of BW, and the damage was detected from the opposite side, i.e., the outside surface of BW.

The vibration of specimen was detected by a LDV (ONOSOKKI, LV-1710) with a He-Ne laser (633 nm, 1 mW), owning a detecting frequency range of 1 Hz–3 MHz. Signals captured by the LDV were stored by using an oscilloscope with a sampling rate of 12.5 MHz. Q-switch signals were used for the triggers of recording LDV signals. Each vibration waveform was averaged by 180 signals to reduce the background noise. A microscope lens focused the He-Ne laser beam to a spot with 3.5 μm in diameter on the specimen surface to obtain a high spatial resolution. The distance between the center point of the Nd-YAG laser beam spot and the He-Ne laser beam spot was fixed to be ca. 5 mm.

2.2. Specimen & Cavitation-induced erosion damage

Figure 3 shows a photograph of the specimen with distributed cavitation damage. The material of specimen is type 316 SS. The damage was imposed on the specimen by using an electro-Magnetic Impact Testing Machine (MIMTM). The detailed characterizations of MIMTM and the distributed cavitation damage was stated in reference [12], and only brief descriptions will be given presently. MIMTM was adopted to simulate the pressure change through the pulsed mechanical impacts of a striker against the HLM. In the MIMTM, mercury is adopted instead of LBE for two main considerations: 1) mercury is in the liquid state at room temperature, whereas the melting point of LBE is ca. 298 K, so that the experiments can be performed more conveniently by using mercury, and 2) some physical properties of mercury, which closely relating to the formation of cavitation bubbles and the released impact force of bubble collapse, are similar to LBE, e.g., density, surface tension. Though the vapor pressure of mercury at room temperature (2.6×10^{-7} MPa) is much higher than that of LBE at the nominal operation temperature (5.8×10^{-13} MPa @ 600 K), both the vapor pressures are far lower than the standard air pressure (0.1 MPa), so the effects of the differences of vapor pressures on cavitation bubble formation can be ignored. The specimen was divided into five areas, and damages with various degrees were conducted at each area, respectively. The maximum peak-to-peak depth of the damage at each area was: (I) 45 μm, (II) 35 μm, (III) 25 μm, (IV) 15 μm, (V) as received. Then the damage at each area was evaluated by the laser ultrasonic systems.

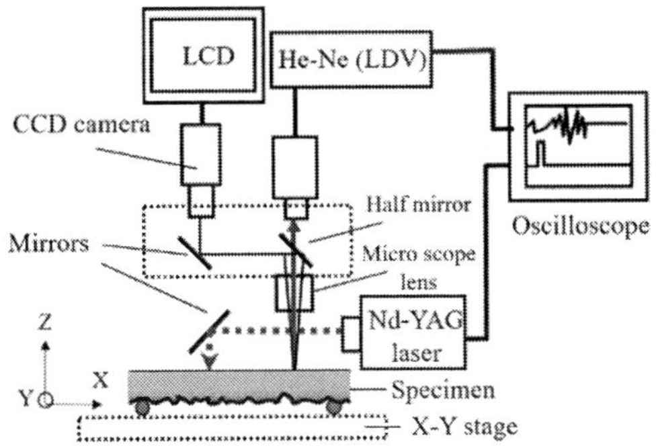

Figure 2. Schematic drawing of the laser ultrasonic system.

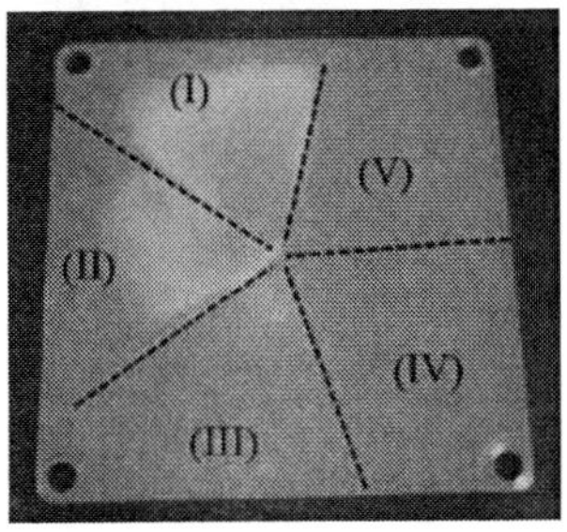

Figure 3. Specimen with distributed cavitation damage: maximum depth of cavitation damage in each area: (I) 45 μm, (II) 35 μm, (III) 25 μm, (IV) 15 μm, (V) as received.

3. Numerical simulations

The numerical simulations were performed using a commercial Finite Difference Method (FDM) code Wave3000 (Cyber Logic). The specific acoustic equation used in Wave3000 is given by [13]:

$$\rho \frac{\partial^2 w}{\partial^2 t} = \left[\mu + \eta \frac{\partial}{\partial t}\right] \nabla^2 w + \left[\lambda + \mu + \phi \frac{\partial}{\partial t} + \frac{\eta}{3} \frac{\partial}{\partial t}\right] \nabla(\nabla \cdot w), \tag{1}$$

where ρ is the material density, kg/m^3; λ is the first Lame constant, N/m^2; μ is the second Lame constant, N/m^2; η is the shear viscosity, N·s/m^2; ϕ is the bulk viscosity, N·s/m^2; ∇ is the gradient operator; $\nabla \cdot$ is the divergence operator; t is the time, s; w is a three-dimensional column vector whose components are the x, y, and z components of the displacement of the medium at location (x, y, z).

To systematically investigate the dependency of ultrasonic wave propagation on number and depth of damage, numerical simulation model with simulated pits was made, as shown in Figure 4(a). The 3D model owns a size of 30×30×2.5 mm^3. The mesh size is 0.1 mm. The maximum Z surface possess an infinite boundary, whereas the other surfaces are "free-free". The source on the upper surface is a pulsed longitudinal sinusoidal wave with a size of 3.2 mm in diameter, a frequency of 1 MHz and a duration of 1 μs. The vibration velocity in the Z-direction was detected by the receiver with 0.2 mm in diameter. Distance of source and receiver is 5.0 mm. Simulated pentahedral pits with various numbers N (ranges from 0 to 900), and depths D (ranges from 0 to 1.4 mm), were established at the bottom of the model. The aspect ratio (depth/width) of simulated pits was 0.5.

It should be mentioned that according to reference [14], wave3000 is a "memory-hungry" program. The approximating required memory is to multiply the number of finite difference grid voxels, say M, by 42. In this study, the maximum depth of the damage is 45 μm. To maintain the structure details of such indent and obtain accurate results, the grid size would be required to be approximately 5 μm. For an object of 30 mm × 30 mm × 2.5 mm, the memory required is about 4400 GB. Obviously, such numerical simulations are impossible to be carried out on a PC. Therefore, to reduce the calculation cost, the depth of damage was selected as 0.4 mm when changing the number of simulated pits, while the number of damage was 900 when varying the depth of simulated pits.

In addition, damages with random profile were also established at the bottom of the model, as shown in Fig. 4(b). The damaged area occupies 90% of the bottom surface, while the maximum depth of random damage was 0.4 and 0.9 mm, respectively.

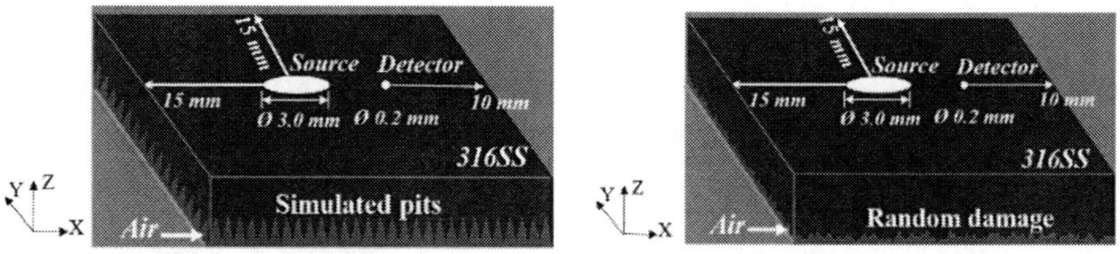

(a) Simulated pits (b) Random damage

Figure 4. Schematic drawings of numerical simulation models.

4. Wavelet differential analysis

The wavelet differential process [15] for the vibration signals was conducted to clearly indicate the differences between signals. The process is described by Eq. (2) as follows:

$$WDI_D = |Img_D - Img_0|, \tag{2}$$

where WDI is the wavelet differential image, Img is the wavelet image of the vibration signals, and the subscripts D and 0 represents the damage with a depth of D μm and no damage (or D and 0 represents the damage with a number of D and no damage), respectively. Then, the average intensity (I_a) of the wavelet differential image (WDI) was calculated according to equations (3) – (5):

$$I_a = \int^{\Delta f} \int^{\Delta t} V(t, f) \mathrm{d}t \mathrm{d}f, \tag{3}$$

$$\Delta f = f_2 - f_1 , \tag{4}$$

$$\Delta t = t_2 - t_1 , \tag{5}$$

where the measurement frequency ranges from f_1 to f_2 (Hz), and the measurement time ranges from t_1 to t_2 (s); $V(t, f)$ is the *WDI* intensity (dB) at time t and frequency f, and I_a is the average intensity of the differential image.

5. Results and Discussions

5.1. Time responses of displacement velocity

Figure 5 shows the examples of time responses of the displacement velocity detected from different areas on the specimen with cavitation damage by using the LDV system. The results obtained by locate the detection laser in the opposite side of each damage area. It can be seen from the results that there are two parameters related to the damage. First one is the amplitude of the first positive peak, referred to as V_p (marked in Fig. 5); the second one is the attenuation coefficient (α, marked in Fig. 5) of waveforms, which was defined by the following equation.

$$V_t = V_P e^{-\alpha \cdot t}, \tag{6}$$

where α is the attenuation coefficient, μs^{-1}; V_t is the vibration velocity at time t, m/s, and V_p is the vibration velocity of the first positive peak, m/s.

In the case of the vibrational signal obtained from the damaged specimen, the amplitude of the largest positive peak and the attenuation coefficient of the waveforms increased. As discussed in a previous study, the positive peak was caused by the shear waves that were reflected from the bottom of the specimen, whereas the attenuation was caused by the longitudinal waves that propagate between the upper and bottom surface of the specimen [11].

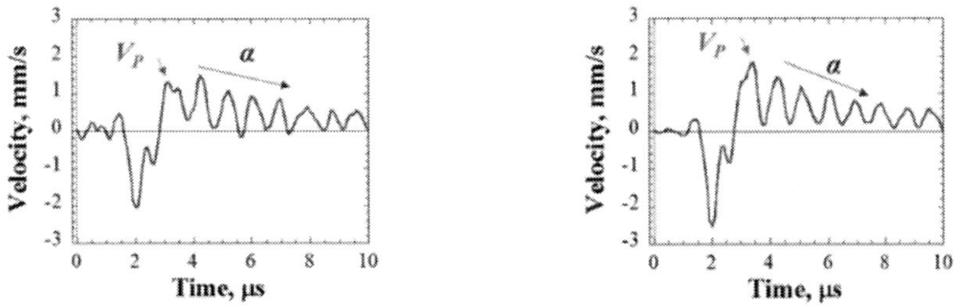

(a) W/O damage (b) W/ damage

Figure 5. Time responses of displacement velocity detected from different areas on the specimen with cavitation damage by using the LDV system.

5.2. Damage evaluation by the amplitude of positive peak

In this subsection, the distributed cavitation damage was evaluated using the LDV system by the positive peak. It is expected to evaluate the maximum depth of damage through the amplitude of positive peak. The numerical simulations were also carried out to evaluate simulated pits and random damage with various depths.

Figure 6 shows the examples of numerical simulation results without damage and with damage. The experimental results shown in Fig. 5 and numerical simulation results shown in Fig. 6 were plotted as functions of the maximum depth of cavitation damage, the results were shown in Fig. 7. From the results it can be seen that the tendency of experimental and the numerical simulation result is coincidence with each other, and the maximum depth of damage can be expressed by the following equation:

$$D = \mathrm{a} \times \left(\frac{V_P - V_{P,0}}{V_{P,0}}\right)^b, \tag{7}$$

where D is the maximum depth of damage; V_P is the positive peak of reflected waveforms with damage; $V_{P,0}$ is the positive peak of reflected waveforms without damage; a, b are constants related to the material property.

The cavitation damage reduces the equivalent thickness of the specimen [11], and thus the ultrasonic wave propagation distance from the source to the receiver is affected by the damage. When the damage depth trends to larger, the ultrasonic propagation distance becomes shorter. Therefore, the attenuation of ultrasonic waves due to spreading becomes weaker in the process of propagating from the source to the receiver, so that more energy will be reflected to the receiver and lead to a stronger amplitude of vibration. The variation of positive peak is mainly dominated by the spreading mechanism in the present study.

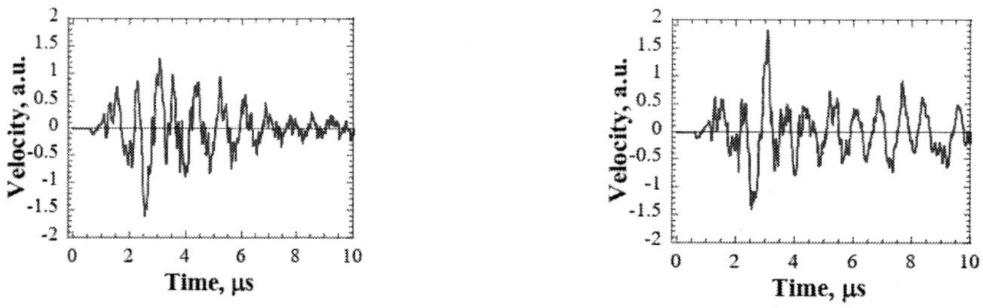

(a) W/O damage (b) W/ damage

Figure 6. Time responses of displacement velocity obtained by the numerical simulations.

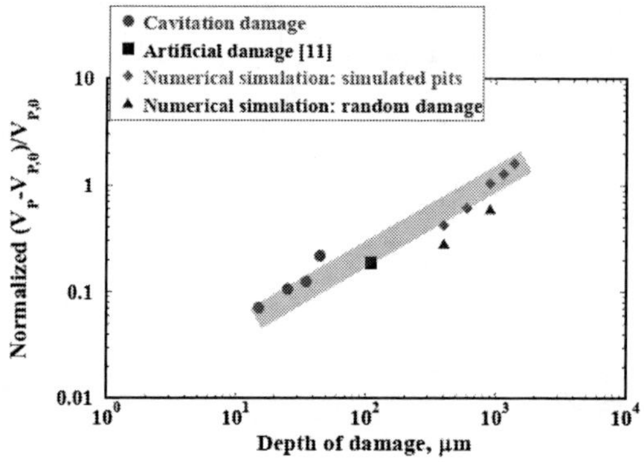

Figure 7. Damage depth as a function of normalized differential positive peak velocity.

5.3. Damage evaluation by WDA

It can be seen from Fig. 7 that the differences is not very obvious to recognize the damage dependency in cases of the damage depths are considerably small and are close to each other in the experiments. This may because the positive peaks just give the damage information at a time point. It is expected to involve all the information of ultrasonic wave attenuation in both time and frequency domains. The WDA technique provides such an approach. For this purpose, the WDA was used to quantitatively evaluate the damage dependency.

Figure 8 shows the typical examples of *WDI*s for cavitation damages with various maximum depths. Obvious differences can be observed from the *WDI* results. To quantitatively evaluate the differences of *WDI* results, the average intensity of *WDI*s was calculated. Figure 9 shows the average intensity of *WDI*s as a function of damage depth. Clear dependency on damage depth can be recognized from the average intensity results. The results indicates that the average intensity of *WDI*s is very sensitive to damage.

When the maximum depth of the damage increases up to approximately 15 μm, the variation in *WDI* results is mainly indicated in the relatively higher frequency range. If the maximum depth of damage keeps increasing, more obvious differences were shown in the relatively lower frequency range. Cavitation damage develops from isolated micro localized damage to area erosion when the maximum depth increases up to approximately 15μm [16, 17]. At this stage, the number of damage changed obviously while the maximum depth of damage almost keeps a constant. With the further accumulation of cavitation damage, the maximum depth of area erosion keeps increasing. It is assumed that the differences showed in the relatively higher frequency components are mainly attribute to the ultrasonic wave scattering from damage with various numbers, whereas the differences in the relatively lower frequency range is mainly due to the ultrasonic waves scattering from damage with various depths. Detailed discussion will be given in the next sub-section.

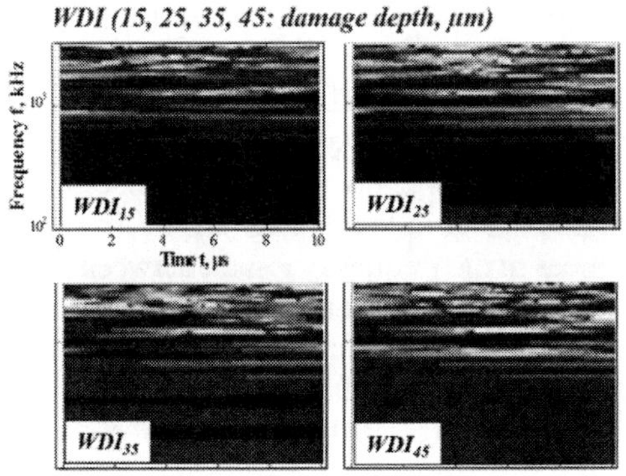

Figure 8. *WDI* results for the area distributed cavitation damage.

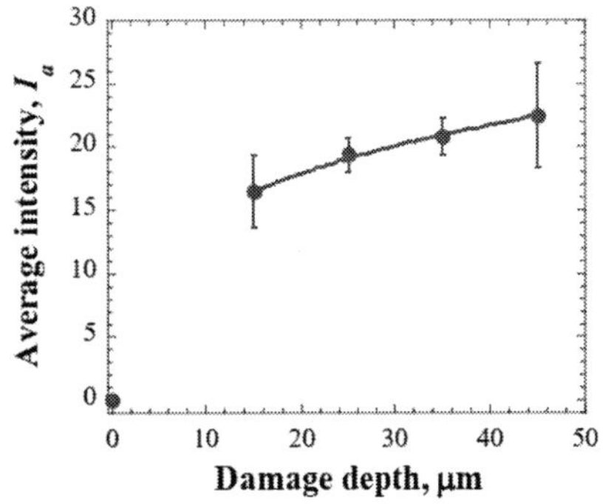

Figure 9. Average intensity of *WDI*s as a function of damage depth.

5.4. Effects of number and depth of damage on WDI results

Numerical simulations were carried out to investigate the effects of number and depth of damage on the *WDI* results. The model with simulated pits was employed for this study. Mesh size of the model was 0.05 mm to provide enough resolution for studying the propagation of ultrasonic waves with relative higher frequencies.

Figure 10 shows the effects of number of damage on the *WDI* results. When the number of damage increases, the differences in the relatively higher frequency region trends more obvious. However, differences can hardly be observed in the relatively lower frequency region. Figure 11 shows the average intensity of *WDI* as a function of each frequency component for various numbers of simulated pits. The damage depth was fixed as 0.4 mm when varying the number of simulated pits. For all the cases, the results can be divided into three regimes: (1) in the region that frequencies are higher than ca. 2 MHz, significantly differences of average intensity appear in the results, which is caused by intense scattering of ultrasonic waves from the damage; (2) when the frequencies locates between ca. 0.8 MHz and 2 MHz, the average intensity shows differences but not so significant as that above ca. 2 MHz. This is considered to be caused by the moderate scattering of ultrasonic waves from the damage; (3) if the frequencies of ultrasonic waves below ca. 0.8 MHz, differences can hardly be observed. Weak scattering of ultrasonic waves from damage occurred in this region.

The phenomenon is related to the relationship between the wavelength of ultrasonic waves and the size of damage. Here, the frequency components of waveforms obtained by simulations are from ca. 0.3 to 19.2 MHz. Assuming that the propagation velocity of ultrasonic waves in stainless steel is 5.7 m/s at room temperature, then the wavelength of the waveforms ranges from ca. 0.30 mm to 19.0 mm. Ultrasonic waves with a frequency of 2 MHz own a wavelength of 2.35 mm, which is ca. 6 times of that of damage depth of 0.4 mm. Ultrasonic waves with such a wavelength or shorter will be strongly scattered by the damage, and the diffusion field distributed in all the scattering directions according to the Kirchhoff scattering theory [18, 19]. Therefore, if the frequency above 2 MHz, almost all the energy of ultrasonic waves will be diffused due to the strong scattering. It is easy to understand that scattering of ultrasonic waves with frequency above 2 MHz will be enhanced due to the increase of number of damage.

If ultrasonic waves with frequencies locate between ca. 2 MHz and 0.8 MHz encounter the damage, the wavelength of ultrasonic waves is ca. 6 to 16 times of that of damage size. In this case, moderate scattering occurred but the diffusion field in the forward scattering direction is dominant [20], which can be detected by the receiver, and thus *WDI* results show difference in this frequency region but not obvious as the first regime.

If the frequency of ultrasonic wave bellows ca. 0.8 MHz, then the wavelength of the ultrasonic waves keep increasing and trend to far larger than the damage size. As a result, the ultrasonic waves would be weakly scattered by the damage. In this situation, the reflection of ultrasonic waves at the bottom of specimen will obey the Snell's law. Most of the ultrasonic waves will be reflected to the receiver direction. Therefore, the *WDI* results showed tiny differences in this frequency region.

Figure 12 shows the effects of depth of damage on the *WDI* results. The differences were mainly observed in the relatively lower frequency range. Figure 13 shows the average intensity of *WDI* as a function of frequency for various depths of simulated pits. The lower boundary of intense scattering zone shifts from approximately 5.5 MHz to 1 MHz with increasing in the damage depth from 0.2 mm to 1.0 mm. Therefore, *WDI* results display obvious differences in the frequency region between 5.5 MHz and 1 MHz as shown in Fig. 12. On the contrast, ultrasonic waves with frequency higher than 5.5 MHz will not be strongly affected due to the increase of damage depth, because they have already been sufficiently diffused due to the strong scattering when the damage size is 0.2 mm. Therefore, the *WDI* results showed slight differences in the frequency region higher than 5.5 MHz

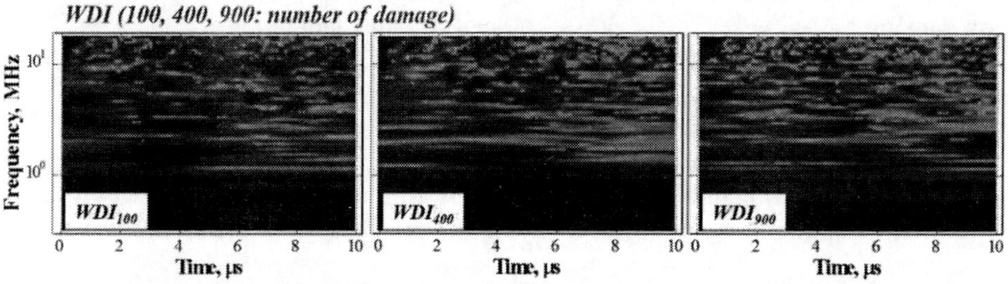

Figure 10. Effects of number of damage on the *WDI* results.

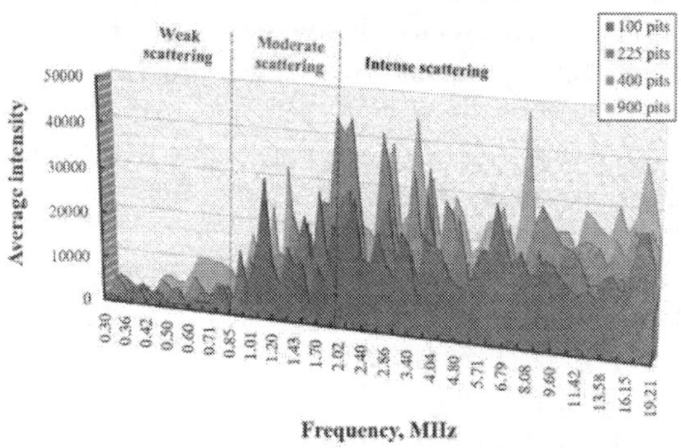

Figure 11. Average intensity of *WDI* as a function of each frequency component for various numbers of simulated pits.

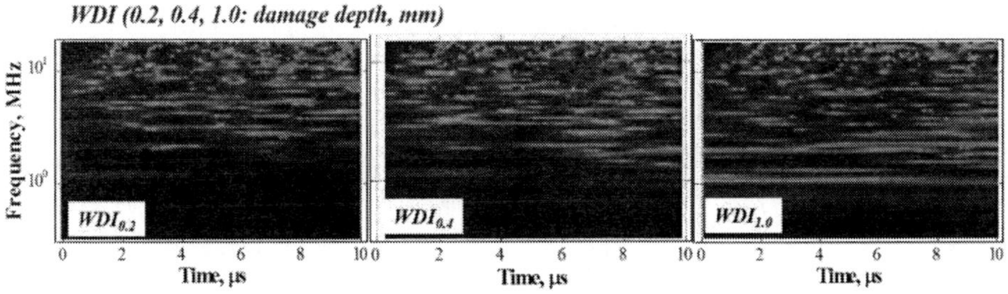

Figure 12. Effects of depth of damage on the *WDI* results.

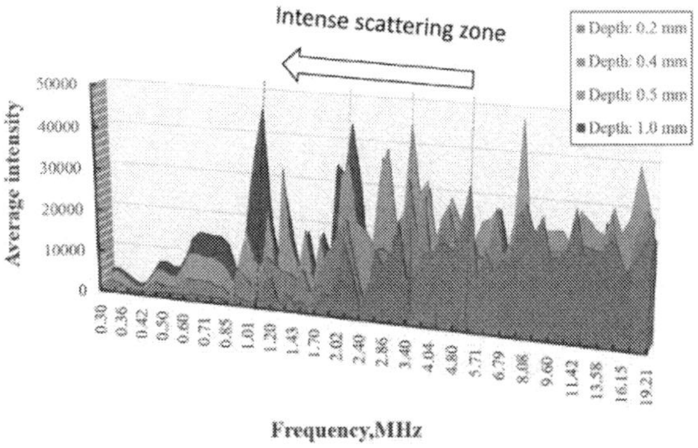

Figure 13. Average intensity of *WDI* as a function of each frequency component for various depths of simulated pits.

6. Conclusion

In the present study, the cavitation-induced erosion damage was evaluated by the laser generation and detection of ultrasonic waves. The WDA technique was developed to clearly indicate the damage dependency. The influences of damage number and depth were studied. The results are concluded as the followings:

(1) Damage depth dependency can be quantitatively and clearly indicated by using laser ultrasonic technique integrated with the WDA technique, the developed techniques are very sensitive to the damage with small size.

(2) The variation of number of damage mainly impose effects on scattering of ultrasonic waves with relatively higher frequency, while the variation of depth of damage mainly affects scattering of ultrasonic waves with relative lower frequency.

(3) Scattering of ultrasonic waves from cavitation damage can be divided into three regimes depends on the correlation between ultrasonic wavelength and damage size: 1) ultrasonic waves with wavelength less than 6 times of damage size could be intensely scattered, and the scattering field distributed in all the scattering directions; 2) ultrasonic waves with wavelength approximately 6 to 16 times of damage size could be moderately scattered, and the forward scattering field is dominant; 3) ultrasonic waves with wavelength far larger than damage size could be weakly scattered.

In the future, loop experiments will be carried out, to investigate the dependency of vibration of target vessel filled with liquid on erosion damage, by adopting the techniques presented in this study.

Acknowledgements

The authors thank to Mr. T. Ejiri, a student of Ibaraki University, for carrying out the laser experiments. This work was partly supported by the Japan Society for the Promotion of Science through a Grant-in-Aid for Scientific Research C (No. 23360088).

References

[1] Oigawa H *et al* 2011 *J. Nucl. Mater.* **415** 229
[2] Tsujimoto K *et al* 2007 *J. Nucl. Sci. Technol.* **44** 483
[3] Sugawara *et al* 2013 *Ann. Nucl. Energy* **55** 238
[4] Obayashi H *et al* 2015 *JPS Conf. Proc.* **8** 041002
[5] Sasa T and Oigawa H 2014 *Plasma and Fusion Research* **9** 4401113
[6] Sasa T 2015 *Prog. Nucl. Energ.* **82** 64
[7] Futakawa M, Kogawa H and Hino R 2000 *J. Phys. VI France.* **10** 237
[8] Futakawa M *et al* 2000 *Nucl. Instrum. Meth. A* **439** 1
[9] Teshigawara T *et al* 2010 *J. Nucl. Mater.* **398** 23
[10] Wan T, Naoe T and Futakawa M 2016 J. *Nucl. Mater.* **468** 321
[11] Wan T *et al* 2014 *Mater. Trans.* **55** 1024
[12] Wan T *et al* 2013 *Int. J. Mater. Prod. Tec.* **46** 141
[13] Delsanto P P, Schechter R S and Mignogna R B 1997 *Wave Motion* **26** 329
[14] CyberLogic Wave3000 manual
[15] Wan T *et al* 2014 *Appl. Mech. Mater.* **566** 629
[16] Futakawa M *et al* 2003 *J. Nucl. Sci. Technol.* **40** 895
[17] Futakawa M *et al* 2005 *J. Nucl. Mater.* **343** 70
[18] J.A. Ogilvy 1988 *J. Phys. D: Appl. Phys.* **21** 260
[19] J.A. Ogilvy 1987 *Rep. Prog. Phys.* **50** 1553
[20] J.A. Ogilvy 1989 *Ultrasonics* **27** 69

IOP Conf. Series: Journal of Physics: Conf. Series **842** (2017) 012011 doi:10.1088/1742-6596/842/1/012011

Application of vision measurements for modal analysis of wires for the purpose of overhead transmission lines monitoring

Krzysztof Mendrok, Ziemowit Dworakowski, Krzysztof Holak, Piotr Kohut

AGH University of Science and Technology, Al. Mickiewicza 30, 30-059 Kraków, Poland

e-mail: mendrok@agh.edu.pl

Abstract. Overhead transmission power lines are still one of the crucial elements of electro-energetic system. There are obvious advantages of using overhead transmission in the distribution of electricity. The amount of energy transported through a power line is determined by the distance between the wire and the ground or other objects placed beneath it (eg. trees). This distance is not fixed and depends on the overhang of the wire. This, in turn, is determined by many factors such as ambient temperature, humidity, precipitation, the value of current flowing through the wire. In order to optimize the wires electrical load, the monitoring of that overhang is required. One way to measure it is the non-contact measurement by vision system. It has the advantage, that using high-speed cameras respectively it also allows for vibration measurement and analysis of dynamic performance. That is very important while the wires are susceptible to the influence of wind, and the resulting vibrations interfere with the correct measurement of the overhang. The paper presents the results of vision measurements of the system vibrations and modal analysis carried out on their basis. The study was conducted on a specially made laboratory stand.

1. Introduction

No one needs convincing about the dominant role of electricity in the life of modern man. Electricity is produced in power plants and distributed by overhead power lines of high and highest voltage. However transmission of electricity over long distances is burdened with power losses which are proportional to the square of the electric current flowing through the conductor. Therefore, the voltage of transmitted electricity is increased and its electrical current is decreased. But one cannot arbitrarily increase the voltage in the wire, and thus increase the amount of transported energy, because it is limited by the risk of surges. To obtain the optimum transmission in overhead power lines one should constantly monitor the distance between the cable and the objects located beneath it. This distance is variable and depends on the wire overhang (assuming that the infrastructure under the transmission line does not change). Therefore, in modern transmission lines of high and highest voltages sensors measuring wire overhang are increasingly used (e.g. inclinometers mounted on the line). Unfortunately, the work of inclinometer is distorted when the cable starts to vibrate and then the readings are unreliable. Therefore, apart from measurements of the wire overhang, monitoring system should also be equip with the ability to measure vibration.

Vibrations can be measured by many methods, which in most of the cases are the contact one, such as accelerometers, strain gauges or FBG sensors. Application of these types of sensors requires them to be connected to the cable. It rises many complications related to the electricity flowing through the wire and its temperature and it makes such a system very expensive and inconvenient. No operator would be willing to turn off the voltage on their lines for mounting the sensors. It is better then, to develop non-contact measurement technique. There are many such methods presented in the literature [1, 2]. They can be based on geodesic or laser tools, but they rises another complications. That is why vision systems seem to be the best solution.

There are some examples of application of vision systems for overhead transmission lines monitoring. The main fields of application can be grouped as follows: an extraction of transmission line cables for later analysis, an estimation of ice cover thickness and damage detection, measurement of displacements of selected wire points. Although only third group is important from the paper topic point of view, for complete picture, examples of all application groups will be presented below. Paper [3] presents an example of the first group. Authors developed an algorithm for image segmentation to extract cables from the background. The automatic curve detection was done using a UAV. In order to process low contrast images, the Otsu method was applied [4]. In the article [5], segmentation was performed using genetic algorithms and particle swarm optimization. The impact of changing lighting conditions was removed by the use of homomorphic filters [6]. The automatic detection of transmission wires and classification of different types of poles was presented in the paper [7]. The classification of towers was based on edges extraction. The article [8] represents the second group of applications. A vision system was applied to measure the thickness of ice on the cable based on a comparison between the edge images of ice-covered cables and ones without an ice. Two methods of edge detection were implemented: wavelet transform and image morphology. The authors of [9] also presented a system for ice cover monitoring. The image processing consists of many steps, with the main ones being background subtraction, edge extraction and classification of ice cover types. The paper [10] shows an application of 3D vision methods to measure ice cover thickness. Manitoba Hydro presented "the Ice Vision System" [11] dedicated to the early detection of icing, measurement of ice accumulation and classification of the type of icing. Damage detection [12] methods for transmission line cables are also being developed. The paper [13], on the other hand, presents a vision system for lightning-induced cable damage detection. The system detects arc-shaped damage as well as cuts and cracks of cable bundles. The authors of the article [14] present an example of an application belonging to the third group – related thematically to the presented paper. The measurement of a wire's vibration was carried out in order to estimate the parameters of the structure and to detect the occurrence of damage. The algorithm consists of edges extraction, vertical displacements of measurement points reconstruction and the Fourier transform of the obtained displacement amplitude signal computation. The paper [15] describes a tool which aids vision-based modal analysis of transmission line cables. The image sequence was captured by a camera influenced by external vibrations. Therefore, the frames were blurred due to relative camera and object motion. The authors proposed multi-pattern matching based on DIC, which reconstructs vibration signals. The phenomenon of forcing wire vibrations due to the shedding of an ice cover is described in the paper [16]. A vision-based system is applied to measure the amplitude, frequency and damping factor of a vibrating wire. A ball marker was used as a measurement point. The articles [17,18] show an application of computer vision for geometry estimation and localization of measurement points in the preliminary steps of modal experiments, for the measurement of vibration signals, vision-based modal analysis and damage detection and localization based on vision data. An extension of the proposed method to a full 3D system is proposed in [19]. The paper [20] presents a non-contact method for the measurement of tensile forces in the suspension cables of bridges. A DIC algorithm is applied for displacement computation. The compensation of the random motion of a camera due to wind or vibration is also proposed. The correction is provided by the observation of a fixed object on the scene, e.g. a distant building. Tension in the cables was estimated based on dynamical response and natural frequencies. A damage detection method based on mode shapes was also introduced.

In the next paragraphs the algorithm used for points displacements calculation from recorded video data, experimental setup and performed modal analysis are presented.

2. Measurement algorithm

The vision system consists of two high-speed digital cameras, halogen lamps for lighting and a set of crash test markers mounted on the analyzed wire. The field of view can be adjusted using different lenses, depending on the desired accuracy. Applied method uses a tracking algorithm and three-dimensional reconstruction of the markers' trajectories. Trajectory reconstruction is based on the positions of corresponding markers on two image planes of the camera system [21]. The system measures the amplitude of vibrations in a reference frame from one of the cameras. This requires the calibration of external and internal parameters [21].

The measurement procedure can described as follows:

1) calibration of the system,
2) measurement setup preparation,
3) image sequence acquisition,
4) tracking and reconstruction of three-dimensional trajectories.

After arrangement of the camera and lighting system, calibration has to be performed. The internal calibration of the system was done using a flat checkerboard pattern. The parameters obtained in the calibration step were used to remove distortions introduced by the optics. The external calibration, necessary for three-dimensional reconstruction algorithms, was performed based on a set of markers and one known dimension on the observed scene. A set of crash test markers was mounted at points of interest on the wire. After acquisition of an image sequence, the image trajectories of markers were tracked across a sequence using Tema's algorithm for quadrant symmetry. The three-dimensional trajectories of markers are reconstructed using a stereo-vision technique. The vibration signal is then scaled to metric units using data obtained in the external calibration step.

3. Performed measurements

In the following paragraph, detailed descriptions of the laboratory measurements are given. The first subsection presents a laboratory stand used for testing an experiment performed to evaluate the sag of a transmission line cable. The second subsection provides information on the experiment in which the dynamic response of a wire was measured.

3.1. Laboratory stand

A laboratory stand for testing contactless measuring methods of the wire type objects was designed and constructed. The modular design of the structure makes it easy to assemble and disassemble with a variable configuration of poles supporting the wire. The cable can be hung 0.8 m above the floor. The tension in the cable can be adjusted continuously. The modularity of the setup makes it possible to analyze the load acting on a given section, as well as the propagation of a disturbance across sections, with one of the sections subjected to a time-dependent force. The variable tension of the cable provides the capability of testing the static states and dynamic characteristics of the system. Figure 1 presents general view of designed stand in the exemplary four-pole configuration.

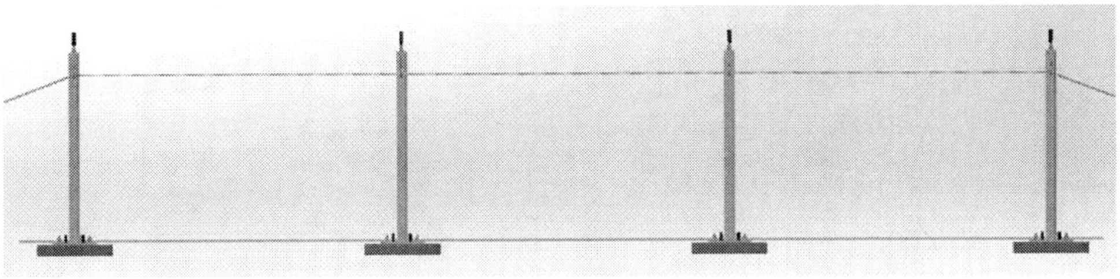

Figure 1. The general view of laboratory stand

IOP Conf. Series: Journal of Physics: Conf. Series **842** (2017) 012011 doi:10.1088/1742-6596/842/1/012011

Figure 2 shows the way of wire suspension and adjustable tension mechanism.

Figure 2. The wires routing on the test bench

3.2. Vision-based measurement of the cable's vibrations

In the performed test, the vibrations of markers attached to the wire were investigated. The measurement was carried out by a stereo-vision system equipped with two high-speed digital cameras. The examined object was excited by burst random noise signal (50 % ratio) using an electrodynamic shaker. The test setup consisted of the following elements:

1) a stereo-vision system with two high-speed Phantom v9.1 cameras of a maximum resolution of 1632x1200, with a fixed focal length Carl Zeiss lens of f = 25 mm. The system was placed on a stereo-vision head and mounted on a tripod at a distance of 1.3 m from the wire. The baseline of the system was 661.0059 mm. The relative position and orientation of the cameras was obtained in an external calibration as it was said before. The parameters of the cameras were set as follows: resolution – 1632x800, frame rate – 200, exposure time – 0.004 s, acquisition time – 23.7 s, Camera 9645 – Master mode, Camera 9644 – Slave mode.
2) an electrodynamic shaker exciting the analyzed wire with a burst white noise signal.
3) a laboratory stand was set in two-pole configuration (distance between poles – 2.25 m). The wire was covered with 51 evenly spaced crash test markers.

Entire test bench with basic parameters is shown in Figure 3.

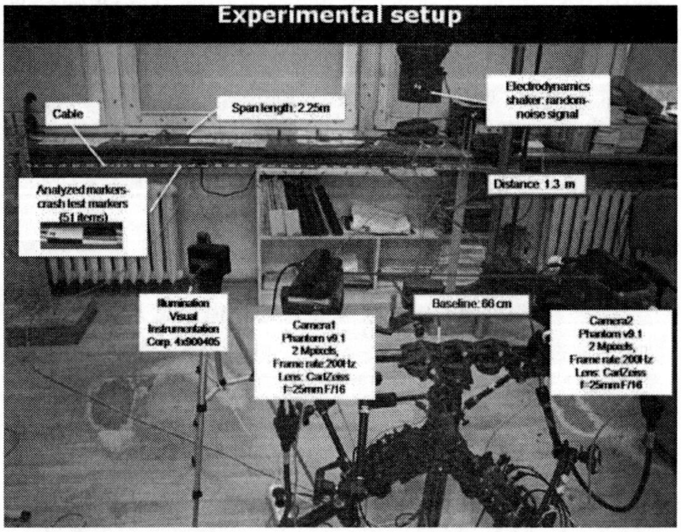

Figure 3. Laboratory stand for vibration measurement

During the measurements, two sessions were carried out, one referential case (Reference1) and one involving an extra mass for further tests of the damage detection algorithms (Mass1). In each session, five measurement series were performed. The positions of markers on the lab setup are shown in Figure 4. To increase the spatial resolution of the vision system, motion analysis was conducted on three segments of the analyzed overhead line stand. They correspond to the three smaller fields of view of the stereo-vision system, as demonstrated in Figure 4. Each measurement area (field of view) included 21, 26 and 19 crash-test markers, respectively. For example, at a distance of 1.3 m from the analyzed object, the field of view of the stereo-system amounted 880 mm.

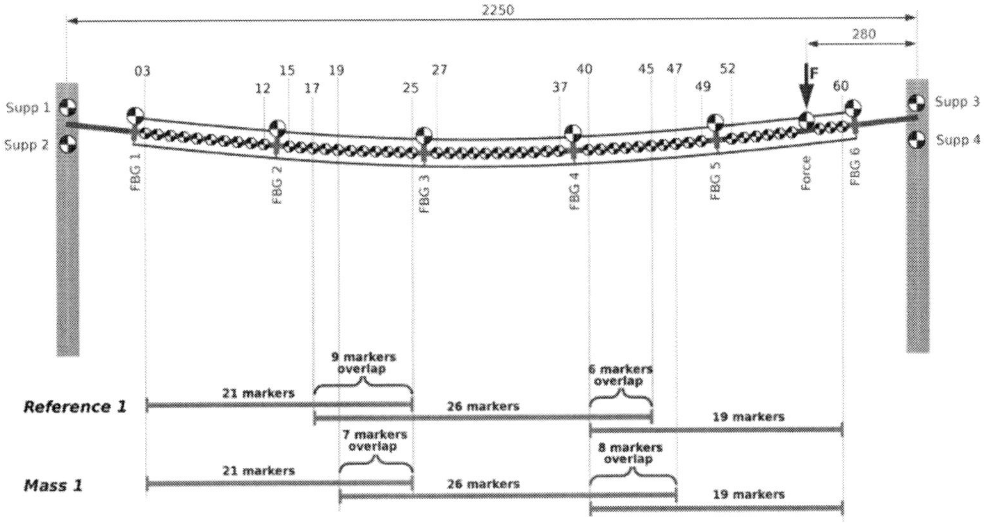

Figure 4. Schematic layout of the markers on the test bench.

The high-speed camera system's data was recorded and stored. Next, the data was loaded in TEMA Automotive software [21] in order to extract the positions of each of the markers and track their trajectories across frames of the sequence. The tracking method chosen for the data was Quadrant Symmetry. The "Quadrant Symmetry" tracker with a radius of 4 pixels was used. Figure 5 presents example of configuration of one measurement point in the software.

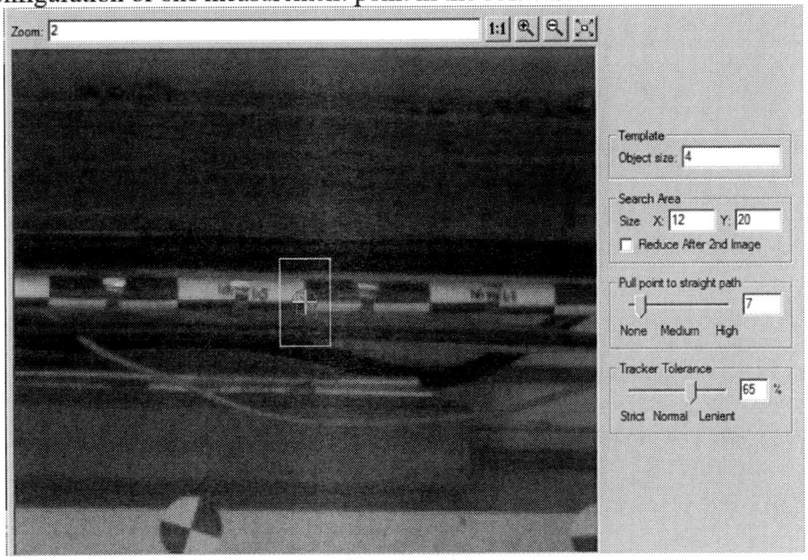

Figure 5. Configuration example of the measurement point.

Configuration of markers was controlled during the tracking of points. Errors in positions of the points were detected by observing the position of all points in graphs in a function of time (See Figure 6). Sudden changes in individual points in comparison to others was treated as a prerequisite for an error of the tracking algorithm. All such errors were removed by adjusting the configuration of the point and re-examine the problematic frames of the film.

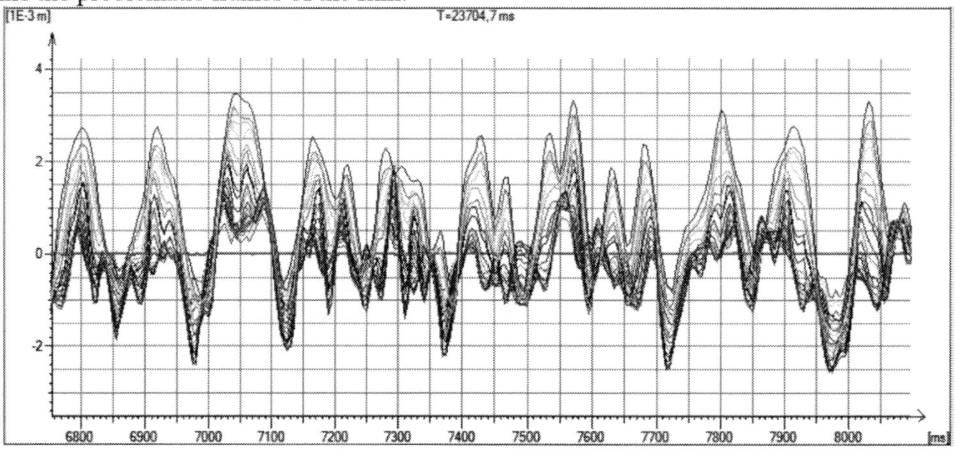

Figure 6. Time history fragment for the z axis

Data concerning the three-dimensional trajectories for each of the movie sequences were exported to .xls files and then to the MATLAB programming environment, where the frequency response of the system was estimated. An example of the obtained frequency spectrum is presented in Figure 7.

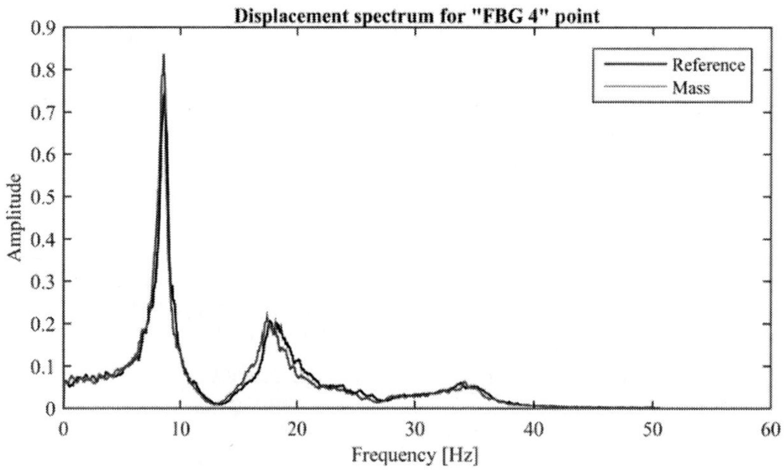

Figure 7. Displacement spectrum for point FBG4

4. Modal analysis

The following paragraph presents the procedure and results of experimental modal analysis performed with use of characteristics measured by the vision system.

Using response frequency spectra obtained from vision measurements, the frequency response functions (FRFs) were calculated with respect to the force signal measured simultaneously with vision recording. Next experimental modal analysis was performed. For the estimation of modal parameters VIOMA toolbox was used. It is a Matlab toolbox created at Department of Robotics and Mechatronics AGH [22]. For analyses Least Square Frequency Domain (LSFD) algorithm was selected [23 -]. Calculations were carried for frequency band 0 - 50 Hz. As a result of these analyses five first natural frequencies of the wire were estimated. The results are shown in Table 1.

Table 1. Modal parameters of tested wire

No.	Natural freq. [Hz]	Modal damp. [%]
1	8.03	8.23
2	16.37	8.58
3	21.95	8.59
4	28.73	3.58
5	35.12	3.49

Since the analysis was conducted based on data collected in the three exposures, in order to properly visualize mode shapes (MS), it was necessary to merge together these three pieces. For proper scaling of consecutive fragments, overlapped points were used - those that were found in two neighboring scenes. Scaling was performed in two stages. First the phase of consecutive points was unified. Then the amplitude of the common points was leveled and the resulting scaling factor was used for the remaining points of the exposure. Figure 8 shows the first MS for which at one of the merging points the scaling was done (right side), and at the other was not (left side).

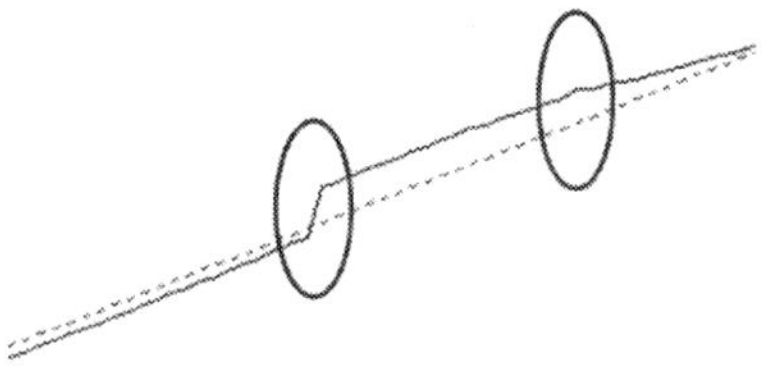

Figure 8. Mode shape at first natural frequency

5. Conclusions

The article presents the results of a modal analysis performed using vision measurements. Discussed analyses are part of bigger research to develop a system for dynamic energy transfer management of the transmission grid and the monitoring system of that grid. The method used to track the motion of markers (measuring points) on the video sequence was shown and the way to visualize MS obtained in several exhibitions was described.

References

[1] Malhara S, Vittal V, Mechanical state estimation of overhead transmission lines using tilt sensors, IEEE Transactions on Power Systems, 2010; Vol. 25 No. 3 pp. 1282-1230.
[2] Seppa T.O, A practical approach for increasing the thermal capabilities of transmission lines, Power Delivery, 1993; Vol. 8 No. 3 pp. 1536-1550.
[3] Wei-Guo Tong, Bao-Shu Li, Jin-Sha Yuan, Shu-Tao Zhao, Transmission Line Extraction And Recognition From Natural Complex Background, Proceedings Of The Eighth International Conference On Machine Learning And Cybernetics, Baoding, 12-15 July 2009
[4] Sun Fengjie, Wang He, Fan Jieqing, 2D Otsu Segmentation Algorithm Based on Simulated Annealing Genetic Algorithm for Iced-Cable Images, Information Technology and Applications, 2009. IFITA '09. International Forum, Vol, 2, Digital Object Identifier: 10.1109/IFITA.2009.171, pp. 600 - 602

[5] Sun Feng-jie, Tian Ye, Transmission Line Image Segmentation Based GA and PSO Hybrid Algorithm, Computational and Information Sciences (ICCIS), International Conference 17 - 19 Dec. 2010 pp. 677 - 680
[6] Yuqin Yao, Wenbin Wang, Feixiang Chen, Yuanpeng Yu, Nondestructive testing of transmission lines based on the image method, Electronics, Communications and Control (ICECC), 2011 International Conference, Digital Object Identifier: 10.1109/ICECC.2011.6067716, pp. 2143 - 2146
[7] Jun Zhao, Shu-tao Zhao and Qi-neng Jiang, Power Transmission Tower Type Determining Method Based on Aerial Image Analysis, Electricity Distribution (CICED), 2010 China International Conference, pp. 1 - 5
[8] Xiaopeng Wang, Jianlin Hu, Bin Wu, Lin Du, Caixin Sun, Study on Edge Extraction Methods for Image-based Icing On-line Monitoring on Overhead Transmission Lines, International Conference on High Voltage Engineering and Application, Chongqing, China, November 9-13, 2008
[9] Geng Xin, Hu Xiaoguang, On-line Monitoring Method and Observation System of Transmission Line Icing, Proceeding of IEEE 4th Conference on Industrial Electronics and Applications, pp. 1265-1269, 2009
[10] Chunsheng Yu, Qingjin Peng, Randy Wachal, and Pei Wang, An Image-Based 3D Acquisition of Ice Accretions on Power Transmission Lines, CECE, 2006, pp. 2005-2008. IEEE,
[11] Wachal R., Stoezel J., Peckover M., Godkin D., A computer vision early-warning ice detection system for the Smart Grid, Transmission and Distribution Conference and Exposition (T&D), 2012 IEEE PES , 7-10 May 2012, pp.1 - 6 , ISSN 2160-8555
[12] Alberto Vallan, Filippo Molinari, Image Analysis of Metallic Ropes for the Measurement of the Lay Length, IEEE International Conference on Instrumentation and Measurement Technology , Victoria, Vancouver Island, Canada, May 12-15, 2008
[13] Ishino R., Tsutsumi F., Detection system of damaged cables using video obtained from an aerial inspection of transmission lines, Power Engineering Society General Meeting, 2004. IEEE, Digital Object Identifier: 10.1109/PES.2004.1373201,Vol.2, pp: 1857 - 1862
[14] Busca G., Cigada A., Manenti A., Zappa E., Vision-based measurements for slender structures vibration monitoring, Environmental, Energy, and Structural Monitoring Systems, 2009. EESMS 2009. IEEE Workshop, 10.1109/EESMS.2009.5341309, pp: 98 - 102
[15] Byeong Hwa Kim, Extracting modal parameters of a cable on shaky motion pictures, Mechanical Systems and Signal Processing, In Press
[16] Xiaobo Meng, Lei Hou, Liming Wanga, Mark MacAlpine, Guanjun Fu, Baoqiang Sun, ZhichengGuan, Wei Hud, Yong Chen, Oscillation of conductors following ice-shedding on UHV transmission lines, Mechanical Systems and Signal Processing, 2012, Vol. 30, pp. 393–406
[17] Kohut P., Kurowski P., Application of Vision for Modal Experiment, Machine Dynamics Problems, Vol. 29, No 2, pp.81-90,2005
[18] Kohut P., Kurowski P., The Integration of Vision System and Modal Analysis for SHM Application, A Conference&Exposition on Structural Dynamics, IMAC XXIV, January – February 02, St. Louis, 2006
[19] Kohut P., Kurowski P., The 3D vision-based measurements for modal analysis, 3rd Workshop on Optical and Measurement Techniques for Structures and Systems, Belgium, Leuven 28-30 May, Proc of OPTIMESS, 2007
[20] Sung-Wan Kim, Nam-Sik Kim, Dynamic characteristics of suspension bridge hanger cables using digital image processing, NDT&E International, 2013, Vol. 59 pp. 25–33
[21] TEMA User's guide, © Image Systems AB, 2011
[22] Uhl T., Lisowski W., Kurowski P., In-operation modal analysis and its application, AGH Publishers, 2001.
[23] M. Mergeay. Least squares complex exponential method and global system parameter estimation used by modal analysis. In Proceedings of the 8th International Seminar on Modal Analysis, Leuven, Belgium, 1983.

Data driven innovations in structural health monitoring

M J Rosales and R Liyanapathirana

Mailing Address: Western Sydney University, School of Computing, Engineering and Mathematics, Locked bag 1797, Penrith, NSW Australia 2751

Tel: +61247360082

E-mail: r.liyanapathirana@westernsydney.edu.au

Abstract. At present, substantial investments are being allocated to civil infrastructures also considered as valuable assets at a national or global scale. Structural Health Monitoring (SHM) is an indispensable tool required to ensure the performance and safety of these structures based on measured response parameters. The research to date on damage assessment has tended to focus on the utilization of wireless sensor networks (WSN) as it proves to be the best alternative over the traditional visual inspections and tethered or wired counterparts. Over the last decade, the structural health and behaviour of innumerable infrastructure has been measured and evaluated owing to several successful ventures of implementing these sensor networks. Various monitoring systems have the capability to rapidly transmit, measure, and store large capacities of data. The amount of data collected from these networks have eventually been unmanageable which paved the way to other relevant issues such as data quality, relevance, re-use, and decision support. There is an increasing need to integrate new technologies in order to automate the evaluation processes as well as to enhance the objectivity of data assessment routines. This paper aims to identify feasible methodologies towards the application of time-series analysis techniques to judiciously exploit the vast amount of readily available as well as the upcoming data resources. It continues the momentum of a greater effort to collect and archive SHM approaches that will serve as data-driven innovations for the assessment of damage through efficient algorithms and data analytics.

Key words: Structural Health monitoring; Wireless Sensor networks, Time series analysis, data analytics

1. Introduction

Throughout the world, civil infrastructures such as buildings, bridges and tunnels involve substantial investments as they have become significant pillars of modern civilizations. These structures will eventually deteriorate or disintegrate as they are incessantly subjected to weathering effects and operational loads. Defective and unstable operation of these infrastructures can cause economic encumbrance and pose major threats to public health and safety. Hence, the serviceability of these assets should be consistently and accurately monitored and evaluated for risk minimization. For example, localization and detection of structural damage should be performed at the onset of its emergence by using precision equipment. Damage is defined as deviations introduced into a system that can potentially cause harmful effects on both the existing or future performance of that system [1]. It has also been suggested that damage is the occurrence of an imperceptible quantity of micro cracks on a structural element [2]. Traditionally, visual inspections are done to assess the health of the civil structures. However, this method proved to be labor-intensive, costly, and time-consuming. Also, visual assessments can be highly subjective as some damages may be below the surface and less visible producing varying results even by trained inspectors [3].

Structural Health Monitoring (SHM) is a complimentary approach to visual inspections providing precise real-time information about the condition of the civil structure based on quantifiable response parameters such as humidity, temperature, strain and acceleration then converting these into digital format to be delivered to a central data collection station [4]. Central to the discipline of SHM is to monitor and assess the health of the structure based on measurable response parameters as these can ultimately become unstable or faulty due to weathering effects, excessive loads and natural corrosion [3]. The deployment of SHM in monitoring civil infrastructures can prolong the serviceability of the structure and guarantee life-safety standards over their operational lives [5]. Evolving wireless technological advancements and applications in conjunction with micro-electromechanical systems (MEMs) laid the foundation of wireless sensor networks (WSNs). Interest in using WSNs for structural health monitoring has increased as more research demonstrates successful application to increasingly complex structures.

According to [20], accurate and efficient damage detection in long-term health monitoring still encounters many difficulties, and recorded big-data requires efficient new damage detection algorithms. In a novel algorithm described in [19,20], transmissibility along with Mahalanobis distance and Hotelling T-square were used for a study involving a numerically simulated beam and an experimentally tested laboratory structure. A more advanced damage detection and localization technique based on the BAT algorithm is found in [21]. This technique enables not only single but also the detection of multiple damage positions along with the rate of damage. Two recent studies on composite beam structures based on vibration analysis can be found in [22,23].

A common view is that the interpretation of the monitoring data in terms of the integrity of the structure is the key function of the SHM system. The data management section involves the filtering out of extraneous data, direct and remote collection of data, conversion of the data into responses, interpretation of these responses to state the condition of the structure and recommendation for future data collection and interpretation [6]. The captured data, from a structural monitoring perspective, constitute a rich resource for the detection and assessment of the imminent behavior, damage and deterioration within the infrastructure under regular or extreme loadings. The massive amount of data related to long-term SHM systems presents many issues such as data quality, relevance, re-use, and decision support. Additional challenges relate to data scrubbing and robust data management, as well as the requirements for increased memory, storage and computing power [7].

Over the past decade, numerous SHM systems have become increasingly versatile and capable of handling larger data capacities and achieving speedier transmission, measurement and storage of data acquired from the infrastructure system being monitored. The integration of data analytics in SHM will bring about the following gains which include reduction of the analysis time, increase in objectivity, automation of pre-processing, identification of meaningful datasets and utilization of the knowledge gained from historical datasets [8]. Fundamentally, damage will not be considered as significant without comparing two different states of the system, one being the undamaged or baseline that will represent the initial state of that system [9]. Thus, the need for a methodical approach to optimally maximize the information at hand is self-evident.

This paper is focused on the determination of the methodological approach to apply data science with the intention of productively managing and analyzing the huge sensor dataset generated from an existing WSN setup. The feature extraction techniques specifically used in this paper are the outlier detection techniques based on time-domain statistical analysis such as Auto Regression (AR) and Auto Regression with exogenous input method (ARX) [15, 16]. The Mahalanobis distance [15] is the feature discrimination algorithm employed in this study to demonstrate the signal deviations occurring in the damaged states against the undamaged conditions.

2. Experimental study

2.1 Experimental setup

The study is primarily designed to quantify and experiment on accessible data collected from a prevailing wireless sensor networking system within a building. The paper emphasis is on the simulation work done for SHM of an existing laboratory test data obtained from an instrumented laboratory three-story structure reported in [10]. The basic dimensions and the schematic depiction of the test structure setup are shown in Figure 1.

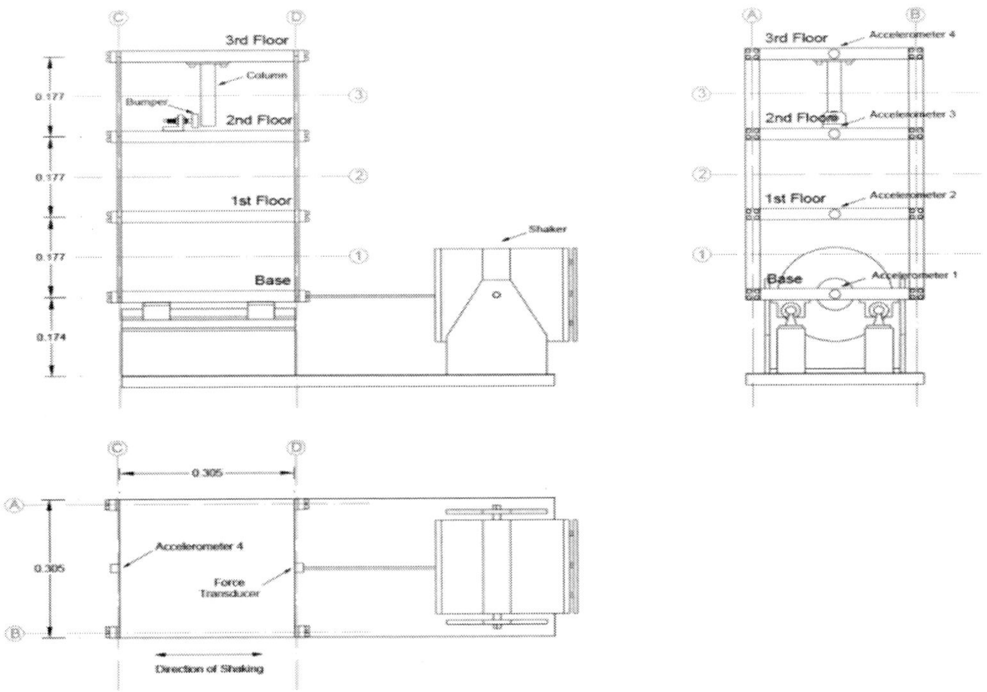

Figure 1. Basic dimensions of the three-story building structure [10].

2.2 Data acquisition

Along the centre line of the structure, an electro-dynamic shaker was installed to provide the lateral excitation. Both the shaker and the test structure are grounded on an aluminium baseplate (76.2 x 30.5 x 2.5 cm) [10]. The entire structural setup is mounted on rigid foam which will be in place to reduce the effect of the certain extraneous excitation sources that may be introduced to the system through its base. The input force from the shaker is measured using a load cell, and the overall system response is measured using four accelerometers. The analog sensor signals output from the data acquisition system was discretized and sampled producing a total of 8,192 data points sampled to 320 Hz frequency at 3.125 ms intervals. Time histories at approximately 25.6s periods were yielded from these sampling parameters. The yielded spectra in the frequency domain produced 3,600 lines demonstrating the data up to a frequency of 140.6 Hz at 0.0391 Hz resolution. The structure was excited using a band-limited random excitation in the 20-150 Hz range. The excitation signal chosen for experimentation was set to above 20 Hz to prevent the effects of the rigid body modes of the structures that are typically existent below 20 Hz [10].

2.3 Dataset Description

Table 1 shows the different states considered in the experiment and the collection of the force- and acceleration histories for the different structural state conditions. A case in point, state #4 describes a state condition that has "87.5% stiffness reduction in column 1BD." This shows that the stiffness in the column situated between the intersection of plane B and D located and between the first floor and the base will be reduced by 87.5% [10].

The experiment setup was conducted following four main structural state conditions divided into groups. The first category was called the baseline condition or state #1 which denotes the reference undamaged structural state. The second group, states #2 to #9, comprises of the circumstances where either mass or stiffness properties of the columns were altered. The second category is intended to simulate these scenarios through varying stiffness levels and mass change. The mass change in undamaged states #2 and #3, denoted by m, involved adding 1.2 kg which is approximately 19% of the overall mass of each floor [10].

The changes made with the stiffness, from states #4 to #9, was carried out by reducing the rigidity of one or more columns in the test-bed by 87.5%. The ensuing state conditions (from state #10 to #14) comprise the damaged state conditions replicated through the nonlinearities initiated into the structure through the suspended column and the bumper and varying gap levels (0.05, 0.10, 0.13, 0.15 and 0.20 mm) for a given level of excitation between them. Lastly, the fourth group involves the damaged states #15 to #17 which involve differing gaps between the bumper and the suspended column with the addition of the mass and stiffness deviations to represent the environmental and operational aspects in real world structures.

The variability in the data is taken into account by performing ten tests for every condition state which follows that ten time histories were also measured corresponding to each of the five transducers as well. Furthermore, the associated coherence functions and frequency response functions (FRFs) at the accelerometers (from Channel 2 to 5) were also measured relative to the input generated from the load cell channel #1 were recorded in the data acquisition system.

Table 1. Structural state conditions used in the experimental set up in [10].

Data Label	State Condition	Description
State # 1	Undamaged	Baseline condition
State # 2	Undamaged	m = 1.2 kg at the base
State # 3	Undamaged	m = 1.2 kg on the 1st floor
State # 4	Undamaged	87.5% reduction of stiffness in column 1 BD
State # 5	Undamaged	87.5% reduction of stiffness in column 1 BD and 1 AD
State # 6	Undamaged	87.5% reduction of stiffness in column 2 BD
State # 7	Undamaged	87.5% reduction of stiffness in column 2 BD and 2 AD
State # 8	Undamaged	87.5% reduction of stiffness in column 3 BD
State # 9	Undamaged	87.5% reduction of stiffness in column 3 BD and 3 AD
State # 10	Damaged	Gap = 0.20 mm
State # 11	Damaged	Gap = 0.15 mm
State # 12	Damaged	Gap = 0.13 mm
State # 13	Damaged	Gap = 0.10 mm
State # 14	Damaged	Gap = 0.05 mm
State # 15	Damaged	m = 1.2 kg at the base and Gap = 0.20 mm
State # 16	Damaged	m = 1.2 kg on the 1st floor and Gap = 0.20 mm
State # 17	Damaged	m = 1.2 kg on the 1st floor and Gap = 0.10 mm

3 Damage detection using time series analysis

The underlying principle of vibration-based damage detection is that the presence of damage will considerably change the properties of the system such as energy dissipation, mass and stiffness that will sequentially change the measured dynamic response of that system [9]. A damage detection algorithm should be able to satisfy the following general features: detect even the smallest of damages, be robust to environmental conditions, and necessitate minimal wireless data transmission. It would also be desirable if the damage detection algorithm is less computationally intensive and can accommodate energy saving strategies such as sleep/awake states [11].

A time series is a succession of data points, measured at continuous and regularly spaced time intervals. Time-series analysis is concerned with the fact that the seasonal trend, variation and autocorrelation of the data points taken over time can be utilized to extract the Damage Sensitive Features (DSFs) that can directly be applied with the measured data for damage detection.

Regression analysis is based on time-series analysis and is a form of supervised learning classification that can be employed in the damage detection process. Some of the regression models that have been successfully applied so far include radial basis functions and neutral networks [12]. Regression techniques have been proven as capable of detecting the existence, quantify and localize the damages present in the structure through the use of modal data. To determine the strengths and limitations of the different statistical procedures used for distinguishing damages within the civil structures, the data set attained from the test structure assembled in the laboratory will be subjected to the LANL's statistical pattern recognition paradigm.

3.1 Autoregression (AR) model

In the AR model formulation, the model order is defined by the largest time lag resulting from the AR models regressing the response at one location onto a delayed version of itself [13]. To reliably deduce that damage is indeed present, it is desirable to define the threshold of how far apart the AR coefficients must be [14]. It would be necessary to employ some form of supervised learning where the damage state is known beforehand. The damage detection algorithm in supervised learning will use training data to recognize and discriminate between undamaged and damaged data sets in order to determine a threshold value. Damage is present in excess of this threshold value. In an autoregressive model, the DSFs are the AR coefficients and the structural response measurements are the acceleration time histories. The summarized AR model is described via the equations below:

Let $x_i(t)$ represent the acceleration data from sensor i. This will be subdivided into segments $x_{ij}(t)$ where the sensor number is denoted by i and the segments are symbolized by j. These segments will be subjected to data pre-processing. Data pre-processing is a process of standardization and normalization of the data set such as the acceleration time series. This is done to minimize or eliminate dependence on variable environmental conditions [15]. The estimation of the AR order will be done subsequent to the completion of the pre-processing procedure [13]. The time signals will be denoted as:

$$x_i = \sum_{j=1}^{p} \emptyset_j \, x(i-j) + e_i \qquad (1)$$

where,

x_i= Signal measured at discrete time index i

$\emptyset_j$ = k^{th} AR coefficient, used as DSFs

p = AR model order

e_i= residual term or unobservable random error

Reference [15] proposed the Burg algorithm that is used in conjunction with the above equation by [13] for the estimation of the AR coefficients. The AR model will be run in each segment that will result in N vectors of AR coefficients that is also equivalent to the overall number of segments from one sensor having a length

$$N = \frac{total\ signal\ length}{segment\ length} \quad (2)$$

The mean and standard deviation of the AR coefficients at various segment lengths will be calculated to determine the optimal segment length and identify at which point these started to stabilize.

In general, the optimum AR model order is unspecified. A higher order model may entirely represent or match the data, but may lose generality when applied to other datasets whereas a lower order model may not necessarily capture the fundamental system response of the physical entity being observed [17].

3.2 Autoregressive with Exogeneous (ARX) Inputs Model

In a study conducted by Lei et al. (2003), they have modified the damage detection approach to take into account the excitation variability and factors of the prediction model orders on the originally obtained damage detection feature. Signals generated from the undamaged structure are compared against the residual error of a new signal coming from an unknown structural condition referenced from the prediction model. The ARX is constructed from the selected reference database as

$$x_k\,(t) = \textstyle\sum_{i=1}^{na} \Sigma a_{ki}\, x_k\,(t-i) + \sum_{i=0}^{nb} \Sigma b_{ki}\, e_x^k\,(t-i) + \varepsilon_x^k\,(t) \quad (3)$$

where, the ARX model order numbers are given by *na* and *nb* , the AR coefficients and the exogenous input are represented by a_{ki} and b_{ki} ,while the $\varepsilon_x^k\,(t)$ signifies the ARX(*na, nb*) residual error, correspondingly. The resultant model will then be employed for the prediction of the new signal designated as $y(t)$ given in

$$y\,(t) = \textstyle\sum_{i=1}^{na} \Sigma a_{ki}\, y(t-i) + \sum_{i=0}^{nb} \Sigma b_{ki}\, e_y(t-i) + \varepsilon_y(t) \quad (4)$$

wherein a_{ki} and b_{ki} are the coefficients associated with $x_k(t)$ derived from Eq. (3) and the residual error of the new signal is symbolized by $\varepsilon_y(t)$.

In the formulation of this model, the responses of two locations having different time lags are employed to execute the regression analysis. The ARX algorithm is used to obtain DSFs, model identification and filtering by means of a least-square approach. The data is fitted in a defined ARX model to determine the relationship between the input and output parameters. This procedure is intended to enhance the damage detection process through the exploitation of the information associated with an identified input specified by the sensing system [13].

4. Damage Localization

The statistical model used to perform the feature discrimination and quantify the extent of damage is the Mahalanobis distance. This model was introduced to isolate the outliers for the purpose of damage detection in a signal [18]. The primary goal of this algorithm is to categorize the extracted features from the normal condition versus the potentially damaged states that recorded the outliers. The feature discrimination training phase involves the learning of the mean and the covariance matrix of the AR model coefficients, serving as the feature vectors, to train a Mahalanobis distance-based detector functionality. The normalization for the features are done through the machine learning algorithm by reducing each feature vector to a score. In multivariate statistics, the Mahalanobis distance is used to perform both identification and quantification of outliers. The striking difference between the Mahalanobis distance and the Euclidean distance is that the latter is not dependent on the scale of the observations but rather reliant on the correlation between the variables. Given the group of *m p*-dimensional real-valued patterns classified under $\Re^p$ having the multivariate mean vector $\mu = (\mu_1, \mu_2, \dots, \mu_p)^T$ and covariance matrix, Σ, the Mahalanobis distance between the new pattern $x = (x_1, x_2, \dots, x_p)^T$ and that group will be represented by

$$D = \sqrt{(x-\mu)^T\,\Sigma^{-1}(x-\mu)} \tag{5}$$

where the normal operational condition is represented by the covariance matrix (Σ) and the mean vector (μ). The potentially damaged condition will be symbolized by x.

The key events in the order in which they happened in this study are as follows. Initially, a series of five-second blocks of acceleration time histories was acquired from the sampled data set for each undamaged state condition. This was followed by the extraction of the AR-based feature vectors done through the fitting of the linear autoregressive model to each of the time histories. The next step involved the training of a Mahalanobis distance-based detector of the sampled dataset by learning the covariance matrix and the mean of the feature vectors from the undamaged structure. On completion, the trained sample runs several test cases one at a time within the structure in different undamaged and damaged states wherein the resultant of which will aid in the determination of the threshold based on 99% confidence to classify the extent of damage in the structure [10].

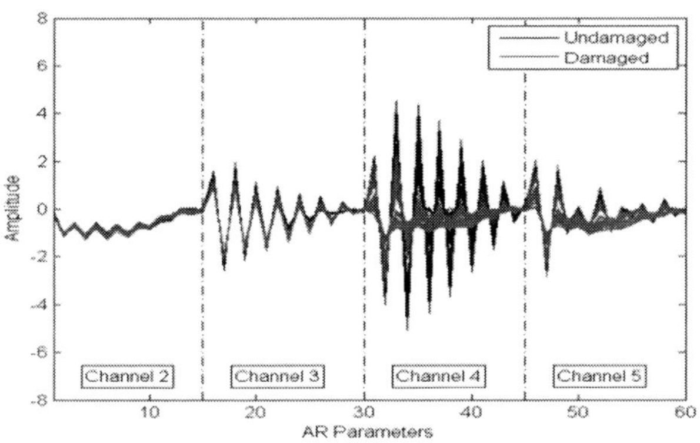

Figure 2. Concatenated AR(15) parameters for all instances

5. Results and discussion

This paper implemented two major stages of data processing to undertake supervised learning statistical modelling analysis: training and live testing. After acquiring the data, this will be undergoing the training phase to learn a model of the baseline or undamaged system. This will then be segmented into smaller fragments of data that will be processed in sequence in the live testing stage to be classified as either "undamaged" or "damaged" referenced from the training or learned model. The threshold will then be derived statistically.

5.1 *AR Method*

The auto-regressive (AR) model is an outlier/novelty detection method that can be used to localize the source of damage in the structure. As mentioned before, the coefficients of the AR model operate as the feature vectors from an array of sensors and will be used as DSFs damage-sensitive features. Changes have been observed on the values of the AR coefficients for acceleration time histories when there is a reduction in the stiffness of the structure under ambient vibrations caused by damage, thus, can be considered as appropriate DSFs. Fig.2 reveals that the AR coefficients are capable of detecting changes in the structure and that different types of structural damage will have varying effects on the AR coefficients implying that the algorithm can potentially be used to correlate the damage measures to the type of damage. AR parameters appear to be potential DSFs (damage-sensitive features). The source of damage seems to induce changes in those parameters related to the level of damage and therefore, this feature can also be used to evaluate the severity of the damage in the structure.

5.2 ARX Method

ARX damage detection algorithm can be utilized for the detection, localization and quantification of damage that is present in the structure. ARX is another statistical outlier/novelty detection method wherein the input parameters are used as damage-sensitive features as well through the use of the least square approach. The ARX analysis and determination of the model order is carried out for one-acceleration-time history of the reference condition (State #1) from Channel 5. Fig.3 shows that the ARX parameters perform better discrimination of the state conditions, especially in channel #3 than the AR parameters considering the same order number of 15. The location of the damage can be identified by comparing the number of outliers per channel. It has been apparent from the plot that the magnitude of the outliers tends to increase and more likely is able to indicate the location where the damage source may either be present or at least proximate.

A comparison of AR and ARX models in Figure 4 reveals noteworthy and highly significant differences. One of the striking dissimilarities is based on their computational costs to execute the damage detection feature generation. The AR modelling required less computational time whereas the ARX recorded higher magnitudes exceeding nearly three-fold than the magnitudes of the AR modelling especially for escalating model orders. On the other hand, however, the application of ARX outperformed AR modelling in damage localization [13].

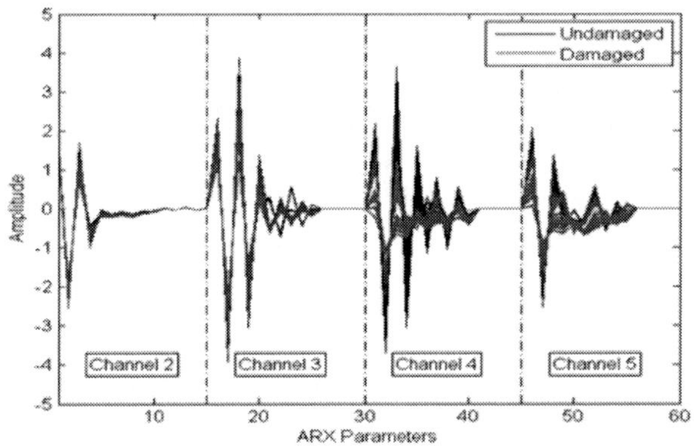

Figure 3. ARX(10,5) parameters versus the amplitude plot for all instances.

The tested data consists of one-time history acquired from each condition. The damage indicators recorded from the undamaged condition (black) for the feature vectors is constant contrary with the feature vectors from the damaged states which displayed an escalating trend. Given the Mahalanobis distance plot, it can be seen that the incremental change in the damage measure will capture the change from one damage state to the next.

Also it can be seen that the extracted features from Channels 2 to 3 are lesser sensitive from Channels 4 and 5 to discriminate the undamaged (1-9) and damaged (10-17) state conditions. This is an indication that the source of damage is located near Channels 4 and 5.

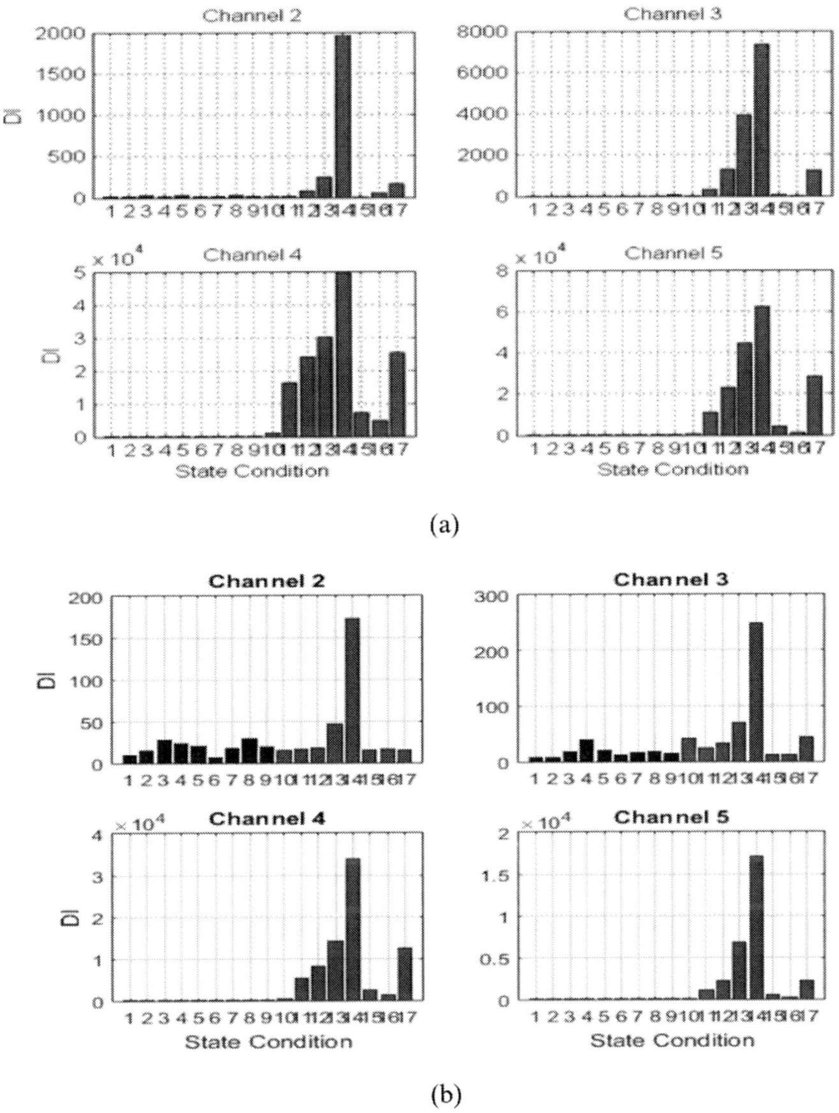

Figure 4. ARX (a) and AR (b) damage indices using Mahalnobis distance algorithm.

6. Conclusions

In summary, a comprehensive literature review was carried out to discuss the theoretical background and importance of SHM, prevalence of WSNs in SHM, and the motivation of the study. The applicability of data analytics in the context of SHM was also examined following a succinct discussion on the existing statistical classifiers and feature discrimination used in the damage detection algorithms. The methodology used in this paper utilized the large amount of experimental data collected from a suitable simulation model designed by LANL. The laboratory test structure attempts to duplicate the dynamic response of actual structures by subjecting the test bed to controlled damage tests. Damages were successfully detected in the simulated models through the use of the damage detection algorithms founded on statistical pattern recognition techniques.

In reference to the damage detection methods, the ARX model provided a better discrimination of the state conditions than the AR model as viewed on the magnitude of the outliers per channel to locate the damage that incurred in the infrastructure. The ARX model also outperformed and returned more distinguishing features than AR using the Mahalanobis learning distance algorithm to accomplish the

IOP Conf. Series: Journal of Physics: Conf. Series **842** (2017) 012012 doi:10.1088/1742-6596/842/1/012012

feature classification procedure. The dependability of the Mahalanobis distance learning algorithm was also demonstrated in this study as a function of both AR and ARX damage detection algorithms.

If appropriately utilised, it is believed that damage detection technologies will extend the lifetime of buildings by allowing deterioration or damage to be recognized earlier and thereby permitting comparatively inconsequential remedial actions to be taken before the deterioration or damage matures to a state where major actions will be mandatory. There are still a number of challenges and opportunities worth pursuing. Future work will involve the experimental endeavors and focus on using statistical pattern recognition approaches to identify the outliers or the different levels of damage within the structure. New data analysis and datamining tools must be developed to address the issue or the existing tools may be revisited. Future data mining research should also focus on providing another level of damage identification to include estimating the remaining service life of a structure.

7. References

[1] Doebling S, Farrar C, and Prime M 1998 *Shock Vib. Digest* A Summary Review of Vibration based Damage Identification Methods **30** pp 91 -105

[2] Delicato F, dos Santos I, Pirmez L, Pires F and de Farias C 2014 *Large Scale Network-Centric Distributed Systems* Challenges in the Use of Wireless Sensor Networks for Monitoring the Health of Civil Structures edited by H. Sarbazo and A. Zomaya, (New York: John Wiley & Sons) pp 148-64

[3] Wang Y, Lynch J and Law K, 2007 *Structure and Infrastructure Engineering,* A wireless structural health monitoring system with multithreaded sensing devices: design and validation **3** pp 103-20

[4] Nagayama T and Spencer, Jr B 2007 *Structural Health Monitoring using Smart Sensors,* (Illinois, USA, Department of Civil and Environmental Engineering, University of Illinois)

[5] Lynch J 2007 *Mathematical, Physical and Engineering Sciences* An Overview of Wireless Structural Health Monitoring for Civil Structures", Philosophical Transactions of the Royal Society A **365** pp 345-72

[6] Rivera E, Mufti A and Thomson D 2006 *The Arabian Journal of Science and Engineering,* Civionics Specifications for Fibre Optic Sensors for Structural Health Monitoring pp 241-48

[7] Mahadevan S, Adams D and Kosson D 2014 *Proc. Annual Conference of the Prognostics and Health Management Society, Department of Civil and Environmental Engineering, Vanderbily University, Tennessee, USA* Challenges in Concrete Structures Health Monitoring

[8] Forstner E and Wenzel H, *22nd International Workshop on Database and Expert Systems Applications (DEXA), 2* The Application of Data Mining in Bridge Monitoring Projects : Exploiting Time Series Data of Structural Health Monitoring **011** pp 297-301

[9] Farrar C, Duffey T, Doebling S and Nix D 1999 *2nd International Workshop on Structural Health Monitoring, Stanford, CA* A Statistical Pattern Recognition Paradigm for Vibration-Based Structural Health Monitoring

[10] Figueireido E, Park G, Figueiras J and Worden K 2009 *Ph.D. Portugal : University of Porto* Structural Heath Monitoring Algorithm Comparisons Using Standard Data Sets

[11] Uy B. Zhu X, Xiang Y, Zou J, Liyanapathirana R, Gunawardana U, Pap L and Perera H. 2011 *Structural Health Monitoring in Australia,* Civionics-An Emerging Interdisciplinary Research Area in Australia (Nova Science Publishers) ISBN 978 1 61728860 9 pp 145-76

[12] Kanapady R, and Lazarevic A 2009 *Encyclopedia of Data Warehousing and Mining, 2nd edn., New York, USA : IGI Global* Data Mining in Structural Health Monitoring pp 450-57

[13] Matarrazzo T, Shahidi G, Chang M and Pakzad S 2015 *Structural health monitoring and damage detection* Are today's SHM procedures suitable for tomorrow's BIGDATA?, In C. Niezrecki, ed.1 pp 59-64

[14] Cheung A, Cabrera C, Sarabandi P, Nair K, Kiremidjian, A and Wenzel H 2008 *Smart materials and structures* The Application of statistical pattern recognition methods for damage detection to field data **17** pp 1-12

[15] Jayawardhana M 2014 *Decentralised structural health monitoring using sensor networks, PhD thesis* Western Sydney University, Australia

[16] Brockwell R and Davis R 2003 *Introduction to Time Series and Forecasting* 2nd edition, Berlin: Springer p 456

[17] Noman A, Deeba F and Bagchi A 2013 *Health Monitoring of Structures Using Statistical Pattern Recognition Techniques, Performance of Constructed Facilities* **27** pp 575-584

[18] Worden K, Manson G and Fieller N 1999 *Sound Vib. Digest* Damage Detection using Outlier Analysis **229** pp 647-67

[19] Zhou Y-L, Maia N and Abdel Wahab M 2016 Damage detection using transmissibility compressed by principal component analysis enhanced with distance measure, *Journal of Vibration an Control* doi:10.1177/1077546316674544

[20] Zhou Y-L and Abdel Wahab M. 2016 Rapid early damage detection using transmissibility with distance measure analysis under unknown excitation in long-term health monitoring, *Journal of Vibroengineering* **18** (7) pp 4491-9

[21] Khatir S, Belaidi I, Serra R, Abdel Wahab M and Khatir T 2016 Numerical study for single and multiple damage detection and localization in beam-like structures using BAT algorithm, *Journal of Vibroengineering* **18**(1) pp 201-13

[22] Khathir S, Belaidi I, Serra R, Abdel Wahab M and Khathir T 2015 Damage detection and localization in composite beam structures based on vibration analysis *Mechanika* **21**(6) pp 472-9

[23] Gillich GR, Praisach ZI, Abdel Wahab M and Vasile O 2014 Localization of transversal cracks in sandwich beams and evaluation of their severity. *Shock and Vibration* http://dx.doi.org/10.1155/2014/607125

A new frequency matching technique for FRF-based model updating

Xiuming Yang[1], **Xinglin Guo**[1], **Huajiang Ouyang**[1,2], **Dongsheng Li**[1]
[1]State Key Laboratory of Structural Analysis for Industrial Equipment, Dalian University of Technology, Dalian, P. R. China
[2]University of Liverpool, Liverpool L69 3BX, United Kingdom

E-mail: allalongyxm@163.com, xlguo@dlut.edu.cn, h.ouyang@gmail.com, dsli@dlut.edu.cn

Abstract. Frequency Response Function (FRF) residues have been widely used to update Finite Element models. They are a kind of original measurement information and have the advantages of rich data and no extraction errors, etc. However, like other sensitivity-based methods, an FRF-based identification method also needs to face the ill-conditioning problem which is even more serious since the sensitivity of the FRF in the vicinity of a resonance is much greater than elsewhere. Furthermore, for a given frequency measurement, directly using a theoretical FRF at a frequency may lead to a huge difference between the theoretical FRF and the corresponding experimental FRF which finally results in larger effects of measurement errors and damping. Hence in the solution process, correct selection of the appropriate frequency to get the theoretical FRF in every iteration in the sensitivity-based approach is an effective way to improve the robustness of an FRF-based algorithm. A primary tool for right frequency selection based on the correlation of FRFs is the Frequency Domain Assurance Criterion. This paper presents a new frequency selection method which directly finds the frequency that minimizes the difference of the order of magnitude between the theoretical and experimental FRFs. A simulated truss structure is used to compare the performance of different frequency selection methods. For the sake of reality, it is assumed that not all the degrees of freedom (DoFs) are available for measurement. The minimum number of DoFs required in each approach to correctly update the analytical model is regarded as the right identification standard.

1. Introduction

Damage identification is an important tool that protects people from the loss of property and even life caused by structure failure. Methods based on changes in vibration characteristics of the structure have attracted the attention of many researchers as they provide quantitative global damage assessment to complex structures [1-3]. Among them model updating techniques, which correct the finite element (FE) model by experimental data, have been developed extensively in recent decades [4-9].

The frequency-domain model updating method can be roughly divided into three categories based on different dynamic characteristics used: eigenvalue based, mode shapes based and FRF based methods. One of them, The FRF based model updating method, directly uses the measured FRF data to update the structure's parameters. Imregun, et al. [10] proposed an early version of this method. Lin and Zhu [11] presented a method which employed measured response data under base excitation for model updating. Esfandiari, et al. [12] used the FRF data of the measured displacement, velocity or acceleration of the damaged structure to update the model. Kwon and Lin [13] developed a frequency selection method for efficient FRF based model updating. Garcia-Palencia and Santini-Bell [14] utilized a two-step algorithm to update the unknown mechanical properties and damping ratios of linear elastic damped structures.

The main idea of these methods is to establish the equation of the residuals between analytical and experimental FRF data with the sensitivity matrix of updating parameters and uses the least-square method to solve the over-determined equation. The updated parameters that bring the theoretical response closest to the actual response are considered to be the actual parameters and the reduction in their initial state, if big enough, is considered to be the occurrence of damage. Using FRF data directly does not require thorny modal identification work and hence the results will not be affected by the errors it produces. Moreover, available test data are broadened to the whole frequency domain of interest beyond measured frequencies. The abundant information provided by the FRF data not only allows more parameters to be updated but also can be used to reduce the effects of noise by choosing more precise regions of FRF. Also the reliability of the updating results can be checked by using different combinations of the experimental data.

However, there are some problems which the FRF based model updating methods have to face. The first is the incompleteness of the experimental data of FRF. Nearly all FRF based model updating methods require at least a complete row or column of the FRF matrix while it is inevitable that some DoFs cannot be measured in actual tests. To solve this problem Imregun used the analytically-generated FRFs for the unmeasured responses [10]. The reduction of the theoretical model and the expansion of the measured FRFs by dynamic reduction method [15] were also suggested. Similarly, Gang, et al. [16] proposed a new iteration formulation for the reduced model using pseudo master DoFs to improve the convergence of that technique.

Besides, the magnitude of FRF date changes dramatically in the vicinity of a natural frequency while most FRF based model updating methods use the theoretical and experimental FRF data at the same frequency to establish the sensitivity equation without the control of the magnitude difference between them. If this difference is big, the test and modelling errors will be enlarged and the level of ill-conditioning of the sensitivity matrix will increase. To overcome this situation, Pascual [17, 18] presented the Frequency Domain Assurance Criterion (FDAC) to localize the right special theoretical frequency to be used by determining the frequency shift at all measured frequencies. Asma and Bouazzouni [19] selected the theoretical frequency at which the distance between theoretical and measured FRFs is shortest. Both methods use the residuals of theoretical and experimental FRFs at different frequencies. However, the distance between these relative frequencies is not easy to control. If the whole frequency domain of interest is searched, it is not only time-consuming but also easy to make bad selections when using these methods per se may select inappropriate frequency values (whose FRFs are required) that fit these criterions best.

This paper presents a new frequency selection method which directly finds the theoretical frequency that minimizes the difference of the order of magnitude between the theoretical and experimental frequency responses in a certain range of measured frequencies. Since the frequency searching range is limited, it is not time-consuming to apply this frequency matching technique and the robustness of the algorithm is improved remarkably, as proved by a simulated truss structure.

2. Basic Theory

The essential purpose of the FRF based model updating method is to use the parameterized theoretical FRFs to approximate the measured ones. A brief description of the fundamental theory is given below which begins with a simple equation:

$$\mathbf{H}_{\mathrm{a}}(\omega)-\mathbf{H}_{\mathrm{x}}(\omega)=\mathbf{H}_{\mathrm{a}}(\omega)-\mathbf{H}_{\mathrm{x}}(\omega) \tag{1}$$

where $\mathbf{H}_{\mathrm{a}}(\omega)$ and $\mathbf{H}_{\mathrm{x}}(\omega)$ are $\mathrm{n}\times\mathrm{n}$ theoretical and experimental FRF matrices at frequency ω, respectively. Equation (1) can be rewritten as

$$\mathbf{H}_{\mathrm{a}}(\omega)\left(\mathbf{H}_{\mathrm{x}}^{-1}(\omega)-\mathbf{H}_{\mathrm{a}}^{-1}(\omega)\right)\mathbf{H}_{\mathrm{x}}(\omega)=\mathbf{H}_{\mathrm{a}}(\omega)-\mathbf{H}_{\mathrm{x}}(\omega) \tag{2}$$

Replacing $\mathbf{H}^{-1}(\omega)$ by the dynamic stiffness matrix $\mathbf{Z}(\omega)$

$$\mathbf{H}_{\mathrm{a}}(\omega)\Delta\mathbf{Z}\mathbf{H}_{\mathrm{x}}(\omega)=\mathbf{H}_{\mathrm{a}}(\omega)-\mathbf{H}_{\mathrm{x}}(\omega) \tag{3}$$

where

$$\Delta\mathbf{Z}=\mathbf{Z}_{\mathrm{x}}(\omega)-\mathbf{Z}_{\mathrm{a}}(\omega)=-\omega^2\Delta\mathbf{M}+\mathrm{j}\omega\Delta\mathbf{C}+\Delta\mathbf{K} \tag{4}$$

The basic assumption here is that the damage causes a reduction of the mass, damping and stiffness matrices of the model at element level. So the difference of the 3 global matrices can be expressed as

$$\Delta\mathbf{M}=\sum_{i=1}^{\mathrm{n_e}}p_{\mathrm{m}i}\mathbf{M}_i^{\mathrm{e}},\quad \Delta\mathbf{C}=\sum_{i=1}^{\mathrm{n_e}}p_{\mathrm{c}i}\mathbf{C}_i^{\mathrm{e}},\quad \Delta\mathbf{K}=\sum_{i=1}^{\mathrm{n_e}}p_{\mathrm{k}i}\mathbf{K}_i^{\mathrm{e}} \tag{5}$$

where $\mathrm{n_e}$ is the number of updating elements, $\mathbf{M}_i^{\mathrm{e}}$, $\mathbf{C}_i^{\mathrm{e}}$ and $\mathbf{K}_i^{\mathrm{e}}$ are respective contributions of the ith element to the global mass, damping and stiffness matrices and $p_{\mathrm{m}i}$, $p_{\mathrm{c}i}$ and $p_{\mathrm{k}i}$ are the corresponding scalar multipliers representing the proportional changes from their values in the intact state and are also regarded as updating parameters. Substituting equations (5) and (4) into (3) and assuming only one column vector $\mathbf{h}_{\mathrm{x}}(\omega)$ of the measured FRF matrix $\mathbf{H}_{\mathrm{x}}(\omega)$ is available, one obtains

$$\begin{bmatrix}\mathbf{S}_{\mathrm{m}}(\omega) & \mathbf{S}_{\mathrm{c}}(\omega) & \mathbf{S}_{\mathrm{k}}(\omega)\end{bmatrix}\begin{Bmatrix}\mathbf{P}_{\mathrm{m}}\\ \mathbf{P}_{\mathrm{c}}\\ \mathbf{P}_{\mathrm{k}}\end{Bmatrix}=\mathbf{h}_{\mathrm{a}}(\omega)-\mathbf{h}_{\mathrm{x}}(\omega) \tag{6}$$

where $\mathbf{S}_{\mathrm{m}}(\omega)$, $\mathbf{S}_{\mathrm{c}}(\omega)$ and $\mathbf{S}_{\mathrm{k}}(\omega)$ are the sensitivity matrices with n_{e} submatrices arranged in a row and $\mathbf{P}_{\mathrm{m}}$, $\mathbf{P}_{\mathrm{c}}$ and $\mathbf{P}_{\mathrm{d}}$ are the column vectors of the corresponding updating parameters. The ith submatrix of the 3 sensitivity matrices is calculated as

$$\begin{aligned}\mathbf{S}_{\mathrm{m}}^{i}(\omega)&=-\omega^{2}\mathbf{H}_{\mathrm{a}}(\omega)\mathbf{M}_{i}^{\mathrm{e}}\mathbf{h}_{\mathrm{x}}(\omega)\\ \mathbf{S}_{\mathrm{c}}^{i}(\omega)&=\mathrm{j}\omega\mathbf{H}_{\mathrm{a}}(\omega)\mathbf{C}_{i}^{\mathrm{e}}\mathbf{h}_{\mathrm{x}}(\omega)\\ \mathbf{S}_{\mathrm{k}}^{i}(\omega)&=\mathbf{H}_{\mathrm{a}}(\omega)\mathbf{K}_{i}^{\mathrm{e}}\mathbf{h}_{\mathrm{x}}(\omega)\end{aligned} \tag{7}$$

Equation (6) is the final form of most FRF based methods. However, as mentioned before, choosing good theoretical frequencies different from their counterparts can improve the robustness of the method. To achieve this, equation (6) and (7) are rebuilt where two different frequencies are considered: ω_{a} as the analytical frequency and ω_{x} as the experimental frequency. They turn into

$$\begin{bmatrix}\mathbf{S}_{\mathrm{m}}(\omega_{\mathrm{a}},\omega_{\mathrm{x}}) & \mathbf{S}_{\mathrm{c}}(\omega_{\mathrm{a}},\omega_{\mathrm{x}}) & \mathbf{S}_{\mathrm{k}}(\omega_{\mathrm{a}},\omega_{\mathrm{x}})\end{bmatrix}\begin{Bmatrix}\mathbf{P}_{\mathrm{m}}\\ \mathbf{P}_{\mathrm{c}}\\ \mathbf{P}_{\mathrm{k}}\end{Bmatrix}=\mathbf{h}_{\mathrm{a}}(\omega_{\mathrm{a}})-\mathbf{h}_{\mathrm{x}}(\omega_{\mathrm{x}})-\mathbf{e}_{\omega}(\omega_{\mathrm{a}},\omega_{\mathrm{x}}) \tag{8}$$

and

$$\begin{aligned}\mathbf{S}_{\mathrm{m}}^{i}(\omega_{\mathrm{a}},\omega_{\mathrm{x}})&=-\omega_{\mathrm{x}}^{2}\mathbf{H}_{\mathrm{a}}(\omega_{\mathrm{a}})\mathbf{M}_{i}^{\mathrm{e}}\mathbf{h}_{\mathrm{x}}(\omega_{\mathrm{x}})\\ \mathbf{S}_{\mathrm{c}}^{i}(\omega_{\mathrm{a}},\omega_{\mathrm{x}})&=\mathrm{j}\omega_{\mathrm{x}}\mathbf{H}_{\mathrm{a}}(\omega_{\mathrm{a}})\mathbf{C}_{i}^{\mathrm{e}}\mathbf{h}_{\mathrm{x}}(\omega_{\mathrm{x}})\\ \mathbf{S}_{\mathrm{k}}^{i}(\omega_{\mathrm{a}},\omega_{\mathrm{x}})&=\mathbf{H}_{\mathrm{a}}(\omega_{\mathrm{a}})\mathbf{K}_{i}^{\mathrm{e}}\mathbf{h}_{\mathrm{x}}(\omega_{\mathrm{x}})\end{aligned} \tag{9}$$

where the additional item $\mathbf{e}_{\omega}(\omega_{\mathrm{a}},\omega_{\mathrm{x}})$ comes from equation (4) representing the difference of theoretical dynamic stiffness matrices between the two frequencies

$$\mathbf{e}_{\omega}(\omega_{\mathrm{a}},\omega_{\mathrm{x}})=(-\omega_{\mathrm{x}}^{2}+\omega_{\mathrm{a}}^{2})\mathbf{H}_{\mathrm{a}}(\omega_{\mathrm{a}})\mathbf{M}\mathbf{h}_{\mathrm{x}}(\omega_{\mathrm{x}})+\mathrm{j}(\omega_{\mathrm{x}}-\omega_{\mathrm{a}})\mathbf{H}_{\mathrm{a}}(\omega_{\mathrm{a}})\mathbf{C}\mathbf{h}_{\mathrm{x}}(\omega_{\mathrm{x}}) \tag{10}$$

By now the relationship between the difference of the FRFs and the updating parameters has been established. The next task is to find the best corresponding theoretical frequencies for every experimental frequency. A well-known method for doing this is the FDAC method [17] which matches the measured frequency with the theoretical frequency whose FDAC value is the greatest. The FDAC value can be regarded as the correlation of the two frequencies and its expression is

$$FDAC(\omega_{\mathrm{a}},\omega_{\mathrm{x}})=\frac{\{\mathbf{h}_{\mathrm{a}}(\omega_{\mathrm{a}})\}^{\mathrm{T}}\{\mathbf{h}_{\mathrm{x}}(\omega_{\mathrm{x}})\}}{\left|\{\mathbf{h}_{\mathrm{a}}(\omega_{\mathrm{a}})\}_{i}\right|\left|\{\mathbf{h}_{\mathrm{x}}(\omega_{\mathrm{x}})\}_{i}\right|} \tag{11}$$

The experimental FRF can be considered as the shift of theoretical FRF caused by the change of the updating parameters. The key point of FDAC method is to find the corresponding theoretical frequency to the experimental one. However, this method requires searching for the whole frequency range, which is very time-consuming.

Since the main idea of this method is to avoid a big difference of the theoretical and experimental FRFs, a new index is proposed to find the theoretical frequency with a minimum order of magnitude difference (MOMD) from the measured frequency and is expressed as:

$$MOMD(\omega_a,\omega_x)=\sum_{i=1}^{n}\left|\log_{10}(\mathbf{momd}_i)\right|$$
$$\mathbf{momd}_i=\begin{cases}\left|\dfrac{\{\mathbf{h}_a(\omega_a)\}_i}{\{\mathbf{h}_x(\omega_x)\}_i}\right|, & \{\mathbf{h}_a(\omega_a)\}_i\cdot\{\mathbf{h}_x(\omega_x)\}_i\geq 0\\ \left|\dfrac{\{\mathbf{h}_a(\omega_a)\}_i-\{\mathbf{h}_x(\omega_x)\}_i}{\min\left(\left|\{\mathbf{h}_a(\omega_a)\}_i\right|,\left|\{\mathbf{h}_x(\omega_x)\}_i\right|\right)}\right|, & \{\mathbf{h}_a(\omega_a)\}_i\cdot\{\mathbf{h}_x(\omega_x)\}_i<0\end{cases} \tag{12}$$

where $\mathbf{momd}_i$ represents the degree of closeness of every pair of elements in the two vectors.

$\mathbf{momd}_i$ has different expressions to reduce the influence of anti-phase vector with the same modulus value. The closer its value is to 1, the smaller *MOMD* is. After calculating all the *MOMD* values in a certain interval around the measured frequencies, the one with smallest *MOMD* value is chosen as the theoretical frequency. By using the MOMD index, the robustness of the algorithm has been improved significantly, which is verified in the next section. For simplicity the FRF based model updating methods without and with different frequency matching techniques under comparison are referred to as the original method, the MOMD method and the FDAC method, respectively.

To deal with the problem that in practice usually not all DoFs are measurable and incomplete measured data cannot be directly substituted into equations (8), (9) and (10), the dynamic reduction method is considered which generates the slave DoFs by the master DoFs if a harmonic excitation is applied in the master DoFs. The transformation is given as

$$\begin{bmatrix}\mathbf{x}_m\\ \mathbf{x}_s\end{bmatrix}=\begin{bmatrix}\mathbf{I}\\ -\left(-\omega^2\mathbf{M}_{ss}+j\omega\mathbf{C}_{ss}+\mathbf{K}_{ss}\right)^{-1}\left(-\omega^2\mathbf{M}_{sm}+j\omega\mathbf{C}_{sm}+\mathbf{K}_{sm}\right)\end{bmatrix}\left[\mathbf{x}_m\right]=\mathbf{T}_d\mathbf{x}_m \tag{13}$$

where subscripts m and s stand for the master and slave DoFs, respectively, and displacement vector $\mathbf{x}$ and mass, damping and stiffness matrices $\mathbf{M}$, $\mathbf{C}$ and $\mathbf{K}$ are divided into sub vectors and sub matrices relating to these DoFs. By suitably dividing the measured DoFs into the master DoFs and the slave DoFs, the theoretical model is reduced by

$$\mathbf{M}_r=\mathbf{T}_d^T\mathbf{M}\mathbf{T}_d,\quad \mathbf{C}_r=\mathbf{T}_d^T\mathbf{C}\mathbf{T}_d,\quad \mathbf{K}_r=\mathbf{T}_d^T\mathbf{K}\mathbf{T}_d \tag{14}$$

where $\mathbf{M}_r$, $\mathbf{C}_r$ and $\mathbf{K}_r$ are the reduced mass, damping and stiffness matrices.

It should be noted that the fewer the measured DoFs there are, the less information the experimental FRF data contains and the more model errors the theoretical model produces, which causes the updating procedure to diverge. Thus, the method which updates parameters correctly with fewer measured DoFs is considered to be more robust.

3. Case study using simulated FRF data

A six-bay truss structure with 25 rods of identical cross-section and 21 DoFs as shown in figure 1, adapted from [12], is used in simulation. The structure is modelled with 2D truss elements. The properties of the structure are Young's modulus of 200GPa, mass density of 7800kg/m^3 and cross section area of 1800mm^2. The DoFs are arranged in the order of the node numbers and for each node the horizontal displacement is placed ahead of the vertical displacement.

In this study, two damage scenarios listed in Table 1 are considered to gauge the performance of the proposed method. Both damage scenarios are represented by the reduction factor of the mass and stiffness matrices of some elements and for the sake of simplicity, the 'experimental' FRF data from only the first column of the FRF matrices directly obtained by the finite element simulation, are used to update the original intact structure.

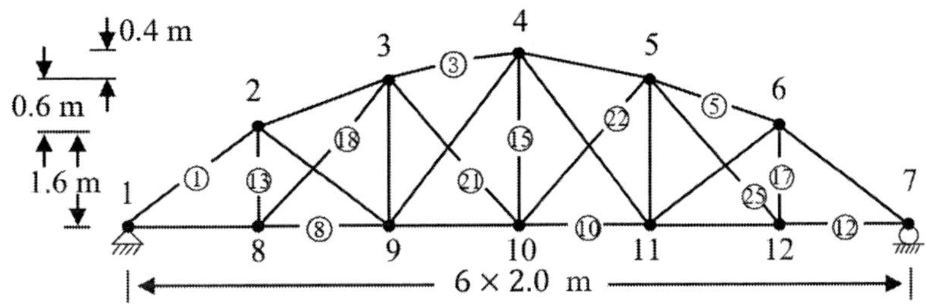

Figure 1. Geometry of a truss model

Table 1. Damage scenarios of the truss structure

Scenario 1			Scenario 2		
Element no.	M (%)	K (%)	Element no.	M (%)	K (%)
7	15	20	7	50	70
12	30	40	12	60	80
21	25	32	21	40	60

First, scenario 1 which represents a moderate damage case is used to get experimental FRF data without noise. Figure 2 shows the results of damage identification using both methods when all DoFs are measurable. It is clear that the FRF based model updating method is able to update so many parameters at the same time if enough and accurate test data are available.

Scenario 1 is also used to compare the identification performance of the three methods. The minimum number of the required measurable DoFs (MD) is considered a criterion. For the sake of fair comparison, for both methods the measurable DoFs are removed gradually from the end of the DoF sequence and the experimental frequencies (whose FRFs are measured) are selected at a regular interval in the range of 0 to 900Hz controlled by the total number of measured frequency points, denoted by NF, which is from 20 to 60. Table 2 lists the results of this comparison which show that the MOMD method can get acceptable updating results using far fewer measured DoFs than the FDAC and the original method. The FDAC method is slightly better than the Original method as it has more acceptable NFs with the same MD. But it takes the longest time to get the final solution of updating parameters.

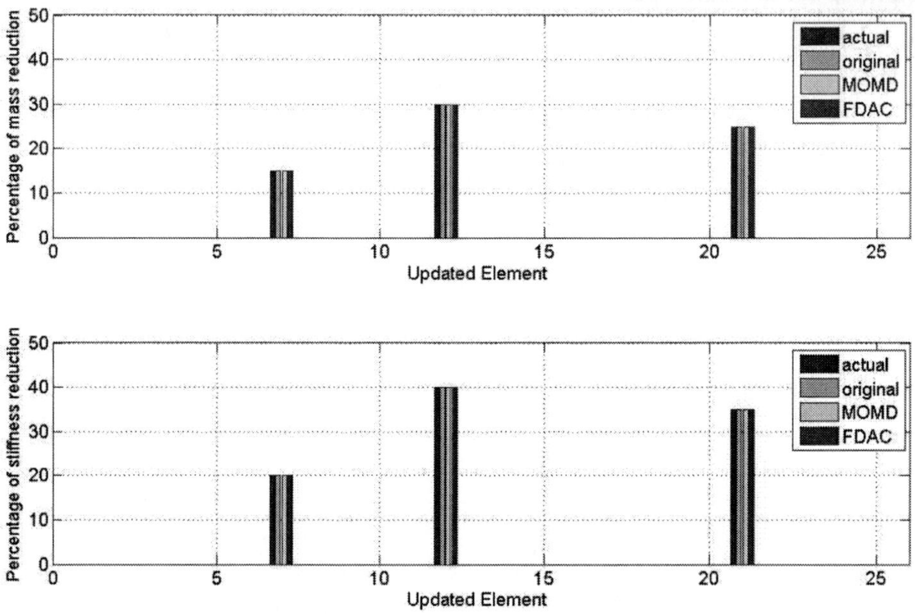

Figure 2. Actual and updating results for damage scenario 1 (top graph for mass and bottom graph for stiffness)

Table 2. Updating limits of the three methods

Method	minimum of MD	Acceptable NF	Single operation time(s) (MD=15,NF=30)
Original	8	21, 24, 32, 38, 40, 58	1.75
MOMD	4	33, 59	4.26
FDAC	8	20, 22, 28, 37, 40, 41, 42, 44, 47, 48, 49, 57, 59	83.86

Secondly, scenario 2 which represents a more severe damage case than scenario 1 is used to compare the performance of the three methods in extreme circumstances. This time the experimental FRF data is contaminated by 1% white Gaussian noise which is introduced though function awgn in MATLAB. Other parameter values are the same as before. The numbers of measured DoFs are from 12 to 18; and the corresponding numbers of acceptable NF, for which an acceptable result can be obtained by the three FRF-based methods are illustrated in figure 3. It is clear that for each number of measured DoFs the MOMD method is more likely to achieve convergence than the original method.

This time the FDAC method performs poorly. A reasonable explanation of this phenomenon is that the severe locations of damage cause a large frequency shift and the big difference between theoretical and experimental frequencies may overwhelm the FRF residuals. That is why the MOMD method targets the theoretical frequency in the vicinity of the experimental frequency and thus has a better performance.

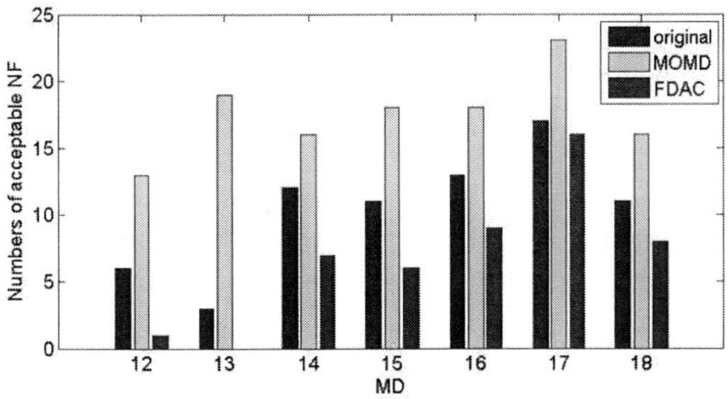

Figure 3. Numbers of acceptable NF for the three methods with different numbers of measured DoFs

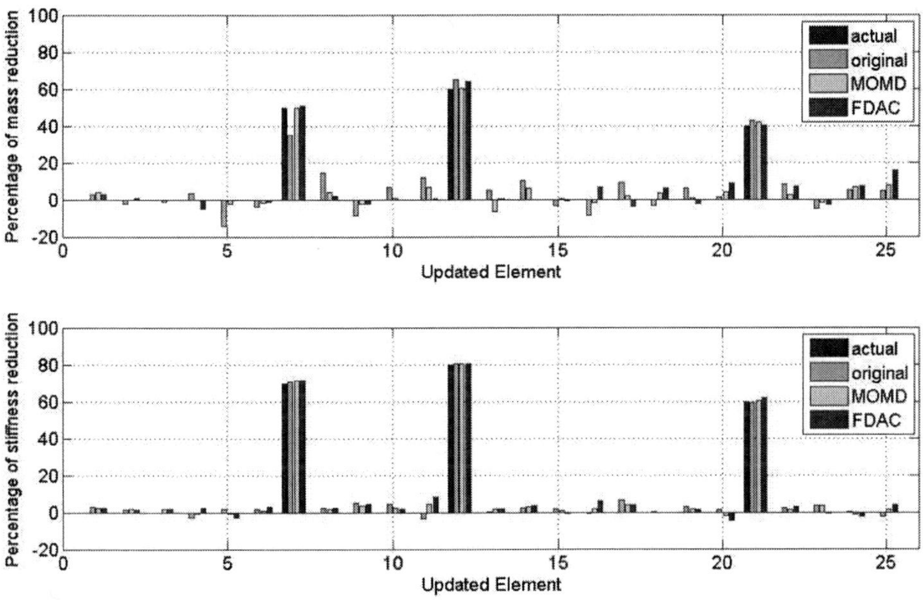

Figure 4. Actual and updating results for damage scenario 2 with 3% white Gaussian noise polluted data (top graph for mass and bottom graph for stiffness)

Then, the proportion of white Gaussian noise is increased to 3% to compare the noise robustness of the three methods. The MD and NF are set to 17 and 30 respectively, by which an acceptable updating results can be obtained by all methods. After enough iterations, the results are considered convergent solutions as shown in figure 4 which illustrates that although all methods are able to identify the damage, the MOMD method makes the best predictions.

4. Conclusions

In this paper, a new frequency matching technique is proposed which can be used for the FRF based model updating method. With a small additional amount of computational time, a best theoretical frequency whose FRF has the minimum order of magnitude difference (MOMD) from the experimental one is selected in the vicinity of the related experimental frequency. The results of the simulated truss structure show that the MOMD method makes a big improvement in the performance of convergence over conventional FRF-based identification method.

Acknowledgements

The research work is supported by the National Natural Science Foundation of China (Grant No. 51121005) and by the Fundamental Research Funds for the Central Universities (DUT16RC(3)027)

References

[1] Li J C, Dackermann U, Xu Y L and Samali B 2011 Damage identification in civil engineering structures utilizing PCA-compressed residual frequency response functions and neural network ensembles *Struct Control Health Monit* **18** 207-26.

[2] Lee U and Shin J 2002 A frequency response function-based structural damage identification method *Comput Struct* **80** 117-32.

[3] Doebling S W, Farrar C R and Prime M B 1998 A summary review of vibration-based damage identification methods *Shock and Vibration Digest* **30** 91-105.

[4] Jaishi B and Ren W X 2006 Damage detection by finite element model updating using modal flexibility residual *Journal of Sound and Vibration* **290** 369-87.

[5] Goge D and Link M 2003 Results obtained by minimizing natural frequency and mode shape errors of a beam model *Mech Syst Signal Proc* **17** 21-7.

[6] Bohle K and Fritzen C P 2003 Results obtained by minimising natural frequency and mac-value errors of a plate model *Mech Syst Signal Proc* **17** 55-64.

[7] Thonon C and Golinval J C 2003 Results obtained by minimising natural frequency and mac-value errors of a beam model *Mech Syst Signal Proc* **17** 65-72.

[8] Xia Y, Hao H, Brownjohn J M W and Xia P Q 2002 Damage identification of structures with uncertain frequency and mode shape data *Earthq Eng Struct Dyn* **31** 1053-66.

[9] Teughels A, Maeck J and De Roeck G 2002 Damage assessment by FE model updating using damage functions *Comput Struct* **80** 1869-79.

[10] Imregun M, Visser W J and Ewins D J 1995 Finite element model updating using frequency response function data: I. Theory and initial investigation *Mech Syst Signal Proc* **9** 187-202.

[11] Lin R M and Zhu J 2007 Finite element model updating using vibration test data under base excitation *Journal of Sound and Vibration* **303** 596-613.

[12] Esfandiari A, Bakhtiari-Nejad F, Rahai A and Sanayei M 2009 Structural model updating using frequency response function and quasi-linear sensitivity equation *Journal of Sound and Vibration* **326** 557-73.

[13] Kwon K S and Lin R M 2004 Frequency selection method for FRF-based model updating *Journal of Sound and Vibration* **278** 285-306.

[14] Garcia-Palencia A J and Santini-Bell E 2013 A Two-Step Model Updating Algorithm for Parameter Identification of Linear Elastic Damped Structures *Comput-Aided Civil Infrastruct Eng* **28** 509-21.

[15] Friswell M and Mottershead J E 1995 *Finite element model updating in structural dynamics* ed Gladwell G M L (Dordrecht: Kluwer Academic), pp: 65-6.

[16] Gang X Y, Chai S, Allemang R J and Li L J 2014 A new iterative model updating method using incomplete frequency response function data *Journal of Sound and Vibration* **333** 2443-53.

[17] Pascual R, Razeto M, Golinval J C and Schalchli R 2002 A robust FRF-based technique for model updating *International Conference on Noise and Vibration Engineering* pp: 1037-45.

[18] Pascual R, Golinval J C, Razeto M and Soc Exptl M 1997 A frequency domain correlation technique for model correlation and updating *Proceedings of the 15th International Modal Analysis Conference* pp: 587-92.

[19] Asma F and Bouazzouni A 2005 Finite element model updating using FRF measurements *Shock Vib* **12** 377-88.

IOP Conf. Series: Journal of Physics: Conf. Series **842** (2017) 012014 doi:10.1088/1742-6596/842/1/012014

Application of proper orthogonal decomposition and radial basis functions for crack size estimation using particle swarm optimization

B Benaissa[1], M Köppen[2], M Abdel Wahab[3] and S Khatir[3]

[1] LEMI, Department of Mechanical Engineering, University M'hamed Bougara of Boumerdes. Algeria.

[2] Faculty of Computer Science and Systems Engineering, Department of Creative Informatics, Kyushu Institute of Technology. Japan.

[3] Department of Electrical Energy, Systems and Automation, Faculty of Engineering and Architecture, Ghent University. Belgium.

E-mail: Brahim.benaissa@hotmail.fr

Abstract. Complex engineering problems require simulations, which are computationally expensive in cases of inverse identification tasks since they commonly requires hundreds of thousands of simulations. This paper propose a method based on model reduction for crack size estimation, combining the proper orthogonal decomposition method with radial basis functions. The reduced model is validated by comparing the obtained boundary displacements with the corresponding results from a finite element model. This inverse procedure is formulated as the minimization of the difference between the measured and computed values of displacement at selected boundary nodes, called sensor points, using particle swarm optimization algorithm. Convex and a non-convex specimens have been considered for investigations of crack presence, and identification of its size, different crack sizes have been tested to demonstrate the efficiency of the proposed approach.

1. Introduction

Crack initiation and propagation is an omnipresent fact in all structures undergoing cyclic loads due to the fatigue phenomenon. In most cases, cracks are engaged in a predictable location. Thus maintenance measures give big importance to the crack size, trying to follow its state to prevent reaching the dangerous level. Damage detection using vibration data has been the focus of a large number of studies in the literature [1-8]. There are several numerical methods for crack detection [9-11], that employ different theoretical bases. Many of these methods are dedicated to the use of parameters which are not accessible experimentally.

Inverse problems are defined as the problems, in which the output is known and the input or source of output remains to be determined. They are opposite to the direct problems, in which output or response are determined using information from input [12]. In the case of the Inverse Elastostatics Problem of internal flaw detection, the geometric parameters of the flaw are unknown, but the displacements along the boundaries are known. In order to analyze this kind of problems, the boundary displacements, usually called "experimental data", are obtained under known boundary conditions and compared with the calculated ones.

Inverse crack identification problems can be stated as an optimization task [4]. There are several optimization techniques summoned in [13]. The particle swarm optimization (PSO) is a very popular algorithms used in diverse range of applications. The most utilized methods in the calculation of the mechanical behaviour of structures are boundary element method (BEM), the finite element method (FEM) [14-22] and recently isogeometric analysis (IGA) [23-29], mainly used to obtain the displacement field. The FEM and BEM was employed to solve inverse methods in structural analysis [30, 31]. Generally speaking, the weak point of FEM based inverse methods is their very high computational cost. Model reduction is an alternative to solve this FEM difficulty.

The proper orthogonal decomposition (POD) is a model reduction techniques proceed by the approximation of the problem solution using the appropriate set of approximation functions [32], which contributes to the huge acceleration of the procedure since, once a trained model is built, the system response are computed by means of it, in a time shorter by about five orders of magnitude compared to FEM [33]. This leads to a very quick alternative in inverse problems, which provides simplicity and a considerably lower computational time.

Boundary measurement are employed to determine cracks since several decades [34]. The proposed crack size estimation procedure uses the same principle of classical approaches. Its main contribution lies in the way the structural response is obtained, which opposed to existing methods that employs simulation methods, they are calculated through a reduced model. Section 2 and 3 present the basic blocks of this approach, which are respectively the model reduction method POD-RBF and optimization method PSO, section 4 is dedicated to the formulation of the inverse problem by combining the POD-RBF and PSO, section 5 studies the ability of model reduction in the estimation of boundary displacement in both convex and non-convex specimen. Finally in section 6, the efficiency of this approach have been tested in identifying small, medium and large cracks in both specimens.

2. POD-RBF as a model reduction method

POD is a powerful statistical method for data analysis, used as model order reduction technique in different fields [35, 36]. In our study, the POD was applied to determine the boundary displacement field of a two dimensional elastic structure containing an unknown crack. Used input data were finite element boundary displacements corresponding to various known cracks, called snapshots. They were first stored in matrix U, which is expressed in equation 1, as shown in figure 1.

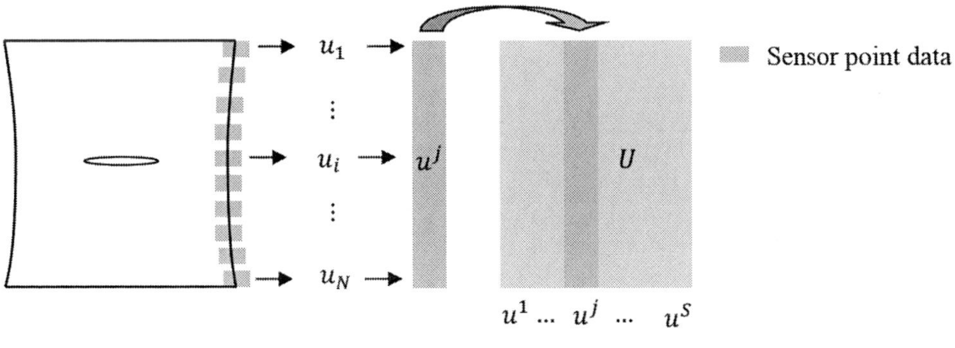

Figure 1. Illustration of the snapshot concept

$$\mathrm{U} = \begin{bmatrix} u_1^1 & u_1^2 & \cdots & u_1^S \\ u_2^1 & u_2^2 & & u_2^S \\ \vdots & \vdots & \ddots & \vdots \\ u_N^1 & u_N^2 & \cdots & u_N^S \end{bmatrix}, \quad (1)$$

Where N is the number of sensor points and S represents the number of snapshot vectors u^j that represent the boundary displacement field of each crack configuration. The matrix P stores the crack parameter sets P_i of all simulations, considered in our study as the crack size.

The main purpose of the POD method is to propose a set of orthogonal vectors Φ called POD basis vectors, to reassemble the snapshot matrix U in an optimal way. Φ is related to U by the following linear relationship:

$$U = \Phi \cdot A\,, \tag{2}$$

In Equation (2), A is the amplitude matrix collecting the coefficients of the new basis combination and, according to the orthogonality of Φ, it can be computed from:

$$A = \Phi^T \cdot U\,, \tag{3}$$

Optimal basis vectors are defined by the performance of the POD method, also known as the singular value decomposition operation [37, 38]:

$$\Phi = U \cdot V \cdot \Lambda^{-1/2}, \tag{4}$$

Where V is the matrix storing the normalized eigenvectors of the covariance matrix C , and Λ the diagonal matrix storing its eigenvalues. The matrix C is given by the following equation:

$$C = U^T \cdot U, \tag{5}$$

Due to the optimality of the new system Φ constructed as a POD basis, a low dimensional approximation $\hat{\Phi}$ of high accuracy is extracted from it by preserving only K (K $\ll$ S) columns that correspond to the largest eigenvalues. Since the eigenvalues of the covariance matrix C , called the energy of the system, are stored in a descending order, POD directions that hold little information are then discarded without influencing the accuracy of the representation. This is known as the truncation of the POD basis, and is accomplished by choosing the fraction of system energy that will be neglected in later calculations. Consequently, the amplitude matrix $\hat{A}$ is given by:

$$\hat{A} = \hat{\Phi}^T \cdot U\,, \tag{6}$$

since:

$$U = \hat{\Phi} \cdot \hat{A}. \tag{7}$$

To determine the boundary displacement field of a two dimensional elastic structure containing an unknown crack, RBF interpolation was used. This method can generate different sets of parameters, which were not included in the initial selection in the matrix P. The amplitude matrix A is defined as a multiplicative form of the function G, defined as the matrix of interpolation parameters, and the matrix B containing the unknown coefficients:

$$A = B \cdot G. \tag{8}$$

The interpolation functions are expressed by [32, 37, 39]:

$$g_i = g_i(|P-P_i|) = \frac{1}{\sqrt{|P-P_i|^2+c^2}}\,, \tag{9}$$

P_i is the parameter corresponding to U_i (i=1,2,...,S). The argument of the i-th RBF is the distance $|P-P_i|$, P and Pi being respectively current and reference parameters. c is the RBF smoothing factor defined in the range from 0 to 1. If all or some of the the knot points P_i are relatively close to each other, the matrix G could be singular, which is circumvented by reducing the c value. After the evaluation of the coefficient matrix B , a low-dimensional model issued from (8) is written in the following vector form:

$$a(P) = B \cdot g(P), \tag{10}$$

Equation (7) can be expressed as an approximation of the snapshot u corresponding to a new parameter vector P :

$$u(P) = \hat{\Phi} \cdot a(P), \tag{11}$$

This model will now be referred as the trained POD-RBF network. It is capable of reproduce the unknown boundary displacement field of the structure that corresponds to any set of crack parameters P . It must be noted that extrapolation outside the range of P leads probably to poor precision of the model.

Increasing the value of the smoothing factor c leads to a better interpolation. But it can make the matrix G singular, depending on the closeness of knot points. In the present work, the parameter c was chosen to be constant for all functions, and equal to the mean value of normalized parameters.

3. Particle swarm optimization

The Particle Swarm Optimization (PSO) is a population based optimization method inspired from the behaviour of bird flocks that is characterized by distinct social and psychological principles. Large attention has been paid to this method in few last decades. The algorithm was initially proposed by Kennedy and Eberhart [40]. Its implementation requires a small number of parameters, which facilitates its application and reduces the computational cost.

The main idea of PSO is to consider the potential solution as a particle moving through the space, looking for the global optimum position. Initiated as a group of random particles, each particle is characterized by its position in the multidimensional space and by its movement speed. These particles, each other, cooperate to achieve the solution, based on their personal previous experience and the experience of other particles. The speed and the position of the particles are calculated, respectively, as follows:

$$\{v^i(t+1)\} = w\{v^i(t)\} + c_1\{r_1\}\cdot(\{x^{Pb,j}\} - \{x^j(t)\}) + c_2\{r_2\}\cdot(\{x^{Gb}\} - \{x^j(t)\}) \quad (12)$$

and

$$\{x^i(t+1)\} = \{x^i(t)\} + v^i(t+1). \quad (13)$$

The weight inertia w is multiplied by the particle speed value at every iteration to maintain the particle acceleration in its original direction. c_1 is a positive constant, called cognitive parameter and controlling the step size toward the particle's personal best position. c_2 is social parameter that controls the step size toward the global best position. $\{r_1\}$ and $\{r_2\}$ are vectors containing random numbers within the interval [0,1]. $\{x^j(t)\}$ is the vector of the current positions of particles. $\{x^{Pb,j}\}$ is the vector of the personal best position found by the particle j and $\{x^{Gb}\}$ is the vector of the global best position found by the entire swarm.

4. Inverse problem formulation

The existence of a crack changes the behavior of the plate when put under traction, therefore the deformation of the structure, which is also affected by the changes of the crack length. Benefiting from this fact, the deformation of the structure's border is measured using deformation sensors for inverse crack size estimation.

The crack estimation algorithm consists of two main stages. In the first stage, the identification problem is defined, and the response data corresponding to the unknown crack is chosen. In the second stage, the optimization algorithm is executed. All fitness function values are obtained from calculation on the reduced model, unlike classical methods where this operation needs a full analysis of the whole structure. Figure 2 summarizes the proposed approach. The fitness function was evaluated from the following equation:

$$\begin{cases} f(P) = \dfrac{\|u(P_0) - u(P)\|^2}{\|u(P_0)\|^2} \\ f(P_{optimal}) = \min[f(P)] \end{cases} \quad (14)$$

By introducing the crack parameters, corresponding to each possible solution, in the trained POD-RBF network, the resulting boundary displacement vector $u(P)$ is generated. Then, the fitness function value is the norm error between this vector and the reference displacement vector $u(P_0)$ caused by the real crack parameters.

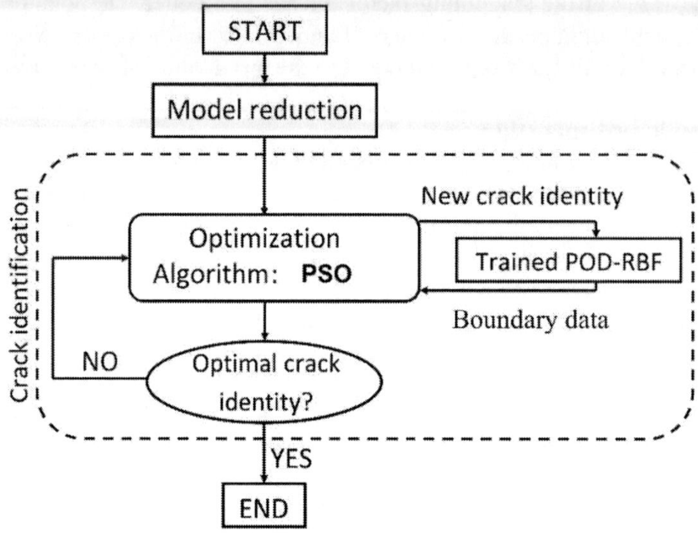

Figure 2. Algorithm of the crack identification approach.

The PSO algorithm is operating in the following manner:
Step 1. Initialization of the algorithm by randomly generating the position vectors of the particles within the design space $(\overrightarrow{x_1}, \overrightarrow{x_2}, \dots, \overrightarrow{x_{NP}})$ and calculating their corresponding speed vectors $(\overrightarrow{v_1}, \overrightarrow{v_2}, \dots, \overrightarrow{v_{NP}})$, where NP the number of particles.
Step 2. Analysis and evaluation of the fitness value for current positions of the particles $(\mathbb{F}(\overrightarrow{x_1}), \mathbb{F}(\overrightarrow{x_2}), \dots, \mathbb{F}(\overrightarrow{x_{NP}}))$, where NP the number of particles. Equation (14) in our study.
Step 3. Personal best calculation, if the current fitness value is better than the best fitness value (Pbest) in the particle's history then this value is set as the new Pbest and the current position as the new $x^{Pb,j}$ for each j particle.
Step 4. Global best calculation, the best fitness value of all the particles Pbest is set as Gbest and the corresponding position as the new x^{Gb}.
Step 5. Updating particle velocity from Eq. (2.40) and update particle position from Eq. (2.41).
Step 6. Check for any dimension i $x_i \leq x^L$ or $x_i \geq x^U$ set $x_i = x^L$ or $x_i = x^U$respectively and $v_i = 0$.
Step 7. If the maximum number of generations or a defined fitness level is reached, the algorithm is terminated; else, the steps 2-6 are repeated.

5. POD-RBF for the computation of boundary displacement field

The POD in this stage has been used to build a reduced model describing the effect of crack size on the vertical boundary displacements, this model is based on a snapshot matrix containing boundary displacement vectors collected from 11 scenarios, corresponding to crack length s varying in the range: 0 (no crack) to 2.5 mm, in two structures where the largest value of the crack (2.5 mm) is half the width of the specimen.

The first specimen (Specimen 1) consists of a rectangular plate, subjected to traction force from the upper and lower sides simulated as a to displacement of 0.1 mm. Young modulus and Poisson coefficient of the material are $E = 210$ GPa and $\nu = 0.3$, respectively. The crack is located at the center of the plate. The displacement data considered for validating the POD-RBF model is obtained from the higher half of the vertical boundary as describes in Figure 3, which contains 40 nodes. The second (Specimen 2) specimen is a non-convex structure, with same material properties and subjected to the boundary conditions as described in Figure 4. The displacement results obtained from the 49 boundary nodes, highlighted by red color in descriptive figure, are considered for validating the POD-RBF.

The choice of the best K value is important, as a too low value will lead to poor precision, and too large value will cost more computational time. To test the effect of the chosen truncation point K on the accuracy of the model, we compared the POD-RBF results, corresponding to crack with size equal to 1.4 mm, with equivalent results from FEM. Figure 5 and Figure 6 study respectively the convex and the non-convex examples by means of the ratio error of the boundary displacement field, produced by POD-RBF model, issue from models constructed based on different K values, and equivalent boundary displacement field from FEM.

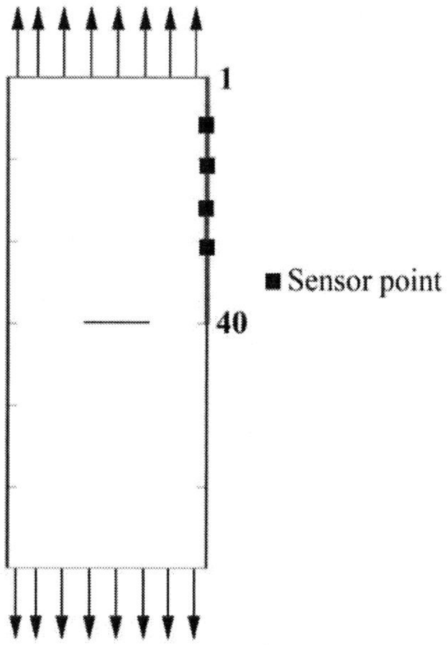

Figure 3. Specimen 1

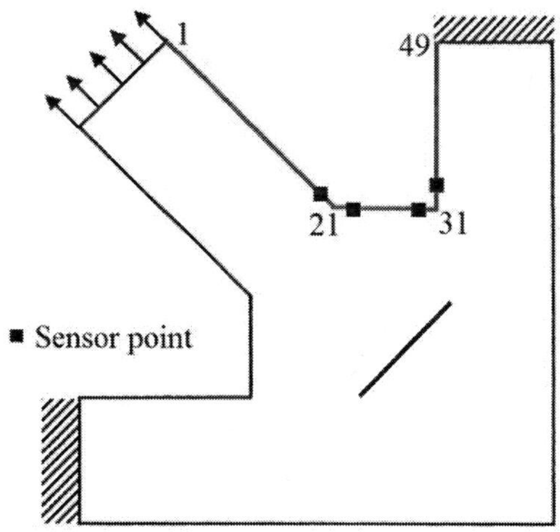

Figure 4. Specimen 2

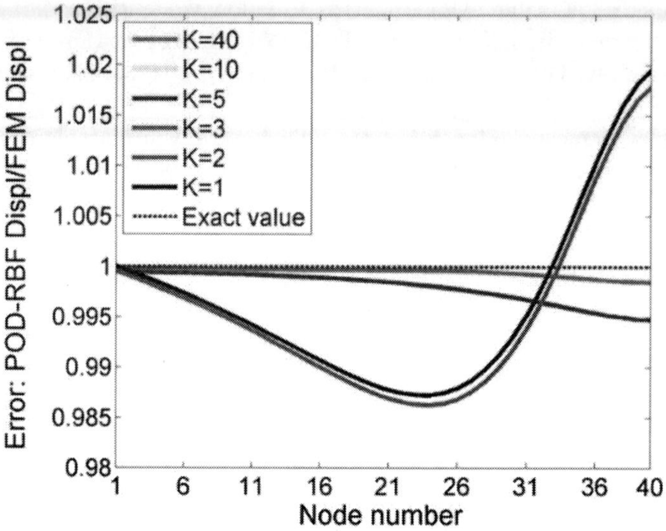

Figure 5. Effect of k value on the boundary displacement error for Specimen 1

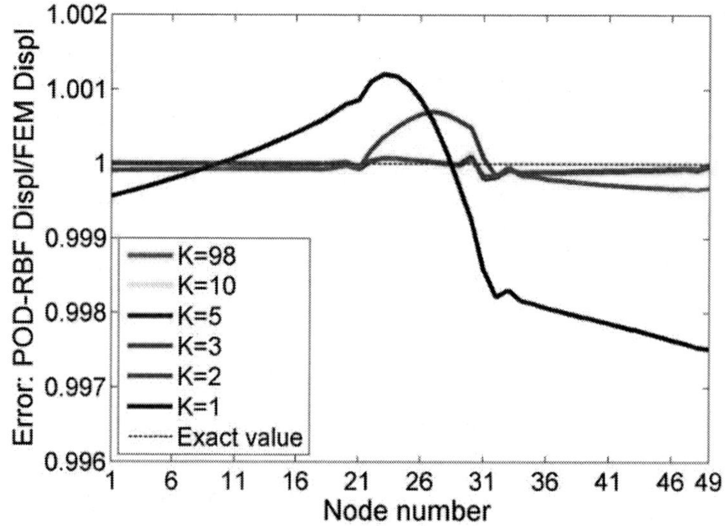

Figure 6. Effect of K values on the boundary displacement error for Specimen 2

From Figure 5, we notice that maximum error for all K values is located at node number 40. This can be explained by the fact that this node is located on the centre of the structure, where the displacement is equal to zero. Therefore, a small difference between estimated results and FEM results affect largely the error value (POD-RBF/FEM). We can also see that the largest error value for all K values is about 2% found for K=1 and K=2. On the other hand, more than 4 modes ($K \geq 4$) have led to better results with error around 1%. Moreover, when the first three modes are used, the results were even better with error less than 0.5%.

Figure 6 shows that node 49 has the largest error value for all modes, similarly to the convex problem, this is due to the fact that displacement value in this node is close to null. It is also noticed that error in the interval between nodes 21 and 31 has a considerable error values for K=1 and K=2, which is not present for K values larger than 3 with error less than 0.1%. Therefore, the first three modes are chosen to represent the full model in the inverse calculations.

IOP Conf. Series: Journal of Physics: Conf. Series **842** (2017) 012014 doi:10.1088/1742-6596/842/1/012014

6. Crack size estimation

The inverse problem solved by PSO, minimizing the cost function, which is the norm error between the vertical displacement vector caused by the crack that we want to estimate and the one proposed by PSO. The displacement data considered in the fitness evaluation is obtained from only 4 sensor point in both cases, convex and non-convex, where their position, as described in Figures 3 and 4, are chosen based on early study [40], in which, it has been found that better results are acquired when the sensors are dispersed uniformly respecting a near distance between every sensor point.

Tables 1 summarize the PSO parameters, and Table 2 and 3 present crack size estimation results, for convex and non-convex specimen, respectively, the results presented in both table represent the best over 10 runs. In each table, three levels of crack size are studied (small, medium and large), as well as, the case of the absence of the crack, which is represented by crack size value equal to zero. The fitness value of each case is shown along with the error of the estimated results.

Table 1. PSO parameters

Parameter	value
Number of particles N	10
Number of iterations M	100
weight inertia w	0.9
cognitive parameter c1	2
social parameter c2	2

Table 2. Crack size estimation in specimen 1

Real (mm)	Estimated (mm)	fitness value	error
0	0.0421	0.000029	_
0.35	0.3687	0.000002	5.3%
1.4	1.4131	0.000010	0.93%
2.45	2.4506	0.000046	0.02%

Table 3. Crack size estimation in specimen 2

Real (mm)	Estimated (mm)	fitness value	error
0	0.0263	0.000045	_
0.35	0.3304	0.000050	5.6%
1.4	1.3951	0.000034	0.35%
2.45	2.4507	0.000059	0.02%

From Tables 2 and 3, we can see that the presented approach could estimate the absence of the crack in both specimens with high accuracy, leading, for specimen 1, to a very small crack of 0.04 mm. It is noted that the fitness value does not follow any order. This is due to using only four sensors, as the displacement results in those sensors varies from one crack size to another. For the three levels of crack sizes, the maximum error value is found in the smallest crack (0.35mm), with a difference between the estimated and real crack equal to 0.02 mm, not the error is large because is calculated compared to a value which is near zero, the larger cracks (1.4 mm and 2.45 mm) have been identified with better precision, due their larger effect of the boundary displacement.

From both tables, we can see the estimation results are very close, which indicates that geometry of the specimen doesn't affect the inverse estimation method. The average computation time in PC with Intel Dual-Core Processor 3.0 GHz and 2 GB RAM, is about 4 seconds.

For the sake of illustration, figures 7 and 8 depict the crack size evolution and the fitness convergence, for the crack size equal to 1.4 in the case of specimen 2.

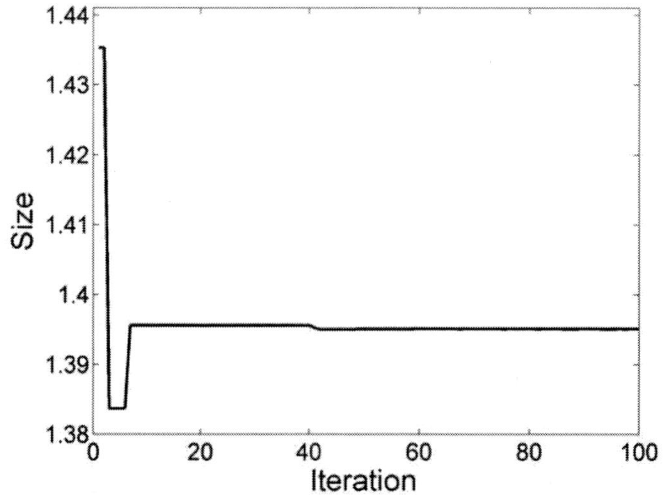

Figure 7. Crack size evolution through iterations for example of real size equal to 1.4 mm in non-convex specimen

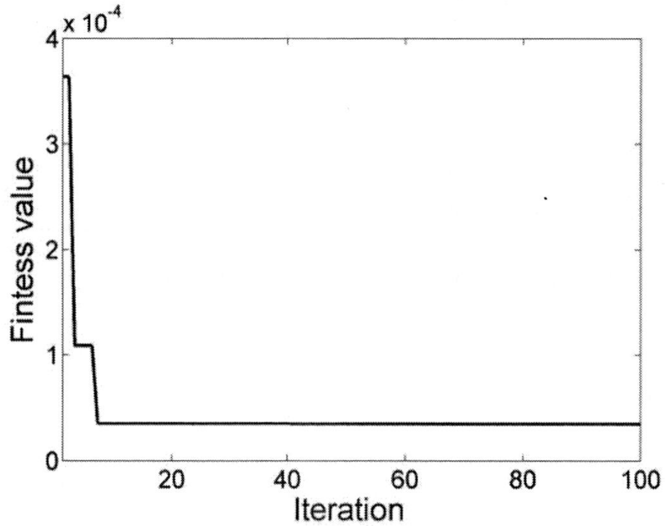

Figure 8. Fitness convergence for example of real size equal to 1.4 mm in non-convex specimen

Figures 7 and 8 shows that the algorithm approached the final result very early, i.e. before the 10th iteration, and reach the optimal result around the 40th iteration. This means that good precision can be conserved even though a low number of iteration is adopted as stopping criteria.

7. Conclusion

In this numerical study, we presented POD-RBF method as model reduction of cracked specimens under traction. The finite element model of the structure was created for different crack lengths, and the results were obtained from the new model was compared to the original ones to insure the efficiency. Crack size was investigated based on vertical boundary displacement data using the PSO for the inverse calculation. The results have clearly shown that the developed algorithm was capable of estimating crack size accurately, and proved its effectiveness even with a very low number of sensors, in both convex and non-convex structures.

The crack size estimation using PSO showed high accuracy, even with a very low number of sensors, and showed that the inverse calculation on the reduced model by POD-RBF was practical and helpful in avoiding computational time problems typical for the simulation based inverse problems.

References

[1] Khatir S, Belaidi I, Serra R, Abdel Wahab M and Khatir T 2015 *Mechanika* **21**(6) 472-479

[2] Zhou Y-L and Abdel Wahab M 2016 *Journal of Vibroengineering* **18**(7) 4491-4499

[3] Gillich G-R, Praisach Z-I, Abdel Wahab M, Gillich N, Mituletu IC and Nitescu C 2016 *SHOCK AND VIBRATION*

[4] Khatir S, Belaidi I, Serra R, Abdel Wahab M and Khatir T 2016 *Journal of Vibroengineering* **18**(1) 202-213

[5] Zhou Y-L and Abdel Wahab M 2017 *Engineering Structures* **141** 175-183

[6] Khatir A, Tehami M, Khatir S and Abdel Wahab M 2016 *Journal of Vibroengineering* **18**(8) 5063-5073

[7] Zhou Y-L, Maia N and Abdel Wahab M 2016 *Journal of Vibration and Control* doi: 10.1177/1077546316674544

[8] Zhou Y-L, Maia NMM, Sampaio R and Wahab MA 2016 *Structural health monitoring* **DOI: https://doi.org/10.1177/1475921716680849**

[9] Li B, Chen X, Ma J and He Z 2005 *Journal of Sound and Vibration* **285**(4) 767-782

[10] Nikolakopoulos P, Katsareas D and Papadopoulos C 1997 *Computers & Structures* **64**(1) 389-406

[11] Chinchalkar S 2001 *Journal of Sound and Vibration* **247**(3) 417-429

[12] Kubo S 2006 Proc. 12th A-PCNDT 2006–Asia-Pacific Conference on NDT ed editors)

[13] Venter G 2010 *Encyclopedia of aerospace engineering*

[14] Yue T and Abdel Wahab M 2014 *WEAR* **321** 53-63

[15] Yue T and Abdel Wahab M 2017 *Tribology International* **107** 274-282

[16] Pereira K, Yue T and Abdel Wahab M 2017 *Tribology International* **110** 222-231

[17] Martínez JC, Vanegas Useche LV and Wahab MA 2017 *International Journal of Fatigue* **100, Part 1** 32-49

[18] Kumar D, Biswas R, Poh LH and Abdel Wahab M 2017 *Tribology International* **109** 124-132

[19] Bhatti NA and Abdel Wahab M 2017 *Tribology International* **109** 552-562

[20] Yue T and Abdel Wahab M 2016 *Materials* **9** 597; doi:510.3390/ma9070597

[21] Resende Pereira KdF, Bordas S, Tomar S, Trobec R, Depolli M, Kosec G and Abdel Wahab M 2016 *Materials* **9** 639; doi:610.3390/ma9080639

[22] Ferjaoui A, Yue T, Abdel Wahab M and Hojjati-Talemi R 2015 *International Journal of Fatigue* **73** 66-76

[23] X. Nguyen H, N. Nguyen T, Abdel Wahab M, Bordas SPA, Nguyen-Xuan H and P. Voa T 2017 *Computer Methods in Applied Mechanics and Engineering* **313** 904-940

[24] Phung-Van P, Qui LX, Nguyen-Xuan H and Wahab MA 2017 *Composite Structures* **166** 120–135

[25] Phung-Van P, Ferreira AJM, Nguyen-Xuan H and Abdel Wahab M 2017 *Composites Part B: Engineering* **118** 125-134

[26] Tran LV, Phung-Van P, Lee J, Wahab MA and Nguyen-Xuan H 2016 *Composite Structures* **140** 655-667

[27] Thai CH, Ferreira AJM, Abdel Wahab M and Nguyen-Xuan H 2016 *Acta Mechanica* **227**(5) 1225-1250

[28] Thai C, Zenkour AM, Abdel Wahab M and Nguyen-Xuan H 2016 *Composite Structures* **139** 77-95

[29] Phung-Van P, Tran LV, Ferreira AJM, Nguyen-Xuan H and Abdel-Wahab M 2016 *Nonlinear Dynamics* 1-16; doi:10.1007/s11071-11016-13085-11076

[30] Burczynski T and Beluch W 2001 *Engineering Analysis with Boundary Elements* **25**(4) 313-322

[31] Gavrus A, Massoni E and Chenot J 1996 *Journal of Materials Processing Technology* **60**(1) 447-454

[32] Rogers CA, Kassab AJ, Divo EA, Ostrowski Z and Bialecki RA 2012 *Inverse Problems in Science and Engineering* **20**(5) 749-767

[33] Buljak V 2011 Inverse analyses with model reduction: proper orthogonal decomposition in structural mechanics ed editors. Springer Science & Business Media

[34] Friedman A and Vogelius M 1989

[35] Schilders WH, Van der Vorst HA and Rommes J 2008 Model order reduction: theory, research aspects and applications ed editors. Springer

[36] Chatterjee A 2000 *Current science* **78**(7) 808-817

[37] Buljak V and Maier G 2011 *Engineering Structures* **33**(2) 492-501

[38] Benaissa B, Aït Hocine N, Belaidi I, Hamrani A and Pettarin V 2016 *Structural and Multidisciplinary Optimization* 1-10

[39] Ostrowski Z, Białecki R and Kassab A 2008 *Inverse Problems in Science and Engineering* **16**(1) 39-54

[40] Kennedy J and Eberhart RC 1997 Systems, Man, and Cybernetics, 1997. Computational Cybernetics and Simulation., 1997 IEEE International Conference on ed editors) IEEE p. 4104-4108

IOP Conf. Series: Journal of Physics: Conf. Series **842** (2017) 012015 doi:10.1088/1742-6596/842/1/012015

Influence of Torsional Excitation on Dynamic Responses of Rotors with a Breathing Slant Crack

Zhiwen Lu, Chunrong Hua, Dawei Dong, Bing Yan, Kang Fan

School of Mechanical Engineering, Southwest Jiaotong University, Chengdu 610031

Email: hcrong@swjtu.cn

Abstract. Focusing on a rotor-bearing system with a breathing slant crack in the power transmission machine, influence of torsional excitations on the coupled nonlinear responses of the system is studied in this work. The slant crack element stiffness matrix is derived based on energy principal and the crack breathing phenomenon is simulated by the Crack Closure Line Position (CCLP) model; and the time-varying coupled dynamic equation of a rotor with a slant breathing crack considering the eccentricity of static unbalance is established using the finite element method and is solved by the NEWMARK method; then the influences of static torque and periodic torsional excitations on rotor dynamic responses in transverse and torsional directions are discussed. Results show that with the increment of static torque, cracks will become open gradually and the nonlinearity degree of rotors will increase firstly and then decrease. For periodic torsional excitation, the torsional excitation frequency and its rotating frequency combination can be found in transverse vibration response, and the larger is the amplitude of excitation, the larger are the combinational frequency components. Then a crack monitoring method for power transmission machines can be suggested by monitoring the coupled response characteristics and their variation from transverse responses of rotors before and after the loads change.

1. Introduction

Rotors are the critical parts of power transmission machines, the shaft easily produces fatigue crack and induces structural bearing failure under long-term complicated and severe operating condition. The slant crack is the most common form of crack owing to the torque transmitting characteristic of power transmitting rotor, and the crack will appear open and close breathing phenomenon under the rotor gravity, eccentricity, and external excitation when the rotor is rotating. Therefore, it is significant for crack detection to investigate the influence of torsional excitation on the dynamic characteristic of the power transmission rotor with slant breathing crack.

Though much research has been carried out on the cracked rotor dynamics [1-4], most of which aim at rotors with transverse cracks and assume the gravity dominant, yet the slant crack is seldom studied at present. However, the cracks in rotors are often slant cracks due to transmitted torque. Moreover, the static torque applied on rotors will affect the dynamic characteristics of rotor systems, which has not been paid enough attention at present.

Based on the above problems, the coupled vibration characteristic of a rotor-bearing system with a slant breathing crack under torsional excitation including periodic and static torsional excitation is investigated in this paper.

A model that reflects the essential behaviour of a crack is vitally important. There are many methods to model a crack. Nonlinear 3D finite element method was adopted to model a breathing crack in a rotor in [5-6], which may be the most accurate model, but the computation workload is very heavy. In [7], rigid finite element method was put forward to model a cracked rotor, which also had good accuracy to model a breathing crack. By considering gravity dominant, rotation dependent breathing model such as switching model [8], cosine model [9], truncated Fourier series model [10] etc. are promoted for breathing crack modelling. Ref. [3] reviewed the crack model in rotors based on strain energy release rate (SERR), and showed that the method put forward by Darpe et al. [11, 12] could model a breathing crack in rotors more accurately, and it had the advantages of allowing general excitations without assuming that gravitational force is dominant, and the behaviour of the breathing crack was response dependent instead of being rotation dependent.

As gravity dominant is not always the case for rotors of power transmission equipment, it is unreasonable to simulate the crack model of these rotors with rotation dependent breathing model. Considering synthetically the model accuracy and computation complexity, the Crack Closure Line Position (CCLP) model proposed by Darpe et al. is adopted in this paper, the slant crack element stiffness matrix is derived based on energy principal, and then the time-varying coupled dynamic equation of a rotor with a slant breathing crack considering the rotor eccentricity is established; the NEWMARK method is used to solve the motion equations in time domain, and the influences of static torque and periodic torsional excitations on rotor dynamic responses in transverse and torsional directions are discussed. The research will provide theoretical basis for crack detection in rotors of power transmission equipment.

2. Stiffness matrix of a shaft with a breathing slant crack

2.1. The slant crack element stiffness derivation based on energy principle

Fig.1 shows a slant cracked shaft element of length l and radius R. P_1–P_{12} are the loads acting on the 12 degrees-of-freedom of the two nodes in the element coordinate system $x-y-z$. The local coordinate system $x'-y'-z'$ is defined on the flat crack face to describe the crack cross-section. θ_c is the crack angle between the crack face and the shaft centreline (formed by the negative z' axis turning to the negative x axis in the counter-clockwise direction) , x_L is the location of the crack centre in the element coordinate system. CCL (crack closure line) is an imaginary line that separates the open and closed parts of the crack which will be used to simulate the breathing of crack. The hatched area corresponds to the open area of the crack.

Based on Castigliano theorem [13], the flexibility coefficient of a crack element can be represented as:

$$g_{ij}=\frac{\partial u_i}{\partial P_j}=\frac{\partial^2}{\partial P_i\partial P_j}(U^0+U^c)=\frac{\partial^2 U^0}{\partial P_i\partial P_j}+\frac{\partial^2 U^c}{\partial P_i\partial P_j}=g_{ij}^0+g_{ij}^c \tag{1}$$

where u_i is the generalized displacement of the ith freedom, P_i is the node force in the ith freedom, U is the total strain energy, U^0 is the strain energy of the un-cracked element; U^c is the additional strain energy induced from crack, g_{ij} and g_{ij}^0 are flexibility coefficients of the cracked and un-cracked element respectively; g_{ij}^c is the additional flexibility coefficient caused by crack.

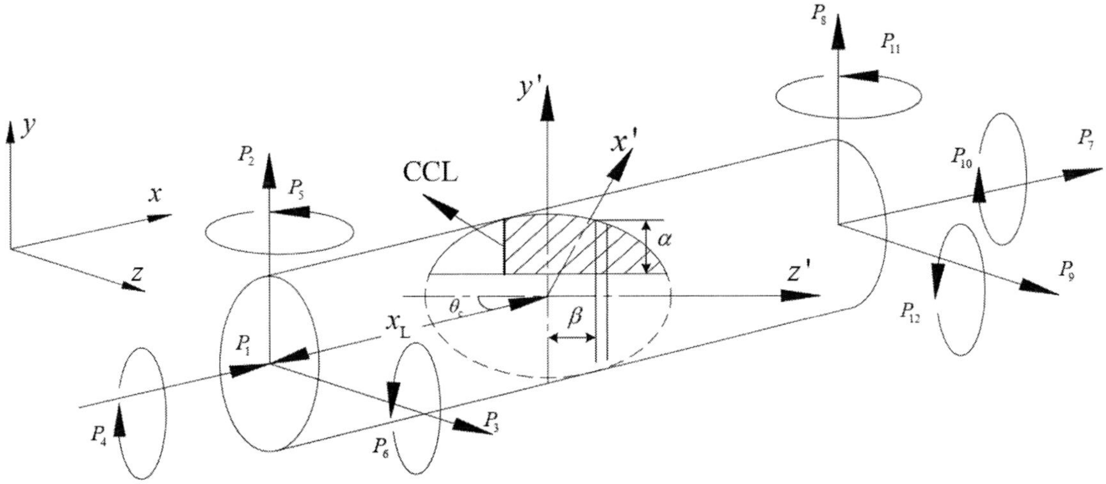

Fig. 1 Element with slant crack and diagram of the crack area

The strain energy of the un-cracked shaft element in Fig.1 is:

$$U^0 = \frac{1}{2}\int\left[\frac{P_1^2}{EA}+\frac{\alpha_s P_2^2}{GA}+\frac{\alpha_s P_3^2}{GA}+\frac{P_4^2}{EI_0}+\frac{(P_3x_L+P_5)^2}{EI_y}+\frac{(P_6-P_2x_L)^2}{EI_z}\right]\mathrm{d}x \tag{2}$$

where E is elastic modulus, G is shear modulus, A is cross section of the shaft;α_s is shear coefficient, $\alpha_s = (1+\nu)/(1.305+1.273\nu)$; I_0 is polar moment of inertia; I_x, I_y are area moment inertias.

According to the strain energy release rate theory [3], the additional strain energy of the shaft resulted from crack is:

$$U^c = \frac{1}{E'}\iint_{A^c}\left[(\sum_{i=1}^{6}K_{\mathrm{I}i})^2+(\sum_{i=1}^{6}K_{\mathrm{II}i})^2+(1+\nu)(\sum_{i=1}^{6}K_{\mathrm{III}i})^2\right]\mathrm{d}A^c \tag{3}$$

where $E' = E/(1-\nu^2)$, ν is Poisson's ratio; $K_{\mathrm{I}i}$ $K_{\mathrm{II}i}$ $K_{\mathrm{III}i}$ are the opening, sliding and tearing mode SIFs respectively; A^c is the open area of crack.

Substituting equation (2) and (3) into equation (1), and flexibility matrix of the cracked and un-cracked shaft element $(\mathbf{G}^0)_{6\times6}$ and $(\mathbf{G}^c)_{6\times6}$ can be derived, and the expression of each element of the matrix can be obtained from [11].

Then the total flexibility matrix of the cracked element is:

$$\mathbf{G}_{ce} = \mathbf{G}^0 + \mathbf{G}^c \tag{4}$$

Considering static equilibrium of the shaft element in Fig.1:

$$\{P_1, P_2, \ldots, P_{12}\}^{\mathrm{T}} = \mathbf{T}\{P_1, P_2, \ldots, P_6\}^{\mathrm{T}} \tag{5}$$

where

$$\mathbf{T}^{\mathrm{T}} = \begin{bmatrix} 1 & 0 & 0 & 0 & 0 & 0 & -1 & 0 & 0 & 0 & 0 & 0 \\ 0 & 1 & 0 & 0 & 0 & 0 & 0 & -1 & 0 & 0 & 0 & l \\ 0 & 0 & 1 & 0 & 0 & 0 & 0 & 0 & -1 & 0 & -l & 0 \\ 0 & 0 & 0 & 1 & 0 & 0 & 0 & 0 & 0 & -1 & 0 & 0 \\ 0 & 0 & 0 & 0 & 1 & 0 & 0 & 0 & 0 & 0 & -1 & 0 \\ 0 & 0 & 0 & 0 & 0 & 1 & 0 & 0 & 0 & 0 & 0 & -1 \end{bmatrix} \tag{6}$$

According to the definition of flexibility

$$\mathbf{T}^{\mathrm{T}}\{u_1,u_2,...,u_{12}\}^{\mathrm{T}} = \mathbf{G}_{\mathbf{ce}}\{P_1,P_2,...,P_6\}^{\mathrm{T}} \tag{7}$$

From equation (5) and (7), one can get:

$$\{P_1,P_2,...,P_{12}\}^{\mathrm{T}} = \mathbf{T}(\mathbf{G}_{\mathbf{ce}})^{-1}\mathbf{T}^{\mathrm{T}}\{u_1,u_2,...,u_{12}\}^{\mathrm{T}} \tag{8}$$

So the stiffness matrices of the cracked and un-cracked shaft element are:

$$\mathbf{K}_{\mathbf{ce}} = \mathbf{T}(\mathbf{G}_{\mathbf{ce}})^{-1}\mathbf{T}^{\mathrm{T}} \tag{9}$$

$$\mathbf{K}_{\mathbf{uce}} = \mathbf{T}(\mathbf{G}^{\mathbf{0}})^{-1}\mathbf{T}^{\mathrm{T}} \tag{10}$$

2.2. The breathing slant crack model

The crack will represent open and close breathing state when the rotor is rotating, so the equation (9) will change with the crack state. Using the breathing crack model based on the Crack Closure Line Position (CCLP) [11], the crack front is subdivided into m part, K_{I} can be calculated from equation (11), which determines the position of the crack closure line, the positive K_{I} represents the corresponding position of the crack front is open and vice versa.

$$K_{\mathrm{I}} = \sum_{i=1}^{6} K_{\mathrm{I}\,i} \tag{11}$$

The SIFs of the slant cracked element under node forces from each freedom in Fig.1 are:

$$\begin{cases} K_{\mathrm{I}\,1} = P_1/(\pi R^2)\sin^2\theta_c\sqrt{\pi\alpha}F_1; K_{\mathrm{I}\,2} = 0; K_{\mathrm{I}\,3} = kP_3/(\pi R^2)\sin 2\theta_c\sqrt{\pi\alpha}F_1 \\ K_{\mathrm{I}\,4} = 2P_4/(\pi R^4)\sqrt{R^2-\beta^2\sin^2\theta_c}\sin 2\theta_c\sqrt{\pi\alpha}F_2 \\ K_{\mathrm{I}\,5} = 4(P_5+P_3(x_{\mathrm{L}}+\beta\cos\theta_c))/(\pi R^4)\beta\sin^3\theta_c\sqrt{\pi\alpha}F_1 \\ K_{\mathrm{I}\,6} = 4(P_2(x_{\mathrm{L}}+\beta\cos\theta_c)-P_6)/(\pi R^4)\sqrt{R^2-\beta^2\sin^2\theta_c}\sin^2\theta_c\sqrt{\pi\alpha}F_2 \end{cases} \tag{12}$$

where, θ_c is crack angle, F_1 and F_2 are empirical coefficients of SIF, and the definition of other parameters are shown in Fig.1. From equation (12), it can be seen that the torsional excitation will affect the $K_{\mathrm{I}\,4}$, then influence the open and close state of crack and thus affect the system responses.

3. The motion equation of the cracked rotor and its time domain solution

For the slant crack rotor-bearing system in Fig. 2(a), supposing the crack only influence the local stiffness, and the system motion equation will be derived once the breathing slant crack element is integrated into the rotor-bearing system based on the finite element method.

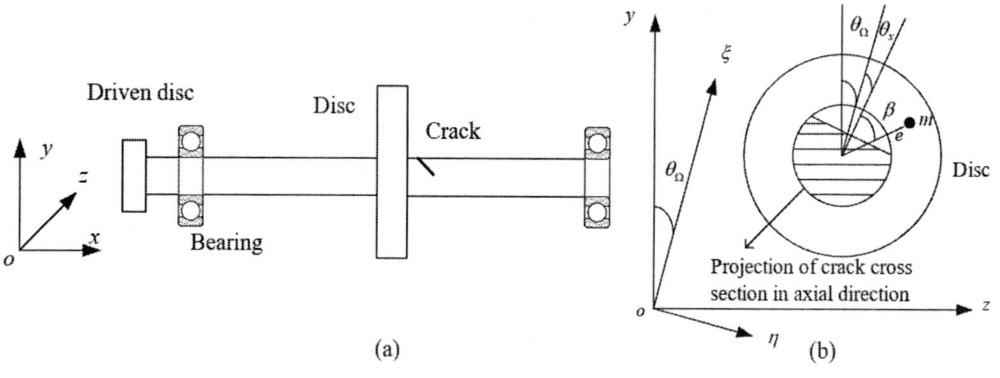

Fig. 2 (a) Diagram of slant cracked rotor-bearing system in power transmission (b) Rotating and stationary coordinates of the rotor

Setting the displacement vector of node i is $\mathbf{q}_i$, and 3 translational and 3 rotational degrees of freedoms are considered in each node,

$$\mathbf{q}_i = \{x_i, y_i, z_i, \theta_{xi}, \theta_{yi}, \theta_{zi}\}^{\mathrm{T}} \tag{13}$$

Therefore, the equation of motion of the cracked rotor in global static coordinate system can be represented as:

$$\mathbf{M}\ddot{\mathbf{q}} + (\mathbf{D} + \Omega \mathbf{D}_{\mathrm{g}})\dot{\mathbf{q}} + \mathbf{K}(t)\mathbf{q} = \mathbf{F}_{\mathrm{e}} + \mathbf{F}_{\mathrm{g}} + \mathbf{F}_{\mathrm{ex}} \tag{14}$$

$$\mathbf{q} = \{\mathbf{q}_1, \mathbf{q}_2, \cdots \mathbf{q}_i, \cdots, \mathbf{q}_n\}^{\mathrm{T}} \tag{15}$$

where, $\mathbf{M}$ is system mass matrix; $\mathbf{K}$ is system stiffness matrix (the stiffness matrix of the cracked rotor are composed from un-cracked and cracked element stiffness matrix according the node position, as the stiffness matrix of breathing crack element is response-dependent nonlinear, so does the system stiffness matrix). $\mathbf{D}$ is system damping matrix, considering Rayleigh damping, $\mathbf{D} = a\mathbf{M} + b\mathbf{K}$; Ω is rotating speed of the rotor; $\mathbf{D}_{\mathrm{g}}$ is Gyroscopic matrix; $\mathbf{F}_{\mathrm{g}}$ is gravity vector; $\mathbf{F}_{\mathrm{e}}$ is eccentric excitation; $\mathbf{F}_{\mathrm{ex}}$ is external excitation; $\mathbf{q}$ is generalized displacement vector.
The gravity vector of the disc in node i:

$$\mathbf{F}_{gi} = \{0, -mg, 0, 0, 0, 0\}^{\mathrm{T}} \tag{16}$$

Eccentric excitation of the rotating disc is [14]:

$$\mathbf{F}_{ei} = \{F_{exi}, F_{eyi}, F_{ezi}, M_{exi}, M_{eyi}, M_{ezi}\}^{\mathrm{T}} \tag{17}$$

$$\begin{cases} F_{exi} = 0; M_{eyi} = 0; M_{ezi} = 0 \\ F_{eyi} = me[(\Omega + \dot{\theta}_{xi})^2 \cos(\Omega t + \theta_{xi} + \beta) + \ddot{\theta}_{xi} \sin(\Omega t + \theta_{xi} + \beta)] \\ F_{ezi} = me[(\Omega + \dot{\theta}_{xi})^2 \sin(\Omega t + \theta_{xi} + \beta) - \ddot{\theta}_{xi} \cos(\Omega t + \theta_{xi} + \beta)] \\ M_{exi} = me[\ddot{y} \sin(\Omega t + \theta_{xi} + \beta) - (\ddot{z} + g)\cos(\Omega t + \theta_{xi} + \beta)] \end{cases} \tag{18}$$

where, $\theta_{\Omega}(t)$ is the relative angular displacement between the rotating and stationary coordinate system; $\theta_x(t)$ is the torsional angular displacement fluctuation of the rotor; β is unbalanced orientation angle. These parameters are shown in Fig. 2(b).
From equation (18), one can see that the eccentric excitation of rotating disc is bending and torsional coupling which will result in more complicated rotor responses.
In this paper the torsional excitation is mainly considered as external excitation, including the static load torque T_s applied into the torsional freedom of the output node and the periodic torsional excitation T_p applied into the torsional freedom of the rotating disc node.
As the breathing state of a crack is decided by the nodal forces in the cracked element, and the rotor responses are related to the eccentric excitation, therefore the motion equation of the cracked rotor has time-variable and nonlinear characteristic. NEWMARK method [15] is adopted to solve the equations in time-domain, and updating the system stiffness matrix, damping matrix and coupling excitation in each time step, and the next time step starts as the last time step converges.

4. The influence of torsional excitation on the dynamic response of rotor

IOP Conf. Series: Journal of Physics: Conf. Series **842** (2017) 012015 doi:10.1088/1742-6596/842/1/012015

The main influence paths of torsional excitation on dynamic responses of a cracked rotor are: the static torque produces static torsional deformation influencing the opening mode SIF then the crack breathing state, leading to variation of system stiffness and responses; the periodic torsional excitation changes system inputs on one hand, and affects the crack state on the other hand, resulting in changes of the dynamic responses. Furthermore, the above factors act on a nonlinear cracked rotor with stiffness coupling which will make the vibration responses in each direction more complicated.

Parameters of the rotor in Fig. 2 (a) are in Tab.1. The rotor is divided into 15 two-node Timoshenko beam elements. Each node includes 3 translational and 3 rotational degrees-of-freedom. The two discs are considered rigid bodies and simplified as three translational and three rotational inertias, which are added to the mass matrix elements at the corresponding degrees-of-freedom. Gyroscopic effect of the two discs is also included. The ball bearings are simplified as isotropic linear springs and dampers, one of which constrains the axial degree-of-freedom. The torsional degree-of-freedom in the driven disc of the rotor is also constrained and the connection misalignment is neglected.

The slant crack is in the 8th element, the ratio of crack depth to diameter of the rotating shaft is 0.2, the crack angle θ_c is **45°** , the eccentricity is 10% of the maximum static deformation of the rotor, the unbalance orientation angle is **0°** . The rotating speed is set as 1/10 of the critic speed to investigate the coupled vibration responses of the rotor at subcritical speed.

Tab. 1 Parameters of the rotor

Parameter	Value(units)	Parameter	Value(units)
Shaft length	0.75m	Young's modulus	$2.1\text{x}10^{11}$Pa
Shaft diameter	0.016m	Poisson's ratio	0.3
Disc mass	2.17Kg	Rotating speed	162r/min
Polar moment of inertia of the disc	$7\text{x}10^{-3}$Kg.m^2	Rayleigh damping coefficient (a)	1.3186
Area moment of inertia of the disc	$3.5\text{x}10^{-3}$ Kg.m^2	Rayleigh damping coefficient (b)	$1.905\text{x}10^{-4}$
Driven disc mass	0.96Kg	Bearing stiffness	$2.5\text{x}10^{5}$N/m
Polar moment of inertia of the driven disc	$1.2\text{x}10^{-3}$ Kg.m^2	Bearing damping	100Kg/s
Area moment of inertia of the driven disc	$6.33\text{x}10^{-4}$ Kg.m^2	First bending frequency	26.9Hz
Maximum static deformation of rotor	$2.66\text{x}10^{-4}$m	Second bending frequency	64.1Hz
Gravitational acceleration	9.8m/s^2	First torsional frequency	68.3Hz
Density of rotor	$7.8\text{x}10^{3}$kg/m^3	First axial frequency	1018.7Hz

The NEWMARK method is adopted to solve the system in time domain, and the initial condition is the initial static deformation of the un-cracked rotor, the NEWMARK constant is set as 0.25 and 0.5, and the convergence precision is $\mathbf{10^{-11}}$, the sampling frequency is **1000Hz**.

4.1. The influence of the static torque

When the rotor has no crack, the transmitted static torque does not affect the dynamic response of the rotor system, as the rotor appears crack, the static torque will change the extension direction of the crack, and change the stress intensity factor of the front crack, which will influence the respiratory state of the crack.

In order to analyze the influence of static torque, vibration responses of each direction of the rotor are compared, when only the static torque and gravity force are applied into the rotor. The value of the static torque is provided by the ratio between the static torque and gravity $r_{S/G}$.

Fig.3 is the torsional and vertical vibration responses of the rotor as $r_{S/G}$ are 0, 0.18 and 0.3. It shows that the transmitted static torque not only affects the static torsional deformation, but also influences the torsional and transverse vibrations of the rotor system. As to the torsional vibration, the static torsional deformation increases with the transmitted static torque, and doubling frequency components appear in the spectra, which increase first and decrease and then diminish with increasing the static torsional torque. As to the vertical vibration, with the increase of the static torque, the amplitude of fundamental frequency and double frequency components increase gradually, the triple frequency and more high frequency components increase first and decrease and then diminish. The influence of torsional excitation on transverse vibration is similar to that on vertical vibration.

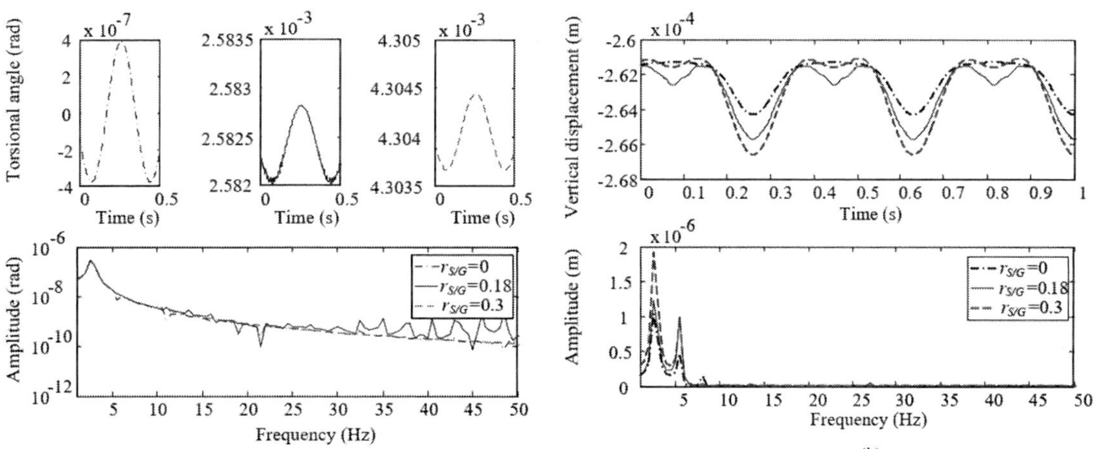

Fig.3 Vibration responses of the rotor with $r_{S/G}$ =0, 0.18, 0.3 (a) Torsional response; (b) Vertical response

In order to explain the influence of static torque on each direction vibration of the rotor, the variation of crack element stiffness is recorded as $r_{S/G}$ are 0, 0.1, 0.18 and 0.3 (see Fig.4). It shows that, with the increase of the static torque, the crack is more largely in open state, and when the static torque is not large enough, the crack will experience close to open completely, but the developing speed of the process will become faster, so the nonlinearity will be increased, leading to the increase of the higher order components. However, when the static torque increases continually, the crack will experience partially open to full open instead of full close to full open, so the crack is developing to an open crack, making the system stiffness variation smaller and smaller, and resulting in reduction in the nonlinearity of the cracked rotor system and decrease of the higher order frequency components. As the static torque exceeds some value, the crack is in full open state, the nonlinear system is developed to a linear system, as the stiffness of the cracked rotor is asymmetric, just the fundamental frequency and double frequency components are remained as well as other higher frequency components being vanished.

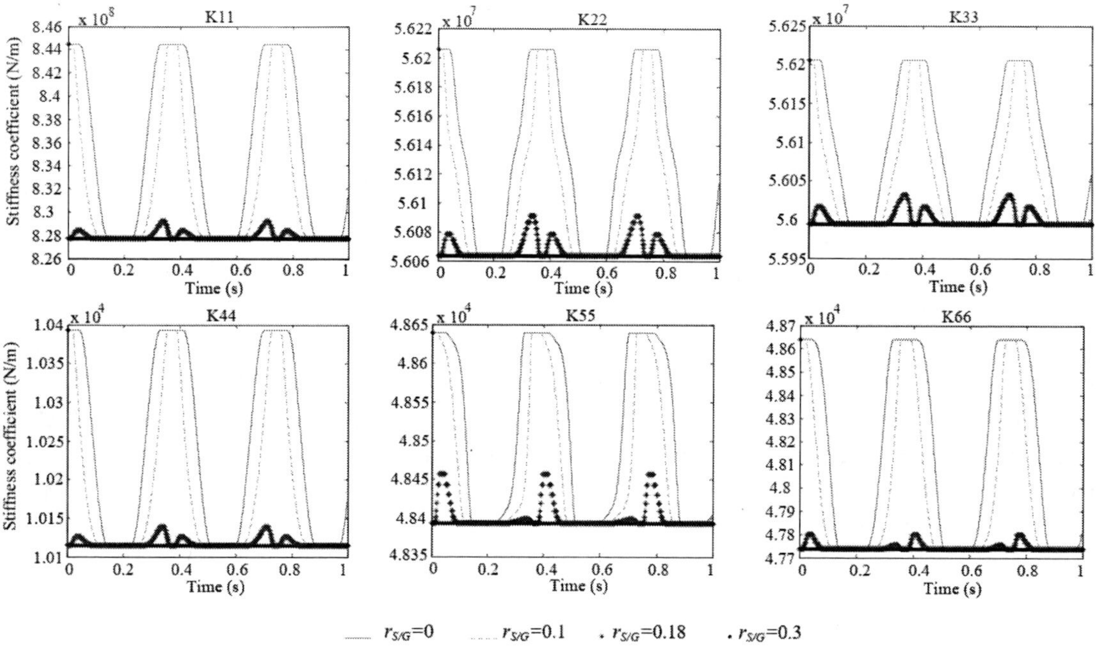

Fig.4 Main stiffness coefficients of crack element with $r_{S/G}$ =0, 0.1, 0.18, 0.3

4.2. The influence of periodic torsional excitation

The power transmission rotor may transmit periodic torque besides the static torque. Supposing the rotating disc is under periodic torsional excitation in different magnitudes, the exciting frequency (ω_T) of which is the first bending frequency (ω_b). The amplitude of the periodic torsional excitation is given by the ratio between torsional excitation to the rotor gravity $r_{P/G}$.

Fig.5 is the vibration responses in each direction of the slant cracked rotor as $r_{P/G}$**=5%, 10%, 15%** .It shows that, torsional vibration presents torsional single frequency response, and frequency components $\Omega, 2\Omega$, torsional exciting frequency ω_T and its combined frequencies $\omega_T \pm \Omega$ appear in vertical vibration response, indicating there are bending and torsional vibration coupling, and the amplitude of various frequencies increase with excitation amplitude; frequency components $\Omega, 2\Omega$ and the combined torsional frequencies $\omega_T \pm \Omega, \omega_T \pm 2\Omega$ appear in the transverse vibration response, the amplitude of various frequencies also increase with the excitation amplitude. The frequency components and change rule of the system response are in accordance with Ref. [11].

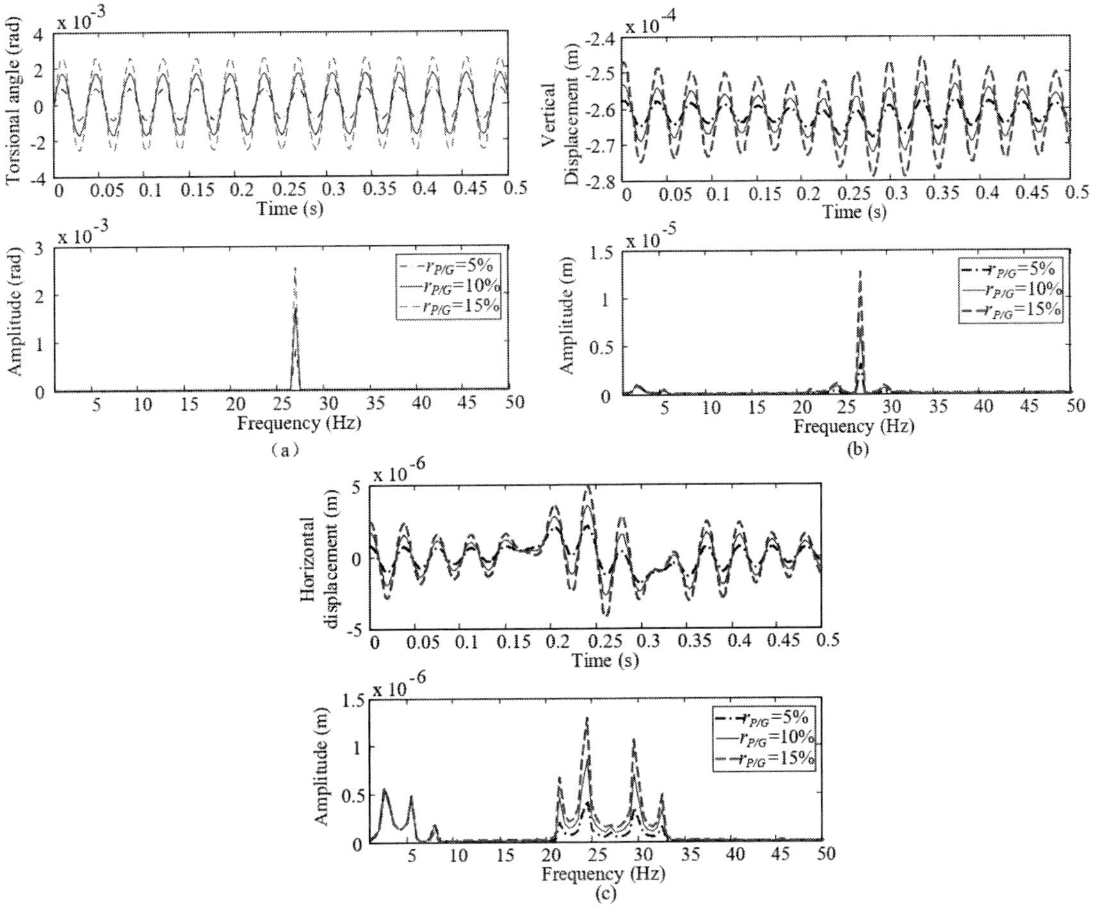

Fig.5 Vibration responses of the rotor with $r_{P/G}$=5%、10%、15% (a)Torsional response; (b) Vertical response; (c) Horizontal response

As to the torsional vibration, the torsional vibration excited by periodic torsional excitation is far greater than that by coupling, so the frequency components by coupling is submerged, and the torsional vibration shows single frequency response. As to the transverse vibration, the larger is the amplitude of the periodic torsional excitation, the greater is the influence on fluctuation of the crack stiffness. Though the crack is in the unchanged breathing rule from full open to full close, more local fluctuation in partial close state is added from open to close state of the crack, therefore, the greater is the nonlinearity of the rotor system, resulting in greater higher order frequency components and combined frequency components.

4.3. The synthesized effect

The static torque, periodic torsional excitation and eccentricity always simultaneously exist in power transmission machines, so their synthesized effect is investigated. Supposing the amplitude of the periodic torsional excitation is 10% of the transmitted static torque. Fig.6 shows the vibration responses of the rotor under synthesized action with $r_{S/G}$ **=0.1, 0.15, 0.3**.

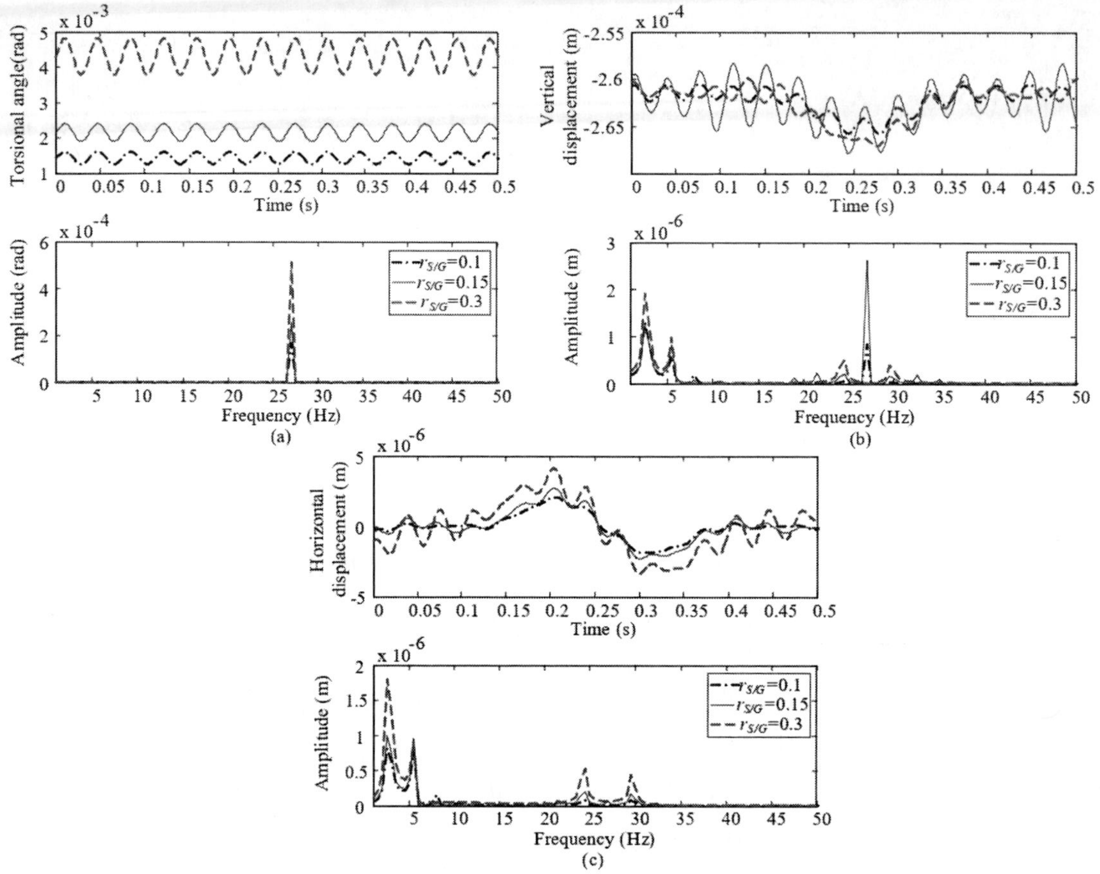

Fig.6 Vibration responses of the rotor with $r_{S/G}$=0.1, 0.15, 0.3 under comprehensive factors (a) Torsional response; (b) Vertical response; (c) Horizontal response

It can be seen from Fig.6 that, the torsional vibration mainly represents the frequency component of periodic torsional excitation, and the amplitude of which increases with the static torque while the double frequency components is submerged. The frequency component Ω and combined frequency components $\omega_T \pm \Omega$ in vertical and transverse responses all increase with the static torque; the frequency components $2\Omega, 3\Omega$ and ω_T increase first and then decrease with the static torque. However the frequency component ω_T in vertical vibration increase with the periodic torsional excitation when there is no static torque (see Fig. 5), which indicates that the static torque changes the breathing state of the crack, leading to variation of the nonlinearity and coupling behaviour of the rotor.

Hence, as the transmitted static torque will make the crack open and change the nonlinearity of the system, the crack detection can be carried out by loading and unloading the power transmission machines first, and then testing the vibration characteristics before and after load change, furthermore, the load should be used as an important monitoring parameter in condition monitoring on rotor systems.

5. Conclusions

By studying the influence of the static torque and periodic torsional excitation on the transverse and torsional dynamic response of an unbalanced rotor with breathing slant crack, conclusions are can be drawn as follows:

(1) Static torque has great effect on vibration responses of each direction of breathing slant cracked rotor. Static torque makes the breathing crack become open crack, and the nonlinearity of the cracked rotor increase first and then decrease with the static torque. Thus the influence of static toque should be considered in crack detection.
(2) Owing to the coupling of bending and torsional vibration of slant cracked rotor, the torsional excitation frequency and the combined frequency components of rotating frequency will appear in transverse vibration response of the rotor under the action of periodic torsional excitation, and the higher the exciting amplitude is, the larger the combined frequency components are.
(3) As the transmitted static torque will make the crack in normally open state and change the nonlinearity of the system, the crack detection can be carried out by loading and unloading the power transmission, and comparing the vibration characteristics before and after load change; furthermore, the load should be used as an important monitoring parameter in condition monitoring on rotor systems.

References

[1] Bovsunovsky A, Surace C 2015. Non-linearities in the vibrations of elastic structures with a closing crack: A state of the art review. *Mech. Syst. Signal. Pro.*, **62**: 129-148.

[2] Kumar C, Rastogi V 2009. A Brief Review on Dynamics of a Cracked Rotor. *Int. J. Rotating Machinery*, **2009**: 1-6.

[3] Papadopoulos C A 2008. The strain energy release approach for modeling cracks in rotors: A state of the art review. *Mech. Syst. Signal. Pro.*, 22(**4**): 763-789.

[4] Friswell M I, Penny J E 2002. Crack modeling for structural health monitoring. *Struct. Health Monit.*, 1(**2**): 139-148.

[5] Giannopoulos G I, Georgantzinos S K, Anifantis N K 2015. Coupled vibration response of a shaft with a breathing crack. *J. Sound Vib.*, **336**: 191-206.

[6] Bachschmid N, Pennacchi P, Tanzi E 2008. Some remarks on breathing mechanism, on non-linear effects and on slant and helicoidal cracks. *Mech. Syst. Signal. Pro.*, 22(**4**): 879-904.

[7] Kulesza Z, Sawicki J T 2012. Rigid finite element model of a cracked rotor. *J. Sound Vib.*, 331(**18**): 4145-4169.

[8] Jun O S, Eun H J, Earmme Y Y, et al 1992. Modelling and vibration analysis of a simple rotor with a breathing crack. *J. Sound Vib.*, 155(**2**): 273-290.

[9] Sinou J J, Lees A W 2007. A non-linear study of a cracked rotor. *Eur. J. Mech - A/Solid*, 26(**1**): 152-170.

[10] Sekhar A S, Prabhu B S 1994. Transient analysis of a cracked rotor passing through critical speed. *J. Sound Vib.*, 173(**3**): 415-421.

[11] Darpe A K 2007. Coupled vibrations of a rotor with slant crack. *J. Sound Vib.*, 305(**1-2**): 172-193.

[12] Darpe A K, Gupta K, Chawla A 2004. Coupled bending, longitudinal and torsional vibrations of a cracked rotor. *J. Sound Vib.*, 269(**1-2**): 33-60.

[13] Tada H, Paris P C, Irwin G R 2000. *The Stress Analysis of Cracks Handbook, Third Edition*. Pennsylvania: Paris Productions.

[14] Yuan Z, Chu F, Lin Y 2007. External and internal coupling effects of rotor's bending and torsional vibrations under unbalances. *J. Sound Vib.*. 299(**1**), 339-347.

[15] Newmark N M 1959. A method of computation for structural dynamics. *J. Eng. Mech. Division*, 85(**3**), 67-94.

Review on structural damage assessment via transmissibility with vibration based measurements

Yun-Lai Zhou[1], Cao Hongyou[2], Ni Zhen[1], Magd Abdel Wahab[3]

[1]Department of Civil and Environmental Engineering, National University of Singapore, 2 Engineering Drive 2, Singapore 117576
[2]School of civil engineering and architecture, Wuhan University of Technology, Wuhan 430070, China
[3]Soete Laboratory, Faculty of Engineering and Architecture, Ghent University, Technologiepark Zwijnaarde 903, B-9052 Zwijnaarde, Belgium

Email: zhouyunlai168168@gmail.com

Abstract. In this study, transmissibility based damage assessment techniques with vibration measurement are reviewed with highlighting the recent advancements since damage might induce severe changes and cause huge economic losses in both civil and mechanical engineering structures. In recent years, transmissibility underwent booming and divergent application for damage assessment both in experimental model and engineering application, and this review provides a fundamental understanding for transmissibility based damage assessment by summarizing those research outputs, which can serve as useful reference for further investigations.

1. Introduction

Damage assessment, as the main issue in structural health monitoring (SHM), underwent more than centuries since maintenance and repairing were indispensable after the long-term serviceability of almost all constructed structures such as buildings and bridges. However, from the wide use of civil infrastructures like oil pipeline, long span bridge, tall buildings and so on, and mechanical structures like rotational machines, damage assessment has become increasingly essential as the failure of such large scale or highly integrated structures may lead to huge economic loss. During last decades, damage assessment underwent booming development, thus, leading to abundant research outputs with applications in engineering including in time and frequency domain techniques, empirical and model-based approaches. Critical reviews on vibration based damage identification can be found in [1, 2], where vibration based damage identification techniques were summarized in a thorough perspective [1], while the review focused on identifying damages in composite structures with vibration testing [2].

For SHM/damage assessment, damage is the key and fundamental conception that should be clearly defined, which includes deterioration and degradation, and so on. A widely accepted general definition is given by Sohn et al. [1] as "*changes introduced into a system that adversely affect its current or future performance. Implicit in this definition is the concept that damage is not meaningful without a comparison between two different states of the system, one of which is assumed to represent the initial, and often undamaged, state.*" In [2], the authors summarized the different damage types like stiffness change, boundary condition change, mass change or connectivity change. Then the key idea for damage identification would be seeking the difference between undamaged pattern and damaged pattern often via some structural characteristics like mechanical, electrical, or magnetic properties, and thus predicting the occurrence of possible damage. From this perspective, those newly developed techniques like eddy current, ultrasonic testing, and X ray, are applied.

1.1. Schemes of damage identification

In SHM, damage identification/assessment is the key issue [3-10]. And for damage identification, the damage state of a system can be described as a five-step process along the lines of the process discussed in [11]. The damage state is described by answering the following questions [1, 11]: (1) Is there damage in the system (existence)? (2) Where is the damage in the system (location)? (3) What kind of damage is present (type)? (4) How severe is the damage (extent)? (5) How much useful life remains (prognosis)?

In [3], the authors summarized damage identification methods into two categories, namely physical model based, and statistical/data model based, where physical model based approaches aim to detect defects/deteriorations via the difference between real structural response and its numerical prediction, where finite element (FE) model requires well understanding of the physical model as prior. As finite element analysis (FEA) should be well conducted, it may require experience and skills. Finite element analysis has been widely used in the literature for many engineering applications that have been recently published [12-30] . The woSystem identification can also applied to extract the structural dynamic characteristics. FE model updating is extended to damage identification while initially designed for optimize the FE model. As one can see, the demanding of physical model immediately sets an overwhelming conundrum and largely restricts its potential application, since in actual engineering the obtaining of exact physical model constantly encounters suffering from boundary conditions, loads, environmental parameters and so on. In addition, the FE model for large-scale structures will require high cost in computation, which also implies the necessity in seeking cost-effective approaches for damage assessment. In such approaches, techniques involve ARMA family models in time domain, dynamic stiffness, and algorithms like Radial Basis Function (RBF), BAT algorithm [6]. Note that damage not only means the damage in reinforced concrete structures, bridges, composite beams, but also damage in fatigue-induced damage.

On the other hand, statistical/data model based methods try to unveil the damages according to the response analysis utilizing pattern recognition, machine learning and so on, which will avoid the measurement of structural excitations, and thus simplifying the damage identification procedure. In [1], the authors described damage assessment problem as one of statistical pattern recognition paradigm, which consists of a four-part process: (1) Operational Evaluation; (2) Data Acquisition, Fusion, and Cleansing; (3) Feature Extraction and Information Condensation; (4) Statistical Mode Development for Feature Discrimination. Note that the key step is to extract a proper feature to interpret the corresponding damage. One feature might be function of only one kind, or also function of several kinds of damage. Features like modal frequencies, damping, and so on are commonly used in the previous investigation for SHM. Pattern recognition methodologies are then adopted to better discriminate the damaged patterns from the undamaged ones. Machine learning algorithms, discriminant analysis, pattern recognition and so on have been introduced into damage identification for facilitating SHM analysis particularly in large-scale or complex structures.

1.2. Significance of this study

This study tries to illustrate the main developments of damage assessment techniques via the adoption of transmissibility in the past years, and thus to give a valuable reference for further investigation. Conventional techniques may require the excitation to be white noise, or demand the measurement of excitation, while the use of transmissibility solved these problems. And booming and diverge developments might be better to be summarized for several categories in order to give a clear and efficient understanding.

2. Damage assessment via transmissibility

For vibration based damage assessment techniques, modal testing is the commonest tool, which might be divided into Experimental Modal Analysis (EMA) and Operational Modal Analysis (OMA). The key difference is that EMA normally takes the testing object into laboratory and apply excitation, and then conduct modal analysis, while OMA directly measures the dynamic response during operating condition of the tested object, and tries to unveil the structural characteristics from the measured data. The advantage of OMA in comparison with EMA is that there is no restriction to the testing object. This gives the possibility in applying the technique to large scale structures like wind blades and suspension bridges. Another merit of OMA is that no necessity is required for the excitation. And transmissibility based technique is a typical one based on OMA.

Transmissibility, a conception raised decades ago, has been applied in damage assessment during the past years. Some general reviews can be found in [31-33] and a detailed review is given in [34]. Transmissibility definition is thoroughly reviewed and discussed in [31], and application of transmissibility in force identification is illustrated in [32]. The application of transmissibility in damage assessment is generally reviewed in [3, 33], while a more general review of transmissibility referring to theoretical development and engineering application is given in [34]. Conventional damage assessment techniques via transmissibility can be found in [35-40], while from 2010, transmissibility has been increasingly widely applied in damage assessment and will be illustrated hereinafter.

2.1. Transmissibility definition.

A general definition for transmissibility is the ratio between two structural dynamic responses, which can be indicated as:

$$T_{(i,j)} = \frac{X_i}{X_j} \tag{1}$$

where i, j means the nodes of two outputs, and X_i, X_j mean the frequency spectrum of dynamic response x_i, and x_j. Note transmissibility is also extended into nonlinear part holding the same philosophy [41].

Similar to coherence in modal analysis, transmissibility coherence (TC) [34, 42] is defined as:

$$TC_{(i,j)} = \left| \frac{G_{ij}^2}{G_{ii} G_{jj}} \right| \tag{2}$$

where G means the cross- or auto- spectrum.

2.2. Damage assessment with transmissibility.

To all transmissibility-based techniques, the net change can be the easiest one in identifying damage [36], and later transmissibility is extended with curvature by analog with FRF curvature [43]. The net transmissibility change is considered in percentage with respect to the undamaged baseline with accumulating in a specific frequency range [33]. Outlier analysis is also introduced in [44]. Neural networks are another alternative, which try to predict damage after being trained by taking transmissibility as a feature [38]. Autoregressive model is also introduced and control chart and factor analysis is delivered [45]. Euclidean distance and cosine similarity measure are introduced to detect the damages in steel platforms [46]. A detection methodology combining transmissibility with Mahalanobis distance and principal component analysis (PCA) are applied in damage localization in real aircraft wing [47].

Unlike pattern recognition algorithms, an indicator might be easier and simpler in adoption. From this idea, modal assurance criterion (MAC) is introduced for damage detection and damage localization and damage quantification [48]. This idea also extends to a specific frequency range of transmissibility since some parts might be contaminated by noise [49]. Since the key idea of MAC is to estimate the similarity between two vectors, it can also be applied to transmissibility extended vectors, like inverse subtraction transmissibility function [50], transmissibility coherence [3, 34, 42, 51]. For instance, in [2, 42], the authors considered that the peaks of transmissibility subtraction were in consistence with the poles of structures, vectors were constructed around the peaks with all transmissibilities of the structural system, and later MAC was applied to identify the damages. Similar idea is extended to transmissibility coherence [51, 52].

For damage assessment, since the key idea is to unveil the difference between undamaged and damaged conditions, distance measure is the commonest way to be considered [8, 53]. Mahalanobis distance and Euclidean distance are compared for transmissibility and FRF, with drawing out that Mahalanobis distance has more noise tolerance than Euclidean distance [54]. Cosine distance is adopted after condensing the transmissibility with PCA, with comparing with distance measure [55]. In they compared city block distance, Chebyshev distance, Minkowski distance, Mahalanobis distance with taking PCA in condensing the transmissibility. Transmissibility based damage assessment has also considered in nonlinear part [56], metro tunnel structure [57], bridges [58], and so on. Certainly other transform like wavelet transform is also discussed [59].

3. Conclusions

This study illustrated the development of transmissibility based damage assessment techniques and also generally discussed the conventional vibration based damage assessment techniques, which may give a valuable reference for further investigation. Conventional techniques may require the excitation to be white noise, or demand the measurement of excitation, while the use of transmissibility solved these problems. And booming and diverge developments were summarized for several categories in order to give a clear and efficient understanding.

Acknowledgement

The first author thanks CWO (Commissie Wetenschappelijk Onderzoek), Faculty of Engineering and Architecture, Ghent University for providing financial support for a research stay at Soete Laboratory.

References

[1] Sohn H, Farrar CR, Hemez FM, Shunk DD, Stinemates SW, Nadler BR and Czarnecki JJ 2004 *Technical Report LA-13976-MS* Los Alamos National Laboratory

[2] Montalvao D, Maia NMM and Ribeiro AMR 2006 *The Shock and Vibration Digest* **38**(4) 295-324

[3] Zhou Y-L, Maia N and Abdel Wahab M 2016 *Journal of Vibration and Control* doi: 10.1177/1077546316674544

[4] Gillich G-R, Praisach Z-I, Abdel Wahab M, Gillich N, Mituletu IC and Nitescu C 2016 *Shock and Vibration* **2016**(Article ID 2086274) 10 pages; http://dx.doi.org/10.1155/2016/2086274

[5] Khatir S, Belaidi I, Serra R, Abdel Wahab M and Khatir T 2015 *Mechanika* **21**(6) 472-479

[6] Khatir S, Belaidi I, Serra R, Abdel Wahab M and Khatir T 2016 *Journal of Vibroengineering* **18**(1) 202-213

[7] Zhou Y-L, Maia NMM, Sampaio R and Wahab MA 2016 *Structural health monitoring* **DOI: https://doi.org/10.1177/1475921716680849**

[8] Zhou Y-L and Abdel Wahab M 2016 *Journal of Vibroengineering* **18**(7) 4491-4499

[9] Zhou Y-L and Abdel Wahab M 2017 *Engineering Structures* **141** 175-183

[10] Khatir A, Tehami M, Khatir S and Abdel Wahab M 2016 *Journal of Vibroengineering* **18**(8) 5063-5073

[11] Rytter A 1993 *PhD thesis*

[12] X. Nguyen H, N. Nguyen T, Abdel Wahab M, Bordas SPA, Nguyen-Xuan H and P. Voa T 2017 *Computer Methods in Applied Mechanics and Engineering* **313** 904-940

[13] Tran Vinh L, Lee J, Nguyen-Van H, Nguyen-Xuan H and Abdel Wahab M 2015 *International Journal of Non-Linear Mechanics* **72** 42-52

[14] Tran LV, Phung-Van P, Lee J, Wahab MA and Nguyen-Xuan H 2016 *Composite Structures* **140** 655-667

[15] Thai CH, Ferreira AJM, Abdel Wahab M and Nguyen-Xuan H 2016 *Acta Mechanica* **227**(5) 1225-1250

[16] Thai C, Zenkour AM, Abdel Wahab M and Nguyen-Xuan H 2016 *Composite Structures* **139** 77-95

[17] Phung-Van P, Tran LV, Ferreira AJM, Nguyen-Xuan H and Abdel-Wahab M 2016 *Nonlinear Dynamics* 1-16; doi:10.1007/s11071-11016-13085-11076

[18] Phung Van P, Nguyen LB, Tran Vinh L, Dinh TD, Thai CH, Bordas SPA, Abdel Wahab M and Nguyen-Xuan H 2015 *International Journal of Non-Linear Mechanics* **76** 190-202

[19] Phung Van P, De Lorenzis L, Thai CH, Abdel Wahab M and Nguyen-Xuan H 2015 *Computational Materials Science* **96** 495-505

[20] Phung Van P, Abdel Wahab M, Liew KM, Bordas SPA and Nguyen-Xuan H 2015 *Composite Structures* **123** 137-149

[21] Yue T and Abdel Wahab M 2017 *Tribology International* **107** 274-282

[22] Yue T and Abdel Wahab M 2016 *Materials* **9** 597; doi:510.3390/ma9070597

[23] Resende Pereira KdF, Bordas S, Tomar S, Trobec R, Depolli M, Kosec G and Abdel Wahab M 2016 *Materials* **9** 639; doi:610.3390/ma9080639

[24] Noda N-A, Chen X, Sano Y, Wahab MA, Maruyama H, Fujisawa R and Takase Y 2016 *MATERIALS & DESIGN* **96** 476-489

[25] Ferjaoui A, Yue T, Abdel Wahab M and Hojjati-Talemi R 2015 *International Journal of Fatigue* **73** 66-76

[26] Tran Vinh L, Lee J, Ly HA, Abdel Wahab M and Nguyen-Xuan H 2015 *International Journal of Mechanical Sciences* **96-97** 65-78

[27] Yue T and Abdel Wahab M 2016 *Materials* **9**(7)

[28] Junyan Ni and Wahab MA 2017 *Computers & Structures* **186** 35–49

[29] Phung-Van P, Qui LX, Nguyen-Xuan H and Wahab MA 2017 *Composite Structures* **166** 120–135

[30] Phung-Van P, Ferreira AJM, Nguyen-Xuan H and Abdel Wahab M 2017 *Composites Part B: Engineering* **118** 125-134

[31] Maia NMM, Urgueira APV and Almeida RAB 2011 *Vibration Analysis and Control - New Trends and Developments* ISBN 978-953-307-433-977, 364 pages, Publisher: InTech. pp. 197-216.

[32] Maia NMM, Lage YE and Neves MM 2012 *Advances in Vibration Engineering and Structural Dynamics* ISBN 978-953-951-0845-0840, 0378 pages, Publisher: InTech, pp.0103-0132.

[33] Chesne S and Deraemaeker A 2013 *Mechanical Systems and Signal Processing* **38**(2) 569-584

[34] Zhou YL 2015 *PhD thesis*

[35] Worden K 1997 *Journal of Sound and Vibration* **201**(1) 85-101

[36] Sampaio RPC, Maia NMM, Ribeiro AMR and Silva JMM 1999 *Proceedings of six international congress on sound and vibration* Copenhagen, Denmark, 5-8 July 1999.

[37] Johnson TJ and Adams DE 2002 *Transactions of the ASME* **124** 634-641

[38] Chen Q, Chan YW and Worden K 2003 *Computers & Structures* **81**(22-23) 2165-2172

[39] Allarangaye MD, Traore O, Traore EVS, Millogo RJ and Konate G 2006 *Journal of Plant Pathology* **88**(3) 309-315

[40] Kess HR and Adams DE 2007 *Mechanical Systems and Signal Processing* **21**(6) 2394-2405

[41] Lang ZQ, Park G, Farrar CR, Todd M, Mao Z, Zhao L and Worden K 2011 *International Journal of Non-Linear Mechanics* **46** 841-853

[42] Zhou YL, Figueiredo E, Maia N and Perera R 2015 *Shock and Vibration*

[43] Sampaio RPC, Maia NMM, Silva JMM and Ribeiro AMR 2000 *Proceedings of European COST F3 Conference on System Identification and Structural Health Monitoring* Madrid, Spain, June 2000.

[44] Worden K, Manson G and Fieller NRJ 2000 *Journal of Sound and Vibration* **229**(3) 647-667

[45] Kullaa J and Heine T 2008 *Shock and Vibration* **15**(3-4) 207-215

[46] Surace C and Worden K 2010 *Mechanical Systems and Signal Processing* **24** 1114-1128

[47] Papatheou E, Manson G, Barthorpe RJ and Worden K 2010 *Journal of Sound and Vibration* **329** 2349-2366
[48] Maia NMM, Almeida RAB, Urgueira APV and Sampaio RPC 2011 *Mechanical Systems and Signal Processing* **25**(7) 2475-2483
[49] Zhou Y-L, Wahab MA and Perera R 2015 *International Journal of Fracture Fatigue and Wear* **3** 254-259
[50] Zhou YL, Perera R and Sevillano E 2012 *Proceedings of the 6th European Workshop on Structural Health Monitoring* Dresden, Germany, July 2012
[51] Zhou Y-L, Yang XB and Wahab MA 2016 *International Journal of Fracture Fatigue and Wear* **4** 200-203
[52] Zhou Y-L, Figueiredo E, Maia NMM and Perera R 2015 *Proceedings of the International Conference on Structural Engineering Dynamics (ICEDyn 2015)* Lagos, Algarve, Portugal, June 2015
[53] Zhou Y-L, Figueiredo E, Maia NMM, Sampaio R and Perera R 2015 *Proceedings of ISMA 2014 - International Conference on Noise and Vibration Engineering and USD 2014 - International Conference on Uncertainty in Structural Dynamics (ISMA 2014)* Leuven, Belgium, September 2014
[54] Zhou Y-L, Figueiredo E, Maia NMM, Sampaio R and Perera R 2015 *Structural Control and Health Monitoring* **22** 1209-1222
[55] Zhou Y-L, Maia NMM, Sampaio R, Qian XD and Wahab MA 2016 Proceedings of the 2016 Leuven Conference on Noise and Vibration Engineering (ISMA 2016) ed editors)
[56] Lang ZQ, Park G, Farrar CR, Todd MD, Mao Z, Zhao L and Worden K 2011 *International Journal of Non-Linear Mechanics* **46**(6) 841-853
[57] Feng L, Yi XH, Zhu DP, Xie XY and Wang Y 2015 *Mechanical Systems and Signal Processing* **60-61** 59-74
[58] Li J and Hao H. Damage Detection of Shear Connectors Based on Power Spectral Density Transmissibility. In: Basu B, editor. Damage Assessment of Structures X, Pts 1 and 2. Key Engineering Materials. 569-5702013. p. 1241-1248.
[59] Fan Z, Feng X and Zhou J 2013 *Smart Structures and Systems* **12**(3-4) 291-308

IOP Conf. Series: Journal of Physics: Conf. Series **842** (2017) 012017 doi:10.1088/1742-6596/842/1/012017

Damages detection in cylindrical metallic specimens by means of statistical baseline models and updated daily temperature profiles

Rodolfo Villamizar-Mejia[1], Luis-Eduardo Mujica-Delgado[2], Magda-Liliana Ruiz-Ordóñez[2], Jhonatan Camacho-Navarro[2], Gustavo Moreno-Beltrán[1]

[1]Universidad Industrial de Santander (UIS), Escuela de Ingenierías Eléctrica, Electrónica y de Telecomunicaciones (E3T), Bucaramanga, Colombia
[2]Universitat Politécnica de Catalunya (UPC) BARCELONATECH, Department of Mathematics, Escola d'Enginyeria de Barcelona Est. (EEBE), Campus Diagonal Besòs, Edifici A, C. Eduard Maristany, 10-14, 08019, Barcelona, Barcelona, Spain

rovillam@uis.edu.co

Abstract. In previous works, damage detection of metallic specimens exposed to temperature changes has been achieved by using a statistical baseline model based on Principal Component Analysis (PCA), piezodiagnostics principle and taking into account temperature effect by augmenting the baseline model or by using several baseline models according to the current temperature. In this paper a new approach is presented, where damage detection is based in a new index that combine Q and T^2 statistical indices with current temperature measurements. Experimental tests were achieved in a carbon-steel pipe of 1m length and 1.5 inches diameter, instrumented with piezodevices acting as actuators or sensors. A PCA baseline model was obtained to a temperature of 21° and then T^2 and Q statistical indices were obtained for a 24h temperature profile. Also, mass adding at different points of pipe between sensor and actuator was used as damage. By using the combined index the temperature contribution can be separated and a better differentiation of damages respect to undamaged cases can be graphically obtained.

1. Introduction

Early damage detection in metallic structures is of main interest for different type of industries such as aviation, hydrocarbon, aeronautic, among others, in order to avoid potential future accidents **[1]**. Several approaches have been reported in literature for structural health monitoring (SHM) of metallic structures by using several techniques such as optical, vibration, guide waves **[2]-[6]**. In recent years, the use of guided waves for damage assessment in pipeline structures has been reported as successful **[7]**. Thus, by taking advantage of piezoelectric properties and guided waves, piezoelectric PZT devices are a cheaper technology with promising results for pipeline damage assessment **[8]**. In addition, the effectiveness of principal component analysis for structural damage detection algorithms has been demonstrated (for example in pipeline crack detection algorithms **[9]**). Although piezoelectric (PZT) devices have been reported as a promising technique to detect damages in metallic structures such as aircraft wings, pipelines and train rails, some limitations are presented during the travel of the generated wave, for example external disturbances such as temperature changes.

Some structural damage detection methodologies that account temperature variations have been reported in literature **[10]**, **[11]**, **[12]**, where for example several or augmented models are created for different temperature levels. However, they have taken into account temperature changes in steady state but not dynamic response of the wave travel during the transient changes or along a daily temperature profile.

By taking into account the above exposed limitation, this work proposes to detect damages in metallic cylindrical structures by combining environmental temperature measurements with statistical indices T^2 and Q obtained from a statistical baseline model for a reference temperature. First, a baseline statistical model is obtained by using Principal Component Analysis at a reference environmental temperature based on PZT responses at the end of a pipe section when a wave is generated by PZTs at the other end. Then, PZT responses are recorded during one or more days where temperature changes are presented. Finally, damages are emulated consisting of mass adding at different points of structure between actuation and sensing points. A first order model that relates statistical indices T^2 and Q with temperature is obtained, what is used to estimate undamaged statistical indices. They are subtracted from the current indices and the temperature effect is then compensated.

2. Damage Detection Methodology

The proposed methodology consists of detecting structural damages in metallic cylindrical specimens based on guided waves generated by PZTs and statistical indices obtained from a statistical PCA baseline model. Updated T^2 and Q statistical indices are computed by using both current and estimated, where estimation is based on current environmental temperature measurements. Figure 1 presents the main steps followed by the proposed methodology.

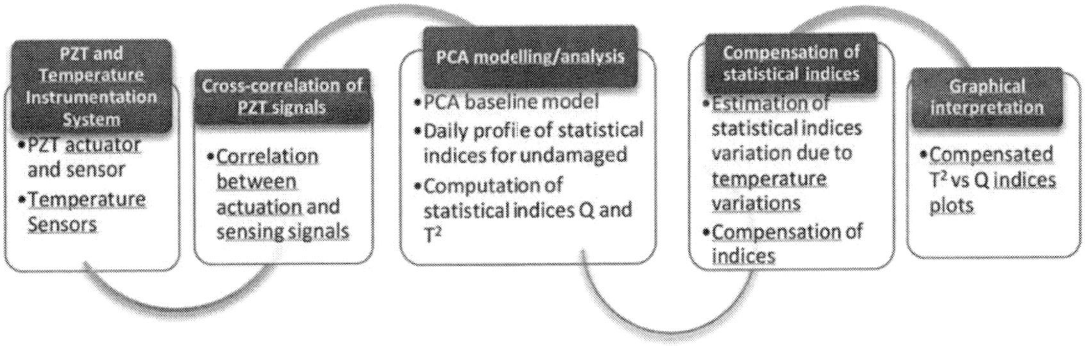

Figure 1. Damages detection methodology

2.1. *PZT and Temperature Instrumentation System*

PZTs should be attached to the monitored structure such that at least one (1) should be actuator and at least one (1) as sensor. Actuator(s) should be located at the end of the cylindrical specimen and sensor(s) at the other end such that any damage between both of them can be detected. At least one temperature sensor should be installed near to the structure (guide wave path should not be affected) to record environmental temperature at each experiment. Additionally, a signal recorder with capacity of storing 300 experiments (at least two signals recorded during at least 2 ms at a frequency greater than 1MHz, repeated approximately each 8s during at least 24 hours.

2.2. *Cross-correlation of actuating and sensing PZT signals*

The cross correlation analysis is used to exclude common external noise signals and then a better separation boundaries for damage conditions is obtained. Thus, actuation and sensing piezo-signals are correlated previously to apply the principal component analysis (Figure 2). The cross-correlation function between two signals $X(t)$ and $Y(t)$ is defined as:

$$R_{XY}(t,t+\tau) = \lim_{N\to\infty}\frac{1}{N}\sum_{k=1}^{N} X_k(t)Y_k(t+\tau) \quad (1)$$

Where N is the number of signal samples and τ is the lag time interval used to compute the cross-correlation function.

2.3. PCA modelling and analysis

PCA is commonly used to obtain a reduced space representation for multidimensional, however in this approach it is used into two situations: Obtaining a statistical baseline model (modeling) and computing T^2 and Q statistical indices (validation). According to **[16]** PCA follow the next general procedure:

i. Normalize the original variables X by means of standard deviations ($\hat{\sigma}_j$) and mean values ($\hat{\mu}_j$)

$$\hat{X} = \frac{X-\hat{\mu}_j}{\hat{\sigma}_j} \quad (2)$$

Where $\hat{X}$ corresponds to normalized data.

ii. Compute the covariance matrix of centered data matrix.

$$C_{\hat{X}} = \frac{1}{m-1}\hat{X}'\hat{X} \quad (3)$$

iii. Estimate the singular value decomposition for covariance matrix.

$$C_{\hat{X}} = \varphi\Sigma\varphi' \quad (4)$$

iv. Transform the original variables onto the orthogonal space defined by the Eigenvectors (φ) (loading matrix) of the covariance matrix.

$$Z = \varphi \cdot \hat{X} \quad (5)$$

v. Keep only the first r components in order to obtain a reduced representation for original variables. The variance for each new variable corresponds to their respective r eigenvalues.

$$Z_r = \varphi_r \cdot \hat{X} \quad (6)$$

Statistical baseline model corresponds to the loading matrix or eigenvectors obtained in iii. T2 and Q statistics are used to measure deviations of each experiment respect to the PCA baseline model (detection of damages), which are defined by equations 7 and 8.

$$T^2 = X^T\varphi_{und}(\Sigma^T\Sigma)^{-1}(\varphi_{und.})^T X \quad (7)$$

$$Q_i = \tilde{x}_i^T\tilde{x}_i, \quad \tilde{x}_i = [I - \varphi_{und} * (\varphi_{und})^T]X_i \quad (8)$$

Where $\tilde{x}_i$ is the residual projection for each experiment. T^2 statistics index is a variation measurement of each experiment within the statistical PCA model (undamaged model) and the Q statistics index is the squared 2-norm that measures deviations of the experiment respect to the lower-dimensional PCA representation **[20]**.

2.3.1. Modelling. Two models are obtained in this step:

i) Statistical baseline model. It consists of obtaining the loading matrix φ (Equation 4), by using 300 repeated experiments at a reference environmental temperature (T_0), for undamaged structure (no mass added). A $n \times m$ measurements matrix is organized, where m is the number of experiments (300) and n the samples number of correlated signal, which is used to obtain the mentioned loading matrix.

ii) $\Delta Q/\Delta T$ and $\Delta T^2/\Delta T$ transfer functions. They relate statistical indices variations ($\Delta Q = Q_k - Q_0$) and $\Delta T^2 = T^2_k - T^2_0$) due to environmental temperature variations ($\Delta T=T_k-T_0$). For this purpose T^2 and Q statistical indices are computed by using the baseline model and from experiments periodically achieved (each 8s) during at least 24 hours for the undamaged structure and environmental temperature. Based on the daily indices and temperature profiles, process identification technique is used to obtain the mentioned transfer functions parameters. Then, by using discrete transformation of continuous system, a discrete linear equation that computes statistical indices variation at sample k+1 in function of index at instant k and environmental temperature variation at instant k.

2.3.2. Validation. During validation, this stage consists of: i) Obtaining the Q and T^2 indices by projecting the correlated signal of current experiment on the principal component space by means of statistical baseline model and then to compute the mentioned indices, according to PCA analysis. ii) Computing the variation of the statistical indices due to the environmental temperature variation by using the obtained discrete linear equation.

2.4. Compensation of statistical indices

This step is used during the damage detection. It consists of subtracting the indices variation previously computed from the computed T^2 and Q indices. Thus a new compensated statistical indice are obtained to distinguish between undamaged and damage conditions of monitored structure.

3. Experimental Results

3.1. Experimental Setup

Experimental tests were achieved on a metallic cylindrical specimen consistent of a carbon-steel pipe of 1m length, 1.5 inches diameter and 0.003 m thickness (see figure 2). A PZT as actuator is attached at one end of the pipe while another one as sensor at the other end. A metallic piece that can be attached and displaced along the pipe section is used to simulate mass adding at different distances respect to the sensor (Figure 3 depicts the simulated damages). The experiment consisted of applying a 9-pulse 100 KHz burst signal from the actuator and to record the travelled wave at the opposite end during 2ms at a sample rate of 3.47MHz. Also, environmental temperature is recorded for each experiment by means of two (2) LM35 sensors.

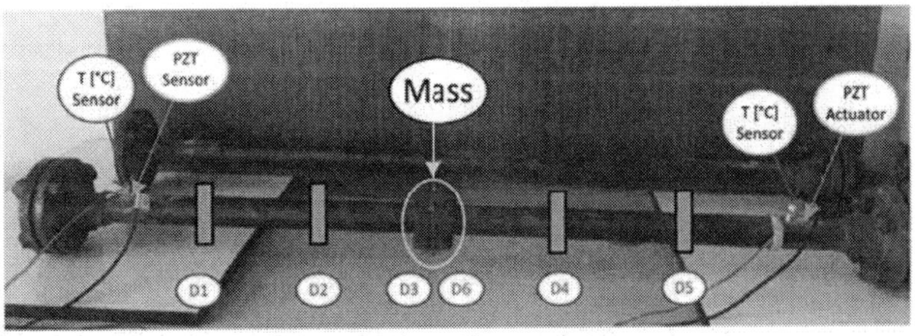

Figure 2. Experimental set-up

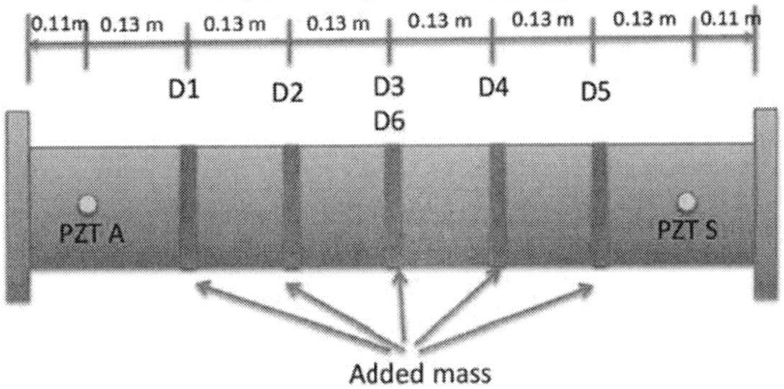

Figure 3.Graphical description of simulated damages

3.2. Set of experiments

Three sets of experiments were achieved.

i) Obtaining of statistical baseline model. A set of 300 repetitions of the experiment was achieved at a reference environmental temperature of 23.3°C and for the undamaged structure. Stages 2.2 to 2.3.1 were achieved and loading matrix or statistical baseline model was obtained.
ii) Statistical indices and environmental temperature daily profile. Experiments were achieved each 8.84s during 24 hours for the undamaged condition (no mass addition), thus vectors of 9763 elements for Q, T^2 and T were obtained and are presented in figure 4. Based on these curves, step 2.3.1 to obtain two first order transfer functions (*$\Delta Q(s)/\Delta T(s)$ and $\Delta T^2(s)/\Delta T(s)$*) of the form:

$$G(s) = \frac{K}{\tau s + 1} \tag{9}$$

Where parameters for $G_Q(s) = \frac{\Delta Q(s)}{\Delta T(s)} = \frac{k_Q}{\tau_Q s+1}$ are $k_Q = 1.4$, $\tau_Q = 4450$ s and for $G_{T^2}(s) = \frac{\Delta T^2(s)}{\Delta T(s)} = \frac{k_{T2}}{\tau_{T^2}s+1}$ are $k_{T^2} = 3500$, $\tau_{T^2} = 2450$ s. The differential equation resulting from the first order transfer function is of the form:

$$\tau_Q \Delta\dot{Q}(t) + \Delta Q(t) = k_Q \Delta T(t) \; ; \; \tau_{T^2} \Delta\dot{T}^2(t) + \Delta T^2(t) = k_{T^2} \Delta T(t) \tag{10}$$

The equivalent discrete representation of the continuous firs order transfer function is of the form:

$$\Delta Q(k+1) = e^{-\frac{Ts}{\tau_Q}} \Delta Q(k) + k_Q \left(1 - e^{-\frac{Ts}{\tau_Q}}\right) \Delta T(k) \tag{11}$$

$$\Delta T^2(k+1) = e^{-\frac{Ts}{\tau_{T^2}}} \Delta T^2(k) + k_{T^2}(1 - e^{-\frac{Ts}{\tau_{T^2}}}) \Delta T(k) \tag{12}$$

Where Ts is the sample time (8.84s). The resulting discrete equations are:

$$\Delta Q(k+1) = e^{-\frac{Ts}{\tau_Q}} \Delta Q(k) + k_Q \left(1 - e^{-\frac{Ts}{\tau_Q}}\right) \Delta T(k) \tag{13}$$

$$\Delta T^2(k+1) = e^{-\frac{Ts}{\tau_{T^2}}} \Delta T^2(k) + k_{T^2}(1 - e^{-\frac{Ts}{\tau_{T^2}}}) \Delta T(k) \tag{14}$$

Figure 5 presents the comparison of the measured against estimated T^2 and Q statistical indices by using the above discrete equations, where it can be observed an acceptable error for compensation purposes.

iii) Validation experiments set. A new set of experiments for about 2.5 hours was achieved, where mass was eventually added during 0.1 to 0.8 hours at different locations, time instants and environmental temperatures. In total six damages (explained in figure 3) that correspond to the same mass added at different locations between actuator and sensor, joint to undamaged conditions were registered during this step, whose statistical indices Q and T^2, and environmental temperature are presented in figure 6. By applying step 2.4 statistical indices were compensated by subtracting the statistical indices variations obtained by using the discrete equations obtained previously. New compensated indices are presented in figure 7, were it can be observed that new indices can differentiate undamaged respect to damage conditions. Finally figure 8 presents a graphical interpretation of new T^2 against Q indices, where a clear distinction of damage respect to undamaged case can be achieved. It can be observed that by using only compensated T^2 index D1 and D3 can be clearly distinguished from the undamaged condition and other damages, while D2, D4 to D6 are confused with undamaged condition. On other hand, by using only the compensated Q index, each damage can be differentiated respect to the undamaged case. Thus, for only detection purposes, compensated Q index could be effective and enough. However, if location of damage is necessary, a combination of both indices could be more effective. For example, damages D4 and D5 contain similar Q values, however by including T^2 index a separation of damages could be obtained for classification purposes.

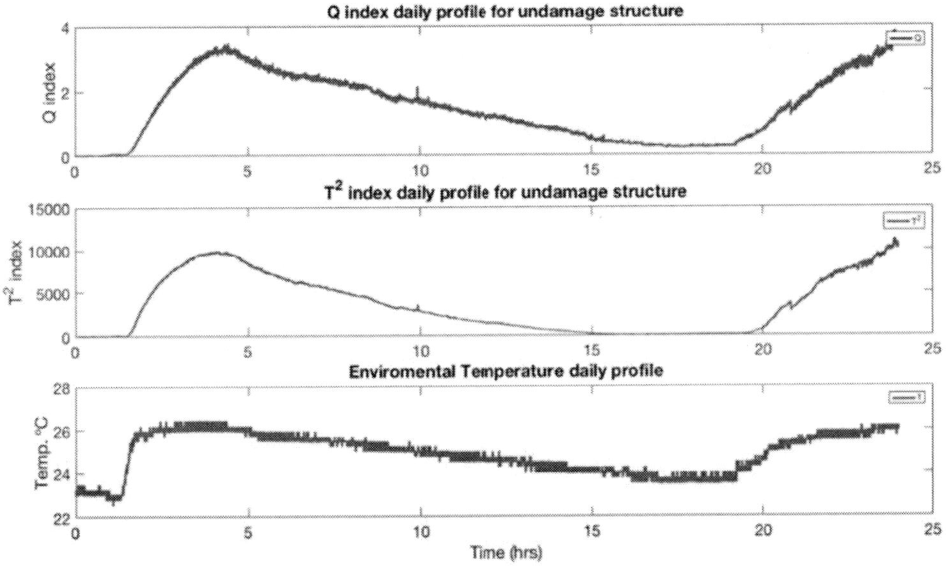

Figure 4. Temperature and statistical indices daily profile

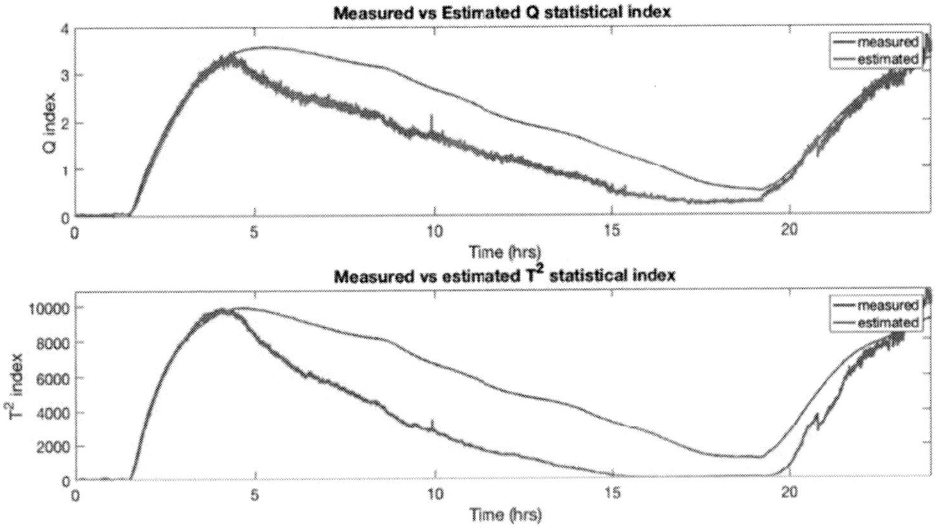

Figure 5. Estimated vs measured statistical indices for a daily profile

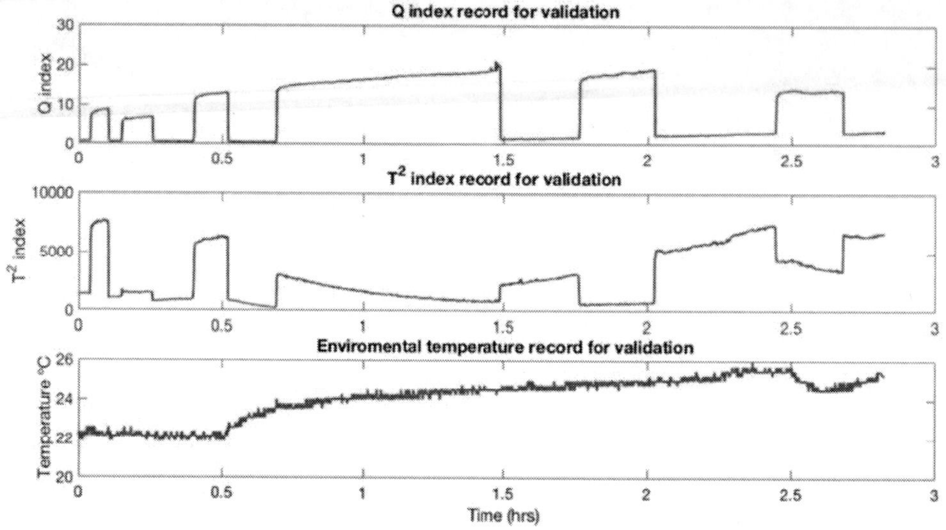

Figure 6. Q, T^2 and T record for validation purposes

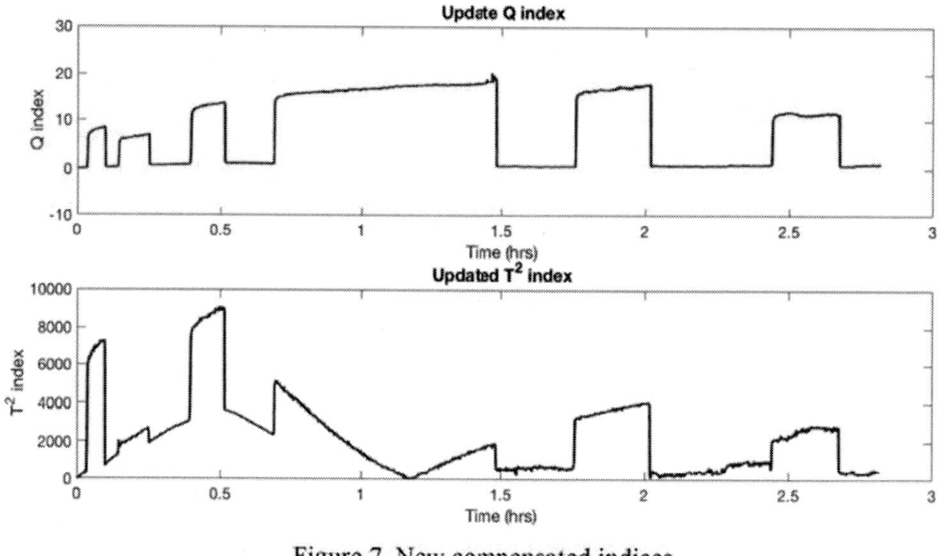

Figure 7. New compensated indices

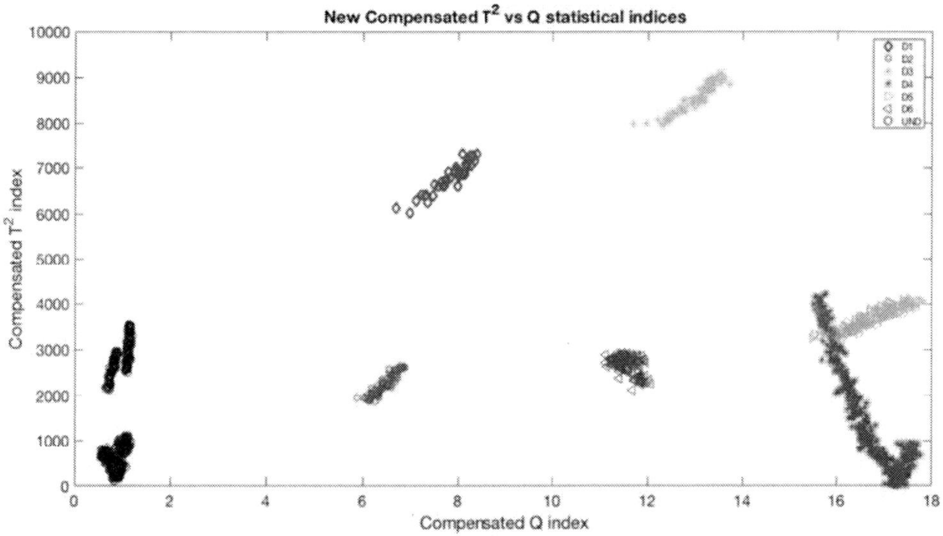

Figure 8. New compensated T^2 vs Q indices for undamaged and damage cases

4. Conclusion

In this approach compensated statistical index is proposed for detecting damages in metallic cylindrical specimens by means of piezo-diagnosis principle and based on PCA. The novelty here was to compute the contribution of environmental temperature changes on the traditional T^2 and Q statistical indices and to exclude their effect for damage detection. The temperature variations corresponded to natural environmental temperature variations during a day and two first order models were identified and used to compute the mentioned contributions. Thus the proposed damage detection methodology becomes robust to detect damages front to temperature variations by using a compensated Q index. Additionally, by taking into account the graphical results is easy to distinguish types of damages (different location of added mass), thus at least a qualitative location of damage could be inferred by using a combination of both compensated indices T^2 and Q.

References

[1] Mavi G S 2013 Feasibility of using ultrasonic guided waves for under water pipe inspection, *Thapar University, Patiala.*

[2] Lowe M J, Alleyne D N, Cawley P 1998 Defect detection in pipes using guided waves. *Ultrasonics* **36(1)** pp 147-154A

[3] Khatir A, Tehami M, Khatir S and Abdel Wahab M 2016 Multiple damage detection and localization in beam-like and complex structures using co-ordinate modal assurance criterion combined with firefly and genetic algorithms Journal of Vibroengineering 18(8) 5063-5073

[4] Zhou Y-L, Maia N and Abdel Wahab M 2016 Damage detection using transmissibility compressed by principal component analysis enhanced with distance measure Journal of Vibration and Control doi: 10.1177/1077546316674544

[5] Khatir S, Belaidi I, Serra R, Abdel Wahab M and Khatir T 2015 Damage detection and localization in composite beam structures based on vibration analysis Mechanika 21(6) 472-479

[6] Gillich G-R, Praisach Z-I, Abdel Wahab M, Gillich N, Mituletu IC and Nitescu C 2016 Free vibration of a perfectly clamped-free beam with stepwise eccentric distributed masses Shock and Vibration 2016(Article ID 2086274)

[7] Gharibnezhad F, Mujica L E, Rodellar J, Fritzen C P 2013 Damage detection using robust fuzzy principal component analysis *Proceedings 6th European Workshop on Structural Health Monitoring (Dresden)* pp 1-6.

[8] Park G, Sohn H, Farrar C R, Inman D J 2003 Overview of piezoelectric impedance-based health monitoring and path forward *Shock and Vibration Digest,* **35(6)** pp 451-464.

[9] Mujica L E, Rodellar J, Fernandez A, Guemes A 2011 Q-statistic and T2-statistic PCA-based measures for damage assessment in structures *Structural Health Monitoring* **10**(5) pp 539–553.

[10] Anaya M, Tibaduiza DA, Torres MA, Pozo F, Ruiz M, Mujica LE 2014 Data-driven methodology to detect and classify structural changes under temperature variations. *Smart Materials and Structures* **23(4)** p 045006

[11] Mujica LE, Rodellar J, Vehí J, Worden K, Staszewski W 2009 Extended PCA visualisation of system damage features under environmental and operational variations *Proc. of SPIE Smart Struct and Mat Nondestr Eval and Health Monitoring (San Diego, USA)*

[12] Gharibnezhad F, Rodellar J, Mujica LE 2014 Robust damage detection in smart structures. *Materia(s)* **7** p 04

Acknowledgments
This work has been partially funded by the Spanish Ministry of Economy and Competitiveness through the research projects DPI2014-58427-C2-1-R and by Universidad Industrial de Santander through the postdoctoral stay commission for Rodolfo Villamizar Mejía at the Department of Mathematics of the UPC, Barcelona, Spain.

Structural damage continuous monitoring by using a data driven approach based on principal component analysis and cross-correlation analysis

Jhonatan Camacho-Navarro[1], Magda Ruiz[1], Rodolfo Villamizar[2], Luis Mujica[1], Gustavo Moreno-Beltrán[2], Jabid Quiroga[1]

[1]Universitat Politécnica de Catalunya (UPC) BARCELONATECH, Department of Mathematics, Escola d'Enginyeria de Barcelona Est. (EEBE), Campus Diagonal Besòs, Edifici A, C. Eduard Maristany, 10-14, 08019, Barcelona, Barcelona, Spain.

[2]Universidad Industrial de Santander (UIS), Escuela de Ingenierías Eléctrica, Electrónica y de Telecomunicaciones (E3T), Bucaramanga, Colombia

jhonatan.camacho@estudiant.upc.edu, magda.ruiz@upc.edu, rovillam@uis.edu.co luis.eduardo.mujica@upc.edu, Gustavo.moreno4@correo.uis.edu.co, jabib@uis.edu.co

Abstract. Continuous monitoring for damage detection in structural assessment comprises implementation of low cost equipment and efficient algorithms. This work describes the stages involved in the design of a methodology with high feasibility to be used in continuous damage assessment. Specifically, an algorithm based on a data-driven approach by using principal component analysis and pre-processing acquired signals by means of cross-correlation functions, is discussed. A carbon steel pipe section and a laboratory tower were used as test structures in order to demonstrate the feasibility of the methodology to detect abrupt changes in the structural response when damages occur. Two types of damage cases are studied: crack and leak for each structure, respectively. Experimental results show that the methodology is promising in the continuous monitoring of real structures.

1. Introduction

"SHM is the integration of sensing and possibly also actuation devices to allow the loading and damaging conditions of a structure to be recorded, analysed, localized, and predicted in a way that non-destructive testing (NDT) becomes an integral part of the structure and a material" [1]. Thus, one of the main concerns in the field of condition monitoring is the implementation of systems with the capability to continuously evaluate the health of a structure. It is desirable that SHM system satisfy characteristics regarding to reliability, accuracy, robustness, and high sensitivity to presence of damage [2]. Additional requirements in order to achieve continuous and efficient SHM systems are related to hardware and software resource consumption when big data are recorded [3]. Therefore, the use of algorithms with low computational cost facilitates the implementation of diagnostic assessment approaches. In this sense, Principal Components Analysis (PCA) based methodology for structural damage detection becomes in a promising technique to be considered for online monitoring [4]. PCA is a mathematical tool widely used for feature extraction and pattern recognition that consists of well-defined matrix operations [5], which can be programmed by means of optimized procedures considering balanced memory/processor performance [6].

PCA in combination with piezo-diagnostics principle have been demonstrated be useful for structural health monitoring, with applicability for damage detection in aluminum plates, composite structures, aircraft sections and pipe work structures [7]. Damage detection is achieved by the use of PCA for obtaining a baseline of the pristine structure taking advantage of guided waves dispersion, which is statistically analyzed through squared prediction error indices. Thus, abnormal states are identified by means of scatter plots.

In this work, a data-driven approach for structural damage detection based on PCA technique is studied for online monitoring. The main contribution is focused on applying cross-correlation analysis as pre-processing technique. The performance of this approach is evaluated by analyzing data from two structures, a laboratory tower and a pipe section. The feasibility to detect cracks and leaks is demonstrated by processing online measurements.

2. Methods and Procedure for Damage Detection

The approach used in this work to detect structural damages is summarized in **Figure 1**. It consists of four main stages: Instrumentation system, Data pre-processing, statistical data-driven modelling and condition monitoring.

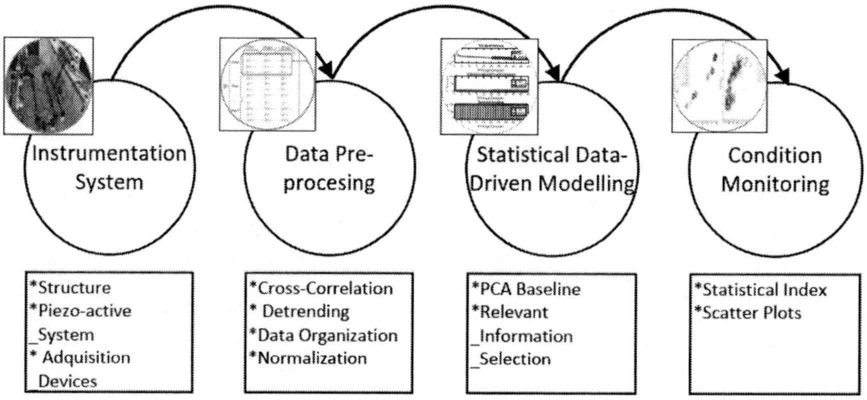

Figure 1. Damages detection methodology

2.1. *Instrumentation System*

In this work, the piezo-diagnostic principle is used as main approach for damage identification, which is based on analyzing guided waves propagation through the structure [8]. Guided waves are generated by means of piezo-electrical devices in order to find patterns with high sensitivity to structural damages. In this way, taking advantage of piezoelectric effect, information about the scattering, reflection, and mode conversion from elastic wave travelling caused by discontinuities is examined.

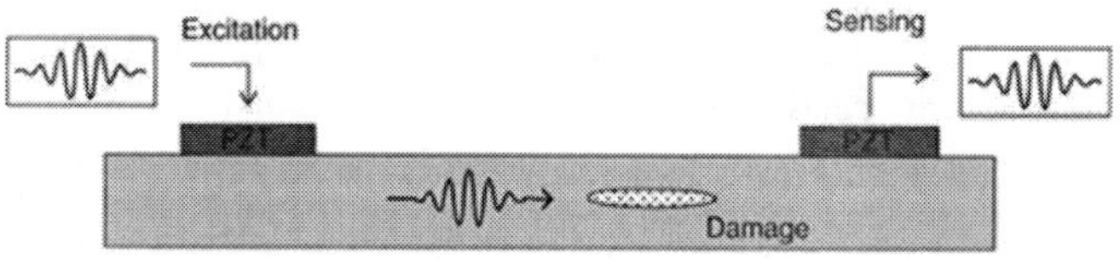

Figure 2. Piezo-diagnostic principle [8]

According to **Figure 2**, the instrumentation system belongs to a piezoelectric active scheme, where one of the PZT mounted on the surface structure operates as actuator, and the remaining PZT works as sensors in a pitch–catch mode. Other components of the instrumentation equipment consist of fine-tuning filters, high wide-band amplifiers and acquisition elements, among others.

2.2. Data pre-processing

The aim of data pre-processing stage is to minimize the presence of outliers and atypical data, as well as the application of methods intended for data cleansing and normalization. It also includes data organization in order to facilitate data fusion, which allows the implementation of statistical techniques. For this purpose, cross-correlation functions, *detrending* analysis, and *Group Scaling (GS)* procedure are applied.

Initially, linear trends are removed from piezo-electrical measurements due to artefacts noise and low frequency disturbances. Then cross-correlation is computed between actuation and sensing piezo-signals, with the objective to exclude common external noise signals, outliers filtering and as a tool for improving separation boundaries for damage conditions. A data matrix (*X*) is built using cross-correlated functions corresponding to each PZT sensors, and considering only measurements from pristine condition. Several repetitions from undamaged state are organized in this unfolded matrix as shown in Figure 3.

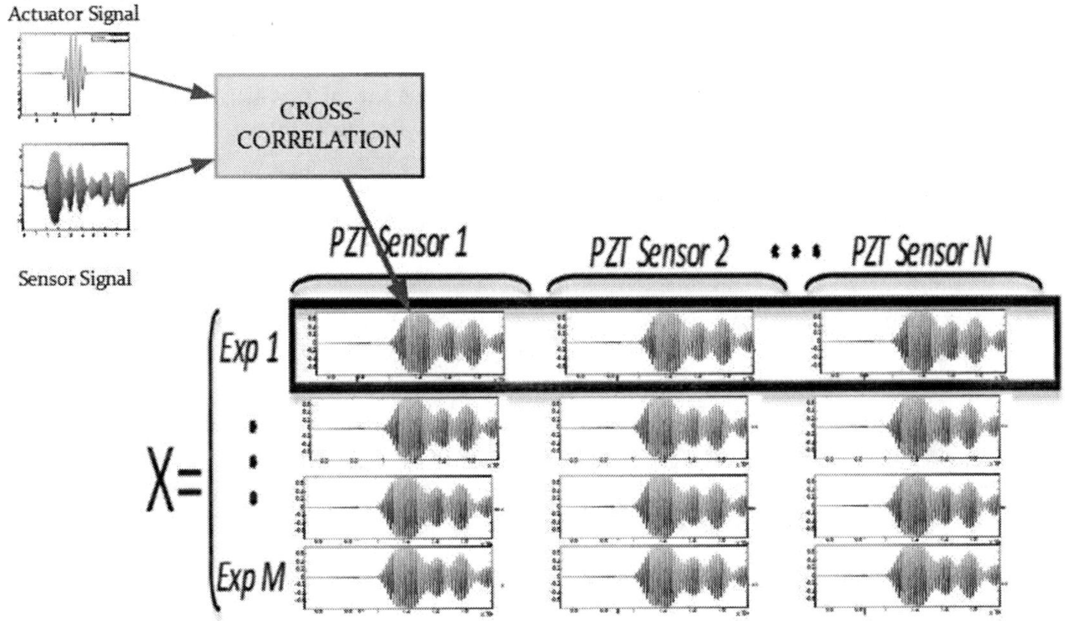

Figure 3. Undamaged cross-correlated baseline Matrix

The cross-correlation function between two signals $X(t)$ and $Y(t)$ is used to obtain elements in the undamaged cross-correlated baseline matrix, it is defined as:

$$R_{XY}(t,t+\tau) = \lim_{N\to\infty} \frac{1}{N} \sum_{k=1}^{N} X_k(t) Y_k(t+\tau), \qquad (1)$$

where N is the number of signal samples and τ is the lag time interval used to compute the cross-correlation function.

Finally, the undamaged cross-correlated baseline matrix is normalized by applying *Group Scaling (GS)* procedure to avoid scaling and bias issues. Normalization reduces the influence of different source of variability due to changes in environmental and operational conditions. In *Group Scaling* each data-point from the undamaged cross-correlated baseline matrix (*X*) is scaled by considering changes between sensors and the nature of data by estimating standard deviation for each block of piezo measurements [9]. The standardization of undamaged cross-correlated baseline matrix X is computed by using the mean of each time sample for every experiment and the standard deviation of each sensor sample vector, which result in the normalized data matrix $\bar{X}$:

$$\bar{x}_{ij} = \frac{x_{ij}-\hat{\mu}_j}{\hat{\sigma}_j}, \quad (2)$$

where $\hat{\sigma}_j$ is the standard deviation per piezo sensor and $\hat{\mu}_j$ is the mean value per column of undamaged cross-correlated baseline matrix X.

2.3. *Statistical Data-Driven modelling*

The baseline model is built by applying PCA to the standardized undamaged baseline matrix $\bar{X}$. Thus, cross-correlation functions from undamaged measurements are represented in the reduced space of principal components according to equation (3).

$$\bar{X} = TP' + E = model + noise \quad \textbf{(3)}$$

The model in (3) belongs to a new reduced space of coordinates with minimal redundancy, based on the variance–covariance of the original data. T is the projection to the reduced space, P is a linear transformation matrix that relates the normalized data matrix $\bar{X}$ in the new coordinates and denotes the principal components (eigenvectors). The noise E-matrix is the part of $\bar{X}$, which is not explained by the model and describes the residual variance neglected by the statistical model. The estimation of the baseline model involves the following steps [10]:

1. Compute the cross-correlation between actuating signal and measurements from PZT sensors when the structure operates in healthy condition.
2. Organize cross-correlation functions in the unfolded data matrix X as is shown in Figure 3.
3. Normalize the undamaged cross-correlated baseline matrix by using *Group Scaling* procedure, which produce mean values ($\hat{\mu}_j$), and standard deviations ($\hat{\sigma}_j$) from experiment repetitions in the nominal condition.
4. Compute the singular value decomposition of the standardized undamaged cross-correlated baseline matrix by applying PCA. Thus, the eigenvectors and eigenvalues are obtained.
5. Keep only the first $\boldsymbol{r}$ components in order to obtain a reduced representation for original variables. The variance for each new variable corresponds to their respective $\boldsymbol{r}$ eigenvalues.

As a result, the statistical baseline model corresponds to mean values ($\hat{\mu}_j$), standard deviations ($\hat{\sigma}_j$), and the r eigenvectors (P) and eigenvalues (λ).

2.4. *Condition Monitoring*

New PZT measurements representing the current structural state are compared with the baseline representation in order to identify structural condition, where differences between baseline model and current state are attributed to damage. Thus, the cross-correlation between actuation and sensing signals of these measurements are organized in a row vector and standardized by applying *Group Scaling* with the mean values and standard deviations of the undamaged cross-correlated baseline matrix ($\hat{\mu}_j, \hat{\sigma}_j$) with the help of eq. (2). This normalized row vector is projected onto the reduced space by using the statistical model (3). In order to compare the signature structure respect to undamaged condition two statistical indexes are computed: T-squared statistics (T^2) and squared prediction error (Q):

$$Q = \sum_j \left(e_j \right)^2 \quad (4)$$

$$T^2 = T' \lambda^{-1} T \quad (5)$$

Where, e_j is the residual error for each $j-th$ principal component used to reconstruct the trial experiment and λ (singular values) are the respective variances of the reduced-space.

The whole methodology can be summarized in two stages: Training and Monitoring. Figure 4 shows the concept of the system:

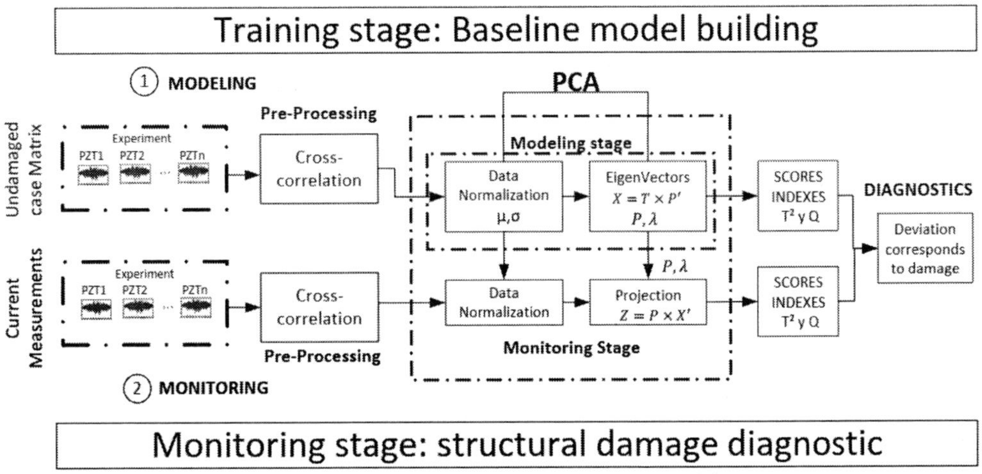

Figure 4: Two staged approach based on PCA for structural damage assessment.

3. Experimental Results

Experimental tests were conducted on two structural lab models in order to evaluate the capability of the methodology for continuous monitoring: A carbon steel pipe section and a laboratory tower. In both cases, the piezoelectric actuator device is excited with a periodic high frequency (~100 KHz) burst type signal in order to induce a guide wave. Each damage scenario includes 100-experiment repetitions during 1s of periodic excitation signal.

3.1. Carbon steel pipe section

The first specimen used as test structure is a carbon-steel pipe section of dimensions 100x 2.54 x0.3 cm (length, diameter, thickness). It was conditioned with piezoelectric devices in order to induce guided waves along the surface structure. Operational conditions are controlled through bridles, a valve, a manometer and a compressor. The air pressure is settled from a compressor in 80 psi at one of the ends. The test structure is depicted in Figure 5, where several sections can be identified. However, only measurements from the first section were considered in this study. In addition, bolts and other elements used to recreate leak damages are included in the nominal state of the structure and consequently in the statistical baseline model.

IOP Conf. Series: Journal of Physics: Conf. Series **842** (2017) 012018 doi:10.1088/1742-6596/842/1/012018

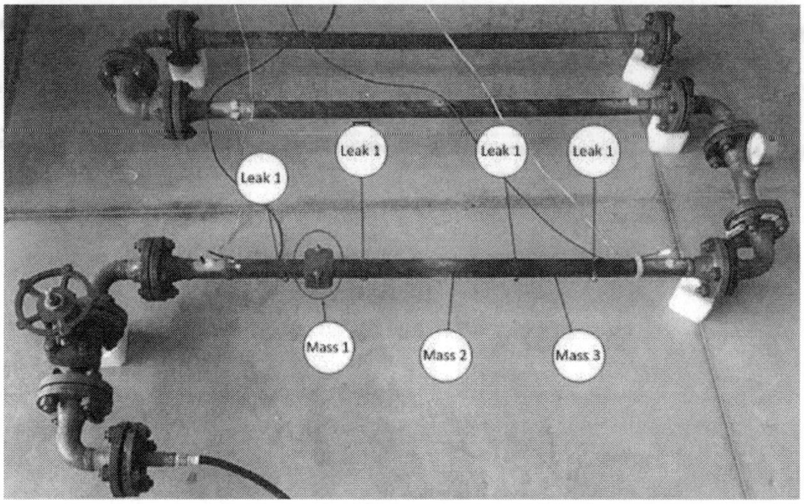

Figure 5. Pipeline experimental set-up

As illustrated in Figure 5, leaks were induced by a full opening of a hole between the PZT devices (Actuator-Sensor) and located at different points along the structure. These kinds of leaks are recreated by means of four ¼-inch holes which were drilled along the pipe section wall. Thus adjustable screws were used to control the leak magnitude to be produced. Likewise, a special shaped accessory was added to the surface pipe section in order to simulate damage types corresponding to adding masses at different locations of the surface.

The T-squared and Q-statistical indices are depicted in Figure 6. The performance without cross-correlation is also presented in order to evaluate the influence of using cross-correlation analysis in the damage detection scheme.

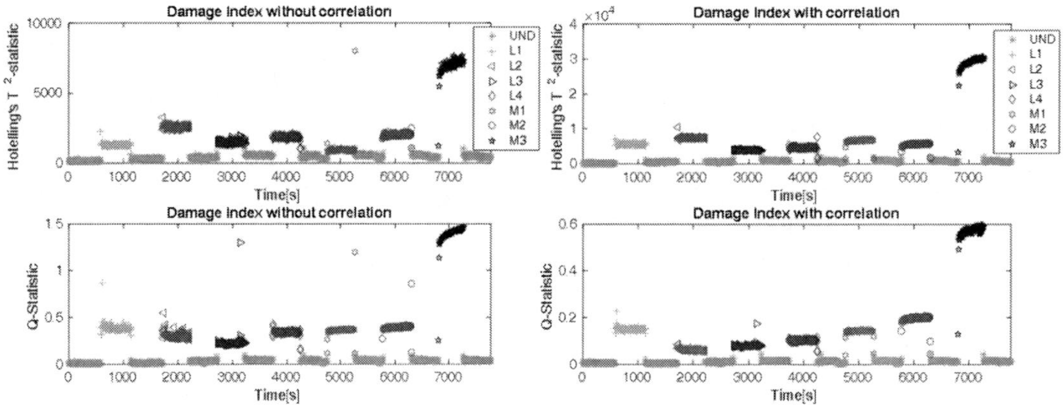

Figure 6. Statistical indexes for pipeline experiment. Left: without cross-correlation. Right: cross-correlated piezo-diagnostic approach

According to results in Figure 6, it can be observed that more dispersion appears without cross-correlation analysis. Also, the transient response when damage appears can be captured and it is possible to identify the time occurrence of damage, which facilitate decision making. Additionally, if correlation analysis is included in the damage detection approach, then some atypical data-cases are filtered. In order to emphasize the advantage of using cross-correlated functions, the Q Vs T^2 scattered plots are shown in Figure 7, where more clear boundaries between damage conditions can be observed when cross-correlation are included.

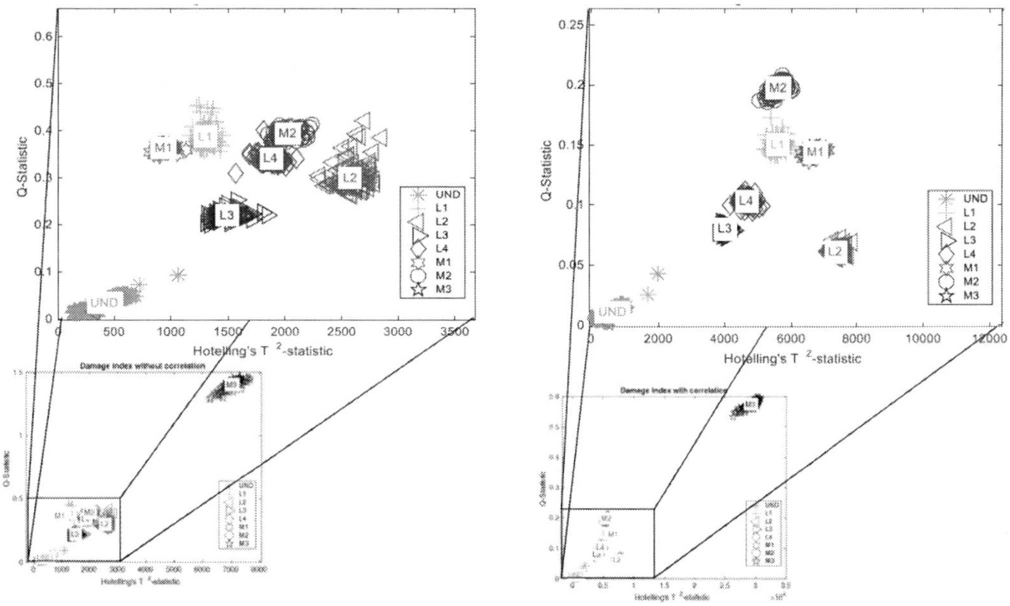

Figure 7. Q vs T^2 scatter plots for pipeline experiment. Left: without cross-correlation. Right: cross-correlated piezo-diagnostic approach

3.2. Laboratory tower

The second test structure is a tower model, representing a wind turbine model [11] (Figure 8). The structure (2.7 m high) is composed by three components (Figure 8a): jacket, tower and nacelle. Five PZT sensors were installed in the jacket (Figure 8a, red markers correspond to PZT devices) in order to record 50-experiment repetitions from guided wave structural responses produced by the PZT actuator using a sample time Ts = 32.0 [ns].

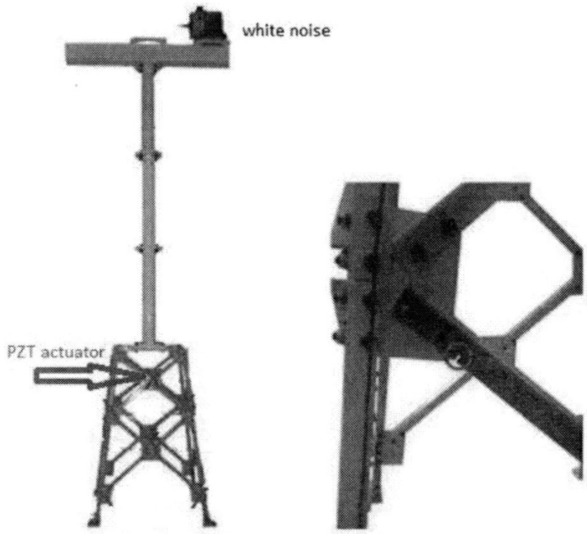

(a) Sensor Location (b) Damage in structure

***Figure 8.** Laboratory tower structure.*

Two kinds of damages were studied in the tower test bench. The first one corresponds to crack condition, which was induced by replacing one of the undamaged sections in the jacket with a 5-mm cracked section (Figure 8b). The second one consists of full/partial unbolted screws. **Figure 9** shows the evolution for index values when damages are caused in the structure. The tag specification in Figure 9 corresponding to each damage is as follow: unbolt two screws in one section (D1), unbolt all screws in the section (D2), replacing cracked element (D3) and repetition of cracked damage (D4). All damages are generated in the same section.

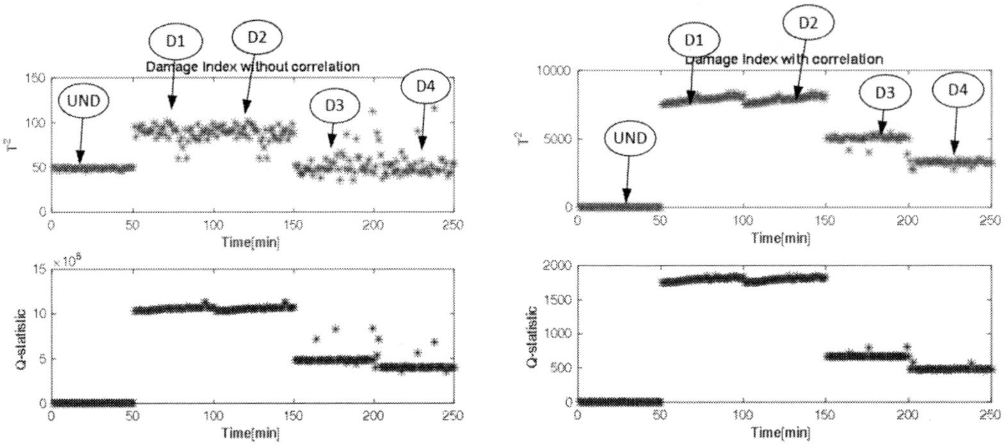

Figure 9: Statistical indexes for the Laboratory tower.

According to results in Figure 9, it can be distinguished the damage state from undamaged one (UND). However, the transient dynamics it is no clearly observable. It is possible establish meaning differences regarding to the undamaged state and a low dispersion for all experiments. Also, clear thresholds are identified between both damage types (unbolt and crack). Nevertheless, a local low linear trend is noted in the index values, which can be produced for sensor drifts.

4. Conclusion

The feasibility of using a data driven approach based on principal component analysis and cross-correlation analysis for monitoring cracks, leaks and mass aggregation in structures was experimentally validated in a pipe loop and in a laboratory tower. In this sense, the squared prediction error (Q-statistic) and T-squared indexes were proved to be suitable as damage index for continuous monitoring with good capability to differentiate unhealthy from undamaged structural conditions. It was demonstrated the applicability of using cross-correlation to reduce outlier data, minimization of damage type group dispersion and for obtaining better boundary separation damage classes. It is required future studies to include environmental and operational variations as well as methods to separate sensor faults such a bias and drift issues.

5. Acknowledgement

The results in this paper were obtained as a collaborative work between the *"Universidad Industrial de Santander –UIS- (Colombia)"* and *"Universitat Politécnica de Catalunya -UPC- (Spain)"*. It has been partially funded by the Spanish Ministry of Economy and Competitiveness through the research projects DPI2014-58427-C2-1-R and also it corresponds to a continuation of the *"Monitorización y detección de defectos en estructuras usando algoritmos expertos embebidos"* research project, financed by the *"Departamento administrativo de ciencia y tecnología Francisco José de Caldas – COLCIENCIAS"* and *"Banca Mundial"*. Finally, authors are grateful to the research center IKERLAN-IK4, for their collaboration in some experimentation.

References

[1] Boller, C., Chang, F. K., & Fujino, Y. (Eds.) 2009 *Encyclopedia of structural health monitoring. Wiley.*

[2] Bennouna, O. & Roux, J.P. 2013. Real Time Diagnosis & Fault Detection for the Reliability Improvement of the Embedded Systems, *Journal of Signal Processing Systems,* **73(2)** pp. 153-160

[3] Zhou, Y. L., & Wahab, M. A. (2016). Rapid early damage detection using transmissibility with distance measure analysis under unknown excitation in long-term health monitoring. Journal of Vibroengineering, **18(7)**.

[4] Zhou, Y. L., Maia, N. M., & Wahab, M. A. (2016). Damage detection using transmissibility compressed by principal component analysis enhanced with distance measure. Journal of Vibration and Control, 1077546316674544.

[5] Jolliffe, I., 2002 Principal component analysis. *John Wiley & Sons, Ltd.*

[6] Liang, Y. C., Lee, H.P., Lim, S.P., Lin, W.Z., Lee, K.H., & Wu, C.G. 2002 Proper orthogonal decomposition and its applications—Part I: Theory. *Journal of Sound and vibration,* **252(3)** 527-544.

[7] Mujica, L. E., Rodellar, J., Fernandez, A., & Guemes, A. 2010. Q-statistic and T2-statistic PCA-based measures for damage assessment in structures. *Structural Health Monitoring.*

[8] An, Y. K., Kim, M. K., & Sohn, H 2014. Piezoelectric transducers for assessing and monitoring civil infrastructures. *Sensor Technologies for Civil Infrastructures: Sensing Hardware and Data Collection Methods for Performance Assessment,* **86**.

[9] Mujica, L. E., Vehi, J., Ruiz, M., Verleysen, M., Staszewski, W., & Worden, K 2008. Multivariate statistics process control for dimensionality reduction in structural assessment. *Mechanical Systems and Signal Processing,* **22(1)** 155-171.

[10] Maćkiewicz, A., & Ratajczak, W 1993. Principal components analysis (PCA). *Computers & Geosciences,* **19(3)** 303-342.

[11] Zugasti, E., Mujica, L. E., Anduaga, J., & Martinez, F 2013 Feature selection-Extraction methods based on PCA and mutual information to improve damage detection problem in offshore wind turbines. *Key Engineering Materials,* **569** 620-627.

Ensemble learning as approach for pipeline condition assessment

Jhonatan Camacho-Navarro[1], Magda Ruiz[1], Rodolfo Villamizar[2], Luis Mujica[1], Gustavo Moreno-Beltrán[2]

[1]Universitat Politécnica de Catalunya (UPC) BARCELONATECH, Department of Mathematics, Escola d'Enginyeria de Barcelona Est. (EEBE), Campus Diagonal Besòs, Edifici A, C. Eduard Maristany, 10-14, 08019, Barcelona, Barcelona, Spain

[2]Universidad Industrial de Santander (UIS), Escuela de Ingenierías Eléctrica, Electrónica y de Telecomunicaciones (E3T), Bucaramanga, Colombia

jhonatan.camacho@estudiant.upc.edu, magda.ruiz@upc.edu, rovillam@uis.edu.co luis.eduardo.mujica@upc.edu, Gustavo.moreno4@correo.uis.edu.co

Abstract. The algorithms commonly used for damage condition monitoring present several drawbacks related to unbalanced data, optimal training requirements, low capability to manage feature diversity and low tolerance to errors. In this work, an approach based on ensemble learning is discussed as alternative to obtain more efficient diagnosis. The main advantage of ensemble learning is the use of several algorithms at the same time for a better proficiency. Thereby, combining simplest tree decision algorithms in bagging scheme, the accuracy of damage detection is improved. It takes advantage by combining prediction of preliminary algorithms based on regression models. The methodology is experimentally validated on a carbon steel pipe section, where mass adding conditions are studied as possible failures. Data from an active system based on piezoelectric sensors are stored and characterized through the T^2 and Q statistical indexes. Then, they are the inputs to the ensemble learning. The proposed methodology allows determining the condition assessment and damage localizations in the structure. The results of the studied cases show the feasibility of ensemble learning for detecting occurrence of structural damages with successful results.

1. Introduction

In the last years, a special interest is noted for the condition monitoring of structures such as wings, bridges, oil pipes, towers and sea platforms among others, which are widely used in tasks related to mechanical, civil and aeronautical applications. As a result, new techniques have been developed in the field of Non-Destructive Testing (NDT), in order to detect structural damages due to aging, overloads, fatigue or external disturbances. Thus, when damage is early detected, proper actions can be conducted to repair or reinforce structural elements that minimize accidents risk, economical losses, catastrophic events, and avoid possible human deaths.

The importance of NDT methods has promoted new trends for "*the integration of sensing and possibly also actuation devices to allow the loading and damaging conditions of a structure to be recorded, analysed, localized, and predicted in a way that non-destructive testing (NDT) becomes an integral part of the structure and a material*", which is named as Structural Health Monitoring (SHM) [1]. However, the implementation of SHM demands the management of uncertainty caused by operational and environmental conditions when continuous action is considered [2] and in this way, the erroneous diagnostic is minimized. In addition, damage localization strategies that facilitate the failure identification and reduce maintenance costs are included. Examples of SHM techniques comprises statistical analysis with high computational costs and sensitivity to external disturbances. For instance, envelope signals have been used to obtain an approximation of structural damage location [3]. Also, a PZT sensor network has been proposed as an alternative to locate the damage [4], [5].

This work proposes the use of ensemble learning as approach for structural damage assessment, since one of the main requirements to achieve reliable continuous monitoring systems is to minimize false alarms and missing reports. The philosophy of *Ensemble Learning* (EL) is the combination of machine learning algorithms in order to improve the probability of damage identification taking advantage of statistical performance [6]. Also, EL is suitable to manage some issues when large or little data volumes are available by the adaptation of resampling techniques and decision averaging. Additionally, EL is useful to integrate features from diverse information sources, which serves as a simple data fusion scheme. Therefore, the efficiency and accuracy of expert systems for SHM are maximized by using EL.

Since, the main purpose of this work is to evaluate the feasibility of using Ensemble Learning as approach for SHM tasks an EL architecture implemented through a bagging configuration of decision trees is evaluated. The whole methodology combines features obtained from piezo-diagnostics principle in order to build an ensemble with capability for detecting mass adding in a pipe section. As a result, the proposed system is able to condition assessment and predict the location of possible damages using information from guided waves along the structure which is characterized by means of statistical indexes and supervised learning. Experimental tests were conducted over a pipe section with masses emulating potential damages. Thus, the section 2 discuss the methodology framework of the approach used as damage identification system, while experimental results is described in section 3. Finally, section 4 corresponds to the main conclusion of this work. In summary, a promising approach for pipeline damage identification based on ensemble decisions was successful validated with promising results for real world application in structural assessment.

2. Damage condition monitoring methodology

The approach used in this work to detect and locate structural damages is based on piezo-diagnostics principle and ensemble learning approach. The main components of the whole expert system are detailed in next sections. Specifically, pre-processing techniques, data organization and instrumentation scheme, as well as theoretical framework are described.

2.1. Ensemble Learning as approach for SHM

The basic idea of ensemble learning is to combine multiple models to improve prediction performance. They are considered meta-algorithms designed to work on top of existing learning algorithms. Ensemble Learning is also known in the literature with several keywords: ensembles, ensemble methods, ensemble learning methods, model combination, combining models, combining classifiers, multiple classifiers, multiple classifier systems, majority voting or mixtures of experts. This diversity of keywords related to EL hinder the identification of previous works in any application field [6], [7], [8]. Figure 1 shows the concept of the expert system based on Ensemble Learning.

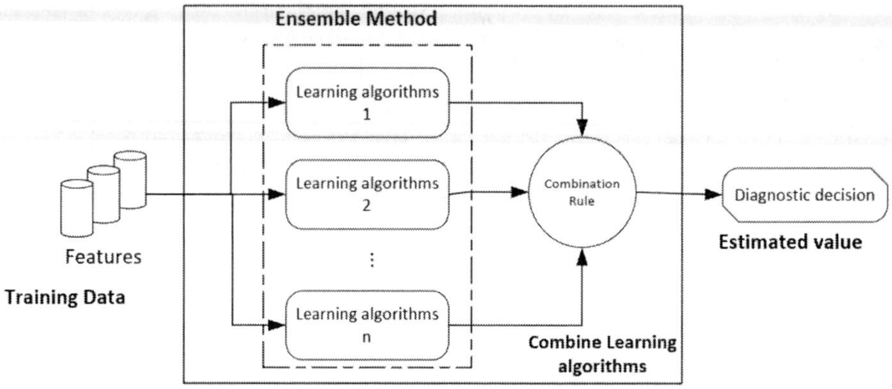

Figure 1. Expert System approach based on Ensemble Learning

According to figure 1, the main components of ensemble approach correspond to multiple learning algorithms organized in a parallel scheme and a combination function (fuser). Thus, the feature inputs are processed and manipulated with the methods selected to construct ensembles in order to obtain the final diagnostic decision. Although figure 1 specifies a parallel architecture, others such a cascading, hierarchical or hybrid can be used as ensemble scheme.

The most popular approaches used to generate ensembles are *Bagging* and *Boosting*. Bagging approach corresponds to Bootstrap Aggregation and it consists on multiple models which are built from different subsamples of the training dataset. In this sense, in bagging a set of models is generated each of which are trained on a random sampling of the data by means of bootstrap resampling where sample of instances with replacement are used. The final output prediction in Bagging ensemble is averaged across the predictions of all of the sub-models, or produced on the concept of voting or weighted average according to the performance of ensemble models [9], [10]:

$$f_{ens}(x) = \sum_{i=1}^{M} w_i \hat{f}_i(x) \tag{1}$$

In equation (1), the combined prediction $f_{ens}(x)$ is obtained from the M models of the ensemble, and the output $\hat{f}_i$ of model *i* on an input x. The weights w_i can be seen as the relative confidence in the correctness of the models predictions, and are thus constrained to sum to 1 ($w_i > 0, \sum_{i=1}^{M} w_i = 1$) [11].

Instead, boosting method constructs models in an iterative manner, where new models learns to fix the prediction errors of a prior model in the chain. Thus, *boosting* ensemble algorithms creates a sequence of models that attempt to correct the mistakes of the models before them in the sequence. There are several variants of boosting algorithm such a AdaBoost, LogitBoost, GentleBoost, and LPBoost. Specifically, LSBoost (least squares boosting) is used to fit regression ensembles by minimizing mean-squared error [12]. In this case, the ensemble fits new learners to the difference between the observed response and the aggregated prediction of all learners grown previously.

On the other hand, a typical example of ensemble machine learning algorithms is the decision trees [10]. A tree model is a set of rules for predicting the target, which are constructed by recursively partitioning a learning sample of data and represented by a node in the tree. For instance, a linear tree model implements a general linear function by means of piecewise linear approximation to the target function. Therefore, the final tree model consists of a tree with linear regression functions at the leaves. In the same way, the prediction for an instance is obtained by sorting it down to a leaf and using the prediction of the linear model associated with that leaf [11].

EL has not been applied sparingly in SHM despite the advantages of its use for task related to pattern recognition. As an illustration, support vector machines and neural networks have been used in combination to detect and classify two damage types in the aircraft fuselage [13]. Other example is detailed in [14], where a classifier is proposed as fusion strategy to manage different sources of information from acoustic emissions, ultrasound tests and flows measurements. The aim of this system is to improve leak detection in pipeline structures. A similar application is described in [15], where independent classifiers are used to combine measurements from different type of sensors in order to obtain higher level diagnostic responses. In this sense, a vote majority scheme is implemented to identify pipe damages as an approach in nondestructive testing. Other example is discussed in [16], where the authors present results of full-scale fatigue test (FSFT), which are analyzed by means of an ensemble of Artificial Neural Networks (ANNs). This last approach allows compensating temperature and environmental effects present during acquisition of Lamb-wave signals.

2.2. Piezo-diagnostic feature extraction

The features input for the structural damage identification methodology previously described are obtained by means of piezo-diagnostics principle. Hence, guided waves travelling along the structure are statistically characterized and implemented through piezoelectric devices. The suitability of piezo-diagnostic principle to find patterns with high sensitivity in structural damages has been previously demonstrated. Therefore, it has been successfully implemented for analyzing experimental data from aircraft wings, pipe bench tests, and plate structures, among others [17], [18]. The main components of feature extraction methodology based on piezo-diagnostics are summarized in figure 2.

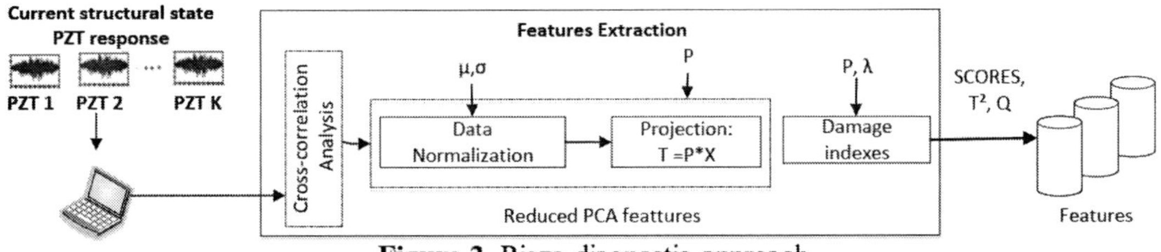

Figure 2. Piezo-diagnostic approach

According to figure 2, an active piezoelectric scheme is implemented in order to acquire the signature response of the structure to the scattering, reflection, and mode conversion from elastic wave travelling, caused by the presence of discontinuities [19]. Also, a pre-processing stage by using cross-correlation analysis and data normalization is included in order to minimize the influence of outliers and atypical data, as well as to facilitate data cleansing and common external noise filtering.

The data-driven strategy illustrated in figure 2 is implemented in two phases. In the first phase, a baseline model is obtained by processing data from repetitions of undamaged structural condition. As a result, the data matrix (*X*) corresponding to cross-correlated functions from PZT sensor measurements are represented in a reduced space according to Principal Component Analysis (PCA) procedure:

$$X = TP' + E = \text{model} + noise \quad (2)$$

The baseline model expressed in equation (2) is a representation of the pristine structural condition with minimal redundancy. In equation (2), T is the projection of X to the reduced space (SCORES), P is a linear transformation matrix which denotes the conserve r principal components and E corresponds to the noise matrix describing the residual variance neglected by the reduced baseline model.

PCA procedure is an efficient data reduction method, also known as Karhunen-Loμeve decomposition or Proper Orthogonal Decomposition. If data X processed by using PCA lie in the Hilbert space, then it can be approximated by orthonormal basis functions φ_i as stated in equation (3) [20]:

$$X = \sum_{i\in\mathbb{I}}(X,\varphi_i)\varphi_i \tag{3}$$

In theorem (3) the real numbers (X,φ_i) are referred as the *Fourier coefficients* of X in the basis $\{\varphi_i\}_{i\in\mathbb{I}}$ [20], where $\mathbb{I}$ is a countable index set and the relative contribution of each term decreases with increasing i. Such series will be referred as a Fourier expansion, used to approximate X data, which corresponds to a similar compact representation of statistical model in (2) [21].

On the other hand, the second phase consists of determining the current structural condition by processing new PZT measurements. For this purpose, the row vector of new PZT measurements is projected onto the reduced space by using the statistical model (2). Then, two statistical indexes are computed in order to identify differences between baseline model and current state: T-squared statistic (T^2) and squared prediction error (Q) [5].

$$T^2 = T'\lambda^{-1}T \tag{4}$$

$$Q = \sum_j \left(e_j\right)^2 \tag{5}$$

Where, e_j is the residual error for each $j-th$ principal component used to reconstruct the trial experiment and λ are the respective variances of the space determined by principal components.

3. Experimental Results

A carbon steel pipe section of 100 length, 2.54 diameter and 0.3 cm thickness, was used as structural lab model in order to conduct experimental tests. The test specimen was instrumented with two piezoelectric devices, which work in a pitch–catch mode. Other components comprising the instrumentation equipment consist of fine-tuning filters, high wide-band amplifiers and acquisition elements, among others. The test structure is depicted in figure 3, where PZT actuator and sensor can be identified.

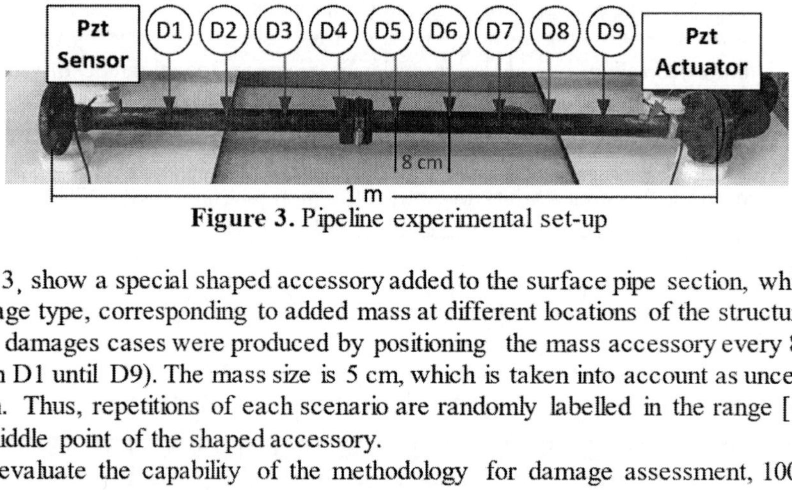

Figure 3. Pipeline experimental set-up

Also, figure 3, show a special shaped accessory added to the surface pipe section, which is used to simulate a damage type, corresponding to added mass at different locations of the structure surface. In this sense, nine damages cases were produced by positioning the mass accessory every 8 cm (tagged with labels from D1 until D9). The mass size is 5 cm, which is taken into account as uncertainty in the damage position. Thus, repetitions of each scenario are randomly labelled in the range [-2.5, 2.5] cm respect to the middle point of the shaped accessory.

In order to evaluate the capability of the methodology for damage assessment, 100 experiment repetitions of each damage scenario were acquired, while 1500 repetitions of undamaged condition (tagged as *'orig'*) were used to build the baseline model. The piezoelectric actuator device is excited with a 1s periodic high frequency (~100 Khz) burst type signal to induce guided waves. Linear trends and bias issues are removed from piezo-electrical measurements in order to reduce artefacts noise and low frequency disturbances. The cross-correlation between sensing and actuation signals obtained from the 1500 piezoelectric measurements used to build the baseline model are depicted in figure 4.

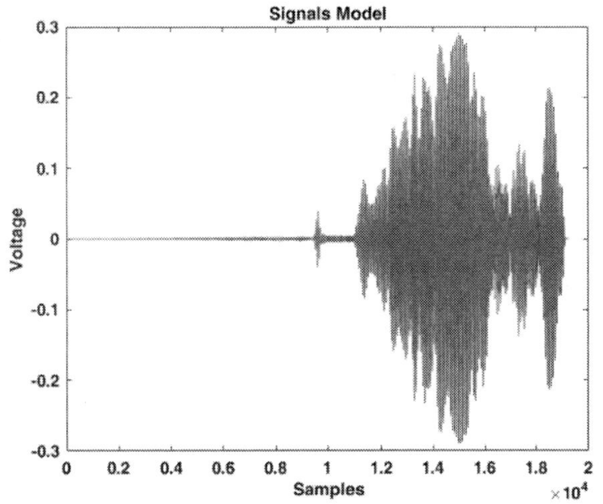

Figure 4. Cross-correlation signals used for baseline modelling.

According to figure 4, the most significant differences are regarding to amplitude variations and possible mode conversions between guided wave packages, which represent the structural response of pristine condition.

Then, the baseline model was built by estimating 25 principal components and considering as healthy structure, the pipe section including bolts and other elements attached to the pipe structure. The input feature vector is formed by concatenating 8 of the first scores with the statistical indexes (T^2 and Q). In this augmented feature vector, only scores from 2 to 8 are included (the first is omitted since it contains amplitude influence), thus, scaling issues are minimized by excluding the first major component. Above influence is demonstrated according to results in figure 5, where baseline model variances for cross-correlation signals in figure 4 as well as contribution of each component are detailed.

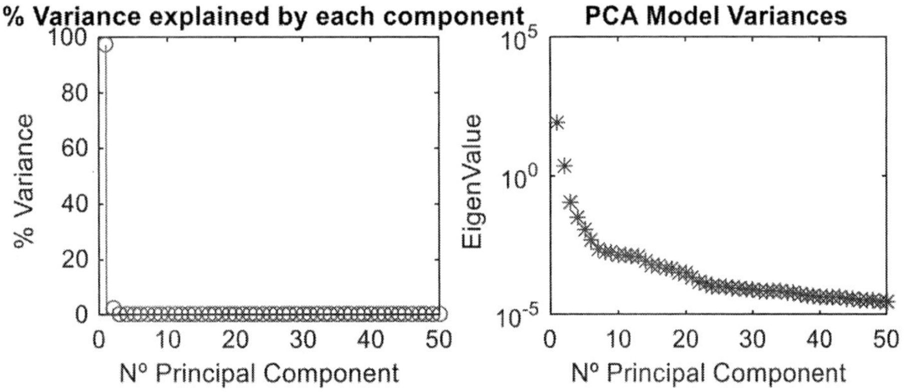

Figure 5. Variance of baseline modelling.

Figure 5a (left) shows that first component mask the contribution of remaining principal components and additionally, percentage variance (figure 5a-left) is not a good indicator of the principal component importance. While, cumulative variance does not allow establishing a meaning significance, there exists a relative variance that can be comparable through the eigenvalues magnitude in a semi-log scale (figure

5b-rigth). Above behaviour indicates that by excluding the first principal component from the augmented feature vector, structural response can be suitable to be represented. Plots of the first fourth Fourier modes are given in figure 6.

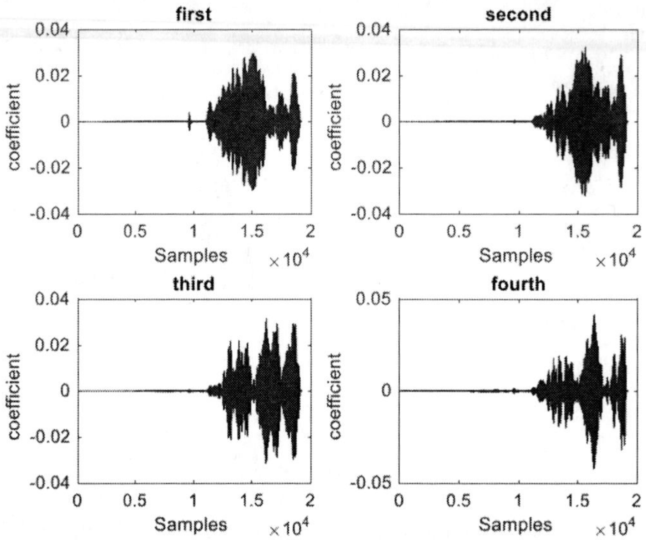

Figure 6. Fourier modes of baseline model.

In figure 6 it is observed that first component is a smoothed-roughly approximation of signals used to build the baseline model (figure 4). Besides, the other principal components capture high variation of baseline model signals. It confirms that first component is highly influenced by amplitude and scaling variations as was previously mentioned.

On the other hand, an iterative analysis was performed to examine the performance of Mean Square Error (MSE) of the estimated damage location. In this way, a proper number of ensembles are used to identify a suitable method to construct ensemble are performed. Figure 7 shows the evolution of the MSE for the Bag (Bagging) and LSBoost methods.

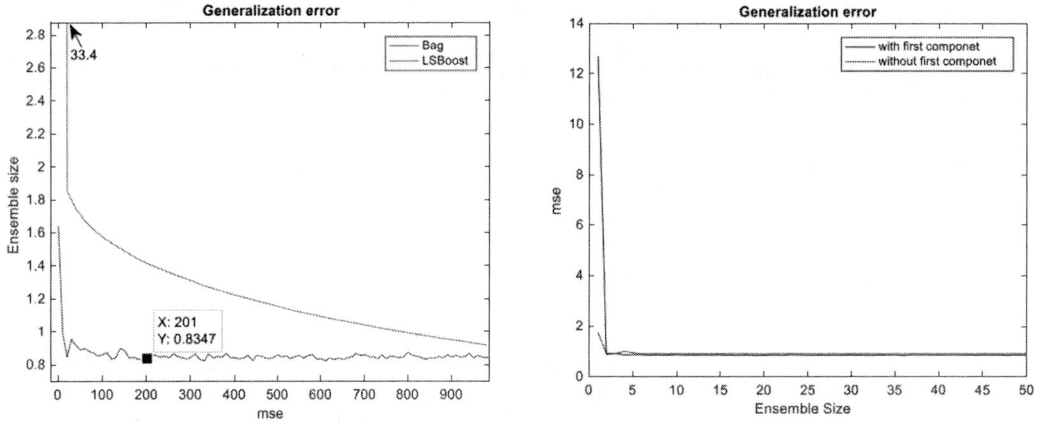

Figure 7. MSE evolution for ensemble methods

According to figure 7, the Bag method show better performance and more stable behaviour with a lower ensemble size. Consequently from now, the results are computed by using Bagging method with 200 ensembles.

In addition, the T-squared and Q-statistical indices as well as the variance of the PCA baseline model are depicted in figure 8.

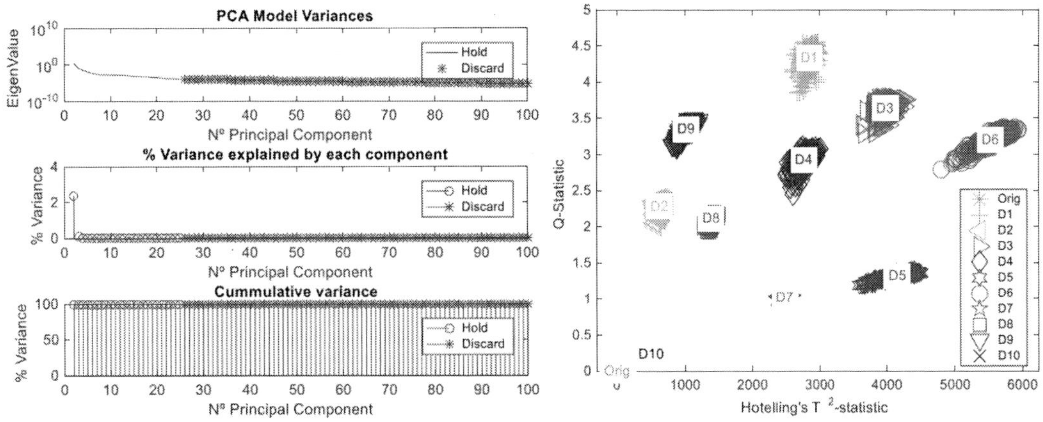

Figure 8. PCA results a) Baseline model variance b) Damage statistical indexes

Figure 8a (left) shows that the first 25 components retain more than 99% of cumulative variance, which is statistically significant to describe the normal (healthy) condition of the structure. Also, different damage types (locations) are clearly separated by boundaries between clusters generated by the T^2 and Q statistical indexes (see figure 8b - right). Thus, damage detection is achieved in a simple way by the examination of scatter plot of damage indexes, while automation of this task can be implemented through proper thresholds definition.

However, in order to solve the damage location problem, complementary information is required. It can be supplied using the first scores computed by the baseline model. Hence, information about deviations respect to healthy condition is included in the augmented feature input vector, then the ensemble tree is trained with supervised learning. Figure 9 shows the performance of training and validation errors, where black dots correspond to original data used to evaluate the regression algorithm and color points to the respective ensemble response.

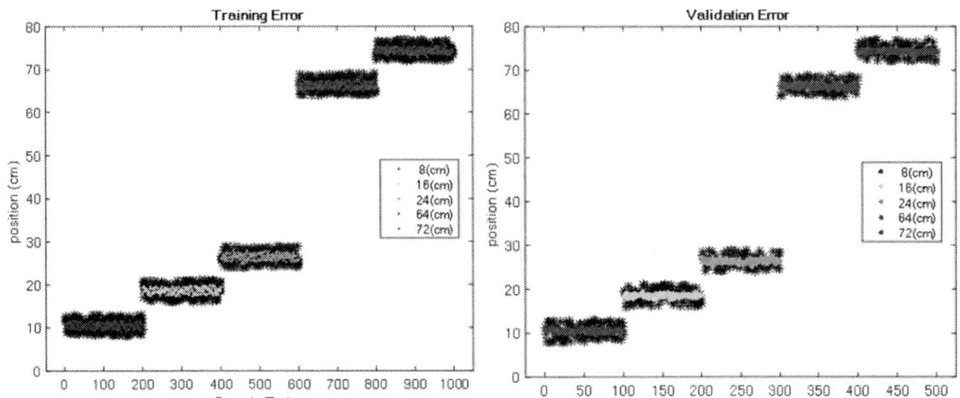

a) Training error. b) Validation error

Figure 9. Ensemble tree performance

Besides, in figure 9 is observed the results obtained within the uncertainty range. Small errors and low dispersion were obtained since data used for training (Figure 9a) and validation (Figure 9b) purposes, corresponds to same damage location biasing the capacity of the algorithm. In this way for a new set of data (Prediction) the error estimate is calculated and illustrated in figure 10.

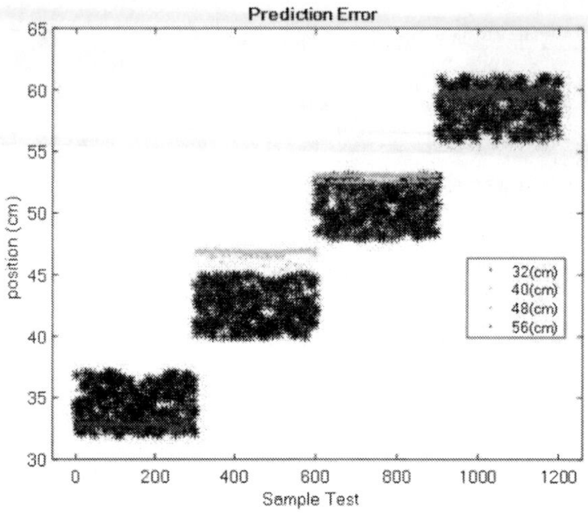

Figure 10. Ensemble prediction error.

Figure 10 shows the position of intermediate damage locations predicted by the ensemble. Although, error can be considered as minimum since it is lower than the size of the element used to simulate damages (damage quantification < 5cm).

4. Conclusions

In this paper an ensemble machine learning approach for improving the performance of models on SHM problem was detailed. The feasibility of combining different learning algorithms, in an ensemble scheme, for monitoring mass aggregation damages in pipe like structures was experimentally validated.

The damage detection is properly performed by the statistical indexes and the scores of PZT measurements. Nevertheless, with the methodology developed in this work, statistical modelling through PCA, piezo-diagnostics approach for damage characterization and bagging method to construct ensembles, the damage location is performed with mean square errors lower than 0.8.

The capability of prediction could be limited to be reliable only in the range within the experimental uncertainty. However, in the Predictive step is recommended to perform additional experiments that do not match with those used during the validation step. On the contrary, the capability of prediction could be limited.

In general, the decision making process for the experiment conditions and the approach based on ensemble learning proposed in this paper has been suitable. As future work, environmental and operational variations should be included, as well as methods to manage unsupervised algorithms, which probably would improve the generalization errors.

5. Acknowledgement

The results in this paper were obtained as a collaborative work between the "Universidad Industrial de Santander –UIS- (Colombia)" and "Universitat Politécnica de Catalunya -UPC- (Spain)". It has been partially funded by the Spanish Ministry of Economy and Competitiveness through the research projects DPI2014-58427-C2-1-R and also it corresponds to a continuation of the "Monitorización y detección de defectos en estructuras usando algoritmos expertos embebidos" research project, financed by the "Departamento administrativo de ciencia y tecnología Francisco José de Caldas – COLCIENCIAS" and "Banca Mundial".

References

[1] Boller, C., Chang, F. K., & Fujino, Y. (Eds.) 2009 *Encyclopedia of structural health monitoring. Wiley*.

[2] Sohn, H. (2007). Effects of environmental and operational variability on structural health monitoring. Philosophical Transactions of the Royal Society of London A: Mathematical, Physical and Engineering Sciences, **365(1851)**, 539-560.

[3] Ebrahimkhanlou, A., Dubuc, B., & Salamone, S. (2016). Damage localization in metallic plate structures using edge-reflected lamb waves. Smart Materials and Structures, **25(8)**, 085035.

[4] Mujica, L. E., Ruiz, M., Pozo, F., Rodellar, J., & Güemes, A. (2013). A structural damage detection indicator based on principal component analysis and statistical hypothesis testing. Smart materials and structures, **23(2)**, 025014.

[5] Tibaduiza, D. A., Mujica, L. E., Rodellar, J., & Güemes, A. (2016). Structural damage detection using principal component analysis and damage indices. Journal of Intelligent Material Systems and Structures, **27(2)**, 233-248

[6] Polikar, R. (2006). Ensemble based systems in decision making. IEEE Circuits and systems magazine, **6(3)**, 21-45.

[7] Opitz, D., & Maclin, R. (1999). Popular ensemble methods: An empirical study. Journal of Artificial Intelligence Research, **11**, 169-198.

[8] Rokach, L. (2010). Ensemble-based classifiers. Artificial Intelligence Review, **33(1-2)**, 1-39.

[9] Breiman, L. (1996). Bagging predictors. Machine learning, **24(2)**, 123-140.

[10] Breiman, L. (2001). Random forests. Machine learning, **45(1)**, 5-32.

[11] Ikonomovska, E. (2012). Algorithms for learning regression trees and ensembles on evolving data streams (Doctoral dissertation).

[12] Mathworks.com. (2017). Ensemble Methods - MATLAB & Simulink. [online] Available at: https://www.mathworks.com/help/stats/ensemble-methods.html [Accessed 10 Ene. 2017].

[13] Candelieri, A., Sormani, R., Arosio, G., Giordani, I., & Archetti, F. (2013). A hyper-solution framework for SVM classification: Improving damage detection on helicopter fuselage panels. AASRI Procedia, **4**, 31-36.

[14] Co-pi, R. P. (2006). A Data Fusion System for the Nondestructive Evaluation of Non-Piggable Pipes Final Report Dr . Shreekanth Mandayam (PI), Dr . John C . Chen (Co-PI) February 2006 DE-FC26-02NT41648 Rowan University 201 Mullica Hill Road Glassboro , NJ 08028 Name and Ad

[15] Dworakowski, Ziemowit, Krzysztof Dragan, and Tadeusz Stepinski. 2016. "Lamb-Wave-Based Monitoring of the Aircraft during Full-Scale Fatigue Experiment - Results and Conclusions." 8th European Workshop on Structural Health Monitoring (EWSHM) (July 2016): 5–8.

[16] Parikh, Devi et al. 2004. "Combining Classifiers for Multisensor Data Fusion." Conference Proceedings - IEEE International Conference on Systems, Man and Cybernetics 2: 1232–37.

[17] Mujica, L. E., Rodellar, J., Fernandez, A., & Guemes, A. 2010. Q-statistic and T2-statistic PCA-based measures for damage assessment in structures. Structural Health Monitoring.

[18] Zugasti, E., Mujica, L. E., Anduaga, J., & Martinez, F 2013 Feature selection-Extraction methods based on PCA and mutual information to improve damage detection problem in offshore wind turbines. Key Engineering Materials, 569 620-627.

[19] An, Y. K., Kim, M. K., & Sohn, H 2014. Piezoelectric transducers for assessing and monitoring civil infrastructures. Sensor Technologies for Civil Infrastructures: Sensing Hardware and Data Collection Methods for Performance Assessment, 86.

[20] Astrid, P. (2004). Reduction of process simulation models: a proper orthogonal decomposition approach. Ph. D.-thesis: Technische Universiteit Eindhoven.

[21] Huisman, L. (2005). Control of glass melting processes based on reduced CFD models. Technische Universiteit Eindhoven.

Structural system identification by utilizing transmissibility coherence from limited measured data

Yun-Lai Zhou[1], Cao Hongyou[2], Ni Zhen[1], Li Zhang[3], Magd Abdel Wahab[4]

[1] Department of Civil and Environmental Engineering, National University of Singapore, 2 Engineering Drive 2, Singapore 117576
[2] School of civil engineering and architecture, Wuhan University of Technology, Wuhan 430070, China
[3] Xinjiang Laboratory of Phase Transitions and Microstructures in Condensed Matters, College of physical Science and Technology, Yili Normal University, 835000, China
[4] Soete Laboratory, Faculty of Engineering and Architecture, Ghent University, Technologiepark Zwijnaarde 903, B-9052 Zwijnaarde, Belgium

E-mail: zhouyunlai168168@gmail.com

Abstract. System identification performs as a core issue in structural dynamic analysis. In this study, transmissibility coherence is introduced for system identification with recalling the existing techniques based on transmissibility. Unlike previous approaches that require four-point measurement, the proposed methodology in this study only requires two-point measurement. The merit behind this approach is that the transmissibility coherence can be employed to estimate the subtraction of transmissibility between two reference points, by using auto- and cross-spectrum analysis. Verification using experimental data proves the feasibility of the proposed technique.

1. Introduction

System identification, one core issue in structural dynamic analysis, has appealed lots of attention during last decades. The key idea is to extract the structural dynamic characteristics, namely resonant frequency, mode shape and so on. This also underwent two decades using experimental modal analysis (EMA) and operational modal analysis (OMA). For each direction, methods can be summarized into two categories, time domain and frequency domain. And in recent years, due to the booming development of algorithms, algorithms such as Expectation Maximum-likehood (EM) are also introduced into system identification [1].

System identification holds an essential role in structural analysis, which may also be considered as the basis for damage diagnosis [2-6], structural health monitoring (SHM) [7-13], prognosis health monitoring (PHM), condition monitoring and so on. Even the structural characteristics generated by system identification are not the only ones, but they are of great importance compared with other techniques in SHM [14-21]. Modal parameters are widely used in damage detection, localization, and quantification [4]. Frequency response functions (FRFs) might be one of the most essential functions in EMA based system identification. The importance of identifying modal parameters of structures can be seen in the literature for many types of applications [22-28]. In SHM, damage detection can be evaluated using direct methods or indirect methods. For indirect methods, numerical techniques such as Finite

Element Analysis [29-41] and isogeometric analysis [22, 24-27, 42-45] are used along with experimental data.

Previous research on system identification gradually change from EMA to OMA, and during OMA, transmissibility has been proved to be feasible for extracting natural frequencies by using the inverse subtraction between two reference points, which experienced for various verification during past years. A general review can be found in [2]. However, it normally requires four points of measurements in order to fulfill the demanding of constructing the subtraction equation.

This study tries to compress the number of four points to be two via utilizing the transmissibility coherence [2, 15, 16, 21], which will reduce the testing time, and thus leading to a more efficient alternative in system identification.

2. Transmissibility background

2.1. Transmissibility

Transmissibility is defined as ratio between two structural dynamic outputs, and it can be denoted as

$$T_{(i,j)} = \frac{X_i}{X_j} \tag{1}$$

where i, j represent the output nodes, while X_i, X_j indicate the frequency spectrum of dynamic response x_i, and x_j in time domain. Different approaches can be used to estimate transmissibility, and further detail can be found in previous reviews [2, 15, 17].

2.2. Transmissibility coherence

During EMA, frequency response functions are commonly employed in modal analysis, while coherence is simultaneously applied to check the correlation between excitation and dynamic response, thus leading to determine whether or not the experiment is well conducted. Herein, by analog of coherence in EMA, transmissibility coherence (TC) [2, 11, 15, 46] is defined as

$$TC_{(i,j)} = \left| \frac{G_{ij}^2}{G_{ii} G_{jj}} \right| \tag{2}$$

where G means the cross- or auto- spectrum. Note that TC is defining the T_1, T_2 like the FRF estimation H_1 and H_2, and is denoted as

$$T_{1(i,j)} = \frac{X_i}{X_j} = \frac{X_i \leftrightarrow X_j}{X_j \leftrightarrow X_j} = \frac{G_{ij}}{G_{jj}} \tag{3}$$

$$T_{2(i,j)} = \frac{X_i}{X_j} = \frac{X_i \leftrightarrow X_i}{X_j \leftrightarrow X_i} = \frac{G_{ii}}{G_{ji}} \tag{4}$$

Note that TC is firstly raised in [15, 16] as for damage detection and quantification, and it is also applied in small nonlinearity detection and quantification. Later, this idea is extended to system identification in [2] and [21], where the idea is taking advantage of the previous equation of inverse subtraction of transmissibility between two reference nodes. This will be illustrated in sections hereinafter.

2.3. System identification scheme

As proved previously [2, 47], the peaks of inverse subtraction of transmissibility for two reference points are consistent with the natural frequencies, thus, one can use peak picking (PP) method to extract the natural frequencies with the following equation:

$$Fun(i, j, R_1, R_2) = \Delta^{-1} T_{(i,j)}^{R_1 R_2} = \frac{1}{\Delta T_{(i,j)}^{R_1 R_2}} = \frac{1}{T_{(i,j)}^{R_1} - T_{(i,j)}^{R_2}} = \frac{G_{(j,R_1)} G_{(j,R_2)}}{G_{(i,R_1)} G_{(j,R_2)} - G_{(i,R_2)} G_{(j,R_1)}} \tag{5}$$

From Equation (5), if choosing i, j as reference points of R_1 and R_2, then recalling Equations (2) and (5), one can derive the following equation:

$$Fun(i,j) = \Delta^{-1}T^{i,j}_{(i,j)} = \frac{1}{T^{i}_{(i,j)} - T^{j}_{(i,j)}} = \frac{1}{T_{2(i,j)} - T_{1(i,j)}} = \frac{1}{T_{2(i,j)}} \times \frac{1}{1 - \frac{T_{1(i,j)}}{T_{2(i,j)}}} = \frac{1}{T_{2(i,j)}} \times \frac{1}{1 - TC^{2}_{(i,j)}} \quad (6)$$

Note that Equation (6) is firstly proposed in [2], and detailed in [11]. The key idea for Equation (6) is to shorten the number of points measured. Comparing Equation (5) and Equation (6), one can find the merit of TC based frequency extraction method, which only relies on two-point measurement, and this gives a possibility in simplifying the extraction of frequencies, and an alternative apart from the previous techniques.

3. Experimental verification

For verifying the proposed natural frequency extraction methodology, a three-floor aluminum structure is considered [48], as shown in Figure 1. For some further studies, the reader may refer to [49-51]. A three-story aluminum structure is excited by a shaker, and four accelerometers are installed at each floor and the base as well to record the responses. Further details can be found in [15, 16, 49-51].

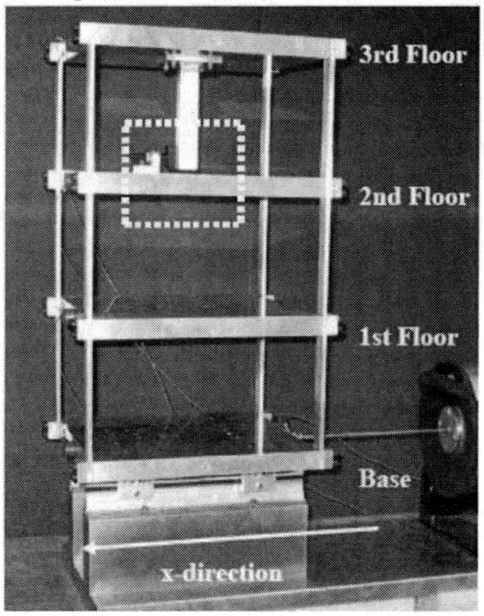

Figure 1. Three-story building structure testing setup [49-51].

4. Results

Results for the three-story building structure are calculated according to the aforementioned equations, and are discussed hereinafter. Figure 2 illustrates FRF (5, 1) and FRFC (5, 1). Note that FRFC herein means the coherence for corresponding FRF. From Figure 2, one can find that the FRFC holds an value adjacent to the value '1', which means that the experiment is well conducted. And note that the peak of FRFC is not corresponding to the peak of FRF.

Figure 3 shows T(5, 2) and TC(5, 2), where one can find that TC also holds a value close to the value of '1', which also implies that the experiment is well conducted. And one can also find that the peak of TC(5, 2) corresponds to the peaks of T(5, 2), this implies the potential interrelation between them.

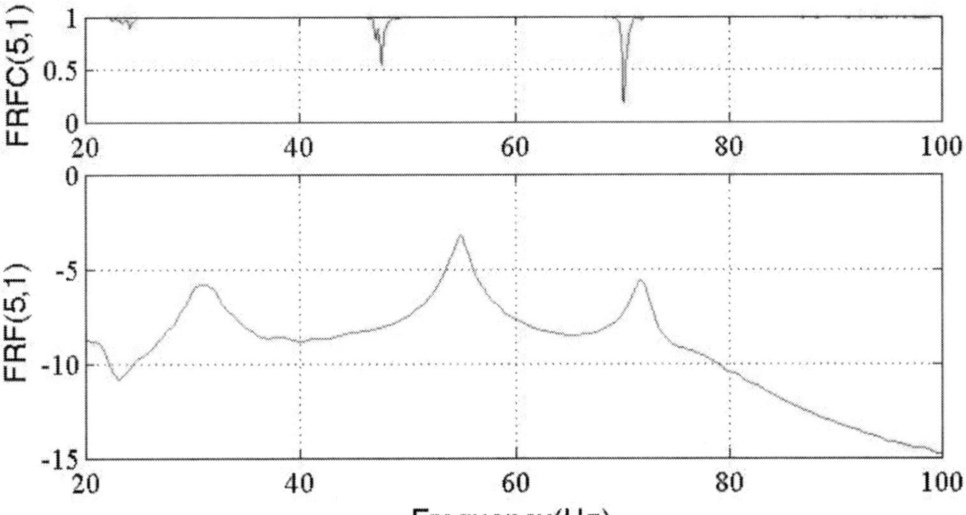

Figure 2. FRF (5, 1) and FRFC (5,1).

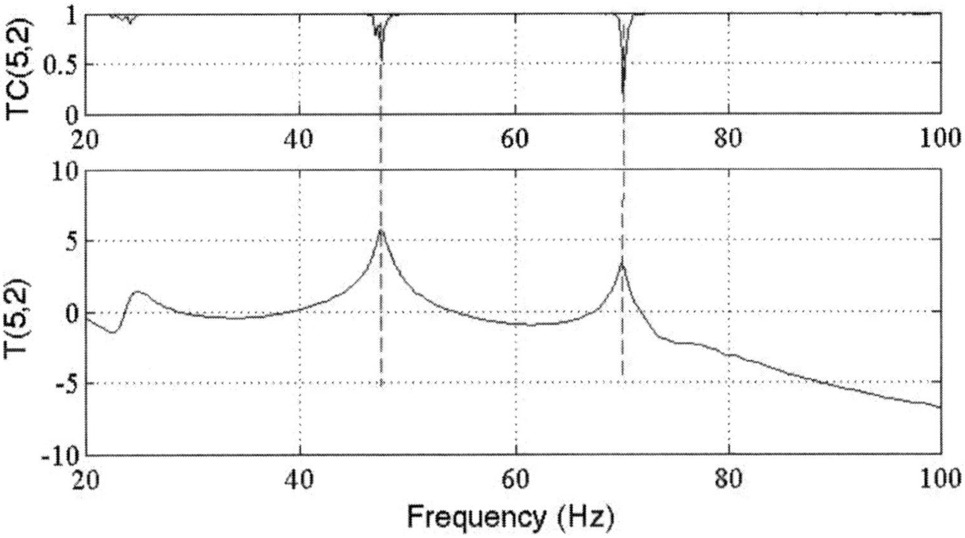

Figure 3. T(5, 2) and TC(5, 2).

Figure 4 shows a comparison between the proposed method in comparison with the conventional technique. One can find that both techniques agree well, while the proposed technique of Equation (6) only relies on two-point measurement. This will give a progress especially applicable for limited measurement condition. Another phenomenon can be found is that unlike theoretical derivation that peaks are corresponding to the natural frequencies, abundant peaks are encountered, note in order to extract the real natural frequencies, this method needs to be combined with other technique or engineering experience to finally fix the natural frequencies. The highest peak does not exactly correspond with the natural frequency; there might be some shift.

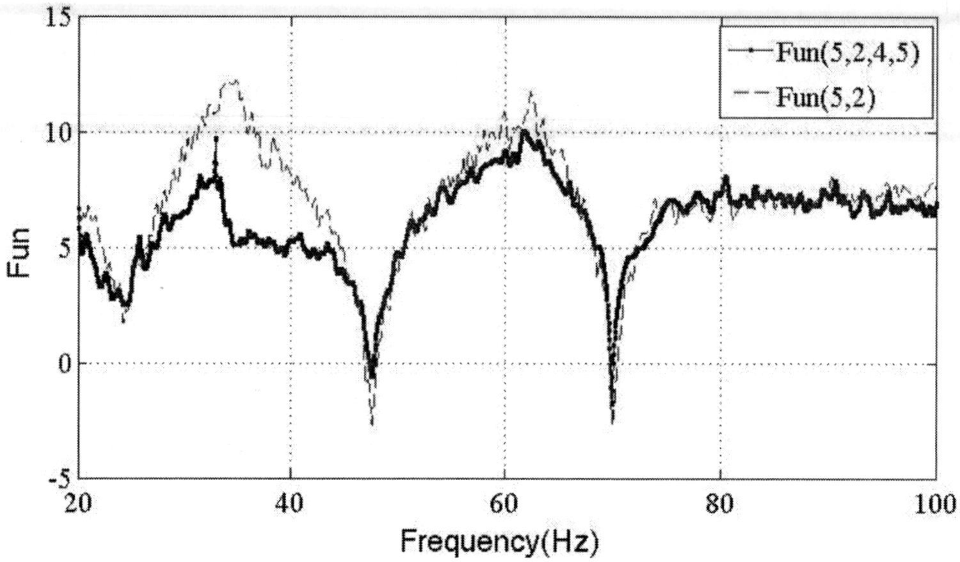

Figure 4. Fun(5, 2, 4, 5) and Fun(5, 2).

Table 1 lists the natural frequencies extracted by the proposed technique in comparison with the natural frequencies derived in the literature [48]. From Table 1, one can find that the natural frequencies agree well with the previous method, and this implies that the proposed technique is feasible and applicable.

Table 1. Natural frequencies extracted.

Mode	Experiment [48]	Equation (6)	Error (%)
1	30.70	30.8230	0.4000
2	54.20	54.1016	-0.0018

5. Conclusions

This study illustrated a new technique for frequency extraction by using transmissibility coherence, which only relies on two-point measurement. A three-story aluminum structure was used as verification, and the proposed technique agreed well with the previous method, while it worked well in frequency extraction. One shortcoming of transmissibility based frequency extraction technique is that more small peaks exist in the derived curve. This requires engineering experience and/or combined with further technique as natural frequency might slightly be shifted. For complex structure, the applicability of the technique requires further investigation to unveil a widely applicable scheme.

Acknowledgement

The first author thanks CWO (Commissie Wetenschappelijk Onderzoek), Faculty of Engineering and Architecture, Ghent University for providing financial support for a research stay at Soete Laboratory.

Reference

[1] Cara FJ, Carpio J, Juan J and Alarcon E 2012 *Mechanical Systems and Signal Processing* **31** 109-129

[2] Zhou YL 2015 *PhD thesis*

[3] Zhou YL and Perera R 2012 *Proceedings of the 1st International Conference on Structural Health Monitoring and Integrity Management (ICSHMIM 2012)* Beijing, China, November 2012

[4] Zhou YL and Perera R 2013 *Proceedings of the 5th European-American Workshop on Reliability of NDE* Berlin, Germany, October 2013
[5] Zhou YL, Perera R and Sevillano E 2012 *Proceedings of the 6th European Workshop on Structural Health Monitoring* Dresden, Germany, July 2012
[6] Zhou Y-L, Wahab MA, Figueiredo E, Maia NMM, Sampaio R and Perera R 2015 *Journal of Physics Conference Series* **628** 012036
[7] Zhou Y-L, Maia N and Abdel Wahab M 2016 *Journal of Vibration and Control* doi: 10.1177/1077546316674544
[8] Gillich G-R, Praisach Z-I, Abdel Wahab M, Gillich N, Mituletu IC and Nitescu C 2016 *Shock and Vibration* **2016**(Article ID 2086274) 10 pages; http://dx.doi.org/10.1155/2016/2086274
[9] Khatir S, Belaidi I, Serra R, Abdel Wahab M and Khatir T 2015 *Mechanika* **21**(6) 472-479
[10] Khatir S, Belaidi I, Serra R, Abdel Wahab M and Khatir T 2016 *Journal of Vibroengineering* **18**(1) 202-213
[11] Zhou Y-L, Maia NMM, Sampaio R and Wahab MA 2016 *Structural health monitoring* **DOI: https://doi.org/10.1177/1475921716680849**
[12] Zhou Y-L and Abdel Wahab M 2017 *Engineering Structures* **141** 175-183
[13] Khatir A, Tehami M, Khatir S and Abdel Wahab M 2016 *Journal of Vibroengineering* **18**(8) 5063-5073
[14] Zhou YL and Perera R 2014 *Proceedings of the 9th European Conference on Structural Dynamics (EURODYN 2014)* Oporto, Portugal, June 2014
[15] Zhou Y-L, Figueiredo E, Maia NMM and Perera R 2015 *Shock and Vibration* **2015**(Article ID 290714)
[16] Zhou Y-L, Figueiredo E, Maia NMM and Perera R 2015 *Proceedings of the International Conference on Structural Engineering Dynamics (ICEDyn 2015)* Lagos, Algarve, Portugal, June 2015
[17] Zhou Y-L, Figueiredo E, Maia NMM, Sampaio R and Perera R 2015 *Structural Control and Health Monitoring* **22** 1209-1222
[18] Zhou Y-L, Figueiredo E, Maia NMM, Sampaio R and Perera R 2015 *Proceedings of ISMA 2014 - International Conference on Noise and Vibration Engineering and USD 2014 - International Conference on Uncertainty in Structural Dynamics (ISMA 2014)* Leuven, Belgium, September 2014
[19] Zhou Y-L and Wahab MA 2016 *Journal of Vibroengineering* **18**(7) 4491-4499
[20] Zhou Y-L, Wahab MA and Perera R 2015 *International Journal of Fracture Fatigue and Wear* **3** 254-259
[21] Zhou Y-L, Yang XB and Wahab MA 2016 *International Journal of Fracture Fatigue and Wear* **4** 200-203
[22] Phung Van P, Nguyen LB, Tran Vinh L, Dinh TD, Thai CH, Bordas SPA, Abdel Wahab M and Nguyen-Xuan H 2015 *International Journal of Non-Linear Mechanics* **76** 190-202
[23] Tran Vinh L, Lee J, Ly HA, Abdel Wahab M and Nguyen-Xuan H 2015 *International Journal of Mechanical Sciences* **96-97** 65-78
[24] Tran Vinh L, Lee J, Nguyen-Van H, Nguyen-Xuan H and Abdel Wahab M 2015 *International Journal of Non-Linear Mechanics* **72** 42-52
[25] Phung Van P, Abdel Wahab M, Liew KM, Bordas SPA and Nguyen-Xuan H 2015 *Composite Structures* **123** 137-149
[26] Phung Van P, De Lorenzis L, Thai CH, Abdel Wahab M and Nguyen-Xuan H 2015 *Computational Materials Science* **96** 495-505
[27] Phung-Van P, Tran LV, Ferreira AJM, Nguyen-Xuan H and Abdel-Wahab M 2016 *Nonlinear Dynamics* 1-16; doi:10.1007/s11071-11016-13085-11076
[28] Zhou Y-L and Abdel Wahab M 2016 *Journal of Vibroengineering* **18**(7) 4491-4499
[29] Gadala I, Abdel Wahab M and Alfantazi A 2016 *MATERIALS & DESIGN* **97** 287-299
[30] Yue T and Abdel Wahab M 2017 *Tribology International* **107** 274-282
[31] Yue T and Abdel Wahab M 2016 *Materials* **9** 597; doi:510.3390/ma9070597
[32] Yue T and Abdel Wahab M 2014 *Wear* **321** 53-63

[33] Wang C, Sun Q, Abdel Wahab M, Zhang X and Xu L 2015 *Waste Management* **43** 19-27
[34] Vanegas-Useche L, Abdel Wahab M and Parker G 2015 *Waste Management* **43** 28-36
[35] Resende Pereira KdF, Bordas S, Tomar S, Trobec R, Depolli M, Kosec G and Abdel Wahab M 2016 *Materials* **9** 639; doi:610.3390/ma9080639
[36] Noda N-A, Chen X, Sano Y, Wahab MA, Maruyama H, Fujisawa R and Takase Y 2016 *MATERIALS & DESIGN* **96** 476-489
[37] Hojjati Talemi R and Abdel Wahab M 2013 *Tribology International* **60** 176-186
[38] Hafiz TA, Abdel Wahab M, Crocombe AD and Smith PA 2013 *International Journal of Adhesion and Adhesives* **40** 188-196
[39] Ferjaoui A, Yue T, Abdel Wahab M and Hojjati-Talemi R 2015 *International Journal of Fatigue* **73** 66-76
[40] Chen X, Noda N-A, Abdel Wahab M, Sano Y, Maruyama H, Wang H, Fujisawa R and Takase Y 2016 *Journal of the Chinese Society of Mechanical Engineers* **37**(1) 1-10
[41] Chen X, Noda N-A, Abdel Wahab M, Akaishi Y-I, Sano Y, Takase Y and Fekete G 2015 *Acta Polytechnica Hungarica* **12**(8) 61-79
[42] X. Nguyen H, N. Nguyen T, Abdel Wahab M, Bordas SPA, Nguyen-Xuan H and P. Voa T 2017 *Computer Methods in Applied Mechanics and Engineering* **313** 904-940
[43] Tran LV, Phung-Van P, Lee J, Wahab MA and Nguyen-Xuan H 2016 *Composite Structures* **140** 655-667
[44] Thai CH, Ferreira AJM, Abdel Wahab M and Nguyen-Xuan H 2016 *Acta Mechanica* **227**(5) 1225-1250
[45] Thai C, Zenkour AM, Abdel Wahab M and Nguyen-Xuan H 2016 *Composite Structures* **139** 77-95
[46] Zhou Y-L, Maia NMM, Sampaio R, Qian XD and Wahab MA 2016 Proceedings of the 2016 Leuven Conference on Noise and Vibration Engineering (ISMA 2016) ed editors)
[47] Devriendt C and Guillaume P 2007 *Mechanical Systems and Signal Processing* (21) 2689-2696
[48] Figueiredo E, Park G, Figueiras J, Farrar CR and Worden K 2009 *Los Alamos National Laboratory Report LA-14393* Los Alamos National Laboratory
[49] Figueiredo E 2010 Porto University, Portugal
[50] Figueiredo E and Flynn E 2013 *http://wwwlanlgov/projects/national-security-education-center/engineering/ei-software-download/thanksphp*
[51] http://www.lanl.gov/projects/national-security-education-center/engineering/ei-software-download/thanks.php

IOP Conf. Series: Journal of Physics: Conf. Series **842** (2017) 012021 doi:10.1088/1742-6596/842/1/012021

Identification of Bouc-Wen hysteretic parameters based on enhanced response sensitivity approach

Li Wang, Zhong-Rong Lu*

Department of applied mechanics and engineering, Sun Yat-sen University, Guangzhou, 510006, P.R. China

E-mail: wangli75@mail.sysu.edu.cn (Li Wang), lvzhr@mail.sysu.edu.cn (Zhong-Rong Lu)

Abstract. This paper aims to identify parameters of Bouc-Wen hysteretic model using time-domain measured data. It follows a general inverse identification procedure, that is, identifying model parameters is treated as an optimization problem with the nonlinear least squares objective function. Then, the enhanced response sensitivity approach, which has been shown convergent and proper for such kind of problems, is adopted to solve the optimization problem. Numerical tests are undertaken to verify the proposed identification approach.

1. Introduction

The hysteretic behavior is frequently observed in many physical systems [1,2], such as mechanical systems, structural dampers, oscillatory circuits and so on. Mathematically, hysteresis is characterized as a special type of memory-based relation between the input and the output, that is, the output at a given time instant depends not only on the instantaneous input but also on its past history. Various mathematical models [1,3–5], including Preisach, Ishlinskii, Bouc-Wen models, have been proposed to describe the hysteretic behavior. In this paper, the focus is on the Bouc-Wen model since it can represent a wide variety of softening/hardening, smoothing-varying/nearly-bilinear hysteretic behavior [6,7].

Identifying the Bouc-Wen hysteretic parameters is an important and nontrivial task in practical application due to the inherent nonlinearity and memory nature. Over the years, two main kinds of identification methods have been developed. On the one hand, introducing the time discretization and treating the parameters as state variables, the problem becomes a typical discrete state equation and then, parameters are obtained through state estimate using the extended/unscented Kalman filter [8–10] or the wavelet multiresolution analysis [11]. The whole procedure is very simple, however, in order to make the discrete state equation well approximate the original continuous problem, the sampling time step should be small enough. What's worse, the extended/unscented Kalman filter is not an unbiased estimator for such strongly nonlinear problems and in wavelet analysis, the exponential parameter of the Bouc-Wen model should be known *a priori*.

On the other hand, identification of the hysteretic parameters from the measured data belongs to a class of inverse problem and is typically formulated as an optimization problem. The objective function is often established as least squares of the error between the measured data and the derived data and then, actual model parameters should minimize the objective function.

Different kinds of the measured data have led to different identification procedures. Ni, Ko and Wong [12] proposed to use the frequency-domain displacement data from periodic vibration experiment and then, the Levenberg-Marquardt algorithm was adopted to solve the nonlinear least squares optimization problem. However, getting the frequency-domain data of nonlinear dynamic systems is not so straightforward as of linear systems and considerable efforts should be taken for large amount of experimental frequency-domain data. In contrast, the time-domain data is much more easily accessed. Yar and Hammond [7] used the displacement data as well as the restoring force data in time domain to get the parameters. With both displacements and restoring forces obtained, the hysteretic loop graph can be directly plotted and then, parameters are easily derived from the graph. Sues, Roberts and Loh *et al* [13–15] identified the parameters using the time-domain data through a multi-stage estimate scheme. Though satisfactory accuracy as well as convergence was gained, the multi-stage approach is usually time-consuming.

In the present work, the enhanced response sensitivity approach [16–18] which has been widely used in structural damage identification is applied to estimate the Bouc-Wen hysteretic parameters. There are several remarkable features. Firstly, it is within the framework of the least squares optimization and therefore, there is no strict constraint on sampling time step. Secondly, the easily available time-domain data is adopted and arbitrarily single kind or combinatory of the data— the acceleration, the velocity, the displacement or even the restoring force can be used to identify the parameters. Thirdly, the proposed approach can be easily extended to parameter identification of other hysteretic models.

2. Problem statement

2.1. Bouc-Wen hysteretic model

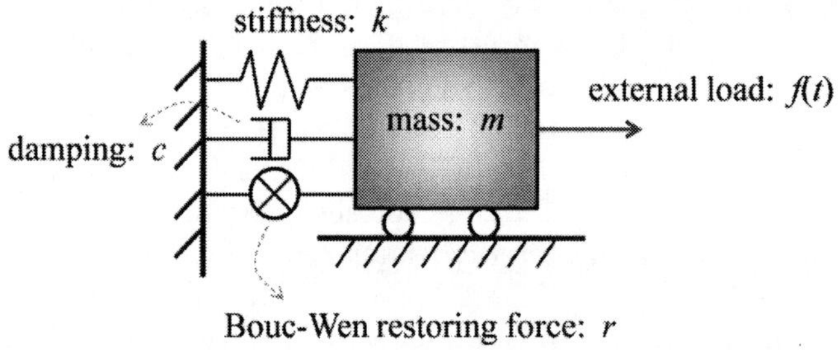

Figure 1. SDOF system with Bouc-Wen hysteresis

Consider a single degree-of-freedom (SDOF) system with the Bouc-Wen hysteretic/restoring force (see Figure 1). The governing equation for motion of the mass is given by

$$\begin{cases} m\ddot{x} + c\dot{x} + kx + r = f(t), t \geq 0 \\ x(0) = x_0, \dot{x}(0) = \dot{x}_0 \end{cases} \tag{1}$$

where x is the displacement, $x_0, \dot{x}_0$ are initial displacement and velocity of the system and m, c, k, r are respective mass, damping, stiffness and restoring force of the system as shown in Figure 1. In case of the Bouc-Wen hysteretic model, the restoring force r with initial value $r(0) = r_0$ pertains to the following equation [19, 20],

$$\dot{r} = \frac{1}{\eta}[A\dot{x} - \nu(\beta r|r|^{n-1}|\dot{x}| + \gamma|r|^n\dot{x})] \tag{2}$$

where $\eta, A, \nu, \beta, \gamma$ and n are design parameters for the hysteretic behavior. In order to render the parameters identifiable and without loss of generality, the parameters η, ν, which are typically used to control degrading and pinching behavior [10], respectively, are both set to unity herein. The exponential parameter verifies $n \geq 1$ and in order to fulfill the thermodynamic admissibility [20], another constraint $|\gamma| \leq \beta$ should be enforced. In this paper, the linear system parameters m, c, k as well as the Bouc-Wen hysteretic parameters A, β, γ, n are to be identified and the admissible space for all parameters are set to be

$$\mathcal{A} = \{\boldsymbol{\alpha} := (1/m, c/m, k/m, A, \beta, \gamma, n) : 1/m > 0, c/m \geq 0, k/m \geq 0, A > 0, |\gamma| \leq \beta, n \geq 1\}. \tag{3}$$

For practical computation, the problem (1) and (2) is reformulated as

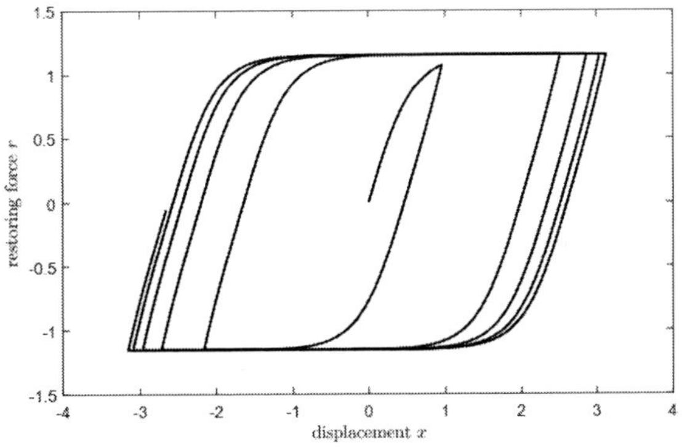

Figure 2. Graph of Bouc-Wen hysteretic loop with parameters $A = 2, \beta = 1, \gamma = 0.5, n = 2$

$$\dot{\boldsymbol{x}} = \boldsymbol{F}(\boldsymbol{x}, \boldsymbol{\alpha}, t) \Leftrightarrow \begin{cases} \dot{x}_1 = x_2 \\ \dot{x}_2 = \alpha_1(f(t) - x_3) - \alpha_2 x_2 - \alpha_3 x_1 \\ \dot{x}_3 = \alpha_4 x_2 - (\alpha_5 x_3 |x_3|^{\alpha_7 - 1}|x_2| + \alpha_6 |x_3|^{\alpha_7} x_2) \end{cases} \tag{4}$$

where $\boldsymbol{x} = [x; y; r]$ and from here on, v_k denotes the kth element of the vector $\boldsymbol{v}$. Then, given parameters $\boldsymbol{\alpha} \in \mathcal{A}$ of the system, the displacement x, the velocity $\dot{x} = y$, the acceleration $\ddot{x} = \dot{y}$ and the restoring force r of the forward problem are solved from (4) by the Runge-Kutta methods, for which one can refer to the ode45 function of MATLAB for details. To visual, the hysteretic loop graph under parameters $m = 1, c = 0.2, k = 1, A = 2, \beta = 1, \gamma = 0.5, n = 2$ and loads $f(t) = 2\cos(t)$ for $t \in [0, 30]$ is shown in Figure 2.

2.2. Inverse problem

Now, assume that some measured data at the time points $t_1 < t_2 < ... < t_l$ has been obtained, which is designated as a column vector $\hat{\boldsymbol{R}} = [\hat{R}_1; ...; \hat{R}_l]$ with $\hat{R}_j := \hat{R}(t_j)$ and then, the parameters are to be identified from the data; this forms a typical inverse identification problem. Practically, the measured quantity R can be one or more of the four quantities—the displacement x, the velocity $\dot{x}$, the acceleration $\ddot{x}$ and the restoring force r and for convenience purpose, it is designated that $R = \mathcal{L}\boldsymbol{x}$ with $\mathcal{L}$ a linear operator, e.g., $\mathcal{L} = [1, 0, 0]$ for $R = x$ and $\mathcal{L} = [0, \frac{d}{dt}, 0]$ for $R = \ddot{x}$. Special attention should be paid to the case when only the restoring force r is measured since then the parameters $\mathrm{a}m, \mathrm{a}c, \mathrm{a}k, \mathrm{a}A, \mathrm{a}\beta, \mathrm{a}\gamma, n$ can lead to the same restoring force solution for all positive constants $\mathrm{a} > 0$. Thus, when only the restoring force r is measured, one shall additional set, for instance, the mass to be known and then, other parameters are well identifiable.

Mathematically, one can deduce from the forward problem (4) that the measured quantity at the time points $t_1 < t_2 < ... < t_l$ is an implicit function of the parameters and therefore, is expressed as $\boldsymbol{R}(\boldsymbol{\alpha}) := [R(\boldsymbol{\alpha}, t_1); ...; R(\boldsymbol{\alpha}, t_l)]$. As usual, the inverse identification procedure can be formulated as: find the actual parameters $\boldsymbol{\alpha}^* \in \mathcal{A}$ such that the objective function $g(\boldsymbol{\alpha})$—defined as the weighted least squares of the error between the measured data $\hat{\boldsymbol{R}}$ and the derived data $\boldsymbol{R}(\boldsymbol{\alpha})$—is minimized, or, more precisely,

$$\boldsymbol{\alpha}^* = \arg\min_{\boldsymbol{\alpha}\in\mathcal{A}} \left\{ g(\boldsymbol{\alpha}) := ||\hat{\boldsymbol{R}} - \boldsymbol{R}(\boldsymbol{\alpha})||^2_{\mathbf{W}} \right\} \tag{5}$$

where $\mathbf{W}$ is a user-defined (positive definite) weight matrix and $||\boldsymbol{v}||^2_{\mathbf{W}} = \boldsymbol{v}^T\mathbf{W}\boldsymbol{v}$. Under the above setting, identifying the parameters has become an strongly nonlinear optimization problem. In what follows, how to efficiently and robustly get the minimizer of the objective function $g(\boldsymbol{\alpha})$ by an enhanced response sensitivity approach is elaborated.

3. Enhanced response sensitivity for parameter identification

Usually, $\boldsymbol{R}(\boldsymbol{\alpha})$ is a nonlinear function of $\boldsymbol{\alpha}$ and therefore, the objective function $g(\boldsymbol{\alpha})$ is in the nonlinear least squares form. For such a nonlinear least-squares optimization problem (5), iterative methods should be used for which the key ingredient is to determine an update $\Delta\boldsymbol{\alpha}$ from a given $\bar{\boldsymbol{\alpha}}$ so that $g(\bar{\boldsymbol{\alpha}} + \Delta\boldsymbol{\alpha})$ becomes as small as possible. The Newton method is costly and improper [18] for such a problem. A simple yet feasible strategy is to linearize $\hat{\boldsymbol{R}} - \boldsymbol{R}(\boldsymbol{\alpha})$ at $\bar{\boldsymbol{\alpha}}$ [18,21], i.e.,

$$\hat{\boldsymbol{R}} - \boldsymbol{R}(\boldsymbol{\alpha}) = \Delta\boldsymbol{R}(\bar{\boldsymbol{\alpha}}) - \mathbf{S}(\bar{\boldsymbol{\alpha}})\Delta\boldsymbol{\alpha}; \Delta\boldsymbol{R}(\bar{\boldsymbol{\alpha}}) := \hat{\boldsymbol{R}} - \boldsymbol{R}(\bar{\boldsymbol{\alpha}}), \mathbf{S}(\bar{\boldsymbol{\alpha}}) = \nabla_{\boldsymbol{\alpha}}\boldsymbol{R}(\bar{\boldsymbol{\alpha}}), \Delta\boldsymbol{\alpha} = \boldsymbol{\alpha} - \bar{\boldsymbol{\alpha}}. \tag{6}$$

Specifically, the key matrix $\mathbf{S}(\boldsymbol{\alpha})$ is computed as

$$\mathbf{S}(\boldsymbol{\alpha}) = [\nabla_{\boldsymbol{\alpha}}R(\boldsymbol{\alpha}, t_1); ...; \nabla_{\boldsymbol{\alpha}}R(\boldsymbol{\alpha}, t_l)]; \nabla_{\boldsymbol{\alpha}}R(\boldsymbol{\alpha}, t_k) = \mathcal{L}\nabla_{\boldsymbol{\alpha}}\boldsymbol{x}(\boldsymbol{\alpha}, t_k). \tag{7}$$

Note that the key term $\nabla_{\boldsymbol{\alpha}}\boldsymbol{x}$ is just the (first-order) sensitivity of the response $\boldsymbol{x}$ to the parameters $\boldsymbol{\alpha}$ and this initiates the name '*response sensitivity approach*'. Obviously, $\nabla_{\boldsymbol{\alpha}}\boldsymbol{x}$ is time-dependent and pertains to the following ordinary differential equation by applying the differential chain rule to (4),

$$\frac{d(\nabla_{\boldsymbol{\alpha}}\boldsymbol{x})}{dt} = \nabla_{\boldsymbol{x}}\boldsymbol{F}(\boldsymbol{x}, \boldsymbol{\alpha}, t) \cdot \nabla_{\boldsymbol{\alpha}}\boldsymbol{x} + \nabla_{\boldsymbol{\alpha}}\boldsymbol{F}(\boldsymbol{x}, \boldsymbol{\alpha}, t). \tag{8}$$

For the SDOF system (4) with Bouc-Wen hysteresis, there are

$$\nabla_{\boldsymbol{x}}\boldsymbol{F}(\boldsymbol{x}, \boldsymbol{\alpha}, t) = \begin{pmatrix} 0 & 1 & 0 \\ -\alpha_3 & -\alpha_2 & -\alpha_1 \\ 0 & \frac{\partial r}{\partial x_2} & \frac{\partial r}{\partial x_3} \end{pmatrix}; \begin{array}{l} \frac{\partial r}{\partial x_2} := \alpha_4 - (\alpha_5 x_3|x_3|^{\alpha_7-1}\mathrm{sgn}(x_2) + \alpha_6|x_3|^{\alpha_7}) \\ \frac{\partial r}{\partial x_3} := -\alpha_7(\alpha_5|x_3|^{\alpha_7-1}|x_2| + \alpha_6|x_3|^{\alpha_7-1}x_2\mathrm{sgn}(x_3)) \end{array} \tag{9}$$

and

$$\nabla_{\boldsymbol{\alpha}}\boldsymbol{F}(\boldsymbol{x}, \boldsymbol{\alpha}, t) = \begin{pmatrix} 0 & 0 & 0 & 0 & 0 & 0 & 0 \\ f(t) - x_3 & -x_2 & -x_1 & 0 & 0 & 0 & 0 \\ 0 & 0 & 0 & \frac{\partial r}{\partial \alpha_4} & \frac{\partial r}{\partial \alpha_5} & \frac{\partial r}{\partial \alpha_6} & \frac{\partial r}{\partial \alpha_7} \end{pmatrix};$$
$$\frac{\partial r}{\partial \alpha_4} := x_2, \frac{\partial r}{\partial \alpha_5} := -x_3|x_3|^{\alpha_7-1}|x_2|, \tag{10}$$
$$\frac{\partial r}{\partial \alpha_6} := -|x_3|^{\alpha_7}x_2, \frac{\partial r}{\partial \alpha_7} := -(\alpha_5 x_3|x_3|^{\alpha_7-1}|x_2| + \alpha_6|x_3|^{\alpha_7}x_2)\ln(|x_3| + \epsilon)$$

where sgn$(\cdot)$ is the usual sign function and $\epsilon > 0$ is a small positive constant, e.g.,=1.0e-32, to guarantee the positiveness of $|x_3|+\epsilon$ for $\ln(\cdot)$ function. Combining (4) with (8) and setting $\boldsymbol{\alpha} = \bar{\boldsymbol{\alpha}}$, the response $\boldsymbol{R}(\bar{\boldsymbol{\alpha}})$ and the response sensitivity $\mathbf{S}(\bar{\boldsymbol{\alpha}})$ are obtainable. Using the linearization (6), an approximate objective function $\hat{g}(\bar{\boldsymbol{\alpha}}, \Delta\boldsymbol{\alpha}) = ||\Delta \boldsymbol{R}(\bar{\boldsymbol{\alpha}}) - \mathbf{S}(\bar{\boldsymbol{\alpha}})\Delta\boldsymbol{\alpha}||^2_{\mathbf{W}}$ is obtained which is in the linear least squares form.

The update $\Delta\boldsymbol{\alpha}$, if computed as minimizer of the linear least squares objective function $\hat{g}(\Delta\boldsymbol{\alpha}, \bar{\boldsymbol{\alpha}})$, can be easily and directly obtained; however, $\hat{g}(\Delta\boldsymbol{\alpha}, \bar{\boldsymbol{\alpha}})$ is only an approximation of the original objective function $g(\bar{\boldsymbol{\alpha}} + \Delta\boldsymbol{\alpha})$ through linearization. Note that linearization can give good approximation of the original nonlinear nonlinear function only when $\Delta\boldsymbol{\alpha} + \bar{\boldsymbol{\alpha}}$ is in the vicinity/trust-region of $\bar{\boldsymbol{\alpha}}$, in other words, the trust-region constraint $||\Delta\boldsymbol{\alpha}|| \leq \eta$ should hold for some small positive number η. On considering this, a reliable way to get the update $\Delta\boldsymbol{\alpha}$ is clarified: for a certain $\eta > 0$, one has

$$\Delta\boldsymbol{\alpha} = \arg \min_{\delta\boldsymbol{\alpha} \in \mathcal{A} - \bar{\boldsymbol{\alpha}}, ||\delta\boldsymbol{\alpha}|| \leq \eta} ||\Delta \boldsymbol{R}(\bar{\boldsymbol{\alpha}}) - \mathbf{S}(\bar{\boldsymbol{\alpha}})\delta\boldsymbol{\alpha}||^2_{\mathbf{W}}. \tag{11}$$

Moreover, to measure how well the linearized version $\hat{g}(\Delta\boldsymbol{\alpha}, \bar{\boldsymbol{\alpha}})$ agrees with the original nonlinear version $g(\bar{\boldsymbol{\alpha}} + \Delta\boldsymbol{\alpha})$, an *agreement indicator* $\rho(\Delta\boldsymbol{\alpha}, \bar{\boldsymbol{\alpha}})$ [18, 21] is introduced,

$$\rho(\Delta\boldsymbol{\alpha}, \bar{\boldsymbol{\alpha}}) := \frac{g(\bar{\boldsymbol{\alpha}}) - g(\bar{\boldsymbol{\alpha}} + \Delta\boldsymbol{\alpha})}{\hat{g}(\mathbf{0}, \bar{\boldsymbol{\alpha}}) - \hat{g}(\Delta\boldsymbol{\alpha}, \bar{\boldsymbol{\alpha}})} = \frac{||\Delta \boldsymbol{R}(\bar{\boldsymbol{\alpha}})||^2_{\mathbf{W}} - ||\Delta \boldsymbol{R}(\bar{\boldsymbol{\alpha}} + \Delta\boldsymbol{\alpha})||^2_{\mathbf{W}}}{||\Delta \boldsymbol{R}(\bar{\boldsymbol{\alpha}})||^2_{\mathbf{W}} - ||\Delta \boldsymbol{R}(\bar{\boldsymbol{\alpha}}) - \mathbf{S}(\bar{\boldsymbol{\alpha}})\Delta\boldsymbol{\alpha}||^2_{\mathbf{W}}}. \tag{12}$$

Good agreement requires $\rho(\Delta\boldsymbol{\alpha}, \bar{\boldsymbol{\alpha}}) \geq \rho_{cr}$ where ρ_{cr} often verifies $\rho_{cr} \in [0.25, 0.75]$. To conclude, in view of (11), different ηs result in different updates $\Delta\boldsymbol{\alpha}$ and then, one natural idea is to properly choose the η such that the resulting update verifies the *agreement condition* $\rho(\Delta\boldsymbol{\alpha}, \bar{\boldsymbol{\alpha}}) \geq \rho_{cr}$, which constitutes just the core of trust-region algorithms [21].

In the above derivation, the response sensitivity approach is enhanced with the trust-region restriction, leading to the *enhanced response sensitivity approach.* Nonetheless, solving the inequality-constraint problem (11) is not easy and straightforward at all. Herein, an equivalent but much simpler manner [18] to deal with the trust-region consideration by the Tikhonov regularization is adopted. The procedure reads: find a proper Tikhonov regularization parameter $\lambda > 0$ such that the following update

$$\begin{aligned} \Delta\boldsymbol{\alpha}_\lambda &= \arg \min_{\delta\boldsymbol{\alpha} \in \mathcal{A} - \bar{\boldsymbol{\alpha}}} \{||\Delta \boldsymbol{R}(\bar{\boldsymbol{\alpha}}) - \mathbf{S}(\bar{\boldsymbol{\alpha}})\delta\boldsymbol{\alpha}||^2_{\mathbf{W}} + \lambda ||\delta\boldsymbol{\alpha}||^2\} \\ &= [\mathbf{S}^T(\bar{\boldsymbol{\alpha}})\mathbf{W}\mathbf{S}(\bar{\boldsymbol{\alpha}}) + \lambda\mathbf{I}]^{-1}\mathbf{S}^T(\bar{\boldsymbol{\alpha}})\mathbf{W}\Delta \boldsymbol{R}(\bar{\boldsymbol{\alpha}}) \end{aligned} \tag{13}$$

satisfies the agreement condition $\rho(\Delta\boldsymbol{\alpha}_\lambda, \bar{\boldsymbol{\alpha}}) \geq \rho_{cr}$. On the one hand, (13) is still posed to be a linear least squares problem and therefore, is directly solved. On the other hand, the proper Tikhonov regularization parameter $\lambda > 0$ can be simply found as follows: (a) set the regularization parameter $\lambda_L(\bar{\boldsymbol{\alpha}})$ obtained by the L-curve method [22, 23] as the initial regularization parameter, i.e., $\lambda := \lambda_L(\bar{\boldsymbol{\alpha}})$ since such a choice has already been shown to work well for many linear system parameter identification problems [16, 17]; (b) compute the update $\Delta\boldsymbol{\alpha}_\lambda$ as (13); (c) calculate the agreement indicator $\rho(\Delta\boldsymbol{\alpha}_\lambda, \bar{\boldsymbol{\alpha}})$; (d) if the agreement condition is satisfied, terminate the procedure and both the proper λ and the proper update $\Delta\boldsymbol{\alpha}_\lambda$ are obtained, otherwise, increase λ up to a factor $a > 1$, i.e., $\lambda := a\lambda$ and return to step (b). Up till now, a general way to get a proper update from a certain $\bar{\boldsymbol{\alpha}}$ has been presented and based on this, an iterative yet convergent algorithm [18] to get the minimizer of $g(\boldsymbol{\alpha})$ or identify the parameters is naturally established as given in Table 1.

4. Numerical tests

The SDOF system with Bouc-Wen hysteretic force in Figure 1 is elaborately studied herein to testify the proposed parameter identification approach. Actual system parameters are

Table 1. Algorithmic details for enhanced response sensitivity approach

- set initial parameters $\boldsymbol{\alpha}_1 \in \mathcal{A}$ and define weight matrix $\mathbf{W}$,
- define error tolerance tol (e.g., $= 10^{-6}$) for convergence criterion,
- fix the maximum number of iterations $Nmax$ (e.g., =1000),
- fix trust-region parameters $\rho_{cr} \in [0.25, 075]$ (e.g., =0.5) and $a > 1$ (e.g., =2),
- set the maximum number of steps for trust-region procedure Ntr, (e.g.,=20),
- load the measured response data $\hat{\boldsymbol{R}}$,
- for $k = 1 : Nmax$
  - solve (4) and (8) to get response $\boldsymbol{R}(\boldsymbol{\alpha}_k)$ and response sensitivity $\mathbf{S}(\boldsymbol{\alpha}_k)$,
  - compute response change $\Delta\boldsymbol{R} = \hat{\boldsymbol{R}} - \boldsymbol{R}(\boldsymbol{\alpha}_k)$,
  - use L-curve method to get the initial regularization parameter $\lambda_L(\boldsymbol{\alpha}_k)$,
  - for $i = 1 : Ntr$ *% Trust-region procedure*
    - $\lambda = a^{i-1}\lambda_L(\boldsymbol{\alpha}_k)$,
    - compute the update $\Delta\boldsymbol{\alpha} = [\mathbf{S}^T(\boldsymbol{\alpha}_k)\mathbf{W}\mathbf{S}(\boldsymbol{\alpha}_k) + \lambda\mathbf{I}]^{-1}\mathbf{S}^T(\boldsymbol{\alpha}_k)\mathbf{W}\Delta\boldsymbol{R}$,
    - if $\boldsymbol{\alpha}_k + \Delta\boldsymbol{\alpha} \notin \mathcal{A}$ continue,
    - solve (4) to get response $\boldsymbol{R}(\boldsymbol{\alpha}_k + \Delta\boldsymbol{\alpha})$,
    - compute new response change $\Delta\boldsymbol{R}_{new} = \hat{\boldsymbol{R}} - \boldsymbol{R}(\boldsymbol{\alpha}_k + \Delta\boldsymbol{\alpha})$,
    - calculate the agreement indicator $\rho = \frac{||\Delta\boldsymbol{R}||^2_{\mathbf{W}} - ||\Delta\boldsymbol{R}_{new}||^2_{\mathbf{W}}}{||\Delta\boldsymbol{R}||^2_{\mathbf{W}} - ||\Delta\boldsymbol{R} - \mathbf{S}(\boldsymbol{\alpha}_k)\Delta\boldsymbol{\alpha}||^2_{\mathbf{W}}}$,
    - if $\rho \geq \rho_{cr}$ break,
  - end for
  - update stiffness parameters $\boldsymbol{\alpha}_{k+1} = \boldsymbol{\alpha}_k + \Delta\boldsymbol{\alpha}$,
  - if $||\Delta\boldsymbol{\alpha}||/||\boldsymbol{\alpha}_{k+1}|| \leq tol$ break.
- end for

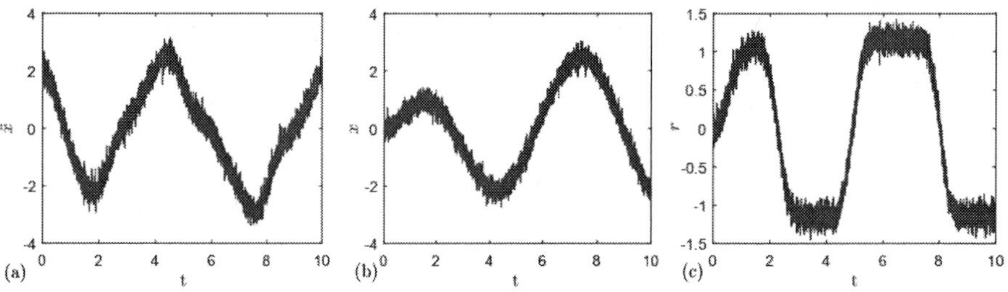

Figure 3. Measured data with 5% noise for parameter identification: (a) acceleration $\ddot{x}$, (b) displacement x and (c) restoring force r

$\boldsymbol{\alpha}^* = (1, 0.2, 1, 2, 1, 0.5, 2)$. A load $f(t) = 2\cos(t)$ is enforced and the initial states are zero. Herein, the measured data is obtained from numerical simulation over the time domain $[0, 10]$ at the sampling rate of 500Hz and possible measurement noise is added to simulated solution as follows,

$$\hat{\boldsymbol{d}} = \boldsymbol{d} + e_{noise}\boldsymbol{R}_{and}var(\boldsymbol{d}) \tag{14}$$

where $\boldsymbol{d} = [d(t_1); ...; d(t_l)]$, $\hat{\boldsymbol{d}}$ represents the respective simulated and polluted solutions, e_{noise} is the noise level (e.g., =5%), $\boldsymbol{R}_{and}$ is a standard normal distribution vector of length l and $var(\boldsymbol{d})$ is the variance over the time history $\boldsymbol{d}$. To visual, the simulated acceleration, displacement and restoring force with measurement noise $e_{noise} = 5\%$ are shown in Figure 3.

Four identification scenarios as detailed in Table 2 are considered. The algorithm in Table 1 is adopted to derive the parameters from the measured data in all four scenarios, for which,

Table 2. Four identification scenarios for SDOF system with Bouc-Wen hysteresis

scenario	measured quantities	measurement noise
1	acceleration $\ddot{x}$	0%
2	acceleration $\ddot{x}$	5%
3	displacement x and restoring force r	0%
4	displacement x and restoring force r	5%

Table 3. Four identification scenarios for SDOF system with Bouc-Wen hysteresis

scenario	identified results
1	(1.0000,0.2000,1.0000,1.9999,1.0001,0.4997,2.0005)
2	(0.9985,0.1933,0.9989,2.0128,1.0202,0.4678,1.9716)
3	(1.0000,0.2000,1.0000,1.9998,1.0002,0.4996,2.0006)
4	(1.0001,0.2003,1.0001,2.0028,0.9990,0.5000,2.0241)

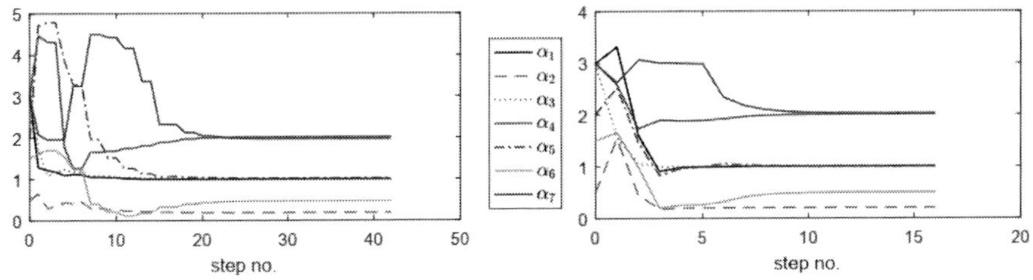

Figure 4. Identification procedure for scenario 2 (left) and scenario 4 (right)

the initial parameters are set to be $\boldsymbol{\alpha}_1 = (3, 0.5, 3, 3, 2, 1.5, 3)$, the weight matrix $\mathbf{W}$ equals to identity matrix and other parameters including $tol, Nmax, \rho_{cr}, a, N_{tr}$ take the default values in corresponding brackets of Table 1. Using this algorithm, the parameters can be identified and results are displayed in Table 3. Moreover, to see how the results converge in the iterative procedure, the results in each step of the algorithm are recorded and those for scenarios 2 and 4 are plotted in Figure 4. In case of no noise, either the measured $\ddot{x}$ or the measured x, r can give rise to quite accurate results. While in case of 5% noise, the maximum relative error for measured $\ddot{x}$ (scenario 2) reaches 6% but is only 1.2% for measured x, r (scenario 4) and the results are identified in 42 steps for scenario 2 and 16 steps for scenario 4, this is somehow reasonable since more data often leads to more accurate and faster identification, nevertheless, the identified results are satisfactory.

5. Conclusions

A general way to identify the Bouc-Wen hysteretic parameters has been presented in this paper. It follows the general inverse identification procedure, i.e., transforming the identification to an optimization problem and therefore, can be easily generalized to other hysteretic model parameter identification problems. Then, a recently-proposed enhanced response sensitivity approach is used to solve the nonlinear least squares optimization problem. Numerical tests on four identification scenarios, using single acceleration measurement or displacement and restoring force measurements, with 0% or 5% measurement noise are conducted. All parameters are well identified, highlighting the proposed identification approach.

Acknowledgement
The present investigation was performed under the support of the National Natural Science Foundation of China (No. 11172333) and Guangdong Province Natural Science Foundation (No. 2015A030313126).

References
[1] Macki JW, Nistri P, Zecca P. Mathematical models for hysteresis. *SIAM Review* 1993, 35(1): 94-123.
[2] Ismail M, Ikhouane F, Rodellar J. The hysteresis Bouc-Wen model, a survey. *Archives of computational methods in engineering* 2009, 16: 161-188.
[3] Wen YK. Method of random vibration of hysteretic systems. *ASCE Journal of the Engineering Mechanics Division* 1976, 102: 249-263.
[4] Belbas SA, Mayergoyz ID. Optimal control of dynamical systems with Preisach hysteresis. *International journal of nonlinear mechanics* 2002, 37: 1351-1361.
[5] Ikhouane F., Rodellar J. On the hysteretic Bouc-Wen model. Part I: Forced limit cycle characterization. *Nonlinear dynamics* 2005, 42: 63-78.
[6] Rochdi Y, Giri F, Ikhouane F, Chaoui FZ. Parametric identification of nonlinear hysteretic systems. *Nonlinear dynamics* 2009, 58: 393-404.
[7] Yar M, Hammond JK. Parameter estimation for hysteretic systems. *Journal of sound and vibration* 1987, 117: 161-172.
[8] Lin JS, Zhang Y. Nonlinear structural identification using extended Kalman filter. *Computers and structures* 1994, 52: 757-764.
[9] Corigliano A, Mariani S. Parameter identification in explicit structural dynamics: performace of the extended Kalman filter. *Computer methods in applied mechanics and engineering* 2004, 193: 3807-3835.
[10] Wu M, Smyth A. Real-time parameter estimation for degrading and pinching hysteretic models. *International jounral of nonlinear mechanics* 2008, 43: 822-833.
[11] Chang CC, Shi Y. Identification of time-varying hysteretic structures using wavelet multiresolution analysis. *International journal of nonlinear mechanics* 2010, 45: 21-34.
[12] Ni YQ, Ko JM, Wong CW. Identification of nonlinear hysteretic isolators from periodic vibration tests. *Journal of sound and vibration* 1998, 217(4): 737-756.
[13] Sues RH, Mau ST, Wen YK. Systems identification of degrading hysteretic restoring forces. *ASCE Journal of the Engineering Mechanics* 1988, 114: 833-846.
[14] Roberts JB, Sadeghi AH. Sequential parameter identification and response of hysteretic oscillators with random excitation. *Structural safety* 1990, 8: 45-68.
[15] Loh CH, Chung ST. A three-stage identification approach for hysteretic systems. *Earthquake engineering and structural dynamics* 1993, 22: 129-150.
[16] Lu ZR, Law SS. Features of dynamic response sensitivity and its application in damage detection. *Journal of sound and vibration* 2007, 303: 305-329.
[17] Fu YZ, Lu ZR, Liu JK. Damage identification in plates using finite element model updating in time domain. *Journal of sound and vibration* 2013, 332: 7018-7032.
[18] Lu ZR, Wang L. An enhanced response sensitivity approach for structural damage identification: convergence and performance. *International journal for numerical methods in engineering* 2017, DOI: 10.1002/nme.5502.
[19] Brewick PT, Masri SF. An evaluation of data-driven identificaton strategies for complex nonlinear dynamic systems. *Nonlinear dynamics* 2016, 85: 1297-1318.
[20] Erlicher S, Point N. Thermodynamic admissibility of Bouc-Wen type hysteresis models. *Comptes rendus mecanique* 2004, 332: 51-57.
[21] More JJ. The Levenberg-Marquardt algorithm: implementation and theory. Chapter Numerical analysis, volume 630 of the series *Lecture Notes in Mathematics* 1978, pp: 105-116.
[22] Hansen PC. Analysis of discrete ill-posed problems by means of the L-curves. *SIAM Review* 1992, 34(4): 561-580.
[23] Hansen PC. Regularization tools—A matlab package for analysis and solution of discrete ill-posed problem. *Numerical algorithms* 1994, 6(1): 1-35.

Multi-damage identification based on joint approximate diagonalisation and robust distance measure

S Cao[1] and H Ouyang[1,2]

[1] Centre for Engineering Dynamics, School of Engineering, The University of Liverpool, Liverpool, L69 3GH, U.K.
[2] State Key Laboratory of Structural Analysis for Industrial Equipment, Dalian University of Technology, Dalian 116023, China

E-mail: h.ouyang@liverpool.ac.uk

Abstract. Mode shapes or operational deflection shapes are highly sensitive to damage and can be used for multi-damage identification. Nevertheless, one drawback of this kind of methods is that the extracted spatial shape features tend to be compromised by noise, which degrades their damage identification accuracy, especially for incipient damage. To overcome this, joint approximate diagonalisation (JAD) also known as simultaneous diagonalisation is investigated to estimate mode shapes (MS's) statistically. The major advantage of JAD method is that it efficiently provides the common Eigen-structure of a set of power spectral density matrices. In this paper, a new criterion in terms of coefficient of variation (CV) is utilised to numerically demonstrate the better noise robustness and accuracy of JAD method over traditional frequency domain decomposition method (FDD). Another original contribution is that a new robust damage index (DI) is proposed, which is comprised of local MS distortions of several modes weighted by their associated vibration participation factors. The advantage of doing this is to include fair contributions from changes of all modes concerned. Moreover, the proposed DI provides a measure of damage-induced changes in 'modal vibration energy' in terms of the selected mode shapes. Finally, an experimental study is presented to verify the efficiency and noise robustness of JAD method and the proposed DI. The results show that the proposed DI is effective and robust under random vibration situations, which indicates that it has the potential to be applied to practical engineering structures with ambient excitations.

1. Introduction

Vibration-based structural damage identification and health monitoring attract more and more attention recently. This is an inverse problem and the basic approach to addressing this issue is to monitor damage-induced changes in vibration responses [1-2]. Basically, modal parameters and vibration response statistics which are extracted from vibration responses are used to identify the damage. However, a noticeable drawback of detecting damage through the estimated damage features is that they are easily compromised by noise, which generally comes from four sources: operational, environmental, measurement and computational [3]. To overcome this noise problem, a current trend is to estimate spatial modal parameters based on the common Eigen-structure of correlation/covariance matrices or power spectral density (PSD) matrices, which are functions of vibration responses [4-5]. From a statistical point of view, the identified averaged spatial modal parameters will be much more accurate and robust for structural damage identification.

For a standard second-order blind identification (SOBI), two steps are traditionally adopted: vibration data is pre-whitened and then joint approximate diagonalisation (JAD) is applied. It is noticed that the pre-whitening procedure is possible to introduce bias or error to SOBI, which cannot be corrected in the JAD step [6]. Moreover, SOBI has limitations in detecting the presence of spatially close modes or repeated frequencies. In order to solve these problems, applying the JAD algorithm to PSD matrices for mode shape (MS) estimation is investigated. The PSD matrix possesses the auto- and cross- spectral correlations of output vibration responses and provides an average energy distribution of a random process. Different from traditional frequency domain decomposition (FDD) method, the proposed JAD diagonalises a set of PSD matrices to obtain the common eigenvectors, which minimises the effect of the leakage error and statistically increase the noise robustness of estimated mode shapes. In this paper, MS's are estimated individually based on JAD method using a narrow frequency band around each resonant frequency without the pre-whitening procedure.

The mathematical approach of applying JAD to PSD matrices was presented by the authors in [5]. In this study, a new damage index (DI) is proposed as the squared Euclidean distance of MS local distortions and the distortions of different MS's are weighted by their corresponding vibration participation factors. In addition, this damage index has the ability to indicate the relative damage severity. In the numerical study, two linear cracks are modelled using fracture mechanics approach [7-8] to validate the feasibility of the proposed DI and coefficient of variation (CV) of MS's over 1000 Gaussian noise realisations is adopted to better demonstrate the noise robustness of JAD method. Then an experimental study of a beam with two cracks tested by PSV500 Scanning Laser Vibrometer is presented to verify noise robustness of MS estimation and the proposed DI. Moreover, the physical meaning and the roles of the real part and imaginary part of MS's in damage identification are discussed.

2. Robust MS estimation based on JAD method

The stochastic vibrations of structures under operational conditions are essentially uncertain in nature and statistical models should be utilised to analyse the random processes. In this study, measured random velocity responses $\mathbf{Y}(t) \in \mathbb{R}^{N_o \times N}$ collected at N_o locations with N samples are used to formulate PSD matrices and estimate MS's according to JAD method. In terms of modal expansion theorem of structural response, measured vibration responses can be expressed using mode shapes and modal coordinates as

$$\mathbf{y}(t_j) = \mathbf{\Phi}_\mathrm{y}\mathbf{q}(t_j) + \mathbf{v}(t_j),\ j = 1,2,\cdots,N \tag{1}$$

where $\mathbf{q}(t) \in \mathbb{R}^n$ is the modal coordinate vector (the total number of modes is n) , $\mathbf{\Phi}_\mathrm{y} \in \mathbb{R}^{N_o \times n}$ represents the mode shape matrix at the measured degrees of freedom and $\mathbf{v}(t_j) \in \mathbb{R}^{N_o}$ denotes a vector of measurement noise. Under the general assumption that $\mathbf{v}(t)$ and $\mathbf{\Phi}_\mathrm{y}\mathbf{q}(t)$ are uncorrelated, correlation matrix should be computed as

$$\mathbf{R}(\tau) = \mathbf{\Phi}_\mathrm{y}\mathbf{R}_\mathrm{qq}(\tau)\mathbf{\Phi}_\mathrm{y}^\mathrm{T} + \mathbf{R}_\mathrm{vv}(\tau) \tag{2}$$

where $\tau\ (= 0,1,\cdots,N-1)$ implies time delay. $\mathbf{R}_\mathrm{qq} \in \mathbb{R}^{n \times n}$ and $\mathbf{R}_\mathrm{vv} \in \mathbb{R}^{N_o \times N_o}$ indicate correlation matrices of modal coordinates and noise, respectively. Taking discrete Fourier transform of equation (2), the PSD matrix is obtained as

$$\mathbf{S}(\omega) = \mathbf{\Phi}_\mathrm{y}\mathbf{S}_\mathrm{qq}(\omega)\mathbf{\Phi}_\mathrm{y}^\mathrm{H} + \mathbf{S}_\mathrm{vv}(\omega) \tag{3}$$

where ω denotes the discrete frequency of excitation and superscript H indicates Hermitian transpose. In experiment, PSD matrix $\mathbf{S}(\omega)$ is calculated directly from the output vibration responses as

$$\mathbf{S}(\omega)=\begin{bmatrix} s_{y_1y_1}(\omega) & s_{y_1y_2}(\omega) & \dots & s_{y_1y_{N_o}}(\omega) \\ s_{y_2y_1}(\omega) & s_{y_2y_2}(\omega) & \cdots & s_{y_2y_{N_o}}(\omega) \\ \vdots & \vdots & \ddots & \vdots \\ s_{y_{N_o}y_1}(\omega) & s_{y_{N_o}y_2}(\omega) & \cdots & s_{y_{N_o}y_{N_o}}(\omega) \end{bmatrix} \quad (4)$$

where $\mathbf{S}(\omega)$ is a Hermitan and positive definite matrix. According to FDD method, the singular value decomposition (SVD) of $\mathbf{S}(\omega)$ is

$$\mathbf{S}(\omega)=\mathbf{U}(\omega)\mathbf{D}(\omega)\mathbf{U}(\omega)^{\mathrm{H}} \quad (5)$$

The columns of $\mathbf{U}(\omega)$ are complex unitary singular vectors and $\mathbf{D}(\omega)$ is a real nonnegative diagonal matrix with its diagonal entries in a descending order: $\sigma_1 \geq \sigma_2 \geq \cdots \geq \sigma_{N_o} \geq 0$, which represent the energy contribution factors of corresponding singular vectors in $\mathbf{U}(\omega)$ to vibration responses at this frequency. At a resonant frequency ω_r, the estimated mode shapes of measured degrees of freedom from $\mathbf{U}(\omega_r)$ could be biased due to the SVD orthogonal/unitary criterion. Fortunately, the mode shapes of weak modes at this resonant frequency are mainly affected by the bias whilst the estimated dominant mode shape is still good. Thus, the mode shapes are obtained individually using their corresponding resonant frequencies.

In FDD method, the dominant singular vector in $\mathbf{U}(\omega_r)$ is taken as the corresponding MS. From a statistical point of view, it is not accurate and robust just using a single PSD matrix at each resonant frequency. Moreover, the dominant mode shape of each resonant frequency does not change much around the resonant frequency. Therefore, the PSD matrices of several adjacent frequencies at each resonant frequency can be decomposed simultaneously for robust MS estimation. Here, a simultaneous diagonalisation technique is proposed by applying JAD method, which estimates the MS's based on a PSD matrix set according to the least squares criterion or maximum likelihood approach. Equation (6) demonstrates the problem of diagonalising a set of PSD matrices around a resonant frequency to find the common unitary matrix $\mathbf{U}(\omega_r)$:

$$\mathbf{S}(\omega_{r+k})=\mathbf{\Phi}_{\mathrm{y}}\mathbf{S}_{\mathrm{qq}}(\omega_{r+k})\mathbf{\Phi}_{\mathrm{y}}^{\mathrm{H}}+\mathbf{S}_{\mathrm{vv}}(\omega_{r+k})=\mathbf{U}(\omega_r)\mathbf{D}(\omega_{r+k})\mathbf{U}^{\mathrm{H}}(\omega_r)+\mathbf{E}(\omega_{r+k}) \quad (6)$$

where r indicates the resonant frequency ω_r, $k\ (=-K,-K+1,\cdots,K)$ denotes the adjacent frequency around ω_r and $\mathbf{E}$ represents the error matrix. The least-squares criterion is used and the over determined diagonalisation is now equivalent to a minimization problem of variables $\mathbf{U}(\omega_r)$ and $\mathbf{D}(\omega_{r+k})$:

$$J\big(\mathbf{U}(\omega_r),\mathbf{D}(\omega_{r+k})\big)=\textstyle\sum_{k=-K}^{K}\left\|\mathbf{S}(\omega_{r+k})-\mathbf{U}(\omega_r)\mathbf{D}(\omega_{r+k})\mathbf{U}^{\mathrm{H}}(\omega_r)\right\| \quad (7)$$

It is worth noting that the identified MS's are unitary complex vectors. According to the Hermitain properties of PSD matrix, the real part of MS's corresponds to the in-phase vibration whilst their imaginary part is related to the out-phase vibration. For damage identification, the damage-induced local stiffness reduction can be detected from the amplitude change of in-phase vibration or absolute value of MS's whereas the damage-induced phase changes are sensitively reflected by the imaginary part of MS's.

3. Damage identification index

With the estimated MS's $\mathbf{\Phi}^{\mathrm{d}}=\left(\boldsymbol{\varphi}_1^{\mathrm{d}},\boldsymbol{\varphi}_2^{\mathrm{d}},\cdots,\boldsymbol{\varphi}_L^{\mathrm{d}}\right)$ of damaged structures, damage identification is traditionally accomplished by comparing with the MS's $\mathbf{\Phi}=(\boldsymbol{\varphi}_1,\boldsymbol{\varphi}_2,\cdots,\boldsymbol{\varphi}_L)$ of healthy structures [9]. The difference or distance between $\mathbf{\Phi}$ and $\mathbf{\Phi}^{\mathrm{d}}$ at measurement point l is measured by

$$D_l\left(\mathbf{\Phi}\parallel\mathbf{\Phi}^{\mathrm{d}}\right)=\textstyle\sum_{r=1}^{L}w_{lr}\left|\varphi_{lr}-\varphi_{lr}^{\mathrm{d}}\right|^2 \quad (8)$$

where $\mathbf{W}$ denotes the weighting coefficient matrix and r indicates the r-th mode shape such as $\boldsymbol{\varphi}_r$. In the case that mode shape matrix of healthy structures is not available, $\mathbf{\Phi}$ is obtained by polynomial smoothing approach of $\mathbf{\Phi}^{\mathrm{d}}$ based on the assumption that the mode shapes of healthy structures are smooth [10]. In this study, gapped smoothing method (GSM) is used to estimate the undamaged mode shape matrix $\widehat{\mathbf{\Phi}}$

$$\hat{\varphi}_{lr} = c_3 x_l^{\,3} + c_2 x_l^{\,2} + c_1 x_l + c_0 \tag{9}$$

where x_l indicates the location of point l and $\mathbf{c} = [c_0, c_1, c_2, c_3]$ are coefficients of the cubic polynomial. GSM is sensitive to the local peaks and valleys of MS's associated with higher frequencies. However, the higher modes are normally more sensitive to local damage. Hence, for different mode shapes, a compromise between damage identification accuracy and sensitivity should be made when using GSM. The proposed damage index (DI) is computed as

$$DI(\widehat{\mathbf{\Phi}} \parallel \mathbf{\Phi}^{\mathrm{d}}) = \mathrm{diag}\left(\left|\widehat{\mathbf{\Phi}} - \mathbf{\Phi}^{\mathrm{d}}\right|\mathbf{W}\left|\widehat{\mathbf{\Phi}} - \mathbf{\Phi}^{\mathrm{d}}\right|^{\mathrm{T}}\right) \tag{10}$$

Here, $\mathbf{W} = \mathrm{diag}(w_1, w_2, \cdots, w_L)$ is a diagonal matrix and the diagonal terms are defined by normalised vibration contribution factors of MS's in terms of

$$w_r = \frac{\sigma_{r1}}{\sum_{r=1}^{L}\sigma_{r1}}, \quad \sum_{r=1}^{L} w_r = 1 \tag{11}$$

The basic steps to calculate the proposed DI are summarised as: (1) Construction of the response PSD matrices. (2) Joint approximate diagonalisation of a set of PSD matrices and (3) Compute the proposed damage index according to equation (10).

4. Numerical study

A cantilever beam with two open cracks (marked in red in figure 1) is simulated to demonstrate the validity of the proposed DI. This damaged beam is modelled according to Euler-Bernoulli beam theory with Rayleigh damping, $\mathbf{C} = \alpha\mathbf{M} + \beta\mathbf{K}$ ($\alpha = 4.0136$ and $\beta = 5.0850 \times 10^{-6}$), using 40 elements in MATLAB. Its geometrical and material properties are tabulated in table 1. In addition, the configurations of the cracks are presented in table 2 and the cracks are modelled according to fracture mechanics approach. Random excitation is applied at point 20 and velocity time series are 'measured' at the prescribed 20 points along the beam with an equal distance of 0.035m as shown in figure 1.

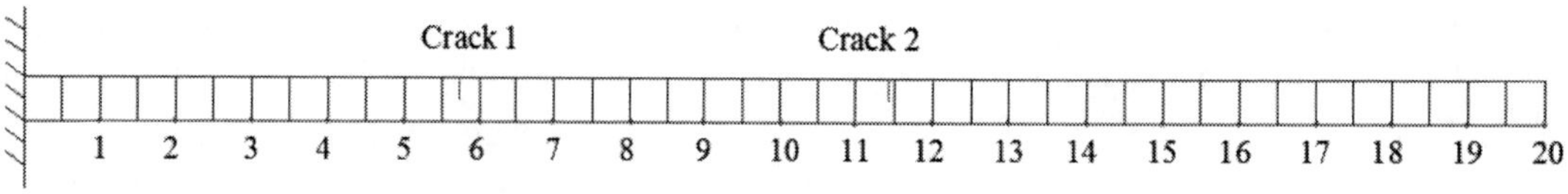

Figure 1. Cantilever beam with two open cracks.

Table 1. Material properties of steel beam.

Property	Value
Length (m)	0.7
Cross-section (m × m)	0.02×0.02
Young's modulus (Gpa)	210
Mass density ($\mathrm{kg/m^3}$)	7850
Poisson ratio	0.33

Table 2. Crack information of numerical study.

Cracks	Location	Measurement points	Depth
Crack 1	0.199m	5~6	0.002m
Crack 2	0.399m	11~12	0.002m

In order to compare the noise robustness of computed MS's and DI between JAD method and FDD method, Gaussian white noise is generated to contaminate the velocity responses in terms of

$$\hat{\mathbf{Y}}_l = \mathbf{Y}_l + \mathbf{d}n_{\mathrm{level}}\sigma(\mathbf{Y}_l), l = 1,2,\cdots,N_o \tag{12}$$

where $\mathbf{d} \in \mathbb{R}^{1\times N}$ is a vector containing normally distributed random values with a zero mean and variance being 1, n_{level} is the noise level range of [0 1] and $\sigma(\mathbf{Y}_l)$ represents standard deviation of vibration responses at lth point. Noise is added independently to $\mathbf{Y}$ 1000 times of the same level 3%. With each noise realization, MS's of peak singular value points in figure 2(a) are calculated by JAD and FDD, respectively. The mean absolute value and coefficient of variation (CV) of MS's over 1000 noise realisations are given in figure 3.

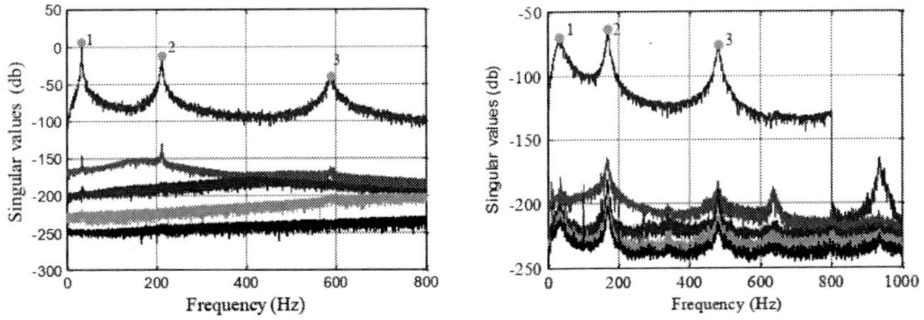

(a) (b)

Figure 2. Singular value spectrum plot: (a) numerical study (b) experimental study.

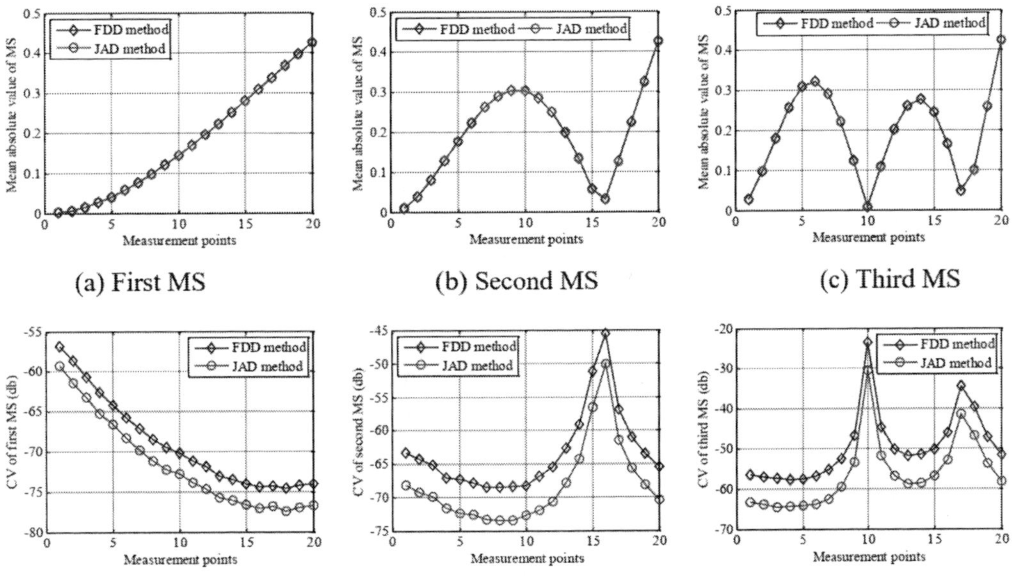

(a) First MS (b) Second MS (c) Third MS

(d) First MS (e) Second MS (f) Third MS

Figure 3. Mean absolute MS values and CV of MS's.

It is obvious in figures 3 (d)-(f) that the identified MS's of JAD method are more robust to noise than those of FDD method due to their smaller CV. The absolute mode shapes in figures 3 (a)-(c) demonstrate some irregular shape features in MS's of higher frequencies, which introduce difficulties of applying GSM method. To overcome this, the signed absolute value of MS's (having the same sign as the real part of MS's) is used to compute the proposed DI and figure 4 illustrates the damage identification results using the real part and signed absolute part of MS's, respectively.

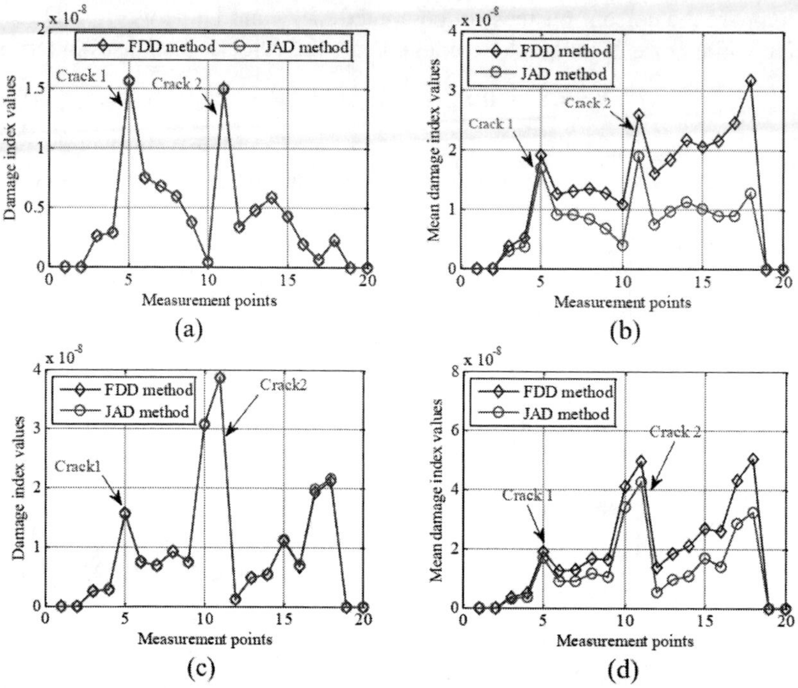

Figure 4. Damage identification results: (a) real part of MS's without noise (b) real part of MS's with noise (c) signed absolute value of MS's without noise (d) signed absolute value of MS's with noise.

It is demonstrated in figure 4 that the two open cracks can be both identified using the real part and signed absolute value of MS's and the overall damage identification performance of JAD method is better than FDD method. Furthermore, the real part of MS's of JAD method provides the most noise robust damage identification results. To further validate the efficiency and feasibility of the proposed DI, an experimental study is conducted next.

5. Experimental study

Experimental test is performed on a cantilever beam of dimensions $700 \times 20 \times 20$ mm^3 with two open cracks. Detailed material information about this beam is given in table 1. Figure 5(a) shows the experimental set-up using a PSV-500 Scanning Laser Vibrometer. Pseudo-random excitation in the frequency range of 0-800 Hz is generated by the PSV-500 system and applied to the cantilever beam by a shaker (LDS V406) as shown in figure 5(b).

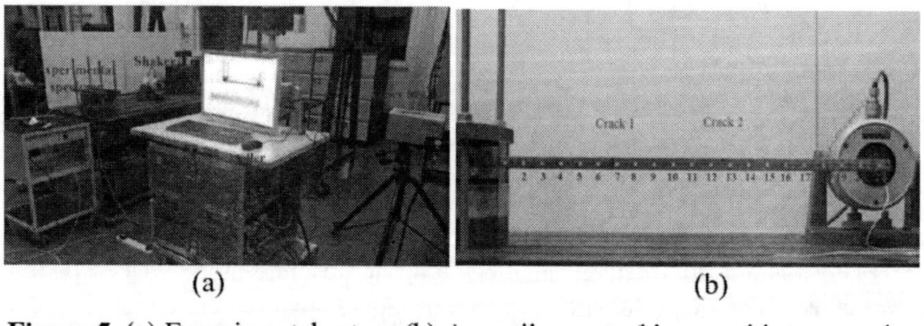

Figure 5. (a) Experimental set-up (b) A cantilever steel beam with two cracks.

Damage is simulated in terms of open cracks by cutting slots of uniform depth and width. In addition, the information of the two cracks is shown in table 3. They are located on the rear surface and are marked as blue lines on the front surface in figure 5(b). Surface velocity data at prescribed 21 are collected using Scanning Laser Vibrometer for damage identification.

Table 3. Crack configurations of experiment study.

Cracks	Location	Measurement points	Depth	Width
Crack 1	0.2m	6~7	0.004m	0.001m
Crack 2	0.4m	12~13	0.004m	0.001m

Figure 2(b) presents the singular value spectrum plot computed by SVD of PSD matrices. The estimated MS's and their curvatures are shown in figure 6. The positions of two cracks are hard to be observed from figure 6. In figures 6 (d)-(f), it is worth noting that the identified MS curvatures by JAD method are smoother than those by FDD method.

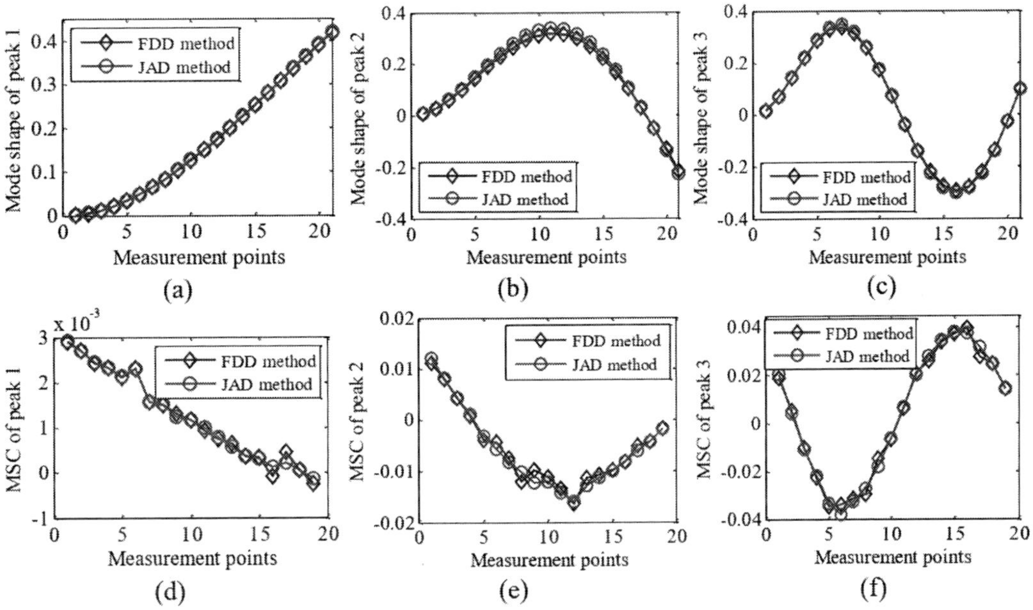

Figure 6. Mode shapes and their curvatures of selected peak points of Figure 2(b).

The damage identification results are presented in Figure 7 and a comparison of damage identification performance between JAD method and FDD method is conducted as well.

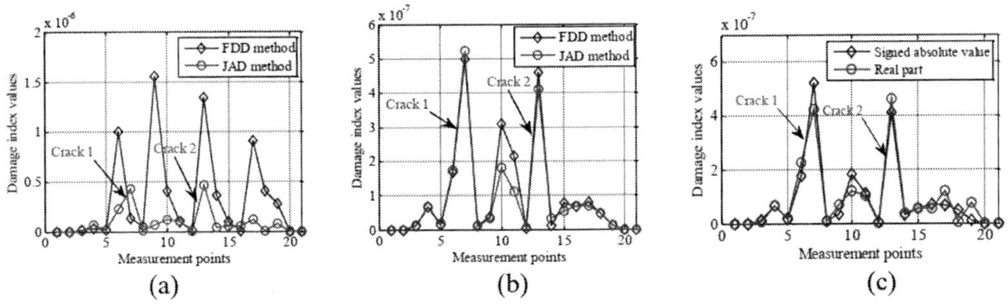

Figure 7. Damage identification results: (a) DI of real part of MS's (b) DI of signed absolute value of MS's (c) DI comparison of JAD method using signed absolute value and real part value of MS's.

Figure 7(a) shows that the real part of MS's of JAD method accurately identifies the locations of two cracks while the real part of MS's of FDD method is impossible to identify the damage locations correctly. Figure 7(b) presents the DI values calculated using the signed absolute value of MS's, which implies that the absolute value of MS's can effectively improve the damage identification accuracy of FDD method. However, the enhancement for JAD method by using signed absolute value is not obvious which even degrades the damage identification results at measurement point 10 as shown in figure 7(c). Another conclusion from Figure 7 is that overall DI values calculated by JAD method are much more accurate and noise robust than those of FDD method.

6. Conclusions

This study proposes a mode shape-based damage identification method without requiring the baseline information of healthy structures. Two aspects of improvements have been made. First, a new algorithm based on joint approximate diagonalisation is proposed to statistically estimate the dominant mode shape at each resonant frequency by diagonalising a set of PSD matrices. Secondly, local polynomial fitting method is used to extract the damage-caused local shape distortions of MS's and several MS's are combined according to their vibration participation factors to compute the proposed damage index. Numerical simulation with Gaussian white noise is used to demonstrate the noise robustness of MS's and damage index of JAD method. Furthermore, form an experimental study, the obtained MS's and damage index of JAD method are illustrated to be more accurate and noise robust than those of FDD method. Consequently, JAD method is demonstrated to be effective in addressing the noise effects in MS estimation and damage identification. In addition, the proposed PSD-based damage index has a good potential to be applied to practical engineering applications with ambient excitations.

Another conclusion is that JAD is effective to enhance the accuracy and noise robustness of mode shapes extracted from experiments. As a result, it should be used to improve accuracy of the PSD based modal analysis.

Acknowledgments

The authors are grateful to the joint University of Liverpool-China Scholarship Council Scholarship and the Fundamental Research Funds for the Central Universities (DUT16RC(3)027).

References

[1] Ahmadian H, Mottershead J and Friswell M 2000 Damage location indicators from substructure mode Shapes *Inverse Probl. Eng.* **8** 309-23
[2] Zhou Y, Maia N, Sampaio R and Wahab M 2016 Structural damage detection using transmissibility together with hierarchical clustering analysis and similarity measure *Struct. Health Monit.* doi: 10.1177/1475921716680849
[3] Kess H R and Adams D E 2007 Investigation of operational and environmental variability effects on damage detection algorithms in a woven composite plate *Mech. Syst. Signal Pr.* **21** 2394-2405
[4] Zhou W and Chelidze D 2007 Blind source separation based vibration mode identification *Mech. Syst. Signal Pr.* **21** 3072-87
[5] Cao S and Ouyang H 2017 Robust structural damage detection and localization based on joint approximate diagonalization technique in frequency domain *Smart Mater. Struct.* **26** 015005
[6] McNeill S and Zimmerman D 2008 A framework for blind modal identification using joint approximate diagonalization *Mech. Syst. Signal Pr.* **22** 1526-48
[7] Lee Y S and Chung M J 2000 A study on crack detection using eigenfrequency test data *Comput Struct.* **77** 327-42
[8] Peng Z, Lang Z and Chu F 2008 Numerical analysis of cracked beams using nonlinear output frequency response functions *Comput. Struct.* **86** 1809-18
[9] Zhou Y, Maia, N and Wahab, M 2016 Damage detection using transmissibility compressed by principal component analysis enhanced with distance measure, *J. Vib. Control* doi: 10.1177/1077546316674544
[10] Xu Y and Zhu W 2016 Non-model-based damage identification of plates using measured mode shapes *Struct. Health Monit.* **16** 3-23

Structure damage identification under ambient excitation based on wavelet packet analysis

Qingzhou Li[1] and Dongsheng Li[1]

[1]State Key Laboratory of Coastal and Offshore Engineering, Dalian University of Technology, Dalian, Liaoning, 116023, People's Republic of China

E-mail: dsli@dlut.edu.cn

Abstract. There are significant changes in vibration responses when various damages appear in structures. Therefore, structural vibration responses contain rich structure damage information. In this paper, virtual impulse response of structure under ambient excitation is obtained through the inversion of a transfer function, and is decomposed into components by wavelet packet transform. Then the energy of these wavelet packet components is calculated to obtain the wavelet packet energy spectrum. And a novel damage index i.e. Energy Spectrum Anomaly Measure is developed to identify the existence of damage. Damage extent can also be inferred from the relative value of Energy Spectrum Anomaly Measure at various locations. Numerical results are presented to evaluate the effectiveness and reliability of the proposed index. It is shown that the proposed damage index works well in identifying structure damages free from the effect of various excitation and there is a positive correlation between the index and the damage severity.

1. Introduction

Various damages will appear in engineering structures or structure components during its service life due to material degradation, initial design defect and environmental factor changes. And the occurrence of damages in structure will finally lead to the reduction of physical properties i.e. stiffness, damping or mass. The accumulation of damages will lead to out-of-service conditions, and even dangerous collapses. Thus, it is significant to identify structure damages timely and accurately. And much more attention should be paid to the damage identification in structures to avoid or reduce accidents in the future.

Most nondestructive damage identification methods can be categorized as either local or global damage identification techniques [1]. Local damage identification methods, including visual, acoustic, ultrasonic and X-ray methods, require that the location of damage should be known in advance and that the inspected portion is accessible for test. However, these requirements cannot be guaranteed in most cases in civil engineering structures. Hence, the vibration-based global damage identification methods are developed to overcome these drawbacks based on dynamic characteristics of structure.

Vibration-based damage identification methods have been widely applied in civil engineering to evaluate the state of structure in the past few decades. The fundamental theory [2] for vibration-based damage identification is that the damage-induced changes in the physical properties (mass, damping, and stiffness) will cause detectable variation in modal properties such as natural frequencies, mode shapes, mode shape curvature, modal flexibility, and modal strain energy etc. Therefore, it is evident that damage can be identified by examining the variation of modal properties. Natural frequency-based methods [2] choose the natural frequency as the damage index which can be easily extracted from vibration responses obtained by just a few sensors on the structure and are usually insensitive to experimental noise. However, Natural frequencies are less sensitive to damage because in comparison to the changes caused by environmental and operational conditions, the damage-induced frequency changes are relatively small. By contrast, the advantage of methods using mode shapes as the damage index is quite obvious. Mode shapes and their derivatives with local information are more sensitive to local damages and can be used directly in multiple damage detection [3]. But its disadvantages are also apparent, that is, they are more sensitive to noise contamination and more sensors are required to obtain the mode shapes and their derivatives.

Compared with aforementioned methods, the recent vibration-based damage identification method based on wavelet analysis with better sensitivity and robust to structural damage has been widely used in structure detection [4-7]. As is known, with its good performance of simultaneous time-frequency localization, the method based on wavelet analysis has great advantages in demonstrating some hidden features of the vibration responses. Hou et al. [8] proposed a wavelet-based method to identify the moment and location of damage occurrence using a simple structural model with breakage springs. It is shown that the wavelet transform (WT) can identify both abrupt and accumulative damages successfully. Wu and Wang [9] used the spatial wavelet transform with Gabor wavelet to analyze the deflection profile of a cracked aluminum cantilever beam. Based on the experimental study, the spatial wavelet transform is proven to be effective to identify the damage area even when the crack depth is around 26% of the beam thickness. Jiang, Ma, and Ren [10] introduced complex continuous wavelet transform to detect crack from the angle characteristics of complex wavelet coefficients. This method is easier to detect the exact locations of singularities even when the signals are contaminated by noise. Chen et al. [11] presented a novel multiple-mode wavelet-index based on biorthogonal wavelet using discrete wavelet transform of the modal frequency curves to identify the structural damage. Finite element model and aluminium samples have been studied to demonstrate the performance of the proposed method. The results suggest that the developed indicator provides a robust and unambiguous damage identification. However, the rather poor resolution of WT in the high-frequency region makes it face difficulties in dealing with discriminating signals which contain high frequency components.

The wavelet packet transform (WPT) is the extension of the WT and provides complete level-by-level decomposition. Many investigators have reported the implementation of WPT to identify damage in structures. Sun and Chang [12] proposed an integrated method with wavelet packet transform and neural network for assessment of structures based on numerical simulation of a three-span continuous bridge under the impact excitation. It is concluded that the WPT-based component energy is a sensitive condition index for structural damage assessment and insensitive to measurement noise. Yam et al. [13] extracted the structural damage feature based on energy variation of structure vibration responses decomposed by wavelet packet transform before and after the occurrence of structural damage. This method can be applied to online structural damage detection and health monitoring. Zabel [14] used the wavelet packet component energies as the damage index for reinforced concrete structure damage assessment. Law et al. [15] developed a method to identify damage in structures using wavelet packet sensitivity. Han, Ren and Sun [16] proposed "wavelet packet rate index (WPERI)" for steel beam damage identification. Ding and Li [17] proposed "energy ratio variation deviation (ERVD)" to identify the state of reinforced concrete slab. Yu et al. [18] applied the Wavelet Packet Energy Curvature Difference (WPECD) method to a real replaced girder at the Ziya River New Bridge in Cangzhou to verify the validity of WPECD in structural damage identification. Shahverdi et al. [19] used "Detail Signal Energy Rate Index (DSERI)" for damage assessment of Jacket type Offshore Platform. Zhu and Sun [20] developed a novel damage detection index i.e. the sum square of wavelet packet energy change rate (WPERSS) for bridge damage assessment through a simply supported beam numerical model. All these indexes mentioned above can identify the structure damage extent and location validly. However, what was applied to models was the constant excitation instead of various ambient one. These methods have much dependence on environmental factors. That is to say the effect of damage may be masked by the effect of ambient excitation, experimental noise and uncertainty, etc.

Based on the discussion above, impulse response functions and virtual impulse response functions that contain all of the structural dynamic properties are taken into account [21, 22]. In this paper two node responses of the structure under ambient excitation are directly used to calculate the virtual impulse responses. Virtual impulse response is decomposed into components based on wavelet packet transform. The energy of these wavelet packet components is calculated to obtain the wavelet packet energy spectrum, and the change of which pre and post damage is used to obtain proposed index i.e. Energy Spectrum Anomaly Measure. Numerically simulated data with different damage scenarios show that the suggested index works well in identifying structural damage.

This paper is organized as follows. The next section provides basic theory of wavelet packet energy spectrum, virtual impulse response and the novel damage index i.e. Energy Spectrum Anomaly Measure. In section 3, numerical data is used to verify the effectiveness and reliability of the proposed index. Then conclusions are stated finally.

2. Theoretical background

Virtual impulse response of structure under ambient excitation is obtained through the inversion of a transfer function, and is decomposed into components based on wavelet packet transform. The energy of these wavelet packet components is calculated to obtain the wavelet packet energy spectrum. A novel damage index i.e. Energy Spectrum Anomaly Measure (ESAM) is then developed to identify the existence of damage.

2.1. Virtual Impulse Response Function [21]

Frequency response function $H(\omega)$ can be expressed as:

$$H(\omega)=\frac{Y(\omega)}{U(\omega)} \tag{1}$$

Where $Y(\omega)$ means Fourier transform of response signals of structure; and $U(\omega)$ means Fourier transform of excitation signals of structure. From the equation (1), $Y(\omega)$ can be expressed as

$$Y(\omega)=H(\omega)U(\omega) \tag{2}$$

Vibration responses at a reference point j are regarded as virtual excitation in a multi-degree-of-freedom (MDOF) structure. Then the auto-power spectral density $Guu(\omega,j)$ of the virtual excitation $x_j(t)$ can be expressed as

$$Guu(\omega,j)=Y^{*}(\omega,j)*Y(\omega,j) \tag{3}$$

Where $Y(\omega,j)$ is the Fourier Transform of the virtual excitation $x_j(t)$ and $Y^{*}(\omega,j)$ is the complex conjugate of $Y(\omega,j)$.

The cross-power spectral density $Gyu(\omega,i,j)$ between the virtual excitation $x_j(t)$ and the response $x_i(t)$ of calculating point i can be expressed as

$$Gyu(\omega,i,j)=Y^{*}(\omega,j)*Y(\omega,i) \tag{4}$$

Where $Y(\omega,i)$ is the Fourier transform of the response $x_i(t)$ of calculating point i and $Y^{*}(\omega,i)$ is the complex conjugate of $Y(\omega,i)$.

From the equation (2) $Y(\omega,i)$ and $Y(\omega,j)$ can be respectively expressed as

$$Y(\omega,i)=H(\omega,i)U(\omega) \qquad Y(\omega,j)=H(\omega,j)U(\omega) \tag{5}$$

As a result, the virtual frequency response function $Hyu(\omega,i,j)$ between $x_j(t)$ and $x_i(t)$ can be expressed as

$$\begin{aligned} Hyu(\omega,i,j) &= \frac{Gyu(\omega,i,j)}{Guu(\omega,j)} = \frac{Y^*(\omega,j)*Y(\omega,i)}{Y^*(\omega,j)*Y(\omega,j)} \\ &= \frac{(H^*(\omega,j)U^*(\omega))*(H(\omega,i)U^*(\omega))}{(H^*(\omega,j)U^*(\omega))*(H(\omega,j)U^*(\omega))} \\ &= \frac{H^*(\omega,j)*H(\omega,i)}{H^*(\omega,j)*H(\omega,j)} \end{aligned} \tag{6}$$

After the inverse Fourier transform, VIRF $Hyu(t,i,j)$ between $x_j(t)$ and $x_i(t)$ can be obtained. In summary the virtual impulse responses calculated by using two node responses of MDOF structures contain all of the structural inherent dynamic properties free from the effect of various ambient excitation. Thus, this method is robust to damage identification.

2.2 Wavelet packet component energy and Energy Spectrum Anomaly Measure

The method based on wavelet packet transform (WPT) for vibration signal analysis provides complete level-by-level decomposition leading to good time-frequency localization. The structural dynamic responses can be decomposed into wavelet packet components to obtain the signal energy distribution at different scales. The signal $x(t)$ decomposed by jth level WPT can be expressed as

$$x(t) = \sum_{i=0}^{2^j-1} x_j^i(t) \tag{7}$$

$$x_j^i(t) = \sum_k c_{j,k}^i \psi_{j,k}^t \tag{8}$$

The WPT coefficient $C_{j,k}^i$ is defined as

$$C_{j,k}^i = \int_{-\infty}^{\infty} x(t)\psi_{j,k,i}(t)dt \tag{9}$$

Since the family $\psi_{j,k,i}(t)$ is an orthonormal basis for $L^2(\mathbb{R})$. If $m \neq n$, it has

$$\psi_{j,k}^m \psi_{j,k}^n = 0 \tag{10}$$

The signal energy [23] in the jth level and the ith frequency band can be expressed as

$$E_j^i = \sum_k \left|C_{j,k}^i\right|^2 \tag{11}$$

The total energy of the original signal $x(t)$ can be written as

$$ES = \sum_{i=0}^{2^j-1} E_j^i \qquad (j = 0,1,2,...,2^i-1) \tag{12}$$

The energy spectrum ratio in the jth level and the ith frequency band can be expressed as

$$I_i = \frac{E_j^i}{\sum_{i=0}^{2^j-1} E_j^i} \quad (j = 0,1,2,...,2^i - 1) \tag{13}$$

ESV (Energy Spectrum Variation) can be obtained

$$ESV = \{ESV_i\} = \{|I_{u,i} - I_{d,i}|\} \tag{14}$$

In which $I_{u,i}$ and $I_{d,i}$ are the ESV of the undamaged and damaged structures in the jth level and the ith frequency band.

Finally, a novel damage index i.e. Energy Spectrum Anomaly Measure ($ESAM$) is presented as

$$ESAM = \|ESV\| \times \sqrt{\sum_{i=0}^{2^j-1}(ESV_i - \overline{ESV})^2} \quad (j = 0,1,2,...,2^j - 1) \tag{15}$$

$$\overline{ESV} = \frac{1}{2^j} \times \sum_{i=0}^{2^j-1} ESV_i \quad (i = 0,1,2,...,2^j - 1) \tag{16}$$

Where $\overline{ESV}$ is the average for ESV. $ESAM$ can be employed to describe the state of structure validly.

3. Numerical Studies

A chain like structure with 3 DOF, shown in Fig. 1, is used as an example to illustrate the proposed damage index i.e. Energy Spectrum Anomaly Measure. The properties of the structure are as follows. Mass matrix $M = \begin{bmatrix} 1 & 0 & 0 \\ 0 & 1.5 & 0 \\ 0 & 0 & 2 \end{bmatrix}$; Stiffness matrix $K = \begin{bmatrix} 1 & -1 & 0 \\ -1 & 3 & -2 \\ 0 & -2 & 5 \end{bmatrix} * 600$; The natural frequencies of modes $w = [14.522 \quad 31.048 \quad 46.50]^T$ and the damping ratios are assumed to be 5% for all modes. The Gaussian stationary white noise, shown in Fig. 2, is assumed to be an earthquake excitation applied at the foundation of the structure. The inertia force vector $E = [1 \; 1 \; 1]^T$. The initial velocity and displacement of the lumped mass are zero. Fifty seconds responses at each lumped mass of the 3-DOF chain-like structure are recorded, then calculated and decomposed using the WPT. The sampling frequency for all signals is 100 Hz. The structure is assumed to be a time-variant mechanical structure subjected to abrupt stiffness degradation that is used to simulate the damage at particular locations. A total of 7 damaged cases, shown in Table 1, are used to identify damage.

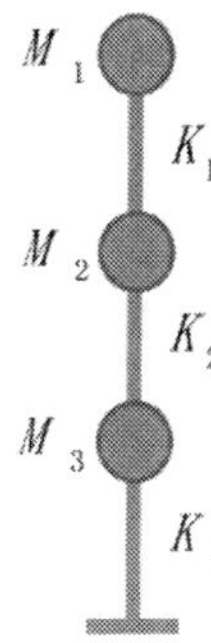

Figure 1 A chain like structure with 3 degrees of freedom

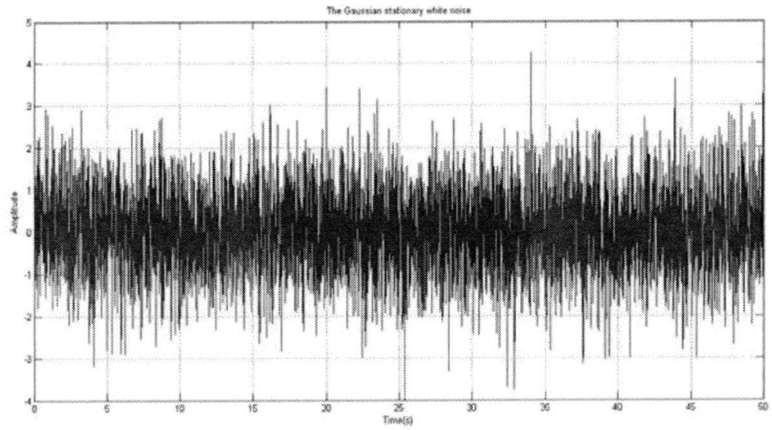

Figure 2 The Gaussian stationary white noise

Table 1. The deterioration of stiffness K_1

Case No.	1	2	3	4	5	6	7
Damage severity	0	5%	10%	15%	20%	25%	30%

3.1 The damage detection procedure

It is worth mentioning that the input Gaussian excitation is not kept constant in each case to evaluate the robustness of the method to various excitations. And a new random excitation is generated in each case.

1. Vibration responses at a reference point 3 are regarded as virtual excitation in the 3-DOF chain-like structure. Then two node responses of the structure under ambient excitation are directly used to

IOP Conf. Series: Journal of Physics: Conf. Series **842** (2017) 012023 doi:10.1088/1742-6596/842/1/012023

calculate the virtual impulse responses.

2. Virtual impulse response is decomposed into components based on wavelet packet transform. The energy of these wavelet packet components is calculated to obtain the wavelet packet energy spectrum.

3. Generate Energy Spectrum Anomaly Measure.

4. Localize damage by comparing ESAM of different measured points.

As will be illustrated, Gaussian white noise with SNRs of 40 and 30 dB are added to the responses to study the noise influence on the proposed damage detection method.

3.2 Simulation results

Numerical simulation has been induced for investigating the influence on ESAM caused by damage. The deterioration of stiffness K_1 is introduced, and ESAM is obtained in each case (Figure 3). It is worth mentioning that numerical simulation is induced for 10 times repeatedly and the initial state is kept constant. There are five wavelet packet decomposition layers, and the Daubechies 30 function is selected as the wavelet function. The deterioration of stiffness K_1 is deliberately selected for this investigation as damage had the lowest impact on the global stiffness of the structure compared with stiffness K_2 and K_3. As is seen in Figure 3, ESAM has been plotted for cases 1-7. And the early part of the graph, cases 1–2, is almost invisible. However, after case 2, the change in the ESAM is abrupt and rapid. On the basis of Figure 3, an increasing trend in the ESAM is obtained after case 2. Case 2 corresponds to 5% reduction of stiffness K_1 (Table 1). As deterioration evolves, higher values of ESAM are obtained, which means the structure is getting far from its initial case. The higher the ESAM is, the less closeness will become between the current case and the initial case. The change in the ESAM is an indicator of damage presence. On the basis of this demonstration, the proposed method can reliably detect abrupt stiffness deterioration in the structure with minor damage severity (0–10%), moderate damage severity (10–20%), and severe damage severity (20–30%). The other interesting point is that the method is robust to changes in excitations. As is said earlier, in each case of deterioration, the response of the structure is obtained under a newly generated random excitation.

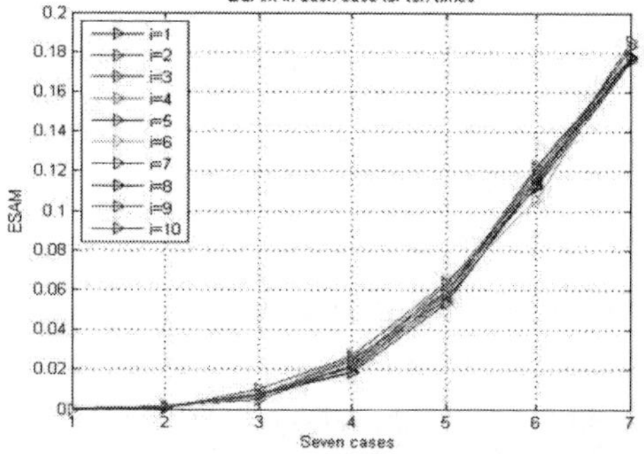

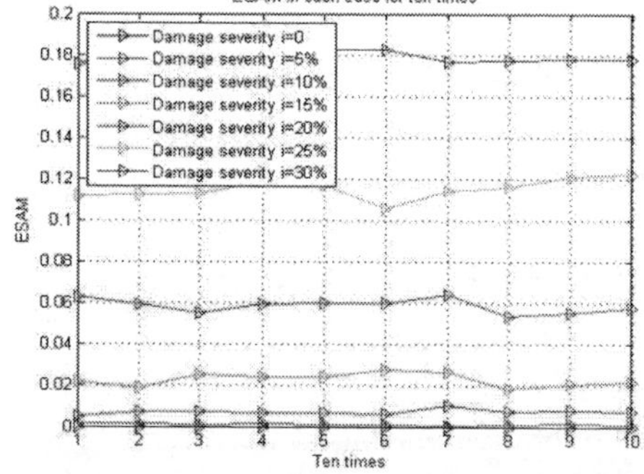

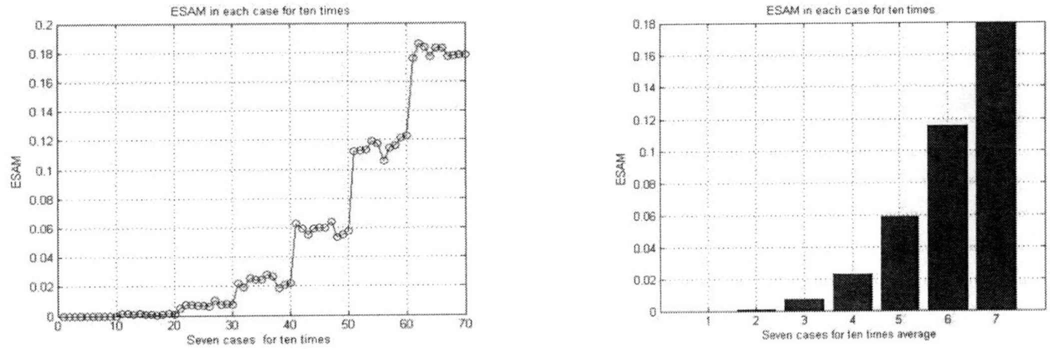

Figure 3 (a, b, c and d) ESAM in each case for ten times (without noise)

As is said earlier, to study the noise influences on the proposed damage detection method, Gaussian white noise with SNRs of 40 and 30 dB are again added to the responses. Figure 4 and 5 show the change of ESAM for these two noise levels respectively. It is shown that the presence of noise does not seem to affect the value of ESAM significantly. The results suggest that ESAM is able to identify the damage extent and location validly even under noisy conditions.

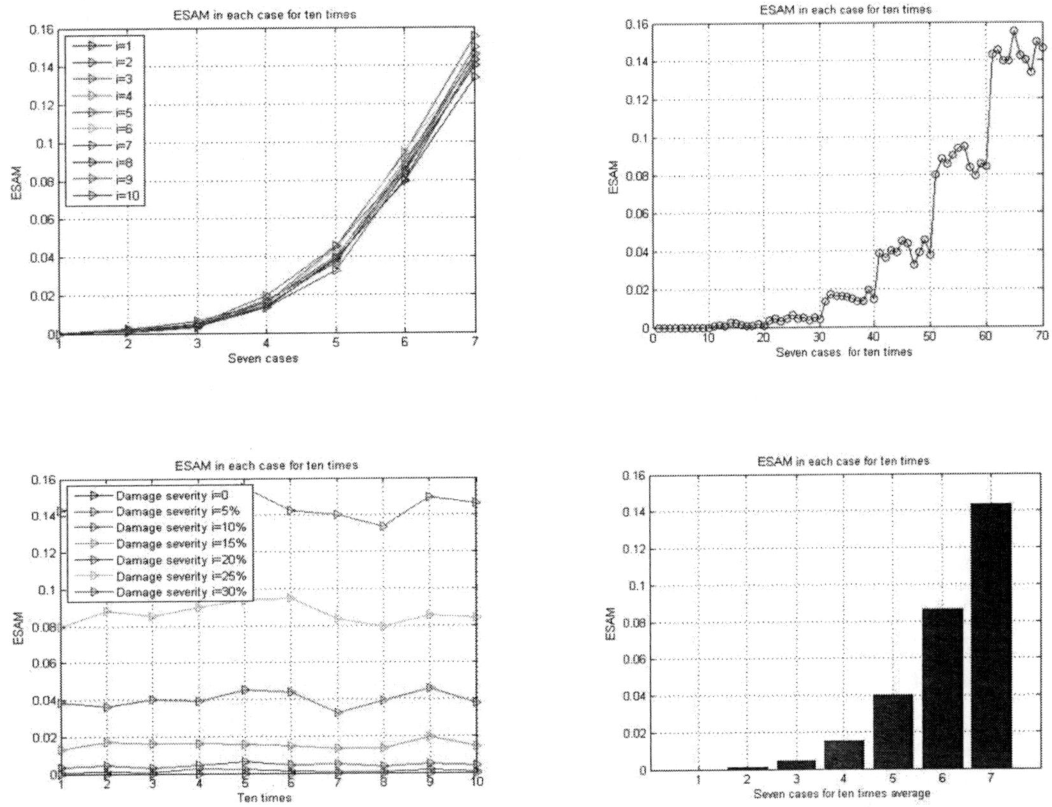

Figure 4 (a, b, c and d) ESAM in each case for ten times (SNR=40 dB)

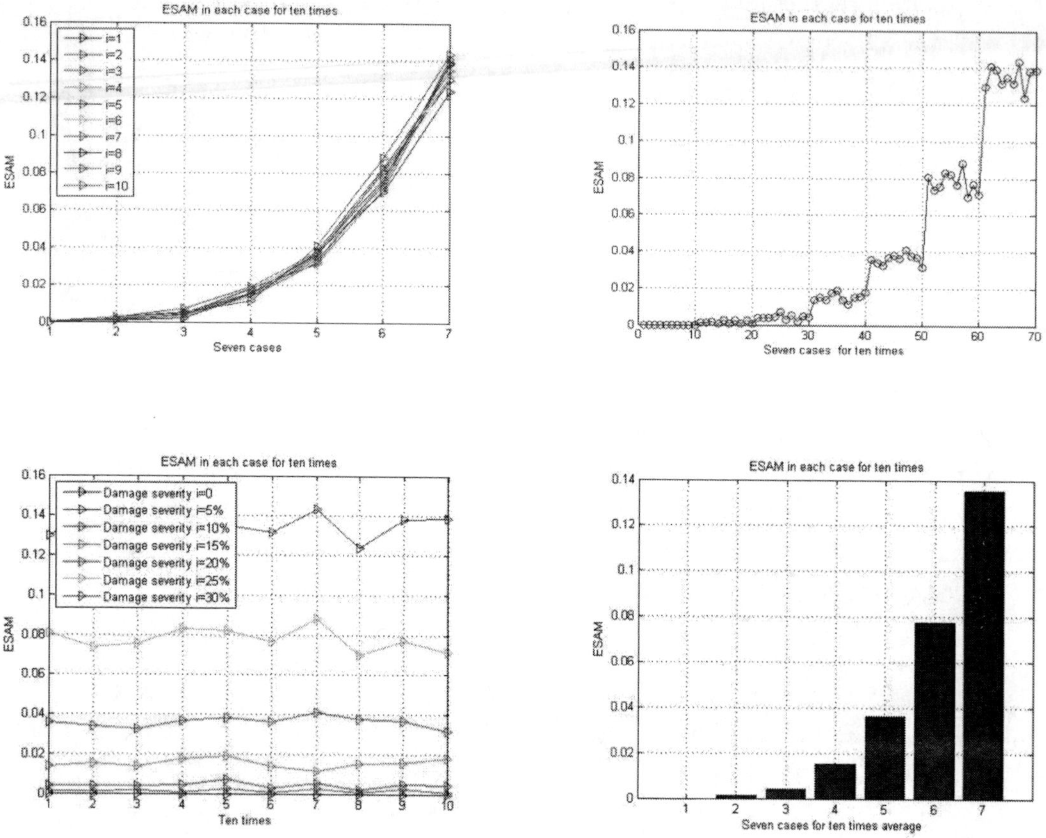

Figure 5 (a, b, c and d) ESAM in each case for ten times (SNR=30 dB)

4. Conclusions

Wavelet packet transform has been widely applied recently as a powerful mathematical tool to identify changes in structural properties induced by damage. In this paper, the use of the virtual impulse response, wavelet packet transform and Energy Spectrum Anomaly Measure for damage detection of civil engineering structures were investigated. And the novel damage index i.e. Energy Spectrum Anomaly Measure is proposed to identify the existence of damage. The results of numerical simulation, using chain-like 3DOF model, demonstrate that the combination of these techniques, reflecting all of the structural inherent dynamic characteristics, have much better sensitivity and reliability to the damage detection of civil engineering structures free from the effect of various excitation. In addition, the results suggest that ESAM is able to identify the damage extent validly even under noisy conditions. Consequently, this damage identification method may be used to identify damage within civil engineering structures, which is of great significance for disaster prevention and mitigation.

Acknowledgements
This research work was jointly supported by the National Natural Science Foundation of China (Grant No. 51121005, and 51578107), and 973 Project (Grand No. 2015CB057704).

References

[1] S. W. Doebling, C. R. Farrar, M. B. Prime, and D. W. Shevitz 1996 Damage identification and health monitoring of structural and mechanical systems from changes in their vibration characteristics: A literature review *Shock & Vibration Digest* **30** 2043-2049.

[2] F. Wei and Q. Pizhong 2010 Vibration-based Damage Identification Methods: A Review and Comparative Study *Structural Health Monitoring* **10** 83-111.

[3] C. R. Farrar and G. H. J. Iii 1997 SYSTEM IDENTIFICATION FROM AMBIENT VIBRATION MEASUREMENTS ON A BRIDGE *Journal of Sound & Vibration* **205** 1-18.

[4] X. Jiang and S. Mahadevan 2011 Wavelet spectrum analysis approach to model validation of dynamic systems *Mechanical Systems & Signal Processing* **25** 575-590,.

[5] D. H. Xie and H. W. Tang 2012 Application of Wavelet Packet Transform and Neural Network to Detect Damage of Elastic Thin Plate *Advanced Materials Research* **594-597** 1105-1108.

[6] M. Solís, M. Algaba, and P. Galvín 2013 Crack Location in Beams Using Wavelet Analysis *Key Engineering Materials* **569-570** 1021-1028.

[7] S. H. Mahdavi and H. A. Razak 2015 A wavelet-based scheme for optimum measurement/monitoring of dynamic responses of structures *Int. Conf., 11th on Damage Assessment of Structures (DAMAS 2015), Vol 628* (Beijing) *pp* 012-024.

[8] Z. K. Hou, M. N. Noori, and R. S. Amand 2000 Wavelet-Based Approach for Structural Damage Detection *Journal of Engineering Mechanics* **126** 677-683.

[9] N. Wu and Q. Wang 2011 Experimental studies on damage detection of beam structures with wavelet transform *International Journal of Engineering Science* **49** 253-261.

[10] X. Jiang, Z. J. Ma, and W.-X. Ren 2012 Crack Detection from the Slope of the Mode Shape Using Complex Continuous Wavelet Transform *Computer-Aided Civil and Infrastructure Engineering* **27** 187-201.

[11] Chen, Y., & Oyadiji, S. O. 2017 Damage detection using modal frequency curve and squared residual wavelet coefficients-based damage indicator *Mechanical Systems & Signal Processing,* **83**, 385-405.

[12] Z. Sun and C. C. Chang 2002 Structural Damage Assessment Based on Wavelet Packet Transform *Journal of Structural Engineering* **128** 1354-1361.

[13] L. H. Yam, Y. J. Yan, and J. S. Jiang 2003 Vibration-based damage detection for composite structures using wavelet transform and neural network identification *Composite Structures* **60** 403-412.

[14] D. I. V. Zabel 2004 A Wavelet-Based Damage Detection Indicator for Reinforced Concrete Structures in *Pro. Int. Conf., 22nd on Modal Analysis* (Dearborn: Detroit) USA.

[15] J.-G. Han, W.-X. Ren, and Z.-S. Sun 2005 Wavelet packet based damage identification of beam structures *International Journal of Solids and Structures* **42** 6610-6627.

[16] S. S. Law, X. Y. Li, X. Q. Zhu, and S. L. Chan 2005 Structural damage detection from wavelet packet sensitivity *Engineering Structures* **27** 1339-1348.

[17] Y. L. Ding, L. I. Ai-Qun, C. Q. Miao, and T. Liu 2006 Structural Damage Alarming Indices for Long-Span Bridges Based on Wavelet Packet Energy Spectrum *China Journal of Highway &*

Transport **19** 34-40.

[18] Z. Yu, H. Xia, J. M. Goicolea, and J. W. Zhan 2013 Experimental study on bridge damage identification based on wavelet packet energy curvature difference method *Zhendong Yu Chongji/journal of Vibration & Shock* **32** 20-25.

[19] S. Shahverdi, M. A. Lotfollahi-Yaghin, and B. Asgarian 2013 Reduced wavelet component energy-based approach for damage detection of jacket type offshore platform *Smart Structures & Systems* **11** 589-604.

[20] Z. Jin-song and S. Ya-dan 2015 Study of a novel wavelet packet energy based damage detection index for bridges *Journal of Vibration, Measurement&Diagnosis* **35** 7.

[21] A Li and Y Ding 2006 *Engineering Structures Damage Alert Theory and Application* (Beijing: Science Press)

[22] Y. S. Diao, Q. L. Zhang, and D.M.Meng 2012 Damage localization of off shore platform based on the virtual impulse response function *Advanced Materials Research* **368 - 373** 1676 - 1680.

[23] O. A. Rosso, M. T. Martin, A. Figliola, K. Keller, and A. Plastino 2006 EEG analysis using wavelet-based information tools *Journal of Neuroscience Methods* **153** 163-182.

Research on the spatial analysis method of seismic hazard for island

Jing Jia [1*], Jitong Jiang [1], Qiuhong Zheng [1] and Huiying Gao [1]

[1] Civil Engineering Department, College of Engineering, Ocean University of China, No.238 Songling Road, Qingdao 266100, P. R. China

*E-mail: jingjia@ouc.edu.cn, TEL: 86-13156383701, Fax: 86-0532-66781550

Abstract. Seismic hazard analysis(SHA) is a key component of earthquake disaster prevention field for island engineering, whose result could provide parameters for seismic design microscopically and also is the requisite work for the island conservation planning's earthquake and comprehensive disaster prevention planning macroscopically, in the exploitation and construction process of both inhabited and uninhabited islands. The existing seismic hazard analysis methods are compared in their application, and their application and limitation for island is analysed. Then a specialized spatial analysis method of seismic hazard for island (SAMSHI) is given to support the further related work of earthquake disaster prevention planning, based on spatial analysis tools in GIS and fuzzy comprehensive evaluation model. The basic spatial database of SAMSHI includes faults data, historical earthquake record data, geological data and Bouguer gravity anomalies data, which are the data sources for the 11 indices of the fuzzy comprehensive evaluation model, and these indices are calculated by the spatial analysis model constructed in ArcGIS's Model Builder platform.

Keywords: Seismic hazard analysis, spatial analysis method of seismic hazard for island, fuzzy comprehensive evaluation model

1. Introduction

Island is the land area which is surrounded by water and above water at high tide, according to the definition in "United Nations convention on the law of the sea", and such definition is also referenced in "People's Republic of China island protection act" [1]. Since the majority islands are quite small, island is defined as a small piece of land scattered in the ocean whose area is more than 500 square meters in "National Standard Oceanographic Terms Marine Geology". According to "The bulletin of island statistic survey in 2015", there are 11000 islands in China, accounting for 0.8% of the total land area in China. Chinese island resource is relatively abundant, since the world island area does not exceed 1/15 of the world's total land area.

Among Chinese islands, there are more than 400 islands with permanent residents, which accounts for more than 98% of the Chinese total island area. For the uninhabited islands, the first uninhabited islands exploitation list is published, including 187 islands, in 2011, following the promulgation of “Island Protection Act”. The list shows that the dominant uses are tourist & recreation, transportation & industry. Facing the increasing islands exploitation projects, it is a fundamental work to carry out the island Seismic hazard analysis (SHA).

Chinese island laws and regulations systems are gradually improving, and it is put forward in “Island Protection Act” that Island Management Information System should be constructed [2]. With the fast development of spatial science & technology and information technology, spatial analysis tools and spatial decision support models are facilitating this work. In the exploitation and construction process of both inhabited and uninhabited islands, SHA is a key component of island engineering’s earthquake disaster prevention work, whose result could provide parameters for seismic design microscopically and also is the requisite work for the island conservation planning’s earthquake and comprehensive disaster prevention planning macroscopically. Under this background, the research content of this paper is designed as constructing a specialized spatial analysis method of seismic hazard for island (SAMSHI) which is suitable for island and serves as the basement for island earthquake resistance and disaster prevention planning, under the application analysis of the existing SHA methods and island feature.

2. Application analysis of the existing SHA method

SHA is the scientific evaluation work of the probable earthquake influencing degree of the engineering construction site in the differently defined following years, which is under the foundation of the detailed research and analysis of seismic and geologic environment, and providing the probability level of the site earthquake influencing degree in the differently defined following years, considering all seismic influences in and surrounding the site, using rigorous mathematical statistics method [3]. The influencing degree is finally indicated by acceleration peak, acceleration response spectrum, duration of ground motion, etc.

SHA works can be Zoning Map of Ground Motion Parameters (ZMGMP) of China, Seismic Zoning (SZ), and Seismic safety evaluation (SSE) for different application area, and these applications all adopt Probabilistic Method. For the SHA works of island, the most convenient way is referencing GB18306-2015, i.e. ZMGMP of China. But the accuracy and scale could not satisfy the requirement for island [4]. The specialized SZ and SSE works could satisfy the accuracy while the expense is high and the field exploration equipment is not convenient in island for transportation difficulties [5]. The SHA method for island, researched in this paper, is used directly in the phase of earthquake resistance and disaster prevention planning, but not engineering construction phase, in which SZ and SSE methods are more appropriate. Since three existing methods are all not suitable for the research purpose of this paper, an island SHA method need be explored for the phase of disaster prevention planning.

At present, some scholars began to study the application of soft computing in disaster risk zoning to express the fuzzy feature of disaster probability [6]. In this paper, fuzzy mathematics theory and spatial analysis technology, based on geographic information system (GIS), are adopted to develop the spatial analysis method of seismic hazard for island (SAMSHI). The SAMSHI mainly serves for the disaster prevention planning, included the models of fuzzy mathematics, statistics, etc., bases on the spatial database of faults data, historical earthquake record data, geological data and Bouguer gravity anomalies data, uses the software platform of GIS and spatial analysis modeling tools, and finally generates the island's fuzzy identification result for each seismic hazard degree.

3. Spatial database for SAMSHI

3.1 Data processing

Like other SHA methods, the basic data for SAMSHI include faults data, geological data and seismic data. All these data needed for SAMSHI are stored in a basic spatial database, Geodatabase in the Platform of ArcGIS 9, as vector data. The coordinate system (geographic coordinate system and projection) are uniformed [7]. The geographic coordinate system select the frequently-used GCS_Beijing_1954, and projection coordinate system select Beijing_1954_GK_zone_20N, because the research areas are mainly located on the longitude range of 114-120 which is the 20th zone of the 6 degree-zoning system. The parameters for all SAMSHI's spatial data are unified as:

Horizontal coordinate system
Projected coordinate system name: Beijing_1954_GK_Zone_20N
Geographic coordinate system name: GCS_Beijing_1954
Map Projection Name: Transverse Mercator
Scale Factor at Central Meridian: 1.000000
Longitude of Central Meridian: 117.000000
Latitude of Projection Origin: 0.000000
False Easting: 500000.000000
False Northing: 0.000000
Planar Coordinate Information
Planar Distance Units: meters
Coordinate Encoding Method: coordinate pair
Coordinate Representation
Abscissa Resolution: 0.000362
Ordinate Resolution: 0.000362
Geodetic Model
Horizontal Datum Name: D_Beijing_1954
Ellipsoid Name: Krasovsky_1940
Semi-major Axis: 6378245.000000
Denominator of Flattening Ratio: 298.300000

Even after the uniformed setting of coordinate system, some maps may not overlap each other well, then Geo-referencing is needed to rectify them. For example, 3 times of affine transforming and 2 times of projection transforming are used to adjust the geologic data to match the fault map.

IOP Conf. Series: Journal of Physics: Conf. Series **842** (2017) 012024 doi:10.1088/1742-6596/842/1/012024

3.2 The fault data

The Chinese Fault Map is shown in figure 1. In the table of content column, there are 27 categories of faults, i.e. compressive torsional fault, measured detachment fault, measured right-lateral strike-slip fault, measured composite fault zone, measured active fault, measured thrust fault, etc. there are 32522 features in this map, whose spatial and attribute data are stored in the Geodatabase.

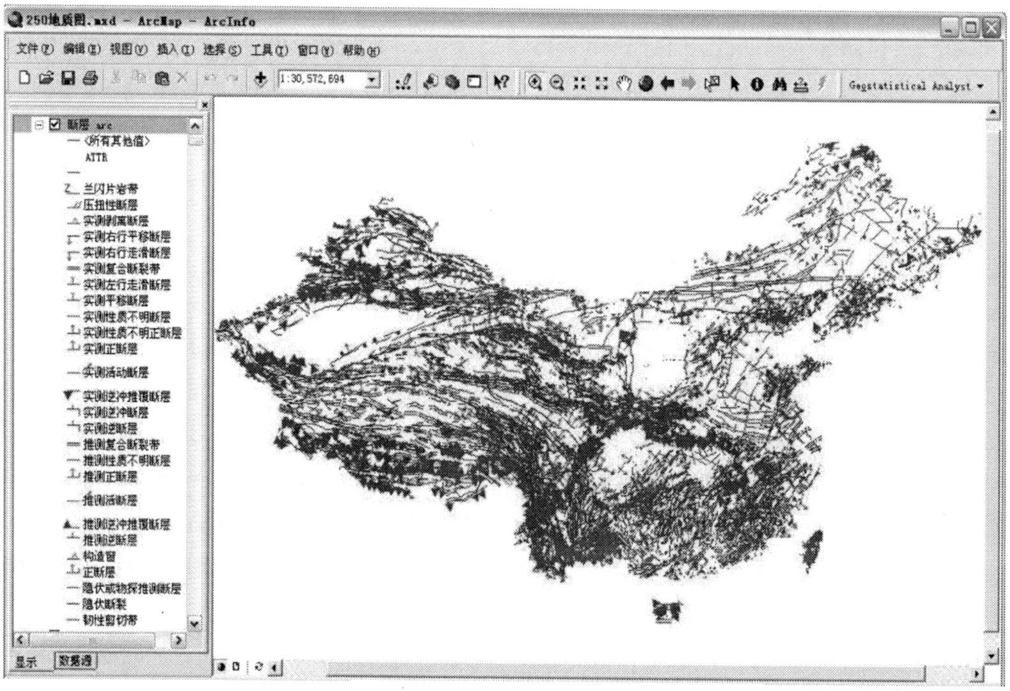

Figure 1. Chinese fault map.

3.3 The fault data

The Chinese Geological Map is shown in figure 1. In this map, there are 61825 geological units. In the table of content column, the displaying symbols and labels of geological units reference Stratigraphic Code of China [8]. Zooming to Shandong Province and Taiwan Province, the Geological Map can be shown more clearly. There are 61825 features in this map, whose spatial and attribute data are stored in the Geodatabase.

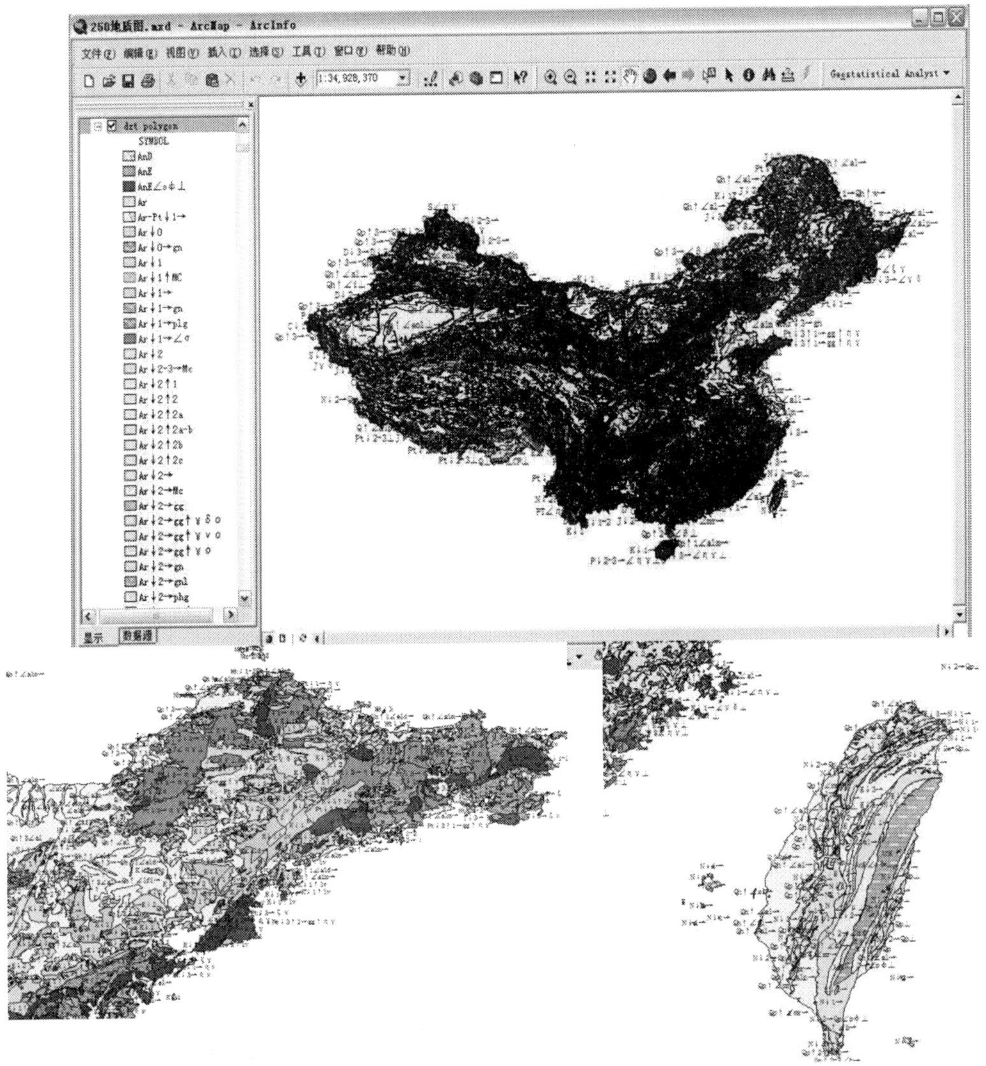

(a) Chinese Shandong Province. **(b)** Chinese Taiwan Province.

Figure 2. Chinese geological map

3.4 The earthquakes data

Parts of Chinese historical earthquakes data are shown in figure 3 as Chinese earthquakes (above magnitude 6) map, and the histograms based on year and magnitude are also appended. The attribute data include earthquake time, location, magnitude, focal depth, etc. There are six feature classes of earthquake data, i.e. modern earthquake swarm, modern sequence, historical earthquake swarm, historical sequence, modern seismic data and historical seismic data in the Geodatabase.

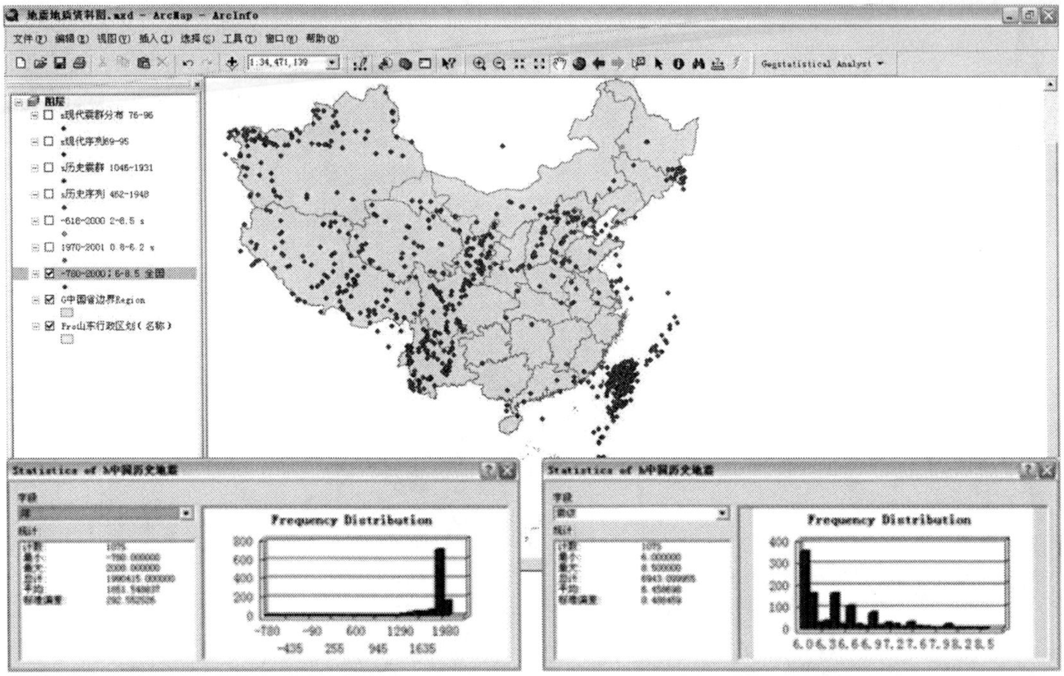

Figure 3. Chinese earthquakes (above magnitude 6) map

4. Fuzzy comprehensive evaluation model of SAMSHI

The ultimate calculation model of SAMSHI is based on fuzzy comprehensive evaluation method (FCEM). The SHA is an essentially comprehensive evaluation process of multiple factors, i.e. converting statistical index of the evaluation object in different aspects and dimensions to dimensionless relative evaluation values and synthesizing these relative evaluation values.

4.1 Evaluation set of FCEM

The seismic hazard degrees of SAMSHI are defined as large, medium and small, which form the evaluation set of FCEM for SAMSHI. The Evaluation set and its corresponding ground motion parameters are shown in table 1.

Table 1. Evaluation set of FCEM for SAMSHI.

	Seismic hazard degrees	Earthquake magnitude beyond the probability of 10% for the next 50 years
v_1	Small	$M < 6$
v_2	Medium	$6 \le M \le 7$
v_3	Large	$M > 7$

4.2 Evaluation indices system of FCEM

The domestic and foreign research scopes of seismic hazard analysis generally include the following four aspects: analysis of uncertainty of earthquake location, analysis of uncertainty of earthquake occurrence time, analysis of uncertainty of earthquake magnitude and analysis of uncertainty of earthquake motion propagation [9]. The construction of evaluation indices system of FCEM for SAMSHI should consider and combine these scopes.

Following the systematic, concise, independent and operable principles, this paper uses system analysis, frequency statistics and expert consultation method to select evaluation index, and finally form the evaluation indices system of FCEM for SAMSHI, which is shown in table 2. It should be noted that all of these indices selected are quantitative, so as to avoid the subjectivity of qualitative indices and ensure the objectivity of the evaluation.

Table 2. Evaluation indices system of FCEM for SAMSHI.

	Indices	Weight
B_1	Within 150 km of the evaluation area, the area of Bouguer gravity anomalies slope grid map in set threshold.	0.0472
B_2	Within 150 km of the evaluation area, the average value of geological parameter.	0.0265
B_3	Within 150 km of the evaluation area, the earthquakes number above M6.	0.2252
B_4	Within 150 km of the evaluation area, the maximum earthquake magnitude.	0.0805
B_5	Within 150 km of the evaluation area, the maximum earthquake magnitude in recent 200 years.	0.1218
B_6	Within 150 km of the evaluation area, the M-lnY curve's area integral ratio of the recent 200 years to the 200 years before last	0.0572
B_7	Within 150 km of the evaluation area, the seismic frequency ratio of the recent 200 years to the 200 years before last	0.026
B_8	Within 150 km of the evaluation area, the intercept of Gutenberg curve based on the seismic frequency statistics.	0.1143
B_9	Within 150 km of the evaluation area, the absolute value of the slope of Gutenberg curve based on the seismic frequency statistics.	0.1143
B_{10}	Within 150 km of the evaluation area, the annual average incidence rate of earthquakes above magnitude 4 in recent 100 years.	0.1275
B_{11}	Within 150 km of the evaluation area, the total length of fault	0.0595

4.3 Weights of indices

The Analytic Hierarchy Process method, which is put forward by T. L. Saaty, is adopted to calculate the weights of indices [9]. The decision matrix for evaluation indices system of FCEM for SAMSHI is shown in table 3.

Table 3. Decision matrix for evaluation indices.

	B1	B2	B3	B4	B5	B6	B7	B8	B9	B10	B11
B1	1	3	1/5	1/4	1/4	1	3	1/3	1/3	1/3	1
B2	1/3	1	1/5	1/3	1/4	1/3	1	1/3	1/4	1/5	1/3
B3	5	5	1	3	2	3	5	3	3	2	4
B4	4	3	1/3	1	1/2	2	3	1/2	1/2	1/2	1
B5	4	4	1/2	2	1	2	4	1	1	1	2
B6	1	3	1/3	1/2	1/2	1	2	1/2	1/2	1/2	1
B7	1/3	1	1/5	1/3	1/4	1/2	1	1/5	1/4	1/5	1/3
B8	3	3	1/3	2	1	2	5	1	1	1	2
B9	3	4	1/3	2	1	2	4	1	1	1	2
B10	3	5	1/2	2	1	2	5	1	1	1	3
B11	1	3	1/4	1	1/2	1	3	1/2	1/2	1/3	1

According to the consistency test result, the consistency ratio value 0.0251 can satisfy the consistency test demand of being smaller than 0.1, and the weights set result is:

Wi (i =1, 2, …, 11)= (0.0472, 0.0265, 0.2252, 0.0805, 0.1218, 0.0572, 0.0260, 0.1143, 0.1143, 0.1275, 0.0595), which is shown in the column weight in table 2.

4.4 Spatial analysis models of the 11 indices

The spatial analysis models for the 11 indices of the evaluation indices system of FCEM for SAMSHI are all constructed in the Model Builder of ArcGIS 9 which is embedded in Arcmap and can use all the spatial analysis tools in ArcToolbox. An island in Yantai of Shandong Province is taken as a sample, and its spatial analysis process of index B1 is elaborated as an example.

Bouguer Gravity Anomalies (BGA) is the observation result of gravimeter, a gravity difference after the rectification of latitude, height, interface layer and terrain, reflecting the depth change of Moho surface, i.e. the change of crust depth [10], [11].

The original BGA map is contour map. After the spatial analysis process of "contour→TIN→DEM→slope", the slope grid map of BGA is generated and shown in figure 4. It also reveals that the value distribution of the slope grid has strong correlation with the earthquake distribution, with the overlapping and further comparison of the distribution map of historical earthquakes above magnitude 6.

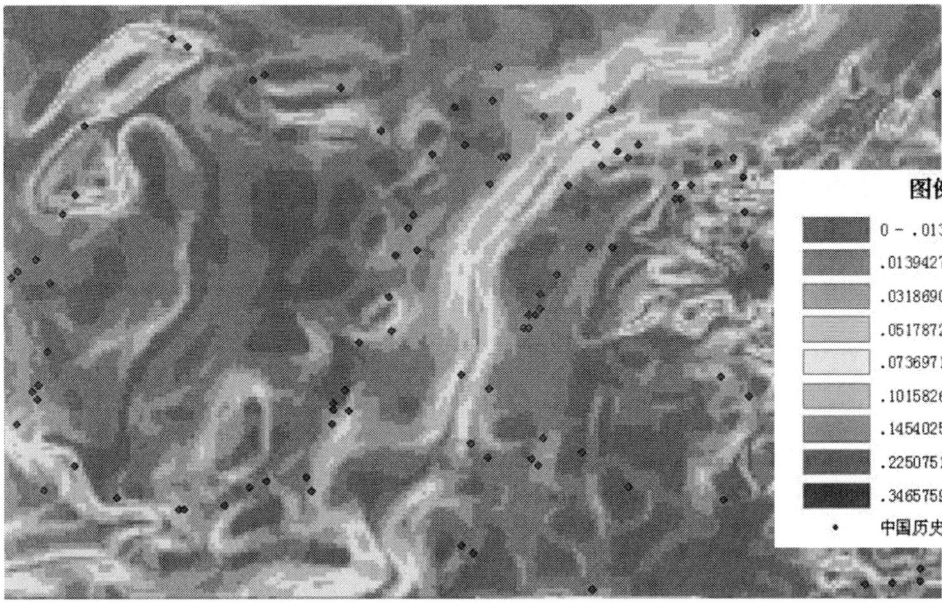

Figure 4. Slope grid map of BGA

According to the statistical analysis of the slope grid, the threshold is set to be 0.086, and the area of slope grids whose values are smaller than 0.086 is 7124.83 square kilometers, in the slope grid map of BGA, within 150 km of the evaluation area, i.e. an island in Yantai of Shandong Province.

Finally, the other 10 indices of an island in Yantai of Shandong Province are all calculated, and another sample of an island in Qingdao of Shandong Province's analyzed for comparison. These two samples' calculated results are shown in table 4.

Table 4. 11 indices value of FCEM for two samples

	B_1	B_2	B_3	B_4	B_5	B_6	B_7	B_8	B_9	B_{10}	B_{11}
An island of Yantai	7124.83	0.474	4	7.4	7.4	435/162.2 2.682	126/42 3	3.875	0.778	0.2	638119
An island of Qingdao	14362.16	0.342	3	7	6.2	293/95.2 3.078	114/26 4.38	3.252	0.668	0.07	1519401

5. Membership function of the indices

The membership function of all the 11 indices for v_I fuzzy subset uses the descending half trapezoid distribution function. The membership function expression is:

$$\frac{\widetilde{A}(x)}{v_1}=\begin{cases}1 & 0\le x\le a_{i1},\\ \dfrac{a_{i2}-x}{a_{i2}-a_{i1}}, & a_{i1}<x\le a_{i2}, \quad i=1,2,\cdots,11, \text{ indicating 11 indices.}\\ 0 & a_{i2}<x,\end{cases}$$

The membership function of all the 11 indices for v_3 fuzzy subset uses the ascending half trapezoid distribution function. The membership function expression is:

$$\frac{\widetilde{A}(x)}{v_3}=\begin{cases}1 & 0\le x\le a_{i3},\\ \dfrac{a_{i4}-x}{a_{i4}-a_{i3}}, & a_{i3}<x\le a_{i4},\\ 0 & a_{i4}<x,\end{cases}$$ i=1,2,···,11, indicating 11 indices.

The membership function expression of all the 11 indices for v_2 fuzzy subset is ：

$$\frac{\widetilde{A}(x)}{v_2}=1-\frac{\widetilde{A}(x)}{v_1}-\frac{\widetilde{A}(x)}{v_3}.$$

The parameter values of a_{i1}、a_{i2}、a_{i3}、a_{i4}, the membership degrees for the evaluation set of the two samples are shown in table 5.

Table 5. Membership degrees for the evaluation set of the two samples.

B_i	a_{ij}	r_{mn}(An island in Yantai)	r_{mn}(An island in Qingdao)	Transformation of x
B_1	a_{11}=0.03 a_{12}=0.08 a_{13}=0.12 a_{14}=0.22	r_{11}=0 r_{12}=1 r_{13}=0	r_{11}=0 r_{12}=0.17 r_{13}=0.83	$x/\pi \cdot 150^2$
B_2	a_{21}=0.3 a_{22}=0.35 a_{23}=0.39 a_{24}=0.55	r_{21}=0 r_{22}=0.47 r_{23}=0.53	r_{21}=0.14 r_{22}=0.86 r_{23}=0	no
B_3	a_{31}=2 a_{32}=5 a_{33}=7 a_{34}=10	r_{31}=0.33 r_{32}=0.67 r_{33}=0	r_{31}=0.67 r_{32}=0.33 r_{33}=0	no
B_4	a_{41}=6.5 a_{42}=7.0 a_{43}=7.5 a_{44}=8.5	r_{41}=0 r_{42}=1 r_{43}=0	r_{41}=0 r_{42}=1 r_{43}=0	no
B_5	a_{51}=6.0 a_{52}=6.5 a_{53}=7.0 a_{54}=7.5	r_{51}=0 r_{52}=0.2 r_{53}=0.8	r_{51}=0.6 r_{52}=0.4 r_{53}=0	no
B_6	a_{61}=0.5 a_{62}=1.5 a_{63}=2.0 a_{64}=3.0	r_{61}=0 r_{62}=0.32 r_{63}=0.68	r_{61}=0 r_{62}=0 r_{63}=1	no
B_7	a_{71}=1.0 a_{72}=3.0 a_{73}=3.5 a_{74}=5.0	r_{71}=0 r_{72}=1 r_{73}=0	r_{71}=0 r_{72}=0.41 r_{73}=0.59	no
B_8	a_{81}=2.5 a_{82}=3.0 a_{83}=3.5 a_{84}=4.0	r_{81}=0 r_{82}=0.25 r_{83}=0.75	r_{81}=0 r_{82}=1 r_{83}=0	no
B_9	a_{91}=1.25 a_{92}=1.32 a_{93}=1.4 a_{94}=1.5	r_{91}=0.5 r_{92}=0.5 r_{93}=0	r_{91}=0 r_{92}=0.03 r_{93}=0.97	x^{-1}
B_{10}	$a_{10,1}$=0.03 $a_{10,2}$=0.08 $a_{10,3}$=0.15 $a_{10,4}$=0.25	$r_{10,1}$=0 $r_{10,2}$=0.5 $r_{10,3}$=0.5	$r_{10,1}$=0.2 $r_{10,2}$=0.8 $r_{10,3}$=0	no
B_{11}	$a_{11,1}$=5 $a_{11,2}$=5.5 $a_{11,3}$=6.0 $a_{11,4}$=6.5	$r_{11,1}$=0 $r_{11,2}$=1 $r_{11,3}$=0	$r_{11,1}$=0 $r_{11,2}$=0.64 $r_{11,3}$=0.36	log(x)

Note: i=1,2,···,11; j=1,2。 r_{mn}(m=1,2,···,11; n=1,2)

Weighted average fuzzy operator is used here, i.e. $B=W \circ R$, to get the fuzzy comprehensive evaluation result, which is shown in table 6.

Table 6. The fuzzy comprehensive evaluation result of SAMSHI for two samples.

vi	An island in Yantai	An island in Qingdao
v1	0.1315	0.2532
v2	0.5687	0.5028
v3	0.2998	0.2440
result	v2	v2

Although Seismic hazard degrees of the two samples are all evaluated as being medium, the membership degrees for each subset are totally different, and the seismic hazard degrees of an island in Yantai is obviously higher.

6. Conclusion

The existing seismic hazard analysis methods are not applicable for island seismic hazard analysis, a specialized spatial analysis method of seismic hazard for island (SAMSHI) is given to fit the island feature and support the further related work of earthquake disaster prevention planning. The basic spatial database, the fuzzy comprehensive evaluation model, and the spatial analysis process of the indices are introduced, taking two islands in Yantai and Qingdao of Shandong province as samples.

7. References

[1] Zelin W 2006 Analysis on island system in the law of the sea stipulates of united nations *Journal of Huazhong University of Science and Technology* **1** pp 51-55

[2] Yang G 2013 *The comparative study of island management system between China and foreign countries* (Qingdao: The ocean university of China) pp 65

[3] Huiteng C 2006 *The seismic risk analysis in downtown areas of Chongqing* (Chongqing: Chongqing University) pp 134

[4] Mengtan G and Shoude L 2006 The discussion on principles of seismic zonation of the next generation *Technology for earthquake disaster prevevtion* **1** pp 1-6

[5] Xingxian H, Yuming Z, Mengtan G, Rongyu T , Guoxing C, Xiaojun L and Fengxin Z 2005 *Evaluation of seismic safety for engineering sites* (Beijing: Standards Press Of China)

[6] Hemetsberger M, Klinger G, Niederer S and Benedikt J 2002 Risk assessment of a avalanches - a fuzzy GIS application *Proc. of the 5th int. FLINS Conf.* pp 395-402

[7] Banggu L and Xuepei H 2004 A study on coordinate transformation of multi-spatial data in GIS *Geomatics & Spatial Information Technology* **1** pp 26-30

[8] LiwenX, Zhaoling Z and et al 1999 *Stratigraphical lexicon of China* (Beijing: Seismological Press) p 95

[9] Saaty T L 1980 *The Analytic Hierarchy Process* (New York: McGraw-Hill)

[10] Shengming F and Rui F 1997 Seismicity and characteristics of anomalous isostatic gravity in central Asia *Acta Seismologica Sinica* **6** pp 650-654

[11] Shaoan S, Aimin X and Hui L 1999 Research on evolution of western Yunnan and Beijing regional gravity fields and their relationship with earthquake *Earthquake* **1** pp 97-106

Assessment of vocal cord nodules: a case study in speech processing by using Hilbert-Huang Transform

M Civera[1], C M Filosi[2], N M Pugno[3,4,5], M Silvestrini[2], C Surace[1], K Worden[6]

[1] Politecnico di Torino, Department of Structural, Building and Geotechnical engineering, Corso Duca degli Abruzzi, 24, 10129, Turin, Italy
[2]Azienda Provinciale per i Servizi Sanitari - Provincia Autonoma di Trento, 38123, Trento, Italy
[3] Laboratory of Bio-Inspired and Graphene Nanomechanics, Department of Civil, Environmental and Mechanical Engineering, University of Trento, 38123, Trento – Italy;
[4]School of Engineering and Materials Science, Queen Mary University of London, Mile End Road, E1 4NS London - United Kingdom;
[5]Italian Space Agency, Via del Politecnico snc, 00133 Rome, Italy.
[6] Dynamics Research Group, Department of Mechanical Engineering, University of Sheffield, Mappin Street, Sheffield S1 3JD, UK.

E-mail: marco.civera@studenti.polito.it

Abstract. Vocal cord nodules represent a pathological condition for which the growth of unnatural masses on vocal folds affects the patients. Among other effects, changes in the vocal cords' overall mass and stiffness alter their vibratory behaviour, thus changing the vocal emission generated by them. This causes dysphonia, i.e. abnormalities in the patients' voice, which can be analysed and inspected via audio signals. However, the evaluation of voice condition through speech processing is not a trivial task, as standard methods based on the Fourier Transform, fail to fit the non-stationary nature of vocal signals. In this study, four audio tracks, provided by a volunteer patient, whose vocal fold nodules have been surgically removed, were analysed using a relatively new technique: the Hilbert-Huang Transform (HHT) via Empirical Mode Decomposition (EMD); specifically, by using the CEEMDAN (Complete Ensemble EMD with Adaptive Noise) algorithm. This method has been applied here to speech signals, which were recorded before removal surgery and during convalescence, to investigate specific trends. Possibilities offered by the HHT are exposed, but also some limitations of decomposing the signals into so-called intrinsic mode functions (IMFs) are highlighted. The results of these preliminary studies are intended to be a basis for the development of new viable alternatives to the softwares currently used for the analysis and evaluation of pathological voice.

1. Introduction.

Healthy condition of vocal cords is of paramount importance, as their mucosal vibration is the source of the human vocal emissions [1]. Thus, studies about the vibratory mechanisms of vocal folds have been conducted for over twenty years [2][3], as well as research about the frequency content of the acoustic output generated by them [4]. Both of these are nontrivial tasks, as human speech represents a complex issue, mainly due to its well-known nonlinearity and non-stationarity [5][6]. Also, the acoustics of human speech has been studied and modelled at least since the 1960s [7] and progress is still in development nowadays, especially for the realisation of naturally-sounding synthesised voices [8].

The Discrete Fourier Transform (DFT), which is traditionally the preferred and most used signal processing tool, is limited by its assumption of stationarity of the given signal; in the case of human voice, this supposition does not hold true, as the mean, variance and other statistical parameters change with time [9]. Algorithms currently used for the extraction of fundamental frequencies in voice recordings generally assume the signal to be generated by a linear source (which is not), and to be locally stationary [10]. These assumptions are too oversimplified in the case of pathological vocal emissions, which, even more than normal speech signals, are characterised by nonlinearities due to incomplete closure of the vocal cords and aperiodic vibrations. As a result, alternative methods should be tested, in order to verify their capacity to properly address the topic.

In this study, four voice recordings, originated by an unique patient before and after surgical remove of vocal cord nodules, have been analysed by applying a technique of signal processing - the Hilbert-Huang Transform (HHT) - to Intrinsic Mode Functions (IMFs), obtained through the Complete Ensemble EMD with Adaptive Noise (CEEMDAN). This algorithm has been proposed by Colominas, Schlotthauer and Torres [11] in its current form and applied to electroglottograms, electrocardiograms and electroencephalograms [12]. However, previous versions of the algorithm have been studied by the same authors since 2011 [13]. The EMD has also been applied to the identification of pathological voices and for the classification of patients with Adductor Spasmodic Dysphonia (AdsD) and Muscular Tension Dysphonia (MTD) [14], as a tool for differential diagnosis. Here, the viability of this relatively novel approach is tested for the analysis of healing processes during convalesce and for the estimation of the immediate effects of surgery removal of nodules. In more detail, the HHT disclosed some well-defined trends during the convalescence time, even if it showed also some limitations, mainly due to its being an empirical decomposition. The investigated method could bring to light new non-invasive tools for healing process monitoring, if the identified trends could be systematically observed in a large sample of patients.

The paper is organised as follows: in Section 2, vocal cords nodules are briefly described. Some hint about epistemology, symptoms, signs and treatment are presented, too, in order to provide a global background of this research. An overview of the techniques currently used for the diagnosis concludes this part. In Section 3, the case report is introduced. In Sections 4, the investigated methods, based on Hilbert-Huang Transform, are employed to disclose trends of the frequency response of the given signals; Hilbert Spectra for Time-Frequency Analysis and Marginal Hilbert Spectra for Energy Content Analysis, as obtained from the input signals, are presented and commented. In Section 5 obtained results are shown. Finally, in Section 6 the paper concludes with some overall discussion.

2. Vocal Cords Nodules.

Vocal cords nodules are localised, benign (i.e., noncancerous) and callous-like masses, present on both folds although not necessary symmetrical, and located within the lamina propria. Their onset, development and – if medically treated – removal and/or regression alter the mechanical and vibratory properties of the vocal folds. In most of the cases, no other functions than phonation are compromised; also, for milder cases, lesions may resolve naturally by reducing voice use.

2.1. Causes, Early Signs and Symptoms

Growth of pathological masses in vocal folds is generally caused by excessive and repeated mechanical stress. In fact, vocal cords are subject to collision forces at each vibratory cycle. Moreover, the air forced through the small gap between the folds during voice modulation causes also drying. Therefore, nodules arise from vocal cords tissue trauma, which in turn is due to chronic vocal overuse or misuse. Over time, these vocal abuses generate firstly soft and swollen spots, which then evolve into nodules and become bigger and stiffer if the incorrect vocal use persists.

Ordinarily, the first symptoms noticed by people affected are difficulties to produce sounds belonging to the upper vocal pitch range [15] [16]. Nevertheless, the definition of "healthy" vocal range is still not well defined, and changes according to the field of interest. Conventionally, the limits of "regular" vocal emissions reach as low as 50 Hz (at least) and up to 20 kHz [17]; differences in vocal fold size, due to gender, genetics, age and other causes, makes the definition of "physiological" vocal range quite vague.

However, since nodules interfere more or less markedly – depending on their size – with the vibrational behaviour of the vocal cords, differences between frequency responses of healthy vs. pathological conditions are known and documented in the literature [18], even if the vibrational behaviour of vocal folds is not at all easy to understand nor to reproduce, mainly due to its known nonlinearity [5] [19]. Generally, nodules cause frequency and intensity ranges to be reduced, but fundamental frequencies and intensity often do not undergo any dramatic change when these masses develop [15] [16]. Other prominent signs of vocal fold nodules are breathiness and hoarseness. The latter is a sign due to aperiodic vibrations of vocal fold, while the former results from the incomplete closure of folds upon phonation [15]. The patient's voice may also be perceived as more harsh and rough than usual.

Another common symptom that may arise is a sensation of pain or soreness in the neck, lateral to the larynx. This generally happens because of the increased effort required to produce the voice [15] [16]. As will be discussed in more detail in Sections 4 and 5, this point represents also a great limitation to the usefulness of audio recordings as they are currently performed for vocal cord nodules estimation. Indeed, the patients, more or less involuntarily, arrange their vocal emission in order to provide the requested tone; speech production is a closed loop process – and so patients will adjust in order to produce the correct sound if possible. This means that some parameters, especially volume (and so, energy content of the signal) become meaningless when compared between different audio tracks, as the input (i.e., air pressure as produced by the lungs) is out of control.

2.2. Diagnostic Methods

In the current state of knowledge, a kaleidoscope of techniques is available for the diagnosis of vocal fold nodules. By simplifying, they can be sorted into two main groups: methods that depend on the direct observation of vocal cords and the so-called "non-invasive" ones, based on vocal emissions analysis. It should be remember that "vocal emissions" include all kinds of vocal output – voiced, unvoiced and plosive sounds. Furthermore, non-invasive methods can be divided between the ones based on acoustic measurements – therefore, objective and quantitative approaches – and the ones that rely on perceptual evaluation (a subjective measure of voice, performed by specialists). These two approaches dominate the current state of clinical evaluation of voice quality, not only regarding nodules, but also for all kinds of voice disorders. Even more, perceptual methods can be further divided into clinician-based (e.g., GRBAS/GIRBAS and CAPE-V) and patient-based ones, such as V-RQOL and IPVI. Attempts have been made in order to link objective and subjective estimations [20], while the effectiveness and reliability of subjective tests has been amply discussed in literature [21][22].

In this case, the pre-operative conditions and the follow-ups over convalescence time were investigated through the application of a computerised tool for acoustic voice analysis, the Multi-Dimensional Voice Program (MDVP™). First introduced in 1993, MDVP™ software has been applied in several contexts [23]; its validity has been analysed and compared to other available software in recent years [24]. At least, more than 33 acoustic parameters are currently inspected in a quantitative way thanks to MDVP™ [25]. This has been made possible by the introduction in last decades of new digital instruments, such as the first digital spectrograph (DSP Sonograph), introduced by Kay Elemetrics ® in the late 70s.

Also perceptual evaluation was performed as a first assessment of patient's condition, according to the GIRBAS scale. This method works simply by rating six parameters of the voice, ranging from 1 (non-pathological) to 5 (strongly affected). In its first definition [26], the scale was known as GRBAS, where the five elements considered were the Grade, Roughness, Breathiness, Aesthenia and Strain of vocal emissions. The "I" parameter stands for the Instability of voice and was added at a later stage by [27], making it in its current definition. The GRBAS/GIRBAS scale is currently accepted as standard by the European Group on the Larynx and by the Japanese Society of Logopedics and Phoniatrics and represents the most commonly used perceptual methodology.

2.3. Treatments

In some cases, surgery is not needed, nor recommended, for the treatment of vocal cord nodules [28] [29]. Non-surgical techniques, such as behavioural voice therapy, ordinarily performed by speech-language pathologists, are generally able to produce a reduction in the size and severity of nodules, even if traumatic injuries are unlikely to heal completely without any aftermath [28] [30]. However, removal surgery may be necessary when behavioural interventions are not effective. Nevertheless, it is not impossible that the vocal range will be permanently altered post surgery [31]. In the particular case reported here, the patient underwent a clinical surgery removal, as voice therapy alone proved to be insufficient.

3. Case Report.

The patient, an Italian male adult (one of the Authors), started to suffer from voice disorders in April 2013, probably due to the overuse of voice. The clinical picture showed an upper phlogosis and was mainly characterised by hoarseness worsening as a result of vocal strain; an endoscopic evaluation indicated a haemorrhagic vocal cord polyp and some signs of vocal stress.

Two months later, on 11th June 2013, the patient underwent the removal of the right vocal cord polyp, under analgosedation with remifentanil, according to the technique of Target Controlled Infusion (TCI) and local anaesthesia. The procedure was conducted in an awake setting and the patient was dismissed on the same day.

The patient received one session of preoperative counselling, targeting vocal hygiene instruction and surgery preparation; he also followed a course of postoperative therapy.

On perceptual evaluation, GIRBAS decreased from 1-1-1-1-2-1 to all zeros 5 months postoperatively. A voice handicap index (VHI) was also administered before and 5 months after the operation to evaluate the patients' perceived satisfaction with his voice and it went from mild deficit (score 12) to normal results (score 1 point).

In the acoustic analysis carried out by MDVP™ model, an improvement of the parameters (jitter, shimmer, NHR, high pitch range) were also obtained, as will be explained in more depth later.

Since treatment, the patient has not presented any dysphonia recurrence even if he continues to use the voice intensively, mainly due to his work, which is highly demanding in terms of voice use.

The recordings have been all realised at ENT Unit, Santa Chiara Hospital, Trento, Italy. The microphone and instrumentations are produced by KAYPENTAX, provided by default for MDVP™ analysis.

3.1. Speech Records

Each speech record comes from the same patient, is 3.75 seconds long and is composed by a single emission of a sustained vowel 'a' according to Italian pronunciation (/a/ as defined by the International Phonetic Alphabet). Sampling frequency (i.e., the samples of sounds per second to represent the speech recorded digitally) is 44100 Hz, resulting in 165375 elements inside each digital record. This

IOP Conf. Series: Journal of Physics: Conf. Series **842** (2017) 012025 doi:10.1088/1742-6596/842/1/012025

sampling rate has been set in order to cover the entire 20 – 20000 Hz range of human hearing. The audio tracks have been labelled this way: (I) track #1: 11th June 2013, pre-operative (day of surgery, immediately before operation with pathological voice); (II) track #2: 25th September 2013 (103 days since surgical intervention – i.e. circa three months and a half later); (III) track #3: 26th November 2013 (166 days or circa five months and a half later); (IV) track #4: 24th June 2014 (378 days – i.e. circa one year since surgery; 210 days since previous record). They can be seen in Figure 1. By comparing the first two records, it is possible to have direct insight into the surgical intervention results and the immediate aftermath, while evolution of specific trends between tracks #2, #3 and #4 have also provided knowledge about follow-ups and healing process of voice condition during convalescence.

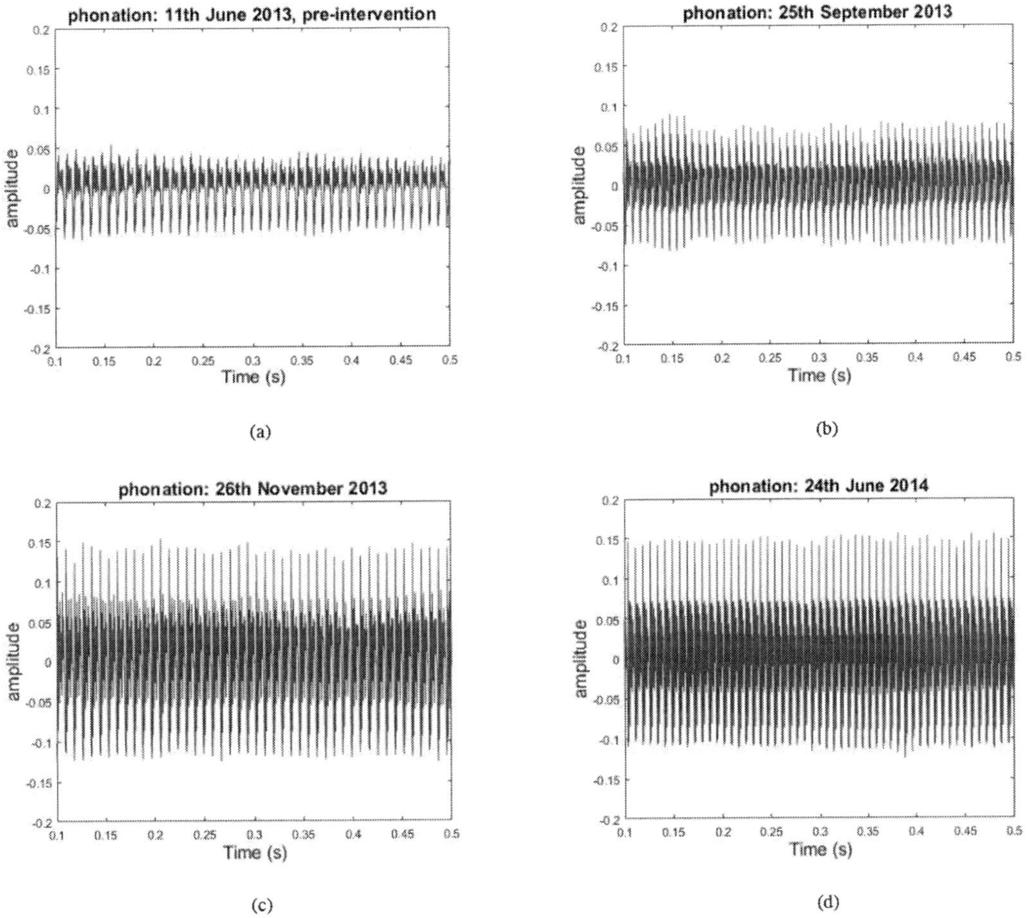

Figure 1. Audio tracks of the patient vocal emission (sustained /a/). (a) track #1 (pathological voice). (b) track #2 (circa three months after removal surgery). (c) track #3 (circa five months since operation). (d) track #4 (one year since operation). For all figures, the signal is stopped at 0.5 seconds.

3.2. MDVP Clinical Reports

Four medical reports have been filed from the analysis of all tracks via MDVP™. Among all the parameters considered, average fundamental frequency (*F0*) is the most commonly used for the evaluation of voice disorders; ordinarily, fundamental frequencies will fall between 85 to 180 Hz for male adults, and between 165 to 255 Hz for the same-aged females [32]. *F0* can be automatically tracked by using peak picking strategies, autocorrelation techniques or other equivalent methods [33].

It represents, obviously, the inverse of the fundamental period *T0*, that is to say, the elapsed time between two successive laryngeal pulses [10]. Nevertheless, as stated before, *T0* is defined in a speech signal according to the assumptions of linearity and local stationarity, which can be a problem if used to approximate a pathological voice. As it will be discussed later in this paper, HHT-based frequency related parameters bypass these limitations.

It results from these reports that pathological voice is not subject to any impressive shift in fundamental frequency, as expected ([15] [16]).

In Figure 2, some results from MDVP™ Reports are reproduced. The 11 indicated parameters are (clockwise moving from the top) the Jitter percent (Jitt); the Fundamental Frequency variation (v*F0*); the Shimmer percent (Shim); the Peak-to-Peak Amplitude Variation (vAm); the Noise-to-Harmonic Ratio (NHR); the Voice Perturbation Index (VTI); the Soft Phonation Index (SPI); the F0-Tremor Intensity Index (FTRI); the Amplitude Tremor Intensity Index (ATRI); the Degree of Voice Break (DVB); and the Degree of Sub-Harmonics (DSH). The green circle encloses the threshold of healthy conditions (every parameter has a different scale).

These parameters are not the only ones produced by MDVP™ analysis, but represent those generally most taken into account, as they are considered the most important objective measures for assessment of several voice disorders [23].

On this particular case, the parameters of interest – the ones which exceed the respective thresholds – are the Jitt, the v*F0*, the Shim and vAm. In more detail, v*F0* exceeds its threshold (1.100 %) before removal, but decreases and stabilises since then; Also Jitt falls drastically just after operation (from 1.655 %, with a threshold of 1.040 %, to 0.379 %) and remains inside the limits afterwards. On the other hand, Shim, which is just slightly over the limit before the intervention (3.851 % respect to 3.810 %), increases in the first months of convalescence (to 5.139 %) and then starts to decrease (to 4.003 % and 2.467 %, respectively for track #3 and #4). Also vAm, which was not above the limit before – for pathological conditions –, went beyond the 8.200 % threshold after operation, reaching a maximum of 10.427 % (track #2) and then declining to 6.341 % (track #3) and to 5.255 % (track #4).

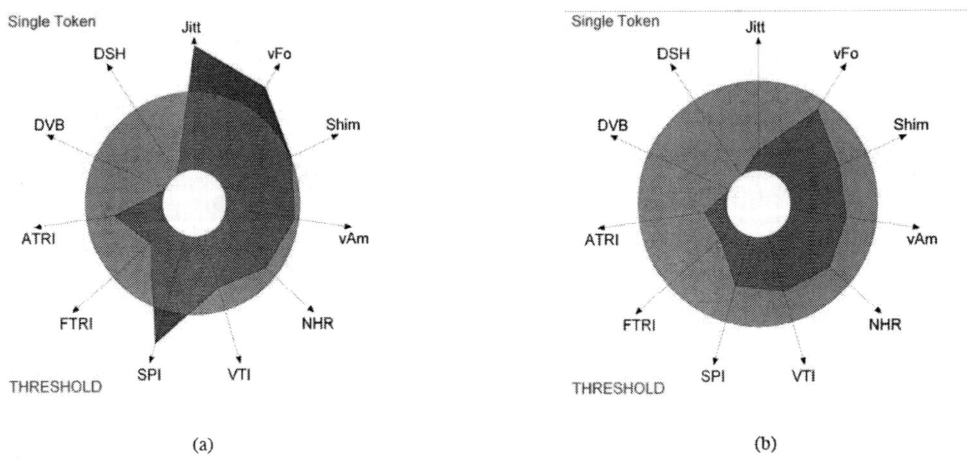

Figure 2. MDVP Clinical Reports. (a) 11th June 2013 (track #1) and (b) 24th June 2014 (track #4).

By considering these data, it is possible to deduce that a transient effect of removal surgery was a temporary increase in variability of the amplitude, which affected both the parameters linked with it, ie. Shimmer and vAm (that is to say, the amplitudes of consecutive periods, divided by the average amplitude, and the Peak-to-Peak Amplitude). Instead, Jitter, which represents the cycle-to-cycle variation of *F0*, and other *F0*-related parameters were positively affected by the surgery and suffered no transient worsening. However, these reports do not provide any information about which frequencies were afflicted the most by the nodules' presence and removal. This information was provided by the EMD-based analysis described here in the next Section.

4. Hilbert-Huang Transform for Speech Processing.

The Hilbert-Huang Transform (HHT) was proposed for the first time in its current form by Huang et al [6]; it represents a suitable option for representing data from nonlinear, non-stationary processes without losing any time-domain information. Essentially, the HHT is made up by two parts: Empirical Mode Decomposition (EMD) and Hilbert Spectral Analysis (HSA). Although the HHT is being used more and more often in the signal-processing context, some background theory will be provided here in order to make this paper a little more self-contained. Much deeper explanations can be found in Huang's original papers [34][6][35], as well as in his recent book [36].

For an arbitrary time series, we can define the Hilbert Transform as

$$HT[g(t)] = \tilde{g}(t) = \frac{1}{\pi}\int_{-\infty}^{+\infty} \frac{g(\tau)}{(t-\tau)} d\tau \qquad (1)$$

This way, the generic time series g(τ) is convoluted with the function 1/πt. This emphasises the local properties of the signal analysed. The tilde ˜ is used here to denote the transformed function, which is still a function of time, as the HT maps function of time or frequency into the same domain, in contrast to the DFT. It should be remarked that the HT, if applied to some dataset, could return physically meaningless results – that is to say, negative frequencies. In order to avoid these problems, two conditions must be applied to the input data: (1) the function must be symmetrical respect to the local zero mean (2) function must have the same number of zero crossing and extrema, or differ at most by one.

Empirical Mode Decomposition provides "modes" that satisfy both these restrictions. These modes – the so-called Intrinsic Mode Functions (IMFs) – can be regarded as the oscillations embedded in the original signal [37]; differently from the harmonics obtained through Fourier Transform, their amplitude and frequency is not constant over time. The process by which they are extracted from the signal can be found in [12]; this step-by-step method is also known as the *sifting process*.
The process is quite straightforward:

(1) for $k = 0$, all the extrema (local maxima and minima) of the analysed data (i.e., of $r_0 = x$) are identified;

(2) local maxima of r_k are connected by a cubic spline line, defining $\mathrm{e_{MAX}}(t)$; likewise, local minima of the same function are linked revolving to the same kind of spline interpolation, thus obtaining $\mathrm{e_{MIN}}(t)$.These two lines form, respectively, the upper and the lower envelops for the given data, which all stand between them;

(3) local mean $\mathrm{m}(t)$ is evaluated as the mean between $\mathrm{e_{MAX}}(t)$ and $\mathrm{e_{MIN}}(t)$;

(4) IMF candidate $\mathrm{d}_{k+1}(t)$ is extracted as $\mathrm{d}_{k+1}(t) = r_k(t) - \mathrm{m}(t)\ \forall t$;

(5) the properties of $\mathrm{d}_{k+1}(t)$ are checked: if $\mathrm{d}_{k+1}(t)$is an IMF, for that instant ' t ' $\mathrm{x}(t)$ is replaced by $\mathrm{d}_{k+1}(t)$, the residue $r_{k+1} = x - \sum_{\mathrm{i}=1}^{k} d_{\mathrm{i}}$ is computed, flag k is increased by one ($k = k + 1$) and r_k is treated as input data for step (2). If $\mathrm{d}_{k+1}(t)$ is not an IMF, $\mathrm{d}_{k+1}(t)$ itself is treated as input data for step (2).

Iteration process ends when the residual r_k satisfies a predefined stopping criterion. In this work, the decomposition has been stopped when, for the *n*-th iteration, the residual $\mathrm{r}_n(t)$ became a monotonic function, from which no more IMFs can be extracted, or when the set number of maximum iterations was reached, whichever came first. Once computed, the IMFs form a complete and nearly orthogonal basis for the original signal; each group is formed by data which have, at any point, zero mean for both the maxima and the minima envelopes [38], as they are – by definition – monocomponent signals.

For this study, the MatLab ® script 'ceemdan.m' has been used. This code has been developed by Marcelo Colominas and was introduced in the current version in [12]. CEEMDAN, or the Complete Ensemble EMD with Adaptive Noise algorithm, is an improvement of the basic EMD algorithm, which is affected by the problem of the so-called "*mode mixing*", the overlap between different modes that have so small differences that they can led to misaddressing of their components. This causes an alias in the frequency-time distribution, leading to a loss of physical meaning of the decomposed data. To solve this issue, Ensemble EMD was first proposed. To keep the discussion brief, the idea is to add white Gaussian noise at each stage of decomposition; then, the generic k-th IMF is computed as the mean over an ensemble of trials ($\overline{IMF_k}$) of the corresponding IMF_k obtained via EMD. Step-by-step, the algorithm can be defined so:

(1) departing from the original signal $x[n]$, $x^i[n]$ is generated, as $x^i[n] = x[n] + w^i[n]$, where $w^i[n]$ are different realisations of white Gaussian noise for $i = (1, \dots, I)$;

(2) each $x^i[n]$ is decomposed by classic EMD, obtaining the modes $IMF_k^i[n]$ for the i-th Gaussian noise and the k-th mode, where $k = (1, \dots, K)$;

(3) k-th IMF is assigned as $\overline{IMF_k}[n] = \frac{1}{I}\sum_{i=1}^{I} IMF_k^i[n]$.

A much more detailed description of the EEMD can be found in [13] and [14]. CEEMDAN, instead, uses each mode for the computation of the next one, sequentially, in a deflationary scheme; basically, in CEEMDAN the several modes are computed as the difference between the current residual and the average of its local means (considering also the noise added to the signal), while in EEMD each $x^i[n]$ is decomposed independently from the other, thus generating I different residuals $r_k^i[n]$ for each mode.

Again, it is possible to describe also the CEEMDAN algorithm in subsequent steps:

(1) I realisation of white Gaussian noise ($w^i[n]$) are used to define $x^i[n] = x[n] + \epsilon_0 w^i[n]$, with ϵ_0 representing the arbitrary value of noise standard deviation for the first step. Then, first modes are computed exactly as for EEMD: $\widetilde{IMF_1}[n] = \overline{IMF_1}[n] = \frac{1}{I}\sum_{i=1}^{I} IMF_1^i[n]$;

(2) for $k = 1$, the first residual is calculated as $r_1[n] = x[n] - \widetilde{IMF_1}[n]$;

(3) departing from the first residual $r_1[n]$, $r_1{}^i[n]$ is generated, as $r_1{}^i[n] = r_1[n] + \epsilon_1 E_1(w^i[n])$, where $w^i[n]$ are different realisations of white Gaussian noise for $i = (1, \dots, I)$ and the operator $E_j(\cdot)$ indicate the whole process that, given a signal, produces the j-th mode by EMD. Then, the second mode is defined as $\widetilde{IMF_2}[n] = \frac{1}{I}\sum_{i=1}^{I} E_1(r_1[n] + \epsilon_1 E_1(w^i[n]))$;

(4) for $k = 2$ onwards, the k-th residuals are computed as $r_k[n] = r_{k-1}[n] - \widetilde{IMF_k}[n]$;

(5) $r_k^i[n] = r_k[n] + \epsilon_k E_k(w^k[n])$ realisations are decomposed. Then, the $(k + 1)$-th mode is defined as $\widetilde{IMF}_{k+1}[n] = \frac{1}{I}\sum_{i=1}^{I} E_1(r_k[n] + \epsilon_k E_k(w^i[n]))$;

(6) steps (4) and (5) are reiterated until $k = K$.

The process is relatively time-consuming, as a large number of iterations are generally required, but reduces substantially the risk of mode mixing. In this work, the initial noise standard deviation ϵ_0 has been set to 0.2; the number of realisations (NR) to 15; and the maximum number of sifting iterations allowed to 3000. The code was also required to automatically increase the SNR for every stage.

It is also important to state that all the data from the four audio tracks have been filtered before being decomposed into IMFs. In more detail, a Butterworth low-pass filter of order 10 has been applied in both the forward and reverse directions, to ensure zero-phase distortion. Filtering was needed in order to reduce the effects of background noise from the audio recordings.

From each track, a set of IMFs has been obtained. From these four sets, in order to speed up the comparison process, four subsets have been extracted, considering only IMFs 7 to 11 (Figure 3). The IMFs chosen for the subset have their Mean Frequencies, f_M, close to the estimated fundamental frequencies of each track, as supplied by the MDVP™ clinical reports.

It must be remarked that the four tracks did not provide the same number of IMFs when decomposed, and that this represents a great limitation to the analysed technique, as explained before. In fact, EMD is a purely empirical decomposition – as the name itself states, obviously – and performs blind signal separation. Hence, no physical interpretation can be provided to justify the obtained number of IMFs. In detail, decomposition of track #1 and track #3 produced both 18 Intrinsic Mode Functions, while 19 functions were extracted from track #2 and track #4. In this particular case, the Authors were able to state that IMFs 7 to 11 include, with small differences, the same data for all the tracks; this was only possible by supervising the EMD operations at any iteration, keeping track of the whole process. However, this is a serious limit to any further attempt to generalise the method, as this distinction can be not always enough evident or easy to oversee.

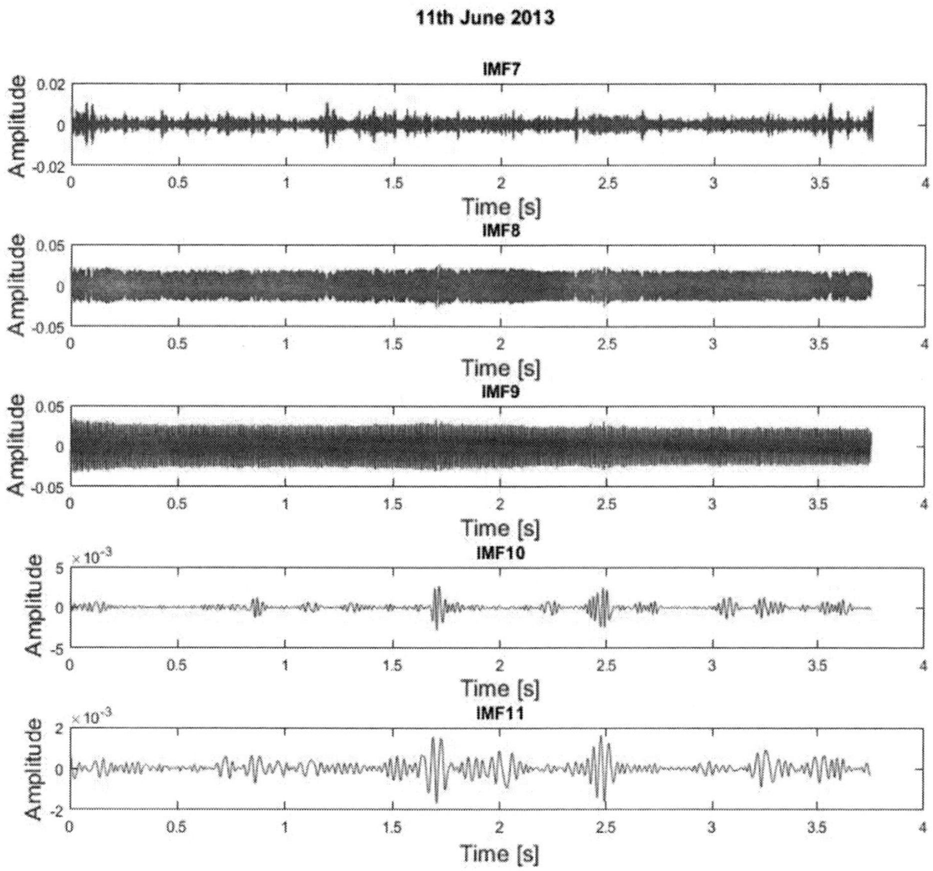

Figure 3. Selected subset (IMFs #7 to #11) for track #1 (11th June 2013).

5. Results.

From each one of the subsets previously described, four parameters have been considered and analysed, in order to test their capability as indices of pathological conditions and/or improvement in healing: the IMF Mean Frequency (f_M), the IMF Standard Deviation (SD), the Total Energy Content (E_{TOT}) and the IMF Content Energy (E_i). The first two will be addressed hereinafter as time-frequency parameters, while the latter two as energetic parameters. These features can be defined as follow:

$$f_M[n] = \sum_{i=1}^{N} f_i\,[n] \tag{3}$$

where N stands for the number of elements inside the investigated signal (here 165375, as the signal is 3.75 seconds long with a sampling rate of 44100 Hz) and $f_i[n]$ are the several instantaneous frequencies of the n-th mode.

$$SD[n] = \sqrt{\frac{\sum_{i=1}^{N}(f_i[n] - f_M[n])^2}{n-1}} \tag{4}$$

where $SD[n]$ simply represents the standard deviation between the several instantaneous frequencies f_i and the mean frequency f_M of the n-th mode.

$$E_i[n] = \sum_{i=1}^{N} c^2\,[n] \tag{5}$$

where $c[n]$ is the amplitude, and hence the energy, of the n-th mode, computed for any element, as it is not constant in time.

$$E_{TOT} = \sum_{n=1}^{M} E_i[n] \tag{6}$$

where M is the total of the IMFs that compose the original signal (18 for tracks #1 and #3, 19 for tracks #2 and #4).

5.1. The Hilbert Spectrum and Time-Frequency Analysis

The Hilbert Spectrum is the graphical representation of the instantaneous frequency over time, computed separately by applying the Hilbert Transform at each one of the IMFs included into the four subsets defined previously. As one can see in Figure 4, the signal taken before removal surgery shows much more often, high frequency peaks, than in the tracks recorded post-operation.

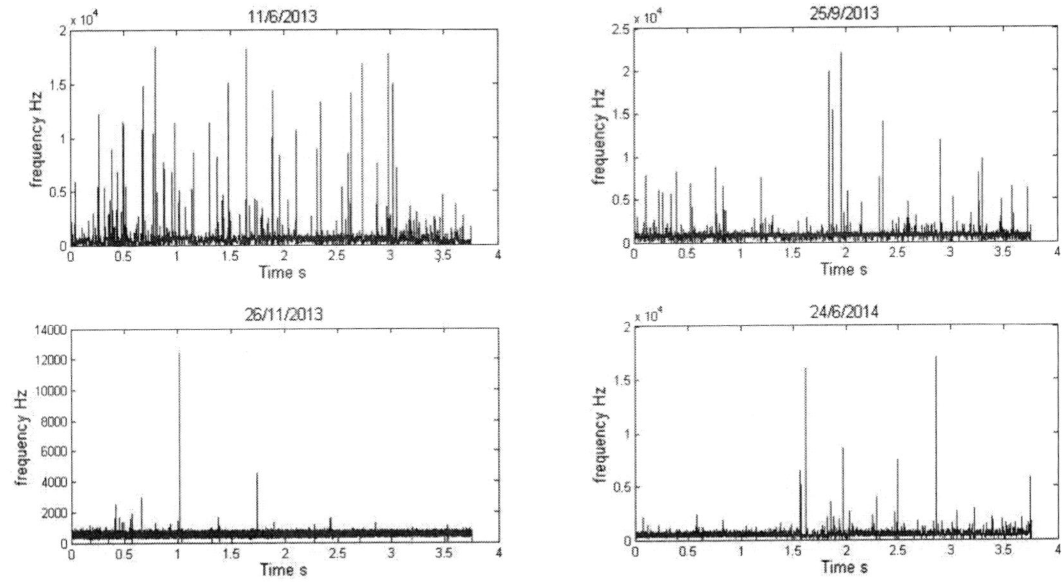

Figure 4. Hilbert Spectra of IMF #7 for all tracks.

By tracking the evolution of both f_M and SD over time, it is possible to recognise some distinct trends. These results, described later, can be seen in the graphs shown in Figure 5. f_M shows a peak corresponding to the post-operative record closest to the surgical intervention (25th September 2013). This reflects the trend of the average fundamental frequency $F0$ between track #1 and track #2, discussed before in Section 3. Nevertheless, $F0$ rebounded slightly after a local minimum occurred corresponding to track #3 and reassessed to values close to the peak for track #4 (mostly the same, circa 134 Hz). Instead, f_M keeps on decreasing for all IMFs, with the sole exception of IMF 8. However, even the 8th mode does not reach again its peak value; overall, all modes seems to tend to stabilising. The exact cause of this behaviour cannot be ascertained with absolute sureness, but could be explained by the fact that surgery is a traumatic event for vocal fold tissues, which change their vibratory characteristics suddenly. A plausible explanation would be scarification and/or swelling. What is certain, is that, since natural frequency is directly proportional to the stiffness and inversely proportional to the mass, any explanation for the immediate post-operative peak will be related to (1) an obvious decrease in mass, due to the nodules' removal; (2) a sudden increase in stiffness, maybe linked to immediate effects of scarification; (3) a combination of these two factors and/or other effects, maybe not related directly to the vocal folds, but to other components of the voice-production mechanisms. Successively, the observed assessment could be most probably due to the healing processes, with scarified tissues reabsorbed over time and a reduction in overall stiffness.

The standard deviation (SD) shows a trend that is strongly related with the one of $F0$. Indeed, since in track #4 the mean frequencies of IMFs 7, 9, 10 and 11 (just to cite the ones reported here) tend to stabilise, while instead $F0$ tends to increase, a higher value of variability is understandable. As mentioned several times earlier, non-pathological voice is supposed to be more coherent and to have a larger range of frequency; both these factors contribute to increase variability of the frequencies contained into the IMFs. Noteworthy is the very high value of SD for IMF 7 in track #1. Most probably, this mode is much more sensitive to the pathological conditions of the pre-operative voice, somehow. A possible explication is that non-homogeneity induced by nodules' presence affects more the modes that are related to higher frequencies, like IMF 7; after surgical intervention, this disturbance is greatly attenuated and this mode starts to behave as its companions do.

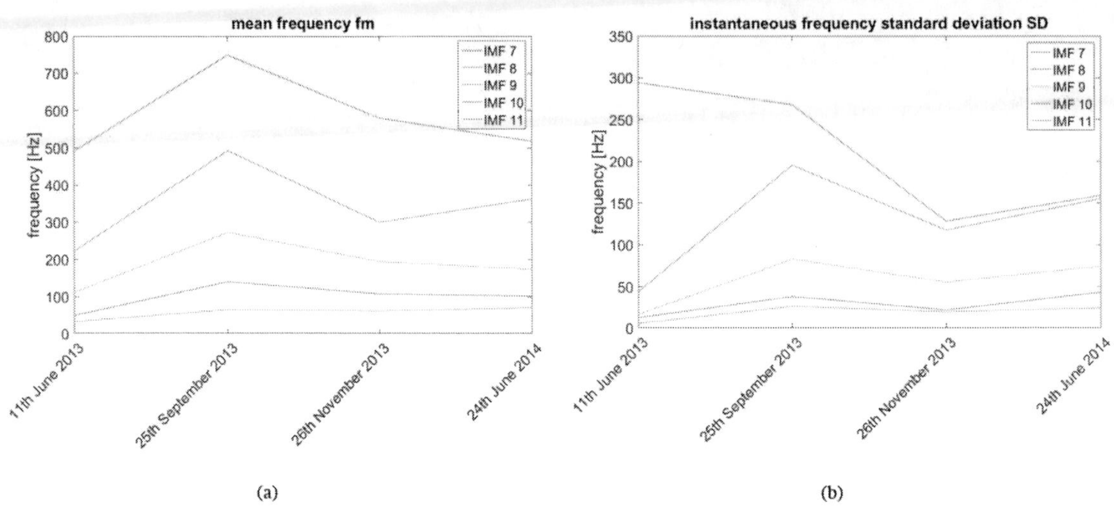

Figure 5. (a) Trends of mean frequencies along time. (b) Trends of instantaneous frequency standard deviations along time.

5.2. Marginal Hilbert Spectra and Energy Content Analysis

Given the Hilbert Spectrum as $H(\omega, t)$ – being, by definition, a portrayal of instantaneous frequency over time – Marginal Hilbert Spectrum can be written as

$$h(\omega) = \int_0^T H(\omega, t) \tag{7}$$

Thus, Marginal Hilbert Spectra (MHS), as shown in Figure 6, permit an immediate visual inspection of the amplitude (i.e., energy) contribution of each frequency.

As can be seen also in Figure 7, each IMF evolves along time according to its own manner, but all of them can be related to the trend of Total Energy E_{TOT}.

The pre-intervention record (track #1) presents a spectrum generally composed of frequencies lower than those preeminent in the post-operative tracks. This evidence agrees with two of the symptoms generally associated with vocal cord nodules, the shrinking of voice range extension and the difficulty to perform the highest frequencies.

It is also possible to notice in Table 1 that no comparison between the IMFs is possible in terms of Energy Content. The values of E_i fluctuate from one record to the other in a fashion that gives no noticeable trends. Only two prominent results can be clearly seen here. Firstly, if IMF 11 is excluded, all the other modes seem to reproduce, broadly speaking, the behaviour of the total energetic content, E_{TOT}. As mentioned before, it has been noticed that nodules seem to affect more the IMFs that contain higher frequencies. Thus, IMF 11 should be the least affected of them all; furthermore, its contribution to the E_{TOT} is negligible (always less than 1%), both in general and when compared to the other modes included in the subset. Secondly, it is possible to observe that before surgery, the Energy Content was very different between different modes. These differences results strongly attenuated nine months after the operation; the vertical bands in Figure 10 show this plainly. This can be seen as evidence that, when voice is healed and back to physiological condition, the energy is distributed between modes in a more uniform fashion. However, it has been evaluated that E_i, as well as E_{TOT} and any other possible energy-related index, is too affected by variations of the voice volume to produce any reliable result when different tracks are compared; mean frequencies and standard deviations of instantaneous frequencies provided results that look much more reliable.

IOP Conf. Series: Journal of Physics: Conf. Series **842** (2017) 012025 doi:10.1088/1742-6596/842/1/012025

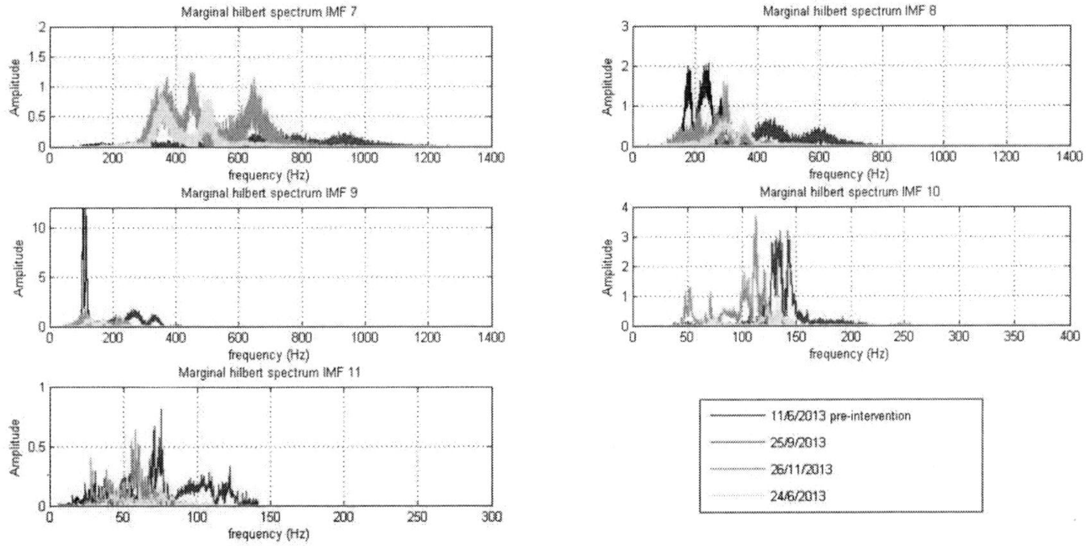

Figure 6. Marginal Hilbert Spectra, IMFs #7 to #11, for all tracks.

Table 1. E_i for IMFs 7 to 11, all tracks. E_{TOT} reported for comparison.

	Track #1 (11th June 2013)	Track #2 (25th September 2013)	Track #3 (26th November 2013)	Track #4 (24th June 2014)
IMF 7 [-]	0.970	12.809	54.026	19.361
IMF 8 [-]	25.583	22.236	60.544	7.301
IMF 9 [-]	50.930	32.406	36.188	10.488
IMF 10 [-]	0.053	39.246	25.023	3.133
IMF 11 [-]	0.032	1.22	0.578	0.186
E_{TOT} [-]	88.795	139.972	477.692	312.09

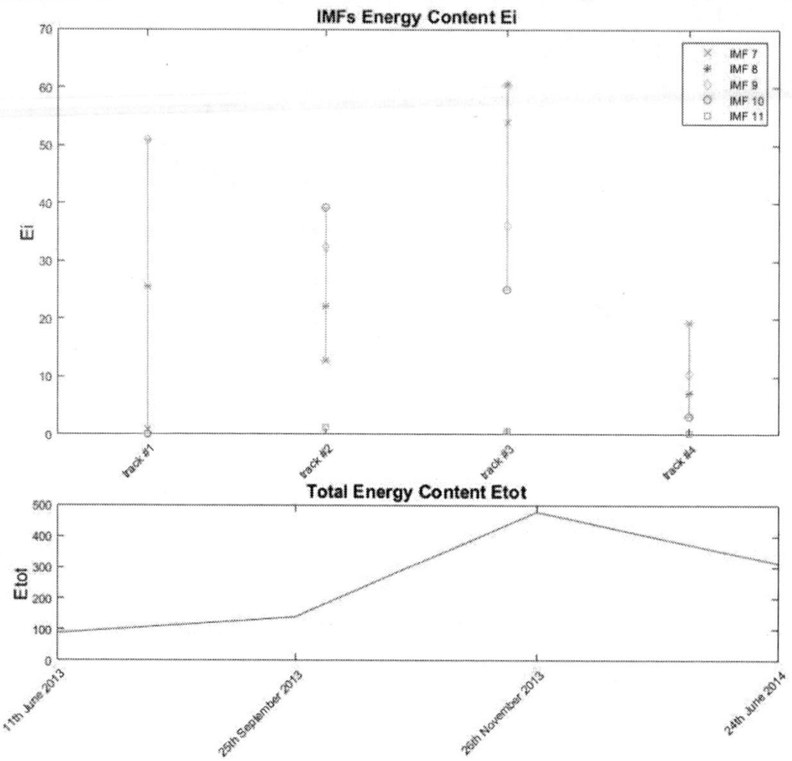

Figure 7. Trends of IMFs Energy Content (Total Energy Content shown for comparison).

6. Conclusions.

The main aim of this research was to investigate the viability of the HHT and CEEMDAN algorithms to track the healing process of vocal cord tissues over convalescence time, as well as to define the immediate aftermath of the removal surgery. A subset has been extracted by the decomposition of the audio tracks into IMFs, specifically IMFs 7 to 11; evolution of these five modes has been studied along the four audio tracks provided. In particular, four parameters have been investigated: the IMF Mean Frequency (f_M), the IMF instantaneous frequency Standard Deviation (*SD*), the Total Energy Content (E_{TOT}) and the IMFs Content Energy (E_i). Some speculations have been drawn from the observed results.

Regarding the trend of f_M over time, all IMFs produced similar result: peak values just after surgery, followed by a stabilisation to values not distant from the pre-operative ones. This proves that vocal cords nodules do not affect substantially the frequency content of voice, as expected, while surgery does, even if for a limited period of time. Higher-numbered modes, which are related to lower frequencies, show to be less affected by the surgery in the short run. Post-operative peaks can be related to a decrease in mass, due to the nodules' removal; to an increase in stiffness, due to tissue scarification; or to a combination of both these effects plus some other mechanisms. The assessment of f_M in the long run, after one year from surgery removal, is almost surely due to an overall reduction of stiffness, which can be related to the healing of scars.

IMFs 8, 9, 10 and 11 showed a higher value of Standard Deviation *SD* in healed conditions than in pathological ones. This increment is sensibly less marked for higher-numbered modes, as they show again to be less influenced by both nodules' presence and surgery aftermaths. This larger variability of the instantaneous frequencies could be a consequence of the broader range of frequency of the healed voice. IMF 7 counterintuitive behaviour may be a result of the non-homogeneous effects of nodules, as lower-numbered modes proved to be more affected and this "deviation" from the otherwise regular trend is limited only to the pre-operative first record.

It was observed that the range of E_i values among the five IMFs reduced noticeably after convalescence, demonstrating how the energy is more uniformly distributed between modes in healthy vocal conditions. Moreover, the Marginal Hilbert Spectrum of track #1 showed also a composition made mostly by frequencies much lower than those prominent in the following records, thus highlighting the reduction of vocal range to lower-than-usual frequencies that ordinarily occurs with the onset of vocal fold nodules. However, apart these two speculations, energy-related features were considered not to be reliable. Overall amplitude of the four signals resulted to be too heavy influenced by voice volume, a parameter that was not taken in account during the several recording operation. Thus, E_{TOT} and E_i cannot be completely trustworthy parameters. By comparing *SD* (which rebounds slightly or stabilises for all the investigated IMFs between track #3 and track #4) with E_{TOT} and E_i, which instead decrease in all cases, it is also possible to state that removing nodule masses from the vocal cords allows them to vibrate much freely, producing a wider range of frequencies, spending less energy.

To sum up, some points become clear from the investigation performed:

(1) The Hilbert-Huang Transform works, as expected, for the analysis of non-stationary data originated by a nonlinear source, as in the case of human voice; theoretically speaking, the tool is perfectly suited for the non-invasive analysis of vocal fold conditions through speech processing. Different to other current methodologies, the HHT performs time-frequency analysis without requiring any assumption of stationarity of the signal or linearity of the system, thus avoiding the risks inherent in the oversimplifications that other techniques are restrained by.

(2) The EMD-based approach allows analysis of the different frequencies separately, even if the decomposition itself, being empirical, is not free from issues. Since there are no guarantees that the decomposition will produce the same amount of IMFs, if the content of this Mode Functions is not – by chance – clearly similar, as in this particular case, the Hilbert-Huang Transform would be itself viable, but results would be much more difficult, or even impossible, to compare.

(3) Energy Content, both for the overall signal (E_{TOT}) and for each one IMF inspected (E_i), is too much influenced by the volume of the patient's voice. Some information can nonetheless be extracted from their trends, but it seems that this results must be handled with great care and not uncritically.

(4) IMF mean frequency (f_M) and instantaneous frequency standard deviation (*SD*) showed clear trends, both during post-operative convalescence time (long run, tracks #2, #3 and #4) and immediately after surgery (short run, tracks #1 and #2). These results are by far the most interesting ones and explain all the possibilities offered by time-frequency analysis for health monitoring the follow-ups after surgical intervention through speech processing.

With all the difficulties and limitations encountered, the authors find that the HHT proved to be feasible, technically speaking, for the proposed aim. f_M and *SD* can be easily included as additional parameters to MDVP™or to any other software for Acoustic Voice Analysis. Nevertheless, the technical obstacles given by the need to have comparable IMFs between different tracks may be just too extended to make the method advantageous in economical terms. Even more, all the results should be identified in a large, statistically valid population before being accepted definitely. Comparing them with records of patients affected by other voice disorders will also help to discern if the patterns encountered are typical of the pathology studied or detectable for other kinds of disease, too. Even so, the results are encouraging, as trends in time-frequency analysis are evident, and exhort the authors to test other options, especially Wavelets, which could probably overcome the technical problematics that afflict HHT and EMD and provide more crystalline answers.

Acknowledgements

The Authors would like to acknowledge Mr. Jacopo Fiorino, M.Sc. in Biomedical Engineering, for his help in setting the basis of this work. N. M. Pugno is supported by the European Research Council PoC 2015 "Silkene" No. 693670, by the European Commission H2020 under the Graphene Flagship Core 1 No. 696656 (WP14 "Polymer Nanocomposites") and under the Fet Proactive "Neurofibres" No. 732344.

References

[1] Timcke R, von Leden H and Moore P 1958 Laryngeal vibrations: measurements of the glottic wave: part I. The normal vibratory cycle *AMA arch. of otolaryngol.* **68(1)** 1-19

[2] Berke G S and Gerratt B R 1993 Laryngeal biomechanics: an overview of mucosal wave mechanics *J. Voice* **7(2)** 123-128

[3] Titze I R, Jiang J J and Hsiao T Y 1993 Measurement of mucosal wave propagation and vertical phase difference in vocal fold vibration *Ann. Otol. Rhinol. Laryngol.* **102(1)** 58-63

[4] Kasuya H *et al* 1986 Normalized noise energy as an acoustic measure to evaluate pathologic voice *J. Acoust. Soc. Am.* **80(5)** 1329-34

[5] Giovanni A *et al* 1999 Nonlinear behavior of vocal fold vibration: the role of coupling between the vocal folds *J. Voice* **13(4)** 465-476

[6] Huang N E *et al* 1998 The empirical mode decomposition and the Hilbert spectrum for nonlinear and non-stationary time series analysis *Proc. Math. Phys. Eng. Sci.* **454(1971)** 903-995

[7] Davis S B 1979 Acoustic characteristics of normal and pathological voices *Speech and language: advances in basic research and practice* **1** 271-335

[8] Zen H and Senior A 2014 Deep mixture density networks for acoustic modeling in statistical parametric speech synthesis *2014 IEEE International Conference on Acoustics, Speech and Signal Processing (ICASSP)* 3844-48

[9] Henríquez P *et al* 2009 Characterization of healthy and pathological voice through measures based on nonlinear dynamics *IEEE transactions on audio, speech, and language processing (TASLP)* **17(6)** 1186-95

[10] Hess W J 2008 Pitch and voicing determination of speech with an extension toward music signals *Springer Handbook of Speech Processing*, ed Springer Berlin Heidelberg pp 181-212

[11] Colominas M A *et al* 2012 Noise-assisted EMD methods in action *Advances in Adaptive Data Analysis* **4(4)** 1250025-1–1250025-11

[12] Colominas M A, Schlotthauer G and Torres M E 2014 Improved complete ensemble EMD: a suitable tool for biomedical signal processing *Biomed. Signal Proces.* **14** 19-29

[13] Torres M E *et al* 2011 A complete ensemble empirical mode decomposition with adaptive noise *2011 IEEE International Conference on Acoustics, Speech and Signal Processing (ICASSP)* 4144-47

[14] Schlotthauer G, Torres M E and Rufiner H L 2010 Pathological voice analysis and classification based on empirical mode decomposition *Development of multimodal interfaces: active listening and synchrony*, ed Springer Berlin Heidelberg pp 364-381

[15] Verdolini K, Rosen C A and Branski R C 2014 Vocal Fold Nodules (Nodes, Singer's Nodes, Screamer's Nodes". *Classification Manual for Voice Disorders-I*, ed Psychology Press pp 37–40

[16] Prater R J and Swift R 1984 *Manual of voice therapy*, ed Little Brown and Company

[17] McKinney J C 1994 *The diagnosis and correction of vocal faults: a manual for teachers of singing and for choir directors (Revised ed.)*, ed Genevox Music Group

[18] Deliyski D D 1993 Acoustic model and evaluation of pathological voice production *Eurospeech 1993* **93** 1969-72

[19] Jiang J J, Zhang Y and Stern J 2001 Modeling of chaotic vibrations in symmetric vocal folds *J. Acoust. Soc. Am.* **110(4)** 2120-28

[20] Bhuta T, Patrick L and Garnett J D 2004 Perceptual evaluation of voice quality and its correlation with acoustic measurements *J. Voice* **18(3)** 299-304
[21] Dejonckere P H *et al* 1993 Perceptual evaluation of dysphonia: reliability and relevance *Folia Phoniatrica et Logopaedica* **45(2)** 76-83
[22] Nemr K *et al* 2012 GRBAS and Cape-V scales: high reliability and consensus when applied at different times *J. Voice* **26(6)** 812-e17–812-e22
[23] Christmann M K *et al* 2015 Use of the program MDVP in different contexts: a literature review *Revista CEFAC* **17(4)** 1341-1349
[24] Manfredi C *et al* 2012 Perturbation measurements in highly irregular voice signals: performances/validity of analysis software tools *Biomed. signal proces.* **7(4)** 409-416
[25] Nicastri M *et al* 2004 Multidimensional Voice Program (MDVP) and amplitude variation parameters in euphonic adult subjects. Normative study *Acta Otorhinolaryngol. Ital.* **24(6)** 337-341
[26] Hirano M 1981 Psycho-acoustic evaluation of voice: GRBAS scale. Clinical examination of voice. ed Springer Verlag
[27] Dejonckere P H *et al* 1995 Differentiated perceptual evaluation of pathological voice quality: reliability and correlations with acoustic measurements *Revue de laryngologie-otologie-rhinologie* **117(3)** 219-224.
[28] McCrory E 2001 Voice therapy outcomes in vocal fold nodules: a retrospective audit *Int. J. Lang. Comm. Disord.* **36(sup 1)** 19-24.
[29] Pedersen M and McGlashan J 2001 Surgical versus non-surgical interventions for vocal cord nodules *Cochrane Libr.* 1-12
[30] Stemple J C and Hapner E R 2014 *Voice therapy: clinical case studies,* ed Plural Publishing
[31] Benninger M S *et al* 1996 Vocal fold scarring: current concepts and management *Otolaryng. Head Neck* **115(5)** 474-482
[32] Titze I R and Daniel W M 1998 Principles of voice production *J. Acoust. Soc. Am.* **104(3)** 1148-48
[33] Gerhard D 2003 Pitch extraction and fundamental frequency: history and current techniques *Tech. Rep., Department of Computer Science, University of Regina, Regina, Canada* 0-22
[34] Huang N E, Long S R and Shen Z 1996 The mechanism for frequency downshift in nonlinear wave evolution *Adv. Appl. Mech.* **(32)** 59-117C
[35] Huang N E, Shen Z and Long S R 1999 A new view of nonlinear water waves: the Hilbert Spectrum 1. *Annu. Rev. Fluid Mech.* **31(1)** 417-457
[36] Huang N E 2014 Hilbert-Huang transform and its applications, ed World Scientific
[37] Kizhner S *et al* 2004 On the Hilbert-Huang transform data processing system development *2004 IEEE Aerospace Conference Proceedings* **(3)** 1961-79
[38] Fontugne R *et al* 2012 Empirical mode decomposition for intrinsic-relationship extraction in large sensor deployments *Workshop on Internet of Things Applications, IoT-App* **(12)**

Minimum constitutive relation error based static identification of beams using force method

Jia Guo[1], **Izuru Takewaki**[1]

[1]Dept of Architecture and Architectural Engineering, Kyoto University, Japan

E-mail: kaku.ka.25x@st.kyoto-u.ac.jp, takewaki@archi.kyoto-u.ac.jp

Abstract. A new static identification approach based on the minimum constitutive relation error (CRE) principle for beam structures is introduced. The exact stiffness and the exact bending moment are shown to make the CRE minimal for given displacements to beam damages. A two-step substitution algorithm—a force-method step for the bending moment and a constitutive-relation step for the stiffness—is developed and its convergence is rigorously derived. Identifiability is further discussed and the stiffness in the undeformed region is found to be unidentifiable. An extra set of static measurements is complemented to remedy the drawback. Convergence and robustness are finally verified through numerical examples.

1. Introduction

This work focuses on the damage identification method which is performed based on static test data through a minimum constitutive relation error (CRE) procedure [1]. For structures such as beams, static tests are easily performed and accurate enough to give useful results as dynamic tests [2-10]. According to the recent variational theory of Geymonat and Pagano [11], constitutive parameters and stresses can be identified through minimization of the CRE as long as full-field displacement measurements are readily acquired under given loads and this idea has been proposed very recently by Florentin and Lubineau [12-16] for parameter identification in plane elasticity problems. This method is also quite tailored for beam structures, to which this paper is just devoted, for the reason that global displacement measurements are easily available and the standard force method [17] based on the minimum complementary energy principle is easily conducted and it often requires substantially less degrees of freedom than conventional displacement-based finite element models. Practical problems such as algorithm implementation, convergence, identifiability and effect of measurement noises will also be discussed in detail.

The remainder of the paper is organized as follows. The minimum CRE principle for inverse identification of a Bernoulli-Euler beam model is simply introduced and the convergence and identifiability of the proposed approach are also discussed in Section 2. Numerical tests are performed in Section 3 and final conclusions are drawn in Section 4.

2. Formulations of the problem

2.1. Reference problem

Let us consider a one-dimensional Bernoulli-Euler beam defined in interval $X=[x_s,x_e]$. The flexural stiffness of the beam is assumed to be $k(x)=EI(x)$. Loading conditions are distributed loads $q(x)$ and concentrated loads F^{ex},M^{ex} along with boundary displacements w_p,θ_p on the prescribed displacement boundary. Then the beam model can be described as a combination of the following three parts:

- Kinematic constraints

$$w\in\mathcal{U};\tag{1}$$

- Equilibrium equations

Find bending moment $M\in\Im$ with

$$\Im=\{M:V=M',M''=q(x),V\text{ and }M\text{ are in equilibrium with }F^{ex}\text{ and }M^{ex}\},\tag{2}$$

where V is a shear force and x^{ex} is the position where concentrated loads F^{ex},M^{ex} are enforced;

- Constitutive relation

$$M=kw''\tag{3}$$

with $k\in\mathcal{C}=\{k\in L_2(X):\exists 0<c_0<C_0,\text{ such that }c_0\le k(x)\le C_0\}$ representing a positive stiffness field.

Generally, $\mathcal{U}$, $\Im$, $\mathcal{C}$ that will be used frequently in what follows are called the spaces of (kinematically) admissible displacement field, (statically) admissible bending moment field and (constitutively) admissible elastic stiffness field, respectively.

In this paper, identification of damage in beams is performed with the help of *static measurements* which contain the information of static loads and full-field displacement measurements. The objective is to identify the damaged stiffness $k(x)\in\mathcal{C}$, given the displacement measurements $w\in\mathcal{U}$ and the respective static loading conditions $\Im$; that is to say, the inverse identification problem reads: find $(k,M)\in\mathcal{C}\times\Im$ with given $w\in\mathcal{U}$ which is a reverse of the forward problem: find $(w,M)\in\mathcal{U}\times\Im$ with known $k\in\mathcal{C}$. Evidently, by solving the inverse problem, one can not only identify the damage, but also reconstruct the stiffness and bending moment distribution.

2.2. The minimum-CRE principle and identification algorithm

Consider an admissible solution trio $(\tilde{w},\tilde{M},\tilde{k})$ which satisfies

$$\tilde{w}\in\mathcal{U},\tilde{M}\in\Im,\tilde{k}\in\mathcal{C}.\tag{4}$$

To measure the distance of the admissible solution trio to the exact solution trio in energy product, an error in constitutive relation is introduced

$$e_{CRE}(\tilde{w},\tilde{M},\tilde{k})=\frac{1}{2}\int_X\frac{(\tilde{M}-\tilde{k}\tilde{w}'')^2}{\tilde{k}}dx\tag{5}$$

which is termed the constitutive relation error (CRE) [1] (or constitutive equation gap (CEG) [12]).

For the present inverse problem, the displacement field $w\in\mathcal{U}$, the spaces $\mathcal{C}$ and $\Im$ for search of the stiffness k and the bending moment M, respectively, are already known. Minimization of the CRE can then be interpreted as the following procedure,

$$(M,k)=\arg\min_{\tilde{M}\in\Im,\tilde{k}\in\mathcal{C}}F(\tilde{M},\tilde{k});\ F(\tilde{M},\tilde{k}):=e_{CRE}(w,\tilde{M},\tilde{k}),\tag{6}$$

which is known as *the minimum CRE principle* [11] for inverse identification. For practical implementation, a two-step substitution algorithm is developed.

Firstly, minimization of the CRE function over $\tilde{M}$ yields a minimum complementary energy problem that should be solved by the force method. Therefore, the procedure is simply designated as

$$\begin{aligned}\tilde{M} &= Force\_Method(\tilde{k},\Im,\mathcal{U}) \\ &:= \arg\min_{\tilde{M}\in\Im}\{\frac{1}{2}\int_X \frac{\tilde{M}^2}{\tilde{k}}dx - \sum_p(\theta_p\tilde{M}|_p + w_p(\tilde{M})'|_p)\}\end{aligned} \tag{7}$$

and this step is called the *force-method* step.

Secondly, minimization of the CRE over $\tilde{k}$ gives

$$\int_X\left\{\frac{\tilde{M}^2}{\tilde{k}^2} - (w'')^2\right\}\delta\tilde{k}dx = 0. \tag{8}$$

Practically, it is sufficient to assume that the stiffness $\tilde{k}$ is piecewise constant, that is to say,

$$\begin{aligned}&X = \bigcup_{i=1}^N X_i, \forall 1 \le i \ne j \le N, X_i \text{ and } X_j \text{ are non-overlapping,} \\ &\tilde{k}(x) = \{\tilde{k}_i, x \in X_i\}\end{aligned} \tag{9}$$

where N is the number of non-overlapping pieces, or *elements*. Under this circumstance, equation (8) is reduced into

$$\int_{X_i}\left\{\frac{\tilde{M}^2}{\tilde{k}_i^2} - (w'')^2\right\}\delta\tilde{k}_i dx = 0, i = 1,2,...,N \tag{10}$$

and then, one has

$$\tilde{k}_i = \sqrt{\frac{\int_{X_i}\tilde{M}^2 dx}{\int_{X_i}(w'')^2 dx}}, i = 1,2,...,N. \tag{11}$$

It turns out that equation (11) looks like the direct use of the constitutive relation in equation (3). Thus, this step is named the *constitutive-relation* step. After integration of both equations (7) and (11), a two-step iterative algorithm for inverse identification of the damaged stiffness could be established.

2.3. Identifiability

In fact, convergence of the algorithm requires the uniqueness of the minimizer and therefore, the algorithm may not work well in the case of $\mu(w''(x) = 0, x \in X) > 0$. Let us assume that the measured displacements do not deform on a certain region, e.g., $w'' = 0$ on X_l with $l \in \{1,2,...,N\}$ being a number. Thus, the CRE is slightly modified into

$$e_{CRE}^{\text{mod}}(\tilde{w},\tilde{M},\tilde{k}) = \frac{1}{2}\int_{X\backslash X_l}\frac{(\tilde{M}-\tilde{k}\tilde{w}'')^2}{\tilde{k}}dx \tag{12}$$

that is to say, to identify the stiffness of a beam at a certain region, practical loads must render this region deformable, i.e. $w'' \ne 0$. If the stiffness of an undeformed region must be identified, another static load should be enforced on the beam so that this region is deformed and another set of static measurements is obtained. To be distinct, denote the extra measured displacement field by $w_1 \in \mathcal{U}_1$ and the corresponding bending moment is sought in $\tilde{M}_1 \in \Im_1$. Then, the inverse identification procedure can proceed in an analogous way to equation (13),

$$(M_1, M, k) = \arg\min_{\tilde{M}\in\Im,\tilde{M}_1\in\Im_1,\tilde{k}\in C_h}\left\{re_{CRE}(w,\tilde{M},\tilde{k}) + r_1 e_{CRE}(w_1,\tilde{M}_1,\tilde{k})\right\} \tag{13}$$

with $r, r_1 \geq 0$ and $r + r_1 = 1$. The parameters r and r_1 are determined by the reliability of practical measurements. In a word, the strategy to introduce additional sets of static measurements can help improve and enhance the identifiability of the algorithm, and provide remedy for possible unidentifiability of the proposed approach accordingly. In fact, the strategy is also effective for other inverse identification problems [17].

3. Numerical tests

To show the effectiveness of the proposed damage identification approach, two beams are studied. They are a simply supported beam and a propped cantilever beam. A Gaussian noise is added to the simulated response as

$$\text{noise}(x) = randn \cdot a \cdot u(x) \tag{14}$$

where $randn$ is the standard Gaussian random distribution with zero mean and unit standard deviation, a is the applied noise level and $u(x)$ is the simulated displacement at a certain point x. For application of the two-step substitution algorithm, the convergence tolerance is practically set to $\text{TOL} = 0.001$.

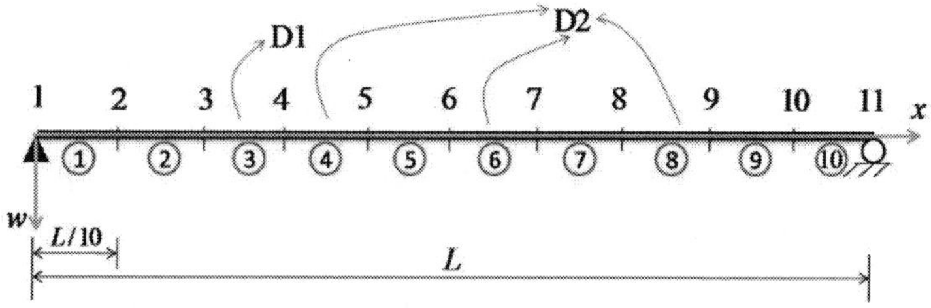

Figure 1. Geometry and damage location of simply supported beam

3.1. Example 1—a simply supported beam

Damage identification of a simply supported beam is studied herein. There are eleven uniformly distributed measurement points which divide the beam into ten elements as enumerated in **Figure 1**. Parameters of the intact beam are: length $L = 6$ m, Young's modulus $E = 1 \times 10^{10}$ Pa and the rectangle section is of size $0.1\text{m} \times 0.1\text{m}$; that is to say, inertial moment of the beam is $I = 0.08333 \times 10^{-4} \text{m}^4$ and the bending stiffness of the intact beam is $k^0 = EI = 8.33 \times 10^4 \text{N} \cdot \text{m}^2$ over the whole beam.

Two damage cases D1 and D2, having single damage region and multiple damage regions, respectively, are considered:

- Case D1: stiffness reduced to 50% in element 3;
- Case D2: stiffness reduced to 50%, 30% and 70% in elements 4, 6 and 8, respectively.

Moreover, two different load cases are enforced separately as

- Case L1: concentrated force $F = 1$ kN at the middle point (point 6); and
- Case L2: uniformly distributed load $q = 0.5$ kN/m over the whole beam,

such that the effect of load types on damage identification is examined. The combinatory of the two damage cases D1 and D2 with the two load cases L1 and L2 provides four damage identification scenarios (I1:D1+L1; I2:D1+L2; I3:D2+L1; I4:D2+L2). The displacements under the four scenarios (I1, I2, I3 and I4) are measured numerically by finite element computations, since for an elastic Bernoulli-Euler beam with piecewise constant stiffness, finite element analysis with cubic Hermit shape functions can give rise to the exact nodal displacements if every element is of constant stiffness [18].

For this statically determinate beam, the exact bending moment is directly known, a single step in equation (11) would complete the identification of the damaged stiffness. For the sake of convenience, the damage index DAI is introduced to represent the scaled damage in the beam, i.e.

$$\text{DAI}_i = k_i^{damage} / k_i^0 \tag{15}$$

for the i th element where k_i^{damage} is the damaged stiffness in element i and k_i^0 is the undamaged initial stiffness. Then, damage indices for the four scenarios are computed and listed in **Table 1**. Obviously, the damage is perfectly identified and reproduced for all the four scenarios. Nevertheless, it is seen that the concentrated load (in scenarios I1 and I3) can lead to slightly better damage identification results than the uniformly distributed load (in scenarios I2 and I4); this is reasonable since Hermit interpolation of the exact pointwise displacements can lead to the exact displacement field for a concentrated load, but only gives an approximate displacement field w for a uniformly distributed load.

Table 1. Identified damage indices of simply supported beam

Element number		1	2	3	4	5	6	7	8	9	10
Damage identification scenarios	I1	1.0000	1.0000	0.5000	1.0000	1.0000	1.0000	1.0000	1.0000	1.0000	1.0000
	I2	1.0001	1.0000	0.5000	1.0000	1.0000	1.0000	1.0000	1.0000	1.0000	1.0001
	I3	1.0000	1.0000	1.0000	0.5000	1.0000	0.3000	1.0000	0.7000	1.0000	1.0000
	I4	1.0001	1.0000	1.0000	0.5000	1.0000	0.3000	1.0000	0.7000	1.0000	1.0001

In addition, to investigate the robustness of the proposed approach, measurement noises of levels 1%, 2%, 5% and 10% are taken into consideration for scenarios I1 and I4. The identified results are displayed in **Figure 2**. It is seen that when measurement noise is not greater than 5%, the damage can be well identified; while if noises rise to 10%, the proposed approach may not work well for a statically determinate beam structure. To examine how identified results are affected by the number of measurement points, the damage case D1—50% stiffness reduction over interval $1.2\text{m} \le x < 1.8\text{m}$ of the beam— under concentrated moments M_0= -1kNm at the left end and M_L=1kNm at the right end is further considered for 11 and 21 uniformly distributed measurement points and under measurement noises of levels 0%, 2% and 5%. As shown in **Figure 3**, the identified results are almost of the same quality under either 10 or 20 measurement elements. Thus, it is deduced that the quality of the identified results depends mainly on the accuracy of measurement rather than the number of measurement points.

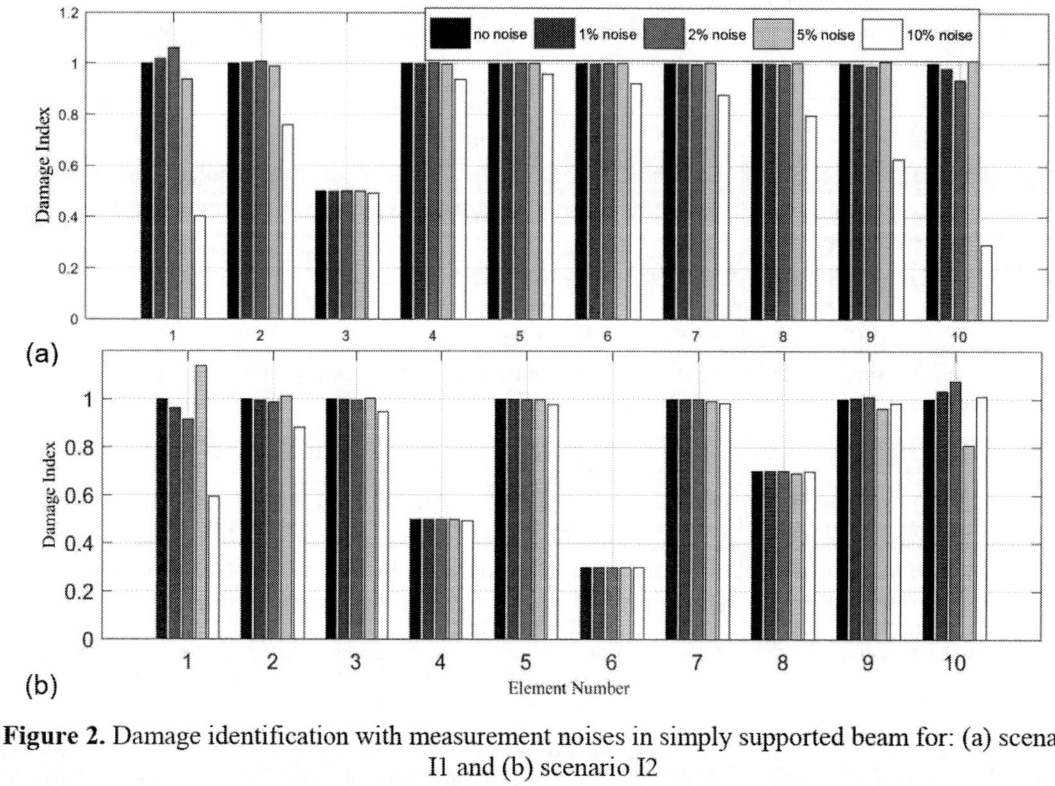

Figure 2. Damage identification with measurement noises in simply supported beam for: (a) scenario I1 and (b) scenario I2

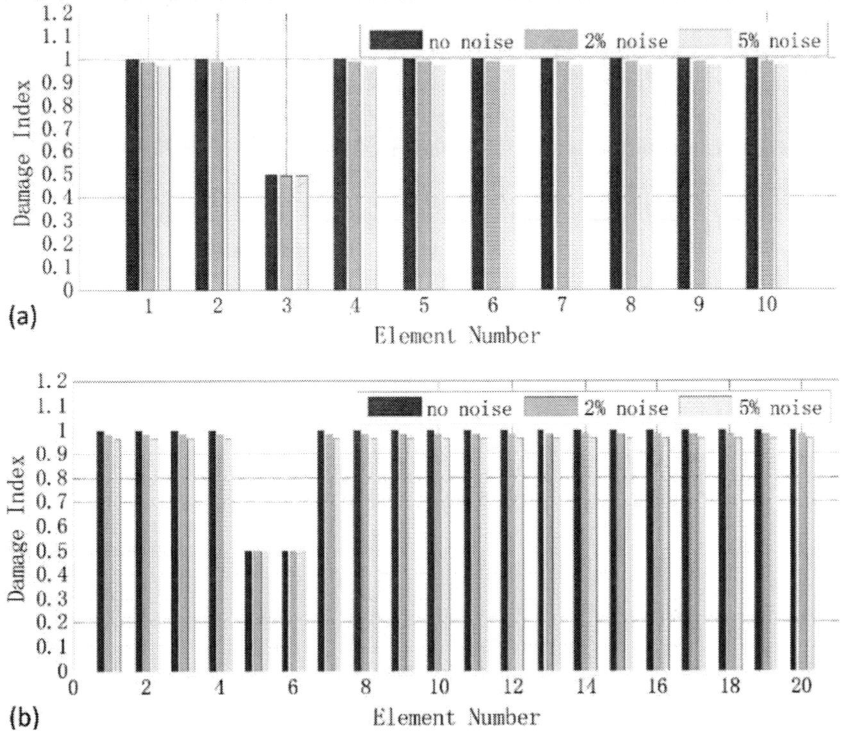

Figure 3. Damage identification with measurement noises in simply supported beam under (a) 10 measurement elements and (b) 20 measurement elements

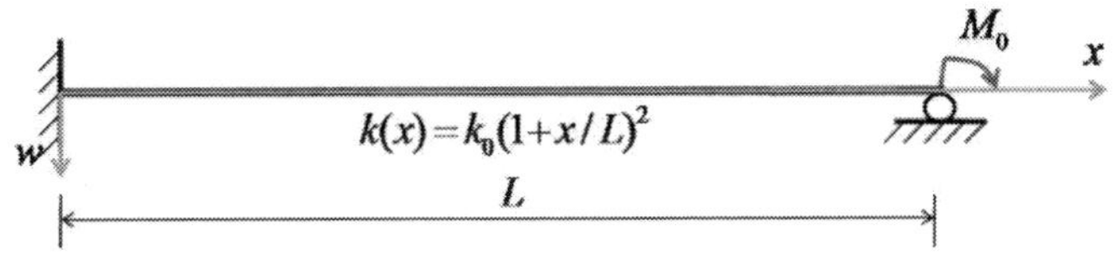

Figure 4. Geometry and loads for a propped cantilever beam

3.2. Example 2—a propped cantilever beam

A propped cantilever beam of continuously varying stiffness $k(x)=k_0(1+x/L)^2$ with $k_0=1\text{kN}\cdot\text{m}^2$ (see **Figure 4)** is studied. The geometric parameter is $L=1\text{m}$ and the load is a concentrated moment $M_0=1\text{kN}\cdot\text{m}$ at the right end. For practical measurements, points are uniformly distributed and pointwise displacements are obtained through refined finite element computation. Three sets of static measurements for 11, 51 and 101 measurement points are studied to see the effect of the number of measurement points on damage identification. To measure the error, an error index ERI is introduced as

$$\text{ERI}=\max_{x\in X}\left|k^{id}(x)-k^{ex}(x)\right| \tag{16}$$

with k^{id} and k^{ex} denoting the identified and exact stiffness. Eventually, the piecewise constant stiffness is identified for all the three sets of measurements as displayed in **Figure 5**. It is found that the increase of number of measurement points from 11 to 51 makes ERI decrease substantially and the stiffness is

IOP Conf. Series: Journal of Physics: Conf. Series **842** (2017) 012026 doi:10.1088/1742-6596/842/1/012026

better identified; however, when the number continues to rise to 101, ERI increases, being even larger than that under 11 measurement points. To explain the unexpected result under 101 measurement points, the bending moment shown in **Figure 5**(d) is examined. The largest error is found to occur in the vicinity of zero bending moment where $M \approx 0$ and hence $w'' \approx 0$; that is to say, the nearby elements hardly deform and therefore, the perturbation and deviation in **Figure 5**(c) is reasonable due to the limited identifiability of the proposed approach.

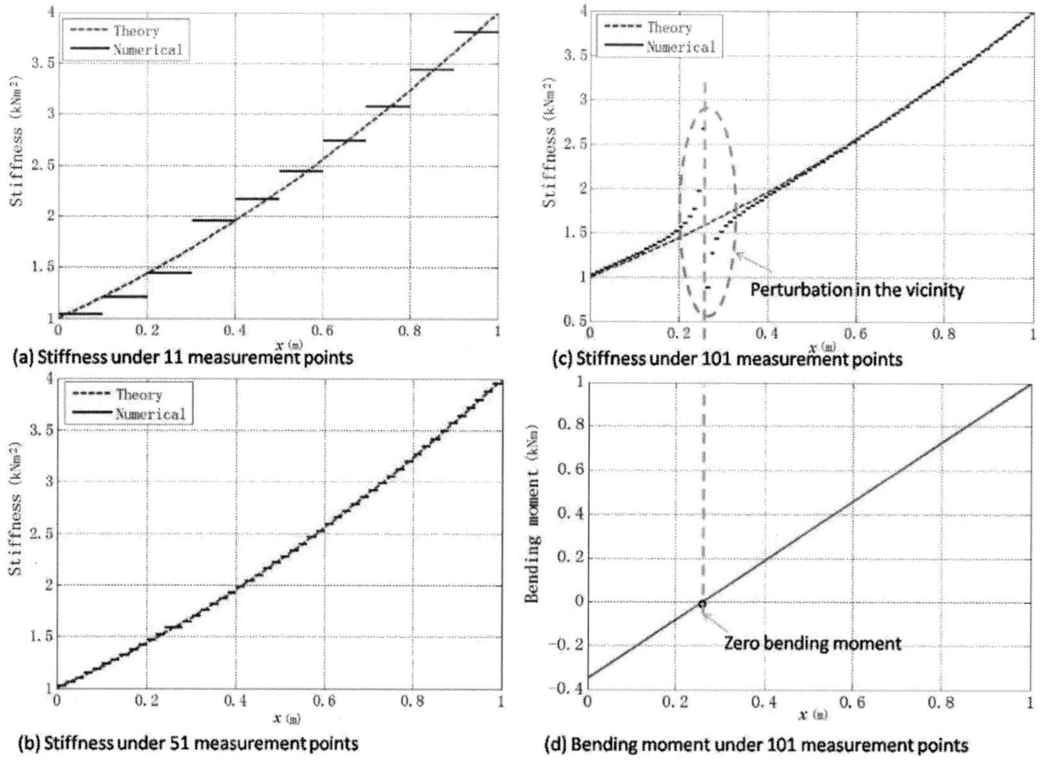

Figure 5. Damage identification in propped cantilever beam under right-end concentrated moment

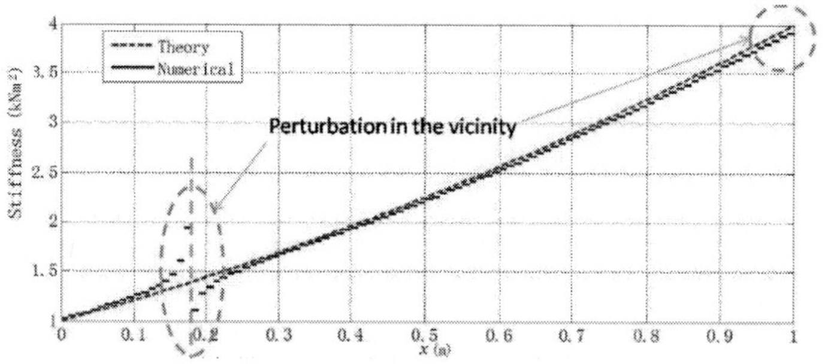

Figure 6. Stiffness identification in propped cantilever beam under concentrated force at x=0.3

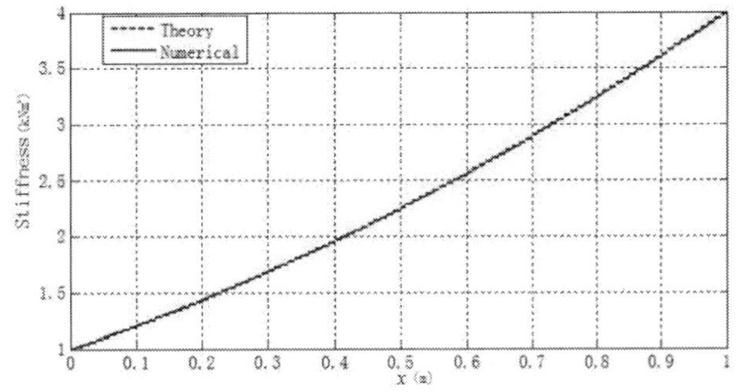

Figure 7. Damage identification in propped cantilever beam using two sets of static measurements

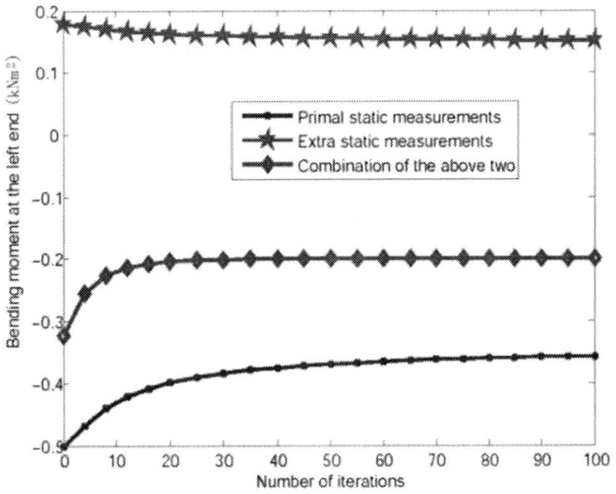

Figure 8. Convergence of bending moment at left end of propped cantilever beam

To identify the stiffness of the beam with 101 measurement points accurately, the strategy presented in Section 2.3 is adopted. Another set of static measurements is required. The load is selected as a concentrated force $F = 1\text{kN}$ at $x = 0.3\text{m}$. Under the single extra set of static measurements, the stiffness can be identified similarly and the bending moment is also obtained as exhibited in **Figure 6**. As expected, the identified stiffness perturbs in the vicinity of the zero bending moment. Fortunately, it is found that the location of the zero bending moment is different from that in the previous case (see **Figure 5**(d)) and it is deduced that after combination of the extra measurements with the previous measurements, the stiffness can be well identified. Then, the identified results are shown in **Figure 7**. Evidently, the stiffness is perfectly identified.

In addition, to show the convergence of the proposed two-step substitution algorithm, the bending moment at the left end of the beam $M(x=0)$ is observed at each iteration step for all the three cases: the single primal set of static measurements, the single extra set of static measurements and combination of the former two. Detailed results are displayed in **Figure 8**. It is seen that the bending moment converges after 80 iterations for all the three cases, verifying the convergence of the two-step substitution algorithm. Furthermore, the combination case becomes convergent after only 20 iterations, being obviously faster than other two individual cases. This is reasonable since the introduction of an extra set of static measurements can enhance identifiability and stability of the algorithm.

4. Conclusions

A new approach based on the minimum constitutive relation error (CRE) principle has been proposed for damage identification in Bernoulli-Euler beams. A two-step substitution algorithm has been established to fulfil the practical implementation and its convergence has been proved. The flaw of the proposed approach for the stiffness in zero bending moment regions can be remedied by introducing another set of static measurements. Numerical tests have been carried out to verify the approach. The sound performance of the proposed approach in the following aspects has been observed:

- It is well applicable to both statically determinate and indeterminate beam structures;
- The proposed remedy for identifiability limits performs well;
- It can well identify single damage as well as multiple damage, under concentrated or uniformly distributed testing loads;
- Damage can be well identified under a noise up to 5%, verifying the robustness of the approach.
- The two-step substitution algorithm converges well in practice.

Therefore, it is believed that the proposed approach can serve as an effective tool for practical structural damage identification.

Acknowledgements:

The authors are grateful to Professor Hongzhi Zhong of Tsinghua University and Professor Li Wang of Sun Yat-sen University for the valuable comments for revision. The present investigation was performed under the support of the National Natural Foundation of China (Nos. 51178247 and 51378294) and the MEXT scholarship of Japan.

References

[1] Ladevèze P and Leguillon D 1983 Error estimate procedure in the finite element method and application *SIAM journal on numerical analysis* **20(3)** 485-509
[2] Banan MR and Hjelmstad KD 1994 Parameter estimation of structures from static response I: Computational aspects *Journal of structural engineering* **120** 3243-58
[3] Banan MR and Hjelmstad KD 1994 Parameter estimation of structures from static response II: Numerical simulation studies *Journal of structural engineering* **120** 3259-83
[4] Viola E and Bocchini P 2015 Non-destructive parametric system identification and damage detection in truss structures by static tests *Structure and infrastructure engineering: Maintenance, Management, Life-Cycle Design and Performance* **9** 384-402
[5] Ghrib F, Li L and Wilbur P 2012 Damage identification of Euler-Bernoulli beams using static responses *Journal of engineering mechanics* **138** 405-415
[6] Avril S and Pierron F 2007 General framework for the identification of constitutive parameters from full-field measurements in linear elasticity *International journal of solids and structures* 44 4978-5002
[7] Yang QW and Sun BX 2011 Structural damage localization and quantification using static test data *Structural health monitoring* **10** 381-9
[8] Rezaiee-Pajand M, Kazemiyan MS and Aftabi SA 2014 Static damage identification of 3D and 2D frames *Mechanics based design of structures and machines: an international journal* **42** 70-96
[9] Caddemi S and Greco A2006 The influence of instrumental errors on the static identification of damage parameters for elastic beams *Computers and structures* **84** 1696-1708.
[10] Bakhtiari-Nejad F, Rahai A and Esfandiari A 2005 A structural damage detection method using static noisy data *Engineering structures* **27** 1784-1793
[11] Geymonat G and Pagano S 2003 Identification of mechanical properties by displacement field measurement: a variational approach *Meccanica* **38** 535-545
[12] Florentin E and Lubineau G 2010 Identification of the parameters of an elastic material model using the constitutive equation gap method *Computational mechanics* **46** 521-531
[13] Lubineau G 2009 A goal-oriented field measurement filtering technique for the identification of

material model parameters *Computational mechanics* **44** 591-603

[14] Florentin E and Lubineau G 2011 Using the constitutive equation gap method for identification of elastic material parameters: technical insights and illustrations *International journal on interactive design and manufacturing* **5** 227-234

[15] Moussawi A, Lubineau G, Florentin E and Blaysat B 2013 The constitutive compatibility method for identification of material parameters based on full-field measurements *Computer methods in applied mechanics and engineering* **265** 1-14

[16] Sedaghati R and Suleman A 2003 Force method revisited *AIAA journal* **41(5)** 957-966

[17] Widlak T and Scherzer O 2015 Stability in the linearized problem of quantitative elastography *Inverse problems* **31** 035005

[18] Tong P 1969 Exact solution of certain problems by finite element method *AIAA journal* **7** 178-180

IOP Conf. Series: Journal of Physics: Conf. Series **842** (2017) 012027 doi:10.1088/1742-6596/842/1/012027

Damage detection and quantification in composite beam structure using strain energy and vibration data

S Tiachacht[1], A Bouazzouni[1] , S Khatir[2], M A behtani[1],Y -L Zhou[3] and M Abdel Wahab[2]

[1]Laboratory of Mechanics, Structure and Energetics (LMSE), Mouloud Mammeri University of Tizi-Ouzou, B.P.N°17 RP, 15000, Algeria.
[2]Department of Electrical Energy, Systems and Automation, Faculty of Engineering and Architecture, Ghent University, Belgium
[3]Department of Civil and Environmental Engineering, National University of Singapore, 2 Engineering Drive 2, Singapore 117576

E-mail: Magd.AbdelWahab@UGent.be

Abstract. In this research paper, the damage in Carbon Fibre Reinforced Polymer (CFRP) laminate beams under free vibration and simply supported conditions, is investigated numerically by Finite Element Method (FEM) using Matlab program. The developed Cornwell Indicator, which is based on strain energy fraction, is used for damage detection and quantification in the considered composite beams. The data was acquired by developing a software that performs dynamic analysis of composite beams based on FEM. The results show the efficiency of the developed indicator to detect and quantify damage for single and multiple damage scenarios.

1. Introduction

The damage detection methods based on vibration analysis have received considerable attention in literature in recent years. Since there is a considerable scientific and technical interest in the resolution of the structural damage identification problem. Not only extension of techniques that are based upon structural linear vibration analysis, but also the emergence of non-linear methodology and analysis have been investigated [1]. It is generally admitted that Rytter gave the four principal damage stages of structural health monitoring:

1- The determination of the presence of damage in the structure,
2- The determination of the damage location in the structure,
3- The quantification of the severity of the damage,
4- The prognosis of the remaining service life of the damaged structure.

In most investigations of damage identification based on modal parameters, the combination of natural frequency and mode shape were widely utilized by [2, 3]. In references [4, 5], the authors used the Chebyshev pseudo spectral modal curvature formulation for damage detection in beam-like composite structures.

An extensive literature review [6] of the state of the art of damage detection and health monitoring from vibration characteristics has recently been published. This interest is attested by the large number of bibliographic reviews [7] dedicated for damage detection. The changes in lumped strain energy was proposed by Carrasco et al. [8] as a damage detection and localization parameter, since differences in modal deformations were observed in the vicinity of a damaged element. This reasoning is also the foundation of Doebling et al. [9] method, who showed that a strategy of mode selection, based on the damaged structure maximum strain energy, produced better damage location results than a strategy based on the undamaged structure maximum strain energy or on minimum frequency of analyzed truss-like structures [10]. The acceleration responses energy based damage detection approach, numerical analysis on long-span cable stayed bridge was performed by using the proposed method and the traditional mode shape curvature strategy, and at the same time damage quantification analysis and robustness analysis for noise pollution are carried out. Transmissibility functions have been proposed to detect damage in structures using only output data [11-14]. Khatir et al. [15-17] presented an approach based on inverse damage detection and localization of composite beam structures based on model reduction and Finite Element Method (FEM) coupled with different optimization methods. FEM has been widely used in the literature for many engineering applications that have been recently published [18-32]. The inverse approach, based on an Finite Element model of the structure, are pattern recognition and signal processing techniques [33, 34]. These methods determine whether or not damage has occurred, based on feature vectors, which encode the important dynamic properties of the structure. Localization is usually performed by determining which candidate set of sub-structures is damaged.

In this article, the developed Cornwell Indicator is used for damage detection and severity in Carbon Fibre Reinforced Polymer (CFRP) composite beam structures with different damages locations, the results shown that the efficiency of the developed indicator.

2. Damage indicator using strain energy

2.1 Cornwell indicator: Indicatorβ_I

Cornwell et al. [1] proposed a damage indicator based on the variation of the potential deformation energy of damaged and undamaged structure. Considering the deformation energy of each element j and the total deformation energy of the beam for the *ith* eigenmodes, the strain energy fraction FU_{ij} is written as:

$$\Gamma_{ij} = \frac{U_{ij}}{U_i} \tag{1}$$

Cornwell damage indicator is defined as follows:

$$\beta_{j1} = \frac{Z_j - \bar{Z}_j}{\sigma_Z} \tag{2}$$

and

$$Z_j = \frac{\Gamma^*_{ij}}{\Gamma_{ij}} \tag{3}$$

where:

Γ_{ij} : Fraction of deformation energy for the undamaged structure.

Γ^*_{ij}: Fraction of deformation energy for the damaged structure.

$\bar{Z}$: Average value of Z .

σ_Z : The standard deviation of ZU .

2.2 Proposed Indicator : Indicator β_{II}

The proposed indicator of damage called, indicator β_{II} for the *jth* element and the *ith* mode, is to calculate

IOP Conf. Series: Journal of Physics: Conf. Series **842** (2017) 012027 doi:10.1088/1742-6596/842/1/012027

the difference between the deformation energy fraction of the largest values minus the smallest of the undamaged and damaged structures, which will be normalized to their greatest value, i.e.

$$\beta_{jII} = \frac{\Gamma_{ij} - \Gamma_{ij}^{*}}{\left(\Gamma_{ij} - \Gamma_{ij}^{*}\right)_{\max}} \tag{4}$$

The Methodological approach for damage detection used in this paper is illustrated in Figure 1.

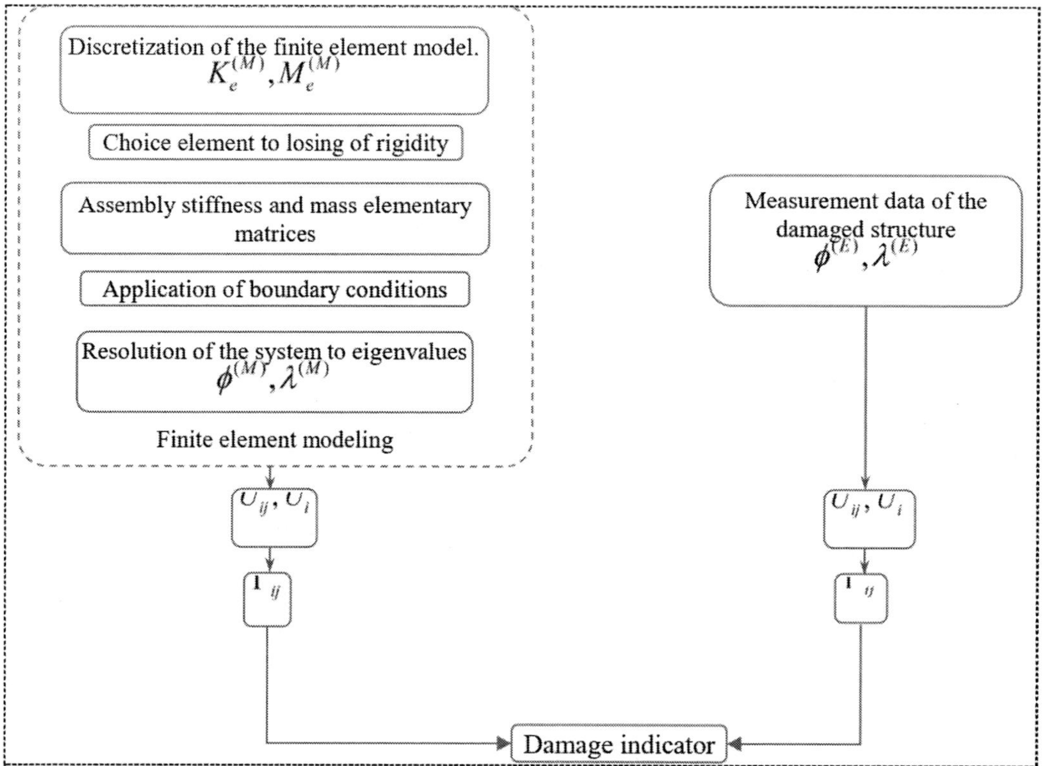

Figure 1. Methodological approach

3. Applications and results

3.1. Beam like structure

A Carbon Fibre Reinforced Polymer (CFRP) material simply supported discretized into 11 elements. The characteristics mechanics and geometrics are given in table 1.

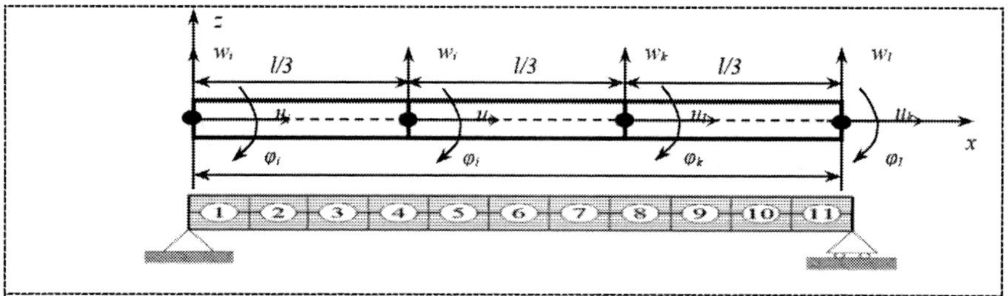

Figure 2. The FE model for the simply supported CFRP laminate beams.

We consider two types of planar structures: unidirectional and laminate beam. We model the structure into SI12 finite elements. Each node of the finite element has three degrees of freedom: displacement w normal to the beam, longitudinal displacement u and rotation φ around the y-axis [35].

$$Q_{11} = b\sum_{k=1}^{K} A_{11}^{(k)}\left[Z_k - Z_{k-1}\right]$$
$$Q_{55} = b\sum_{k=1}^{K} A_{55}^{(k)}\left[Z_k - Z_{k-1}\right]$$
$$B_{11} = \frac{1}{2} b\sum_{k=1}^{K} A_{11}^{(k)}\left[Z_k^2 - Z_{k-1}^2\right]$$
$$D_{11} = \frac{1}{3} b\sum_{k=1}^{K} A_{11}^{(k)}\left[Z_k^3 - Z_{k-1}^3\right] \tag{5}$$

k :shear correction factor

$A_{11}^{(k)} = E_x^{(k)}; \qquad A_{11}^{(k)} = G_{xz}^{(k)}$

$E_x^{(k)}$: Young's modulus of the *k-th* layer in the x direction. $G_{xz}^{(k)}$: Transverse shear modulus of the *k-th* layer.

The generalized mass densities ρ_0 , ρ_1 and ρ_2 are given by the following equations:

$$\rho_0 = b\sum_{k=1}^{K} \rho_k\left[Z_k - Z_{k-1}\right]$$
$$\rho_1 = \frac{1}{2} b\sum_{k=1}^{K} \rho_k\left[Z_k^2 - Z_{k-1}^2\right]$$
$$\rho_2 = \frac{1}{3} b\sum_{k=1}^{K} \rho_k\left[Z_k^3 - Z_{k-1}^3\right] \tag{6}$$

b :width of the beam. z_k: the coordinate of the *k-th* layer

The elementary stiffness matrix $[K_e]$ can be written as:

$$K_e = \frac{1}{2}\int_0^L B^T D B dx \tag{7}$$

Where

$$D = \begin{bmatrix} Q_{11} & 0 & 0 \\ 0 & D_{11} & 0 \\ 0 & 0 & kQ_{55} \end{bmatrix}; \quad B = LN \quad ; \quad L = \begin{bmatrix} \frac{\partial}{\partial x} & 0 & 0 \\ 0 & 0 & \frac{\partial}{\partial x} \\ 0 & \frac{\partial}{\partial x} & 1 \end{bmatrix}$$

The elementary mass matrix $[M_e]$ is defined by the following relationship:

$$M_e = \int_0^l N^T R_0 N dx \tag{8}$$

Where N is the matrix of shape functions and R_0 is the generalized matrix densities.

IOP Conf. Series: Journal of Physics: Conf. Series **842** (2017) 012027 doi:10.1088/1742-6596/842/1/012027

Table 1. Geometric and mechanical parameters of CFRP laminate[36].

Width $b(mm)$	Thickness t (mm)	Total length $L_t(mm)$	Length $L(mm)$	Young's modulus $E(Ns^2/mm^2)$	Density $\rho(Ns^2/mm^4)$
35	2.4	400	350	93850	1.95×10^{-9}

Table 2. Theoretical and experimental frequency values of undamaged CFRP beam.

	E [36]	*F* [36]	*M*	*Er1*(%)	*Er2*(%)	*Er3*(%)
f_1 **(Hz)**	69,4	73,61	73,62	6,07%	6,08%	0,01%
f_2 **(Hz)**	258,4	294,72	294,46	14,06%	13,96%	0,09%
f_3 **(Hz)**	564,2	663,95	662,57	17,68%	17,44%	0,21%

E: Experiment [12].
F: FEM [12].
M: FEM
$Er1=100\times|E-F|/E; Er2=100\times|E-M|/E; Er3=100\times|F-M|/F$

Table 3. Damage scenarios.

Scenario	Damages elements	Loss of rigidity D1	D2	D3	D4	D5	D6
1	1	20%	-	-	-	-	-
2	6	50%	-	-	-	-	-
3	2 and 6	30%	30%	-	-	-	-
4	6 and 11	10%	10%	-	-	-	-
5	3, 5, 6 and 8	30%	40%	50%	45%	-	-
6	4, 5, 6, 7 and 8	35%	40%	45%	40%	35%	-
7	2, 3, 6, 7, 9 and 10	50%	40%	35%	35%	40%	50%

3.2. Case 1: Damage case induced at element 2

Table 4. Theoretical, experimental and FEM frequency values of undamaged and damaged CFRP beam.

	E [36]	*F* [36]	*M*	*D*	*Er1*(%)	*Er2* (%)	*Er3* (%)
f_1 **(Hz)**	69,4	73,61	73,62	73,32	5,65%	0,39%	0,40%
f_2 **(Hz)**	258,4	294,72	294,46	290,72	12,51%	1,36%	1,27%
f_3 **(Hz)**	564,2	663,95	662,57	650,06	15,22%	2,09%	1,89%

E: Experiment [12].
F: FEM [12].
M: FEM.
D: FEM damaged.
$Er1=100\times|E-D|/E$; $Er2=100\times|F-D|/F$; $Er3=100\times|M-D|/M$

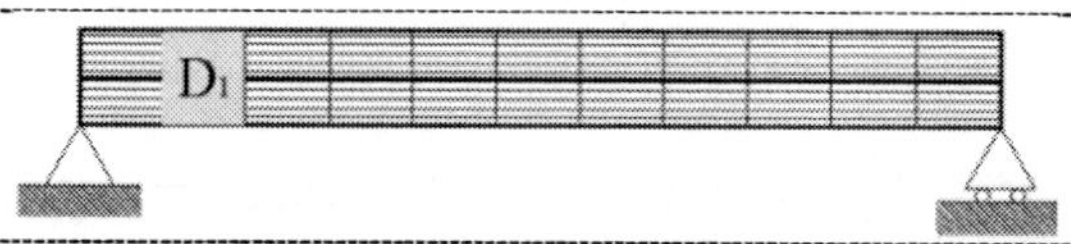

Figure 3. CFRP laminate beams having a damage case induced at element 2

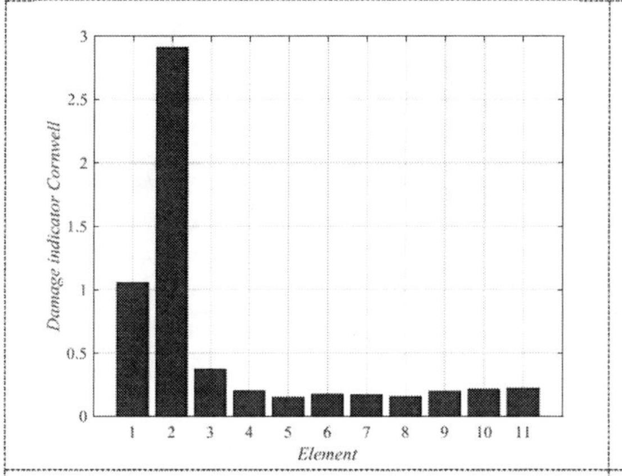

Figure 4. Damage detection from Cornwell method.

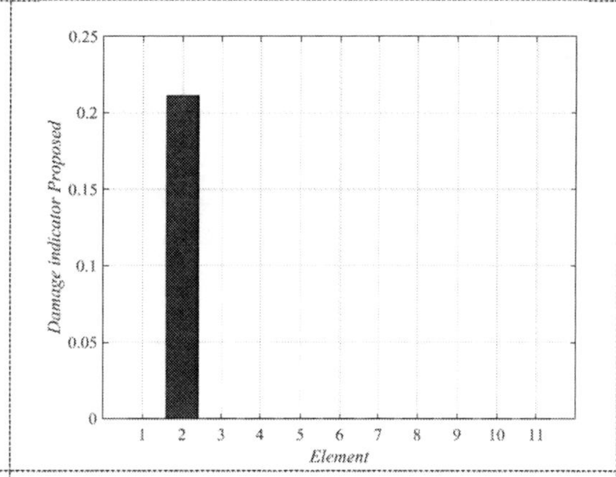

Figure 5. Damage detection from methode proposed.

3.3. Case 2: Damage case induced at element 6.

Table 5. Theoretical, experimental and FEM frequency values of undamaged and damaged CFRP beam.

	E [36]	*F* [36]	*M*	*D*	*Er1*(%)	*Er2* (%)	*Er3* (%)
f_1 **(Hz)**	69,4	73,61	73,62	67,74	2,39%	7,97%	7,98%
f_2 **(Hz)**	258,4	294,72	294,46	293,74	13,68%	0,33%	0,24%
f_3 **(Hz)**	564,2	663,95	662,57	618,48	9,62%	6,85%	6,65%

E: Experiment [12].
F: FEM [12].
M: FEM.
D: FEM damaged.
$Er1=100\times|E-D|/E$; $Er2=100\times|F-D|/F$; $Er3=100\times|M-D|/M$

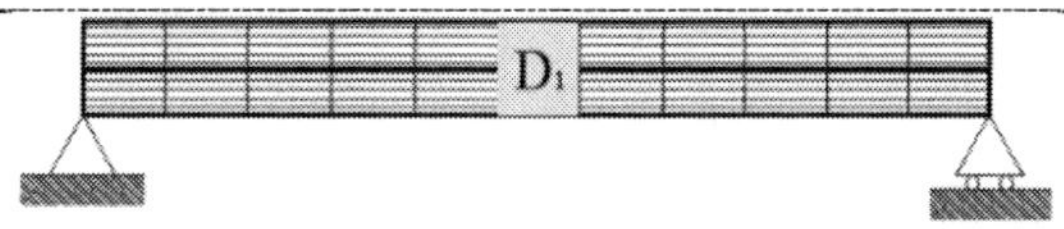

Figure 6. CFRP laminate beams having a damage case induced at element 6.

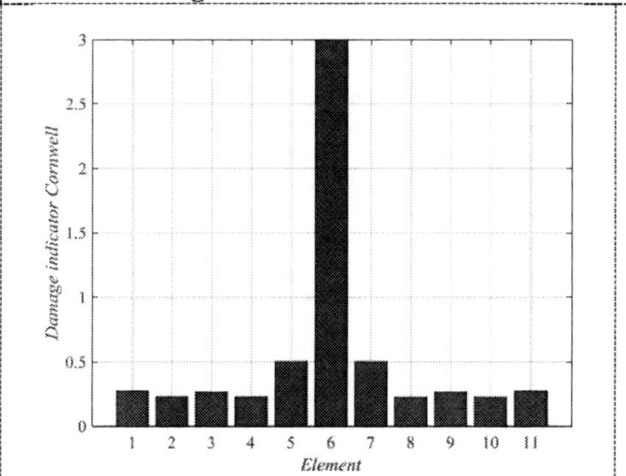

Figure 7. Damage detection from Cornwell method.

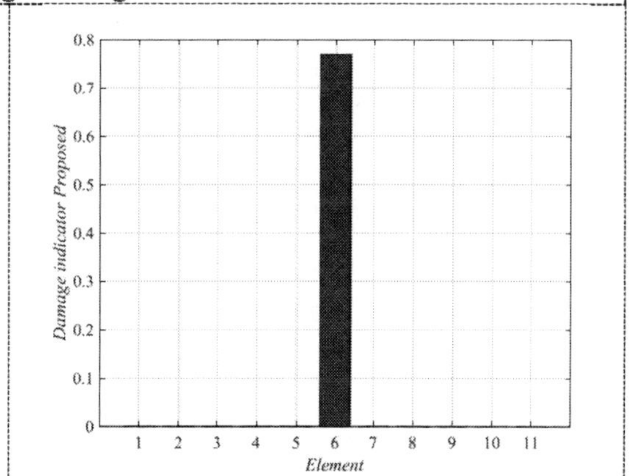

Figure 8. Damage detection from methode proposed.

3.4. Case 3: Damage case induced at element 2 and 6

Table 6. Theoretical, experimental and FEM frequency values of undamaged and damaged CFRP beam.

	E [36]	*F* [36]	*M*	*D*	*Er1*(%)	*Er2* (%)	*Er3* (%)
f_1 **(Hz)**	69,4	73,61	73,62	70,47	1,55%	4,26%	4,27%
f_2 **(Hz)**	258,4	294,72	294,46	287,82	11,39%	2,34%	2,25%
f_3 **(Hz)**	564,2	663,95	662,57	621,23	10,11%	6,43%	6,24%

E: Experiment [12].
F: FEM [12].
M: FEM.
D: FEM damaged.
Er1=100×|*E-D*|/*E*; *Er2*=100×|*F-D*|/*F*; *Er3*=100×|*M-D*|/*M*

Results:

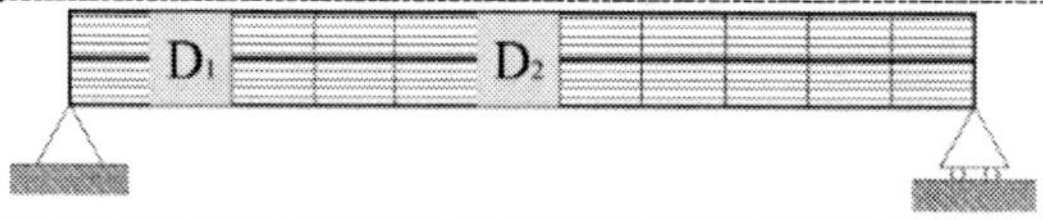

Figure 9. CFRP laminate beams having a damage case induced at elements 2 and 6

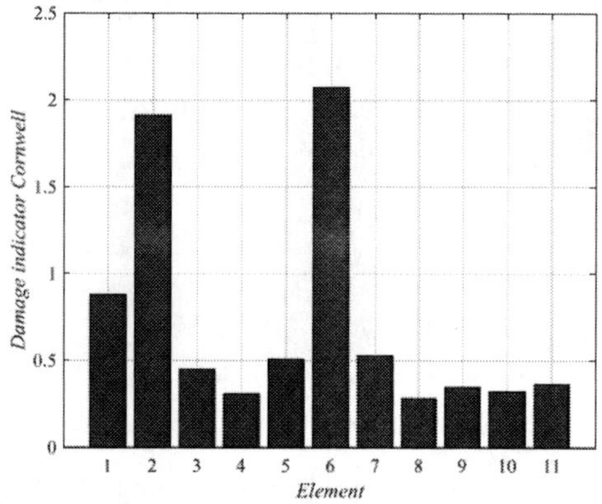

Figure 10. Damage detection from Cornwell method.

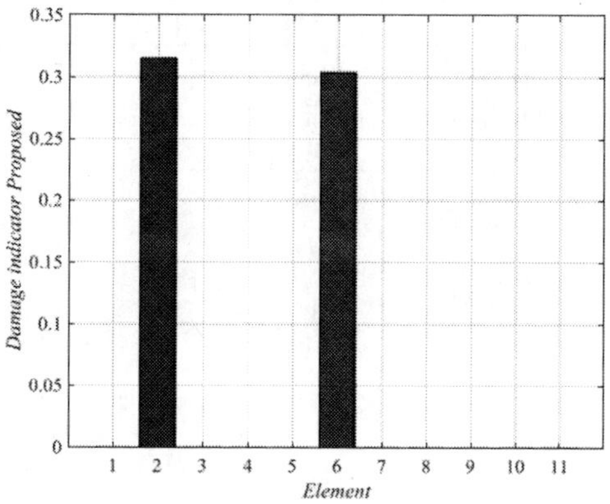

Figure 11. Damage detection from method proposed.

3.5. Case 4: Damage case induced at elements 6 and 11.

Table 7. Theoretical, experimental and FEM frequency values of undamaged and damaged CFRP beam.

	E [36]	*F* [36]	*M*	*D*	*Er1*(%)	*Er2* (%)	*Er3* (%)
f_1 **(Hz)**	69,4	73,61	73,62	72,87	5,00%	1,00%	1,02%
f_2 **(Hz)**	258,4	294,72	294,46	294,08	13,81%	0,22%	0,13%
f_3 **(Hz)**	564,2	663,95	662,57	655,08	16,11%	1,34%	1,13%

E: Experiment [12].
F: FEM [12].
M: FEM.
D: FEM damaged.
Er1=100×|*E-D*|/*E*; *Er2*=100×|*F-D*|/*F*; *Er3*=100×|*M-D*|/*M*

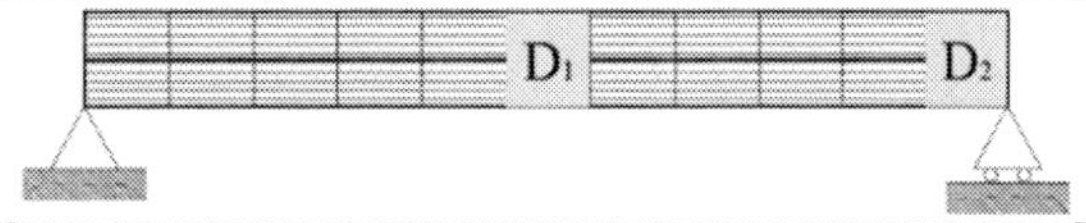

Figure 12. CFRP laminate beams having a damage case induced at elements 6 and 11.

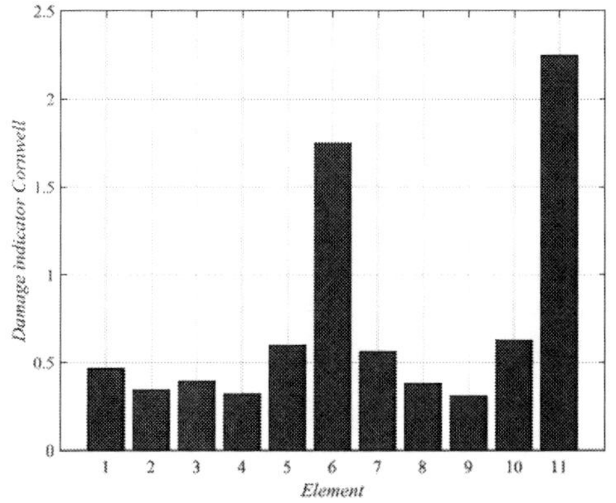

Figure 13. Damage detection from Cornwell method.

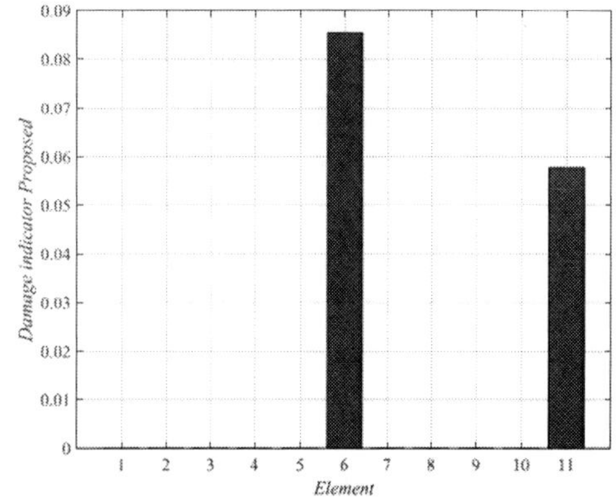

Figure 14. Damage detection from method proposed.

3.6. Case 5: Damage case induced at elements 3, 5, 6 and 8.

Table 8. Theoretical, experimental and FEM frequency values of undamaged and damaged CFRP beam.

	E [36]	*F* [36]	*M*	*D*	*Er1*(%)	*Er2* (%)	*Er3* (%)
f_1 **(Hz)**	69,4	73,61	73,62	61,56	11,30%	16,37%	16,38%
f_2 **(Hz)**	258,4	294,72	294,46	264,67	2,43%	10,20%	10,12%
f_3 **(Hz)**	564,2	663,95	662,57	587,61	4,15%	11,50%	11,31%

E: Experiment [12].
F: FEM [12].
M: FEM.
D: FEM damaged.
$Er1=100\times|E\text{-}D|/E$; $Er2=100\times|F\text{-}D|/F$; $Er3=100\times|M\text{-}D|/M$

Results:

Figure 15. CFRP laminate beams having a damage case induced at elements 3, 5, 6 and 8.

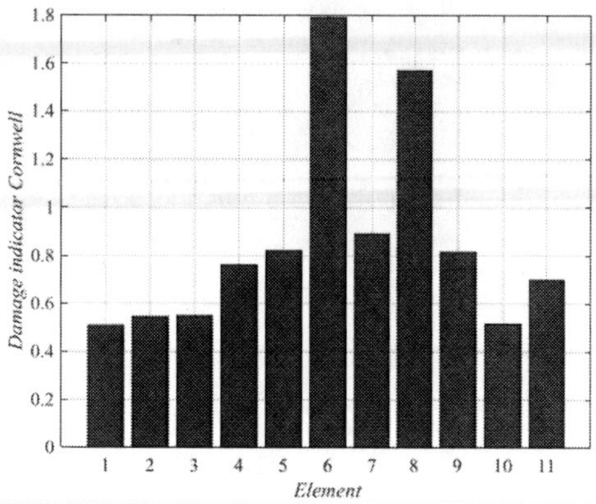

Figure 16. Damage detection from Cornwell method.

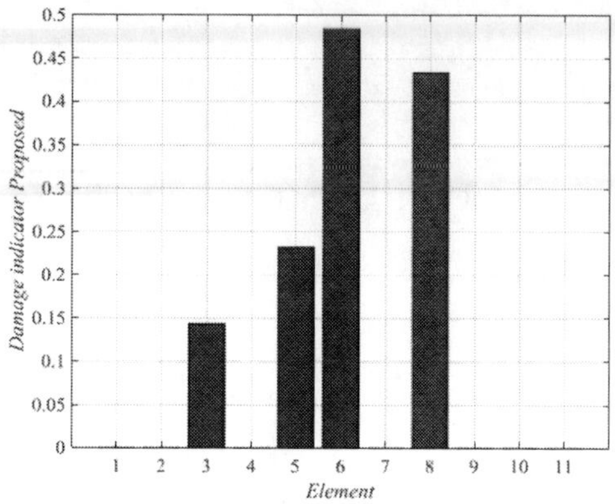

Figure 17. Damage detection from method proposed.

3.7. Case 6: Damage case induced at elements 4, 5, 6,7 and 8.

Table 9. Theoretical, experimental and FEM frequency values of undamaged and damaged CFRP beam.

	E [36]	*F* [36]	*M*	*D*	*Er1*(%)	*Er2* (%)	*Er3* (%)
f_1 **(Hz)**	69,4	73,61	73,62	59,95	13,62%	18,56%	18,57%
f_2 **(Hz)**	258,4	294,72	294,46	264,49	2,36%	10,26%	10,18%
f_3 **(Hz)**	564,2	663,95	662,57	600,73	6,47%	9,52%	9,33%

E: Experiment [12].
F: FEM [12].
M: FEM.
D: FEM damaged.
$Er1=100\times|E-D|/E$; $Er2=100\times|F-D|/F$; $Er3=100\times|M-D|/M$

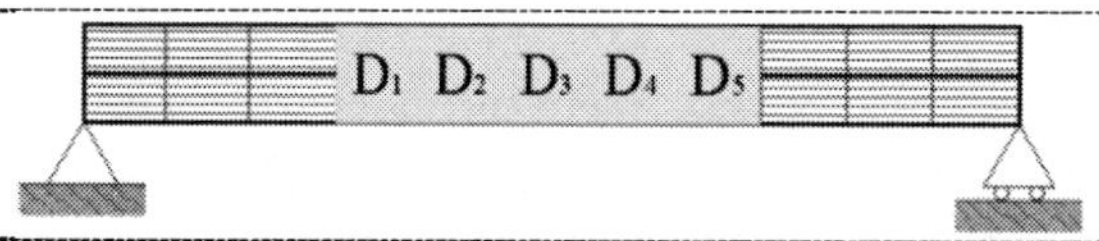

Figure 18. CFRP laminate beams having a damage case induced at elements 4, 5, 6, 7 and 8.

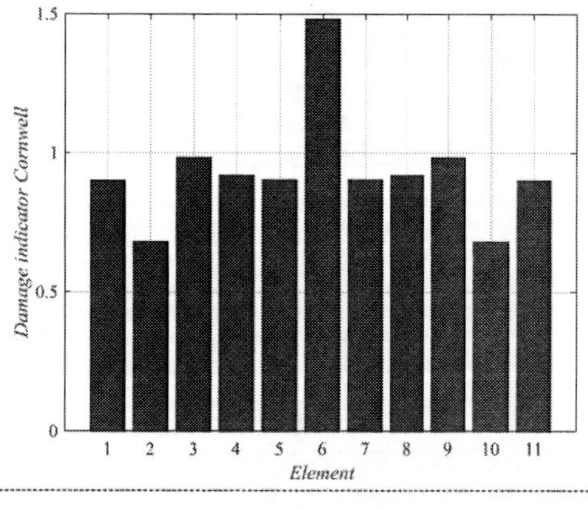

Figure 19. Damage detection from Cornwell method.

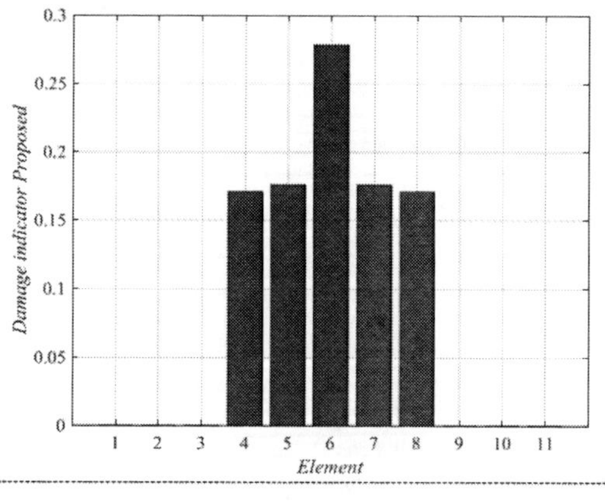

Figure 20. Damage detection from method proposed.

3.8. Case 7: Damage case induced at elements 2, 3, 6, 7, 9 and 10

Table 10. Theoretical, experimental and FEM frequency values of undamaged and damaged CFRP beam.

	E [36]	*F* [36]	*M*	*D*	*Er1*(%)	*Er2* (%)	*Er3* (%)
f_1 **(Hz)**	69,4	73,61	73,62	63,23	8,89%	14,10%	14,11%
f_2 **(Hz)**	258,4	294,72	294,46	242,75	6,06%	17,63%	17,56%
f_3 **(Hz)**	564,2	663,95	662,57	522,43	7,40%	21,31%	21,15%

E: Experiment [12].
F: FEM [12].
M: FEM.
D: FEM damaged.
$Er1=100\times|E-D|/E$; $Er2=100\times|F-D|/F$; $Er3=100\times|M-D|/M$

Figure 21. CFRP laminate beams having a damage case induced at elements 2, 3, 6, 7, 9 and 10.

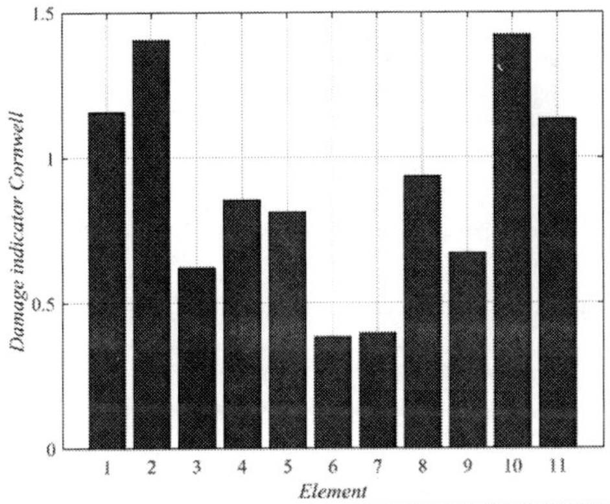

Figure 22. Damage detection from Cornwell method.

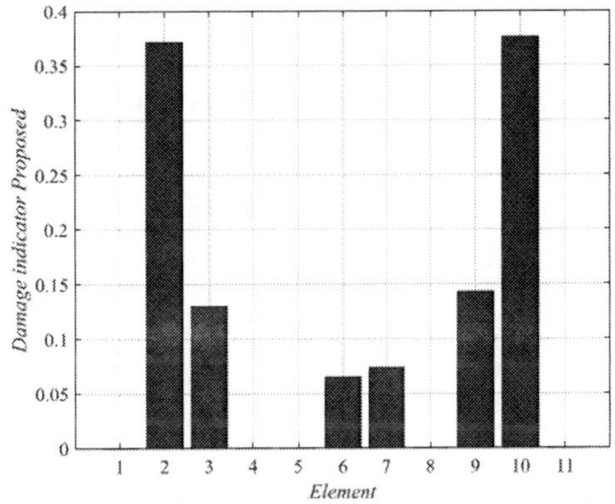

Figure 23. Damage detection from method proposed.

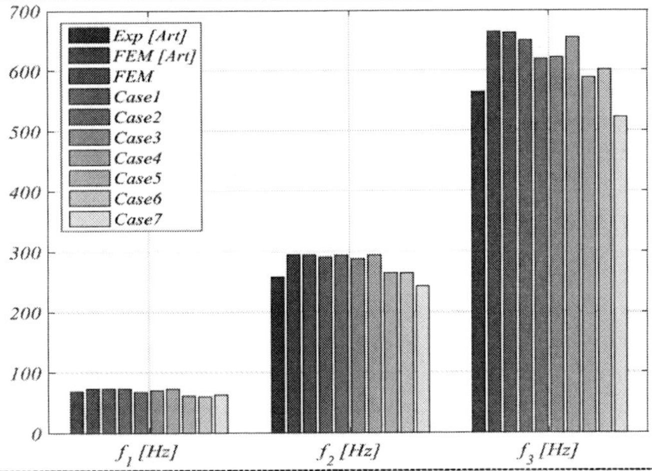

Figure 24. Frequencies of all cases.

In the Cornwell indicator method, a threshold that separates healthy and damaged components must be fixed (see Figure 25). We notice the existence of several elements which are not damaged, for example, apart from the damaged elements 5, 6 and 8, we note the existence of other damaged elements which are 4, 7, 9, 11.The use of the Cornwell indicator to locate several damages present in the structure studied is conditioned by the determination of a certain threshold separating the damaged state from the healthy state. This threshold is determined by numerical simulations and probabilistic statistical analysis. The proposed indicator facilitates the location of damage, since its concept is qualitative: when the indicator of damage is non-zero, it indicates the presence of damage.

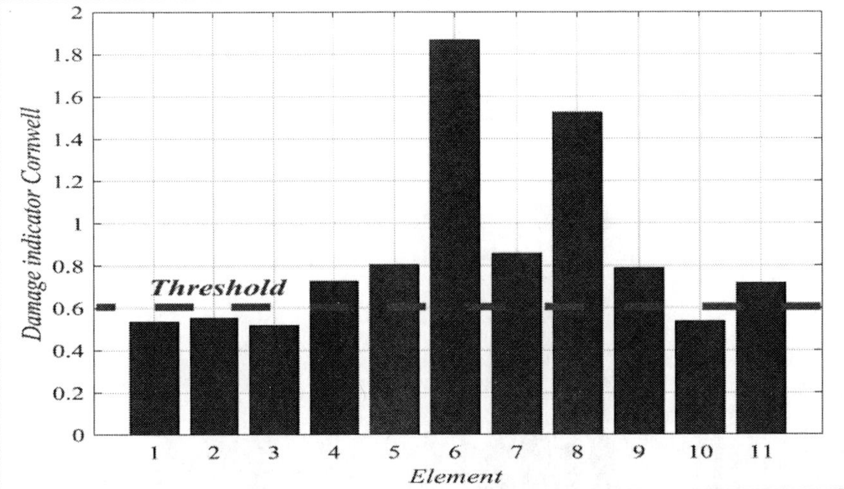

Figure 25. Threshold representations that separates the healthy elements from the damaged ones

Through the results found by two indicators in all case, we notice that the first indicators have difficulties to locate a damage compared with proposed indicator.

4. Conclusion

This study presented a developed approach for damage identification based on Cornwell Indicator. A comparison with the Cornwell Indicator, which is based on strain energy fraction, and developed indicator demonstrated the efficiency and reliability of the proposed approach. The damage detection and quantification of Carbon Fiber Reinforced Polymer (CFRP) beams structure based on Finite Element Method (FEM) using Matlab program are used. The results show that the efficiency of the developed indicator to quantify damage compared with original method for single and multiple damage.

References

[1] Rytter A. 1993.
[2] Alvandi A, Cremona C. *Journal of Sound and Vibration*. 2006;**292**(1):179-202.
[3] Choi S, Park S, Park N-H, Stubbs N. *Journal of Sound and Vibration*. 2006;**297**(3):865-79.
[4] Yang Z-B, Radzienski M, Kudela P, Ostachowicz W. *Composite Structures*. 2017;**168**:1-12.
[5] Yang Z-B, Radzienski M, Kudela P, Ostachowicz W. *Composite Structures*. 2017;**168**:372-83.
[6] Doebling SW, Farrar CR, Prime MB, Shevitz DW. 1996.
[7] Montalvao D, Maia NMM, Ribeiro AMR. *Shock and vibration digest*. 2006;**38**(4):295-324.

[8] Carrasco CJ, Osegueda RA, Ferregut CM, Grygier M, editors. Damage localization in a space truss model using modal strain energy. Proceedings-SPIE the international society for optical engineering; 1997: SPIE INTERNATIONAL SOCIETY FOR OPTICAL.
[9] Doebling SW, Hemez FM, Peterson LD, Farhat C. *AIAA journal*. 1997;**35**(4):693-9.
[10] Xu Z-D, Wu Z. *Engineering structures*. 2007;**29**(4):609-17.
[11] Zhou Y-L, Abdel Wahab M. *Engineering Structures* 2017;**141**:175-83.
[12] Zhou Y-L, Maia NMM, Sampaio R, Wahab MA. *Structural health monitoring*. 2016;**DOI: https://doi.org/10.1177/1475921716680849**
[13] Zhou Y-L, Maia N, Abdel Wahab M. *Journal of Vibration and Control* 2016:doi: 10.1177/1077546316674544
[14] Zhou Y-L, Abdel Wahab M. *Journal of Vibroengineering*. 2016;**18**(7):4491-9.
[15] Khatir S, Belaidi I, Serra R, Wahab MA, Khatir T. *Mechanics*. 2016;**21**(6):472-9.
[16] Khatir S, Belaidi I, Serra R, Wahab MA, Khatir T. *Journal of Vibroengineering*. 2016;**18**(1).
[17] Khatir S, Belaidi I, Serra R, Benaissa B, Saada AA, editors. Genetic algorithm based objective functions comparative study for damage detection and localization in beam structures. Journal of Physics: Conference Series; 2015: IOP Publishing.
[18] Junyan Ni, Wahab MA. *Computers & Structures*. 2017;**186**: 35–49.
[19] Yue T, Abdel Wahab M. *Tribology International* 2017;**107**:274-82.
[20] Pereira K, Yue T, Abdel Wahab M. *Tribology International* 2017;**110**:222-31.
[21] Martínez JC, Vanegas Useche LV, Wahab MA. *International Journal of Fatigue* 2017;**100, Part 1**:32-49.
[22] Kumar D, Biswas R, Poh LH, Abdel Wahab M. *Tribology International* 2017;**109**:124-32.
[23] Bhatti NA, Abdel Wahab M. *Tribology International* 2017;**109**:552-62.
[24] Yue T, Abdel Wahab M. *Materials*. 2016;**9**:597; doi:10.3390/ma9070597
[25] Resende Pereira KdF, Bordas S, Tomar S, Trobec R, Depolli M, Kosec G, Abdel Wahab M. *Materials*. 2016;**9**:639; doi:10.3390/ma9080639.
[26] X. Nguyen H, N. Nguyen T, Abdel Wahab M, Bordas SPA, Nguyen-Xuan H, P. Voa T. *Computer Methods in Applied Mechanics and Engineering*. 2017;**313**:904-40.
[27] Phung-Van P, Qui LX, Nguyen-Xuan H, Wahab MA. *Composite Structures* 2017;**166**:120–35.
[28] Phung-Van P, Ferreira AJM, Nguyen-Xuan H, Abdel Wahab M. *Composites Part B: Engineering*. 2017;**118**:125-34.
[29] Tran LV, Phung-Van P, Lee J, Wahab MA, Nguyen-Xuan H. *Composite Structures* 2016;**140**:655-67.
[30] Thai CH, Ferreira AJM, Abdel Wahab M, Nguyen-Xuan H. *Acta Mechanica* 2016;**227**(5):1225-50.
[31] Thai C, Zenkour AM, Abdel Wahab M, Nguyen-Xuan H. *Composite Structures* 2016;**139**:77-95.
[32] Phung-Van P, Tran LV, Ferreira AJM, Nguyen-Xuan H, Abdel-Wahab M. *Nonlinear Dynamics* 2016:1-16; doi:0.1007/s11071-016-3085-6.
[33] Trendafilova I, editor Pattern recognition methods for damage diagnosis in structures from vibration measurements. Key Engineering Materials; 2001: Trans Tech Publ.
[34] Worden K, Manson G, Fieller NR. *Journal of sound and vibration*. 2000;**229**(3):647-67.
[35] Rikards R. *Riga: Riga Technical University*. 1991.
[36] Capozucca R, Bonci B. *Composite Structures*. 2015;**122**:367-75.

Damage localization and quantification of composite stratified beam Structures using residual force method

A Behtani[1], A Bouazzouni[1] , S Khatir[2], S Tiachacht[1], Y-L Zhou[3] and M Abdel Wahab[2]

[1]Laboratory of Mechanics, Structure and Energetics (LMSE), Mouloud Mammeri University of Tizi-Ouzou, B.P.N°17 RP, 15000, Algeria
[2]Department of Electrical Energy, Systems and Automation, Faculty of Engineering and Architecture, Ghent University, Belgium
[3]Department of Civil and Environmental Engineering, National University of Singapore, 2 Engineering Drive 2, Singapore 117576

E-mail: magd.abdelwahab@ugent.be

Abstract. In this paper, the problem of using measured modal parameters to detect and locate damage in beam composite stratified structures with four layers of graphite/epoxy [0°/902°/0°] is investigated. A technique based on the residual force method is applied to composite stratified structure with different boundary conditions, the results of damage detection for several damage cases demonstrate that using residual force method as damage index, the damage location can be identified correctly and the damage extents can be estimated as well.

1. Introduction

The damage assessment of structure using vibration analysis based data has been receiving significant attention for the last three decades. A large number of proposed methods for detecting damage based on vibration characteristics by Doebling et al. [1, 2]. The most existing damage detection study was based on modal curvature and investigated the changes of an indicator between the damaged and undamaged structures [3]. Rytter [4] proposed a comparison between different techniques, which consists of four levels. The first level is based on detection, the second level is based on localization, the third level on assessment, and finally the fourth level, which is the consequence of damage, predicts the remaining life of the structure in a certain state of damage. The mode shape expansion method based on the best eigenvector concept to solve the incomplete measurement problem was presented by Yang and Liu [5]. Recently, transmissibility functions, which are based on using only output data, have been proposed to detect damage in structures [6-9]. Numerical techniques such as Finite Element Analysis [10-22] and isogeometric analysis [23-31] have been used along with experimental data to detect and quantify damage in structures. Khatir and al. [32] used finite element method to build snapshot matrix based on proper orthogonal decomposition with radial basis function (POD-RBF) combined with optimization method for damage identification of composite beam structure using vibration data. The finite element method combined with optimization methods for detecting and locating damage of composite structure based on objective function using several objective functions was introduced by Khatir and al. [33-35].

The modal data in damage detection in the form of the response showed that the use of vibration at higher frequencies is better that the identification of delamination in the cantilever composite beams introduced by Valdes and Soutis [36]. In the procedure for detecting and locating the variability in structural stiffness using the data obtained by damaged structure, the mode shapes of undamaged structures were smoothed without irregularity and a curve fitting technique was proposed. The first few mode shapes were approximated using this procedure by Yoon et al. [37]. Chiang and Lai [38] combined the residual forces method with the method of simulated evolution for damage identification. The damage detection and localization approach based on the residual forces method was used successfully to locate structural damage based on analytical model for damage identification. Kahl and Sirkis [39] proposed to use the residual forces concept along with the subspace rotation damage identification approach.

This work presents a force residual method for damage detection and localization of stratified beam structure with four layers of graphite/epoxy [0°/90$_2$°/0°] discretised in 20 elements with different boundary conditions using FEM based on Matlab programming.

2. The residual forces method

The damage index of the jth element is here expressed as the change of the rigidity of a finite element:

$$\Delta[K]_j^e = \left([K]_j^e - [K]_{dj}^e\right) = \alpha_j [K]_j^e \tag{1}$$

Where $[K]_j^e$ and $[K]_{dj}^e$ are the jth element of the elementary matrix of the damaged and undamaged structure, respectively. $\Delta[K]_j^e$ represents the variation of stiffness. $\alpha \in [0,1]$ indicates a loss of rigidity of jth element; i.e. for undamaged element $\alpha = 0$ and for damaged element $\alpha = 1$. We consider that the mass matrix of the damaged structure is not affected by damage, and the rigidity matrix of the damaged element change as given below:

$$[\Delta M] = 0 \tag{2}$$

$$\Delta[K]_j^e = \alpha \Delta[K]_j^e \tag{3}$$

The modal residual force vector can be written as:

$$\{R\}_i = [\Delta K]\{\phi\}_{di} = \{\Delta f\}_i = \left[\{f\}_1^e \{f\}_2^e \cdots \{f\}_m^e\right] \begin{Bmatrix} \alpha_1 \\ \alpha_2 \\ \cdot \\ \cdot \\ \cdot \\ \alpha_m \end{Bmatrix} = [F]_i\{\alpha\} \tag{4}$$

Equation (4) can be written in matrix form as:

$$[F]\{\alpha\} = \{R\} \tag{5}$$

Where matrix $[F]$ is:

$$\{F_{ij}\} = [K]_j^e [\phi]_{dij}^e \tag{6}$$

The modal residual force vector can be written as:

$$\{R\}_i = ([K] - \lambda_{di}[M])\{\phi\}_{di} \tag{7}$$

Equation (5) can be rewritten as:

$$\begin{bmatrix} \{F\}_{11} & \{F\}_{12} & \cdots & \{F\}_{1m} \\ \{F\}_{21} & \{F\}_{22} & \cdots & \{F\}_{2m} \\ \cdot & \cdot & \cdot & \cdot \\ \cdot & \cdot & \cdot & \cdot \\ \{F\}_{n1} & \{F\}_{n21} & \cdots & \{F\}_{nm} \end{bmatrix} \begin{Bmatrix} \alpha_1 \\ \alpha_2 \\ \cdot \\ \cdot \\ \cdot \\ \alpha_m \end{Bmatrix} = \begin{Bmatrix} \{R\}_1 \\ \{R\}_2 \\ \cdot \\ \cdot \\ \cdot \\ \{R\}_n \end{Bmatrix} \tag{8}$$

Where n is the number of modes, while m is the number of elements. The resolution of equation (8) allows us to determine the values of the damage indicators as:

$$\begin{Bmatrix} \alpha_1 \\ \alpha_2 \\ \cdot \\ \cdot \\ \cdot \\ \alpha_m \end{Bmatrix} = \begin{bmatrix} \{F\}_{11} & \{F\}_{12} & \cdots & \{F\}_{1m} \\ \{F\}_{21} & \{F\}_{22} & \cdots & \{F\}_{2m} \\ \cdot & \cdot & \cdot & \cdot \\ \cdot & \cdot & \cdot & \cdot \\ \{F\}_{n1} & \{F\}_{n21} & \cdots & \{F\}_{nm} \end{bmatrix}^{+} \begin{Bmatrix} \{R\}_1 \\ \{R\}_2 \\ \cdot \\ \cdot \\ \cdot \\ \{R\}_n \end{Bmatrix} \tag{9}$$

The methodological approach used in this paper is illustrated in Figure 1.

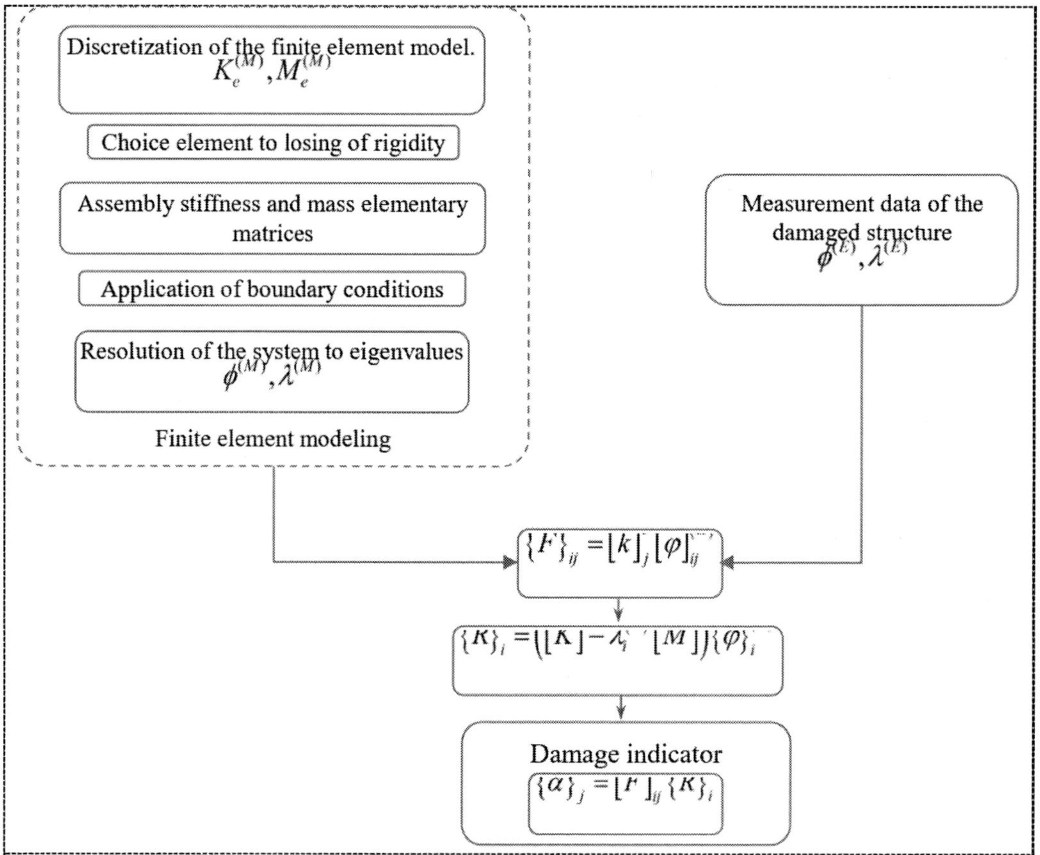

Figure 1. Methodological approach

3. Numerical applications

To evaluate the performance of the proposed residual force method, we consider a stratified composite beam Graphite/epoxy [0°/90$_2$°/0°] discretized into 20 finite elements SI12 as shown in Figure 2 [40]. The mechanical and geometrical properties are presented in Table 1.

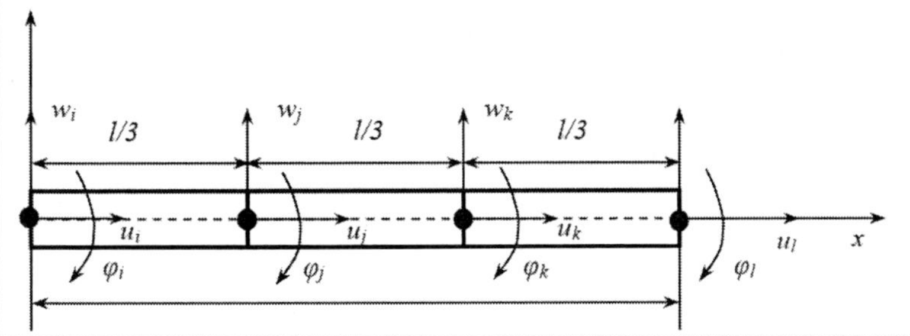

Figure 2. SI12 FEM.

Table 1. The material properties and beam structure composite Graphite / epoxy [0°/90$_2$°/0°].

Length (m)	Width (m)	Thickness (m)	Young modulus E_x (GPa)	Young modulus E_y (GPa)	G_{yz} (GPa)	$G_{xy} = G_{xz}$ (GPa)	ρ (Kg/m^3)	ν_{xy}
1	0.05	0.04	14.48	0.965	0.345	0.414	1390	0.25

3.1 Case 1. Clamped free beam

In first section, we consider clamped – free a stratified beam structure with four layers of graphite/epoxy [0°/90$_2$°/0°].

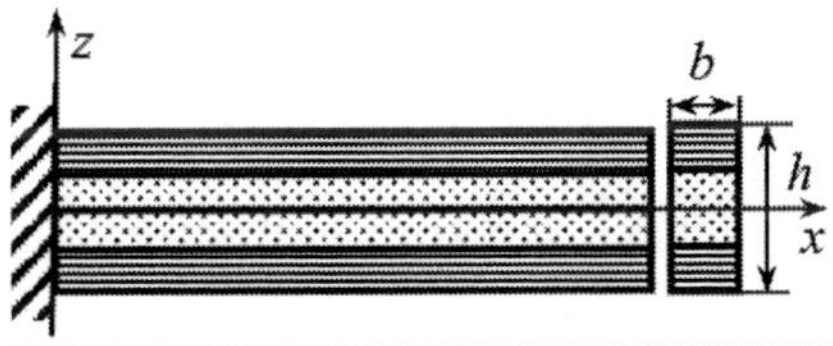

Figure 3. Clamped free beam stratified graphite/epoxy [0°/90$_2$°/0°].

3.1.1 Single damage

The results of single damaged element 2, 7 and 13, having 20% loss of rigidity, are presented in Figure 4 and the results of a single damaged element 10 with different loss of rigidity 30%, 40% and 50% are presented in the Figure 5.

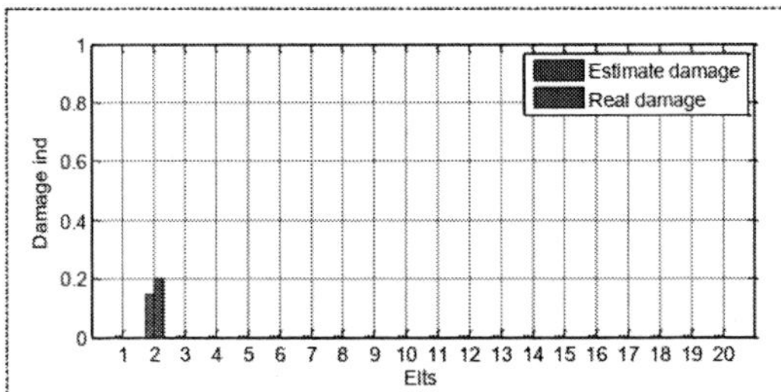

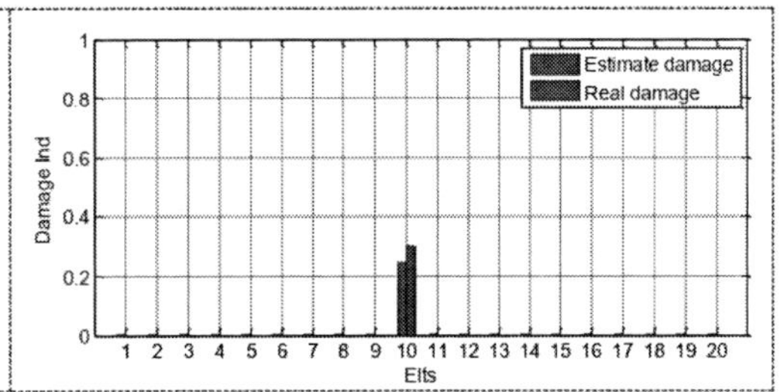

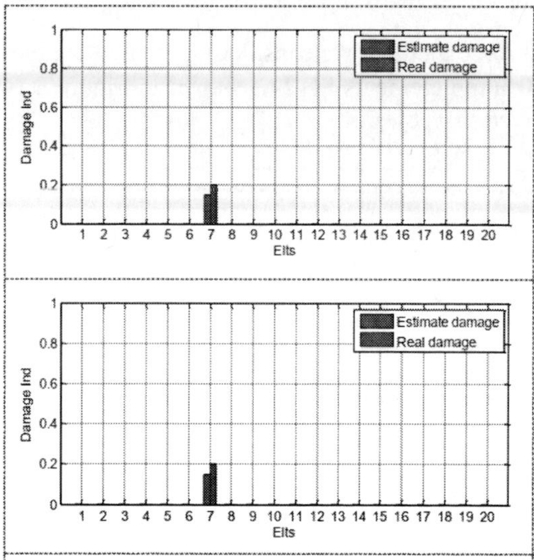

Figure 4. The variation of the damage indicator of clamped–free stratified beam structure with single damaged elements 2, 7 and 13 by 20% loss of rigidity

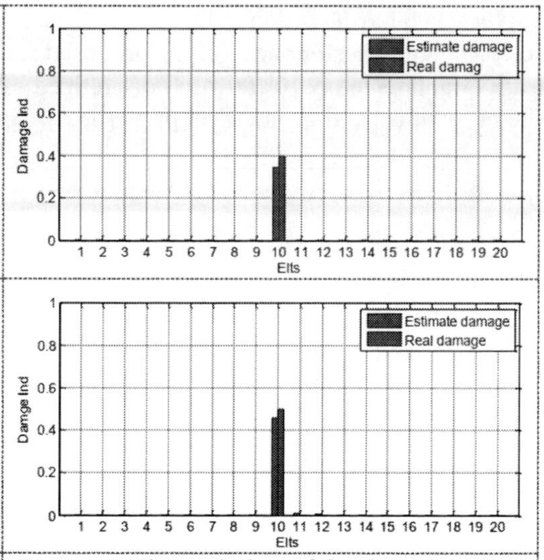

Figure 5. The variation of the damage indicator of clamped–free stratified beam structure with different loss of rigidity 30%, 40% and 50% of element 10

According the results sited in the figures 4-5 the value of the damage indicator of the damaged element affected is largest and that whatever the position and level of damage.

3.1.2 Multiple damage

The results of multiple damaged elements (3, 12), (5, 16) and (10, 15) with loss of rigidity 30% are presented in Figure 6.

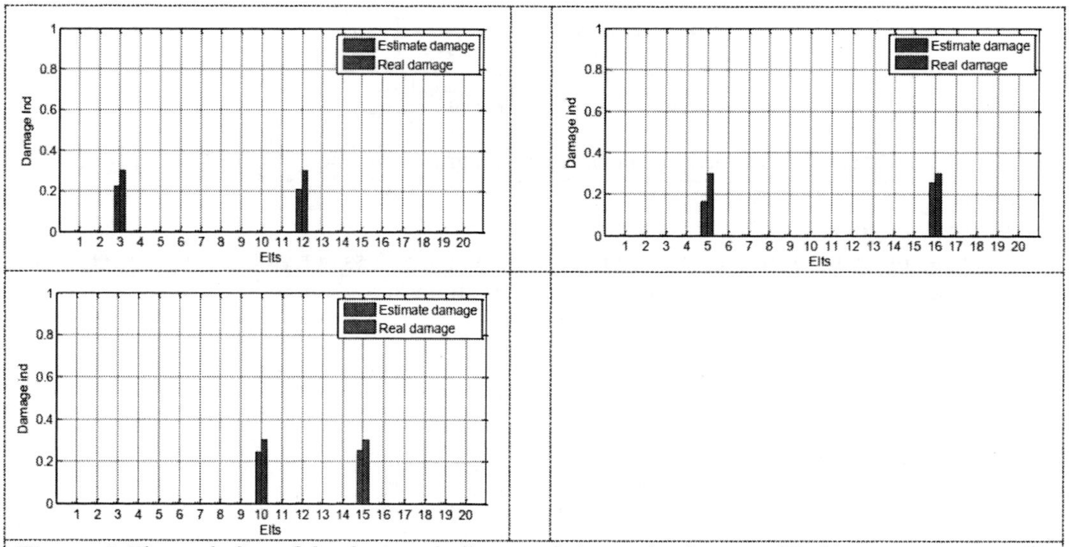

Figure 6. The variation of the damage indicator of clamped – free stratified beam structure with multiple damaged elements (3, 12), (5, 16) and (10, 15) with loss of rigidity 30%

The results of multiple damaged elements (2, 11, 17), (3, 9, 11, 17) and (2, 6, 11, 13, 16) with loss of rigidity 30%, 35% and 40%, respectively, are presented in Figure 7.

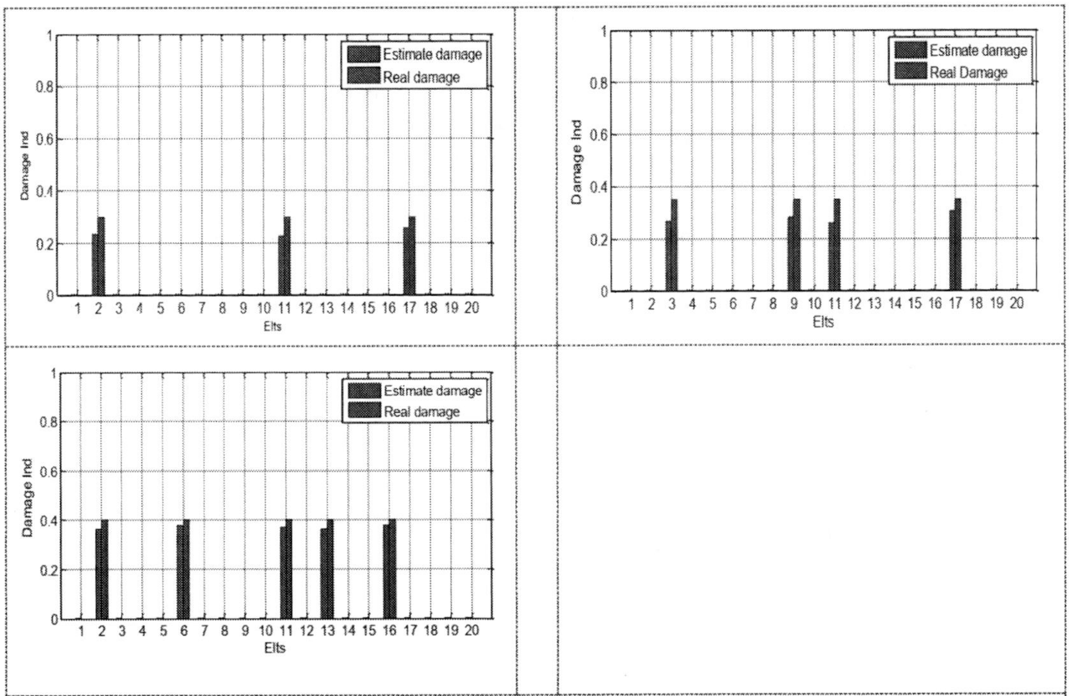

Figure 7. The variation of the damage indicator of clamped – free stratified beam structure with multiple damaged elements (2, 11, 17), (3, 9, 11, 17) and (2, 6, 11, 13, 16) with loss of rigidity 20%, 30% and 35% Successively.

From the results obtained in this application, we can conclude that the method based on the modal residual force method it possible to locate the damage in case of a beam with multiple damages. The frequencies of damaged and undamaged beam stratified structure are presented in the Table 2.

Table 2

		Frequency (Hz)				
		f_1	f_2	f_3	f_4	f_5
FEM		62.20	357.70	911.80	1599.90	1863.40
	2 (20%)	60.00	354.90	910.00	1599.50	1840.80
	7 (20%)	60.70	356.30	906.00	1599.60	1845.90
	13 (20%)	61.10	354.50	904.30	1599.60	1856.20
	3 and 12 (30%)	59.40	349.60	904.50	1589.30	1811.90
Damaged	5 and 16 (30%)	60.00	355.90	891.90	1563.40	1824.70
	10 and 15 (30%)	60.80	348.70	895.50	1563.20	1835.30
	2, 11 and 17 (20%)	59.90	350.60	904.50	1578.30	1828.90
	3, 9, 11 and 17 (30%)	58.90	343.50	900.00	1547.40	1784.30
	2, 6, 11, 13, and 16 (35%)	57.30	344.90	863.60	1541.50	1741.10

3.2 Case 2. Clamped – clamped beam

In second section, we consider clamped-clamped stratified beam structure with four layers of graphite/epoxy [0°/90_2°/0°] a shown in Figure 8.

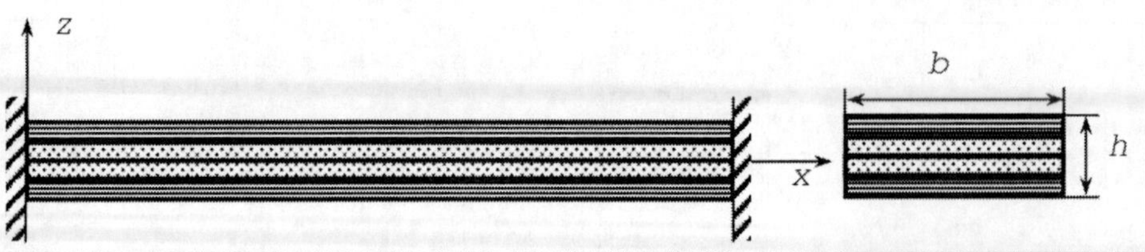

Figure 8. Clamped – clamped a stratified beam structure with four layers of graphite/epoxy [0°/90$_2$°/0°].

3.2.1 Single damage

The results of single damaged elements 2, 7 and 13, with 20% loss of rigidity clamped–clamped stratified beam structure are presented in Figure 9, and the results of single damaged element 10 with different loss of rigidity 30%, 40% and 50% are presented in Figure 10.

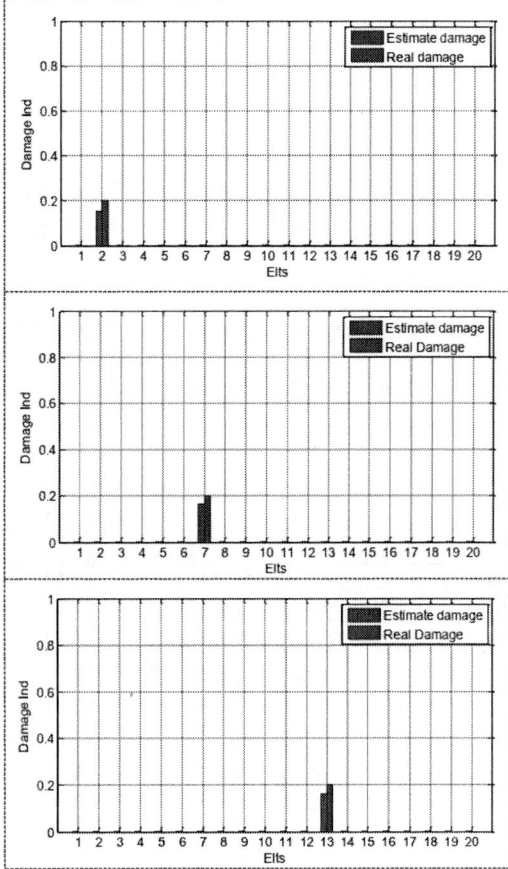

Figure 9. The variation of the damage indicator of clamped – clamped stratified beam structure with single damaged elements 2, 7 and 13 by 20% loss of rigidity.

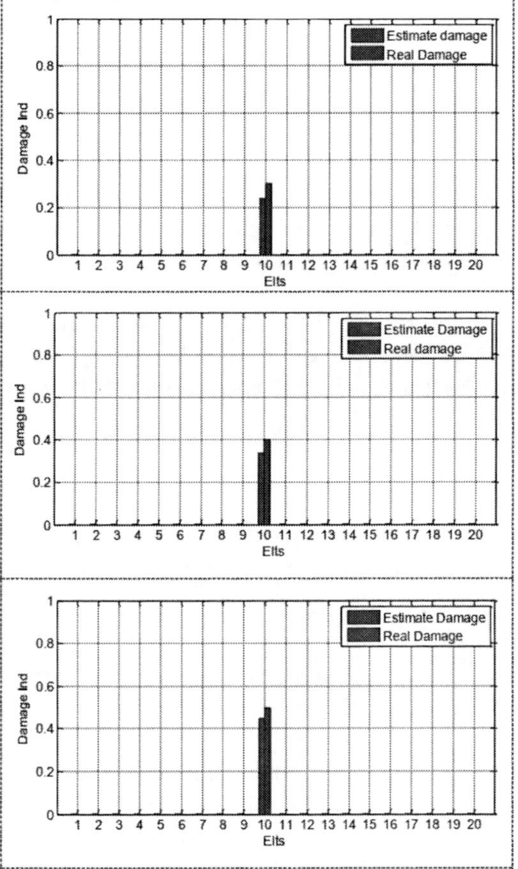

Figure 10. The variation of the damage indicator of clamped – clamped stratified beam structure with single damaged elements 10 with different loss of rigidity 30%, 40% and 50%

IOP Conf. Series: Journal of Physics: Conf. Series **842** (2017) 012028 doi:10.1088/1742-6596/842/1/012028

3.2.2 Multiple damage

The results of multiple damaged elements (3, 12), (5, 16) and (10, 15) with loss of rigidity 30% of clamped–clamped stratified beam structure are presented in Figure 11.

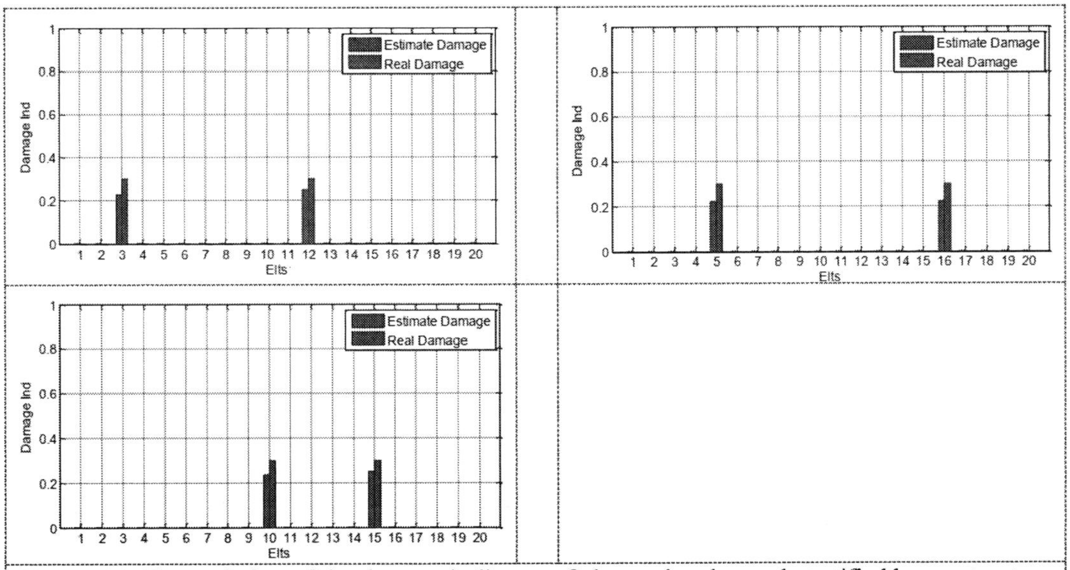

Figure 11. The variation of the damage indicator of clamped – clamped stratified beam structure with multiple damaged elements (3, 12), (5, 16) and (10, 15) with loss of rigidity 30%.

The results of multiple damaged elements (2, 11, 17), (3, 9, 11, 17) and (2, 6, 11, 13, 16) with loss of rigidity 20%, 30% and 35% , respectively, of clamped – clamped stratified beam structure are presented in Figure 12.

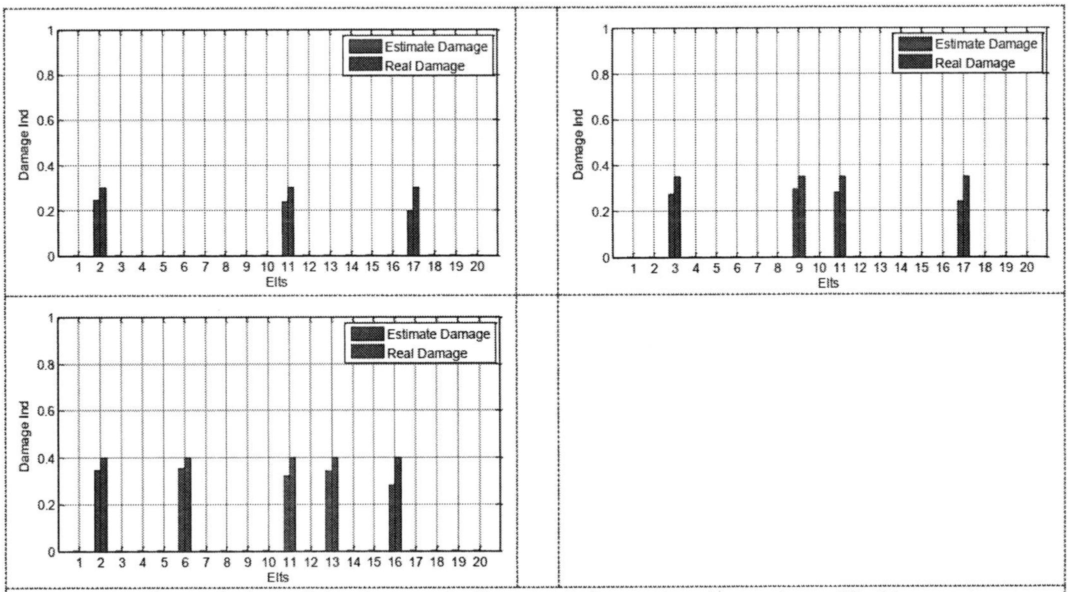

Figure 12 The variation of the damage indicator of clamped – clamped stratified beam structure with multiple damaged elements (2, 11, 17), (3, 9, 11, 17) and (2, 6, 11, 13, 16) with loss of rigidity 20% ,30% and 35% Successively.

From the results obtained in Figures 9 to 12, we can conclude that the method based on the modal residual force method is successful to locate the damage in the case with single and multiple damages. The frequencies of damaged and undamaged beam stratified structure are presented in the Table 3.

Table 3

		Frequency (Hz)				
		f_1	f_2	f_3	f_4	f_5
FEM		350.10	860.60	1502.50	2217.90	2976.30
Damaged	2 (20%)	347.20	858.80	1502.10	2216.50	2972.20
	7 (20%)	349.20	854.80	1502.00	2210.40	2963.50
	13 (20%)	348.60	856.40	1501.30	2205.10	2972.90
	3 and 12 (30%)	344.50	857.10	1488.60	2190.60	2958.60
	5 and 16 (30%)	350.00	847.70	1477.80	2209.90	2969.70
	10 and 15 (30%)	345.60	850.40	1478.40	2213.20	2939.30
	2, 11 and 17 (20%)	344.50	857.10	1485.00	2207.10	2958.00
	3, 9, 11 and 17 (30%)	340.00	854.20	1461.80	2174.40	2935.30
	2, 6, 11, 13, and 16 (35%)	335.90	828.20	1453.00	2179.00	2909.40

4. Conclusion

The case study demonstrates a methodology for damage detection and localization in beam composite stratified structures with four layers of graphite/epoxy [0°/902°/0°]. Several boundary conditions discussed for the robustness of this method for detecting and locating damage in a structure.The presence of damage in a composite structure based on changes in the dynamic properties of the structure. The results show that the residual force method can be detecting single and multiple damage scenarios.

5. References

[1] Doebling SW, Farrar CR, Prime MB and Shevitz DW 1996
[2] Zou Y, Tong L and Steven GP 2000 *Journal of sound and vibration* **230**(2) 357-378
[3] Wang Y-l 2014 *Structural Engineering And Mechanics* **51**(6) 1037-1046
[4] Doebling SW, Farrar CR and Prime MB 1998 *Shock and vibration digest* **30**(2) 91-105
[5] Yang Q and Liu J 2007 *Journal of sound and vibration* **305**(1) 298-307
[6] Zhou Y-L and Abdel Wahab M 2017 *Engineering Structures* **141** 175-183
[7] Zhou Y-L, Maia NMM, Sampaio R and Wahab MA 2016 *Structural health monitoring* DOI: https://doi.org/10.1177/1475921716680849
[8] Zhou Y-L, Maia N and Abdel Wahab M 2016 *Journal of Vibration and Control* doi: 10.1177/1077546316674544
[9] Zhou Y-L and Abdel Wahab M 2016 *Journal of Vibroengineering* **18**(7) 4491-4499
[10] Gadala I, Abdel Wahab M and Alfantazi A 2016 *MATERIALS & DESIGN* **97** 287-299
[11] Yue T and Abdel Wahab M 2017 *Tribology International* **107** 274-282
[12] Yue T and Abdel Wahab M 2016 *Materials* **9** 597; doi:510.3390/ma9070597
[13] Yue T and Abdel Wahab M 2014 *Wear* **321** 53-63
[14] Wang C, Sun Q, Abdel Wahab M, Zhang X and Xu L 2015 *Waste Management* **43** 19-27
[15] Vanegas-Useche L, Abdel Wahab M and Parker G 2015 *Waste Management* **43** 28-36
[16] Resende Pereira KdF, Bordas S, Tomar S, Trobec R, Depolli M, Kosec G and Abdel Wahab M 2016 *Materials* **9** 639; doi:610.3390/ma9080639
[17] Noda N-A, Chen X, Sano Y, Wahab MA, Maruyama H, Fujisawa R and Takase Y 2016 *MATERIALS & DESIGN* **96** 476-489
[18] Hojjati Talemi R and Abdel Wahab M 2013 *Tribology International* **60** 176-186
[19] Hafiz TA, Abdel Wahab M, Crocombe AD and Smith PA 2013 *International Journal of Adhesion and Adhesives* **40** 188-196
[20] Ferjaoui A, Yue T, Abdel Wahab M and Hojjati-Talemi R 2015 *International Journal of Fatigue* **73** 66-76

[21] Chen X, Noda N-A, Abdel Wahab M, Sano Y, Maruyama H, Wang H, Fujisawa R and Takase Y 2016 *Journal of the Chinese Society of Mechanical Engineers* **37**(1) 1-10
[22] Chen X, Noda N-A, Abdel Wahab M, Akaishi Y-I, Sano Y, Takase Y and Fekete G 2015 *Acta Polytechnica Hungarica* **12**(8) 61-79
[23] X. Nguyen H, N. Nguyen T, Abdel Wahab M, Bordas SPA, Nguyen-Xuan H and P. Voa T 2017 *Computer Methods in Applied Mechanics and Engineering* **313** 904-940
[24] Tran Vinh L, Lee J, Nguyen-Van H, Nguyen-Xuan H and Abdel Wahab M 2015 *International Journal of Non-Linear Mechanics* **72** 42-52
[25] Tran LV, Phung-Van P, Lee J, Wahab MA and Nguyen-Xuan H 2016 *Composite Structures* **140** 655-667
[26] Thai CH, Ferreira AJM, Abdel Wahab M and Nguyen-Xuan H 2016 *Acta Mechanica* **227**(5) 1225-1250
[27] Thai C, Zenkour AM, Abdel Wahab M and Nguyen-Xuan H 2016 *Composite Structures* **139** 77-95
[28] Phung-Van P, Tran LV, Ferreira AJM, Nguyen-Xuan H and Abdel-Wahab M 2016 *Nonlinear Dynamics* 1-16; doi:10.1007/s11071-11016-13085-11076
[29] Phung Van P, Nguyen LB, Tran Vinh L, Dinh TD, Thai CH, Bordas SPA, Abdel Wahab M and Nguyen-Xuan H 2015 *International Journal of Non-Linear Mechanics* **76** 190-202
[30] Phung Van P, De Lorenzis L, Thai CH, Abdel Wahab M and Nguyen-Xuan H 2015 *Computational Materials Science* **96** 495-505
[31] Phung Van P, Abdel Wahab M, Liew KM, Bordas SPA and Nguyen-Xuan H 2015 *Composite Structures* **123** 137-149
[32] Khatir S, Belaidi I, Serra R, Wahab MA and Khatir T 2016 *Mechanics* **21**(6) 472-479
[33] Khatir S, Belaidi I, Serra R, Wahab MA and Khatir T 2016 *Journal of Vibroengineering* **18**(1)
[34] Khatir S, Belaidi I, Serra R, Benaissa B and Saada AA 2015 Journal of Physics: Conference Series ed editors) IOP Publishing p. 012035
[35] Khatir A, Tehami M, Khatir S and Abdel Wahab M 2016 23rd International Congress on Sound and Vibration (ICSV23) ed editors) International Institute of Acoustics and Vibration
[36] Valdes SD and Soutis C 1999 *Journal of sound and vibration* **228**(1) 1-9
[37] Yoon M, Heider D, Gillespie J, Ratcliffe C and Crane R 2005 *Journal of sound and vibration* **279**(1) 119-139
[38] Chiang D-Y and Lai W-Y 1999 *AIAA journal* **37**(10) 1331-1333
[39] Kahl K and Sirkis J 1996 *AIAA journal* **34**(12) 2609-2614
[40] Rikards R 1991 *Riga: Riga Technical University*

Sub-surface defects detection of by using active thermography and advanced image edge detection

Peter W. Tse[1,*] and **Gaochao Wang**[2]

Croucher Optical Nondestructive Testing Laboratory (CNDT)
Department of Systems Engineering and Engineering Management,
City University of Hong Kong, Tat Chee Avenue, Hong Kong, China

[1]*Email of corresponding author: meptse@cityu.edu.hk, [2]Email: gcwang2@cityu.edu.hk

Abstract. Active or pulsed thermography is a popular non-destructive testing (NDT) tool for inspecting the integrity and anomaly of industrial equipment. One of the recent research trends in using active thermography is to automate the process in detecting hidden defects. As of today, human effort has still been using to adjust the temperature intensity of the thermo camera in order to visually observe the difference in cooling rates caused by a normal target as compared to that by a sub-surface crack exists inside the target. To avoid the tedious human-visual inspection and minimize human induced error, this paper reports the design of an automatic method that is capable of detecting sub-surface defects. The method used the technique of active thermography, edge detection in machine vision and smart algorithm. An infrared thermo-camera was used to capture a series of temporal pictures after slightly heating up the inspected target by flash lamps. Then the Canny edge detector was employed to automatically extract the defect related images from the captured pictures. The captured temporal pictures were preprocessed by a packet of Canny edge detector and then a smart algorithm was used to reconstruct the whole sequences of image signals. During the processes, noise and irrelevant backgrounds exist in the pictures were removed. Consequently, the contrast of the edges of defective areas had been highlighted. The designed automatic method was verified by real pipe specimens that contains sub-surface cracks. After applying such smart method, the edges of cracks can be revealed visually without the need of using manual adjustment on the setting of thermo-camera. With the help of this automatic method, the tedious process in manually adjusting the colour contract and the pixel intensity in order to reveal defects can be avoided.
Keywords: Active thermography; Edge detection; Quality inspection, Automation, Image processing

1. Introduction

Active thermography has been proven to be a reliable tool to defect internal cracks. A great deal of research work has been carried out to implement qualitative and quantitative analysis [1]. One of the most popular applications in thermography is the capability in detecting internal cracks in various structures. Most of the cracks are invisible from external view. However, ignoring these cracks will ultimately cause serious consequence as they may continuously deteriorating and finally fatal breakdown occurred. To reveal these invisible cracks, active thermography has been proposed to predict, detect and quantify the cracks [2]. The advantages of active thermography include non-contact and non-intrusive inspection, possibility in detecting internal cracks and a fast and portable method. Recently, active thermography has also been applied to detect invisible damages occurred in composite materials [3] [4] [5] as the existence of internal cracks may cause uneven density which affected the internal heat flow in time [6].

In active thermography, an external stimulus is needed to apply to an inspected specimen causing non-stationary heat flow. The stimulus could be an optical stimulation method, such as flash light, that is used to deliver heat energy to the specimen. The heat energy propagates as thermal waves inside the specimen. The propagation of heat depends on the specimen's material properties, such as thermal conductivity or density [2]. During the propagation, when the heat flow encountered a defect inside the specimen, it will change the heat propagation rate and produce thermal contrast. Among different optical stimulation methods, optical pulsed thermography, optical lock-in thermography, and optical step heating thermography are the most common ones. Compared with the other two methods, the optical pulsed thermography is less time consuming. Since only the temperature decay is of interest, the signal-to-noise ratio (SNR) of defects obtained during the raw cooling down period may not be sufficient high. Conventionally, various signal enhancement methods would be employed to enhance the SNR. They include the thermal contrast [7][8][9], the pulsed phase thermography[10], the principal component thermography [11], the logarithmic polynomial fit and derivatives [12], the dynamic thermal tomography [13] etc. In summary, a fully automatic defect detection method for active thermography has been designed and reported here. It can detect internal cracks in a pipe without any human intervention. Moreover, an improved edge detection method for enhancing the SNR of image and a packet of canny edge detectors for removing non-uniform background have been embedded with the automatic defect detection method. Hence, the automatic method is able to convert the temperature signal, which has low thermal contrast between the defective and normal areas of the specimen, into detectable edges and distinguishable cracks automatically.

2. Automatic detection

The proposed approach for detecting a damage from a sequence of image signals is broken down into several steps which are outlined in the following sections. Figure 1 shows a summary of the steps. The detailed procedure will be introduced in the following sections.

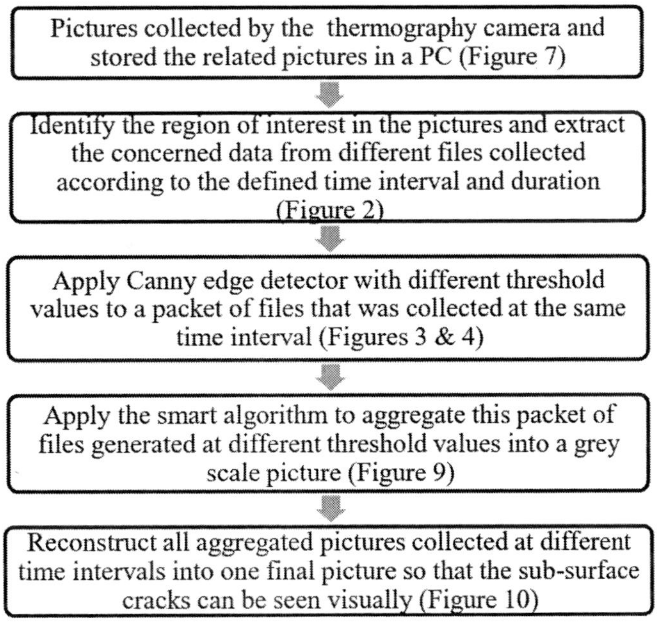

Figure 1. The flow chart of proposed smart algorithm for detecting internal cracks.

2.1 Active thermographic inspection

Experiment is applied using two flash lamps of 4800 watts at 50% capacity. The lamps are turned on for a moment and triggering the source at milliseconds. Before and after the lamps flash, the temperature of specimen is recorded for a total of 10 seconds. Infrared images are acquired using a FILR SC3000 thermal camera. This FILR SC3000 operates in 8 to 9 μm waveband and produces thermal images of 320*240 pixels. Before applying the reconstruction method, we could extract the region of interest and from the temperature time history of the thermographic data. In particular, the stimulating period is too short for analysis, so we will only count the cooling down period as shown in Figure 2. The number of frames starting from the stimulating one to last one of the interest region is set as 60 here, which means 2s infrared video signals at the 30*Hz*.

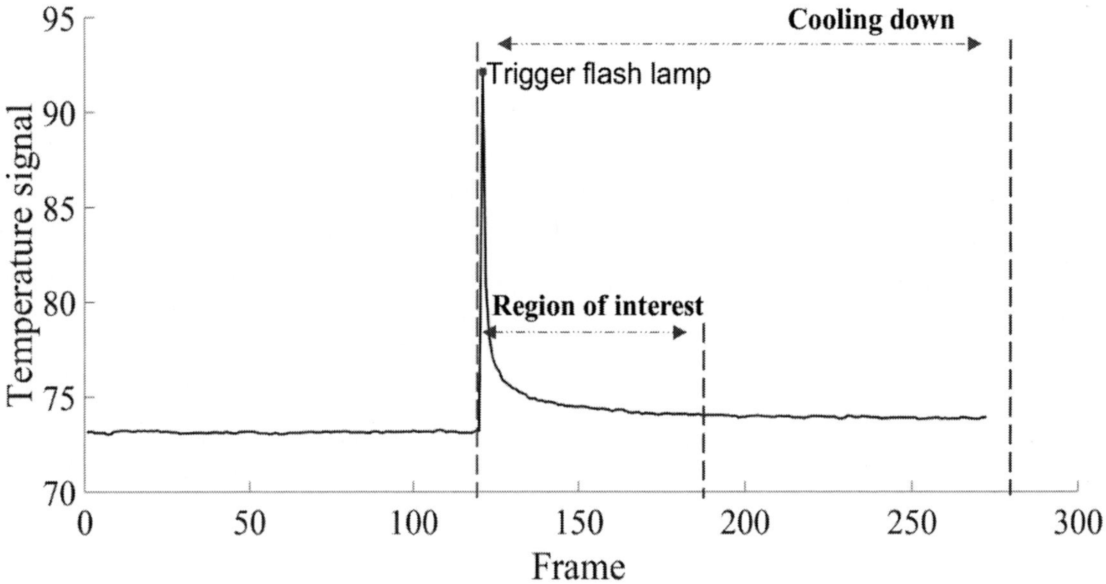

Figure 2. Temperature distribution of the inspected object after triggering the flash lamp at frame 122, the selected region of interest and during the cooling down period.

2.2 Canny edge detector

The Canny edge detection algorithm is known to many as the optimal edge detector [14]. The process of the implementation of canny detector is below:

a. Smoothing the image with an appropriate Gaussian filter to reduce the irreverent details of image.

$$d(i,j,\sigma) = \frac{1}{2\pi\sigma^2} e^{-\frac{i^2+j^2}{2\sigma^2}} \quad (1)$$

$$g(i,j) = d(i,j,\sigma) * O(i,j) \quad (2)$$

where $O(i,j)$ is the original image signal and $g(i,j)$ is signal after the image was smoothed.

b. Determine gradient magnitude and gradient direction at each pixel.
Compute a and b partial derivative and:

$$g'_a(a,b) \approx G_a = [g(a+1,b) - g(a,b) + g(a+1,b+1) - g(a,b+1)]/2 \quad (3)$$

$$g'_b(a,b) \approx G_b = [g(a,b+1) - g(a,b) + g(a+1,b+1) - g(a+1,b)]/2 \quad (4)$$

$$M[a,b] = \sqrt{G_a(a,b)^2 + G_b(a,b)^2} \quad (5)$$

$$\theta[a,b] = \arctan(\frac{G_a(a,b)}{G_b(a,b)}) \quad (6)$$

where $M[a,b]$ stands for edge intensity, and $\theta[a,b] = arg\,max(M[a,b])$ is the direction of edge.

By comparing each $M[a,b]$ in four directions, if the gradient magnitude at a pixel is larger than those at its two neighbours in the gradient direction, mark the pixel as an edge. Otherwise, mark the pixel as the background. The weak edges were removed by doubling the threshold values. The low threshold value t_l equal to the high threshold value t_h times 0.4 by default. If the edge pixel's gradient value is higher than the high threshold value, they are marked as strong edge pixels. If the edge pixel's gradient value is smaller than the high threshold value and larger than the low threshold value, they are marked as weak edge pixels. If the pixel value is smaller than the low threshold value, they will be suppressed. Then a group Canny edge detectors with different threshold values were applied the images. The threshold values were started from 0.001 to 0.1, a total of 100 numbers (n =100), decided based on the edge map results, and the maxima threshold value must turn out to be sufficient large after testing. The last frame of infrared video signals, which could be regarded as a fully cooling down infrared image, was pro-processed by Canny edge detector with the scale from 0.001 to 0.999. The result of total edge numbers in edge map is shown in Figure 3. When the threshold values are between 0.012 and 0.1, the total edge numbers are decaying immediately. Then after the value of 0.1, it starts to converge. Four critical threshold values were chosen and compared by their edge maps as shown in Figure 4. Here, the value of 0.1 is the upper limit of threshold scale at a confidence level of 95%. The parameter selected for Gaussian filer is consistent and decided based on experience.

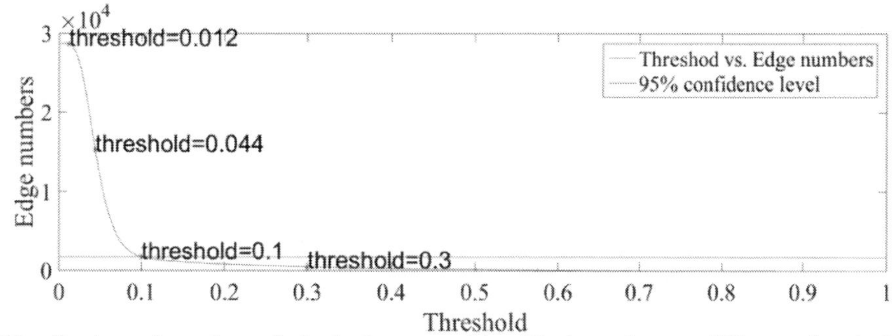

Figure 3. Distribution of number of pixels that were identified as edges at different threshold values.

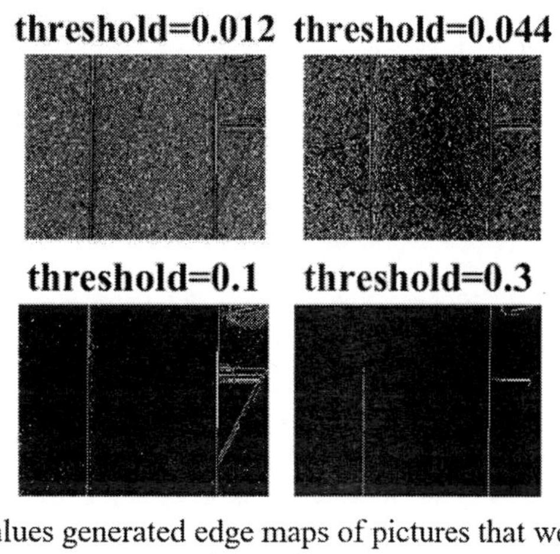

Figure 4. Different threshold values generated edge maps of pictures that were captured at the same time interval.

2.3 The smart algorithm for crack detection

The smart algorithm is to apply a group of different threshold values for Canny edge detector to each infrared picture and then reconstruct the raw video signals to a grey scale picture. The smart processes are described in Figure 5. Every single infrared image captured by the thermal camera was processed by a packet of canny edge detector with different threshold values. The number of detectors was set as 200. Then a reconstructed the image was formed by computing the sum of edge map value of each pixel and made it as a new sequence of image signal as shown in Figure 5(d). Figure 5(e) shows the comparison of raw defect signals and normal signals that has no significant difference between them. Figure 5(f) shows the reconstructed signals by the smart algorithm. There is a large difference in edge indexes between the defect and normal signals.

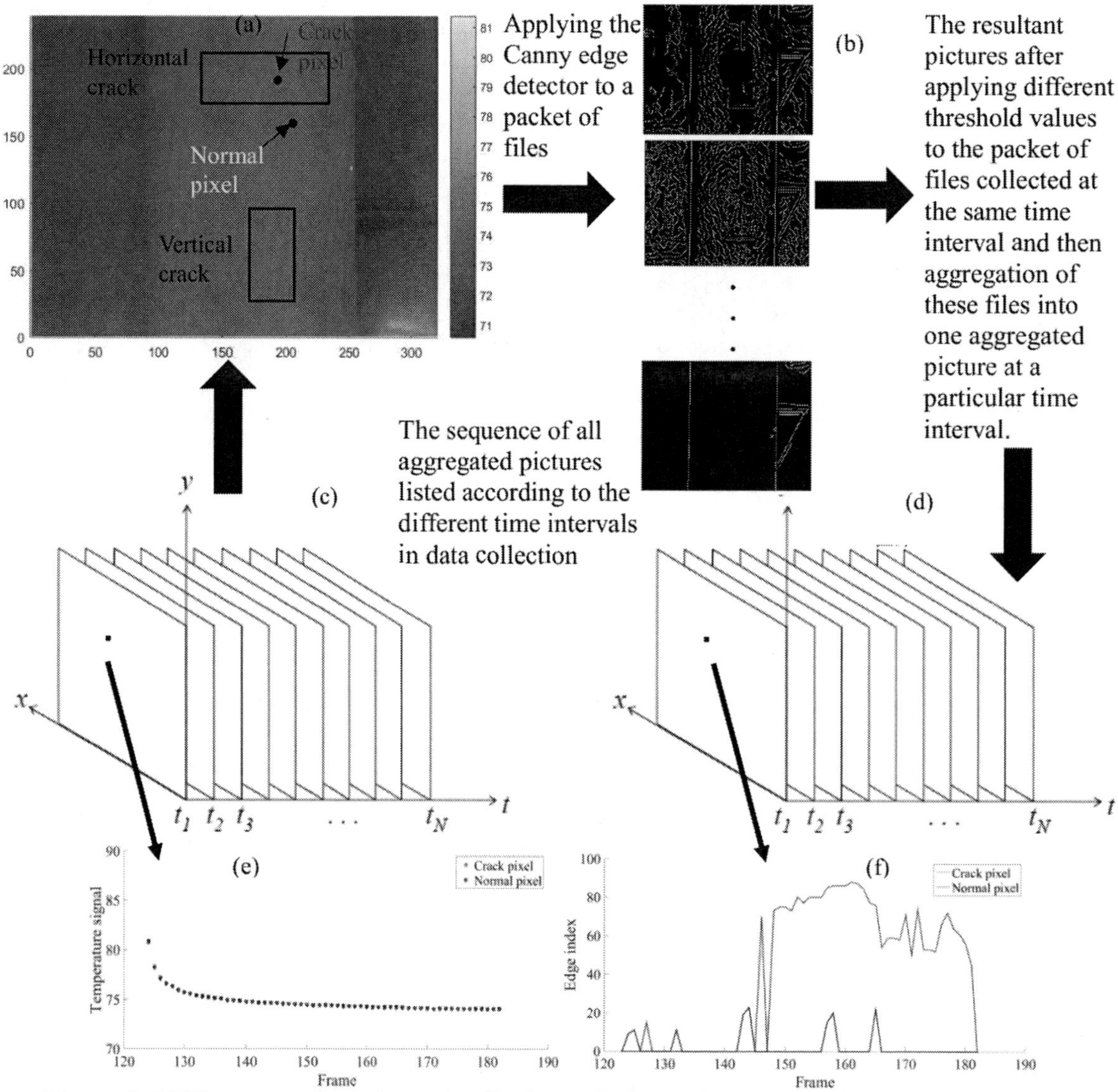

Figure 5. (a) The captured raw image hardly shows the internal cracks; (b) the pictures after applying Canny edge detector at different threshold values; (c) the time sequence of captured raw pictures; (d) the time sequence of all aggregated pictures; (e) the comparison of frame number vs. temperature distribution

IOP Conf. Series: Journal of Physics: Conf. Series **842** (2017) 012029 doi:10.1088/1742-6596/842/1/012029

between images with and without cracks (the difference can hardly be seen); (f) the comparison of frame number vs. edge index between images with and without cracks (the difference is obvious).

3. Experimental setup and results

The proposed procedure to detect damage automatically has been applied to a specimen (Figure 6) which contain two artificial cracks inside with different directions. The device consists of a thermal camera and two flash lamps as external stimulus, as can be seen in Figure 7. The distance between thermal camera and specimen is more than 1m long. Figure 8 shows some infrared pictures in the different period of the whole experiment and the result obtained from the proposed procedure for internal cracks detection. We can see from Figure 9 that the contrast of the edge of cracks area in enhanced compared to its raw image signal in Figure 8. The proposed detection procedure is able to detect both the two defects and locate the edge of these cracks accurately from a long distance. In Figure 10, the shape of pipe can be seen clear that is easy to be classified as the background of the pipe specimen.

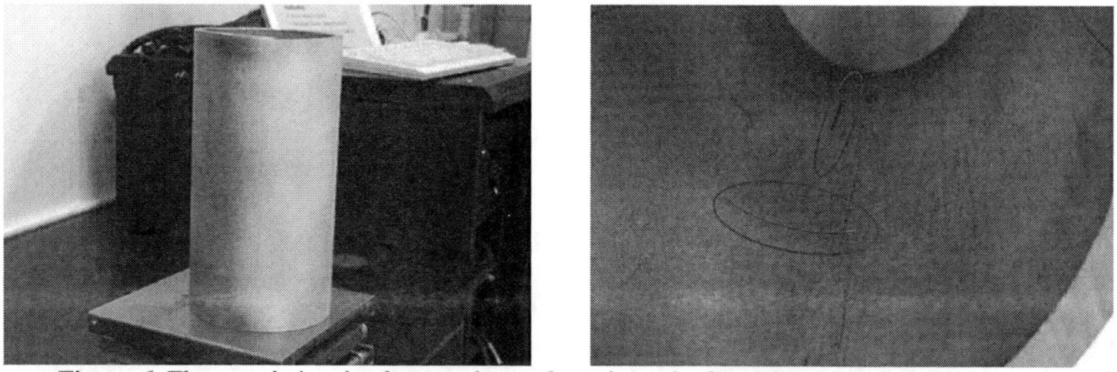

Figure 6. The tested pipe that has two internal cracks - a horizontal crack and a vertical crack.

Figure 7. The experimental setup with the thermal camera, the flash lamps and the PC.

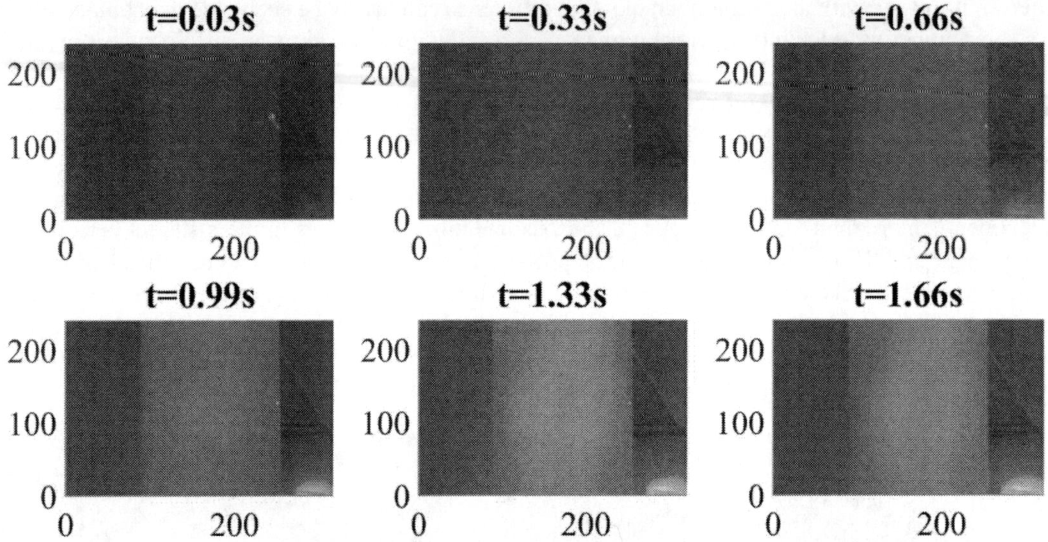

Figure 8. The raw pictures captured by the thermal camera at different time intervals.

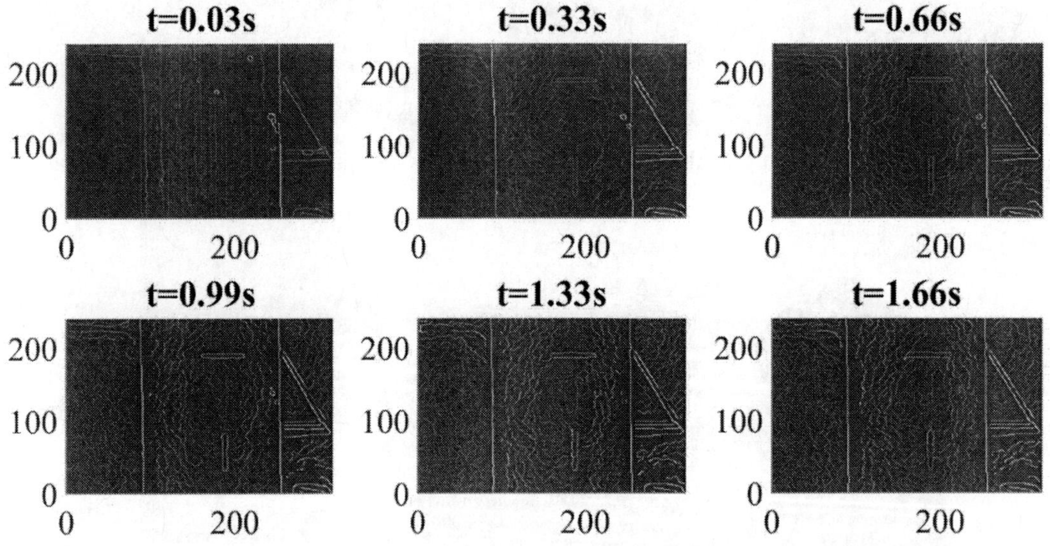

Figure 9. The aggregated edge maps after applying our method to the raw pictures that were collected at different time intervals.

Figure 10. The reconstructed picture after by calculating the average of all pixels from all aggregated pictures and then used these averaged pixels to reconstruct the final picture of which the two sub-surface cracks can be revealed visually.

4. Conclusion

This paper presents an automatic system for detecting internal/sub-surface cracks from a sequences of infrared pictures captured by using the method of active/pulsed thermography. A pipe with two internal cracks was chosen for testing the effectiveness of the proposed method that includes the active thermography and Canny edge detection algorithm. An external stimulus in terms of triggering two flash lamps was applied to the pipe in order to induce relevant thermal contrasts between the defective and non-defective areas of the pipe. During the inspection processing, the suspended region of interest was selected in the captured thermo-pictures. The image pixel values within the region of interest were extracted from different temporal files collected in respective to the defined time interval and duration. Then different Canny edge detectors with different threshold values were applied to the captures pictures to eliminate the effect caused by non-uniform heating as well as enhance the local difference between the pixels that contained the normal and crack information. Subsequently, all of the resultant pictures treated by having different threshold values but captured at the same time interval were aggregated together. Eventually, the pixel values were averaged among all these files and then reconstructed into one grey scale picture. Experimental results prove that the edge of both two internal cracks can be visually observed from this final grey scale picture. All the above processes were conducted automatically. Hence, the proposed method introduced in this paper is superior to the conventional way which uses the tedious and manual adjustment to tune the temperature/colour intensity so that the sub-surface cracks can be finally identified. With the help of this method, a fully automatic defect detection method can be developed to reveal sub-surface defects by combining the active thermography and Canny edge detector and smart algorithm.

5. References

[1] Ibarra-Castanedo, Clemente, and Xavier Maldague. "Pulsed phase thermography reviewed." Quantitative Infrared Thermography Journal 1.1 (2004): 47-70.

[2] Usamentiaga, R., Venegas, P., Guerediaga, J., Vega, L., & López, I. (2013). Automatic detection of impact damage in carbon fiber composites using active thermography. Infrared Physics & Technology, 58, 36-46.

[3] Vavilov, V. P., Almond, D. P., Busse, G., Grinzato, E., Krapez, J. C., Maldague, X., & Wu, D. (1998). Infrared thermographic detection and characterisation of impact damage in carbon fibre composites: results of the round robin test. Proc. QIRT 98 (Paris), 43-52.

[4] Ball, R. J., & Almond, D. P. (1998). The detection and measurement of impact damage in thick carbon fibre reinforced laminates by transient thermography. NDT & e International, 31(3), 165-173.
[5] Montanini, R., & Freni, F. (2012). Non-destructive evaluation of thick glass fiber-reinforced composites by means of optically excited lock-in thermography. Composites Part A: Applied Science and Manufacturing, 43(11), 2075-2082.
[6] Maldague, X. P. (2002). Introduction to NDT by active infrared thermography. Materials Evaluation, 60(9), 1060-1073.
[7] Vavilov, V. (1992, July). Thermal non destructive testing: short history and state-of-art. In QIRT (Vol. 92, pp. 179-194).
[8] Benítez, H. D., Ibarra-Castanedo, C., Bendada, A., Maldague, X., Loaiza, H., & Caicedo, E. (2008). Definition of a new thermal contrast and pulse correction for defect quantification in pulsed thermography. Infrared Physics & Technology, 51(3), 160-167.
[9] Balageas, D. L. (2012). Defense and illustration of time-resolved pulsed thermography for NDE. Quantitative InfraRed Thermography Journal, 9(1), 3-32.
[10] Maldague, X., Galmiche, F., & Ziadi, A. (2002). Advances in pulsed phase thermography. Infrared physics & technology, 43(3), 175-181.
[11] Rajic, N. (2002). Principal component thermography for flaw contrast enhancement and flaw depth characterisation in composite structures. Composite Structures, 58(4), 521-528.
[12] Shepard, S. M., Lhota, J. R., Rubadeux, B. A., Wang, D., & Ahmed, T. (2003). Reconstruction and enhancement of active thermographic image sequences. Optical Engineering, 42(5), 1337-1342.
[13] Vavilov, V., Maldague, X., Picard, J., Thomas, R. L., & Favro, L. D. (1992). Dynamic thermal tomography: new NDE technique to reconstruct inner solids structure using multiple IR image processing. In Review of progress in quantitative nondestructive evaluation (pp. 425-432). Springer US.
[14] Maini, Raman, and Himanshu Aggarwal. "Study and comparison of various image edge detection techniques." International journal of image processing (IJIP) 3.1 (2009): 1-11.

IOP Conf. Series: Journal of Physics: Conf. Series **842** (2017) 012030 doi:10.1088/1742-6596/842/1/012030

On the physical significance of the Effective Independence method for sensor placement

Yaoguang Jiang, Dongsheng Li and Gangbing Song

State Key Laboratory of Coastal and Offshore Engineering,Dalian University of Technology, Dalian, Liaoning, 116023, People's Republic of China

E-mail:dsli@dlut.edu.cn

Abstract: Optimally deploy sparse sensors for better damage identification and structural health monitoring is always a challenging task. The Effective Independence(EI) is one of the most influential sensor placement method and to be discussed in the paper. Specifically, the effect of the different weighting coefficients on the maximization of the Fisher information matrix(FIM) and the physical significance of the re-orthogonalization of modal shapes through QR decomposition in the EI method are addressed. By analyzing the widely used EI method, we found that the absolute identification space put forward along with the EI method is preferable to ensuring the maximization of the FIM, instead of the original EI coefficient which was post-multiolied by a weighting matrix. That is, deleting the row with the minimum EI coefficient can't achieve the objective of maximizing the trace of FIM as initially conceived. Furthermore, we observed that in the computation of EI method, the sum of each retained row in the absolute identification space is a constant in each iteration. This potential property can be revealed distinctively by the product of target mode and its transpose, and its form is similar to an alternative formula of the EI method through orthogonal-triangular(QR) decomposition previously proposed by the authors. With it, the physical significance of re-orthogonalization of modal shapes through QR decomposition in the computation of EI method can be obviously manifested from a new perspective. Finally, two simple examples are provided to demonstrate the above two observations.

Key Word: Effective Independence method; Fisher information matrix; sensor placement method; structural health monitoring; QR decomposition

1. Introduction

Sensor placement as an essential issue in dynamic testing of large structures has been widely concerned by many fields in the last decade, with the increasing number of instrumented large structures for health monitoring[1]. Due to the reasons of economy and structural accessibility limitations, the sensors installed in structures are sparse and far less than the actual degrees of freedom in these structures. Furthermore, the identification, feature extraction, and even the health state estimation of a structure are mainly based on the data that obtained from these deployed sensors[2]. Consequently, the optimization of sensor placement scheme is an crucial task for the objective of better structural identification with deploying limited number of sensors.

In the past, a vast amount of literatures on this issue has been produced. And various of methods have been proposed, such as, the modal kinetic energy method selects those sensors with high amplitude of responses[3], the Guyan reduction method which deploys the sensors on master degrees of freedom of the reduced matrix[4], damage sensitivity based methods[5], neural networks and genetic algorithms[6], observability based methods[7], etc. And this paper is based on an influence and commonly used method .i.e. the effective independence (EI) method, which is an iterative method proposed by D.C.Kammer in 1991. This method selects candidate sensor positions with the aim that optimizing the linearly independence of the mode shapes and making the mode shapes contain the most information of the target mode[8,9]. To achieve this objective, Kammer maximized the trace[8,10] of the Fisher information matrix (FIM) as the criterion. Although the EI method has been investigated by many scholars and developed very mature, some inner problems are confused yet, for instance, the physical significant of the re-orthogonalization of modal shapes through orthogonal-triangular (QR) decomposition[11] previously proposed by the authors.

In this paper, the EI method is analyzed from the viewpoint of theory. It is found that the traditional EI coefficients which is post-multiolied by a weighting matrix can't achieve the objective of maximizing the trace of FIM, as a substitution the absolute identification space performs better. Meanwhile, it arrives at the physical significant of the re-orthogonalization of modal shapes through QR decomposition by analyzing the effect of different weighting coefficients on the maximization of the FIM and the property of the absolute identification space in detail. Finally, Two simple examples are employed to demonstrate the above analysis.

2. Theory

In this section, the traditional EI method and the potential re-orthogonalization of target modes in the computation of EI method are presented initially. Subsequently, the effect of different weighting coefficients on the maximization of the FIM is discussed and the physical significance of the re-orthogonalization of target modes can be obviously manifested from a new perspective through the discussion.

2.1. Effective independence method

As mentioned previously in the introduction, the EI method is one of the most influential sensor placement algorithms, Kammer simplified and extended the method proposed by Udwadia[12] which is based on sensitivity analysis from the viewpoint of estimation theory, and thus put forward the EI method. The theoretical analysis of the EI method starts from an efficient unbiased estimator of the modal coordinate q, the covariance matrix of the estimate error can be formulated as follow:

$$E\left[(q-\hat{q})(q-\hat{q})^T\right]=\left[\left(\frac{\partial y}{\partial q}\right)^T\left[\Psi_0^2\right]^{-1}\left(\frac{\partial y}{\partial q}\right)\right]^{-1} \tag{1}$$

in which E denotes the expected value, $\hat{q}$ represents an efficient unbiased estimator of the vector $\boldsymbol{q}$, $\boldsymbol{y}$ is a measurement column vector indicating which positions of the structure are measured, and the Ψ_0^2 is the variance of the stationary Gaussian measurement white noise ε.

In this formulation, because of the assumption that the measurement noise is uncorrelated and possesses identical statistical properties of each sensor , which means the variance of stationary Gaussian measurement is a constant. Therefore, the equation (1) can be simplified as follow:

$$E\left[(q-\hat{q})(q-\hat{q})^T\right]=\left[\frac{1}{\Psi_0^2}\Phi^T\Phi\right]^{-1}=A^{-1} \tag{2}$$

where A is the FIM with the definition of the product of the mode shape matrix and its transpose: $A=\Phi^T\Phi$.

Obviously, maximizing the FIM will results in the minimum of the covariance matrix of the estimate error and achieves the best state estimate of q. Therefore, the trace of the FIM is provided as original criterion to maximize the FIM. And the analysis begins by solving the eigenvalue equation:

$$\left[\Phi^T\Phi-\lambda I\right]\Psi=0 \tag{3}$$

where Ψ denotes the orthogonal eigenvectors, λ are the corresponding eigenvalues. The eigenvalues of the FIM can be expressed distinctly by forming the absolute identification space which will be called the eigenvalue contribution matrix in this paper. In this matrix each row contains the square of the product of the mode shape matrix Φ and the eigenvector matrix Ψ.

$$G=[\Phi\Psi]\otimes[\Phi\Psi] \tag{4}$$

in which G denotes the eigenvalue contribution matrix, i.e. the absolute identification space, $\otimes$ represents a term-by-term matrix multiplication. Postmultiplying G by the inverse of the matrix of eigenvalues λ, the EI coefficients of the candidate sensors are computed as follow:

$$E_D=[\Phi\Psi]\otimes[\Phi\Psi]\lambda^{-1}\cdot 1 \tag{5}$$

in which, E_D is the EI index.

The traditional EI method computational process is deleting the row with the smallest EI coefficients at each iteration, then repeating the same process until the remained number of the sensors equals to the preconceived value. And those remained sensor locations are regarded as the measurement ones. The details of the EI method is presented in reference [8].

2.2. *The re-orthogonalization of target mode*

According to the formula of the EI method, target modes Φ and its reduced form are employed apparently in the iterative computation of the EI coefficients. Nevertheless, a potential re-orthogonalization of the target modes exists in each iteration of the computation, which essentially changes the directions of the target mode vectors. For a better understanding of the re-orthogonalization of target modes, an alternative method is provided[13], in which the EI index are computed as the diagonal of the projection matrix.

$$E_D = diag\left[\Phi\left(\Phi^T\Phi\right)^{-1}\Phi^T\right] \tag{6}$$

where the matrix $\Phi\left(\Phi^T\Phi\right)^{-1}\Phi^T$ represents the projection matrix of vector space spanned by target modes.

For the objective to reveal the connection between the modal kinetic energy method and the EI method, the authors Li, etc[11,14] disposed the target mode matrix Φ through the shinny orthogonal-triangular(QR) decomposition.

$$\Phi = QR \tag{7}$$

Then the equation (6) can be written as

$$E_D = diag[QR\left(R^TQ^TQR\right)^{-1}R^TQ^T] = diag(QQ^T) \tag{8}$$

in which Q denotes an $n\times m$ orthonormal matrix with the same dimensions as Φ , and R is an $m\times m$ upper triangular matrix.

After the first iteration, a new reduced system taken shape. The reduced target mode matrix Φ_1 should be re-normalized through the QR decomposition as the original system, and the EI index is computed as follow:

$$E_{D1} = diag[\Phi_1(\Phi_1^T\Phi_1)^{-1}\Phi_1^T] = diag[Q_1R_1(R_1^TQ_1^TQ_1R_1)^{-1}R_1^TQ_1^T] = diag(Q_1Q_1^T) \tag{9}$$

Obviously, in the first iteration the target mode matrix is disposed by QR decomposition, then in the next iteration an re-orthogonalization through QR decomposition is applied to the reduced target mode matrix Φ_1 which is not strictly orthogonal anymore after the certain row of the Φ is deleted in the proceeding iteration[15]. The QR decomposition of the reduced matrix Φ_1 can be regarded as a procedure to re-extract an orthogonal subspace from the matrix Φ_1 . Comparing equation (8) to equation (9), the potential re-orthogonalization of the target modes is manifested distinctly.

2.3. *The physical significance of the re-orthogonalization of target modes*

As mentioned in the first section of the theory, the aim of the EI method is to select the candidate sensor locations which optimize the linear independence of mode shapes. From the sensitivity analysis of the modal coordinate with estimation theory and equation (2), the objective of optimizing the linear independence of mode shapes can be simplified as the maximization of the FIM. And the trace of the FIM was applied by Kammer as the suitable norm of the FIM to be maximized. For instance, the trace of the FIM can be formulated as[16]:

$$tr(FIM) = \lambda_1 + \lambda_2 + \cdots\cdots + \lambda_n \tag{10}$$

where λ_i is the eigenvalue of the FIM.

Therefore, the analysis begins by obtaining the eigenvalues through equation (3). As the eigenvalues of the FIM are positive, and the eigenvectors Ψ are orthonormal. Then according to the concept of similarity in matrix analysis[16], the FIM is similar to the diagonal matrix of eigenvalues as follow:

$$\Psi^T(\Phi^T\Phi)\Psi = \lambda \tag{11}$$

equation (11) can also be written as:

$$(\Phi\Psi)^T(\Phi\Psi) = \lambda \quad \text{or} \quad \lambda = diag[(\Phi\Psi)^T(\Phi\Psi)] \tag{12}$$

According to equation (12), the elements of the diagonal matrix of eigenvalues are exactly the corresponding norms of the row vectors from the matrix $(\Phi\Psi)^T$. For distinguishing the contribution of each row of the target modes to the total eigenvalues of the FIM, Kammer made a transformation of equation (12) and arrived at the formula as equation (4), where the matrix G will be called the eigenvalue contribution matrix here. Equation (4) will be rewritten below with its detailed form:

$$G=[\Phi\Psi]\otimes[\Phi\Psi]=\begin{bmatrix} g_{11} & g_{12} & \cdots & g_{1n} \\ g_{21} & g_{22} & \cdots & g_{2n} \\ \vdots & \vdots & \ddots & \vdots \\ g_{n1} & g_{n2} & \cdots & g_{nn} \end{bmatrix} \tag{13}$$

where g_{ij} is the *ith* row element of the *jth* column of the eigenvalue contribution matrix G. And the sum of each column denotes the corresponding eigenvalue of the FIM, meanwhile each row of the matrix G sums to the contribution of the corresponding sensor to the total eigenvalues of the FIM.

$$\lambda_i = g_{1i} + g_{2i} + \cdots\cdots + g_{ni} \tag{14a}$$

$$\lambda_{isensor} = g_{i1} + g_{i2} + \cdots\cdots + g_{in} \tag{14b}$$

where λ_i denotes the *ith* eigenvalue, $\lambda_{isensor}$ is the contribution of the *ith* sensor. And as presented by the original EI method, the eigenvalue contribution matrix G should be postmultiplied by the inverse of the diagonal matrix of eigenvalues λ. The formula will be rewritten with its detailed form:

$$F_E=[\Phi\Psi]\otimes[\Phi\Psi]\lambda^{-1}=\begin{bmatrix} \lambda_1^{-1}g_{11} & \lambda_2^{-1}g_{12} & \cdots & \lambda_n^{-1}g_{1n} \\ \lambda_1^{-1}g_{21} & \lambda_2^{-1}g_{22} & \cdots & \lambda_n^{-1}g_{2n} \\ \vdots & \vdots & \ddots & \vdots \\ \lambda_1^{-1}g_{n1} & \lambda_2^{-1}g_{n2} & \cdots & \lambda_n^{-1}g_{nn} \end{bmatrix} \tag{15}$$

From the above formula, the F_E matrix is equivalent to multiply each column of G with different weighting coefficients. That is, postmultiplying the inverse of the matrix λ can be regarded as weighting matrix G by a matrix with different coefficients. The F_E matrix can be analyzed in the same way, the sum of each column of the F_E matrix represents the fraction rank of the FIM, and each row sums to the contribution of the corresponding sensor position to the rank of the FIM.

$$1=\lambda_i^{-1}(g_{1i} + g_{2i} + \cdots\cdots + g_{ni}) \tag{16a}$$

$$r_{isensor} = \lambda_1^{-1}g_{i1} + \lambda_2^{-1}g_{i2} + \cdots\cdots + \lambda_n^{-1}g_{in} \tag{16b}$$

For achieving the aim that maximizing the FIM, the row of the matrix F_E with the minimum summation is removed as originally proposed. However it is found that retaining the rows of the matrix F_E with large summation in the iterative calculation can't reflect the eigenvalues well, meanwhile can't guarantee the maximization of the FIM either. And the matrix G reflects the eigenvalues more directly in comparison with the matrix F_E by analyzing equation (14b) and equation (16b). Element of the matrix F_E represents the proportion of *ith* sensor position to the *jth* eigenvalue, while the element of the matrix G, i.e. g_{ij}, is the contribution of the *ith* sensor position to the *jth* eigenvalue. For a more profound understanding, a special situation is discussed which the FIM is a 2×2 matrix with two eigenvalues.

$$G=\begin{bmatrix} g_{11} & g_{12} \\ g_{21} & g_{22} \end{bmatrix} \quad \text{and} \quad G*1=\begin{bmatrix} g_{11}+g_{12} \\ g_{21}+g_{22} \end{bmatrix} \tag{17a}$$

$$F_E=\begin{bmatrix} \lambda_1^{-1}g_{11} & \lambda_2^{-1}g_{12} \\ \lambda_1^{-1}g_{21} & \lambda_2^{-1}g_{22} \end{bmatrix} \quad \text{and} \quad F_E*1=\begin{bmatrix} \lambda_1^{-1}g_{11}+\lambda_2^{-1}g_{12} \\ \lambda_1^{-1}g_{21}+\lambda_2^{-1}g_{22} \end{bmatrix} \tag{17b}$$

where 1 denotes a 2×1 column vector with all elements of 1. From above equations, it is assumed that $(g_{11}+g_{12})>(g_{21}+g_{22})$ in the matrix G which means the contribution to the total eigenvalues of the 1st sensor is larger than that of the 2nd sensor. However, if the proportion of the 1st sensor to the eigenvalue λ_1 and λ_2 is smaller than that of the 2nd sensor, i.e. $(\lambda_1^{-1}g_{11}+\lambda_2^{-1}g_{12})<(\lambda_1^{-1}g_{21}+\lambda_1^{-1}g_{22})$, the above two criteria arrive different conclusion about maximizing the trace norm of the FIM. According to the matrix F_E*1 the 2nd sensor position is remained, oppositely, the matrix $G*1$ remains the 1st sensor position. And obviously the matrix $G*1$ reflects the trace of the FIM better.

With the analyzing above, another interesting phenomenon is that each element of the matrix $G*1$ is a constant in each iteration. When recalculating the new eigenvalue contribution matrix $G'*1$ after the preceding iteration, the only different between $G'*1$ and $G*1$ is the removed element which is corresponding to the sensor position selected by using the EI coefficients. For the objective to explain this property of the matrix G, an analysis is made as follow. Because of the matrix G is related to the eigenvalues of the FIM, and is obtained by equation (13). A transformation of equation (13) is provided as:

$$G*1=[\Phi\Psi]\otimes[\Phi\Psi]*1=diag([\Phi\Psi][\Phi\Psi]^T) \tag{18}$$

the eigenvectors Ψ of the FIM have the property that $\Psi^T\Psi=I$, therefore,

$$G*1=diag(\Phi\Psi\Psi^T\Phi^T)=diag(\Phi\Phi^T) \tag{19a}$$

$$G*1=diag(\Phi\Phi^T)=[\phi_1\phi_1^T,\phi_2\phi_2^T,\cdots\cdots,\phi_2\phi_2^T]=[\|\phi_1\|_2,\|\phi_2\|_2,\cdots\cdots,\|\phi_n\|_2] \tag{19b}$$

where ϕ_i denotes the *ith* row of the target mode matrix Φ. From the equation above, it is obviously that the element of the matrix $G*1$ equals to the row norm of the target mode matrix Φ. Therefore, the elements is constant.

Distinctly, the form of equation (19a) is similar to the equation (9). And the two formulas are compared as follow:

$$E_D=F_E*1=[\Phi\Psi]\otimes[\Phi\Psi]\lambda^{-1}*1=diag(QQ^T) \tag{20a}$$

$$G*1=[\Phi\Psi]\otimes[\Phi\Psi]*1=diag(\Phi\Phi^T) \tag{20b}$$

The physical significance of the re-orthogonalization of the target mode in each iteration can be regarded as weighing the target mode matrix with the inverse of the eigenvalue matrix λ of the FIM. The effect of different weighting coefficients on the maximization of the FIM and the physical significance of the re-orthogonalization of target modes are demonstrated in the following examples.

3. Two Simple Numerical Examples

A 4-degrees of freedom beam model with concentrated quality is used primarily to illustrate the effect of different weighting coefficients on the maximization of the FIM and the physical significance of the re-orthogonalization of target modes as deduced above. The beam is a simplified model with 4 elements as shown in **Figure 1**. For the purpose of simplifying the computation, the parameters are provided in a simple form. The total length of the beam is 4 and each element is quartered. The mass of the 4 elements are, m_1=1.5, m_2=1.5, m_3=1.25, m_4=0.5. The other parameters of the beam are provided as, the cross section is 1, the density of the beam is 1,and the E=1, I=1.

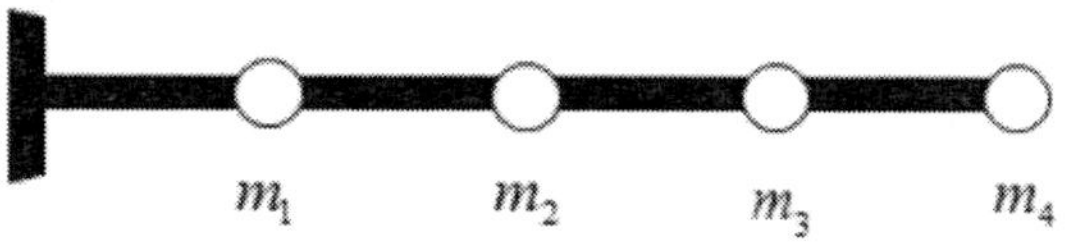

Figure 1. The 4-dofs beam model with concentrated quality

The first two orders of the mode shapes are used as target modes to participate in calculation. In the first iteration, computing the FIM and its eigenvector matrix Ψ and the eigenvalue matrix λ according to the equation (3). Then the matix $G*1$ and the matrix F_E*1, i.e. the matrix of EI index, can be obtained from equation (13) and equation (15) as follow:

$$G*1=\begin{bmatrix}0.3739\\0.3243\\0.4494\\0.7818\end{bmatrix}\begin{matrix}\\-removed\\\\\\\end{matrix} \quad \text{and} \quad E_D=F_E*1=\begin{bmatrix}0.5255\\0.4785\\0.3741\\0.6220\end{bmatrix}\begin{matrix}\\\\-r\ e\ m\ o\\\\\end{matrix}$$

$$trace_{G*1} > trace_{F_E*1}$$

Obviously, the row with the minimum element of the above two matrix are different. If the 2nd row of the target mode is deleted in this iteration according to the matrix $G*1$, the trace of the FIM in the next iteration is 1.6051. However, if the 3rd row of the target mode is removed according to the matrix F_E*1, the trace in the next iteration is 1.4800. That is, $trace_{G*1}>trace_{FE*1}$ after the first iterative computation. And as discussed above in equation (17), it is found that the value of the 2nd row surpasses that of the 3rd row by postmultiplying the inverse of the matrix of eigenvalues λ. Analyzing the 2nd and the 3rd rows of the target modes in details as follow:

$$G*1=[\Phi\Psi]\otimes[\Phi\Psi]_{2^{nd}and3^{rd}}*1=\begin{bmatrix}0.3176 & 0.0067\\0.0242 & 0.4252\end{bmatrix}*1=\begin{bmatrix}0.3243\\0.4494\end{bmatrix}$$

$$F_E*1=G*\lambda^{-1}*1=\begin{bmatrix}\lambda_1^{-1}*0.3176 & \lambda_2^{-1}*0.0067\\\lambda_1^{-1}*0.0242 & \lambda_2^{-1}*0.4252\end{bmatrix}*1=\begin{bmatrix}0.4785\\0.3741\end{bmatrix}$$

It demonstrates the deduction of the equation (17) that the large proportion of the 2nd row of the eigenvalues can't guarantee its contribution to the trace of the FIM is large as well. Therefore, the matrix $G*1$ is preferable to maximizing the trace of the FIM as well as the FIM itself, instead of the matrix F_E*1 proposed as original.

To illustrate the physical significance of the re-orthogonalization of target modes, the matrix $G*1$ is analyzed in each iteration. The initial matrix $G*1$ is provided as:

$$G*1=\begin{bmatrix}0.3739\\0.3243\\0.4494\\0.7818\end{bmatrix}$$

Then in the next iteration with the assumption that one row of the target modes is removed, the new reduced matrix $G'*1$ is computed as follow:

$$G'*1=\begin{bmatrix} \\ 0.3243 \\ 0.4494 \\ 0.7818 \end{bmatrix}\begin{matrix} -1st.removed \\ \\ \\ \\ \end{matrix} \qquad G'*1=\begin{bmatrix} 0.3739 \\ \\ 0.4494 \\ 0.7818 \end{bmatrix}\begin{matrix} \\ -2nd.removed \\ \\ \\ \end{matrix}$$

$$G'*1=\begin{bmatrix} 0.3739 \\ 0.3243 \\ \\ 0.7818 \end{bmatrix}\begin{matrix} \\ \\ -3rdremoved \\ \\ \end{matrix} \qquad G'*1=\begin{bmatrix} 0.3739 \\ 0.3243 \\ 0.4494 \\ \\ \end{bmatrix}\begin{matrix} \\ \\ \\ -4th.removed \end{matrix}$$

Obviously in the next iteration, comparing to the reduced matrix $G'*1$, the removed element of the initial matrix $G*1$ is the row corresponding to the one deleted from the target modes in the preceding iteration. And the other rows of the matrix $G*1$ are constant. In the follow iterations the property of the matrix $G*1$ is stayed as well. Furthermore, the value of the removed row of the matrix $G*1$ equals to the reduction of the trace of the FIM. Therefore, the physical significance of the re-orthogonalization of the target modes deduced by equation (20) is demonstrated.

To demonstrate the deduction is generalized, another beam with distributed mass is provided, as shown in the **Figure 2**. The beam with the length 4 is divided into 4 equal elements. And the simplified parameters are provided as well, whose the cross section is 1, E=1 and I=1. The density of the 2nd element is 2 while the others are 1.

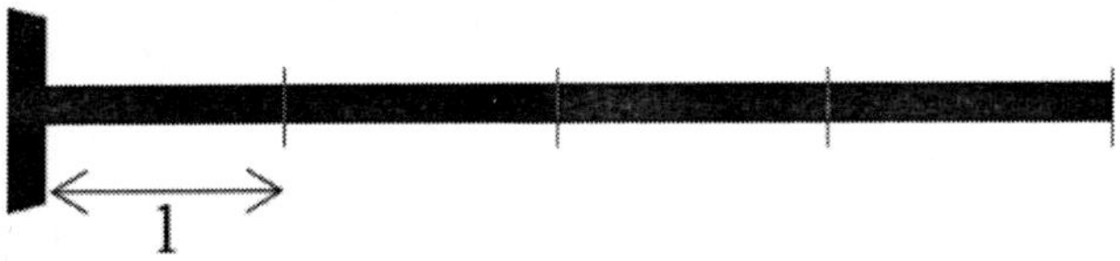

Figure 2. Beam model with distributed mass

The first 4 orders of the mode shapes are remained as the target modes. According to equation (13) and equation (15) the corresponding matrix $G*1$ and F_E*1 can be obtained and the values of each row of the two matrices are listed below in **Table1.**:

Table 1. Values of the matrix $G*1$ and F_E*1

	G*1	F_E*1
1st row	0.6585	0.5932
2nd row	1.1617	0.3663
3rd row	0.7624	0.4140
4th row	1.7599	0.5244
5th row	1.1239	0.4236
6th row	1.4843	0.4056
7th row	3.7408	0.5573
8th row	10.4685	0.7157

From the Table 1 above, it is obviously that the row with the minimum value of the two matrices are different. Furthermore, the sequence of the row according to the matrix F_E*1, i.e. the EI index, is greatly different from that according to the matrix $G*1$. Therefore, the removed rows of the target modes in the following iterations according to the two matrices are different either. And the trace of the corresponding FIM in each iteration based on different criterion which defined by the two matrices are shown in **Figure 3.** as follow,

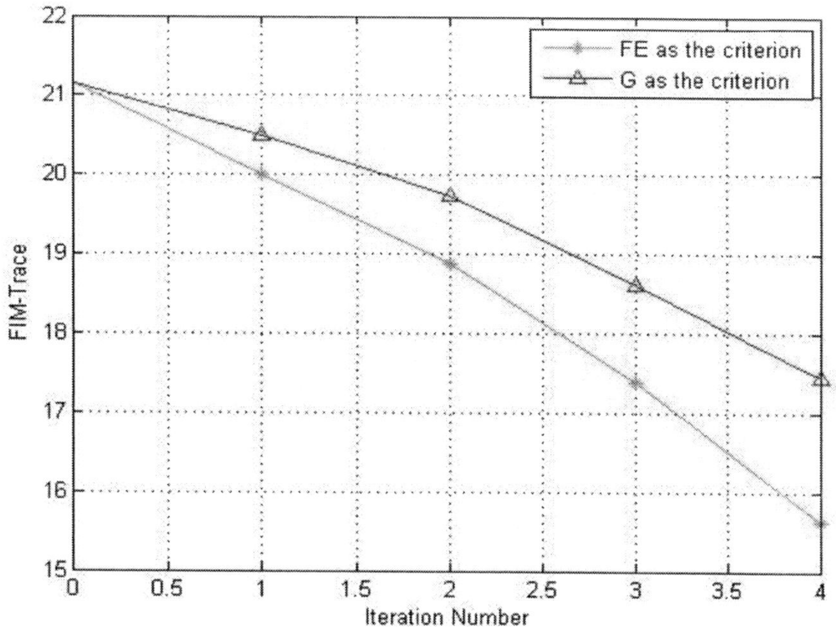

Figure 3. The trace of FIM in each iteration according to different criterion

It is observed that the trace of the FIM according to the criterion defined by the matrix $G*1$ are larger. Therefore, the deduction is proved that the matrix $G*1$ is preferable to maximizing the trace of the FIM as well as the FIM itself, other than purposed originally.

Then the matrices $G*1$ in each iteration corresponding to the reducing sequences of the target mode are provided as follow in **Table 2**. The specific row of the target mode is removed in each iterative computation.

After the 1st iteration, the 1st row of the matrix $G*1$ is removed because of the target modes are reduced with the 1st row is removed, and the other rows are remained as initial. So are the following iterations. Therefore the conclusion deduced through equation (20) that the physical significance of the re-orthogonalization of the target mode can be regarded as weighing the target mode matrix with the inverse of the eigenvalue matrix λ of the FIM is demonstrated once again.

Table 2. The matrix G*1 in each iteration.

0.6585			
1.1617	1.1617	1.1617	1.1617
0.7624	0.7624		
1.7599	1.7599	1.7599	1.7599
1.1239	1.1239	1.1239	
1.4843	1.4843	1.4843	1.4843
3.7408	3.7408	3.7408	3.7408
10.4685	10.4685	10.4685	10.4685
$G*1$	$G'*1$	$G''*1$	$G'''*1$

4. Conclusions

Based on the influential sensor placement method, i.e. EI method, the effect of different weighting coefficients on the maximization of the FIM is discussed. It is found that the sum of each row of eigenvalue contribution matrix *G*, i.e. the matrix *G*1*, is preferable to maximizing the trace of the FIM as well as the FIM itself instead of the EI index purposed originally. Furthermore, the physical significance of the re-orthogonalization of target mode in the EI method is derived by analyzing the eigenvalue contribution matrix *G* in details. The procedure of re-orthogonalizing the target modes in each iteration can be regarded as weighing the target mode matrix with the inverse of the eigenvalue matrix λ of the FIM. And the two simple examples are provided eventually, which demonstrating the theory numerically. Furthermore, an alternative form to compute the matrix *G*1* is shown in equation (19a).

Acknowledgements

This research work was jointly supported by the National Natural Science Foundation of China (Grant No. 51121005, and 51578107), and 973 Project (Grand No. 2015CB057704).

References

[1] Deraemaeker K and Worden K 2010 *New trends in vibration based structural health monitoring* (Wien: Springer)
[2] Balageas D, Fritzen C P and Guemes A 2006 Structural health monitoring (London: ISTE)
[3] Papadopoulos M and Garcia E 1998 Sensor placement methodologies for dynamic testing *J. AIAA.* **36** 256-63
[4] Penny J E T, Friswell M I and Garvey S D 1994 The automatic choice of measurement locations for dynamic testing *J. AIAA.* **32** 407-14
[5] Coote J E, Lieven N A J and Skingle G W 2005 *Sensor placement optimisation for modal*

testing of a helicopter fuselage Proc. 24th Int. Model Analysis Conf. (Orlando) p 07

[6] Guo H Y, Zhang L, Zhang LL and Zhou J X 2004 Optimal placement of sensor for structural health monitoring using improved genetic algorithms *J. Smart Mater.Struct.* **13** 528-34

[7] Skelton R E 1998 *Dynamic systems control* (New York: Wiley)

[8] Kammer D C 1991 Sensor placement for on-orbit modal identification and correlation of large space structures *J. Guidance, Control and Dynamics.* **14** 251-59

[9] Kammer D C and Tinker M L 2004 Optimal placement of tri-axial accelerometers for modal vibration test *J. Mechancical systems and signal processing.* **18** 29-41

[10] Yu T K and Sein J H 1973 Observability and optimal measurement locations in linear distributed parameter systems *Int. J. of Control* **18** 785-99

[11] Li D S, Li H N and Fritzen C P 2007 The connection between effective independence and modal kinetic enery methods for sensor placement *J.Sound and Vibration* **305** 945-55

[12] Udwadia F E and Garba J A 1985 *Optimal seensor Locations for structural identification Proc. Of workshop on identification and control of flexible space structures* pp247-61

[13] Heo G, Wang M L and Satpathi D 1997 Optimal transduceer placement for health monitoring of long span bridge *J. Soil Dynamics and Earthquake Engineering.* **16** 495-502

[14] Li D S, Li H N and Fritzen C P 2009 A note on fast computation of effective independence through QR downdating for sensor placement *J. Mechanical System and Signal Processing.* **23** 1160-68

[15] Li D S, Li H N and Fritzen C P 2006 *On the physical significance of the norm based sensor placement method Proc. 3rd European workshop on structural health monitoring* (Granada Spain) 1542-50

[16] Horn R A and Johnson C R 1990 *Matrix analysis* (London: Cambridge university press)

Condition Assessment of Corrosion-damaged Bridge Girders Strengthened with Post-tensioned Composite Strips

Yail J. Kim[1], **Jae-Yoon Kang**[2], **Jong-Sup Park**[2], and **Woo-Tai Jung**[2]

[1] Department of Civil Engineering, University of Colorado Denver, Denver, CO 80217, USA
[2] Structural Engineering Research Division, SOC Research Institute, Korea Institute of Civil Engineering and Building Technology, Gyeonggi, Korea

woody @kict.re.kr

Abstract. This paper presents the condition evaluation of bridge girders upgraded using post-tensioned near-surface-mounted (NSM) carbon fiber reinforced polymer (CFRP) strips subjected to corrosion damage. Computational models are developed to predict the behavior of the girders over a 100-year service period. Dynamic analysis exhibits that damage localization takes place, in conjunction with various mode shapes. As the extent of damage rises, the effectiveness of the NSM CFRP increases and the flexural stiffness of the girders decreases.

1. Introduction

Among many detrimental factors affecting the performance of existing bridge structures, corrosion may be a critical attribute because of a reduction in steel area, concrete cover failure, and deteriorated concrete-steel interface [1,2]. On account of the pores in concrete, steel bars in a reinforced concrete member can be exposed to detrimental chemicals such as chlorides [3], which generate multiple stages in steel corrosion: initiation and progression. Recent research on corrosion damage for concrete structures involves deterministic and probabilistic investigations [4-6]. Structural rehabilitation with carbon fiber reinforced polymer (CFRP) composites is promising to upgrade the functionality of existing concrete bridges at affordable costs. Stallings et al. [7] repaired a decrepit bridge using CFRP sheets, which had been in service for over 40 years. The deflection of girders was measured under the operation of known-weight trucks, and was compared against the case without CFRP-strengthening. Badawi and Soudki [8] conducted a laboratory test to examine the behavior of corrosion-damaged concrete beams repaired with CFRP composites. The repair resulted in an increase in the beams' load-carrying capacity. Kasan et al. [9] proposed a strand-splicing repair method in tandem with CFRP sheets. An analytical model was constructed to evaluate the load rating of bridge girders.

A strengthening technique called near-surface-mounted (NSM) CFRP is emerging to enhance the performance of bridges. Along the tensile soffit of a concrete member, grooves are saw-cut to insert CFRP composites that serve as an additional reinforcement to the member. The CFRP-positioned grooves are filled with a structural epoxy. De Lorenzis and Teng [10] states that the NSM method is better than conventional externally bonded CFRP applications in terms of interfacial behavior between the concrete and CFRP. Alkhrdaji et al. [11] carried out a strengthening project for an in-situ bridge deck. The focus was on flexural capacity and failure characteristics. A predictive model was developed to verify the adequacy of test results. Dalfre and Barros [12] performed slab-strengthening with several NSM CFRP types, and studied crack progression and moment redistribution. NSM CFRP may be post-tensioned to upgrade the efficacy for repair or strengthening.

Although extensive endeavors have been made to investigate a number of technical aspects related to NSM CFRP, scarce information is available on the behavior of bridge members upgraded with post-tensioned NSM CFRP when subjected to corrosion damage. In this research, computational investigations are conducted to examine the time-dependent response of bridge girders strengthened with post-tensioned NSM CFRP strips covering a service period of 100 years. The study approach includes three-dimensional finite element modeling from static and dynamic perspectives, in accordance with the degree of chloride-based corrosion.

2. Model Development

2.1. Bridge girders and CFRP-strengthening

Three benchmark prestressed concrete bridge girders were designed and employed for a numerical study. The girders heights and length varied from 1,750 mm to 2,200 mm and from 25 m to 35 m, respectively, as shown in Fig. 1(a). The specified compressive strength of the girder concrete was f'_c = 40 MPa. The low-relaxation prestressing tendons (12.7 mm in diameter) were draped and pretensioned at about 65%f_{pu}, where f_{pu} is the tendon's ultimate capacity of 1,860 MPa with an elastic modulus of E_p = 190 GPa. Figure 1(b) illustrates a strengthening method for these girders, which needs the following steps:

a) a narrow groove is cut along the concrete substrate and CFRP is inserted
b) jacking apparatus is mounted to post-tension the CFRP
c) the post-tensioning force is permanently transferred to the girder
d) the groove is grouted to improve aesthetics

The CFRP tendons used had a cross sectional area of A_f = 700 mm^2 with an ultimate capacity of f_{fu} = 2,500 MPa and a modulus of E_f = 165 GPa.

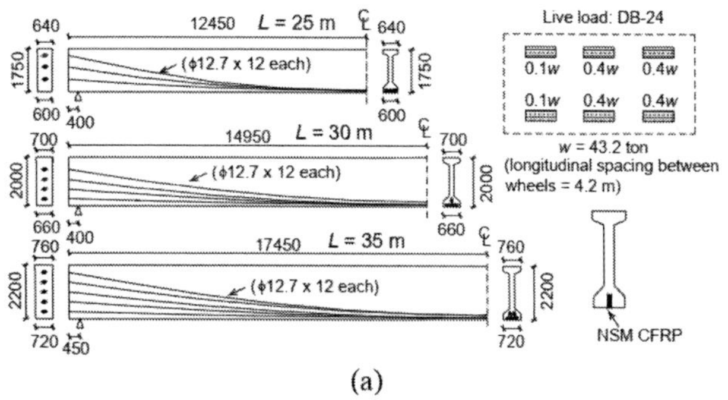

(a)

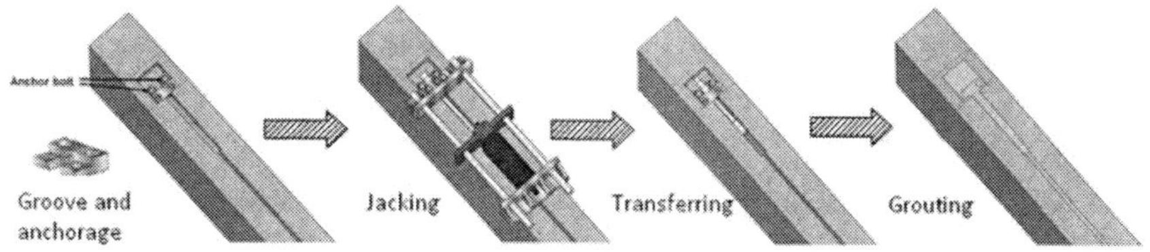

(b)

Figure 1. Benchmark girders and strengthening method: (a) prestressed concrete girders; (b) proposed strengthening method (patent numbers: 10-0653632, 10-1005347, and 10-1083626)

Figure 2. Three-dimensional finite element model
Computational model

2.1.1. Static model
A commercial finite element program was employed to predict the behavior of the strengthened and unstrengthened girders (Fig. 2). Shell and link elements represented the concrete girders and the prestressing tendons, respectively. The NSM CFRP was also modeled using the link elements. Perfect bond was assumed between the concrete and reinforcements, so that displacement compatibility was accomplished. The center of gravity of the prestressing tendons was determined and a single representative tendon was included in the model. Initial strains were applied to the CFRP and steel to simulate prestressing effects. Since the live load was in a service condition, material nonlinearity was not necessary in the girder models. Boundary conditions were involved by constraining nodes.

2.1.2. Dynamic model
The static model was expanded to investigate the dynamic characteristics of the bridge girders. Of interest was the analysis for mode shapes and damage localization. The mass of the individual materials was obtained from their densities: concrete = 2,400 kg/m^3, steel = 7,810 kg/m^3, and CFRP = 1,500 kg/m^3. The girders' modes were acquired by the Block Lanczos method. A sparse matrix solver was exploited to obtain dynamic results. Figure 3 reveals model validation against experimental frequencies reported by Saiidi et al. [13] and Xia and Brownjohn [14]. Figure 3(a) shows the first frequency of laboratory-scale concrete beams (100 mm in width by 125 mm in depth by 3,600 mm in length) consisting of four reinforcing steel bars (284 mm^2) and a prestressing strand (99 mm^2). The average error of margin between the experiment and model was 6%. The other validation shown in Fig. 3(b) was concerned with a beam-slab assembly having dimensions of 250 mm deep by 1,000 mm wide. Good agreement with an average discrepancy of 19% was observed. It is worth noting that structural members with post-tensioned NSM CFRP were not able to be modeled and assessed, because no experimental work had been conducted previously.

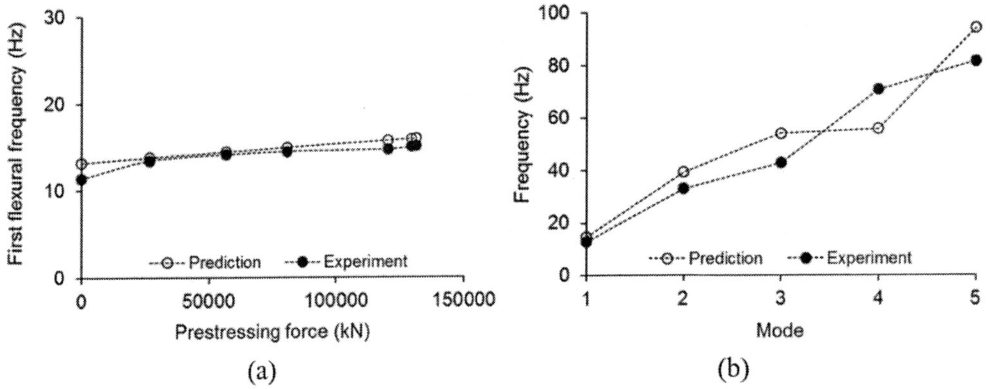

Figure 3. Model validation: (a) prestressed concrete beam; (b) beam-slab assembly

2.1.3. *Damage model*
Corrosion initiation was estimated by [15]:

$$t_i = \frac{C^2}{4D_c}\left[erf^{-1}\left(\frac{C_{cr}-C_0}{C_i-C_0}\right)\right]^{-2} \tag{1}$$

where t_i is the corrosion initiation time in years; C is the concrete cover in cm; D_c is the diffusion coefficient in cm^2/sec; *erf* is the Gauss error function; C_{cr} is the critical chloride concentration; C_0 is the equilibrium chloride concentration on the concrete surface; and C_i is the initial chloride concentration (C_i = 0). In accordance with others' research [15,16], the following can be taken: $D_c = 2.0\times 10^{-8}$ cm^2/s; C_{cr} = 0.4%, and C_0 = 1.6%. The loss of steel area was predicted using Faraday's law [17,18]:

$$\phi(t_p) = \phi_0 - r_{corr}t_p \text{ with } r_{corr} = C_c\frac{W_a i_{corr}}{n\rho} \tag{2}$$

$$i_{corr}(t_p) = \frac{37.8(1-w/c)^{-1.64}}{C}\alpha t_p^{\beta} \tag{3}$$

where $\phi(t_p)$ is the diameter of the prestressing steel strands in mm; t_p is the time of corrosion in year; ϕ_0 is the initial diameter of the strands in mm; r_{corr} is the corrosion rate in mm/year; C_c is a conversion factor (0.00327 for mm/year); W_a is the atomic weight (55.9 g/mol for steel); i_{corr} is the corrosion current density in μA/cm^2; n is the number of electrons involved (n = 2 for steel since $Fe \rightarrow Fe^2 + 2e$); ρ is the density (7 g/cm^3); $i_{corr}(t_p)$ is the corrosion density in μA/cm^2; w/c is the water cement ratio (w/c = 0.4 was used); and α and β are constants (α = 1 and β = 0 up to one year after corrosion initiation and α = 0.85 and β = -0.3 beyond the one year). The corrosion of the girders was assumed to be uniform without considering the contribution of the NSM CFRP. This assumption is justified by the fact that the narrow NSM CFRP does not contribute to inhibiting chloride ingress into the girder concrete. The long-term prestress loss of the CFRP was represented by [19]:

$$f_{fn}(\%) = 102.0 - 3.2Log(t) \tag{4}$$

where f_{fn} is the normalized CFRP stress at year t.

3. Results

3.1. Long-term behavior of reinforcement
The long-term behavior of the CFRP and steel is provided in Fig. 4. It was predicted that the steel diameter gradually decreased after the initiation of corrosion at 12.8 years. The relatively rapid reduction in steel diameter between 12.8 years and 20 years may be attributed to a rapid reaction in chloride diffusion when the corrosion process began. The variation of the post-tensioned CFRP was not significant for a 100-year service period, including a decrease of approximately 4%.

3.2. Static deflection
Figure 5(a) exhibits the development of girder deflections with service year. For conciseness, the deflections of the control and 100-year cases (L = 30 m) are only provided for the 30 m girders.

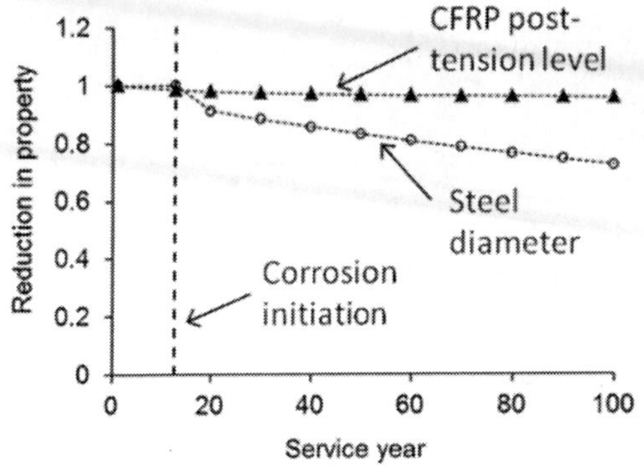

Figure 4. Long-term response predicted

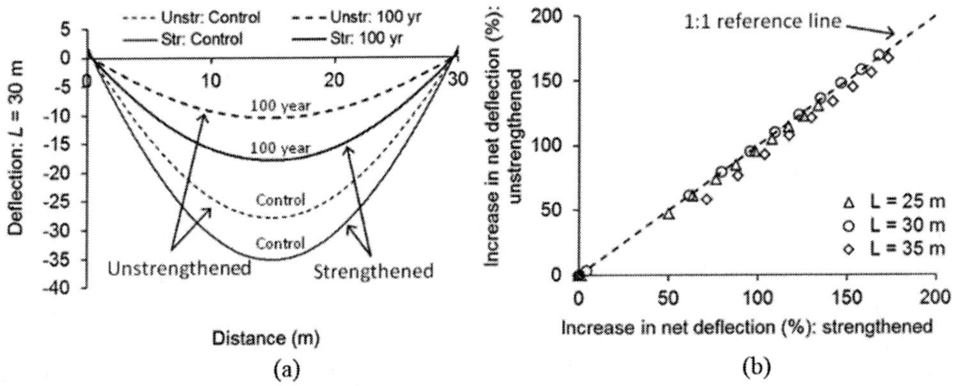

Figure 5. Static deflection: (a) development; (b) deflection due to live load

Upon strengthening, the midspan deflection of the control increased by 26% owing to post-tensioning. As the girders were corrosion-damaged, the upward deflection decreased. Irrespective of CFRP-strengthening, the girder deflection induced by live load did not change (Fig. 5(b)).

3.3. Dynamic mode

The five modes of the strengthened girder at 100 years (L = 35 m) are shown in Fig. 6. The first mode was in-plane single curvature bending (Fig. 6(a)), whereas the second mode was double curvature bending (Fig. 6(b)). The third mode was also single curvature bending (Fig. 6(c)), followed by lateral sway (Fig. 6(d)) and twisting (Fig. 6(e)).

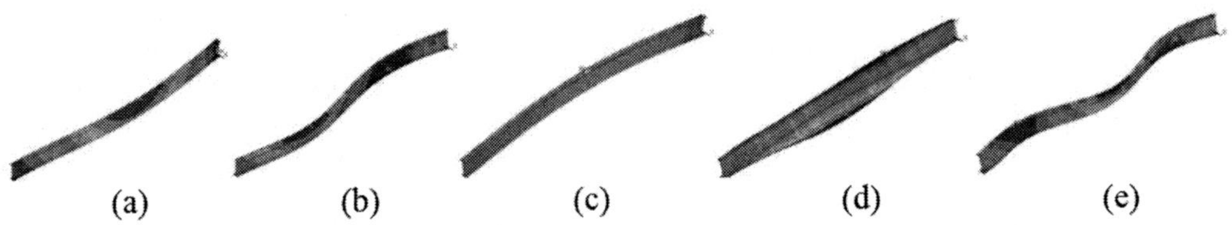

Figure 6. Mode shape of 35 m girder at 100 years:
(a) mode1; (b) mode2; (c) mode 3; (d) mode 4; (e) mode 5

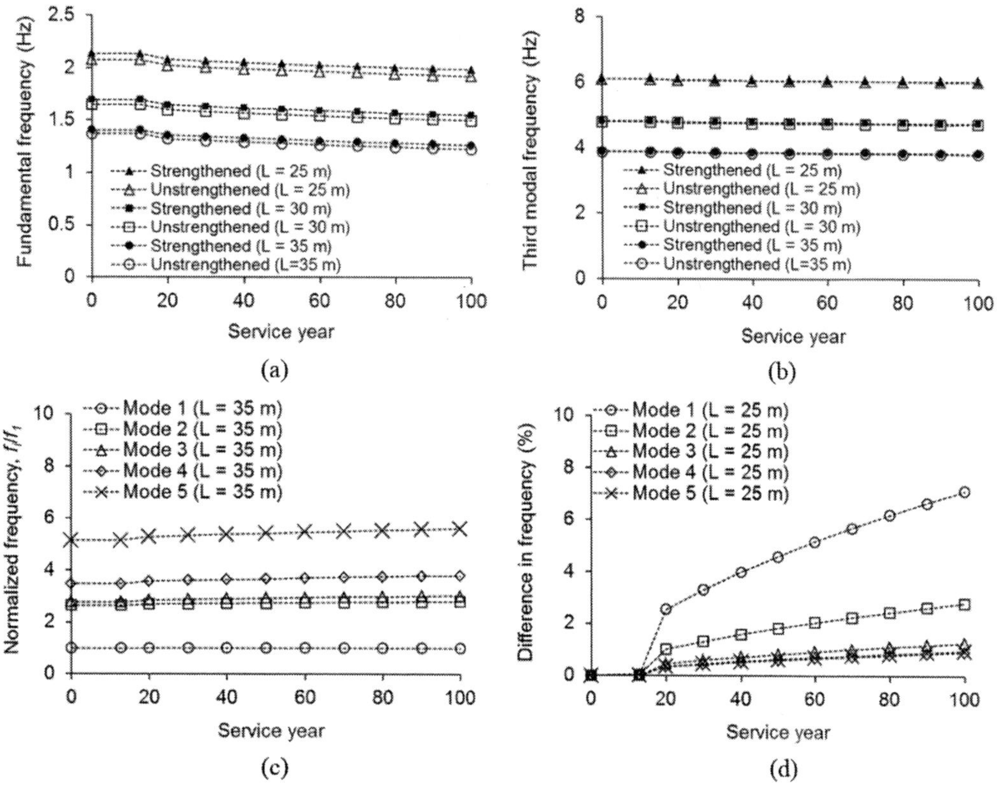

Figure 7. Frequency variation: (a) fundamental frequency; (b) third frequency (c) normalized frequency; (d) frequency difference

3.4. Frequency

Figure 7 shows the frequency variation of the strengthened girders. The fundamental frequencies of the strengthened girders increased in comparison with those of the unstrengthened ones owing to the CFRP's contribution (Fig. 7(a)). This tendency, however, was not apparent as the mode rose (Fig. 7(b)). The individual frequencies of the strengthened girders (L = 35 m) were normalized by the fundamental frequency and their time-dependent responses were assessed (Fig. 7(c)). At an intact state (service year = 0), the second and third modes were almost the same; however, the fourth and fifth modes were markedly higher. Frequency differences with and without corrosion damage (L = 25 m) are plotted in Fig. 7(d). The fundamental frequency of the girder showed an increase of 7% at 100 years, while the difference was reduced in higher modes.

4. Conclusions

The long-term response of prestressed concrete bridge girders strengthened with post-tensioned NSM CFRP composites was discussed when subjected to chloride-induced corrosion damage. The length of the girders varied from 25 m to 35 m. To predict the behavior of the girders over a 100-year service period, computational models were developed. The following is concluded:

- When corrosion damage initiated, the diameter of the steel tendons rapidly decreased. The time-dependent CFRP properties were relatively stable, including a 5% reduction in post-tensioning force for 100 years.

IOP Conf. Series: Journal of Physics: Conf. Series **842** (2017) 012031 doi:10.1088/1742-6596/842/1/012031

- The efficacy of the NSM CFRP became more pronounced with an increase in service year. The girders' fundamental frequency was influenced by corrosion damage within a variation range of 7%.

References

[1] Clifron J R 1993 Predicting the service life of concrete *ACI Mat. J.* 90 611-617.

[2] O'Connor A J, Sheils E, Breysse D, and Schoefs F 2013 Markovian bridge maintenance planning incorporating corrosion initiation and nonlinear deterioration, *J. Bridge Eng.*, 18(3) 189-199.

[3] Dyer T 2014 *Concrete durability*, CRC Press, Boca Raton, FL.

[4] Vu K, Stewart M G, and Mullard J 2005 Corrosion-induced cracking: experimental data and predictive models *ACI Struct. J.* 102(5) 719-726.

[5] Du Y G, Chan A H C, Clark L A, Wang X T, Gurkalo F, and Bartos S 2013 Finite element analysis of cracking and delamination of concrete beam due to steel corrosion, *Eng. Struct.* 56 8-21.

[6] Li S, Xu Y, Li H, and Guan X 2014 Uniform and pitting corrosion modeling for high-strength bridge wires *J. Bridge Eng.* 19(7) 04014025

[7] Stallings J M, Tedesco J W, El-Mihilmy M, and McCauley M 2000. Field performance of FRP bridge repairs *J. Bridge Eng* 5(2) 107-113.

[8] Badawi M and Soudki K 2010 CFRP repair of RC beams with shear-span and full-span corrosions *J. Compos. Constr.* 14(3) 323-335.

[9] Kasan J L, Harries K A, Miller R, and Brinkman R J 2014. Repair of prestressed-concrete girders combining internal strand splicing and externally bonded CFRP techniques *J. Bridge Eng.* 19(2) 200-209.

[10] De Lorenzis L and Teng J G 2007 Near-surface mounted FRP reinforcement: an emerging technique for strengthening structures *Compos. Part B* 38 119-143.

[11] Alkhrdaji T, Nanni A, and Chen G 1999. *Destructive and non-destructive testing of bridge J857 Phelps County Missouri* Technical Report CIES 99-08A.

[12] Dalfre G M and Barros J A O 2013. NSM technique to increase the load carrying capacity of continuous RS slabs *Eng. Struct.* 56 137-153.

[13] Saiidi M, Douglas B, and Feng S 1994 Prestress force effect on vibration frequency of concrete bridges *J. Struct. Eng.* 120(7) 2233-2241.

[14] Xia P -Q and Brownjohn M W 2004. Bridge structural condition assessment using systematically validated finite element model *J. Bridge Eng.* 9(5) 418-423.

[15] Thoft-Christensen P, Jensen F M, Middleton C R, and Blackmore A 1996. Assessment of the reliability of concrete slab bridges *7th IFIP WG 7.5 Working Conf. Boulder CO* 1-8.

[16] Stewart M G and Rosowsky D V 1998 Time-dependent reliability of deteriorating reinforced concrete bridge decks *Struc. Safe.* 20 91-109.

[17] Vu K A T and Stewart M G 2000 Structural reliability of concrete bridges including improved chloride-induced corrosion models *Struc. Safe.* 22 313-333.

[18] Ahmad Z 2006 *Principles of corrosion engineering and corrosion control* Butterworth-Heinemann Oxford UK.

[19] El-Hacha R, Wight R G, and Green M F 2003 Innovative system for prestressing fiber-reinforced polymer sheets *ACI Struc. J.* 100(3) 305-313.

Damage Detection Sensitivity of a Vehicle-based Bridge Health Monitoring System

Ayaho Miyamoto[1], Akito Yabe[2] and Válter J. G. Lúcio[3]

[1] Emeritus Professor, Dr. Eng., Yamaguchi University, Ube, Japan
[2] Dr. Eng., Sustainable Solutions Dept., Kozo Keikaku Engineering Inc, Tokyo, Japan
[3] Professor, PhD, Faculty of Sci. and Eng., NOVA de Lisboa Univ., Lisbon, Portugal

E-mail: miya818@yamaguchi-u.ac.jp

Abstract. As one solution to the problem for condition assessment of existing short and medium span (10-30m) reinforced/prestressed concrete bridges, a new monitoring method using a public bus as part of a public transit system (called "Bus monitoring system") was proposed, along with safety indices, namely, "characteristic deflection", which is relatively free from the influence of dynamic disturbances due to such factors as the roughness of the road surface, and a structural anomaly parameter. In this study, to evaluate the practicality of the newly developed bus monitoring system, it has been field-tested over a period of about four years by using an in-service fixed-route bus operating on a bus route in the city of Ube, Yamaguchi Prefecture, Japan. In here, although there are some useful monitoring methods for short and medium span bridges based on the qualitative or quantitative information, the sensitivity of damage detection was newly discussed for safety assessment based on long term health monitoring data. The verification results thus obtained are also described in this paper, and also evaluates the sensitivity of the "characteristic deflection", which is a bridge (health) condition indicator used by the bus monitoring system, in damage detection. Sensitivity of "characteristic deflection" is verified by introducing artificial damage into a bridge that has ended its service life and is awaiting removal. Furthermore, the sensitivity of "characteristic deflection" is verified by 3D FEM analysis.

1. Introduction

In the field of bridge management engineering, a great deal of decision making often depends on the assessment and experience of the domain experts in related fields, such as professional experience, knowledge on bridge management, etc. Then an important parameter in management for existing bridges is the remaining life assessment that is crucial as a kernel of the systematization for bridge maintenance throughout service life including economic analyses referring to initial cost and assessment technologies. In bridge structures there are many different unforeseen conditions because bridges are larger and often need to serve longer more than 100 years, compared with the other products such as electrical, mechanical and systems engineering fields. And they are subjected to diverse types of deterioration mechanism such as corrosion, fatigue, carbonation, alkali aggregate reactions, etc. Therefore it is probable to have some risk for not fulfilling the complete standards of safety so some failure probability is possible. Structural aging, environmental conditions, and reuse are examples of circumstances that could affect the reliability and the life of a bridge structure. Engineers have been visually inspecting, monitoring and proof testing bridges for Centuries. However, presently health and performance are described based on subjective measures that are not universal. In addition, defects, deterioration and damage are not discovered until it is possible to visually observe the signs they exhibit at which time these would have taken their toll on health. These shortcomings impact the timeliness, effectiveness and the reliability in any management decision irrespective of any sophistication in the management process. Moreover, even experienced engineers may find visual signs of defects, deterioration and damage and still not be able to diagnose the causative mechanisms, or their impact on the reliability of the system and global health. The global health of an entire bridge as a system, inclusive of the performance criteria corresponding to every one of the limit-states is actually what is needed for effective management decisions. There are needs of periodic inspections to detect deterioration resulting from normal operation and environmental attack or inspections following extreme events, such as strong-motion earthquakes or hurricanes. To quantify these system performance measures requires some means to monitor and evaluate the integrity of civil structures while in service [1-8]. Then, since the necessity of developing a practical health monitoring system has been point out for detecting deterioration phenomenon as early as possible for bridge maintenance.

The authors have been developing a long term health monitoring method using a public bus as part of a public transit system (called "Bus monitoring system") [9-11] to be applied mainly to short and medium span bridges. As a safety index from the system, "characteristic deflection" which is relatively free from the influence of dynamic disturbances due to such factors as the roughness of the road surface, and a structural anomaly parameter was proposed. The advantages of bus monitoring system are as follows: a) If a large vehicle about 10 m long is used for measurement, it is highly likely that when the vehicle crosses a short and medium span bridge, that is the only vehicle in the same lane on the bridge, b) If a short and medium span bridge, which is has relatively high flexural stiffness, is to be vibrated, it is necessary to use a relatively heavy vehicle, c) If a fixed-route bus is used as a source of bridge excitation, it is easy to reproduce measuring conditions such as the time of passage, route, frequency and velocity, d) Since a fixed-route bus equipped with a sensor makes the rounds, it is possible to monitor main short and medium span bridges in a particular area on a regular basis. As a result, substantial cost reduction can be achieved because there is no need to install sensors to all bridges to be monitored, e) The electric power for the measuring instruments used can be supplied by the power supply of the bus.

This paper described not only the validation results obtained from the long-term monitoring and discusses the usefulness of the system but also the problems of the conventional observation method based on the "characteristic deflection", which is a bridge condition indicator that makes possible efficient detection of structural anomalies of the bridge being monitored, are identified, and a new observation method that enhances the damage detection sensitivity of the system is evaluated. This study also examines the influence of artificial damage (guardrail removal) on the characteristic deflection to evaluate the sensitivity of the system in detecting damage given to the field test bridge. The sensitivity of the characteristic deflection is verified by 3D FEM analysis. Finally, various study results as mentioned above are put together to systematically discuss the practical scope of application, damage detection accuracy and remaining problems of the system.

2. Basic Concept of the Bus Monitoring System

The bus monitoring system has been developed by the method of monitoring short and medium span bridges by using fixed-route buses operated as part of public transport systems [9-11]. **Fig. 1** illustrates a basic concept and damage detection flow of the bus monitoring system. Main reasons why the use of fixed-route buses, which are large and heavy vehicles, was considered are as follows: (1) If a large vehicle about 10 m in length is used for on-the-move measurement on a short or medium span bridge, that vehicle is likely to be the only vehicle passing over the bridge at point in time, (2) In order to make a short or medium span bridge, whose stiffness is relatively high, it is necessary to use a reasonably heavyweight vehicle, (3) Large vehicles used to vibrate bridges make it easier to reproduce

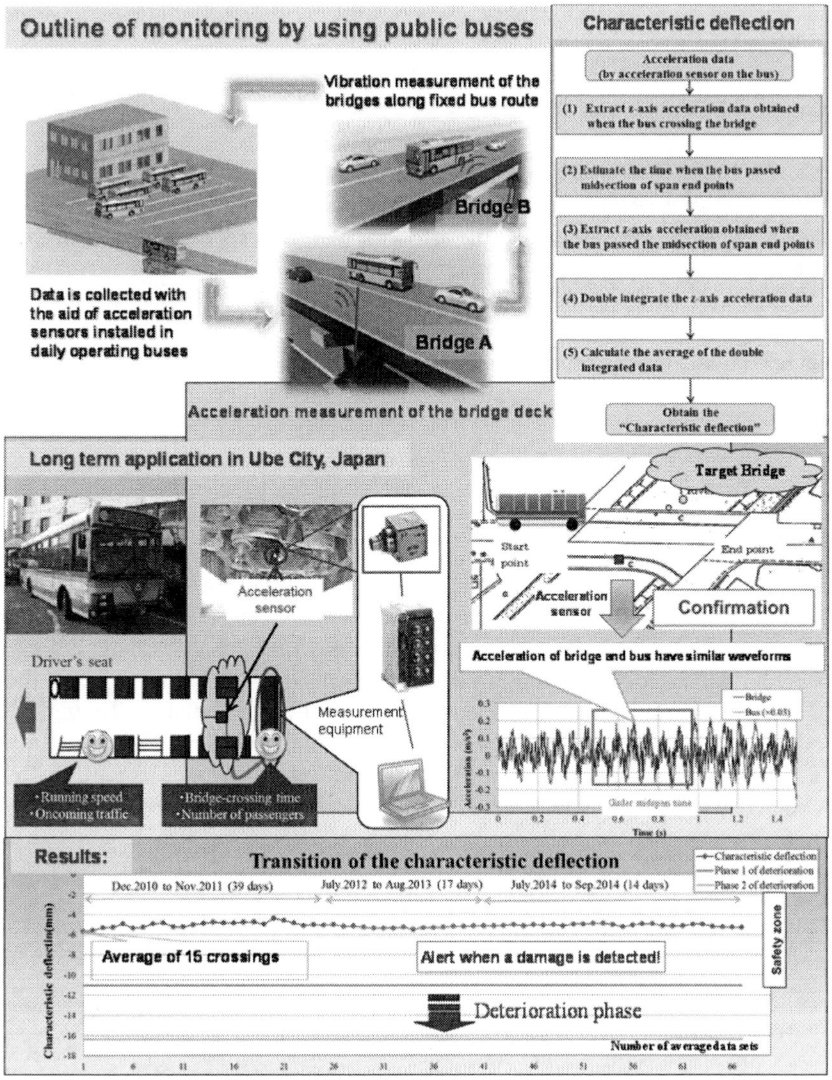

Figure 1. Basic concept and damage detection flow of the bus monitoring system

measuring conditions such as travel time, route, frequency of passages and vehicle speed, (4) By using a fixed-route bus equipped with acceleration sensors, main short and medium span bridges in the area of interest can be monitored on a regular basis. By so doing, monitoring cost would be substantially less than in the case where sensors were installed at all bridges, and (5) Power for the measuring devices can be supplied from the power supply available in the bus.

As shown in **Fig. 1**, the vertical displacement of the bridge and the bus under its rear wheel springs is an estimated deflection obtainable by integrating the vertical acceleration waveform twice. The displacement can be expressed, as the sum of (1) the static displacement dependent on bridge stiffness and bus weight, (2) a non-steady-state vibration component having a stochastic nature characterized as a Gaussian process attributable to a rough road surface and having a mean value of 0, (3) vibration components governed by the equation of motion such as the bridge and the vehicle, and (4) external disturbance factors due to differences in the operating conditions of the bus and differences in data extraction and processing methods. For the purpose of this study, the characteristic deflection [9],[10] which is relatively free from the influence of dynamic disturbances due to the roughness of road surface, etc., is defined as an indicator that may be useful in efficiently detecting a structural anomaly of existing bridges [11].

3. Examination and Discussion of Damage Detection Sensitivity

3.1 Examination by an Existing Bridge with Artificial Damage

To evaluate the sensitivity of the characteristic deflection used by the bus monitoring system as a serious damage indicator, the influence of artificial damage (bridge guardrail removal) on the characteristic deflection was evaluated by using a decommissioned bridge (a real old bridge to be removed shortly). This chapter deals with the field test results thus obtained. A simulation analysis using an analysis model allowing for the coupling between the bus and the bridge is also performed to calculate changes in the characteristic deflection of the artificially damaged bridge, and damage detection sensitivity is compared and evaluated from the analytical point of view.

3.1.1 Overview of the test

The field test was carried out by using a 72-year-old (as of the time of the study) reinforced concrete bridge to be dismantled and removed shortly ("Sakae(SK) Bridge" [12]; see **Fig. 2**). In view of factors such as seasonal (temperature) changes and the presence or absence of bridge guardrails, it was decided to conduct measurement using the bus monitoring system a total of four times, namely, in September 2012 and in January, February and March 2013. As in the case of the long-term field test mentioned earlier, the vertical (z-axis) acceleration response and "characteristic deflection" occurring when the municipal bus (vehicle) running at a constant speed crosses the bridge in the in-bound and out-bound lanes (15 in-bound runs and 15 out-bound runs) are calculated. Taking advantage of the absence of other road traffic (ideal condition) because the bridge was already in the process of demotion and removal, the measurements were conducted under various conditions by varying such conditions as vehicle weight and vehicle speed (constant speed).

3.1.2 Types and characteristics of vehicles used for measurement

The vehicles used for the measurement purpose are Ube-city's sightseeing bus having a gross weight of about 15 tf and a mini vehicle having a gross weight of about 1.3 tf. As in the long-term field test mentioned earlier, an acceleration sensor was installed at a similar position (under the rear wheel spring) of each vehicle for continuous measurement. Besides this sensor, three other acceleration sensors of comparable performance were also installed [on the right side, at the center (normal location) and on the left side under the rear wheel spring] to evaluate the influence of sensor locations. **Fig. 3(a)** and **(b)** show the general views of the two types of vehicles used and the acceleration sensor locations. **Table 1(a)** and **(b)** show the specifications of the vehicles used and data on the performance and other details of the sensors.

Figure 2. General view of SK-Bridge in service

3.1.3 Outline of the bridge

IOP Conf. Series: Journal of Physics: Conf. Series **842** (2017) 012032 doi:10.1088/1742-6596/842/1/012032

The SK-Bridge used for the field measurement is a 168.3-meter-long, 11.0-meter-wide eight-span simple cantilever reinforced-concrete T-girder bridge completed in 1941 (managed by the Ministry of Land, Infrastructure, Transport and Tourism). For reconstruction, the bridge was demolished and removed by stages over a period of two years from fiscal year 2012. **Table 2** and **Fig. 4** show the structural specifications and pre-removal configuration and dimensions of the SK-Bridge and the bridge section where the field test was conducted. Since there was no other road traffic on the bridge at the time of measurement, only seasonal (temperature) changes were taken into consideration as an external disturbance factor affecting characteristic deflection in evaluating the influence of the presence or absence of bridge guardrails (regarded as artificial damage). For the purpose of that evaluation, a field test was carried out on four occasions (once in autumn, twice in winter and once in

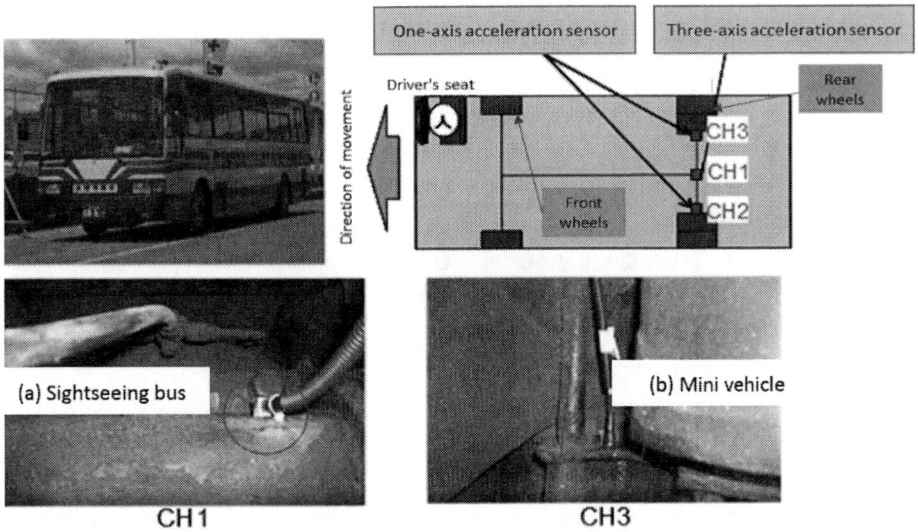

(a) Sightseeing bus

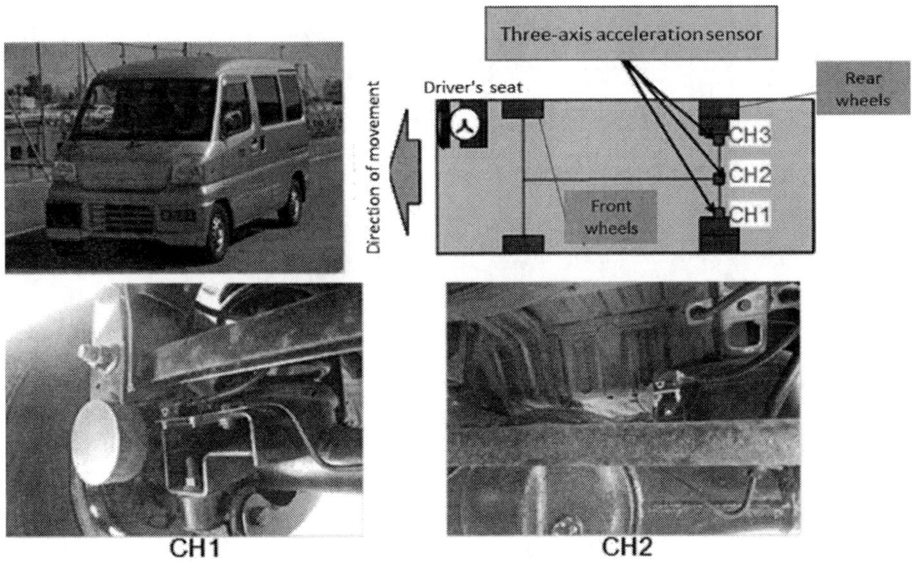

(b) Mini vehicle

Figure 3. Measurement vehicles and acceleration sensor locations(Sightseeing bus & Mini-van)

Table 1. Specifications of measurement vehicles and acceleration sensors

(a) Measurement vehicles

Items	Bus	Mini vehicle
Riding capacity	57 persons	2(4)
Length	1,194 cm	339 cm
Width	249 cm	147 cm
Height	330 cm	189 cm
Vehicle weight	11,810 kg	860 kg
Gross weight	14,945 kg	1,320 kg
Front axle weight	4,030 kg	440 kg
Rear axle weight	7,780 kg	420 kg

(b) Acceleration sensors

Items	Model No.	Serial No.	Channel	Axis	Sensitivity	Unit	Location
Bus	SA11ZSC-TI	5692	CH1	Z	0.99	mV/ms^{-2}	Center
	M353B16	106161	CH2	Z	1.058	mV/ms^{-2}	Right
	M353B12	106410	CH3	Z	0.483	mV/ms^{-2}	Left
Mini vehicle	2422	1236	CH1	Z	794.0	mV/G	Center
		1237	CH2	Z	795.1	mV/G	Right
		1238	CH3	Z	797.3	mV/G	Left

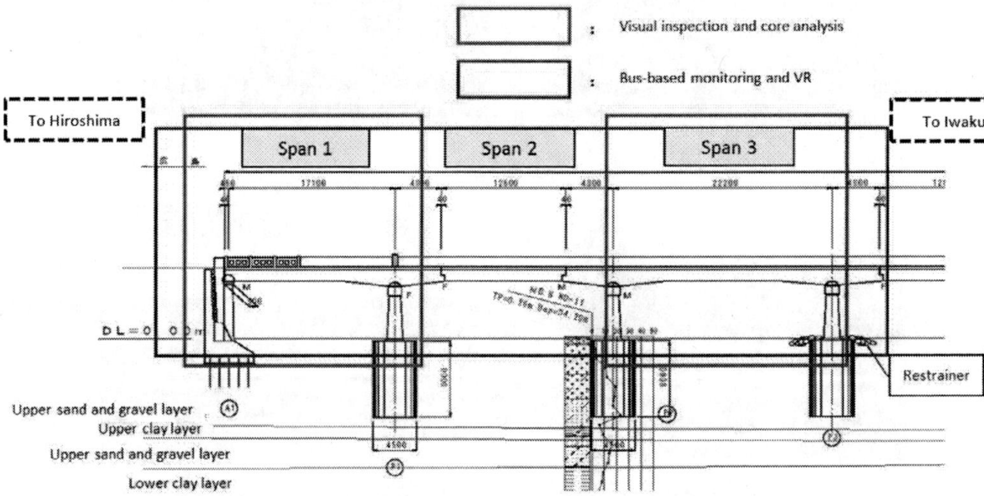

Figure 4. Configuration and dimensions of SK-Bridge and spans used for field testing

spring). For the evaluation of changes in the characteristic deflection before and after bridge guardrail removal, the data obtained in January and February 2013 were compared. **Fig. 5** shows the bridge before and after the guardrail removal.

3.1.4 Test results and discussion

This section puts together and discusses the results obtained from the field test conducted by using the decommissioned bridge as mentioned above.

Table 2. Structural specifications of field test bridge (SK-Bridge)

Length	L= 168.29 m
Width	W= 11.0m(two lanes + sidewalks)
	W= 2.5 m(sidewalks)
Span	Eight spans
Super structure	Cantilever reinforced concrete(RC) T-girder bridge

1) Differences in sensor location

Changes in characteristic deflection depending on the location of the acceleration sensor installed under the rear wheel spring were examined. **Fig. 6** compares the acceleration response waveforms obtained from the different sensor locations under the rear wheel spring of the sightseeing bus(see **Table 1(a)**). As shown in **Fig. 6**, acceleration response can be measured at any sensor location with accuracy good enough to make characteristic deflection calculation possible. From this, it can be said that an acceleration sensor needed by the bus monitoring system to calculate the characteristic deflection as a serious damage indicator may be installed at any location under the rear wheel spring.

2) Differences in axle weight due to vehicle type

To evaluate the effects of vehicle axle weight on the characteristic deflection, the characteristic deflection was calculated from the acceleration responses of the sightseeing bus and the mini vehicle, which have a gross weight difference of more than 10 tf, measured while they were moving. As examples, **Figs. 7(a)** to **(d)** show an under-rear-wheel-spring (center) acceleration response waveform of the sightseeing bus and an under-rear-wheel-spring (right) acceleration response waveform of the mini vehicle, along with acceleration response waveforms of the sightseeing bus and the mini vehicle obtained from the acceleration sensor installed to the bridge when they crossed the bridge. Comparison of **Figs. 7(a)** and **(b)** reveals that the response waveform obtained from the acceleration sensor installed to the mini vehicle tends to be larger, in both the noise portion and the signal portion, than the waveform obtained from the sightseeing bus. The reason for this is thought to be that since the mini vehicle is lighter than the sightseeing bus (smaller axle weight), the former is more easily affected by the roughness of the road surface when in motion so that vertical vibration of the mini vehicle becomes greater. Comparison of **Figs. 7(c)** and **(d)** reveals that the response waveform obtained from the bridge acceleration sensor when the mini vehicle cross the bridge is significantly smaller than the response waveform recorded when the sightseeing bus crossed the bridge. From this, it is presumed that the mini vehicle, whose gross weight is small, is not heavy enough to cause the bridge to vibrate and that external disturbance factors such as road surface roughness prevent the under-spring structure of the vehicle and the bridge from vibrating together (i.e. the similarity between the vehicle and the bridge does not hold). As a result, it is presumed, the requirements of the bus monitoring system are not met (i.e. not suitable for the calculation of the characteristic deflection).

3) Comparison of characteristic deflections before and after guardrail removal

In this section, 15 calculated values of the characteristic deflection derived from the measurement results obtained at four different times of year by using the sightseeing bus having a gross weight of about 15 tf are used to examine the effects of the artificial damage (guardrail removal) introduced into the bridge ("SK-Bridge"). As an example, **Fig. 8** shows the characteristic deflection calculation results for Span 2(cantilever part) and Span 3 (see **Fig. 4**) of the bridge obtained by using the sightseeing bus (moving at 40 km/h). The concrete guardrail removal was carried out between the second measurement (January 2013) and the third measurement (February 2013). The characteristic deflection

(a) Before guardrail removal

(b) After guardrail removal

Figure 5. Bridge before and after guardrail removal

values obtained before and after the guardrail removal are compared below.

First, turning attention to Span 3 (see **Fig. 4**), we notice that the two characteristic deflection values (−2.97, −2.89) obtained after the guardrail removal are about 10% larger than the values (−2.67, −2.61) obtained before the guardrail removal. This is thought to be because the decrease in the flexural stiffness of the bridge resulted in an increase in the characteristic deflection. In Span 2, which includes a cantilever structure, however, the characteristic deflection value (−2.24) obtained after the fourth measurement is slightly smaller than the value obtained before the guardrail removal. Since Span 2 includes a cantilever structure, it is thought likely that the cantilever structure somehow influenced the

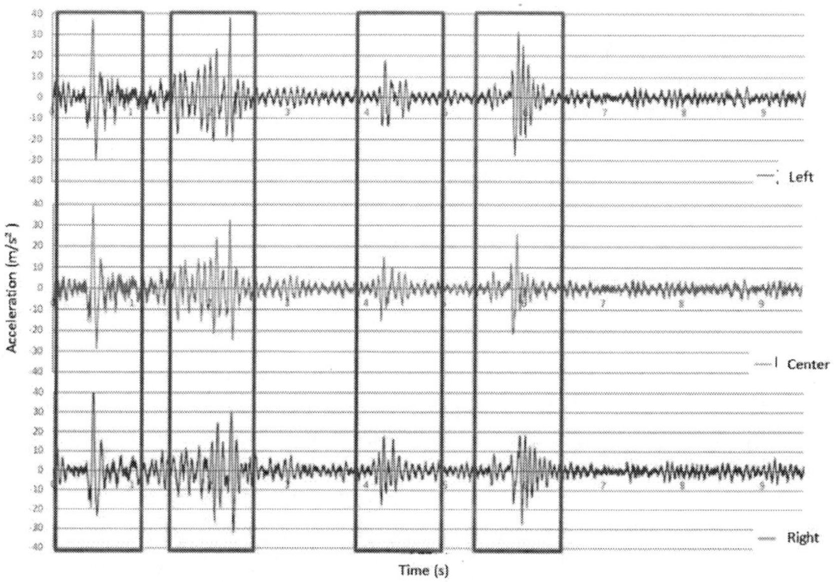

Figure 6. Comparison of acceleration response waveforms obtained from sensors installed at different under-rear-wheel-spring locations of the sightseeing bus

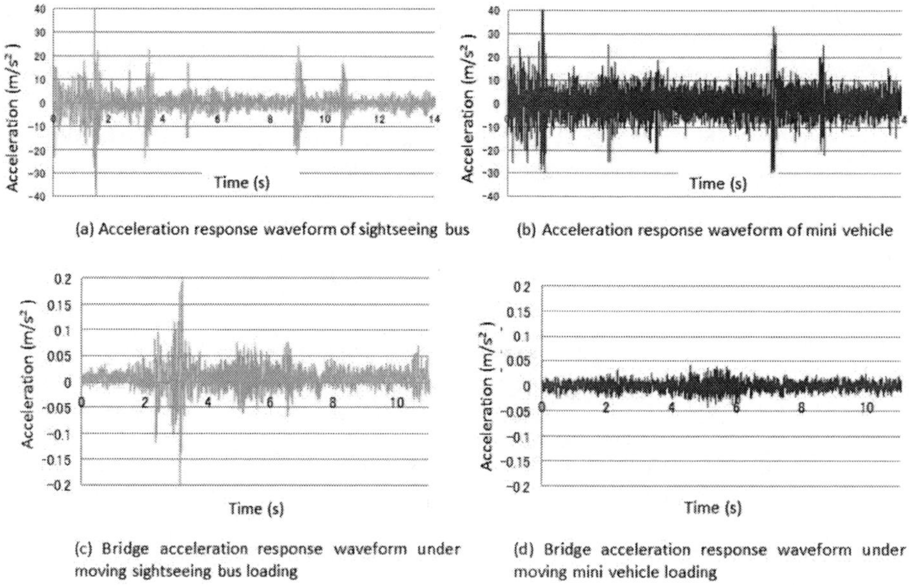

(a) Acceleration response waveform of sightseeing bus

(b) Acceleration response waveform of mini vehicle

(c) Bridge acceleration response waveform under moving sightseeing bus loading

(d) Bridge acceleration response waveform under moving mini vehicle loading

Figure 7. Effects of vehicle type on vehicle and bridge acceleration response waveforms characteristic deflection. When calculating characteristic deflection of a bridge with a cantilever structure, therefore, it is necessary to perform a simulation analysis and compare the calculated values with the analytical results.

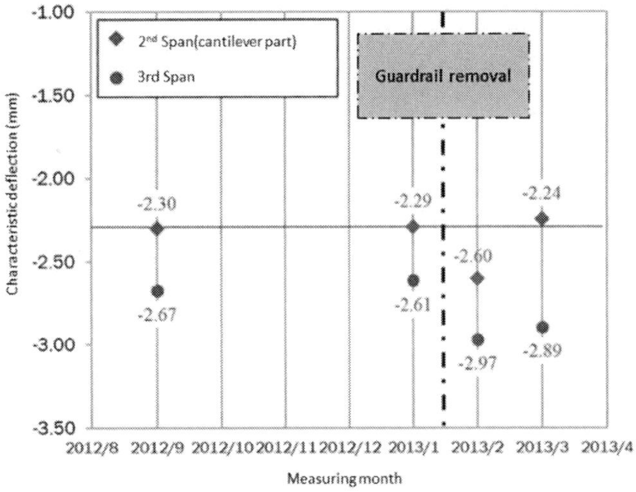

Figure 8. Changes in characteristic deflection due to railing removal

3.2 Examination by Model-based Simulation Analysis

This section deals with a model-based simulation analysis of the bridge measurement results obtained from the bus monitoring system performed to analytically identify the bridge behavior before and after the guardrail removal.

In the analysis, a static FEM analysis [13] was conducted of the simple girder structure excluding Span 2 on the Hiroshima side to calculate the maximum deflection ratio and evaluate the influence of the bridge guardrails on girder stiffness. The analysis, by using the three-dimensional finite analysis method, calculates and compares the midspan deformation in the cases where a midspan vertical downward unit concentrated load (1 kN) is applied. Static analyses were conducted of the four cases listed below:

Case 1: With guardrails, no damage
Case 2: Without guardrails, no damage
Case 3: With guardrails, damaged
Case 4: Without guardrails, damaged

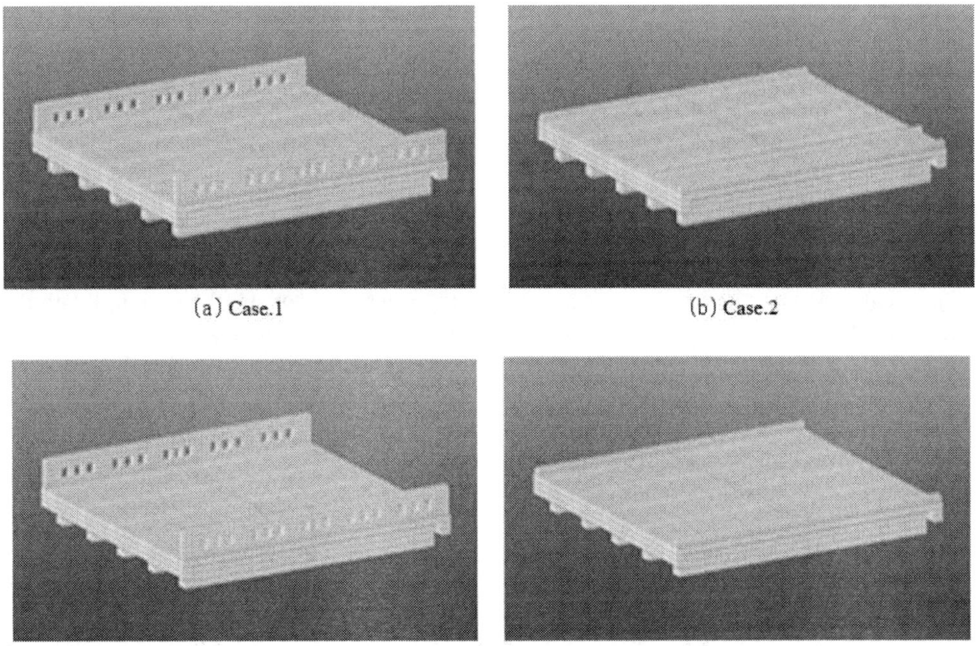

(a) Case.1 (b) Case.2

(c) Case.3 (d) Case.4

Figure 9. Model used for analysis

The analysis used MIDAS-GEN (MIDAS-IT Co.).

3.2.1 Analysis model and analysis conditions

Fig. 9 shows how the analysis model used looks. **Fig. 10** shows an enlarged view of an assumed damage region in the "damaged" case shown in **Fig. 9**. The static analysis conditions are shown in **Table 3**. An eigenvalue analysis showed that the first mode natural frequency is 16 Hz. The first mode frequency determined from theoretical solutions and other study findings [14],[15] is also 16 Hz. It is therefore thought that the model mentioned above is suitable for use in this study.

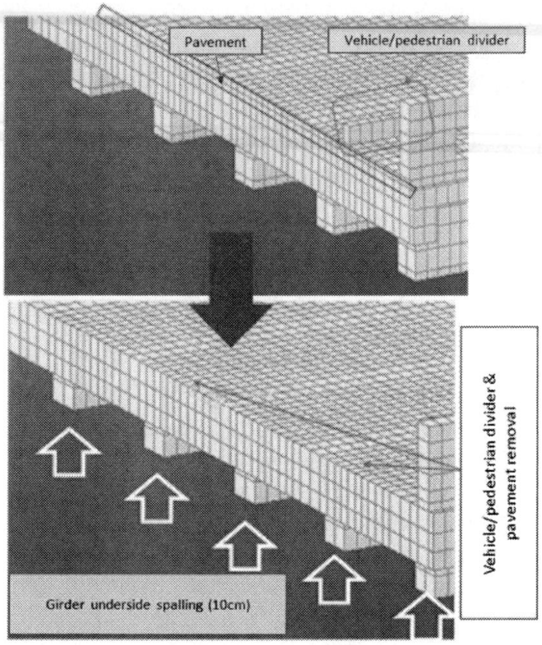

Figure 10. Assumed damage

3.2.2 Comparison between analytical results and characteristic deflection values

Table 4 shows the obtained analytical results. The change ratio in the "with guardrails and no damage" case under the assumed conditions is 1.04, indicating that the change amounts only to about 4%. The change ratio in the case where general delamination of the underside of the girder is assumed is about 1.05, indicating a change of about 5%. The geometrical moment of inertia was also calculated in both the "with guardrails" and "without guardrails" cases. The second moments of area thus obtained are shown in **Table 5**. As shown in **Table 5**, the geometrical moment of inertia shows a decrease of about 40% as a result of the guardrail removal. This translates to a deflection ratio of 1.64.

Table 3. Analysis conditions

Items	Conditions
Analysis method	Static elasticity analysis using 3D solid elements
Material constants	JIS (RC) Fc24 or equivalent
	Modulus of elasticity : 2.2668E7
	Poisson's ratio : 0.2
	Unit weight : 24.0
Load	Midspan vertical downward unit concentrated load (1 kN)
Boundary conditions	Simple beam; restrained at bottom of girder end

The results of the 3D FEM analysis mentioned earlier show that the change ratio ranges from 1.04 to 1.05(see **Table 4**), and the effects of the guardrail removal are very small compared with the values obtained from the geometrical moment of inertia. This is thought to indicate that stress transmission paths extend two-dimensionally under the midspan loading due to the moving bus. This does not agree with the result of stiffness evaluation based on the assumption, for cross-sectional calculation, of Navier's hypothesis (of plane sections remaining plane) and bending in the bridge axis direction.

Comparison with the characteristic deflection reveals that the calculated values of the characteristic deflection for Span 2 show decreases after the guardrail removal.

Table 4. Analysis results (midspan deflection)

Case	Deflection (≠ characteristic deflection)	Ratio to "with guardrail" case
1	-2.89×10^{-6} m	
2	-3.00×10^{-6} m	1.04 (4% increase)
3	-3.29×10^{-6} m	
4	-3.47×10^{-6} m	1.05 (5% increase)

3.2.3 Discussions

According to the characteristic deflection calculation

results for Span 3, the characteristic deflection values obtained after the guardrail removal (average: 2.93 mm) indicate a decrease of about 10% compared with the characteristic deflection values (average: 2.64 mm) obtained before the guardrail removal. In view of the fact that the change ratio obtained through the static analysis in the case assuming general delamination of the underside of the girder was about 5%, it can be said that the characteristic deflection is reasonably sensitive to the guardrail removal. For Span 2, which includes a cantilever structure, however, the characteristic deflection after the guardrail removal showed a slight increase from the value obtained before the guardrail removal. This result is not consistent with the results of 3D FEM analysis, either. One likely reason for this is that the model used for the analysis consists of a simple span without a cantilever structure, while the calculated values of the characteristic deflection have been influenced by the cantilever structure. Changes in measurement results due to seasonal factors are possible, but that seems unlikely in view of the results for Span 3. Span 2 requires further study, and it is also necessary to examine other factors such as the possible influence of the type of bridge structure on the characteristic deflection.

These results have shown, at least at this stage of study in which the usefulness and validity of the characteristic deflection as an evaluation indicator for use by the bus monitoring system is being verified, that the method of using a three-dimensional model for comparison and evaluation is useful when measuring a bridge in which stress transmission paths extend two-dimensionally.

Table 5. Geometrical moment of inertia calculation results

Items	With guardrails	Without guardrails	Decrease ratio
Cross-sectional area (m^2)	6,650	5,948	10.5%
Neutral axis location (m)	0.974	0.877	
Geometrical moment of inertia (m^4)	1.612	0.980	39.2%

4. Concluding remarks

This paper has discussed the results of damage detection sensitivity of the bus monitoring system for short and medium span bridges, that is the characteristic deflection used as an evaluation indicator. Damage detection performance of the characteristic deflection has been verified systematically by introducing artificial damage (guardrail removal) into a decommissioned bridge that was in the process of demolition. Main results of this study are summarized as follows:

1) In the field test conducted by using a bridge that was decommissioned and was undergoing demolition, the influence (sensitivity) of artificial damage (guardrail removal) on the characteristic deflection was evaluated. The results obtained for Span 3 of the bridge confirmed that the characteristic deflection is reasonably sensitive to a decrease in the flexural stiffness of the entire bridge. In the case of Span 2, which includes a cantilever structure, however, the characteristic deflection before the guardrail removal was greater than the characteristic deflection after the guardrail removal. This result is not consistent with the 3D FEM analysis results, either. It is likely that the calculated values of the characteristic deflection have been influenced by the cantilever structure. Further study is needed, therefore, particularly on the influence of the type of bridge structure on the characteristic deflection.
2) At present, the usefulness and validity of the characteristic deflection as an evaluation indicator are being evaluated and verified. It can be concluded, at least at this stage, that the method of using a three-dimensional model for comparison and evaluation is useful when measuring a bridge in which stress transmission paths extend two-dimensionally.

Finally, the long-term field test of the bus monitoring system has produced useful results associated with the practical application of the system although a number of problems still remain to be solved. Important challenges for the bus monitoring system include the automation and rationalization of measurement. The structural health evaluation of bridges can be made simpler and more efficient by analyzing and solving the problems identified as a result of this study.

Acknowledgments
The authors would like to express our sincere thanks to Dr. Hisao Emoto, Kouzoubutsu Clinic Co., Ltd. for his great support on this research work, and also to the Traffic Division of Ube-city's Transportation Bureau for the generous cooperation in connection with the long-term field test conducted for the purposes of this study including bus sensor installation and bus operation arrangements.

References
[1] Miyamoto A and Emoto H 2015 Performance evaluation and remaining life prediction of an existing RC bridge by J-BMS *fib Symposium Proceedings on Concrete – Innovation and Design* fib Copenhagen 1-18.

[2] Khatir A, Tehami M, Khatir S and Abdel Wahab M 2016 Multiple damage detection and localization in beam-like and complex structures using co-ordinate modal assurance criterion combined with firefly and genetic algorithms *Journal of Vibroengineering* **18**(8) 5063-5073.

[3] Zhou Y-L, Maia NMM, Sampaio R and Wahab MA 2016 Structural damage detection using transmissibility together with hierarchical clustering analysis and similarity measure *Structural Health Monitoring* doi: 10.1177/1475921716680849.

[4] Zhou Y-L, Maia N and Abdel Wahab M 2016 Damage detection using transmissibility compressed by principal component analysis enhanced with distance measure *Journal of Vibration and Control* doi: 10.1177/1077546316674544.

[5] Zhou Y-L and Abdel Wahab M 2016 Rapid early damage detection using transmissibility with distance measure analysis under unknown excitation in long-term health monitoring *Journal of Vibroengineering* **18**(7) 4491-4499.

[6] Khatir S, Belaidi I, Serra R, Abdel Wahab M and Khatir T 2016 Numerical study for single and multiple damage detection and localization in beam-like structures using BAT algorithm *Journal of Vibroengineering* **18**(1) 202-213.

[7] Khatir S, Belaidi I, Serra R, Abdel Wahab M and Khatir T 2015 Damage detection and localization in composite beam structures based on vibration analysis *Mechanika* **21**(6) 472-479.

[8] Gillich G-R, Praisach Z-I, Abdel Wahab M, Gillich N, Mituletu IC and Nitescu C 2016 Free vibration of a perfectly clamped-free beam with stepwise eccentric distributed masses *Shock and Vibration* **2016**(Article ID 2086274) 10 pages; http://dx.doi.org/10.1155/2016/2086274.

[9] Miyamoto A and Yabe A 2012 Development of practical health monitoring system for short and medium-span bridges based on vibration responses of city bus *Journal of Civil Structural Health Monitoring* ISHMII **2**(1) 47-63.

[10] Yabe A, Miyamoto A and Nagata M 2015 Verification tests for practical application of a health monitoring system for short- and medium-span bridges based on public bus vibrations *Journal of Civil Structural Health Monitoring* ISHMII **5**(1) 67-95.

[11] Miyamoto A, Lúcio V.J.G, Emoto H and Tanaka H 2015 Long term application of bus monitoring system to short and medium span bridges and damage detection *Journal of Physics: Conference Series* **628**(2015)012037, IOP Publishing, 1-14 doi:10.1088/1742-6596/628/1/012037.

[12] Miyamoto A, Emoto H, Takahashi J and Yabe A 2015 Structural health evaluation and remaining life estimation of a decommissioned bridge based on site investigation: Sakae Bridge investigation results *Infrastructure Management Series No. 22* The Research Center for Environmental Safety Yamaguchi University Ube Japan **22** 1–104.

[13] Yabe A 2011 Fundamental study on development of bridge anomaly detection system using in-service fixed-route bus *Doctoral Dissertation* Yamaguchi University Ube Japan.

[14] Omi H, Mori S and Twayan R P 2013 Detectability of natural frequency of concrete girder bridge by vehicle loading and hammering *Proceedings of the Annual Conference of JSCE Shikoku Branch*, **I-4**.

[15] Tamada K 2013 Experimental study on damage and dynamic characteristics of bridges managed by local governments Japan Institute of Countryology and Engineering *Proceedings of the 13th JICE Annual Presentation of Subsidized Research*, **I-5**.

IOP Conf. Series: Journal of Physics: Conf. Series **842** (2017) 012033 doi:10.1088/1742-6596/842/1/012033

Damage detection in multi-span beams based on the analysis of frequency changes

G R Gillich[1], J L Ntakpe[1,2], M Abdel Wahab[3], Z I Praisach[1] and M C Mimis[1]

[1]NDT Laboratory, Faculty of Engineering and Management, "Eftimie Murgu" University of Resita, Romania
[2]Alphabet International GmbH, Munich, Germany
[3]Laboratory Soete, Faculty of Engineering and Architecture, Ghent University, Belgium

E-mail: gr.gillich@uem.ro

Abstract. Crack identification in multi-span beams is performed to determine whether the structure is healthy or not. Among all crack identification methods, these based on measured natural frequency changes present the advantage of simplicity and easy to use in practical engineering. To accurately identify the cracks characteristics for multi-span beam structure, a mathematical model is established, which can predict frequency changes for any boundary conditions, the intermediate supports being hinges. This relation is based on the modal strain energy concept. Since frequency changes are relative small, to obtain natural frequencies with high resolution, a signal processing algorithm based on superposing of numerous spectra is also proposed, which overcomes the disadvantage of Fast Fourier Transform in the aspect of frequency resolution. Based on above-mentioned mathematical model and signal processing algorithm, the method of identifying cracks on multi-span beams is presented. To verify the accuracy of this identification method, experimental examples are conducted on a two-span structure. The results demonstrate that the method proposed in this paper can accurately identify the crack position and depth.

1. Introduction

Even if beams with multiple supports are frequently used in practice, the research about such structures in the damaged state is not broadly present in the literature. In this research, uniformly spaced spans are commonly considered. An example is the study of a beam with two identical spans traversed by a train moving at a constant velocity, which had the aim to highlight the resonance characteristics [1]. The Euler-Bernoulli beam theory was used to model the two beam segments, and constraints and compatibility conditions were imposed at the beam center. The effect of the random deviation of the span lengths from the ideal design was studied in [2], using the perturbation method. Transverse vibrations of a slender beam with infinite length mounted on discrete elastic supports are studied in [3]. To describe the dynamic behavior of multi-span beams, the supports are replaced by forces acting on the support locations. The deflection is described by a single function, the reactions at the support locations being included in the eigenvalue solution [4].

In the damage detection methods, the multi-span structures with transverse cracks are treated as an assembly of intact sub-segments. The cracks are replaced with massless rotational spring, while instead of the supports, constraints and compatibility conditions are imposed [5]. The models become complicated, and, by replacing the intermediate supports or the crack, a new set of equations has to be solved. Several theoretical studies and experiments aiming to study and analyze the changes of the natural frequencies and mode shapes occurring due to damage in general terms [6-13] and due to a crack in particular [14-17], were presented in the literature.

Previous publications, [18-20], present a mathematical relation that predicts the frequency changes occurred in single span beams due to cracks with known depth and position. The input data is limited to the curvature achieved by the intact beam in the location where damage is expected, and the absolute damage severity derived from the deflections of the intact and damaged beam. Based on this relation a database was build for a multitude of damage scenarios and differently supported beams, therefore damage assessment becomes an inverse problem.

This paper presents the exact solution for the frequencies of beams with two unequal spans, for both healthy and damaged state. Any support and crack position can be taken into consideration; the crack depth is limited to around 50% by the linearity condition. The results are implemented in a database and used to demonstrate the robustness of the proposed damage assessment method.

2. Modal parameters for the intact two-span beam

The study is performed on a long slender beam with two spans, as shown in figure 1, having the length L, the width b and the thickness h, and consequently the second moment of area of the cross section I and the cross section area A. At the right end the support is a hinge and the left end is clamped; supplementary, a hinge at distance L_1 is intercalated. The relevant material characteristics are: the Young Modulus E, the volumetric mass density ρ, and the Poisson ratio μ.

If the beam is subjected to a damage, this is an open transversal crack which has the depth a and the width l. Its position is indicated as c.

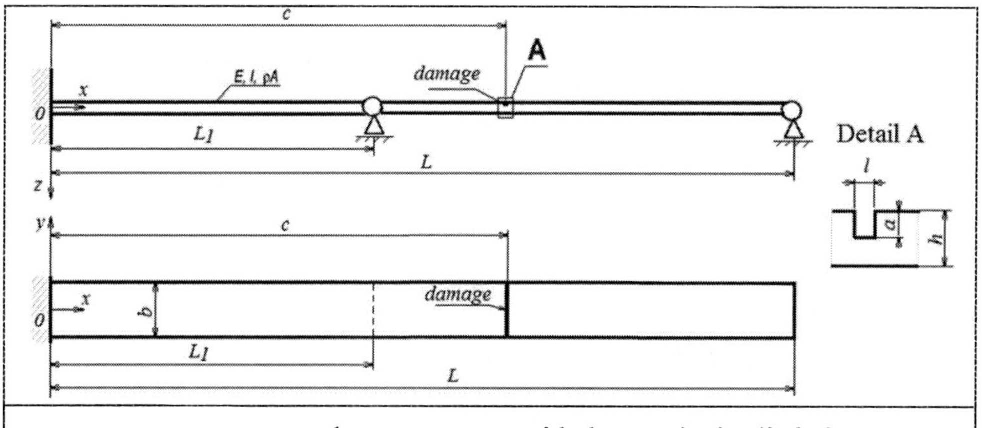

Figure 1. Two-span beam geometry with damage in detailed view.

The bending vibrations of this beam are described by the following equation:

$$\frac{\partial^4 z}{\partial x^4} + \frac{\rho A}{EI} \cdot \frac{\partial^2 z}{\partial t^2} = 0 \tag{1}$$

where z is the vertical beam displacement at distance x measured from the left end. The solution z of equation (1) can be written as a standing wave, $z(x,t) = W(x) \cdot \sin(\omega t + \varphi)$, thus spatial and temporal components are separated. For the spatial component the well-known solution is

$$W(x) = A\sin\alpha x + B\cos\alpha x + C\sinh\alpha x + D\cosh\alpha x \tag{2}$$

where *A, B, C* and *D* and $\alpha^4 = \dfrac{\rho A \omega^2}{EI}$ are coefficients defined by the particular boundary conditions.

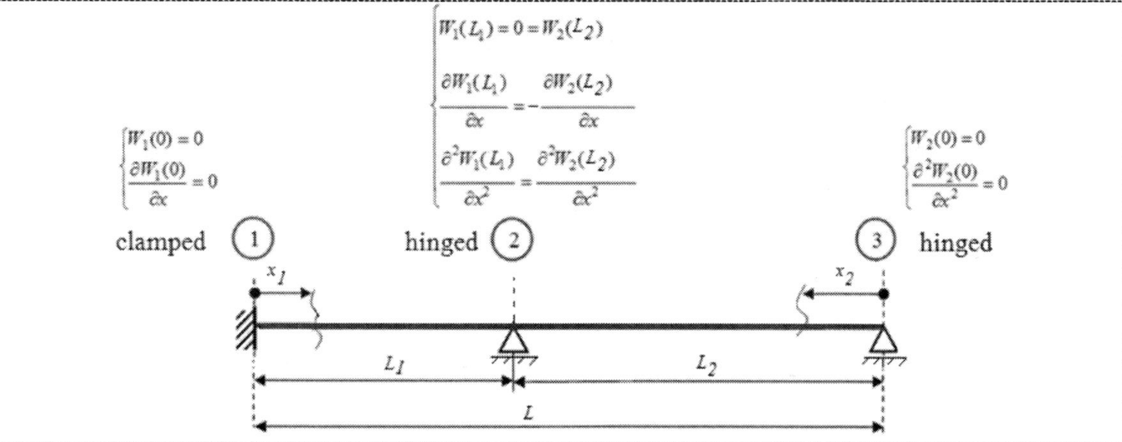

Figure 2. Boundary conditions for the analyzed two-span beam.

In the vibration analysis, each segment between two supports is considered as a separate beam. The left segment has associated the space interval $x_1 \in (0 \ldots L_1)$, and the origin of the reference system is the fixed beam end (point 1 in figure 2). The right segment has associated the space interval $x_2 \in (0 \ldots L_2)$, and the origin of the reference system is the hinge placed at the right beam end (point 3 in figure 2). The two solutions of equation (1) get, for the two segments, the form

$$\begin{cases} W_1(x_1) = A_1 \sin(\alpha x_1) + B_1 \cos(\alpha x_1) + C_1 \sinh(\alpha x_1) + D_1 \cosh(\alpha x_1) \\ W_2(x_2) = A_2 \sin(\alpha x_2) + B_2 \cos(\alpha x_2) + C_2 \sinh(\alpha x_2) + D_2 \cosh(\alpha x_2) \end{cases} \tag{3}$$

Applying the eight constraints and continuity conditions to the solutions in equation (3), it results a system with eight algebraic equations. After removing the eight coefficients $A_1 \ldots D_2$ from the algebraic system, an equation with the variable $\lambda = \alpha / L$ is attained, that is

$$\begin{aligned} &2 \cdot \left[\left(\Omega_1(\lambda L_1)\right)^2 + \Pi_1(\lambda L_1) \cdot \Gamma_1(\lambda L_1) \right] - \\ &- \left[\frac{\Sigma_1(\lambda L_1)\Pi_1(\lambda L_1) - \Omega_1(\lambda L_1)\Gamma_1(\lambda L_1)}{\sin(\lambda L_2)} \right] \cdot \left[\frac{\cos(\lambda L_2)}{\sin(\lambda L_2)} - \frac{\cosh(\lambda L_2)}{\sinh(\lambda L_2)} \right] = 0 \end{aligned} \tag{4}$$

where $\Gamma_1(\lambda L_1) = \sin(\lambda L_1) + \sinh(\lambda L_1)$, $\Pi_1(\lambda L_1) = \sin(\lambda L_1) - \sinh(\lambda L_1)$, $\Sigma_1(\lambda L_1) = \cos(\lambda L_1) + \cosh(\lambda L_1)$ and $\Omega_1(\lambda L_1) = \cos(\lambda L_1) - \cosh(\lambda L_1)$.

By solving equation (4) for the central hinge position removed iteratively along the beam, the evolution of the dimensionless wave number λ in form of variation curves is obtained (see figure 3). From this figure one can easily observe that for $\bar{L}_1 \to 0$ the wave numbers achieve very close values to that of a beam clamped at the left end and hinged at the right end (marked with a red square). If the intermediate hinge is close to the hinged end, i.e. $\bar{L}_1 \to 1$, the wave number values are very similar with that of a beam with both ends clamped (marked with a blue square).

The two-span beam's frequencies can be derived using the well-known frequency relation

$$f_i = \frac{\lambda_i^2}{2\pi} \sqrt{\frac{EI}{\rho A L^4}} \tag{5}$$

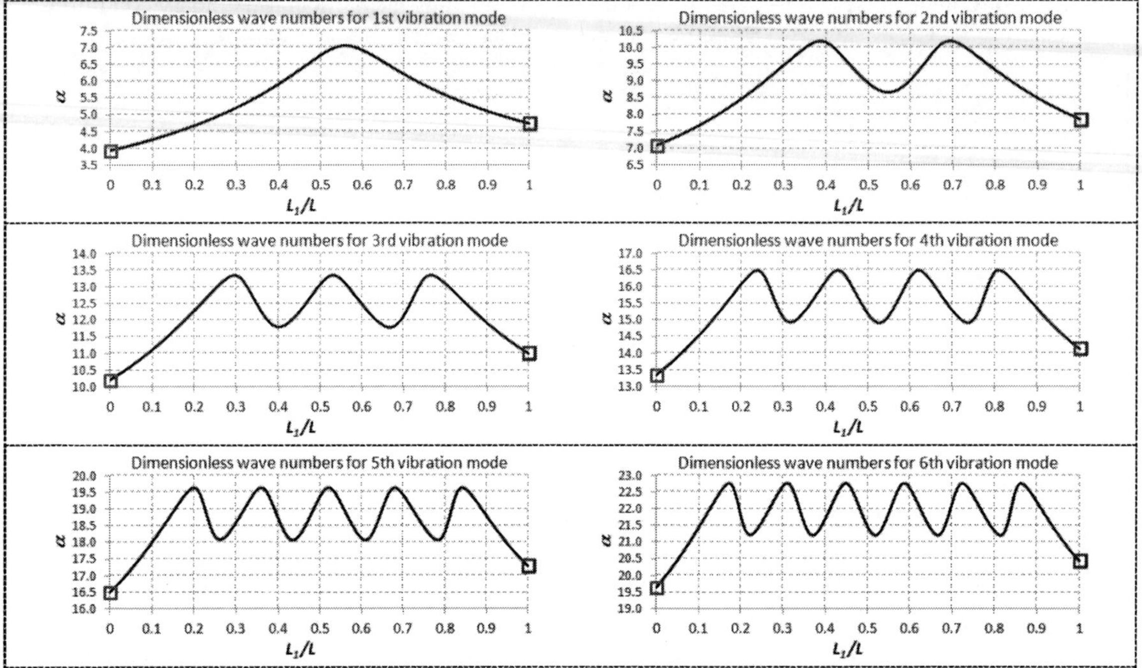

Figure 3. Dimensionless wave number for the first six vibration modes central hinge location.

By replacing the coefficients A_1, B_1, C_1, D_1, A_2, B_2, C_2, D_2 in equations (4), results the mode shape function for the considered continuous beam. For the left segment it is:

$$W_{1-i}(x_1,L_1,\lambda_i)=D_1\left\{\frac{\Omega_1(\lambda_i L_1)}{\Pi_1(\lambda_i L_1)}\cdot\left[\sin(\lambda_i x_1)-\sinh(\lambda_i x_1)\right]-\left[\cos(\lambda_i x_1)-\cosh(\lambda_i x_1)\right]\right\} \quad (6)$$

while for the right segment it is:

$$W_{2-i}(x_2,L_1,\lambda_i)=D_1\frac{\Sigma_1(\lambda L_1)\Pi_1(\lambda L_1)-\Omega_1(\lambda L_1)\Gamma_1(\lambda L_1)}{2\cdot\Pi_1(\lambda L_1)\cdot\sin(\lambda L_2)}\left[\frac{\sin(\lambda_i L_2)}{\sinh(\lambda_i L_2)}\cdot\sinh(\lambda_i x_2)-\sin(\lambda_i x_2)\right] \quad (7)$$

Note that, to obtain normalized mode shapes, D_1 must be determined separately for each location of L_1 and vibration mode.

The analytically achieved results are now compared with that from FEM simulations. A beam with following geometrical and physical-mechanical parameters is involved in the analysis: $L=2\text{m}$, $b=20\text{mm}$, $h=5\text{mm}$, $\rho=7850\text{kg/m}^3$, $E=2\cdot10^{11}\text{N/m}^2$ and $\nu=0.3$. In the first case the central hinge is placed at the normalized distance $L_1/L=0.33$ and in the second case it is located at distance $L_1/L=0.45$. The FEM analysis, performed by means of the ANSYS 14 software, made use of a hexahedral mesh element with imposed dimensions 2 mm.

Table 1 indicates the wave numbers and frequencies derived analytically, as well as the simulation results. The differences between synonym frequencies is less than 5%, except two modes of the beam with central hinge at $L_1/L=0.33$. These differences occur because certain slenderness is necessary to permit applying the Euler-Bernoulli theory. Thus, if one of the beam segments is short, bigger errors are present, imposing a more complex model. In addition, the FEM model does not exactly reflect the conditions induced by the hinges in the analytical model.

Table 1. Wave numbers and frequencies for two position of the intermediate hinge.

Vibration mode i	Hinge at $L_1/L = 0.33$			Hinge at $L_1/L = 0.45$		
	Analytic		FEM	Analytic		FEM
	Wave number	Frequency [Hz]	Frequency [Hz]	Wave number	Frequency [Hz]	Frequency [Hz]
1	5.407	8.475	9.358	6.348	11.681	12.538
2	9.806	27.873	27.334	9.569	26.542	27.789
3	12.908	48.296	53.050	12.264	43.593	43.824
4	15.008	65.292	65.813	16.262	76.662	81.506
5	19.219	107.073	106.538	18.196	95.978	99.190
6	22.358	144.903	150.784	22.773	150.341	151.798

For the hinge located at $L_1/L = 0.33$, the first six bending mode shapes are depicted in figure 4, where those corresponding to the analytic approach are to the left and those resulted from the FEM analysis to the right. One can observe in this figure that curves fit well.

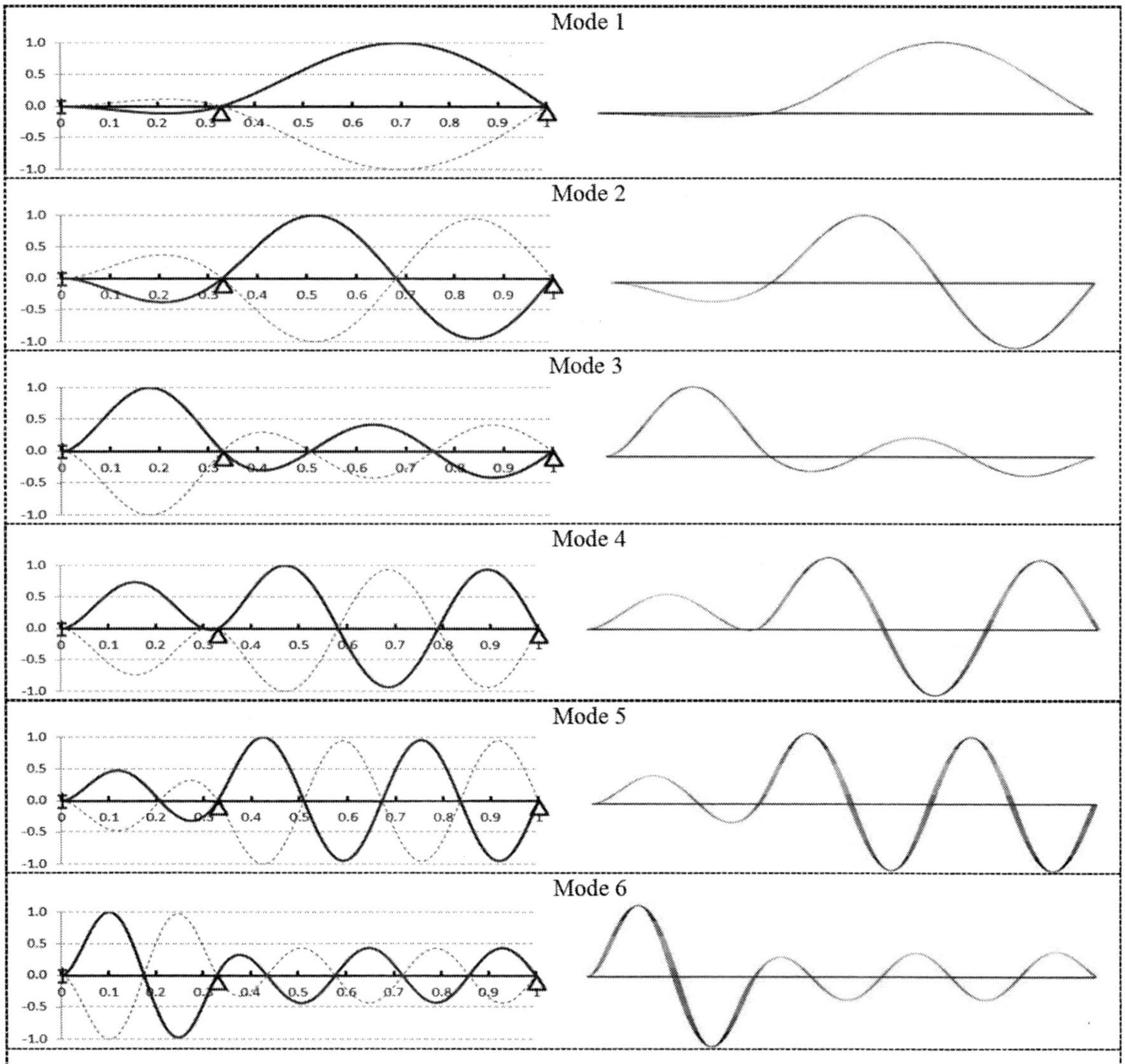

Figure 4. Modes shapes derived analytically and by means of FEM

3. Natural frequencies of the damaged beam

The crack considered in this study is transversal and has the depth a, as shown in figure 1. It is located at distance c from the fixed end. As shown in our previous papers, the energy stored in a slice in the undamaged state can indicate the natural frequency drop if a crack occurs in that slice [19]. The mathematical relation permitting to determine the natural frequencies for the damaged beam, adapted for the particular case of the two-span beam is

$$f_{D-i}\left(c,L_1,\bar{a}\right)=f_i\left\{1-\gamma(\bar{a})\cdot\left[\bar{W}_i''\left(c,L_1,\alpha_i\right)\right]^2\right\} \tag{8}$$

where f_i is the natural frequency for the undamaged beam for the i-th vibration mode and $\gamma(\bar{a})<1$ is the damage severity characterizing the effect of a crack with relative depth [20], which is null for the undamaged beam. The normalized modal curvature at location c, denoted $\bar{W}_i''(c,L_1,\alpha_i)$ in equation (8), is found from equations (9) or (10), depending on which segment the crack is located.

$$W_{1-i}''\left(x_1,L_1,\alpha_i\right)=k_1k_2\left\{-\frac{\Omega_1\left(\alpha_iL_1\right)}{\Pi_1\left(\alpha_iL_1\right)}\cdot\left[\sin\left(\alpha_ix_1\right)+\sinh\left(\alpha_ix_1\right)\right]+\left[\cos\left(\alpha_ix_1\right)+\cosh\left(\alpha_ix_1\right)\right]\right\} \tag{9}$$

$$W_{2-i}''\left(x_2,L_1,\alpha_i\right)=k_1\frac{\Sigma_1(\alpha L_1)\Pi_1(\alpha L_1)-\Omega_1(\alpha L_1)\Gamma_1(\alpha L_1)}{2\cdot\Pi_1(\alpha L_1)\cdot\sin(\alpha L_2)}\left[\frac{\sin(\alpha_iL_2)}{\sinh(\alpha_iL_2)}\cdot\sinh\left(\alpha_ix_2\right)-\sin\left(\alpha_ix_2\right)\right] \tag{10}$$

The functions in equations (9) and (10) are proportional with the modal curvatures of the two segments. Normalization is achieved by involving the proper coefficients k_1 and k_2. Based on equation (8), the frequencies of the damaged beam for any crack location can be calculated. Figure 5 shows the frequency evolution if the crack is replaced along the beam, for the intermediate hinge located at $L_1/L=0.33$. The damage severity is $\gamma(\bar{a})=0.01$; this severity is achieved for the open crack with depth $a=1.25\text{mm}$ and width $l=2\text{mm}$.

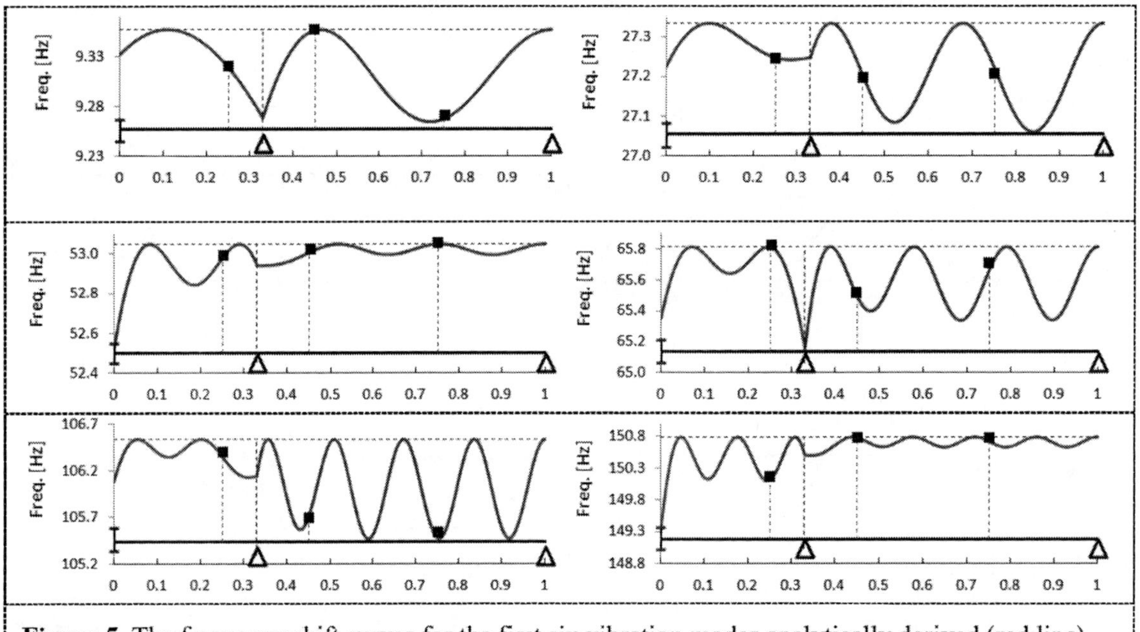

Figure 5. The frequency shift curves for the first six vibration modes analytically derived (red line) and samples attained from the FEM analysis.

Modal analysis using ANSYS was made for three crack positions $c_1 / L = 0.25$, $c_2 / L = 0.45$ and $c_3 / L = 0.75$. The results are indicated with black squares in figure 5. Having a look onto this figure, it is easy to remark the fit between analytically and FEM-based attained results, meaning that equation (8) can be used to define damage patterns which can be further used in the damage detection process. For beams with multiple supports the frequency shifts are slight, making the improvement of frequency readability an important issue in experimental modal analysis [21-22].

4. The damage detection algorithm

The proposed assessment algorithm is based on the particularity that any crack provides a particular behavior change to that beam. As consequence of a crack, frequencies of different bending modes shift differently, in respect to the energy stored in the damaged slice, as shown in previous section. For the i-th bending mode of a two-span beam, the relative frequency shift is derived as:

$$\Delta\bar{f}_i\left(c, L_1, \bar{a}\right) = \frac{f_i - f_{D-i}}{f_i} = \gamma(\bar{a}) \cdot \left[\bar{W}_i''\left(c, L_1, \alpha_i\right)\right]^2 \tag{11}$$

For every damage locations c and relative depths $\bar{a}$ result the relative frequency shifts $\Delta\bar{f}_1 ... \Delta\bar{f}_n$ which constitute the damage pattern. In optimal conditions two frequency shifts are sufficient to identify location and severity [23], but an increased number of analyzed frequencies, we recommend $n > 6$, permit precise assessment because redundancy reduce evaluation or measurement errors.

Normalization of relative frequency shift values for each location c by the highest value of the sequence cancel the influence of damage depth. This happens because both numerator and denominator contain the severity $\gamma(\bar{a})$. In that way the sequence had become independent of the damage severity and its parameters take values between 0 and 1. Such a sequence is named Damage Location Indicator (DLI) and the individual parameters Damage Location Coefficients (DLC). For the crack positioned at distance c_j, the DLCs are

$$\Phi_1(c_j) = \frac{\left(\bar{W}_1''(c_j)\right)^2}{\max\left\{\left(\bar{W}_i''(c_j)\right)^2\right\}}, \ \ldots, \Phi_n(c_j) = \frac{\left(\bar{W}_n''(c_j)\right)^2}{\max\left\{\left(\bar{W}_i''(c_j)\right)^2\right\}} \tag{12}$$

These DLCs can be derived just using information about the beam in its intact state. For asymmetric structures the DLCs uniquely indicate the damage location, while for symmetric structures two mirrored locations provide the same values of DLCs. An infinite number of damage location indicators, denoted $\Phi(c_j)$, can be derived in respect to the chosen crack positions j. From our expertise we recommend limiting to $p = 100$ locations.

For the real beam measured relative frequency shifts $\Delta\bar{f}_1^m ... \Delta\bar{f}_1^m$ can be also obtained from equation (11), if the frequencies in intact state f_i^m and in the presence of damage f_{D-i}^m are known. Normalizing the measured relative frequency shifts after the same rule as that used for the DLCs, the normalized measured frequency shifts NFR are obtained. These are expressed as

$$\Psi_1 = \frac{\Delta\bar{f}_1^m}{\max\left\{\Delta\bar{f}_i^m\right\}}, \ \ldots, \Psi_n = \frac{\Delta\bar{f}_n^m}{\max\left\{\Delta\bar{f}_i^m\right\}} \tag{13}$$

and are also severity-independent. The entire sequence $\boldsymbol{\Psi}$ is nominated as the Damage Signature (DS); it has the same meaning as the DLIs but is derived from experimental data.

Example of two DLIs, for the beam with a hinge at $L_1 / L = 0.33$ and damages at position $c_1 / L = 0.24$, $c_2 / L = 0.46$ and $c_3 / L = 0.76$ are presented in figure 6.

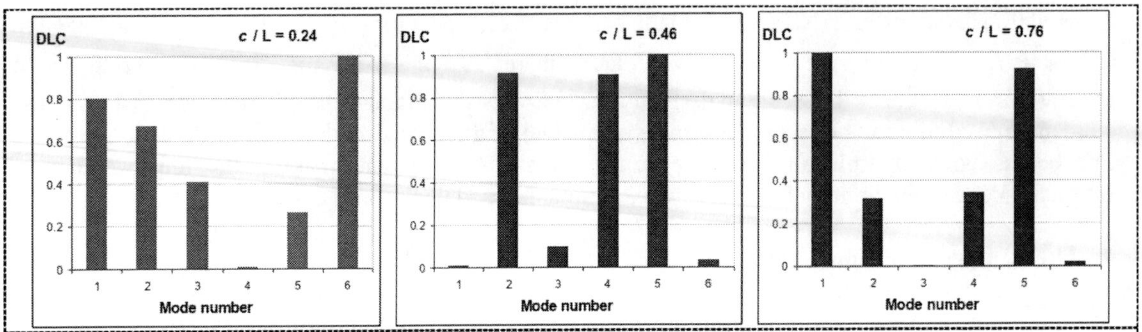

Figure 6. Examples of damage location indicators for the crack position $c_1 / L = 0.24$ (left diagram), $c_2 / L = 0.46$ (right diagram), and $c_3 / L = 0.76$ (right diagram).

If the damage signature $\boldsymbol{\Psi}$ is compared with several damage location indicators $\Phi(c_j)$, it will be found one that best fit. The similitude test between the Damage Signature $\boldsymbol{\Psi}$ and all Damage Location Indicators $\Phi(c_j)$, in fact vectors with n elements, indicate the crack position by designating the index j. Herein the Minkowski distance of second order is uded as a Damage Index, that is

$$DI_j = \left(\sum_{i=1}^{n} \left| \Phi_i(c_j) - \Psi_i \right|^2 \right)^{\frac{1}{2}} \tag{14}$$

is used to find the most similar vectors. The value j for which the objective function DIj achieves the lowest value is searched; it indicates the crack location c_j.

Briefly, the proposed damage assessment algorithm is detailed in the following steps: If the beam is already damaged in the initial state, the assessment algorithm shows the damage evolution from this stage on.

S1 - frequency evaluation for the intact state is performed for the first n weak-axis bending vibration modes. It result a vector $U:\{f_1^m ... f_n^m\}$

S2 - periodical measurements are performed, for each attempt being obtained a sequence $D:\{f_{D-1}^m ... f_{D-n}^m\}$

S3 - frequencies for initial and actual state are compared in order to find the frequency shifts. The values obtained in step S2 are subtracted from those obtained in step S1, resulting $S:\{\Delta f_1^m ... \Delta f_n^m\}$. For insignificant differences the structure is considered intact, else occurrence of damage is presumed.

S4 - if the damage is expected, the relative frequency shifts are determined as $R:\{\Delta\bar{f}_1^m ... \Delta\bar{f}_n^m\}$

S5 - the relative frequency shifts are normalized, resulting the Damage Signature $\boldsymbol{\Psi}:\{\Psi_1 ... \Psi_n\}$

S6 - the normalized squared modal curvatures for the healthy beam are analytically determined for p equidistant locations, resulting p vectors $C_j:\left\{\left[\bar{W}''_{1,2-1}(c_j, L_1, \bar{a})\right]^2 ... \left[\bar{W}''_{1,2-n}(c_j, L_1, \bar{a})\right]^2\right\}$

S7 - the DLIs are derived by dividing the DLCs of each C_j to the highest value of the sequence, resulting $\boldsymbol{\Phi}(c_j):\left\{\Phi_1(c_j) ... \Phi_n(c_j)\right\}$

S8 - the analytically derived $\Phi(c_j)$ vectors are compared to the $\boldsymbol{\Psi}$ vector determined by measurements by means of a similitude estimator (herein the Minkowski Distance) in order to identify the index j ensuring the lowest value, and subsequent the crack position c_j.

S9 - the damage severity can be determined from severity curves traced for different cross-sections after a methodology presented in [24].

IOP Conf. Series: Journal of Physics: Conf. Series **842** (2017) 012033 doi:10.1088/1742-6596/842/1/012033

5. Numerical experiment

The algorithm is tested for the three damages indicated at the end of section 3, with absolute positions $c_1 = 500$ mm, $c_2 = 900$ mm and $c_3 = 1500$ mm. The beam described in figure 1 is analyzed. It has the geometry and the physical-mechanical parameters provided in section 2, and the intermediate hinge position is at the position $L_1 = 660$ mm, resulting $L_1 / L = 0.33$.

The measurements are replaced by FEM simulations, the evaluated frequencies for the intact as well as for the three damage cases being presented in tables 2 to 4. We worked through steps the first 5 steps of the algorithm and found the RFS and the damage signatures $\boldsymbol{\Psi}$ for the three analyzed cases. These are compared in tables 2 to 4 with the DLCs of the closest damage location indicators. Note that, in steps S6 to S8, just 51 crack locations are considered. The reduced number of DLIs assumed in the database increases the degree of difficulty for the identification process. If crack identification is performed successful, it is demonstrated that the damage signature attained from six vibration modes doubtless indicate the crack position.

Table 2. Frequencies and damage patterns for the crack positioned at c_1 = 500 mm.

Vibration mode i	Intact	Damage at c_1 = 500 mm			Closest found DLCs	
	Frequency [Hz]	Frequency [Hz]	RSF	DS	DLC 480	DLC 520
1	9.358	9.316	0.4400	1	0.7987	1
2	27.334	27.245	0.3224	0.7327	0.6740	0.6947
3	53.050	52.991	0.1103	0.2508	0.4067	0.1623
4	65.813	65.800	0.0193	0.0439	0.0065	0.0562
5	106.538	106.326	0.1984	0.4509	0.2643	0.4919
6	150.784	150.160	0.4136	0.9400	1	0.8332

Table 3. Frequencies and damage patterns for the crack positioned at c_2 = 900 mm.

Vibration mode i	Intact	Damage at c_2 = 900 mm			Closest found DLCs	
	Frequency [Hz]	Frequency [Hz]	RSF	DLC	DLC 880	DLC 920
1	9.358	9.357	0.0080	0.0122	0.0309	0.0075
2	27.334	27.186	0.5402	0.8182	0.4323	0.9147
3	53.050	53.014	0.0661	0.1002	0.1134	0.0968
4	65.813	65.458	0.5379	0.8147	0.4504	0.9058
5	106.538	105.834	0.6601	1	1	1
6	150.784	150.758	0.0168	0.0255	0.0001	0.0319

Table 4. Frequencies and damage patterns for the crack positioned at c_3 = 1500 mm.

Vibration mode i	Intact	Damage at c_3 = 1500 mm			Closest found DLCs	
	Frequency [Hz]	Frequency [Hz]	RSF	DLC	DLC 1480	DLC 1520
1	9.358	9.266	0.9838	1	1	0.9681
2	27.334	27.236	0.3582	0.3721	0.3184	0.5139
3	53.050	53.048	0.0034	0.0033	0.0044	0.0001
4	65.813	65.616	0.2987	0.2962	0.3459	0.1447
5	106.538	105.544	0.9327	0.9521	0.9247	1
6	150.784	150.736	0.0312	0.0335	0.0215	0.0669

Similitude tests between the damage signatures and the damage location indicators are performed in step eight. The three vectors Ψ are individually compared with $p = 51$ vectors $\Phi(c_j)$ involving the Minkowski distance of second order. The results are graphically represented in figure 7, in the form of evolution curves. This figure gives direct information about the damage location. The similitude curve traced for the damage position $c_1 = 500$ mm attains the lowest values around the crack position, framing it. A clear minima is achieved for the crack positioned at $c_2 = 900$ mm, while the curve for the crack located at $c_3 = 1500$ mm present two possible solutions. A supplementary test can be performed by comparing the individual elements of the possible DLIs and the DS. Because the DS can fit just one DLI, this will indicate the damage position.

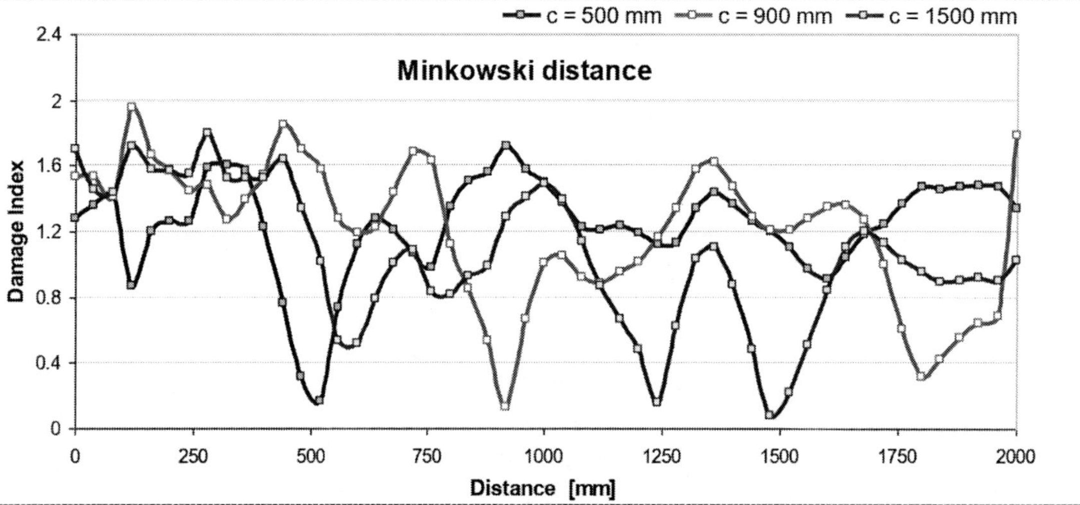

Figure 7. Examples of damage location identification using the Minkowski distance.

Alternatively, the damage position can be found by identifying the index j for which the damage index in equation (14) get the lowest value(s). This process can be automated, no human intervention being necessary. If more DI values qualify as possible solutions, in order to eliminate misinterpretation, again the individual elements of the potential DLIs and the DS can be compared.

In both analysis strategies, the increase of DLIs populating the database leads to an improved precision in locating cracks. This is reflected in damage index values close to null, eliminating doubts.

6. Conclusion

The problem of crack assessment in multi-span beams was considered via a simple and robust vibration-based method. The proposed method is based on the effect that the energy loss in a slice due to damage has upon the magnitude of frequency change. For the intact beam, the energy distribution in various vibration modes was contrived, and a predictive behavioral model developed. This model is used to define DLIs, which are integrated in a database including any possible damage scenarios for the analyzed beam. Adaption of the model, allowing the use of other end supports, can by simply made by changing the boundary conditions.

The method proved effective in detecting cracks and high accuracy is achieved in estimating the crack position. However, the early observation and precise localization depend on the possibility to quantify slight frequency changes and on the density of DLIs involved in the analysis. The similarity estimation strategy plays also a crucial role.

Acknowledgements
The work has been funded by the Sectoral Operational Programme Human Resources Development 2007-2013 of the Ministry of European Funds through the Financial Agreement POSDRU/159/1.5/S/132395.

References

[1] Wang Y, Wei Q C, Shi J and Long X 2010 Resonance characteristics of two-span continuous beam under moving high speed trains *Latin American Journal of Solids and Structures* **7**(2) 185-199
[2] Yang J N and Lin Y K 1964Frequency Response Functions of a Disordered Periodic Beam *Journal of Sound and Vibration* **38**(3) 317–340.
[3] Hosking R J, Husain S A and Milinazzo F 2004 Natural flexural vibrations of a continuous beam on discrete elastic supports *Journal of Sound and Vibration* **272** 169–185.
[4] Khodabakhsh S and Bhat R B 2011 Clustered natural frequencies in multi-span beams with constrained characteristic functions *Shock and Vibration* **18** 698-707.
[5] Lien T V and Hao T A 2014 Determination of the Shape Function of a Multiple racked Beam Element and Its Application for the Free Vibration Analysis of a Multiple Cracked Frame Structure, *American Journal of Civil Engineering and Architecture* **2**(1) 12-15.
[6] Zhou Y-L, Maia NMM, Sampaio R and Wahab MA 2016 Structural damage detection using transmissibility together with hierarchical clustering analysis and similarity measure *Structural health monitoring* DOI: **https://doi.org/10.1177/1475921716680849**
[7] Zhou Y-L, Maia N and Abdel Wahab M 2016 Damage detection using transmissibility compressed by principal component analysis enhanced with distance measure *Journal of Vibration and Control* doi: 10.1177/1077546316674544
[8] Zhou Y-L and Abdel Wahab M 2016 Rapid early damage detection using transmissibility with distance measure analysis under unknown excitation in long-term health monitoring *Journal of Vibroengineering* **18**(7) 4491-4499
[9] Khatir S, Belaidi I, Serra R, Abdel Wahab M and Khatir T 2016 Numerical study for single and multiple damage detection and localization in beam-like structures using BAT algorithm *Journal of Vibroengineering* **18**(1) 202-213
[10] Khatir A, Tehami M, Khatir S and Abdel Wahab M 2016 Multiple damage detection and localization in beam-like and complex structures using co-ordinate modal assurance criterion combined with firefly and genetic algorithms *Journal of Vibroengineering* **18**(8) 5063-5073
[11] Gillich G-R, Praisach Z-I, Abdel Wahab M, Gillich N, Mituletu IC and Nitescu C 2016 Free vibration of a perfectly clamped-free beam with stepwise eccentric distributed masses *Shock and Vibration* **2016**(Article ID 2086274) 10 pages; http://dx.doi.org/10.1155/2016/2086274
[12] Khatir S, Belaidi I, Serra R, Abdel Wahab M and Khatir T 2015 Damage detection and localization in composite beam structures based on vibration analysis *Mechanika* **21**(6) 472-479
[13] Zhou Y-L and Abdel Wahab M 2017 Cosine based and extended transmissibility damage indicators for structural damage detection *Engineering Structures* **141** 175-183
[14] Li S G and Li H 2011 The Fractional Diagnosis of Multi-Span Continuous Bridge's Structural Damage Based on Neural Network and Genetic Algorithm *Applied Mechanics and Materials* **71-78** 1298-1304.
[15] Kullaa J, Santaoja K and Eymery A 2013 Vibration-based structural health monitoring of a simulated beam with a breathing crack *Key Engineering Materials* **569-570** 1093-1100.
[16] Surace C, Ruotolo R, Storer D 2011 Detecting nonlinear behaviour using the volterra series to assess damage in beam-like structures *Journal of Theoretical and Applied Mechanics* **49** 905-926.
[17] Sakellariou J S and Fassois S D 2008 Vibration based fault detection and identification in an

aircraft skeleton structure via a stochastic functional model based method *Mechanical Systems and Signal Processing* **22**(3) 557-573.

[18] Gillich G R and Praisach Z I 2013 Detection and Quantitative Assessment of Damages in Beam Structures Using Frequency and Stiffness Changes *Key Eng. Mat.* **569-570** 1013-1020.

[19] Gillich G R, Praisach Z I and Negru I 2012 Damages Influence on Dynamic Behaviour of Composite Structures Reinforced with Continuous Fibers *Materiale Plastice* **49**(3) 186-191.

[20] Gillich G R, Birdeanu E D, Gillich N, Amariei D, Iancu V and Jurca C S 2009 Detection of damages in simple elements *Annals of DAAAM for 2009 & Proceedings of the 20th International DAAAM Symposium* **20** 623-624.

[21] Onchis-Moaca D, Gillich G R and Frunza R 2012 Gradually Improving The Readability Of The Time-Frequency Spectra For Natural Frequency Identification In Cantilever Beams 2012 Proceedings of the 20th European Signal Processing Conference (EUSIPCO) 809-813.

[22] Gillich G R, Maia N M M, Mituletu I C, Praisach Z I, Tufoi M and Negru I 2015 Early Structural Damage Assessment by Using an Improved Frequency Evaluation Algorithm *Latin American Journal of Solids and Structures* **12**(12) 2311-2329.

[23] Cawley P and Adams R D 1979 The locations of defects in structures from measurements of natural frequencies *Journal of Strain Analysis* **14**(2) 49–57.

[24] Gillich G R, Abdel Wahab M, Praisach Z I and Ntakpe J L 2014 The influence of transversal crack geometry on the frequency changes of beams *Proceedings of International Conference on Noise and Vibration Engineering (ISMA2014) and International Conference on Uncertainty in Structural Dynamics (USD2014)* 485-498.

IOP Conf. Series: Journal of Physics: Conf. Series **842** (2017) 012034 doi:10.1088/1742-6596/842/1/012034

Damage severity estimation from the global stiffness decrease

C Nitescu[1,2], G R Gillich[1], M Abdel Wahab[3], T Manescu[1] and Z I Korka[1]

[1] NDT Laboratory, Faculty of Engineering and Management, "Eftimie Murgu" University of Resita, Romania
[2]BMW AG, Munich, Deutschland
[3] Laboratory Soete, Faculty of Engineering and Architecture, Ghent University, Belgium

E-mail: gr.gillich@uem.ro

Abstract. In actual damage detection methods, localization and severity estimation can be treated separately. The severity is commonly estimated using fracture mechanics approach, with the main disadvantage of involving empirically deduced relations. In this paper, a damage severity estimator based on the global stiffness reduction is proposed. This feature is computed from the deflections of the intact and damaged beam, respectively. The damage is always located where the bending moment achieves maxima. If the damage is positioned elsewhere on the beam, its effect becomes lower, because the stress is produced by a diminished bending moment. It is shown that the global stiffness reduction produced by a crack is the same for all beams with a similar cross-section, regardless of the boundary conditions. One mathematical relation indicating the severity and another indicating the effect of removing damage from the beam. Measurements on damaged beams with different boundary conditions and cross-sections are carried out, and the location and severity are found using the proposed relations. These comparisons prove that the proposed approach can be used to accurately compute the severity estimator.

1. Introduction

Integrity assessment of structures involving vibration-based techniques has got an increased attention in last decades. The core of these techniques is the vibration measurement, which provides information in real time regarding the structure's parameter changes, in this way making possible the detection of damage occurrence [1]. In the last few years, many research reports on vibration-based damage detection techniques have been published [2-9]. In respect to the geometry and orientation, damages can be differently classified. This study is limited to analyze the dynamic behavior of Euler-Bernoulli beams with transverse cracks. The rigidity of such beams is clearly diminished by the cracks, and also the mass is reduced. Among many past studies, two main crack ranges raised up: open cracks assuming the loss of rigidity and mass, and breathing cracks acting just on the rigidity [10].

In the case of open cracks, no contact between the crack edges is present, thus the vibration evolution is considered as linear. In all applications, single degree of freedom (SDOF) models are used for these systems. In the case of breathing cracks, the open and the closed stages alternate. The inertial forces compress the crack and close it. In the next stage, the forces act in the opposite direction and open the crack. This mechanism is known as breathing [11]. Studied papers, see for instance [12] and

[13], presented the SDOF models as always useful for the beams with breathing cracks and proposed bilinear model for nonlinearity. As a consequence, the vibration is considered nonlinear.

Severity estimation for both damage types is usually performed by means of the fracture mechanics approach [14-17]. For each damage geometry, cross-section shape combinations are necessary an individual analysis. This paper deals with the limitations of the above-mentioned approaches and proposes a new behavioral model that globally characterizes the damaged beam's dynamic behavior. In this model, the modal stored energy is introduced as a measure of the natural frequency and the damage severity is found from the decrease of the stored energy due to crack. The energy is evaluated from the beam deflection, which indicates an energy loss when it increases.

2. Actual models of cracked beams

It is largely recognized that, for open cracks, the beam rigidity equally decreases in both senses in transverse directions, e.g; from k in the healthy state to αk in the damaged one. Therefore, the vibration signal is a sinusoid, having decreased frequency in comparison with that of the original intact beam. If the intact and damaged beams are excited with the same amount of energy, a larger deflection is expected for the damaged ones. The behavior of a beam with an open crack close to the beam center is presented in figure 1. The blue line indicates the deflection achieved by the damaged beam, while the dashed line represents the case of the intact beam excited with the same amount of energy.

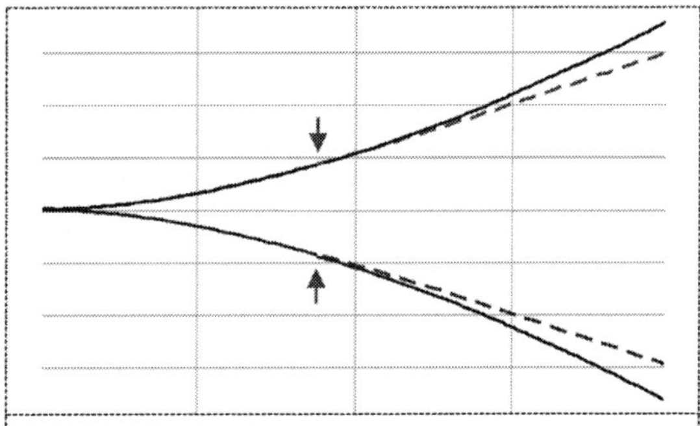

Figure 1. Maximum elongation of a beam with open crack near the center.

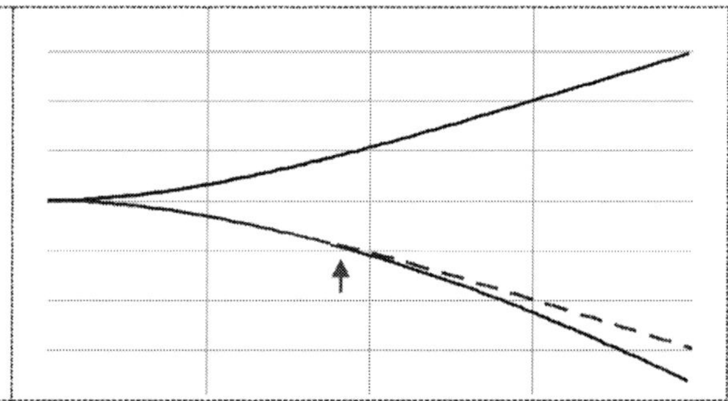

Figure 2. Maximum elongation of a beam with breathing crack near the center.

The behavior of a beam with a breathing crack is described in figure 2. One observes that, for the open stage the beam with a breathing crack behaves similar to the beam with an open crack, while in the closed stage it behaves like the healthy beam. For both stages, the amplitude is marked with blue color, the dashed line again representing the case of the intact beam. As a consequence, for a beam with a breathing crack, the stiffness depends on whether or not the crack is closed. In fact, it takes two different values, depending on the bending moment's sign [11]. Usually, the stiffness function $\tilde{k}(z)$ is defined as:

$$\tilde{k}(z) = \begin{cases} k_C = k & \text{if } z \geq 0 \\ k_O = \alpha k & \text{if } z < 0 \end{cases} \quad (1)$$

where z is the transversal direction.

The parameter α is called the stiffness ratio and takes values in the range of 0 to 1. For the stiffness ratio $\alpha = 1$, the system is linear, else it is stepwise linear. The two states correspond to the stages of the breathing mechanism: a) in the closed stage, indicated by index C, the stiffness ratio is equal to the unit and b) in the open stage, marked with O, the stiffness ratio is less that 1. Figures 3, 4 and 5 present the SDOF models of the intact beam, the beam with open crack and closed crack, respectively.

If involving the stiffness from equation (1) to describe the two distinct stages of breathing, the signal displays a half-period T_C similar to that of the intact beam, followed by an increased half-period T_O, for which the decreased rigidity is responsible. Larger amplitude for the open stage is observed.

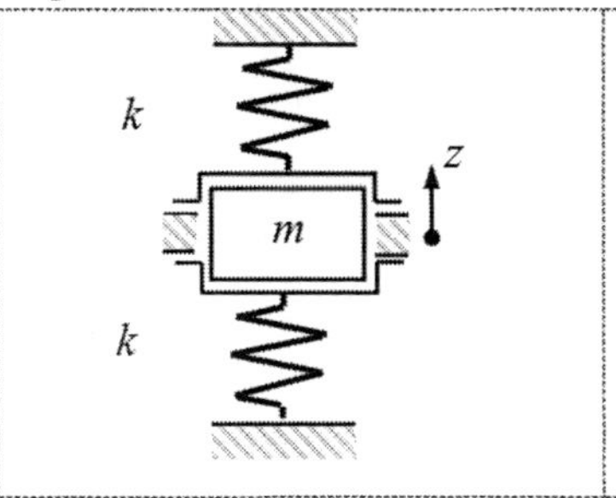

Figure 3. SDOF model for the intact beam.

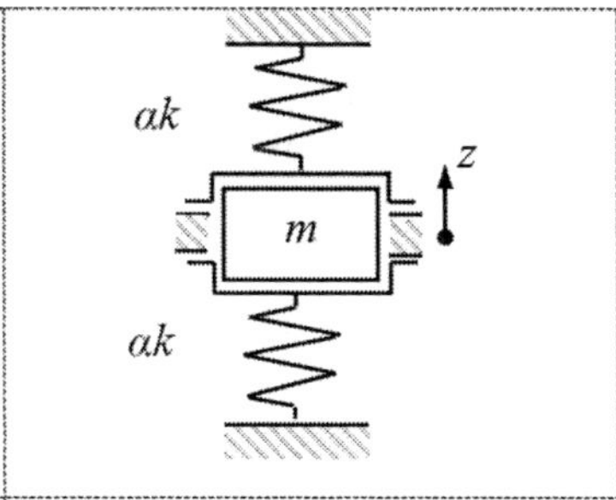

Figure 4. SDOF model for the beam with an open crack.

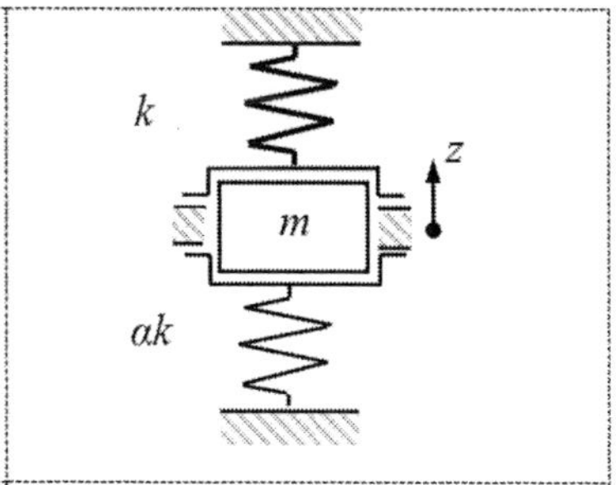

Figure 5. SDOF model for the beam with a breathing crack.

In [12], the solutions for the angular frequencies are: $\omega_C = \sqrt{k_C / m}$ for the closed crack, and $\omega_O = \sqrt{k_O / m}$ for the open crack. The frequency ω_D of the bilinear oscillator in free vibration, the so-called bilinear frequency, satisfies the condition $\omega_O < \omega_D < \omega_C$, and is derived as:

$$\omega_D = \frac{2\omega_C \omega_O}{\omega_C + \omega_O}. \tag{2}$$

Apparently, the phenomenon is well-described, but some inadvertencies are observed by a deeper analysis [18]. These arise if the analysis is made by means of energetic methods and will be demonstrated in the next section.

3. Derivation of damage severity

Given that the law of conservation of energy is valid, the maximum strain energy in the open and closed stage is the same. In addition, the kinetic energy at the rest position should also be equal to these values. This can be formalized as:

$$TE = \max EP_O = \max EK = \max EP_C = \text{const.} \tag{3}$$

Where TE is the total energy, EP_O and EP_C are the strain energy in open and closed stages, respectively, and EK is the kinetic energy.

3.1. The frequencies of the intact cantilever beam

For a uniform cantilever beam, the kinetic energy of a given mode i is derived as:

$$EK_i = \frac{\bar{m}}{2} \int_0^L \left[X_i'(x,t)\right]^2 dx \tag{4}$$

where $X_i'(x,t)$ is the velocity of a point located at distance x from the origin (left beam end), $\bar{m} = \rho A$ is the distributed mass, E is the Young's modulus, I is the moment of inertia and t stays for time. If $X_i(x,t)$ is the beam deflection in transversal direction, and assuming a sinusoidal displacement law, $X_i(x,t)$ can be written as:

$$X_i(x,t) = \phi_i(x) \cdot \sin(\omega_i t) \tag{5}$$

Here $\phi_i(x)$ denotes the mode shape at maximum elongation and ω_i the angular frequency of vibration mode i. Thus, the highest kinetic energy at the rest position is given as:

$$\max EK_i = \frac{\bar{m}}{2}\int_0^L [\phi_i'(x)]^2 dx = \frac{\omega_i^2 \cdot \bar{m}}{2}\int_0^L [\phi_i(x)]^2 dx \tag{6}$$

The potential energy is:

$$EP_i = \frac{EI}{2}\int_0^L [X_i''(x,t)]^2 dx \tag{7}$$

Thus, the pick value is obtained for the extreme positions of the beam as:

$$\max EP_i = \frac{EI}{2}\int_0^L [\phi_i''(x)]^2 dx \tag{8}$$

From equations (6) and (8), the angular frequency is calculated as:

$$\omega_i^2 = \frac{EI\int_0^L [\phi_i''(x)]^2 dx}{\bar{m}\int_0^L [\phi_i(x)]^2 dx} \tag{9}$$

It is known that the mode shape for the cantilever beam is:

$$\phi_i(x) = \cosh\alpha_i x - \cos\alpha_i x - \frac{\cos\alpha_i L + \cosh\alpha_i L}{\sin\alpha_i L + \sinh\alpha_i L}(\sinh\alpha_i x - \sin\alpha_i x) \tag{10}$$

and the modal curvature is:

$$\phi_i''(x) = \alpha_i^2\left[\cosh\alpha_i x + \cos\alpha_i x - \frac{\cos\alpha_i L + \cosh\alpha_i L}{\sin\alpha_i L + \sinh\alpha_i L}(\sinh\alpha_i x + \sin\alpha_i x)\right] \tag{11}$$

where α is a dimensionless parameter reflecting the boundary conditions. Substituting these expressions in equation (9), one obtains the angular frequency of the *i*-th mode, as:

$$\omega_i^2 = \frac{\alpha_i^4 EI\int_0^L \left[\cosh\alpha_i x + \cos\alpha_i x - \frac{\cos\alpha_i L + \cosh\alpha_i L}{\sin\alpha_i L + \sinh\alpha_i L}(\sinh\alpha_i x + \sin\alpha_i x)\right]^2 dx}{\bar{m}\int_0^L \left[\cosh\alpha_i x - \cos\alpha_i x - \frac{\cos\alpha_i L + \cosh\alpha_i L}{\sin\alpha_i L + \sinh\alpha_i L}(\sinh\alpha_i x - \sin\alpha_i x)\right]^2 dx} \tag{12}$$

In [19], it is proved that the functions under the two integrals provide the same result $0.25L\delta^2$, where δ is the free end deflection. After substitution and reduction, the angular squared frequency becomes:

$$\omega_i^2 = \frac{\alpha_i^4 EI \cdot 0.25L}{\bar{m} \cdot 0.25L} = \alpha_i^4 \frac{EI}{\bar{m}} \tag{13}$$

Hence, the well-known expression of the natural frequency of a beam for the bending modes is obtained as:

$$f_i = \frac{\omega_i}{2\pi} = \frac{\lambda_i^2}{2\pi}\sqrt{\frac{EI}{\bar{m}L^4}} \tag{14}$$

with $\lambda = \alpha / L$ derived from the beam's characteristic equation. For the cantilever, this equation is:

$$\cos\lambda\cosh\lambda + 1 = 0 \tag{15}$$

One can easily observe in Eq. (14) that the term under the square is inversely proportional to the beam deflection at the free end if only dead weight is acting, i.e.:

$$\delta = \frac{\overline{m} g L^4}{8EI} \tag{16}$$

where g denotes the gravitational acceleration. From equations (14) and (16), the natural frequencies in respect to the free end deflection are achieved under dead weight as:

$$f_i = \frac{\lambda_i^2}{2\pi}\sqrt{\frac{g}{8\delta(L)}} \tag{17}$$

3.2. The frequencies of a cantilever beam with a breathing crack

Let us now consider the cracked beam vibrates, and in open stage, the free end achieves amplitude δ^* similar to the deflection of the intact beam under dead weight. Because of energetic equilibrium, a smaller amplitude δ is achieved in the closed stage. In figure 6, both open and closed states are depicted on the same side, in order to permit a facile comparison. The shape in the open state is marked with a blue line and that in the closed state with a gray dashed line. Note that the same quantity of strain energy is stored in both states since the larger deflection occurs just due to a supplementary rotation in a narrow slice subjected to damage. In all other damaged beam slices, the same energy is stored as in the intact beam. This is reflected in figure 7 by the squared modal curvature. The strain energy for the open and closed stages are both represented with bold blue line, the single difference between the two distribution consisting in the small hill present in the slice where supplementary rotation is possible.

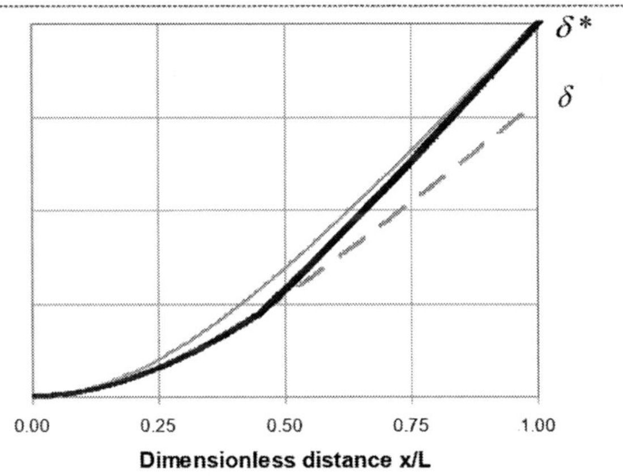

Figure 6. Mode shape of the beam with breathing crack in open state (blue line) and closed state (gray dashed line) and substituted intact beam (gray line).

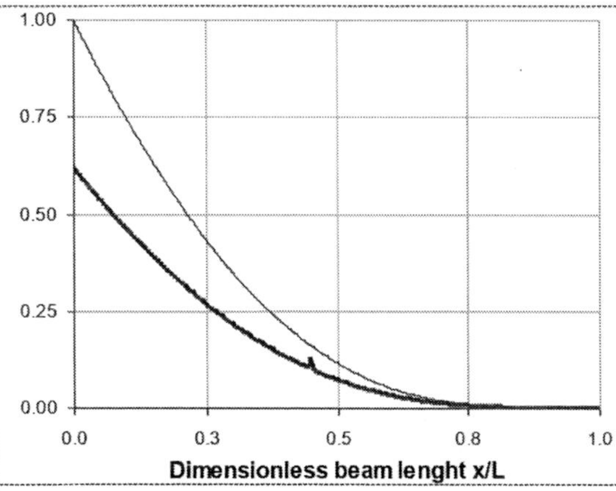

Figure 7. Square of the mode shape curvature for the damaged beam in open and closed stage (bold line) respectively the surrogate intact beam (thin line).

The above presented example shows that stored energy and deflection ratio must be considered in the analysis. Now, we can replace the damaged beam with a healthy one with decreased rigidity and achieving similar deflection as the damaged in open stage. This will store, according to Castigliano's second theorem, which is applicable to the stepwise linear system under analysis, the same energy as the damaged beam. Let this beam, nominated herein as surrogate beam, have the pick deflection δ^* and the subsequent rigidity $EI_D(c)$. It is worth to mention that the distributed mass $\overline{m}$ is maintained similar with that of the intact original beam. Also, the surrogate rigidity $EI_D(c)$ is influenced by the damage position c; the closer the damage to the fixed end, the lower the surrogate rigidity is.

The squared modal curvature for the surrogate beam, represented in figure 7 with thin blue line, achieves larger amplitude as the damaged beam in closed stage, due to the decreased rigidity. The ratio is defined by deflections δ^* and δ, their values being deductible from equation (19).

The rigidity $EI_D(c)$ is derived from the deflections of the original and surrogate beams at the free end, which are:

$$\delta^* = \frac{\bar{m}gL^4}{8EI_D(c)} \quad \text{and} \quad \delta = \frac{\bar{m}gL^4}{8EI} \tag{18}$$

thus

$$EI_D(c) = EI\frac{\delta}{\delta^*} \tag{19}$$

For the intact and surrogate beam, respectively, the angular frequencies ω_i and ω_{i-D} are deductible from:

$$\omega_i^2 = \frac{EI\int_0^L [\phi_i''(x)]^2 dx}{\bar{m}\int_0^L [\phi_i(x)]^2 dx} = \frac{EI}{\bar{m}} \quad \text{and} \quad \omega_{i-D}^2 = \frac{EI_D(c)\int_0^L [\phi_{i-D}''(x)]^2 dx}{\bar{m}\int_0^L [\phi_{i-D}(x)]^2 dz} = \frac{EI_D(c)}{\bar{m}} \tag{20}$$

Hence, the frequency of a cracked beam in closed state is

$$f_{i-D} = f_i\sqrt{\frac{EI_D(c)}{EI}} = f_i\sqrt{\frac{\delta}{\delta^*}} \tag{21}$$

and the dimensionless frequency is

$$\bar{f}_{i-D} = \frac{f_{i-D}}{f_i} = \sqrt{\frac{\delta^*}{\delta}} \tag{22}$$

The deflection ratio, $r = \delta^* \cdot \delta^{-1}$, presented in equations (21) and (22) merits a special remark. Since just the inertial forces act on the beams and no large deflections occur (introducing nonlinearities), this ratio depends only on the damage position c and its depth a.

Now, we change the initial condition. The analysis is made for the free end of the cracked beam removed with a distance δ in such a way that the crack closes. By releasing the free end, the beam vibrates, in the opposite position achieving a larger deflection δ^*. The case becomes similar with that of the closed crack presented in subsection 3.2. The same surrogate beam results, and the same frequency as indicated in equations (21) and (22) are obtained.

3.3. Some considerations regarding the vibration of cracked beams

A conclusion from this study is that the two half-periods T_O and T_C are identical for the free vibration of beams subjected to breathing cracks. Also, it results that the frequency of the damaged beam can be found involving equation (21), basing on the intact beam frequency and the beam deflection in intact respectively damaged state.

If a given damage is located in a slice where the curvature achieves maxima, i.e. the fixed end for the cantilever beam, the lowest value for ratio r is achieved. Here, the lowest deflection ratio and surrogate rigidity are achieved. This beam is nominated herein as equivalent beam and the rigidity as equivalent rigidity EI_{eq}.

For a beam with an open crack, the damage expansion on the longitudinal axis increases the deflection due to the weak segment between the crack edges. This has as a consequence an additional frequency decrease in comparison with that produced by a breathing crack with a similar depth. The frequency is still deductible from equation (21).

Because the mode shapes and modal curvatures are reduced in equation (20), the equation (21) can be used for deriving the frequencies for damaged beams with other boundary conditions. Obviously, the intact beam's frequencies for the given boundary conditions must be involved.

4. A simple damage severity estimator

As shown in previous section, the damage depth is one of the parameters affecting the natural frequencies. The second parameter is the damage position. Let us firstly consider the damage at the fixed end, and take as variable the damage depth a. In open stage, the deflection at the free end is $\delta_O(0,\bar{a})$, where 0 indicates the damage location, i.e. $c=0$, and $\bar{a}=a/h$ is the dimensionless damage depth. If the deflection of the intact beam is δ, the frequency shift for a beam with rigidity decrease can be calculated as:

$$\Delta f_{i-D} = f_i - f_{i-D} = \frac{\lambda_i^2}{2\pi}\sqrt{\frac{8}{g\delta}} - \frac{\lambda_i^2}{2\pi}\sqrt{\frac{8}{g\delta_O(0,\bar{a})}} = \frac{\lambda_i^2}{2\pi}\left(\sqrt{\frac{8}{g\delta}} - \sqrt{\frac{8}{g\delta_O(0,\bar{a})}}\right) \tag{23}$$

Hence, the relative frequency shift, obtained by normalizing the frequency of the intact beam f_i, becomes:

$$\Delta\bar{f}_{i-D} = \frac{f_i - f_{i-D}}{f_i} = \frac{\frac{\lambda_i^2}{2\pi}\left(\sqrt{\frac{8}{g\delta}} - \sqrt{\frac{8}{g\delta_O(0,\bar{a})}}\right)}{\frac{\lambda_i^2}{2\pi}\sqrt{\frac{8}{g\delta}}} = \frac{\sqrt{\delta_O(0,\bar{a})} - \sqrt{\delta}}{\sqrt{\delta_O(0,\bar{a})}} = \gamma(0,\bar{a}) \tag{24}$$

The absolute damage severity $\gamma(0,\bar{a})$ indicates the biggest effect, which a given damage can produce on the cantilever beam, being valid for all bending vibration modes. It can be simply calculated from the deflection of the beam with a breathing crack in intact and damaged state. Based on this damage evaluator, the frequency of a damaged beam with a breathing crack at the fixed end can be found as:

$$f_{i-D} = f_i - f_i \cdot \Delta\bar{f}_{i-D} = f_i\left[1 - \frac{\sqrt{\delta_O(0,\bar{a})} - \sqrt{\delta(L)}}{\sqrt{\delta_O(0,\bar{a})}}\right] = f_i\left[1-\gamma(\bar{a})\right] \tag{25}$$

Let us now consider the crack located elsewhere on the beam, for instance at normalized distance $\bar{c}=c/L$ from the fixed end. Obviously, the deflection of the damaged beam will be smaller. We take now advantage of introducing the surrogate beam. This is concretized in distributing the real rigidity decrease of one beam element onto all elements, as depicted in figure 8 for the second vibration mode. In figure 8.a, the element near the fixed end has a reduced rigidity due to the crack. If this is located on another element, the effect will be decreased in function of the energy stored in that element, thus in respect to $\left[\bar{\phi}_i''(\bar{c})\right]^2$. In this way, the effect of damage location upon the severity is introduced. Now, we can write the damage severity as:

$$\gamma(\bar{c},\bar{a}) = \gamma(0,\bar{a}) \cdot \left[\bar{\phi}_i''(\bar{c})\right]^2 \tag{26}$$

$$\gamma(\bar{c},\bar{a})=\frac{\sqrt{\delta_O(\bar{c},\bar{a})}-\sqrt{\delta}}{\sqrt{\delta_O(\bar{c},\bar{a})}}=\frac{\sqrt{\delta_O(0,\bar{a})}-\sqrt{\delta}}{\sqrt{\delta_O(0,\bar{a})}}\cdot\left[\bar{\phi}_i''(\bar{c})\right]^2 \quad (27)$$

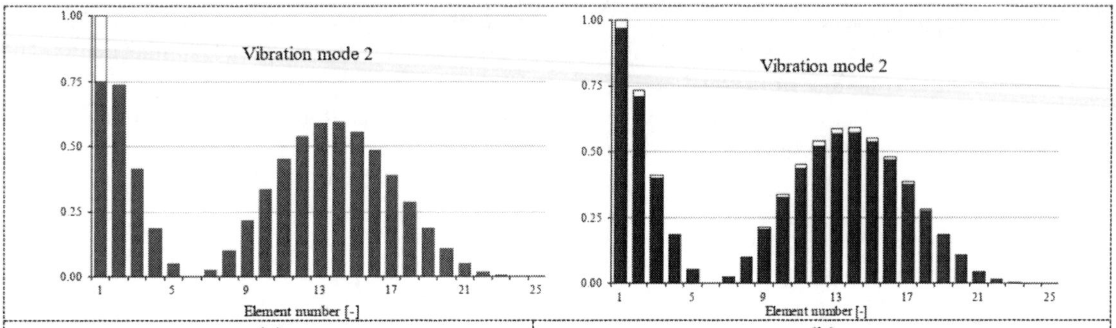

(a) (b)

Figure 8. Normalized square of the modal curvature corresponding to bending vibration mode two: (a) damaged beam; (b) equivalent beam.

The natural frequency of a beam with a breathing crack with depth $\bar{a}$ and position $\bar{c}$ is

$$f_{i-D}=f_i\left\{1-\gamma(\bar{a})\cdot\left[\phi_i''(\bar{c})\right]^2\right\} \quad (28)$$

This simple mathematical relation can be used to develop damage scenarios in order to create databases usable in damage detection procedures. For open cracks, in case of a fine tuning, the mass loss should also be taken into consideration [20].

5. Numerical examples

The robustness of the proposed damage severity indicator was proved by numerical experiments on steel beams with length $L=600\text{mm}$ and three different cross-sections shapes: square, hexagonal and circular. The cross-section dimensions were chosen as following: the square edges $h_S=12\text{mm}$, the circle diameter $d=12\text{mm}$ and the distance between the parallel sides of the hexagon $h_H=12\text{mm}$. All simulations were made by means of the ANSYS package, the mechanical characteristics being that of the structural steel from the dedicated library. The medium size of the tetrahedral elements, used for meshing, was fixed to 0.5mm.

The damage, in form of a breathing crack, was positioned near the fixed end. Several crack depths were used in the simulations. First a static analysis in order to find the deflections was performed. Afterward, the modal analysis was used to find out the natural frequencies. The achieved results are presented separately for each cross-section shape in tables 1 to 3. Percentile error was contrived, to evaluate the damage severity precision.

Table 1. Results achieved for the hexagonal beam. The absolute damage severity and the relative frequency shift are indicated together with the contrived error.

a [mm]	δ_O [mm]	γ [%]	f_1 [Hz]	RFS [%]	Error [%]
0	0.62271	0	24.82598	0	0
1	0.62572	0.240813	24.76592	0.241942	0.466644
2	0.63387	0.884216	24.60571	0.887279	0.345255
3	0.64781	1.956435	24.33850	1.963586	0.364182
4	0.66976	3.576406	23.93435	3.591538	0.42134
5	0.70426	5.967841	23.33672	5.998789	0.515904
6	0.76214	9.608927	22.42499	9.671274	0.644665

Table 2. Results achieved for the circular beam. The absolute damage severity and the relative frequency shift are indicated together with the contrived error.

a [mm]	δ_O [mm]	γ [%]	f_1 [Hz]	RFS [%]	Error [%]
0	0.69271	0	23.53858	0	0
1	0.69431	0.115289	23.51134	0.115736	0.386
2	0.70192	0.658224	23.38283	0.661669	0.520
3	0.71569	1.618542	23.15626	1.624223	0.349
4	0.73936	3.206153	22.78038	3.221107	0.464
5	0.77817	5.650743	22.20134	5.681056	0.533
6	0.84435	9.423721	21.30601	9.484736	0.643

Table 3. Results achieved for the prismatic beam. The absolute damage severity and the relative frequency shift are indicated together with the contrived error.

a [mm]	δ_O [mm]	γ [%]	f_1 [Hz]	RFS [%]	Error [%]
0	0.51890	0	27.19549	0	0
1	0.52218	0.314563	27.10972	0.315372	0.256585
2	0.53077	1.124509	26.88894	1.127184	0.237287
3	0.54527	2.448032	26.52740	2.456623	0.349687
4	0.56779	4.402185	25.99316	4.421045	0.426585
5	0.60257	7.202114	25.22600	7.241957	0.550169
6	0.65810	11.20349	24.12754	11.28109	0.687906

To calculate the damage severity, the equation (24) was used, where the input terms are the beam deflections. Also equation (24) is used to find the relative frequency shifts, herein the input data consists of the first natural frequencies of the beams.

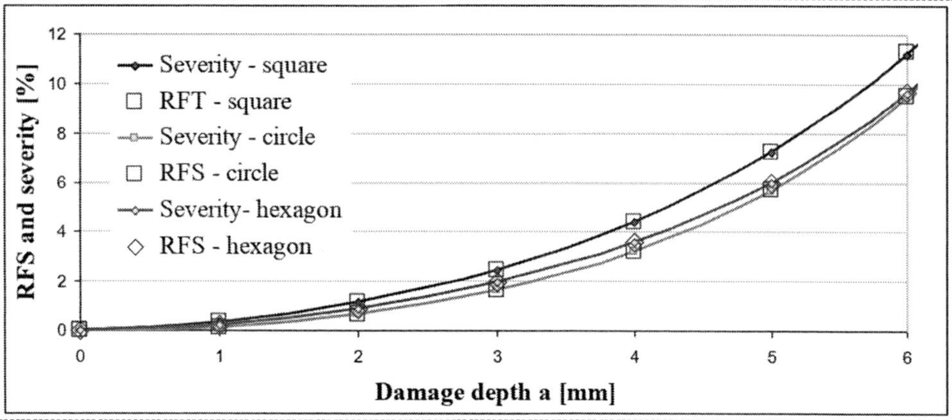

Figure 9. Damage severity and the relative frequency shift for damage depth evolution.

The curves plotted in figure 9 present good similarity with that presented in the literature, see [14-17]. Even more significant is the extremely low error level found between the damage severity and relative frequency shifts. As it was calculated in percentage, the values are indicated in the last column of tables 1 to 3. One can observe that the errors are less than 1%, meaning that the proposed relation to contrive the damage severity is reliable. This can be also concluded from figure 9, where the perfect fit between the compared values is observable.

6. Conclusion

This paper introduces a robust damage severity evaluator that predicts the frequency decrease occurred in a cantilever beam due to a breathing crack. It can be simply calculated using the deflections produced by the dead mass to the intact beam and the beam with a crack in the open state respectively. This introduces the important advantage that evaluator can be used for any cross-section shape. A particular case of this feature is the absolute damage severity, which is derived for the crack located in a slice where the bending moment achieves maxima. Thus, it indicates the biggest effect, which a given damage can produce on the frequencies, and it is unique for all bending vibration modes. In contrary, the damage severity takes, for a given crack, different values in respect to the position on the beam and the vibration mode number. The ratio between the damage severity and the absolute damage severity is actually the normalized squared modal curvature of the intact beam at the crack location. An important conclusion resulted from this study, which permits a correct interpretation of the vibration of beams with breathing cracks, is that the two half-periods T_O and T_C are identical. This happens because the frequencies are proportional to the stored energy which is the same in the open and closed stage. Therefore, the estimator can be involved for a beam with an open crack as well. The single difference here is the damage expansion on the longitudinal axis, producing an increased deflection due to the weak segment between the crack edges. This has as result an additional frequency decrease in comparison with that produced by a breathing crack with similar depth; it is proportional to the deflection increase. In the mathematical relations used to calculate the deflection for different boundary conditions just the numerical value is different. Thus, the deflection ratio r is identical for beams with any support type, because this value is reduced in the ratio. As a consequence, the deflection ratio r for the cantilever beam can be used for any beam configuration. It is the most convenient case, since here the largest deflections are expected and so reduced errors are presumed. Remarkable is the fact that the absolute damage severity estimator permits predicting the frequencies of the damaged beam by involving only the intact beam frequency and the damage depth and its location (in order to calculate the normalized square of the modal curvature for the intact beam). This means that the frequency for damaged beams with any boundary conditions is derivable knowing the crack location and depth.

Acknowledgements

The work has been funded by the Sectoral Operational Programme Human Resources Development 2007-2013 of the Ministry of European Funds through the Financial Agreement POSDRU/159/1.5/S/132395.

References

[1] Gillich G R, Praisach Z I, Abdel Wahab M and Vasile O 2014 Localization of transversal cracks in sandwich beams and evaluation of their severity, *Shock Vib.* Article ID 607125

[2] Zhou Y-L, Maia NMM, Sampaio R and Wahab MA 2016 Structural damage detection using transmissibility together with hierarchical clustering analysis and similarity measure *Structural health monitoring* doi: 10.1177/1475921716680849

[3] Zhou Y-L, Maia N and Abdel Wahab M 2016 Damage detection using transmissibility compressed by principal component analysis enhanced with distance measure *Journal of Vibration and Control* doi: 10.1177/1077546316674544

[4] Zhou Y-L and Abdel Wahab M 2016 Rapid early damage detection using transmissibility with distance measure analysis under unknown excitation in long-term health monitoring *Journal of Vibroengineering* **18**(7) 4491-4499

[5] Khatir S, Belaidi I, Serra R, Abdel Wahab M and Khatir T 2016 Numerical study for single and multiple damage detection and localization in beam-like structures using BAT algorithm *Journal of Vibroengineering* **18**(1) 202-213

[6] Khatir A, Tehami M, Khatir S and Abdel Wahab M 2016 Multiple damage detection and localization in beam-like and complex structures using co-ordinate modal assurance criterion

combined with firefly and genetic algorithms *Journal of Vibroengineering* **18**(8) 5063-5073

[7] Gillich G-R, Praisach Z-I, Abdel Wahab M, Gillich N, Mituletu IC and Nitescu C 2016 Free vibration of a perfectly clamped-free beam with stepwise eccentric distributed masses *Shock and Vibration* **2016**(Article ID 2086274) 10 pages; http://dx.doi.org/10.1155/2016/2086274

[8] Khatir S, Belaidi I, Serra R, Abdel Wahab M and Khatir T 2015 Damage detection and localization in composite beam structures based on vibration analysis *Mechanika* **21**(6) 472-479

[9] Zhou Y-L and Abdel Wahab M 2017 Cosine based and extended transmissibility damage indicators for structural damage detection *Engineering Structures* **141** 175-183

[10] Shen M H H and Chu Y C 1992 Vibrations of beams with a fatigue crack, *Comput. Struct.* **1**(45) 79–93

[11] Rivola A and White P R 1998 Bispectral analysis of the bilinear oscillator with application to the detection of fatigue cracks, *J. Sound Vib.* **5**(216) 889-910.

[12] Yan G, De Stefano A, Matta E and Feng R 2013 A novel approach to detecting breathing-fatigue cracks based on dynamic characteristics *J. Sound Vib.* **2**(332) 407–422.

[13] Bovsunovsky A and Surace C 2015 Non-linearities in the vibrations of elastic structures with a closing crack: A state of the art review *Mech. Syst. Signal Pr.* **62-63** 129-148.

[14] Bilello C 2001 Theoretical and Experimental Investigation on Damaged Beams under Moving Systems. Ph.D.Thesis, Universita` degli Studi di Palermo, Italy.

[15] Chondros T J, Dimarogonas A D and Yao J 1998 A continuous cracked beam vibration theory *J. Sound Vib.* **1**(215) 17–34.

[16] Ostachowicz W M and Krawczuk C 1991 Analysis of the effect of cracks on the natural frequencies of a cantilever beam *J. Sound Vib.* **2**(150) 191–201.

[17] Liebowitz H and Claus Jr W D S 1968 Failure of notched columns *Eng. Frac. Mech.* **2**(1) 379–383.

[18] Gillich G R and Praisach Z I 2012 Robust method to identify damages in beams based on frequency shift analysis *Proc. SPIE 8348* p. 83481D.

[19] Gillich G R, Praisach Z I, Iancu V, Furdui H and Negru I 2015 Natural Frequency Changes due to Severe Corrosion in Metallic Structures *Stroj. Vestn.-J. Mech. E.* **61**(12) 721-730

[20] Gillich G R and Praisach Z I 2013 Detection and Quantitative Assessment of Damages in Beam Structures Using Frequency and Stiffness Changes *Key Eng. Mat.* **569-570** 1013-1020.

Research of Ancient Architectures in Jin-Fen Area Based on GIS&BIM Technology

Jing Jia[1], Qiuhong Zheng [1*], Huiying Gao[1] and Hai Sun[1]
1 Civil Engineering Department, College of Engineering, Ocean University of China, No.238 Songling Road, Qingdao 266100, P. R. China

[*]E-mail: zhengqiuhongouc@163.com, TEL: 86-1576227290, Fax: 86-0532-66781550

Abstract. The number of well-preserved ancient buildings located in Shanxi Province, enjoying the absolute maximum proportion of ancient architectures in China, is about 18418, among which, 9053 buildings have the structural style of wood frame. The value of the application of BIM (Building Information Modeling) and GIS (Geographic Information System) is gradually probed and testified in the corresponding fields of ancient architecture's spatial distribution information management, routine maintenance and special conservation & restoration, the evaluation and simulation of related disasters, such as earthquake. The research objects are ancient architectures in JIN-FEN area, which were first investigated by Sicheng LIANG and recorded in his work of "Chinese ancient architectures survey report". The research objects, i.e. the ancient architectures in Jin-Fen area include those in Sicheng LIANG's investigation, and further adjustments were made through authors' on-site investigation and literature searching & collection. During this research process, the spatial distributing Geodatabase of research objects is established utilizing GIS. The BIM components library for ancient buildings is formed combining on-site investigation data and precedent classic works, such as "Yingzao Fashi", a treatise on architectural methods in Song Dynasty, "Yongle Encyclopedia" and "Gongcheng Zuofa Zeli", case collections of engineering practice, by the Ministry of Construction of Qing Dynasty. A building of Guangsheng temple in Hongtong county is selected as an example to elaborate the BIM model construction process based on the BIM components library for ancient buildings. Based on the foregoing work results of spatial distribution data, attribute data of features, 3D graphic information and parametric building information model, the information management system for ancient architectures in Jin-Fen Area, utilizing GIS&BIM technology, could be constructed to support the further research of seismic disaster analysis and seismic performance simulation.

Key word: Ancient architectures in Jin-Fen Area, BIM components library for ancient buildings, BIM model of Mahavira hall in Guangsheng temple

1. Introduction

The number of well-preserved ancient buildings located in Shanxi Province, enjoying the absolute maximum proportion of ancient architecture in China, is about 18418, among which, 9053 buildings have the structural style of wood frame. Four existing Tang Dynasty wooden structures and 75% wooden structures of the Five Dynasties are located in the Shanxi Province. The number of the National Important Historical Monuments under Special Preservation, being located in Shanxi, ranks first in China. In 1985, Sicheng LIANG has visited the research area of this paper to explore the Chinese ancient architectures and wrote "Record of the pre-investigation of the ancient buildings in the JIN-FEN area" [1] to record the characteristics of ancient architectures in JIN-FEN area. In this paper, the ancient buildings' data in JIN-FEN area are collected as the sample group to research the protection of ancient building by using the BIM & GIS technology. The author built components models to form BIM components library for ancient buildings by using Revit software according to the "Yingzao Fashi" [2], an architectural practice book of Song dynasty, and tried to use the BIM library to build a whole ancient building. During this research process, the spatial distributing Geodatabase is established utilizing GIS to manage all the information collected for the ancient architectures in the research area. A building of Guangsheng temple in Hongtong County is then selected as an example to elaborate the BIM model construction process based on the BIM components library for ancient buildings.

2. Application of BIM technology in ancient building protection

BIM technology is widely used in the construction industry. Since it was invented for modern building modeling, it was primarily used in the planning, design and construction phases of buildings. Recently, BIM technology has begun to be used in the maintenance, renovation and management of existing buildings throughout their lifecycle. It is primarily used in the imitation and design of ancient buildings and combined with 3D scanning technology. The main process is using scanning equipment to three-dimensionally scan the whole building and the data will be uploaded to the cloud database, which are downloaded into the BIM software later to generate 3D building model. The advantage of this method is that it is possible to restore the appearance of the ancient building with high accuracy. Nevertheless, the accuracy of the model is greatly related to the performance of the scanner, and a 3D scanner costs a lot. BIM technology combined with 3D scanning technology for model construction is the method from the macro to the details, i.e. from the construction of the overall building model to the refined internal component models. In this paper, the author proposes a method of constructing ancient building models from micro to macro. Models of ancient wooden fabrications are built first, and then the whole ancient building model is assembled and adjusted.

In the latest research, Xiang LUO and Guohua JI introduced the application of Revit Architecture in parametric modeling of ancient buildings [3]. Weichao SUN introduced the detailed method of building ancient building information models with Revit Architecture [4]. Basing on the research of'Gongcheng Zuofa Zeli" written in Qing Dynasties, Weixin SUN studies the model parameterization of the ancient building's components in Ming and Qing Dynasties [5]. Carlo Biagini et al. used the BIM technique to simulate the renovation and restoration of historical buildings, and the construction process is simulated, basing on a typical case study of the Nome di Maria church [6]. Maurice Murphy,

IOP Conf. Series: Journal of Physics: Conf. Series **842** (2017) 012035 doi:10.1088/1742-6596/842/1/012035

Eugene McGovern and Sara Pavia initially demonstrated the potential of HBIM in protecting historic buildings and the environment [7]. The ancient architectures in research area of Jin-Fen are partly built in the Song Dynasty, which has more referencing value for the study and protection of ancient buildings before the Qing Dynasty. Constructing BIM models of ancient building components is the main challenge, by referencing the rules and model illustrations of "Yingzao Fashi".

In recent years, Historic Building Information Modelling was brought forward abroad. Historic building information modelling (HBIM) is proposed as a new system for modelling historic structures. The HBIM process uses a combination of a ground laser scanner and a digital camera to collect data remotely, and then uses a series of software programs to combine images and scanned data [7].

3. Ancient buildings of JIN-FEN area

The number of well-preserved ancient buildings located in Shanxi Province is more than 9000, enjoying the absolute maximum proportion of ancient architecture in China. Almost all types of buildings from Tang Dynasty to Qing Dynasty can be found here. Among them, 106 buildings were built before the Song Dynasty and 4 buildings were built in the Tang Dynasty. The number of ancient buildings, which were recorded in "Record of the pre-investigation of the ancient buildings in the JIN-FEN area" is 20. These buildings include a building of Song Dynasty, JIN ancestral temple, ten buildings of the Yuan Dynasty and four buildings of the Ming Dynasty. Therefore, it is a very meaningful work to follow the footprint of Sicheng LIANG to explore the current situation of the ancient buildings in JIN-FEN area, which is especially beneficial for the maintenance of existing ancient buildings. The Current situations of ancient buildings in JIN-FEN area are shown in Table 1.

Table 1. Current situations of ancient buildings in JIN-FEN area.

Name	Current situation	The cause of the damage
Tianlong temple	Disappear	Unknown
Chongsheng temple	Disappear	Unknown
Lingyan temple	Pagoda and caves preserved	War
Guoning temple	Disappear	Unknown
Shengmu temple of Wenshui country	Disappear	Military
Wen temple of Wenshui country	Disappear	Unknown
Dongyue temple	Disappear	Road widening
Taiqing temple	Disappear	Unknown
Wen temple of Huozhou country	Disappear	Unknown
Zhusheng temple	Some preserved	Occupied by flour factory
Xifuchang temple	Disappear	Unknown
Shengmu temple of Huozhou country	Disappear	Unknown
County yamen of Huozhou country	Well preserved	
Bridge and iron ox	Disappear	Flood
Nvwa temple	Disappear	Unknown
Guangsheng temple	Well preserved	

Mingyingwang temple	Well preserved	
Zhongzhen temple	Disappear	War
Jinci temple	Well preserved	

Unfortunately, Sicheng LIANG's visiting did not arouse local people's awareness of the protection of these buildings. Since the reform and opening up, only seven ancient buildings survive, i.e. Jinci temple (Song Dynasty), Guangsheng Higher and Lower temples (Yuan Dynasty), Mingying temple, Yamen of Huozhou County, Zhushen temple of Huozhou County and Lingyan temple (Ming Dynasty), and the rest are destroyed due to war or other reasons. Now, some existing architectures, such as Jinci temple, have become famous tourist attractions and received enough investigation for maintenance and protection from the government every year. Because of the raising awareness of the protection of ancient buildings, some of the ancient buildings were renovated and rebuilt by the local villagers. However, methods of refurbishment and reconstruction are very inappropriate, which make ancient buildings become cement constructions and lose historical value. Such kind of behavior causes irreparable loss to the valuable ancient architectures. The Protection of ancient buildings should be the protection of their cultural value and the maintenance of their original appearances.

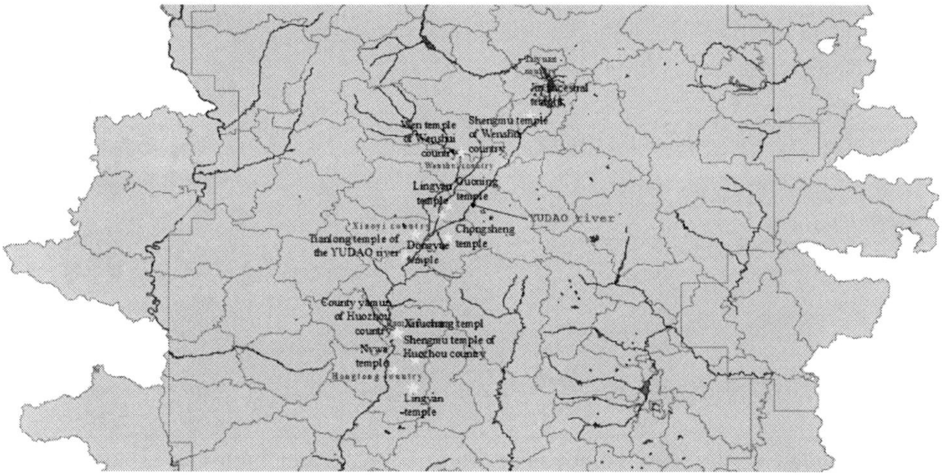

Figure 1. Spatial distribution map of ancient buildings in JIN-FEN area

As shown in Figure 1, JIN-FEN area is actually around the Yudao river, which includes Wenshui county, Xiaoyi county, Huozhou city, Hongtong county and Taiyuan city. Most of the ancient architectures are located in villages, where the protection measure is weak and only simple and original means are adopted to keep them from collapsing. It is very urgent to take steps to protect these disappearing ancient architectures.

4. BIM models of ancient buildings

Chinese ancient buildings are mainly composed of large wood components, and can be separated into scattered components. The sizes and construction sequence of these wooden components must obey detailed rules. Hence models of ancient wooden components should be built first, and then the whole ancient building model could be built according to the structure system of the building. The BIM components library for ancient buildings is an important basis for the construction of the whole

IOP Conf. Series: Journal of Physics: Conf. Series **842** (2017) 012035 doi:10.1088/1742-6596/842/1/012035

building model, which should contain abundant BIM models of ancient building components. In the construction process of the BIM components library for ancient buildings, "Yingzao Fashi", the official code for constructional engineering in Song Dynasty, is referenced, which is the most complete architectural technology book in Chinese history. It is an indispensable reference book for the study of the architectures built in and after Song Dynasty. For better understanding of this ancient book, "Notes of Yingzao Fashi", which is written by Sicheng LIANG, is referenced too. Every BIM model of ancient building components is constructed by Revit software.

4.1 Revit models of ancient building components

Chinese ancient buildings abide by the strict hierarchy. The scales of residential buildings (area and high) are based on the owner's official position, and the sizes for the wood components (beam, column, etc.) are correspondingly different. The authors use Revit software to build the components' models, in addition to the basic door, window and column, the other particular models are shown in table 2.

Table 2. Models of components.

Name	Type	Diagram
NI-DAO Gong	Gong	
MAN Gong	Gong	
LING Gong	Gong	
SAN Dou	Dou	
QIXIN Dou	Dou	
LU Dou	Dou	
JIAOHU Dou	Dou	

The construction of the BIM components library for ancient buildings is mainly through the method of constructing BIM families in Revit software, which can be controlled by the set family parameters. Each BIM family, representing a particular component of ancient buildings, shares the same shape and feature, and the specified size for different buildings can be adjusted for the corresponding building. The Completion of the library of BIM components will greatly speed up the establishment of ancient models of buildings.

4.2 BIM model of Mahavira hall in Guangsheng temple

Guangsheng temple is located in Hongtong country, and buildings of the temple were mainly built in Yuan Dynasty (year 1309). This temple is a National Important Historical Monuments under Special Preservation. Mahavira hall of Guangsheng temple, which was also built in the Yuan Dynasty, is taken as an example to elaborate the construction process of BIM model. The main structure contains seven halls, and the doors are set in the central three halls. High mullioned windows are set by the right and left side of the doors. The roof style uses suspended hill. The construction of Dougong is five sets, heavy Gong, and only used in column cap, without Dougong sets between columns. The size of Dougong is bigger than usual, and the comprehensive design is relatively concise. The distribution of columns adopted the method of moving and minus columns, which is popular in Song Dynasty, to reduce the number of inner columns and increase inner space.

The construction of the BIM model of Mahavira hall of Guangsheng temple mainly contains three steps. The ultimate BIM model of Mahavira hall of Guangsheng temple is shown in Figure 3, comparing with the photo of the hall.

4.2.1. The overall structure of Mahavira hall.

According to the "Chinese ancient architectures survey report", the basic information and architectural plans of the building are obtained, it is 25 meters long and 14.6 meters wide with a total area of 365 square meters. It is clearly shown in this picture that the eave columns and inner columns were not centered, which reveals the method of moving and minus columns.

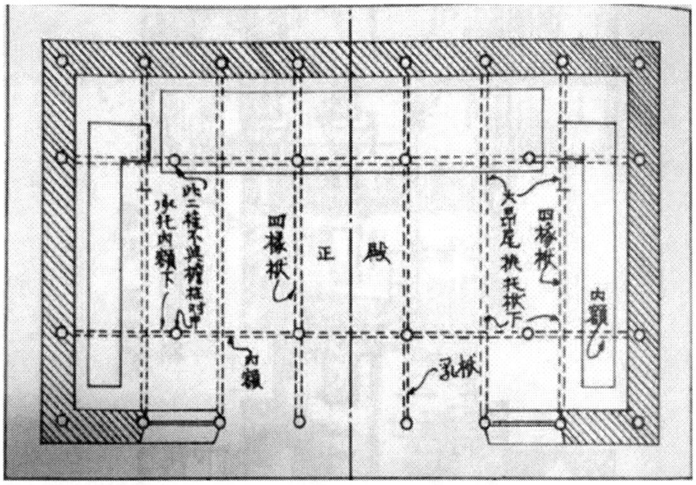

Figure 2. Architectural plan of Mahavira hall.

4.2.2. Analysis of the components type.

The types of ancient architecture components, being used in Mahavira hall and construction mode, are summarized by on-site investigation and literature searching & collection.

4.2.3. Construction of the whole building model.

The Revit models needed for Mahavira hall are selected from the library of BIM components. For the Dougong area, a complete Dougong is firstly constructed to form a new family file by combining Dou, Gong and other components. Then the columns, beams, Dougongs are placed according to the architectural plan. Finally, the roof is built by the built-in models in Revit.

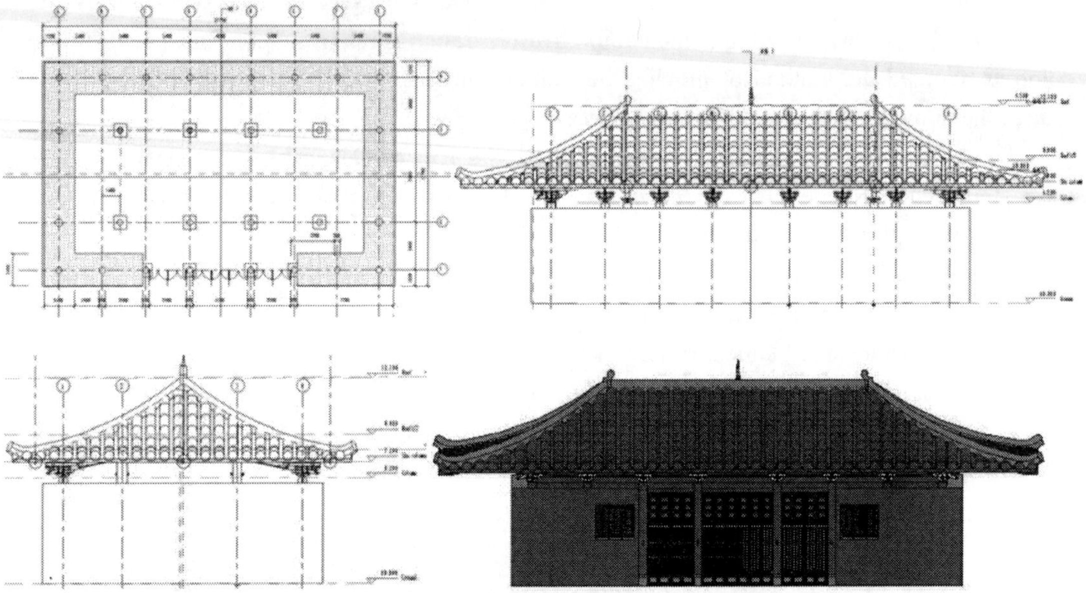

Figure 3. BIM model of Guangsheng temple.

Figure 4. Mahavira hall.

4.3 Conclusion of the BIM modeling process of Mahavira hall

The method of building the entire model based on models of components, according to "Yingzao Fashi", is proved to be feasible and has satisfactory accuracy, and the BIM components library for ancient buildings can be reused for the building of new ancient architectures models in the future. Through a typical case study of a building in Guangsheng temple, some suggestions on using BIM software for the construction of ancient architectures models are puts forward.

1. About the heights of the ancient architectural columns. "Yingzao Fashi" can be used as a reference for us, but the actual size of the component should come from on-site investigation. The height of columns of the Mahavira Hall, according to "Yingzao Fashi", should be ten meters. However, it more meets the standard high of eight-Yuanjian hall, about 4.83 meters.

2. The ancient walls are thick. For example, the wall of Mahavira hall is 2.4 meters in thickness. The built-in model method in Revit software is used to establish inclined walls and their surface

coating.

3. The problems about the open and overlapping order of Dougong should be noticed. Constructing complete Dougong BIM families, using the existing components models form the library, could facilitate the following work.

4. Roof is the main component of an ancient building, which has many decorative objects with Chinese characteristics. Hence it is better to use the built-in model method in the new file to build a roof, and then use the link to the original Revit file to combine two documents.

5. The whole model is usually made up of various components, and the model accuracy should be set differently considering the character of these components. This point will be elaborated in the following section.

4. 4 Accuracy of ancient building model

In order to make the BIM models of ancient buildings more credibility and have more historical value, certain accuracy standards should be obeyed to construct the BIM models. The accuracy for the main structure of the ancient buildings are differently defined, considering the features of components, and these components are classified to be Dougong, roof, joinery work (doors, windows, railings, etc.), carpentry work (columns, beams, purlins, etc.), and masonry structures for corresponding accuracy needs.

The accuracy of the model depends on the purpose of it. Referencing the analysis of surveying and mapping accuracy of relic building, Zhewen LUO, a specialist of ancient building, divided the purposes of surveying and mapping of relic buildings into three levels, i.e. forming “scientific records”, “fundamental drawing information for repairing design and construction” and “basic information for the research of Chinese architectural history and architectural theory” [8]. Therefore, the accuracy of the ancient building model can be divided into three levels. The first level is suitable for the model construction of the important ancient buildings, which requires the complete construction of Dougong and decorations on roof and carpentry work, and refined joinery work for the models. The second level is applicable to the establishment of scientific management archives of ancient buildings, which requires that the important decorations on roof, carpentry work, and joinery work should be constructed, and Dougongs should be constructed to satisfy the requirements. The third level is suitable for the analysis and survey of the ancient buildings, and the construction of the model is only required to express the information of size, quantity and materials.

5. Conclusion

The destruction and extinction of ancient buildings in JIN-FEN area in the past few decades represent the situation of the ancient architectures of China. Now it is critical to take steps to protect their original states and slow down the disappearing speed. GIS and BIM technology can be an effective tool for protecting, retrofitting, renovating and managing of existing ancient buildings. With the development of science and technology, the protection of ancient architecture should also develop with advance of the times.

Based on the foregoing work results of spatial distribution data, attribute data of features, 3D graphic information and parametric building information model, the information management system for ancient architectures in Jin-Fen Area, utilizing GIS&BIM technology, could be constructed, which is beneficial to the protection of cultural heritage, the management and preservation of ancient

IOP Conf. Series: Journal of Physics: Conf. Series **842** (2017) 012035 doi:10.1088/1742-6596/842/1/012035

building and also supports the further research of seismic disaster analysis and seismic performance simulation combining with the structural analysis software and seismic analysis software, such as PKPM, SAP2000 and Autodesk Robot Structural Analysis.

References

[1] Sicheng L. 2011 *Chinese ancient architectures survey report* (Beijing: SDX Joint Publishing Company).

[2] Sicheng L. 2013 *The annotation of "Yingzao Fashi"* (Beijing: SDX Joint Publishing Company).

[3] Xiang L. Guohua J. 2009 On parametric modeling of traditional Chinese architecture based on family component of Revit Architecture *Chinese & Overseas Architecture*, **8** pp42-44.

[4] Weichao S. 2012 *An approach to the system of ancient building information modeling based on Revit architecture* (Tianjin: Tianjin University).

[5] Weixin S. 2013 *Research on parametric information model of ancient buildings of Ming and Qing based on BIM* (Xian:Xi`an University of Architecture and Technology).

[6] Carlo B, Pietro C, Vincenzo D, Nora F. 2016 Towards the BIM implementation for historical building restoration sites. *Automation in Construction.* **71** pp 74–86.

[7] Maurice M, Eugene M, Sara P. 2013 Historic building information modelling – adding intelligence to laser and image based surveys of European classical architecture. *ISPRS Journal of Photogrammetry and Remote Sensing*, **76** pp 89–102.

[8] Chengjun B, Cong W, Yajing D. 2015 Surveying and mapping of cultural relic building based on the material system accuracy analysis *Journal of Xi'an University of Architecture & Technology (Natural Science Editi on)*.**47** pp 560-564

IOP Conf. Series: Journal of Physics: Conf. Series **842** (2017) 012036 doi:10.1088/1742-6596/842/1/012036

Identification of Delamination in Concrete Slabs by SIBIE Procedure

M. Yamada[1], Y. Yagi[1] and M. Ohtsu[2]

[1] Fuji P. S. Corporation, Tokyo, JAPAN
[2] Graduate School of Engineering, Kyoto University, Kyoto, JAPAN

m.yamada@fujips.co.jp

Abstract. The Impact-Echo method is known as a non-destructive testing for concrete structures. The technique is based on the use of low-frequency elastic waves that propagate in concrete to determine the thickness and to detect internal flaws in concrete. The presence and locations of defects in concrete are estimated from identifying peak frequencies in the frequency spectra, which are responsible for the resonance due to time-of-flight from the defects. In practical applications, however, obtained spectra include so many peak frequencies that it is fairly difficult to identify the defects correctly. In order to improve the Impact-Echo method, Stack Imaging of spectral amplitudes Based on Impact-Echo (SIBIE) procedure is developed as an imaging technique applied to the Impact-Echo data, where defects in concrete are identified visually at the cross-section. In this study, the SIBIE procedure is applied to identify the delamination in a concrete slab. It is demonstrated that the delamination can be identified with reasonable accuracy.

1. Introduction

The Impact-Echo method is well known as a non-destructive testing for concrete structures [1]. The method has been applied to thickness measurement of a slab, grouting performance and void detection in a post-tensioning tendon duct, identification of surface-opening crack depth, location of delamination and determination of material properties.

The Impact-Echo method has been widely applied to identification of shallow delamination [1, 2]. In principle, the location of delamination is estimated by identifying peak frequencies in the frequency spectrum. However, the frequency spectrum cannot always be interpreted successfully, because many peaks are often observed in the spectrum. In order to circumvent this fact, Stack Imaging of spectral amplitudes Based on Impact-Echo (SIBIE) procedure is developed [3, 4]. In this study, the SIBIE procedure is applied to identify the delamination in concrete slab.

2. Principle

2.1. Impact-Echo method

The principle of void detection in the Impact-Echo is shown in Figure 1. When the elastic wave is driven, the paths of the elastic wave are shown. Frequency spectrum is analyzed by Fast Fourier transform (FFT), applying to an obtained waveform detected at the output point. As shown in the figure, peak frequencies could appear at f_T and f_{void}, which correspond to resonant frequencies due to the thickness of a plate and the depth of a void, and are calculated as,

$$f_T = C_p / 2T, \tag{1}$$
$$f_{void} = C_p / 2d, \tag{2}$$

where C_p = velocity of P-wave, T = plate thickness, and d = covered depth. The plate thickness and the cover depth are obtained by substituting f_T or f_{void} with C_p into these equations.

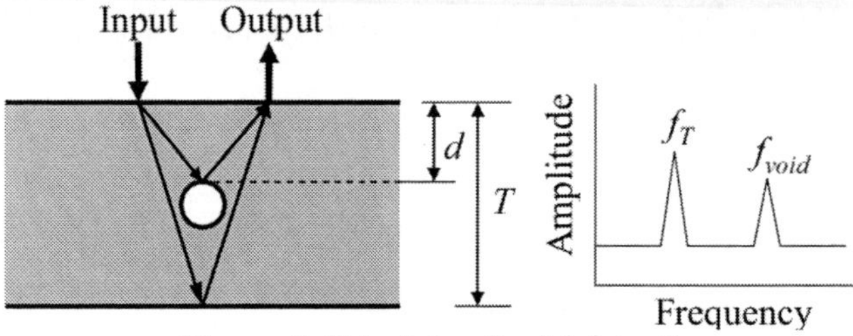

Figure 1. Principle of void detection.

2.2. *SIBIE procedure*

SIBIE procedure is a post-processing technique to Impact-Echo data. This is an imaging technique for detected waveforms in the frequency domain. In the procedure, first, a cross-section of concrete is divided into square elements as shown in Figure 2[5]. Then, resonance frequencies due to reflections at the elements are computed. The travel distance from the input location to the output via the element is calculated as,

$$R = r_1 + r_2. \tag{3}$$

The resonance frequency due to reflection at each element is calculated from,

$$f_R = C_p / R, \tag{4}$$

Spectral amplitudes corresponding to the resonance frequencies in the frequency spectra are summed at the elements. Thus, reflection intensity at each element is estimated as a stacking image as shown in Figure 3[5]. The minimum size of the square element, Δx, for the SIBIE analysis is approximately equal to $C_p\Delta t / 2$.

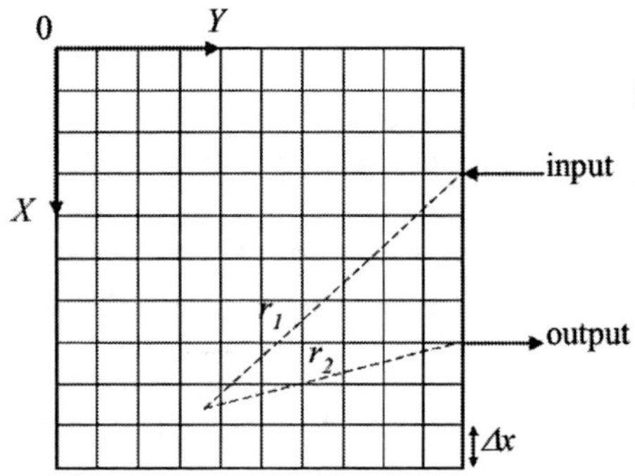

Figure 2. SIBIE imaging model.

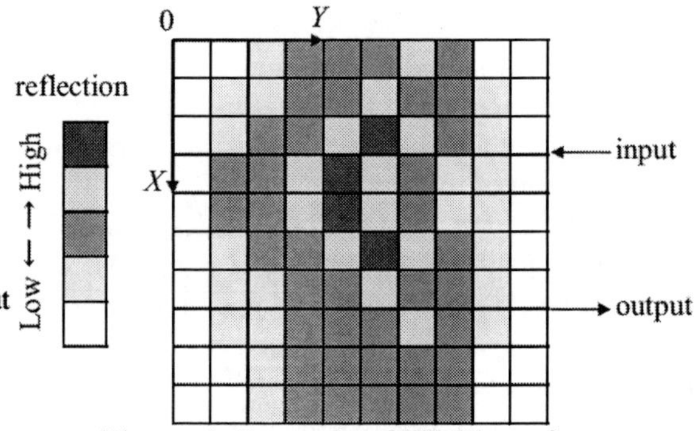

Figure 3. Example of SIBIE result.

3. Experiments

3.1. *Test site*

Piture1 shows a deck slab, which was removed from a bridge. In this deck slab, delaminated areas were found at some locations. One delamination was located at 160 mm depth from the top of the slab. Tests were conducted at three area of the slab which are selected as shown in Picture 2, 3, and 4. Picture 2 shows the normal area of the deck which has the pavement of 70 mm depth and an existing slab of 250 mm thickness. Picture 3 and Picture 4 show the overlaid areas of the deck with slabs of 250 mm thickness. The former has the pavement of 70 mm depth and the overlaid slab of 90 mm thickness without a delamination below. In the latter, a delaminated area is observed between the overlaid slab 90 mm thick and the existing slab 160 mm thick.

Picture1. Deck slab

Picture 2. Normal area (unit: mm)

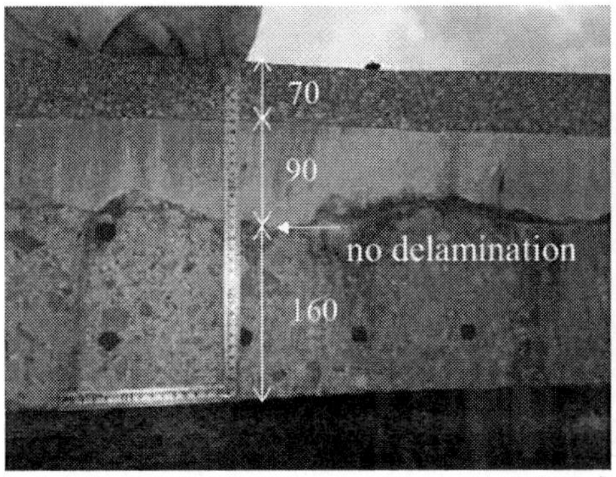

(mm)

Picture 3. Overlaid area (no delamination)

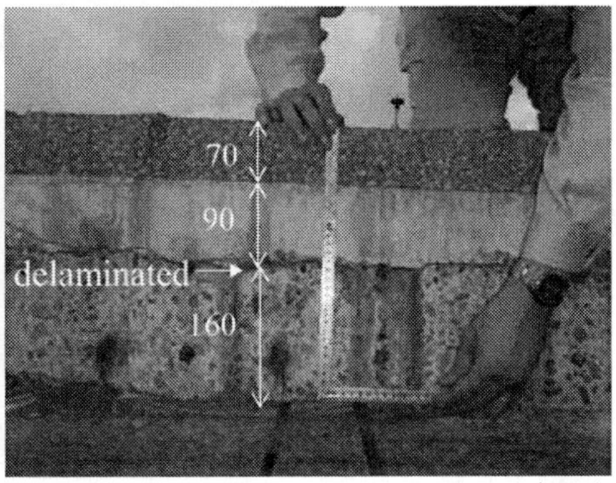

(mm)

Picture 4. Overlaid area (delamination)

3.2. Impact test

Picture 5 shows the impactor, which is equipped with accelerometers. It is confirmed that the upper-bound frequency due to the impact could cover up to 30 kHz, by using an accelerometer system. Fourier spectra of accelerations were analyzed by Fast Fourier Transform (FFT).

Impact test was done once at each point, normal area, overlaid area, delaminated area, respectively. The impactor was shot by spring with 100 N to generate elastic waves. Sampling time of the wave recorded in Picture 6 was 4μsec and the number of digitized data for each waveform was 2048.

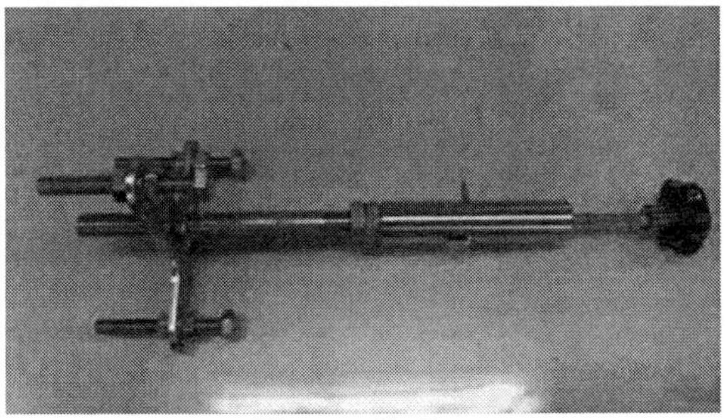

Picture 5. Impactor

Picture 6. Waveform recorder .

The locations of impact and two detection points are also shown in Picture 7. Two accelerometers were placed at the detection points to record surface motions caused by reflections of the elastic waves. Here, the interval between the impact point and the detection point is 50mm. P-wave velocity is a very important parameter in using SIBIE procedure. P-wave velocity of the test deck slab was obtained as 4500 m/s by the Impact-Echo method [6].

Picture 7. Impact point

4. Results and Discussion

4.1. Normal area

The result of Impact-Echo method is given in Figure 4. Figures 4 is the spectrum of the impact test at the Normal area. Calculated values of the resonance frequencies due to thickness, $f_T = C_p/2T$ is indicated with a block line. It can be seen from the frequency spectrum that there is the peak at f_T, however, it is difficult to identify the peak only from the specrum.

The SIBIE result at the normal area is given in Figure 5, which shows a cross-section of the deck (Picture 2). The size of cross-section is 320 mm × 500 mm. The red color regions indicate the highest reflection of the elastic wave. The impact point and the detection points are indicated by a downward arrow and upward arrows, respectively.

It is clearly seen that there exists high reflection zone only at the bottom of the deck. There are no other high reflections observed at the cross-section

.

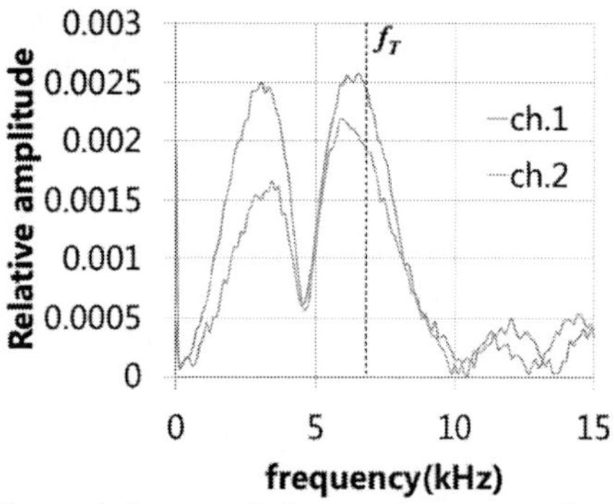

Figure4. Impact-Echo result at normal area

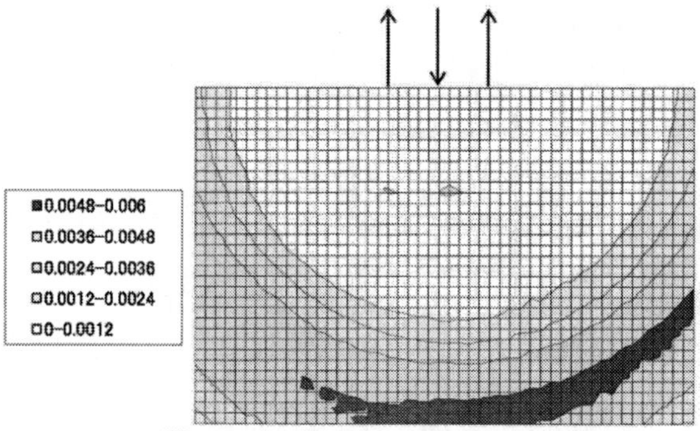

Figure5. SIBIE result at normal area

4.2. Overlaid area (no delamination)

In Figure 6, frequency spectrum obtained by the impact-test is shown. The resonance frequency f_T is observed at 7.03 kHz, although it is not a strong peak.

SIBIE result of the impact test for the case at the overlaid area (Picture 3) is shown in Figure 7. Red and yellow color of high reflection are clearly observed at the bottom of the deck again. There are no other high reflections observed at the cross-section. Thus, no differences are observed between the SIBIE result at the normal area and that of the overlaid area.

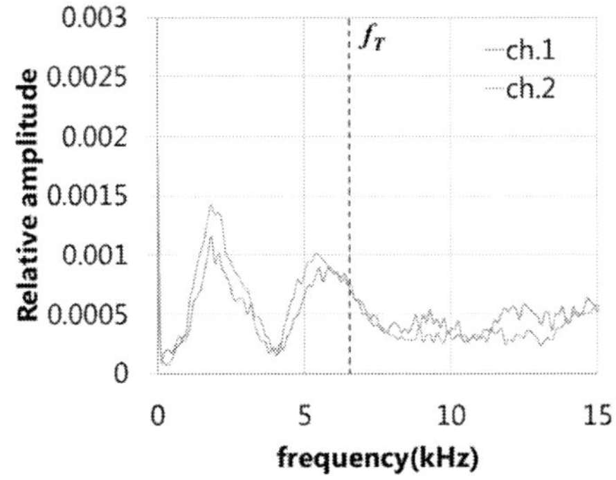

Figure6. Impact-Echo result at overlaid area

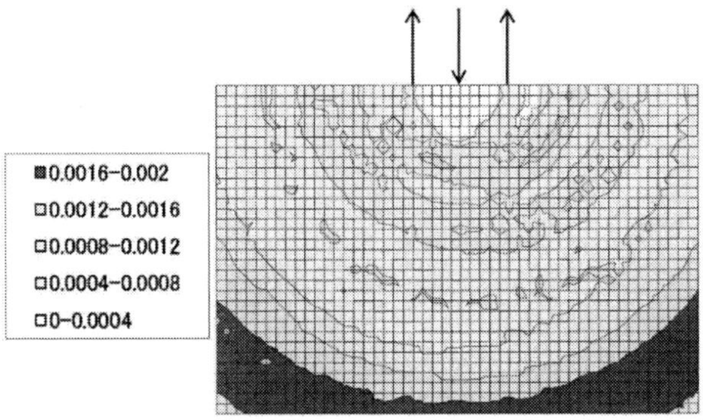

Figure7. SIBIE result at overlaid area

4.3. Overlaid area (delamination)

Figure 8 shows the result of Impact-Echo method at the delaminated area. The resonance frequency of delamination at 160 mm depth is calculated as 14.1 kHz is observed, and it's amplitude is higher than the resonance frequency f_T at 7.03 kHz.

In Figure 9, SIBIE result of the impact test for the case at the delaminated area is observed. The high intense region due to the delaminaion is observed at 160 mm from the top of the deck, while the high intense red regions can not be observed at the bottom. From this result, it is confirmed that the SIBIE procedure is available for void detection .

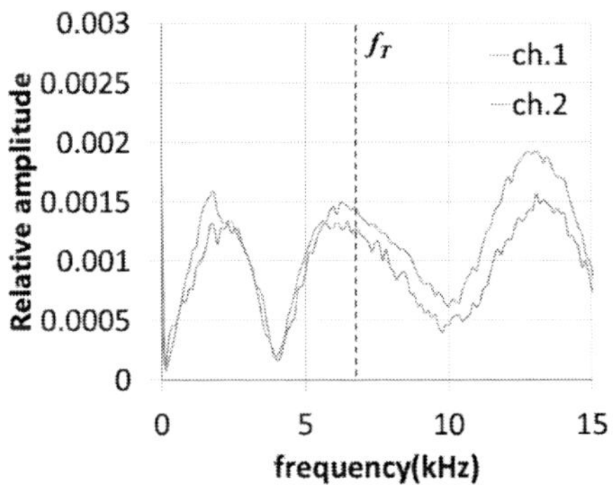

Figure8. Impact-Echo result at delaminated area

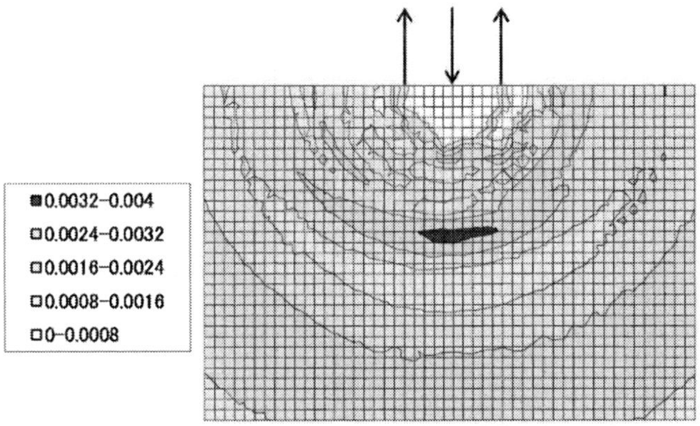

Figure9. SIBIE result at delaminated area

5. Conclusions

Based on the study, conclusions can be drawn, as follows:

(1) In the concrete slab taken out of a bridge, it was found to be not easy to identify the resonant frequencies of delamination and thickness only from the spectra obtained by the Impact-Echo method.

(2) According to the SIBIE results of normal area and overlaid area, there exist high reflection zones only at the bottom of the deck. On the SIBIE result of delaminated area, in contrast, the high intense region of reflection due to the delaminaion is observed clearly at 160 mm from the top of the deck. Therefore, it is confirmed that the location of the delamination can be clearly identified using the SIBIE procedure.

Thus, it is demonstrated that the delamination between an existing part and an overlaid part can be identified with the reasonable accuracy by the SIBIE procedure.

IOP Conf. Series: Journal of Physics: Conf. Series **842** (2017) 012036 doi:10.1088/1742-6596/842/1/012036

References

[1] Sansalone M.J. & Streett W.B. 1997a. Impact-echo, Ithaca, NY Bullbrier Press.
[2] Sansalone M. 1997b. Impact-echo: The complete story, ACI Structural Journal, 94(6), 777-786
[3] Ohtsu M. & Watanabe T. 2002. Stack imaging of spectral amplitudes based on impact-echo for flaw detection, NDT&E International, 35, 189-196.
[4] Ohtsu M. 2016. Innovative AE and NDT Techniques for On-Site Measurement of Concrete and Masonry Structures, RILEM State-of-the-Art Reports, Vol.20, 89-103
[5] Yamada M., Ohtsu M. 2010. Case studies on ungrouted tendon-duct in prestressed concrete models by SIBIE, Construction and Building Materials, Vol.24, Isuue12, 2376–2380
[6] Japanese Society for Non-Destructive Inspection Standard, NDIS 2426-2: Non-destructive testing of concrete : elastic wave method : Part 2: Impact elastic wave method

IOP Conf. Series: Journal of Physics: Conf. Series **842** (2017) 012037 doi:10.1088/1742-6596/842/1/012037

Fatigue damage assessment of electric roads based on probabilistic load models

R Ceravolo[1] G Miraglia[1] C Surace[1]

[1] *Politecnico di Torino, Corso Duca degli Abruzzi 24, Torino, Italy*

Abstract. The electro-mobility is becoming an increasingly present reality in recent years. The most important drawback of this technology is known to be limited battery autonomy. In an attempt to overcome this problem, for specific studies and testing, a number of roads have been implemented with coil systems in order to transfer power to electric vehicles, as described in this article. While on the one hand this could solve the problem of charging, on the other hand the introduction of a technology within an existing infrastructure could result in further structural issues. Since little or no information on the possible structural effect of the introduction of a charging system in the road is currently available, this study has focused on the long-term fatigue analysis of an electric road infrastructure in which an inductive wireless charging system has been introduced into the road structure. To perform the fatigue analysis, a recursive procedure defined within a probabilistic framework was developed and applied to a benchmark case study. The results obtained from the analysis represent an initial database for the definition of strategies and protocols for the monitoring, maintenance and operations of future electric roads infrastructures.

1. Introduction

The electric road (termed 'e-road'), an infrastructure able to provide power to vehicles, is potentially one of the key components of future smart cities. Electric power can be transferred to the vehicles operating on the e-road using a Wireless Power Transmission (WPT), a system that transmits electric power through an air-gap between vehicle and road without direct contact. In particular, the Inductive Power Transfer (IPT) technology is a near-field WPT technology suited for the contactless charging solution of electric vehicles in e-roads, [1]. In order to optimise the WPT between the e-road and the vehicle, the wear layer of the pavement needs to be very thin. Typical reference values of its thickness are 40-50 mm, [2].

The use of e-road would result in several advantages including the possibility of realising lighter and cheaper electric vehicles through the use of smaller batteries which, being recharged continually, need to store a reduced amount of energy, [3]. In addition, in a wider vision for the future, electric vehicles could store energy and return it to the grid (vehicle to e-road energy transfer phase) in times of greatest need. Furthermore, the energy could be transferred to the network even with the use of innovative solutions applied to the field of renewable resources. Although the benefits of such an infrastructure are very attractive, their development depends on the actual feasibility and viability of realisation. In this sense, a technical feasibility study (http://www.fabric-project.eu/), represents the first step towards the definition and development of a suitable technology which is at the same time accessible and reasonably easy to use. The technical feasibility and sustainability of e-roads touches many themes and research fields, such as Power Engineering, Telecommunications Engineering, Structural Engineering. As concerns structural feasibility, the lack of information regarding the long-terms structural response of e-roads is a major issue due to its impact on maintenance and the potentially high related costs. An estimate of the fatigue endurance of e-roads infrastructure is thus needed.

This consideration has led the authors to investigate ways for predicting the long-term effects that the introduction of recharging devices within the e-road pavement can cause. Correspondingly, this article presents a methodology for the fatigue damage assessment of e-roads based on data from numerical models to simulate the existence of a monitoring system typical for e-roads.

The numerical models take into account some specific features of e-roads such as, for example, a small misalignment of the vehicle during travel. For recharging reasons, in fact, the maximum allowed misalignment for power transfer is about 20 cm, [4]. This means that the vehicle is forced to follow a specific trajectory, determining a possible acceleration of the damage evolution as well as premature rutting formation, leading to high driveability problems that could potentially compromise also the recharging phase. The methodology was applied for two solution of e-road: one that uses Cement Asphalt Mortar (CAM) to fill the Charging Unit (CU), and one that uses asphalt (see Figure 1).

To evaluate the long-term effects (such as damage evolution and fatigue resistance) within a numerical framework, the choice of the appropriate damage law is of paramount importance. One useful method to identify the life of a road pavement is the definition of the number of load cycles to failure. There are different ways to evaluate the number of load cycle of an asphalt layer based on strains, stresses, temperatures or on the dissipated energy, [5]. The strain-based models are most general models to assess the fatigue damage accumulated by the pavement which are related to the tensile strain at the bottom of the asphalt layer. In the present study, a generalised Maxwell constitutive model was used to represent the road materials and the damage evolution law chosen, [6] taking into account the generalised Maxwell type on several terms with inclusion of a viscous strain. The damage is then related to the energy density release rate considering both the elastic and viscous effects.

In this paper, Section 2 contains a description of the case study analysed. Section 3 reports the definition of the simulated monitoring system to provide the input data for the analysis. In Section 4, the probabilistic methodology used to evaluate the lifetime of e-roads is briefly presented before the damage evolution law applied to the methodology is deeply analysed. Finally, Section 5 reports the entire recursive procedure followed by the simulation results.

2. Benchmark case study description

The e-road solution analysed involves the use of a coil box which contains the recharging technology. The coil-box is covered with a bituminous material (CAM or Asphalt) in the first layer of the e-road. Figure 1 shows the layout of the case study investigated.

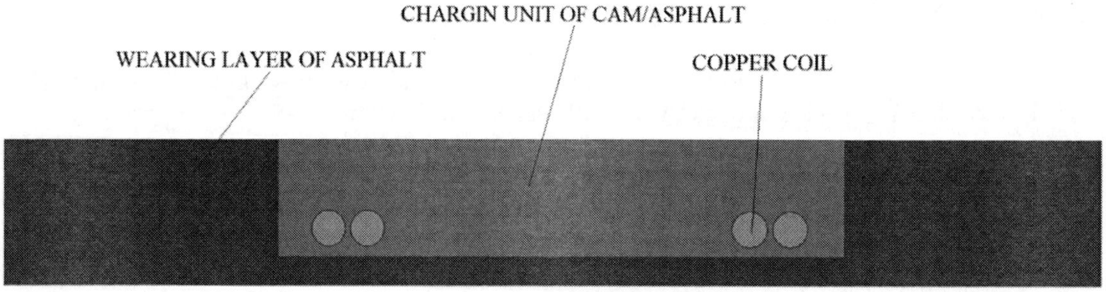

Figure 1. Layout of the charging solution installed in the first layer of the e-road in the middle of the lane.

The recharging equipment is directly inserted under the surface of the road in a hole cut with a depth of 60-70 cm. The power electronics and control equipment are placed in manholes at the track sides, optimising the installation of the Charging Unit (CU). Successively the wires are embedded and the hole is covered with asphalt or Cement Asphalt Mortar (CAM). The use of CAM as an alternative to the asphalt is due mainly to its impact load absorption properties, [7]. In addition, CAM does not need to be compacted, an action that could potentially affect the technology.

3. Monitoring System

3.1. *Traffic data simulation*

In the present study, the variation ranges for the simulation of the time series are imposed as follows. The speed is assumed to vary between a minimum of 50 km/h (below which the charging system is switched off for safety reasons) and a maximum of 130 km/h. At the same time, the footprint area is assumed to vary from 0.1 m to 0.6 m. As regards the vehicle load and number of vehicle axles, it is assumed that between 8 kN and 35 kN the vehicle has 2 axles and can be considered as a 'light'. Above 35 kN the vehicle is classified as 'heavy' with 2 axles up to 180 kN, 3 axles between 180 kN and 260 kN, 4 axles between 260 kN and 400 kN and 5 axles over 400 kN, the upper boundary being set at 440 kN. As regards the axle width, for light vehicles the boundary conditions were set between 1.2 m and 1.9 m, while for heavy vehicles, the random generation of the axle width has affected values from 1.9 up to 2.6 m.

The transverse misalignment of the vehicle was simulated, as described in, [8]. Assuming a Poisson process, [9] to simulate the number of axles in a reference period (set equal to 20 years in this analysis), the distribution parameter was calibrated so as to obtain values of the inter-arrival time of the axles within a reasonable range (e.g. 0.8-12 s). The dynamic amplification of the gross weight load is assumed to be 1.4. Finally, the lane width is assumed to be 3.75 m.

Figure 2 shows the simulated axle load series, while in Figure 3 the generated transverse position of the vehicle axle is shown for both e- and t-road infrastructure.

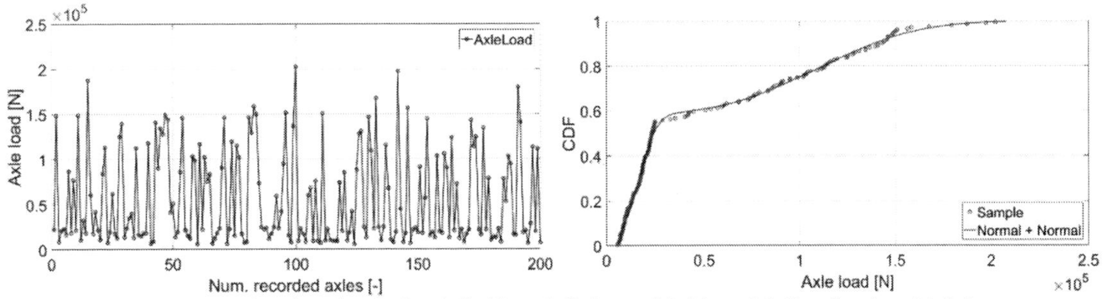

Figure 2. Simulated axle load (left) and fitting with bimodal distribution (right).

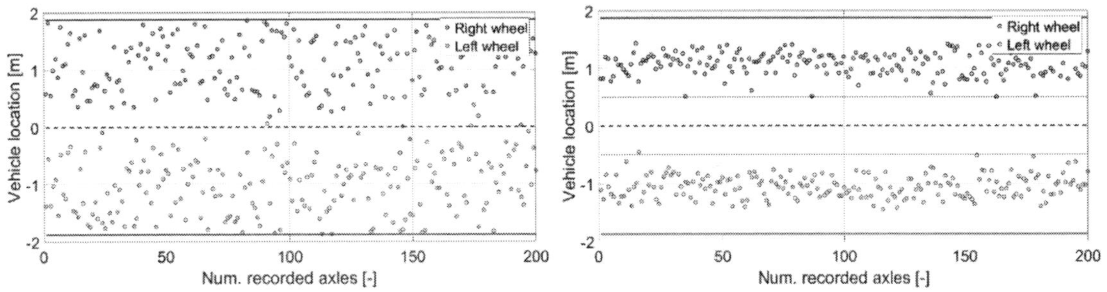

Figure 3. Simulated vehicle misalignment for both traditional (left) and e-road (right) infrastructures.

4. Basic theory

4.1. The probabilistic methodology

In order to evaluate the lifetime of the e-road infrastructure, a strategy that takes into account four elements has been followed. These elements can be summarised as:

- The equivalent static load approaches, expressed by the generalised Maxwell model modified with an equivalent viscosity, [4];
- The statistical evaluation of the load intensity, q, and the inter-arrival time between two axles $T_a(x)$;
- The fatigue damage law, [6];
- Finite Element Model.

For the recursive procedure, reference was made to, [8], which is based on the following assumption: starting from a reference value of lifetime, t_j, the axle load intensity q_a and the redistributed inter-arrival time $T_a(x)$ is calculated. Then it is possible to calculate the correction factor for the viscous effect $C(x)$ that takes into account the actual cyclic nature of the vehicle loads, [10].

$$C(x) = \frac{T_a(x)}{\tau_a} \quad (1)$$

where τ_a is the impulse duration of the load.
Successively, the Finite Elements (FEs) analysis can be performed with the load intensity, q_a, and equivalent viscosity, obtained by multiply the actual viscosity for the correction factor of Equation (1). From the analysis, the horizontal tensile strains at the bottom of the asphalt layers (commonly related to the failure of the surface layers), is considered in order to calculate the number of cycle to failure at constant strain by following Equation (2):

$$N_f = k_A\left(\frac{1}{\varepsilon_t}\right)^{k_B} \quad (2)$$

with, k_A=1.29e-6, and k_B=3.02, experimental parameters that were set to the values obtained in, [11], with Accelerated Pavement Testing (APT), which simulate the road conditions as closely as possible by considering vehicle 'wandering', resulting in a longer life estimation.
ε_t is the transverse strain at the bottom of the asphalt layers.

Finally, the damage accumulation law can be applied. This study assumed the law identified by the Equation (14). When the damage, D, reaches the value of 1, the number of applied load cycles, N (corresponding to the ratio between the equivalent time used in the generalised Maxwell model and the inter-arrival time T_a), is assumed to have reached the number of fatigue cycles to failure, N_f. The new reference time, t_{j+1}, can be thus updated within all related probabilistic models (i.e. load, inter arrival time, vehicle misalignment). The recursive procedure continues until the estimate for the lifetime converges to a value $|t_{j+1} - t_j| < \varepsilon$.
In the present study, the residual ε, is assumed to be ε=604,800 s (1 week).

4.2. Damage law description

A framework based on the continuum damage mechanics and thermodynamics of irreversible processes with internal state variables is used to characterize the distributed damage in viscoelastic asphalt materials. This is obtained by simulating damage in the form of micro-crack initiation and accumulation. A model based on a non-associated damage evolution law is capable to describe the temperature-dependent processes of micro-crack initiation, evolution and macro-crack formation, respectively, in good agreement with the material response in the Superpave Indirect Tensile (IDT) strength test, [12].

With this type of model, the energy density release rate, *Y*, can be written, without loss of generality, in terms of the characteristic relaxation function and the total strain as, [6]:

$$Y(t) = -\frac{\partial \psi}{\partial D} = \frac{1}{2} E(t)\varepsilon_t^2 \tag{3}$$

where, ε_t, is the transverse strain field, and ψ represents the Helmholtz free energy, which, for the described generalized Maxwell model, can be expressed as:

$$\psi = (1 - D)(\psi_\infty + \sum_{i=1}^{n} \psi^i{}_{neq}) \tag{4}$$

with $\psi_\infty$and $\psi^i{}_{neq}$ representing the free energy functions of the long-term equilibrium (time independent) and the non-equilibrium (time dependent) parts respectively, *D* the value of damage, and *E*(*t*) the relaxation modulus of Maxwell model at reduced time *t*.

The energy-based damage law unifies many particular damage modes such as ductile, creep, fatigue, quasi-brittle damage with the critical damage densities related to mesocrack initiation. The energy-based damage model in, [13][14][15], is adapted and further extended for the damage characterization in viscoelastic asphalt concrete material. The micro-crack initiation criterion can be expressed in the form:

$$f^d = \varrho_1{}^*(Y) - \varrho_{1,c}{}^*(S_0) - R(r) = 0 \tag{5}$$

where $\varrho_1{}^*(Y)$ is the micro-crack initiation potential function, $\varrho_{1,c}{}^*(S_0)$ is the critical micro-crack damage threshold and *R*(*r*) is the isotropic damage hardening function. The micro-crack initiation potential function $\varrho_1{}^*(Y)$, which is used to obtain the point at which micro-crack initiates, is a function of the linear viscoelastic energy *Y*, and can be expressed as:

$$\varrho_1{}^*(Y) = \frac{S_0}{k_2 + 1}\left(\frac{Y}{S_0}\right)^{k2+1} \frac{1}{1 - D} \tag{6}$$

where k_2, S_0 are material parameters. The parameter S_0 is the critical energy when the micro-crack initiates.

When *Y*=S_0, the linear viscoelastic energy reaches the value for which the damage mechanism starts to evolve. In this situation, the value of damage is supposed to be $D = 0$. Incorporating these assumptions in Equation (6), it is possible to get the value of the potential function corresponding to the damage activation mechanism, expressed as:

$$\varrho_{1,c}{}^*(S_0) = \frac{S_0}{k_2 + 1} \tag{7}$$

Using a non-associated micro-crack evolution rule, a micro-crack propagation criterion is introduced and expressed as:

$$F_D = \varrho_2^*(Y) - \frac{k_1}{k_2}\varrho_{1,c}^*(S_0) - R(r) = 0 \tag{8}$$

where $\varrho_2^*(Y)$ is the micro-crack propagation potential, expressed in the form:

$$\varrho_2^*(Y) = \frac{k_1}{k_2}\frac{S_0}{k_2+1}(\frac{Y}{S_0})^{k2+1} \tag{9}$$

The evolution of micro-crack, can be obtained with respect to the dissipative micro-crack potential F_D by taking the derivative of the dissipation potential, which reads [6]:

$$\dot{D} = \dot{\lambda}\frac{\partial F_D}{\partial Y} = \frac{k_1}{k_2}(\frac{Y}{S_0})^{k2}\,\dot{\lambda} \tag{10}$$

The damage (micro-crack) hardening variable, r, can be obtained as a function of the Lagrange multiplier λ and expressed as:

$$\dot{r} = -\dot{\lambda}\frac{\partial F_D}{\partial R} = \dot{\lambda} \tag{11}$$

With, $\dot{r} = \dot{p}(1-D)$, where $\dot{p}$ denotes the cumulative plastic strain, [14]. The resulting damage evolution law is given as:

$$\dot{D} = \frac{k_1}{k_2}(\frac{Y}{S_0})^{k2}\dot{p}(1-D) \tag{12}$$

Finally, the damage law results from the solution of the following differential equation:

$$\frac{dD}{dt} = \frac{k_1}{k_2}\left(\frac{Y}{S_0}\right)^{k2}\frac{d\varepsilon_t}{dt}(1-D) \tag{13}$$

For simplicity, Y was assumed as constant within each interval $\Delta\varepsilon$

$$D = 1 - \exp(-\frac{k_1}{k_2}\left(\frac{Y}{S_0}\right)^{k_2}\Delta\varepsilon_t + \frac{k_1}{k_2}\Delta\varepsilon_t) \tag{14}$$

The damage model coefficients used to model the damage evolution in the asphalt mixtures are listed hereinafter, [6]:

- k_1 =27.43
- k_2 =0.29
- S_0 = 0.437 $[\frac{kJ}{m^3}]$

Finally, the full damage constitutive equation, which was implemented successively in the algorithm for evaluating the time life of e-roads, can be summarised as:

$$\begin{cases} if\ \varrho_1^* < \varrho_{1,c}^* & \text{no microcrack initiation} \\ if\ \varrho_1^* = \varrho_{1,c}^* & \text{micrcrack initiation} \\ if\ \varrho_2^* > \dfrac{k_1}{k_2}\varrho_{1,c}^* & \text{microcrack evolution} \end{cases} \qquad (15)$$

5. Recursive procedure for e-road lifetime estimation

5.1. Algorithm description

In order to calculate the lifetime, an algorithm was specifically created in the Matlab environment to implement a recursive procedure. The possibility of working in batch mode optimises the calculation process and allows to calculate the infrastructure's lifetime by exploiting the potential of FE analysis.
In more detail, the Matlab 'script' generates input data for the FE model, as shown previously. Having defined the load intensity and the correction factor of Prony's serie, the algorithm starts the batch mode analysis of the FE model, which produces a time history of parameters such as strain and energy. These values are used as data input for the cumulative damage law. The final outcome is an estimate for the lifetime, which corresponds to the attainment of D=1. If the damage does not reach its maximum value, the lifetime is set >20 years.
The algorithm can take into account different aspects.

- Transverse position of vehicles loads. In this case 2 critical positions for the load are considered (see Figure 5), but the analysis can be virtually conducted for any position of load in the transverse direction;
- Any type of cumulative damage law can be assumed;
- For the CU, the fatigue damage is supposed to be reached when the asphalt under it reaches its t_{f0}.

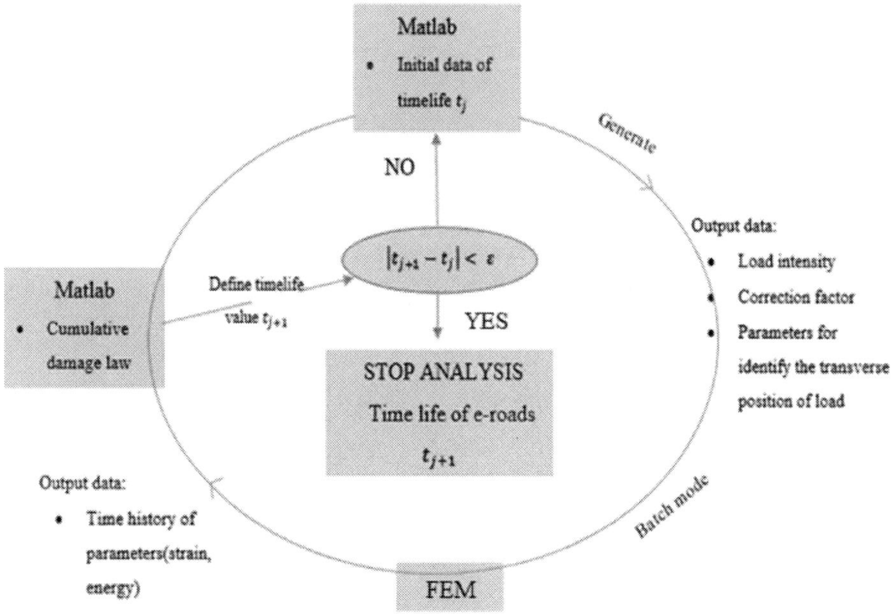

Figure 4. Layout of the updating recursive procedure.

The cycle of Figure 4 evaluates the lifetime of e-roads referring to a single coupled of p_s and p_w, which characterise the percentage of passage of standard and heavy vehicles, respectively.

5.2. Results

With the aim to assess both the fatigue endurance of e-roads and the use of CAM as alternative material to fill the Charging Unit (CU), two load conditions were defined. The first one, for the estimation of the e-road lifetime, with centred load, and a second one for the estimation of the damage in the CU, with a load placed on it. These load conditions are represented in Figures 5.

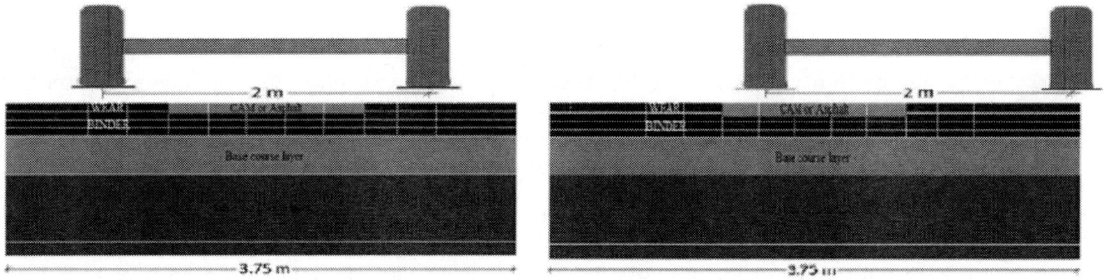

Figure 5. Load conditions for the estimation of the lifetime of e-roads (left), and the damage in the CU (right).

The results show that the fatigue life increases with increasing inter-arrival time of heavy vehicles, following a power law. Negligible differences were found between CAM or asphalt filling solutions in terms of fatigue endurance of the entire e-road. Consequently, outside the CU, the material used to fill the coils does not affect significantly the fatigue life. In the analysis, the average value of the characteristic load at 95% is found to be, for standard and heavy vehicles respectively, equal to 25 kN and 150 kN.

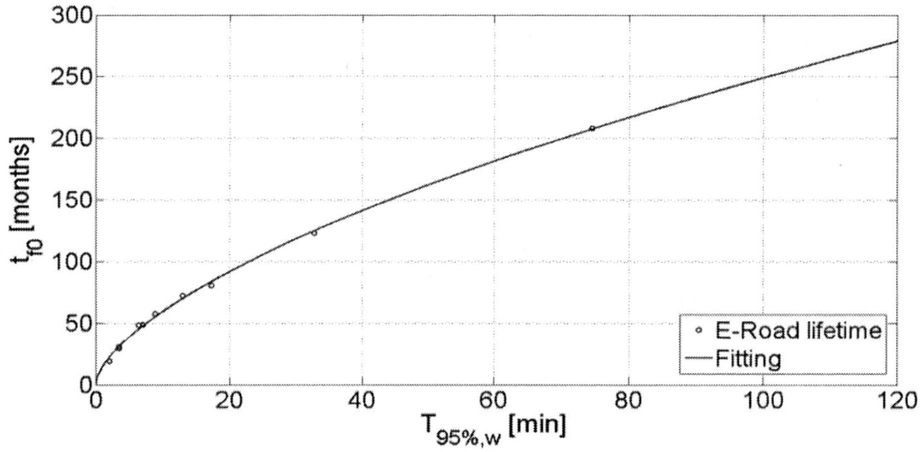

Figure 6. E-roads lifetime function of $T_{95\%,w}$.

The lifetime decreases with decreasing inter-arrival time (95 percentile) of heavy vehicles, $T_{95\%,w}$, accelerating for small values. Figure 6 shows the trend of the fatigue life, t_{f0}, as approximated by Equation (16).

$$t_{f0} = 14.26 \cdot T_{95\%,w}{}^{0.6209} \tag{16}$$

In the absence of heavy vehicles, it would be sufficient to particularise the analysis by referring to a t_{f0} associated to high values of $T_{95\%,w}$. Accordingly, the lifetime associated with the highest $T_{95\%,w}$ would correspond to the value searched. In this case, with a characteristic load of standard vehicles of 25 kN, the fatigue in presence of a lowest number of heavy vehicles ($T_{95\%,w}$ = 75 min) is equal to about 17 years.

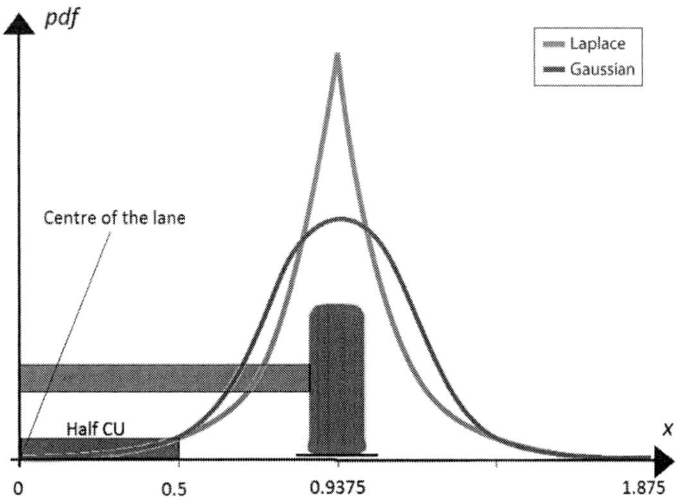

Figure 7. Transverse distribution of loads.

To estimate the damage near the CU, a local assessment was also performed on the endurance of the portion of infrastructure closed to it. The CU's fatigue endurance is supposed to be reached when the asphalt under it reaches its t_{f0}. In fact, once the link between the CU and infrastructure fails, any further transit of vehicles could irremediable damage the CU and the whole e-road infrastructure.

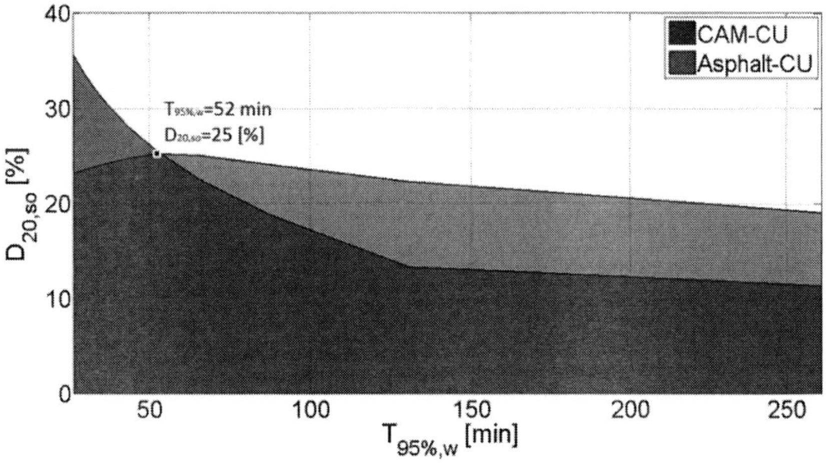

Figure 8. Damage function of $T_{95\%,w}$.

Accordingly, the CU's endurance was estimated to exceed 20 years in both CAM and asphalt solutions. Figure 8 reports the damage value rated to 20 years, in the case of both asphalt- and CAM-CU. The damage is less than 40% in both solutions, which ensures a considerable safety margin from the limit condition. However, it can be seen for $T_{95\%,w}$ exceeding 52 minutes that the damage is greater in an asphalt-CU. Under this value, there is a reversal of the behaviour. More specifically, for a CAM-CU the damage increases with decreasing inter-arrival time of heavy vehicles. By contrast, under

$T_{95\%,w}$=52 min, the asphalt-CU solution damage decreases with decreasing of $T_{95\%,w}$. This is due to the greater importance that takes the Laplace distribution with respect to the Gauss one. In fact, when the number of heavy vehicles increases, the transverse distribution tends to a Laplace [8], which has a smaller second order moment (see Figure 7), this limiting the damage to the CU.

6. Conclusions

The results of the analysis can be summarised as follows.

- The fatigue analysis revealed a high dependence of the lifetime on the inter-arrival time of heavy vehicles, being the characteristic load assumed in this study equal to 25 kN and 150 kN, for standard and heavy vehicles respectively.
- The fatigue endurance of the CU can be set greater of 20 years for both asphalt or CAM-filling solutions. However, when the inter-arrival time of heavy vehicles is less than 52 min, better results in terms of damage are found with asphalt. For longer inter-arrival times the CAM gives a better performance (less damage).

The overall procedure can be easily applied to any configuration of different e-road solutions, providing indications to define an optimal maintenance plan for the entire lifetime of the e-road. During the service life, the procedure also allows for updating the maintenance plan, based on the observed levels of damage in the e-road structure, and for scheduling interventions.

Acknowledgements This research has been supported by the European Commission within the FP7 projects UNPLUGGED (Grant No. 314126) and FABRIC (Grant No. 605405).

References

[1] Chen F, Taylor N and Kringos N 2015 Electrification of roads: Opportunities and challenges, Applied Energy, 150, 109–119. doi:10.1016/j.apenergy.2015.03.067More references.
[2] Viktoria Swedish ICT 2013 Slide-in Electric Road System, Gothenburg, Sweden.
[3] Boys JT and Covic GA 2013 Magnetic Circuits for Powering Electric Vehicles, IPT Fact Sheet Series: No. 2.
[4] Ceravolo R, Miraglia G, Pinotti E, Surace C and Zanotti Fragonara L 2015, Modelling of a road infrastructure adapted for dynamic inductive recharging for maintenance and monitoring, 7th International Conference on Structural Health Monitoring of Intelligent Infrastructure, SHMII-7, Turin.
[5] Baburamani P 1999 Asphalt fatigue life prediction models: A literature review, ARRB Group Transport Research, ARR 334, 0158-0728.
[6] Onifade I, Birgisson B, Balieu R 2015 Energy-based damage and fracture framework for viscoelastic asphalt concrete.
[7] Wang Q, Yan PY, Ruhan A, Yang JB and Kong XM 2011 Strength mechanism of cement-asphalt mortar, Journal of Materials in Civil Engineering, 23(9):1353-1359.
[8] Ceravolo R, Miraglia G, Surace C and Zanotti Fragonara L 2016 A Computational Methodology for Assessing the Time-Dependent Structural Performance of Electric Road Infrastructures, Computer-Aided Civil and Infrastructure Engineering.
[9] Chengming L, Hui L and Jinping O 2011 Traffic load modelling based on structural health monitoring data, Structure and Infrastructure Engineering, 7:5, 379-386, DOI: 10.1080/15732470902726809.
[10] Ceravolo R, Miraglia G, Surace C 2016 Strategies for Assessing the Structural Performance of Electric Road Infrastructures, 6th Civil Structural Health Monitoring (CSHM-6) annual ISHMII workshop.
[11] Pais JC, Pereira PAA, Minhoto MJC, Fontes LPTL, Kumar D and Silva BTA 2009 The prediction of fatigue life using the k1-k2 relationship, In 2nd Workshop on Four Point Bending (pp. 39–

46).
[12] AASHTO T 321 2014 Edition
[13] Krajcinovic D, Lemaıtre J 1987 Continuum damage mechanics: theory and applications, Springer-Verlag Kg.
[14] Lemaitre J, Desmorat R 2005 Engineering damage mechanics: ductile, creep, fatigue and brittle failures, Berlin; New York: Springer.
[15] Lemaitre J, Sermage JP and Desmorat R 1999 A two scale damage concept applied to fatigue, Int J Fract;97:67–81. http://dx.doi.org/10.1023/A:1018641414428.

IOP Conf. Series: Journal of Physics: Conf. Series **842** (2017) 012038 doi:10.1088/1742-6596/842/1/012038

Energy dissipation and absorption capacity influence on experimental modal parameters of a PC girder

O S Luna Vera[1], C W Kim[1] and Y Oshima[2]

[1] Infrastructure Innovation Engineering Laboratory, Department of Civil and Earth Resources Engineering, Graduate School of Engineering, Kyoto University, JAPAN
[2] Center for Advanced Engineering Structural Assessment and Research (CAESAR), Public Works Research Institute (PWRI), Tsukuba, JAPAN

E-mail: kim.chulwoo.5u@kyoto-u.ac.jp

Abstract. This experimental investigation focuses on collecting vibration data after different loading scenarios in a PC girder while measuring the performance at the same time. The vibration data is used for modal analysis and since several samples were taken in each scenario with different impact hammer tips, a comparison of the outcomes is also possible. The performance is based on recorded displacements along the bottom surface of the girder, information that makes possible the calculation of energy absorbed before cracking and the energy dissipation that takes place after cracking. Furthermore, two major damage events, a tendon breakage and cracking, were induced and monitored respectively. This study also clarified the influence of change in energy on changes in modal parameters.

1. Introduction

Obtaining the current performance of concrete bridges typically implies exhausting non-destructive tests for each specific structure and also takes long periods of time with qualified engineers. Structural health monitoring has made a great progress in damage detection by using sensors that can collect information easily once installed. However, using monitored data to obtain performance parameters, such as degradation, serviceability and decision making is still under discussion. Also, dealing with the materials uncertainties is a big challenge for engineers because these uncertainties are merged with the monitored data and might be the reason of abnormal deviations.

The natural frequencies from deteriorated concrete beams have shown to be sensitive features for the strength reduction when evaluating the residual strength in bridges with 50 years of service life. At the same time, it rises the likelihood of using them as a symptom to be monitored in order to estimate the structural reliability [1]. Literature has also shown that the modal damping ratio can be a damage sensitive feature for reinforced and prestressed concrete structures [2]. Moreover, researches indicate that a trend of getting higher value of modal damping ratios with the age of the bridges could be a sign of possible deterioration, which is a strong reason to investigate the relationship between damping parameter and degree of damage [2]. Last but not least, discrepancies between the results from experimental testing and theoretical models come up due to variety of reasons that needs to be faced [3].

Prestressed concrete (PC) girders are essential elements in bridge stocks, and efficient inspection of the PC bridges is a keen technical issue. This study, thus, investigates vibration-based structural health monitoring for PC bridges especially focusing on the relation between energy absorption capacity and modal parameters, as well as the effect of energy dissipation.

2. Experimental setup and plan

The PC girder used for this experiment comes from an earlier investigation about the evaluation of grout failure process with tendon rupture by means of the acoustic emission [4]. Regardless of its previous use, the concrete element is still suitable for the purpose of the research. The girder has a total length of 4.5 meters, 0.5 meters height and 0.25 meters width. The free span or distance between the steel roller supports is 4.1 meters. Seven servo-accelerometers (A1-A7) and seven displacement transducers, as shown in Figure 1, were allocated. Two more displacement gauges were deployed at the ends of the concrete specimen to observe longitudinal displacements. The girder has four post-tensioned tendons, of which two were already severed as shown in Figure 2, but leaving the possibility of severing one more tendon for the experiment aim.

The displacement transducers are Tokyo Sokki Kenkyujo type CDP-50 and CDP-25. The loading actuator has a maximum capacity of 400 KN and it was connected to the PC girder through a steel beam that is in contact in two points separated by a distance of one meter. Both, the displacement transducers and the actuator were connected to a data logger TDS-530 from Tokyo Sokki Kenkyujo. The accelerometers are piezoelectric type ONOSOKKI NP-2120 with a maximum acceleration of 8000 m/s^2 and a sensitivity of 5pC/ (m/s2) +/-2dB.

The experiment plan is described in Figure 3 with a timeline. Along all the experiment, ten vibration tests were performed. Also, the static loading plan consists of eight total load-unload stages, from which the first four are the same load level (150 kN) that is below of the cracking load predicted. These four loads allowed to monitor the response behavior under analogous conditions. Besides, the tendon severing could be introduced in between to test the effect of this event on the response. Later, three more load-unload stages were introduced with an increase of 50 kN each time, where cracking begins and propagates subsequently, and finally a last load step until failure was performed.

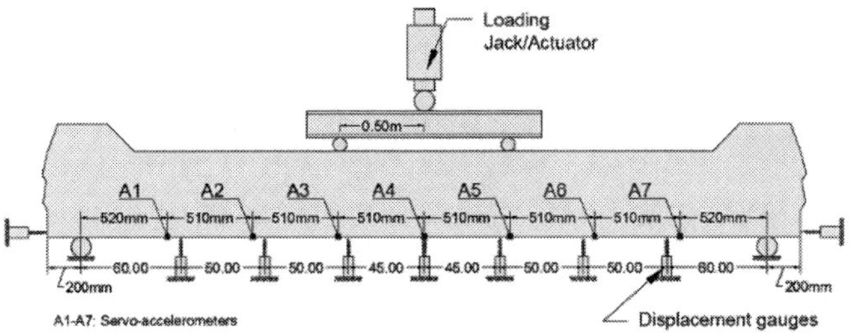

Figure 1. Sensors location in the PC girder (displacement transducers and servo-accelerometers)

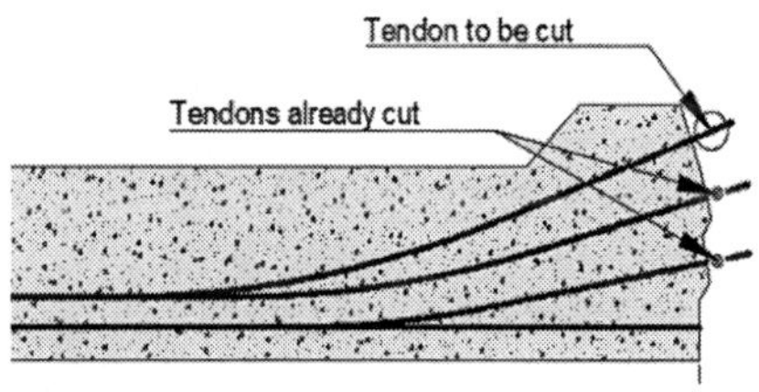

Figure 2. Tendons condition and breakage plan for the experiment

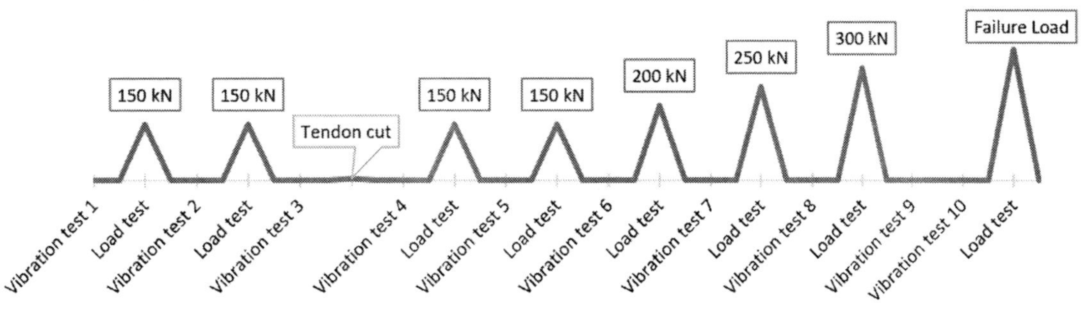

Figure 3. Loading stages and timeline of the experiment

3. Static load testing and energy calculation

Besides the tendon severing, the static load tests change the intrinsic parameters of the girder by producing cracks when increasing the load levels. These changes are to be monitored by examining the energy absorbed and dissipated in each load stage.

3.1. Cracks monitoring and failure

The beginning of cracking was detected roughly in 170 kN (by visual inspections every 5 kN of load increase), but it was noticed internal cracks sound in 165 kN. Also, the load-displacement curves show a change in the slope from approximately 166 kN. After that, the cracks were monitored and marked by hand in each stage (see Figure 4). Then, the calculation of the lengths was achieved by using a CAD software, and the summary is in Table 1.

The failure of the girder was reached by exceeding the compressive strength of the concrete in the top surface of the specimen. From the records it can be seen that one section of the girder started yielding around 330KN and then failed in 351KN. Locating and determining the exact load failure can be easily done by inspecting the displacement records in the final stage (see Figure 5).

Table 1. Total lengths of cracks in the PC girder after each stage

	Cracks after 200 kN	Cracks after 250 kN	Cracks after 300 kN	Cracks after failure
Front face	666.49 mm	2121.05 mm	3602.75 mm	5359.94 mm
Back face	657.96 mm	2030.60 mm	3695.59 mm	4877.01 mm
Average	662.23 mm	2075.83 mm	3649.17 mm	5118.48 mm

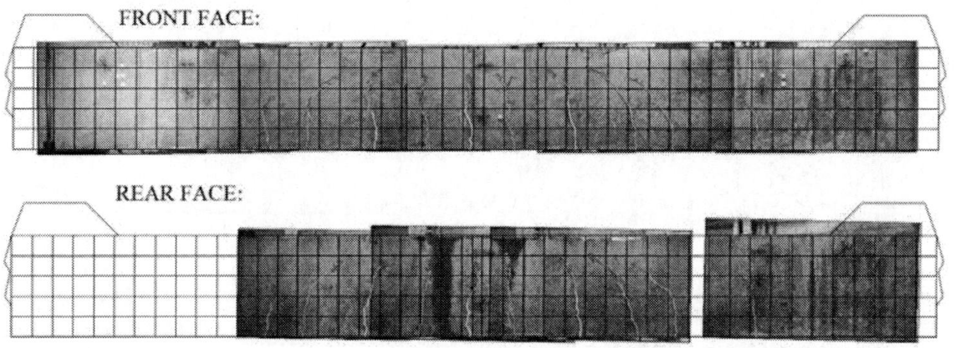

Figure 4. Cracks identification after different load stages: after 200KN (yellow); after 250KN (Blue); after 300KN (Green); after ultimate load (Red)

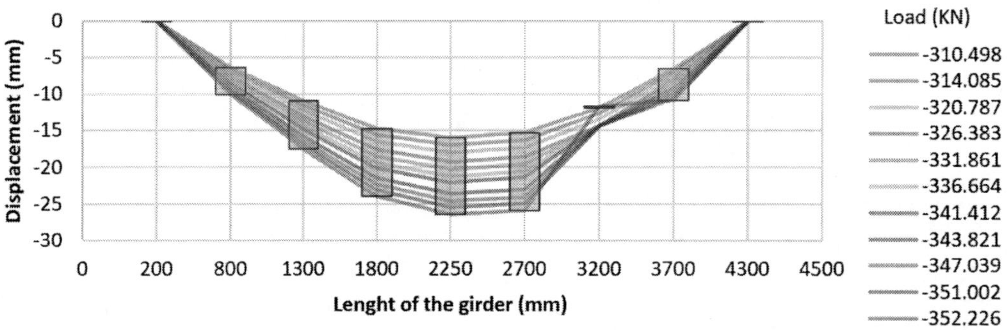

Figure 5. Displacement progress of a specific load range in the ultimate load stage: Green bars are displacements increase (down); Red bar is displacement jump (up)

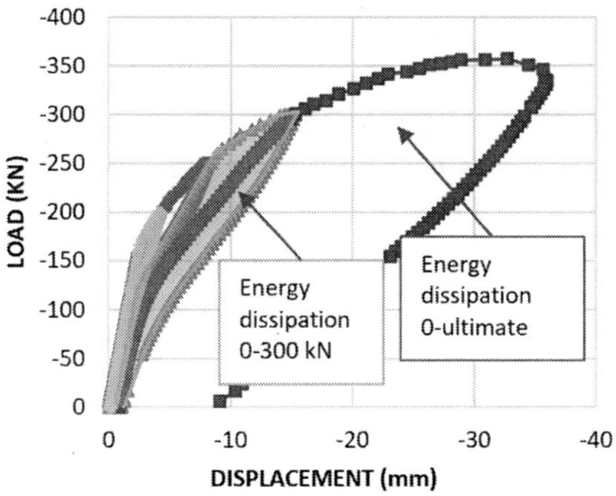

Figure 6. Energy dissipation in different stages from displacement transducer located in the span center

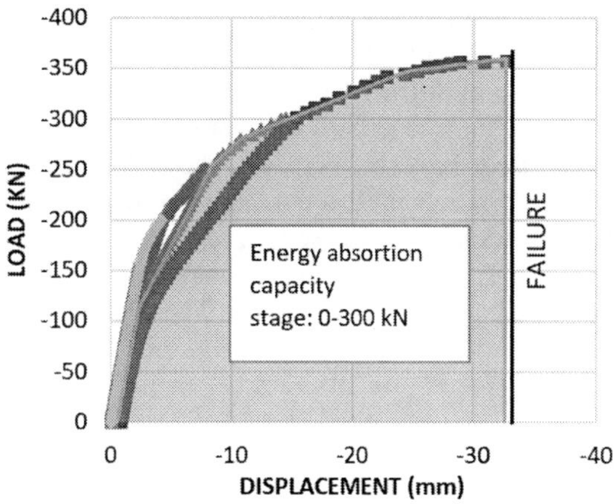

Figure 7. Energy absorption capacity of a specific stage from the displacement transducer at the span center

Table 2. Energy dissipated and energy absorption capacity from all loading stages

	0-150 kN (1)	0-150 kN (2)	0-150 kN (3)	0-150 kN (4)
Energy dissipated (Joules)	15.95	8.67	10.78	9.77
Energy absorption capacity (Joules)	3892.47	3888.18	3890.81	3891.28
	0-200 kN	0-250 kN	0-300 kN	0-ultimate
Energy dissipated (Joules)	67.27	191.80	574.08	2841.17
Energy absorption capacity (Joules)	3886.86	3863.84	3797.60	3531.60

3.2. Energy calculation
The energy dissipation is the area within the load-unload curve vs. displacement (graphically described in Figure 6) and the energy absorption capacity is the area beneath the load-displacement curve under the assumption of reaching the failure load (see Figure 7). The sum of all transducers data allows the calculation of the total work in the entire girder and this is summarized in Table 2.

4. Dynamic testing and vibration monitoring
The equipment used for the dynamic testing consists of an impact hammer Bruel & kjaer type 8208 – Modal sledge hammer 3-pound head with a maximum compressive force of 44.4 kN and a sensitivity of 0.225 mV/N. The accelerometers and the impact hammer were connected to a Multi-input Data Acquisition System Keyence Series NR-600.

4.1. Impact Hammer test and FRF
The impact roving technique was performed to obtain the modal parameters where the impact points (seven) are located in the opposite face (top of the girder) of the accelerometers location (bottom part). Two impact tips were used to produce different amplitudes of excitation (soft tip and medium tip) and 30 hits with each tip were carried out in each impact point of every stage.

Transfer functions were easily obtained and since the hammer and accelerometers were synchronized, the single-input multiple-output (SIMO) approach was also used for estimating the frequency response functions (FRF). Each value of the modal parameters represents the mean and standard deviation of 210 samples per tip of every stage (4200 samples in total for the whole experiment).

IOP Conf. Series: Journal of Physics: Conf. Series **842** (2017) 012038 doi:10.1088/1742-6596/842/1/012038

4.2. Frequency monitoring

Structural damage is interpreted as a decay of the mechanical properties of the structure, which means a decrease of stiffness. However, the trend of natural frequencies to the progress of damage shows an anomalous increase from one damage configuration to the next one even experimentally [5]. The first natural frequency (see Figure 8) shows a typical decrease as severity of damage increases, but there is a particular increment that occurs without any load step in between and the reason is because of changes in displacements since the elastic curve of the girder is recovering after the last load-unload step. The second frequency has diverse changes from the moment that cracks started and the location of these ones in the girder (see Figure 9). The third modal frequency might be the only one that shows a change due to the tendon cut and it also agrees with the increase of the first frequency cause of the elastic recovery (see Figure 10).

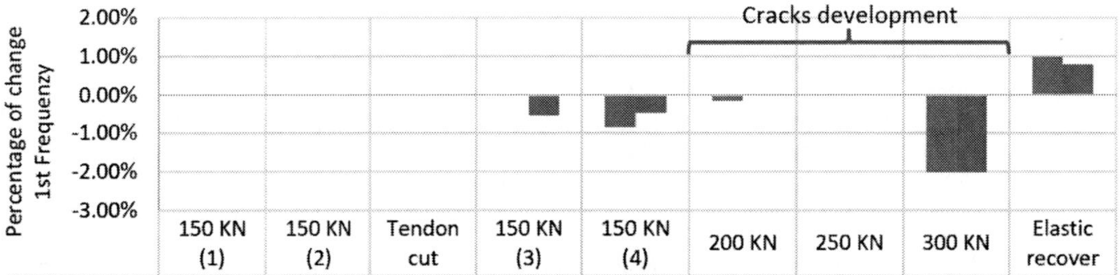

	150 KN (1)	150 KN (2)	Tendon cut	150 KN (3)	150 KN (4)	200 KN	250 KN	300 KN	Elastic recover
■ Soft Tip	0.00%	0.00%	0.00%	0.00%	-0.85%	-0.15%	0.00%	-2.02%	0.99%
■ Medium Tip	0.00%	0.00%	0.00%	-0.54%	-0.46%	0.00%	0.00%	-2.02%	0.80%

Figure 8. Percentage changes in the first bending mode frequency along specific events

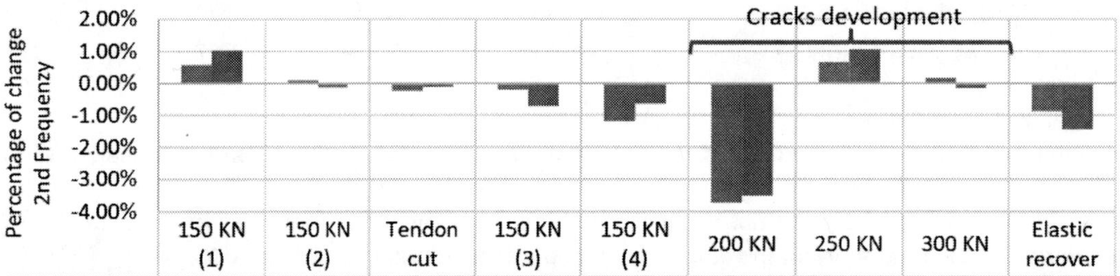

	150 KN (1)	150 KN (2)	Tendon cut	150 KN (3)	150 KN (4)	200 KN	250 KN	300 KN	Elastic recover
■ Soft Tip	0.57%	0.10%	-0.23%	-0.20%	-1.19%	-3.73%	0.66%	0.17%	-0.86%
■ Medium Tip	1.03%	-0.13%	-0.11%	-0.72%	-0.64%	-3.51%	1.06%	-0.15%	-1.45%

Figure 9. Percentage changes in the second bending mode frequency along specific events

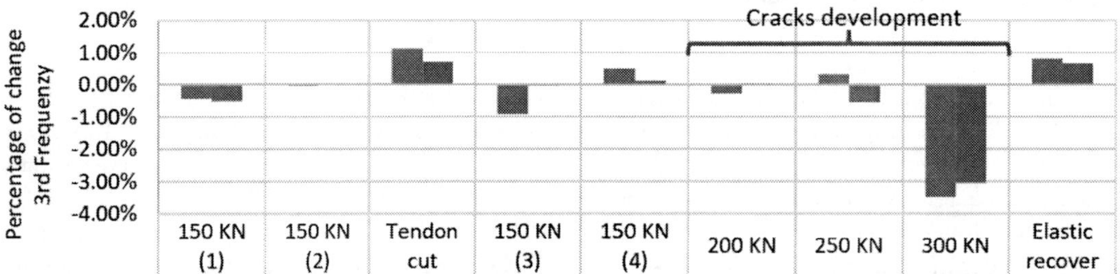

	150 KN (1)	150 KN (2)	Tendon cut	150 KN (3)	150 KN (4)	200 KN	250 KN	300 KN	Elastic recover
■ Soft Tip	-0.44%	-0.05%	1.12%	-0.92%	0.50%	-0.29%	0.32%	-3.49%	0.81%
■ Medium Tip	-0.52%	0.01%	0.71%	-0.03%	0.13%	0.03%	-0.55%	-3.06%	0.66%

Figure 10. Percentage changes in the third bending mode frequency along specific events

4.3. Damping ratio monitoring

The damping ratio was calculated by identifying the half power (-3dB) points of the magnitude of the frequency response function and once having the frequency bandwidth between the two half power points, the damping ratio is the ratio of the bandwidth with two times the frequency resonance. Figure 11 shows the first modal damping ratio development with no clear trend of specific damage. The second modal damping ratio, as shown in Figure 12, seems to change after cracks begin. Figure 13 shows how the third damping ratio might be altered by the tendon cut once again as shown in the third bending mode frequency.

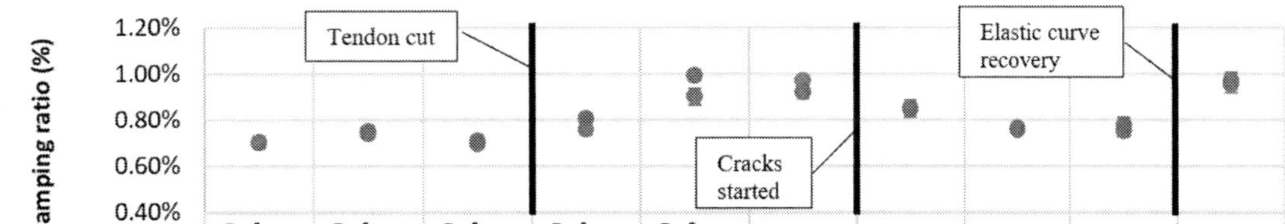

	Before 150 KN (1)	Before 150 KN (2)	Before tendon cut	Before 150 KN (3)	Before 150 KN (4)	Before 200 KN	Before 250 KN	Before 300 KN	After 300 KN	Before ultimate
● Soft Tip	0.70%	0.74%	0.70%	0.76%	0.90%	0.97%	0.85%	0.77%	0.75%	0.95%
● Medium Tip	0.70%	0.75%	0.71%	0.81%	0.99%	0.92%	0.85%	0.76%	0.78%	0.97%

Figure 11. Change in the first modal damping ratio along all experiment stages

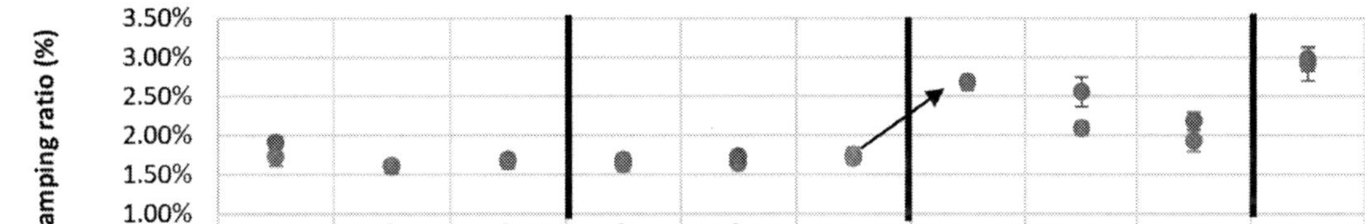

	Before 150 KN (1)	Before 150 KN (2)	Before tendon cut	Before 150 KN (3)	Before 150 KN (4)	Before 200 KN	Before 250 KN	Before 300 KN	After 300 KN	Before ultimate
● Soft Tip	1.73%	1.60%	1.66%	1.63%	1.64%	1.75%	2.68%	2.09%	1.93%	2.91%
● Medium Tip	1.91%	1.61%	1.69%	1.69%	1.73%	1.71%	2.67%	2.55%	2.18%	2.98%

Figure 12. Change in the second modal damping ratio along all experiment stages

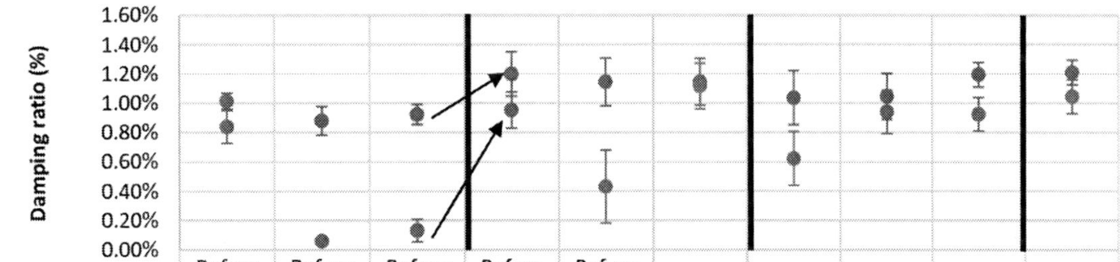

	Before 150 KN (1)	Before 150 KN (2)	Before tendon cut	Before 150 KN (3)	Before 150 KN (4)	Before 200 KN	Before 250 KN	Before 300 KN	After 300 KN	Before ultimate
● Soft Tip	0.84%	0.06%	0.13%	0.95%	0.43%	1.12%	0.62%	0.94%	0.92%	1.04%
● Medium Tip	1.01%	0.88%	0.92%	1.20%	1.14%	1.15%	1.04%	1.05%	1.19%	1.21%

Figure 13. Change in the third modal damping ratio along all experiments stages

IOP Conf. Series: Journal of Physics: Conf. Series **842** (2017) 012038 doi:10.1088/1742-6596/842/1/012038

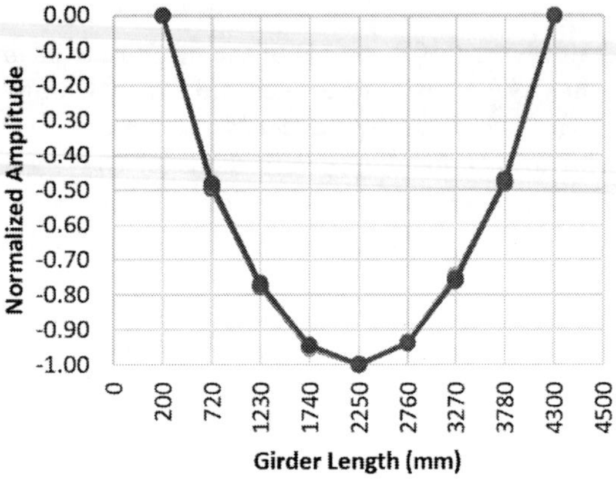

Figure 14. First bending mode shapes overlapped from all stages

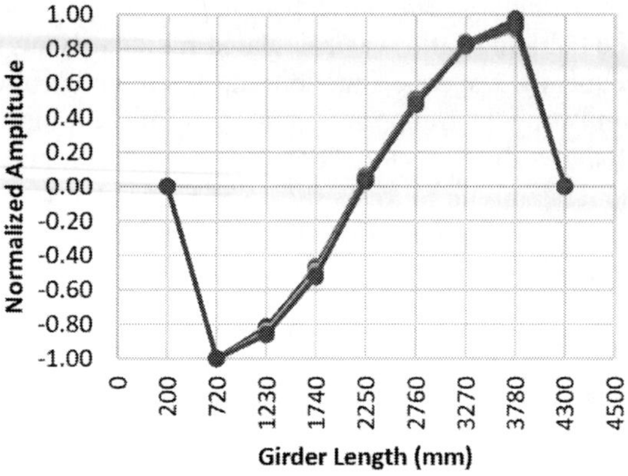

Figure 15. Second bending mode shapes overlapped from all stages

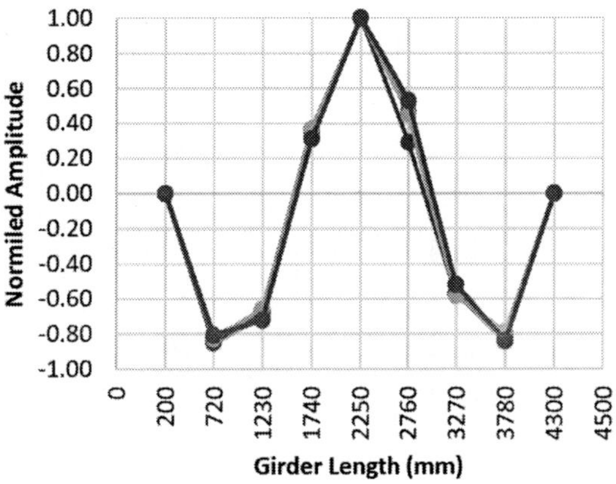

Figure 16. Third mode shapes overlapped from all experiment stages

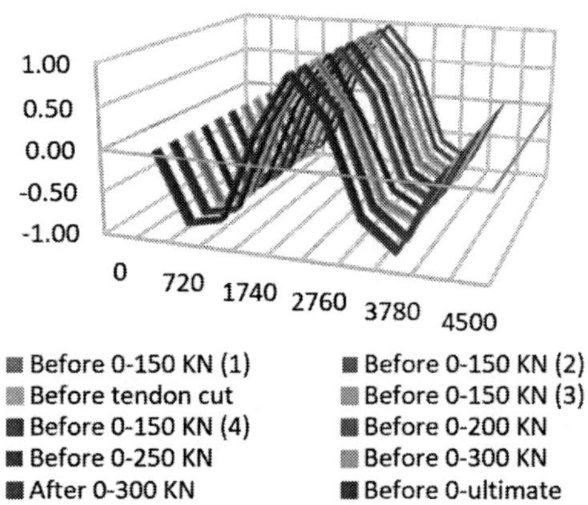

Figure 17. 3D view of third mode shapes overlapped from all experiment stages

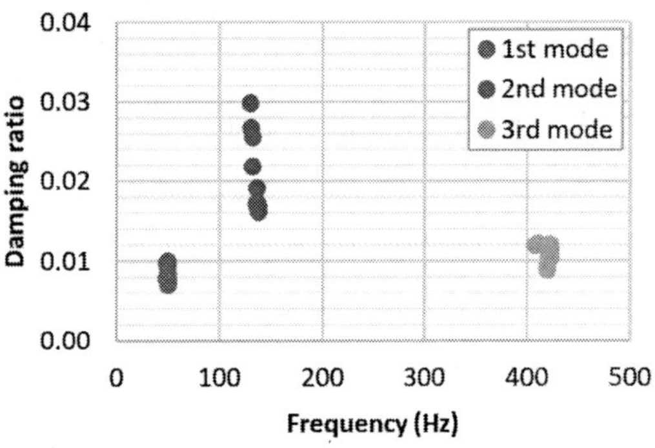

Figure 18. Variation of natural frequencies vs. modal damping ratios

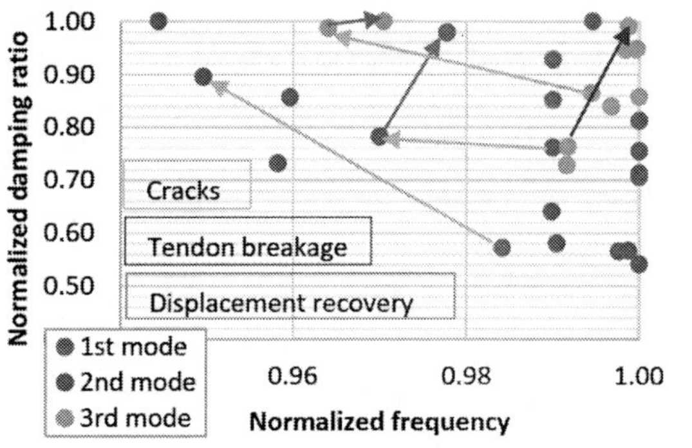

Figure 19. Correlation between normalized frequency and normalized damping ratio

4.4. Mode Shapes monitoring

Mode shapes were extracted from the imaginary part of the frequency response functions and as shown in Figure 14, Figure 15, Figure 16 and Figure 17; after being normalized it seems that there are no drastic changes, except for the third mode shape that showed a change due to cracking.

IOP Conf. Series: Journal of Physics: Conf. Series **842** (2017) 012038 doi:10.1088/1742-6596/842/1/012038

5. Discussions on experimental results

Similar results were obtained in Figure 18 to the ones in a previous investigation of concrete bridges [2], but this time it can be identified the mode that is more sensitive to damages, that is the second mode. In Figure 19, a correlation between normalized frequency and normalized damping ratio was investigated, but the only mode that has a strong fit is the second bending mode. Although, some clear data deviations were identified in all of them when specific events occurred. The effect of the tendon damage did not have relevant impact on the first two frequencies as happened in previous and similar studies [6], [7] and it also verifies that prestressing measurement force is not viable through frequencies only [8], but higher frequencies and damping ratios might be able to detect some changes as shown before.

Figure 20 might be an evidence that permanent deformations has a strong influence in modal parameters since the only ordinate that stores two stages data is when the displacement recovery happens. Changes in different modes at different times might be due to the crack state and its location. Figure 21 is the description of the girder health only cause of cracks, where 100% means no cracks and 0% is considering the total length of the cracks in the final stage. The resemblance of the changes with Figure 22 is a clear evidence that the second mode is the more sensitive to cracks and a corroboration of using frequency data for crack detection in beam-type structures [9].

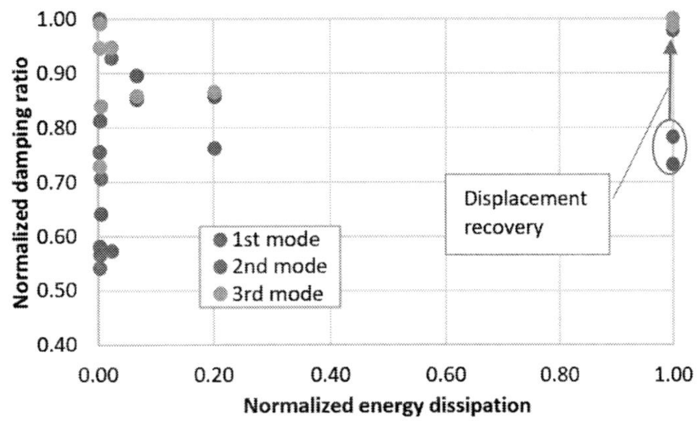

Figure 20. Correlation between energy dissipation and damping ratio

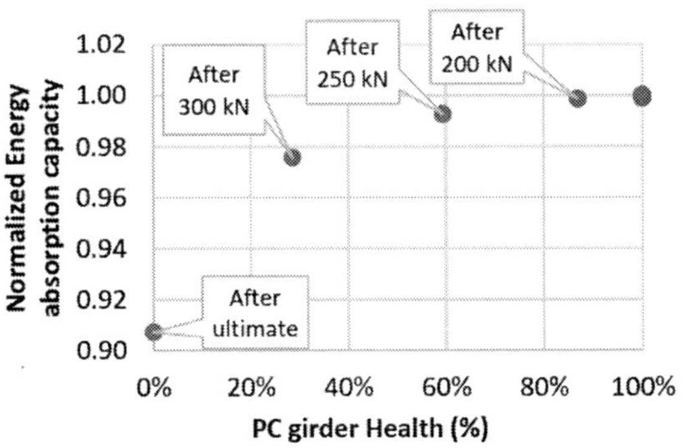

Figure 21. Effect of energy absorption capacity on the girder health due to cracks occurrence

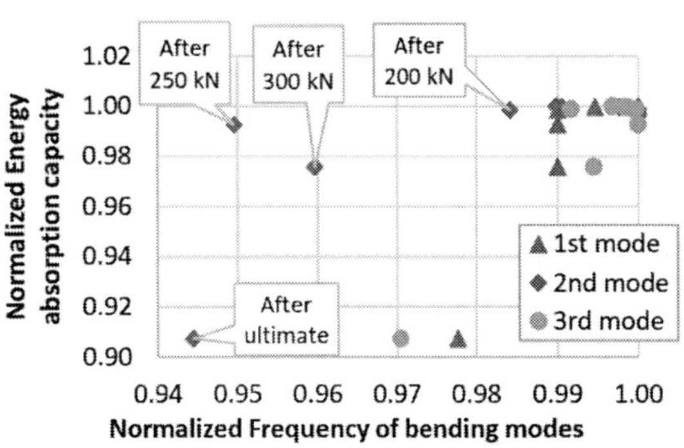

Figure 22. Correlation between frequency and energy absorption capacity

IOP Conf. Series: Journal of Physics: Conf. Series **842** (2017) 012038 doi:10.1088/1742-6596/842/1/012038

6. Conclusions

The amplitude of excitations when performing vibration tests has an effect on the accuracy of higher modes. There should be some precautions when interpreting the deviations of modal parameters because the deviation could be caused by damage or inadequate excitation. The elastic curve of the girder has a clear impact on modal parameters, which demonstrates that some attention must be paid to assess the permanent displacements, correlation with frequencies and damping ratio. There is a clear decrease in frequency as damage occurs but also an increase or decrease depending of the relative permanent displacements. The energy correlations with modal parameters are evidence that the last one can be developed and used for performance assessment.

References

[1] Quattrone, A. et al., 2012. Vibration tests on dismounted bridge beams and effects of deterioration. *Journal of Physics: Conference Series,* **382**

[2] Dammika, A. J. et al., 2015. *Proc. Int. Symp. on Life-Cycle Civil Engineering.* (London, Taylor & Francis Group), pp. 404-410

[3] Udwadia, F. E., 2005. Structural Identification and Damage Detection from Noisy Modal Data. *Journal of Aerospace Engineering,* **18**(3), pp. 179-187.

[4] Shiotani, T., Oshima, Y., Goto, M. & Momoki, S., 2013. Temporal and spatial evaluation of grout failure process with PC cable breakage by means of acoustic emission. *Construction and Building Materials,* **48**, pp. 1286-1292.

[5] Dilena, M. & Morassi, A., 2011. Dynamic testing of a damaged bridge. *Mechanical Systems and Signal Processing,* **25** (5), pp. 1485-1507.

[6] Limongelli, M. P. et al., 2016. Damage detection in a post tensioned concrete beam - Experimental investigation. *Engineering Structures,* Volume **128**, pp. 15-25.

[7] Döhler, M., Hille, F., Mevel, L. & Rücker, W., 2014. Structural health monitoring with statistical methods during progressive damage test of S101 Bridge. *Engineering Structures,* Volume **69**, pp. 183–193.

[8] Noble, D., Nogal, M., O'Connor, A. J. & Pakrashi, V., 2016. The effect of prestress force magnitude and eccentricity on the natural bending frequencies of uncracked prestressed concrete beams. *Journal of Sound and Vibration,* Volume **365**, pp. 22-44.

[9] Kim, J.-T. & Stubbs, N., 2002. Crack detection in beam-type structures using frequency data. *Sound and Vibration,* **259**(1), pp. 145-160.

IOP Conf. Series: Journal of Physics: Conf. Series **842** (2017) 012039 doi:10.1088/1742-6596/842/1/012039

Developments in damage assessment by Marie Skłodowska-Curie TRUSS ITN project

A González

College Lecturer and TRUSS ITN Project Coordinator, Civil Engineering School, University College Dublin, Dublin, Ireland

Email: arturo.gonzalez@ucd.ie

Abstract. The growth of cities, the impacts of climate change and the massive cost of providing new infrastructure provide the impetus for TRUSS (Training in Reducing Uncertainty in Structural Safety), a €3.7 million Marie Skłodowska-Curie Action Innovative Training Network project funded by EU's Horizon 2020 programme, which aims to maximize the potential of infrastructure that already exists (http://trussitn.eu). For that purpose, TRUSS brings together an international, inter-sectoral and multidisciplinary collaboration between five academic and eleven industry institutions from five European countries. The project covers rail and road infrastructure, buildings and energy and marine infrastructure. This paper reports progress in fields such as advanced sensor-based structural health monitoring solutions – unmanned aerial vehicles, optical backscatter reflectometry, monitoring sensors mounted on vehicles, ... – and innovative algorithms for structural designs and short- and long-term assessments of buildings, bridges, pavements, ships, ship unloaders, nuclear components and wind turbine towers that will support infrastructure operators and owners in managing their assets.

1. Introduction

Buildings, energy and transport infrastructure are key elements for supporting society in their day-to-day activities. The infrastructure network is ageing and deteriorating rapidly under an increasing demand in operational and environmental loads. While an efficient infrastructure network provides economic and social benefits, infrastructure chaos in terms of capacity or reliability can involve economic costs and lower quality of life. For infrastructure to remain effective and structurally safe, a management strategy that guarantees proper maintenance and best use of the resources available is needed. However, this is a complex task due to uncertainties associated to the structural capacity and to the demand on a structure. TRUSS (Training in Reducing Uncertainty in Structural Safety) is a 4-year Marie Skłodowska-Curie Innovative Training Network (ITN) project [1], which started in January 2015, aiming to improve infrastructure management via advanced structural reliability and monitoring methods. Figure 1 illustrates the TRUSS concept. TRUSS consortium (Figure 2) is composed by:

- 5 Universities (University College Dublin –UCD– and Trinity College Dublin –TCD– in Ireland, Universitat Politecnica de Catalunya –UPC– in Spain, University of Nottingham –UNOTT– in UK and Université de Nantes in France).
- 11 Industry participants (Lloyd's Register EMEA, Full Scale Dynamics Ltd, AECOM and Microlise in UK, Equipos Nucleares S.A. –ENSA–, COMSA and COTCA in Spain, Arup and Burgmann Packings in Ireland, Phimeca Engineering in France, and Greenwood Engineering in Denmark).

- 1 Research institute (Transport Research Laboratory –TRL– in UK).

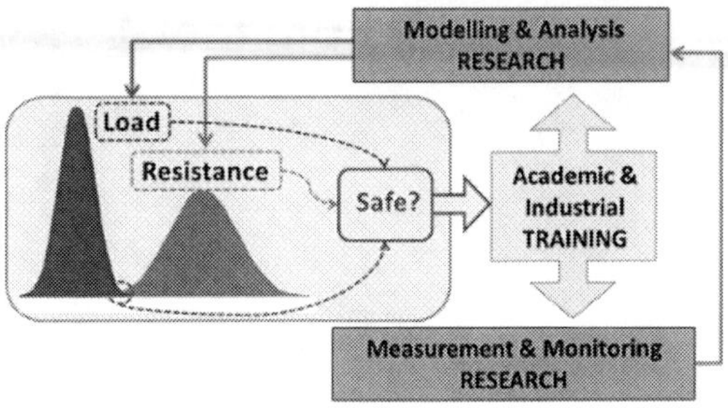

Figure 1. Concept of TRUSS.

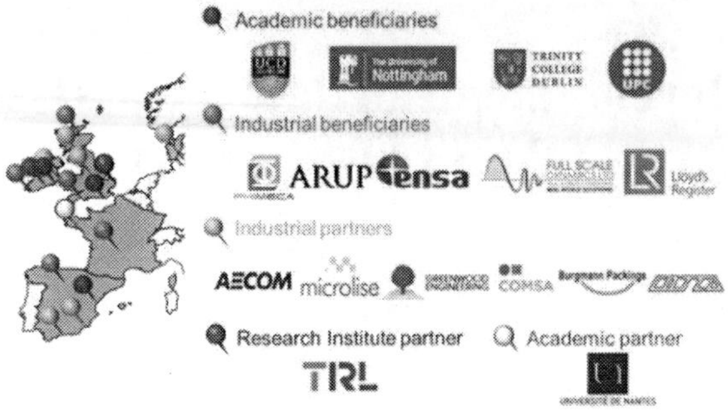

Figure 2. TRUSS consortium.

The main objectives of TRUSS are twofold:

- To develop reliable monitoring systems and structural, material and loading models to be achieved through research that will contribute to: (a) more efficient infrastructure design, assessment, monitoring and management, (b) maintain current infrastructure stock in operation while minimizing risks, and (c) reduce infrastructure costs and demand for non-renewable and carbon intensive resources while maintaining or improving safety levels.
- To offer a high-quality intersectoral and multidisciplinary training in structural safety to a new generation of Early Stage Researchers (ESRs) through network-wide and local activities. The training programme combines taught modules with original research supported by secondments, which will allow ESRs gaining experience and enhancing their career prospects in both industrial and academic sectors.

In order to meet these objectives, 14 ESRs have been recruited to carry out research in 14 different individual projects. While moving forward the state of the art in the topics of these projects, TRUSS prepares ESRs for dealing with the challenges faced at the assessment and management stages of aging large scale structures. TRUSS is structured into five Work Packages (WPs): WP1 on Management (led by UCD), WP2 on Dissemination and Outreach (led by UPC), WP3 on Structured Training (led by UCD), and two research WPs: WP4 on Buildings, Energy and Marine (led by UCD) and WP5 on Rail and Road infrastructures (led by UNOTT). The following sections report on the progress of the project in the period up to December 2016.

2. Dissemination and outreach

Following a recruitment period, ESRs joined TRUSS ITN between September and December 2015. In their first year, ESRs have published a total of 25 technical papers: 2 in peer reviewed journals and 23 in conferences, that are included in the list of references. 16 of these papers include academic and industrial co-authors, which shows evidence of the narrow collaboration between both sectors. Dissemination by TRUSS (WP2) is keenly aware of the importance of not only producing and presenting research outputs for the scientific community and key stakeholders (i.e., via conferences, workshops, publications and reports), but also engaging the general public in line with the Innovation Union objectives [2]. TRUSS activities contribute to make citizens aware that:

- infrastructure is aging and failing, and funding has been insufficient to repair and replace it,
- the important role of the Marie Skłodowska-Curie Actions [1] in forming 21st century engineers that will have the skills to face the formidable challenge of modernizing the fundamental infrastructure that supports civilization.

The website (http://trussitn.eu) has 135 pages of content with the top 8 countries with more visitors to the website being United States, United Kingdom, Ireland, Spain, India, France, Germany and Italy. TRUSS social media profiles include Facebook (https://www.facebook.com/trussitn.eu), Google+ (https://plus.google.com/+TrussITN), LinkedIn (https://www.linkedin.com/in/trussitn), Twitter (https://twitter.com/TRUSSITN), YouTube (https://www.youtube.com/c/TrussITN), Research Gate (https://www.researchgate.net/profile/Truss_Itn), and Blogger with TRUSS ESRs that have delivered 81 posts so far. The fact that the research topics impact directly on the day-to-day lives of people gives the research a relevance that captures the interest and inspire younger students with an interest in engineering. With this in mind, TRUSS has carried out outreach activities that include activities in Junior and High Schools, research exhibitions and Open Days at Universities. Therefore, TRUSS researchers have been featured in press releases in mainstream Italian and Spanish newspapers, in Italian and Irish magazines and on Irish television.

3. Training

Figure 3 illustrates the TRUSS training programme, which it is structured in two ways:

- Network-wide and local training activities, and
- Supervised research towards a doctoral award.

TRUSS holds network-wide meetings approximately every 6 months. During these meetings, ESRs have an opportunity to practise their communication skills, and to receive academic and industrial training and feedback. The network-wide training programme is extended by a range of advanced research methods, project management, language courses, transferable skills and communication modules available for the ESR at each host. Another significant form of training is on-the-job training, which ESRs have received from the local research group at their host, and at the institution/s where they are seconded.

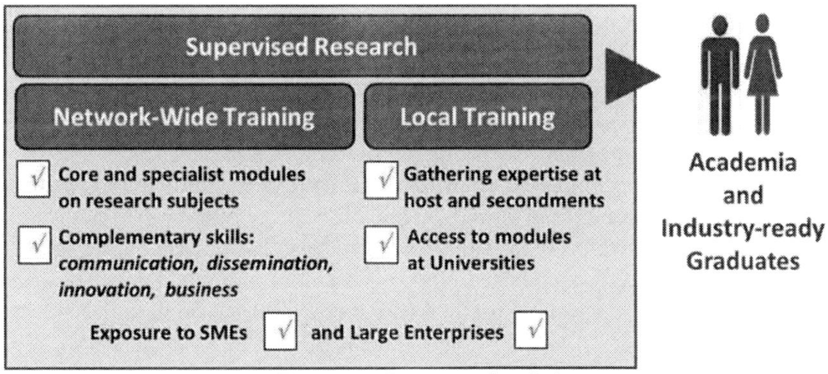

Figure 3. Structured training.

Each ESR is registered to a PhD programme in one of the academic beneficiaries. The primary support to help ESRs to conduct the high-level research necessary to achieve a PhD is their main supervisor and Doctoral Studies Panel (DSP) composed by consortium experts. In addition to the formal supervision structure, each ESR is required to maintain a Personal Career Development Plan (PCDP). Following presentations by the ESRs at the network-wide meetings, the consortium provides feedback and then, each ESR meets with his/her DSP in a focus group that discuss the PCDP and future plans of action. This structure plays a major part in informing the direction of the research and training and empowers ESRs to take ownership of their individual projects.

4. Research on buildings, energy and marine infrastructure

Table 1 lists the TRUSS projects falling within WP4, which is characterized by the very aggressive environments that the infrastructure is subjected to (corrosive, radioactive, non-linear structural responses) or relatively high uncertainties regarding materials and modelling.

Table 1. Individual research projects in WP4.

	Title	Host	Secondment
ESR1	Reliability of Concrete Structures Reinforced with Braided FRP	University College Dublin (Ireland)	Burgmann Packings (Ireland)
ESR2	Reduction of Uncertainty in Assessing Concrete Strength of Existing Structures	Arup (Ireland)	University College Dublin (Ireland)
ESR3	Reduction of Uncertainty in Design of Free Standing Nuclear Spent Fuel Rack	Equipos Nucleares, SA (Spain)	University College Dublin (Ireland)
ESR4	Probabilistic Optimization of the Design of Offshore Wind Turbine Towers	Trinity College Dublin (Ireland)	Lloyd's Register (UK)
ESR5	Integrity Management of Ship Structures	Lloyd's Register EMEA (UK)	University College Dublin (Ireland)
ESR6	Residual Life Assessment and Management of Ship Unloaders	Lloyd's Register EMEA (UK)	University College Dublin (Ireland)

4.1. Materials and buildings

Uncertainty in material strength is addressed via reliability models of concrete structures reinforced with Basalt Fiber Reinforced Polymer (BFRP) (ESR1) and assessment and testing of concrete strength of existing structures (ESR2).

4.1.1. Reliability of concrete structures reinforced with braided FRP. This project by Antonopoulou et al [3-5] aims to assess the reliability of structures manufactured using braided BFRP reinforcement. Basalt fibres and epoxy resin are used as reinforcement and matrix respectively (Figures 4, 5 and 6). Bars are manufactured using braiding and a vacuum assisted resin infusion technique. The work conducted so far can be summarised into the following tasks. The first includes the design and manufacture of braided BFRP preforms while changing key parameters (angle, no of layers). Ten different configurations in various diameters (5, 8, 10 mm) have been produced in Burgmann Packings Ltd. Then, the suitability of manufactured BFRP samples for epoxy resin impregnation in both regular and spiral configurations was explored in SuperTEX Austria, Burgmann Packings Ltd and UCD. Finally, a theoretical numerical approach based on the Classical Laminate Theory has been developed for the prediction of the elastic properties of braided BFRP rebars based on the influence of geometrical factors and processing conditions.

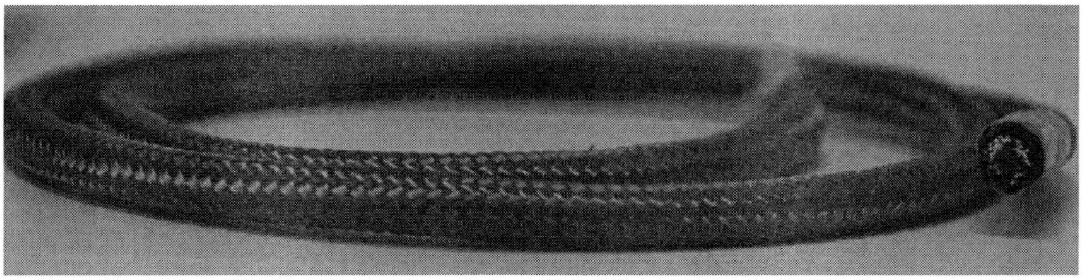

Figure 4. Braided preform.

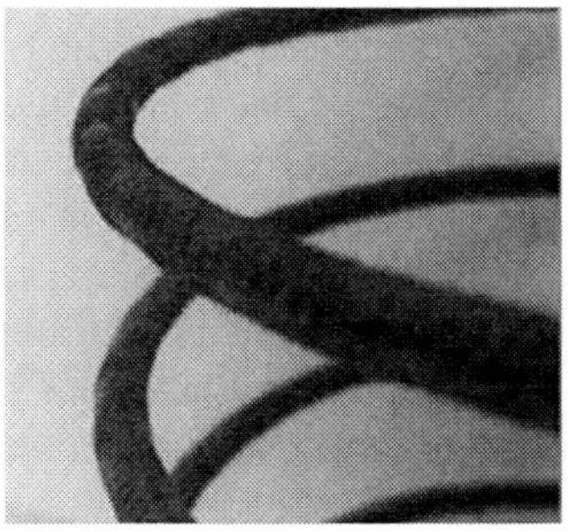

Figure 5. Resin impregnated braided BFRP.

Figure 6. BFRP sample.

4.1.2. Reduction of uncertainty in assessing concrete strength of existing structures. With the goal of reusing and extending the life of structures, assessment of concrete strength in existing structures is a must. Non-destructive tests (NDTs) allows quick collection of data at a limited cost for such assessment. As most NDTs indirectly measure concrete strength, they introduce uncertainty in the overall assessment. Existing NDTs are best suited in assessing the uniformity and quality of concrete in the structures. High variability in results have been observed following some preliminary studies using rebound hammer (Figure 7) and ultrasonic pulse velocity tests. Complete pullout failure during the pullout of post-installed concrete screw has potential to measure compressive strength of concrete as threads of screw installed in the drilled hole cut into concrete and exert bearing force on the surrounding concrete. Research into the application of post-installed screw for concrete strength assessment is being conducted and evaluated for practical application by Sourav et al [6,7].

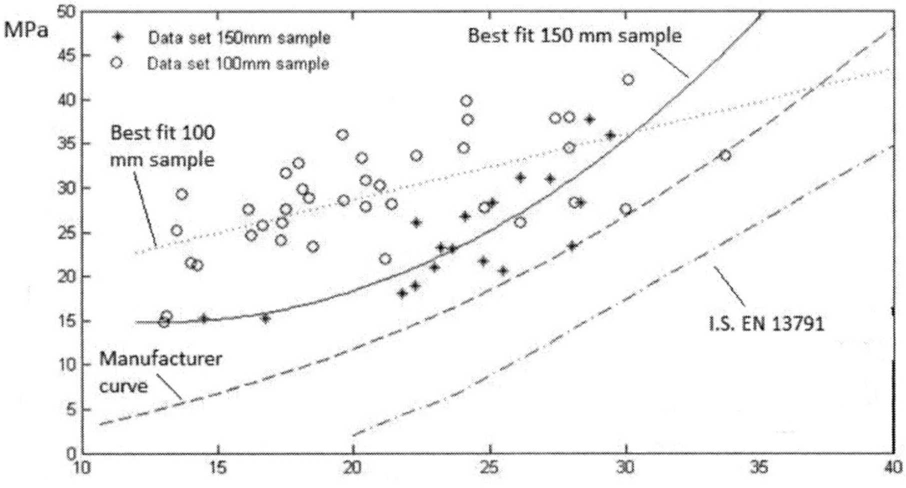

Figure 7. Compressive strength versus rebound number.

4.2. Energy infrastructure

Uncertainty in the design of energy infrastructure is addressed via different response models for a free standing nuclear spent fuel rack (ESR3) and probabilistic optimisation for wind turbine towers (ESR4).

4.2.1. Reduction of uncertainty in design of free standing nuclear spent fuel rack. Spent fuel racks rest in free-standing conditions submerged in water at 12 m depth (Figure 8). Their seismic response exhibits a highly geometrical nonlinear behaviour with dynamic contacts and large displacements influenced by the fluid-structure interaction. An ad-hoc methodology implements the added mass concept to represent the water effect in a cost-effective way. The transient analysis is therefore carried out in a Finite Element (FE) structural model (Figure 9) with direct integration of the equation of motion. However, some dispersion of results still exists and several sources of uncertainty have been identified. Modelling issues have been introduced and a parametrical analysis of the stochastic input variables has been launched. Moreover, a 2-rack physical model has been designed to study the water coupling forces between units. As a result, a better understanding of the rack seismic behaviour is being gathered and current safety margins will be quantified more accurately. A direct application of the results can lead to increase the storage capacity of the existing fuel storage pools, which would bring an increase in the operation span of nuclear power plants without fuel reprocessing or dry cask storage. The dynamic analysis of other submerged sliding structures can also benefit from the research undertaken here by Gonzalez et al [8-10].

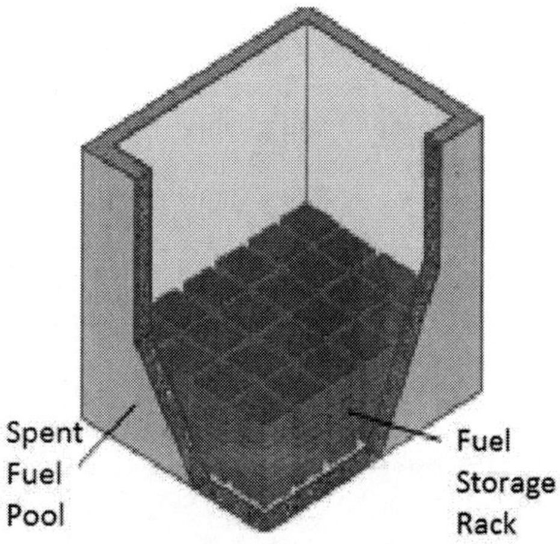

Figure 8. Fuel storage racks resting on the bottom of spent fuel pool

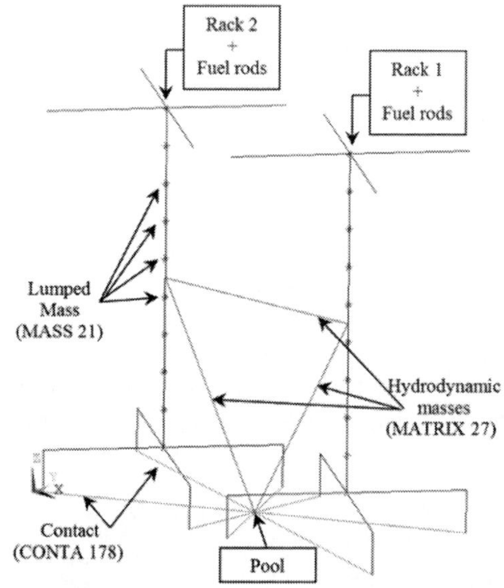

Figure 9. Finite element model of free standing rack.

4.2.2. Probabilistic optimization of the design of offshore wind turbine towers. In the particular case of Offshore Wind Turbines (OWT), despite a relatively new sector, the implemented practices still disregard the need of a fully comprehensive description of the system's behavior. In such a complex system, the space of variables to analyze grows fast, making the approach to create robust designs very resource intensive. Quantifying OWT long-term uncertainty on a practical basis has been therefore a major topic of research in the presented framework. It is of major importance in the field of OWT engineering to establish techniques to extrapolate extreme events and quantify their uncertainty. Substantial work has been developed in this topic and on how to accurately and efficiently quantify the long term uncertainty in both; the external loading variables (e.g., wave characterization) and on the OWT itself. Results show that new reliability approaches can efficiently approximate the long term occurrences based on short-term data, e.g., Figure 10. This project by Teixeira et al [11,12] connects research and its real application via the partnership with Lloyd's. A main line of research connected with improving the efficiency of the probabilistic design of offshore wind turbines using surface models, such as the Kriging models, is being exploited. These, known also as Gaussian process models, are of interest for the topic of reliability analysis due to their interpolation capacity, the flexibility to approximate arbitrary functions with a high level of accuracy and the capability of accounting for a local uncertainty measure. Currently, an integrated model of an offshore wind turbine with monopole foundation is being built and the model response being fitted to a Kriging surface.

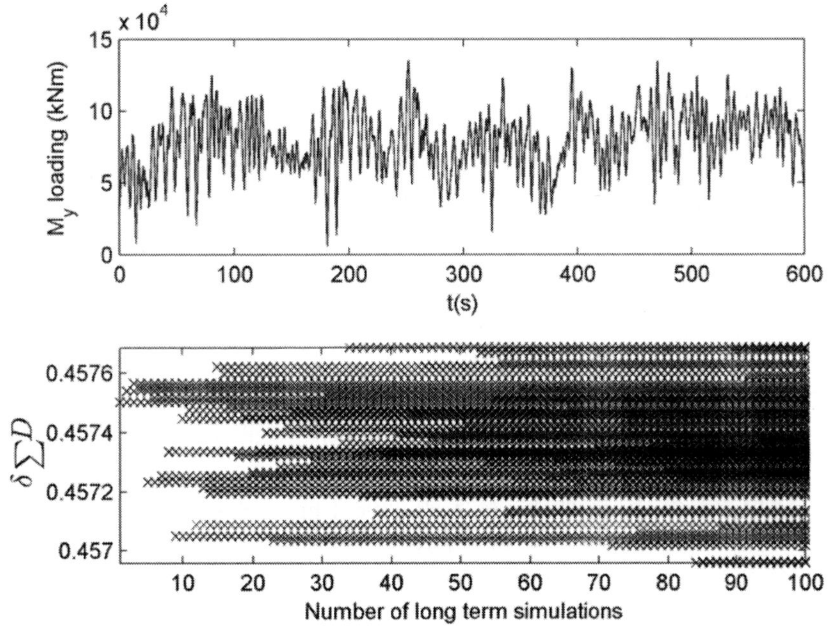

Figure 10. Example of statistical characterization from short-term system description to long term response.

4.3. Infrastructure in a marine environment
Uncertainty in the assessment and residual life of infrastructure in a marine environment is addressed via an integrity management based on long-term monitoring data for ageing marine structures (ESR5) and ship unloaders (ESR6).

4.3.1. Integrity management of ship structures. Numerical investigations have been carried out on: 1) fatigue reliability with S-N model and fracture mechanics (FM) model); 2) Crack initiation and reliability-based inspection planning; 3) Calibration of equivalent initial crack size (EIFS); 4) Calibration of time-to-crack-initiation (TTCI). This project by Guang et al [13,14] proposes to use a fracture mechanics based reliability analysis method that takes the crack initiation stage into account via the concept of TTCI. The optimum inspection plan for a fatigue prone ship structural component (i.e., Figure 11) is derived by the new approach and compared to the commonly-used method that only considers crack propagation life. Two inspection planning approaches are tested to investigate the influence of incorporating crack initiation period: (i) target reliability approach and, (ii) equidistant inspection times approach. With each planning approach, two inspection methods are adopted: close visual and magnetic particle inspection. Recommendations on the inspection method and planning approach to adopt are provided depending on the crack initiation stage (Figure 12) being considered or not. The following step is to develop a normalized calibration procedure of FM model to S-N curves, followed by research on probabilistic crack management and maintenance planning.

Primarily, this project will contribute to life-cycle fatigue management methods for ship and marine structures, and will lead to optimum decisions for the design, inspection and maintenance of structures in an integrated framework. Secondly, the project will facilitate the establishment of a crack database in marine engineering by providing the least data that need to be measured and stored, and showing how historical data can be utilized in decision-making for inspection and maintenance actions. Lastly, the project will bring awareness and facilitate the use of reliability methods to the marine community, by demonstrating their strong capacities in incorporating uncertainties associated with wave loads, material properties, structural geometries, analysis, inspection and maintenance methods.

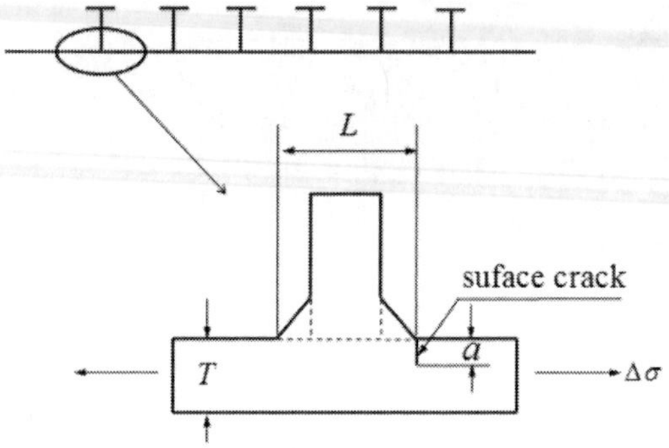

Figure 11. A fillet welded joint in a stiffened plate.

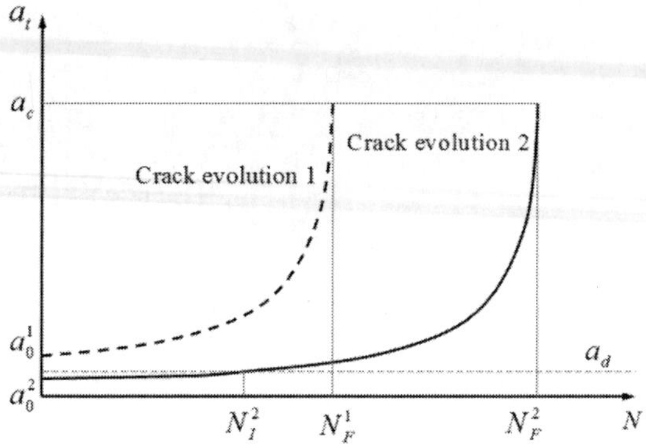

Figure 12. Schematic representations of crack evolution.

4.3.2. Residual life assessment and management of ship unloaders. This project by Milana et al [15-17] focuses on residual life assessment of ship unloader. Incidences related to a potential failure on these cranes or their parts can have catastrophic consequences in terms of both fatalities and economic impact. In addition, the deterioration process for ship unloaders is generally higher than for other infrastructures, since they are exposed to both an extremely aggressive environment and continuous alternating loadings. Following an initial phase of the state-of-art review, the standard procedure for fatigue life assessment in Figure 13 has been identified. New concepts and additional analyses, shown in red, represent some ideas formulated in order to reduce uncertainties arising from assumptions involved in the process. Data provided by a monitoring system have been processed and used to calibrate a FE model of the structure. Comparing results from the FE model to the recorded stresses, it has been noticed that the dynamic amplification factor currently used is very conservative for some locations. Therefore, location-specific DAFs has been introduced to improve the assessment of the structure while keeping calculations relatively simple. It is planned to investigate next the impact of end pins not free to rotate on fatigue life. As a result of this research, the residual life of an existing ship unloader will be assessed in a more reliable way with a significant impact on operational capacity and safety of ports.

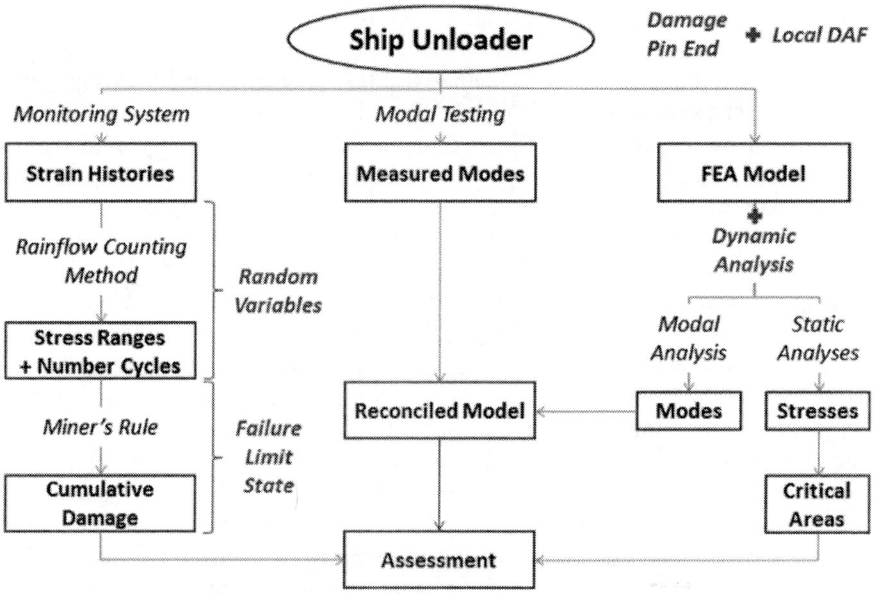

Figure 13. Procedure for fatigue life assessment.

5. Research on rail and road infrastructure

The infrastructures covered in WP5 are characterized by a variable traffic load, and they include individual research projects ESR7 to ESR14.

Table 2. Individual research projects in WP5.

	Title	Host	Secondment
ESR7	Railway Bridge Safety and Condition Assessment	Full Scale Dynamics Ltd (UK)	University College Dublin (Ireland)
ESR8	Probabilistic Modelling of Bridge Damage based on Damage Indicators	Phimeca Engineering (France)	University College Dublin (Ireland)
ESR9	Railway Bridge Condition Monitoring and Fault Diagnostics	University of Nottingham (UK)	AECOM (UK)
ESR10	Assessment of Bridge Condition and Safety based on Measured Vibration Level	Universitat Politecnica de Catalunya (Spain)	COMSA (Spain)
ESR11	Development of Optical Fibre Distributed Sensing for SHM of Bridges and Large Scale Structures	Universitat Politecnica de Catalunya (Spain)	COTCA (Spain)
ESR12	Bridge Damage Detection Using an Instrumented Vehicle	University College Dublin (Ireland)	Greenwood Engineering (Denmark)
ESR13	Using Truck Sensors for Road Pavement Performance Investigation	University of Nottingham (UK)	Microlise (UK) TRL (UK)
ESR14	Reduction of Uncertainty through Regularized, Automated Road Inspection	University College Dublin (Ireland)	Arup (Ireland)

5.1. Bridges

Uncertainty in bridge safety is addressed a new damage detection algorithm based on rotations due to a moving load (ESR7), reliability analysis of damage indicators (ESR8), condition monitoring and fault detection methods (ESR9), assessment of the damage sensitivity of various vibration parameters to develop a condition index, with a vibration parameter range that could serve as safety limits (ESR10), testing of different adhesives for the deployment of fiber optic sensing (ESR11) and application of an instrumented vehicle for bridge monitoring purposes (ESR12).

5.1.1. Railway bridge safety and condition assessment. In this research by Huseynov et al [18,19], numerical and experimental analyses are carried out in an effort to develop damage detection methods for bridge structures using rotations as the main parameter. Initially, an algorithm is developed in MATLAB, that employs the Influence Line (IL) and Moment – Area theorems, to simulate a moving vehicle loading across the hypothetical 2-D bridge structure. Formulations involved enable to define different levels of damage severity in terms of change in stiffness. A 3 m long simply supported beam structure is built in the lab to test the proposed damage detection concepts (Figure 14). The beam structure is instrumented with 9 inclinometers (at equal distances) to record rotation while a point load is moved (statically) across the length of the structure several times. Later, the structure is strengthened and the foregoing damage detection concepts are applied to successfully identify the location and severity of change in stiffness across the length of the beam structure (Figure 15).

Figure 14. Lab test set-up.

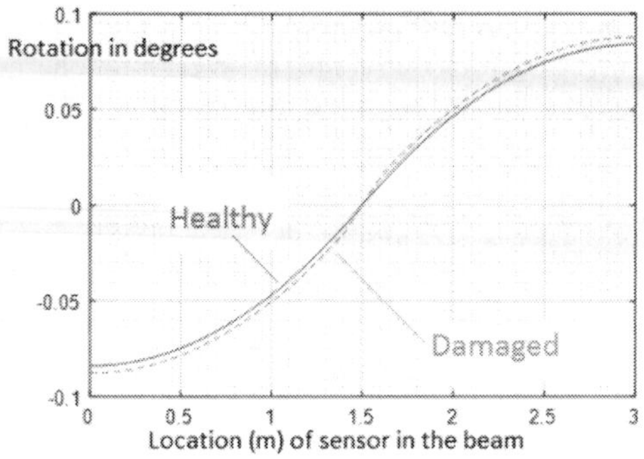

Figure 15. Rotation versus sensor location.

5.1.2. Probabilistic modelling of bridge damage based on damage indicators. This project by Heitner et al [20,21] is developing a bridge safety model in a probabilistic context. It involves Weigh-in-Motion data on the traffic load side, while it relies on results reported in the literature on the dead load and resistance sides. One of the main aims of the project is to improve this initial model using Bayesian updating and incorporating damage indicators (DIs). So far the focus has been on DIs based on deflection and rotation measurement data. It is important to note that most of the available sensors or health monitoring systems, which target to observe the actual performance of the entire structure, are able to provide information related to the stiffness of the bridge and therefore it has also been a crucial issue to find an appropriate way of connecting information of stiffness change to resistance change. Figure 16 reveals both the strong and obvious connection between the DI and the bar area loss; and the great amount of uncertainty related to the DI, which needs to be addressed appropriately.

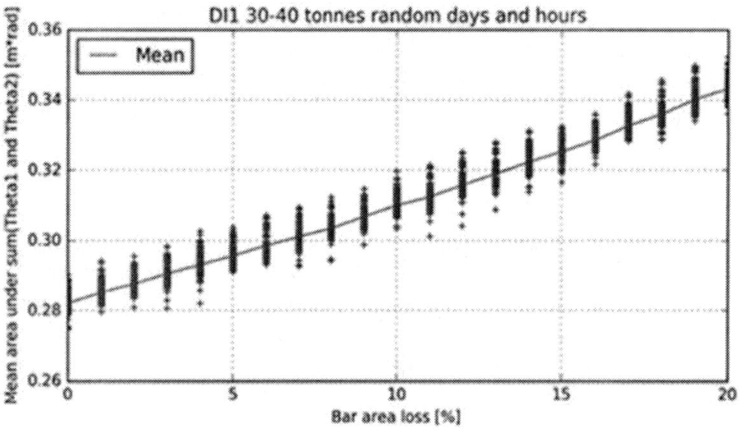

Figure 16. Relationship between reinforcing bar area loss (due to corrosion) and a DI based on rotation measurements for hourly random (30-40 tonnes) truck traffic flow in a 20 metres long simply supported reinforced concrete slab bridge.

5.1.3. Railway Bridge Condition Monitoring and Fault Diagnostics. Railway bridges are key elements for the reliability, availability and safety of the whole railway network, as they are pushed to their physical limits due to continuously changing environmental conditions, such as increasing of traffic and climate change that produces extreme events. In order to maximise the reliability, availability and safety of the railway network, bridge condition-monitoring strategies can help the railway industry by providing rapid and reliable information on the health state of each bridge. With this aim, a Bayesian Belief Network (BBN) method is developed by Vagnoli et al [22] in order to monitor in real-time the health state of a steel truss railway bridge. A FE model is perturbed by simulating the degradation mechanisms of the bridge elements, and the data collected from these simulations is used to as input of the BBN. As soon as a new measurements of the bridge behaviour are available, the BBN is able to assess in real-time the health state of the steel truss bridge and to identify the most degraded element of the bridge (Figure 17). This research will allow bridge managers to know the health condition of the bridge in real time and, consequently, the maintenance schedule will be optimized leading to efficient usage of the available budget.

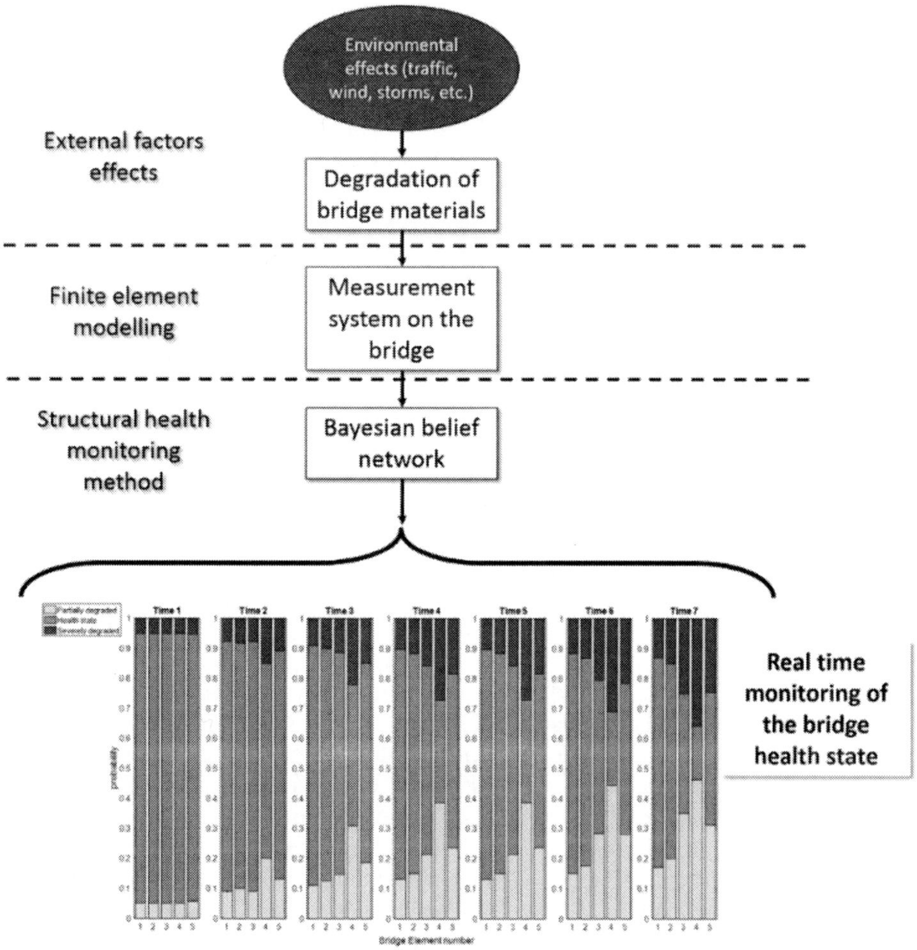

Figure 17. Bayesian Belief Network for structural health monitoring.

5.1.4. Assessment of bridge condition and safety based on measured vibration level. The primary technical achievements of the project by Moughty et al [23-26] thus far centre about the damage sensitivity assessment of numerous vibration parameters. This work was completed on data obtained from two real bridges in Austria and Japan, which were intentionally damaged incrementally and subjected to different loading regimes: ambient vibration and truck passage over the bridge, respectively. Results from this phase shall advance the shared knowledge regarding the performance of some of the more under-utilised damage features in the world of bridge SHM. Figure 18 demonstrates the degree of damage sensitivity and resolution that can be achieved using one such vibration parameter (vibration intensity). Additional technical achievements have been accomplished regarding the mitigation of uncertainty surrounding sources of ambient vibration, while present focus now surrounds the currently trending problem of accurately analysing non-linear, non-stationary vibration signals recorded from damaged bridges. This research has the potential to develop an empirical relationship between vibration level and bridge condition for various bridge types, resulting in a reduction of safety uncertainty. If successful, the standard protocol of bridge inspection may be changed from the compulsory 2 year period to a condition-based inspection philosophy for sufficiently instrumented bridges. This would improve bridge inspection efficiency and accuracy, while also drawing closer attention to bridges that present signs of deterioration. The knock-on socio-economic impact of this would be that bridge owners could save money through increased efficiencies and fewer bridges would deteriorate unnoticed. This could prevent possible collapses / closures resulting in lost revenues to communities and reduced access to important services.

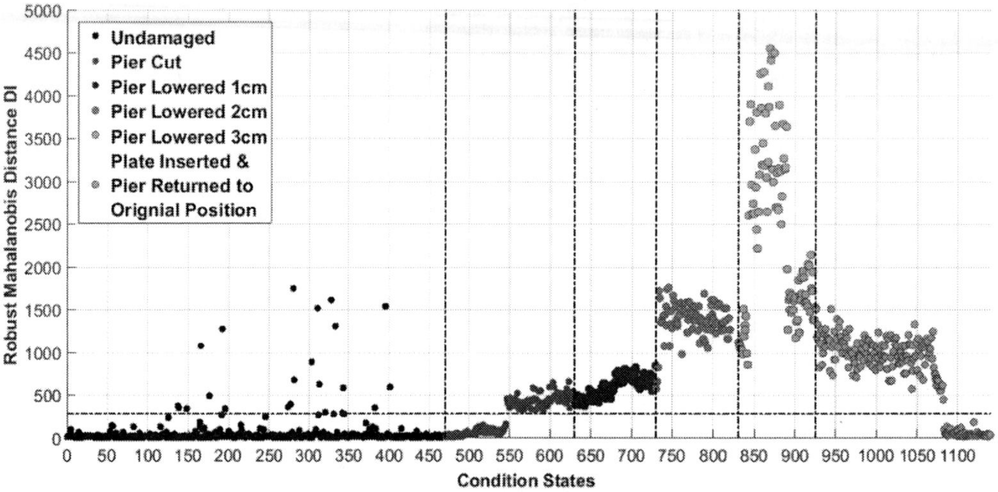

Figure 18. Robust Mahalanobis distance versus condition state.

5.1.5. Development of optical fibre distributed sensing for SHM of bridges and large scale structures. This project by Barrias et al [27-31] investigates the applicability of distributed optical fiber sensing (DOFS) on the monitoring of concrete structures. The main and predetermined goals of this study are the analysis of the possible spatial resolution, strain accuracy and long-term reliability of the measurements performed with this technology. In this way, a thorough and comprehensive literature review was conducted where the theoretical background of this technology and applications already conducted on civil engineering structures were analysed. As part of the secondment at COTCA S.A., it was possible to participate in experiments conducted at the moment, including an active monitoring procedure of a real-world structure (Sarajevo Bridge at Barcelona) where DOFS are being used (Figure 19). Important takeaways are taken from this experience by witnessing in first-hand the main challenges associated with the use of this technology out of controlled environments. Finally, the first laboratory experiments with innovative implementations of this technology (i.e., regarding bonding adhesives and techniques of attachment) to reinforced concrete elements have been started and their results are currently being assessed (Figure 20). This research will have an important and favourable socio-economic impact since it will bring additional and more accurate information for bridge and large-structures owners and in this way enable better structural health monitoring systems to prevent possible collapses and closures resulting on important savings for the users.

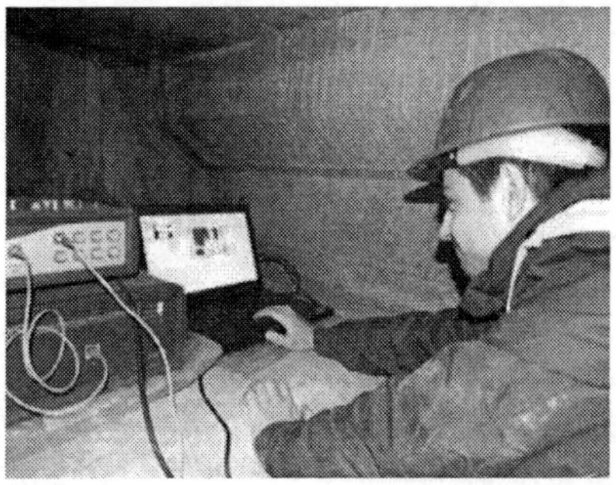

Figure 19. Structural health monitoring using DOFS.

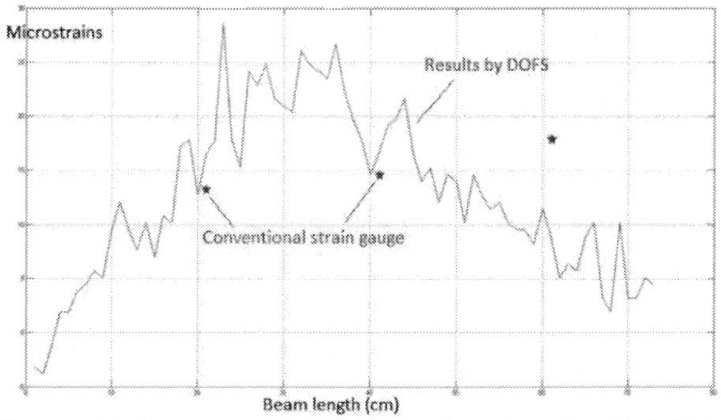

Figure 20. Strain across the beam length using DOFS in a lab test.

5.1.6. Bridge damage detection using an instrumented vehicle. Figure 21 shows the Traffic Speed Deflectometer (TSD). The TSD is a vehicle designed for pavement engineering, where vertical deflection can be calculated through the measured data from the trailer with high levels of resolution. The characteristics of the TSD have the potential to be employed as a damage detection tool for bridges. The curvature of the bridge is traditionally considered to be a highly sensitive damage indicator. Considering that the response of the vehicle-bridge system is time dependent, two types of curvature can be defined using TSD measurements: (1) Instantaneous Curvature (IC) when curvature is calculated at an instant in time using three different positions damage is detected and located from the relationship between curvature and stiffness; (2) Moving Reference Curvature (MRC) is a curvature calculated at different time instants and the information that can be extracted from the damage is limited. Damage can be quantified but not located with MRC. Figure 22 shows variation of IC curvature with the position of the vehicle using simulations for the case of a bridge with two damaged zones. A quick increase in curvature is noticeable at the damaged portions of the bridge. Although MRC is not able to locate and quantify damage to the same extent than IC, it appears to be less sensitive to noisy scenarios. First results using numerical simulations are promising, although the method still needs to be validated on the field. This research by Martinez et al [32-37] has a lot of potential for efficiently monitoring short span bridges which constitute the largest proportion of the bridge stock and where the cost of instrumenting many bridges may result prohibitive.

Figure 21. Traffic Speed Deflectometer.

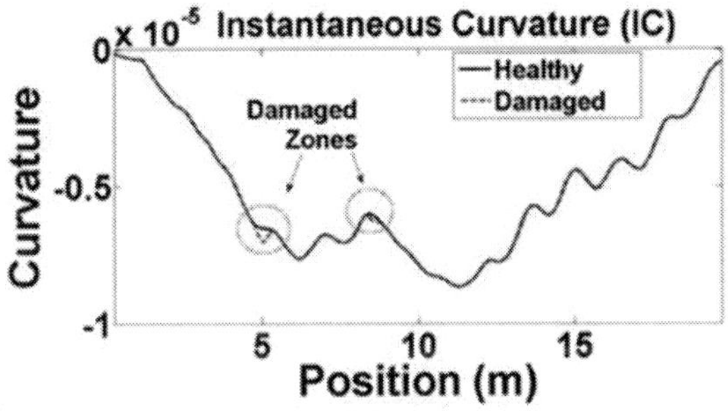

Figure 22. Impact of damaged on curvature.

5.2. Road infrastructure
Uncertainty in pavement safety is addressed via the use of production vehicle sensors (i.e., available engine performances data) for road pavement performance investigation (ESR13) and Unmanned Aerial Vehicles (UAVs) for data collection, using reconstructed 3D data with new algorithms for image-based point cloud damage detection (ESR14).

5.2.1. Using truck sensors for road pavement performance investigation. In a first phase, this project by Perrotta et al [38] focused on the definition of a general methodology for to assess the impact of road surface conditions on truck fleet fuel economy. Using this approach, real time data can be used and any generated model can be progressively updated with time, following improvements in vehicle and road technology. First results show that approximately 3 and 5% of the whole fuel consumption can be, respectively, a result of the evenness and macrotexture of the road. This shows that road maintenance policies can have a very significant impact of fuel costs and greenhouse gas emissions. In a second phase, a validation of the obtained results, extending this approach to a wider range of vehicle and road types, will improve confidence in including the pavement-vehicle interaction component into road pavement Life Cycle Assessment (LCA) studies increasing the applicability of the results. Figure 23 shows predicted fuel consumption versus real measurements referred to articulated trucks with 6 axles in total and Euro 6 engine – 1290cc. The result of these investigations will lead to the definition of more accurate models for the LCA of pavements.

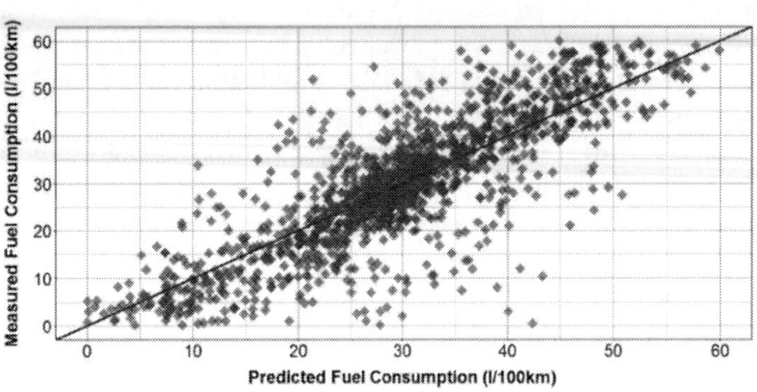

Figure 23. Linear regression: predicted fuel consumption vs real measurements.

5.2.2. Reduction of uncertainty through regularized, automated road inspection. The achievements of this project by Chen et al [39,40] are reflected in several areas. The first represents a state-of-the art review of the numerous UAVs currently available through commercial and research efforts. Based on this work, a UAV platform customized for inspection data acquisition was developed. This platform is capable of capturing high-quality images and real-time videos for offline and online analysis. Next, the workflow of most popular images-based 3D reconstruction platforms, such as the VisualSFM (open source), PhotoScan (commercial) and Pix4D (commercial), was investigated. These were applied to a wide range of highly varied data sets of buildings, roads and bridges (Figure 24) to test their usefulness, robustness, and scalability. Their respective advantages and disadvantages were compared, from which a data processing strategy was developed. Finally, a framework was to determine the accuracy of various 3D reconstruction efforts, which was then used to optimize data efforts. As part of this, a laboratory experiment was devised to exploring the interaction of camera angles and shooting distance with respect the final accuracy level achieve. The current work is now mainly focused on damage detection. This project contributes to develop a safer, cheaper and more efficient way for road infrastructure inspections using UAV technology.

Figure 24. The Boyne bridge reconstruction.

6. Conclusions

This paper has provided a summary of the progress and expected impact of each individual project in TRUSS. The individual projects in TRUSS are quite diverse, focusing on a wide array of application structures. Nonetheless, the need to overcome uncertainty in material, load and structural performance represents a core thread that ties the projects together. This has led to interactions between researchers, and to generate innovations that forms the basis of their PhD research. The objectives of reducing uncertainty and improving infrastructure management are achieved via the development of new material/sensor/monitoring technologies (ESR1, ESR2, ESR11, ESR12, ESR13, ESR14) that allow more efficient data collection, and new algorithms (ESR3, ESR4, ESR5, ESR6, ESR7, ESR8, ESR9, ESR10) that process the data collected from the structure to estimate its safety more accurately than current approaches.

IOP Conf. Series: Journal of Physics: Conf. Series **842** (2017) 012039 doi:10.1088/1742-6596/842/1/012039

With the help of innovative health monitoring, damage detection, structural simulations and tests, use of new materials and probabilistic assessments altogether, TRUSS will have an impact on: (a) economic activities, by avoiding unnecessary repair works and optimising structures in terms of their entire life-cycle; (b) sustainability, by reducing the waste materials during construction and rehabilitation works and by utilizing innovative and environment-friendly inspection, maintenance and rehabilitation methodologies; and (c) social terms by avoiding unnecessary repairs or replacements, that can affect the public in a variety of forms, i.e., road closures in the case of bridge repairs or failure that will lead to longer travel times and will increase costs in many economic sectors. TRUSS outputs provide tools that will ensure a more durable infrastructure capable of successfully supporting society needs and expectations over the next decades.

7. References

[1] European Commission 2014 *Horizon 2020 Work Programme 2014-2015 Marie Skłodowska-Curie Actions* European Commission Decision C (2014) 4995 of 22 July 2014

[2] European Commission 2014 *State of the Innovation Union, Taking stock 2010-2014* Commission Staff Working Document Acoompanying the Document "Communication from the Commission to the European Parliament, The Council, The European Economic and Social Committee and the Committee of the Regions" Research and innovation as sources of renewed growth COM(2014) 339

[3] Antonopoulou S, McNally C and Byrne G 2016 Developing braided FRP reinforcement for concrete structures *Proc. of Civil Engineering Research in Ireland Conference (CERI 2016)* (Galway, Ireland, August 29-30)

[4] Antonopoulou S, McNally C and Byrne G 2016 Development of braided basalt FRP rebars for reinforcement of concrete structures *Proc. of 8th International Conference on Fibre-Reinforced Polymer (FRP) Composites in Civil Engineering (CICE 2016)* (Hong Kong, China, December 14-16)

[5] Antonopoulou S and McNally C 2017 Reliability assessment of braided BFRP reinforcement for concrete structures *Proc. European Safety and Reliability ESREL2017* (Portorož, Slovenia, June 18-22)

[6] Sourav SNA, Al-Sabah S and McNally C 2016 Strength assessment of in-situ concrete for the evaluation of structural capacity: State of the art *Proc. of Civil Engineering Research in Ireland Conference (CERI 2016)* (Galway, Ireland, August 29-30)

[7] Sourav SNA, Al-Sabah S and McNally C 2017 Post-installed concrete screws for in-situ assessment of concrete strength *Proc. European Safety and Reliability ESREL2017* (Portorož, Slovenia, June 18-22)

[8] Gonzalez Merino A, Costas L and Gonzalez A 2016 Uncertainties in seismic design of free-standing HDSFS racks *Proc. of the ninth International Youth Nuclear Congress (IYNC2016)* (Hangzhou, China, July 24-30)

[9] Gonzalez Merino A, Costas L and Gonzalez A 2016 Dynamic analysis of the nonlinear response of high density fuel storage racks, *Proc. of Civil Engineering Research in Ireland Conference (CERI 2016)* (Galway, Ireland, August 29-30)

[10] Gonzalez Merino A, Costas L and Gonzalez A 2017 Sensitivity analysis of a finite element model for the seismic analysis of free-standing spent fuel racks *Proc. European Safety and Reliability ESREL2017* (Portorož, Slovenia, June 18-22)

[11] Teixeira R, O'Connor AJ and Nogal M 2016 Comparative analysis of the probabilistic methods to estimate the probability of failure of offshore wind turbine towers *Proc. of Civil Engineering Research in Ireland Conference (CERI 2016)* (Galway, Ireland, August 29-30)

[12] Teixeira R, O'Connor AJ, Nogal M, Nichols J and Spring M 2017 Structural probabilistic assessment of Offshore Wind Turbine operation based on Kriging response interpolation, *Proc. European Safety and Reliability ESREL2017* (Portorož, Slovenia, June 18-22)

[13] Zou G, Banisoleiman K and Gonzalez A 2016 Methodologies for Crack Initiation in Welded

Joints Applied to Inspection Planning *International Journal of Civil, Environmental, Structural, Construction and Architectural Engineering* **10(11)** 1406-1413

[14] Zou G, Banisoleiman K and Gonzalez A 2017 Reliability-based inspection planning in view of both crack initiation and propagation *Proc. European Safety and Reliability ESREL2017* (Portorož, Slovenia, June 18-22)

[15] Milana G, Banisoleiman, K and Gonzalez A 2016 Fatigue life assessment methods: the case of ship unloaders *Proc. of the 1st International Conference on Natural Hazards & Infrastructure (ICONHIC2016)* (Chania, Crete Island, Greece, June 28-30)

[16] Milana G, Banisoleiman K and Gonzalez A 2016 Sources of structural failure in ship unloaders *Proc. of the European Safety and Reliability Conference (ESREL2016)* (Glasgow, UK, September 25-29)

[17] Milana G, Banisoleiman K and Gonzalez A 2017 Field characterization of location-specific dynamic amplification factors towards fatigue calculations in ship unloaders *Proc. European Safety and Reliability ESREL2017* (Portorož, Slovenia, June 18-22)

[18] Huseynov F, Brownjohn JMW, O'Brien EJ and Hester D 2016 Analysis of load test on composite I-girder bridge *Proc. of the CSHM-6: Structural health monitoring of new and ageing infrastructure* (Queens University, Belfast, Northern Ireland, United Kingdom, May 26-27)

[19] Xu Y, Huseynov F, Brownjohn JMW, O'Brien EJ and Hester D 2017 Tracking deflection in the field using optical system: a case study *Proc. European Safety and Reliability ESREL2017* (Portorož, Slovenia, June 18-22)

[20] Heitner B, OBrien E, Schoefs F, Yalamas T, Decatoire R and Leahy C 2016 Probabilistic modelling of bridge safety based on damage indicators *Procedia Engineering* **156** 140-147

[21] Heitner B, OBrien E, Schoefs F, Decatoire R, Yalamas T and Leahy C 2016 Evaluation of bridge safety based on Weigh-in-Motion data *Proc. of Civil Engineering Research in Ireland Conference (CERI 2016)* (Galway, Ireland, August 29-30)

[22] Vagnoli M, Remenyte-Prescott R and Andrews J 2017 A fuzzy-based Bayesian Belief Network approach for railway bridge condition monitoring and fault detection *Proc. European Safety and Reliability ESREL2017* (Portorož, Slovenia, June 18-22)

[23] Moughty JJ and Casas JR 2016 Vibration based damage detection techniques for small to medium span bridges: A review and case study *Proc. of the 8th European Workshop on Structural Health Monitoring (EWSHM 2016)* (Bilbao, Spain, July 5-8)

[24] Moughty JJ and Casas JR 2016 Advancements of vibration based damage detection techniques for small to medium span bridges *Proc. of Civil Engineering Research in Ireland Conference (CERI 2016)* (Galway, Ireland, August 29-30)

[25] Moughty JJ, Casas JR and Rodrigues F 2016 Assessment of vibration intensity as a condition and safety indicator for short-span bridges *Proc. of the Fifth International Symposium on Life-Cycle Civil Engineering (IALCCE 2016)* (Delft, The Netherlands, October 16-19)

[26] Casas JR and Moughty JJ 2017 Bridge damage detection based on vibration data: past and new developments *Frontiers in the Built Environment* DOI: 10.3389/fbuil.2017.00004

[27] Rodríguez G, Casas JR Villalba S and Barrias A 2016 Monitoring of shear cracking in partially prestressed concrete beams by distributed optical fiber sensors *Proc. of the 8th International Conference on Bridge Maintenance, Safety and Management (IABMAS 2016)* (Foz do Iguaçu, Brazil, June 26-30)

[28] Barrias A, Casas JR and Villalba 2016 A Review of Distributed Optical Fiber Sensors for Civil Engineering Applications *Sensors* **16(5)** 748

[29] Barrias A, Casas JR and Villalba S 2016 Review of civil engineering applications with distributed optical fiber sensors *Proc. of the 8th European Workshop on Structural Health Monitoring (EWSHM 2016)* (Bilbao, Spain, July 5-8)

[30] Barrias A, Casas JR, Villalba S and Rodriguez G 2016 Health monitoring of real structures by distributed optical fiber *Proc. of the Fifth International Symposium on Life-Cycle Civil Engineering, IALCCE 2016* (Delft, The Netherlands, October 16-19)

[31] Barrias A, Casas JR, Villalba S and Rodriguez G 2017 UPC - BarcelonaTech experience on the use of Rayleigh based distributed optical fiber sensors for SHM of concrete structures *Proc. European Safety and Reliability ESREL2017* (Portorož, Slovenia, June 18-22)
[32] Sevillano E, OBrien EJ and Martinez D 2016 Instantaneous curvature in bridge damage detection, *Proc. of the CSHM-6: Structural health monitoring of new and ageing infrastructure* (Queens University, Belfast, Northern Ireland, United Kingdom, May 26-27)
[33] OBrien EJ, Sevillano E and Martinez D 2016 Monitoring the condition of a bridge using a traffic speed deflectometer vehicle travelling at highway speed *Proc. of the 3rd International Balkans Conference on Challenges of Civil Engineering* (Tirane, Albania, May 19-21)
[34] Martinez D, OBrien EJ and Sevillano E 2016 Drive-by bridge damage detection using curvatures in uncertain environments *Proc. of Civil Engineering Research in Ireland Conference (CERI 2016)* (Galway, Ireland, August 29-30)
[35] Martinez D, OBrien EJ and Sevillano E 2016 Damage Detection by Drive-by Monitoring Using the Vertical Displacements of the Bridge *Proc. Structural Engineering, Mechanics and Computation (SEMC 2016)* (Cape Town, South Africa, September 5-7)
[36] Malekjafarian A, Martinez D and OBrien EJ 2017 Pavement Condition Measurement at High Velocity using a TSD *Proc. European Safety and Reliability ESREL2017* (Portorož, Slovenia, June 18-22)
[37] OBrien EJ, Martinez D and Malekjafarian A 2017 A New Damage Indicator for Drive-by Monitoring using Instantaneous Curvature *Proc. AustRoads Bridge Conference ABC2017* (Melbourne, Australia, April 3-6)
[38] Perrotta F, Parry T and Neves L 2017 Using truck sensors for road pavement performance investigation *Proc. European Safety and Reliability ESREL2017* (Portorož, Slovenia, June 18-22)
[39] Chen S, Laefer DF and Mangina E 2016 State of Technology Review of Civilian UAVs *Recent Patents on Engineering* **10(3)** 160-174
[40] Chen S, Laefer DF, Byrne J and Natanzi AS 2017 Angle affect on image-based 3D reconstruction for brick structure *Proc. European Safety and Reliability ESREL2017* (Portorož, Slovenia, June 18-22)

Acknowledgments

The TRUSS ITN project (http://trussitn.eu) has received funding from the European Union's Horizon 2020 research and innovation programme under the Marie Skłodowska-Curie grant agreement No. 642453.

An anisotropic damage model for concrete structures under cyclic loading-uniaxial modeling

Yuchuan Long[1] **and Yuming He**[2]

[1] Key laboratory of New Technology for Construction of Cities in Mountain Area(Chongqing University), Ministry of Education, Chongqing, China 400045

[2] School of Civil Engineering, Chongqing University, Chongqing, China 400045

E-mail: longyc02@cqu.edu.cn

Abstract. An anisotropic damage model is developed based on conventional rotating crack approach. It uses nonlinear unloading/linear reloading branches to model the hysteretic behavior of concrete. Two damage variables, determined by the ratio of accumulated dissipating energy to fracture energy, are introduced to represent the stiffness degradation in tension and compression. Three cyclic tests are simulated by this model and sensitivity analyses are conducted as well. The numerical responses calculated by the damage model are consistent with those obtained from the experiments. The numerical results reflect the nonlinear behavior observed in those tests, such as the damage-induced stiffness degradation, accumulation of residual deformation, energy dissipation caused by hysteretic behavior and stiffness recovery effect due to crack closure. Sensitivity analyses show that the damage exponents have significant influence on the computational accuracy. It is concluded that the anisotropic damage model is applicable to the nonlinear analyses of concrete structures subjected to cyclic loading.

1. Introduction

Analyzing the mechanical response of concrete structures subjected to general loading conditions is an essential yet challenging task for structural engineering, not to mention predicting the failure patterns of complex structures. The distinct failure mechanisms due to tensile cracking and compressive crushing also complicate the constitutive modeling of concrete. The failure process is accompanied by the initiation and propagation of cracks, which may be modeled macroscopically as softening behavior of the material.

The stiffness degradation is conveniently modeled by continuum damage mechanics. Mazars [1], [2] and Cervera [3] introduced a scalar damage factor to describe the relationship between stress and effective stress based on the assumption of isotropic damage. Papa[4], Al-Gadhib[5], Li Qingbin[6], Challamel[7],Badel[8] and Pröchtel[9] utilized a second-order or fourth-order damage tensor to model the anisotropic properties of concrete damage. Their models reflected the stiffness degradation. However, it is insufficient for those models to represent the irreversible deformation and inelastic volumetric expansion in compression.

Simo[10], Lubliner[11],Luccioni[12], Faria[13], Lee[14], Grassl[15], Wu Jianying[16], Voyiadjis[17] and Poh[18] developed plastic-damage models by combining the advantages of classical plasticity and damage mechanics. In this approach, the irreversible deformation and volumetric expansion in multiaxial compression is represented using the flow theory of plasticity and the stiffness degradation is modeled by introducing scalar damage variable or damage tensor. However, the plastic-damage models are very complicated and only a few simple ones have sufficient efficiency to conduct dynamic or cyclic analysis of complex concrete structure[19], [20]. Moreover, these models utilize linear stress branches to simulate the unloading/reloading responses, which is difficult to represent the hysteretic behavior of concrete.

Palermo[21] have proposed a cyclic concrete model which implements nonlinear unloading/linear reloading branches into rotating crack approach for modeling hysteretic behavior. In this model, stress branches are defined as functions of strain and stress along the damage axis and the key parameters are statistically derived from the concrete tests. Sima[22] developed a cyclic constitutive model using independently tensile and compressive damage parameters to simulate the stiffness degradation. The compressive hysteretic behavior is therein modeled by exponential unloading/linear reloading branches, whereas the tensile one is neglected.

An anisotropic damage model for cyclic loading was developed by introducing nonlinear unloading/linear reloading stress branches into conventional rotating crack approach[23]. The stiffness degradation was modeled by tensile and compressive damage variables which were determined by the ratio of accumulated dissipating energy to fracture energy. The stiffness recovery due to crack closure was also taken into account by using the approach proposed by Lee[14].

2. The constitutive relation of anisotropic damage model

Following presuppositions are used to model the nonlinear behavior of concrete.

(i) The strain is decomposed into elastic part and inelastic part for any material point.

(ii) There are three mutually orthogonal axes of material which always keep aligned with the directions of principal strain (and stress) during loading histories.

(iii) The damage constitutive relations are mutually decoupled and formulated as functions of principal stress and inelastic strain on the rotating damage axes.

2.1. Mathematical formulation of the damage model in incremental theory

As shown in Fig.1, y_1 axis is the principal direction on which damage occurs. The strain array $\boldsymbol{\varepsilon}$ in global coordinates $ox_1x_2x_3$ is decomposed into elastic part $\boldsymbol{\varepsilon}^{\mathrm{e}}$ and inelastic part $\boldsymbol{\varepsilon}^{\mathrm{in}}$ and linear elasticity is given by:

$$\boldsymbol{\varepsilon} = \boldsymbol{\varepsilon}^{\mathrm{e}} + \boldsymbol{\varepsilon}^{\mathrm{in}}, \qquad \boldsymbol{\varepsilon}^{\mathrm{e}} = \mathbf{E}_0^{-1}\boldsymbol{\sigma} \tag{1}$$

where $\boldsymbol{\sigma}$ denotes stress array and the initial elastic stiffness $\mathbf{E}_0$ is a 6×6 matrix.

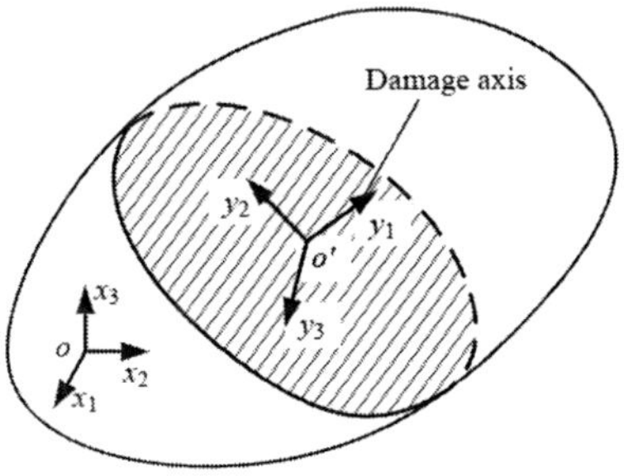

Figure 1. Local coordinates determined by the principal directions of damaged concrete

The orthotropic axes of material, represented by the local coordinates $o'y_1y_2y_3$ in Fig.1, keep aligned with the directions of principal strain and stress. Thus, the shear components of stress and strain array vanish in the coordinate $o'y_1y_2y_3$. Additionally, the strain $\boldsymbol{e}$, elastic part $\boldsymbol{e}^{\mathrm{e}}$ and inelastic part $\boldsymbol{e}^{\mathrm{in}}$ in the local coordinates $o'y_1y_2y_3$ observe the strain decomposition assumption as well.

$$\boldsymbol{e} = \boldsymbol{e}^{\mathrm{e}} + \boldsymbol{e}^{\mathrm{in}} \tag{2}$$

Herein the relation between local and global inelastic strains follows:

$$\varepsilon^{\text{in}} = \mathbf{T}_{\text{r}}\, \boldsymbol{e}^{\text{in}} \tag{3}$$

where $\mathbf{T}_{\text{r}}$ is the transformation matrix of strain from the global coordinate $ox_1x_2x_3$ to the local one $o'y_1y_2y_3$. Similarly, the relation between local (or principal) stress s and global one σ satisfies:

$$s = \mathbf{T}_{\text{r}}^{\text{T}}\, \sigma \tag{4}$$

The relation between principal stress s and inelastic strain e^{in} in $o'y_1y_2y_3$ is formulated as follows:

$$\begin{aligned} e^{\text{in}} &= \mathbf{G}\, s \\ \boldsymbol{G} &= \begin{bmatrix} g_1(e_1^{\text{in}}) & 0 & 0 \\ 0 & g_2(e_2^{\text{in}}) & 0 \\ 0 & 0 & g_3(e_3^{\text{in}}) \end{bmatrix} \end{aligned} \tag{5}$$

in which $\mathbf{G}$ is the flexibility matrix of damage in local coordinates $o'y_1y_2y_3$ and $g_i(e_i^{\text{in}})$ $i = 1,\cdots,3$ denotes the secant flexibility on damage axis i.

Therefore, the constitutive relations are derived as follows from equation (1)~(5):

$$\begin{aligned} \sigma &= \mathbf{D}\,\varepsilon \\ \mathbf{D} &= \left(\mathbf{E}_0^{-1} + \mathbf{T}_{\text{r}}\mathbf{G}\mathbf{T}_{\text{r}}^{\text{T}}\right)^{-1} \end{aligned} \tag{6}$$

where $\mathbf{D}$ is the constitutive matrix between stress and strain after damage occurs. The transformation matrix $\mathbf{T}_{\text{r}}$ and damage flexibility one $\mathbf{G}$ are determined by current strain and its inelastic part respectively. $\mathbf{T}_{\text{r}}^{\text{T}}$ is the transposed matrix of $\mathbf{T}_{\text{r}}$.

Because the constitutive relation in Equation (6) is nonlinear, the linearization concept is employed to derive the incremental form of constitutive relation. The deriving process is similar to that of the rotating crack model which may refer to the work of [24] and [25]. Finally, the constitutive relation in incremental form is formulated as follows:

$$\Delta\sigma = \mathbf{D}_{\tan}\Delta\varepsilon = \mathbf{T}^{\text{T}}\mathbf{D}'\mathbf{T}\Delta\varepsilon \tag{7}$$

in which $\mathbf{D}_{\tan}$ and $\mathbf{D}'$ denote the tangent material stiffness matrices in global and local coordinates, respectively; and $\mathbf{T}$ is the transformation matrix of engineering strain from $ox_1x_2x_3$ to $o'y_1y_2y_3$.The formulation of matrix $\mathbf{D}'$ is given as follows:

$$\begin{aligned} \mathbf{D}' &= \begin{bmatrix} \tilde{\mathbf{D}}_{\text{n}} & \mathbf{0} \\ \mathbf{0} & \tilde{\mathbf{D}}_{\text{s}} \end{bmatrix} \\ \tilde{\mathbf{D}}_{\text{n}} &= \left(\tilde{\mathbf{C}}_{\text{e}} + \tilde{\mathbf{C}}_{\text{c}}\right)^{-1} \\ \tilde{\mathbf{C}}_{\text{e}} &= \frac{1}{E}\begin{bmatrix} 1 & -\nu & -\nu \\ -\nu & 1 & -\nu \\ -\nu & -\nu & 1 \end{bmatrix} \\ \tilde{\mathbf{C}}_{\text{c}} &= \begin{bmatrix} \left(\dfrac{\mathrm{d}s_1}{\mathrm{d}e_1^{\text{in}}}\right)^{-1} & 0 & 0 \\ 0 & \left(\dfrac{\mathrm{d}s_2}{\mathrm{d}e_2^{\text{in}}}\right)^{-1} & 0 \\ 0 & 0 & \left(\dfrac{\mathrm{d}s_3}{\mathrm{d}e_3^{\text{in}}}\right)^{-1} \end{bmatrix} \end{aligned} \tag{8}$$

$$\tilde{\mathbf{D}}_s = \begin{bmatrix} \frac{s_1 - s_2}{2(e_1 - e_2)} & 0 & 0 \\ 0 & \frac{s_1 - s_3}{2(e_1 - e_3)} & 0 \\ 0 & 0 & \frac{s_2 - s_3}{2(e_2 - e_3)} \end{bmatrix}$$

where E is the initial elastic modulus and ν denotes Poisson ratio. Equation (7) and (8) show that the constitutive relations are only dependent on the function $g_i(e_i^{\text{in}})$ and the derivative $\mathrm{d}s_i/\mathrm{d}e_i^{\text{in}}$ in the local coordinates. Consequently,establishing the mathematical expression of cyclic stress branches is the key issue of the model.

2.2. Formula of stress branches on the rotating damage axis

Stress branches under cyclic tension or compression consist of loading, unloading, reloading, partial unloading and partial reloading. The stress branches under uniaxial cyclic loading are shown in Fig.2, with $s_1 > 0, e_1 > 0$ for tensile response and $s_1 < 0, e_1 < 0$ for compressive one.

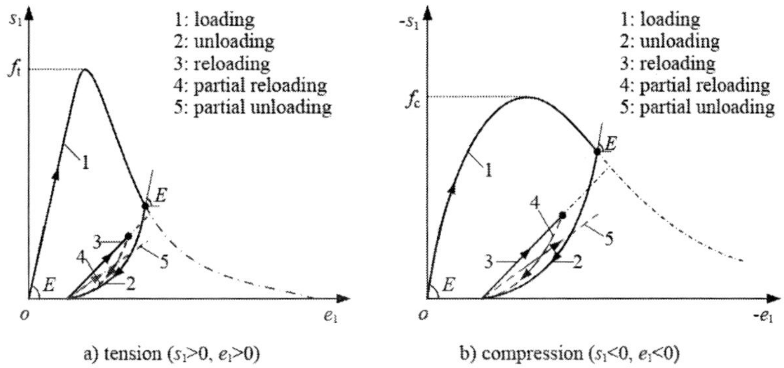

Figure 2. Stress branches under uniaxial cyclic loading: a) tension; b) compression

According to Equation (5)~(7), the relations between principal stress and inelastic strain are decoupled for all material axes. Furthermore, it is suitable to define stress branches in the form of principal stress and inelastic strain because there is no energy dissipation caused by the relations between principal stress and elastic strain.

2.2.1. Loading branch A piecewise linear function which connects characteristic data points of experimental response is utilized to define the loading stress branch.

$$|s_1| = \begin{cases} \sigma_i^\kappa + k_i^\kappa\left(\left|e_1^{\text{in}}\right| - a_i^\kappa\right) & a_i^\kappa \leqslant \left|e_1^{\text{in}}\right| < a_{i+1}^\kappa \ (i = 0, n-1) \\ \sigma_n^\kappa & a_n^\kappa \leqslant \left|e_1^{\text{in}}\right| \end{cases} \tag{9}$$

$$k_i^\kappa = \frac{\sigma_{i+1}^\kappa - \sigma_i^\kappa}{a_{i+1}^\kappa - a_i^\kappa}, \qquad \kappa = \begin{cases} \text{t} & \text{if } e_1^{\text{in}} > 0,\ s_1^{\text{in}} > 0 \\ \text{c} & \text{if } e_1^{\text{in}} < 0,\ s_1^{\text{in}} < 0 \end{cases}$$

where $\kappa \in (\text{t}, \text{c})$ is state variable with $\kappa = \text{t}$ for tension and $\kappa = \text{c}$ for compression, $(a_i^\kappa, \sigma_i^\kappa)$ are the absolute values of inelastic strain and corresponding stress of the ith point. Fig.3 shows the loading stress branch defined by equation (9). However, equation (9) is merely applicable to uniaxial loading or the conditions with low lateral stress because it doesn't account for the influence of multiaxial stress state on tensile or compressive strength. Therefore, it is necessary to develop a general function of loading branch under multiaxial stress state in the further study.

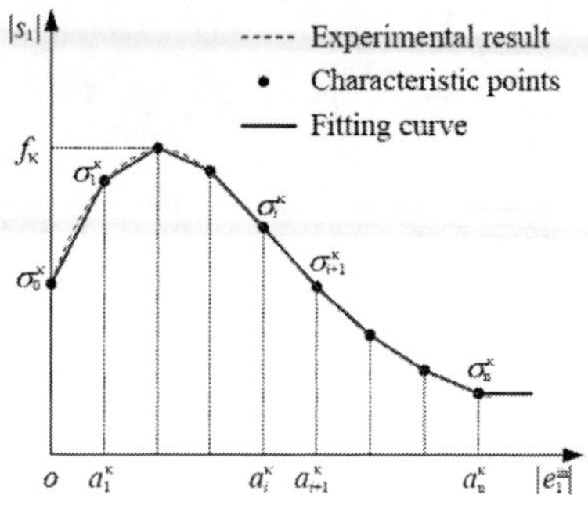

Figure 3. The stress branch of loading defined by principal stress and inelastic strain

2.2.2. *Unloading branch* Concrete tests conducted by Karsan[26], Buyukozturk[27], Bahn[28], Gopalaratnam[29] and Yankelevsky[30] show that irrecoverable deformation occurs even completely removing external loads. The results of confined tests from Buyukozturk[27] also indicate that the irrecoverable deformation was not remarkably affected by confining stresses or strains. The residual strain, called plastic offset strain by Palermo[21], is essentially the amount of irrecoverable damage resulting from concrete crushing, compressing of internal voids and micro-crack opening.

The unloading stress branch shown in Fig. 4 is given as follows.

$$\begin{aligned} |s_1| &= |\sigma_m| + B^\kappa \left|e_m^{in} - e_1^{in}\right| + C^\kappa \left|e_m^{in} - e_1^{in}\right|^{\alpha^\kappa} \\ B^\kappa &= \beta^\kappa \left[\alpha^\kappa |\sigma_m| \cdot \left|e_m^{in} - e_p^\kappa\right|^{\alpha^\kappa - 1} - H_t^\kappa \left|e_m^{in} - e_p^\kappa\right|^{\alpha^\kappa}\right] \\ C^\kappa &= \beta^\kappa \left[H_t^\kappa \left|e_m^{in} - e_p^\kappa\right| - |\sigma_m|\right] \\ \beta^\kappa &= \frac{1}{(1-\alpha^\kappa)\left|e_m^{in} - e_p^\kappa\right|^{\alpha^\kappa}} \end{aligned} \tag{10}$$

in which e_1^{in} and s_1 are the present inelastic strain and principal stress on damage axis y_1, e_m^{in} and σ_m denote the previously maximum inelastic strain and corresponding stress at the onset of unloading, e_p^κ is the plastic offset strain, H_t^κ represents the tangent modulus at a zero stress and exponent α^κ $(0 < \alpha^\kappa < 1)$ is material constant determined by experimental response. Here $\kappa \in (t, c)$ denotes tensile or compressive state respectively.

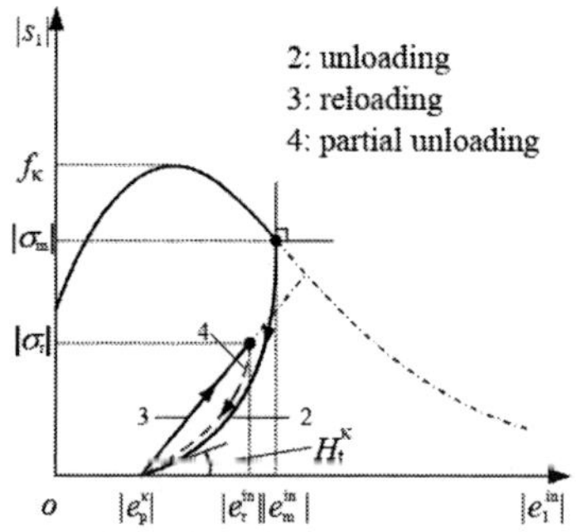

Figure 4. The nonlinear unloading branch defined in the form of principal stress and inelastic strain

2.2.3. Reloading branch In this model, reloading is modeled using a linear response combined with a damage factor d representing stiffness degradation.

$$\begin{aligned} |s_1| &= |\sigma_a| + H\left|e_1^{in} - e_a^{in}\right| \\ H &= \frac{1-d}{d}E, \quad 0 < d < 1 \end{aligned} \tag{11}$$

where e_a^{in} and σ_a are the inelastic strain and corresponding stress at the onset of reloading, H is the slope of reloading branch. Damage factor d is defined as a function of tensile and compressive damage variables, which is formulated in section 2.4.

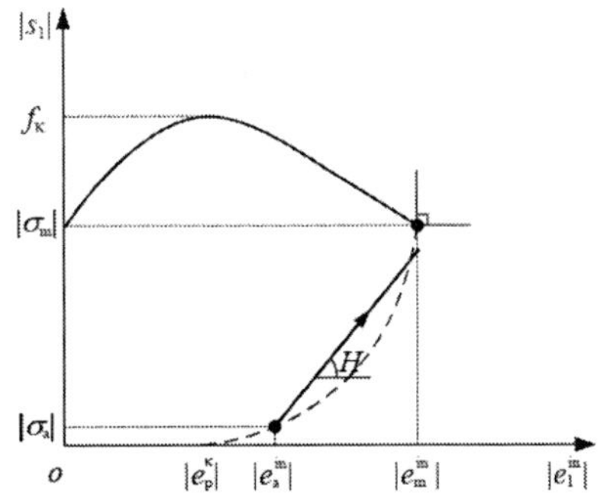

Figure 5. The linear reloading branch defined in the form of principal stress and inelastic strain

Fig. 5 shows the linear reloading branch defined by equation (11). It should be noted that reloading concludes once the stress updated on the reloading branch exceeds the envelope defined by equation (9). This means that concrete undergoes loading again and therefore the principal stress s_1 and its derivative $\mathrm{d}s_i/\mathrm{d}e_i^{in}$ should be calculated according to equation (9).

2.2.4. Partial unloading branch Most of the models available in the literature ignores the behavior of concrete for the case of partial unloading/reloading because there is a lack of experimental information considering the general case where partial unloading is followed by partial reloading to strains less than the previously maximum unloading strain. Herein the proposed rule for the partial unloading response is similar to that assumed for full unloading. Nevertheless, the previously maximum inelastic strain e_m^{in} and corresponding stress σ_m at the onset of full unloading are replaced by the inelastic strain e_r^{in} and stress σ_r at the onset of partial unloading,

$$\begin{aligned} |s_1| &= |\sigma_r| + B_r^\kappa\left|e_r^{in} - e_1^{in}\right| + C_r^\kappa\left|e_r^{in} - e_1^{in}\right|^{\alpha^\kappa} \\ B_r^\kappa &= \beta_r^\kappa\left[\alpha^\kappa|\sigma_r|\cdot\left|e_r^{in} - e_p^\kappa\right|^{\alpha^\kappa-1} - H_t^\kappa\left|e_r^{in} - e_p^\kappa\right|^{\alpha^\kappa}\right] \\ C_r^\kappa &= \beta_r^\kappa\left[H_t^\kappa\left|e_r^{in} - e_p^\kappa\right| - |\sigma_r|\right] \\ \beta_r^\kappa &= \frac{1}{(1-\alpha^\kappa)\left|e_r^{in} - e_p^\kappa\right|^{\alpha^\kappa}} \end{aligned} \tag{12}$$

in which parameter H_t^κ and α^κ are identical to those used in equation (10). Fig. 4 also shows the partial reloading branch defined by equation (12).

2.2.5. Partial reloading branch The stress branch of partial reloading is also assumed as a linear response similar to that of full reloading:

$$\begin{aligned} |s_1| &= |\sigma_b| + H\left|e_1^{in} - e_b^{in}\right| \\ H &= \frac{1-d}{d}E, \quad 0 < d < 1 \end{aligned} \tag{13}$$

where e_b^{in} and σ_b are the inelastic strain and stress at the onset of partial reloading. The slope H and damage factor d are identical to those used in equation (11).

Based on the derivation of anisotropic damage model in section 2.1 and the descriptions of stress branches in section 2.2, it is convenient to calculate the stress and tangent material stiffness matrix. Then, the nonlinear equations of equilibrium can be solved using the Newton-Raphson method.

2.3. The definitions of damage variables
Herein, the definition of damage variable d_κ, $\kappa \in (t, c)$ depends on the density of accumulated dissipating energy Ψ^κ, $\kappa \in (t, c)$ and that of fracture energy g_f^κ, $\kappa \in (t, c)$ in pure tension or compression. As shown in Fig. 6a, the density of fracture energy is defined as the area of the shadow region:

$$g_f^\kappa = \int_0^{\varepsilon_f^\kappa} s_1 de_1^{in} = \frac{G_f^\kappa}{l^\kappa} \tag{14}$$

in which ε_f^κ, $\kappa \in (t, c)$ denotes the inelastic strain corresponding to complete damage in tension or compression, G_f^κ is a material property representing the tensile fracture energy or its counterpart in the compressive state, and l^κ is the size of localization zone determined by the characteristic length of element to maintain objective results at the structural level. However, it is still an open issue whether the counterpart of the fracture energy in compressive failure is a material property.

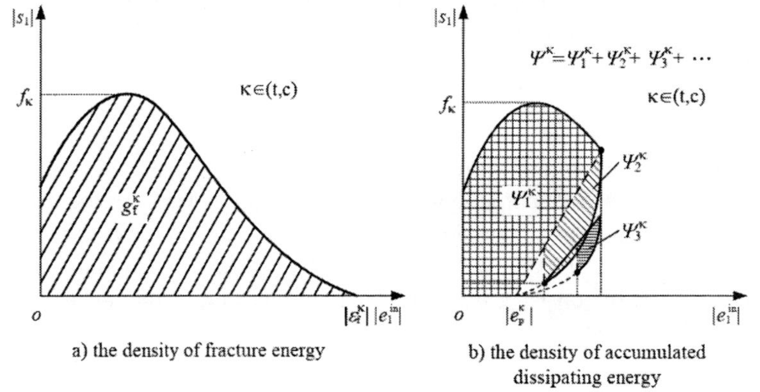

Figure 6. The definition of fracture energy and accumulated dissipating energy under pure tension or compression: a) the density of fracture energy; b) the density of accumulated dissipating engergy

The accumulated dissipating energy per unit volume Ψ^κ is defined as the summation of energy dissipation during previously cyclic loading. As illustrated in Fig.6b, the Ψ^κ is calculated as:

$$\Psi^\kappa = \Psi_1^\kappa + \Psi_2^\kappa + \Psi_3^\kappa + \cdots \tag{15}$$

where Ψ_1^κ, Ψ_2^κ and Ψ_3^κ denote the energy dissipation corresponding to loading, unloading/reloading and partial unloading/reloading responses respectively. Thus, the damage variable d_κ is defined as:

$$d_\kappa = (\kappa_\kappa)^{\theta^\kappa}, \quad \kappa_\kappa = \frac{\Psi^\kappa}{g_f^\kappa}, \quad \kappa \in (t, c) \tag{16}$$

in which κ_K is a dimensionless variable defined as the ratio of accumulated dissipating energy to fracture energy, and θ_K denotes damage exponent which is material property. Obviously, $\kappa_K = 0$ represents intact state with $d_K = 0$ while $\kappa_K = 1$ denotes completely damage with $d_K = 1$. Moreover, the damage factor d used in equation (11) and (13) equals the damage variable d_K for pure tension or compression.

2.4. *The effect of stiffness recovery*

The approach proposed by [14] is used to model the effect of stiffness recovery due to crack opening/closing. Herein, the damage factor d on the rotating damage axis is defined as follows by combining the tensile damage variable d_t with the compressive one d_c:

$$d = 1 - (1 - S_t d_c)(1 - S_c d_t) \tag{17}$$

where S_t and S_c are functions of the stress state which are introduced to model stiffness recovery effects associated with stress reversals. They are defined as

$$\begin{aligned} S_t &= 1 - w_t r^*(s_1) \qquad 0 \le w_t \le 1 \\ S_c &= 1 - w_c[1 - r^*(s_1)] \quad 0 \le w_c \le 1 \\ r^*(s_1) &= \begin{cases} 1 & \text{if } s_1 > 0 \\ 0 & \text{if } s_1 < 0 \end{cases} \end{aligned}$$

The weight factors w_t and w_c are assumed to be material properties which controls the recovery of tensile and compressive stiffness upon load reversal. Concrete tests show that the compressive stiffness is recovered upon crack closure as the load changes from tension to compression, whereas the tensile one is not recovered as the load reverses from compression to tension. Therefore, $w_t = 0$ and $w_c = 1$ representing the above-mentioned behavior are the default values used in this model.

This approach is an approximate modeling of stiffness recovery effect because it assumes that the recovery of stiffness accomplishes immediately at point $(e_p^K, 0)$ once stress reversal occurs. In fact, concrete tests conducted by [31] and [32] show that complete closing of cracks requires a certain amount of compression and the stiffness is not affected by accumulated damage in tension once the cracks are closed. Therefore, its accuracy should be carefully investigated in the further study.

3. Numerical simulations of cyclic concrete tests

3.1. *Cyclic compressive test conducted by Karsan and Jirsa*

The specimens tested by [26] were 82.6 × 82.6 × 82.6 mm in dimension.Material properties are: Young's modulus $E = 3.17 \times 10^4$ MPa, Poisson's ratio $\nu = 0.2$, compressive strength $f_c = -27.6$ MPa, the density of compressive fracture energy $g_f^c = 0.20\,\text{N/mm}^2$ and the exponent of unloading branch $\alpha^c = 0.5$. Fig.7 shows the finite element mesh and loading branch used to model the test.

The compressive plastic offset is given as

$$\frac{e_p^c}{\varepsilon_c^0} = c_c + k_c\left(\frac{e_1^{in}}{\varepsilon_c^0}\right) + m_c\left(\frac{e_1^{in}}{\varepsilon_c^0}\right)^2 \tag{18}$$

where ε_c^0 denotes the strain at the peak stress under uniaxial monotonic compression. Based on the experimental data of Karsan's test, $\varepsilon_c^0 = 2.5 \times 10^{-3}$, $c_c = -0.1175$, $k_c = 0.5060$, $m_c = 0.0989$ and the damage exponenent $\theta_c = 0.2678$ are obtained by an inverse analysis.

The stress-strain response calculated by the model is compared with that obtained by the test in Fig.8a. Numerical responses such as residual deformation, nonlinear unloading branches and degraded stiffness are in close agreement with the experimental ones. The hysteretic behavior of concrete is also represented via the nonlinear unloading/linear reloading branches. Moreover, the numerical reloading branch approaches the experimental one and models the stiffness degradation. This indicates that it

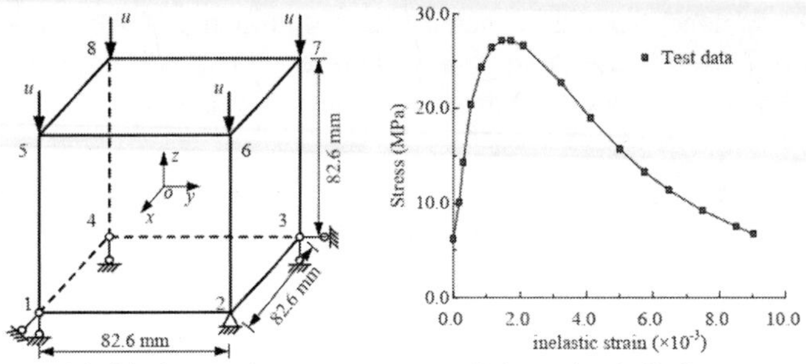

a) the finite element mesh b) the stress branch of loading

Figure 7. Numerical modeling of Karsan's test: a)the finite element mesh; b)the stress branch of loading

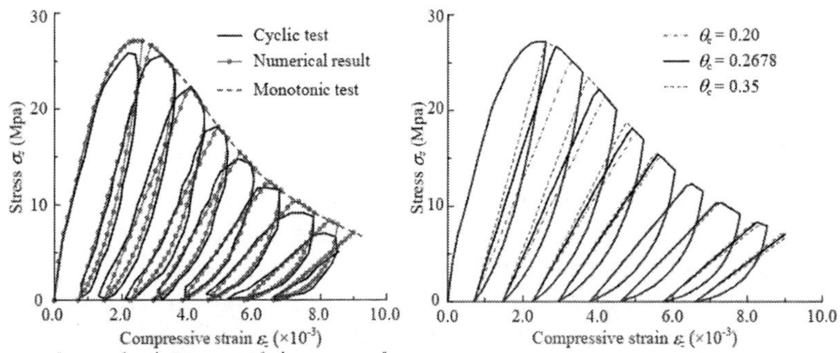

a) comparison between numerical respsonse and experimental one b) sensitivity analysis for θ_c

Figure 8. The stress-strain response: a) comparison between numerical response and experimental one; b) sensitivity analysis for θ_c

is feasible to define the damage factor with the accumulated dissipating energy and fracture energy. However, linear reloading branch is approximate because the experimental reloading appears remarkable nonlinearity in the vicinity of the loading branch.

Fig.8b shows the stress-strain responses and corresponding relations between d_c and κ_c obtained with $\theta_c = 0.20, 0.2678$ and 0.35. These analyses obtain identical unloading branches for each load cycle because the displacement loading conditions provide the same coefficients which determine the unloading response. However, the reloading branches are significantly influenced by θ_c. The results show that the damage exponent θ_c has substantial influence on the accuracy of the present model.

3.2. Cyclic tensile test conducted by Gopalartnam and Shah

Cyclic tests conducted by [29] are analyzed by the present model to verify its capability for simulating the nonlinear behavior of concrete in tension. The material properties are given as follows: Young's modulus $E = 3.10 \times 10^4$ MPa, Poisson's ratio $\nu = 0.2$, tensile strength $f_t = 3.48$ MPa, tensile fracture energy density $g_f^t = 6.80 \times 10^{-4}\,\mathrm{N/mm^2}$ and the exponent of unloading branch $\alpha^t = 0.75$. Fig.9 shows the finite element mesh and loading branch used in the simulation.

The tensile plastic offset is given as

$$\frac{e_p^t}{\varepsilon_t^0} = c_t + k_t \left(\frac{e_1^{in}}{\varepsilon_t^0} \right) \tag{19}$$

IOP Conf. Series: Journal of Physics: Conf. Series **842** (2017) 012040 doi:10.1088/1742-6596/842/1/012040

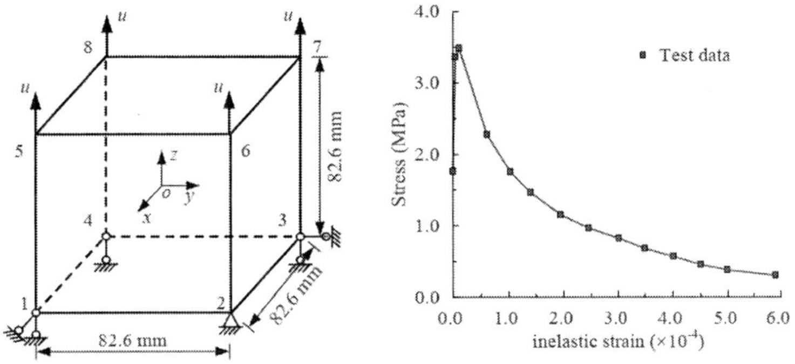

a) the finite element mesh b) the stress branch of loading

Figure 9. Numerical simulation of Gopalaratnam's test: a) the finite element mesh; b) the stress branch of loading

where ε_t^0 denotes the strain at the peak stress under uniaxial monotonic tension. $\varepsilon_t^0 = 1.12 \times 10^{-4}$, $c_t = -0.119$, $k_t = 0.824$and the damage exponenent $\theta_t = 0.824$ are calculated from an inverse analysis.

The numerical result is shown in Fig.10, compared with the experimental one. The stress-strain response is comparable to that obtained in test, such as the stiffness degradation, irrecoverable deformation and hysteretic behavior. The effect of θ_t on the response is also investigated by modeling the test with $\theta_t = 0.75, 0.8242$ and 0.90. Under displacement loadings, the unloading branches obtained in these three cases are identical and the reloading ones are slightly different.

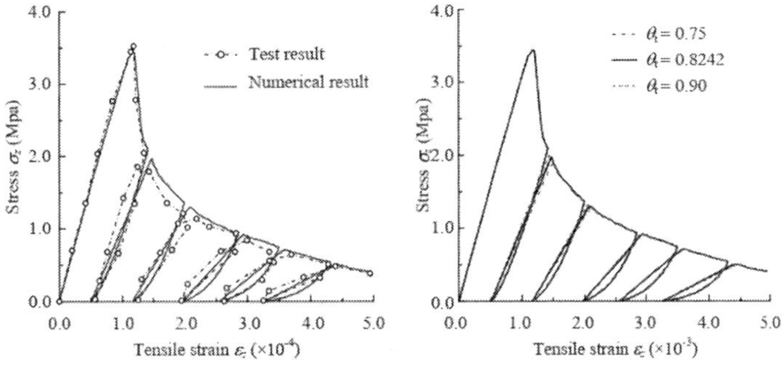

a) comparison between numerical respsonse and experimental one b) sensitivity analysis for θ_t

Figure 10. Comparison between numerical results and experimental ones: a) comparison between numerical and experimental results; b) sensitivity analysis for θ_t

3.3. Cyclic tensile-low compressive test conducted by Reinhardt

The test conducted by conducted by [31] is simulated herein. The material properties are as follows: Young's modulus $E = 3.90 \times 10^4$ MPa, Poisson's ratio $\nu = 0.20$, tensile strength $f_t = 3.20$ MPa, compressive strength $f_c = -40.0$ MPa, the density of tensile fracture energy $g_f^t = 1.13 \times 10^{-3}\,\text{N/mm}^2$, its counterpart under compression $g_f^c = 0.294\,\text{N/mm}^2$, and the tensile and compressive exponents of unloading branch $\alpha^t = 0.75$ and $\alpha^c = 0.50$. The tensile plastic offset is given by Equation (19), with $\varepsilon_t^0 = 0.82 \times 10^{-4}$, $c_t = 0.2019$, $k_t = 0.6098$ obtained from an inverse analysis. Fig.11 shows the finite element mesh and loading branch used in this simulation.

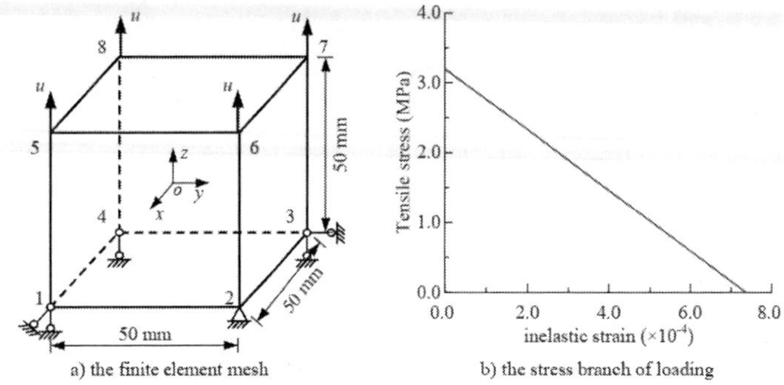

Figure 11. Numercial model of Reinhardt's test: a) the finite element mesh; b) the loading branch

The recovery of compressive stiffness is modeled with w_c = 1.0, 0.5, 0.3 and 0.0 and that of the tensile stiffness is assumed as w_t = 0.0. Fig.12 shows the stress-strain responses obtained from the numerical simulations and experiment.

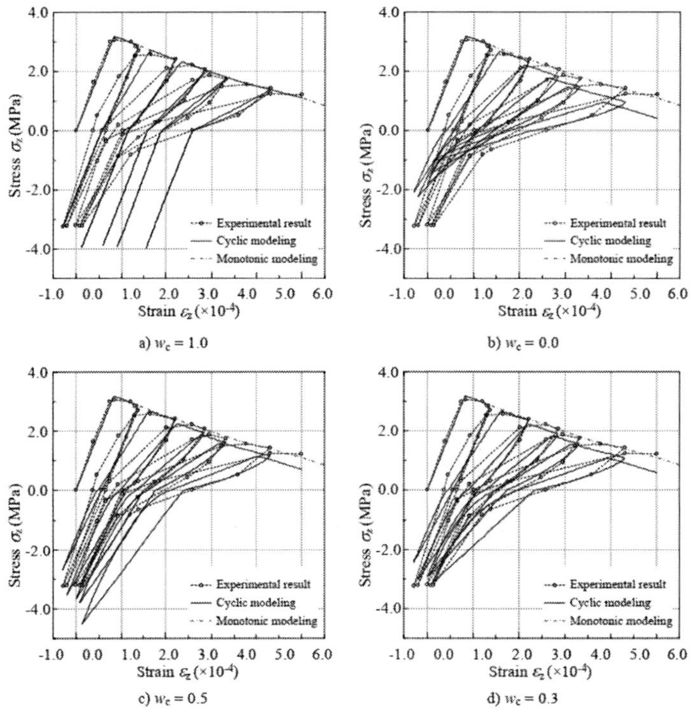

Figure 12. The comparison between numerical and experimental results : a) w_c = 1.0; b) w_c = 0.0; c) w_c = 0.5; d) w_c = 0.3

Compared with the experimental result, the four cases obtain similar responses in tension which reflect the stiffness degradation, accumulation of residual deformation and hysteretic behavior of concrete under cyclic loading. However, the variation of w_c causes significant difference in the numerical response, especially in the compressive domain.

In the case of w_c = 1.0, the compressive stiffness is fully recovered once the load reverses from tension to compression and is not influenced by the tensile damage factor d_t. Linear loading/unloading branches, whose stiffness is equal to the initial elastic stiffness E, are obtained in the compressive domain because the stress is much lower than the compressive strength of concrete. As shown in Fig.12a, the compressive stress-strain response obtained with $w_c = 1.0$ is overestimated compared to the experimental result and it does not reflect the hysteretic behavior in compression.

In the case of $w_c = 0.0$, the effect of stiffness recovery is completely ignored and the compressive stiffness simultaneously takes into account the tensile and compressive damage. Due to the degraded stiffness, the linear reloading branch yields inelastic strain which causes the occurrence of plastic offset strain and nonlinear unloading behavior in compression. Thus, this case reflects the hysteretic behavior in the compressive domain based on the linear reloading/nonlinear unloading branches. However, the result analyzed with $w_c = 0.0$ remarkably underestimates the compressive stress-strain response observed in the cyclic test.

In the cases of $w_c = 0.5$ and 0.3, the compressive stiffness is partially recovered to $(1 - d_c)\,[1 - (1 - w_c)d_t]\,E$ and the stiffness of reloading branch increases with the value of w_c. As shown in Fig.12, the compressive response calculated with $w_c = 0.3$ is in the closest agreement with the experimental result while that obtained with $w_c = 0.5$ is slightly overestimated.

These results show that the stiffness recovery factor w_c has significant influence on the computational accuracy of the present model. With appropriate w_c, this model can obtain consistent response with the experimental one, which reflect the stiffness degradation, accumulation of residual deformation and hysteretic behavior in tension and compression. Therefore, it is necessary to study the calibration of w_c.

4. Conclusions

An anisotropic damage model is developed for analyzing the response of concrete structures subjected to cyclic loading. Nonlinear unloading/linear reloading branches are implemented into the model to represent the hysteretic behavior. Two independent damage variables, one for tensile and the other for compressive load, are determined by the ratio of the accumulated dissipating energy to the fracture energy to model the stiffness degradation. In addition, this study presents the calibrating procedures of the key parameters which govern the constitutive behavior. Using this model, cyclic tests are numerically analyzed and the results draw the following conclusions:

(i) The stress-strain response calculated by the numerical analysis is similar to that obtained by the experiment. The numerical response reflects the damage-induced degradation of stiffness, the accumulation of residual deformation, energy dissipation due to hysteretic behavior and the recovery of stiffness upon crack closure. From an engineering point of view, the proposed damage model is applicable to simulating the nonlinear behavior of concrete under cyclic loading.

(ii) The calculated responses of unloading and reloading are consistent with the experimental ones. This indicates that the hysteretic behavior can be approximately modeled using the nonlinear unloading/linear reloading branches and it is feasible to define the tensile or compressive damage variable d_κ as a function of the ratio of the accumulated dissipating energy to the fracture energy.

(iii) The plastic offset strain e_p^κ, tangent modulus at a zero stress of unloading H_t^κ and damage exponent θ_κ are the key parameters which govern the accuracy of the model. The response calculated with the calibrated parameters is in close agreement with that obtained by the experiment, which means that the calibrating procedures proposed in this study is suitable to determine these key parameters.

(iv) The effect of stiffness recovery is approximately modeled using a sudden recovery of stiffness with weight factor w_c. This approach oversimplifies the simulation of crack closure and the value of w_c has significant influence on the computational accuracy. A refined approach which reasonably represents the crack closure mechanism should be implemented into the present model in the further study.

Acknowledgments
Financial support for this study has been provided by the National Natural Science Foundation of China under project 51578088 and 50909109 and the Xinjiang Production and Construction Corps under project 2012BB013, which are gratefully acknowledged.

References
[1] Mazars J., 1986. A description of micro- and macroscale damage of concrete structures. Eng. Fract. Mech. **25**(5-6), 729–37.
[2] Mazars J., Pijaudier-Cabot G., 1989. Continuum damage theory – application to concrete. ASCE J. Eng. Mech. **115**(2), 345–65.
[3] Cervera M., Oliver J., Faria R., 1995. Seismic evaluation of concrete dams via continumm damage models. Earthq. Eng. Struct. D. **24**(9), 1225–45.
[4] Papa E., Taliercio A., 1996. Anisotropic damage model for the multiaxial static and fatigue behavior of plain concrete. Eng. Fract. Mech. **55**(2), 163–79.
[5] Al-Gadhib A.H., Baluch M.H., Shaalan A., Khan A.R., 2000. Damage model for monotonic and fatigue response of high strength concrete. Int. J. Damage Mech. **9**(1), 57–78.
[6] Li Q.B., Zhang L.X., Ansari F. 2002. Damage constitutive for high strength concrete in triaxial cyclic compression. Int. J. Solids Struct. **39**(15), 4013–25.
[7] Challamel N., Lanos C., Casandjian C. 2005. Strain-based anisotropic damage modelling and unilateral effects. Int. J. Mech. Sci. **47**(3), 459–73.
[8] Badel P., Godard V., Leblond J.B., 2007. Application of some anisotropic damage model to the prediction of the failure of some complex industrial concrete structure. Int. J. Solids Struct. **44**(18-19), 5848–74.
[9] Pröchtel P., Häußler-Combe U., 2008. On the dissipative zone in anisotropic damage models for concrete. Int. J. Solids Struct. **45**(16), 4384–406.
[10] Simo J.C., Ju J.W., 1987. Strain- and stress-based continuum damage models–I.Formulation. Int. J. Solids Struct. **23**(7), 821–40.
[11] Lubliner J., Oliver J., Oller S., Onate E. 1998. A plastic-damage model for concrete. Int. J. Solids Struct. **25**(3), 299–326.
[12] Luccioni B., Oller S., Danesi R., 1996. Coupled plastic-damaged model. Comput. Methods Appl. Mech. Eng. **129**(1-2), 81–9.
[13] Faria R., Oliver J., Cervera M. 1998. A strain-based plastic viscous-damage model for massive concrete structures. Int. J. Solids Struct. **35**(14), 1533–58.
[14] Lee J., Fenves G.L., 1998. Plastic-damage model for cyclic loading of concrete structure. ASCE J. Eng. Mech. **124**(8), 892–900.
[15] Grassl P., Jirásk M., 2006. Damage-plastic model for concrete failure. Int. J. Solids Struct. **43**(22-23), 7166–96.
[16] Wu J.Y., Li J., Faria R., 2006. An energy release rate-based plastic-damage model for concrete. Int. J. Solids Struct. **43**(3-4), 583–612.
[17] Voyiadjis G.Z., Taqieddin Z.N., Kattan P.I., 2008. Anisotropic damage-plasticity model for concrete. Int. J. Plasticity **24**(10), 1946–65.
[18] Poh L.H., Swaddiwudhipong S., 2009. Over-nonlocal gradient enhanced plastic-damage model for concrete. Int. J. Solids Struct. **46**(25-26), 4369–78.
[19] Lee J., Fenves G.L., 1998. Plastic-damage concrete model for earthquake analysis of dams. Earthq. Eng. Struct. D. **27**(9), 937–56.
[20] Faleiro J., Oller S., Barbat A.H., 2008. PlasticCdamage seismic model for reinforced concrete frames. Comput. Struct. **86**(7-8), 581–97.
[21] Palermo D., Vecchio F.J., 2003. Compression field modeling of reinforced concrete subjected to reversed loading: formulation. ACI Struct. J. **100**(5), 616–25.
[22] Sima J.F., Roca P., Molins C., 2008. Cyclic constitutive model for concrete. Eng. Struct. **30**(3), 695–706.
[23] Rots J., Blaauwendraad J., 1989. Crack models for concrete: discrete or smeared? Fixed, multi-directional or rotating? Heron **34**(1), 3–59.
[24] Bazant, Z.P., 1983. Comment on orthotropic models for concrete and geomaterials. ASCE J. Eng. Mech. **109**, 849–65.
[25] Willam K., Pramono E., Sture S., 1987. Fundamental issues of smeared crack models. In: Shah S.P., Swartz S.E.(Eds.), Proceedings SEM/RILEM International Conference of Fracture of Concrete and Rock. Springer-Verlag, Heidelberg, Berlin, New York, pp. 142–157.
[26] Karsan I.K., Jirsa J.O., 1969. Behaviour of concrete under compressive loadings. ASCE J. Struct. Div. **95**(12), 2543–63.
[27] Buyukozturk O., Tseng T.M., 1984. Concrete in biaxial cyclic compression. ASCE J. Struct. Eng. **110**(3), 461–76.
[28] Bahn B.Y., Hsu C.T.T., 1998. Stress-strain behaviour of concrete under cyclic loading. ACI Mater. J. **95**(2), 178–93.
[29] Gopalaratnam V.S., Shah S.P., 1985. Softening response of plain concrete in direct tension. J. ACI **82**(2), 310–23.

[30] Yankelevsky D.S., Reinhardt H.W., 1989. Uniaxial behaviour of concrete in cyclic tension. ASCE J. Struct. Eng. **115**(1), 166–82.
[31] Reinhardt H.W., 1984. Fracture mechanics of an elastic softening material like concrete. Heron **29**(2), 1–42.
[32] Ramtani S., Berthaud Y., Mazars J, 1992. Orthotropic behavior of concrete with direction aspects: Modeling and experiments. Nucl. Eng. Des. **133**(1), 97–111.

Fatigue Analysis of Proton Exchange Membrane Fuel Cell Stacks Based on Structural Stress Distribution

C.W. Wu [1], B. Liu [1], M.Y. Wei [2], L.F. Liu [1]

[1] State Key Laboratory of Structure Analysis for Industrial Equipment, Department of Engineering Mechanics, Dalian University of Technology, Dalian 116024, China.
[2] Department of Compression Systems, Aero Engine Corporation of China Commercial Aircraft Engine Co., LTD, Shanghai 200241, China.
Email: cwwu@dlut.edu.cn

Abstract. Proton exchange membrane fuel cell (PEMFC) stack usually undergoes various vibrations during packing, transportation and serving time, in particular for those used in the automobiles and portable equipment. Based on the Miner fatigue damage theory, the fatigue lives of the fuel cell components are first assessed. Then the component fatigue life contours of the stack are obtained under four working conditions, *i.e.* the three single-axial (in X-, Y- and Z-axis separately) and multi-axial random vibrations. Accordingly, the component damage under various vibrations is evaluated. The stress distribution on the gasket and PEM will greatly affect their fatigue lives. Finally, we compare the fatigue lives of 4-bolt- and 6-bolt-clamping stacks under the same total clamping force, and find that increasing the bolt number could improve the bolt fatigue lives.

1. Introduction

Proton exchange membrane fuel cell (PEMFC) stacks are expected to be widely used in vehicles. In general, the requirement of stack serving life in vehicles should be more than 5000 hours [1]. However, they often encounter such problems as short life and poor durability [2]. These problems may be caused by the structural damage resulting from rigorous working conditions, especially the vibrations during the serving time [3]. Rajalakshmi et al. [4] carried out a sine sweep and random vibrating test on a 3-axis-vibrating platform and observed that the torque applied to some tightening bolt was reduced. The bolt torque release may be due to either structural failure or joint wear under vibration [5-6]. Hou et al. [7-8] finished a vibration test for PEMFC stack under strengthened road vibration and found the external leakage increased by 1.73 times. In other vibration tests, it was found that the polarization curve was changed at high current density [9-10]. Banan et al. [11] found that the increase of vibration magnitude and frequency obviously aggravates the MEA damage. To accurately study the effects of cyclic stress caused by vibration on the fatigue life, Khorasany et al. [12] conducted fatigue experiments for the PEM specimens. They used the elastic-plastic finite element method (FEM) to predict the PEM fatigue life [13]. Although their studies focus mainly on the single components, the large experimental data become the basis of our simulation. However, they left interactions between components and the structure details (such as the channels) out of consideration.

In our published work, the effects of the clamping configuration and the clamping force magnitude on vibration modes were discussed [14-15]. However, there was still no visual and quantitative analysis on the fatigue life of stack components in service. Based on the Miner fatigue damage theory [16], the fatigue life and component damage (such as bolts, gaskets and PEMs) are evaluated in the present paper.

2. Fatigue Analysis Method of PEMFC Stacks

The stacks studied here are typical bolt-clamping stacks used in vehicles [17, 18]. 10-cell-stacks (1 kW) are set up as shown in Figure 1. Along the positive Z-axis, the unit cells numbered from 1 to 10 are clamped in series along Z-axis by the endplates, bolts (M10) and nuts. The dimensions of the bipolar plate in X, Y and Z directions are 85 mm,165 mm and 4 mm, respectively. On the bipolar plate, each parallel gas channel is 1 mm in width and 0.6 mm in depth. The cooling channel of the bipolar plate has a rectangular cross-section of 3 mm×2 mm. MEA is composed of two gas diffusion layers (GDLs) and one PEM in the middle. Here the thicknesses of GDL and PEM are 0.275 mm and 0.05 mm, respectively. The MEA sides are sealed with the gaskets. The stack rigid-body motion should be constrained. Specifically, the right endplate is fully fixed (see Figure 1), but the Z-axial translation of the left endplate is not fixed. The sizes of the endplate, collector plate and insulator plate in X- and Y-axis are 180 mm and 250 mm, and their thicknesses in Z-axis are 20 mm, 20 mm and 4 mm, respectively. The bolt positions for configurations with 4 and 6 bolts are shown in Figure 1. A torque of 12 Nm and 8 Nm are respectively applied to each bolt to provide the same total clamping force, about 22 kN [18].

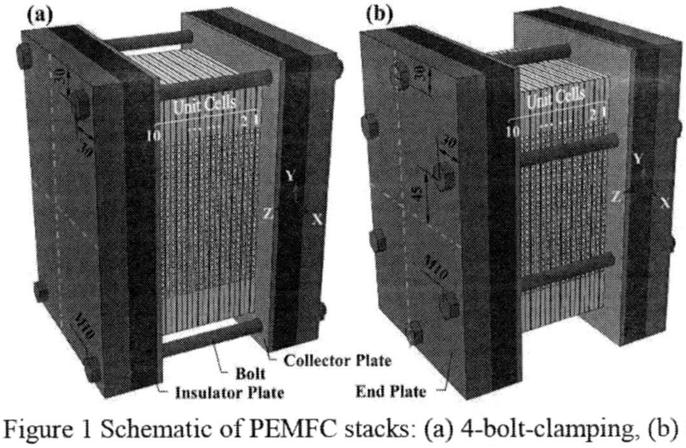

Figure 1 Schematic of PEMFC stacks: (a) 4-bolt-clamping, (b) 6-bolt-clamping.

GDL is made of carbon paper (TGP-H-90). This material shows isotropy and linear elasticity in XY-plane but shows nonlinear elasticity in Z-axis. All the other components are assumed to be isotropic. The operating temperature of PEM is 70 °C and relative humidness is 90% .

The *S-N* curves of the bolt, gasket and PEM materials are straight line in log-log coordinate, i.e., the cyclic number (N_s) increases exponentially with the decrease of cyclic stress(S), as the following:

$$S^{\alpha} N_s = C \quad (1)$$

where α and C are obtained by fitting material fatigue test data. For bolt, α equals 2.10 and C equals 3.40×10^9 [19]; for PEM, α equals 7.57 and C equals 3.16×10^9; for gasket, α equals 2.26 and C equals 3.53×10^5 [20].

In the present paper, a four hour random vibration is truncated. After periodic extension, a virtual infinite random vibration is acquired for fatigue analysis. The stacks studied under four conditions, including the single-axial (in X-, Y- and Z-axis separately) and multi-axial random vibrations. Here the multi-axial random vibration is the mixture of single-axial vibration in X-, Y- and Z-axis, whose weight coefficients are 1, 1 and 1, separately. The acceleration power spectral density (PSD) of random vibration in each direction is kept as the same with the PSD-curve. According to the stack vibration test criterions and corresponding literatures [21], the magnitude of acceleration usually ranges from 1 g to 20 g. Therefore we set the maximum acceleration to 20 g. The FEM model of each stack contains about 1 million elements. The fatigue life can be evaluated based on Miner fatigue damage theory.

3. Fatigue Analysis Results

3.1 Fatigue Life Contours of Bolts

The cyclic stress magnitude in stack components is one main influence factor to the fatigue lives. The endplate bending under clamping force makes the bolts bend to one side of the unit cells. In this condition, when the 4- and 6-bolt-clamping stacks are exposed to vibration, the bolts will be subjected to the coupled stress resulting from bending moment and tension. As a result, under the vibration in X- or Y-axis, the bolts have high stress near the nuts and thus have relatively short fatigue life in these regions, as shown in Figure 2. Under the vibration in Z-axis, the bolts will bear the inertia load of the whole stack in Z-axis, and thus will be subjected to a large cyclic tension stress. Due to the cyclic stress, the fatigue lives in most regions of the bolts decrease to about 1000 hours, as shown in Figures 2c and 2g. In addition, the bolts bending to one side of the unit cells will cause larger cyclic stretching stress to this side, giving rise to shorter fatigue life. As the bolt damage caused by the multi-axial vibration is the damage accumulation resulting from each single-axial vibration, fatigue lives in the most regions of the bolt are less than 1000 hours, as shown in Figures 2d and 2h.

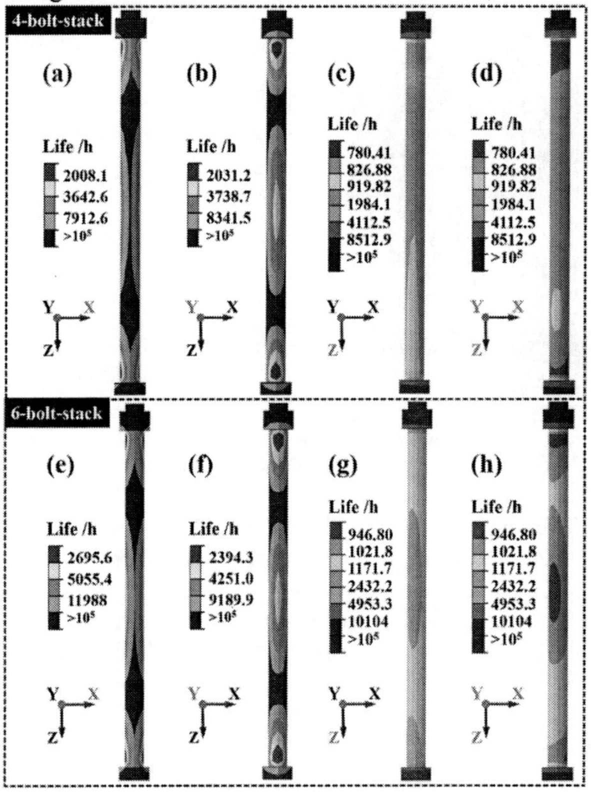

Figure 2. Fatigue life contours of the bolts: bolts in (a-d) belong to 4-bolt-clamping stack, (e-h) belong to 6-bolt-clamping stack, applying the vibrations in (a, e) X-axis, (b, f) Y-axis, (c, g) Z-axis, and (d, h) multi-axes.

The fatigue life is usually short in the bolt regions near the nuts and endplates. When the life limit is reached, the bolt threads in those regions may give rise to fretting wear, the possible damage that may cause the torque release. These phenomena are also observed in the experiments [4]. Furthermore, the possible torque loss may also cause decreasing of the inner pressure and increasing of hydrogen leakage rate and contact resistance. In mechanical design, there are many methods to avoid torque release, for examples, spring washers applied between nuts and endplate, rough contact surfaces between nuts, washers and endplate, and etc.
Since the 6-bolt- and 4-bolt-clamping stacks share the same total clamping force, each bolt in 6-bolt-clamping stack will be subjected to a smaller tension stress. This makes their fatigue lives longer than those in 4-bolt-clamping stack under all the conditions. Especially under the vibration in Z-axis, since the bolts in 6-bolt-clamping stack are also subjected to much less inertia load due to the vibration, the minimum bolt fatigue life in 6-bolt-clamping stack is 27.4% longer than the 4-bolt-clamping stack, as shown in Figures 2c and 2g. For the same reason, when multi-axial vibration is applied to the 6-bolt-clamping stack, the minimum fatigue life is 21.3% longer than the 4-bolt-clamping stack, as shown in Figures 2d and 2h.

3.2 Fatigue Life Contours of PEMs and Gaskets

The numerical results show that some regions in the gasket and PEM will be subjected to relative larger cyclic stress, leading to the localized failure or short fatigue life. The fatigue life contours are greatly affected by the stress distribution. Based on the color histogram analysis of the fatigue life contours (in XY-plane), the ratio of the area with fatigue life less than 5000 hours to the total area can be obtained. Hereafter this "ratio" will be called "failure area ratio". Generally, a high gasket failure area ratio will lead to a high leakage rate. A high PEM failure area ratio will result in performance degradation even a stack shutdown. As the gasket and PEM in the No.10 unit cell have relatively short fatigue life, Figures 3 and 4 demonstrate the fatigue life contours in this unit cell.

Under the vibration in X-axis, the stack will be subjected to bending. Correspondingly, warpage may occur in both left and right sides of the gasket, as shown in Figures 3a and 3e, which causes large coupled cyclic stress resulting from shearing and compression. Therefore the fatigue lives in these regions in stacks clamped with 4 and 6 bolts are both short. The fatigue lives are 3858.6 hours and 3367.7 hours, and the failure ratios are 14.0% and 15.4%, respectively. Analogously, under the vibration in Y-axis, the upwards and downwards sides of gasket have shorter fatigue lives of 4342.1 hours and 3701.6 hours, and the failure ratios are 6.7% and 11.4% respectively, as shown in Figures 3b and 3f. Thus it can be seen that under transverse (X- and Y-axis) vibration, the outside gasket region in 6-bolt stack has shorter fatigue life and larger failure ratio than that in 4-bolts region. Under vibration in Z-axis, the inertia load of unit cells in Z-axis is mostly applied to the gaskets. As a result, the gasket failure area ratio is 100% in the 4-bolt-clamping stack and 30% in the 6-bolt-clamping stack, as shown in Figures 3c and 3g. Under the clamping force and vibration in Z-axis, the outside of endplates will show bending deformation. Therefore, the outside region in the gasket bears a larger cyclic stress, whereas the internal region only needs to bear relative small stress. This would lead to a longer fatigue life in the internal gasket region. Under multi-axial vibration, the combination of the reasons mentioned above will result in the fatigue lives in the most gasket regions that are less than 2000 hours and in the outside gasket regions that are even less than 1100 hours. This may cause a severe gas leakage in the stack during the long time serving.

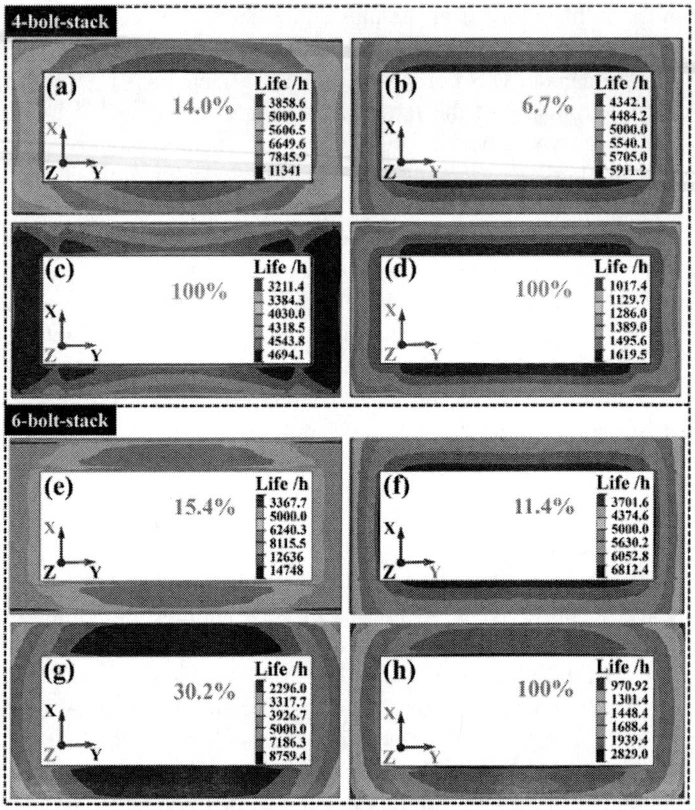

Figure 3 Fatigue life contours of the gaskets in the No.10 unit cell: gaskets in (a-d) belong to 4-bolt-clamping stack; (e-h) belong to 6-bolt-clamping stack, applying the vibrations in (a, e) X-axis, (b, f) Y-axis, (c, g) Z-axis, and (d, h) multi-axes.

With the service time increasing, the adding failure ratio of PEM means the shrinkage of electrochemical reaction region and the degradation of stack performance. Under vibrations in X- and Y-axis separately, the PEM in 4-bolt-clamping stack has failure area ratios of 1.1% and 0.9%, as shown in Figures 4a and 4b. Whereas, as shown in Figures 4e and 4f, the PEMs in 6-bolt-clamping stack have smaller failure area ratios, i.e. 1.0% and 0.3%. Similarly to gasket, the failure mainly occurs in the outside region of PEM. Under vibration in Z-axis, the inertia load of unit cells in Z-axis mostly applies to the gasket, which makes the PEM has a fatigue life more than 1 million hours, as shown in Figures 4c and 4g. This indicates that the single-axial vibrations may lead to very small PEM failure area ratios. Under multi-axial vibration, the PEM failure area ratios in 4-bolt- and 6-bolt-clamping stack are 13.5% and 2.6% respectively, as shown in Figures 4d and 4h. As discussed above, only small cyclic stress is generated in the PEM, resulting in the much smaller failure area ratio in PEM than that in the gasket. It is also worth mentioning that stress concentration exists in the contact areas of MEA and channel ribs, giving rise to a short fatigue life in this area. Generally speaking, most of the regions (more than 80%) of the stacks analyzed in the present paper can work for a long time. However the outside PEM regions give short fatigue lives usually less than 1000 hours.

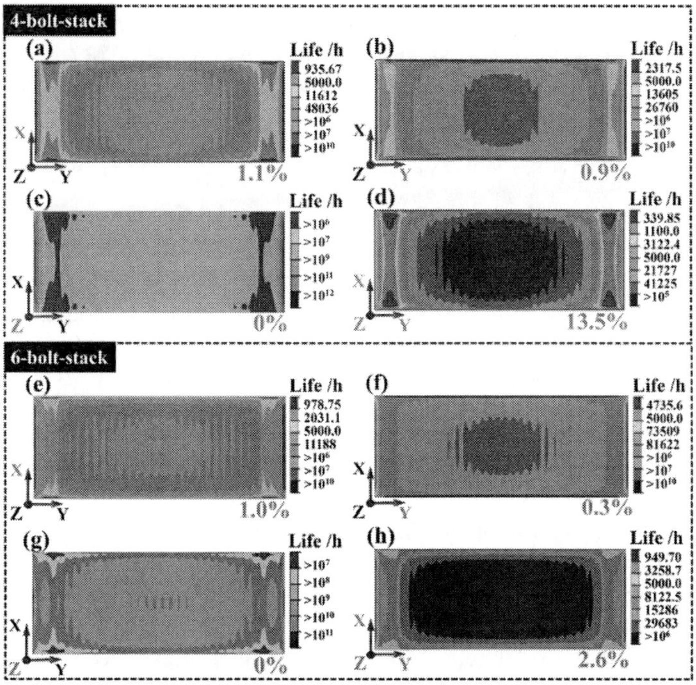

Figure 4 Fatigue life contours of PEMs in the No.10 unit cell: PEMs in (a-d) belong to 4-bolt-clamping stack; (e-h) belong to 6-bolt-clamping stack, applying the vibrations in (a, e) X-axis, (b, f) Y-axis, (c, g) Z-axis, and (d, h) multi-axes.

4. Conclusions

(1) Under the four vibration conditions studied in the present paper, the cyclic stress distribution on the stack components will greatly affect their fatigue lives. According to the fatigue contours, the fatigue lives in the bolt regions close to nuts and the outside regions of gasket and PEM are relatively short. This may cause severe hydrogen leakage under long term vibration.

(2) Comparing the 4- and 6-bolt-clamping stacks, as the total clamping force and the inertial load along Z-axis can be shared by more bolts, the bolts in 6-bolt-clamping stack have longer fatigue lives than 4-bolt-clamping stack. The quantity and configuration of bolts will also affect the stress distribution on gasket and PEM.

(3) The structure design for the stack has a great effect on the component fatigue life. For example, increasing rigidity of endplate could decrease the magnitude of cyclic stress on the outside regions of gasket and PEM, and thus increase the fatigue lives. A different thickness design of MEA and gasket can make the stress distribution more properly and improve both the fatigue lives of gasket and PEM.

Acknowledgements

This work was supported by National Basic Research Program of China (2015CB057306), and the NSFC (11572080).

References

[1] Sharaf O Z *et al* 2014 An overview of fuel cell technology fundamentals and applications *Renew. Sustain. Energy Rev.* Vol 32 pp 810-853

[2] Wu J *et al* 2008 A review of PEM fuel cell durability degradation mechanisms and mitigation strategies *J. Power Sources* Vol 184 pp 104-119

[3] Shi W Y *et al* 2007 The effect of H_2S and CO mixtures on PEMFC performance *Int. J. Hydrogen Energy* Vol 32 pp 4412-4417

[4] Rajalakshmi N *et al* 2009 Reliability evaluation of an open-cathode PEMFC at operating state and longtime vibration by mechanical loads *Int. J. Hydrogen Energy* Vol 34 pp 3833-3837

[5] Jiang Y *et al* 2004 An Experimental study of self-loosening of bolted joints *J. Mechanical Design* Vol 126 pp 925-931

[6] Basava S *et al* 1998 Bolted joint clamping force variation due to axial vibration *J. Sound & Vibration* Vol 210 pp 255-265

[7] Hou Y *et al* 2016 Effect of strengthened road vibration on performance degradation of PEM fuel cell stack *Int. J. Hydrogen Energy* Vol 41 pp 5123-5134

[8] Hou Y *et al* 2011 Experimental investigation of gas-tightness and electrical insulation of fuel cell stack under strengthened road vibrating conditions *Int. J. Hydrogen Energy* Vol 36 pp 13763-13768

[9] Betournay M C *et al* 2004 The effects of mine conditions on the performance of a PEM fuel cell *J. Power Sources* Vol 134 pp 80-87

[10] Hou Y *et al* 2012 An investigation of characteristic parameter variations of the polarization curve of a proton exchange membrane fuel cell stack under strengthened road vibrating conditions *Int. J. Hydrogen Energy* Vol 37 pp 11887-11893

[11] Banan R *et al* 2013 Effect of mechanical vibrations on damage propagation in polymer electrolyte membrane fuel cells *Int. J. Hydrogen Energy* Vol 38 pp14764-14772

[12] Khorasany R M H *et al* 2015 Mechanical degradation of fuel cell membranes under fatigue fracture tests *J. Power Sources* Vol 274 pp 1208-1216

[13] Khorasany R M H *et al* 2015 Simulation of ionomer membrane fatigue under mechanical and hygrothermal loading conditions *J. Power Sources* Vol 279 pp 55-63

[14] Liu B *et al* 2016 Effect of impact acceleration on clamping force design of fuel cell stack *J. Power Sources* Vol 303 pp 118-125

[15] Wu C W *et al* 2015 Mechanical response of a large fuel cell stack to impact a numerical analysis *Fuel Cells* Vol 15 pp 344-351

[16] Lee Y L *et al* 2005 Fatigue testing and analysis: theory and practice *Elsevier Butterworth-Heinemann*

[17] Lin P *et al* 2011 Multi-objective topology optimization of end plates of proton exchange membrane fuel cell stacks *J. Power Sources* Vol 196 pp 1222-1228

[18] Wen C Y *et al* 2009 Experimental study of clamping effects on the performances of a single proton exchange membrane fuel cell and a 10-cell stack *J. Power Sources* Vol 192 pp 475-485

[19] Warhadpande A *et al* 2010 A new finite element fatigue modeling approach for life scatter in tensile steel specimens *Int. J. Fatigue* Vol 32 pp 685-697

[20] Tabar R J *et al* 1983 Soft heat and fatigue resistant elastomeric articles *US4419480-A*

[21] Paclisan D *et al* 2013 Real time modelling of the dynamic mechanical behaviour of PEMFC thanks to neural networks *Eng. Appl. Artif. Intel.* Vol 26 pp 706-713

IOP Conf. Series: Journal of Physics: Conf. Series **842** (2017) 012042 doi:10.1088/1742-6596/842/1/012042

Novel tool wear monitoring method in milling difficult-to-machine materials using cutting chip formation

P P Zhang, Y Guo and B Wang

School of Mechatronics Engineering, University of Electronic Science and Technology of China, Chengdu, 611731, Sichuan, CHINA

E-mail: peipei.zhang@uestc.edu.cn, 201421080149@std.uestc.edu.cn, 1016765655@qq.com

Abstract. The main problems in milling difficult-to-machine materials are the high cutting temperature and rapid tool wear. However it is impossible to investigate tool wear in machining. Tool wear and cutting chip formation are two of the most important representations for machining efficiency and quality. The purpose of this paper is to develop the model of tool wear with cutting chip formation (width of chip and radian of chip) on difficult-to-machine materials. Thereby tool wear is monitored by cutting chip formation. A milling experiment on the machining centre with three sets cutting parameters was performed to obtain chip formation and tool wear. The experimental results show that tool wear increases gradually along with cutting process. In contrast, width of chip and radian of chip decrease. The model is developed by fitting the experimental data and formula transformations. The most of monitored errors of tool wear by the chip formation are less than 10%. The smallest error is 0.2%. Overall errors by the radian of chip are less than the ones by the width of chip. It is new way to monitor and detect tool wear by cutting chip formation in milling difficult-to-machine materials.

1. Introduction

Difficult-to-machine materials have gained widespread applications in aerospace, marine and biomedical industries due to their following favorable properties: they are light weight, possess high strength, have excellent fatigue performance and offer high resistance to an aggressive environment [1]. However, the main problems in milling difficult-to-machine materials are the high cutting temperature and the rapid tools wear. Rapid tool wear serious affect the machining efficiency and surface quality. Thus, how to monitor and control tool wear is a pressing problem for milling difficult-to-machine materials.

Controlling tool wear was realized by selection and optimization of cutting parameters (spindle speed, feed rate, and depth of cut) [2]. However, it is impossible to obtain the real-time degradation process by selecting or optimization of cutting parameters. In order to obtain or monitor the degradation process of tool or insert, many work or researches have been done on it. To monitor tool wear, vibration signal is most commonly used to analyze status of tool wear because it is direct and easy to collect vibration signal. For example, average energy was used to monitoring tool wear for turning Ti-6Al-4V [3]. Ref. [4] monitored tool wear using classifier fusion based on acoustic emission signal. Tool wear was correlated with vibration in the milling process of hardened steel in ref. [5].

However, vibration signal can not accurately reflect the states of tool wear because there are many vibration sources. At this time, during the whole processing, it is impossible to measure tool wear by constantly interrupted machine. Milling chip contacted directly with tool is taken to study tool wear because it is a representation of tool wear. Chip morphology was modeled to monitor tool wear for ultra-precision raster milling in ref. [6]. Ref. [7] used multiple sensors in turning to monitor the tool wear and chip formation. For ultra-precision raster milling, chip morphology was studied well to monitor or predict tool wear. Ref. [8] proposed a new method to monitor tool wear for raster milling using cutting chip. This method was realized to monitor tool wear by developing cutting chip geometry model. The effect of cAN/TiAIN coating on tool wear and chip morphology in face milling of Ti6Al4V was studied in Ref. [9]. However, ref. [9] had not mentioned the relationship between tool wear and chip morphology. There are fewer studies on tool wear and chip formation for milling difficult-to-machine materials. Thus, this paper aims to develop the relationship between tool wear and chip morphology to monitor the tool wear.

In this paper, section 2 proposes a novel method based on the problems of developing the relationship between tool wear (obtained un-easily) and chip (obtained easily). Section 3 introduces the experiment for collecting data of tool wear and chip morphology. Section 4 analyses these data and discusses the results which are gotten with the proposed method in section 2. Section 5 gives a conclusion of this paper.

2. Proposed method

Following the introduction, the machining efficiency and process can be accessed timely by monitoring related characterizations such as tool wear, vibration, surface quality, chip and so on. And these characterizations are effected or controlled by same cutting parameters. Some characterizations are obtained easily. However, some characterizations are not obtained easily and even unavailable. It is necessary to develop the relationships or mapping models between characterizations obtained easily and characterizations obtained non-easily if characterizations obtained non-easily are very important for researches or applications. In other word, the characterizations with same variables can be replaced or represented each other. So, in this paper, tool wear (obtained by stopping the machine) will be represented by chips (obtained easily by collecting them after finishing the cutting). This idea is described in mathematic form as following.

Functions with same variables can be replaced each other. Group functions F have i functions f which have same variables **x**,

$$\mathbf{F}_i = f_i(\mathbf{x}) \quad i = 1, 2, \ldots, n \tag{1}$$

i^{th} function $\mathbf{F}_i$ can be replaced by j^{th} function $\mathbf{F}_j$. The ith function $\mathbf{F}_i$ becomes equation (2).

$$\mathbf{F}_i = f_i\left(f_j^{-1}(\mathbf{F}_j)\right) \tag{2}$$

Here, take linear equations as example. There are two functions $\mathbf{F}_1$ and $\mathbf{F}_2$: $\mathbf{F}_1 = a_1x + b_1$, $\mathbf{F}_2 = a_2x + b_2$. $\mathbf{F}_2$ is rewritten with $\mathbf{F}_1$ as equation (3).

$$\mathbf{F}_2 = a_2\left(\frac{\mathbf{F}_1 - b_1}{a_1}\right) + b_2 \tag{3}$$

The calculation results are precise if functions have exact expressions. However, if the function is developed by fitting from the experiments data, there are going to have errors or deviation. Figure 1 shows the schematic diagram of errors forming process. Function f_1 is fitted based on round black point. Function f_2 is fitted based on star shape point. Function f_1 is used to monitor or predict function f_2. Real value (experiment data) of f_1 at point x_1 is put in function f_1. The point x_2 is obtained in function f_1 (blue round point in the line of $\mathbf{F}_1$). After x_2 is put in the f_2, $f_2(x_2)$ is the monitored value (blue star point in the line of $\mathbf{F}_2$). The Δy_2 is the monitored error or deviation of function f_2 by the function f_1. In contrast, Δy_1 is the monitored error or deviation of function f_1 by the function f_2 (red round point and red star point).

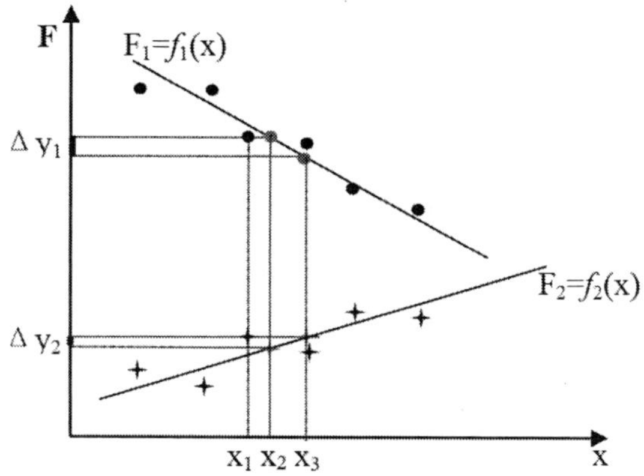

Figure 1. Schematic diagram of errors forming process.

3. Description of experiment

3.1. Work-piece material

The tested material was an alpha-beta titanium alloy Ti-6AI-4V (TC4). Titanium alloy is an attractive material due to their unique high strength, which is classified as difficult-to-machine materials. The main problems in machining titanium alloy are the high cutting temperature and rapid tools wear. The chemical compositions of the titanium alloys (in wt.%) is given in table 1. The mechanical properties of tested material are 912MPa (tensile strengthσ_b) , 839 MPa (Yield Strength $\sigma_{0.2}$), 10% (elongation) and around 27 HRC (hardness). The work-piece used in the experiment was a block with the size 200 x 200 x 80 (mm).

Table 1. Chemical Compositions (in wt.%)

Element	**Ti**	**AL**	**V**	**Fe**	**C**	**N**	**H**	**O**	**other**
Proportion	remainder	6.4	3.7	0.24	0.085	0.023	0.011	0.17	<0.10

3.2. Cutting tool material

The tool (inserts) used for the machining experiments were cemented carbide CG10 SEKT1204AFFN-HL. The shape of the insert is square with four angles 45 degree cut (figure 2 (a)). Five inserts were fixed in the tool handle at the same time (figure 2 (b)). That means five measured values of tool wear can be obtained at the same time.

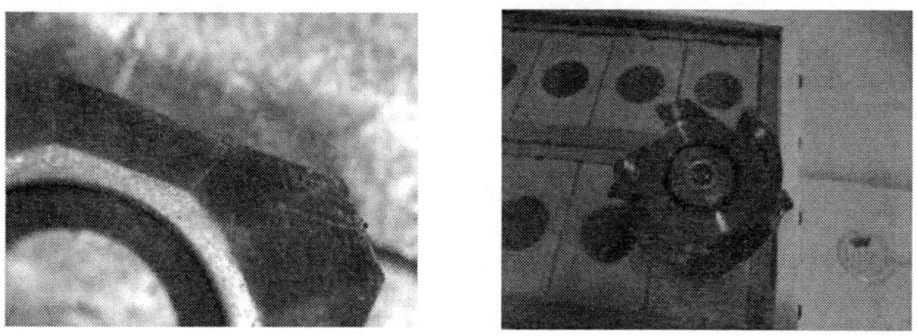

(a) (b)

Figure 2. Tool handle with five inserts.

3.3. Machining experiments

All the machining experiments were carried out on a vertical CNC milling machine XH714 which was controlled by FANUC controller (figure 3). The CNC milling has a continuously variable spindle speed. Throughout the whole experiments, the width of cutting was kept constant at 40 mm. and the spindle speeds (*n*) were set at 700, 1000 and 1300 rpm. The feed rates (*vf*) were set at 80, 100 and 120 mm/min. The depths of cut (*dp*) were set at 0.1, 0.3 and 0.5 mm. The machining experiments were carried out with cutting fluid. The cutting conditions used in this paper are shown in table 2.

Figure 3. Worktable.

Table 2. Design of experiments

	n rpm	*vf* mm/min	*dp* mm
1	700	100	0.3
2	1300	80	0.5
3	1300	100	0.1

3.4. Tool wear and cutting chip measurement

The tool wear and chip formation were measured at the same time after a surface was milled (twice straight cutter path; width of cut is 40 mm; width of the block is 80 mm). The number of measurement of tool wear and chip formation is 10. It means that the processing is interrupted by 10 times. The time of each processing depended on the feed rate. The maximum wear lands were measured in the longest and the widest by the toolbox of software in microscope (figure 4). The width chip and radian of chip were chosen to be the parameters of chip because they can typically display the chip formation. Figure 5 shows the measurements of cutting chip formation. Cutting process was terminated after 10 processing was finished no matter the inserts were failed or not. Otherwise, cutting was forced to terminate when catastrophic fracture of the edge was observed (appeared a lots of sparks). The cutting process for second set parameters (*n*=1300 rpm, *vf*=80 mm/min, *dp*=0.5mm) is forced to stop at the 7th processing due to catastrophic fracture. Thus, in the section 4, there are only 7 groups' data for the second set parameters.

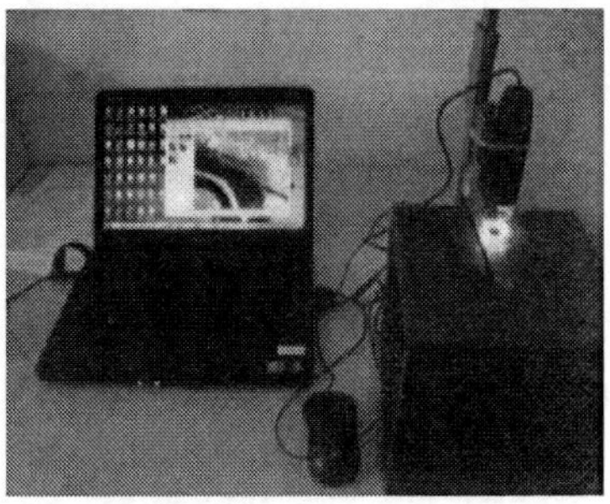

(a) Digital microscope

(b) Measurement of tool wear region

Figure 4. Equipment and measurement.

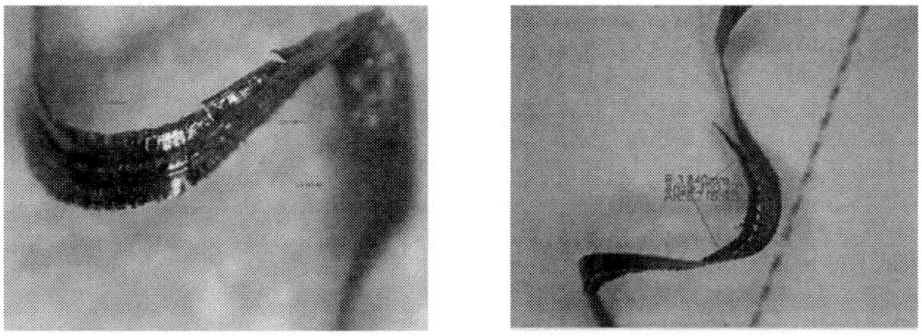

(a)Width of chip (b)Radian of chip

Figure 5. Cutting chip formation measurements.

4. Results and discussion

As discussed before, the relationship between tool wear and chip formation is developed. Figure 6, figure 8 and figure 10 show the tools wear values, width of chip and radian of chip for three sets parameters (table 2) to cutting times. It is easy to find that tool wear gradually increase along with cutting time. In contrast, width of chip and radian of chip are gradually decreased.

By method proposed in section 2, the monitoring values of tool wear by chip formation is shown in the table 3, table 4 and table 5. The errors are also listed in the three tables at the same time. So three tables list from left to right monitored values of tool wear by width of chip (W-by-W), monitored values of tool wear by radian of chip (W-by-R), monitored values of tool wear by average values of width of chip and radian of chip (W-by-WR), real tool wear (R-W), monitored errors by width of chip (E-by-W), monitored errors by radian of chip (E-by-R) and monitored errors by average value of width of chip and radian of chip (E-by-WR). Figure 7, figure 9 and figure 11 show the monitored errors by width of chip, radian of chip and average value of them.

Table 3 and figure 7 show that the monitored errors by radian of chip are less than the ones by width of chip for first set parameters (n=700rpm, vf=100mm/min, dp=0.3mm). The maximum monitored error by radian of chip is 18.95%; the minimum monitored error by radian of chip is 0.29%. The errors less than 10% are 70%. All the errors are less than 20%. However, the maximum monitored error by width of chip is 26.95%. The minimum monitored error by width of chip is 0.24%. The monitored errors by width of chip less than10% are 50%. It means that the monitored errors by radian of chip are less than the one by width of chip. All errors by average value of width of chip and radian of chip are less than 17%.

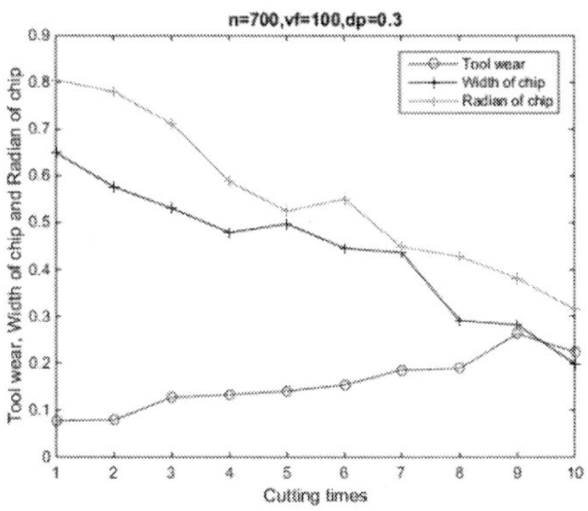

Figure 6. Tool wear, width of chip and radian of chip for the first set parameters.

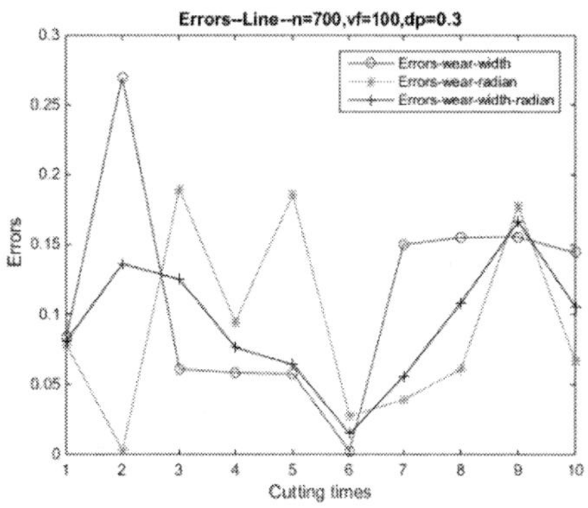

Figure 7. Errors of monitored tool wear by width of chip, radian of chip and average value of width and radian of chip for the first set parameters.

Table3. Monitored tool wears and errors by chip formation for the first set parameters

	W-by-W	W-by-R	W-by-WR	R-W	E-by-W	E-by-R	E-by-WR
1	0.0706	0.0701	0.0708	0.0770	8.35	7.81	8.08
2	0.1003	0.0792	0.0898	0.0790	26.95	0.29	13.62
3	0.1193	0.1029	0.1111	0.1270	6.08	18.95	12.52
4	0.1407	0.1455	0.1431	0.1330	5.83	9.41	7.62
5	0.1329	0.1672	0.1500	0.1410	5.74	18.55	6.41
6	0.1544	0.1582	0.1563	0.1540	0.24	2.75	1.49
7	0.1581	0.1933	0.1757	0.1860	15.01	3.90	5.55
8	0.2184	0.2005	0.2094	0.1890	15.53	6.07	10.80
9	0.2221	0.2163	0.2192	0.2630	15.56	17.77	16.67
10	0.2563	0.2389	0.2476	0.2240	14.43	6.67	10.55

Table 4 and figure 9 show that the monitored errors by radian of chip are less than the ones by width of chip for second set parameters (*n*=1300rpm, *vf*=80mm/min, *dp*=0.5mm). The maximum monitored error by radian of chip is 39.59%; the minimum monitored error by radian of chip is 3.36%. However, the monitored errors by width of chip less than 6% are 71% even though the maximum error is 46.43%. All of errors by average of width of chip and radian of chip are less than 25%.

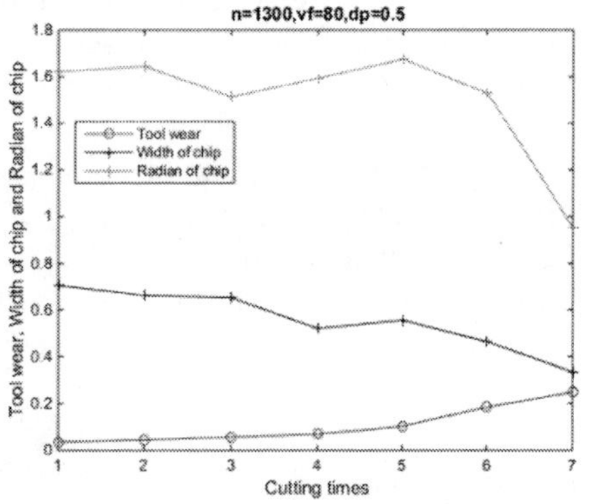

Figure 8. Tool wear, width of chip and radian of chip for the second set parameters.

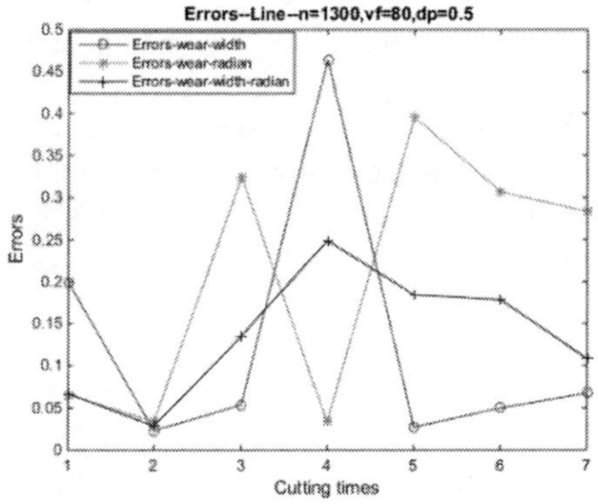

Figure 9. Errors of monitored tool wear by width of chip, radian of chip and average value of width and radian of chip for the second set parameters.

Table 4. Monitored tool wears and errors by chip formation for the second set parameters

	W-by-W	W-by-R	W-by-WR	R-W	E-by-W	E-by-R	E-by-WR
1	0.0881	0.1172	0.1026	0.1100	19.92	6.54	6.69
2	0.1094	0.1082	0.1088	0.1120	2.30	3.40	2.85
3	0.1145	0.1603	0.1374	0.1210	5.37	32.45	13.54
4	0.1816	0.1282	0.1549	0.1240	46.43	3.36	24.89
5	0.1633	0.0961	0.1297	0.1590	2.70	39.59	18.45
6	0.2100	0.1532	0.1816	0.2210	4.96	30.67	17.82
7	0.2761	0.3799	0.3280	0.2960	6.73	28.35	10.81

Table 5 and figure 11 show that the monitored errors by radian of chip are less than the ones by width of chip for the third set parameters (*n*=1300rpm, *vf*=100mm/min, *dp*=0.1mm). The maximum monitored error by radian of chip is 26.33%; the minimum monitored error by radian of chip is 3.55%. The monitored errors less than 15% are 70%. However, the monitored errors by width of chip are about 40% which are not useful for the real engineering. The results of first set parameter show the same phenomenon. All of errors by average value of width of chip and radian of chip are less than 34%.

The overall errors monitored by radian of chip are less than the ones by width of chip for first set parameters and third set parameters. For the second set parameters, the overall errors monitored by radian of chip are equal to the ones by width of chip. The overall errors monitored by average value of width of chip and radian of chip are smaller than the ones by the radian of chip or width of chip. So width of chip is not suitable for monitoring tool wear. Radian of chip or average value of radian and width of chip can be chosen to monitor tool wear.

In addition, width of chip varies from big to small following that the depth of cut narrows. With the fixed parameters, width of chip varies from big to small along with the tool wear getting more serious. Width of chips is proportional with the radian of chips.

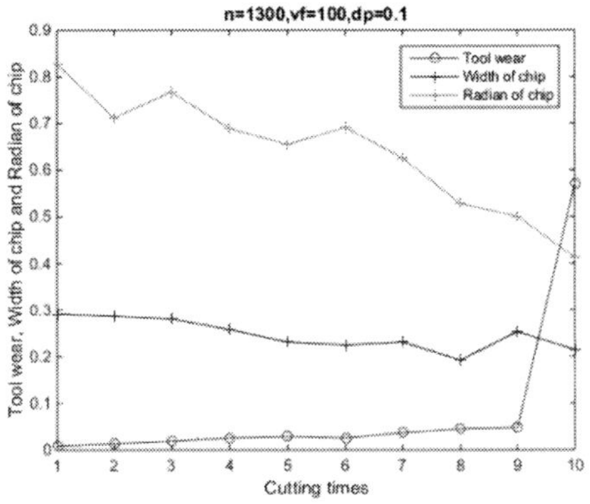

Figure 10. Tool wear, width of chip and radian of chip for the third set parameters.

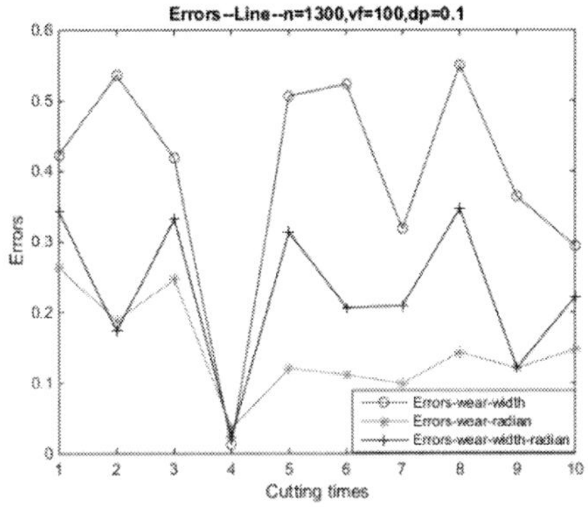

Figure 11. Errors of monitored tool wear by width of chip, radian of chip and average value of width and radian of chip for the third set parameters.

Table 5. Monitored tool wears and errors by chip formation for the third set parameters

	W-by-W	W-by-R	W-by-WR	R-W	E-by-W	E-by-R	E-by-WR
1	0.0035	0.0044	0.0039	0.0060	42.3	26.33	34.32
2	0.0046	0.0119	0.0083	0.0100	53.62	18.89	17.37
3	0.0064	0.0083	0.0073	0.0110	41.81	24.68	33.24
4	0.0132	0.0135	0.0133	0.0130	1.24	3.55	2.40
5	0.0211	0.0157	0.0184	0.0140	50.69	12.07	31.38
6	0.0229	0.0133	0.0181	0.0150	52.40	11.13	20.64
7	0.0211	0.0176	0.0193	0.0160	31.86	9.93	20.89
8	0.0326	0.0240	0.0283	0.0210	55.05	14.33	34.69
9	0.0146	0.0258	0.0202	0.0230	36.39	12.08	12.15
10	0.0261	0.0315	0.0288	0.0370	29.48	14.74	22.11

5. Conclusions

The research explored the method for monitoring tool wear by cutting chip formation. From the study, specific conclusions are as follows: tool wear gradually increase along with cutting time. In contrast, width of chip and radian of chip are gradually decreased along with the cutting time. For a fixed cutting parameters, width of chip decrease along with that the tool wear is getting more serious. It is possible and useful to monitor tool wear in milling difficult-to-machine materials by cutting chip formation. The monitored errors by radian chip and average value of width of chip and radian of chip are less than the ones by the width of chip.

The method proposed in this paper can be used to monitor or predict the related two representations for milling the difficult-to-machine materials. The following work is going to focus on developing the models of cutting parameters and each representation (vibration, surface quality, chips formation and tool wear). Thereby the balance of quality and efficiency for milling difficult-to-machine materials is found by multi-disciplinary optimization.

Acknowledgements

The work described in this paper was partially supported by a grant from the National Natural Science Foundation of China (NSFC) (Project No. 51405062) and The Fundamental Research Funds for the Central Universities (Project No.ZYGX2014J075).

References

[1] https://www.researchgate.net/post/Why_is_difficult_to_machining_titanium_and_titanium_alloys

[2] Zhang P P and Guo Y 2015 Mapping time-variant modeling of tool wears and cutting parameters on difficult-to-machine materials, *Proceedings of the 10th world congress on Engneering Asset Management*, 28-30 september 2015 (Tampere, Finland)

[3] Srinivasa Pai P and D'Mello G 2015 Vibration signal analysis for monitoring tool wear in high speed turning of Ti-6Al-4V, *Indian Journal of Engineering & Materials Science* 22 pp 652-660

[4] Kannatey-Asibu E, Yum J and Kim T H 2017 Monitoring tool wear using classifier fusion, *Mechanical Systems and Signal Processing* 85 pp 651-661

[5] Mendes de Aguiar M, Eduqrdo Diniz A and Pederiva R 2013 Correlating surface roughness, tool wear and tool vibration in the milling process of hardened steel using long slender tools, *International Journal of Machine Tools & Manufacture* 68 pp 1-10

[6] Zhang G Q, To S and Xiao G B 2014 The relation between chip morphology and tool wear in ultra-precision raster milling, *International Journal of Machine Tools & Manufacture* 80-81 pp 11-17

[7] Bhuiyan M S H, Choudhury I A and Dahari M 2014 Monitoring the tool wear, surface roughness and chip formation occurrences using multiple sensors in turning, *Journal of Manufacturing Systems* 33 pp 476-487

[8] Zhang G Q, To S and Xiao G B 2014 Novel tool wear monitoring method in ultra-precision raster milling using cutting chips, *Precision Engineering* 38 pp 555-560

[9] Çalışkan H and Küçükköse M 2015 The effect of cAN/TiAIN coating on tool wear, cutting force, surface finish and chip morphology in face milling of Ti6Al4V superalloy, *International Journal of Refractory Metals and hard Materials* 50 pp 304-312

IOP Conf. Series: Journal of Physics: Conf. Series **842** (2017) 012043 doi:10.1088/1742-6596/842/1/012043

Tuned Liquid Column Damper based Reduction of Dynamic Responses of Scaled Offshore Platforms in Different Ocean Wave Basins

Deirdre O'Donnell[1,2] Jimmy Murphy[3,4], Cian Desmond[4], Vesna Jaksic[5] and Vikram Pakrashi[1,2]

[1] Dynamical Systems and Risk Laboratory, School of Mechanical and Materials Engineering, University College Dublin, Ireland

[2] Marine and Renewable Energy Ireland (MaREI), University College Dublin (UCD), Ireland

[3] Department of Civil, Structural and Environmental Engineering, University College Cork (UCC), Ireland

[4] Marine and Renewable Energy Ireland (MaREI), Environmental Research Institute, University College Cork (UCC), Ireland

[5] Department of Civil, Structural and Environmental Engineering, Cork Institute of Technology (CIT), Ireland

vikram.pakrashi@ucd.ie

Abstract. Control of dynamic responses of renewable energy device platforms is important for their performance, safe operation and efficiency over their lifetime under regular and extreme wave conditions. Tuned Liquid Column Dampers (TLCDs) have been recently considered as a viable passive control mechanism in this regard but limited information is available in relation to their experimental performance. This paper compares scaled experiments conducted in two different ocean wave basins where floating offshore platforms were retrofitted with multiple TLCDs (MTLCDs). Performance of such MTLCDs in these scaled ocean wave basins are evaluated and compared considering control of dynamic responses for a specific objective. This paper shows the potential of MTLCDs to reduce motions in offshore platforms for different designs and platforms of MTLCDs and provides a comparison of the levels of reduction of dynamic responses achieved. The performance of MTLCDs in different wave basins create an experimental evidence base behind the potential use of such solutions, the objectives of such use and highlight related challenges and limitations.

1. Introduction

Offshore wind developments offer many benefits, such as increased wind speeds as they are placed at further distances from the shore, less turbulence and less opposition from the public, as is seen toward onshore turbines [1]. In deep water, floating platforms must be employed to harness wind energy in an economically viable way [2], and with floating platforms comes a greater need for solutions to ensure more stable operating conditions [3]. A platform with a higher tolerance of high sea states, and a lower wave sensitivity will likely be a more economically favourable design [4]. Damper technologies have been employed in civil infrastructure as a means of structural control [5]. They have been applied to the stabilising of tall buildings [6] and as method of earthquake protection [7]. Technologies include tuned-mass dampers, fluid-viscous dampers and tuned liquid column dampers (TLCDs). More recently, these systems have been applied to offshore platforms; spring dampers [8], buoyant mass dampers [9], tuned liquid dampers [10,11], and tuned-mass dampers[12,13] have all been investigated both numerically and experimentally, to varying success. Tuned Liquid Column Dampers (TLCDs) are structural control devices which use the energy dissipated by the movement of water in liquid columns, typically U-shaped, of tuned lengths which dissipate excitations within the structure. They are a low-cost, low-maintenance passive damping solution, which can be easily installed. Parameters of TLCDs affecting their performance have been investigated [14], and systems can be design to meet specific requirements. The application of TLCDs to offshore systems has been well investigated numerically [15, 16, 17, 18] but, though examples exist [19], experimental comparisons in this topic are not adequately well-represented in literature. Dynamic responses and related control can be used for monitoring through markers [20].

In this paper, a recent experimental regime of a new MTLCD design retrofitted to a model of a floating wind platform is presented. The retrofitted platform was tested under various wave conditions in a deep ocean wave basin, while displacement data was recorded, from which acceleration data can be derived. These results are compared with results from previous testing of a different model, in a different ocean wave basin. It is hoped that this paper successfully demonstrates that results obtained in these two different cases, where the tank, the structure and the parameters of the TLCD all differed, are still comparable in scale and offer a benchmark for scaled tank testing of such systems.

2. Experimental Details

Experiments were carried out in the newly commission ocean wave basin in the Lir National Ocean Test Facility in Cork, on a scaled semi-submersible structure, designed by Iberdrola, which is part of the European Union EU FP7 LEANWIND project. TLCDs were retrofitted to the device, their dimensions tuned to reduce hub height accelerations for a set of wave conditions.

Figure 1. Semisubmersible floating wind platform model, fitted with TLCDs, on site at the Lir National Ocean Test Facility, prior to wave testing

2.1. Scaled Model
The platform (Figure 1) is a floating, semi-submersible hollow steel structure, moored by 3 cable-spring systems designed to represent the behaviour of a catenary line at full scale. The bow leg holds a mast on which a fan, used to simulate a turbine at model scale, can be mounted. During testing, a scaled mass was added to the top of the mast, to simulate the weight of a turbine, and so the effect of aerodynamic rotor thrust was not examined during testing.

2.2. Wave tank
The wave tank is 35m long and 12m wide, with a moveable floor which can go up to 3m deep, seen in figure 2 during testing. The platform was subjected to 1:36 Froude scaled waves, and the Bretschneider spectrum was used to simulate ocean conditions. The wave profiles used are defined by significant wave height (Hs) and wave period (Tp), and the time series of the waves are generated by from the frequency domain representation of wave spectra.

Figure 2. The wave tank in operation at the Lir facility.

The wave cases chosen to test the platform were Bretschneider Spectrum waves, of 2.5m significant wave height (*Hs*) and periods (*Tp*) of 6, 9, 11, 12, 15 and 17 seconds respectively. A previous paper by Jaksic et al. [21], showed results with good consistency across different wave heights, so in this paper the focus was on different wave periods and the effect they might have on the effectiveness of a TLCD's tuning to the wave frequency. The wave cases were chosen to cover a wide range of wave frequencies, each for a period of 14 minutes which allowed time to establish the wave, a 10-minute period of continuous steady data acquisition, and time at the end for the oscillations to settle. The wave cases were each run for the three different platform arrangements; no TLCD (to create baseline from which change could be measured), 'TLCD1' in active state (valve open and water free to move), and 'TLCD2', also in its active state.

2.3. Instrumentation
The choice of sensors for wave basin testing has been previously investigated [22], and a high-resolution camera tracking system is employed here. The Qualisys motion capture system, with 4 Oqus3 cameras, was used to record the position of the platform by tracking 4 markers attached to different parts of the platform. This positional data was used to derive acceleration data at hub height.

2.4. TLCD Design

The peak frequency of the set of sea states selected for this analysis was between .08 and .13 Hz. This range was chosen as the frequency to which to tune the TLCD. Comprehensive equations summarising their behaviour were presented by Yalla & Kareem [23], but the key governing design decision taken is based on finding the length of liquid column (L_d) obtained from the frequency to which TLCD is tuned (ω_{TLCD}).

Using this equation, the required length of water column was calculated for the frequency being designed to. Other optimum parameters for TLCDs are investigated by Yalla & Kareem [23], and some of their findings instructed the design featured here. The horizontal to vertical length ratio was kept at a minimum, within practical boundary conditions.

Mass ratios were between 1 and 5%. Two TLCD combinations were designed and tested. The first, referred to as 'TLCD1', seen in figure 3 (a), consists of two symmetrically mounted TLCDs both tuned to the peak input frequency by liquid column lengths of 1.1m. The TLCDs weighed 1.8% of the total platform mass. Ballast was removed from the front leg of the platform to re-level the platform, and the difference in draft, and the change in mooring tension that would result, was accounted for by adjusting the movable tank floor accordingly.

(a)

(b)

Figure 3. (a)'TLCD1' configuration, two 1.1m liquid columns. **(b)**'TLCD2' configuration, four liquid columns of different lengths.

The second, 'TLCD2', shown in figure 3 (b), consists of four TLCDs, each of different liquid column lengths (1.1m, 1.2m, 1.3m and 1.4m, respectively), which corresponded to values bounding the peak frequency value, but covering a greater range of frequency values. The TLCDs in this case weighed 3% of the total platform mass. Additional ballast was removed and the floor was lowered to account for further draft change. The TLCDs were constructed from clear PVC piping, of 25mm and 32mm external diameters. A valve was added to each as a method to control whether the device was 'active' (valve open, water free to move) or 'inactive' (valve closed). They were mounted on custom made steel brackets, which suspended them above the wave surface, to avoid extra drag being introduced to the system.

3. Results and Discussion

As top-of-mast accelerations are often a design limit state criterion, both raw displacement data at the top of the mast and acceleration data derived from this were analysed. Using this data, comparisons were made between different wave cases and different platform configurations, shown in figure 6. Figure 4 (a) shows the comparison of mean surge displacement, and shows a small (approx. 1%) but consistent reduction in mean displacement for both active MTLCD systems, with 'TLCD2' being the most effective. A similar situation is observed in figure 4 (b), where a 1.9% reduction in peak displacement was achieved. The results are reasonably consistent across all wave periods.

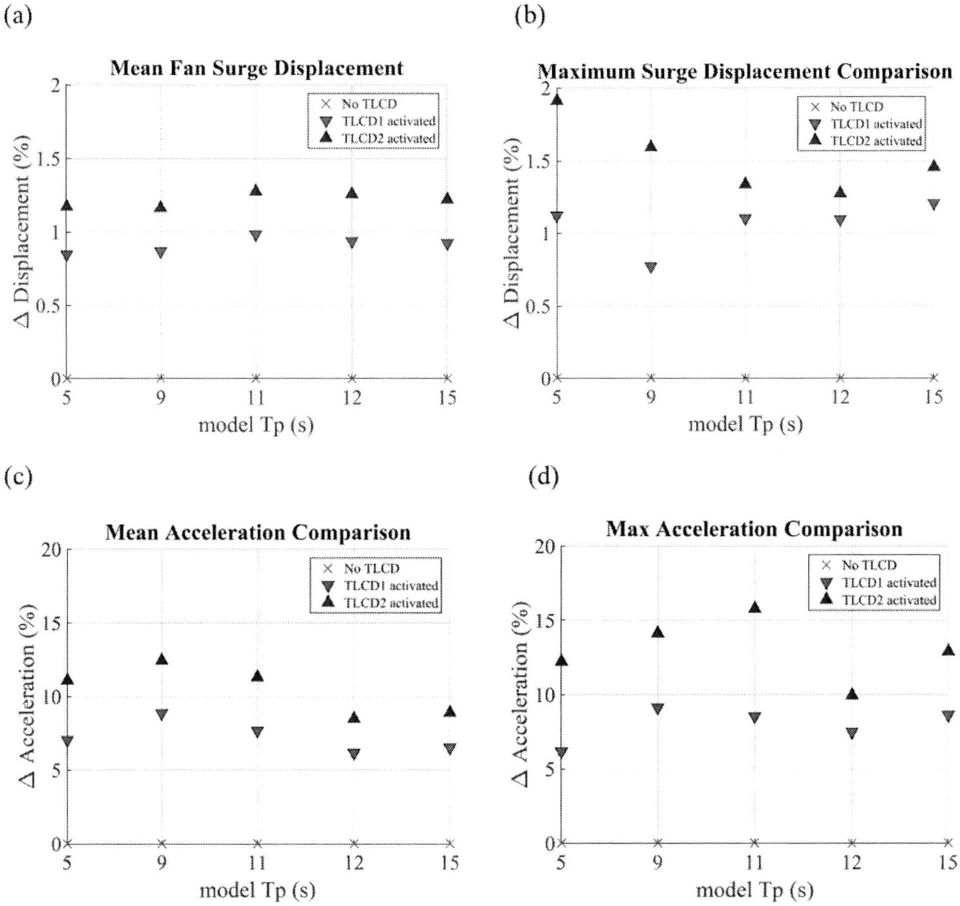

Figure 4. (a) Reduction in mean displacement in the surge direction, **(b)** reduction in maximum displacement in surge direction, **(c)** reduction in mean acceleration, **(d)** reduction in maximum acceleration, all for top of mast.

Figure 4(c) shows a significant percentage reduction in mean acceleration at the top of the tower due to the addition of the MTLCD systems, with 'TLCD2' proving again to be most effective with a maximum of 13% reduction achieved. The peak acceleration sees a maximum of 16% reduction, seen in figure 4 (d), and once again 'TLCD2' is consistently the most effective across all wave periods. Of great interest during this testing was the reduction of hub height accelerations, a given value of which is often an important limit state criterion in the design of such structures.

A Weibull probability density fit (figure 5) was used on the data as a means of assessing the change in data distribution. For the TLCD1 and TLCD2 systems, respectively, the peak of the distribution is closer to the y-axis, making the data more right skewed towards lower acceleration values. A 'flattening' of the tail shows a reduction in the probability of the occurrence of maxima.

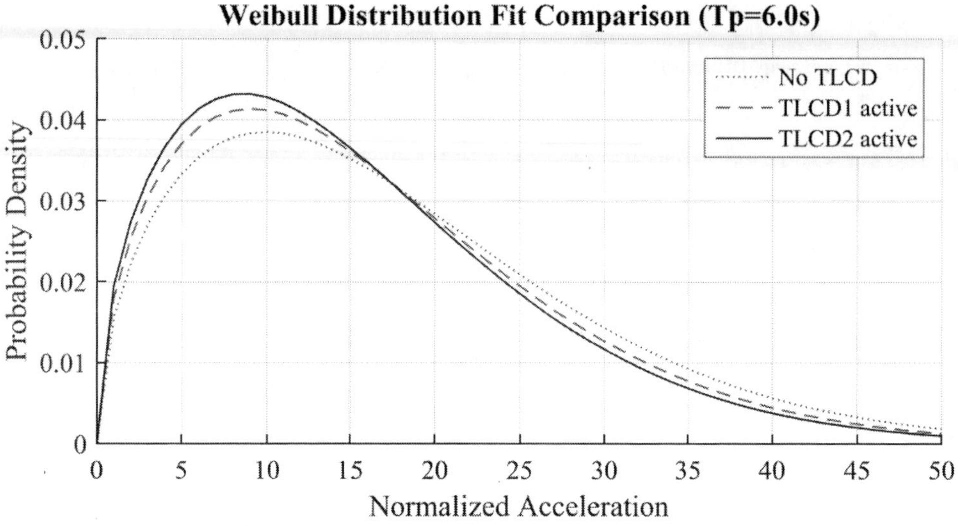

Figure 5. Weibull probability density function comparison.

The work by Jaksic et al. [21] assessed reductions due to three different MTLCD designs when retrofitted to a tension leg platform (TLP) at 1:50, and testing was carried out in a different ocean wave basin.

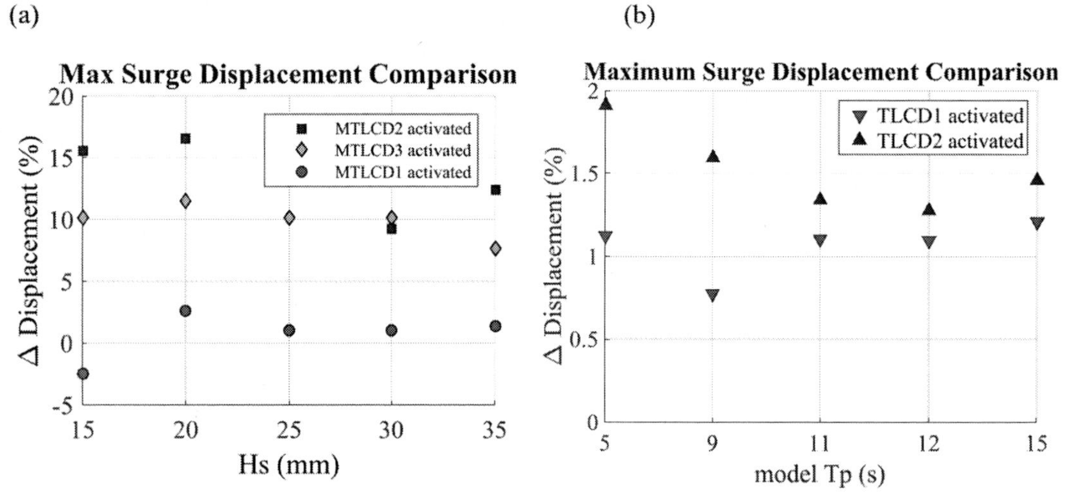

Figure 6. Comparison of maximum surge displacement for **(a)**TLP model **(b)**semisubmersible model.

Figure 6 (a) shows a reduction in surge displacement of up to 16% for TLCD systems of 15, 20 and 30% mass ratios. The current work resulted in maximum displacement reduction of 1.9% (figure 6 (b)), however this was for devices of 1.8 and 3% mass ratios. The differences in testing parameters, such as the TLP platform in the 1m deep wave basin compared to the semisubmersible catenary-moored platform in the 3m deep basin, are to be considered even in this dimensionless side-by-side comparison.

4. Conclusions

This work details the retrofitting a system of multiple tuned liquid column dampers to a scaled semisubmersible floating wind platform model for testing in an ocean wave basin. Investigation into reduction of accelerations at the top of the platform's mast is an important objective for these tests. The installation of TLCDs had a significant, positive effect on reducing hub height accelerations of the mast. TLCD2 was the most effective configuration, the 4 different lengths of water column, designed to control for a range of frequencies, having the desired effect. Results showed the favourable effect of the TLCD on the platform's displacement was small (approx. 1%). Reductions of up to 16% were achieved in maximum acceleration values, and an average of 8% and 11% reduction in mean acceleration values for TLCD1 and TLCD2, respectively. These results were consistent across input wave periods, and TLCD2 was the more effective design in all cases. The load cases were chosen to test the peak frequencies to which the device was designed, and results show that both designs were effective in the design objective of damping the motions these waves induced. The results are comparable to previous scaled tests on a different platform and basin, and this implies a level of repeatability is achievable across different wave tanks in terms of demonstration on control off accelerations using TLCDs at a scaled model level.

Acknowledgements

This work was carried out with the MaREI SFI Centre as part of the LEANWIND project which has received funding from the European Union Seventh Framework Programme under the agreement SCP2-GA-2013-614020. The PhD candidate is supported by the Irish Research Council 2016 Postgraduate Scholarship.

References

[1] Henderson AR, Morgan C, Smith B, Sørensen HC, Barthelmie RJ, and Boesmans B, "Offshore wind energy in europe - A review of the state-of-the-art," *Wind Energy*, vol. **6**, no. 1, pp. 35–52, 2003.

[2] Musial W, Butterfield S, and Ram B, "Energy From Offshore Wind," *Offshore Technol. Conf.*, pp. 1–11, 2013.

[3] European Wind Energy Association, *Deep water*, July. 2013.

[4] Butterfield S, Musial W, Jonkman J, and Sclavounos PP, "Engineering Challenges for Floating Offshore Wind Turbines" Copenhagen Offshore Wind Conference, Copenhagen, Denmark, 2005.

[5] Housner GW, Bergman LA, Caughey TK, and Chassiakos AG, "Structural control: past, present, and future," *J. Eng. Mech.*, vol. **123**, no. 9, pp. 897–971, 1997.

[6] Diana G, Resta F, Sabato D, and Tomasini G, "Development of a methodology for damping of tall buildings motion using TLCD devices," *Wind Struct.*, vol. **17**, no. 6, pp. 629–646, 2013.

[7] Gowda KK and Kiran KK, "Earthquake resistance of structures using dampers - a review," *Int. J. Adv. Struct. Geotech. Eng.*, vol. **2**, no. 1, pp. 31–35, 2013.

[8] Wright C, O'Sullivan K, Murphy J, and Pakrashi V, "Experimental Comparison of Dynamic Responses of a Tension Moored Floating Wind Turbine Platform with and without Spring Dampers," *J. Phys. Conf. Ser.*, vol. **628**, p. 12056, 2015.

[9] Moharrami M and Tootkaboni M, "Reducing response of offshore platforms to wave loads using hydrodynamic buoyant mass dampers," *Eng. Struct.*, vol. **81**, pp. 162–174, Dec. 2014.

[10] Jin Q, Li X, Sun N, Zhou J, and Guan J, "Experimental and numerical study on tuned liquid dampers for controlling earthquake response of jacket offshore platform," *Mar. Struct.*, vol. 20, no. **4**, pp. 238–254, 2007.

[11] Jaksic V, Wright C, Chanayil A, Ali SF, and Murphy J, "Performance of a Single Liquid Column Damper for the Control of Dynamic Responses of a Tension Leg Platform," *J. Phys.*, vol. **12058**, pp. 1–8, 2015.

[12] Wu Q, Zhao X, and Zheng R, "Experimental Study on a Tuned-Mass Damper of Offshore for

Vibration Reduction," *J. Phys. Conf. Ser.*, vol. **744**, pp. 2–10, 2016.

[13] Stewart GM and Lackner MA, "The impact of passive tuned mass dampers and wind–wave misalignment on offshore wind turbine loads," *Eng. Struct.*, vol. **73**, pp. 54–61, Aug. 2014.

[14] Gao H, Kwok KCS, and Samali B, "Optimization of tuned liquid column dampers," *Eng. Struct.*, vol. **19**, no. 6, pp. 476–486, 1997.

[15] Colwell S and Basu B, "Tuned liquid column dampers in offshore wind turbines for structural control," *Eng. Struct.*, vol. **31**, no. 2, pp. 358–368, 2009.

[16] Coudurier C, Lepreux O, and Petit N, "Passive and semi-active control of an offshore floating wind turbine using a tuned liquid column damper," *IFAC-PapersOnLine*, vol. **28**, no. 16, pp. 241–247, 2015.

[17] Chatterjee T and Chakraborty S, "Vibration mitigation of structures subjected to random wave forces by liquid column dampers," *Ocean Eng.*, vol. **87**, pp. 151–161, 2014.

[18] Roderick C, "Vibration Reduction of Offshore Wind Turbines Using Tuned Liquid Column Dampers," Masters Thesis, 2012.

[19] Jaksic V, Wright C, Mandic DP, Murphy J, and Pakrashi V, "A Delay Vector Variance based Marker for an Output-Only Assessment of Structural Changes in Tension Leg Platforms," *J. Phys. Conf. Ser.*, vol. **628**, no. 1, p. 12059, 2015.

[20] Pakrashi V, O'Shea R, Jaksic V, and Murphy J, "The Hurst Exponent as an Indicator of the Behaviour of a Model Monopile in an Ocean Wave Testing Basin," *J. Phys. Conf. Ser.*, vol. **628**, no. 1, p. 12057, 2015.

[21] Jaksic V, Wright CS, Murphy J, Afeef C, Ali SF, Mandic DP, and Pakrashi V, "Dynamic response mitigation of floating wind turbine platforms using tuned liquid column dampers," *Philos. Trans. R. Soc. A Math. Phys. Eng. Sci.*, vol. **373**, no. 2035, pp. 20140079–20140079, 2015.

[22] O'Donnell D, Srbinovsky B, Murphy J, Popovici E, and Pakrashi V, "Sensor Measurement Strategies for Monitoring Offshore Wind and Wave Energy Devices," *J. Phys. Conf. Ser.*, vol. **628**, no. 1, p. 12117, 2015.

[23] Yalla SK and Kareem A, "Optimum Absorber Parameters for Tuned Liquid Column Dampers," *J. Struct. Eng.*, vol. **126**, no. 10, p. 1187, 2000.

IOP Conf. Series: Journal of Physics: Conf. Series **842** (2017) 012044 doi:10.1088/1742-6596/842/1/012044

Potentials of Optical Damage Assessment Techniques in Automotive Crash-Concepts composed of FRP-Steel Hybrid Material Systems

M Dlugosch[1*], B Spiegelhalter[1], T Soot[1], D Lukaszewicz[2], J Fritsch[1], S Hiermaier[1]

[1] *Fraunhofer Institute for High-Speed Dynamics, Ernst-Mach-Institut, EMI, Eckerstrasse 4, 79104 Freiburg, Germany*

[2] *BMW Group, Vehicle Safety, Knorrstrasse 147, 80788 Munich, Germany*

[*] Corresponding author, Email: michael.dlugosch@emi.fhg.de

Abstract. With car manufacturers simultaneously facing increasing passive safety and efficiency requirements, FRP-metal hybrid material systems are one way to design lightweight and crashworthy vehicle structures. Generic automotive hybrid structural concepts have been tested under crash loading conditions. In order to assess the state of overall damage and structural integrity, and primarily to validate simulation data, several NDT techniques have been assessed regarding their potential to detect common damage mechanisms in such hybrid systems. Significant potentials were found particularly in combining 3D-topography laser scanning and X-Ray imaging results. Ultrasonic testing proved to be limited by the signal coupling quality on damaged or curved surfaces.

1. Introduction

Hybrid material systems composed of advanced composite materials (CFRP / GFRP) and metals are one possible way to enhance multi material lightweight design by synergistically exploiting beneficial properties of the individual discrete material phases. Several authors have studied the crash behavior of such hybrid material systems on a component level with different specifications (see for example [1–3]). General results indicate, that hybrid material systems offer – although strongly dependent on the type and the architecture / arrangement of the parent materials – significant weight saving potentials of up to 38 % (compared to conventional materials) in energy absorbing structures [4] and advantages in other crashworthiness performance metrics such as the load uniformity [5]. However, the complexity and the interdependency of parametric effects strictly limit the comparability and transferability between the individual results.

Consequently, based on previous studies on coupon [6] and component [7] levels, the authors have tested composite-intensive hybrid material systems (ratio of cross sectional thicknesses of FRP to metal of higher than 1) in generic automotive structural subsystems under crash conditions as the final experimental program in a building block approach. Main goals of this program were to analyze the damage and failure mechanisms, to assess the potentials regarding performance and weight savings as well as to study the transferability of mechanisms from one stage to the next. Another aim was to generate a comprehensive set of data for the validation of numerical simulation models.

Critical to the performance of such material systems in vehicle-structural applications is the damage behavior of both constituent materials phases and the adhesive interface under crash loading conditions.

This behavior determines the overall crashworthiness of a structural concept composed of hybrid material systems, which in turn can be qualitatively assessed by evaluating the amount of brittle damage in the composite phase, interlaminar and interface delaminations, as well as the residual structural integrity. Non-destructive analysis techniques (NDTs) are commonly used in metal and composite structures for manufacturing and in-service quality or damage assessment [8, 9]. Since the increased research efforts on hybrid material systems such as the ones studied here are a quite recent development, NDT technologies have rarely been applied and evaluated with respect to the particular challenges inherent to such material systems. Droeder et al. have developed an automated ultrasonic device to test hybrid automotive structures for defects in the composite phase [10]. Although the architecture of the hybrid system is not clearly stated, they were able to detect dry areas and defects coupling the waves through an elastomer couplant and a medium to the FRP phase. Gieske and Rumsey used ultrasonic testing to assess the bonding quality in GFRP to metal adhesive interfaces in wind turbine applications [11]. When coupling the signal into the GFRP phase, the bond quality of the adhesive interface as well as defects within the composite phase were detectable. However, their approach using the through transmission technique (other techniques generated ambiguous results) limits the scope of applicability to structures that provide access to the relevant testing positions from both sides. Most studies focusing on X-Ray or computed tomography (CT) with hybrid material systems worked on highly integrative fiber-metal-laminates (FMLs). With composite-metal hybrid systems, the main challenge in X-Ray based analyses is to have enough energy for a total penetration of the metal layers and still obtain a reasonable contrast between the low density materials and defects / voids [12]. Further challenges in CT-analyses of such multi-material systems include severe artefacts stemming from beam hardening or scattering at the metallic parts [13].

In this study, the damage assessment approach is based on visual inspection, followed by a description and categorization of the individual damage mechanisms. In order to enhance this approach, different optical measurement technologies are applied and evaluated regarding their potential to provide deeper insights into the damage mechanisms, a more profound assessment of the overall performance and a larger scope of validation data. Eventually, understanding the damage mechanisms in such hybrid material systems could lead to improved structural designs with respect to crashworthiness and energy absorption.

2. Hybrid material vehicle subsystem crash tests

Chapter two intends to provide an overview of the experiments performed and the specimens tested. Additionally, a brief description of the damage mechanisms usually found in such hybrid structures is derived.

2.1. Subsystem design and crash test results

The test specimens contain three different components: one steel, one CFRP and one GFRP component. For the steel component a dual phase steel (DP-Steel) ("HCT600X+Z100" [14]) was chosen as it is commonly used for energy absorbing structures in BiW designs. For the manufacturing of the composite laminates fiber mats with an areal weight of 300 g/m² made of unidirectional 50k filament industrial grade carbon fiber and standard E-glass fiber, pre-impregnated with epoxy resin ("FT102" [15]) - so called prepregs - have been used. Both prepregs, the carbon fiber ("PREDO PR-UD CS 300/600 FT 102 38" [16]) and the glass fiber prepreg ("PREDO PR-UD EST 300/300 FT 102 35"[17]) were obtained from SGL epo GmbH. They exhibit a stiffness in the 0°-/fiber-direction of 120 GPa/40 GPa and strengths of 1409 MPa/1046 MPa. Joining of these three different material phases was achieved by using a layer of a two component epoxy adhesive ("BETAMATE 2096" [18]).

As shown in Figure 1, the specimen is composed of four beams all of which consist of varying cross sections and material systems. The sill (#4) is made of 12 layers GFRP [0°/±45°] and has a two-sided reinforcement made of 1.5 mm DP-Steel. The cross beam #3 consists of two 2.0 mm DP-Steel U-shaped profiles bonded by adhesive. The square hollow profiles in cross beams #2 is made of 12 layers and #1 is made of 8 layers with a [0°/±45°] layup. Both profiles feature single U-shaped 0.75 mm DP-Steel reinforcements as shown below.

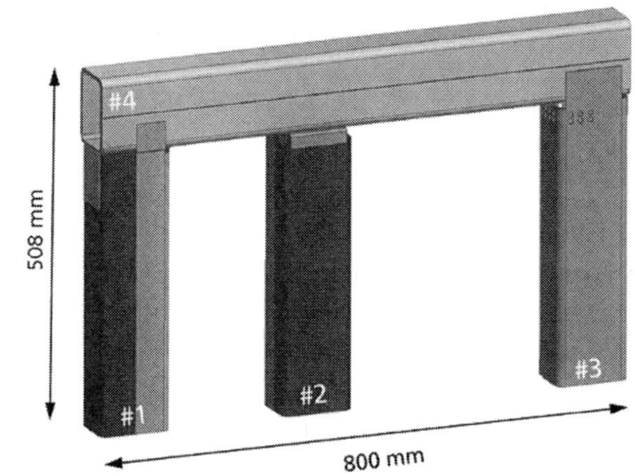

#1: 8 layers CFRP $[-45°/0°/45°/0°]_s$, Reinforced with 0.75 mm DP-Steel

#2: 12 layers CFRP $[-45°/0°/45°/0°/-45°/0°]_s$, Reinforced with 0.75 mm DP-Steel

#3: Pure 2.0 mm DP-Steel

#4: 12 layers GFRP $[-45°/0°/45°/0°/-45°/0°]_s$, Reinforced with 1.5 mm DP-Steel

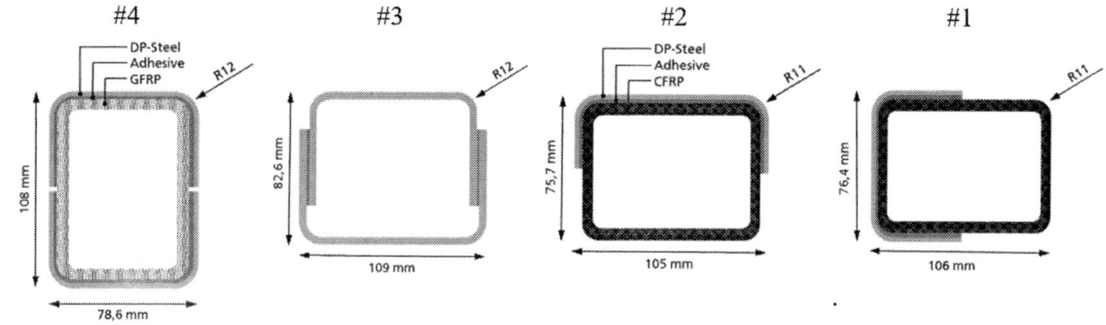

Figure 1. Cross sectional view of the four base parts of the subsystem specimen.

The experiments have been carried out at the Component Crash Test Facility at the Fraunhofer EMI Crash Center. The explosive release of pressure stored in a vessel accelerates a push rod and a sled, guided on a rail system. As the push rod hits a damping device, the sled uncouples from the system and impacts the specimen with a defined speed. The impact energy of 23.9 kJ results of an inpactspeed of 8.6 m/s (±0.1) and an impact mass of 647 kg. The displacement of the specimen and the speed of the sled were measured using a magnetostrictive measuring system. Three 3D-force sensors were installed under each crossbeam and were used to measure the impact force (on each section). Four high-speed video cameras were positioned at varying angles in order to document the damaging process.

Figure 2 shows the force-displacement plot of the central force sensor in impact direction (left) with the corresponding HS-video frames (right) at the indicated time steps.

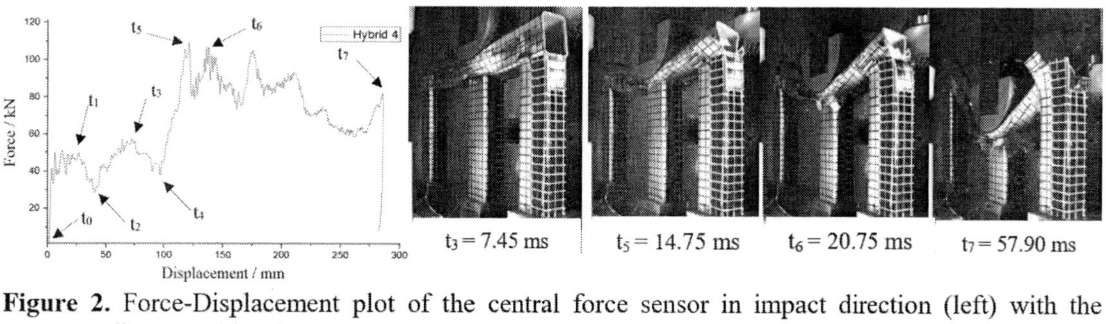

Figure 2. Force-Displacement plot of the central force sensor in impact direction (left) with the corresponding HS-video frames (right) at the indicated time steps.

At t_0 the impactor hits the specimen. The corresponding force signal shows an incline until the sill begins to buckle at t_1 until t_3. Between t_3 and t_4 the sill collapses (coinciding with a drop of the force signal). At t_4 the sill is fully compressed and the impactor reaches the intact CFRP-metal crossbeam #2 underneath the sill. Between t_5 and t_7 the CFRP part of the central crossbeam is crushed and the metal reinforcements are wrinkled. At t_7 the impactor rebounds.

2.2 Qualitative Damage assessment – characteristics and categorization

The damage mechanisms in hybrid material systems cover a broad spectrum defined through the individual constituents' and the respective interface damage mechanisms. Depending on the overall quality and quantity of the damage found in a structure, it is possible to derive more complex measures such as the structural integrity or residual load bearing capacity.

Table 1 depicts the scope of the damage mechanisms found in FRP-metal hybrid material systems under dynamic loading sorted by the respective discrete phases in the hybrid system. Steel is known to yield and buckle when surpassing specific tensile or compressive loads. These two mechanisms can lead to a periodic folding in dynamically crushed structures (see Alexander [19] for a detailed analysis of this mechanism) where yielding and local buckling at the hinges compels the material to fold forming periodic lobes. When the strain to failure is exceeded, ripping with sharp edges occurs.

Table 1. Overview of the scope of damage mechanisms in hybrid material systems under crash loading.

Hybrid Phase	Principal Damage Mechanism	Macro-level Damage
Steel	• Yielding / Necking • Local Buckling • Fracturing / Ripping	• Periodic Folding
Composite	• Interlaminar o Delamination • Intralaminar o Matrix fracture o Fiber fracture	• Progressive Crushing • Splaying
Adhesive	• Adhesive failure • Cohesive failure	• Phase Separation

In the brittle composite phase, the main damage mechanisms are divided into interlaminar delamination and intralaminar matrix and fiber fracture [20]. When subjected to dynamic axial compression, CFRP structures may exhibit a highly energy dissipative damage mechanism called "progressive crushing", which is caused by a complex interaction of the different microscopic damage mechanisms and frictional effects (see also Hull [21]). The adhesive bonding between the composite and the steel phase can fail adhesively at either of the two interfaces or cohesively within the adhesive layer itself. Both types result in a phase separation on a macro-level. Figure 3 depicts hybrid CFRP-steel components after testing under crash loading with an indication of the various damage mechanisms discussed above.

In order to derive integrative damage assessment measures on a structural level, all relevant mechanisms need to be assessed in terms of their quality and quantity. One of the most important structural measures is the structural integrity (SI). However, SI in automotive crash lacks a clear and concise definition. With respect to the field of application, different aspects, such as residual load bearing capacity, the absence of rupture or the integrity of critical components (e.g. fuel tank), are evaluated. According to Sahr, the material strength (no total failure) and overall capability of a component to remain in its original shape mainly define its SI [22]. Since brittle fracture is inevitable with composite structures, this definition does not apply to those material systems tested. Ferenczi et al. presented an approach to the SI assessment of CFRP automotive structures by proving that such structures – in contrast to metallic ones - still exhibit a significant load bearing capacity after major fracture occurs [23].

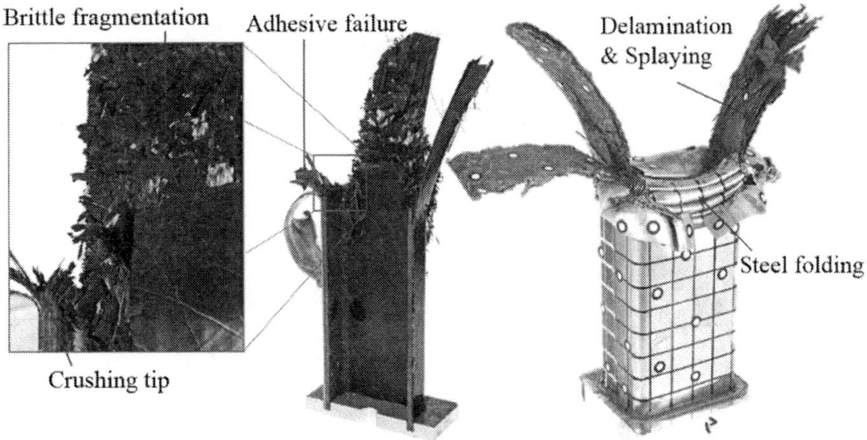

Figure 3. Damage mechanisms in FRP-Metal hybrid systems after dynamic testing.

However, relatively weak formulations such as "splintering should be minimized" and "a certain capacity is maintained" limit the transferability of their interesting work. Furthermore, they propose to apply NDT techniques in order to enhance the damage assessment and SI evaluation. Lukaszewicz extended the discussion on the SI of automotive CFRP structures and emphasized the need of the load paths' functional integrity, as well as all parts of the passenger compartment to be maintained. Local debonding of joints however, is acceptable [24]. Consequently, SI assessment here is also performed by discussing and evaluating several aspects, although the authors are aiming to work on more robust and comprehensive SI assessment methods in the future.

3. Optical Damage Assessment Techniques

The following chapters introduce the assessment NDT-technologies used including a brief technological background.

3.1. 3D-Laser Topography Scanning

Laser scanning is used to create digital three-dimensional (3D) models for applications related to reverse engineering and rapid prototyping by raster scanning of surfaces. A 3D-laser scanner consists of a laser sensor, which determines the distance between the laser source and the surface of the specimen using triangulation and a motion-tracking device, which tracks the orientation and position of the laser sensors itself or that of the object. The 3D-laser scanner used in the course of this work is the handheld device "HandySCAN 3D" (CREAFORM) which uses a stereo-photogrammetric system with optical reflectors for motion tracking. This system enables a dynamic registration of the scans. The software "VXelements 4" is used to control the scanning process and provides a real-time view of the generated 3D-data. Among the advantages of handheld laser scanners in terms of flexibility and scanning speed the performance of a 3D-Scanner is mainly influenced by its accuracy and resolution. In this context, accuracy means the extent of deviation of two physical points from their corresponding digital counterparts. The resolution of the 3D-model is defined by the scanning density, i.e. the number of detectable points in any given surface area. The used system scans with an accuracy of up to 0.030 mm and has a resolution of up to 0.050 mm [25].

3.2. Ultrasonic Analyses

Ultrasonic testing (UT) is based on the propagation and particular manipulation of short ultrasonic pulse-waves through the object or material to be tested. In the reflection or pulse-echo mode – as applied here

- the reflection of the pulses at various internal features, such as voids, defects or imperfections, is received at the same transmitting device. These reflections are caused by jumps in the acoustic impedance of the medium and can be qualified based on the amplitude (intensity of the reflection) and the travel time (depth of the reflecting feature), as schematically depicted in Figure 4. The equipment used is a handheld DolphiCam 1.3 (see [26]), which is a real-time imaging ultrasonic camera with a dry elastomer coupling pad particularly designed to test CFRP-material of up to 16 mm thickness, see Figure 5. The imaging frequency is 2 Hz.

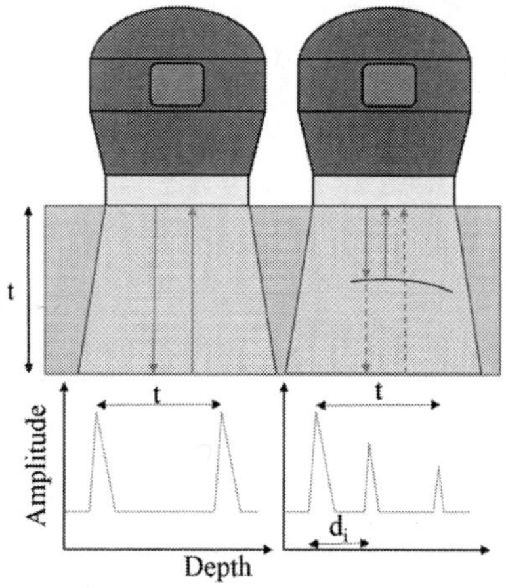

Figure 4. Schematic depiction of the working principle of the UT reflection mode.

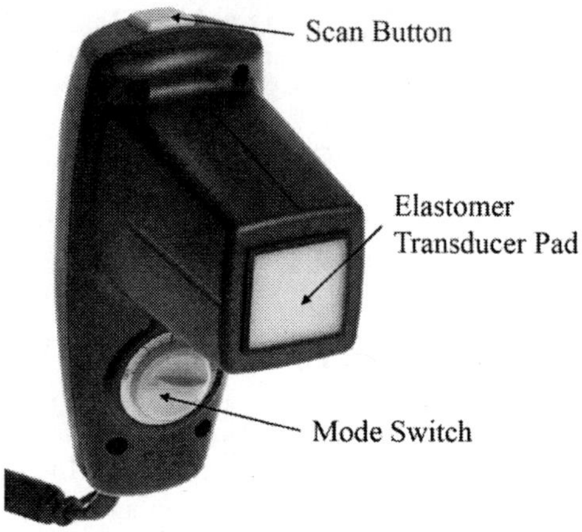

Figure 5. Handheld DolphiCam device used in this study [27].

The transducer is composed of 124 x 124 piezo-electric elements which transmit particularly designed pulses in order to cover a broad spectrum of energies and frequencies and receive the reflected mechanical signals, which in turn are being turned into electrical signals before being processed (e.g. filtered) and visualized. The software is capable to visualize A-Scans, horizontal and vertical B-Scans, C-Scans in the amplitude and the time-of-flight mode as well as 3D-views of the specimen's internal structure. By defining separate amplitude thresholds as "gates", it is possible to inspect particular thickness regions of the material. All values above or below a defined amplitude threshold will be assigned the same color value.

3.3. X-Ray Scanning

X-Ray scanning is a non-destructive high-quality imaging method, which is used to detect irregularities in solid materials. Using a point-shaped radiation source (X-Ray or gamma-ray) differences of the density in a material can be visualized on a digital X-Ray film by means of a projected image. Differences in the material thickness or density are represented by local variations of the film exposure. The thicker or denser the material is, the lower is the amount of radiation that can penetrate through it in a given time and thus the brighter the X-Ray film will appear. The industrial radiography has to deal with different peculiarities e.g. object structures in the direction of transmittance are superimposed as a result of the projective geometric mapping as well as the stronger weakening of X-Ray radiation by metals in comparison to plastics. Generally, the image quality is influence by contrast, blur and noise [28]. The self-constructed X-Ray system used in this project contains an YXLON FXE-225.48 X-Ray tube [29] and an PerkinElmer XRD1621 xN3 ES detector [30]. The X-Ray images were taken at an X-Ray tube voltage of 160 kV, a tube current of 350 mA and an exposure time of 1000 ms.

4. Results and discussion

The following sections introduce the results achieved with the respective damage assessment techniques and discuss the potentials and limitations regarding their application to the hybrid material systems tested.

4.1. 3D-Laser Topography Scanning

Figure 6 shows a photography of the central cross beam of the hybrid specimen taken 1 month after testing (left) and a 3D-Scan of the same specimen recorded directly after testing (right).

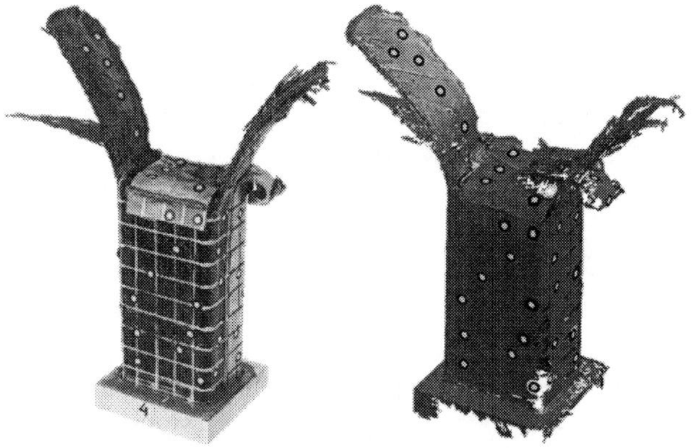

Figure 6. Central Cross beam #2 of the hybrid specimen. Photography of the specimen 1 month after testing (left); 3D-Scan directly after testing (right).

Due to the different recording times, there are small deviations between the photography and the 3D-Scan especially in the angle of the two CFRP fronds. The 3D-Scan has not been digitally altered except for its resolution which was set to its maximum of 0,03 mm. The surface of the digital specimen is shown in blue, the non-scanned backside of this surface in black. Clearly visible are the positioning targets for the 3D-Scanning (white circles with a black rim). There are some dark areas e.g. edges or the black lines on the reflecting steel reinforcement, which have not been captured during scanning. This can be corrected by repeated scans, by adjusting the shutter speed of the scanner during scanning or manually by software-based post processing. Remarkably, the scan captures the exact shape of the steel folds, the fiber orientation patterns on the left CFRP frond (45°) and single ply delaminations.

Figure 7 provides an overview of the performance and handling of 3D-Scans as well as possible applications. 1) shows a photography of the tested hybrid specimen and 2) depicts the related 3D-Scan of the hybrid specimen. The main advantages of this digital model compared to the physical one are the easier handling and the possibility to analyze the surface from different angles and picture sections without much effort. Furthermore, it is possible to directly compare the tested 3D-model with the untested CAD model as shown in 3). This procedure gives a clear visualization of the extent and type of damages in comparison to the untested model. This comparison also allows for a quantitative assessment of the damage by measuring the geometric deviation between the CAD model and the 3D-Scan of the tested specimen in each direction. The method proved particularly suitable for objects with low wall thicknesses where a section cut can provide insightful visualizations of the damage (see 4).

Local damage mechanisms such as the delamination between the GFRP and the reinforced steel or the detached adhesive are shown in 3a). Moreover, in 3b) the classic folding pattern of the steel reinforcement at the central cross beam is detected, as well as severe delaminations of the CFRP phase.

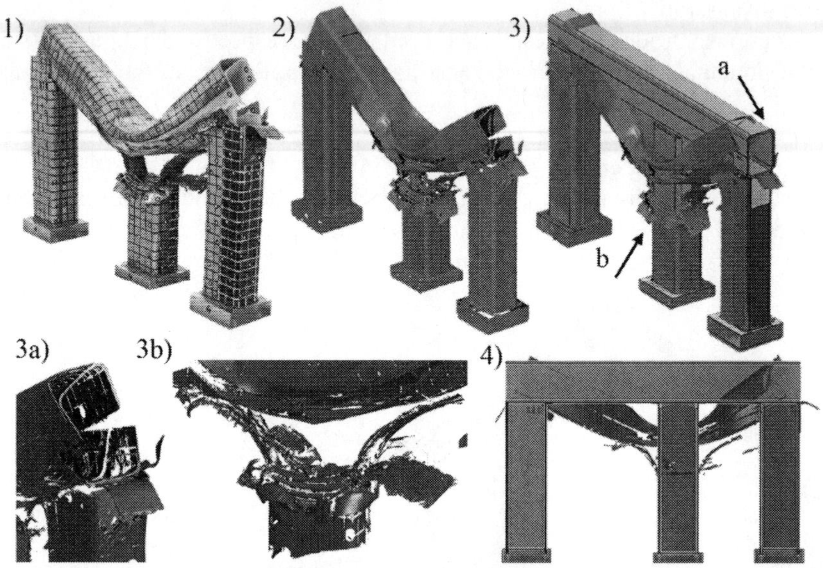

Figure 7. Hybrid Subsystem 1) Photography; 2) 3D-Scan; 3) Comparison between 3D-Scan and the CAD Model; 3a) local enlargement of the 3D-Scan of the damaged sill and torn adhesive connection 3b) local enlargement of the middle crossbeam below the sill 4) Visualization of a section through the symmetry axis of the model (comparison between 3D-Scan and CAD).

4.2. Ultrasonic Analyses

In order to study the sensitivity of the considered ultrasonic device, a representative test specimen is used. The CFRP test specimen consists of 16 plies with a [0/90]-layup and an overall thickness of 5 mm. It contains representative damages realized by blind holes of five different depths and diameters in each case. The residual material thickness varies between 1 mm and 4.5 mm, the blind hole diameters are 1 mm, 3 mm, 6 mm, 9 mm and 12 mm. The material sound velocity is adjusted to 2700 m/s. A setup with three gates, with corresponding threshold values for the amplitude is used (see [26]). Exemplary results are given in Figure 8 for the 3 mm and the 6 mm blind holes with a residual material thickness of 1 mm, 1.5 mm, 2.5 mm, 3.5 mm and 4.5 mm.

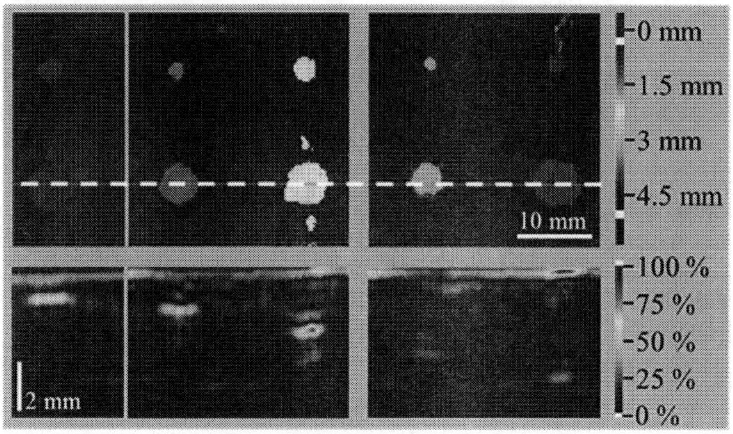

Figure 8. C-Scan / B-Scan of the 3 mm and the 6 mm blind holes.

The analyses show, that the 1 mm blind holes are not reproducibly identifiable with the device. However, all the other damages are determinable and even the slight variation between 4.5 mm residual material thickness and the undamaged area is clearly detected. An analysis beyond the CFRP phase (e.g.

adhesive bond quality) is limited by the echoes from the back surface, which are not distinguishable from the impedance gap towards the adhesive or air (voids). The applicability in the context of the considered subsystems is analyzed in further measurements depicted in Figure 9.

Figure 9. Damaged beam (left) and related C-Scan / B-Scan (right).

For this purpose, the damaged beam is examined in the highlighted delaminated area. The related C-Scan and B- Scan clearly indicate a damaged area (delamination) on the right hand side at a depth of approximately 3.5 mm. This proves, that the pulse-echo based device is generally applicable to the detection of such damage in the hybrid material systems studied here. However, remaining critical issues are the accessibility, the surface finish and shape as well as the resulting coupling. Coupling the signal into the steel phase results in a dominant echo before entering the adhesive phase, which effectively inhibits deeper analyses. This limits the technology to be applied to CFRP surfaces, which in turn need to be accessible to the device. Furthermore, in order to achieve good coupling, the CFRP surface needs to be smooth and without any major curvature.

4.3. X-Ray Scanning

For reasons of limited X-Ray detector size, Figure 10 is composed of several X-Ray images of the tested hybrid specimen separately recorded and post-processed (indicated by the dashed lines). For the optical 3D-scanning approximately 300 round reflecting targets have been glued on the tested specimen. Those targets are also visible in the X-Ray image 1). The highlighted areas 2), 3), and 9) show delaminations within the GFRP and inside the adhesive bond to the steel reinforcement. These delaminations might also be confused with successive folds or indentations which are superimposed along the path of the X-Ray beam. To ensure a sufficiently strong adhesive layer, rectangular metal spacers have been installed between the two components. The metal adhesive spacers are marked with 4). The [0°/±45°]-laminate of the CFRP can be recognized in the non-reinforced crossbeam depicted at 5). The highlighted areas 6) and 8) show severe delamination and splaying of the CFRP composite after crushing. In the marked area 7) the progressively folded metal is shown. In 10) single fibers and composite-fragments/debris inside the square tube are visible. Further analyses are conducted to check whether 11) depicts the adhesive bonding interface or a gap caused by delamination between the two bonded sheets. In terms of structural integrity it can be seen that based on the image in Figure 10, no major voids or gaps can be identified in the transition from beam #1 to #3 and #4. It can thus be assumed that a certain residual load carrying capacity is maintained. However, if the integrity of a loadpath through beam #2 is crucial, the overall SI is clearly violated.

Consequently, X-Ray imaging provides deep insights into the various damage mechanisms described in Table 1, some of which are more accessible than others. Needless to say, a computed tomography scan resulting in a sliceable volume representation could clearly enhance the possibilities of damage assessment and simulation data validation.

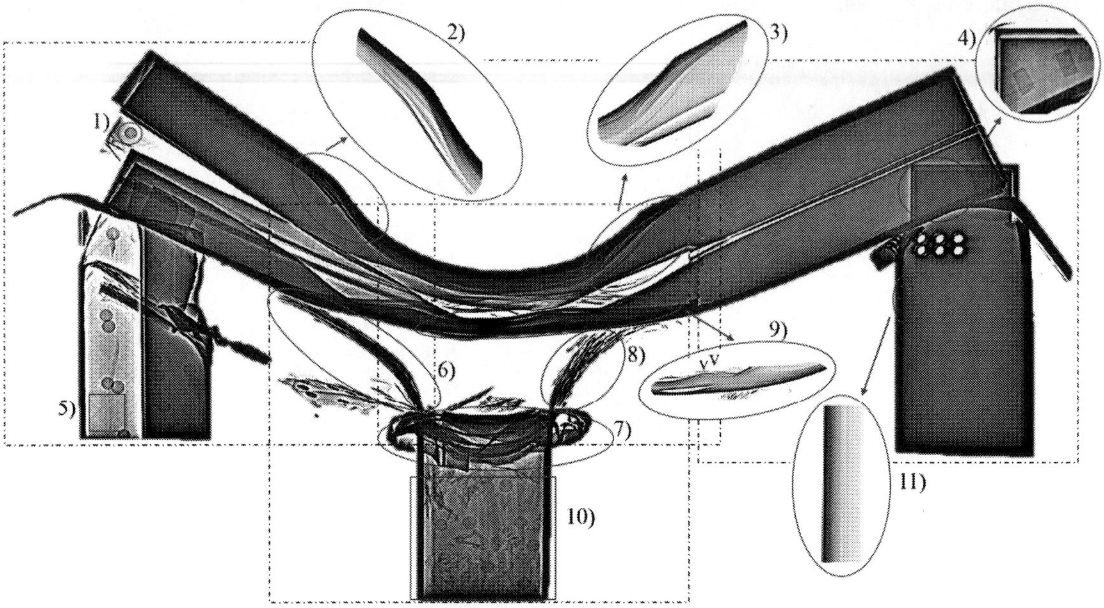

Figure 10. Composed X-Ray image with highlighted areas of hybrid damage mechanisms.

5. Conclusions and future work

Generic automotive structural concepts composed of FRP-metal hybrid material systems have been tested under crash loading conditions and the major damage mechanisms usually found in such systems were discussed. Several NDT techniques have been applied in order to evaluate their potentials to enhance the damage assessment methods regarding such hybrid structures. The 3D-scanning proved to be an effective technique to generate a high-resolution topology representation of the structure covering various outer damaging modes such as FRP delamination, splaying and metal folding. A major advantage is the possibility to virtually validate simulations results interactively using the surface model. Obviously, the technique is limited to superficial damage mechanisms. Ultrasonic testing has shown to be quite difficult, since sufficient coupling of the signal into the structure is merely achieved at smooth and straight FRP surfaces. Damage (such as delamination or voids) within the FRP phase may be detected but the process generally requires good accessibility and a significant amount of expertise. X-Ray scanning as well implies user know-how at setting the filters and post-processing parameters. The resulting damage detection results however, provide significant insights into the structures external and internal damage mechanisms. Various FRP and steel damage types are detected and clearly visible and distinguishable in the composed X-Ray image. Particularly when combining the respective techniques, a good overview of the general state of damage within the structure may be achieved. Based on those results, a proper validation of numerical models (e.g. in terms of damage simulation) and a comprehensive understanding of the damage mechanisms might lead to improved designs of hybrid structural concepts for automotive crash applications.

Future work will focus on the potentials of computed tomography techniques which provide a virtual volume representation of the structure. Given the possibility to slice that volume and analyze internal as well as external damage mechanisms, this might be an approach synergistically combining the advantages of the NDT techniques studied here.

6. Acknowledgements

The authors would like to express their gratitude to their colleagues at the Fraunhofer EMI and at BMW for valuable discussions and support.

7. References

[1] Wang, X. G., Bloch, J. A., and Cesari, D. 1992. Static and dynamic axial crushing of externally reinforced tubes. *ARCHIVE: Proceedings of the Institution of Mechanical Engineers, Part C: Journal of Mechanical Engineering Science 1989-1996 (vols 203-210)* 206, 53, 355–360.

[2] Bambach, M. R. 2013. Fibre composite strengthening of thin-walled steel vehicle crush tubes for frontal collision energy absorption. *Thin-Walled Structures* 66, 15–22.

[3] Mildner, C. 2013. *Numerische und experimentelle Untersuchungen des Crashverhaltens von FVK-verstärkten Metallstrukturbauteilen*. Dissertation, Technische Universität München.

[4] Kim, H. C., Shin, D. K., Lee, J. J., and Kwon, J. B. 2014. Crashworthiness of aluminum/CFRP square hollow section beam under axial impact loading for crash box application. *Composite Structures* 112, 1–10.

[5] Dlugosch, M., Lukaszewicz, D., Fritsch, J., and Hiermaier, S. 2016. Experimental investigation of automotive Experimental investigation of automotive components consisting of hybrid FRP-metal-material systems under dynamic loading. In *Proceedings of the 1st International Conference on Impact Loading of Structures and Materials*.

[6] Dlugosch, M., Lukaszewicz, D., Fritsch, J., and Hiermaier, S. 2016. Experimental investigation of hybrid material systems consisting of advanced composites and sheet metal. *Composite Structures,* 152, 840–849.

[7] Dlugosch, M., Lukaszewicz, D., Fritsch, J., and Hiermaier, S. 2016. Experimental Investigation of Automotive Components consisting of hybrid FRP-Metal-Material Systems under Dynamic Loading. In *ECCM 17. Proceedings of the 17th European Conference on Composite Materials*.

[8] Hellier, C. 2003. *Handbook of nondestructive evaluation*. McGraw-Hill, New York.

[9] Gholizadeh, S. 2016. A review of non-destructive testing methods of composite materials. *Procedia Structural Integrity* 1, 50–57.

[10] Dröder, K., Dietrich, F., Fricke, A., and Gerhard, C. 2016. Automated Non-destructive Testing of Hybrid Structures. In *19th World Conference on Non-Destructive Testing*.

[11] Gieske, J. H. and Rumsey, M. A. 1997. *Nondestructive testing (NDE) of Composite-to-Metal Bond Interface of a Wind Turbine Blade Using An Acousto-Ultrasonic Technique*, Albuquerque, USA.

[12] Leonard, F., Shi, Y., Soutis, C., Withers, P. J., and Pinna, C. 2014. Impact damage characterisation of fibre metal laminates by X-ray computed tomography. In *International Conference on Industrial Computed Tomography (ICT)*.

[13] Gusenbauer, C., Reiter, M., Salaberger, D., and Kastner, J. 2016. Comparison of Metal Artefact Reduction Algorithms from Medicine Applied to Industrial XCT Applications. In *19th World Conference on Non-Destructive Testing*.

[14] Salzgitter Flachstahl. 2013. HCT600XD (HC340XD*/ CR330Y590T-DP**) Mehrphasenstähle zum Kaltumformen – Dualphasenstähle (May. 2013).

[15] Swiss-composite. 2008. Epoxyd Prepreg FT 102 - technical datasheet (Apr. 2008).

[16] SGL Group. 2014. *PREDO PR-UD CS 300/600 FT102 38*. Technical Data Sheet, Willich.

[17] SGL Group. 2014. *PREDO PR-UD EST 300/300 FT102 35*. Technical Data Sheet, Willich.

[18] Dow Automotive AG. Betamate 2096 - Technisches Datenblatt 09.12.2005.

[19] Alexander, J. M. 1960. An approximate analysis of the collapse of thin cylindrical shells under axial loading. *Q J Mechanics Appl Math* 13, 1, 10–15.

[20] Greve, L. 2005. *Damage and Failure Modeling of carbon/epoxy Non Crimp Fabric Composites*. Dissertation, Cranfield University.

[21] Hull, D. 1991. A unified approach to progressive crushing of fibre-reinforced composite tubes. *Composites Science and Technology* 40, 4, 377–421.

[22] Sahr, C. 2011. *Methodische Vorgehensweise der Werkstoffauswahl für die Karosserieentwicklung in Multi-Material-Bauweise*. Dissertation, RWTH Aachen University.

[23] Ferenczi, I., Kerscher, S., and Möller, F. 2013. Energy Dissipation and Structural Integrity in Frontal Impact. In *The 23rd ESV Conference Proceedings*.

[24] Lukaszewicz, D. 2017. *Verbundwerkstoffe im Crash*, Konstanz.
[25] Creaform Inc. Handy Scan 3D - technical datasheet 2016.
[26] Lingvall, F. and Skoglund, E. 2015. *DoplhiCamTM 1.3 Technical Paper*.
[27] Payaro, a., Skoglund, E., and Endrerud, J. O. 2016. *DoplhiCam User Manual*. DolphiTech.
[28] Bauer, N., Ed. 2007. *Handbuch zur industriellen Bildverarbeitung. Qualitätssicherung in der Praxis*. Vision. Fraunhofer-IRB-Verl., Stuttgart.
[29] COMET GmbH. X-Ray Sources - FXE Series - technical data sheet 22.03.2007.
[30] PerkinElmer. Flat Panel X-ray Detector - XRD1621 xN ES - Product note 2014.

Topic Correlation Analysis for Bearing Fault Diagnosis Under Variable Operating Conditions

Chao Chen[1], Fei Shen[1]and Ruqiang Yan[1,*]

[1]School of Instrument Science and Engineering, Southeast University
Nanjing, Jiangsu 210096, China

[*]Email: ruqiang@seu.edu.cn

Abstract. This paper presents a Topic Correlation Analysis (TCA) based approach for bearing fault diagnosis. In TCA, Joint Mixture Model (JMM), a model which adapts Probability Latent Semantic Analysis (PLSA), is constructed first. Then, JMM models the shared and domain-specific topics using "fault vocabulary". After that, the correlations between two kinds of topics are computed and used to build a mapping matrix. Furthermore, a new shared space spanned by the shared and mapped domain-specific topics is set up where the distribution gap between different domains is reduced. Finally, a classifier is trained with mapped features which follow a different distribution and then the trained classifier is tested on target bearing data. Experimental results justify the superiority of the proposed approach over the stat-of-the-art baselines and it can diagnose bearing fault efficiently and effectively under variable operating conditions.

1. Introduction

Rolling bearings, as one of the most important components in rotary machines, are widely used in modern manufacturing systems. However, faults occurring in bearings could cause dreadful consequences such as performance deterioration, costly downtime, production delay or even personnel casualties [1]. Therefore, detecting the faults occurring in bearings as early as possible can not only ensure reliability, safety and good operation of mechanical equipment, but also have the ability to avoid unnecessary losses. With the development of fault diagnosis in rolling bearings, various methods [2-4] have been employed to measure certain physical quantities, including temperature monitoring, oil analysis, vibration signal analysis, and etc., which serve as a crucial preprocessing step for fault diagnosis. Among them, the use of vibration signal analysis has been proven to be quite common, practical and effective. In many existing researches [5-6], the fault diagnosis of bearings can be treated as a problem of pattern recognition solved by machine learning-based algorithms using extracted features from vibration data. However, in many applications, bearings are operated under complex working conditions [7], such as high speed, variable load, high temperature and other complex conditions, effective fault diagnosis for bearings is challenging.

Nowadays, most studies of bearing fault diagnosis focus on a specific working condition(e.g., in laboratory or under controlled conditions), where traditional machine learning methods (e.g., back propagation neural network, transductive support vector machine, K-means) are proved to be effective using labeled, unlabeled data or both of them[8-10]. However, successful utilization of all these methods requires that the training and testing data are drawn from the same distribution (e.g., vibration data for training and testing process are measured from the same set of experiments) [11]. In many real-world applications, bearings are working under variable operating conditions, where there exist various speeds, changing load as well as some unknown jamming, there is no guarantee that the training and testing data are drawn from the same distribution. When the distribution changes, it is common to rebuild the diagnostic model using sufficient newly labeled data. But it is sometimes time-consuming or expensive to obtain vibration data of the target bearing to train a precise classifier, because the installation position of bearing is inside some unable dismantled parts, thus the performance of traditional machine learning methods in terms of bearing fault diagnosis under complex working conditions would come out to be poor. In contrast, transfer learning, a new machine learning approach with ability to recognize and apply knowledge and skills learned in previous tasks to new tasks, has attracted more and more attention since 1995 [12-13]. It is a good solution for fault diagnosis in complex situations, where the training and test data are drawn from different distributions.

Inspired by this, the paper proposes a novel bearing fault diagnosis strategy based on Topic Correlation Analysis (TCA). In TCA, Joint Mixture Model (JMM), a model which adapts Probability Latent Semantic Analysis (PLSA), is constructed first. As JMM is originally designed well for text classification, the following modifications have been taken to ensure success in applying JMM to fault diagnosis: (1) view a time series of vibration data as a document consisted of a number of "fault word", and take energy distribution as well as load, speed, and other parameters related to vibration data as the topics embedded in the document; (2) extract frequency-domain features and time-domain features from vibration data; (3) apply K-means to generate a set of "fault word" using extracted features. Then, JMM models the shared and domain-specific topics using "fault word". After that, the correlations between two kinds of topics are computed and used to construct a mapping matrix. Furthermore, a new shared space spanned by the shared and mapped domain-specific topics is built where the distribution gap between different domains is reduced. Finally, a classifier is trained with mapped features which follow a different distribution and then the trained classifier is tested on target bearing.

The rest of the paper is organized as follows. In section 2, the principle of Topic Correlation Analysis (TCA) along with its modification, applied in bearing fault diagnosis, is introduced. Then experimental study using bearing data set is carried out in Section 3, followed by the conclusions in Section 4.

2. TCA-based transfer learning approach for bearing fault diagnosis

In practice, classification algorithm based on machine learning plays an important role in fault diagnosis. But a basic assumption of traditional machine learning is that the training and testing data follow the same distribution. However, when it comes to bearing fault diagnosis under variable conditions, the distributions of training and testing data do not always satisfy the assumption. In this case, transfer learning will become an ideal tool for bearing fault diagnosis.

Here an improved algorithm, named enhanced TCA, which can reduce the data difference between the two data samples, is put forward for fault diagnosis. TCA is based on the model of Probabilistic Latent Semantic Analysis (PLSA) [14], and the principle of PLSA is introduced first.

2.1. A Brief Review of PLSA

PLSA, originating from latent semantic analysis (LSA), has been widely used for document clustering and related tasks. Given a set of documents $D=\{d_1,d_2,\ldots,d_N\}$, words $W=\{w_1,w_2,\ldots,w_M\}$ and latent topic variables $Z=\{z_1,z_2,\ldots,z_K\}$. A generative model for document co-occurrences, as is shown in Figure 1, can be defined by the following scheme:

- select a document d_i with probability $P(D = d_i)$,
- draw a topic z_k with probability $P(Z = z_k \mid D = d_i)$,
- select a word w_j with probability $P(W = w_j \mid Z = z_k)$.

Where $P(d_i)$ denotes the probability that a word occurrence will be observed in a particular document d_i , $P(w_j \mid z_k)$ denotes the class-conditional probability of a specific word conditioned on latent topic variable z_k , and $P(w_j \mid d_i)$ represents a document specific probability distribution over the latent variable space.

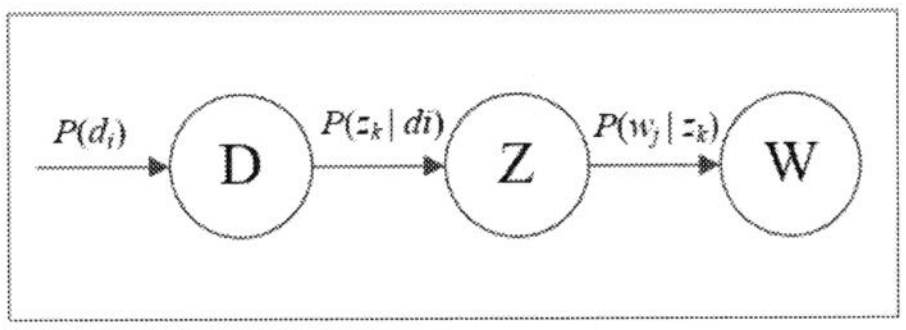

Figure 1. Graphical representation of Probabilistic Latent Semantic Analysis (PLSA) model.

Then $P(d_i, w_j)$ can be obtained according to Bayesian formula

$$P(d_i,w_j)= P(d_i)\, P(w_j \mid d_i) \tag{1}$$

Where

$$P(w_j \mid d_i) = \sum_{k=1}^{K} P(z_k \mid d_i) P(w_j \mid z_k) \tag{2}$$

Then, the Expectation-Maximization (EM) algorithm, which is suitable for estimating model parameters of latent variable model [15], is applied. The aim of EM is to find suitable $P(z_k \mid d_i)$ and $P(w_j \mid z_k)$ by maximizing objective function

$$L = \sum_{i=1}^{N} \sum_{j=1}^{M} n(d_i, w_j) \log P(d_i, w_j) \tag{3}$$

where $n(d_i, w_j)$ denotes the number of times that word w_j occurs in document d_i.

PLSA model is originally designed to address the semantic problem caused by synonyms and polysemy in document clustering task. Similarly, in the field of fault diagnosis, fault features, equivalent to words in document categorization, also have the characteristics of synonyms and polysemy. For example, there must be a number of commonalities between features extracted from the same fault type under different working conditions because of the same root cause of failures, which is similar to synonyms in text. At the same time, due to these commonalities, the extracted features can not only represent their own fault category, but also have a certain contribution to the clustering of the fault type under another working condition, under this circumstance, these fault features can also be seen as polysemy.

In document clustering task, vectors which are represented by word frequency are used to stand for documents. The latent 'bag-of-words' is used to discovery the topics of documents. Thus, in fault diagnosis, if one fault type is viewed as a document consisted of a number of "fault word", and energy distribution as well as load, speed, and other parameters related to vibration data is taken as the latent topics embedded in the document, then the probability distribution of the latent topic can be found with PLSA. Specifically, given a set of vibration signal $D=\{d_1,d_2,\ldots,d_N\}$, some modifications are listed as follows:

(1) Extracting time-domain features and frequency domain features respectively from the vibration signal, then utilizing k-means to generate N fault words

(2) Computing the similarity between each fault word and the features extracted from each vibration signal, then co-occurrence matrix $n(d_i, w_j)$ is obtained, where $n(d_i, w_j)$ denotes the similarity between fault word w_j and vibration signal d_i.

Then, the rest steps are the same as traditional PLSA used in document clustering task.

2.2. TCA model

Before discussing TCA, some basic notations are listed in Table 1.

Table 1. Notations

Symbols	Description
l	domain random variable, $l\in\{t, s\}$
D^l	fault type random variable
Z	topic random variable
π	decision random variable, $\pi\in\{0,1\}$
W	feature word of fault information
d_n^l	nth fault type in domain l
w_j	jth fault word
z_k	kth shared topic
z_r^l	rth domain-specific topic in domain l
$\mu_{d_i^l}$	fault type-level parameter for the decision process
K	number of the shared topics
K^l	number of the domain-specific topics in domain l

The main idea of TCA can be summarized into three parts:

Part I. Modelling both Shared and Domain-Specific Topics

In this part, Joint Mixture Model(JMM), which adapts PLSA, is used to extracting both shared and domain specific topics. In JMM, as is shown in Figure 2 (a), the co-occurrence between a fault sample and a fault word is associated with a latent topic $Z\in\{z_1,...,z_K,z_1^s,...,z_{K^s}^s,z_1^t,...,z_{K^t}^t\}$ which is drawn from either the shared topics, i.e. $\{z_1,...,z_K\}$, or the domain-specific topics, i.e. $\{z_1^s,...,z_{K^s}^s,z_1^t,...,z_{K^t}^t\}$. A latent decision variable $\pi\in\{0,1\}$is used to control which kind of topic is drawn. The generative process for co-occurrence in JMM is defined as:

(1) Selecting a kind of fault sample d_i with probability $P(D = d_i)$

(2) Drawing a decision random variable $\pi \sim \text{Bern}(\mu_{d_i^l})$:if $\pi = 0$, a shared topic z_K with probability $P(Z = z_k| D = d_i, \pi=0)$ is chose, and a fault word w_j with probability $P(W = w_j| Z = z_k)$ is generated; if $\pi = 1$, a domain-specific topic Z_r^l with probability $P(Z=z_r^l\,|\,D^l=d_i^l,\pi=1)$ is chosen, and a fault word w_j with probability $P(W=w_j\,|\,Z=z_r^l)$ is generated.

(3) The joint probability model for the data generative process is described as:

$$P(d_i^l,w_j)=P(d_i^l)P(w_j\,|\,d_i^l) \tag{4}$$

where

$$\begin{aligned}P(w_j\,|\,d_i^l)=&P(\pi=0\,|\,\mu_{d_i^l})\sum_{k=1}^{K}P(z_k\,|\,d_i^l,\pi=0)P(w_j\,|\,z_k)\\&+P(\pi=1\,|\,\mu_{d_i^l})\sum_{r=1}^{K^l}P(z_r^l\,|\,d_i^l,\pi=1)P(w_j\,|\,z_r^l)\end{aligned} \tag{5}$$

The equation (5) related matrix decomposition is illustrated in Figure 2 (b). Then EM algorithm is used to estimate model parameters of the latent variable model by maximizing following objective function

$$L=\sum_{l\in\{s,t\}}\sum_{d_i^l\in D^l}\sum_{w_j\in W}n(d_i^l,w_j)\log P(d_i^l,w_j) \tag{6}$$

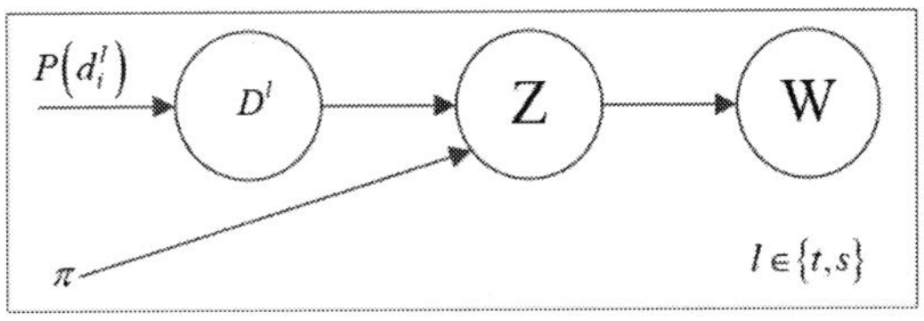

(a)

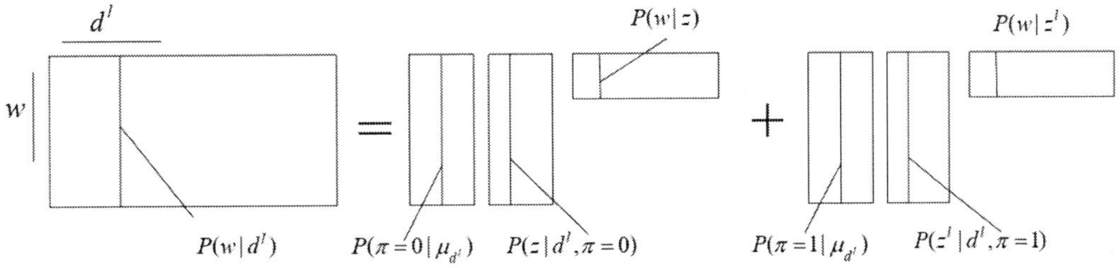

(b)

Figure 2 : (a) JMM graphical model; **(b)** Diagram of matrix decomposition

Part II. Measuring Topic Correlations across Domains

Correlations between domain-specific topics from different domains can be measured as follows:

(1) Calculating the similarity $\theta_{z_i^l,z_k}$ between the shared and the domain specific topics in each domain with the vectorial angle cosine method

(2) The Pearson's Correlation Coefficients (PCCs) [16] is adopted to infer the correlations between the domain-specific topics from different domains. With PCCs, a topic mapping matrix $U=\{\rho(z_i^t,z_j^s)\}$, where $i=1,...,K^t$, $j=1,...,K^s$, $U\in R^{K^t\times K^s}$, can be obtained as

$$\rho(z_i^t,z_j^s)=\frac{\sum_k\left(\theta_{z_i^t,z_k}-\bar{\theta}_{z_i^t}\right)\left(\theta_{z_j^s,z_k}-\bar{\theta}_{z_j^s}\right)}{\sqrt{\sum_l\left(\theta_{z_i^t,z_k}-\bar{\theta}_{z_i^t}\right)^2}\sqrt{\sum_l\left(\theta_{z_j^s,z_k}-\bar{\theta}_{z_j^s}\right)^2}} \tag{7}$$

where $\bar{\theta}_{z_i^l}$ represents the mean similarity between z_i^l and all the shared topics.

Part III. Constructing New Representation for Documents

Here both shared and the mapped domain-specific topics are utilized to construct a new shared space, which can be described as follows:

(1) Representing fault samples $\{d_i^l\}_{l,i}$ from both domains with the extracted topics:

$$\phi(d_i^l)=\left[P(z_1|d_i^l),...,P(z_K|d_i^l),P(z_1^l|d_i^l),...,P(z_{K^l}^l|d_i^l)\right]^T \tag{8}$$

where $\phi(d_i^l)_{[1:K]}$ represents domain-shared part, and $\phi(d_i^l)_{[K+1:K^l]}$ stands for domain-specific part.

(2) Transforming the domain-specific part $\phi(d_i^t)_{[K+1:K+K^t]}$ with the topic mapping matrix U:

$$\varphi(\phi(d_i^t))=\left[\phi(d_i^t)_{[1:K]};U^T\phi(d_i^t)_{[K+1:K+K^t]}\right] \tag{9}$$

The new representation $\varphi(\phi(d_i^t))$ is composed of shared topics and mapped source domain-specific topics. Then, each fault sample in the target domain is mapped into the new feature space spanned by $\varphi(\phi(d_i^t))$.

IOP Conf. Series: Journal of Physics: Conf. Series **842** (2017) 012045 doi:10.1088/1742-6596/842/1/012045

2.3. Overview of Bearing Fault Diagnosis Based on Transfer Learning

The enhanced TCA algorithm can map the feature data that under different working conditions to a new shared space, where the difference between the data will be reduced. Therefore, the bearing fault diagnosis problem under variable conditions can be solved using this transfer learning technology. The complete process of the proposed method is summarized in Algorithm 1.

Algorithm 1: Bearing Fault Diagnosis Based on Enhanced TCA

Input: (1) Source domain labeled vibration signal and target domain unlabeled vibration signal; (2)Number of shared topics K; (3) Number of source/target domain-specific topics K^s, K^t;(4) Number of iterations T
Output: The predicted class label of each target unlabeled vibration signal
Extract time-domain features and frequency-domain features from inputted vibration signal.
Apply K-means to generate a set of "fault word" from the extracted features.
Initialize the parameters of JMM model;
for t←1 to T do
    EM algorithm
end
Measure the topic correlations;
Construct the topic mapping matrix U;
Represent each fault sample with extracted topics;
Transform each fault sample in the target domain;
Train a classifier with the labeled source domain data $D^s = \left\{\left(\phi\left(d_n^s\right), y_n^s\right)\right\}_{n=1}^{N^s}$ and predict the class labels of unlabeled target domain data $D^t = \left\{\left(\phi\left(d_i^t\right)\right)\right\}_{i=1}^{N^t}$

---

3. Experimental study and performance analysis

3.1. Experimental setup

To verify the effectiveness of the presented approach in bearing fault diagnosis, the bearing data set contributed by Western Reserve University Bearing Data Center was analyzed. As is shown in Figure 3, the experiment system consists of a 2 hp motor (left), a torque transducer/encoder (center), a dynamometer (right), and control electronics (not shown). The model of drive end bearing is SKF6205, and the model of fan end bearing is SKF6203.

Vibration signals were measured using a 16 channel DAT recorder with12 kHz sample frequency from the test system under variable speeds and loading conditions. Data were collected at 12,000 samples per second. Speed and horsepower data were collected using the torque transducer/encoder and were recorded by hand. Different bearing working conditions are listed in Table 2.

Table 2. Bearing test condition descriptions.

Experiment Condition	**1**	**2**	**3**	**4**	
Motor Speed (RPM)/ Motor Load (HP)	1797/ 0	1772/ 1	1750/ 2	1730/ 3	——
Fault Diameter (mm)	0.1778	0.3556	0.5334	0.7112	——
Fault Type	Inner Race	Ball	Outer Race Relative to Load Zone (Load Zone Centered at 6:00)		
			6:00	3:00	12:00

Figure 3. Experiment setup

3.2. Experimental results and analysis

Firstly, we extracted 8 time-domain features and 8 frequency-domain features respectively, and then a total of 16 features are obtained. In this study 200 data samples with 16 features from each fault type under each working condition were obtained, thus, the size of original target and source sample feature matrix is 1600*16. It should be pointed out that the training sample and testing sample belong to different working conditions.

To test the effectiveness of enhanced TCA for bearing fault diagnosis, a comparison study is performed between enhanced TCA and TCA_{share}, which only uses the shared topics to represent fault samples for classifier training. Furthermore, the enhanced TCA is also compared with principal component analysis (PCA), kernel principal analysis (KPCA), factor analysis (FA) and locally linear embedding (LLE) for feature selection. The mapped source data set are fed into SVM classifier for training, then the target data are used to test the trained model.

Two cases are studied in this paper: transfer among drive end bearings, and transfer between fan end bearing and drive end bearing. The classification results are shown in Tables 3 and 4, where the notation, for example D1797R1-D1797R2, is explained as follows: the left represents source data (for training only), the right is target data(used for testing), and 1797 represents the speed, R1, R2, R3, R4 stand for fault diameter with 0.1778mm, 0.3556mm, 0.5334mm, 0.7112mm respectively. The symbols F and D represent fan end bearing and drive end bearing respectively.

Table 3. Comparison among enhanced TCA and other algorithms for diagnosis accuracy under different working conditions using drive end bearing data

	Accuracy (%)					
	D1797R1 - D1797R2	D1797R1 - D1797R3	D1772R1 - D1775R1	D1772R1 - D1730R1	D1772R2 - D1730R2	Average
None	80.75	83.5	72.5	78	75.25	78
PCA	85.13	81	80	80	76.5	80.53
KPCA	81.5	81.5	77	79	75.63	78.93
LLE	77.5	79	82.13	80	79	79.53
FA	84	81.38	78	64	86.5	78.78
TCA_{share}	92	89	85	89	84.88	87.98
TCA	96.25	94.5	92.5	91	93.5	93.55

Table 4. Comparison among enhanced TCA and other algorithms for diagnosis accuracy using fan end bearing and drive end bearing data

	Accuracy (%)					
	F1797R1 - D1797R2	F1797R1 - D1797R3	F1772R1 - D1775R1	F1772R1 - D1730R1	F1797R2 - D1730R2	Average
None	71.63	84.5	71.5	78	71.5	75.43
PCA	86	80	81	81.75	79.88	81.73
KPCA	79.75	77.88	75.75	71	74	75.68
LLE	77.38	77	78	76.13	77.5	77.2
FA	80	80.5	82	74.13	76.5	78.63
TCA_{share}	90.5	85.63	85.5	90.25	80.5	86.48
TCA	92	95.75	96	90	92	93.15

The following conclusions can be obtained from Tables 3 and Table 4:

1) Using original features without any feature selection techniques as the input of SVM has the lowest performance.

2) The classification accuracy of traditional machine learning algorithms (PCA, KPCA, LLE and FA) comes to be only 78.88% on average, which means they are not suitable to train a precise diagnostic model for bearings under different conditions.

3) Compared with machine learning methods, the classification accuracy of transfer learning (TCA_{share}, enhanced TCA) is improved by about 8.31% and 14.47% respectively, and the best classification accuracy of enhanced TCA in both two cases is 96.25%. Thus, the enhanced TCA has the best classification performance.

4. Conclusion

A fault diagnosis method based on enhanced TCA strategy is proposed for bearing fault diagnosis under different working conditions. Here vibration data are viewed as a document consisted of a number of "fault word", and with the K-means method, "fault word" can be generated. Compared with traditional machine learning algorithm and TCA_{share}, the enhanced TCA has the superior performance.

Acknowledgments

This project is supported by National Natural Science Foundation of China (Grant No. 51575102).

References

[1] Van M and Kang H J 2015 *J. IET Science Measurement Technology* **9** 671-680.

[2] Wu S D, Wu P H and Wu C W 2012 *J. Entropy* **14** 2650-54.

[3] Ndez-Francos D, Mart and Nez-Rego D 2013 *J. Computers & Industrial Engineering* **64** 357–365.

[4] Pandya D H, Upadhyay S H and Harsha S P 2013 *J. Expert Systems with Applications* **40** 4137–45.

[5] Inoue E, Ma R and Mitsuoka M 2011 *J. Ama Agricultural Mechanization in Asia Africa & Latin*

America **42** 89-93.
[6] Ali J B, Fnaiech N and Saidi L 2015 *J. Applied Acoustics* **89** 16-27.
[7] Sharma S, Abed W and Sutton R 2015 *J. IFAC-PapersOnLine* **48** 49-54.
[8] Yu Y and Junsheng C 2006 *J. Journal of Sound and Vibration* **294** 269-277.
[9] Shen Z, Chen X and Zhang X 2012 *J. Measurement* **45** 30-40.
[10] Shi Z, Song W and Taheri S 2016 *J. Entropy* **18** 70.
[11] Pan S J and Yang Q 2010 *J. IEEE Transactions on Knowledge and Data Engineering* **22** 1345-59.
[12] La L, Guo Q and Cao Q 2014 *J. Neural Computing and Applications* **24** 807-816.
[13] Pan W and Ming Z 2014 *J. IEEE Intelligent Systems* **29** 48-54.
[14] Hofmann T. 2001 *J. Machine Learning* **42** 177-196.
[15] Zhang Y, Brady M and Smith S 2001 *J. IEEE Transactions on Medical Imaging* **20** 45-57.
[16] Ahlgren P, Jarneving B and Rousseau R 2003 *J. Journal of the American Society for Information Science and Technology* **54** 550-560.

Non-negative Matrix Factorization and Co-clustering: A Promising Tool for Multi-tasks Bearing Fault Diagnosis

Fei Shen[1], Chao Chen[1] and Ruqiang Yan[1,*]

[1]School of Instrument Science and Engineering, Southeast University
Nanjing, Jiangsu 210096, China

[*]Email: ruqiang@seu.edu.cn

Abstract. Classical bearing fault diagnosis methods, being designed according to one specific task, always pay attention to the effectiveness of extracted features and the final diagnostic performance. However, most of these approaches suffer from inefficiency when multiple tasks exist, especially in a real-time diagnostic scenario. A fault diagnosis method based on Non-negative Matrix Factorization (NMF) and Co-clustering strategy is proposed to overcome this limitation. Firstly, some high-dimensional matrixes are constructed using the Short-Time Fourier Transform (STFT) features, where the dimension of each matrix equals to the number of target tasks. Then, the NMF algorithm is carried out to obtain different components in each dimension direction through optimized matching, such as Euclidean distance and divergence distance. Finally, a Co-clustering technique based on information entropy is utilized to realize classification of each component. To verity the effectiveness of the proposed approach, a series of bearing data sets were analysed in this research. The tests indicated that although the diagnostic performance of single task is comparable to traditional clustering methods such as K-mean algorithm and Guassian Mixture Model, the accuracy and computational efficiency in multi-tasks fault diagnosis are improved.

1. Introduction

As one of the critical components in rotary machines, rolling element bearing occupies a significant position in modern mechanical equipment, which promotes a series of researches relating to the bearing fault mechanism, fault detection and fault precaution [1]. To highlight the weak fault information contaminated in the noise or reduce the influence of external factors, some time-frequency analysis methods have been studied to effectively extract features of bearing faults, including wavelet transform [2], matching demodulation transform [3], principal components analysis (PCA)[4], etc. Meanwhile, some modified classifiers were designed with the goal of improving the bearing fault diagnostic performance, such as the manifold learning classifier [5], support vector machine (SVM) [6] as well as neural network methods[7]. Other researchers even explored detection of multiple faults appearing in rotary machines. For example, Tang [8] demonstrated the high potential of kurtosis deconvolution to detect the compound faults of rolling bearings.

However, the approaches mentioned above have to suffer from the inefficiency when facing with those complicated or crossed diagnosis tasks. Aiming at the multi-task issue, a general strategy is to consider different tasks individually (e.g., crack size and crack location), thus producing increased computational loads, which, however, cannot meet the requirement for real-time diagnosis. Although some of multi-task strategies have been applied in other fields, like the diffusion least mean square

(LMS) [9] for network node recognition, the multi-task learning in context classification [10], their availability in bearing fault diagnosis is still unknown.

To introduce the concept of multi-task fault diagnosis, a novel idea which combines the non-negative matrix factorization (NMF) and Co-clustering is proposed in this paper. Different from the global characteristics of vector quantization (VQ) and PCA, NMF gives a good description about the local features, specializing in searching the small scale information from several tasks. Meanwhile, since the concept of Co-clustering is put forward by Higbee [11] in 1996, some outstanding algorithms have been developed, including the CTWC (Coupled Two-way Clustering), the Crossing Minimization, BCCA (Bi-Correlation Clustering Algorithm) [12], to realize the clustering in the row and line of a 2d matrix at the same time to meet some special requirements.

2. The basic principle of multi-task fault diagnosis

A challenge on multi-task bearing fault diagnosis is how to classify both fault locations and fault severity levels simultaneously. To tackle this challenge, two strategies were considered in traditional method: 1) Recognize these two tasks one by one, which does not consider the link between multi-purposes; 2) Subdivide the multi-task as a mixture-types classifying model (for example, 4 types for task 1 and 5 types for task 2 means 20 types for mixture task), which has the adverse effects on diagnostic accuracy. Therefore, a multi-task classifier is carried out in this paper to overcome the weaknesses of methods above.

Preparing for the multi-task classifier, short-time Fourier transform (*STFT*) is adopted for feature extraction because of the non-stationary characteristics of vibration signals.

$$STFT(t,w) = \int_{-\infty}^{+\infty} s(\tau)\gamma^{*}(\tau - t)e^{-jw\tau}\, d\tau \tag{1}$$

where $\gamma(t)$ is the window function; $s(\tau)$ is the collecting vibration signals. The results of STFT reflect the energy distribution both in time and frequency domain.

In the STFT-based feature extraction strategy, the time-frequency power spectrum graph is segmented by $M \times N$ windows, and the maximum value in each window is chosen as a feature. So there will be $M \times N$ features, which are defined as: $F = \{f_{(i,j)}\}, i \in [1,2,\dots,M], j \in [1,2,\dots,N]$. A chirp signal as well as its feature graph is shown in Figure 1 as an example, where selection of M and N values depends on the non-stationarity in the time-frequency domain. More fluctuation in curves requires more segments.

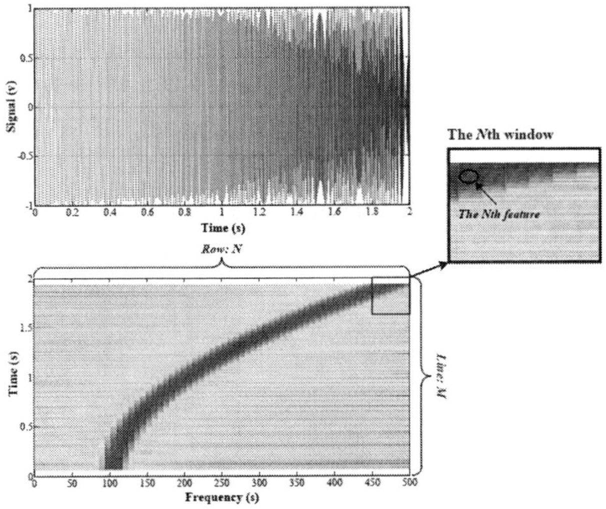

Figure 1. The power spectrum graph of chirp signal.

Then, we assume $CmCnx^2$ samples that are chosen from 2-task sets: 1~Cm category for task 1; 1~Cn category for task 2. x^2 is the number of each sub-category. The process of multitask diagnosis is shown in Figure 2 and described as follows:

1) A group of $M \times N$ feature matrixes is constructed using the STFT features, where each matrix is composed of the identical dimension sub-feature $f_{(i,j)}$, $i \in [1,2,\ldots,M]$, $j \in [1,2,\ldots,N]$ of all $CmCnx^2$ samples:

$$V_{(i,j)} = \begin{bmatrix} f_{(i,j)}^{1,1} & f_{(i,j)}^{1,2} & \cdots & f_{(i,j)}^{1,xCn} \\ f_{(i,j)}^{2,1} & f_{(i,j)}^{2,1} & \cdots & f_{(i,j)}^{2,xCn} \\ \cdots & \cdots & \cdots & \cdots \\ f_{(i,j)}^{xCm,1} & f_{(i,j)}^{xCm,2} & \cdots & f_{(i,j)}^{xCm,xCn} \end{bmatrix} \tag{2}$$

2) A Co-clustering classifier based on information entropy is carried out in each 2D feature matrix after NMF algorithm, and then the clustering matrixes are obtained, where two different types are divided into n and m groups in horizontal and vertical grid respectively. But it is possible that there are a bits of missing categories in each matrix, such as that shown in Figure 2(c):

$$R_{(i,j)} = \begin{bmatrix} C_{(i,j)}^{1,1} & C_{(i,j)}^{1,2} & \cdots & C_{(i,j)}^{1,Cn} \\ C_{(i,j)}^{2,1} & C_{(i,j)}^{2,1} & \cdots & C_{(i,j)}^{2,Cn} \\ \cdots & \cdots & \cdots & \cdots \\ C_{(i,j)}^{Cm,1} & C_{(i,j)}^{Cm,2} & \cdots & C_{(i,j)}^{Cm,Cn} \end{bmatrix}, C_{(i,j)}^{m,n} = \{f_{(i,j)}\}, T_1[f_{(i,j)}] = m \cap T_2[f_{(i,j)}] = n \tag{3}$$

3) The final categories are confirmed by a fuse approach, such as weight fusion, in $M \times N$ result matrixes.

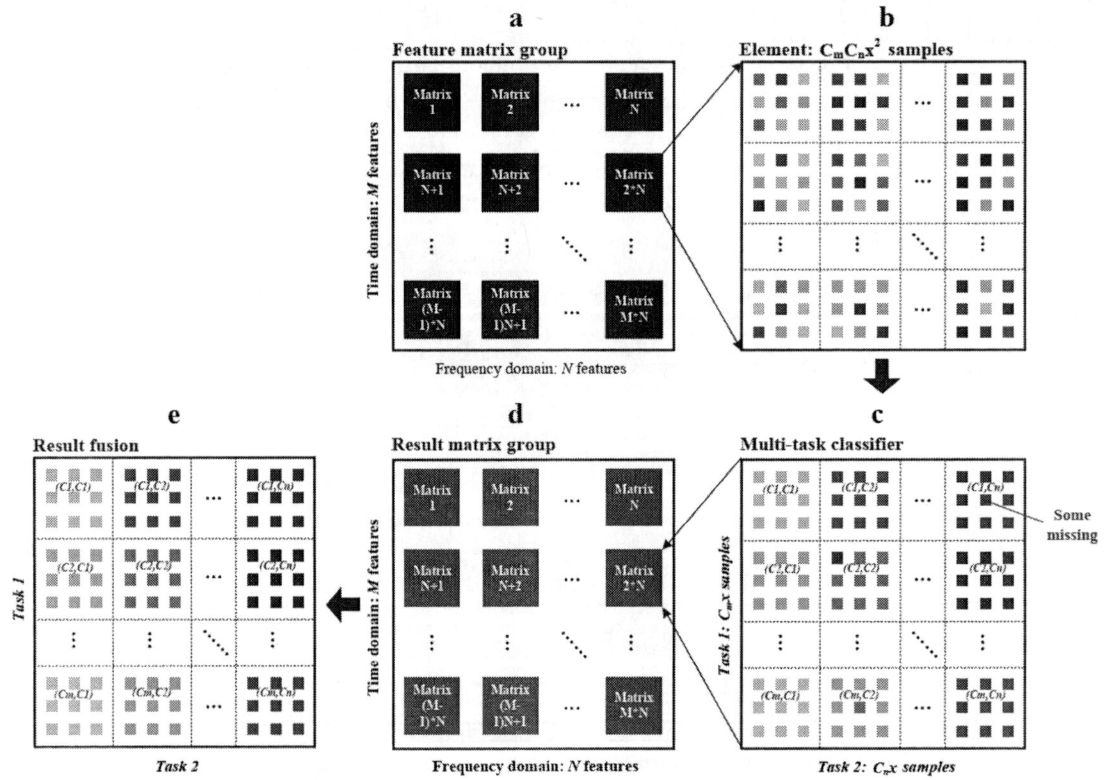

Figure 2. The basic principle of multi-task fault diagnosis

3. The Non-negative Matrix Factorization

Since the non-negative characteristic of energy features, we considered the NMF algorithms for separating two tasks in the same non-negative matrix $\boldsymbol{V}$ **[13].** Two sub-matrixes are represented as $\boldsymbol{W}$ and $\boldsymbol{H}$, respectively.

$$\boldsymbol{V} \approx \boldsymbol{WH} \tag{4}$$

where the dimension of $\boldsymbol{V}$ is $xC_m \times xC_n$, which is then approximately factorized into an $xC_m \times r$ matrix $\boldsymbol{W}$ and an $r \times xC_n$ matrix $\boldsymbol{H}$. Usually r is chosen to be smaller than xC_m or xC_n, so that $\boldsymbol{W}$ and $\boldsymbol{H}$ are smaller than the original matrix. Factorization results mean two compressed versions of the original data matrix.

To find an approximate factorization $\boldsymbol{V} \approx \boldsymbol{WH}$, some cost functions are defined to quantify the degree of approximation. Such a cost function can be constructed using some distance measures between two non-negative matrixes $\boldsymbol{A}$ and $\boldsymbol{B}$:

1) Euclidean distance:

$$D_1(A||B) = \|A - B\|^2 = \sum_{ij} \left(A_{ij} - B_{ij}\right)^2 \tag{5}$$

2) Divergence distance:

$$D_2(A||B) = \sum_{ij} \left(A_{ij} log \frac{A_{ij}}{B_{ij}} - A_{ij} + B_{ij}\right) \tag{6}$$

Equations (5) and (6) are both lower bounded by zero if and only if $\boldsymbol{A} = \boldsymbol{B}$. In addition, the divergence distance is asymmetric between these two matrixes ($D_2(A||B) \neq D_2(B||A)$). Then, the formulation of NMF is considered to be an optimization problem according to the definition of two distances:

$$\min_{\boldsymbol{W},\boldsymbol{H}} D_1(\boldsymbol{V}||\boldsymbol{WH}) \quad \text{s.t. } \boldsymbol{W},\boldsymbol{H} \geq 0 \qquad \textbf{or} \qquad \min_{\boldsymbol{W},\boldsymbol{H}} D_2(\boldsymbol{V}||\boldsymbol{WH}) \quad \text{s.t. } \boldsymbol{W},\boldsymbol{H} \geq 0 \tag{7}$$

To solve the optimization problem, traditional numerical optimization approaches like gradient descent and conjugate gradient are applied to find local minima. But the convergence of the former is slow and the computation of the latter is complex. So a multiplicative update rule is designed in our paper to provide a good compromise between speed and ease of implementation.

1) For $\min_{\boldsymbol{W},\boldsymbol{H}} D_1(\boldsymbol{V}||\boldsymbol{WH})$, the update rule of $\boldsymbol{W}$&$\boldsymbol{H}$ can be designed as [14]:

$$\boldsymbol{H}_{\alpha\mu} \leftarrow \boldsymbol{H}_{\alpha\mu} \frac{\left(\boldsymbol{W}^{\mathrm{T}}\boldsymbol{H}\right)_{\alpha\mu}}{(\boldsymbol{W}^{\mathrm{T}}\boldsymbol{WH})_{\alpha\mu}}$$
$$\boldsymbol{W}_{i\alpha} \leftarrow \boldsymbol{W}_{i\alpha} \frac{\left(\boldsymbol{VH}^{\mathrm{T}}\right)_{i\alpha}}{(\boldsymbol{WHH}^{\mathrm{T}})_{i\alpha}} \tag{8}$$

2) For $\min_{\boldsymbol{W},\boldsymbol{H}} D_2(\boldsymbol{V}||\boldsymbol{WH})$, the update rule of $\boldsymbol{W}$&$\boldsymbol{H}$ can be designed as:

$$\boldsymbol{H}_{\alpha\mu} \leftarrow \boldsymbol{H}_{\alpha\mu} \frac{\sum_i \boldsymbol{W}_{i\alpha}\boldsymbol{V}_{i\mu}/(\boldsymbol{WH})_{i\mu}}{\sum_k \boldsymbol{W}_{k\alpha}}$$
$$\boldsymbol{W}_{i\alpha} \leftarrow \boldsymbol{W}_{i\alpha} \frac{\sum_\mu \boldsymbol{H}_{\alpha\mu}\boldsymbol{V}_{i\mu}/(\boldsymbol{WH})_{i\mu}}{\sum_v \boldsymbol{H}_{\alpha v}} \tag{9}$$

Note that, the divergence is invariant under these updates if and only if $\boldsymbol{W}$ and $\boldsymbol{H}$ are at a stationary point of the divergence. Also, each update consists of multiplication by a factor. In particular, it is straightforward to see that this multiplicative factor is unity when $\boldsymbol{V} = \boldsymbol{WH}$.

As an example of NMF results for two-task matrix, here the full matrix $\boldsymbol{V}$ was defined as a 300×200 task matrix and each 100*100 represent different categories, whatever in horizontal or vertical direction. As shown in Figure 3, the dimensions of sub-matrix $\boldsymbol{W}$ and $\boldsymbol{H}$ are 300×r and r×200, respectively. When the dimension r is set less (e.g. r=10), the classification result is pure and the computational load is also small. However, if the types of full matrix $\boldsymbol{V}$ are not clear, we must increase the dimension of sub-matrix for higher classification accuracy in next classifier.

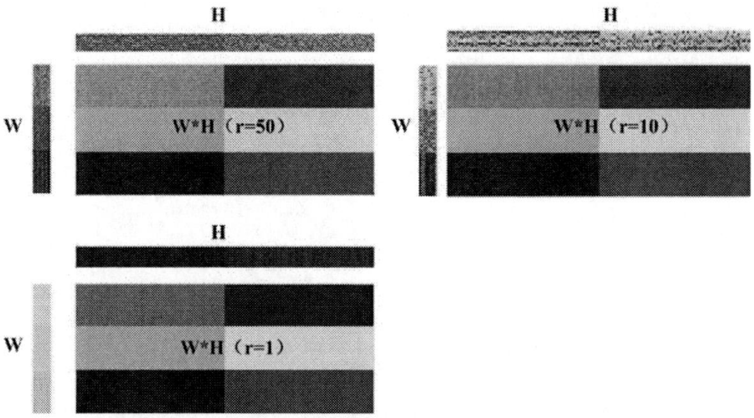

Figure 3. An example of NMF results for two-task matrix.

4. The Co-clustering Based on Information Entropy

After the completion of NMF, the matrix $\boldsymbol{V}$ have been divided into two sub-matrixes $\boldsymbol{W}$ and $\boldsymbol{H}$ in the horizontal and vertical direction respectively. To maximize the gap between different categories and minimize it between same categories, we design an information entropy-based clustering model for co-clustering.

During the model, we assume the classification of task 1 is listed as $\{w_1, w_2 \ldots, w_n\}$, where n is the total number of task 1, and the classification of task 2 is listed as $\{h_1, h_2 \ldots, h_m\}$, where m is the total number of task 2. The mutual entropy $I(W; H)$ between task 1 and task 2 can be calculated as:

$$I(W;H) = \sum_{w}\sum_{h} p(w,h)\log_2 \frac{p(w,h)}{p(w)p(h)} \tag{10}$$

where discrete random variable $\alpha \in \{w_1, w_2 \ldots, w_n\}$, $\beta \in \{h_1, h_2 \ldots, h_m\}$; $p(w,h)$ means the joint probability distribution between α and β; $p(w)$ means the probability distribution of W; $p(h)$ means the probability distribution of H.

Reference [15] proved the optimal formula when the Co-clustering reaches the optimum values:

$$\arg\min_{W^*,H^*}\{I(\alpha;\beta) - I(\alpha^*;\beta^*)\} \tag{11}$$

And in this paper, we choose the Kullback-Leibler (KL) to represent the $I(\alpha;\beta) - I(\alpha^*;\beta^*)$:

$$\begin{gathered} I(\alpha;\beta) - I(\alpha^*;\beta^*) = KL[p(w,h), q(w,h)] \\ q(w,h) = p(w^*,h^*)p(\alpha|w^*)p(\beta|h^*) \end{gathered} \tag{12}$$

where $KL[g,h]$ is the Kullback-Leibler distance between probability distribution function $g(x)$ and $h(x)$, which represents the relative entropy between them and is calculated in equation (13) or (14):

$$KL[p(\alpha,\beta),q(\alpha,\beta)] = \sum_{s^*}\sum_{s\in s^*} p(w)\,KL[p(\beta|w),q(\beta|w^*)] \tag{13}$$

$$KL[p(\alpha,\beta),q(\alpha,\beta)] = \sum_{s^*}\sum_{s\in s^*} p(h)\,KL[p(\alpha|h),q(\alpha|h^*)] \tag{14}$$

Therefore, the smallest mutual information entropy loss can be acquired by minimizing the KL distance between $p(\beta|w)$ and $q(\beta|w^*)$ or the KL distance between $p(\alpha|h)$ and $q(\alpha|h^*)$, and we can find the best mapping function. The detailed procedures can be described as following steps:

1) Initialize the probability density function: the sub-matrix $\boldsymbol{W}$ & $\boldsymbol{H}$ are divided into k & l groups according to the farthest segmentation criteria [16], then calculate the initial probability distribution $p(w,h)$ and $q(w,h)$;
2) Update the row clustering: search a new category label i for each row using the constraint condition (15), to reduce the KL distance of equation (13) as far as possible.

$$i = \arg\min_{S^*,F^*} KL[p(\beta|w),q(\beta|w^*)] \tag{15}$$

Meanwhile, update the probability distribution $q(w,h)$ with (12);

3) Update the column clustering: search a new category label j for each column using the constraint condition (16), to reduce the KL distance of equation (14) as far as possible.

$$j = \arg\min_{S^*,F^*} KL[p(\alpha|h),q(\alpha|h^*)] \tag{16}$$

Meanwhile, update the probability distribution $p(w,h)$ with (12);

4) Compare the mutual entropy $I(\alpha;\beta)$ with a threshold value $I_s(\alpha;\beta)$, if $I(\alpha;\beta)<I_s(\alpha;\beta)$, the program outputs the result of Co-clustering, otherwise, return to step (2) for next update. Particularly, a disturbance search algorithm [16] is adopted during the update process in row clustering or column clustering to avoid the local optimum.

Finally, a result fusion based on the weight of each feature is carried out to obtain the final diagnosis result:

$$i = \frac{1}{M\times N}\sum_{f=1}^{M\times N} \delta_f \times i_f$$
$$j = \frac{1}{M\times N}\sum_{f=1}^{M\times N} \gamma_f \times j_f \tag{17}$$

where M means the number of frequency domain segmentation of STFT power spectrum graphs; N means the number of time domain segmentation of STFT power spectrum graphs; δ_f and γ_f denote the weight of the fth feature in task 1 and task 2, respectively.

5. Experiments and Performance Analysis

The bearing data set from Western Reserve University Bearing Data Center Website were used for the experimental study. As shown in Figure 4, the bearing apparatus consists of a drive motor, a torque transducer, and a dynamometer. Here 16 channel vibration data were collected using accelerometers, which were attached to the drive end and fan end of the motor housing with magnetic bases. Otherwise, speed and horsepower data were collected using the torque transducer/encoder, both with 12 kHz sampling frequency.

The bearing data set from Western Reserve University Bearing Data Center Website were used for the experimental study. As shown in Figure 4, the bearing apparatus consists of a drive motor, a torque transducer, and a dynamometer. Here 16 channel vibration data were collected using accelerometers,

IOP Conf. Series: Journal of Physics: Conf. Series **842** (2017) 012046 doi:10.1088/1742-6596/842/1/012046

which were attached to the drive end and fan end of the motor housing with magnetic bases. Otherwise, speed and horsepower data were collected using the torque transducer/encoder, both with 12 kHz sampling frequency.

Figure 4. The bearing test system.

Table 1. Two fault diagnostic tasks.

Classification	**C1**	**C2**	**C3**	**C4**	**C5**	**C6**
Fault Severity	0	0.178	0.356	0.533	0.712	—
Fault Location	Health	Inner	Ball	Outer 3.00	Outer 6.00	Outer 9.00

Firstly, we extracted and compared the *STFT* features in different diagnostic tasks. Figure 5 illustated the power spectrum graphs of fault severity task (Fault location: inner; Load: 0 horsepower; Rotate speed: 1797Hz). Figure 6 gave the power spectrum graphs of fault location task (Fault severity: 0.533; Load: 0 horsepower; Rotate speed: 1797Hz). From Figures 5 and 6, we can find the power spectrum graphs between different types have a difference in the frequency-axis, but are stable in the time-axis. Obviously, the influence of fault severity and fault location on the frequency domain is larger than that on the time domain. Based on the observed power spectrum graphs, we segmented these graphs using a (100×2) window, where 100 means there being 100 segmentations in the range of 0~500Hz, and 2 is the number of windows from 0s to 120s. Therefore, the dimension of feature vector of each sample is 200, and we chose 100 (10×10) samples in each fault category for NMF algorithm.

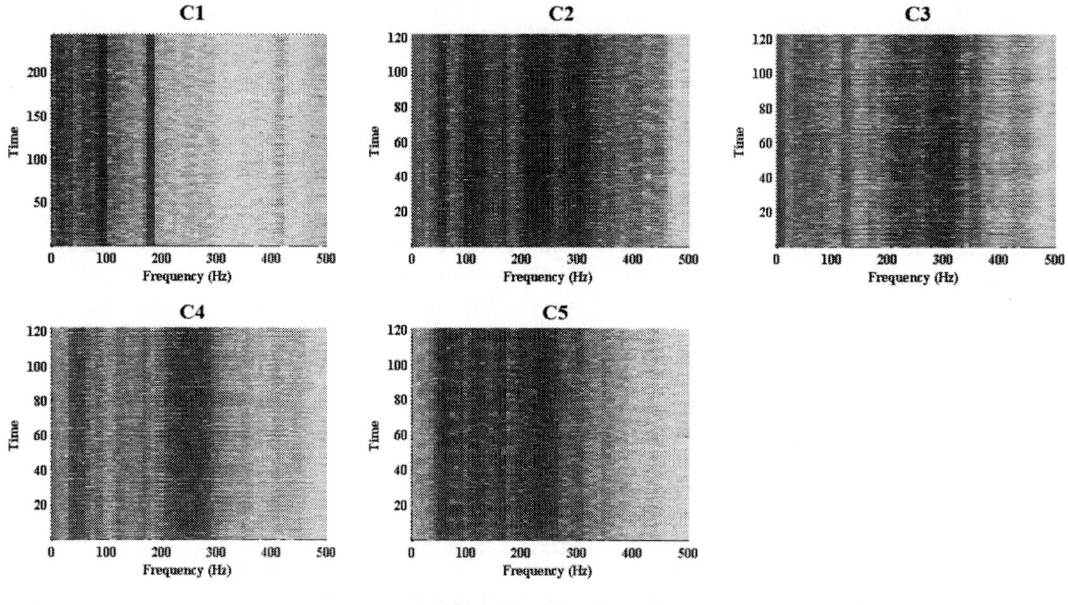

Figure 5. The power spectrum graphs of fault severity task.

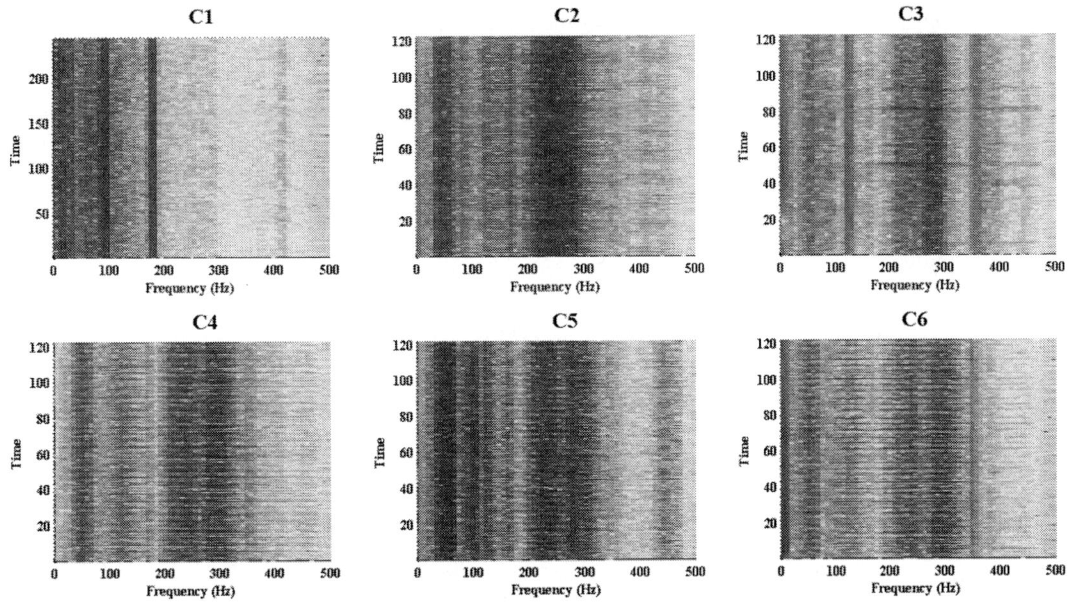

Figure 6. The power spectrum graphs of location task.

Secondly, the NMF and the Co-clustering strategy were carried out for 100×(5+6) samples. We compared the performance of bearing fault diagnosis as well as their computational loads when the r value in W and H increases from 1 to 100, the fault diagnosis accuracy curve along with the relative time cost is shown in Figure 7. It can be seen that the diagnosis accuracy grows from 74.16% (①) to 97.08% (③) when the dimension of NMF sub-matrix rises from 1 to 100, meanwhile, the computational load also appear an exponential increase from approximately 0 to 100. According to Figure 7, a balance point is found in r=39 (②) where the diagnosis accuracy is satisfactory enough (96.04%), also the diagnosis keeps a low level (15.21%). So, the dimension of NMF is designed as 39 when proposed method is applied to the Western Reserve University bearing data set.

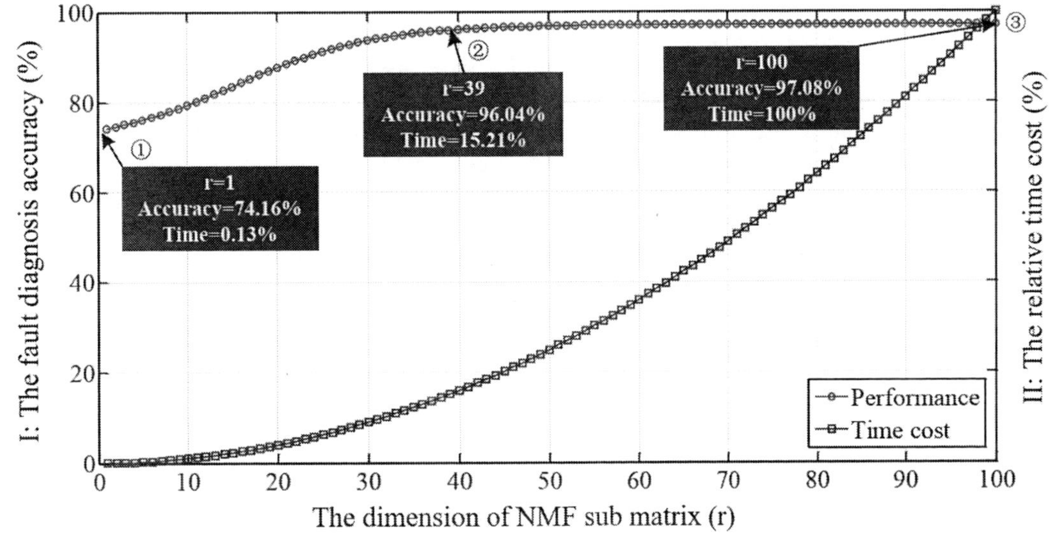

Figure 7. The fault diagnosis accuracy curve as well as the relative time cost curve.

To verify the effectiveness of the proposed approach, we compared its performance with some classical clustering algorithms, such as K-means algorithm and GMM (Guassian Mixture Model) algorithm. The fault diagnosis results are listed in Table 2. Some conclusions can be obtained from this table:

1) The bearing diagnostic accuracy in fault severity (91.16%) is weak smaller than that in fault location (94.08%) in all five strategies, which means that task 1 mainly lies on the time domain features, while task 2 depends on the frequency domain features. So increasing the time domain segmentation number (more than 2) of NMF is a good measure to improve the classification performance of task 1;
2) During classical clustering algorithms, although the accuracy of "Task 1+Task 2" strategy is higher than the "Task 1⊗Task 2" strategy, the time cost of former is also larger than the later (increase by about 30%). Meanwhile, the GMM algorithm requires more time , even though possessing higher performance than K-mean approach;
3) The NMF-based Co-clustering offers good fault diagnosis performance compared with classical clustering algorithms. In addition, task 1 and task 2 are classified at the same time, which guarantees a low computational load, only 70.7% as compared with the K-means.

Table 2. The performance comparison between K-means, GMM, and NMF-based Co-clustering.

Methods	**Accuracy of task 1 (mean±std)**	**Accuracy of task 2 (mean±std)**	**Total accuracy (mean±std)**	**Relative time cost with K-means (%)**
Task 1⊗Task 2: K-means	87.73±4.54	91.46±3.68	90.76±1.40	76.2
Task 1+Task 2: K-means	91.12±4.01	94.55±3.84	95.77±1.56	100.0
Task 1⊗Task 2: GMM	89.44±5.91	92.13±3.50	93.25±2.02	84.1
Task 1+Task 2: GMM	94.26±5.39	95.46±3.77	95.05±2.39	112.3
NMF-based Co-clustering	93.26±4.05	96.82±3.93	95.43±2.13	70.7

Task 1⊗Task 2: an algorithm for the 5×6=30 mixture categories;
Task 1+Task 2: an algorithm for 5 categories of task 1 firstly, then for 6 categories of task 2;

6. Conclusion

A fault diagnosis method based on NMF and Co-clustering strategy is proposed for multi-tasks bearing fault diagnosis. Here the high-dimensional matrixes are constructed using the STFT features, where the dimension of each matrix equals to the number of target tasks. With the NMF method, both fault severity and fault location of bearings can be identified at the same time. The time cost of the proposed method improves about 30% compared with classical clustering algorithms such as K-means and GMM, with acceptable diagnostic performance. Meanwhile, the suitable dimension of sub matrixes by NMF is 39. In that case, both classification accuracy and time cost are satisfactory thus being appealing for the multi-task application in real time bearing fault diagnosis system.

Acknowledgments
This project is supported by National Natural Science Foundation of China (Grant No. 51575102).

References

[1] Yu J 2011 *J. Mechanical Systems and Signal Processing* **25** 2573-88.
[2] Li P, Kong F R and He Q B 2013 *J. Measurement* **46** 497-505.
[3] Wang S, Chen X, Li G, Li X and Liu Z 2014 *J. Instrumentation and Measurement* **63** 1372-83.
[4] Widodo A and Yang B S 2011 *J. Expert Systems with Applications* **38** 8430-37.
[5] Tang B P, Song T and Li F 2012 *J. Renewable Energy* **62** 1-9.

[6] Widodo A and Yang B S 2008 *J. Mechanical Systems and Signal Processing* **21** 2560-74.
[7] Gebraeel N Z and Lawley M A 2008 *J. Automation Science and Engineering* **5** 154-163.
[8] Tang G, Wang X and He Y 2016 *J. Journal of Mechanical Science and Technology* **30** 43-54.
[9] Chen J, Richard C and Sayed A H 2015 *J. Signal Processing* **63** 2733-48.
[10] Xue Y, Liao X and Carin L 2007 *J. Machine Learning Research* **8** 35-63.
[11] Higbee K 1996 *J. Technometrics* **40** 852.
[12] Hochreiter S, Bodenhofer U and Heusel M 2010 *J. Bioinformatics* **26** 1520-27.
[13] Zhongliang F U, Liu J and Jintao L I 2016 *J. Geomatics*.
[14] Lee D D and Seung H S 1999 *J. Nature* **401** 788-791.
[15] Dhillon I S, Mallela S and Modha D S 2003 *J. Kdd* 89-98.
[16] Ke W 2015 *J. International Journal on Digital Libraries* **16** 1-15.

IOP Conf. Series: Journal of Physics: Conf. Series **842** (2017) 012047 doi:10.1088/1742-6596/842/1/012047

A new damage diagnosis approach for NC machine tools based on hybrid Stationary subspace analysis

Chen Gao, Yuqing Zhou*, Yan Ren

College of Mechanical and Electrical Engineering, Wenzhou University, Wenzhou 325035, China

*: Correspondence should be addressed to Yuqing Zhou: zhouyq@wzu.edu.cn

Abstract. This paper focused on the damage diagnosis for NC machine tools and put forward a damage diagnosis method based on hybrid Stationary subspace analysis (SSA), for improving the accuracy and visibility of damage identification. First, the observed single sensor signal was reconstructed to multi-dimensional signals by the phase space reconstruction technique, as the inputs of SSA. SSA method was introduced to separate the reconstructed data into stationary components and non-stationary components without the need for independency and prior information of the origin signals. Subsequently, the selected non-stationary components were analysed for training LS-SVM (Least Squares Support Vector Machine) classifier model, in which several statistic parameters in the time and frequency domains were exacted as the sample of LS-SVM. An empirical analysis in NC milling machine tools is developed, and the result shows high accuracy of the proposed approach.

1. Introduction

Numerical control (NC) machines play important roles in industrial relying on their high automation and stability. However, NC machines generally are subject to failures by inner flimsy tools in practice, which may influence the manufacturing quality, operation safety, et al. Therefore, damage diagnosis for NC machine tools accurately and reliably has attracted considerable interests.

Tool's damage, such as wear, gap, will occur progressively on the tool's face contacting with the chip or on the flank due to the friction between the tool and the workpiece material. A common used effective method is analysing signals gathered by sensors to reveal damage features with the help of effective signal processing techniques. Such as time-domain analysis [1], frequency-domain analysis [2], Wavelet transform [3]. However, how to choose a suitable wavelet basis function to match the machine signal structure remains an open issue despite of many alternative types currently. There is still a lack of generally accepted effective methods to solve the issue [4].

In this paper, a new damage diagnosis approach for NC machine tools were provide, which integrated Stationary subspace analysis (SSA), phase space reconstruction (PSR), and Least Squares Support Vector Machine (LS-SVM). SSA was used to separate multi-dimensional signals into the stationary components and non-stationary components without the need for independency and prior information of the origin signals. PSR was introduced to reconstruct dimensionality of signal to satisfy the constraint of SSA' input data. Subsequently, the selected non-stationary components were analysed for training LS-SVM classifier model, which inputs are from eight statistic parameters in the time and frequency domains.

2. Methods

2.1. Stationary subspace analysis

Stationary Subspace Analysis (SSA) is a blind source separation algorithm which factorizes a multivariate time series into stationary and non-stationary components [5]. According to the SSA model, the observed multivariate time series $X(t) \in R^D$ is assumed to be generated as a linear superposition of stationary components $S^s(t) \in R^d$ and non-stationary components $S^n(t) \in R^{D-d}$.

$$X(t) = AS(t) = (A^s A^n)\binom{S^s(t)}{S^n(t)} \quad (1)$$

where A is an unknown matrix; $A^s$and $A^n$are the basis of the stationary and non-stationary subspace respectively; $S^s(t)$are the stationary components, $S^n(t)$are the non-stationary components.

The aim of SSA is to estimate the inverse mixing matrix $\hat{A}^{-1}$ separating the stationary from non-stationary components in $X(t)$. That is, we want to find a matrix, $\hat{A}^{-1} = \begin{bmatrix} B^s \\ B^n \end{bmatrix}$ with $B^s \in R^{d\times D}$ and $B^n \in R^{(D-d)\times D}$ that consists of the matrices B^s and B^n to the stationary and non-stationary directions respectively. Thus, if we apply such an ideal $\hat{A}^{-1}$ to the observed data $X(t)$,

$$\begin{bmatrix} \hat{S}_t^s \\ \hat{S}_t^n \end{bmatrix} = \hat{A}^{-1}X(t) = \hat{A}^{-1}A\begin{bmatrix} S_t^s \\ S_t^n \end{bmatrix} = \begin{bmatrix} B^sA^s & B^sA^n \\ B^nA^s & B^nA^n \end{bmatrix}\begin{bmatrix} S_t^s \\ S_t^n \end{bmatrix} \quad (2)$$

The SSA algorithm uses the optimal stationary signal recognition criteria, only the stationary source signal S_t^s and n- space can be uniquely determined, and the non-stationary source signal S_t^n can only get its estimate $\hat{S}_t^n$ by maximizing the non-stationary. Among them, the stationary judgment is based on the weak stationary condition, that is, a time series is stationary if its first two moments are constant over time [6]. The specific processs of SSA are as follows:

(1) Divide the observed data $X(t)$ into $\mathrm{N}(\mathrm{N} \geq (\mathrm{D}-\mathrm{d})/2+2)$ continuous time epochs, and calculate the mean $\hat{\mu}_i$ and covariance $\hat{\Sigma}_i$ of each time period, for an arbitrary choice of B^s, get the estimated mean $\hat{\mu}_{s,i} = B^s\hat{\mu}_i$ and covariance matrix $\hat{\Sigma}_{s,i} = B^s\hat{\Sigma}_i$ in each time period.

(2) Measure the difference between the estimated stationary signal distribution and normal distribution in each time period by Kullback-Leibler (KL) divergence. Then sum up all epochs' KL divergences, and construct the objective function $\mathrm{L}(B^s)$ with independent variable B^s. Minimizing the objective function $\mathrm{L}(B^s)$, obtain the optimal stationary projection $\hat{B}^s$, and according to the formula (2) to estimate the stationary source signal $\hat{S}_t^s$.

$$\mathrm{L}(\hat{B}^s) = \min \mathrm{L}(B^s) = \min \textstyle\sum_{i=1}^n D_{KL}\left[Norm(\hat{\mu}_{s,i}, \hat{\Sigma}_{s,i})||Norm(0,\mathrm{I})\right]$$

(3) Construct the objective function $\mathrm{G}(B^n)$ with the above step (1) and (2). Maximizing the objective function $\mathrm{G}(B^n)$, obtain the optimal non-stationary projection $\hat{B}^n$, and according to the formula (2) to estimate the non- stationary source signal $\hat{S}_t^n$.

$$\mathrm{G}(\hat{B}^n) = \max \mathrm{G}(B^n) = \max \textstyle\sum_{i=1}^n D_{KL}\left[Norm(\hat{\mu}_{n,i}, \hat{\Sigma}_{n,i})||Norm(0,\mathrm{I})\right]$$

2.2. Phase space reconstruction

Phase space reconstruction (PSR) is a Chaos technology according to the delay coordinate method proposed by Takens and Packard [7]. The idea of PSR is that the evolution of any component of a system is determined by the interaction with other components, and the time series implies the development process of each component. Considering a component, PSR take some measurement processing in fixed time delay point as a new dimension, the value of delay is considered as a new coordinate. Repeating this process at different time points, many of these points could be generated to a new space. For one-dimensional time series $X(t)$, through different time delay 0, t, 2t,... (m-1)t to construct m- dimensional phase space vector.

$$X_i(t) = \{x(t+i), x(t+i+\tau), x(t+i+2\tau), \dots, x(t+i+(m-1)\tau)\}, i = 0,1,\dots,m-1$$

where t is the time delay and τ is the embedding dimension.

$$\mathrm{P} = \begin{bmatrix} x_1 & x_2 & \cdots & x_{N-(m-1)\tau} \\ x_{1+\tau} & x_{2+\tau} & \cdots & x_{N-(m-2)\tau} \\ \vdots & \vdots & & \vdots \\ x_{1+(m-1)\tau} & x_{2+(m-1)\tau} & & x_N \end{bmatrix}$$

The key of phase space reconstruction is the time delay t and embedding dimension τ. Values of t and τ are obtained by the mutual information method and Cao's method [8], respectively.

2.3. Least squares support vector machine

Support Vector Machines (SVM) is a powerful methodology for solving problems in nonlinear classification, function estimation and density estimation, which has been introduced within the context of statistical learning theory and structural risk minimization. Least squares support vector machine (LS-SVM) proposed by Suykens [9] is one of the most influential achievements in recent years.

Given a training data set $\{(x(t), y(t)), t = 1,2,\dots,N\}$, $x(t) \in R^m$ is the input data, $y(t) \in R$ is the category label corresponding to $x(t)$, and N is the training samples number. In LS-SVM, The estimation model is $y_t = w^T\varphi(x_t) + b + e_t$, where $\varphi(.)$ is a nonlinear transformation that map $x(t)$ to the high feature space, w is weight vector, b is intercept, and e_t is the error term that is assumed to be independent and identically distributed (i.i.d) with zero mean and finite variance.

According to [9,10], the LS-SVM model is expressed in dual form as: $\hat{y}(z) = \sum_{t=1}^{N} \alpha_t K(z, x_t) + b$. Note that, the unknown items w and $\varphi(.)$ are not included in the dual model, the mapping transformation operations are achieved implicitly through the use of kernel function$K(.,.)$. That is to say, it is necessary to find the expression of mapping function $\varphi(.)$. For the kernel function $K(.,.)$, it is not difficult to determine the kernel function, and the function satisfying the Mercer theorem can be used as a kernel function. There are several choices presently, such as linear kernel, spline kernel, and RBF kernel.

2.4. Damage diagnosis approach based on hybrid Stationary subspace analysis

The damage diagnosis approach for NC machine tools in this paper includes two phases. The first phase focuses on model training, sample feature parameters in different tool states are obtained by the proposed method. Firstly reconstructed multi dimensions signal are transformed by PSR, stationary parts and non-stationary parts are extracted from multi-dimensional signals by SSA. Subsequently, some dimensionless statistic time- frequency parameters in non-stationary components are calculated as the sample data of LS-SVM for classification. The second phase is damage detection for unknown working tools. The specific procedure is constructed through the following main steps as shown in Figure 1.

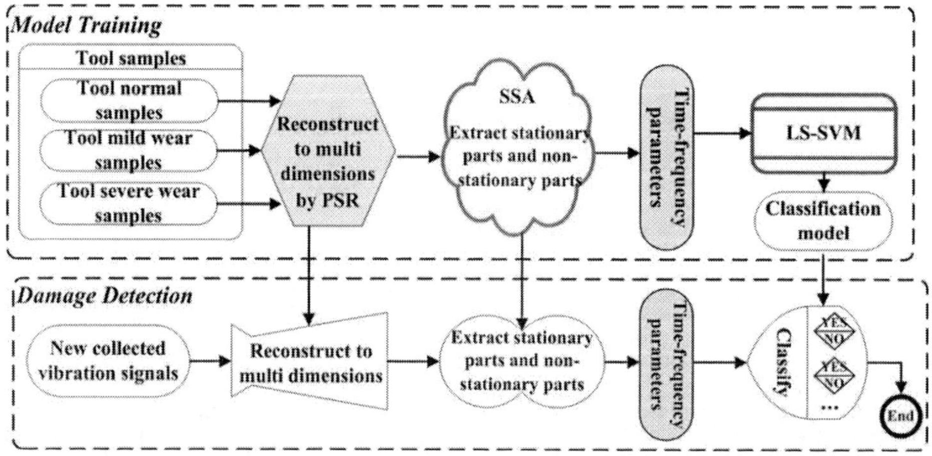

IOP Conf. Series: Journal of Physics: Conf. Series **842** (2017) 012047 doi:10.1088/1742-6596/842/1/012047

Figure 1. Damage diagnosis procedure based on hybrid Stationary subspace analysis

3. Experiment and discussion

3.1. Experiment Design

The GSVM714A NC milling machine is used for the experimental test. The cutting tools used in this test are carbide tools #RPMW1003# in a face milling cutter #EMR 5R160-40-8T# with four teeth. The experimental setup for the in-process detection of tool fault is illustrated in Fig 2. The tool vibration signals are monitored by a laser vibrometer by Polytec Inc. (shown in Fig 3), and sent to a data acquisition instrument (ECON Avant (MI-7016)) and a portable computer.

Tool states in this experiment are alternatively three situations: normal, mild wear, and severe wear (shown in Fig 4). Aluminum alloy samples #7075-T351# of size (150 mm, 100 mm, 500 mm) are fixed on the machining center. In this experiment, cutting speed and depth are 1000 rpm and 1 mm respectively.

3.2. Result and Discussion

In this experiment, three tool damage classes were measured including (1) normal tool, (2) mild wear tool, (3) severe wear tool. It is assumed that all investigated tools have the same damage mechanism. The vibration signals were measured and the sampling frequency was 48,000 Hz. Vibration signals in three tool classes are measured 30 times and divided into two parts: 20 times as training data and 10 times testing data. The training set is used to train model in Matlab R2012.

In the first phase model training, the two parameters t=9 and τ=4 by the mutual information method and Cao's method. The result of PSR are chosen as input for the SSA decomposition, and extracted to 3 (d=3) stationary components and 1 non-stationary component. Figure 5 shows a SSA result of the class of mild wear tool three classes in sample data. Then eight dimensionless time- frequency parameters outlined in [11] were choose in the non-stationary component, which are the training sample data of LS-SVM classifier model. Gaussian RBF kernel was selected in LS-SVM.

The classification results of hybrid SSA proposed are shown in Table 1 for testing data. According to Table 1, 10 normal tool groups can be correctly identified, however, the accuracy of mild wear groups and severe wear groups are 80% and 90%, there are 2 mild tool groups were identified to severe tool class, and 1 severe tool group was identified to mild tool class. The hybrid SSA method proposed was compared with the original LS-SVM. The original LS-SVM method denotes that the original vibration signals measured in experiment are the input of LS-SVM, which do not extract signal by SSA and PSR. As shown in Table 1, the accuracy of the original LS-SVM is less than hybrid SSA. There are 1 normal group, 4 mild wear groups and 2 severe wear groups get error results, and the accuracy of normal, mild wear and severe wear groups are 90%, 60% and 80%. The hybrid SSA proposed in this paper is outperform with original LS-SVM. SSA can extract effectively stationary sources and non-stationary sources, and obtain valuable information to damage diagnosis.

IOP Conf. Series: Journal of Physics: Conf. Series **842** (2017) 012047 doi:10.1088/1742-6596/842/1/012047

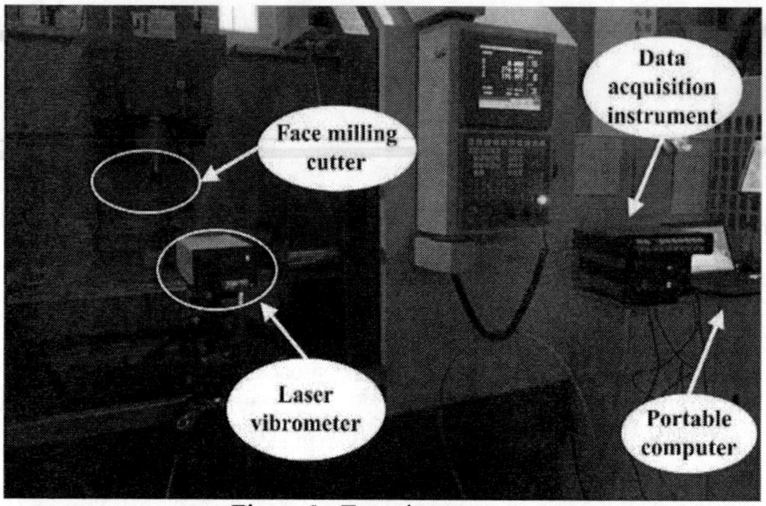

Figure 2. Experiment setup

Figure 3. Measuring vibration signal by a laser vibrometer

(a)Face milling cutter (b) normal tool (c) mild wear tool (d) severe wear tool

Figure 4. Samples of milling cutter shape and different tool states

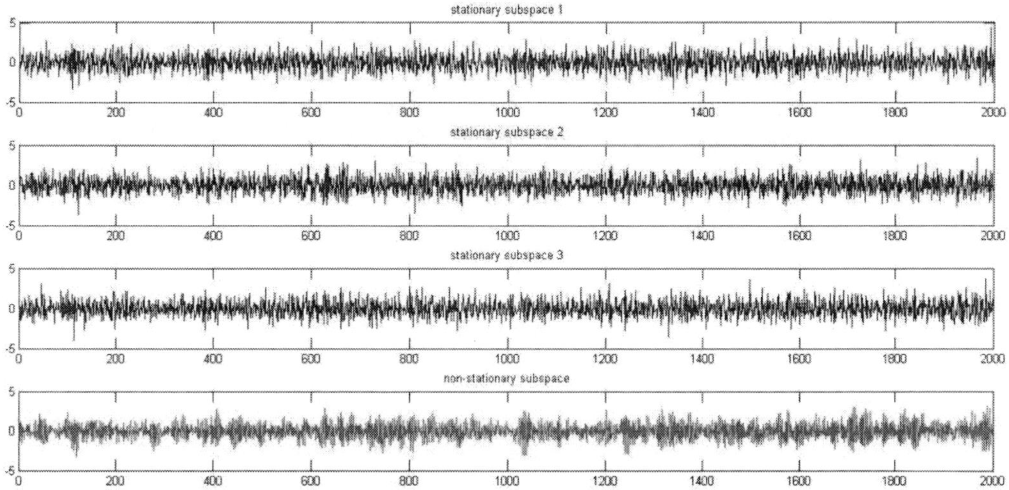

Figure 5. SSA results of mild wear tool sample

Table 1 Compare in result of different methods

Methods	Testing classes	Testing result			Accuracy
		normal tool	mild wear tool	severe wear tool	
Hybrid SSA proposed	normal tool	10	0	0	100%
	mild wear tool	0	8	2	80%
	severe wear tool	0	1	9	90%
Original LS-SVM	normal tool	9	1	0	90%
	mild wear tool	3	6	1	60%
	severe wear tool	0	2	8	80%

4. Conclusion

The present study proposes a damage diagnosis approach to NC machine tools based on hybrid SSA. Firstly, the original data set are transformed single dimensional data into high dimensional manifold signals by PSR. The SSA method is applied to extract the stationary parts and non-stationary parts from multi-dimensional signals without the need for independency and prior information of the source signals. Subsequently, the selected non-stationary components were analyzed for classification by LS-SVM, in which eight dimensionless parameters in the time and frequency domains were exacted as the inputs of LS-SVM. An empirical analysis in NC milling machine tools is developed, and the result shows the proposed approach is outperform with the original LS-SVM approach. Future work can be centered on improvement of the accuracy, efficiency and calculation speed of the proposed approach.

Acknowledgments

This work was supported by the National Nature Science Foundation of China (Grant No. 51405346), the project of Natural Science Foundation of Zhejiang Province in China (No. LY17E050005) and the Public Projects of Wenzhou City (No. G20150014, G20160015).

References

[1] Lv Junjie, Wang Jie, Wang Mei, Wu Yue 2010 Research on tool wear condition monitoring Based on combination of SOM and HMM *China Mechanical Engineering* **21** pp 1531-1535.

[2] Zaghbani, M. Lamraoui, V. Songmene, M. Thomas, and M. ElBadaoui 2011 Robotic high speed machining of aluminum alloys *Advanced Material Research* **188** pp 584-589.

[3] Manpreet Singh and Rajesh Kumar 2013 Thrust bearing groove race defect measurement by wavelet decomposition of pre-processed vibration signal. *Measurement* **46** pp 3508-3515.

[4] Zhipeng Feng, MingLiang, and FuleiChu 2013 Recent advances in time-frequency analysis methods for machinery fault diagnosis: A review with application examples *Mechanical*

IOP Conf. Series: Journal of Physics: Conf. Series **842** (2017) 012047 doi:10.1088/1742-6596/842/1/012047

Systems and Signal Processing **38** pp 165-205.
[5] Von B P, Meinecke F C, Király F C 2009 Finding stationary subspaces in multivariate time series *Physical Review Letters* **103** pp 1-4.
[6] Von Bü nau P 2012 Stationary Subspace Analysis: Towards Understanding Non-Stationary Data Ph.D. Thesis, *Berlin Institute of Technology*, Berlin, Germany.
[7] Packard N H, Crutchfield J P, Farmer J D Geometry from a time series. *Phys Rev Lett* 1980 **45** pp 712-716.
[8] Xia D, Song S, Wang J, Shi J, et al 2012 Determination of corrosion types from electrochemical noise by phase space reconstruction theory *Electrochemistry Communications* **15** pp 88-92.
[9] Suykens J A K 2000 Least squares support vector machines for classification and nonlinear modeling *Neural Network World* **10** pp 29-48.
[10] Feng Y, Jia B, Yan G, et al. 2016 Prediction model of high-speed oblique cutting temperature based on LS-SVM. *The Inter Jour of Adv Man Tec* **85** pp 1-8.
[11] Zhou Yuqing, Liu Xinfang, Li Fengping, Sun Bingtao, Xue Wei 2015 An online damage identification approach for numerical control machine tools based on data fusion using vibration signals *Journal of Vibration and Control* **21** pp 2925-2936.

Study on quantitative risk assessment model of the third party damage for natural gas pipelines based on fuzzy comprehensive assessment

Zeyang Qiu[1], Wei Liang[1*], Xue Wang[2],Yang Lin[1] and Meng Zhang[1]

[1]College of Mechanical and Transportation Engineering, China University of Petroleum-Beijing

[2]China Building Material Test & Certification Group Co., Ltd.

lw@cup.edu.cn

Abstract. As an important part of national energy supply system, transmission pipelines for natural gas are possible to cause serious environmental pollution, life and property loss in case of accident. The third party damage is one of the most significant causes for natural gas pipeline system accidents, and it is very important to establish an effective quantitative risk assessment model of the third party damage for reducing the number of gas pipelines operation accidents. Against the third party damage accident has the characteristics such as diversity, complexity and uncertainty, this paper establishes a quantitative risk assessment model of the third party damage based on Analytic Hierarchy Process (AHP) and Fuzzy Comprehensive Evaluation (FCE). Firstly, risk sources of third party damage should be identified exactly, and the weight of factors could be determined via improved AHP, finally the importance of each factor is calculated by fuzzy comprehensive evaluation model. The results show that the quantitative risk assessment model is suitable for the third party damage of natural gas pipelines and improvement measures could be put forward to avoid accidents based on the importance of each factor.

Key Words. natural gas pipelines; third party damage; quantitative risk assessment model; AHP; FCE.

1. Introduction

As natural gas pipelines have characteristics of continuous operation, cover different sites complex environments, therefore, the safety of gas transmission pipelines not only affects the safety of pipeline running, social energy supply, but also threatens people's life and living environment

The third party damage has been get extensive attention as the main reason of natural gas pipeline accidents. Tom Bajcar presents a refined way to quantify the effects of third party interference on risk that is posed on people by transmission pipelines for natural gas. The main focus is set on the influence of population density on risk[1]; Wei Liang focuses on the application of self-organizing maps(SOMs)to assess the risk of third-party interference and classify their risk patterns[2]; Xing-yu Peng based on fragility and internal & external hazard, an overall reliability model can be established as a theoretical basis for quantitative risk assessment on oil and gas pipeline systems[3]. Jun Li calculates the failure possibility of the third-party to urban gas pipeline based on AHP and FCE. However, the importance of each factor for the third-party has not been calculated. Moreover, in this thesis, the main studying object is urban gas pipeline not Long-distance Gas Pipeline [4].

Quantitative risk assessment model has been established for the third party damage used Analytic Hierarchy Process (AHP) and Fuzzy Comprehensive Evaluation (FCE) in this paper. The model could compare the importance of the various factors for the third party damage and prevent accidents effectively.

2. Risk source identification of third party damage

The third party damage of natural gas pipelines generally refers to the damage caused by peoples' behavior. Primary risk factors of third party damage are divided into six different categories in this paper: legal factors, condition of pipeline patrol, related equipment of pipeline, public relations along the pipeline, peripheral activity along the pipeline and alarm system of Pipeline Company.

The author based on the research of third party damage, determining its corresponding second level risk factors of primary risk factors are shown in figure 1.

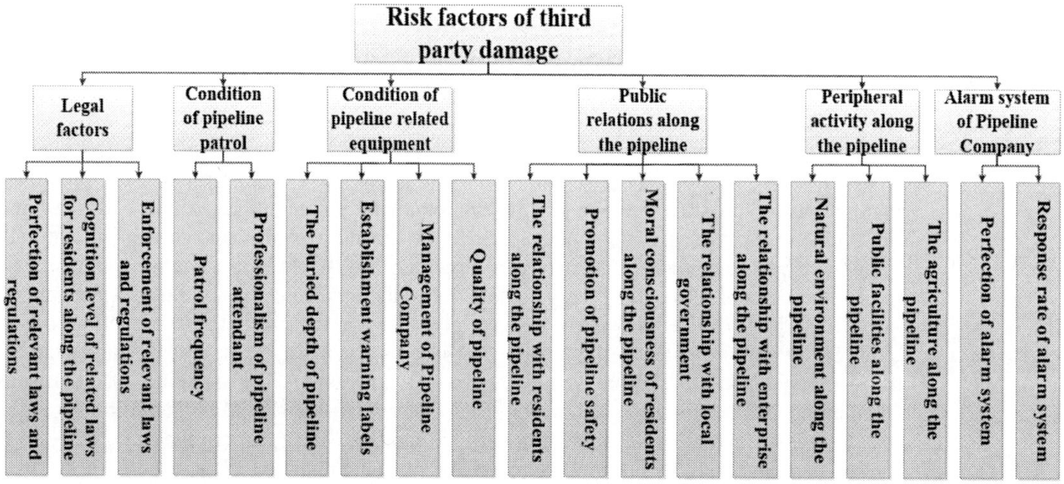

Figure 1. Risk factors of third party damage for natural gas pipelines.

The risk factors of third party damage are analysed to establish quantitative risk assessment model and prevent accidents of natural gas pipelines.

3. The fundamentals of improved AHP

Analytic hierarchy process (AHP) is a structured multi-attribute decision method. The main advantage of AHP is its capability to check and reduce the inconsistency of expert judgments [4].

Generally, the process of applying AHP is divided into three steps. Firstly, establish a hierarchical structure by recursively decomposing the decision problem. Secondly, construct the pairwise comparison matrix to indicate the relative importance of alternatives. A numerical rating including nine rank scales is suggested [5], as shown in Table 1. Thirdly, verify the consistency of pairwise comparison matrix and calculate the priority weights of alternatives. Using the formula: $CR = CI/RI$, $CI = \frac{1}{n-1}(\lambda_{max} - n)$, where, CI is consistency index; RI is average random consistency index; CR is calculate and inspect consistency ratio. If $CR < 0.10$ the constructed pairwise comparison matrix is considered acceptable and the priority weights of alternatives can be obtained through the eigenvector[6]. Otherwise, the comparison matrix needs to be reconstructed, where RI is the random consistency index related to the dimension of matrix[7], listed in Table 2.

Table 1. The 1–9 scales for pairwise comparisons in the AHP.

Definition	Equal importance	Moderate importance	Moderate importance	Moderate importance	Moderate importance	Moderate importance	Moderate importance
Importance intensity	1	3	5	7	9	2,4,6,8	Reciprocals

Table 2. RI value of matrixes with different exponent.

Order: n	1	2	3	4	5	6	7	8	9
RI	0.00	0.00	0.58	0.90	1.12	1.24	1.32	1.41	1.45

For the pairwise comparison matrix $M_{n \times n}$ with acceptable consistency, suppose $\vec{w} = [w_1, w_2 \cdots, w_n]^T$ is the eigenvector of M, whose $w_i (i = 1, 2, \cdots n)$ is indicated as the priority weight of the ith alternative and calculated by $M\vec{w} = \lambda_{\max}\vec{w}$ [8].

The judgment of one expert is calculated and analyzed in traditional AHP. This paper proposes a method combining with Delphi method and AHP in order to calculate the weight of risk evaluation index. [9]。 This method would be more accurate to invite five expert judgment methods, rather than only consider one expert opinion to calculate the weight. The analysis process of improved AHP is shown in figure 2.

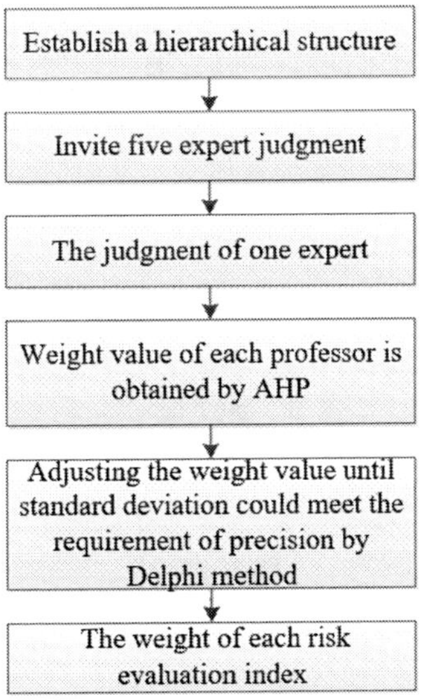

Figure 2. The analysis process of improved AHP.

Weight value of each professor has been processed by Delphi method, values are weighted average and standard deviation is calculated according to expert opinions credibility. If standard deviation could meet the requirement of precision, then the final weight of each risk evaluation index is obtained. Otherwise new data is calculated after experts' adjusting the weight value until standard deviation could meet the requirement of precision, then final result is the weight of each risk evaluation index.

4. The fundamentals of FCE

The fuzzy comprehensive evaluation method, combining fuzzy theory and mathematical model, is a useful method in quantitative risk assessment [10]. The methods for obtaining evaluation index set and weight set have been described above. The fuzzy comprehensive evaluation mainly includes the following procedures:

(1) Factor sets: $\mathbf{U} = \{u_1, u_2, \cdots, u_i, \cdots, u_n\}$

(2) Evaluation sets: $\mathbf{V} = \{v_1, v_2, \cdots, v_j, \cdots, v_n\}$

(3) Weight sets: $\mathbf{A} = \{a_1, a_2, \cdots, a_m\}$. $\sum_{i=1}^{m} a_i = 1$ $a_i \geq 0$.

(4) Fuzzy relation matrix sets: Construct the membership functions, and then adopt the expert scoring method for each single factor scores of factor sets, through membership function obtain membership values rim of each single factor, establish single factor fuzzy evaluation matrix [11].

(5) Comprehensive evaluation [12]. According to the Weight sets and evaluation matrix Calculate factors layer and index layer fuzzy comprehensive evaluation sets. $\mathbf{B} = \mathbf{A} \bullet \mathbf{R}$.

5. Case analysis

5.1. fuzzy comprehensive risk assessment model

Possible destruction styles, influencing factors and the importance of risk assessment index system for third party damage to Shaan-jing pipeline are quantified based on FCE. The analysis process of fuzzy comprehensive risk assessment model for the third party damage is shown in figure 3.

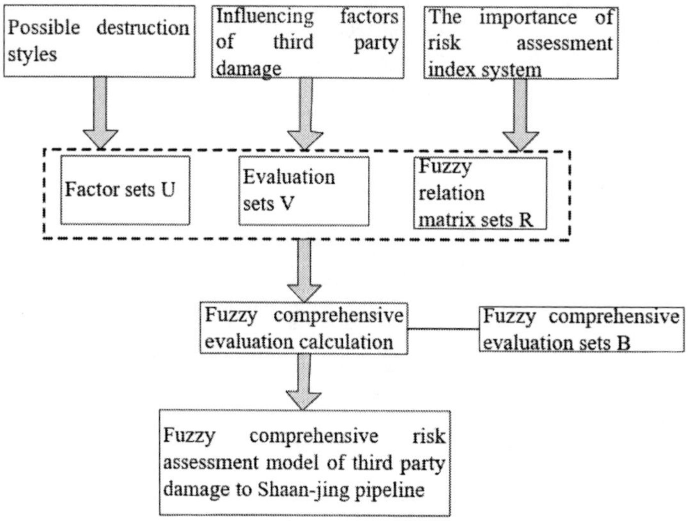

Figure 3. Fuzzy comprehensive risk assessment model of third party damage to Shaan-jing pipeline.

Possible destruction styles, influencing factors and the importance of risk assessment index system for third party damage to Shaan-jing pipeline are graded by five rich field experiences of managers and employees based on risk source identification of third party damage mentioned in section 1. The result of the grading statistics is analysed and fuzzy comprehensive risk assessment model is established based on FCE.

According to the statistical results, possible destruction styles of third party damage: excavating equipment, cultivation of agricultural machinery, artificial malicious damage, transportation vehicle, natural disaster; influencing factors of third party damage: the buried depth of pipeline、overlaying soil properties、Man-made obstacles、natural obstacles、mark stake of pipeline、patrol frequency、the nature of the region passed by pipeline、response of alarm system; the importance of risk assessment index system for third party damage: the buried depth of pipeline、ground activity of pipeline、ground facilities、warning labels、patrol frequency、Public education and publicity、alarming system. Evaluation sets is divided into four grades: low (L), medium (M), between medium and high (MH), high (H). The weight of each factor is calculated based on based on improved AHP mentioned in section 2, factor sets, evaluation sets and weight sets are shown in Table 3. The specific scoring statistics is not present in the paper due to the large number.

Table 3. The content of fuzzy comprehensive evaluation.

Category	Evaluation sets	Factor sets U	Weight of factor sets	Evaluation sets V
Possible destruction styles of third party damage	Excavating equipment B11	{Safety publicity education, The supervision of excavation, The connection of construction unit, Patrol strength}	{0.1, 0.4, 0.2, 0.3}	V1 The possibility of occurrence: {H, MH, M,L}
	Agricultural production B12	{Safety publicity education, The connection of local government, Facility of warning labels, Patrol strength}	{0.2, 0.1, 0.3, 0.4}	
	Artificial malicious damage B13	{Moral consciousness of residents along the pipeline, Safety publicity education, Moral consciousness of residents along the pipeline, Patrol strength}	{0.15, 0.15, 0.2, 0.5}	
	Transportation vehicle B14	{Protection facilities of pipeline, Pipeline route design, Facility of warning labels, Patrol strength}	{0.25, 0.1, 0.35, 0.3}	
	Natural disaster B15	{Pipeline route design, Protection facilities of pipeline, Prediction of natural disasters, Response measurements of natural disasters}	{0.7, 0.05, 0.05, 0.2}	
Influencing factors of third party damage	The buried depth of pipeline B21	{Pipeline crossing situation, The situation of warning sign, The situation of guard plate, The situation of pipe casing}	{0.5, 0.3, 0.1, 0.1}	V2 Influence level: {H, MH, M,L }
	Man-made obstacles B22	{Pipeline crossing situation, Type of protection facilities, Complete situation of protection facilities, Demand situation of protection}	{0.3, 0.2, 0.4, 0.1}	
	Mark stake of pipeline B23	{Complete situation of mark stake, Damage condition of mark stake, Perfection of mark stake, Management of mark stake}	{0.3, 0.3, 0.2, 0.2}	
	Pipeline patrol frequency B24	{Professionalism of pipeline attendant, regulation of pipeline patrol, Quality of pipeline patrol, Management unit of pipeline patrol}	{0.15, 0.15, 0.5, 0.2}	
	Property of region passed by pipeline B25	{Economic development of region, Population density, Development degree, The situation of spiritual civilization}	{0.2, 0.3, 0.2, 0.3}	
	Response of alarm system B26	{ Perfection of alarm system, Response rate, Response mechanism, Personnel assignment }	{0.2, 0.3, 0.3, 0.2}	
The importance of risk assessment index system for third	Buried depth of pipeline B31	{Pipeline crossing situation, The situation of warning sign, The situation of guard plate, The situation of pipe casing}	{0.5, 0.3, 0.1, 0.1}	V3 Importance level : { H,
	Ground activity of pipeline B32	{The ground construction activity, The ground agriculture activity, Natural environment, Pipeline crossing situation }	{0.4, 0.3, 0.1, 0.2}	
	Ground	{Exposure of ground facilities, Protection	{0.5, 0.3,	

IOP Conf. Series: Journal of Physics: Conf. Series **842** (2017) 012048 doi:10.1088/1742-6596/842/1/012048

party damage	facilities B33	facilities, Natural environment, The situation of warning labels}	0.1, 0.1}	MH, M, L }
	Warning labels B34	{Complete situation of warning labels, Damage condition of warning labels, Density of warning labels, Management of warning labels}	{0.3, 0.3, 0.2, 0.2}	
	Patrol frequency B35	{Professionalism of pipeline attendant, Regulation of pipeline patrol, Quality of pipeline patrol, Management unit of pipeline patrol }	{0.15, 0.15, 0.5, 0.2}	
	Public education and publicity B36	{Promotion of pipeline safety, Propaganda way, Contact with local government, Periodic return visits}	{0.2, 0.2, 0.4, 0.2}	
	Alarming system B37	{ Perfection of alarm system, Publicity of alarm system, Response rate, Response mechanism }	{0.2, 0.2, 0.3, 0.3}	

5.2. possible destruction styles of third party damage

(1) Excavating equipment: B_{11}

Weight sets: $A_{11} = [0.1\ 0.4\ 0.2\ 0.3]$

Fuzzy relation matrix sets: $R_{11} = \begin{bmatrix} 0.18 & 0.27 & 0.52 & 0.03 \\ 0.85 & 0.12 & 0.03 & 0 \\ 0.36 & 0.42 & 0.12 & 0.10 \\ 0.73 & 0.24 & 0.03 & 0 \end{bmatrix}$

$$\begin{aligned} B_{11} &= A_{11} \bullet R_{11} \\ &= [(0.1 \cap 0.18) \cup (0.4 \cap 0.85) \cup (0.2 \cap 0.36) \cup (0.3 \cap 0.73), \\ &\quad (0.1 \cap 0.27) \cup (0.4 \cap 0.12) \cup (0.2 \cap 0.42) \cup (0.3 \cap 0.24), \\ &\quad (0.1 \cap 0.52) \cup (0.4 \cap 0.03) \cup (0.2 \cap 0.12) \cup (0.3 \cap 0.03), \\ &\quad (0.1 \cap 0.03) \cup (0.4 \cap 0) \cup (0.2 \cap 0.10) \cup (0.3 \cap 0)] \\ &= [0.1 \cup 0.4 \cup 0.2 \cup 0.3, 0.1 \cup 0.12 \cup 0.2 \cup 0.24, \\ &\quad 0.1 \cup 0.03 \cup 0.12 \cup 0.03, 0.03 \cup 0 \cup 0.10 \cup 0] \\ &= [0.4, 0.24, 0.12, 0.1] \end{aligned}$$

After the normalization: $B_{11} = [0.4651, 0.2791, 0.1395, 0.1163]$

The possibility of third party damage caused by excavating equipment is H and membership is 0.4651 based on maximum membership principle.

Detailed calculation process will not be introduced in the paper; the specific calculation process could be best seen by reference to B_{11}.

(2) Cultivation of agricultural machinery B_{12}.

$$B_{12} = A_{12} \bullet R_{12} = [0.2, 0.1, 0.3, 0.4] \bullet \begin{bmatrix} 0.21 & 0.27 & 0.49 & 0.03 \\ 0.33 & 0.21 & 0.42 & 0.04 \\ 0.42 & 0.42 & 0.16 & 0 \\ 0.79 & 0.21 & 0 & 0 \end{bmatrix} = [0.4255, 0.3139, 0.2128, 0.0426]$$

(3) Artificial malicious damage B_{13}

$$B_{13} = A_{13} \bullet R_{13} = [0.15\ 0.15\ 0.2\ 0.5] \bullet \begin{bmatrix} 0.21 & 0.21 & 0.33 & 0.25 \\ 0.3 & 0.15 & 0.36 & 0.19 \\ 0.33 & 0.21 & 0.42 & 0.04 \\ 0.61 & 0.33 & 0.06 & 0 \end{bmatrix} = [0.4237, 0.2797, 0.1695, 0.1271]$$

(4) Transportation vehicle B_{14}

$$B_{14} = A_{14} \bullet R_{14} = [0.25\ 0.10\ 0.35\ 0.3] \bullet \begin{bmatrix} 0.36 & 0.42 & 0.22 & 0 \\ 0.12 & 0.15 & 0.54 & 0.19 \\ 0.18 & 0.48 & 0.30 & 0.04 \\ 0.24 & 0.76 & 0 & 0 \end{bmatrix} = [0.2709, 0.3552, 0.2248, 0.1491]$$

(5) Natural disaster B_{15}

$$B_{15} = A_{15} \bullet R_{15} = [0.7\ 0.05\ 0.05\ 0.2] \bullet \begin{bmatrix} 0.58 & 0.42 & 0 & 0 \\ 0.09 & 0.06 & 0.81 & 0.04 \\ 0.12 & 0.33 & 0.24 & 0.31 \\ 0.21 & 0.54 & 0.15 & 0.18 \end{bmatrix} = [0.4361\ 0.3158\ 0.1128\ 0.1353]$$

5.3. influencing factors of third party damage

(1) Buried depth of pipeline B_{21}

$$B_{21} = A_{21} \bullet R_{21} = [0.5\ 0.3\ 0.1\ 0.1] \bullet \begin{bmatrix} 0.54 & 0.21 & 0.22 & 0.03 \\ 0.30 & 0.66 & 0.04 & 0 \\ 0.42 & 0.24 & 0.27 & 0.07 \\ 0.12 & 0.09 & 0.51 & 0.28 \end{bmatrix} = [0.4464, 0.2679, 0.1964, 0.0893]$$

(2) Man-made obstacles B_{22}

$$B_{22} = A_{22} \bullet R_{22} = [0.3\ 0.2\ 0.4\ 0.1] \bullet \begin{bmatrix} 0.15 & 0.39 & 0.27 & 0.19 \\ 0.09 & 0.33 & 0.42 & 0.16 \\ 0.24 & 0.66 & 0.10 & 0 \\ 0.21 & 0.27 & 0.48 & 0.04 \end{bmatrix} = [0.2182, 0.3636, 0.2455, 0.1727]$$

(3) Mark stake of pipeline B_{23}

$$B_{23} = A_{23} \bullet R_{23} = [0.3\ 0.3\ 0.2\ 0.2] \bullet \begin{bmatrix} 0.45 & 0.24 & 0.22 & 0.09 \\ 0.54 & 0.21 & 0.25 & 0 \\ 0.30 & 0.27 & 0.39 & 0.04 \\ 0.15 & 0.21 & 0.64 & 0 \end{bmatrix} = [0.3409, 0.2727, 0.2841, 0.1023]$$

(4) Pipeline patrol frequency B_{24}

$$B_{24} = A_{24} \bullet R_{24} = [0.15\ 0.15\ 0.5\ 0.2] \bullet \begin{bmatrix} 0.30 & 0.48 & 0.18 & 0.04 \\ 0.24 & 0.54 & 0.22 & 0 \\ 0.73 & 0.18 & 0.09 & 0 \\ 0.36 & 0.45 & 0.12 & 0.07 \end{bmatrix} = [0.5435, 0.2174, 0.1630, 0.0761]$$

(5) Property of region passed by pipeline B_{25}

$$B_{25} = A_{25} \bullet R_{25} = [0.2\ 0.3\ 0.2\ 0.3] \bullet \begin{bmatrix} 0.09 & 0.54 & 0.28 & 0.09 \\ 0.12 & 0.67 & 0.21 & 0 \\ 0.12 & 0.58 & 0.27 & 0.03 \\ 0.21 & 0.64 & 0.15 & 0 \end{bmatrix} = [0.2593, 0.3704, 0.2593, 0.1111]$$

(6) Response of alarm system B_{26}

$$B_{26} = A_{26} \bullet R_{26} = [0.2\ 0.3\ 0.3\ 0.2] \bullet \begin{bmatrix} 0.15 & 0.57 & 0.15 & 0.13 \\ 0.18 & 0.67 & 0.12 & 0.03 \\ 0.21 & 0.58 & 0.18 & 0.03 \\ 0.15 & 0.48 & 0.33 & 0.04 \end{bmatrix} = [0.2500, 0.3571, 0.2381, 0.1548]$$

5.4. the importance of risk assessment index system for third party damage

(1) Buried depth of pipeline B_{31}

$$B_{31} = A_{31} \bullet R_{31} = [0.5\ 0.3\ 0.1\ 0.1] \bullet \begin{bmatrix} 0.54 & 0.21 & 0.18 & 0.07 \\ 0.30 & 0.61 & 0.04 & 0.04 \\ 0.39 & 0.27 & 0.27 & 0.07 \\ 0.12 & 0.09 & 0.51 & 0.28 \end{bmatrix} = [0.4630, 0.2778, 0.1667, 0.0926]$$

(2) Ground activity of pipeline B_{32}

$$B_{32} = A_{32} \bullet R_{32} = [0.4\ 0.3\ 0.1\ 0.2] \bullet \begin{bmatrix} 0.42 & 0.36 & 0.12 & 0.10 \\ 0.36 & 0.54 & 0.10 & 0 \\ 0.15 & 0.21 & 0.48 & 0.16 \\ 0.15 & 0.30 & 0.45 & 0.10 \end{bmatrix} = [0.3774, 0.3396, 0.1887, 0.0943]$$

(3) Ground facilities B_{33}

$$B_{33} = A_{33} \bullet R_{33} = [0.5\ 0.3\ 0.1\ 0.1] \bullet \begin{bmatrix} 0.54 & 0.21 & 0.21 & 0.04 \\ 0.30 & 0.66 & 0.04 & 0 \\ 0.42 & 0.24 & 0.27 & 0.07 \\ 0.12 & 0.09 & 0.51 & 0.28 \end{bmatrix} = [0.4505, 0.2703, 0.1892, 0.0901]$$

(4) Warning labels B_{34}

$$B_{34} = A_{34} \bullet R_{34} = [0.3\ 0.3\ 0.2\ 0.2] \bullet \begin{bmatrix} 0.45 & 0.24 & 0.22 & 0.09 \\ 0.54 & 0.21 & 0.25 & 0 \\ 0.30 & 0.27 & 0.39 & 0.04 \\ 0.15 & 0.21 & 0.64 & 0 \end{bmatrix} = [0.3409, 0.2727, 0.2841, 0.1023]$$

(5) Patrol frequency B_{35}

$$B_{35} = A_{35} \bullet R_{35} = [0.15\ 0.15\ 0.5\ 0.2] \bullet \begin{bmatrix} 0.30 & 0.48 & 0.18 & 0.04 \\ 0.24 & 0.54 & 0.22 & 0 \\ 0.73 & 0.18 & 0.09 & 0 \\ 0.36 & 0.45 & 0.12 & 0.07 \end{bmatrix} = [0.5435, 0.2174, 0.1630, 0.0761]$$

(6) Public education and publicity B_{36}

$$B_{36} = A_{36} \bullet R_{36} = [0.2\ 0.2\ 0.4\ 0.2] \bullet \begin{bmatrix} 0.24 & 0.45 & 0.18 & 0.13 \\ 0.18 & 0.63 & 0.15 & 0.07 \\ 0.52 & 0.30 & 0.18 & 0 \\ 0.15 & 0.18 & 0.58 & 0.09 \end{bmatrix} = [0.3883, 0.2913, 0.1942, 0.1262]$$

(7) Alarming system B_{37}

$$B_{37} = A_{37} \bullet R_{37} = [0.2\ 0.2\ 0.3\ 0.3] \bullet \begin{bmatrix} 0.24 & 0.21 & 0.39 & 0.16 \\ 0.21 & 0.48 & 0.21 & 0.10 \\ 0.36 & 0.27 & 0.25 & 0.12 \\ 0.58 & 0.24 & 0.15 & 0.03 \end{bmatrix} = [0.3061, 0.2755, 0.2551, 0.1633]$$

5.5. result

The result of fuzzy comprehensive evaluation to Shaan-jing pipeline is summarized in Table 4.

Table 4. The result of fuzzy comprehensive evaluation.

	Factor	Membership	Consequence	Ranking
Possible destruction styles of third party damage	Excavating equipment	0.4651	H	1
	Cultivation of agricultural machinery	0.4255	MH	2
	Artificial malicious damage	0.4237	MH	3
	Natural disaster	0.4361	MH	4
	Transportation vehicle	0.3552	M	5
Influencing factors of third party damage	Pipeline patrol frequency	0.5435	H	1
	Buried depth of pipeline	0.4464	MH	2
	Property of region passed by pipeline	0.3704	M	3
	Man-made obstacles	0.3636	M	4
	Response of alarm system	0.3571	M	5
	Mark stake of pipeline	0.3409	L	6
The importance of risk assessment index system for third party damage	Patrol frequency	0.5435	H	1
	Buried depth of pipeline	0.4630	H	2
	Ground facilities	0.4505	H	3
	Public education and publicity	0.3883	M	4
	Ground activity of pipeline	0.3774	M	5
	Warning labels	0.3409	L	6
	Alarming system	0.3061	L	7

It is shown that factors of possible destruction styles need to be noticed are excavating equipment, agricultural production and artificial malicious damage; factors of influence level for third party damage need to be noticed are pipeline patrol frequency, buried depth of pipeline and property of region passed by pipeline; factors of importance for risk assessment index system need to be noticed are pipeline patrol frequency, pipeline patrol frequency and ground facilities. To solve these basic factors, corresponding improvement measures are put forward from the following several aspects:

1. Pipeline warning belt is founded and mark stakes of pipeline are added in the region of excavating equipment and agricultural production operated frequently.
2. Pipeline patrol path is optimized, the frequency and quality of pipeline patrol is improved and careful patrol maintenance by pipeline attendant is required in the pipe section prone to third party damage.
3. Increasing buried depth in the wrong place and keeping away from the large plant growth soil area after buried depth inspection.

4. Ground facilities are inspected regularly and lost or damaged identification should be maintained and adjusted timely to ensure the accuracy and completeness.

6. Conclusions

Quantitative risk assessment model of third party damage combined with AHP and FCE is established based on fuzzy mathematic theory in this paper. The importance of each factor for third party damage is calculated and valuable prevention measures could be provided, the main conclusions are as follows:
1. In this paper, factors of possible destruction styles, influence level and importance of risk assessment index system for third party damage were considered by improved AHP/FCE. It could be more comprehensive to evaluate the third party damage of natural gas pipelines compared with other evaluation methods.
2. In the process of improved AHP/FCE, the basic factors of third party damage components were analysed, and improvement measures were proposed for the key factors.
3. In the process of improved AHP/FCE, the weight and membership of factors for third party damage is more accurate with traditional AHP/FCE. The process of improved AHP/FCE was shown and the effectiveness of quantitative risk assessment model is verified by an actual applied example.

7. Acknowledgements

This paper is supported by National Science and Technology Major Project of China (No. 2011ZX05055), Science Foundation of China University of Petroleum, Beijing (No. 2462015YQ0406).

References

[1] Bajcar T, Cimerman F, Širok B. Towards more detailed determination of third party impact on risk on natural gas pipelines: influence of population density[J]. *Process Safety & Environmental Protection*, 2014, 94:509–516.
[2] Liang W, Hu J, Zhang L, et al. Assessing and classifying risk of pipeline third-party interference based on fault tree and SOM[J]. *Engineering Applications of Artificial Intelligence*, 2012, 25(3):594-608.
[3] Peng X Y, Yao D C, Liang G C, et al. Overall reliability analysis on oil/gas pipeline under typical third-party actions based on fragility theory[J]. *Journal of Natural Gas Science & Engineering*, 2016, 34:993-1003.
[4] Li J, Zhang H, Han Y, et al. Study on Failure of Third-Party Damage for Urban Gas Pipeline Based on Fuzzy Comprehensive Evaluation[J]. *Plos One*, 2016, 11(11):e0166472.
[5] Aminbakhsh S, Gunduz M, Sonmez R. Safety risk assessment using analytic hierarchy process (AHP) during planning and budgeting of construction projects[J]. *Journal of Safety Research*, 2013, 46(3):99-105.
[6] Kokangül A, Polat U, Dağsuyu C. A new approximation for risk assessment using the AHP and Fine Kinney methodologies[J]. *Safety Science*, 2016, 91:24-32.
[7] Xinyi Zhou, Xinyang Deng, Yong Deng, et al. Dependence assessment in human reliability analysis based on D numbers and AHP[J]. *Nuclear Engineering and Design*, 2017:243–252.
[8] WU Shun-qiu, LIU Gui-fang, YANG Jian-guang. City's Gas Pipeline Safty Based On Analytic Hierarchy Process And Fuzzy Evaluation[J]. *Journal of Hunan City University (Natural Science)*,2013,22(4):51-53.
[9] Xinyi Zhou, Xinyang Deng, Yong Deng, et al. Dependence assessment in human reliability analysis based on D numbers and AHP[J]. *Nuclear Engineering and Design*, 2017:243–252.
[10] Naifang Zhang. Risk assessment of town gas pipeline networks[D]. *Harbin Institute of Technology*,2011.
[11] Jiao J, Ren H, Sun S. Assessment of surface ship environment adaptability in seaways: A fuzzy comprehensive evaluation method[J]. *International Journal of Naval Architecture & Ocean Engineering*, 2016, 8(4):344-359.

[12] Zhou Z, Zhang X, Dong W. Fuzzy Comprehensive Evaluation for Safety Guarantee System of Reclaimed Water Quality[J]. *Procedia Environmental Sciences*, 2013, 18:227–235.

[13] Li Y, Wang W, Liu B X, et al. Research on Oil Spill Risk of Port Tank Zone Based on Fuzzy Comprehensive Evaluation [J]. *Aquatic Procedia*, 2015, 3:216-223.

Dependency of tensile strength of ductile cast iron on strain rate and temperature

Tomohiro IKEDA[1], Takuo UMETANI[1], Nobuhiro KAI[1], Keisaku OGI[1], Nao-Aki NODA[2] and Yoshikazu SANO[2]

[1]HINODE, Ltd., Azaiwasaki, Miyaki-cho, Miyaki-gun, Saga, 849-0101 Japan
[2] Kyushu Institute of Technology, Sensui-cho, Tobata-ku, Kitakyushu, Fukuoka, 804-8550 Japan

E-mail: t_ikeda@hinodesuido.co.jp

Abstract. The dependency of the tensile strength σ_B^{smooth} and the notch strength σ_B^{notch} on strain rate and temperature were investigated for conventional ferrite-pearlite type ductile cast iron (JIS-FCD500) to make clear the applicability of ductile cast iron to components for welded steel structures. High speed tensile tests were conducted on notched and smooth specimens with varying strain rate and temperature. Charpy absorbed energy was also evaluated on notched specimen with varying temperature. It is found that the tensile strength is in a good relationship with strain rate-temperature parameter R for the wide range of strain rate and temperature. With decreasing R parameter, both σ_B^{smooth} and σ_B^{notch} increase even when Charpy absorbed energy starts decreasing. It should be noted that the notch strength σ_B^{notch} is always larger than the tensile strength at room temperature $\sigma_{B,\ RT}^{smooth}$ in the range of R parameter required for the welded structures. Therefore, the tensile strength $\sigma_{B,\ RT}^{smooth}$ is confirmed to be useful for the structural design.

1. Introduction

Ductile cast irons are widely used as structural members for automobile, railway vehicle, machine tool, and so on. In some cases, they have replaced cast steel and forging products [1]. In recent years, ductile cast irons efficiently substitute for several welded components because of their manufacturing and engineering advantages, such as broad range of mechanical properties and easy fabrication of components with complicated shapes. Furthermore, the cast products show better fatigue property than welded parts [1].

Ductile cast iron consists of nodular graphite particles and iron matrix. Spheroidal graphite is obtained by adding 0.035～0.045% Mg to molten iron. Usually, the required strength of ductile cast iron is obtained by controlling the ferrite-pearlite ratio in the matrix. Typically, ferritic ductile cast iron exhibits higher elongation and lower tensile strength, which is approximately 350 MPa, while pearlitic ductile cast iron has lower elongation and higher tensile strength ranging over 800 MPa. So that various strength levels of ductile cast iron castings were utilized in many fields.

However, for designing of structural components, we should take into account the upper limit of strain rate and lower limit of temperature because the toughness and strength of ferrous materials strongly depends on strain rate and temperature. Table 1 shows the example of the strain rate and temperature range required for welded structural members [2, 3]. The values for Charpy impact test is also given in Table 1 [4, 5], indicating Charpy impact test is not suitable for evaluating the property of welded structural components because the impact speed does not correspond to that in real products failure. Therefore, for example, one of the authors studied on high-speed tensile testing with varying tensile speed to investigate the impact properties of engineering plastics [6-8]. In addition, not Charpy absorbed energy but tensile strength and yield strength have been used for designing ductile cast iron products. To design the castings for welded structural members such as for architectures, it is necessary to make clear the influence of strain rate and temperature on the tensile strength of ductile cast iron.

Therefore, high speed tensile tests were conducted on smooth and notched specimens with varying strain rate and temperature to discuss the applicability of ductile cast iron to welded structural components.

Table 1. Example of strain rate and temperature range acting on welded structural components.

	Industrial field	Strain rate (s^{-1})	Temperature (°C)	*R* parameter (K)
Design	Weld toes of beam-column (Architectural structure)	~ 2 [2]	-18 ~ [3]	4522 ~
Test	V-notch Charpy impact test	≒ 10^2 [4, 5]	-18 (Temperature of this table)	3524

2. Experimental procedure

2.1. Test material

Conventional ferrite-pearlite type ductile cast iron having a tensile strength of ≧500MPa (JIS-FCD500) was prepared with 300kW high frequency electric furnace. Ductile cast iron melt was prepared by the Sandwich Method and cast into sand molds for JIS Type II Y-shaped blocks [9]. All tests specimens are taken from the gray part of Y blocks as shown in Figure 1. Table 2 and Figure 2 show the chemical composition and typical microstructures of test specimens. The matrix structure is composed of 47% ferrite and 53% pearlite. Table 3 shows graphite structures of specimen evaluated according to JIS-G5502. Table 4 shows the results of tensile test on JIS No.4 test piece whose diameter = 14 mm and gage length = 50 mm. The test procedure meets JIS-Z 2241 standard [10].

2.2. High speed tensile test

Figure 3 shows the specimens for high speed tensile test. The smooth specimen was prepared to investigate influence of strain rate and temperature on the tensile strength σ_B^{smooth}. Furthermore, to enhance sensitivity to strain rate and temperature, the notched round bar specimen was also prepared to obtain the notch strength σ_B^{notch}. The notch has the same shape and dimension as the Charpy V-notch specimen. The tensile test was carried out in the range of stroke speed, $\upsilon = 8.5 \times 10^{-3} \sim 2.7 \times 10^2$ mm/s, at temperature T = -130 ~ 22 °C by using electrohydraulic servo testing machine. Then, the tensile strength σ_B^{smooth} and the notch strength σ_B^{notch} were calculated by using equation (1).

$$\left.\begin{aligned} \sigma_B^{smooth} &= 4P_{max} / \pi d^2 \\ \sigma_B^{notch} &= 4P_{max} / \pi d^2 \end{aligned}\right\} \qquad (1)$$

Where P_{max} = maximum load (N), d = test specimen diameter (mm). The strain rate $\dot{\varepsilon}^{smooth}$ and $\dot{\varepsilon}^{notch}$ were calculated for the smooth specimen and notched specimen respectively by using equation (2).

Table 2. Chemical composition of test specimens (mass %).

Material	C	Si	Mn	P	S	Cu	Mg
JIS-FCD500	3.75	2.08	0.40	0.021	0.013	0.24	0.039

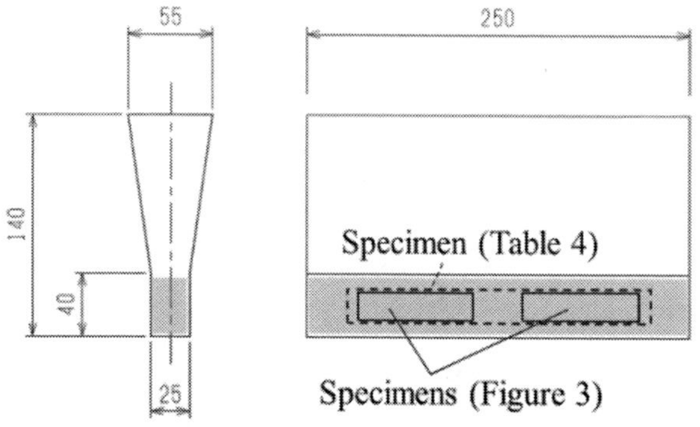

Figure 1. Schematic view of JIS Type II Y-shaped block.

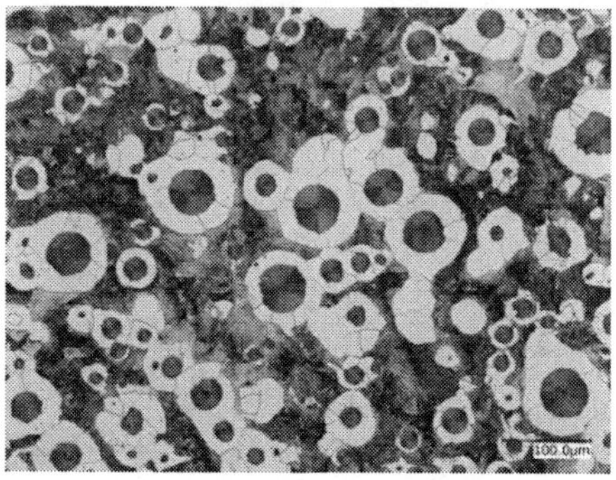

Figure 2. Microstructure of test specimen.

Table 3. Quantitative analysis of graphite nodules structures of as cast sample.

Nodule count (mm^{-2})	Average nodule diameter (μm)	Nodularity (%)	Graphite area fraction (%)
191	27.0	93.2	10.8

Table 4. Tensile property of test specimen.

σ_B (MPa)	$\sigma_{0.2}$ (MPa)	ε_B (%)
566	323	11

σ_B: Tensile strength, $\sigma_{0.2}$: 0.2 % proof stress, ε_B: Fracture strain

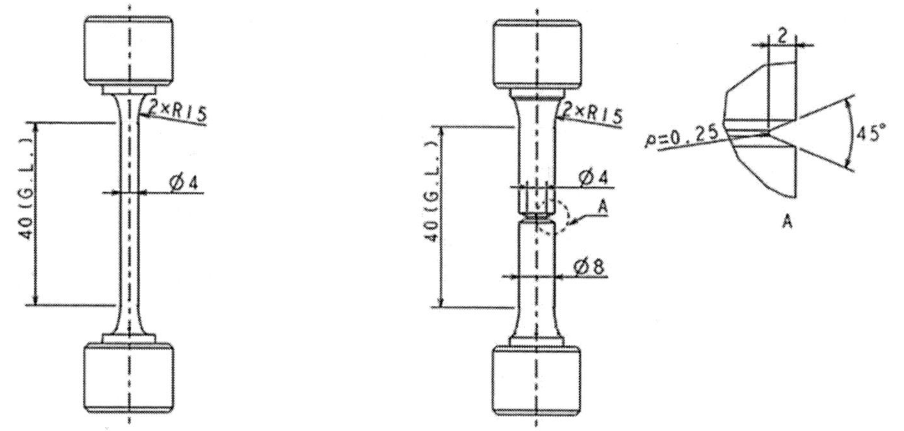

(a) Smooth specimen (d = 4 mm) (b) Notched specimen (d = 4 mm)

Figure 3. Configuration of test specimens (in mm).

$$\left.\begin{aligned}\dot{\varepsilon}^{\text{smooth}} &= u(\text{t}) \;/\; t\,\text{l} \\ \dot{\varepsilon}^{\text{notch}} &= u(\text{t}) \;/\; t\,\text{l} \times K_{\text{t}\dot{\varepsilon}}\end{aligned}\right\} \tag{2}$$

Where u(t) = stroke displacement (mm), t = time (s), l = gauge length, 40 mm. Strain rate concentration factor $K_{\text{t}\dot{\varepsilon}}$ [6-8] was used to calculate $\dot{\varepsilon}^{\text{notch}}$ because measuring the strain rate at the notch root is too difficult. The $K_{\text{t}\dot{\varepsilon}}$ value is 9.49 for the notched specimen in Figure 3.

2.3. *Charpy impact test*

Charpy absorbed energy was evaluated on V-notched specimens at 213 ~ 353 K using a Charpy impact machine with the maximum energy capacity 300 J [11]. The impact speed is estimated as 5.18 × 10^3 mm/s. The total absorbed energy in the fracture process E is determined according to JIS-Z 2242 standard [12].

3. Results and discussion

3.1. *Tensile strength*

Based on Bennett and Sinclair's theory on the influence of strain rate and temperature on yield phenomena of BCC metals [13], strain rate-temperature parameter, R parameter given by the following equation (3), has been introduced to explain the combined influence of strain rate and temperature on the yield stress of steel and ductile cast iron [14-16]. A good correlation between tensile strength and R parameter was also reported on these materials [16, 17].

$$R = T \times \ln(\text{A} \,/\, \dot{\varepsilon}) \tag{3}$$

Where T = temperature (K), A = $10^8\,\text{s}^{-1}$ [13, 16], $\dot{\varepsilon}$ is strain rate (s^{-1}); $\dot{\varepsilon} = \dot{\varepsilon}^{\text{smooth}}$ for smooth specimen, $\dot{\varepsilon} = \dot{\varepsilon}^{\text{notch}}$ for notched specimen. As shown in Figure 4, there is a good correlation between $\sigma_{\text{B}}^{\text{smooth}}$ and R parameter. In Figure 4, the range of R parameter required for the welded structures is also shown as the hatched lines. This required range of R parameter was calculated from the strain rate and temperature for the welded toes of beam-column in Table.1. It is seen that $\sigma_{\text{B}}^{\text{smooth}}$ of JIS-FCD500 continuously increases with the decrease in R parameter over the range of R parameter required for welded structures. Here, decreasing R parameter means increasing strain rate or decreasing temperature, or both. The result of ductile cast iron is consistent with previous research reports on steel [18].

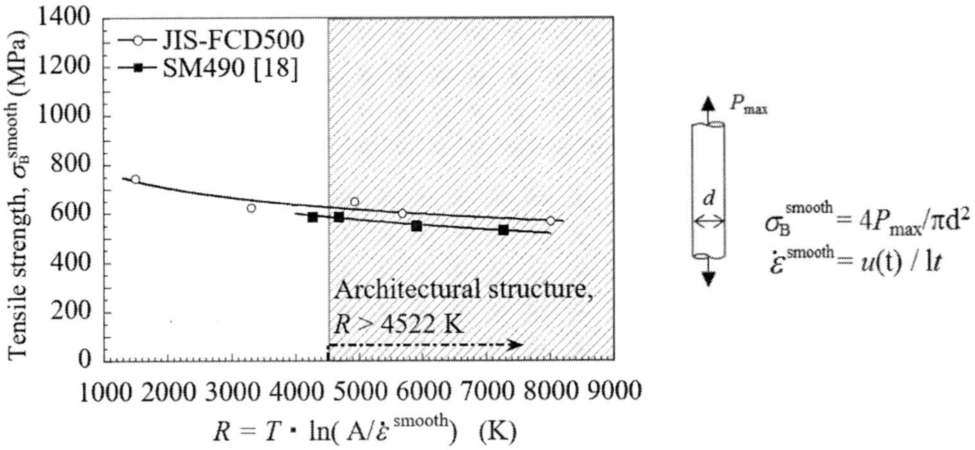

Figure 4. Relationship between tensile strength $\sigma_{\text{B}}^{\text{smooth}}$ and R parameter for JIS-FCD500 and SM490 [18].

IOP Conf. Series: Journal of Physics: Conf. Series **842** (2017) 012049 doi:10.1088/1742-6596/842/1/012049

3.2. Notch strength

The dependency of notch strength σ_B^{notch} of JIS-FCD500 specimens on R parameter is shown in Figure 5. With decreasing R parameter, σ_B^{notch} keeps increasing until R value of 2500, and then it starts decreasing. The σ_B^{notch} values is 100～130 MPa higher than σ_B^{smooth} over the R parameter range required for architectural structure shown by hatched lines.

As the higher strain rate and the lower temperature also give similar effects on the toughness of ductile cast iron [19, 20], we tried to correlate Charpy absorbed energy E to R parameter [11]. Figure 6 shows the relationship between E and R parameter for JIS-FCD500 specimens. The absorbed energy E starts dropping at the critical R parameter indicated by the black arrow. However, it is clear that σ_B^{smooth} and σ_B^{notch} keep increasing even when Charpy absorbed energy starts decreasing.

Figure 7 shows comparison of σ_B^{notch} and the tensile strength at room temperature $\sigma_{B,\ RT}^{smooth}$. It should be noted that σ_B^{notch} is always larger than $\sigma_{B,\ RT}^{smooth}$ in the range of R parameter required for architectural structure. Therefore, tensile strength $\sigma_{B,\ RT}^{smooth}$ is confirmed to be useful in designing ductile cast iron for the welded structural components.

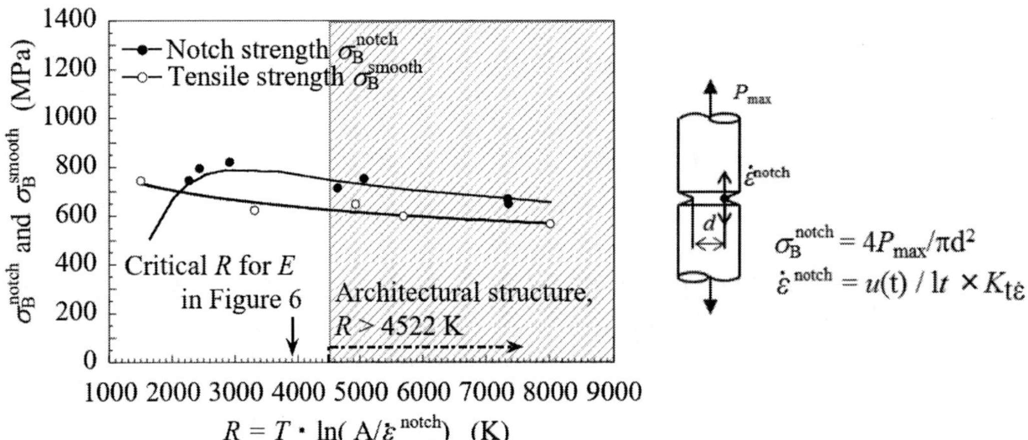

Figure 5. Relationship between notch strength σ_B^{notch} and R parameter for JIS-FCD500.

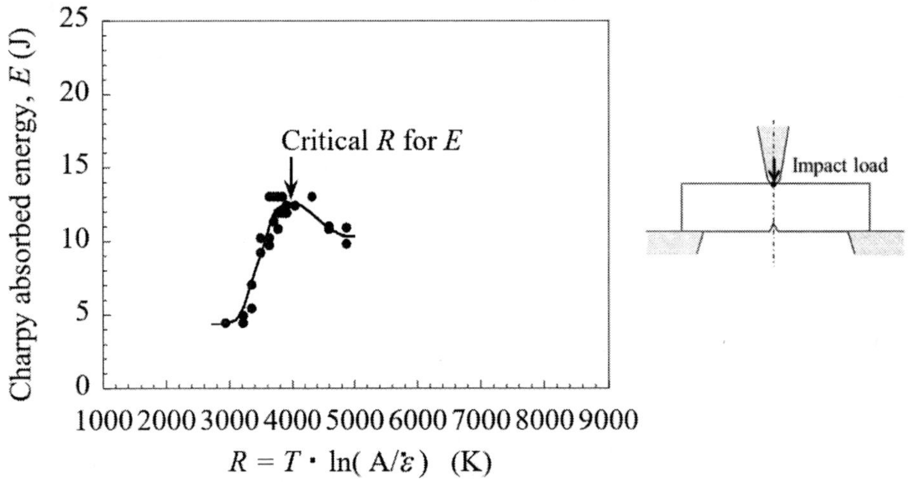

Figure 6. Relationship between Charpy absorbed energy E and R parameter for JIS-FCD500 ($\dot{\varepsilon} = 10^2$ s^{-1} [4, 5]).

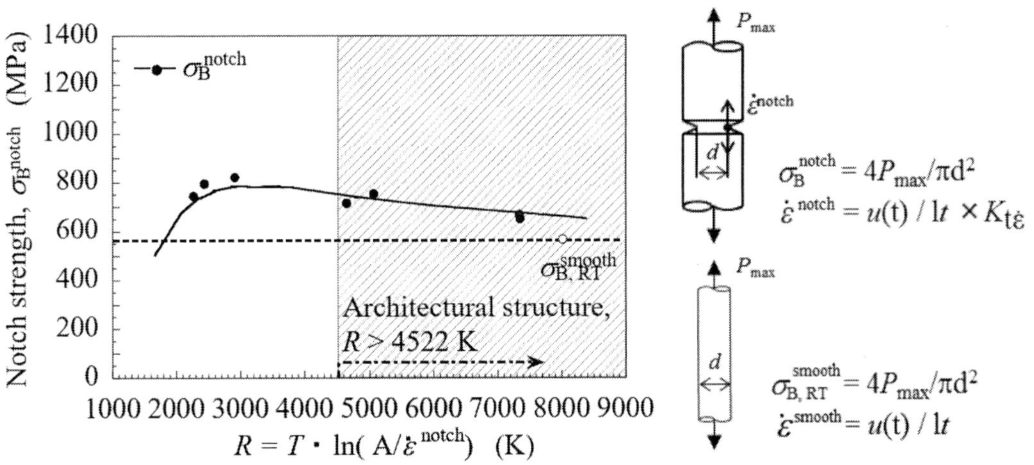

Figure 7. The comparison of σ_B^{notch} and $\sigma_{B,RT}^{smooth}$ in R parameter required from the welded structural design.

4. Conclusions

High speed tensile test was carried out on the notched and smooth ductile cast iron specimens with 47%ferrite-53%pearlite matrix in the range of stroke speed $v = 8.5 \times 10^{-3} \sim 2.7 \times 10^{2}$ mm/s, at temperature $T = -130 \sim 22$ °C. The experimental data were correlated to R parameter, and following conclusions are obtained.

(1) With decreasing R parameter, the tensile strength of both smooth and notched specimens, σ_B^{smooth} and σ_B^{notch}, continuously increases over the range of R parameter required for architectural structures. Therefore, ductile cast iron has wide application potentiality to components for welded steel structures.

(2) The notch strength σ_B^{notch} is consistently larger than the room temperature strength $\sigma_{B,RT}^{smooth}$ in the range of R parameter required for architectural structures. Therefore, $\sigma_{B,RT}^{smooth}$ can be safely used for structural design of architectural components.

References

[1] S.Harada and T.Kobayashi 1999 *Kyujyokokuen-Chutetu no Kyodo Hyoka* (Tokyo: AGNE Gijyutu-Ctnter)
[2] Okashita K, Ohminami R, Michiba K, Yamamoto A, Yomimatsu M, Tanji Y and Miki C 1998 Investigation of the Brittle Fracture at the Corner of P75 Rigid-Frame Pier in Kobe Harbor Highway During the Hyogoken-Nanbu Earthquake *The Journal of Japan Socity of Civil Engineers* **591** pp 243-261
[3] Toyoda M 1996 Lessons Learned from Great Hanshin Earthwuake for Steel Framed Structures with Particular Reference to Materials/Welding *Materia Japan* **35** No.4 pp 370-379
[4] Sakui S, Nakamura T and Ohomori 1963 The Effects of Grain Size and Deformation Rate on the Tensile Proparties of Mild Steel at Low Temperature **49** No.7 pp 996-1003
[5] Sugiura N, Kobayashi T, Yamamoto I, Nishido S and Hayashi K 1995 Comparison of tensile and 3-point bend properties under impact loading in an AC4CH-T6 aluminum casting alloy *Journal of Institute of Lite Metals* **45** No.11 pp 638-642

[6] Noda N.-A, Ohtuka, H, Ando M, Sano Y, Takase Y, Shinozaki T and Guan W 2013 Analysis of dynamic stress concentration and strain rate cocentration for notched specimes used for high speed tensile test *Transactions of the Japan Society of Mechanical Engineers Series A* **79** No.804 pp 1182-1190

[7] Ando H, Noda N.-A, Kuroshima Y, Ishikawa Y and Takeda H 2014 Impact properties of polydimethylsiloxane copolymerized polycarbonate and applicaton of the time-temperature superposition principle *Transactions of the Japan Society of Mechanical Engineers* **80** No.814 DOI: 10.1299/transjsme. 2014smm0149

[8] Noda N.-A, Ohtsuka H, Zheng H, Sano Y, Ando M, Shinozaki T and Guan W 2015 Strein rate concentration and dynamic stress concentration for double-edge-notched specimens subjected to high-speed tensile loads *Fatigue & Fracture of Engineering Materials & Structures* **38** pp 125-138

[9] Japanese standards association (Eds.) 2012 *JIS G 5502 Spheroidal graphite iron castings* (Tokyo: Japanese standards association) pp 1880-1897

[10] Japanese standards association (Eds.) 2011 *JIS Z 2241 Metallic materials –Tensile testing-Method of test at room temperature* (Tokyo: Japanese standards association) pp 321-365

[11] Umetani T, Ikeda T, Sura N, Ashizuka K, Nemoto T, Takada H. and Ogi K 2014 Tensile strength, Fatigue Strength and Impact Strength of Solution Strengthend High Silicon Ferritic Ductile Cast Iron *Journal of Japan Foundry Engineering Society* **86** No.1 pp 36-42

[12] Japanese standards association (Eds.) 2005 *JIS Z 2242 Method for Charpy pendulum impact test of metallic materials* (Tokyo: Japanese standards association) pp 366-377

[13] Bennett P E and Sinclair G M 1966 Parameter Represntation of Low-Temperature Yield Behavior of Body-Centered Cubic Transition Metals *Transactions of the ASME* **65** pp 518-524

[14] Fuji E, Ohkuma Y, Kawaguchi Y and Tsukamoto M 1985 Effects of Temperature and Strain Rate on Dynamic Fracture Toughness of Steel *The society of naval architects of Japan* **158** pp 619-629

[15] Goto K, Hirasawa H and Toyosada M 1944 A Simple Estimating Method of Constiutive Equation for Structural Steel as a Function of Strain Rate and Tmperature *The society of naval architects of Japan* **176** pp 501-507

[16] Yamamoto H, Kobayashi T and Fujita H 1999 Strain Rate-temperature Dependency of Impact Tensile Properties and Ductile Fracture Behavior in Ductile Cast Iron *Tetsu-to-Hagané* **85** No.10 pp 765-770

[17] Minami F, Hashida T, Toyoda M, Morikawa J, Ohmura T, Arimochi K and Konda N 1998 *The society of naval architects of Japan* **184** 453-463

[18] Kurihara, M., Influence of strain rate on tensile properties for steel, Jounal of the Japan Welding Society, Vol.66, No.1(1997), pp.6 (in Japanese)Japanese standards association (Eds.) 2005 *JIS Z 2242 Method for Charpy pendulum impact test of metallic materials* (Tokyo: Japanese standards association) pp 366-377

[19] Yamamoto H, Kobayashi T and Fujita H 2000 Strain Rate Dependency of Ductile-Brittle Transition Behavior in Ductile Cast Iron *Journal of Japan Foundry Engineering Society* **72** No.2 pp 107-112 Minami F, Hashida T, Toyoda M, Morikawa J, Ohmura T, Arimochi K and Konda N 1998 *The society of naval architects of Japan* **184** 453-463

[20] Ikeda T, Umetani T, Kai N, Ogi K, Noda N.-A. and Sano Y 2016 Influence of Silicon Content, Strain Rate and Temperature on Toughness and Strength of Solid Solution Strengthened Ferritic Ductile Cast Iron *MATERIALS TRANSACTIONS* **57** No.12 pp 2132-2138

IOP Conf. Series: Journal of Physics: Conf. Series **842** (2017) 012050 doi:10.1088/1742-6596/842/1/012050

Stress and deformation of ceramic rolls to produce high quality zinc coated steel sheet

Hiromasa SAKAI[1], Nao-Aki NODA[1], Yoshikazu SANO[1], Yasushi TAKASE[1] and Guowei ZHANG [1]

[1]Department of Mechanical and Control Engineering, Kyushu Institute of Technology Sensui-Cho 1-1 Tobata-Ku, Kitakyushu-Shi, Fukuoka, Japan

E-mail: q595103h@mail.kyutech.jp

Abstract. Several ceramic rolls can be used efficiently to produce high quality zinc coated steel sheet used for automobiles. Those ceramics rolls may provide a longer life and reduce the cost for the maintenance because of its large heat resistance and large wear resistance. One example may be seen in sink rolls used in molten zinc bath to manufacture zinc coated steel sheet. Since the rolls are subjected to large thermal stress and mechanical loading, care should be taken for the risk of fracture due to the ceramic brittleness. Moreover, since the sleeve and shafts can be connected only by using small shrink fitting ratio, another failure risk should be considered for the separation of those components [25, 26]. In this paper, therefore, the mechanical and thermal stress and separation condition will be investigated considering the separation of the connected portion. Here, by using the finite volume method the heat transfer coefficient is discussed and by using the finite element method the thermal stress is considered.

1. Introduction

In steelworks, steel manufacturing machinery produces heavy plates, hot-rolled strips, cold-rolled strips, which are used in a broad range of the fields, such as automobiles, electrical appliances, and steel frames for construction. Steel manufacturing machinery includes several types of rolls and rollers, which are used in severely corrosive atmosphere. The first example is found in sink and support rolls used in molten zinc bath to manufacture zinc coated steel sheet. The second example may be seen in conveying rollers used in hot rolling line to transport hot strip. Since those rolls and rollers are used under high temperature and corrosive atmosphere, abrasion and corrosion easily arise on their surfaces in a short period. Therefore, instead of stainless steel rolls and rollers, ceramic rolls and rollers are recently considered to be used [1-8]. Since the ceramics has high heat resistance and abrasion resistance [9], the exchanging cycle of roll and roller can be extended in a large scale. In this paper, we will also focus on the two examples of rolls, first in continuous pickling line [10, 11], second in molten zinc bath [12]. Then, the application of ceramic rolls will be considered in the following sections.

2. Static Strength Analysis and Fatigue Strength Analysis

2.1. Analytical Conditions

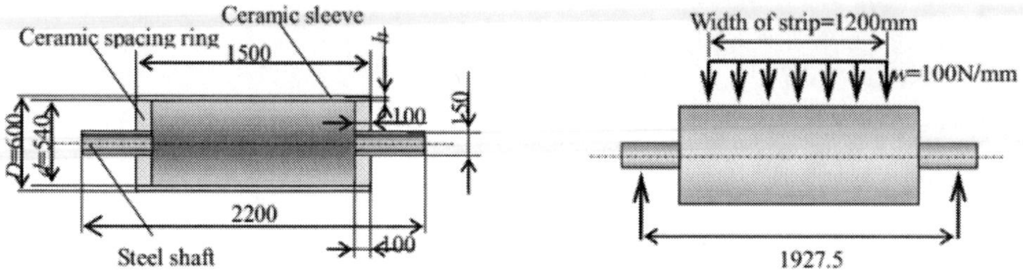

(a) Dimensions (b) Boundary conditions

Figure 1 Ceramics roll dimensions

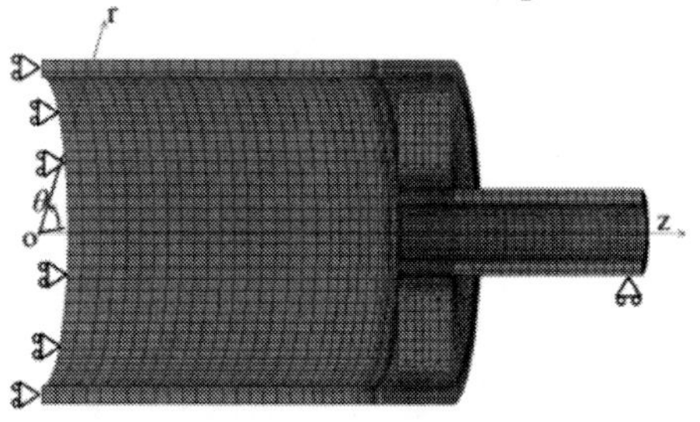

Figure 2 FEM mesh

Table 1 Material properties

	Ceramics (Si_3N_4)	Steel (Hv220)
Young's moduls (MPa)	300	210
Poisson's ratio	0.28	0.3
Tensile strength (MPa)	500	600
Fracture toughness (MPa$\sqrt{m}$)	7	100
Fatigue strength (MPa)	200	300

Define the shrink fitting ratio as δ/d, where δ is the diameter difference with the diameter d. Assume that the roll is subjected to distributed load w=100N/mm and simply supported at both ends as shown in Figure 1. To design the hollow rolls as shown in Figure 1, attention should be paid to the maximum tensile stresses and stress amplitude appearing at the edges of the sleeve [13]. The friction coefficient among the sleeve, the spacer rings and the shafts is assumed as 0.3.

Table 1 shows the material properties of ceramics and steel. Stainless steel is usually used for rolls; however ceramic rolls may provide a longer maintenance span due to their high corrosion resistance and high abrasion resistance.

Figure 2 shows the finite element mesh model of the rolls. The total number of elements is 42920 and the total number of nodes is 53382. The model of 1/4 of the roll is considered due to symmetry.

2.2. *Maximum Tensile Stress*

The shrink fitting ratio when $\delta/d \leq 3.0\times10^{-4}$ will be considered because large values of δ/d are not suitable for ceramics. Figure 3 shows σ_θ under the shrink fitting ratio $\delta/d=3.0\times10^{-4}$ and h=30mm. Figure 3 (a) shows the stress $\sigma_{\theta s}$ due to shrink fitting and Figure 3 (b) shows the stress $\sigma_{\theta max}$ ($=\sigma_{\theta s}+\sigma_{\theta b}$) due to load distribution w=100N/mm after shrink fitting. As shown in Figure 3 (a), the maximum tensile stress at point A and A' are 94.52MPa under the shrink fitting, which becomes 92.70MPa at point A by applying the distribution load after shrink fitting and becomes 93.88MPa at point A' as shown in Figure 3 (b). Maximum stress after applying the load distribution 96.60MPa appears at point B also shown in Figure 3 (b).

2.3. *Effect of Shrink Fitting Ratio and Bending Moment upon the Maximum Tensile Stress* $\sigma_{\theta max}$

Figure 4 (a) shows the relationship between $\sigma_{\theta s}$ and $\sigma_{\theta max}=\sigma_{\theta s}+\sigma_{\theta b}$ vs. δ/d when the load distribution w=100N/mm is applied after shrink fitting. When δ/d is small, the effect of stress concentration at the end of contact area is large and the effect decreases with increasing shrink fitting ratio. From Figure 4 (a) it is found that $\sigma_{\theta max}$ has a minimum value at about $\delta/d=1.0\times10^{-4}$ when h=30mm. When, $\delta/d\geq1.0\times10^{-4}$ $\sigma_{\theta max}$ increases linearly. In addition, $\sigma_{\theta s}$ increases linearly with increasing δ/d at h=30mm. On the other hand, $\sigma_{\theta b}$ decreases with increasing δ/d as shown in Figure 4 (b).

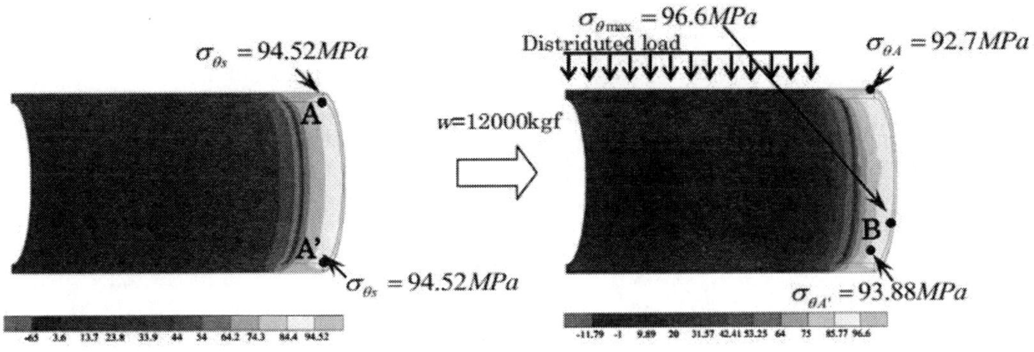

(a) σθ due to shrink fitting (b) σ_θ due to shrink fitting and load distribution

Figure 3 Stress distribution on the sleeve when δ/d=3.0×10^{-4}

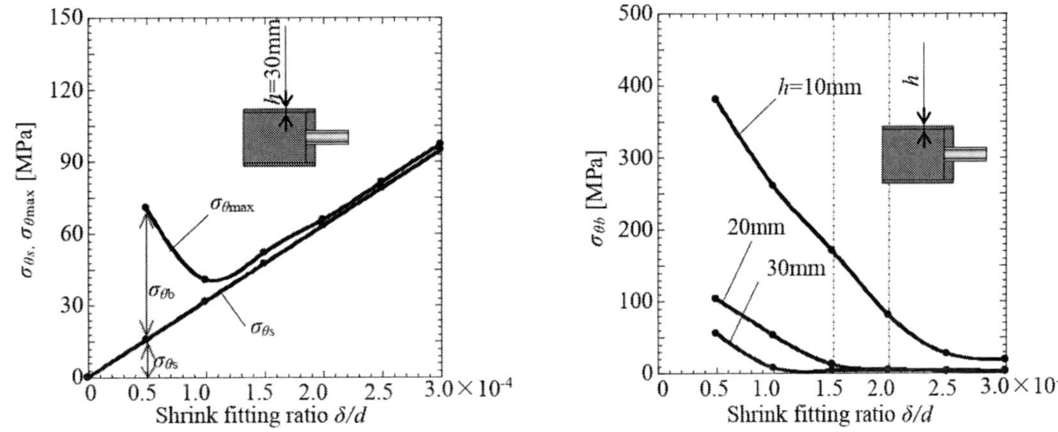

(a) $\sigma_{\theta max}$, $\sigma_{\theta s}$ vs. δ/d (b) $\sigma_{\theta b}$ vs. δ/d

Figure 4 σ_θ vs. δ/d ($\sigma_{\theta max}=\sigma_{\theta s}+\sigma_{\theta b}$)

Here, Figure 4 (b) shows $\sigma_{\theta b}=\sigma_{\theta max}-\sigma_{\theta s}$ vs. δ/d relation for different sleeve thickness h. We assume the thickness as h=10mm, 20mm and 30mm. From Figure 4 (b), it is found that $\sigma_{\theta b}$ decreases with increasing sleeve thickness h. Also, when h=20mm and 30mm, $\sigma_{\theta b}$ becomes constant value over the threshold value of δ/d. That means, when the δ/d is large enough, the spacer rings and the sleeve can be treated as a unit body bonded perfectly [1].

2.4. Fatigue Strength Analysis

Fatigue strength should be also considered for ceramic rolls. Figure 5 shows the mean stress $\sigma_m=(\sigma_{\theta E}+\sigma_{\theta E'})/2$ and the stress amplitude $\sigma_a=(\sigma_{\theta E}-\sigma_{\theta E'})/2$. Due to the rotation of rolls, the values of stress σ_θ at the upper and the lower inside of sleeve are different as shown in Figure 5. Figure 6 shows an endurance limit diagram for h=10mm [14, 15]. In Figure 6, with increasing the shrink fitting ratio δ/d, the stress amplitude σ_a decreases. As shown in Figure 6, the values of stress amplitude σ_a and mean stress σ_m are below the fatigue strengths of ceramics for plain specimen when δ/d>1.0×10^{-4}. In other words, the fatigue strength can be larger by using larger shrink fitting ratio.

2.5. Maximum Stress Appearing at the Outer Surface of Sleeve instead of Inner Surface of Sleeve

In Figure 7 (a) for h=30mm $\sigma_{\theta smax}$ appears at point A (shown in Figure 3 (a)) on the inner sleeve surface. However, in Figure 7 (b) for h=10mm $\sigma_{\theta smax}$ appears at the outer sleeve surface. Two-dimensional shrink fitting systems usually induce tensile stress for outer components and compressive stress for inner components. Then, the maximum tensile stress usually appears at the inner surface of outer components. To understand why the maximum stress appears at the outer surface, the stress

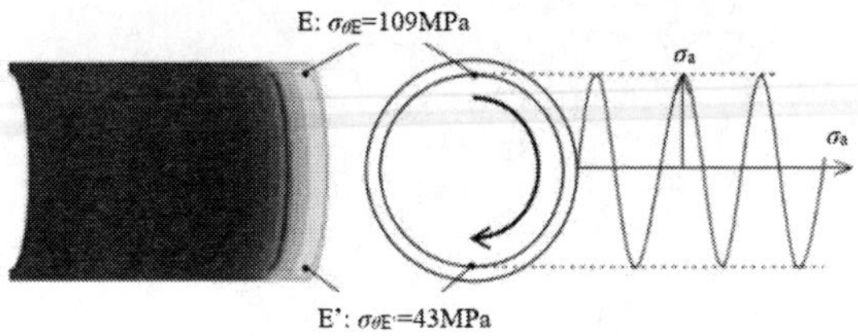

Figure 5 Cycle stress model
(The value of stress when δ/d=3.0×10^{-4})

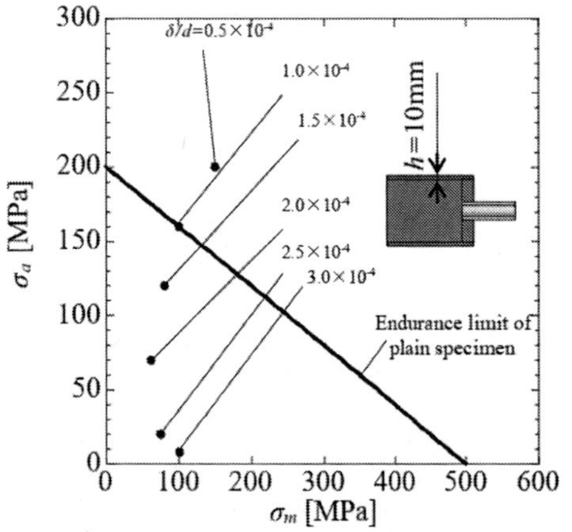

Figure 6 Endurance limit diagram for h=10mm

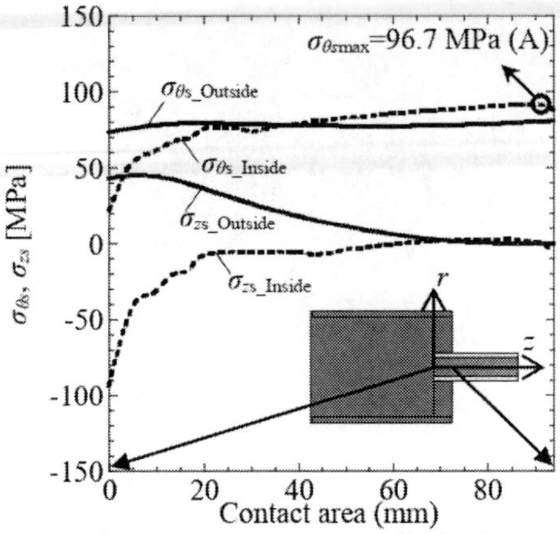

(a) σ_θ, σ_z vs. contact area for h=30mm

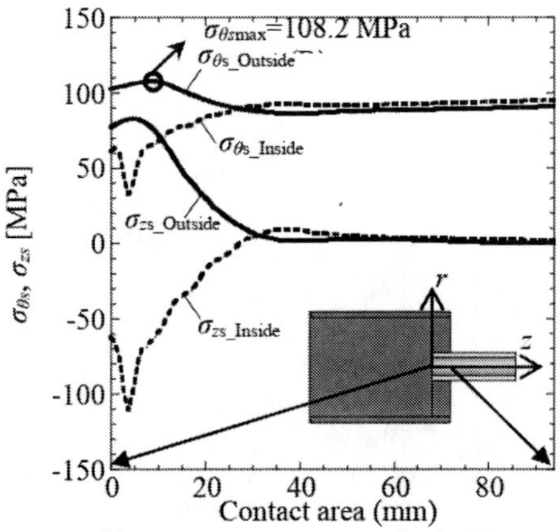

(b) σ_θ, σ_z vs. contact area for h=10mm

Figure 7 Shrink fitting stress distribution on the sleeve when δ/d=3.0×10^{-4}

distributions are compared along the inner and the outer surfaces as shown in Figure 7. Figure 7 (a), (b) indicates that at the left contact end for both h=30mm and h=10mm the outer surface stress is larger than the inner one. It may be concluded that in they three-dimensional partial shrink fitting as shown in Figure 7 the outer stress is always larger than the inner stress. When h=30mm, since the inner stress at the right end is larger than the outer stress at the left end, the maximum stress is on the inner surface similar to 2D shrink fitting system. Care should be taken for the position of the maximum stress to design the ceramic sleeve roll different from the 2D system.

3. Heat-transfer coefficient Analysis

3.1. Analytical Conditions

Besides the ceramic roll, thermal stress should be consider for ceramic stokes dipped into molten metal, which have been used for low-pressure die casting [16]. Usually coefficient of thermal shock has been useful to measure often used to measure the severity [17]. In our previous studies, therefore, simple two-dimensional circular geometries were assumed and the heat transfer coefficients for ceramic structurers dipped into molten aluminum and molten zinc under unsteady flow were analyzed to discuss the thermal stress [3, 7, 18, 19]. Few studies targeting except molten aluminum and molten zinc are available for two-dimensional heat transfer coefficients [20-23]. In this study, two- and three-dimensional cylindrical geometries will be considered and compared with each other to analyze heat-

Table 2 Material properties of molten zinc and aluminum

Physical property	Zinc (420℃)	Aluminum (750℃)
Thermal conductivity, λ (W/m·K)	58.8	112
Kinematic viscosity, ν (mm^2/s)	0.489	0.96
Isobaric specific heat, C_p (kJ/kg·K)	0.505	1.10
Viscosity, η (mPa·s)	3.26	2.20
Prandtl number, P_r	0.028	0.022

Table 3 Material properties of ceramics

Mechanical property	Sialon
Thermal conductivity, (W/m·K)	17
Specific heat, (J/kg·K)	650
Coefficient of linear expansion, (1/K)	3.0×10^{-6}
Young's modulus, (GPa)	294
Specific weight	3.26
Poisson's ratio	0.27
4 Point bending strength, (MPa)	1050
Fracture toughness, (MN/mm$^{3/2}$)	7.5

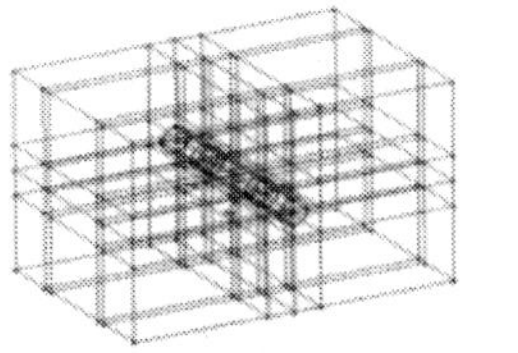

Figure 8 Mesh for 3D model

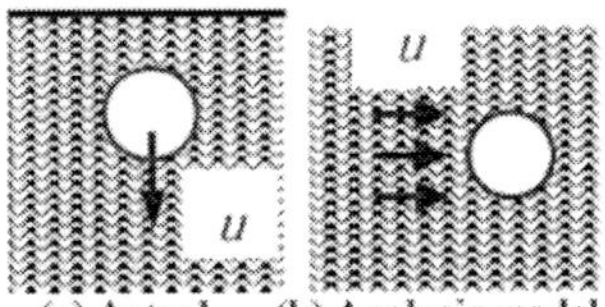

(a) Actual (b) Analysis model

Figure 9 Modeling

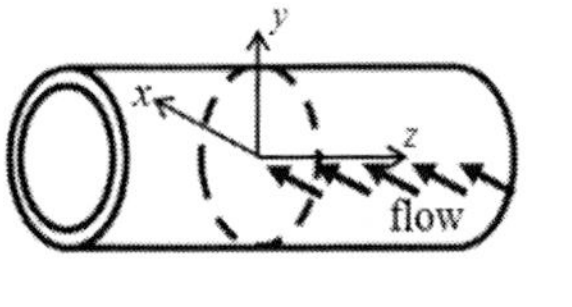

Figure 10 rectangular coordinates

Table 4 Results of heat-transfer coefficient for different models in molten aluminium (α: W/m^2 · K$\times10^3$)

Model	α_{max}, α_{mid}, α_{min}
Circle, 540; $2a$ = 540mm	5.21 (max), 3.07, 1.02 (min)
Circle, 170; $2a$ = 170mm	8.38 (max), 5.63, 1.02 (min)
Circle, 20; $2a$ = 20mm	24.1 (max), 14.6, 0.49 (min)
Ellipse, a=10, b=20	30.3 (max), 9.55, 4.53 (min)
Ellipse, a=10, b=325	39.9 (max), 2.08, 1.61 (min)

Model	α_{max}, α_{mid}, α_{min}
Rectangle, r=20, 650	17.5 (max), 2.45, 1.88 (min)
Rectangle, r=20, 45°	10.6(max), 3.24, 3.05(min); 2.15(max), 1.41, 0.84 (min)
Rectangle, r=20, 90°	4.85(max), 3.51(min), 3.52, 3.51(min); 2.85(max), 1.35(min), 1.30, 1.35(min)

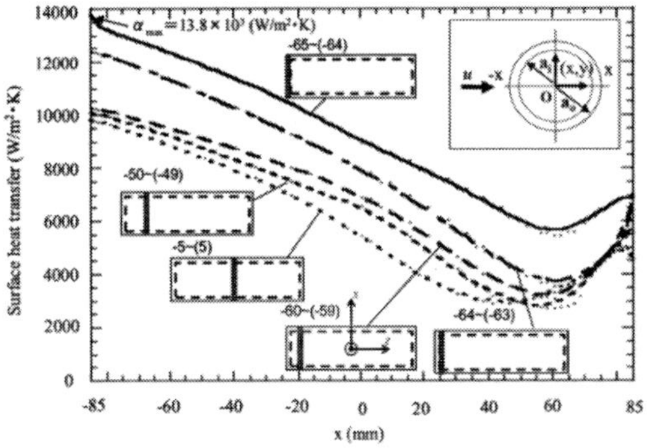

Figure 11 Heat-transfer coefficient as a function of x for cylindrical model in molten aluminum

transfer coefficient when ceramic structures is dipping into molten metal. Here, elliptical and rectangular shapes will be also considered in comparison with the cylindrical shape.

Three-dimensional mesh model shown in Figure 8 was generated by using commercial softwares SOLIDWORKS and GAMBIT2.4.6. The hexahedral mesh is used. The element number is about 3,000,000. Since the dipping of the circular body in Figure 9 (a) is equivalent to the flow problem around the circular body in Figure 9 (b), Figure 9 (b) will be considered. In present study, ANSYS Fluent 6.3.26(Fluent Inc.) was used for calculation with Navier-Stokes equations for a flow of an incompressible viscous fluid being solved using Finite Volume Method (FVM).

Table 2 shows material properties of molten zinc and aluminum. Table 3 shows material properties of ceramics.

To calculate the heat-transfer coefficient more accurately, the finite volume method is applied to the static 2D heat-transfer model. Here, the roll is fixed, and the fluid velocity is assumed as u= 25 mm/s. By using 3Dmodel, the effect of tiriteness the z-direction is also discussed.

3.2. Effects of 2D Models Shape on Heat-transfer coefficient

Table 4 shows the results of the heat-transfer coefficient when the flow rate u=25mm/s of molten aluminum for different ceramic geometries of 2D circle, 3D cylinder, 2D rectangle and 2D ellipse. In Table4, a is miner radius and b is major radius of ellipse. To compare with the rectangle a : b is

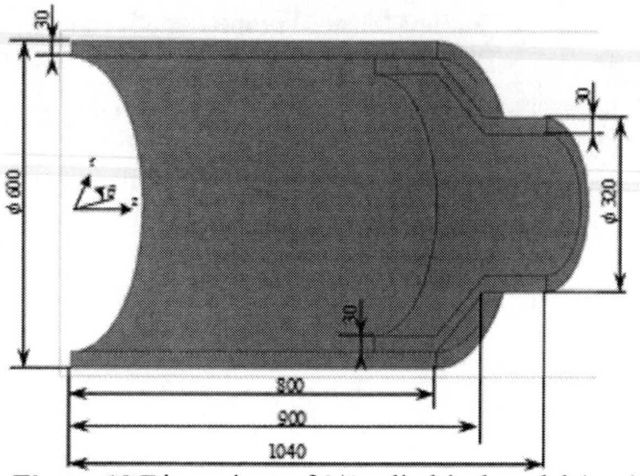

Figure 12 Dimensions of 1/4 cylindrical model (mm)

Table 5 Material properties of Si_3N_4

Thermal conductivity [W/m•K]	65
Specific heat [J/kg•K]	680
Thermal expansion coefficient [1/K]	3.0×10^{-6}
Young's modulus [GPa]	300
Tensile strength [MPa]	500
Mass density [kg/m³]	3200
Poisson's ratio	0.3
Fracture toughness [MPa√m]	7.7

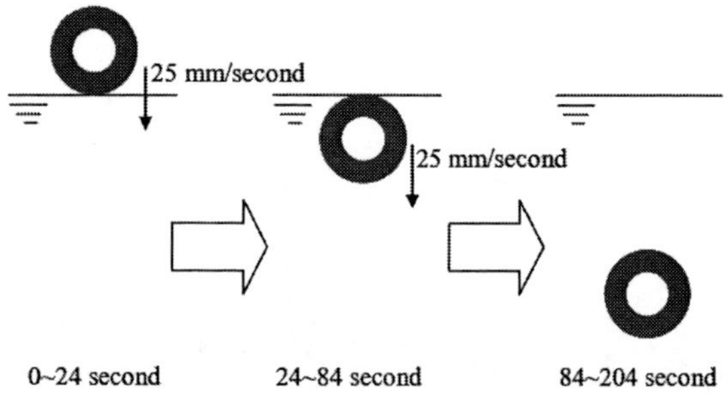

Figure 13 Dipping process

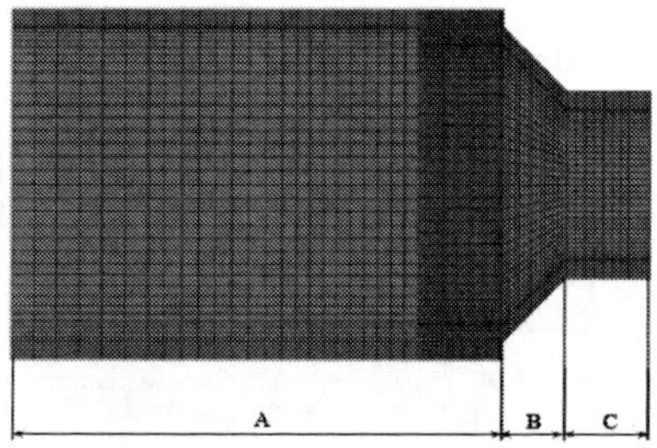

Figure 14 FEM model of ceramics

assumed as 20 : 650. From Table 4, the maximum heat-transfer coefficient appears at the front end of 2D circle and 2D rectangle. With increasing *a*/*b*, the maximum heat-transfer coefficient increases.

3.3. *Heat-transfer coefficient during Dipping in Horizontal Direction*

Figure 11 shows the results of the heat-transfer coefficient of 3D cylinder. Each line shows the results for fixed value of the position z (see Figure 10). From Figure 11, it is found that the results of z=0 almost coincide with the 2D results. In Figure 11, the maximum heat-transfer coefficient α_{max}= 13.8×10^3(W/m² · K) and the minimum heat-transfer coefficient α_{min}=3.0×10^3(W/m² · K) . Therefore, the value of α_{max} is about 4 times larger than the value of α_{min}.

4. Thermal Stress Analysis of Ceramic Rolls Considering Separation

4.1. *Analytical conditions*

In this study, it is necessary to consider the thermal local deformations for the sleeve and the shaft in addition to the heat-transfer analysis for the whole structure shown in Figure12. The thermal-mechanical coupled analysis is therefore performed using three-dimensional hexahedral element with eight nodes. The dimensions are shown in Figure 12. In addition, the temperature of molten metal is 480°C and the initial temperature of roll is 20°C. The sleeve and the shaft are connected by shrink fitting. Define the shrink fitting ratio as δ/d, where δ is the diameter difference with the diameter d. Here, $\delta/d=2.0\times10^{-4}$ is used for investigation. Here, silicon nitride (Si_3N_4) [24] having a high thermal conductivity is assumed for the analysis as shown in Table 5. Moreover, the dipping process is shown in Figure 13. The Roll is gradually dipped into molten metal at 25mm/s from 0s to 84s, while it reaches a steady state from 84s to 204s. Surface heat-transfer coefficient is calculated by FVM software ANSYS Fluent 6.3.26 for the roll dipped into the molten metal, which has a sleeve connected with two shafts by shrink fitting. In actual analysis, the roll is fixed and molten metal flows from the -

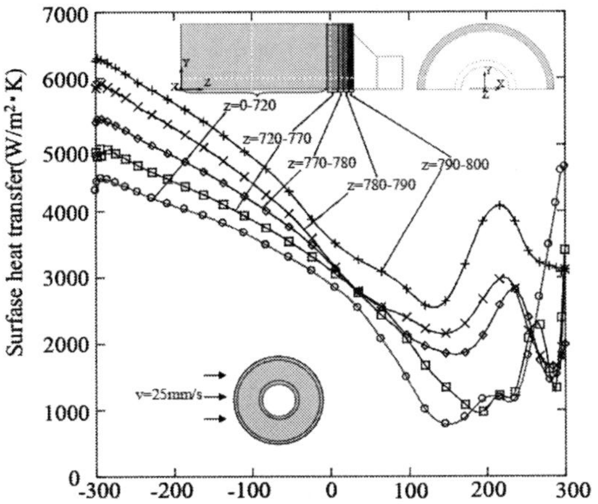

Figure 15 Surface heat transfer coefficient as a function of x for roll in the molten zinc

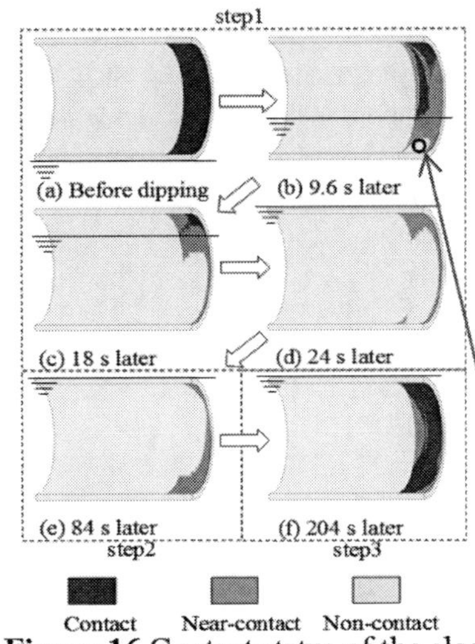

Figure 16 Contact status of the sleeve during dipping

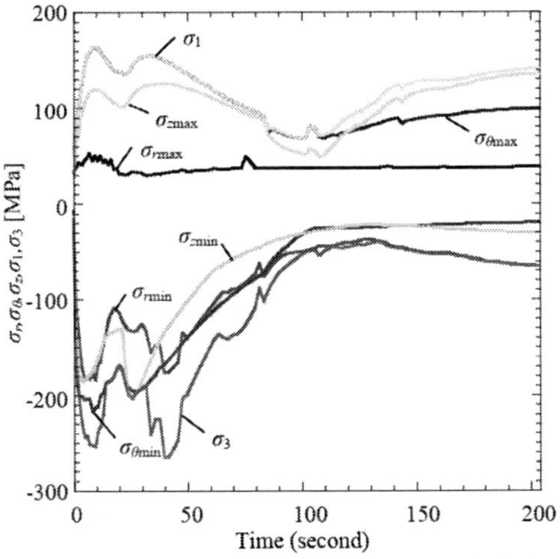

Figure 17 Maximum stress vs. time for sleeve

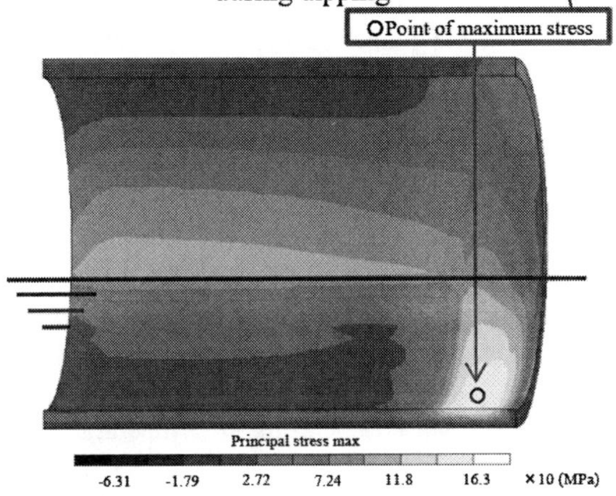

Figure 18 Stress distribution on the sleeve

x-direction to the x-direction at 25 mm/s. Figure 14 shows FEM model. The results of surface heat-transfer coefficients on part A (see Figure 14) of the roll are shown in Figure 15. The maximum of the surface heat-transfer coefficient is 6264 W/m^2 · K at the position (x=-350mm and z= 800 mm) as shown in Figure 15. Here, as the heat-transfer coefficient at the internal surface of the roll α_{min}=1.0×10^3(w/m^2 · K) is assumed, which is the minimum value of the outer surface. After when the roll is installed at the position of 2100mm in Figure 13 the same minimum heat-transfer coefficient α_{min}=1.0×10^3(w/m^2 · K) is used. Thermal stress analysis is performed by FEM in the following.

4.2. Thermal Stress Considering Separation during Dipping

Figure 16 shows the contact status of ceramics roll shown in Figure 12 during dipping. It can be seen that the contact area becomes less during dipping. There is almost no contact at 24s after the roll is dipped into molten metal as shown in Figure 16 (d). However, as dipping goes on, the sleeve and the shaft become contacted again after 204s as shown in Figure 6 (f). Figure 17 shows the maximum and minimum stresses on the sleeve with time during dipping. Here, principal stress (σ_1: Maximum principal stress, σ_3: Minimum principal stress) and the stress components in cylindrical coordinate (σ_r, σ_θ, σ_z) are indicated. Maximum tensile stress σ_{1max}=162.7MPa appears on the sleeve at 9.6s after dipping as shown in Figure 17. The maximum tensile stress σ_{1max}=162.7MPa is smaller than the tensile strength of Si_3N_4 500MPa (Safety factor=3.0). Figure 18 shows the stress distributions for the sleeve when maximum tensile stress appears. It can be seen from Figure 18 that maximum tensile stress on the sleeve appears on the inside surface at the bottom end where is separated from the shaft (see Figure 16). The contact status of the sleeve with the shaft affects the thermal deformation of the sleeve. This is because heat conduction between the sleeve and the shaft is not performed if they are separated each other.

References

[1] N.A. Noda, Hendra, Y. Takase, M. Tsuyunaru: Jounal of Solid Mechanics and Materials Engineering, 2(2008), 11, 1410-1419.
[2] Hendra, M. Tsuyunaru, N.A. Noda, Y. Takase: *Key Engineering Materials*, 385-387(2008), 513-516.
[3] W. Li, N.A. Noda, H. Sakai, Y. Higashi: *Key Engineering Materials*, 452-453(2011), 241-244.
[4] W. Li, Ph.D dissertation, Kyushu Institude of Technology, 2012, 12
[5] N.A. Noda, Y. Yamada, Y. Sano, S. Sugiyama, S. Kobayashi: *Engineering Failure Analysis*, 15(2008), 4, 261-274.
[6] N.A. Noda, Hendra, Y. Takase, W. Li: *Jounal of Solid Mechanics and Materials Engineering*, 3(2009), 10, 1090-1100.
[7] N.A. Noda, Hendra, Y. Takase, H. Ogura, Y. Higashi: *Jounal of Solid Mechanics and Materials Engineering*, 4(2010), 8, 1198-1213.
[8] Y. Takase, W. Li, Hendra, H. Ogura, Y. Higashi, N.A. Noda: *Key Engineering Materials*, 452-453(2011), 233-236.
[9] T. Iwata, H. Mori: *Plant Engineer*, 15(1983), 6, 55-59.
[10] N.A. Noda, Hendra, M. Oosato, K. Suzumoto, Y. Takase, W. Li: *Key Engineering Materials*, 462-463(2011), 1140-1145.
[11] N.A. Noda, Y. Sano, Y. Takase, W. Li, H. Sakai: *International Journal of Engineering Innovation and Management*, 1(2011), 77-82.
[12] S. Hamada, E. Ogawa, K. Shimizu, N.A. Noda, K. Kishi, S. Koga, sokeizai, 51, 12(2010), 54-58.
[13] S. Harada, N.A. Noda, O. Uehara, M. Nagano: *Transactions of the Japan Society of Mechanical Engineering* A, 57(1991), 539, 173-178.
[14] M. Masuda, T. Soma, M. Matsui, I. Oda: *Journal of the Ceramic Society of Japan*, 96(1988), 3, 277-283.
[15] M. Masuda, T. Makino. Y. Nakasuji, M. Matsui: *Materials Science & Technology*, (1993), 55-70.
[16] Y. Harada, K. Komatsu, N. Nomura, Patent Application Publication, No. H3-63565, 1991.
[17] JSME Data Book, Heat Transfer 5th Edition, 2009.
[18] N.A. Noda, M. Yamada, Y. Sano, S. Sugiyama, S. Kobayashi: *Transactions of the Japan Society of Mechanical Engineering* A, 70(2004), 700, 1755-1763.
[19] Hendra, PhD dissertaion, Kyusyu Instutude of Technology, 2010, 9
[20] H. Nakamura, K. Kamemoto: *Transactions of the Japan Society of Mechanical Engineering* B, 67(1996), 662, 2525-2532.
[21] K. Sugiyama, R. Ishiguro: *Transactions of the Japan Society of Mechanical Engineering* B, 51(1985), 464, 1269-1276.
[22] M. Shimizu, T. Nakagawa, Y. Kikuchi: *Transactions of the Japan Society of Mechanical Engineering*., 0.35-1(2003), 203-204.
[23] K. kawakita, N. Takenaka, D, Akie: *Transactions of the Japan Society of Mechanical Engineering*., 02-1(2002), 253-254.
[24] H. Imamura, M. Sohue, S. Hamayoshi, Patent Application Publication, No. H14-293642, 2002.
[25] N.A. Noda, D. Suryadi, S. Kumasaki, Y. Sano, Y. Takase: *Engineering Failure Analysis*, 57(2015), 261-274.
[26] N.A Noda, Y. Xu, D. Suryadi, Y. Sano, Y. Takase: *Journal of ISIJ International*, 56(2016), 2, 303-310.

Coming out prevention by stopper for the shrink fitted sandwiched shaft from the ceramic sleeve

Guowei ZHANG [1],Nao-Aki NODA[1], Yoshikazu SANO[1] , Hiromasa SAKAI[1] and Kazuhiro ODA[2]

[1]Department of Mechanical and Control Engineering, Kyushu Institute of Technology

Sensui-Cho 1-1 Tobata-Ku, Kitakyushu-Shi, Fukuoka, Japan

[2]Department of Mechanical and Energy Systems Engineering, Oita University

700 Dannoharu, Oita-Shi, Oita, Japan

E-mail: q595105c@mail.kyutech.jp

Abstract : Ceramic roller can be used in the heating furnace conveniently because of its high temperature resistance. The roller consists of sleeve and steel shaft connected only under a small shrink fitting ratio because of the brittleness. However, the coming out of the shaft may often happen from the ceramic sleeve under repeated bending load. Therefore, how to prevent the coming out failure becomes an important issue. Based on the previous study, a two-dimensional shrink fitted structure is considered by replacing the shaft with the inner plate and by replacing the sleeve with the outer plate. Then, this research focuses on preventing the inner plate coming out from the outer plate by introducing a newly designed stopper on the outer plate. The simulation results shows that the coming out phenomenon can be prevented effectively due to the contact between the inter plate and the stopper installed on the outer plate. In order to evaluate the contact force between the inner plate and the stopper, the coming out mechanism is clarified. To prevent the coming out by stopper safely, the effects of the magnitude of repeated load and the friction coefficient upon the contact compressive force are investigated under large number of loading cycles by using 2D simulation.

1 Introduction

Steel conveying rollers are used in the heating furnace for producing high-quality steel plates for automobiles. The conventional roller material is steel with ceramic spray coating on the outside of sleeve. To reduce the temperature, inside of the roller is cooled by water. However the thermal expansion mismatch may exceed the adhesion strength of the ceramic layer and causes failure on the roller surface such as crack, peeling, and wearing [1]. Therefore, the life of the roller becomes quite short.

Figure 1 shows a new ceramic roller consisting of steel shaft at both ends and ceramic sleeve having high heat resistance, high corrosion resistance and high wear resistance [2, 3]. Several previous studies suggested that the shrink fitting system may be the most suitable joining method for cylindrical structures to reduce the maintenance time and cost for the shaft replacement [4-8]. However, since only a small shrink fitting ratio can be used because of the ceramic brittleness [9-15], the coming out of the shaft from the sleeve may happen during operation. In our previous studies, therefore, the coming out simulations are performed by using the finite element method [4, 5] for this kind of roll. Suryadi [4] et al built 3D FEM model to simulate the coming out but only until N = 5 loading circles because of the large calculation time. Therefore, Xu [5] et al developed 2D model as shown in Figure 2 to reduce the calculation time and obtained the results until N = 40 circles or more. Those studies [4, 5] have proved that the coming out behavior can be realized in the 2D and 3D simulations. In this paper, how to prevent the shaft from coming out will be newly discussed when the shrink fitting ratio is smaller and the load is larger.

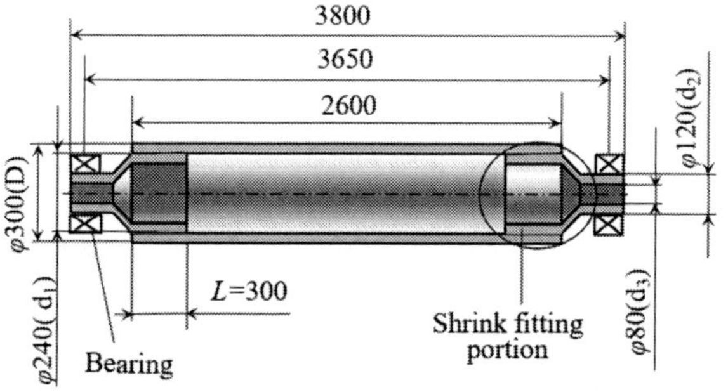

Figure 1.Structure and dimensions of a real ceramic roller (mm)

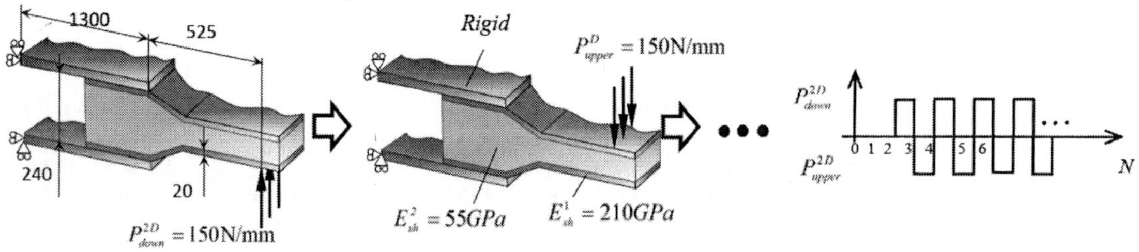

Figure 2. Two-dimensional alternate loading model (Model 2D-alt.)

This paper considers the shrink fitted sandwiched shaft and the newly added stoppers to simulate the prevention of coming out from the roller sleeve as an innovation solution. The two dimensional model will be considered instead of the three dimensional model subjected to rotary bending, which can save much computational time[5]. Here, the inner plate and outer plate will be considered by replacing the shaft and sleeve as shows in Figure 2. The finite element method is applied to simulating the behavior; then several mechanical factors is considered to explain the mechanism of coming out and also to prevent the coming out of the inner plate.

2 Analysis condition

Figure 3 illustrates a new simplified 2D model with the stopper against the old one. Table 1 shows the material properties of the models. For 3D model, the shaft material was steel. For 2D model, the material of 2D shaft assumed to be the composite as described in the literature[5], and the outer plate (Ceramic sleeve) is assumed to be rigid.

Regarding the 2D model, the point load in the 3D model is replaced by the distributed load as shown in Figure 2. Here, the 2D model has a unit thickness which corresponds to 120mm width and the standard load 18KN is represented by 150N/mm considering the diameter of the shaft.

The purpose of this study is to realize the coming out prevention by the stopper for large number of cycles by the numerical simulation to find out the mechanism. In this study, the simulation is considered under room temperature because the coming out occurs more easily. Here, static structural analysis is performed to the models by using MSC. Marc/Mentat 2012 with the Full Newton-Raphson method. A half model is considered due to the symmetry.

Table 1 Material properties

Model	3D−alt., 2D−alt.			3D−rot.	
	Sleeve	Shaft		Sleeve	Shaft
	Rigid	Steel	Composite part	Ceramics	Steel
Young's modulus [GPa]	∞	210	55	300	210
Poisson's ratio	1	0.3	0.3	0.28	0.3
Tensile strength [MPa]	∞	600	−	500	600
Mass density [kg/m^3]	0	7800	7800	3200	7800

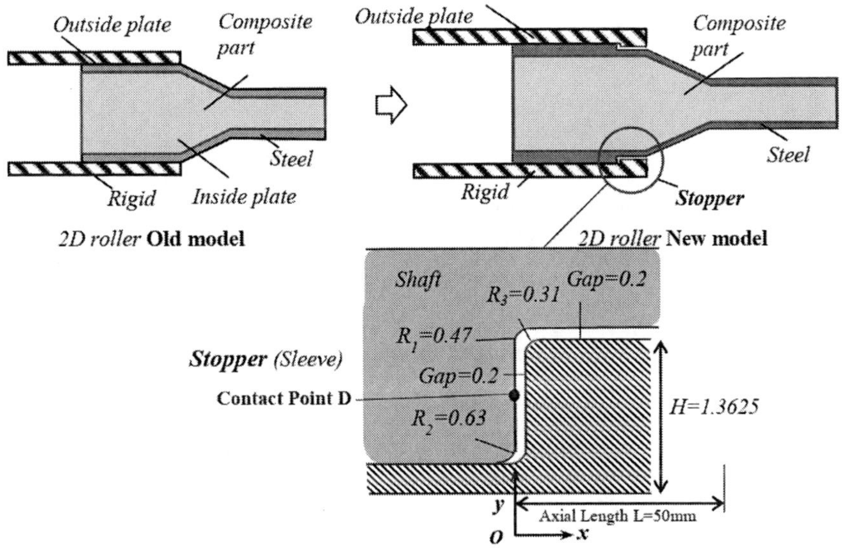

Figure 3. New simplified 2D model with the stopper against the old model of the roller

Equivalent shrink fitting ratios as the roller in the heating furnace may be applied by considering the shaft expansion. The shrink fitting ratio δ/d is defined as the height difference δ divided by height d=240 mm. The shrink fitting is considered in the range $\delta/d=0.01\times10^{-3}\sim0.4\times10^{-3}$. Then, the ratio 0.2×10^{-3} is used as a reference value.

3 Contact condition between the inner plate and the stopper installed on the outer plate under different shrink fitting ratios and bending loads

In order to investigate the contact status between the inner plate and the outer stopper, Figure 4 (a) shows the displacement u_{xD} in the x-direction at Point D when load P=150N/mm is repeated N times for the fixed shrink fitting ratio from $\delta/d=0.01\times10^{-3}\sim0.4\times10^{-3}$. It can be seen that the smaller shrink fitting ratio causes the contact between the inner plate and the stopper easily under small N. If there is no stopper, the shaft continues coming out as have been discussed in the previous research [5]. After the coming out is prevented by the stopper, a constant amplitude of the displacement can be seen when $\delta/d = 0.01\times10^{-3}$and $\delta/d =0.03\times10^{-3}$.

To find out the effect of magnitude of loads on the contact status, Figure 4(b) shows the displacement u_{xD} in the x-direction at Point D for the fixed fitting ratio δ/d=0.2×10^{-3} with fixed friction coefficient μ=0.3. From Figure 4(b), it can be seen that P=600N/mm is the threshold value of the contact and below

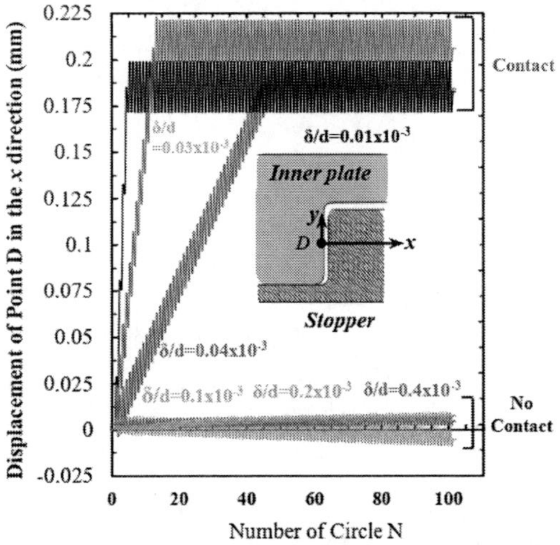

(a)

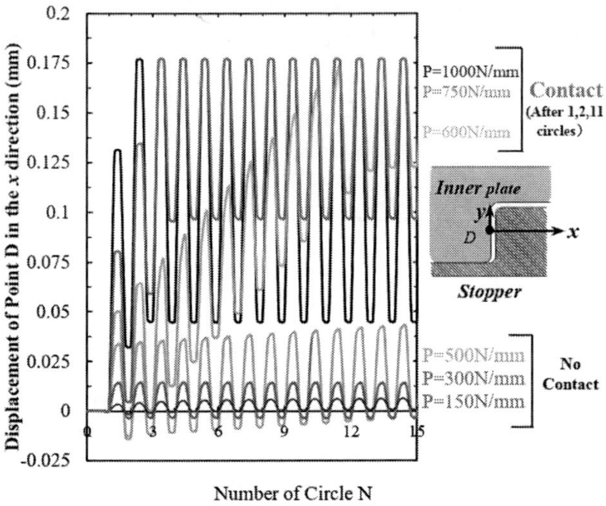

(b)

Figure 4. (a) Displacement u_{xD} vs number of cycle N under fixed load P=150N/mm. (b) Displacement u_{xD} vs number of cycle N under fixed δ/d =0.2×10^{-3}and μ=0.3

which no contact anymore. When load is smaller than the threshold, even after several circles N the displacement u_{xD} does not increases and therefore there is no contact. In other words, the relative position between the inner plate and the stopper is almost steady from the beginning of N. It should be noted that the displacement for P=150, 300, 500N/mm is far less than the gap distance G=0.2mm in Figure 3. On the contrary, when load P= 750N/mm, only after N = 3 circles, the shaft has contacted with the sleeve stopper. The contact appears after for P=600N/mm at N = 11 circles, for P=1000N/m at N = 2 circles, and for P=2000N/mm at N = 1circle. It may be concluded that with increasing load P, the coming out speed increases and the number of cycle until the contact deceases when load P is smaller than 2000N/mm.

4 Coming out mechanism and how to prevent the inner plate from coming out

The simulation is performed to investigate the contact force F_x appearing on the stopper. The contact force F_x may be regarded as the driving force for the coming out. Therefore, to obtain the reaction F_x at the stopper may be useful for understanding the coming out mechanism and finally for preventing from coming out.

Figure 5 shows the contact force F_x when upward load is applied with varying the magnitude of loads P and friction coefficients μ under fixed δ/d =0.2×10^{-3}. In Figure 5, it is seen that with increasing the magnitude of the applied load P, the compressive force F_x increases. It is also seen that with increasing the friction coefficient μ, the compressive force F_x increases.

Due to the shrink fitting and the applied load P, the inner plate is contacting with the outer plate at several portions such as the upside portion, the downside portion and the stopper. It should be noted that the reactions from those contacted portions should be balanced because the coming out behavior is quasi-static. As an example, when the load P =600N/mm is in the upward direction, the shrink fitting ratio δ/d=0.2×10^{-3}, the friction coefficient μ=0.3 and N≥11, the downside inner plate contact with the stopper, the upside does not contact. As shown in Figure 6， from the shear stress distributions , the shearing force at the upward part can be obtained as $F_{\tau u}$=1165.7N, and the shearing force at the downward part can be calculated as $F_{\tau d}$=-931.1N. Then we have:

$$F_s=F_{\tau u}+F_{\tau d} \tag{1}$$

Substituting $F_{\tau u}$ =1165.7N, $F_{\tau d}$=-913.1N into Eq.(1) we have F_s=252.6N as the driving force for coming out. The compressive force F_x on the stopper is calculated by FEM as F_x= -240.3 N[16]. Then we have $F_s+F_x \fallingdotseq 0$ within 5%. It may be concluded that the shear force F_s balanced with the compressive force F_x in the x-direction.

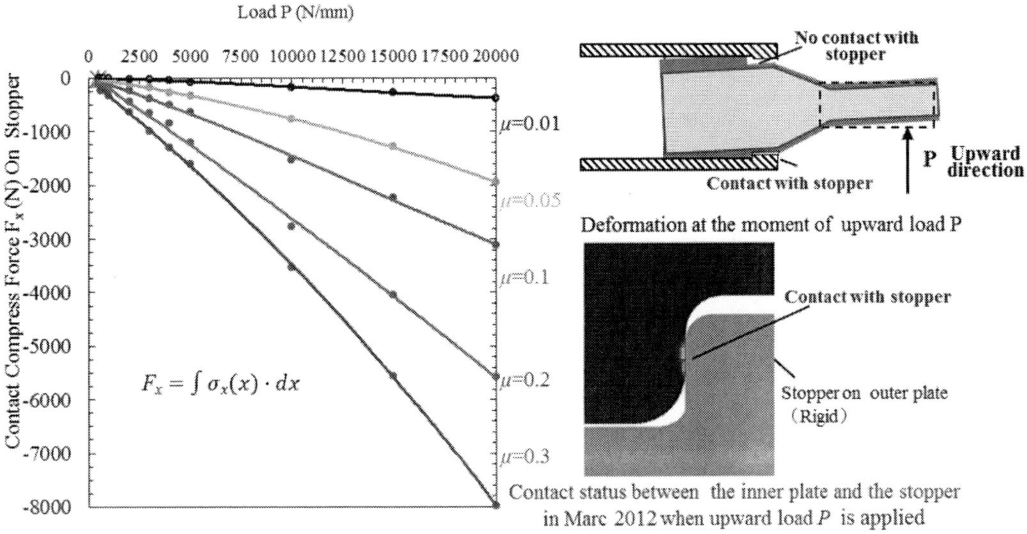

Figure 5. Relation between different load P and compressive contact force F_x on the stopper under different friction coefficient μ when δ/d =0.2×10^{-3}and upward load P is applied

5 Conclusions

In this study, a newly designed stopper is considered to prevent the inner plate from coming out. In order to evaluate the contact status between the inner plate and the stopper, a simulation has been conducted. The effect of the magnitude of applied force on the coming out of the inner plate has been discussed. The conclusion can be summarized in the following way.

(1) The analysis results shows that the several portions of the inner plate contact with the outer plate due to the repeated load P after shrink fitting. At that time, two shearing forces are generated at the upward portion and at the downside portion of the inner plate.

(2) The resultant force $F_s=F_{\tau u}+F_{\tau d}$ can be obtained from the shearing force at the upward portion F_τ and the shearing force at the downside portion $F_{\tau d}$. This resultant force F_s can be regarded as the driving force of coming out. Here, the positive direction is defined as the coming out direction. Due to this positive driving force, the inner plate is coming out.

(3) To prevent the inner plate from coming out can be achieved by applying a negative force to the inner plate. This force can be provided as a contact force F_x appearing at the stopper. Numerical simulation demonstrated that the contact force F_x and the driving force of coming out F_s are almost equal within 5% error. In other words, it is confirmed that the shearing force at the upward portion F_τ and the shearing force at the downside portion $F_{\tau d}$ and the contact force F_x are balanced.

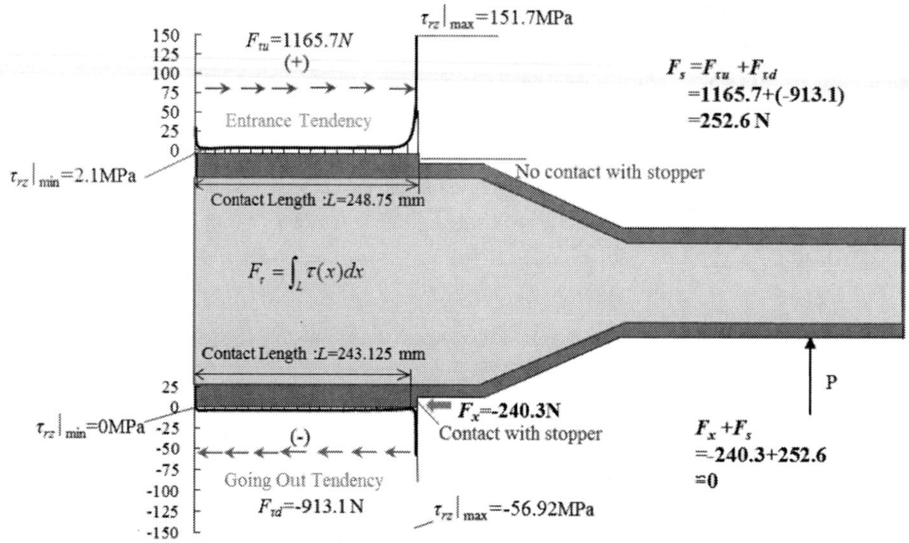

At N =100, P=600N/mm, upward, the downside inner plate contacts with stopper

Figure 6. Shear stress distribution on inner plate at the moment of upward load P=600N/mm is applied when circle N=100

References

[1] M. Fujii, A. Yoshida, J. Ishimaru, S. Shigemura, K. Tani: Trans. Jpn. Soc. Mech. Eng. C, 72(2006), 1354.
[2] C. Liour, T. Mori, H. Kobayashi, T. Mitamura: JCS-Japan, 98(1990), 348.
[3] T. Ono: J. Jpn. Soc. Mech. Eng, 86(1983), 470.
[4] N.A. Noda, D. Suryadi, S. Kumasaki, Y. Sano, Y. Takase Engineering Failure Analysis 57(2015)219-235.
[5] N.A. Noda, Y. Xu, Y. Sano, Y. Takase: ISIJ Int., 56(2016), No. 2, pp. 303–310.
[6] W. Li, N-A. Noda, H. Sakai, Y. Takase: Journal of Solid Mechanics and Materials Engineering, 5(2011), 14.
[7] W. Li, N.A. Noda, H. Sakai, Y. Takase: Key Eng. Material, 452-453(2011), 241.
[8] A. Rusin, G. Nowak, and W. Piecha: Engineering Failure Analysis, 34(2013), 217.
[9] E. Ogawa, K. Shimizu, S. Hamayoshi, N. Kumagai, Y. Ohtsubo, N.A. Noda, Y. Takase, K, Kishi, K. Shobu, T. Tabaru, E, Maeda, S. Koga, T. Matsuda: Hitachi Metals Technical Review, 28(2012), 50.
[10] M. Tsuyunaru, N.A. Noda, Hendra, Y. Takase: Trans. Jpn. Soc. Mech. Eng. A, 74(2008), 919.
[11] N.A. Noda, Hendra. Y. Takase, M. Tsuyunaru: Journal of Solid Mechanics and Materials Engineering, 2(2008), No.2, 1410.
[12] N.A. Noda, M. Yamada, Y. Sano, S. Sugiyama, S. Kobayashi: Engineering Failure Analysis, 15(2008), 261.
[13] N.A. Noda, Hendra, M. Oosato, K. Suzumoto, Y. Takase, W. Li : Key Eng. Material, 462-463(2011), 1140.
[14] S. Matsuda, D. Suryadi, N.A. Noda, Y. Sano, Y. Takase, S. Harada : Trans. Jpn. Soc. Mech. Eng. A, 79(2013), 989.
[15] N-A. Noda, D. Suryadi, S. Matsuda, Y. Sano, Y. Takase: ISIJ Int., 55(2015), 2416.
[16] Marc Mentat team, Theory and User Information. Vol.A, MSC.Software, Tokyo, 2012.531-596.

IOP Conf. Series: Journal of Physics: Conf. Series **842** (2017) 012052 doi:10.1088/1742-6596/842/1/012052

Integrated condition monitoring of a fleet of offshore wind turbines with focus on acceleration streaming processing

Jan Helsen[1], Nicoletta Gioia[1], Cédric Peeters[1], Pieter-Jan Jordaens[2]

[1]Vrije Universiteit Brussel, Pleinlaan 2, 1050 Brussel, Belgium
[2]OWI-lab, Celestijnenlaan 300b, 3001 Leuven, Belgium

E-mail: jan.helsen@vub.ac.be

Abstract. Particularly offshore there is a trend to cluster wind turbines in large wind farms, and in the near future to operate such a farm as an integrated power production plant. Predictability of individual turbine behavior across the entire fleet is key in such a strategy. Failure of turbine subcomponents should be detected well in advance to allow early planning of all necessary maintenance actions; Such that they can be performed during low wind and low electricity demand periods. In order to obtain the insights to predict component failure, it is necessary to have an integrated clean dataset spanning all turbines of the fleet for a sufficiently long period of time. This paper illustrates our big-data approach to do this. In addition, advanced failure detection algorithms are necessary to detect failures in this dataset. This paper discusses a multi-level monitoring approach that consists of a combination of machine learning and advanced physics based signal-processing techniques. The advantage of combining different data sources to detect system degradation is in the higher certainty due to multi-variable criteria. In order to able to perform long-term acceleration data signal processing at high frequency a streaming processing approach is necessary. This allows the data to be analysed as the sensors generate it. This paper illustrates this streaming concept on 5kHz acceleration data. A continuous spectrogram is generated from the data-stream. Real-life offshore wind turbine data is used. Using this streaming approach for calculating bearing failure features on continuous acceleration data will support failure propagation detection.

1. Introduction

Wind energy has been booming for the past decade and a massive amount of turbines have been installed both onshore and offshore. Given the increasing percentage of wind energy in the renewable energy mix, it becomes more important to deliver an as constant as possible output of power; particularly at moments of high electricity demand. For achieving this turbines are clustered in large wind farms. Particularly offshore this is the case, where wind farms even have their own transformer substation. Such a plant will in the future be operated as an integrated power plant that needs to meet electricity demanded by the grid [1]. This will cause the farm operator to be much more responsible for the way in which this energy is produced. Now the wind speed is determining for each turbine how much energy it produces. In the future the wind farm operator will decide how much each turbine in the farm produces. That way he can take into account the price of energy, the condition of the turbines and the other factors.

A second important aspect in farm operation is maintenance planning. Drivetrain reliability is an important aspect in this [2]. Each turbine needs yearly replacement of certain components and fluids.

Additionally if a component fails it needs to be replaced. In ideal conditions the maintenance actions are done during low wind and low electricity demand periods. That way the lost revenue due to downtime is minimal.

In order to be able to achieve this vision it is essential to have a good estimate of the current condition of each individual turbine in the farm. Early detection of failure and insights in the historic loading of each turbine is the basis for such an optimized operation of the farm. This paper suggests an integrated approach to assess the current condition of all turbines in a farm.

2. Methods

A first requirement to be able to assess the current condition of a system is the availability of data in a consistent dataset. This dataset should span all turbines in the farm in order to allow comparison between the behaviour of the different turbines. To be able to deal with the massive amount of measurement data coming from the more than 200 sensors mounted on each turbine in a farm we suggest the use of a big-data storage and streaming approach.

A second requirement is the availability of dedicated data analysis tools to perform anomaly detection. We suggest a combination of dedicated physics based condition monitoring signal processing techniques for high frequency data [3,4] and data-modelling machine learning algorithms for anomaly detection on signals coming from low frequency sampled (<10Hz) sensors embedded in the turbine [5].

The third requirement is a highly capable decision support framework to interpret the alarms from the anomaly detection algorithms. It is important to be able to learn from previous experiences (maintenance actions) how certain alarm sequences can be coupled to certain failures. This then in future allows to use these links to predict initiating failures. Again machine-learning algorithms can bring added value here.

This integrated approach is too broad to discuss in the context of this paper. Therefore one aspect is lifted out in detail: the aspect of high frequency data streaming signal processing. The other aspects are briefly discussed and reference is made to other papers where the details are discussed.

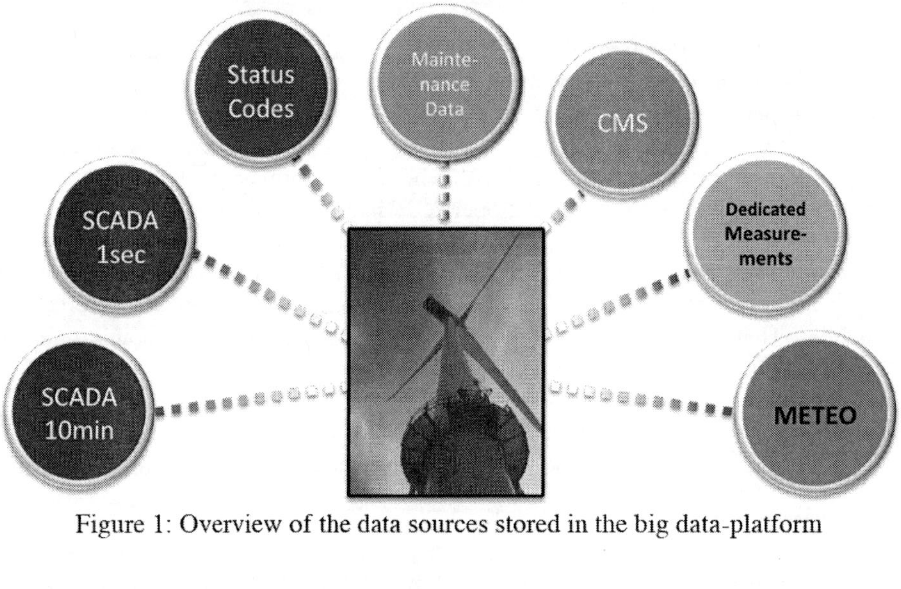

Figure 1: Overview of the data sources stored in the big data-platform

3. Big-data storage

In general there is a large volume of data generated by the different turbines in the farm. In order to be able to perform analysis on this data it is necessary to have it stored in one consistent dataset. This data can be time series or more contextual. For the wind turbine application the most important time series data-sources are: Supervisory Control And Data-Acquisition data (SCADA) either sampled at 10min basis or at 1Hz, turbine status codes [6], data from condition monitoring systems and dedicated measurements of custom monitoring systems for certain components, such as the gearbox. Context information is linked to for example maintenance actions and meteorological information. These data-sources can provide additional insights in the behaviour seen in the different time series of the turbines and link them to failure events.

Given the high variety and high volumes of data we needed to go beyond traditional database approaches. Our automated analysis platform therefore uses a big data no-SQL architecture. Automation is needed in order to be able to process the different sensor channels in parallel. An overview of the different data sources is shown in Figure 1.

In addition to raw data also processed data is stored in the big-data platform. Particularly features linked to anomaly information are useful to keep over long periods of time. Such features can for example be the amplitude value of a ball pass frequency outer ring (BPFO). These features are kept in order to train classifiers that can combine the anomalies found for different sensor signals in a decision support strategy.

In this paper we focus on the high frequency acceleration data, rather than the low frequency SCADA data. This particular data source consists of the data of accelerometers mounted on main bearing, gearbox and generator. It is the data source that signal processing algorithms for condition monitoring will use as input information.

Acceleration data is typically sampled between 500Hz and 10kHz. This high data rate poses a challenge for more traditional processing approaches. In most of those cases the raw data is aggregated into averaged values at a one per minute to one per ten minutes sampling rate by an Extract Transform and Load (ETL) process. The aggregated data is stored permanently in the database and used as basis for further analysis.

However for condition monitoring it is necessary to perform the calculation of different features on each of the high frequency sampled channels and not on the aggregated data. Using the 'store-all' approach this would result in a massive amount of data in the database. Traditionally this is overcome by a reduction in the amount of measurements the condition monitoring system is sending to the database. In most cases key statistics will be sent at many instances, whereas detailed high frequency data is transferred at a limited number of times each day. This raw data is then processed into features, which are again stored in the database.

Having access to a limited number of data points each day allows for alarming, but makes it difficult to detect trends in system degradation. Particularly for the wind turbine application it is necessary to have more long-term data. This in order to have multiple measurement points in time at consistent operating conditions.

In order to overcome this problem we suggest the use of streaming processing approach. In such a data architecture the data is continuously processed as it is generated.

4. High frequency data streaming signal processing

In the streaming context the condition monitoring processing algorithms need to be able to keep up with the data-generation. Therefore, it is necessary to perform the different processing tasks in parallel. Figure 2 illustrates the dataflow.

In this case the input to the data-streaming engine is a stream of files. These files contain the data of one or multiple measurement channels and are generated by the data-acquisition system. Another possibility is the case for which the measurement system is capable of streaming the data directly to the streaming engine.

The streaming engine buffers the data sufficiently long for processing. It splits the data over different parallel data streams tailored to each of the analysis algorithms. Several measurement channels can be repeated in parallel streams if they are used as inputs for different signal processing sequences. The signal features resulting from the signal processing algorithms are stored in the database. All actions are implemented such that they are autonomous and continuous.

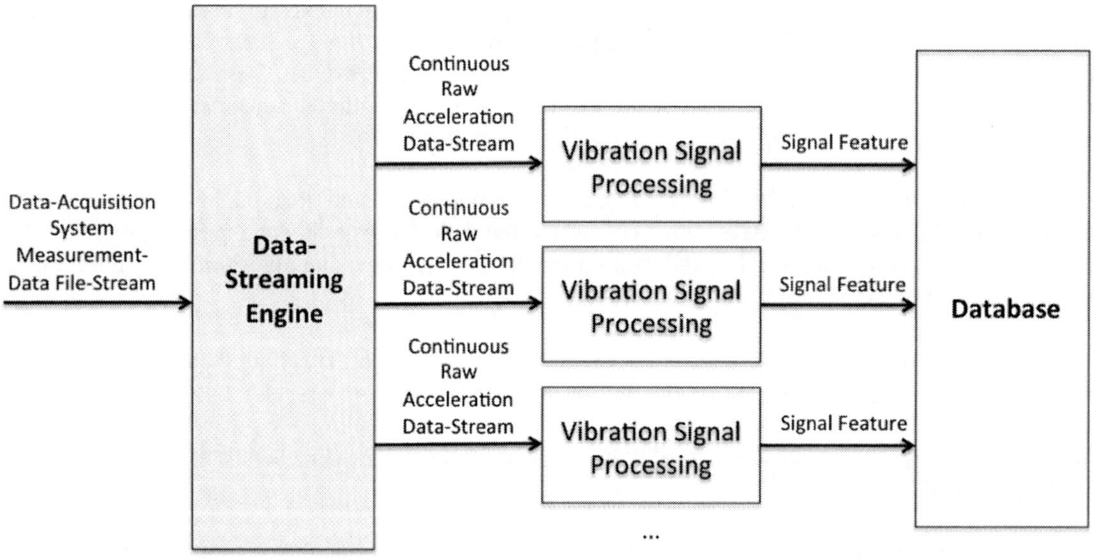

Figure 2: Schematic representation of data-flow for high frequency acceleration data streaming processing

5. Offshore wind turbine gearbox acceleration measurements

To illustrate the working principle of our data streaming approach we show the example of long-term spectrogram processing on an acceleration signal of an offshore wind turbine. The spectrogram is often used as input for condition monitoring analysis and operational modal analysis. It offers insights in the dynamic behaviour over time of a system.

The measurement data that is used to illustrate the concept of the streaming acceleration processing is coming from an offshore wind turbine. A multi-month measurement campaign was performed on the turbine. The instrumentation package consisted of accelerometers mounted on the gearbox. All channels were sampled at 5120Hz.

For rotating machines the spectrogram is typically characterized by the clear presence of harmonics linked to the different rotating shafts and gears. The frequency of these harmonics changes when the speed of the system changes. Hence tracking of the harmonics gives insights in the speed behaviour of the machine. This can for example be used for virtual tachometer extraction using e.g. a Multi-Order Probabilistic Approach (MOPA) [7]. Additionally the spectrogram can be used in the context of Campbell diagram analysis for the detection of resonance frequencies of rotating machines.

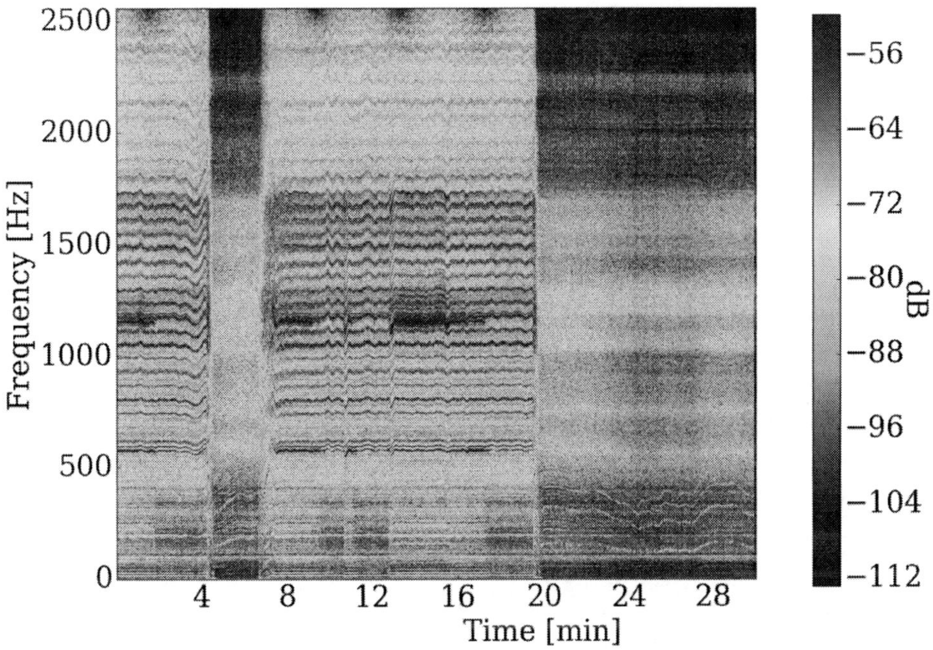

Figure 3: Spectrogram calculated on streaming data of multiple files for a time period of 30 minutes of acceleration data sampled at 5120Hz.

Figure 3 illustrates a streaming spectrogram for one of the acceleration signals on the wind turbine gearbox housing for a stream with duration of 30 minutes. One can clearly see the presence of the orders linked to the different shafts and gear meshes in the gearbox. Since the gearbox has three stages the number of orders is quite high. At about 1200Hz there is clear indication of the presence of a resonance.

During the visualized time period the turbine is continuously changing regime. For this particular case the turbine was operational until minute 4. Then it was manually stopped for 1.5 minutes and restarted. This stop event is shown in Figure 4. The orders clearly mark the braking of the system. During the stop the turbine is still idling very slowly as marked by the remaining low frequency low amplitude orders. At minute 20 the turbine is stopped again. A more detailed view of this stop is shown in Figure 5.

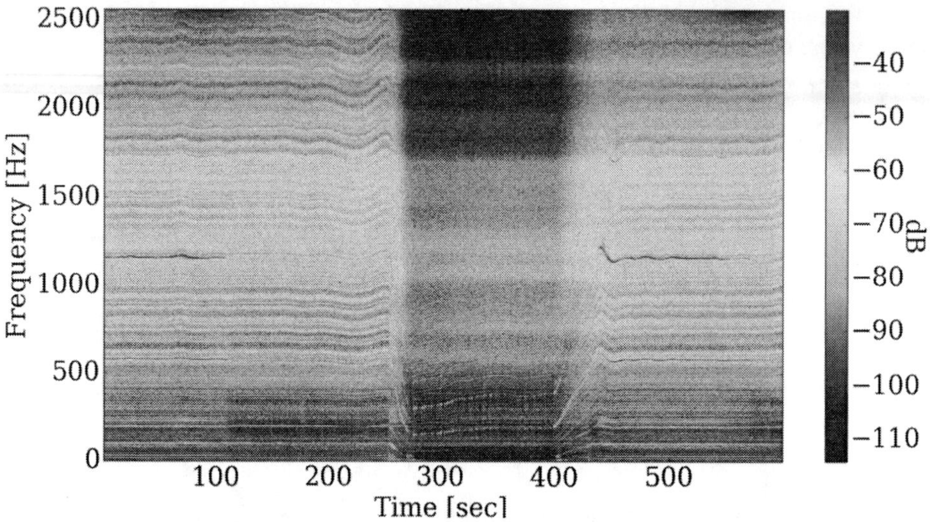

Figure 4: Zoom on the first 10 minutes of the streaming spectrogram of the acceleration signal.

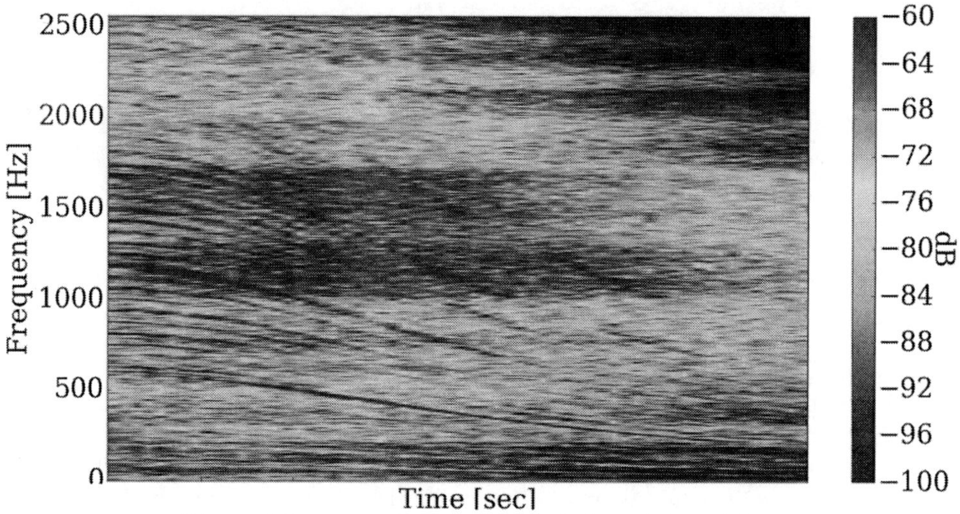

Figure 5: Zoom on the speed degradation of the turbine.

6. Conclusions

This paper illustrated our integrated big-data monitoring approach for a fleet of wind turbines. The general concept was illustrated. Moreover, the concept of streaming processing of high frequency acceleration data was discussed. This approach was illustrated by means of a streaming spectrogram calculation. Performing this analysis on experimental data of an offshore wind turbine it was shown that the dynamic behavior of this system is clearly present and it is a good starting point for streaming modal analysis and condition monitoring analysis. This will be further developed in future work.

Acknowledgements
The authors would like to acknowledge the support of the VLAIO SBO HYMOP and VLAIO VIS OWOME projects. Furthermore the owners of the wind farm are acknowledged for their contributions in the measurement campaign.

References

[1] PAO, L. Y., JOHNSON, K. E.: A tutorial on the dynamics and control of wind turbines and wind farms, 2009, *IEEE American Control Conference, St. Louis, Missouri, USA*
[2] LINK H, LACAVA W, VAN DAM J, MCNIFF B, SHENG S, WALLEN R, MCDADE M, LAMBERT S, BUTTERFIELD S, OYAGUE F.: Gearbox reliability collaborative project report: findings from phase 1 and phase 2 testing (technical report) *NREL/TP-5000-51885, National Renewable Energy Laboratory (NREL), Golden, CO.*
[3] PEETERS, C., GUILLAUME, P., HELSEN, J., 2016, 'Vibration data pre-processing techniques for rolling element bearing fault detection', *Proceedings of the 23rd international conference on sound & vibration*', Athens, Greece
[4] PEETERS, C., GUILLAUME, P., HELSEN, J., 2016, 'Vibration-based bearing fault detection analysis algorithms and diagnostics results on experimental wind turbine gearbox data', *Proceedings of the European Conference of the prognostics and health management society'*, Bilbao, Spain
[5] HELSEN, J., DEVRIENDT, C., WEIJTJENS,. W., GUILLAUME, P., 2015 'SCADA based CMS', *Proceedings of EWEA, Paris, France*
[6] DORO, P., SCERRI, L., GUILLAUME, P., HELSEN, J., 2016, 'Pattern detection in status codes as an optimization tool in offshore wind farms', *Proceedings of the European Conference of the prognostics and health management society'*, Bilbao, Spain
[7] LECLERE, Q., ANDRE, H., ANTONI, J., 2016, 'A multi-order probabilistic approach for instantaneous angular speed tracking debriefing of the CMMNO14' diagnosis contest, *Mechnical Systems and Signal Processing*, DOI: 10.1016/j.ymssp.2016.02.053

Vibration-based angular speed estimation for multi-stage wind turbine gearboxes

Cédric Peeters[1,a], Quentin Leclère[2], Jérôme Antoni[2], Patrick Guillaume[1], Jan Helsen[1]

[1]Vrije Universiteit Brussel, Pleinlaan 2, 1050 Brussel, Belgium
[2]LVA, INSA de Lyon, F-69621, Villeurbanne cedex, France

E-mail: [a]`Cedric.Peeters@vub.ac.be`

Abstract. Most processing tools based on frequency analysis of vibration signals are only applicable for stationary speed regimes. Speed variation causes the spectral content to smear, which encumbers most conventional fault detection techniques. To solve the problem of non-stationary speed conditions, the instantaneous angular speed (IAS) is estimated. Wind turbine gearboxes however are typically multi-stage gearboxes, consisting of multiple shafts, rotating at different speeds. Fitting a sensor (e.g. a tachometer) to every single stage is not always feasible. As such there is a need to estimate the IAS of every single shaft based on the vibration signals measured by the accelerometers. This paper investigates the performance of the multi-order probabilistic approach for IAS estimation on experimental case studies of wind turbines. This method takes into account the meshing orders of the gears present in the system and has the advantage that a priori it is not necessary to associate harmonics with a certain periodic mechanical event, which increases the robustness of the method. It is found that the MOPA has the potential to easily outperform standard band-pass filtering techniques for speed estimation. More knowledge of the gearbox kinematics is beneficial for the MOPA performance, but even with very little knowledge about the meshing orders, the MOPA still performs sufficiently well to compete with the standard speed estimation techniques. This observation is proven on two different data sets, both originating from vibration measurements on the gearbox housing of a wind turbine.

1. Introduction

Wind turbines experience strong dynamical loads during their lifetime that can lead to considerable reduction of the expected life of a wind turbine's drivetrain[1]. Changes in the behavior of the wind and the electricity grid can influence significantly the loads on the drivetrain and thus its lifetime. Gearboxes especially are prone to unexpected premature failures and are costly to replace due to the substantial downtime[2]. Reducing the amount of critical failures and thus increasing wind turbine reliability has the potential to drastically decrease the actual cost of electricity [3].

Current condition monitoring systems often utilize vibration measurements to assess the health of the internals of a gearbox, like the shafts, gears or bearings. Most of these vibration-based systems rely on spectral analysis of the measured vibration signals. Methods based on tracking spectral content however generally assume the rotation speed of the system to be constant or quasi constant. In practical applications like wind turbines, this requirement is unlikely to be met due to the highly dynamic character. To enable the synchronization of the spectral analysis, the instantaneous angular speed (IAS) can be measured by using a tachometer. However the installation and cost of a tachometer is not always feasible or desirable and thus recent developments focus on extracting the IAS directly from the vibration signal. Several approaches already exist and some common techniques include Vold-Kalman filtering [4, 5], Gabor and wavelet transforms [6, 7], band-pass filtering[8] and the Teager-Kaiser energy operator[9]. However most of these techniques rely on the presence of a frequency band that contains only a harmonic of the rotation frequency with a high signal-to-noise ratio (SNR). Other signal components like harmonics that are not related to the rotation speed of interest, or structural resonances are not desirable in the frequency band and usually ruins the chances of successfully extracting the IAS. Since wind turbine gearboxes are usually a complex system consisting of multiple (planetary) stages, the vibration signals measured from such a complex system contain multiple resonances and asynchronous harmonics. As such the assumption of a frequency band containing only one harmonic with high SNR cannot be guaranteed.

The limitation of only taking taking into account one harmonic hints at an approach that utilizes as many harmonics as possible in order to increase the amount of information and correspondingly the accuracy. This paper investigates exactly such a method, namely the multi-order probabilistic approach (MOPA) as first proposed by Leclère et al. [10]. This method uses the available information about the system and does not require to associate one harmonic with the correct order. Knowledge of the gear ratios only is sufficient to extract from the vibration signal an accurate estimation of the IAS.

2. Methodology

This paper compares the IAS estimation accuracy of two methods against each other on two experimental data sets of wind turbine gearboxes. This section brie y describes the theoretical background of both methods.

2.1. Multi-order probabilistic approach

The general idea behind the multi-order probabilistic approach (MOPA) as proposed by Leclère et al. [10] is based on regarding the instantaneous spectrum (which can be obtained through a short time Fourier transform) of the vibration signal as a probability density function (pdf) of the IAS Ω. Consequently, if the spectrum has a high amplitude at frequency f, there is a high probability that the shaft frequency is equal to f/H_i with H_i being the excitation order or for the cases described below the gear ratios. It is important to define a range for the IAS in which the user expects the IAS to reside. This range has a lower bound Ω_{min} and an upper bound Ω_{max}. The following pdf can then be constructed:

$$\begin{cases} [\Omega|H_i] = \frac{1}{\xi_i}A(H_i\omega) & \text{for } \Omega_{min} < \omega < \Omega_{max} \\ [\Omega|H_i] = 0 & \text{for } \omega < \Omega_{min} - \omega > \Omega_{max} \end{cases} \tag{1}$$

with $A(f)$ a whitened version of the vibration signal's spectrum and ξ_i a normalization factor to make sure the pdf has unit area:

$$\xi_i = \int_{\Omega_{min}}^{\Omega_{max}} A(H_i\omega)d\omega. \tag{2}$$

The purpose of the whitening is essentially to reduce the influence of resonances on the generated pdf, since it is undesirable to give a too high probability to a certain part of the spectrum only due to the increased amplitudes because of a resonance.

To improve the IAS estimation and utilize more of the information potential of the spectrum, one has to include more than just one pdf based on one gear ratio or meshing order. Afterwards these different pdfs can be combined together in one pdf by multiplication. Equation1 does not take into account the possibility that a part of the spectrum for a certain harmonic H_i can exceed the Nyquist frequency. In this case the pdf is made uniform above f_{max}/H_i:

$$\begin{cases} [\Omega|H_i] = \frac{1}{\xi_i}A(H_i\omega) & \text{for } \Omega_{min} < \omega < f_{max}/H_i \\ [\Omega|H_i] = \frac{1}{\Omega_{max}-\Omega_{min}}[\Omega|H_i] = 0 & \text{for } \omega < \Omega_{min} - \omega > \Omega_{max} \end{cases} \tag{3}$$

with ξ_i now:

$$\xi_i = \frac{\Omega_{max} - \Omega_{min}}{f_{max}/H_i - \Omega_{min}} \int_{\Omega_{min}}^{f_{max}/H_i} A(H_i\omega)d\omega. \tag{4}$$

The method's inputs are thus an approximate range for the IAS, the meshing orders and the vibration signal. For every order a pdf is then constructed based on the signal's instantaneous spectrum and rescaled to the given range for the IAS. Next, the pdfs are multiplied to combine the information of all the orders so that the main corresponding estimate for the IAS becomes the most dominant peak in the pdf.

Currently the pdfs are still independently generated for each time step and thus do not guarantee any continuity of the IAS, which is a logical assumption for any mechanical system. Due to the inertia of the rotating shafts, strong acceleration or deceleration is improbable. As such, to improve the results further, an a priori of continuity is introduced for the IAS. The concept relies on generating for each time step a pdf that is based on the pdfs of several time steps before and after the central pdf. Appropriate weighting of these pdfs is done by convolving the pdf with a centered Gaussian and the time relationship is introduced by letting the variance depend on the time between the considered pdf and the central pdf. The pdf at time step j generated by the pdf at time step $j + k$ is defined as:

$$[\Omega_j]_{j+k} = \int_{\Omega_{min}}^{\Omega_{max}} [\Omega_j|\Omega_{j+k}]d\omega \propto \exp(\frac{\omega^2}{2\sigma_k^2}) * [\Omega_{j+k}] \tag{5}$$

with $[\Omega_j]_{j+k}$ the pdf at time j that can be obtained by convolution of the pdf at time $j + k$ $[\Omega_{j+k}]$ with a centered Gaussian, and $\sigma_k = |\gamma k \Delta_t|$ with Δ_t being the time step, γ the standard acceleration of the IAS. Similar to the previous step in which the pdfs corresponding to the different orders have to be multiplied for each time step to obtain a single combined pdf, there are now again multiple pdfs for every time step $j + k$ belonging to time steps before and after time step j. Thus the final step is to multiply again all the pdfs for every time step:

$$[\Omega_j]_s \propto \prod_{k=-K}^{K} [\Omega_j]_{j+k}. \tag{6}$$

2.2. Band-pass filtering & instantaneous phase demodulation

The method that is used for comparison of the MOPA method on the experimental data is fairly straightforward and is mainly based on using an ideal band-pass filter around a well-separated, high SNR harmonic of the rotation speed. After defining the optimal lower and upper cutoff frequencies for the band-pass filter, the harmonic is filtered out of the complex spectrum and translated around zero frequency. Next, the complex band-pass filtered spectrum is inverse Fourier transformed to the time domain. The angle is then unwrapped of the complex signal after addition of the demodulation frequency to obtain the instantaneous phase. Finally, the instantaneous angular speed is estimated based on the phase variation.

3. Experimental applications

The two methods described above are investigated on two data sets containing vibration measurements of wind turbine gearboxes. The first data set originates from the diagnosis contest held in light of the International Conference on Condition Monitoring of Machinery in Non-Stationary Operations (CMMNO) in 2014. The second data set was recorded as well on the gearbox housing of a wind turbine but due to confidentiality agreements, the authors cannot disclose the origin of the data.

3.1. CMMNO 2014 diagnosis contest data

The vibration signal used for this analysis was measured on the gearbox housing of a wind turbine near the epicyclic gear train and sampled at 20 kHz. The goal was to estimate the IAS of the high-speed shaft (carrying gear #7 in Fig.1). This estimate was then compared with a reference speed signal measured by an angle encoder. The length of the measurement was approximately 550 seconds.

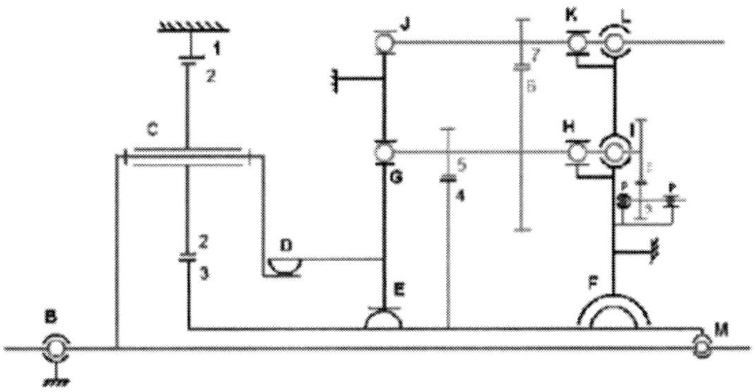

Figure 1. Visualization of the wind turbine gearbox used in the CMMNO 2014 diagnosis contest.

In the first step a spectrogram is generated using a rectangular window with a window length of 0.5 seconds, 1.5 seconds of zero padding, and an overlap of 0.25 seconds. Figure 2a displays the resulting spectrogram, in which there are some clear resonances (*red horizontal lines*) to be seen. Therefore the spectrogram is whitened to remove these resonances as is shown in Fig.2b.

The next step is the generation of the pdfs for every meshing order and combining them in order to obtain a single pdf representing the most likely frequency of the IAS. The result of this operation can be seen in Fig.3a. A total of six meshing orders are exactly known and each of them increases the performance of the MOPA method. The last step is to introduce the continuity condition, which performs a smoothing of the pdfs and significantly improves the results as can be seen in Fig.3b.

Now that the pdfs are constructed, the expected value is extracted for every time step. The result of the IAS estimation of the MOPA method and the band-pass filtering method is shown in Fig.4a. For the latter method, the second harmonic of the rotation speed was chosen as the best candidate for demodulation. Additionally, the IAS was only estimated for windows of around 50 seconds at a time for this method, since otherwise the speed variation would be too large. The relative errors of the MOPA method and the band-pass filtering method to the reference signal are displayed in Fig.4b. The error of the MOPA method is significantly smaller, while the band-pass filter sometimes fails to track the actual speed of the signal.

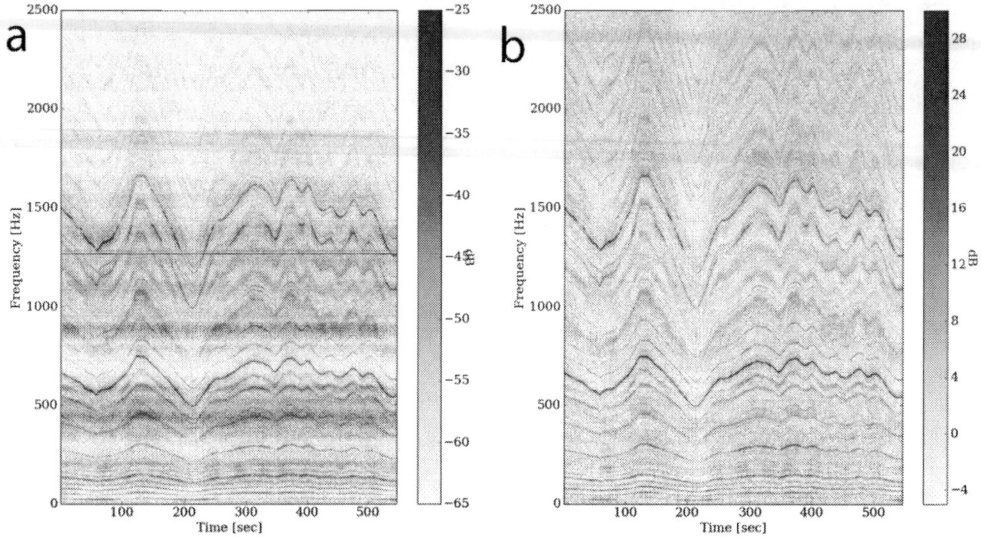

Figure 2. (a) Spectrogram of the CMMNO 2014 vibration signal of the wind turbine gearbox housing. (b) Whitened spectrogram

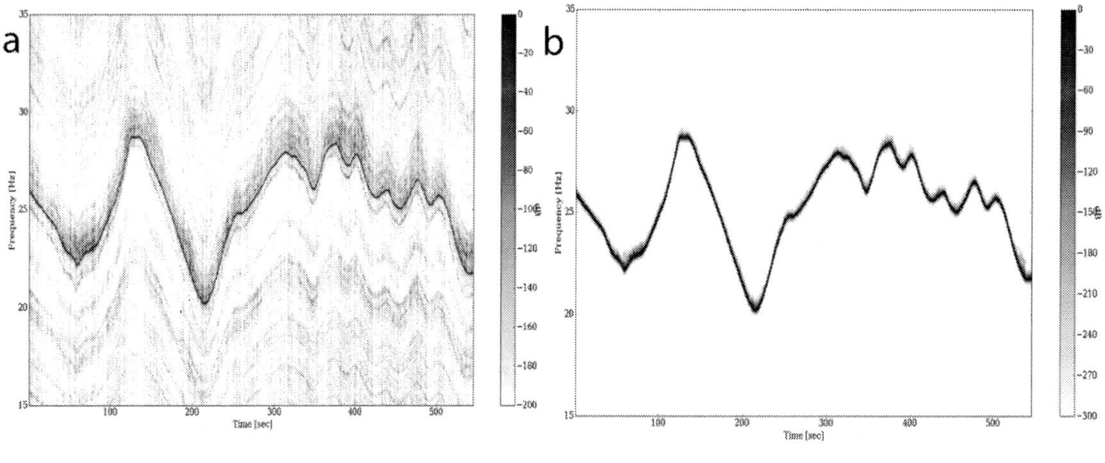

Figure 3. (a) Time-Frequency map of the pdfs for an IAS range from 15 to 35 Hz after combining the information of all meshing orders. (b) Time-Frequency map of the pdfs after the a priori continuity introduction.

3.2. Second wind turbine gearbox case study

To assess the performance of the MOPA method further, the vibration data of another wind turbine gearbox is investigated. Due to a confidentiality agreement however, no details can be given about the wind turbine itself and some of the certain frequency axes are normalized from zero to one in the figures.

In contrast to the previous case, the exact meshing orders present in the system are not known, which forms quite a significant obstacle for the MOPA method since its main input is these meshing orders. However, the spectrogram can be used for estimating the approximate

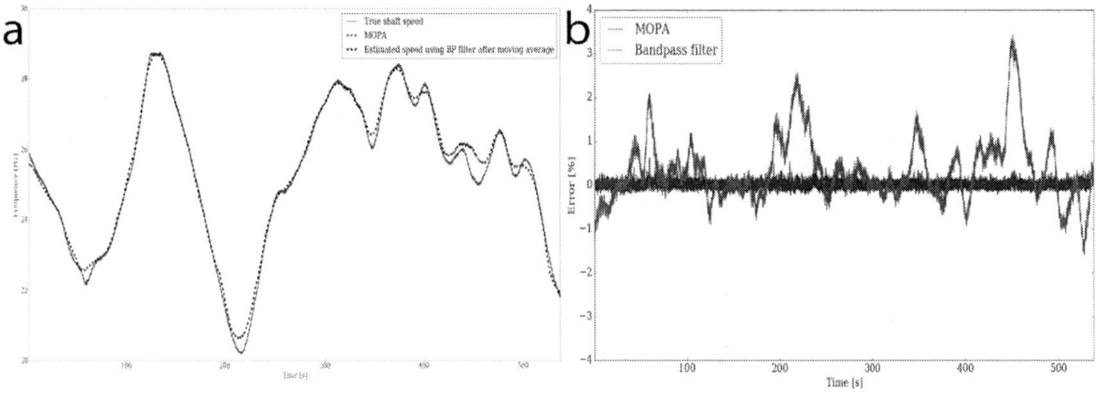

Figure 4. (a) Comparison of the estimated IAS obtained through MOPA and band-pass filtering to the reference speed signal. (b) Relative error (in %) of the two described methods.

meshing orders and is used for this analysis.

The same procedure as in the previous section is followed. First, the spectrogram is generated and whitened as is shown in Fig.5a & Fig.5b. The spectrogram clearly exhibits two harmonics (around 0.35 and 0.7 normalized frequency) that have an increased amplitude due to their presence near a structural resonance of the system. These harmonics are also linked to the fundamental rotation speed (33rd and 66th harmonic) of the high-speed shaft and thus form ideal candidates for band-pass filtering and for inclusion in the MOPA method. It is noted that for the meshing order input to the MOPA method only five meshing orders were used and they are all harmonics of the rotation speed, namely the 1st, 2nd, 3rd, 33rd and 66th harmonic.

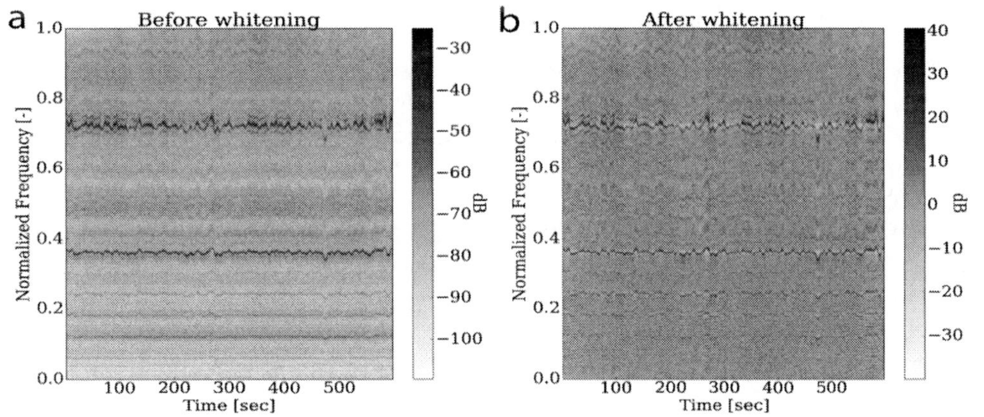

Figure 5. (a) Spectrogram of a wind turbine gearbox vibration signal. (b) Whitened spectrogram.

Figure 6a shows the resulting speed profiles obtained with the MOPA method and through using a band-pass filter around the 33rd harmonic. The blue line displays the reference speed signal which was measured using a tachometer. Both methods fare quite well in tracking the actual reference speed , but it can be seen that around 470s there is a large deviation in speed that the band-pass filter is unable to capture. This is also evidenced in the relative error, shown in Fig.6b.

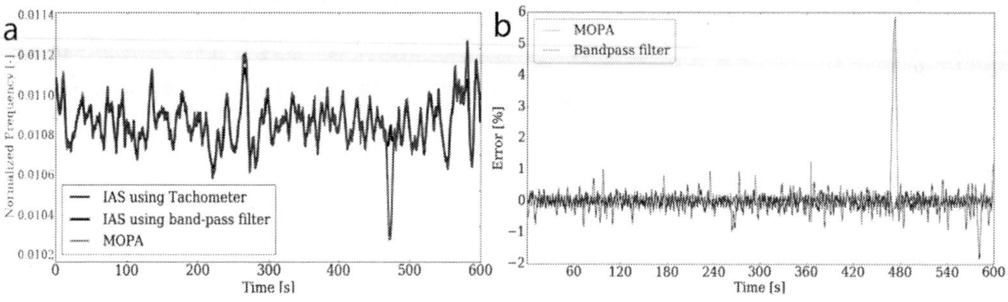

Figure 6. (a) Comparison of the estimated IAS obtained through MOPA and band-pass filtering to the reference speed signal. (b) Relative error (in %) of the two described methods.

4. Discussion & Conclusions

This paper investigates the effectiveness of the multi-order probabilistic approach (MOPA) for estimating the instantaneous angular speed of rotating shafts in wind turbine gearboxes. A brief overview of the theory behind the method is presented in section 2. The method is compared to the commonly used band-pass filtering and phase demodulation approach. Two experimental data sets are examined. The first data set originates from the contest held for the CMMNO 2014 conference and consists of vibration data measured on the wind turbine gearbox housing. The probabilistic approach performs very well in tracking the speed fluctuations for the 10 minute signal. The band-pass filtering approach however has to be limited to processing only 50 seconds at once, since the speed variation is too large in some parts of the signal making it exceed the band-pass limits. Even when taking into account block processing for the band-pass filtering approach, care still has to be taken in making sure that no interfering content crosses the demodulation band. The error of the MOPA method compared to the reference signal is significantly smaller in this case than the one for the second approach. The second case however presents a bigger challenge for the MOPA technique. Since the meshing orders of the system are not exactly known, they have to be derived from the spectrogram which makes it difficult to estimate very accurately these ratios. Additionally the presence of a dominant harmonic due to its vicinity to a resonance and the lack of any interfering content in this frequency band makes this a very well-suited data set for the band-pass filter approach. The results corroborate this and show that both methods have a fairly low discrepancy compared to the reference signal. Even though the vibration signal in this case is more suited towards the band-pass filtering method, the MOPA approach performs equally well and does not show large deviations for sudden speed fluctuations. The multi-order probabilistic approach thus presents itself as a robust and easy to use method for speed estimation of rotating machinery. It manages to perform adequately in low SNR scenarios and with very limited input. More knowledge about the system benefits the MOPA approach as it is able to incorporate this information to improve its estimation process, while other methods generally focus on only part of this information and thus do not use its potential.

5. Acknowledgment

The authors are grateful for the provided data by the CMMNO 2014 Organization Committee. They would also like to thank VLAIO for the support through the VIS-OWOME project.

6. References

[1] Helsen J, Guo Y, Keller J and Guillaume P 2016 *Wind Energy* **19** 22552269
[2] Helsen J, Devriendt C, Weijtjens W and Guillaume P 2016 *Renewable Energy* **85** 259–272
[3] Blanco M I 2009 *Renewable and Sustainable Energy Reviews* **13** 1372–1382
[4] Vold H and Leuridan J 1993 High resolution order tracking at extreme slew rates, using kalman tracking filters Tech. rep. SAE Technical Paper
[5] Stephens D B and Vold H 2014 *Journal of Sound and Vibration* **333** 3818–3830
[6] Shao H, Jin W and Qian S 2003 *IEEE Transactions on Instrumentation and measurement* **52** 754–761
[7] Aller J M, Habetler T G, Harley R G, Tallam R M and Lee S B 2002 *IEEE Transactions on Industry Applications* **38** 1344–1350
[8] Bonnardot F, El Badaoui M, Randall R, Daniere J and Guillet F 2005 *Mechanical Systems and Signal Processing* **19** 766–785
[9] Randall R and Smith W 2016 *PHM Europe conference, Bilbao, Spain*
[10] Leclere Q, André H and Antoni J 2016 *Mechanical Systems and Signal Processing* **81** 375386

Experimental investigations of contact-type damage nonlinearity

A. Klepka[1], K. Dziedziech[1], Ł. Pieczonka[1], M. Adamczyk[1], W.J. Staszewski[1]

[1]Department of Robotics and Mechatronics, AGH University of Science and TechnologyAl. A. Mickiewicza 30, 30–059 Krakow, Poland,

andrzej.klepka@agh.edu.pl

Abstract. The problem of identification of non-linear phenomena associated with contact–type non-linearity is the subject of many research papers. The number of theoretical models describing the non-linear phenomena is constantly increasing. New capabilities of experimental verification and observation of these non-linear effects provide opportunities for a more accurate mathematical description of the non-linear behavior of contact interfaces. Better understanding of the non-linear related contact mechanisms can lead to more accurate numerical models of structures with contact-types defects, damage propagation and prediction algorithms and development of new methods for damage detection. This paper presents research on the non-linear acoustic phenomena in the presence of contact–type damage. Two test samples in contact are subject to vibro-acoustic modulation test. Non-linear spectral components were analysed and compared to the temperature changes generated as a results of the frictional forces.

1. Introduction

Damage detection methods based on structural dynamics are increasingly being used. Among those techniques the nonlinear acoustics is becoming more attractive for contact-type damage. The reasons for this are: high sensitivity, easy of testing and in some cases possibility of damage localization. Problems that until now have not been resolved are: interpretation of the results and the lack of full knowledge about the mechanisms responsible for the nonlinear behavior of the damaged structures.
Nonlinear acoustical phenomena can be related to many defects like imperfections of atomic lattices (intrinsic or material nonlinearity) and/or thermo-elastic behaviour of interfaces (e.g. cracks or contacts) and many others [1,2,3]. The first are related to material nonlinearity and they have been intensively investigated for sixty years [4,5]. The second group are local nonlinearities. Work on the explanation of the phenomena responsible for the local nonlinear effects associated with contact-type damage has been underway for 20 years but in many cases the results allow only for formulating hypotheses related to these phenomena. The reason for that is a variety of proposed contact related nonlinear mechanisms and fact that the similar nonlinear effects can be manifested by different mechanism and vice versa. Additionally it is very difficult or impossible to separate an assigned specific nonlinear mechanism to specific nonlinear effect. As a result, the nonlinear characteristics allow in easy way to determine damage existence when the baseline methods are applied. Due to the lack knowledge of the phenomena occurring in the structures with contact-type damage, it is extremely hard to interpret the results obtained without the knowledge about the states before failure. Physical understanding of nonlinear mechanisms involved is thus very important for implementation and real engineering applications.

The major objective of the paper is to investigate the nonlinear effect related to the contact phenomena during vibro-acoustic modulation test. The case related to friction mode of damage is investigated. The structure of the paper is as follows. Section 2 presents the general information about nonlinear acoustics. In Section 3 the experimental setup and procedures are described. Section 4 concerns on result discussion. Finally, the article is concluded.

2. Nonlinear mechanisms of contact.

The contact nonlinearity phenomenon is investigated for many years. Classically the two main mechanisms caused signal distortion are "clapping" mechanism and nonlinear friction mechanism. The first one comes from asymmetry of stress-strain relation when the two surfaces are driven by acoustic wave [6,7]. The wave has to produce enough stress to move surface interface. During compression phase the stiffness is higher than during tensile phase when the surface are in partial contact or without the contact. This phenomenon can be approximated by a piece-wise/bilinear stress-strain relation. The characteristic is presents in **Figure 1**.

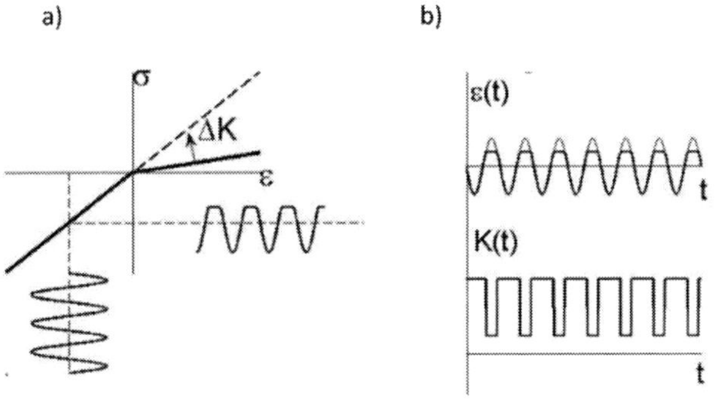

Figure 1. Stiffness characteristic (a) and wave deformation for "clapping" mechanisms.

The clapping mechanism is approximated by equation [8,9]

$$\sigma = K\left[1 - H(\varepsilon)\frac{\Delta K}{K}\right]\varepsilon \tag{1}$$

where $H(\varepsilon)$ is Heaviside function and $\Delta K = K - \frac{\partial \sigma}{\partial \varepsilon}$ for $\varepsilon > 0$. The bi-linear contact caused by a harmonic acoustic wave results in pulse-type modulation of material stiffness and wave deformation. As a result of wave distortion the higher harmonics (even and odd) of the fundamental acoustic wave appear in the response signal spectra.

The second non-linear mechanism which causes changes in stiffness characteristic is related to friction forces between the surfaces. If the amplitude of share wave is low, the damage interfaces are displaced in micro-slip mode between the neighboring roughness areas. This effect is independent of the direction of motion and additionally changes the stiffness characteristics twice (symmetrical non-linearity). In this case only odd harmonics are generated in spectra of the system response. When the amplitude of the acoustic wave increases enough, the contact static friction forces become broken. Then the surfaces of the crack starts sliding relatively to each other in stick and slip mode. This means that first, the asperities coupled by adhesion force deform elastically and then plastically (slip) [10,11]. It causes a cyclic change between static and kinematic friction phase which turns strain–stress characteristic into hysteretic [12,13]. As in the case of micro-slip mode, changes are independent of the direction of displacement and change stiffness twice per one cycle of loading, causing generation of only odd harmonics. **Figure 2** presents stress–strain characteristics for micro-slip as well as stick and slip mode.

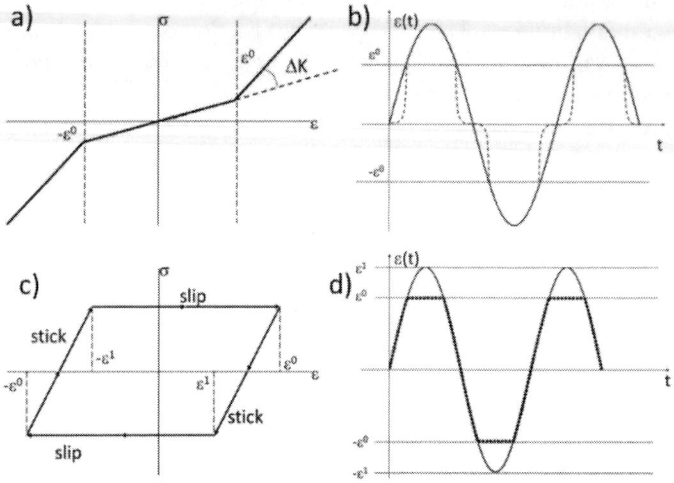

Figure 2. Stress–strain characteristics and wave deformation for micro-slip mode (a,b) and stick and slip mode (c,d)

Besides the classic mechanisms associated with the interaction of the acoustic wave and the surfaces in contact, there are other sources that are not fully explained. Although bilinear stiffness effects a modulation in case of vibro-acoustic modulation test the experimental evidence that the cyclic contact is not the only mechanism involved his effect. Studies have shown that in the case of the wave amplitude is incapable of moving surfaces (cause opening / closing action) of the fatigue crack, the modulation still occurs. Tests have shown that this has to do with the temperature field generated in the contact area [1]. Another neoclassical phenomenon observed for contact type damage is modulation transfer. When the structure is excited by pumping and probing wave, the modulation is transferred from pumping to probing frequency [14]. This phenomenon is compared with Luxembourg-Gorky effect, observed for the first time for radio waves. In work [15] authors concluded that the reason of modulation transfer are dissipative mechanism and strong thermoelectric losses at the defect. Additionally the effect like frequency shift[16], memory effect [17], DC component [18] or hysteretic behavior can be observed for contact-type damage.

3. Experimental setup and procedures.

The experimental work performed focused on the friction forces in case of contact between two surfaces excited simultaneously by low frequency and high frequency acoustic wave. Classical vibro-modulation technique uses these two excitations to induce nonlinear effect in case contact-type damage. It is very important to select the low frequency properly. The most common approach is to select it as a one of the natural frequency of the structure. The natural frequency is related to mode shape. This causes determined moves of surfaces and produce different nonlinear effect. Classically the three crack mode can be defined [1]:

- Crack mode-I – crack faces move directly apart from each other (opening–closing mode);
- Crack mode-II – crack faces slide on each other in the direction perpendicular to the leading edge of the crack (sliding mode);
- Crack mode-III – crack faces move relative to each other and parallel to the leading edge of the crack (tearing mode).

Specified modes are presented in Figure 3.

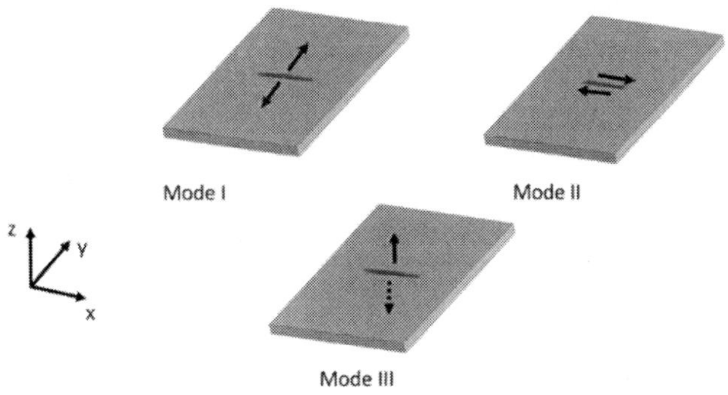

Figure 3. Three basic modes in case contact-type damage.

To find the mode shapes of the investigated structure the modal test is performed. Then after selection proper low frequency excitation, the vibro-acoustic modulation experiment can be carry out.
Two excitation are simultaneously used to excite the structure. One is the low/modal frequency and the second is high frequency acoustic wave. Schematically, the experiment steps are shown in Figure 4.

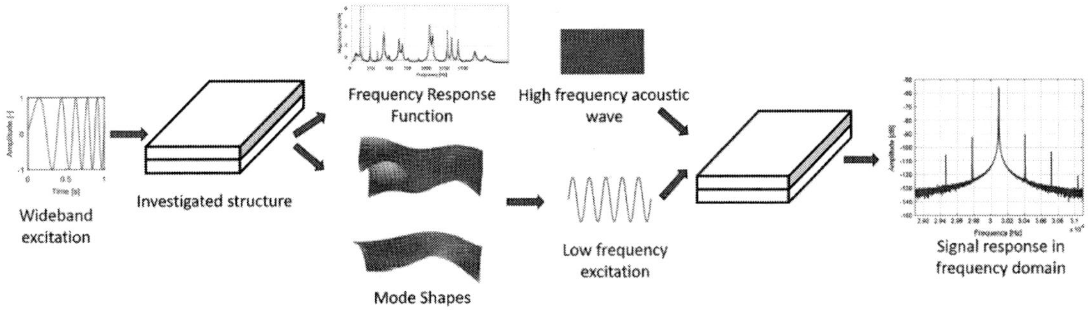

Figure 4. Procedure of vibro-acoustic modulation technique.

As a result of interactions low and high frequency excitations and contact interfaces the response signal can contains a lot of nonlinear components such as higher and subharmonics and modulation. Classical approach assumes that the nonlinear effects are related to closing/opening action caused by low frequency excitation, schematically presented in Figure 5.

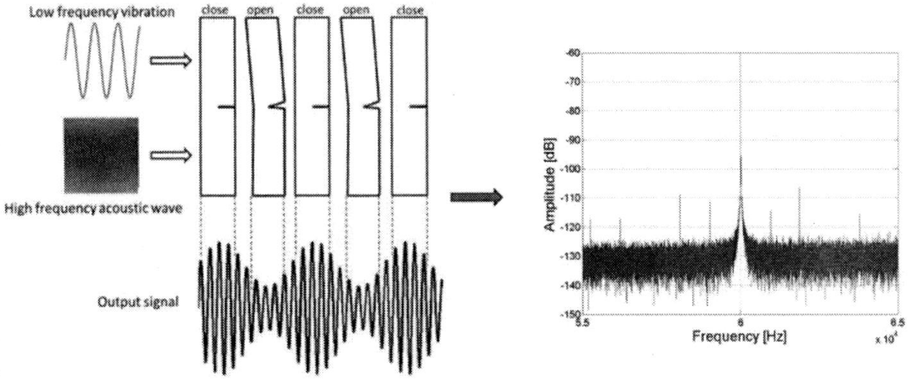

Figure 5. Classical explanation of signal modulation process for contact-type damage.

On the other hand, there are experimental evidence that, besides the classical modulation mechanism, there are others, causing modulation despite the lack of relative displacement between the contact surfaces. In the experiment the frictional nonlinear mechanisms are investigated. Two samples were made from C45 steel. Overall dimensions of the test sample were 40x40x45 mm, with the contact surface of size 10x10 mm. The contact surfaces were grinded with sand paper with the grit size P1500 to obtain smooth surface and remove the trace of milling tool. After sanding, surfaces were cleaned and degreased. The one of the samples was fixed to ground. The second one has been set that the relative motion between contact surfaces occurred in one axis only. The second block was also attached to the electromagnetic shaker. The bottom sample was equipped with piezoelectric transducer for high frequency excitation. The top sample was loaded by extenders with mass from 0 to 500g. The experimental setup is shown in Figure 6.

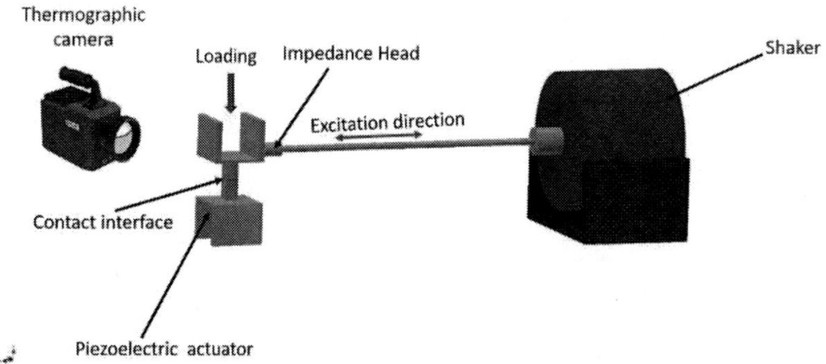

Figure 6. Experimental setup.

4. Result and discussion.

In the first step the classical vibro-acoustic modulation test was performed. The structures were simultaneously excited by different low frequency excitation. In the same time the 45kHz acoustic wave was introduced to the structure. The temperature changes were monitored by high sensitive thermographic camera. Tests were conducted a number of times to check the repeatability of the responses. For different loading level and excitation frequency, the temperature changes were measured. The temperature profiles are presented in Figure 7.

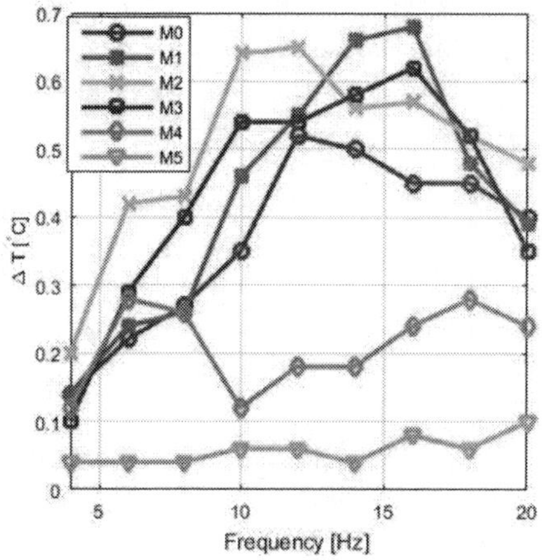

Figure 7. Temperature change profile for different frequencies and loading level.

It can be noticed that temperature changes during test were dependent on loading level and excitation frequency. Their behavior was different for different loading levels. At higher loading levels very small changes of temperatures were observed. Maximal temperature changes occur for loading level M1 and M2.

The examples of response spectra for low frequencies 6 and 16 Hz are presented in Figure 8.

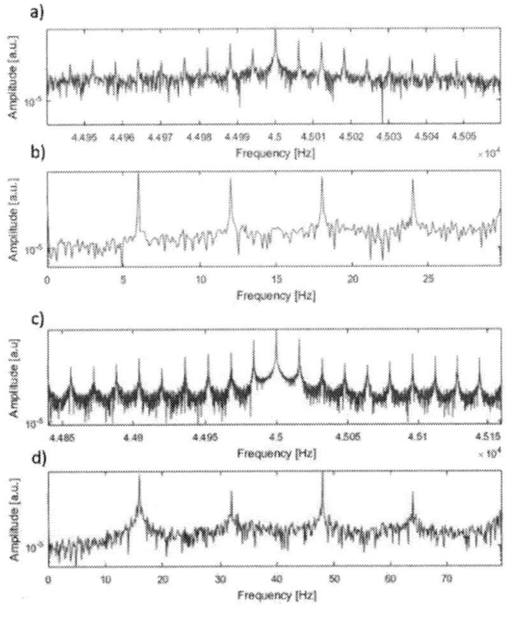

Figure 8. Response spectra focused on vicinity of high and low frequency for 6Hz (a,b) and 16Hz (c,d) low frequency excitation.

Next the 1st and 2nd harmonics analysis was performed. The excitation amplitude was set to 0,4V. The low frequency was changed from 4 to 20Hz with step of 2Hz. The high frequency acoustic wave parameters were kept constant (45kHz@100V). The amplitude of the 1st and 2nd harmonics for different loading are presented in Figure 9.

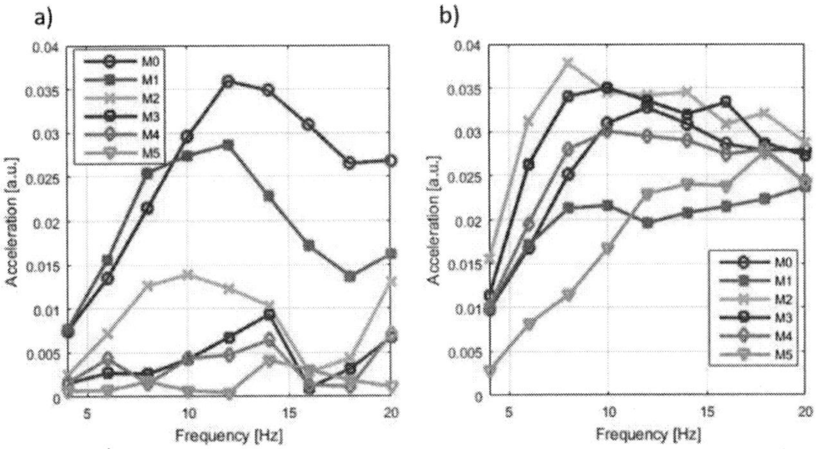

Figure 9. 1st (a) and 2nd (b) harmonics of fundamental frequency for different loading level at different excitation frequency.

As can be noticed the amplitudes of harmonics changed with both excitation frequency and loading level. First the amplitudes increase with frequency values. For different loading levels they reach maximum values at different frequencies and start to decrease. For all cases of loading levels the shape of characteristics is similar.
The next analyses are related to sidebands level. The sidebands amplitude for different frequency excitation and loading level are presented in Figure 10.

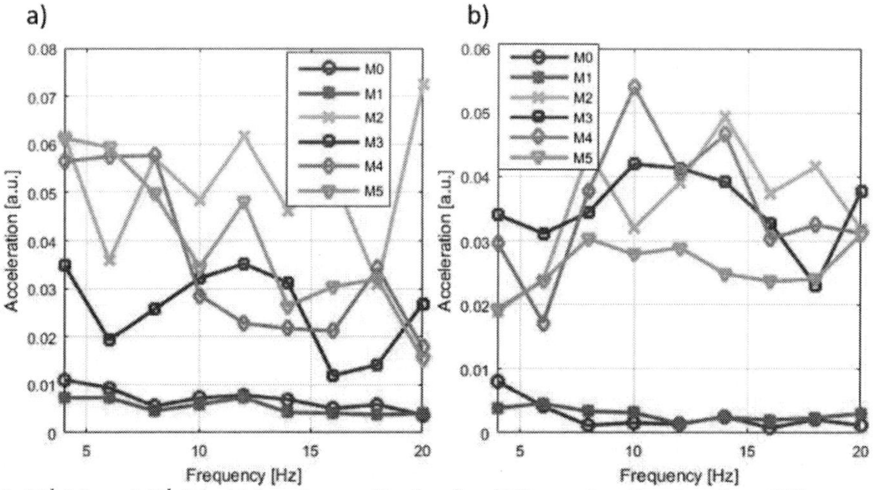

Figure 10. 2nd (a) and 3rd (b) sidebands amplitudes for different loading level at different excitation frequency.

Sidebands levels for M0 and M1 are lowest. It is caused by weak contact between surfaces. Then the high frequency wave is attenuated. Curves related to 2nd sidebands shows different behaviour then 1st harmonics characteristics as shown in Figure 9a (except loading level M3). Regarding the 3rd sidebands for loading levels M2, M3 and M4, the characteristic shape resembles the 2nd harmonic.
In-depth analysis of above presented characteristics leads to conclusion that temperature changes for different loading levels in most cases are related to amplitude changes of given non-linear spectral components (harmonics and sidebands), this could confirm that there is a clear link between temperature generated as results of frictional force acting (temperature field) and non-linear effect related to contact phenomenon.

5. Conclusions.

The non-linear effects related to the contact phenomena in vibro-acoustic modulation test was investigated. Harmonic and sidebands amplitudes were analysed with respect to temperature changes related to friction in contact phenomena. Presented results confirms that temperature filed could correspond to generation of non-linear effects related to contact phenomena. The different behaviour of sidebands and harmonics can be observed for different loading but the shape of the characteristics match to the temperature profile in most cases. This could mean that the distribution of the temperature field generated due to the friction forces have a contribution to generate nonlinear effects observed in the response signal. Finding the relationship between these two phenomena could contribute to understanding of the contact related non-linear mechanisms.
Future works should consider analysis of roughness of the surfaces in contact, as well as different low and high frequency amplitudes of excitations.

ACKNOWLEDGMENT

The work presented in this paper was performed within the scope of the research project 2015/17/B/ST8/03399 financed by the Polish National Science Centre

References

[1]. Klepka A, Staszewski WJ, Jenal RB, Szwedo M, Iwaniec J, Uhl T. Nonlinear acoustics for fatigue crack detection - experimental investigations of vibro-acoustic wave modulations. *Struct Heal Monit.* 2012;11(2):197-211. doi:10.1177/1475921711414236.

[2]. Klepka A, Pieczonka L, Staszewski WJ, Aymerich F. Impact damage detection in laminated composites by non-linear vibro-acoustic wave modulations. *Compos Part B Eng.* 2014;65:99-108. doi:10.1016/j.compositesb.2013.11.003.

[3]. Pieczonka, L., et al. "Impact damage detection in light composite sandwich panels using piezo-based nonlinear vibro-acoustic modulations." Smart Materials and Structures 23.10 (2014): 105021.

[4]. Breazeale MA, Ford J. Ultrasonic studies of the nonlinear behavior of solids. *J Appl Phys.* 1965;36(11):3486-3490. doi:10.1063/1.1703023.

[5]. Bateman T, Mason WP, McSkimin HJ. Third-Order elastic moduli of germanium. *J Appl Phys.* 1961;32(5):928-936. doi:10.1063/1.1736135.

[6]. Solodov IY, Krohn N, Busse G. CAN: An example of nonclassical acoustic nonlinearity in solids. *Ultrasonics.* 2002;40(1-8):621-625. doi:10.1016/S0041-624X(02)00186-5.

[7]. Quan, Li, XiaoZhou Liu, and XiuFen Gong. "Nonlinear nonclassical acoustic method for detecting the location of cracks." *Journal of Applied Physics* 112.5 (2012): 054906.

[8]. Pecorari C, Solodov I. Nonclassical Nonlinear Dynamics of Solid Surfaces in Partial Contact for NDE Applications. *Universality of Nonclassical Nonlinearity.* 2006:309-326. doi:10.1007/978-0-387-35851-2_19.

[9]. Woolfries, S. "The effects of friction between crack faces on nonlinear wave propagation in microcracked media." Journal of the Mechanics and Physics of Solids 46.4 (1998): 621-644.

[10]. Solodov IY. Ultrasonics of non-linear contacts: propagation, reflection and NDE-applications. *Ultrasonics.* 1998;36(1-5):383-390. doi:10.1016/S0041-624X(97)00041-3.

[11]. Awrejcewicz, Jan, and Pawel Olejnik. "Stick-slip dynamics of a two-degree-of-freedom system." *International Journal of Bifurcation and Chaos* 13.04 (2003): 843-861.

[12]. Solodov I. Nonlinear Acoustic Ndt : Approaches , Methods , and Applications. *10th Int Conf Slov Soc Non-destructive Test.* 2009:1-16.

[13]. Awrejcewicz, J., and P. Olejnik. "Analysis of dynamic systems with various friction laws." *Applied Mechanics Reviews* 58.6 (2005): 389-411.

[14]. Trojniar, T., Klepka, A., Pieczonka, L., & Staszewski, W. J. (2014, March). Fatigue crack detection using nonlinear vibro-acoustic cross-modulations based on the Luxemburg-Gorky effect. In SPIE Smart Structures and Materials+ Nondestructive Evaluation and Health Monitoring (pp. 90641F-90641F). International Society for Optics and Photonics.

[15]. Zaitsev V, Gusev V, Castagnede B. Luxemburg-gorky effect retooled for elastic waves: a mechanism and experimental evidence. *Phys Rev Lett.* 2002;89(10):105502. doi:10.1103/PhysRevLett.89.105502.

[16]. Guyer, Robert A., and Paul A. Johnson. "Nonlinear mesoscopic elasticity: Evidence for a new class of materials." *Physics today* 52.4 (1999): 30-36.

[17]. Solodov, Igor Yu, and Boris A. Korshak. "Instability, chaos, and "memory" in acoustic-wave–crack interaction." *Physical review letters* 88.1 (2001): 014303.

[18]. Korshak, B. A., I. Yu Solodov, and E. M. Ballad. "DC effects, sub-harmonics, stochasticity and "memory" for contact acoustic non-linearity." *Ultrasonics* 40.1 (2002): 707-713.

[19]. Moussatov, Alexei, Vitalyi Gusev, and Bernard Castagnede. "Self-induced hysteresis for nonlinear acoustic waves in cracked material." *Physical review letters* 90.12 (2003): 124301.
[20]. Gusev, Vitalyi. "Propagation of acoustic pulses in material with hysteretic nonlinearity." *The Journal of the Acoustical Society of America* 107.6 (2000): 3047-3058.

Fault Features Extraction and Identification based Rolling Bearing Fault Diagnosis

B Qin[1], G D SUN[1], L Y ZHANG[1], J G WANG[1], J HU [2]

[1] School of Mechanical Engineering, Inner Mongolia University of Science & Technology, Baotou 014010
[2] Chungking Pure-Smart & Technology CO. LTD, Chungking 400030

E-mail:nkdqb@163.com

Abstract. For the fault classification model based on extreme learning machine (ELM), the diagnosis accuracy and stability of rolling bearing is greatly influenced by a critical parameter, which is the number of nodes in hidden layer of ELM. An adaptive adjustment strategy is proposed based on vibrational mode decomposition, permutation entropy, and nuclear kernel extreme learning machine to determine the tunable parameter. First, the vibration signals are measured and then decomposed into different fault feature models based on variation mode decomposition. Then, fault feature of each model is formed to a high dimensional feature vector set based on permutation entropy. Second, the ELM output function is expressed by the inner product of Gauss kernel function to adaptively determine the number of hidden layer nodes. Finally, the high dimension feature vector set is used as the input to establish the kernel ELM rolling bearing fault classification model, and the classification and identification of different fault states of rolling bearings are carried out. In comparison with the fault classification methods based on support vector machine and ELM, the experimental results show that the proposed method has higher classification accuracy and better generalization ability.

1. Introduction

Rolling bearing is playing an increasingly important role in rotary mechanical. However, diverse working conditions, including full of dust, high humidity, high temperature, and unpredictable heavy load etc., may cause the power drive system prone to fault, which is also one of the important sources of system vibration, noise, and lead to catastrophic failure or even huge economic loss. Therefore, timely monitoring of bearing and classifying faults into correct types are of great significance.

Today, data-driven fault monitoring and classification methods have attracted much attention[1]. In fact, the diversity and quality of modeling data influence the effectiveness of the fault classification. After collection of data, several crucial points should be discussed: 1) how to extract the fault features from the collected signal; 2) how to improve the classification accuracy of the fault identification model. For the fault-features extraction, Arnaz et al.[2] proposed a method to construct a mixed domain feature set based on wavelet decomposition. However, the time series is only extracted by wavelet decomposition in the frequency domain and time domain features are neglected. On the contrary, Guo et al.[3] extracted fault-features from frequency domain from six aspects, which include original vibration signal and its spectrum, the filtered, and demodulated signal by wavelet packet etc. To better extract fault features, Cui et al.[4] extracted the signal features by wavelet packet decomposition. In time domain, the time delay of original signal was pretreated by wavelet packet decomposition. Then in the frequency domain, threshold of signal was constructed and fault features were extracted through spectrum analysis. However, the selection of wavelet-based functions is difficult and subjective.

Recently, intelligent algorithms have been applied to fault classification of bearing, such as artificial neural network[5], fuzzy logic[6], and least squares support vector machines[7] etc. The traditional back propagation neural network (BPNN) faces the problem of slow convergence rate, difficulty of configuring tunable parameters and easily falling into the local optimum. In comparison with BPNN, the generalization performance of support vector machine (SVM) is improved. However, kernel function and its parameters are usually given according to human experiences. Extreme learning machine (ELM)[8] is a newly developed single hidden layer feed-forward neural network, which does not require continuously to adjust parameters of hidden layer. In ELM, iterative parameter optimization process of traditional neural network is replaced by solving linear equation groups and the outputs of the minimum norm least squares solution are employed as the weights of the network. Therefore, the network is trained one-time without iterations. Compared with BPNN and SVM, ELM greatly improve the training speed and generalization ability, which has been successfully applied in the areas such as pattern recognition[9-14]. The idea of introducing ELM into fault classification for bearing is given in this paper. However, the adjustment of the weights of the input and hidden layer adopts random selection method, which can not guarantee the validity of the weight and easily cause the problem of redundant network structure. Because the structure parameters of fault identification model determine the classification ability, how to obtain the optimal structure parameters is the key to improve the classification accuracy of fault identification model.

To solve the above mentioned problems, a comprehensive and effective fault feature extraction and classification algorithm is proposed in this paper. First, to overcome the influence of disturbances, morphological average filtering algorithm is given to filter the collected signals and then the mode function (MF) is obtained through vibrational mode decomposition (VMD)[15] after de-noising. Mode function presents fault features and MF components that have large correlation coefficients are used to calculate energy index in time domain. Besides, on the other hand, entropy of these MFs are obtained by Permutation Entropy (PE)[16] analysis, which can be used to obtain the singular values. The matrix of PE is reconstructed high dimension feature vector used for classification. Third, the inner product of Gauss kernel function is utilized to express the ELM output function, which adaptively determine the number of the hidden layer nodes. At last, the fault classification model is developed by these feature vectors and the K-ELM algorithm.

The remainder of this paper is organized as follows. In Section 2, the extraction of multiple feature vectors are given. Fault classification method is formulated based on the enhanced ELM algorithm is proposed in Section 3. Experiment results clearly demonstrate the effectiveness and feasibility of the proposed method in the last section.

2. The extraction of feature vectors

The specific processes of feature vectors extraction are shown in Fig. 1. First, rolling bearing vibration signals are de-noised by morphological averaged filter. After de-nosing, feature extraction contains two sections: 1) employ VMD to get intrinsic mode components (MFs) that have large correlation coefficients; 2) calculate permutation entropy using the obtained MFs. Through the above two steps, the calculated feature matrix is treated as feature vectors for bearing fault classification.

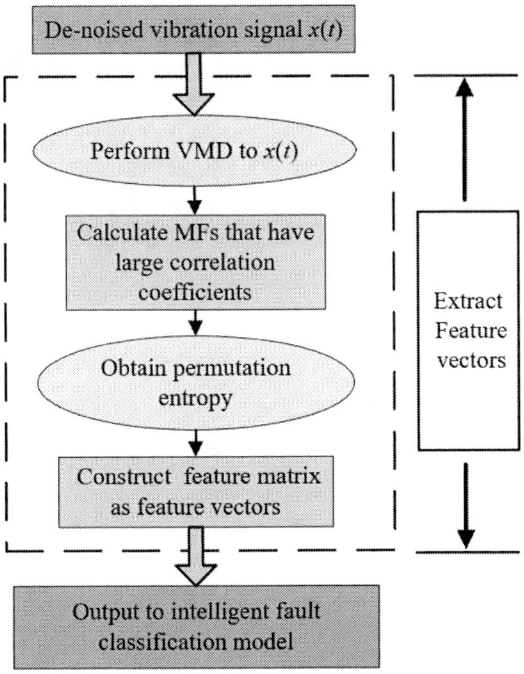

Figure 1 Flow chart of the extraction of feature vectors.

2.1. De-noising of original signal

Because data collected from signal gathering system usually contain noise, here mathematical morphology (MM) and average filtering algorithm[17] are used to filter it. The basic idea of MM is to use some structural elements that have certain shapes to measure and extract images corresponding to the shape and achieve the purpose of image analysis. Based on the geometric characteristics of the signal, MM based average filter can cope with the nonlinear signal noise by morphological operations between structural elements and the original signal. The proposed filter inherits the advantages of MM, including simple operation, processing the local properties of signal in the time domain, does not have amplitude attenuation and phase shift in the waveform. Therefore, it is helpful for the processing of mechanical fault signals.

Opening and closing are the basic operations of MM, as shown in Eqs. (1) and (2), as follows,

$$(f \circ g)(n) = (f \Theta g \oplus g)(n) \tag{1}$$

$$(f \bullet g)(n) = (f \oplus g \Theta g)(n) \tag{2}$$

where Θ is erosion operation presenting the relationship of $(f \Theta g)(n) = \min\left[f(n+m) - g(m)\right] n,m \in [0, N-1]$, $\oplus$ is dilation operation having the relationship of $(f \oplus g)(n) = \max\left[f(n-m) + g(m)\right] n,m \in [0, N-1]$ and $n > m$.

The linear combinations of Eqs. (1) and (2) can be used to construct the average filter (AVG), as shown in Eq. (3). It eliminates positive and negative impulse of the signal and can smooth the signal and reduce the signal noise.

$$\mathrm{AVG}(f) = (f \bullet g + f \circ g)/2 \tag{3}$$

2.2. Extraction of feature vectors

2.2.1. Calculation of MFs components based on VMD

However, harsh operation conditions of rolling bearing, its vibration signal always contains process disturbances, including the resonance, external noise, etc. Then after de-noising using Eq. (3), Variational Mode decomposition[15] is employed to extract the intrinsic characteristics of signals, which attempts to obtain these central frequencies and intrinsic mode functions centered on those frequencies concurrently using alternate direction method of multipliers (ADMM) [18].

Here, note the de-noised signal still as *f* that is represented as Eq. (4) through ADMM,

$$\min_{\{u_k,\omega_k\}}\left\{\sum_k\left\|\partial_t\left[\left(\sigma(t)+\frac{j}{\pi t}\right)u_k(t)\right]e^{-j\omega_k t}\right\|_2^2\right\} \quad s.t.\ \sum_k u_k=f \tag{4}$$

The objective function is defined as an unconstrained optimization problem as Eq. (4) by means of augmented Lagrangian Multiplier method.

$$L(\{u_k\},\{\omega_k\},\lambda)=\alpha\sum_k\left\|\partial_t\left[\left(\sigma(t)+\frac{j}{\pi t}\right)\bullet u_k(t)\right]e^{-j\omega_k t}\right\|_2^2 +\left\|f(t)-\sum_k u_k(t)\right\|_2^2+\left\langle\lambda(t),f(t)-\sum_k u_k(t)\right\rangle \tag{5}$$

In ADMM philosophy, we solve for one variable at a time assuming all others are known. So the formula for updating u_k at the n+1 the iteration is as follows. The convergence of the algorithm is given in[15].

The final algorithm for VMD[15] as follows.

First, initialize $\{\hat{u}_k^1\}$, $\{\omega_k^1\}$, $\hat{\lambda}^1$. $n\leftarrow 0$

Repeat,

$n\leftarrow n+1$

For $k\leftarrow 1{:}K$

Update $\hat{u}_k$ for all ω≥0

$$\hat{u}_k^{n+1}\leftarrow\frac{\hat{f}(\omega)-\sum_{i\neq k}\hat{u}_i(\omega)+\hat{\lambda}(\omega)/2}{1+2\alpha(\omega-\omega_k)^2} \tag{6}$$

Update ω_k:

$$\omega_k^{n+1}\leftarrow\frac{\int_0^\infty\omega|\hat{u}(\omega)|^2\,d\omega}{\int_0^\infty|\hat{u}(\omega)|^2\,d\omega} \tag{7}$$

End

Dual ascent for all $\omega\geq 0$

$$\hat{\lambda}^{n+1}(\omega)\leftarrow\hat{\lambda}^n(\omega)+\tau\left(\hat{f}(\omega)-\sum_k\hat{u}_k^{n+1}(\omega)\right) \tag{8}$$

Until convergence: $\sum_k\left(\left\|u_k^{n+1}-u_k^n\right\|_2^2/\left\|u_k^n\right\|_2^2\right)<\varepsilon$.

2.2.2. Permutation entropy

With the superiority, permutation entropy (PE)[16] has been valid used in many spheres, such as fast calculation, robustness, invariance to nonlinear monotonous transformations, and so on. By comparing the size of the adjacent data, dynamic mutation of the time series are tested by PE. In the process of the above, specific values of data operation will not be involved, which avoid the interference of noise and reduce the computational complexity. When the rolling bearing failure occurs, the nonlinear vibration of fault source aroused, meanwhile the complexity of the bearing vibration signal will be changed. Thus, PE can be used to measure the complexity of the signal. The basic principle of PE is briefly described as follows, more detailed information about PE can be found in Refs. [16] and [33].

For a time series $\{x(k), k=1,2,\ldots,N\}$, an D-dimensional delay embedding vector at time i is defined as:

$$X_i^D = [x(i), x(i+\tau), x(i+2\tau), \cdots x(i+(D-1)\tau)] \tag{9}$$

where $i=1,2,\ldots,N$, $D \geqslant 2$ represents the embedded dimension and τ represents the time delay.

Then, denote the symmetric group of order $D!$ as S_D, which is the group of all the permutations of length D. The symbols in S_D can be set as $\pi_j=(j_1, j_2, \ldots, j_D)$. Then, we say that X_i^D has a permutation π_j, if and only if π_j is unique in S_D and satisfies:

$$x(i+(j_1-1)\tau) \le x(i+(j_2-1)\tau) \le \cdots \le x(i+(j_D-1)\tau) \tag{10}$$

where 1ej_i≤ere T$j_i \neq j_j$.

For each permutation π_j, the relative frequency can be obtained by:

$$P(\pi_j) = \frac{Num\{X_i^D\}}{N-(D-1)\tau} \tag{11}$$

where $1 \le j \le N-(D-1)\tau$.

Then, according to Shannon's entropy theory, the PE is defined as:

$$H_p(D) = -\sum_{\pi_j \in S_D} p(\pi_j)\ln(p(\pi_j)) \tag{12}$$

Obviously, $0 \le H_p(D) \le \ln(D!)$, where the lower bound is attained for an increasing or decreasing sequence of values, and the upper bound for a completely random system where all of the possible $D!$ permutations appear in the time series with the same probability $1/\ln(D!)$. For convenience, Eq. (11) is always normalized onto the interval of [0, 1] by $\ln(D!)$, and denoted as:

$$H_p = H_p(D)/(ln(D!)) = -\frac{1}{ln(D!)}\sum_{i=1}^{k} p(\pi_i) ln(p(\pi_i)) \tag{13}$$

The main idea of the PE algorithm is to match the candidate time series onto a symbolic sequence which describes the relationship between present values and a fixed number of equidistant values at a given past time. To calculate the value of PE only needs relationship between any two sample points in the time series. This is a highly favourable characteristic which makes PE avoid noise interference and has strong robustness with low computational complexity. Moreover, PE measures the departure of the time series from a complete random one: the smaller the value of the PE, the more regular the time series is. This implies that PE value increases with the irregularity or randomness of the time series[20]. The changes of PE value reflect and amplify the tiny changes of the time series.

By processing each group of the signal under different status according to the above steps, we obtain these high dimensional singular value matrixes as feature vectors to classify fault of rolling bearing. And feature vectors are adopted to train classification model of rolling bear based on kernel extreme learning machine, which will be given in the following section.

3. Identification Model Based Kernel Extreme Learning Machine

The accuracy of fault classification depends on the intelligent model used in the process of machine learning methods. Compared with the algorithm of BPNN, SVM, ELM only needs to set the number of nodes for hidden layer to train the network and it has the advantages of high efficiency, fast learning speed and the unique solution. However, two structure parameters of ELM, i.e. input weights and hidden layer threshold, are randomly given and they will inevitably lead to poor accuracy. Having the advantages of dynamic control of global and local search conversion and avoids falling into local optimum, ELM output function is expressed by the inner product of Gauss kernel function to adaptively determine the number of the hidden layer nodes. Therefore, Gauss kernel function conveyed ELM is used in the construction of a rolling bearing fault classification model to improve the precision and generalization ability. The processes of intelligent fault classification are shown in Fig. 2 concretely.

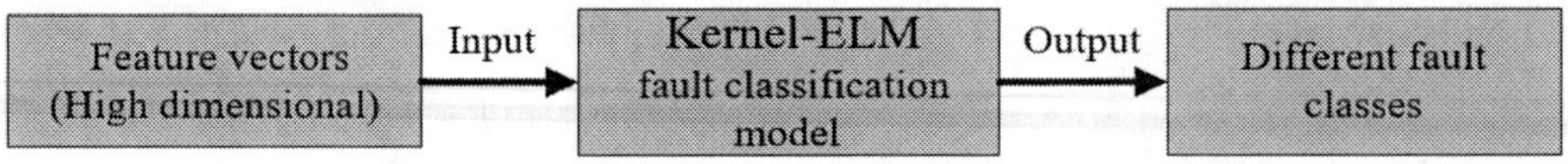

Figure 2 Flow chart of intelligent fault classification.

3.1. The establishment of fault classification model

In this part, the fault classification model is proposed based on Kernel-ELM. Only requiring the number of neurons in hidden layer, ELM randomly generates connection weights and threshold of hidden layer neurons between the input layer and hidden layer and it can obtain the unique optimal solution.

Assuming the number of samples is N, i.e. $(x_i, y_i) i \in [1,2,\ldots N]$, the number of nodes of hidden layer is H and the activation function is $g(x)$, the mathematical model of ELM is below,

$$y_i = \sum_{j=1}^{L} \beta_j g\left(w_j x_i + b_j\right) \tag{14}$$

where $w_j = [k_{j1}, k_{j2}, \ldots, k_{jn}]$ is the connection weights vector between the input node and the jth node of hidden layer, b_j is threshold of the jth node in hidden layer.

In Eq. (14), a feed-forward neural network model of single hidden layer is developed, of which the output is close to zero error,

$$\sum_{i=1}^{N} \left\| y_i - t_i \right\|^2 = 0 \tag{15}$$

Sequentially, parameters w_j, b_j, β_j satisfy the relationship of Eq. (16),

$$\sum_{i=1}^{N} \beta_j g\left(w_j x_i + b_j\right) = t_i, i = 1,2,\ldots,N \tag{16}$$

And Eq. (14) can be further simplified as $\mathbf{H\beta} = \mathbf{T}$, where,

$$\mathbf{H} = \begin{bmatrix} g(k_1, b_1, x_1) & \ldots & g(k_L, b_L, x_1) \\ \ldots & \ldots & \ldots \\ g(k_1, b_1, x_N) & \ldots & g(k_L, b_L, x_N) \end{bmatrix}_{N\times L}, \quad \boldsymbol{\beta} = \begin{bmatrix} \beta_1^T \\ \vdots \\ \beta_L^T \end{bmatrix}, \quad \mathbf{T} = \begin{bmatrix} T_1^T \\ \ldots \\ T_N^T \end{bmatrix}_{N\times m}$$

$\mathbf{H}$ is the output matrix of hidden layer, and $\mathbf{H}(i, j)$ stands for the output of the ith training data in the jth hidden node.

The goal of adjustment is to find a set of optimal parameters w_j, b_j, β_j that make the $\left\|(\mathbf{H\beta})^T - \mathbf{T}\right\|$ minimum.

3.2. The establishment of fault classification model

The weights of input layer and thresholds of hidden layer might be zero which result into some hidden layers might be failed. Thus, the number of hidden layer nodes has to be increased to obtain high classification accuracy. However, it may lead to poor adaptability and low generalization capacity for testing data. The Gauss kernel function as Eq. (17) by means of inner product is employed to express the input weights and threshold of hidden layer of ELM to achieve fast and high accuracy convergence. In this way, the classification accuracy and generalization ability will be improved. The specific process is shown in Fig. 3.

$$k(x,x_i)=exp(-\frac{\|x-x_i\|^2}{\sigma^2}) \tag{17}$$

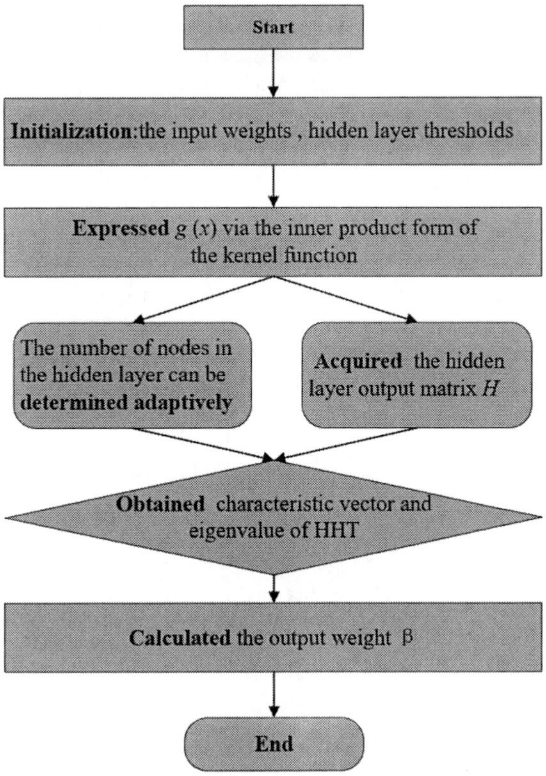

Figure 3 The flow chart of Kernel-ELM.

The hidden node output function $g(x)$ of ELM algorithm is unknown and its kernel function can be used to express $g(x)$ in the form of inner product. However, by using specific kernel functions $K(x,x_i)$ as Eq. (18) we can easily find value of the output function and the hidden layer nodes can be adaptively determined. In the above process, the concrete form on hidden node output function $g(x)$ didn't given directly.

$$K_{ELM}=HH^T:K_{ELM}=g(x_i)\cdot g(x_j)=K(x_i,x_j) \tag{18}$$

To sum up, K-ELM can be summarized as below: Given a containing L samples of the training set (x_i,y_i), x_i represents an input vector, y_i represents the corresponding output, $i=1,2,\ldots,L$, and the kernel function $K(x,x_i)$. And then the output equation as:

$$\begin{aligned} f(x)&=g(x)H^T(1/_C+HH^T)^{-1}T \\ &=\begin{bmatrix} K(x,x_1) \\ \vdots \\ K(x,x_L)\end{bmatrix}^T(1/_C+\Omega_{ELM})^{-1}T \end{aligned} \tag{19}$$

4. Rolling bearing fault classification based VMD ,PE and K-ELM

The actual bearing vibration signals are usually complex, which produced by multiple sources. Although PE can be used to measure the complexity of the signal, it can not identify the bearing failure under different working condition. The vibration signal is adaptive decomposed into a series of different time scales of PE component by VMD. Above all, the entropy of PE component can be used to indicate the intrinsic characteristics of the vibration signals in different time scales, which provided accurate information for fault diagnosis. In addition, Gauss kernel function conveyed ELM is used in the construction of a rolling bearing fault classification model to improve the precision and generalization ability. Process of bearing fault diagnosis is shown in Fig. 4 based on the VMD, PE and K-ELM.

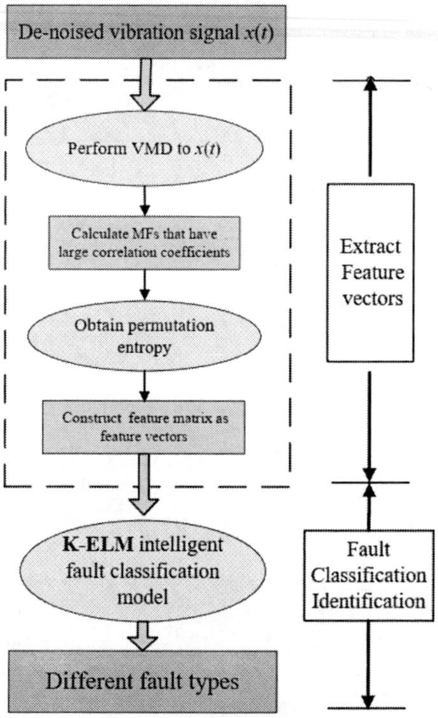

Figure 4 The flow of bearing fault diagnosis based on the VMD,PE and K-ELM.

The specifics are:

(1) Sample k times respectively in accordance with given sampling frequency f_s under normal, inner ring, outer ring and rolling elements of rolling bearing four different conditions, totaling gained 4 k signals eventually. A total of k signals, m signals were randomly selected as training sample of each state, the others (n signals) as testing sample sets.

(2) Decompose above any signal by VMD, and choose four PE component (PE_1,.., PE_4) in front which contains the main failure.

(3) Construct high dimensional feature vector T. First, according to Eq. (9) confirm time delay τ =1 and embedded dimension D=4. And then, calculate normalized PE component PE_1,.., PE_4.

$$\boldsymbol{T}=[PE_1,PE_2,\cdots PE_4] \tag{20}$$

(4) Repeat Steps (2) and (3), construct high dimensional feature vector sets, including 4k signals.

(5) The high dimension feature vector sets are used as the input of K-ELM algorithm to establish the kernel function extreme learning machine rolling bearing fault classification model. Meanwhile, four kinds of status corresponding fault category label is defined. If the output value is 1, it stands for normal status. Similarly, inner ring fault, outer ring fault and rolling body fault are identified as the value of 2, 3 and 4 respectively.

5. Results and discussions

5.1. Data acquisition and extraction of feature vectors

At last, the experimental results on the Case Western Reserve University bearing data set verify the feasibility and validity of the proposed method in this paper. Bearing designation of testing is deep groove ball bearings 6205-2RS. Usually, there are three types of fault: rolling body fault, inner ring fault and outer ring fault. Combined with the normal status, four kinds of signal can be collected. For each operation status, 30 times of experiment are performed. Each experiment contains 5120 data points.

First, morphological average filter is used to de-noise the above signal. The linear structural element is selected, and each structural element value is 0, namely g={0,0,0}. According to the determined structural elements, implement the four states signals' noise reduction by morphological average filter. Then, VMD is used to decompose the state sample under different status. According to the rule given in Subsection 2.2.1, four MFs will be retained. The decomposition of one experiment under inner ring fault status is shown in Fig. 5, and part extraction of de-noising signal feature vectors based VMD and PE in four cases are shown in Table 1.

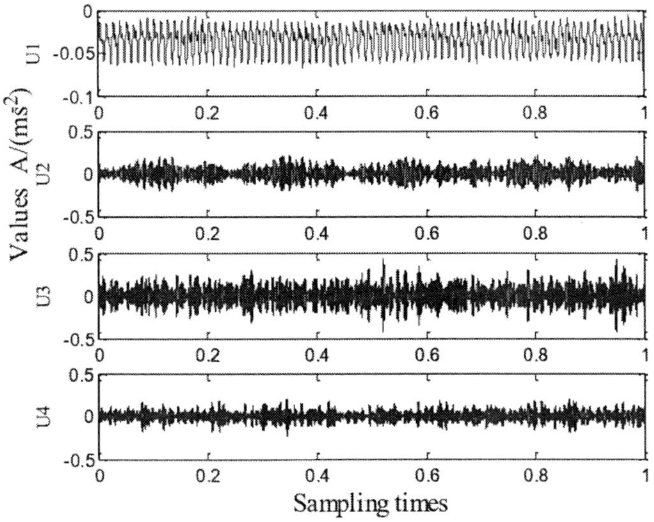

Figure 5 The results of VMD for inner ring fault condition.

Table 1 Part extraction of de-noising signal feature vectors based VMD and PE in four cases.

CONTION	PE_1	PE_2	PE_3	PE_4
Normal	0.2858	0.4668	0.4979	0.6823
	0.2715	0.4475	0.4953	0.6679
	0.2817	0.4655	0.4882	0.6855
	0.2752	0.4509	0.4970	0.6817
Inner ring fault	0.3721	0.4583	0.489	0.6083
	0.3683	0.4558	0.4914	0.5556
	0.3063	0.4587	0.4936	0.5579
	0.3296	0.4686	0.4982	0.5564
Outer ring fault	0.3156	0.4102	0.4756	0.5616
	0.3674	0.4089	0.4826	0.5616
	0.3267	0.4085	0.4959	0.5496
	0.3186	0.4083	0.4921	0.5518
Rolling body fault	0.2054	0.3333	0.4783	0.5703
	0.2135	0.3478	0.4758	0.6298
	0.2510	0.3855	0.4787	0.5888
	0.2210	0.3479	0.4836	0.6742

5.2. The improvement of classification model

Totally, thirty experiment data are conducted under each condition. Twenty of them are used as training data and the remaining ten are used as testing data. Using the value of PE as the input of the inputs of the K-ELM algorithm, the fault classification model is developed.

For the testing samples, fault classification accuracy of K-ELM model is 100%, which is a high accuracy and is shown in Fig. 6. The value of *y*-axis stands for the different operation status. If the value is 1, it stands for normal condition. Similarly, inner ring fault, outer ring fault and rolling body fault are identified as the value of 2, 3 and 4 respectively. To better illustrate the performance of the proposed method, SVM and ELM are employed for comparison. The fault classification model accuracy of SVM is 87.5% and ELM is 95%, which are shown in Figs. 7 and 8. The comparison results are shown in Table 2 and the proposed method has a higher classification accuracy.

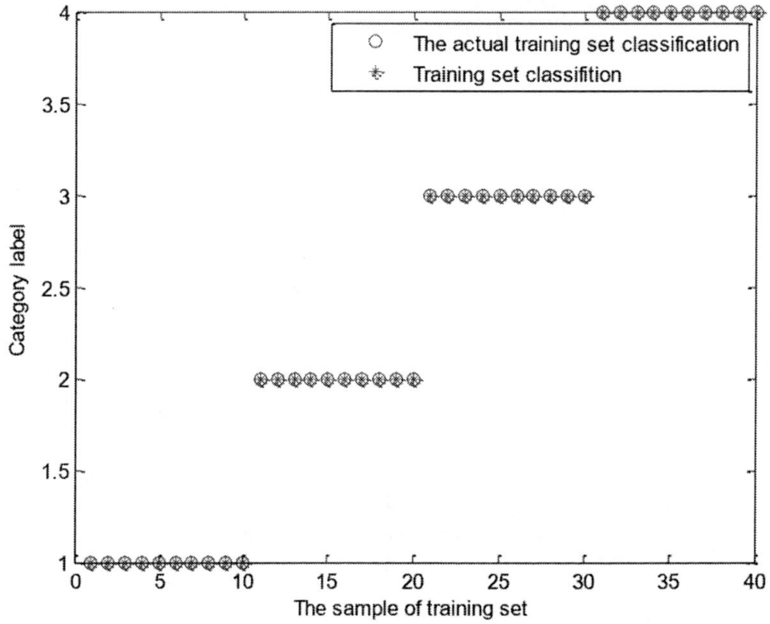

Figure 6 Classification of testing data based on K-ELM.

Table 2 Comparisons of SVM, ELM and K-ELM.

Algorithm	Accuracy (%)			
	Normal	Inner ring fault	Outer ring fault	Rolling body fault
SVM	100	90	60	100
ELM	100	100	80	100
K-ELM	100	100	100	100

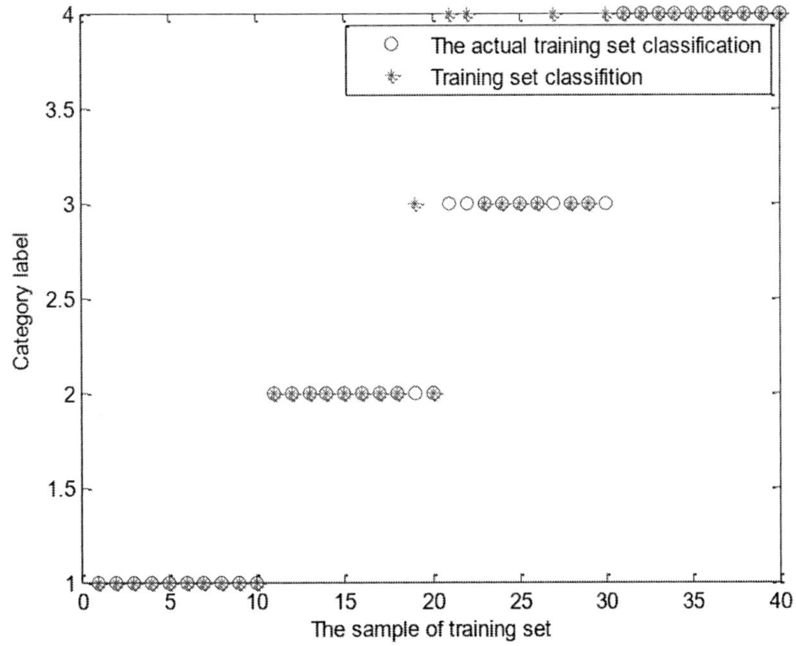

Figure 7 Classification of testing data based on SVM.

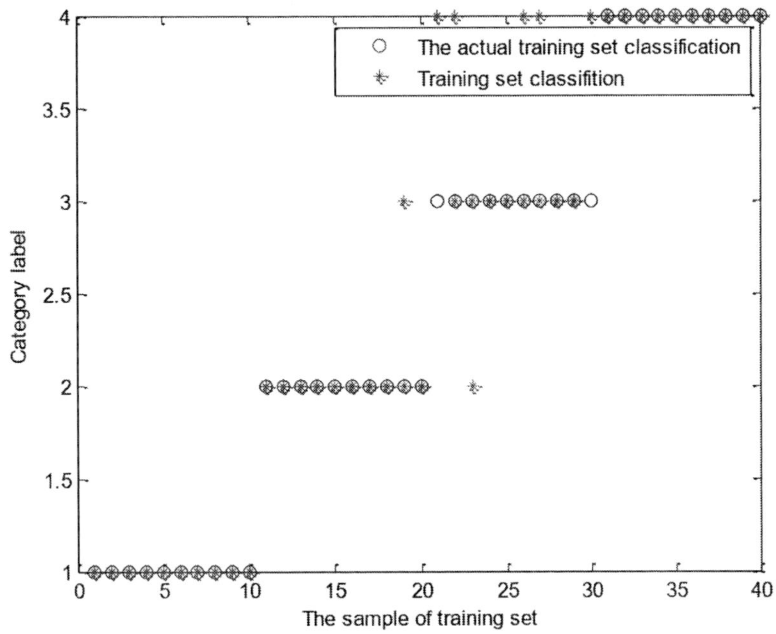

Figure 8 Classification of testing data based on ELM.

6. Conclusions

To solve the existing problems of fault features extraction and fault classification method of rolling bearing, several crucial points are considered in this paper. First, because rolling bearing works under a complex environment, the collected vibration signal needs filter before using. To effectively remove noise, a morphological average filtering algorithm is proposed. Then VMD is performed on the filtered data to obtain high dimensional feature vectors. Then, these feature vector sets are used as inputs for fault modelling. Finally, the fault classification model is developed based on enhanced extreme learning machine, which output function is expressed by the inner product of Gauss kernel function to adaptively determine the number of the hidden layer nodes. In comparison with fault classification methods based on support vector machine and extreme learning machine, the experimental results show that the proposed method has higher classification accuracy and better generalization ability.

References

[1] Yan Qin, Chunhui Zhao, Furong Gao. An Iterative Two-Step Sequential Phase Partition (ITSPP) Method for Batch Process Modeling and Online Monitoring. AIChE Journal, 2016, 62(7): 2358-2373

[2] Arnaz M, Robert X G. PCA-Based Feature Selection Scheme for Machine Defect Classification. IEEE Transactions on Instrumentation and Measurement, 2008, 53(6): 1517-1525

[3] Leiya Guo, Hezheng Jia, Ziyan Yang. Application of an intelligent classification method to mechanical fault diagnosis. Expert Systems with Applications, 2009, 36: 9941-9948.

[4] Cui Lingli, Kang Chenhui. The enhanced resonance demodulation method based on delay correlation and wavelet packet coefficient entropy threshold. Mechanical Engineering, 2010, 46(20): 53-57

[5] González-Grimaldo, Raymundo A,Cuevas-Tello, Juan C. Analysis of time series with artificial neural networks. 7th Mexican International Conference on Artificial Intelligence, MICAI 2008, 131-137

[6] Rubens N O. The Application of fuzzy Logic to the construction of the ranking function information retrieval systems. Computer Modelling and New Technologies, 2006, 10(1): 20-27

[7] Yang Jie, Ma Jun. A sparsity-based training algorithm for Least Squares SVM. 5th IEEE Symposium on Computational Intelligence and Data Mining, CIDM 2014, 345-350

[8] Huang GB, Zhu QY, Siew CK. Extreme learning machine: theory and applications. Neurocomputing, 2006, 70(1): 489-501

[9] Mohammed A A, Minhas R, Jonathan Wu Q M, Sid-Ahmed M A. Human face recognition based on multidimensional PCA and extreme learning machine. Pattern Recognition, 2011, 44(10-11): 2588-2597

[10] Van Heeswijk M, Miche Y, Lindh-Knuutila T, Hilbers P A, Honkela T, Oja E, Lendasse A. Adaptive ensemble models of extreme teaming machines for time series prediction. 2009, Lectures Notes in Computer Science, 5769: 305-314

[11] Tian HX, Mao ZZ. An ensemble ELM based on ModiFed AdaBoost RT algorithm for predicting the temperature of molten steel in ladle furnace. IEEE Transactions on Automation Science and Engineering, 2010, 7(1):73-80

[12] Yuan Q, Zhou W D, Li S F, Cai D M. Epileptic EEG classification based on extreme learning machine and nonlinear features. Epilepsy Research, 2011, 96(1-2): 29-38

[13] Akyildiz I F, Weilian S, Sankarasubramaniam Y, et al. Wireless sensor networks: A survey. Computer Networks, 2002, 38(4): 393-422

[14] Jing Wang, Guanghua Xu, Qing Zhang, et al. Application of improved morphological filter to the extraction of impulsive attenuation signals. Mechanical Systems and Signal Processing. 2009, 23(1): 236-245

[15] Konstantin Dragomiretskiy and Dominique Zosso. Variational Mode Decomposition. IEEE Transactions on Signal Processing. vol. 62, no.3. p.531-544. Feb.2014

[16] X. Zhang, Y. Liang, J. Zhou A novel bearing fault diagnosis model integrated permutation entropy, ensemble empirical mode decomposition and optimized SVM Measurement, 69 (2015), pp. 164-179

[17] Jing Wang, Guanghua Xu, Qing Zhang, et al. Application of improved morphological filter to the extraction of impulsive attenuation signals. Mechanical Systems and Signal Processing. 2009, 23(1): 236-245

[18] Stephen Boyd, Neal Parikh, Eric Chu, Borja Peleato and Jonathan Eckstein. Distributed

Optimization and Statistical Learning via the Alternating Direction Method of Multipliers. Foundations and Trends in Machine Learning. vol. 3, no.1. p.1-122. 2010
[19] S.-D. Wu, P.-H. Wu, C.-W. Wu, J.-J. Ding, C.-C. Wang Bearing fault diagnosis based on multi scale permutation entropy and support vector machine Entropy, 14 (2012), pp. 1343-1356
[20] N. Nicolaou, J. Georgiou, Detection of epileptic electroencephalogram based on permutation entropy and support vector machines, Expert Syst. Appl.39 (2012) 202-209

Acknowledgments
This research is supported in part by the National Natural Science Foundation of China (No. 51565046), Innovation Fund of Inner Mongolia University of Science and Technology (No. 2015QDL12) ,and Natural science Fund of Inner Mongolia (No. 2016MS0543).

IOP Conf. Series: Journal of Physics: Conf. Series **842** (2017) 012056 doi:10.1088/1742-6596/842/1/012056

Study on observation condition and on simple evaluation method of TypeIV creep damage of Mod.9Cr-1Mo steel

F Kawashima[1] , T Masumitsu[2], K Negi[3], K Watanabe[3] and K Fujiwara[1]

[1] Faculty of Advanced Science and Technology, Kumamoto University
2-39-1 Kurokami, Chuoku, Kumamoto 860-8555, Japan
[2] Graduate School of Science and Technology, Kumamoto University
[3] Department of Mechanical Engineering, Faculty of Engineering, Kumamoto University

E-mail: kawashima@mech.kumamoto-u.ac.jp

Abstract. Type IV creep damage is the phenomena that a large number of voids in micron size initiate, grow, coalesce each other and become large cracks. They initiate in welded joints of high chromium steel in power plant. The density of voids, the number of voids per a unit area, is used to evaluate the residual life. The observed density of voids depends on the observation conditions; the observation area and the magnification of observed photograph of metallographic structure, because voids do not distribute uniformly and the small magnification misses the small voids. In previous study, we studied the influence of them with simulated fine-grain HAZ of Mod.9Cr-1Mo steel, and proposed how to determine the appropriate observation area for the temporary allowable error. We also proposed the method to evaluate the start time of initiation of voids, the initiation rate of voids and the growth rate of voids based on the relation between the observed density of voids and the magnification. But the experimental data was short. In this paper we showed new data, but they were yet not sufficient. We used FEM analysis and considered why enough data had been not taken.

1. Introduction

Type IV creep damage is the phenomena that a large number of voids in micron size initiate, grow, coalesce and become large cracks[1]. They initiate in the fine grain heat affected zone (FGHAZ) in the welded joints that have been exposed to stress and high temperature for years. For reliable use of power plants, the method to evaluate the residual life is demanded. The density of voids is defined as the number of voids per a unit area. It is used as the parameter of the life fraction of the component with Type IV creep damage[2]. The observed density of voids depends on the observation conditions[3] which are the observation area and the magnification of observed photograph of metallographic structure, because voids do not distribute uniformly and the small magnification misses the small voids.

Generally Type IV creep damage initiates under high stress triaxiality. However the influence of the stress and the stress triaxiality on the density of voids has been researched[4][5], the quantitative conclusion has not been taken yet. One of the reasons is that the influence of the stress and stress triaxiality is hidden by the influence of the observation conditions. In the previous research[6][7][8], authors proposed the method to determine the appropriate observation area. They showed that the influence of the Heat on the appropriate observation areas was small.

In the previous research[7] the linear relation between the mean density of voids and the reciprocal of the magnification also was taken. There was the probability that the relation represents the distribution of the radius of voids. The initiation rate, the growth rate and the start time of initiation of voids might be evaluated with the distribution of the radius. However, data were short to discuss it. In this research we add data and try to evaluate these parameters and to calculate the density of voids. However, they are not enough, either. We use FEM analysis and consider the reason of short data.

2. Materials, creep test and observation

The creep tests were conducted with simulated FGHAZ and the samples with Type IV creep damage were obtained.

The material was KA SFVAF28, which was forged Mod.9Cr-1Mo steel. Table 1 shows the chemical composition of this material. It was normalized at 1045 C for 3.5 hours and tempered at 760 C for 5.5 hours. After that it was cut to the bars. To reproduce the metallographic structure of FGHAZ, the Heat 1 and the Heat 2 were heated at 910 C for 1 second one time and three times, respectively. After that the test pieces were machined from the bars.

To simulate the Post Weld Heat Treatment (PWHT), and to remove the residual strain, the test pieces were heated for 2 hours at 740 C after the machining. Table 2 shows the Vickers hardness measured with 9.807N and the average grain size after PWHT.

Tests were conducted with notched test pieces and a smooth test piece. Figure 1 (a) shows the notched test piece of Heat 1 with M18 screws. The shape of smooth test piece of Heat 1 was the same as figure 1 except for the notch. Figure 1 (b) shows the notched test piece of Heat 2 with M14 screws. The net section stress of the notched test pieces of the Heat 1, the smooth test piece of Heat 1 and the notched test pieces of Heat 2 were 72.8 MPa, 40.6 MPa and 44.2 MPa, respectively.

The test temperature was 650 C. The fracture time of notched test piece of Heat 1 and that of Heat 2 were 1487 hours and 2302 hours, respectively. The smooth test piece of Heat 1 didn't fracture and the notch factor is unknown. The similar notched test pieces need from 1.3 to 1.4 times larger net section stress to the same fracture time as the smooth test piece[9].

The stress and *TF*, the triaxiality factor, were analyzed with FEM with ANSYS Academic Teaching Introductory. Young's modulus of 157 GPa and Poisson's ratio of 0.314 were used. Equation (1) is the used Norton's law. *TF* is defined with equation (2).

$$\dot{\varepsilon} = 4.3270\times10^{-15}\sigma^{5.8143} \ [\text{hour}^{-1},\ \text{MPa}], \tag{1}$$

where $\dot{\varepsilon}$ is creep rate and σ is stress.

$$TF = \frac{(\sigma_1 + \sigma_2 + \sigma_3)}{\sigma_{mises}}. \tag{2}$$

Table 1 Chemical composition. [%]

	C	Si	Mn	P	S	Ni	Cr	Mo	V	Nb
Heat 1	0.10	0.30	0.45	0.014	0.001	0.16	9.34	0.96	0.21	0.07
Heat 2	0.11	0.30	0.45	0.015	0.001	0.18	9.16	0.96	0.23	0.07

Table 2 Hardness and grain size.

	Hardness [HV][a]	Grain size [μm]
Heat 1	225	2.23[b]
Heat 2	235	4.22[c]

[a] measured with 9.807N
[b] Observed with the magnification of 2520
[c] Observed with the magnification of 2544

IOP Conf. Series: Journal of Physics: Conf. Series **842** (2017) 012056 doi:10.1088/1742-6596/842/1/012056

For the example, figure 2 (a) and (b) show the distributions of stress and *TF* on the cross section of the bottom of the notch, respectively. They distribute uniformly in the area with the radius of less than 1 mm, and we observed this area. The creep damage of the centre of this cross section (radius r=0 mm) and that of the bottom of the notch (r=3.75 mm) were calculated with the time fraction rule shown with equation (3).

$$D_c = \Sigma\left(\frac{t_i}{t_{ri}}\right), \tag{3}$$

where D_c, t_i and t_{ri} are the creep damage, the period during that the principal stress is σ_i and the creep fracture time for σ_i. t_{ri} is calculated with equations (4) and (5).

$$LMP = (T + 273.15)(\log_{10} tr + 20) \ . \tag{4}$$

$$\log_{10}\sigma = -2.143\times10^{-4} LMP + 6.282 \ . \tag{5}$$

The creep damages are shown in figure 3. It was evaluated that the test piece fractures when the damage of bottom of the notch is 1. The creep damage at the centre is approximately 0.3 when the test piece fractures. The reason of the small observed density of voids may be that the creep damage of observed area was small.

Figure 4 shows the observed cross sections. Table 3 shows the test time, the stress and *TF* of the observed cross section. Only the cross section f was on the smooth test piece. Other cross sections were on notched test pieces.

The cross sections were etched with Nital. Figure 5 shows the example of metallographic structure. The observed photographs of metallographic structure were taken and printed with a optical microscope, a digital camera and a printer. The magnification of them, O, were approximately 1000, 500 and 250.

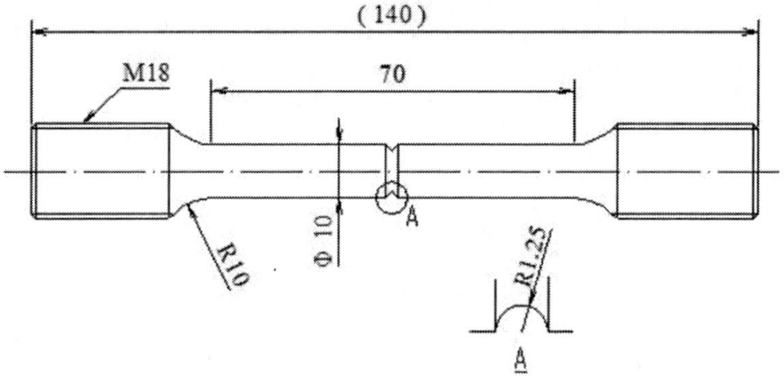

(a)

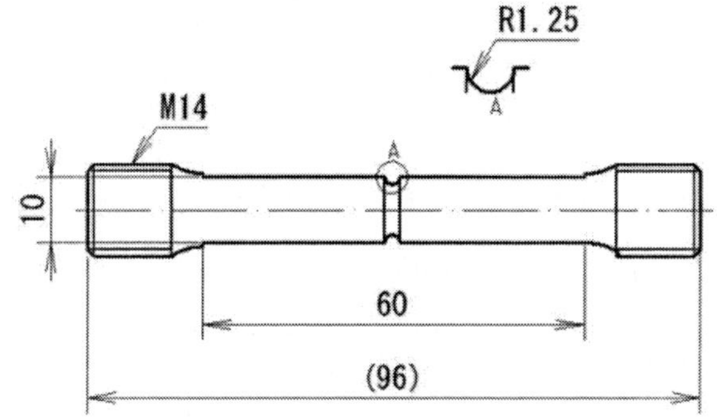

(b)

Figure 1 Notched test piece. (a) and (b) are Heat 1 and Heat 2, respectively.

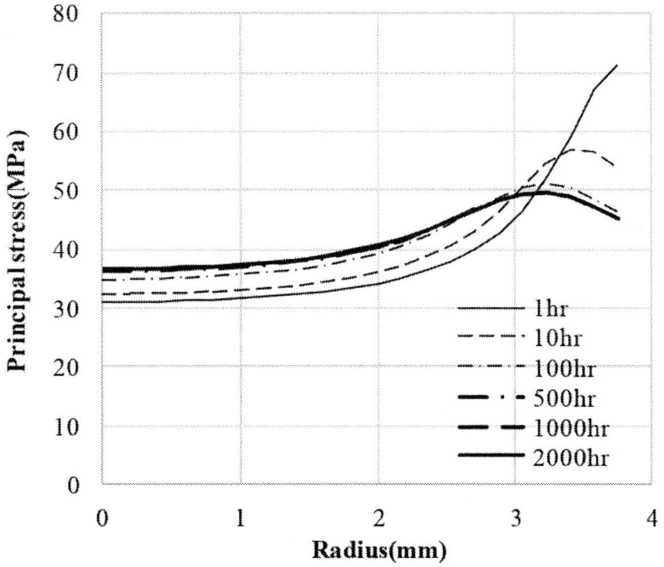

(a)

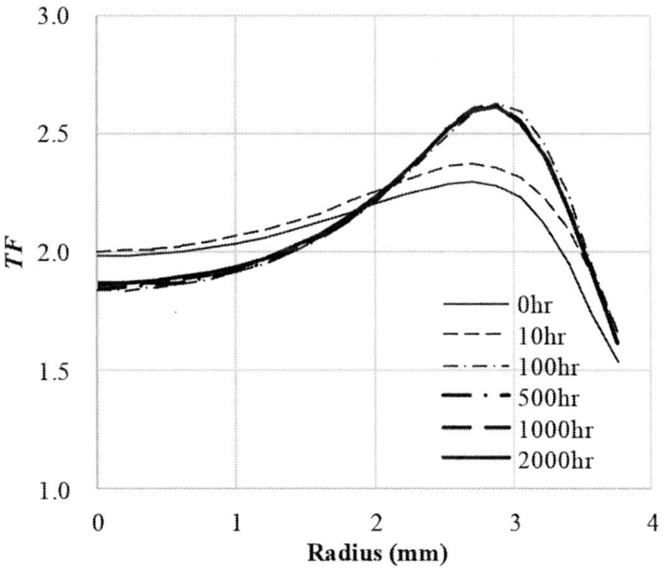

(b)

Figure 2 Stress and *TF* distribution on cross section of the bottom of notch.
(a) and (b) are stress distribution and *TF* distribution, respectively.

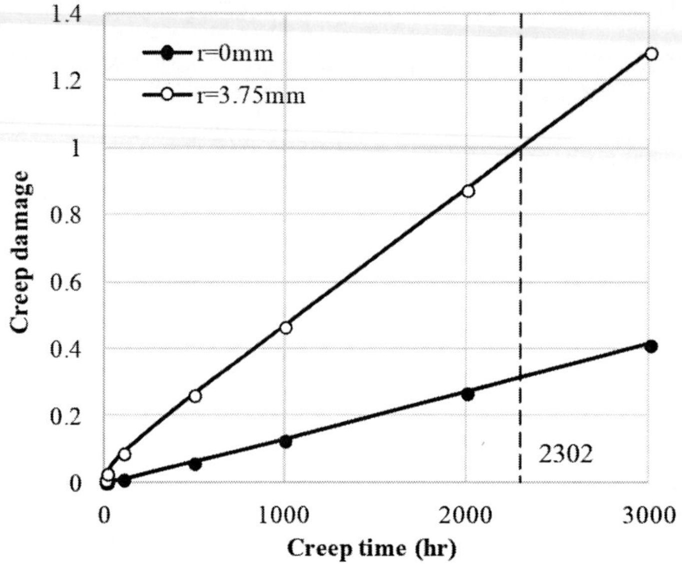

Figure 3 Creep damage of the observed area and the bottom of the notch.

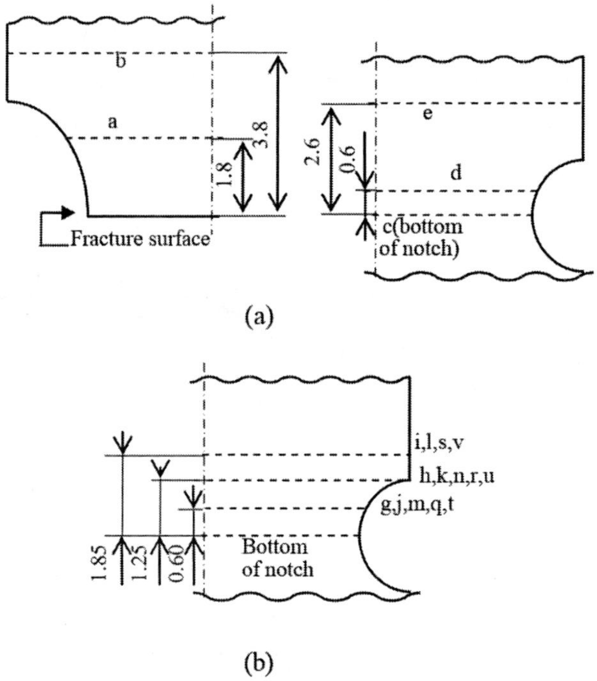

Figure 4 Outline of observed cross sections. (a) and (b) are Heat 1 and Heat 2, respectively.

Table 3 Observed cross section.

Heat	Test Piece	Cross Section	Time (hr)	Y [b] (mm)	S or N[c]	σ_1 (MPa)	*TF*
1	2	a	1487[a]	1.80	N	65.6	1.9
		b		3.80	S	53.9	1.2
	3	c	1353	0	N	62.6	1.9
		d		0.60	N	63.4	2.0
		e		2.60	S	63.2	1.6
	4	f	1242	-	S	40.6	1.0
2	2-3	g	2072	0.60	N	37.4	1.9
		h		1.25	N	38.5	1.9
		i		1.85	S	39.0	1.8
	2-4	j	2181	0.60	N	37.4	1.9
		k		1.25	N	38.5	1.9
		l		1.85	S	39.0	1.8
	2-5	m	2279	0.60	N	37.4	1.9
		n		1.25	N	38.5	1.9
		p		2.06	S	-	-
	2-7	q	2281	0.60	N	37.4	1.9
		r		1.25	N	38.5	1.9
		s		1.85	S	39.0	1.8
	2-8	t	2163	0.60	N	37.4	1.9
		u		1.25	N	38.5	1.9
		v		1.85	S	39.0	1.8

[a] Fracture time
[b] Distance from the cross section of bottom of notch after the test
[c] N: Notched part, S: Smooth part

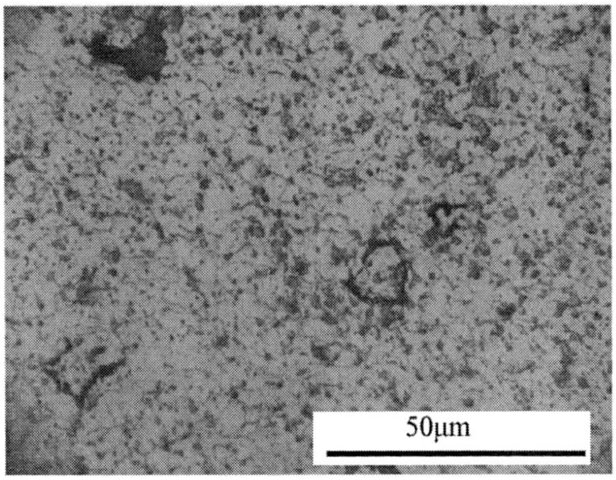

Figure 5 The metallographic structure of cross section.

The pictures were connected to be the circle with the radius of 1mm where the stress distributed uniformly (figure 6). The circle was divided to rectangles. A is the area of each rectangles. Except for cross section a and b, A is 0.04mm^2 when o is 500 or 1000 and 0.08mm^2 when o is 250, respectively. In cross section a, A is 0.10mm^2, 0.07mm^2 and 0.06mm^2 when o is 250, 500 and 1000, respectively. In cross section b, A is 0.09mm^2 when o is 250 and 0.04mm^2 when o is 500 or 1000. Figure 7 shows the examples.

The size of observed area, s, is the same as A, or is twice or four times as large as A. i is the density of voids of each s. The distribution of i was taken. The mean and the standard deviation of i are represented with i_u and i_s, respectively.

Figure 6 Observed metallographic photograph of cross section a.

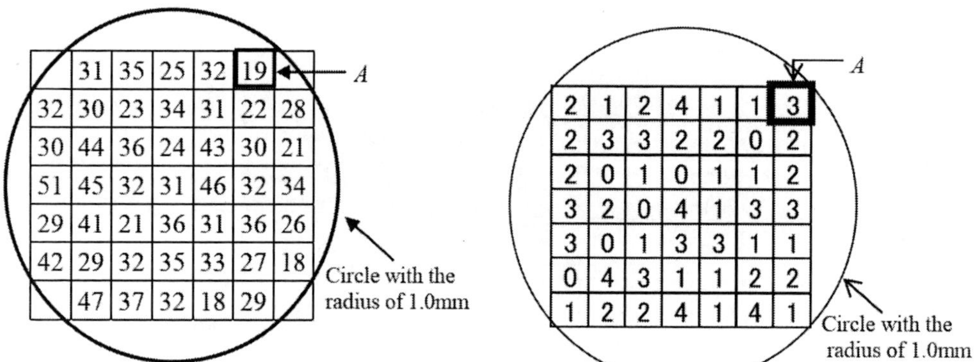

(a) (b)

Figure 7 Distribution of the number of voids.
(a) and (b) are cross section a and cross section n, respectively.

3. Results of observation

Figure 8 shows the example of $F(i)$ that is the cumulative distribution of i. Their approximating curves are normal distributions. i_s is small if s is large.

The allowable error of i should be determined based on the allowable error of the evaluated life fraction. But the relation between i and the life fraction is unknown. We assumed the allowable error of $0.8i_u \le i \le 1.2i_u$. Figure 9 is the relation between the density of voids i and the confidence level $F(1.2i_u) - F(0.8i_u)$ that is the probability of i being in the allowable error. Figure 10 shows that the confidence level increases with the increase of s. The area appropriate for the confidence level of 0.8 is shown in figure 11. It is function of i_u. The grain size of Heat 2 is about twice as large as that of the Heat 1, and the summation of grain boundary length per unit area of Heat 2 is shorter than that of Heat 1. But, in figure 11, there is not the difference between the Heat 1 and Heat 2. The reason may be that at the same i_u the influence of the grain size is cancelled.

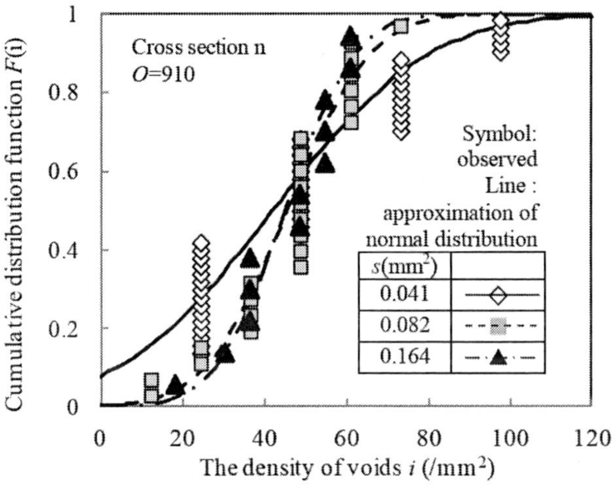

Figure 8 Cumulative distribution Function of density of voids (cross section n, O =910)

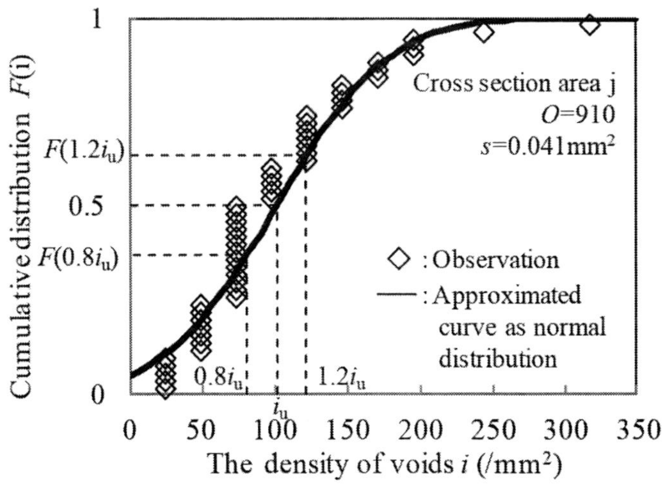

Figure 9 Outline of confidence level (cross section j, O=910).

IOP Conf. Series: Journal of Physics: Conf. Series **842** (2017) 012056 doi:10.1088/1742-6596/842/1/012056

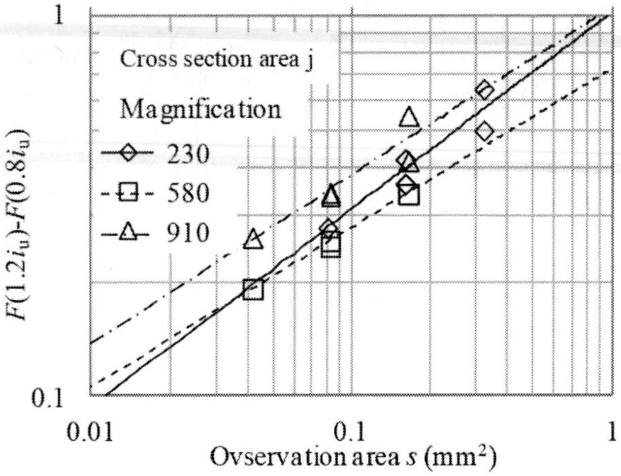

Figure 10 Relation of confidence level $F(1.2i_{\mathrm{u}})-F(0.8i_{\mathrm{u}})$ and observation area s .

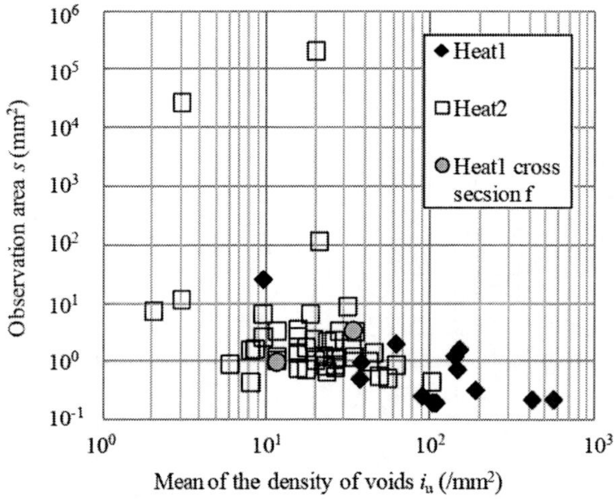

(a)

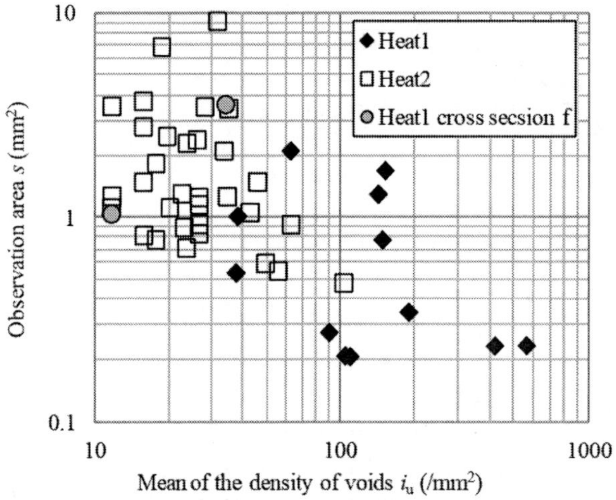

(b)

Figure 11 Range of appropriate observation conditions.
(a) shows all data. (b) shows data in practical range.

4. Evaluation method of initiation and growth of voids

Figure 12 shows the relation between O and i_u. Figure 12 (a) shows the cross sections of Heat 1 observed with 3 magnifications. Figure 12 (b) shows the cross sections of Heat 2 with the distance of 1.85mm from the bottom of notch. i_u seems to be linear to the reciprocal of O except for cross section l. There seemed to be some failures in observation of cross section l.

The minimum radius that one can see by the naked eyes is assumed to be ρ. In the photograph with the magnification of O, the voids with radius larger than ρ / O can be observed. Therefore the relation between O and i_u shows the distribution of the radiuses of voids.

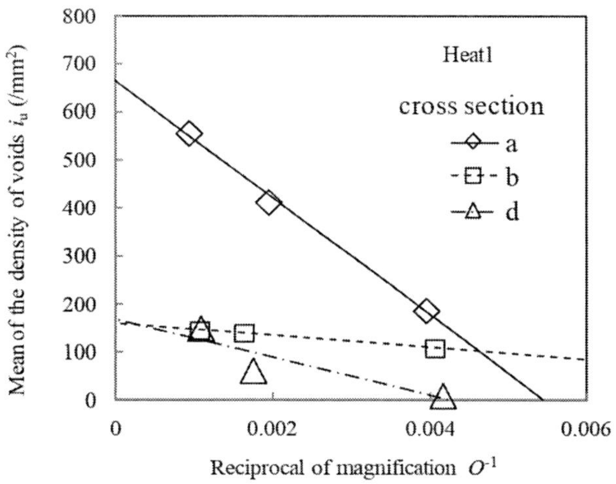

(a)

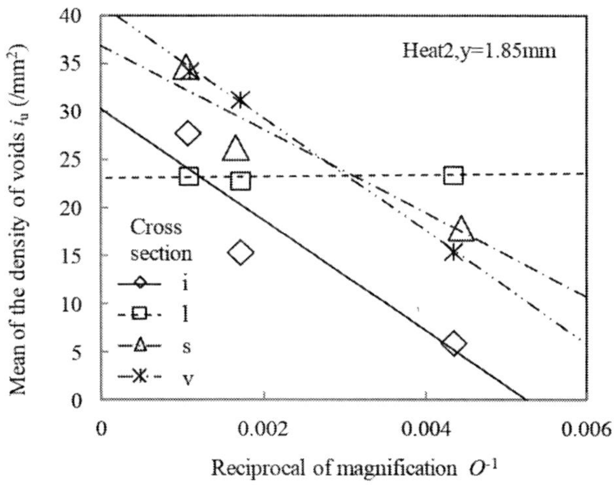

(b)

Figure 12 Relation between mean of density of voids i_u and reciprocal of magnification O^{-1}. (a) shows the cross sections of Heat 1 observed with more than three magnifications. (b) shows the cross sections with the distance of 1.85 mm from bottom of notch.

It is assumed that voids start to initiate at t_1. The initiation rate and the growth rate of voids are assumed to be the functions of time; $G(t)$ mm^{-2}hour^{-1} and $H(t)$ mm hour^{-1}, respectively. The radius of voids that initiates at t_o becomes $r=\int_{t_0}^{t_2} H(t)dt$ at t_2. And the density of voids with radius larger than $r=\int_{t_0}^{t_2} H(t)dt$ is $i=\int_{t_1}^{t_0} G(t)dt$ at t_2.

If the initiation rate and the growth rate are constant, $G(t)=f$ and $H(t)=h$, then the relation become $r=h(t_2-t_0)$ and $i=f(t_0-t_1)$. Equation (6) is made of these two equations. It shows the density of voids $i(r,t)$ whose radius are larger than r at the time of t.

$$i(r,t)=-\frac{f}{h}r+f(t-t_1) \ , \tag{6}$$

where r, t, f, h, and t_1 represent the radius of void (mm), the creep time (hour), the voids' initiation rate (hour^{-1} mm^{-2}), the voids' growth rate (mm hour^{-1}) and the time at that the voids' initiation started (hour), respectively. $r=\rho/O$ is substituted into equation (6), then equation (6) becomes equation (7).

$$i(r,t)=-\frac{f}{h/\rho}O^{-1}+f(t-t_1) \ . \tag{7}$$

In equation (7), i is linear to O^{-1}. It corresponds to the observation. $-\frac{f}{h/\rho}$ and $f(t-t_1)$ in equation (7) are the slope and axial section of figure 12.

Perhaps there are $G(t)$ and $H(t)$ that are not constant [10][11][12][13] [14] [15] and correspond to the observation, but in this research we use the simple assumption.

As shown in table 3 and figure 3, cross section i, s and v are in the same location; 1.85 mm from bottom of the notch. The stress of these cross sections is same. So we assume that f, h and t_1 of these cross sections are same. This assumption is valid because the slopes of them in figure 12 (b) are almost the same. Observed i_u and the creep time of the three cross sections are substituted into $i(r,t)$ and t in equation (7), and f, h and t_1 are calculated with the least squares method. Also f, h and t_1 of the other cross sections of Heat 2 that are 0.6 mm and 1.25 mm from bottom of the notch are calculated. Data of test piece 2-4 (cross section j, k, l) are not used for the calculation because the relations between i_u and O^{-1} are irregular. Table 4 shows the results. The density of voids calculated with equation (7) and the variables in table 4 are compared to the observation in figure 13. Errors of observation are large. The reason of that is thought to be not enough observation area. In this study observation area of i_u is about 2 mm^2, which is the area of the largest square in circle with radius of 1 mm. As shown in figure 11, because i_u of Heat 2 is small, observation area might be need more than 2 mm^2.

Table 4 Estimated start time of initiation t_1, initiation rate f and growth rate h.

y^a (mm)	t_1 (hour)	f (/mm^2/hr)	h/ρ (/hr)
0.6	1317	4.783×10^{-2}	9.214×10^{-6}
1.25	1384	5.220×10^{-2}	6.021×10^{-6}
1.85	1446	4.510×10^{-2}	11.30×10^{-6}

[a] Distance from bottom of notch after the test

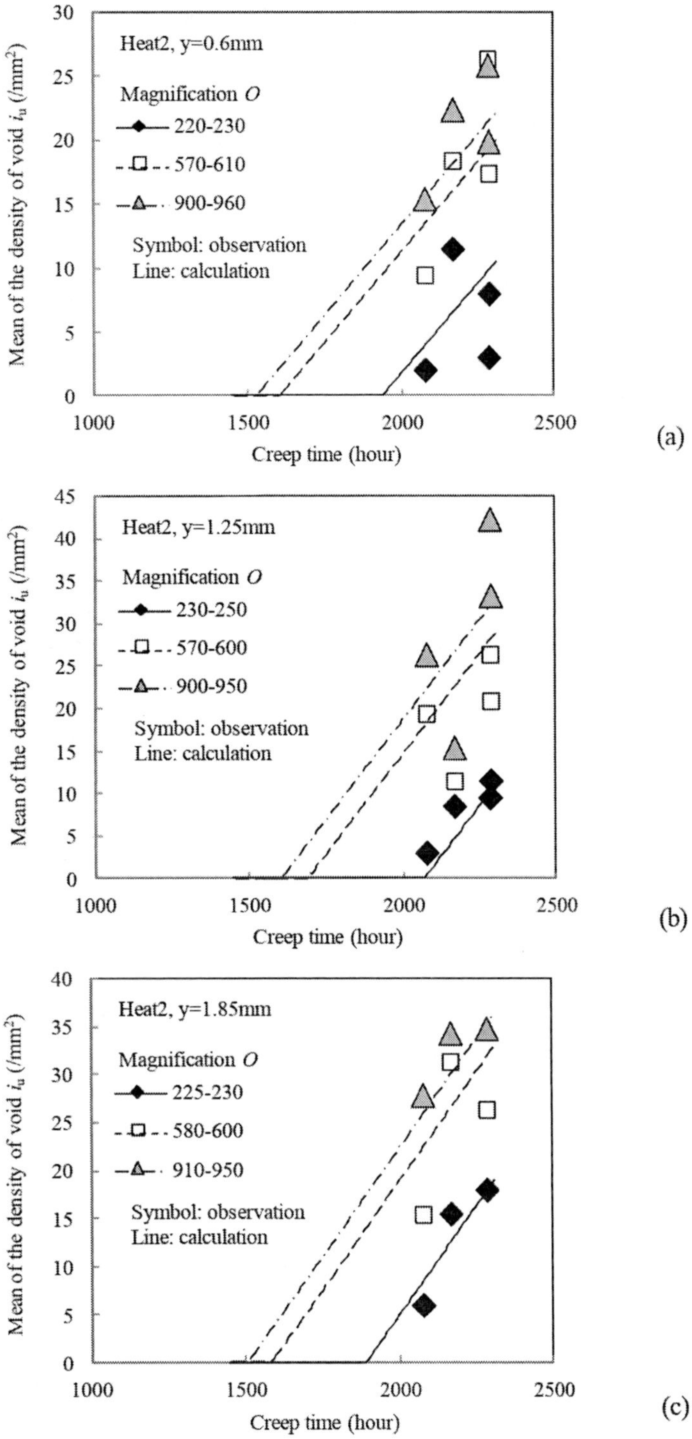

Figure 13 Calculated density of voids compared with the test results. (a), (b) and (c) show the cross sections with y of 0.6 mm, 1.25 mm, and 1.85 mm, respectively.

Though there is the error, equation (7) well expresses the relation of i_u and O^{-1}. So the calculated variables are thought to be valid.

It is not clear whether there are differences among variables of three cross sections. In the next test, we should use the test piece with the observation areas with various *TF* or observation areas of high creep damage (in other words, large density of voids) at fracture time.

In some conditions, i_u might not be liner to O^{-1}. In that case we should study $G(t)$ and $H(t)$ that are not constant but functions of time.

5. Conclusions

The observation area appropriate for the assumed allowable error of $0.8i_u \leq i \leq 1.2i_u$ with the confidence level of 0.8 is determined. The appropriate area is small if the density of voids is large. The influence of Heat on the appropriate area is thought to be small.

The method that estimates the voids' initiation rate, growth rate and the initiation start time is proposed. It needs the observation more than two times with more than two magnifications. This method will be used to estimate the influence of *TF* on Type IV creep damage.

References

[1] Ellis F. V and Viswanathan R 1998 Review of type IV cracking *ASME-PVP* **380** 59-76

[2] Xu Qihua, Xu Qiang, Lu Z and Donnelly S 2013 The development of phisical-base creep damage constitutive equation for P/T22 (2.25Cr-1Mo) steel at low stress *Proc. of 6th International 'HIDA' Conference* S4B-03

[3] The Iron and Steel Institute of Japan 1991 *Manual on evaluation of creep-fatigue damage / lives by replication method* (Tokyo: The Iron and Steel Institute of Japan) Japan 28-33

[4] Fujimoto M, Sakane M, Date S and Yoshida H 2005 Multiaxial creep rupture and damage evaluation for 2.25Cr-1Mo forged steel *Jour. of the Society of Materials Science, Japan* Japan **54** 2 149-54

[5] Himeno T, Chuman Y, Tokiyoshi T, Fukahori T and Igari T 2013 Creep rapture behanior of girth-welded mod. 9Cr-1Mo steel pipe subject to internal pressure and axial load *Proc. of 6th Int. 'HIDA' Conf.* (Nagasaki: 2-4 December, 2013 / European Technology Development) S6-05

[6] Kawashima F, Kinoshita T, Sato A, Enui Y, Fujiwara K and Hata H 2013 Experimental research on the appropriate extent of observation area of creep voids to evaluate the remaining life of Mod.9Cr-1Mo steel with Type IV creep damage *Proc. of 6th Int. 'HIDA' Conf.* (Nagasaki: 2-4 December, 2013 / European Technology Development) S2-02

[7] Kawashima F, Sato A, Irie Y, Yamakawa K, Fujiwara K and Hata H 2015 Study on the observation condition of Type IV creep damage of Mod. 9Cr-1Mo steel *Proc. of the 1st Int. Conf. on Advanced High-Temperature Materials Technology for Sustainable and Reliable Power Engineering (123HiMAT-2015)* (Sapporo: 29 June - 3 July, 2015 / 123 Comitte on Heat Resisting Materials and Alloys, Japan Society for thr Promotion of Science) 82-85

[8] Kawashima F, Sato A, Irie Y, Yamakawa K, Fujiwara K and Hata H 2015 Study on the magnification and area of observation appropriate to measure Type IV creep damage of Mod.9Cr-1Mo steel *Jour. of the Society of Materials Science,Japan* Japan **64** 74-79

[9] Takahashi Y 2009, Creep rapture behavior of modified 9Cr-1Mo steel under multiaxial stress and its modeling, *Jour. of the Society of Materials Science, Japan* Japan **58** 2 115-21

[10] Raj R and Ashby M. F 1975 Intergranular Fracture at elevated temperature *Acta Metallurgica* **23** 653-66

[11] Raj R 1978 Nucleation of cavities at second phase particles in grain boundaries *Acta Metallurgica* **26** 995-1006

[12] Chuang T-J, Kagawa K. I, Rice J. R and Sills L. B 1979 Non-equibrium models for diffuseive cavities of grain intefaces *Acta Metallurgica* **27** 265-84

[13] Pharr G. M and Nix W. D 1979 A nimerical study of cavity growth controlled by surface diffusion *Acta Metallurgica* **27** 1615-31

[14] Rice J. R 1981 Constraints on the diffusive cavitation of isolated grain boundary facets on creeping polycrystals *Acta Metallurgica* **29** 675-81

[15] Nix W. D, Yu K. S and Wang J. S 1983 The effect of segregation on the kinetics of intergranular cavity growth under creep conditions *Metallurgical and Materials Trans.* A **14** 563-70

Fault Diagnosis for Centre Wear Fault of Roll Grinder Based on a Resonance Demodulation Scheme

Liming Wang[1], Yimin Shao[1,3], Lei Yin[1], Yilin Yuan[2] and Jing Liu[1]

[1] State Key Laboratory of Mechanical Transmission, Chongqing University, P.R. CHINA
[2] Ningbo Baoxin Stainless Steel Co., Ltd., Ningbo, Zhejiang, P.R. CHINA
[3] Corresponding author, email address: ymshao@cqu.edu.cn

Abstract. Roll grinder is one of the important parts in the rolling machinery, and the grinding precision of roll surface has direct influence on the surface quality of steel strip. However, during the grinding process, the centre bears the gravity of the roll and alternating stress. Therefore, wear or spalling faults are easily observed on the centre, which will lead to an anomalous vibration of the roll grinder. In this study, a resonance demodulation scheme is proposed to detect the centre wear fault of roll grinder. Firstly, fast kurtogram method is employed to help select the sub-band filter parameters for optimal resonance demodulation. Further, the envelope spectrum are derived based on the filtered signal. Finally, two health indicators are designed to conduct the fault diagnosis for centre wear fault. The proposed scheme is assessed by analysing experimental data from a roll grinder of twenty-high rolling mill. The results show that the proposed scheme can effectively detect the centre wear fault of the roll grinder.
Keywords: fault diagnosis; wear; roll grinder; demodulation

1. Introduction

As one of the most important parts in rolling equipment, roll grinder is utilized to improve or modify the surface quality of the working roll which is used in rolling mill. However, the roll grinder usually contains multiple components, which means multiple vibration sources. Besides, it is difficult for the roll grinder to stay at a stable grinding process. Therefore, some undesirable chatter marks can be found on the roll surface, as shown in Fig. 1, which may have a direct effect on the surface quality of the steel strip [1, 2]. In roll grinder machine, the centre bears the gravity of the roll and alternating stress during grinding process. Therefore, wear or spalling fault is easily observed, which may lead to an unstable vibration and generate chatter marks on the roll surface.

Envelope analysis (*EA*) is widely used in the fault diagnosis of rotating machinery, i.e. bearings and gearboxes [3]. For instance, Yang presented a hybrid method which employed *EA* and empirical mode decomposition to detect different fault patterns of bearings [4]. Krishnappa discussed the development of several parameters based on amplitude and phase modulation method, and showed their applications in the fault diagnosis of helicopter transmissions [5]. Figlus proposed a method based on envelope spectrum for fault diagnosis of gearbox with tooth chipping and wear faults [6]. However, most times it requires the users should have a rich knowledge of the vibration signal and engineering experiences to determine the filter parameters by using the traditional *EA* demodulation method. Therefore, some scholars tried to select the parameters automatically by the application of fast kurtogram based on the principle of maximum kurtosis. This method can yield a sub-band in which the signal should be demodulated [7, 8]. Therefore, based on these techniques it is possible to detect the fault characteristic frequencies in envelope spectrum.

The rest of the paper is organized as follows. Section 2 describes the methodology structure of envelope analysis with fast kurtogram and the computing process of the proposed two health indicators. Section 3 presents the experiment system of the roll grinder. Then the effectiveness of the proposed method is validated in section 4. Conclusions are drawn in section 5.

Fig. 1. Chatter marks on the roll surface

2. Methodology

2.1. A Review of the fast kurtogram

In fault diagnosis of rotating machinery, different sub-band signals carry different amount of fault-related components. Therefore, it is of significance to find the most fault-related sub-band signal for analysis. Firstly introduced by J. Antoni in Ref. [7, 8], fast kurtogram is an advanced technique based on spectral kurtosis (*SK*) that can indicate the presence of series of transients and their locations in the frequency domain through the estimation of kurtosis of each sub-band signal. The calculation of the fast kurtogram is shown in Fig. 2. Firstly signal x is decomposed into k levels through filter banks. In each level C_k^i means the ith (i=0, 1,⋯, 2k-1) sub-band signal in k level and $C_0 \equiv x$ is the original signal. Then the kurtogram is estimated by computing the kurtosis of all sequences, which can be obtained by

$$K_k^i = \frac{\left\langle \left| C_k^i(n) \right|^4 \right\rangle}{\left\langle \left| C_k^i(n) \right|^2 \right\rangle^2} - 2 \tag{1}$$

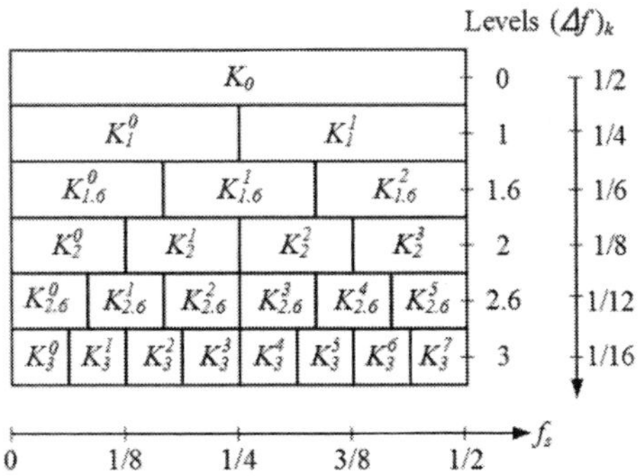

Fig. 2. Paving of the (frequency/frequency resolution) plane in the case of the 1/3-binary tree kurtogram estimator [7]

2.2. Envelope analysis (EA)

When roll fault occurs, impulsive shocks are modulated by high frequencies, i.e. resonance frequencies. This will lead to complex sidebands in frequency spectra, which increases the difficulty of fault detection. *EA* is one of the most effective demodulation techniques to detect these impulses by removing the carrier signal. In order to derive the envelope spectrum of the time domain signal $x(t)$, Hilbert transformation is performed as

$$\hat{x}(t) = -\frac{1}{\pi}\int_{-\infty}^{+\infty}\frac{x(t-\tau)}{\tau}d\tau \tag{2}$$

Then the analytic signal $x_a(t)$ is obtained as

$$x_a(t) = x(t) + j\hat{x}(t), \quad j = \sqrt{-1} \tag{3}$$

The envelope signal is derived by computing the magnitude of $x_a(t)$. After that the envelope spectrum can be obtained by performing fast Fourier transformation to the envelope signal.

2.3. *Two health indicators*

There mainly exist two problems for the detection of wear fault characteristics in the grinding process. One is the time-varying working conditions, i.e. rotating speed and grinding force. Another is that the wear fault signals are contaminated by noise as well as normal grinding signals. Therefore, traditional dimensional indicators may suffer degradation in finding the wear fault in grinding machine, i.e. root mean square (*RMS*), skewness etc. Aiming to overcome these problems, two dimensionless health indicators based on envelope spectrum are proposed: envelope spectrum kurtosis (*ESK*) and envelope spectrum entropy (*ESE*). When there is a local wear fault, some corresponding fault frequency components will be found in envelope spectrum. And the *ESK* and the *ESE* are sensitive to the change of the envelope spectrum structure. Thus they are expected to increase and decrease respectively no matter how the working conditions will be. The *ESK* and *ESE* can be obtained by

$$ESK = \frac{\sum_{m=1}^{M}(S(m)-\bar{S})^4}{M\left(\sum_{m=1}^{M}(S(m)-\bar{S})^2\right)^2} \tag{4}$$

$$ESE = -\sum_{m=1}^{M} P(m)\log_2 P(m) \tag{5}$$

where $S(m)$ is the envelope spectrum for m=1, 2, …, M. M is the number of spectrum components. $\bar{S}$ is the mean value of $S(m)$ and $P(m) = S(m)^2 / \sum_{m=1}^{M} S(m)^2$.

2.4. *Signal processing procedure*

The flow chart of the fault diagnosis model is shown in Fig.3, which mainly consists of 3 stages. First of all the raw data is analysed via fast kurtogram and the most fault-related sub-band is selected. Secondly, based on the Hilbert transform and FFT the envelope spectrum of the filtered signal are obtained. Finally two health indicators: *ESK* and *ESE* are proposed for fault diagnosis of the roll grinder.

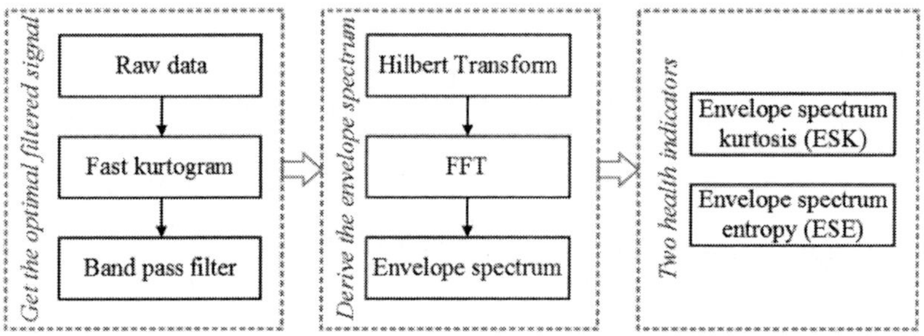

Fig. 3. Flow chart of the fault diagnosis model

3. Experiment

The experimental system consists of two parts: test bench of the roll grinder and the measurement setup, as shown in Fig. 4. The test bench is composed of headstock, tailstock, head centre, tail centre, roll, grinder, support plates and support board. It is noted that the support board will carry the roll to move around in the grinding process, and spalling fault is seeded on the tail centre. Three experiments are conducted to acquire the vibration data in normal condition and centre wear condition respectively. The grinding parameters are listed in table 1, where "*N*" and "*W*" denote normal condition and centre wear condition respectively. For the measurement setup, a PCB ICP (Integrated Circuit Piezoelectric) accelerometer is mounted on the tail centre to acquire vibration signal. NI 9234 is selected for data acquisition. The sampling frequency is chosen as 10k Hz and the signal length is 40 seconds.

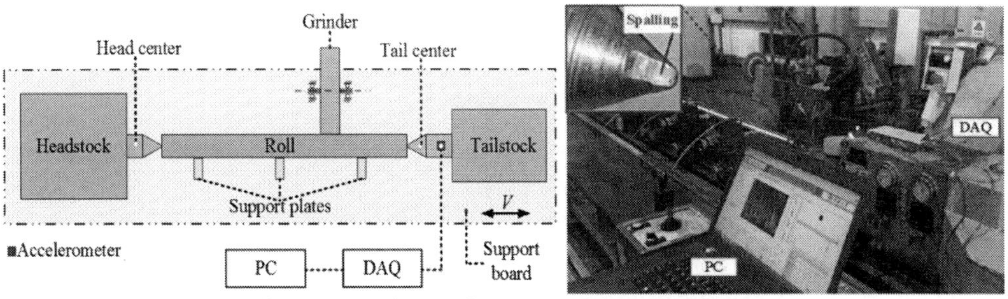

Fig. 4. Experimental system of the roll grinder.

Table 1. Grinding parameters of the 3 experiments

	Grinder speed (m/s)	Roll speed (rpm)	Support board speed (mm/min)	Grinder diameter (mm)	Roll diameter (mm)
Experiment 1	19	100	1000	394.5 (*N*) 450.9 (*W*)	59.22 (*N*) 59.87 (*W*)
Experiment 2	19	100	1000		
Experiment 3	19	110	650 (*N*)/750 (*W*)		

4. Results and discussions

In order to verify the effectiveness of the proposed method, all the datasets of the three experiments are analysed. Figure 5 shows an example of the (a) time domain signal and (b) spectrum in normal and tail centre spalling situation. In Fig. 5 (a), it can be observed that some impulsive components occur, and the time interval between two impulses is equal to the rotating period of the roll. Meanwhile, there are some differences between the normal situation and the seeded centre spalling situation as shown in Fig. 5 (b). Figure 6 shows the fast kurtogram and the envelope spectrum of the example signal. The most fault-related sub-band is selected according to kurtogram with the parameters (frequency centre=1718.75 Hz, bandwidth=312.5 Hz) as marked in Fig. 6 (a). The rotating frequency of the roll and its harmonics "f_r", "$2f_r$" and "$3f_r$" are obvious in the seeded centre spalling situation, which yield higher values than the normal situation, as shown in Fig. 6 (b).

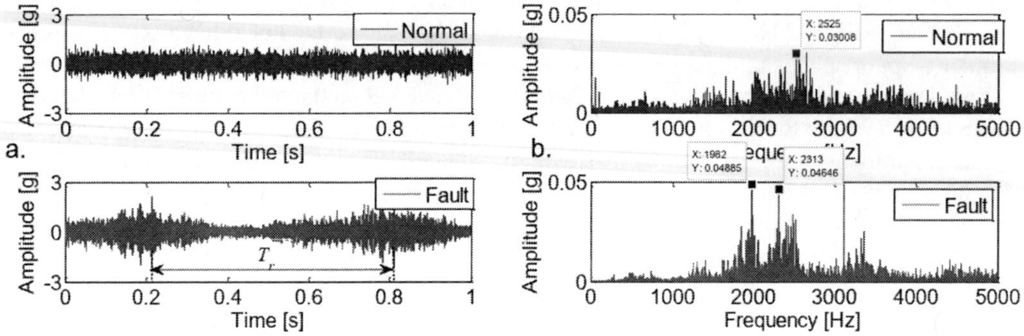

Fig. 5. An example of the (a) time domain signal and (b) spectrum in normal and tail centre spalling situation

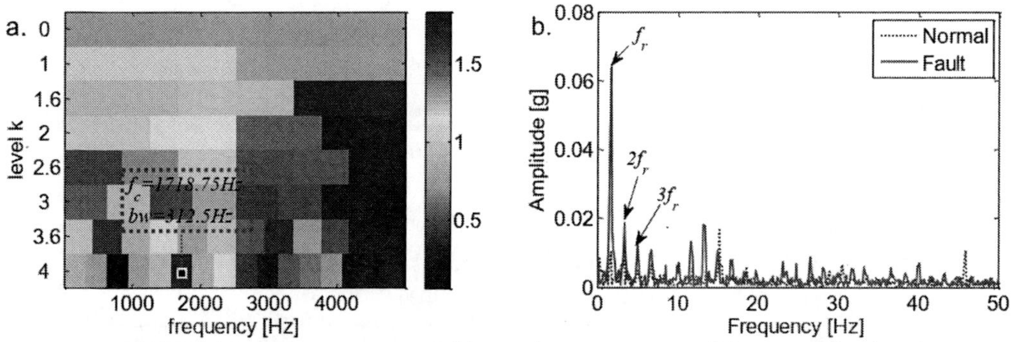

Fig. 6. (a) Fast kurtogram and (b) envelope spectrum of the example signal

Figure 7 to Fig. 9 illustrate two health indicators' plots of the three experiments. The x axis is the feature number and y axis is the amplitude. The dataset in each experiment is divided into 8 segments uniformly. It can be observed that the *ESK* of the vibration signal in seeded centre spalling situation always yields higher value than normal situation, as shown in Fig. 7 (a), Fig. 8 (a) and Fig. 9 (a), while the *ESE* of the vibration signal in seeded centre spalling situation always yields lower value than normal situation, as shown in Fig. 7 (b), Fig. 8 (b) and Fig. 9 (b). Conclusions can be drawn that the *ESK* and *ESE* could be good indicators for the roll grinder with centre spalling fault. Figure 10 shows the scatter plot of the normalized *ESK* and normalized *ESE* of all 3 Experiments. The normal samples are located at the left-up of the scatter plot with a relatively concentrated group, while the seeded centre spalling samples are located at the right-bottom of the scatter plot. Among them 5 samples are near to the normal group and others have a concentrated group.

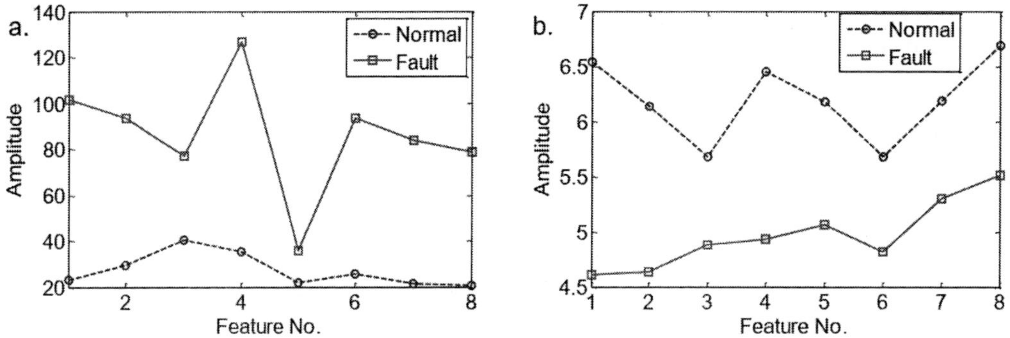

Fig. 7. (a) The *ESK* plot and (b) the *ESE* plot of Experiment 1

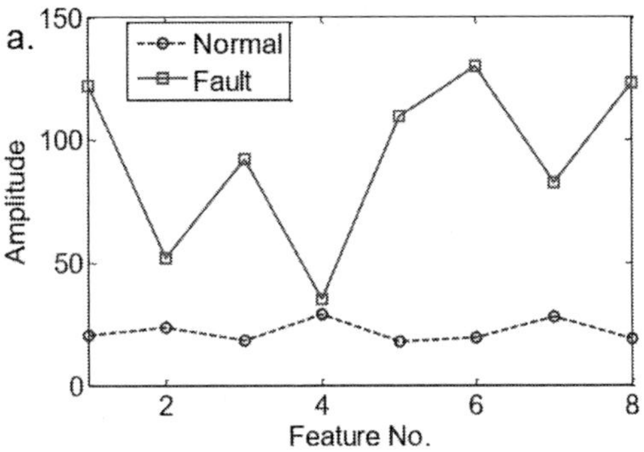

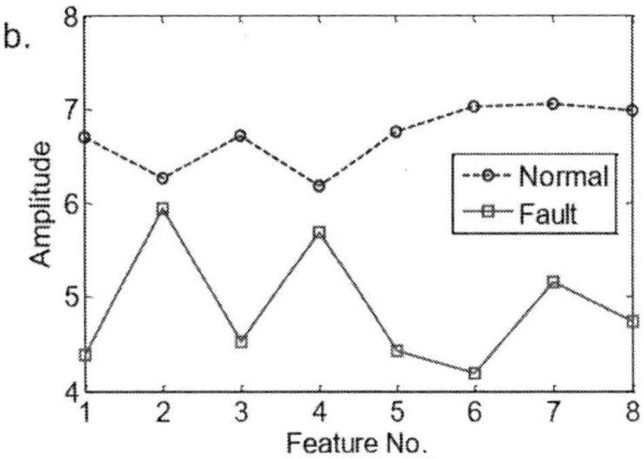

Fig. 8. (a) The *ESK* plot and (b) the *ESE* plot of Experiment 2

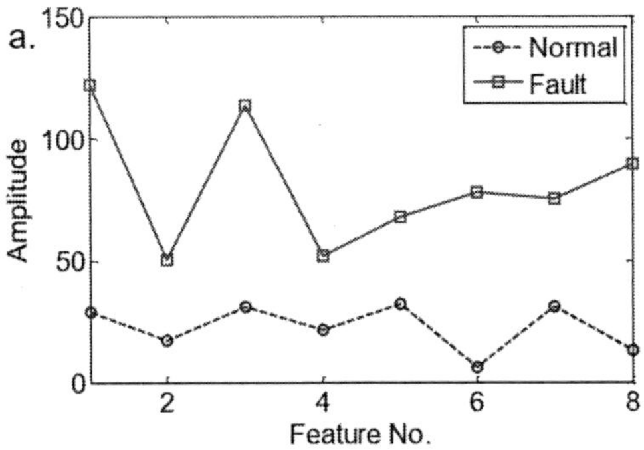

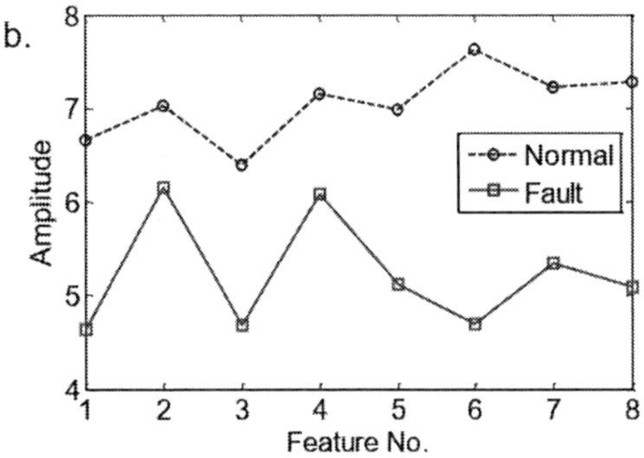

Fig. 9. (a) The *ESK* plot and (b) the *ESE* plot of Experiment 3

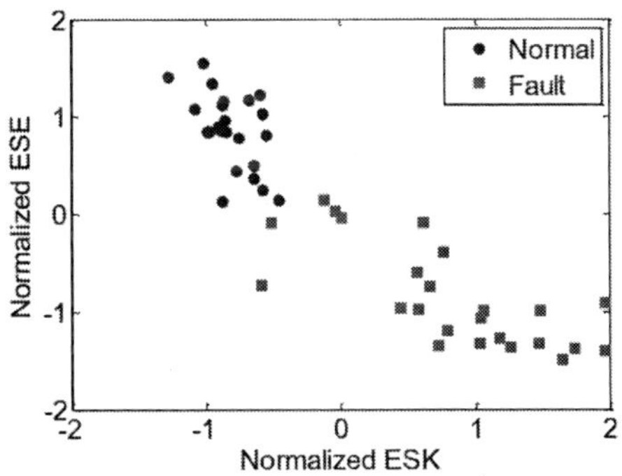

Fig. 10. Scatter plot of the normalized *ESK* and normalized *ESE* of all 3 Experiments

5. Conclusions

In this paper, two health indicators are proposed based on the envelope spectrum to detect the centre spalling fault of roll grinder. Firstly the optimal sub-band signal that contains more fault-related components are selected through fast kurtogram method. Then based on Hilbert transformation, envelope spectrum of the filtered signal is obtained. Finally two dimensionless health indicators are designed based on envelope spectrum to detect the centre spalling fault. It can be summarized from the results that the two health indicators can successfully distinguish the centre spalling situation from normal situation in all 3 experiments, and the scatter plot of all 3 experiments yields a good cluster performance.

Acknowledgement

The authors are grateful for the financial support provided by Ningbo Baoxin Stainless Steel Co., Ltd. and the National Natural Science Foundation of China under Contract No. 51475053.

References

[1] Thompson R A. The Character of Regenerative Chatter in Cylindrical Grinding. Journal of Engineering for Industry, 1973, 95(3):858.
[2] Hardwick B.R. A Technique for the Detection and Measurement of Chatter Marks on Roll Surface. Iron and Steel Technology, 2003, 80(4): 64-70.
[3] Feldman M. Hilbert transform in vibration analysis. Mechanical Systems and Signal Processing, 2011, 25(3): 735-802.
[4] Yang Y, Yu D, Cheng J. A fault diagnosis approach for roller bearing based on IMF envelope spectrum and SVM. Measurement, 2007, 40(9-10):943-950.
[5] Krishnappa G. Gear fault detection parameter development based on modulation techniques//Proceedings of the Fifth International Congress on Sound and Vibration. 1997, 2: 919-926.
[6] Figlus T, Wilk A. Application of Analysis of Envelope's Spectrum for Gearbox Diagnosing. International Review of Mechanical Engineering, 2013, 6(6):1350-1355.
[7] Antoni J, Randall R B. The spectral kurtosis: application to the vibratory surveillance and diagnostics of rotating machines. Mechanical Systems & Signal Processing, 2006, 20(2):308-331.
[8] Antoni J. Fast computation of the kurtogram for the detection of transient faults. Mechanical Systems & Signal Processing, 2007, 21(1):108-124.

An illustration of new methods in machine condition monitoring, Part I: stochastic resonance

K. Worden[1], I. Antoniadou[1], S. Marchesiello[2], C. Mba[2] and L. Garibaldi[2]

[1]Dynamics Research Group, Department of Mechanical Engineering, University of Sheffield, Mappin Street, Sheffield S1 3JD, UK.

[2]Department of Mechanical and Aerospace Engineering, Politecnico di Torino, 10129, Torino, Italy.

E-mail: k.worden@sheffield.ac.uk

Abstract. There have been many recent developments in the application of data-based methods to machine condition monitoring. A powerful methodology based on machine learning has emerged, where diagnostics are based on a two-step procedure: extraction of damage-sensitive features, followed by unsupervised learning (novelty detection) or supervised learning (classification). The objective of the current pair of papers is simply to illustrate one state-of-the-art procedure for each step, using synthetic data representative of reality in terms of size and complexity. The first paper in the pair will deal with feature extraction.

Although some papers have appeared in the recent past considering stochastic resonance as a means of amplifying damage information in signals, they have largely relied on *ad hoc* specifications of the resonator used. In contrast, the current paper will adopt a principled optimisation-based approach to the resonator design. The paper will also show that a discrete dynamical system can provide all the benefits of a continuous system, but also provide a considerable speed-up in terms of simulation time in order to facilitate the optimisation approach.

1. Introduction

The aim of machine condition monitoring is to detect incipient damage in (usually rotating) machinery before it grows to the point where a failure of the machine occurs. This entails detecting the signature of small damage in a component from signals which may be strongly corrupted by the presence of noise from other elements of the machine or environment. When the components of interests are gears or bearings, the 'signature' of damage in vibration data can take the form of a train of impulses. In a gearbox, a crack in a tooth may cause increased flexibility of that tooth, which can manifest as an impulse when the tooth comes into mesh; the period of the impulse train will then be associated with the meshing frequency. In a bearing, a crack or spall in a race will cause an impulse when a ball runs over the damage; the period of the impulse train is then associated with one of the bearing characteristic fault frequencies. There is clearly some benefit in being able to detect the presence of impulses in a noisy background; establishing the period of the impulse train can then give information about which gear in a system, or which race in a bearing, is damaged.

One approach to the problem described above might be to selectively increase the effect of the impulse in the signal relative to the noise background, and this is where the phenomenon of *stochastic resonance* becomes of interest. The phenomenon was first demonstrated in [1] in the context of climatology, although applications in various fields followed [2]. In the context of the problem here, stochastic resonance can be used as a signal processing tool to increase the signal-to-noise ratio (SNR) of a mixture of deterministic signal and noise. This is accomplished by passing the noisy signal through a nonlinear system; for carefully tuned nonlinear systems, the response can show amplification of the deterministic component relative to the noise. Although the technique is most often applied in order to amplify a small harmonic signal in noise, it can also be used for the impulse trains of the condition monitoring context [3, 4, 5]. In a previous article by some of the authors [6], the approach was illustrated on a number of condition monitoring data sets and was shown to be effective, if the parameters of the nonlinear oscillator could be tuned appropriately. Although the results were very good, it was found that parameter tuning required care and expertise on the part of the analysts. The objective of the current paper is to introduce an *optimisation* based tuning scheme, independent of the user, and to validate it via application to two synthetic data sets. (While this paper was in preparation, the authors became aware of the reference [7], which also explores some of the concepts discussed here.)

The layout of the paper is as follows: the following section briefly explains how stochastic resonance can amplify the effect of an impulse train in noise, and how the oscillator parameters might be tuned using an optimisation scheme. Section Three outlines the background theory for the optimisation scheme chosen for this paper - *Self Adaptive Differential Evolution* (SADE). Section Four describes how the synthetic data for the case studies were constructed, and results for the data are presented in Section Five. The paper ends with brief conclusions.

2. Stochastic Resonance

Rather than describe stochastic resonance in general, the discussion here will motivate the approach via the explicit nonlinear oscillator considered in this paper, which, in its simplest form, is a first-order cubic oscillator of the form,

$$\dot{y} - ay + by^3 = p(t) \tag{1}$$

which is a system with a cubic internal restoring force, driven by a mixture of a single impulse component and noise. In the context of stochastic resonance, the amplitude of the impulsive component will usually be small compared to the noise. Now, the potential $V(y)$ associated with the internal force is,

$$V(y) = -\frac{a}{2}y^2 + \frac{b}{4}y^4 \tag{2}$$

which is for $a > 0$ and $b > 0$, a twin-well oscillator as illustrated in Figure 1 (for the values $a = 40$, $b = 1$).

Now, consider a particle moving in this potential. If the level of forcing is low, the particle will remain very close to one of the stable equilibria at $\pm\sqrt{2a/b}$. Without loss of generality, consider the particle to be in the negative well. Suppose that the noise signal is a zero-skewness process; the mixture of the impulse and noise will have a small degree of skewness because of the impulse, however this will usually be imperceptible because A_0 is small compared to X. If one 'turns up' the level of forcing the particle will begin to move between the limits given by $V(y)$, but corresponding to a higher value of y. At the limit where the particle has maximum amplitude, yet remains within the well, the oscillations will be skewed - the signal will be 'drawn out' at the side towards the origin. This 'drawing out' will amplify the skewing effect of the impulse, those indicating the presence of the impulse. By tuning the parameters a and b, one is able to amplify the skewing effect and increase detectability of the impulse; however, one must take care not to choose parameters which allow the particle to pass into the positive well, or the effect will be lost.

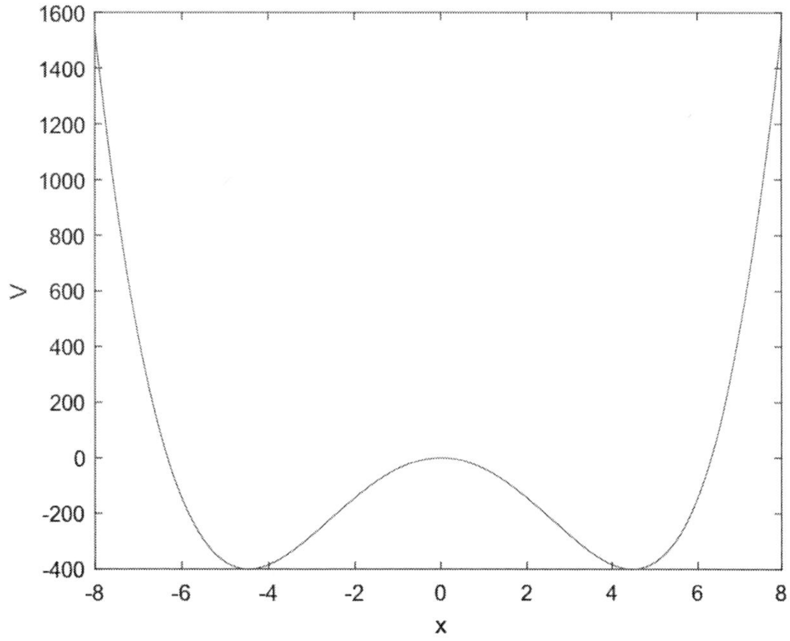

Figure 1. Twin-well potential for the oscillator in equation (1) with $a = 40$ and $b = 1$.

A previous paper by some of the authors here [6], carried out a careful tuning of the parameters in order to demonstrate how effectively impulses could be detected in a noise background. It was clear from the paper that an automated tuning procedure, requiring less intervention from the user, was desirable. The current study will show that this is possible using an optimisation framework. The idea will be to optimise the parameters according to some measure of the 'drawing out' effect, while constraining the particle to a single well. The drawing out or asymmetry could in principle be optimised by maximising the skewness of the oscillator response; however, in [6], it was preferred to use the kurtosis. This approach makes sense because the kurtosis of a signal is a measure of its 'spikiness', and this is increased by emphasising the impulses. A simple penalty term can be used to prevent the particle crossing the origin, or to rule out candidate parameters which make the oscillator unstable[1]. So, the objective function for optimisation involves forcing the nonlinear oscillator with the signal of interest and then computing the kurtosis of the response. Unfortunately, this is time-consuming because it requires solving an initial value problem for each evaluation of the objective function. The solution to this problem is quite simple; rather than using a continuous-time oscillator, one passes to a discrete-time variant. For this case, forward predictions using the oscillator model will be much faster.

Starting from the oscillator model of equation (1), one can approximate the derivative using the discrete difference $(y_{i+1} - y_i)/\Delta t$, yielding,

$$y_{i+1} = (1 + a\Delta t)y_i - b\Delta t y_i^3 + \Delta t p(t) \tag{3}$$

[1] In fact, the kurtosis of the response signal will be increased if the particle occasionally crosses between wells; however, if this happens too often, the desired effect is lost, especially as jumps can occur that are not caused by impulses. For this reason, the well-crossing constraint is applied here.

Incorporating the rescaling ratio $R = \Delta t_r / \Delta t$, discussed in [6], then gives,

$$y_{i+1} = (1 + \frac{a\Delta t_r}{R})y_i - \frac{b\Delta t}{R}y_i^3 + \frac{\Delta_t}{R}p(t) \tag{4}$$

So, with a little relabelling, the discrete oscillator takes the form,

$$y_{i+1} = \alpha y_i - \beta y_i^3 + \gamma p(t) \tag{5}$$

And the stochastic resonance problem is now to maximise the kurtosis of the response, by varying the parameter vector $\underline{\theta} = (\alpha, \beta, \gamma)$.

3. SADE

The standard Differential Evolution (DE) algorithm of reference [9] attempts to transform a randomly generated initial population of parameter vectors into an optimal solution through repeated cycles of evolutionary operations, in this case: *mutation*, *crossover* and *selection*. In order to assess the suitability of a certain solution, a fitness function is needed; the kurtosis of the oscillator response is used here. Figure 2 shows a schematic for the DE procedure for evolving between populations. The following process is repeated with each vector within the current population being taken as a *target vector*; each of these vectors has an associated fitness (kurtosis). Each target vector is pitted against a *trial vector* in a competition which results in the vector with highest fitness advancing to the next generation.

The mutation procedure used in basic DE runs as follows. Two vectors A and B are randomly chosen from the current population to form a vector differential $A - B$. A *mutated* vector is then obtained by adding this differential, multiplied by a scaling factor F, to a further randomly chosen vector C to give the overall expression for the mutated vector: $C + F(A - B)$. The scaling factor, F, is often found have an optimal value between 0.4 and 1.0.

The *trial vector* is the child of two vectors: the target vector and the mutated vector, and is obtained via a crossover process; in this work, uniform crossover is used. Uniform crossover decides which of the two parent vectors contributes to each chromosome of the trial vector by a series of $D - 1$ binomial experiments. Each experiment is mediated by a crossover parameter C_r (where $0 \leq C_r \leq 1$). If a random number generated from the uniform distribution on [0,1] is greater than C_r, the trial vector takes its parameter from the target vector, otherwise the parameter comes from the mutated vector.

This process of evolving through the generations is repeated until the population becomes dominated by only a few high fitness solutions, any of which would be suitable. Like the vast majority of optimisation algorithms, convergence to the global maximum is not guaranteed; however, one of the benefits of the evolutionary approach is that it more resistant to finding a local maximum.

A potential weakness of the standard implementation of the DE algorithm as described above is that it requires the prior specification of a number of *hyperparameters* (parameters which need to be specified before the algorithm can run). Apart from the population size, maximum number of iterations etc., the algorithm needs *a priori* specification of the scaling factor F and crossover probability C_r. The values used are not guaranteed to work as well in all situations and an algorithm which establishes 'optimum' values for these parameters during the course of the evolution is clearly desirable. Such an algorithm is available in the form of the Self-Adaptive Differential Evolution (SADE) algorithm [10, 11]; the description and implementation of the algorithm here largely follows [11].

The development of the SADE algorithm begins with the observation that Storn and Price, the originators of DE, arrived at five possible strategies for the mutation operation [12]:

(i) *rand1*: $M = A + F(B - C)$

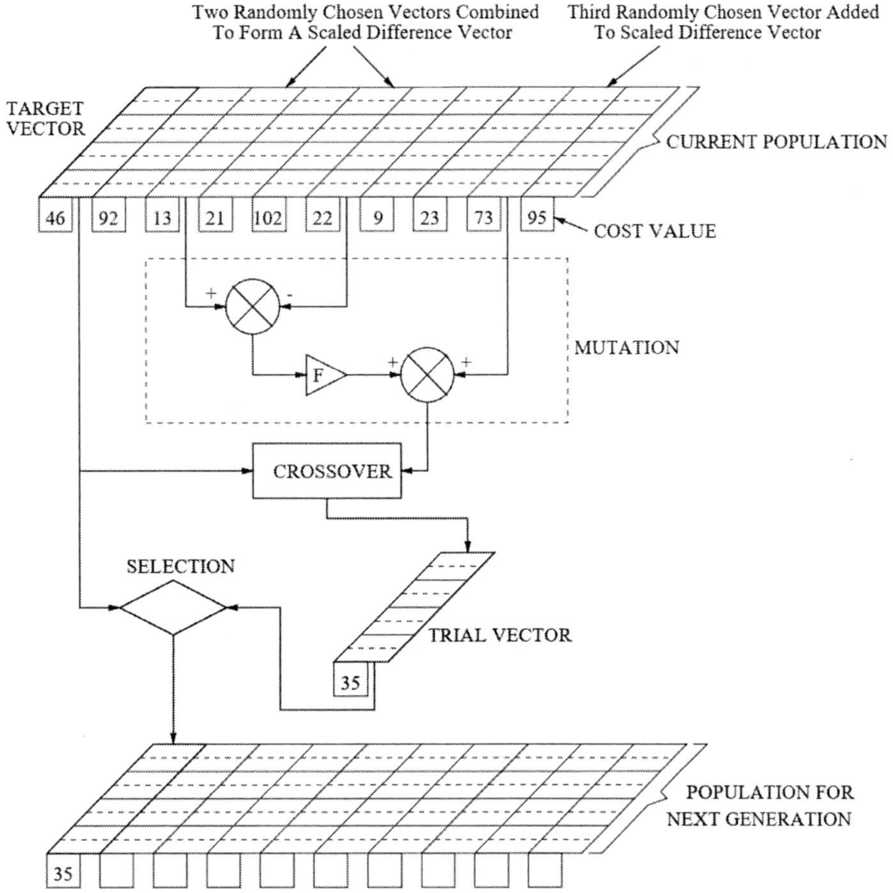

Figure 2. Schematic for the standard DE algorithm.

(ii) *best1*: $M = X^* + F(B - C)$

(iii) *current-to-best*: $M = T + F(X^* - T) + F(B - C)$

(iv) *best2*: $M = X^* + F(A - B) + F(C - D)$

(v) *rand2*: $M = A + F(B - C) + F(D - E)$

where T is the current trial vector, X^* is the vector with (currently) best cost and (A, B, C, D, E) are randomly-chosen vectors in the population distinct from T. F is a standard (positive) scaling factor. The SADE algorithm also uses multiple variants of the mutation algorithm as above; however these are restricted to the following four:

(i) *rand1*

(ii) *current-to-best2*: $M = T + F(X^* - T) + F(A - B) + F(C - D)$

(iii) *rand2*

(iv) *current-to-rand*: $M = T + K(A - T) + F(B - C)$

In the strategy *current-to-rand*, K is defined as a coefficient of combination and would generally be assumed in the range [-0.5,1.5]; however, in the implementation of [11] used here, the prescription $K = F$ is used to essentially restrict the number of tunable parameters. The SADE algorithm uses the standard crossover approach, except that at least one crossover is

IOP Conf. Series: Journal of Physics: Conf. Series **842** (2017) 012058 doi:10.1088/1742-6596/842/1/012058

forced in each operation on the vectors. If mutation moves a parameter outside its allowed (predefined) bounds, it is pinned to the boundary. Selection is performed exactly as in DE; if the trial vector has higher (or equal) fitness to the target, it replaces the target in the next generation.

The adaption strategy must now be defined. First, a set of probabilities are defined: $\{p_1, p_2, p_3, p_4\}$, which are the probabilities that a given mutation strategy will be used in forming a trial vector. These probabilities are initialised to be all equal to 0.25. When a trial vector is formed during SADE, a roulette wheel selection is used to choose the mutation strategy on the basis of the probabilities (initially, all equal). At the end of a given generation, the numbers of trial vectors successfully surviving to the next generation from each strategy are recorded as: $\{s_1, s_2, s_3, s_4\}$; the numbers of trial vectors from each strategy which are discarded are recorded as: $\{d_1, d_2, d_3, d_4\}$. At the beginning of a SADE run, the survival and discard numbers are established over the first generations, this interval is called the *learning period* (and is another example of a hyperparameter). At the end of the learning period, the strategy probabilities are updated by,

$$p_i = \frac{s_i}{s_i + d_i} \tag{6}$$

After the learning period, the probabilities are updated every generation, but using survival and discard numbers established over a moving window of the last N_L generations. The algorithm thus adapts the preferred mutation strategies. SADE also incorporates adaption or variation on the hyperparameters F and C_r. The scaling factor F mediates the convergence speed of the algorithm, with large values being appropriate to global search early in a run and small values being consistent with local search later in the run. The implementation of SADE used here largely follows [10] and differs only in one major aspect, concerning the adaption of F. Adaption of the parameter C_r is based on accumulated experience of the successful values for the parameter over the run. It is assumed that the crossover probability for a trial is normally distributed about a mean $\overline{C}_r$ with standard deviation 0.1. At initiation, the parameter C_r is set to 0.5 to give equal likelihood of each parent contributing a chromosome. The crossover probabilities are then held fixed for each population index for a certain number of generations and then resampled. In a rather similar manner to the adaption of the strategy probabilities, the C_r values for trial vectors successfully passing to the next generation are recorded over a certain greater number of generations and their mean value is adopted as the next $\overline{C}_r$. The record of successful trials is cleared at this point in order to avoid long-term memory effects. The version of the algorithm here adapts F in essentially the same manner as C_r but uses the Gaussian $N(0.5, 0.3)$ for the initial distribution. At this point, the reader might legitimately argue that SADE has simply replaced one set of hyperparameters (F, C_r) with another (duration of the learning period etc.). In fact, because DE and SADE are heuristic algorithms, there is no analytical counter to this argument. However, the transition to SADE is justified by the fact that the algorithm appears to be very robust with respect to the new hyperparameters.

4. Data

The optimisation algorithm will be benchmarked here on the two synthetic data sets examined in [6]. The first set represents a single impulse submerged in noise, while the second considers an impulse train in noise.

4.1. Single Impulse

The first data set was generated by summing a single impulse response function and Gaussian noise, as follows,

$$p(t) = s(t) + n(t) = Ae^{-D(t-T_i)}\sin(2\pi f_0[t - T_i]) + N(0, \sigma) \tag{7}$$

The parameters used were as follows: $A = 0.15$, $D = 12$, $T_i = 1.0$, $f_0 = 16$ Hz and $\sigma = 0.07$. A sampling frequency of 500 Hz was adopted and 1000 samples were generated; this placed the impulse in the centre of the record. As the resampling ratio R is absorbed in the discrete parameters, its value is not important, although it was set at 200 in [6].

4.2. Impulse Train

This is a more realistic simulation, as real machine faults can manifest as impulse trains embedded in noise. For example, faults in bearings can manifest in several ways due to different causes, such as fatigue, wear, poor installation, improper lubrication and occasionally manufacturing faults. As mentioned in the introduction, the period of the impulse train can give indications of the type of fault e.g. different characteristic frequencies arise: ball pass frequency of the outer race, ball pass frequency of the inner race, fundamental train frequency, ball spin frequency.

In order to simulate an impulse train, the individual impulses were set as the impulse responses of an SDOF system with a resonance frequency of 5000 Hz and a damping ratio of 5%; their amplitude was set as $A = 0.30$. The noise was defined as a Gaussian white noise sequence with zero-mean and standard deviation $\sigma = 0.07$. In [6], the resampling factor was set at $R = 20000$. The signal was sampled at a frequency of 96000 Hz and a second of data was collected.

5. Results

The optimisation problem was now investigated on the two data sets. The initial ranges for the SADE algorithm are important so they were initially set here by consideration of equations (3) and (4) which suggest that α will be a number a little greater than unity, and β and γ will be small positive numbers. There is usually a small trial and error exercise with SADE, where one tunes the initial ranges by making a small number of runs and expanding the ranges until no final estimates are pinned at either end of the range. The tuning phase led to inital ranges of $[1.0, 1.4]$, $[0.001, 2]$, $[0.001, 1]$ for α, β and γ respectively. In all runs, the population size was set at 30 individuals, and the number of generations was set at 1000.

The results for the single impulse case will be presented first. Because SADE is a stochastic algorithm based on a randomly-generated initial population, it is usual to make several runs and save the results from the best one. In this case, 10 runs of the algorithm were made and the optimum kurtosis obtained was 13.0. This is not as high as the result in [6]; however, one should bear in mind, that a different oscillator model is being used here. The fitness curves over the 10 runs are shown in Figure 3, with the maximum fitness for each run in red, and the average fitness in blue. The average fitnesses in each run are much lower than the maximum fitnesses because many of the candidate solutions generated in each generation cross the oscillator potential barrier at the origin and are automatically assigned low fitness. The input and ouput signals for the optimum oscillator are given in Figure 4; the impulse is very clearly amplified in the output.

For the second case study - the impulse train, only 5 runs on the SADE algorithm were carried out because the signal comprised 96000 points. The best solution arrived at a kurtosis of 19.3, which is only slightly worse than the 19.5 reported in [6]. The fitness curves for the runs are given in Figure 5, and the input and output signals for the optimised oscillator are given in Figure 6.

The results are very encouraging. During the multiple SADE runs carried out, there were still instances of parameter estimates being pinned to the ends of the initial ranges, so it is possible that improvements can still be made by expanding the initial ranges.

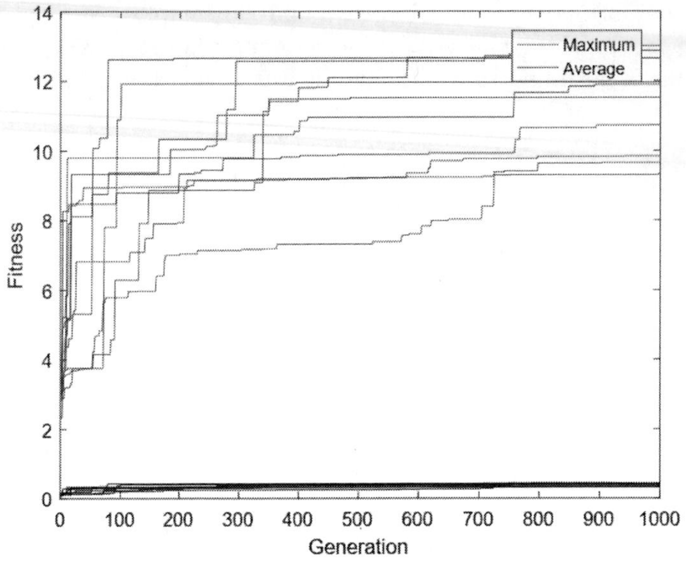

Figure 3. Average and maximum fitness values for 10 SADE runs on example 1: a single impulse in noise.

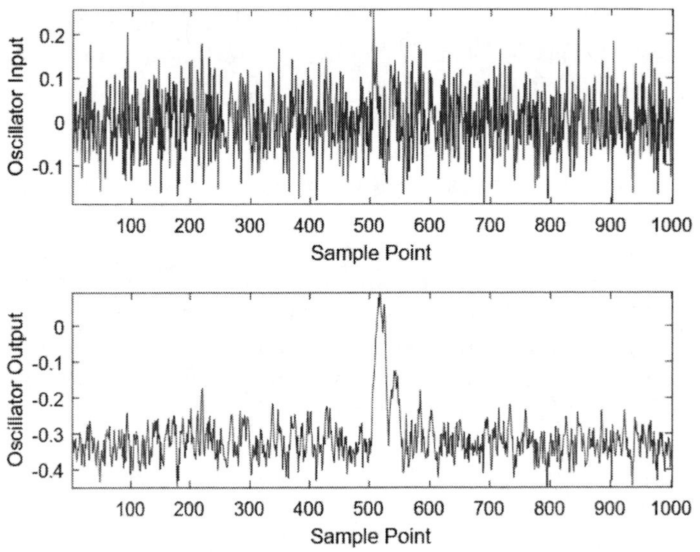

Figure 4. Input and output to SR oscillator for example 1: input kurtosis = 2.83; output kurtosis = 13.0.

6. Conclusions

The paper presents two new ideas in terms of using stochastic resonance for the detection of impulses in background noise. The first is simply to use an evolutionary optimisation scheme

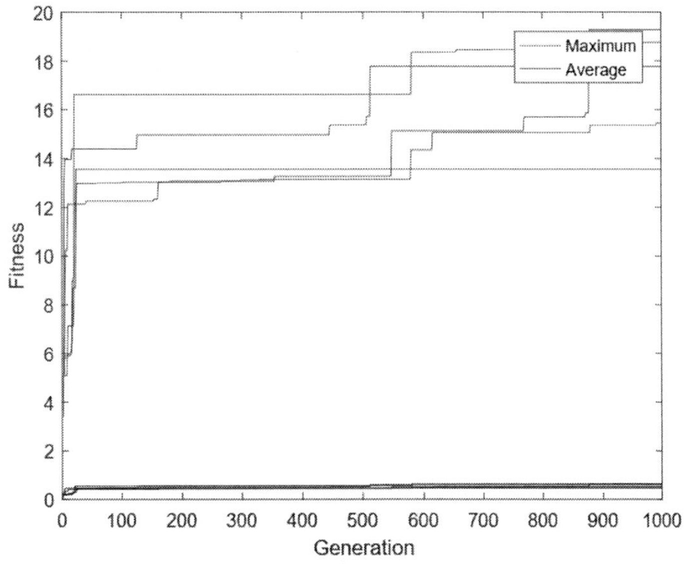

Figure 5. Average and maximum fitness values for 5 SADE runs on example 2: a train of impulses in noise.

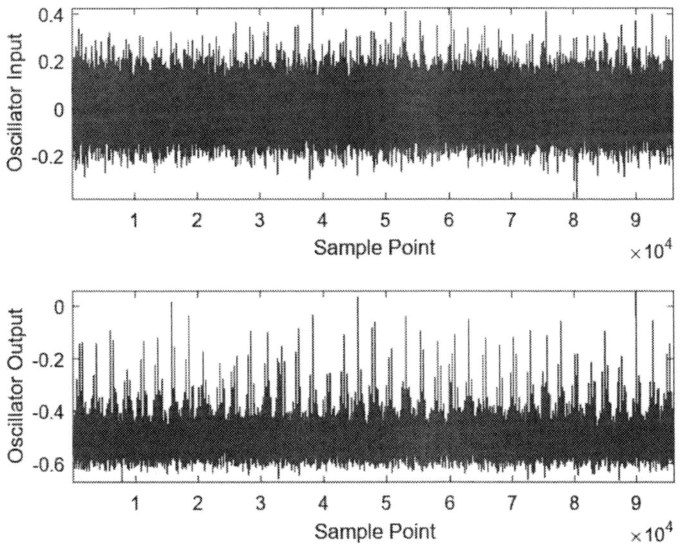

Figure 6. Input and output to SR oscillator for example 2: input kurtosis = 3.38; output kurtosis = 19.3.

to set the oscillator parameters. Because optimisation uses forward prediction through the oscillator model in order to evaluate the objective, it is very slow if one uses the 'traditional' continuous-time representation of the oscillator, so this has been replaced with a discrete-time variant here. The initial results on simulated data are promising, and the next stage will be to evaluate the approach on experimental data. It is anticipated that some small adaptations to the SADE hyperparameters may occur then, so the simulated data may be re-evaluated. The ultimate aim is to freeze all parameters of the algorithm so that it provides a universal solution, independent of problem.

This paper has been concerned with the process of *feature extraction* for condition monitoring; the next paper in the (short) series will consider *anomaly* or *novelty detection*.

Acknowledgments
The authors would like to acknowledge the UK Engineering and Physical Sciences Research Council (EPSRC) for funding this research through the AIMaREM project (EPSRC grant reference number: EP/N018427/1); they would also like to thank Dr Graeme Manson of the Dynamics Research Group, for providing the original code for the SADE algorithm.

References

[1] Benzi R, Sutera A and Vulpiani A 1981 *J. Phys. A* **10** L453.
[2] McDonnell M D and Abbot D 2009 *PloS Comp. Bio.* **5**.
[3] Li J, Chen X and He Z 2013 *J. Sound Vib.* **332** 5999.
[4] Li J, Chen X and He Z 2013 *Mech. Sys. Sig. Proc.* **36** 240.
[5] Lu S, He Q and Kong F 2014 *Mech. Sys. Sig. Proc.* **45** 488.
[6] Marchesiello S, Fasana A, Garibaldi L 2015 *Proc. Int. Conf. on Structural Engineering Dynamics (Lagos, Portugal)*.
[7] Zhou P, Lu S, Liu F, Liu Y, Li and Zhao J 2017 *J. Sound Vib.* **391** 194.
[8] Tan J, Chen X, Wang J, Chen H, Cao H, Zi Y and He Z 2009 *Mech. Sys. Sig. Proc.* **23** 811
[9] Price K and Storn R 1997 *J. Glob. Opt.* **11** 341.
[10] Qin A K and Suganthan P N 2005 *Proc. IEEE Congress on Evolutionary Computation (Edinburgh, Scotland)* 1785.
[11] Huang V L, Qin A K and Suganthan 2006 *Proc. IEEE Congress on Evolutionary Computation (Vancouver, Canada)* 17.
[12] Storn R and Price K 2009 http://www.icsi.berkeley.edu/ storn/code.html: (Accessed 27th October 2009).,

IOP Conf. Series: Journal of Physics: Conf. Series **842** (2017) 012059 doi:10.1088/1742-6596/842/1/012059

An Illustration of New Methods in Machine Condition Monitoring, Part II: Adaptive outlier detection

I. Antoniadou[1], K. Worden[1], S. Marchesiello[2], C. Mba[2], L. Garibaldi[2]

E-mail: I.Antoniadou@shef.ac.uk;
[1]Dynamics Research Group, Mechanical Engineering Department, The University of Sheffield, Mappin Street, Sheffield S1 3JD, United Kingdom
[2]Dipartimento di Ingegneria Meccanica e Aerospaziale Politecnico di Torino Corso Duca degli Abruzzi 24, 10129, Torino, Italy

Abstract. There have been many recent developments in the application of data-based methods to machine condition monitoring. A powerful methodology based on machine learning has emerged, where diagnostics are based on a two-step procedure: extraction of damage-sensitive features, followed by unsupervised learning (novelty detection) or supervised learning (classification). The objective of the current pair of papers is simply to illustrate one state-of-the-art procedure for each step, using synthetic data representative of reality in terms of size and complexity. The second paper in the pair will deal with novelty detection. Although there has been considerable progress in the use of outlier analysis for novelty detection, most of the papers produced so far have suffered from the fact that simple algorithms break down if multiple outliers are present or if damage is already present in a training set. The objective of the current paper is to illustrate the use of phase-space thresholding; an algorithm which has the ability to detect multiple outliers inclusively in a data set.

1. Introduction

The fundamental framework that describes the process of a *Condition Monitoring* (CM) strategy can be outlined in a sequence of steps [1]: operational evaluation, data acquisition, feature extraction and statistical modelling for feature discrimination. It is the last step of this process, i.e. the implementation of algorithms that analyse extracted data features for damage detection and quantification, that the second part of this study is concerned with.

In general, feature discrimination can be performed using machine learning algorithms and statistical methods. These methods can be categorised into the *supervised* and the *unsupervised learning* approaches. *Unsupervised learning* methods for damage detection do not require all the class labels of the acquired data and can provide information of potential abnormality in the data being analysed. On the other hand, although *supervised learning* methods might provide much more informative results, they require training data describing all the damage classes or operational and environmental scenarios of a monitored machine, that in most cases are not available. For this reason, unsupervised learning approaches, might have an advantage over the supervised learning ones in many practical CM applications. For unsupervised learning, one major challenge is that environmental and operational changes, that could affect the potential

IOP Conf. Series: Journal of Physics: Conf. Series **842** (2017) 012059 doi:10.1088/1742-6596/842/1/012059

damage sensitive features, should be taken into account. Although, aspects of this issue are taken into consideration in this study, the analysis presented will not deal with it as its main research objective, therefore the authors refer to papers [2, 3, 5] for further exploration of the problem and its proposed data analysis solutions.

Novelty detection is one of the fundamental unsupervised learning approaches, also known in the statistics field as *outlier analysis*. It works under the assumption that selected features defining a normal condition follow the Gaussian distribution [4]. Distance measures, such as the Mahalanobis squared-distance, can be used as discordancy measures. Procedures, such as the Monte Carlo Method, can be employed for the estimation of a threshold value.

Here, a new adaptive novelty detection method will be discussed for the purposes of damage detection on the synthetic data discussed in Part I. The proposed approach is based on the idea of *spatial adaptation*, a term that, in the statistical literature, describes regression and classification methods that do not require an initially pre-defined fixed basis, because of their ability to create this kernel basis from the data being analysed [6, 7, 8]. The main advantage in this case, is the better recovery of the functions actually occurring within real data [8].

The method proposed here exploits the three following concepts [9]: a) that differentiation enhances the high frequency portion of the signal, b) that the expected maximum of a random series is given by the *universal threshold*, and c) that good data cluster in a dense cloud in phase space or in Poincaré maps. The threshold used in order to identify the outliers in the data, is estimated in an adaptive manner, in the phase space of the dynamic system examined (operational bearing in our case).

The aim is to show that this approach of outlier detection can be a powerful tool for certain CM applications. The layout of the paper is as follows: the following section briefly explains the basic theory on the new adaptive threshold method proposed. Section Three explains the challenges faced in the estimation of thresholds when standard novelty detection methods are not provided with good quality training data. This is demonstrated by comparing results for an outlier detection scheme that uses the Mahalanobis squared-distance and the Monte Carlo process for threshold estimation to the results for the new adaptive threshold method. The paper ends with brief conclusions.

2. Theoretical background: the phase-space outlier detection method

The method iteratively constructs sequential ellipsoids in the dynamical system's three-dimensional phase space: points lying outside the ellipsoids are designated as outliers. The three-dimensional phase space, also known as a Poincaré map or a phase space map, is a plot in which a variable and its derivatives are plotted against each other.

The threshold is defined by the Universal criterion, introduced in the landmark paper given in reference [8], forming an ellipsoid in the phase space that separates inliers from outliers. For n independent, identically distributed, standard, normal random variables ξ_i the expected absolute maximum is:

$$E(|\xi_i|_{max}) = \sqrt{2\log n} = \lambda_U \tag{1}$$

where λ_U is termed the *universal threshold*. For a normal, random variable, that consists of n data points and whose standard deviation is estimated by σ and the mean is zero, the expected absolute maximum is:

$$\lambda_U\sigma = \sqrt{2\log n}\sigma \tag{2}$$

The algorithm consists of a number of sequential iterations that stop when the number of good data becomes constant (or, equivalently the number of new points identified as outliers, peaks in the power diagrams in this case, falls to zero). If u_i is the dataset being analysed, then each iteration has the following steps:

- Calculate *surrogates* Δu_i and $\Delta^2 u_i$ for the first and second derivatives from:

$$\Delta u_i = (u_{i+1} - u_{i-1})/2 \tag{3}$$

$$\Delta^2 u_i = (\Delta u_{i+1} - \Delta u_{i-1})/2 \tag{4}$$

- Calculate the standard deviations of all three variables σ_u, $\sigma_{\Delta u}$ and $\sigma_{\Delta^2 u}$, and then the expected maxima using the *universal criterion.*
- Calculate the rotation angle of the principal axis of the $\Delta^2 u_i$ versus u_i using the cross correlation:

$$\theta = tan^{-1}(\frac{\sum u_i \Delta^2 u_i}{\sum u_i^2}) \tag{5}$$

- Each set of variables $\{u_i, \Delta u_i, \Delta^2 u_i\}$, determines a point in spherical coordinates. For each pair of these variables, an ellipse can be calculated. Therefore, for Δu_i versus u_i the major axis is $\lambda_U \sigma_u$ and the minor axis is $\lambda_U \sigma_{\Delta u}$; for $\Delta^2 u_i$ versus Δu_i the major axis is $\lambda_U \sigma_{\Delta u}$ and the minor axis is $\lambda_U \sigma_{\Delta^2 u}$; and for $\Delta^2 u_i$ versus u_i the major and minor axes a and b respectively, can be shown by geometry to be the solution of:

$$(\lambda_U \sigma_u)^2 = a^2 cos^2\theta + b^2 sin^2\theta \tag{6}$$

$$(\lambda_U \sigma_{\Delta^2 u})^2 = a^2 sin^2\theta + b^2 cos^2\theta \tag{7}$$

- For each projection in space the points that lie outside of the ellipse are identified and replaced with a smoothed estimate in order to perform the next iteration.

At each iteration, replacement of the outliers reduces the standard deviations calculated in step two and thus the size of the ellipsoid reduces until further outlier replacement has no effect.

3. Time-domain outlier detection

3.1. For reference: outlier analysis of the enhanced by stochastic resonance synthetic data

The classic outlier analysis approach, that computes the Mahalanobis squared-distance discordancy measures of the data feature matrix and compares them to a threshold, was initially used to analyse the synthetic data, processed using the optimised stochastic resonance approach, and presented in more detail in Part I.

For multivariate data the *Mahalanobis squared-distance* is estimated from the following equation:

$$D_\zeta^2 = (\{x\}_\zeta - \{\overline{x}\})^T [\Sigma]^{-1} (\{x\}_\zeta - \{\overline{x}\}) \tag{8}$$

where $\{x\}_\zeta$ is the feature vector corresponding to the candidate outlier, $\{\overline{x}\}$ is the sample mean of the normal condition features and $[\Sigma]$ is the normal condition feature sample covariance matrix.

The threshold value that labels the inlier and outlier observations can be estimated through the employment of a *Monte Carlo* method. Briefly, a $p \times n$ matrix (p-observations, n-dimensions, in order that the matrix dimensions match the dimensions of the extracted features) is generated and populated with elements randomly drawn from a zero-mean, unit standard deviation Gaussian distribution. Then the Mahalanobis squared distance is calculated for all elements and the largest value stored. This is repeated a large number of times, each time storing the largest Mahalanobis squared-distance, which are then sorted in order of magnitude. The threshold is assigned as a percentile of the resulting array of Mahalanobis squared-distances.

For the single impulse case, presented in Figure (1), a 5-dimensional feature was defined as a 5-point time window. In total this resulted in a 200×5 feature matrix. A series of 40 features was used as training data. The results are satisfying for this particular case, but restrictions arise regarding the need to define appropriate feature dimensions and choose training data that do not include outliers. The latter restriction, can be significantly challenging for larger and more complex datasets.

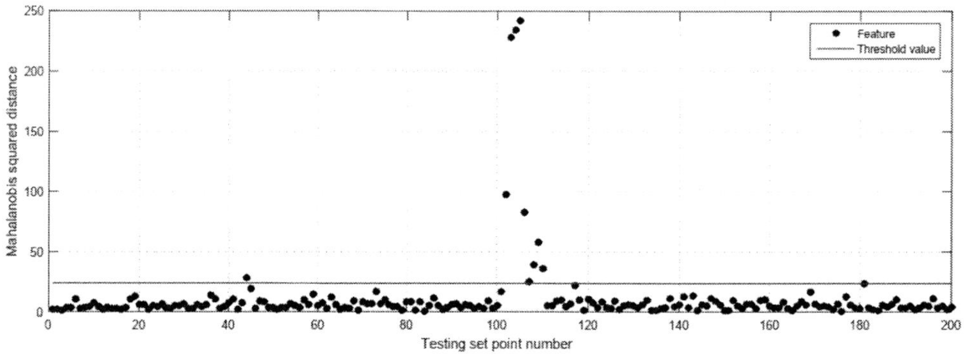

Figure 1. Outlier statistics for single impulse synthetic SR output signal.

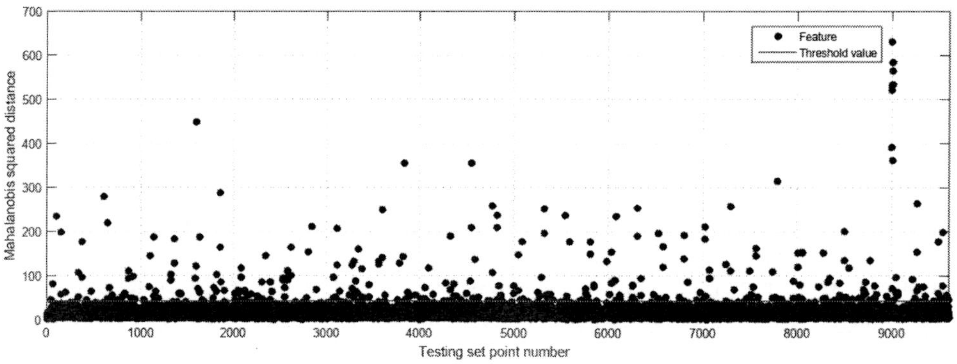

Figure 2. Outlier statistics for multiple impulse synthetic SR output signal.

In Figure 2, the results of the outlier analysis algorithm for the multiple impulse signal are given. The choice of the amount and quality of the algorithm's training data is important for the algorithm's performance. In this case, a 10-dimensional feature was defined as a 10-point time window. In total this resulted in a 9600×10 feature matrix. The first series of 40 features was used as training data. Changing the feature dimensions and the amount of training data results in different outputs and can potentially hinder the robustness of the approach. An additional comment is that the algorithm is not a strong choice for time-domain signal analysis, if the nonstationarities that need to be detected in the signal are impulse-like, due to the short duration of the signal's amplitude variation. In addition, the results can only give approximate information of the time that these changes are happening, since the definition of a feature is a sample point time window. This problem, can be addressed by choosing a frequency domain approach, if the signals analysed are periodic. In many condition monitoring scenarios, this could be a valid assumption. For condition monitoring applications though, where the speed of the rotating machine examined is changing (e.g. wind turbine bearings/gearboxes), this might not be always well founded. Depending on the strong or weak nature of periodicity of the signal's nonstationarities associated with damage, the frequency domain approach could result in a relative quality of performance.

3.2. The phase-space threshold method results and discussion

The phase-space threshold method is used in this section for outlier detection. Figures 3 and 4 show the phase-space threshold analysis results for the single impulse case and Figure 5 shows the multiple impulse case results.

The results demonstrate that the method can be very efficient for detecting outliers and offers the advantage of adaptivity: outliers are detected through an iterative procedure that at each step re-defines the parameters describing the distribution of the analysed data and therefore does not directly depend on the existence of a training set corresponding to healthy state data. This iterative procedure followed for the detection of outliers means in practice that the "threshold" is not necessarily a constant value but is adjusted throughout the dataset according to the output of the algortithm's specific active iteration steps. This may lead to the assumption that variations that may appear in the system monitored, lasting longer than variations taking place for shorter time periods, often linked to common machine damage types in the condition monitoring research, might not inhibit the reliability of the method. This requires further study, but also indicates the potential of the method.

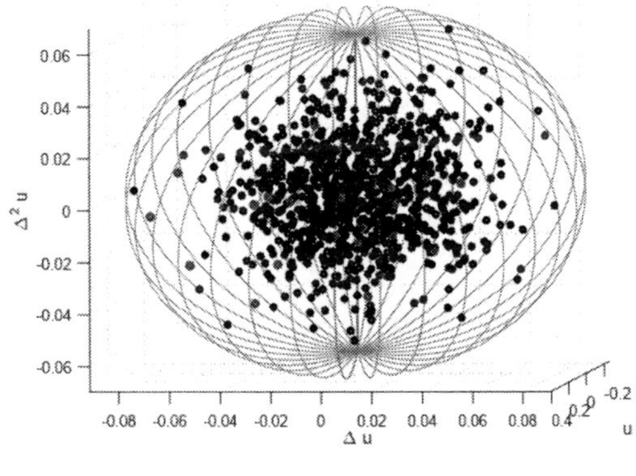

Figure 3. Phase space of SR output data: demonstration of constructed ellipsoid (single impulse case).

In Figure 5, one can see the results of the method for the second case of data analysed (multiple impulse data). The detection of outliers provides an advantage when compared to the results given in Figure 2, in the sense that the outliers can be directly linked to time instants and that the estimated "threshold", changes values throughout the dataset, each time, depending on neighboring data points and the rate of the difference in their values. Further study should be followed in order to confirm the performance of the method in experimental data.

4. Conclusions

The second part of this paper demonstrates how a novel adaptive method for outlier detection can be used in combination with the genetically-optimised stochastic resonance approach, for the

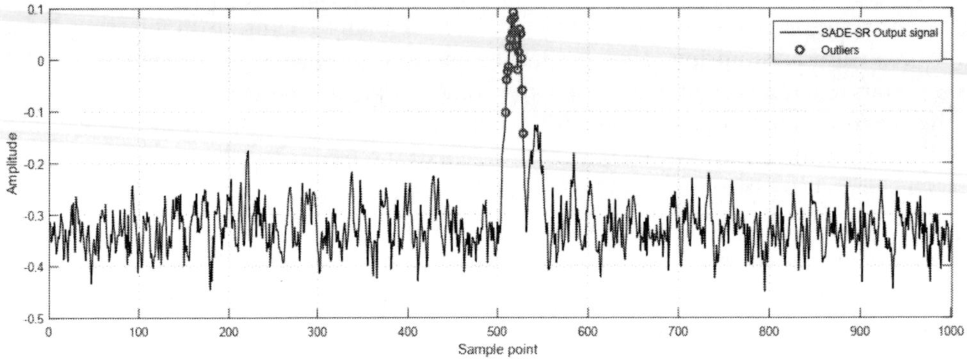

Figure 4. Detected outliers using the phase space thresholding method of the SR output data (single impulse case).

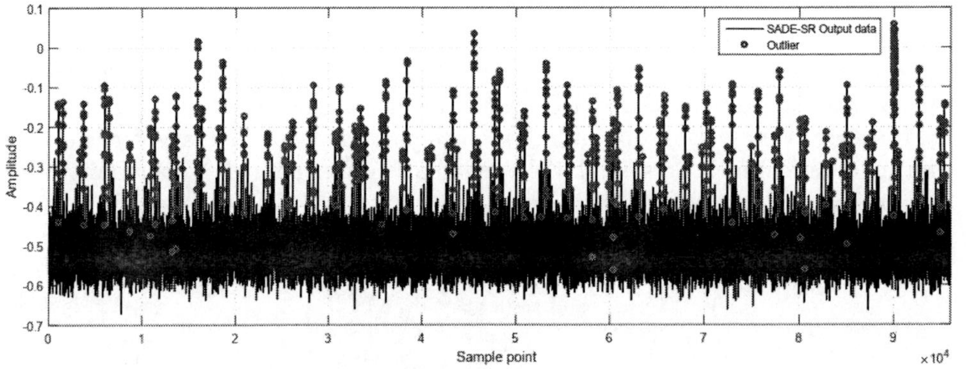

Figure 5. Detected outliers using the phase space thresholding method of the SR output data (multiple impulse case).

purposes of bearing condition monitoring. Standard machine learning or statistics algorithms for novelty detection might under-perform in certain condition monitoring applications due to their training needs and the complexity of the data being analysed. Here, through our analysis of synthetic pre-processed data it is demonstrated that we can potentially avoid this risk by exploiting an adaptive approach. Further research is needed in order to confirm the results on experimental bearing datasets.

Acknowledgments
The authors acknowledge the EPSRC for funding this research through the AIMAREM project (EPSRC grant reference number: EP/N018427/1).

References
[1] Farrar, C., Worden, K.. Structural health monitoring: A machine learning perspective. John Wiley and Sons Ltd 2013;
[2] Cross, E., Worden, K., Chen, Q.. Cointegration: a novel approach for the removal of environmental trends in structural health monitoring data. Proceedings of the Royal Society A: Mathematical, Physical and Engineering Science 2011;467(2133):27122732.
[3] Cross, E., Manson, G., Worden, K., Pierce, S.. Features for damage detection with insensitivity to

environmental and operational variations. Proceedings of the Royal Society A: Mathematical, Physical and Engineering Sciences 2012;:1471 2946.

[4] Worden, K., Manson, G., Fieller, N.. Damage detection using outlier analysis. Journal of Sound and Vibration 2000;229(3):647667.

[5] Antoniadou, I., Manson, G., Staszewski, W.J., Barszcz,T., Worden, K.. A time-frequency analysis approach for condition monitoring of a wind turbine gearbox under varying load conditions. Mechanical Systems and Signal Processing 2015; 64(188-216).

[6] Breiman, L., Friedman, J., Stone, C.J., Olshen, R.. In: Classification and regression trees. Chapman and Hall; 1984.

[7] Friedman, J., Silverman, B.. Flexible parsimonious smoothing and additive modeling. Technometrics 1989;31(1):321.

[8] Donoho, D., Johnstone, I.. Ideal spatial adaptation by wavelet shrinkage. Biometrika 1994;81(3):42555.

[9] Goring, D., Nikora, V.. Despiking acoustic doppler velocimeter data. Journal of Hydraulic Engineering 2002;128:117 126.

A new state evaluation method of oil pump unit based on AHP and FCE

Yang Lin, Wei Liang*, Zeyang Qiu, Meng Zhang and Wenqing Lu

College of Mechanical and Transportation Engineering, China University of Petroleum-Beijing

lw@cup.edu.cn

Abstract. In order to make an accurate state evaluation of oil pump unit, a comprehensive evaluation index should be established. A multi-parameters state evaluation method of oil pump unit is proposed in this paper. The oil pump unit is analyzed by Failure Mode and Effect Analysis (FMEA), so evaluation index can be obtained based on FMEA conclusions. The weights of different parameters in evaluation index are discussed using Analytic Hierarchy Process (AHP) with expert experience. According to the evaluation index and the weight of each parameter, the state evaluation is carried out by Fuzzy Comprehensive Evaluation (FCE) and the state is divided into five levels depending on status value, which is inspired by human body health. In order to verify the effectiveness and feasibility of the proposed method, a state evaluation of oil pump used in a pump station is taken as an example.
Keywords. State evaluation; Oil pump; FMEA; AHP; FCE

1. Introduction

The pipeline system has been serving as a main transportation channel of oil and natural gas, in which the oil pump is a so pivotal component that it guarantees oil and gas transmission safety with the ever-increasing demand of oil [1]. In general, the oil pump driven by motor is the typical form in oil pipeline supporting for long distance transmission of oil. However, unexpected shutdown of such component significantly threatens the production safety and enterprise profits, even causes casualties because of its high running speed and continuous operation for a long period [2]. Therefore, increasing attention has been attracted to the condition assessment of oil pump unit in the last decades.

The first service system for mechanical equipment was designed to consider the age not current condition in the early 90s [3]. With the advancement of condition monitoring instruments and computer technology, running state evaluation of equipment has been paid more attention based on their one or more state feature values such as temperature, vibration and so on [4-6]. However, because the structure of equipment is complex and the relevancy among components of equipment is always close, it's difficult to use some of the feature parameters to evaluate the entire equipment condition [7-9]. Therefore, a state evaluation method for equipment based on multi feature parameters is essential to ensure safety operation.

As an application of fuzzy mathematics, FCE has been widely used to state evaluation in recent years [10]. However, evaluation index and its weighted value in the method usually given by experience are inevitably subjective. Although there isn't an applicable method to establish the evaluation index, the weight vector of evaluation index has been solved by fuzzy AHP, an extension of AHP proposed by

Saaty [11]. Gu et al, calculated the weighted value of all level's evaluation factors in the mechanical movement program evaluation system based on a fuzzy AHP method [12]. Chang et al, proposed a new extent analysis approach for the synthetic extent values of the pair-wise comparison by fuzzy AHP [13]. In order to obtain an objective and comprehensive evaluation index, FMEA is used to establish the evaluation index of oil pump unit in this paper.

The remainder of this paper is organized as follows: the evaluation index of oil pump unit is established based on FMEA in section 2. Section 3 deals with determining weights of parameters in evaluation index. Section 4 presents an application of the proposed multi-parameters state evaluation method based on AHP and FCE. Finally in section 5, conclusions are presented and discussed.

2. Establishment of evaluation index

2.1. *FMEA*

FMEA was first implemented in 1949 by the United States Army and then it has been widely used in various fields with its strength and robustness. This method is an engineering analysis tool of equipment failure, helping engineers to define, identify, and eliminate known or potential failures, problems, errors in different times, including design, process or service. The process of FMEA is showed in figure 1.

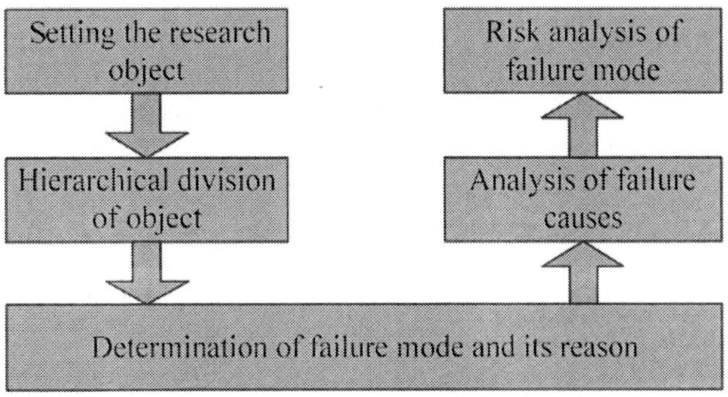

Figure 1. Process of FMEA.

2.2. *Evaluation index*

Oil pump unit driven by motor, the subject in this paper, is composed of two important parts: oil pump and driven motor. And there are so many monitoring parameters that it couldn't be totally contained in evaluation index of oil pump unit. At present, evaluation index of oil pump unit is established based on experience, so it is possible to ignore some important parameters in evaluation index. To establish a comprehensive evaluation index, FMEA is applied and Table 1 shows FMEA results for oil pump. Compared to conventional FMEA, Table 1 adds detection method and it can help select important parameters to establish evaluation index.

According to FMEA results of oil pump unit, an evaluation index of fourteen parameters including driven motor and oil pump is established. Parameters of oil pump in evaluation index include Outlet Pressure (P1), Inlet Pressure (P2), Vibration of Driving End (P3), Vibration of Free End (P4), Temperature of Driving End (P5), Temperature of Free End (P6), Differential Pressure of Oil seal (P7), Temperature of Pump Case (P8). Parameters of driven motor in evaluation index are Temperature of A Phase (M1), Temperature of B Phase (M2), Temperature of C Phase (M3), Current Value (M4), Temperature of Driving End (M5), Temperature of Free End (M6).

Table 1. FMEA for oil pump.

Component	Function	Failure Causes	Failure Effect	Detection Method
Pump Case	Holding fluid	1. Shaft Misalignment 2. Pump vacuum 3. Serious bearing wear 4. Rotor unbalance 5. Anchor bolt loosening 1. Serious bearing wear 2. Excessive discharge 3. Shaft Misalignment 4. Pump wear	1. Excessive mechanical damage 2. Pump shutdown 1. Excessive mechanical damage 2. Shutdown	1. Vibration detection 2. Temperature detection 3. Experience judgment
Rotor (Impeller)	Energy conversion	1. Design and manufacturing defects 2. Poor installation 3. Impeller wear or corrosion 4. Plastic deformation 5. Not concentric between stator and rotor	1. Performance degradation 2. Shutdown	1. Vibration detection 2. Comparison test under different load
Mechanical Seal	Preventing leakage	1. Improper seal installation 2. Severe sleeve wear 3. Rotor eccentricity caused by excessive wear of pump shaft 4. Shaft Misalignment 5. Abnormal seal pressure 6. Poor filler installation 7. Sealing surface damage 8. Pump vacuum in a long time	1. Excessive mechanical damage 2. Shutdown	1. Temperature detection 2. Differential pressure detection
Pump Shaft	Torque transmission and fixing rotor	1. Improper installation of shaft 2. Shaft Misalignment 3. Material defects	1. Excessive mechanical damage 2. Excessive vibration 3. Shutdown	Vibration detection
Bearing	Fixing pump shaft	1. Rotational imbalance 2. Rotor eccentricity caused by excessive wear of pump shaft 3. Impurities or deterioration of lubricating oil 4. Poor installation of bearing 5. Insufficient or excessive lubrication	Excessive bearing wear	1. Vibration detection 2. Temperature detection 3. Experience judgment
Pump Inlet	Fluid inlet	1. No or insufficient pump priming 2. Inlet pipe blockage or small valve opening 3. Fluid gasification 4. Strong viscosity of fluid 5. Suction pipeline leakage	Insufficient flow	Pressure detection
Pump Outlet	Fluid outlet	1. Outlet pipeline leakage	Abnormal flow	Pressure detection

3. Determination of weight vector of evaluation index

3.1. AHP

As a well-known multi-criterion decision-making technique, AHP was firstly used for OHS [14-15]. It is a simple method which is a combination of qualitative and quantitative analysis and its main advantage is checking and reducing the inconsistency of expert judgments. The interaction between different decision variables is expressed by hierarchy in AHP. The hierarchy is established from more general criterion to the particular one and determined by pairwise comparison which is made using a nine-point scale, as shown in Table 2. In this paper, AHP is employed to determine the weight vector of evaluation index of oil pump unit and it is comprised of four main steps: decomposition of structure of evaluation index, structure of judgment matrix, determination of weight vector of evaluation index and coincidence test.

Table 2. Important degree judgment.

f(a, b)	*f*(b, a)	Definition	Verbal explanation
1	1	a and b "equally important"	a and b contribute equally to the property
3	1/3	a and b "slightly important"	The contribution a slightly bigger than b, not obviously
5	1/5	a and b "obvious important"	The contribution a obviously bigger than b, but not extremely obvious
7	1/7	a and b "intense important"	The contribution a extremely obviously bigger than b, but not dominant
9	1/9	a and b "absolutely important"	The contribution a overwhelming superiority bigger than b
2, 4, 6, 8	1/2, 1/4, 1/6, 1/8	a and b situated among various ranks	The assessment falls between two levels

3.2. Weight vector of evaluation index

The hierarchy structure of evaluation index of oil pump unit is established based on evaluation index in section 2 and AHP, as shown in figure 2. And then according to judgment matrix of evaluation index, the weight vector of evaluation index is determined: v= [0.1369 0.0963 0.0474 0.0474 0.0282 0.0282 0.0970 0.0187 0.1553 0.0851 0.0851 0.0851 0.0447 0.0447]. Finally it is proved that weight distribution is reasonable by coincidence test.

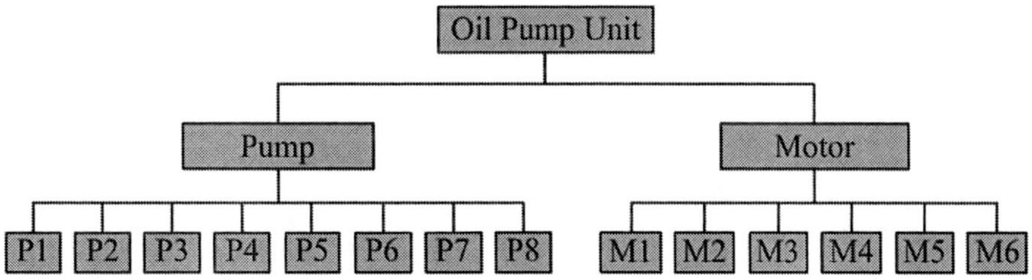

Figure 2. The hierarchy structure of evaluation index of oil pump unit.

Table 3. State grade of oil pump unit.

State grade	Status value	Strategy
Best	[90, 100]	Routine maintenance
Good	[60, 90)	Critical part monitoring
Ill	[40, 60)	Further detection
Stiff	[20, 40)	Urgent maintenance
Worst	[0, 20)	Immediate shutdown and maintenance

4. State evaluation of oil pump unit

4.1. State grade

In order to accurately describe the state of oil pump unit, the running condition is divided into five levels based on status value, which is inspired by human body health. Five states are defined as best, good, ill, stiff, and worst, and status value ranges from 0 to 100. 100 means the best state and 0 means the worst state, as shown in Table 3.

4.2. State evaluation of oil pump unit

Before evaluating the state of oil pump unit, the state of each parameter in evaluation index should be determined. And there is an operating range of each parameter in evaluation index. Therefore, the state of each parameter in evaluation index can be determined based on its operating range and triangular membership. The oil pump unit in a pump station is taken as an example and the membership vector of each parameter to five state grades is calculated: $\mathbf{r}_1$=[0.2, 0.8, 0, 0, 0], $\mathbf{r}_2$=[0.3, 0.7, 0, 0, 0], $\mathbf{r}_3$=[0, 0.4, 0.6, 0, 0], $\mathbf{r}_4$=[0, 0.5, 0.5, 0, 0], $\mathbf{r}_5$=[0.5, 0.5, 0, 0, 0], $\mathbf{r}_6$=[0.4, 0.6, 0, 0, 0], $\mathbf{r}_7$=[0.4, 0.6, 0, 0, 0], $\mathbf{r}_8$=[0.3, 0.7, 0, 0, 0], $\mathbf{r}_9$=[0.4, 0.6, 0, 0, 0], $\mathbf{r}_{10}$=[0.4, 0.6, 0, 0, 0], $\mathbf{r}_{11}$=[0.4, 0.6, 0, 0, 0], $\mathbf{r}_{12}$=[0.4, 0.6, 0, 0, 0], $\mathbf{r}_{13}$=[0.2, 0.8, 0, 0, 0], $\mathbf{r}_{14}$=[0.5, 0.5, 0, 0, 0]. Hence the membership vector of evaluation index to five state grades is obtained: $\mathbf{R}=[\mathbf{r}_1\ \mathbf{r}_2\ \mathbf{r}_3\ \mathbf{r}_4\ \mathbf{r}_5\ \mathbf{r}_6\ \mathbf{r}_7\ \mathbf{r}_8\ \mathbf{r}_9\ \mathbf{r}_{10}\ \mathbf{r}_{11}\ \mathbf{r}_{12}\ \mathbf{r}_{13}\ \mathbf{r}_{14}]^T$.

According to weight vector of evaluation index and membership vector of evaluation index to five state grades, membership vector of oil pump unit to five state grades can be calculated: $\mathbf{A}=\mathbf{vR}$=[0.3216, 0.6264, 0.0521, 0, 0]. Therefore, subordinate values that state grade of oil pump unit corresponds to best, good, ill, stiff, and worst are 0.3216, 0.6264, 0.0521, 0, and 0, respectively. The global state of oil pump unit is "good" based on the principle biggest membership. However, it can be concluded that P3 and P4 are not as good as other parameters by membership vector of evaluation index to five state grades.

Monitor values of P3 and P4 in field are 86 and 72, while their alarm values are 100. Similarly, it can also be concluded that the values of P3 and P4 in field are a little large. Therefore, there is in agreement between actual condition of oil pump unit and results of state evaluation. And these two parameters should be given a careful attention in operation of oil pump unit.

5. Conclusions

1. The evaluation index of oil pump unit should be comprised of important parameters that can reflect the state of oil pump unit. To obtain these important parameters, FMEA for oil pump unit is applied and a comprehensive evaluation index is established based on these parameters.
2. Each parameter in evaluation index has a different influence on the running state of oil pump unit. Based on AHP, the weight vector of evaluation index is calculated and the result is proved to be reasonable by coincidence test.
3. A state grade of oil pump unit is made based on status value, which is inspired by human body health. And an evaluation example of oil pump unit proves that the evaluation method is in agreement with actual state.

However, the good global state can't guarantee that all parts of oil pump unit are good. To evaluate the state of critical part, the relationships between monitoring parameters in evaluation index and the state of critical part should be researched further.

6. Acknowledgements

This paper is supported by National Science and Technology Major Project of China (No. 2011ZX05055), Science Foundation of China University of Petroleum, Beijing (No. 2462015YQ0406).

References

[1] Qiu J, Liang W, Zhang L, et al. The early-warning model of equipment chain in gas pipeline

based on DNN-HMM[J]. *Journal of Natural Gas Science & Engineering*, 2015, 27: 1710-1722.

[2] de León Hijes F C G, Cartagena J J R. Maintenance strategy based on a multicriterion classification of equipments[J]. *Reliability Engineering & System Safety*, 2006, 91(4): 444-451.

[3] Tan Z F, Guo L Z, Yong-Xiu H E. Maintenance theory and optimal design for a compound system under a statistical equilibrium state[J]. *Systems Engineering & Electronics*, 2004, 26(1):125-129.

[4] Yan R, Gao R X. Rotary Machine Health Diagnosis Based on Empirical Mode Decomposition[J]. *Journal of Vibration & Acoustics*, 2008, 130(2):111-120.

[5] Chen Z S, Yang Y M, Hu Z, et al. Detecting and Predicting Early Faults of Complex Rotating Machinery Based on Cyclostationary Time Series Model[J]. *Journal of Vibration & Acoustics*, 2006, 128(5):666-671.

[6] Sun Q, Zhang D, Chen P. Pattern Recognition for Automatic Machinery Fault Diagnosis[J]. *Journal of Vibration & Acoustics*, 2004, 126(2):307-316.

[7] Li X J, Bin G F, Dhillon B S. Model to evaluate the state of mechanical equipment based on health value[J]. *Mechanism & Machine Theory*, 2011, 46(46):305-311.

[8] Gu Y J, Dong Y L, Yang K. Synthetic evaluation on conditions of equipment in power plant based on fuzzy judgment and RCM analysis[J]. *Proceedings of the Csee*, 2004.

[9] Ravi B. Rapid tooling manufacturability evaluation using fuzzy-AHP methodology[J]. *International Journal of Production Research*, 2007, 45(5):1161-1181.

[10] Chen J F, Hsieh H N, Do Q H. Evaluating teaching performance based on fuzzy AHP and comprehensive evaluation approach[J]. *Applied Soft Computing*, 2015, 28(C):100-108.

[11] Saaty T L. The analytic hierarchy process: Planning, priority setting, resource Allocation. McGraw-Hill, NY, USA[M]// The Analytic Hierarchy Process: Planning, Priority Setting, Resource Allocation. 1980.

[12] Gu Y. A Fuzzy AHP Approach to the Determination of Weights of Evaluation Factors in Mechanism Scheme Evaluation Process[J]. *China Mechanical Engineering*, 2007.

[13] Chang D Y. Applications of the extent analysis method on fuzzy AHP[J]. *European Journal of Operational Research*, 1996, 95(3):649-655.

[14] Henderson R D, Dutta S P. Use of the analytic hierarchy process in ergonomic analysis[J]. *International Journal of Industrial Ergonomics*, 1992, 9(4):275-282.

[15] ANDRIS FREIVALDS. Comparison of United States (NIOSH Lifting Guidelines) and European (ECSC Force Limits) Recommendations for Manual Work Limits[J]. *AIHAJ - American Industrial Hygiene Association*, 1987, 48(8):698-702.

Numerical Simulation of cracked orthotropic materials using extended isogeometric analysis

S H Habib[1], I Belaidi[1], S Khatir[2] and M Abdel Wahab[2]

[1] Department of Mechanical Engineering, University of M'hamed Bougara, Boumerdes, Algeria

[2] Department of Electrical Energy, Systems and Automation, Faculty of Engineering and Architecture, Ghent University, Belgium

Email: magd.abdelwahab@ugent.be

Abstract. In the present study, extended isogeometric analysis (XIGA) is used to analyse cracks in orthotropic media. NURBS and T-splines geometric technologies are used to define the geometry and the solution. Knot insertion and order elevation are used in NURBS models, while a new local refinement algorithm is applied to T-spline models. In XIGA, the basic idea of the extended finite element method (X-FEM) is used along with isogeometric analysis for modelling discontinuities by including enrichment functions. Special orthotropic crack tip enrichments are used to reproduce the singular fields near a crack tip, and fracture properties of the models are defined by the mixed mode stress intensity factors (SIFs), which are obtained by means of the interaction integral (M-integral). Results of the proposed method are compared with other available results.

1. Introduction

Composite materials are not immune to manufacture defects such as voids, inclusions and cracks. The crack problems greatly influence in the macroscopic response of composites materials, and this lead to the need for better understanding and analyzing the behavior of these materials under applied loads especially in critical conditions.

Analytical studies based upon the laws of fracture mechanics of composites are available [1-5]. However, the complexity of the analytical solutions even for simple cases requires modelling the mechanical behaviour of this problem using effective numerical methods. Among them we distinguish, boundary element method (BEM) [6], finite element method (FEM) [7-14], which has wide range of applications [8, 11, 15-21], element free Galerkin method (EFGM) [22], extended finite element method (XFEM) [23] and other methods [24, 25]. A few years ago, isogeometric analysis was introduced by Hughes et *al*. [26] as an alternative to other methods. This method is based on the idea of using the same functions adopted in Computer Aided Design (CAD) not only to describe the domain geometry, but also to build the numerical approximation of the problem solution [27-35]. It has been developed and applied in many fields, such as: structural vibrations [33-40], with applications to structural health monitoring (SHM) [41-46], contact problems [47], piezoelectric [33, 34, 38] materials [48], blood flow [49], electromagnetics [50], shape optimization [51] and fracture mechanics [52]. However in fracture mechanics problems, IGA faces certain difficulties that are already exist in finite element method (FEM). This makes IGA a suitable candidate to be extended in the same way that FEM improved.

Recently, IGA was enhanced using the partition unity enrichment, but with local enrichment functions in order to capture the discontinuous without any explicit meshing, this method is known as extended isogeometric analysis (XIGA). For fracture mechanics, XIGA is characterized by the accuracy and the high convergence rate in such problems, as shown in [53]. Many investigations were performed using XIGA, such as in fatigue crack growth [54], thin shell analysis, Helmholtz problems, piezoelectric materials and composite materials [55-61].

The most widely geometric function used in CAD is the NURBS, due to its simplicity and its efficiency in geometrical representations, especially all conical sections, which can be represented exactly, such as, circles, cylinders, spheres and ellipsoids. NURBS basis functions have very useful mathematical properties, like partition of unity, compact support, and high degree of continuity. Also, NURBS are characterized by the multitude of refinement methods (*h*-refinement, *p*-refinement, *k*-refinement). In addition, there are many effective algorithms that can be programmed numerically to generate NURBS objects. As in CAD, NURBS was also used for IGA in most researches, because their mathematical properties are always compatible with analysis conditions regardless of the used geometric shape. Despite all the advantages mentioned there are some disadvantages observed during the analysis, which cannot be avoided even by using multiple patches, where NURBS generates a complicated mesh, which lead to produce superfluous control points.

Sederberg *et al.* [62] proposed T-splines as more generalized tools than NURBS in order to handle their disadvantages. T-splines are characterized by their robustness in the generation of adjacent patches, and also by the local refinement property of the mesh even for a single patch. In addition, T-splines provide several algorithms that can be used to refine geometries locally. Recently, T-splines were introduced as additional modules in some CAD software, like Maya and Rhino. As part of the IGA, T-splines are considered as applicable analytical tools that have proven to be efficient in several studies. From the analysis point of view, T-splines allow to use a minimum number of degrees of freedom compared to the number that can be used in the NURBS, and this leads to simplify the computation. In some domains, T-splines are considered as a better tool for IGA, like in the fracture mechanics, contact mechanics and mechanics of bi-materials, generally in any domain where the property of local refinement is necessary and the geometry is complex. However T-spline bases are not always valid to be used in analysis for different geometric configurations, because the linear independent and the partition unity properties are not always ensured. Li et al. [63] introduced analysis-suitable T-spline, where for any choice of knot vectors the blending functions are linearly independent. Like NURBS bases, analysis-suitable T-spline bases have the properties of the analysis basis functions. Moreover, they provide an efficient algorithm, which allows making highly localized refinement [64].

In this paper, cracks in orthotropic media are analysed using XIGA, whereas NURBS and T-splines are used for modelling the crack and construct the geometry. Orthotropic crack tip enrichment functions [65-67] are implemented to accurately calculate the displacement and stress fields, and mixed mode stress intensity factors (SIFs) are numerically evaluated using the interaction integral (M-integral) [68] to determine fracture properties of the domain. An edge and central crack are investigated by considering the effect of fibre orientation.

2. Fracture mechanics for 2D orthotropic materials

Consider a cracked elastic homogeneous orthotropic body Ω subjected to traction forces f_t applied at Γ_t and displacement conditions applied at Γ_u in the absence of body forces, as shown in Figure 1. The partial differential equation of the stress functions of this problem can be obtained using equilibrium and compatibility conditions [69]:

$$c_{22}\frac{\partial^4\varphi}{\partial x^4}-2c_{26}\frac{\partial^4\varphi}{\partial x^3\partial y}+\left(2c_{12}+c_{66}\right)\frac{\partial^4\varphi}{\partial x^2\partial y^2}-2c_{16}\frac{\partial^4\varphi}{\partial x\,\partial y^3}+c_{11}\frac{\partial^4\varphi}{\partial y^4}=0 \tag{1}$$

where (x,y) is local Cartesian coordinates, φ is the stress function and c_{ij} is the components of the compliance matrix, which is computed in terms of Young moduli (E_1, E_2), Poisson ratios (ν_{12},ν_{13} and ν_{23}) and shear modulus (μ_{12}, μ_{13} and μ_{23}) as follows:

IOP Conf. Series: Journal of Physics: Conf. Series **842** (2017) 012061 doi:10.1088/1742-6596/842/1/012061

$$c=\begin{bmatrix} 1/E_1 & -\nu_{12}/E_1 & -\nu_{13}/E_1 & 0 & 0 & 0 \\ -\nu_{12}/E_1 & 1/E_2 & -\nu_{23}/E_2 & 0 & 0 & 0 \\ -\nu_{13}/E_1 & -\nu_{23}/E_2 & 1/E_3 & 0 & 0 & 0 \\ 0 & 0 & 0 & 1/\mu_{23} & 0 & 0 \\ 0 & 0 & 0 & 0 & 1/\mu_{13} & 0 \\ 0 & 0 & 0 & 0 & 0 & 1/\mu_{12} \end{bmatrix} \tag{2}$$

The general characteristic equation of the partial differential equation is:

$$c_{11}s^4-2c_{16}s^3+\left(2c_{12}+c_{66}\right)s^2-2c_{26}s+c_{22}=0 \tag{3}$$

The four roots of this characteristic equation are complex and conjugated two-two (s_1, $\bar{s}_1$ and s_2, $\bar{s}_2$). These roots were used by [1] to derive the displacement and the stress fields in the vicinity of the crack tip for an infinite domain. The stress components for pure mode I are:

$$\begin{aligned}
\sigma_{xx}^{I}&=\frac{K_I}{\sqrt{2\pi r}}Re\left[\frac{s_1s_2}{s_1-s_2}\left(\frac{s_2}{(cos\theta+s_2sin\theta)^{1/2}}-\frac{s_1}{(cos\theta+s_1sin\theta)^{1/2}}\right)\right]\\
\sigma_{yy}^{I}&=\frac{K_I}{\sqrt{2\pi r}}Re\left[\frac{1}{s_1-s_2}\left(\frac{s_1}{(cos\theta+s_2sin\theta)^{1/2}}-\frac{s_2}{(cos\theta+s_1sin\theta)^{1/2}}\right)\right]\\
\sigma_{xy}^{I}&=\frac{K_I}{\sqrt{2\pi r}}Re\left[\frac{s_1s_2}{s_1-s_2}\left(\frac{1}{(cos\theta+s_1sin\theta)^{1/2}}-\frac{1}{(cos\theta+s_2sin\theta)^{1/2}}\right)\right]
\end{aligned} \tag{4}$$

and for pure mode-II:

$$\begin{aligned}
\sigma_{xx}^{II}&=\frac{K_{II}}{\sqrt{2\pi r}}Re\left[\frac{1}{s_1-s_2}\left(\frac{s_2^2}{(cos\theta+s_2sin\theta)^{1/2}}-\frac{s_1^2}{(cos\theta+s_1sin\theta)^{1/2}}\right)\right]\\
\sigma_{yy}^{II}&=\frac{K_{II}}{\sqrt{2\pi r}}Re\left[\frac{1}{s_1-s_2}\left(\frac{1}{(cos\theta+s_2sin\theta)^{1/2}}-\frac{1}{(cos\theta+s_1sin\theta)^{1/2}}\right)\right]\\
\sigma_{xy}^{II}&=\frac{K_{II}}{\sqrt{2\pi r}}Re\left[\frac{1}{s_1-s_2}\left(\frac{s_1}{(cos\theta+s_1sin\theta)^{1/2}}-\frac{s_2}{(cos\theta+s_2sin\theta)^{1/2}}\right)\right]
\end{aligned} \tag{5}$$

The displacements for pure mode-I are

$$\begin{aligned}
u_x^I&=K_I\sqrt{\frac{2r}{\pi}}Re\left\{\frac{1}{s_1-s_2}\left[s_1d_2(cos\theta+s_2sin\theta)^{1/2}-s_2d_1(cos\theta+s_1sin\theta)^{1/2}\right]\right\}\\
u_y^I&=K_I\sqrt{\frac{2r}{\pi}}Re\left\{\frac{1}{s_1-s_2}\left[s_1e_2(cos\theta+s_2sin\theta)^{1/2}-s_2e_1(cos\theta+s_1sin\theta)^{1/2}\right]\right\}
\end{aligned} \tag{6}$$

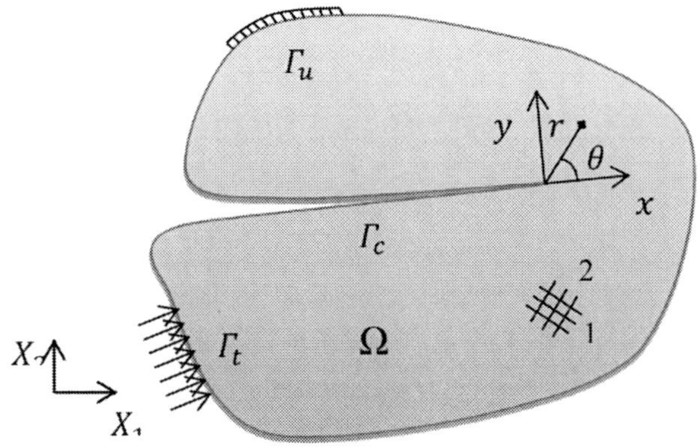

Figure 1. An arbitrary orthotropic cracked body with boundary conditions.

and for pure mode-II:

$$u_x^{II} = K_{II}\sqrt{\frac{2r}{\pi}} Re\left\{\frac{1}{s_1 - s_2}\left[d_2\left(cos\theta + s_2 sin\theta\right)^{1/2} - d_1\left(cos\theta + s_1 sin\theta\right)^{1/2}\right]\right\}$$
$$u_y^{II} = K_{II}\sqrt{\frac{2r}{\pi}} Re\left\{\frac{1}{s_1 - s_2}\left[e_2\left(cos\theta + s_2 sin\theta\right)^{1/2} - e_1\left(cos\theta + s_1 sin\theta\right)^{1/2}\right]\right\} \quad (7)$$

where *Re* represents the real part of a complex number, K_I and K_{II} are the stress intensity factors for modes I and mode II, respectively, and the constant values d_i and e_i are computed as:

$$d_i = c_{11}s_i^2 + c_{12} - c_{16}s_i \quad i = 1,2 \quad (8)$$

$$e_i = c_{12}s_i + \frac{c_{22}}{s_i} - c_{26} \quad i = 1,2 \quad (9)$$

The relation between energy release rate G and stress intensity factors was expressed by Sih et al. [1] for homogeneous composites as follows:

$$G_I = -\frac{\pi}{2}K_I c_{22}\,\mathrm{Im}\left[\frac{K_I(s_1 + s_2) + K_{II}}{s_1 s_2}\right] \quad (10)$$

$$G_{II} = -\frac{\pi}{2}K_{II}c_{11}\,\mathrm{Im}\left[K_I(s_1 + s_2) + K_{II}s_1 s_2\right] \quad (11)$$

3. Non uniform rational B-splines (NURBS)

NURBS entities are constructed by linear combination of NURBS basis functions and their control points. NURBS bases are defined by one knot vector in each parametric direction. Univariate NURBS basis functions R_i of order p are defined by:

$$R_{i,p}(\xi) = \frac{N_{i,p}(\xi)w_i}{\sum_{\hat{i}=1}^{n} N_{\hat{i},p}(\xi)w_{\hat{i}}} \quad (12)$$

where n is the number of basis functions, w is a set of positive weights, ξ is a parametric coordinate of the first direction and N_i are the B-spline basis functions corresponding to the knot vector $\Xi=\{\xi_1,\xi_2,\ldots\ldots,\xi_{n+p+1}\}$, they are given by:

$$N_{i,0}(\xi) = \begin{cases} 1 & \text{if } \xi_i \le \xi < \xi_{n+p+1} \\ 0 & \text{otherwise} \end{cases}, \quad \text{for } p = 0 \quad (13)$$

and

$$N_{i,p}(\xi)=\frac{\xi-\xi_i}{\xi_{i+p}-\xi_i}N_{i,p-1}(\xi)+\frac{\xi_{i+p+1}-\xi}{\xi_{i+p+1}-\xi_{i+1}}N_{i+1,p-1}(\xi);\quad \text{for } p>0 \tag{14}$$

For the two parametric directions ξ and η, bivariate NURBS basis functions are given by:

$$R_{i,j}^{p,q}(\xi,\eta)=\frac{N_{i,p}(\xi)M_{j,q}(\eta)w}{\sum_{\hat{i}=1}^{n}\sum_{\hat{j}=1}^{m}N_{\hat{i},p}(\xi)M_{\hat{j},q}(\eta)w} \tag{15}$$

where M_j are B-spline basis functions of order q corresponding to the second knot vector $\mathrm{H}=\{\eta_1,\eta_2,\ldots\ldots,\eta_{m+q+1}\}$ of which m is the number of basis functions.

4. T-splines

In order to construct a T-spline surface, the T-mesh, which is a mesh of rectangular elements defined by the lines corresponding to knot values, as shown in Figure 2, must be defined according to the basis function orders p and q to extract the local knot vectors Ξ_α and H_α, and compute the blending functions $\mathbf{R}_\alpha$ for each anchor. Then, in a similar way as for NURBS, we can define the T-spline surface by:

$$S=\sum_{\alpha=1}^{k}\mathbf{R}_\alpha \mathrm{P}_\alpha \tag{16}$$

where the blending functions given by:

$$\mathbf{R}_\alpha(\xi,\eta)=\frac{w_\alpha \mathbf{B}_\alpha(\xi,\eta)}{\sum_{\beta=1}^{k}w_\beta \mathbf{B}_\beta(\xi,\eta)} \tag{17}$$

and P is the control points, α is the anchor index, k is the number of the anchors in T-mesh and $\mathbf{B}$ is the bivariate local basis function, which is given by:

$$\mathbf{B}_\alpha=N_{i,p}(\xi)M_{j,q}(\eta) \tag{18}$$

The refinement in T-splines is based on the subdivision of basis functions by insert knots to local knot vector. For one inserted knot z, there are two produced basis functions that are combined linearly to form the original basis function as:

$$N(\xi|\xi_1,\ldots,\xi_5)=XN(\xi|\xi_1,\ldots,\xi_4)+YN(\xi|\xi_2,\ldots,\xi_5) \tag{19}$$

where

$$X=\begin{cases}\dfrac{z-\xi_1}{\xi_4-\xi_1} & \text{for } z<\xi_4\\ 1 & \text{for } z\geq\xi_4\end{cases} \tag{20}$$

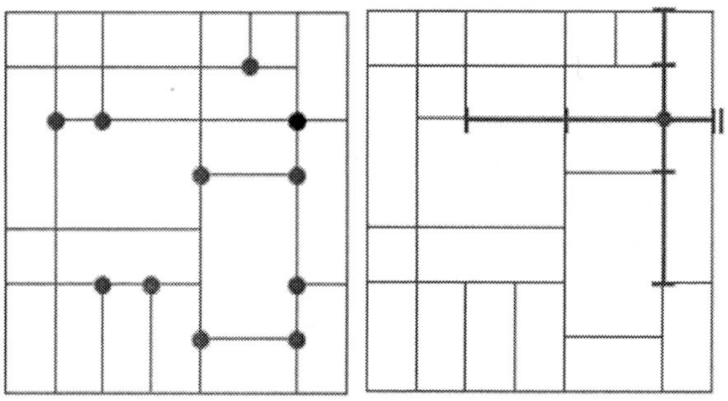

(a) (b)

Figure 2. Shows a T-mesh with: (a) T-junction points (red points) and (b) local support of black point.

$$Y = \begin{cases} \dfrac{\xi_5 - z}{\xi_5 - \xi_2} & \text{for } z > \xi_2 \\ 1 & \text{for } z \leq \xi_2 \end{cases} \tag{21}$$

For more than one inserted knot, the application of these equations is repeated until all coefficients are derived. In refined T-spline space, the relation between the original basis functions $\mathbf{B}^1$ and the generated basis functions $\mathbf{B}^2$ can be expressed in linear system form using the refinement operator $\mathbf{M}$

$$\mathbf{N}_1 = \mathbf{M}\mathbf{N}_2 \tag{22}$$

where $\mathbf{N}_1$ and $\mathbf{N}_2$ are the column vectors of blending functions of the original and the refined T-spline spaces, respectively.

5. Extended isogeometric analysis (XIGA)

In XIGA, the displacement approximation for cracks at a particular point $\zeta=(\xi,\eta)$, can be written in generalized form by extending the IGA approximation:

$$\mathrm{u}(\zeta) = \sum_{i=1}^{n_s} Q_i(\zeta)\,\mathrm{u}_i + \sum_{j=1}^{n_{cf}} Q_j(\zeta) H(\zeta)\,\mathrm{a}_j + \sum_{k=1}^{n_{ct}} Q_k(\zeta)\left(\sum_{\ell=1}^{4} F_\ell(\zeta)\,\mathrm{b}_k^\ell\right) \tag{23}$$

where $\boldsymbol{Q}$ is the T-spline basis function (**R**), Eq. 17 or NURBS basis function (R) Eq. 15. H is the Heaviside function used for modelling the crack face, it takes the value 1 above the crack and -1 below the crack, F are the crack-tip enrichment functions derived from the analytical solution of the displacement field around the crack tip, u_i, a_j and b_k are the displacement vectors correspond to n_s, n_{cf} and n_{ct} control points, respectively.

The crack-tip enrichment functions are obtained from the analytical solution of the displacement field in the vicinity of the crack-tip as [66]

$$\{F_\ell(r,\theta)\}_{\ell=1}^{4} = \begin{Bmatrix} \sqrt{r}\cos\left(\dfrac{\theta_1}{2}\right)\sqrt{g_1(\theta)}, \sqrt{r}\cos\left(\dfrac{\theta_2}{2}\right)\sqrt{g_2(\theta)} \\ \sqrt{r}\sin\left(\dfrac{\theta_1}{2}\right)\sqrt{g_1(\theta)}, \sqrt{r}\sin\left(\dfrac{\theta_2}{2}\right)\sqrt{g_2(\theta)} \end{Bmatrix} \tag{24}$$

where $g_k(\theta)$ and θ_k are given by

$$g_k(\theta) = \sqrt{(\cos\theta + s_{kx})^2 + (s_{ky}\sin\theta)^2}; \quad \theta_k = \arctan\left(\frac{s_{ky}\sin\theta}{\cos\theta + s_{kx}\sin\theta}\right); \quad k = 1,2 \tag{25}$$

6. Numerical simulations

In this section, a circular orthotropic plate with a central crack and a rectangular orthotropic plate with an edge crack are simulated in the plane stress state. The corresponding boundary conditions are shown in Figure 3. First, in the circular plate (E_1=0.1 GPa, E_2=1 GPa, μ_{12}=0.5 GPa, v_{12}=0.03) different inclined cracks are examined, where the orthotropic axes (1,2) coincide with the Cartesian axes (x,y). The problem is analysed using 793 control points and 688 elements (Figure 4). Variation of mode I and II SIF is presented in Figure 6. The obtained values in the case β=30° are compared with other values taken from the literature, as shown in Table 1. In the second part, several orientations of material elastic axes namely, γ=0°, 30°, 45°, 60° and 90° are investigated for the rectangular plate, which is composed of graphite-epoxy material (E_1=114.8 GPa, E_2=11.7 GPa, μ_{12}=9.66 GPa, v_{12}=0.21). The T-spline mesh (1372 DOF) is constructed using 566 control points and 431 elements (Figure 5), and the NURBS mesh (3248 DOF) is constructed using 1624 control points and 1375 elements (Figure 5). The obtained normalized stress intensity factors ($\bar{K} = K / (\sigma\sqrt{\pi a})$) are compared with those of the XFEM, as depicted in Figure 7.

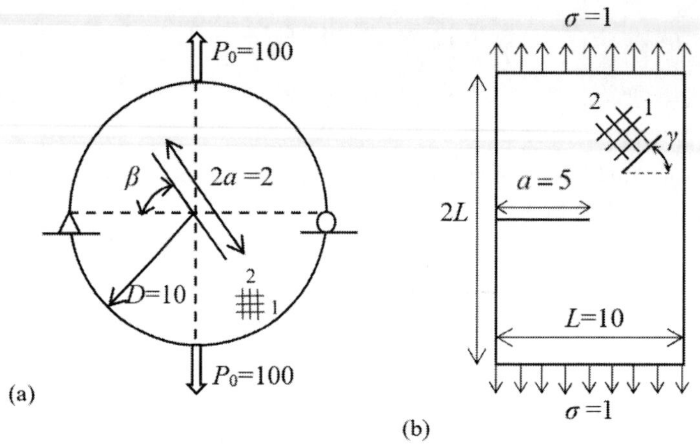

Figure 3. Geometries and boundary conditions for the homogeneous orthotropic examples: (a) a disk with a center crack and (b) a rectangular plate with an edge crack.

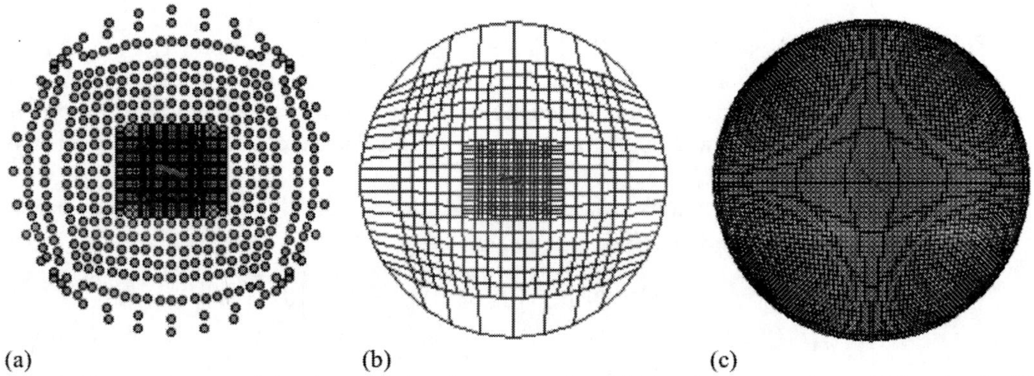

Figure 4. The meshes used for the circular plate example: (a) T-spline control net (793 points), (b) T-spline mesh (688 elements) et (c) NURBS control net (4489 points).

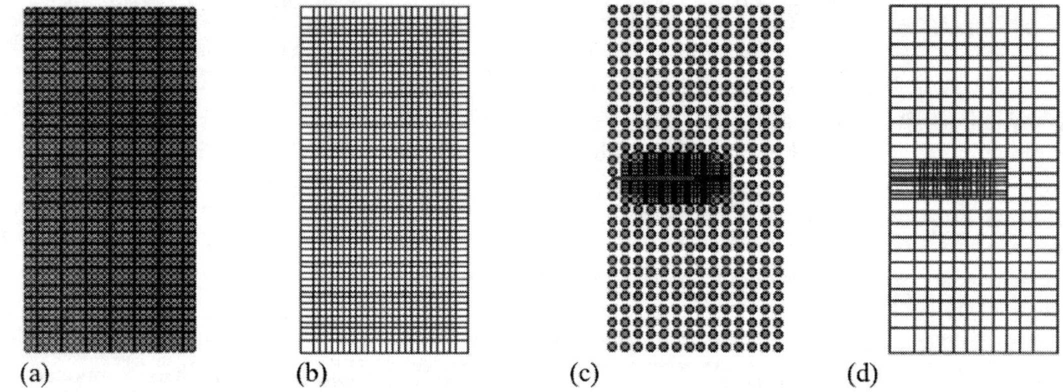

Figure 5. The used meshes of rectangular plate: (a) 1624 control points and (b) 1375 elements, for NURBS representation, (c) 566 control points and (d) 431 elements, for T-spline representation.

In Figure 6 and Table 1 the obtained SIFs are in good agreement with the other methods. T-splines provide a minimal number of nodes due to their ability in describing exactly such geometries. Also, the results of the edge crack problem (Figure 7) in both modes agree well for minimal number of degrees of freedom compared to XFEM and NURBS based XIGA. The mode I SIF changes their trend in 45° and the mode II SIF changes their trend in 30°, so in orthotropic materials the stress intensity factors are not only depend on the crack angle, but they also depend on the direction of elasticity axes. For more details about the results that obtained by other numerical methods for this problem, the reader may refer to [23,69].

Table 1. Comparison of mixed mode stress intensity factors for a central inclined crack (β=30°) in orthotropic disk.

Method	Number of nodes	Number of elements	Number of Degrees of freedom	K_I	K_{II}
MCC [7]	2712	999	5424	16.73	11.33
M-integral [69]	2712	999	5424	16.75	11.38
XEFG [23]	877	-	1507	16.80	11.79
NURBS based XIGA	4489	4096	9266	16.70	11.38
T-splines based XIGA	793	688	1890	16.86	11.78

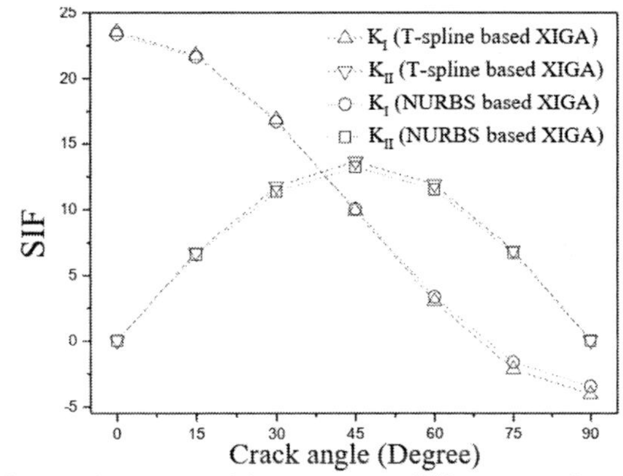

Figure 6. Variation of mode I and II SIFs with respect to different crack angles using analysis-suitable T-splines and NURBS for the circular plate.

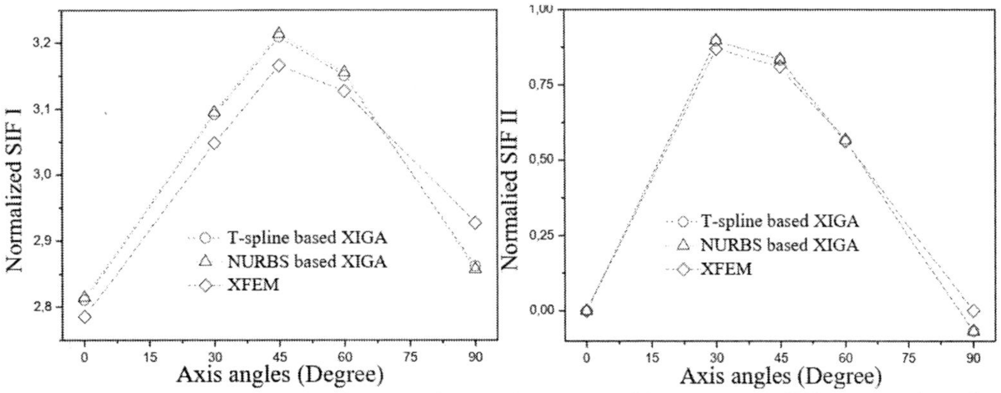

Figure 7. Variations of normalized mode I and II SIFs with respect to different angles of orthotropic axes using analysis-suitable T-splines, NURBS and XFEM for the rectangular plate.

7. Conclusion

Static fracture behaviour for a crack in orthotropic materials is analysed numerically in this study using the Extended Isogeometric Analysis (XIGA). NURBS and T-splines bases are used to define the exact geometry and approximate the solution throughout the analysis. Orthotropic crack-tip enrichment functions, which can be applied to all types of orthotropic materials, are implemented. SIFs for both mode I and II are evaluated using the M-integral to determine fracture properties of the domain. The quality of the obtained results regarding their good agreement with the results of literature demonstrates the accuracy of the present approach and the robustness of the developed code. The inclination angle of the orthotropic axes generates and affects the evolution of mixed mode SIF even if the crack is only subjected to mode I loading, therefore the mode II must also be evaluated.

References

[1] Sih GC, Paris P and Irwin GR 1965 *International Journal of Fracture Mechanics* **1**(3) 189-203
[2] Bowie O and Freese C 1972 *International Journal of Fracture Mechanics* **8**(1) 49-57
[3] Kuo M and Bogy D 1974 *Journal of Applied Mechanics* **41**(1) 197-202
[4] Viola E, Piva A and Radi E 1989 *Engineering Fracture Mechanics* **34**(5) 1155-1174
[5] Hoenig A 1982 *Engineering Fracture Mechanics* **16**(3) 393-403
[6] Aliabadi M and Sollero P 1998 *Composites Science and Technology* **58**(10) 1697-1703
[7] Kim J-H and Paulino GH 2002 *Engineering Fracture Mechanics* **69**(14) 1557-1586
[8] Yue T and Abdel Wahab M 2014 *Wear* **321** 53-63
[9] Wang C, Sun Q, Abdel Wahab M, Zhang X and Xu L 2015 *Waste Management* **43** 19-27
[10] Yue T and Abdel Wahab M 2016 *Materials* **9**(7)
[11] Ferjaoui A, Yue T, Abdel Wahab M and Hojjati-Talemi R 2015 *International Journal of Fatigue* **73** 66-76
[12] Pereira K, Yue T and Abdel Wahab M 2017 *Tribology International* **110** 222-231
[13] Kumar D, Biswas R, Poh LH and Abdel Wahab M 2017 *Tribology International* **109** 124-132
[14] Bhatti NA and Abdel Wahab M 2017 *Tribology International* **109** 552-562
[15] Yue T and Abdel Wahab M 2016 *Materials* **9** 597; doi:510.3390/ma9070597
[16] Vanegas-Useche L, Abdel Wahab M and Parker G 2015 *Waste Management* **43** 28-36
[17] Resende Pereira KdF, Bordas S, Tomar S, Trobec R, Depolli M, Kosec G and Abdel Wahab M 2016 *Materials* **9** 639; doi:610.3390/ma9080639
[18] Noda N-A, Chen X, Sano Y, Wahab MA, Maruyama H, Fujisawa R and Takase Y 2016 *MATERIALS & DESIGN* **96** 476-489
[19] Junyan Ni and Wahab MA 2017 *Computers & Structures* **186** 35–49
[20] Yue T and Abdel Wahab M 2017 *Tribology International* **107** 274-282
[21] Martínez JC, Vanegas Useche LV and Wahab MA 2017 *International Journal of Fatigue* **100, Part 1** 32-49
[22] Ghorashi SS, Mohammadi S and Sabbagh-Yazdi S-R 2011 *Engineering Fracture Mechanics* **78**(9) 1906-1927
[23] Asadpoure A, Mohammadi S and Vafai A 2006 *Thin-Walled Structures* **44**(9) 1031-1038
[24] Asadpoure A, Mohammadi S and Vafai A 2006 *Finite Elements in Analysis and Design* **42**(13) 1165-1175
[25] Atluri SN, Kobayashi AS and Nakagaki M 1975 *International Journal of Fracture* **11**(2) 257-271
[26] Hughes TJ, Cottrell JA and Bazilevs Y 2005 *Computer Methods in Applied Mechanics and Engineering* **194**(39) 4135-4195
[27] X. Nguyen H, N. Nguyen T, Abdel Wahab M, Bordas SPA, Nguyen-Xuan H and P. Voa T 2017 *Computer Methods in Applied Mechanics and Engineering* **313** 904-940
[28] Tran Vinh L, Lee J, Nguyen-Van H, Nguyen-Xuan H and Abdel Wahab M 2015 *International Journal of Non-Linear Mechanics* **72** 42-52

[29] Tran LV, Phung-Van P, Lee J, Wahab MA and Nguyen-Xuan H 2016 *Composite Structures* **140** 655-667

[30] Thai CH, Ferreira AJM, Abdel Wahab M and Nguyen-Xuan H 2016 *Acta Mechanica* **227**(5) 1225-1250

[31] Thai C, Zenkour AM, Abdel Wahab M and Nguyen-Xuan H 2016 *Composite Structures* **139** 77-95

[32] Phung-Van P, Tran LV, Ferreira AJM, Nguyen-Xuan H and Abdel-Wahab M 2016 *Nonlinear Dynamics* 1-16; doi:10.1007/s11071-11016-13085-11076

[33] Phung Van P, Nguyen LB, Tran Vinh L, Dinh TD, Thai CH, Bordas SPA, Abdel Wahab M and Nguyen-Xuan H 2015 *International Journal of Non-Linear Mechanics* **76** 190-202

[34] Phung Van P, De Lorenzis L, Thai CH, Abdel Wahab M and Nguyen-Xuan H 2015 *Computational Materials Science* **96** 495-505

[35] Phung Van P, Abdel Wahab M, Liew KM, Bordas SPA and Nguyen-Xuan H 2015 *Composite Structures* **123** 137-149

[36] Cottrell JA, Reali A, Bazilevs Y and Hughes TJ 2006 *Computer Methods in Applied Mechanics and Engineering* **195**(41) 5257-5296

[37] Tran Vinh L, Lee J, Ly HA, Abdel Wahab M and Nguyen-Xuan H 2015 *International Journal of Mechanical Sciences* **96-97** 65-78

[38] Phung Van P, Tran Vinh L, Ferreira AJM, Nguyen-Xuan H and Abdel Wahab M 2016 *Nonlinear Dynamics* (http://dx.doi.org/10.1007/s11071-016-3085-6)

[39] Tran Vinh L, Phung Van P, Lee J, Abdel Wahab M and Nguyen-Xuan H 2016 *Composite Structures* **140** 655-667

[40] Thai C, Ferreira AJM, Abdel Wahab M and Nguyen-Xuan H 2016 *Acta Mech* **227**(5) 1225-1250

[41] Zhou Y-L, Maia N and Abdel Wahab M 2016 *Journal of Vibration and Control* doi: 10.1177/1077546316674544

[42] Zhou Y-L and Abdel Wahab M 2016 *Journal of Vibroengineering* **18**(7) 4491-4499

[43] Khatir S, Belaidi I, Serra R, Abdel Wahab M and Khatir T 2016 *Journal of Vibroengineering* **18**(1) 202-213

[44] Khatir S, Belaidi I, Serra R, Abdel Wahab M and Khatir T 2015 *Mechanika* **21**(6) 472-479

[45] Gillich G-R, Praisach Z-I, Abdel Wahab M, Gillich N, Mituletu IC and Nitescu C 2016 *Shock and Vibration* **2016**(Article ID 2086274) 10 pages; http://dx.doi.org/10.1155/2016/2086274

[46] Zhou Y-L, Maia NMM, Sampaio R and Wahab MA 2016 *Structural health monitoring* **DOI: https://doi.org/10.1177/1475921716680849**

[47] Temizer I, Wriggers P and Hughes T 2011 *Computer Methods in Applied Mechanics and Engineering* **200**(9) 1100-1112

[48] Willberg C and Gabbert U 2012 *Acta Mechanica* **223**(8) 1837-1850

[49] Bazilevs Y, Calo V, Zhang Y and Hughes TJ 2006 *Computational Mechanics* **38**(4-5) 310-322

[50] Buffa A, Sangalli G and Vázquez R 2010 *Computer Methods in Applied Mechanics and Engineering* **199**(17) 1143-1152

[51] Li K and Qian X 2011 *Computer-Aided Design* **43**(11) 1427-1437

[52] Irzal F, Remmers JJ, Verhoosel CV and Borst R 2014 *International Journal for Numerical Methods in Engineering* **97**(8) 608-628

[53] De Luycker E, Benson D, Belytschko T, Bazilevs Y and Hsu M 2011 *International Journal for Numerical Methods in Engineering* **87**(6) 541-565

[54] Bhardwaj G and Singh I 2015 *Journal of the Brazilian Society of Mechanical Sciences and Engineering* **37**(4) 1065-1082

[55] Nguyen-Thanh N, Valizadeh N, Nguyen M, Nguyen-Xuan H, Zhuang X, Areias P, Zi G, Bazilevs Y, De Lorenzis L, et al. 2015 *Computer Methods in Applied Mechanics and Engineering* **284** 265-291

[56] Peake MJ, Trevelyan J and Coates G 2013 *Computer Methods in Applied Mechanics and Engineering* **259** 93-102

[57] Bui TQ 2015 *Computer Methods in Applied Mechanics and Engineering* **295** 470-509

[58] Jia Y, Anitescu C, Ghorashi SS and Rabczuk T 2015 *IMA Journal of Applied Mathematics* **80**(3) 608-633

[59] Yu T, Bui TQ, Yin S, Doan DH, Wu C, Van Do T and Tanaka S 2016 *Composite Structures* **136** 684-695

[60] Guo Y, Ruess M and Gürdal Z 2014 *Composite Structures* **116** 55-66

[61] Ghorashi SS, Valizadeh N, Mohammadi S and Rabczuk T 2015 *Computers & Structures* **147** 138-146

[62] Sederberg TW, Zheng J, Bakenov A and Nasri A 2003 ACM transactions on graphics (TOG) ed editors) ACM p. 477-484

[63] Li X, Zheng J, Sederberg TW, Hughes TJ and Scott MA 2012 *Computer Aided Geometric Design* **29**(1) 63-76

[64] Scott M, Li X, Sederberg T and Hughes T 2012 *Computer Methods in Applied Mechanics and Engineering* **213** 206-222

[65] Asadpoure A and Mohammadi S 2007 *International Journal for Numerical Methods in Engineering* **69**(10) 2150-2172

[66] Shojaee S, Asgharzadeh M and Haeri A 2014 *International Journal of Applied Mechanics* **06**(06) 1450068

[67] Shojaee S and Daneshmand A 2015 *Engineering Fracture Mechanics* **147** 203-227

[68] Kim J-H and Paulino GH 2003 *International Journal of Solids and Structures* **40**(15) 3967-4001

[69] Lekhnitskii S 1963 Of an anisotropic elastic body ed editors. San Francisco: Holden-Day

IOP Conf. Series: Journal of Physics: Conf. Series **842** (2017) 012062 doi:10.1088/1742-6596/842/1/012062

Development of four-point bending fatigue test method using continuously hydrogen-charging pipe specimen

T Yoshimoto[1] and T Matsuo[2, 3]
[1]Graduate School of Engineering, Fukuoka University, 8-19-1 Nanakuma, Jonan-ku, Fukuoka 814-0180, Japan
[2]Department of Mechanical Engineering, Fukuoka University, 8-19-1 Nanakuma, Jonan-ku, Fukuoka 814-0180, Japan
[3]Institute of Materials Science and Technology, Fukuoka University
8-19-1 Nanakuma, Jonan-ku, Fukuoka 814-0180, Japan

E-mail: tmatsuo@fukuoka-u.ac.jp

Abstract. To evaluate hydrogen embrittlement, the following two types of testing method are available: (i) testing in high-pressure hydrogen gas environment and (ii) testing in ambient air using hydrogen precharged specimen. Testing in high-pressure hydrogen gas environment is technically difficult and expensive because high-pressure gas equipments, such as high-pressure vessel and pipe, have to be installed in the laboratory. On the other hand, in the case of precharging method, outgassing of hydrogen from the specimen occurs during the test. Therefore, hydrogen embrittlement can hardly be evaluated properly, especially, in long-term testing such as high cycle fatigue test at low frequency. In this study, to effectively evaluate the hydrogen embrittlement in fatigue, an experimental method, which was the four-point bending fatigue test system with a mechanism of internal circulation of hydrogen-charging solution in a pipe specimen, was developed. By using this method, the fatigue crack growth properties in the presence of hydrogen were investigated at frequencies of 0.05 Hz and 1 Hz.

1. Introduction

Hydrogen attracts attention as clean energy which does not emit carbon dioxide and air pollutant and the practical use of hydrogen energy is being proceeded in USA, Europe, Japan and so on. A lot of effort has been put on the development of, in particular, hydrogen fuel cell vehicle (HFCV) and hydrogen station. However, in order to ensure the safety of these hydrogen equipments, hydrogen embrittlement: degradation of strength properties of metallic material due to hydrogen penetration into the material, needs to be clarified.

To evaluate hydrogen embrittlement, the following two types of testing method are commonly used: (i) testing in high-pressure hydrogen gas environment and (ii) testing in ambient air using hydrogen precharged specimen. The behavior of materials used in hydrogen equipments can be simulated by the testing in high-pressure hydrogen gas. In recent research [1, 2], the testing in high-pressure hydrogen gas demonstrated that fatigue crack growth properties were affected by hydrogen pressure, test frequency, test temperature, etc. However, this testing method is technically difficult and expensive because high-pressure gas equipments such as a high-pressure vessel and a compressor must be installed in a laboratory satisfying sufficient safety and in addition, the installation cost is of the order of hundred million yen in Japan. On the other hand, the precharging method is easy to handle and inexpensive, while it has a disadvantage that outgassing of hydrogen from the specimen occurs during the test. Figure 1 shows the time variation of hydrogen outgassing from hydrogen precharged carbon steels. Hydrogen content of precharged carbon steels decreased to almost the same level as that of uncharged material during the first 200 h after hydrogen precharging. Thus, hydrogen embrittlement can hardly be evaluated properly, especially in long-term testing such as high cycle fatigue test at low frequency. Investigation of the hydrogen embrittlement in fatigue is essential for assurance of long-term integrity of hydrogen equipment, and therefore, it is important to establish an easy yet effective evaluation method for hydrogen embrittlement in fatigue.

In this study, to effectively evaluate hydrogen embrittlement, an experimental method was developed, which was the four-point bending fatigue test system with a mechanism of circulation of hydrogen-charging solution in a pipe specimen.

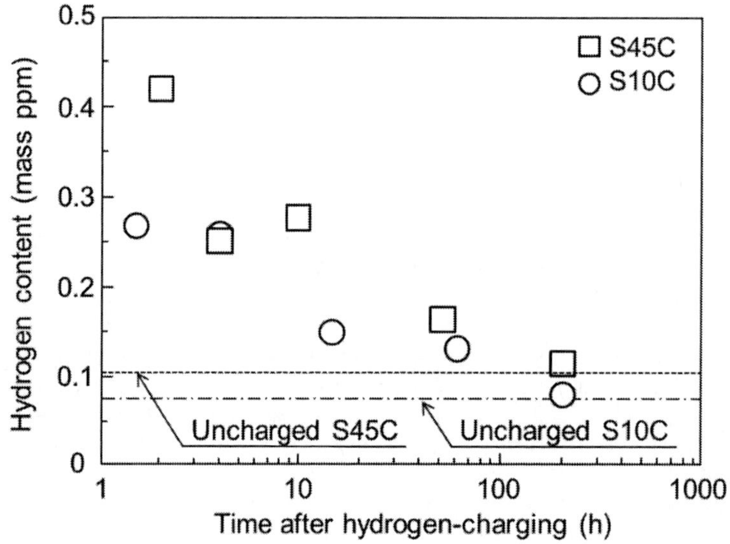

Figure 1. Relationship between hydrogen content and time after hydrogen-charging.

2. Experimental method

2.1 A developed testing setup

Figure 2 shows the concept of testing method developed in this study. The fatigue crack growth test was conducted by the four-point bending fatigue test system with a mechanism of internal circulation of hydrogen-charging solution in a pipe specimen. The hydrogen-charging solution (20 mass% NH_4SCN) was pumped into the pipe specimen through a flexible tube. Two screw holes were introduced on both ends of the specimen, which was connected to a flexible tube by a screw type fitting.

There are the following advantages by employing the four-point bending fatigue test system. The screw holes of both ends of specimen are not subjected to bending moment because the bending moment is zero at the outer side of load points, as indicated by the bending moment diagram shown in Fig. 3. In addition, the through-hole was present on the neutral plane where the bending stress is zero, see Fig. 4. Therefore, the unintentional fracture from the screw holes or the through-hole is avoidable in spite of the direct exposure to hydrogen-charging solution. This specimen is hereafter called the *H-charging specimen*, while the specimen tested without hydrogen-charging is called the *Non-charging specimen*.

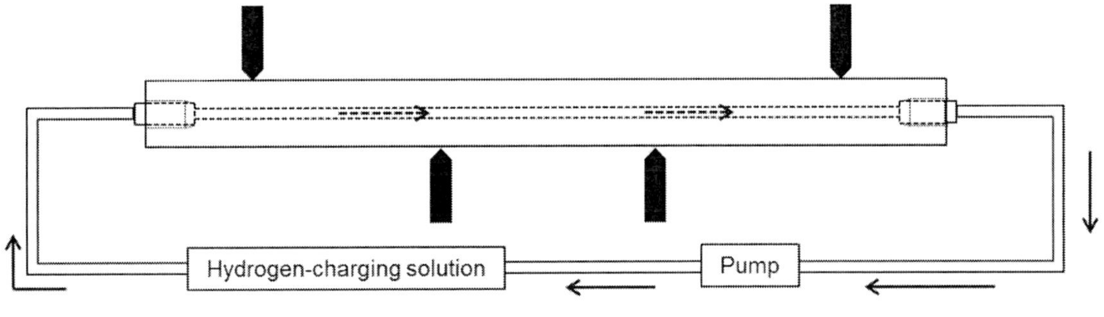

Figure 2. The concept of testing method.
(Four-point bending fatigue test with a mechanism of circulation of hydrogen-charging solution).

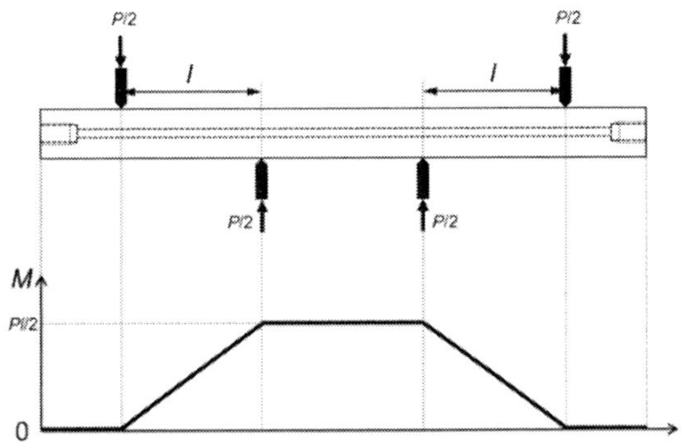

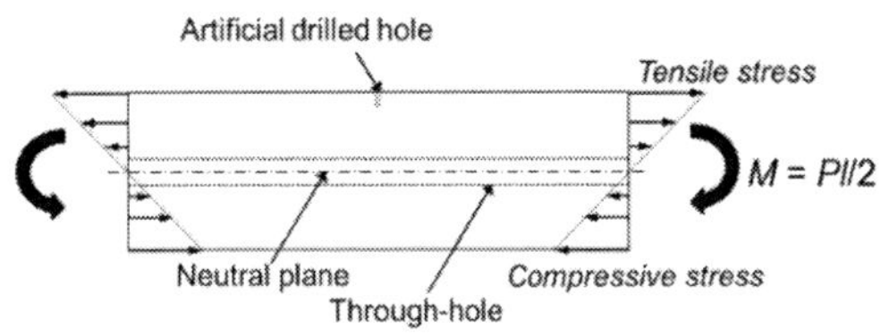

Figure 3. Bending moment diagram (BMD). **Figure 4.** Bending stress distribution.

2.2 Material and specimen

An annealed carbon steel (JIS-S35C) was used in this study. The Vickers hardness of this material is approximately 156. The microstructure is of ferrite-pearlite with an elongated texture in the rolling direction, as shown in Fig. 5.

Figure 6 shows the geometry of pipe specimen. The specimen surface was polished with a #2000 emery paper and subsequently buff-polished using alumina paste with a particle diameter of 1 μm. After polishing, three drilled holes shown in Fig. 7 were introduced on the specimen surface as a crack starter. Then, stress relief annealing in vacuum at 600 ℃ for 1h was conducted.

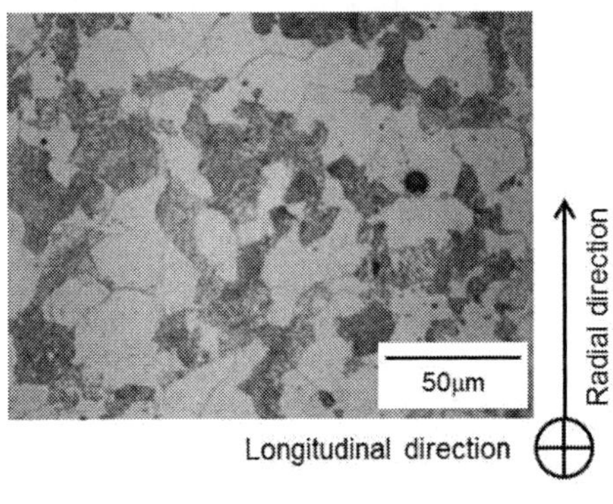

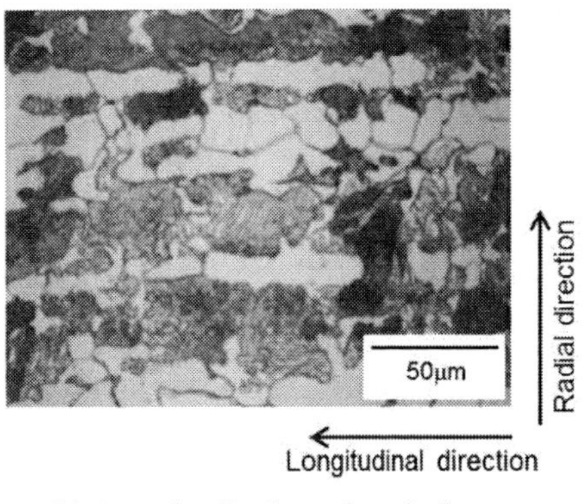

(a) Transverse sectional view (b) Longitudinal sectional view
Figure 5. Microstructures of S35C.

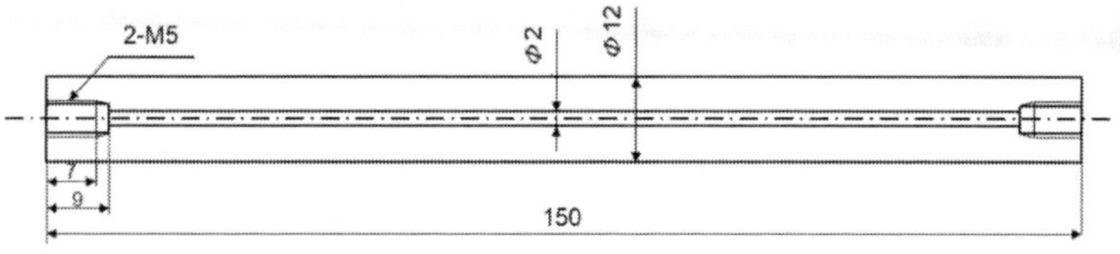

Figure 6. Geometry of pipe specimen (in mm).

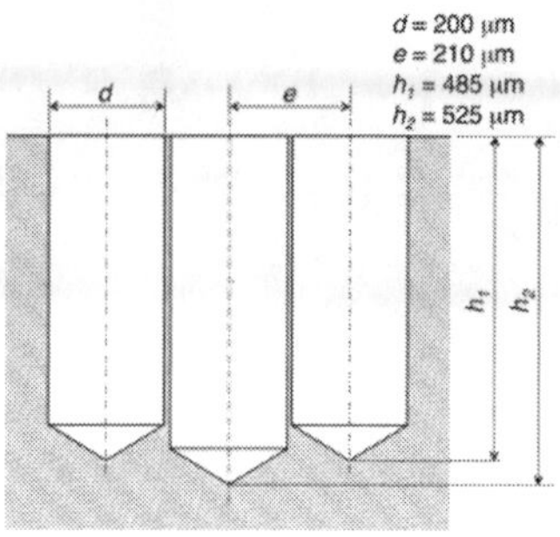

Figure 7. Shape and dimensions of drilled holes.

2.3 Fatigue crack growth test

Fatigue crack growth tests were carried out in the four-point bending fatigue using H-charging specimens in air at room temperature. The stress ratio was 0.1. The test frequency was 0.05 Hz or 1 Hz. The maximum bending stress at specimen surface was controlled to be constant, so that the stress intensity factor (SIF) range increased with crack growth. The crack growth behavior was observed by an optical microscope at the appointed numbers of cycles to measure the crack length. The fracture surface was observed by a scanning electron microscope.

3. Results and discussion

3.1 Effects of continuous hydrogen-charging on the fatigue crack growth rate

The fatigue crack growth test were conducted under the both conditions of continuous hydrogen-charging and of no hydrogen-charging at a stress ratio of $R = 0.1$ and at a test frequency of $f = 1$ Hz. The initial SIF range was about 5.5 MPa·m$^{1/2}$ and the SIF range increased up to about 11 MPa·m$^{1/2}$ with crack growth.

Figure 8 shows the relationship between the crack growth rate, da/dN, and the SIF range, ΔK. At the relatively low ΔK regime ($\Delta K < 7$ MPa·m$^{1/2}$), the crack growth was unstable and the crack growth rate has a large scatter. It appears that there was no noticeable acceleration of crack growth by hydrogen in this low ΔK regime. At the relatively high ΔK regime ($\Delta K > 7$ MPa·m$^{1/2}$), however, the crack growth rate for H-charging specimen was higher than that for Non-charging specimen and the difference in crack growth rate increased with increase in ΔK. As mentioned in the introduction, there are already two conventional methods, in which the test is performed in hydrogen gas or using a hydrogen-precharged specimen. For comparison, in Fig. 8, the present test results are shown together with the conventional test results obtained using CT specimens [10]. Although these results cannot be simply compared owing to the diversity of testing conditions including material, specimen geometry, test frequency, hydrogen content and ΔK-increasing/decreasing loading, no significant difference was found among these three methods. In addition, when the test frequency was dropped from 1 Hz to 0.05 Hz, the crack growth acceleration by hydrogen became more pronounced. At the SIF range of $\Delta K = 10$ MPa·m$^{1/2}$, the acceleration ratio of crack growth rate with hydrogen to that without hydrogen was 4.5 at 0.05 Hz, while it was 2.2 at 1 Hz. This frequency dependence upon hydrogen-induced crack growth acceleration is well known as a feature of hydrogen-related fracture [3-7] and it has been explained by taking the increase in hydrogen accumulation at the crack tip with decrease in test frequency into account.

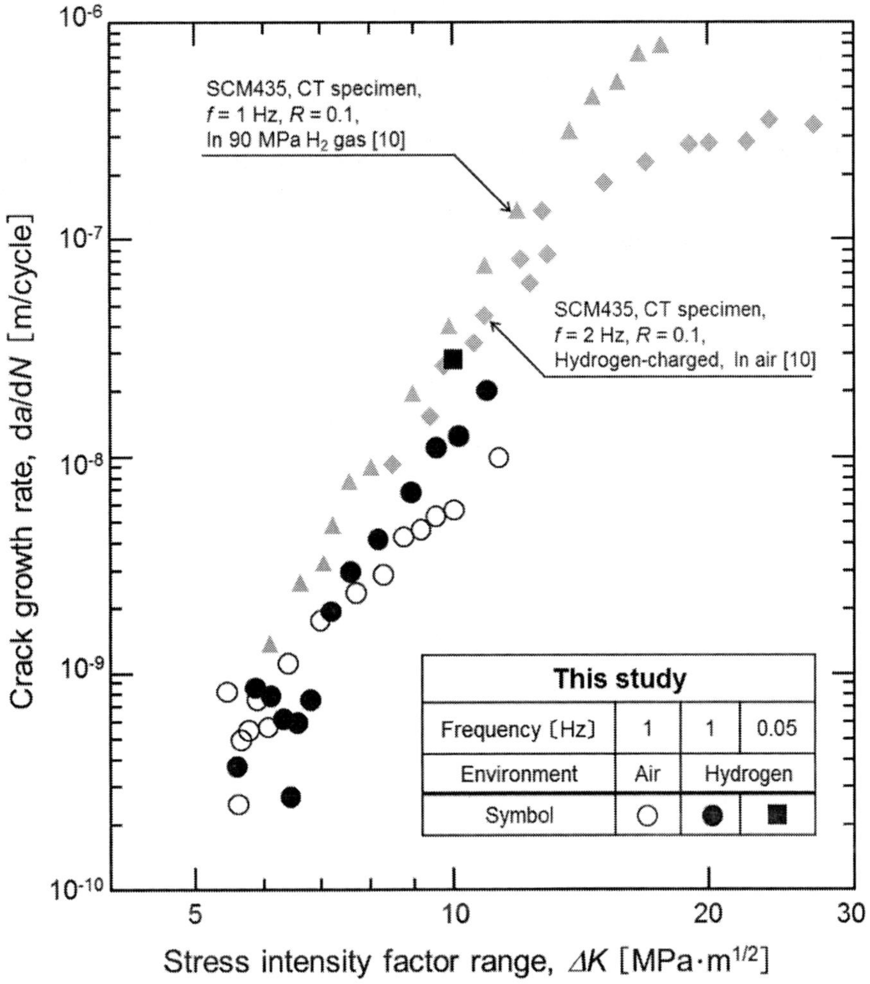

Figure 8. Relationship between crack growth rate and ΔK.

3.2 Fracture surface

Figure 9 (a) shows the fracture surface of Non-charging specimen tested at 1 Hz and Figs. 9 (b) and (c) show the fracture surfaces of H-charging specimens tested at 1 Hz and at 0.05 Hz, respectively. All images were taken at the SIF range of 10 MPa·m$^{1/2}$. At the SIF range of $\Delta K > 7$ MPa·m$^{1/2}$ where the crack growth rate was accelerated by continuous hydrogen-charging, flat fracture facets could be observed partially, as shown in Fig. 9 (b) and (c). These facets are seen to be attributed to intergranular cracking. This facet-type fracture was also observed in conventional testing method in some previous studies [5,7-9]. It is considered that the facet-type fracture induced by continuous hydrogen-charging would account for the acceleration of fatigue crack growth. Focusing on the frequency effect on hydrogen-induced facet formation, there is no marked difference between 1 Hz and 0.05 Hz (Fig. 9 (b) and (c)), though the crack growth acceleration due to hydrogen was more pronounced for $f = 0.05$ Hz.

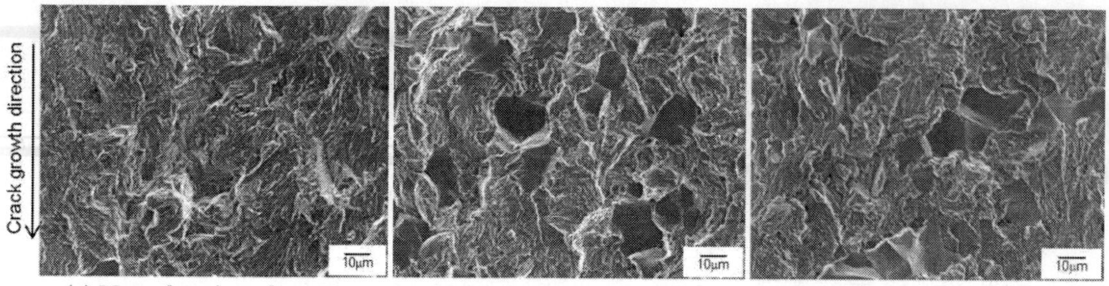

(a) Non-charging, $f = 1$ Hz (b) H-charging, $f = 1$ Hz (c) H-charging, $f = 0.05$ Hz

Figure 9. Fracture surfaces observed at $\Delta K = 10$ MPa·m$^{1/2}$.

3.3 Residual hydrogen content

The residual hydrogen content was measured just after fatigue test using a thermal desorption spectrometer (TDS). The testing sample was a chip cut from the region about 7 mm away from the fracture surface.

The TDS measurement demonstrated that the residual hydrogen content of the chip was about 0.24 mass ppm. Figure 10 shows the hydrogen outgassing properties of the hydrogen-precharged carbon steels (JIS-S10C, S45C) for reference. As seen in this figure, about 200 h after the hydrogen-charging, the hydrogen content of those carbon steels becomes nearly equal to that of uncharged material (nearly equal to or less than 0.1 mass ppm), whereas it is 0.30-0.50 mass ppm just after the hydrogen-charging. The fatigue test conducted in this study spent about 10 days (i.e. 240 h). Nevertheless, much more amount of hydrogen remained in the pipe specimen, compared to precharged carbon steels. It should be noted that a sufficient amount of hydrogen was being supplied into the pipe specimen by the continuous circulation of hydrogen-charging solution over an entire period of testing.

The fatigue test based on the continuous hydrogen-charging demonstrated the influence of hydrogen on the crack growth acceleration, its frequency dependence and the fracture surface morphology. Accordingly, it can be concluded that the developed testing method has an advantage over the conventional precharging method in a long-term fatigue test because of its capability for hydrogen supply over an entire period of testing.

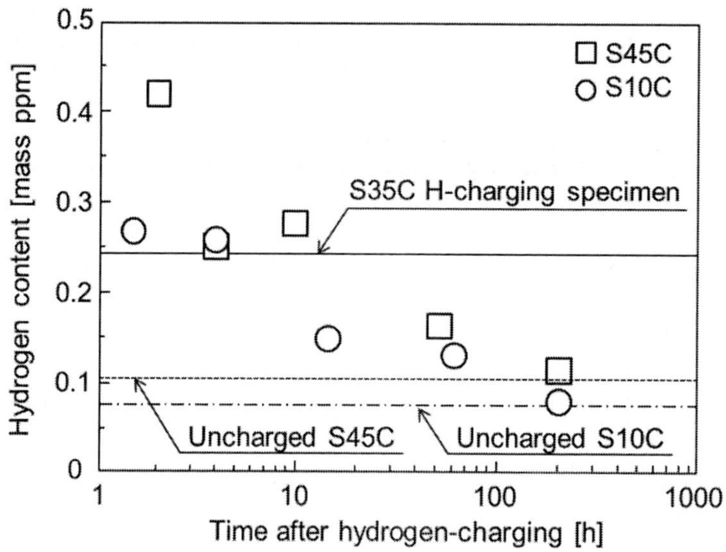

Figure 10. Decrease in hydrogen content in ambient air of hydrogen precharged carbon steel (JIS-S10C, S45C).

4. Conclusions

A fatigue testing method was developed for investigating the effect of hydrogen on the material property, which utilized continuous circulation of hydrogen-charging solution in a pipe specimen. The conclusions can be summarized as follows:

1. The acceleration of fatigue crack growth rate occurred by this method at the relatively high ΔK regime ($\Delta K > 7$ MPa·m$^{1/2}$). The acceleration was associated with flat facet formed by intergranular fracture.
2. Fatigue crack growth acceleration caused by this method became more pronounced with decrease in the test frequency from $f = 1$ Hz to $f = 0.05$ Hz.
3. Sufficient amount of hydrogen was contained in a specimen after the fatigue test of a long period of 10 days.
4. For a long-term fatigue test, the testing method developed in this study has an advantage over the conventional precharging method in terms of its sufficient capability for hydrogen supply over an entire period of testing.

References

[1] Matsuoka S, Tanaka H, Homma N and Murakami Y 2011 *int. J. Fracture* **168** 101
[2] Marchi C S, Somerday B P, Nibur K A, Stalheim D G, Boggess T and Jansto S 2010 *Proc. Asme Pressure Vessels and Piping Conf* **6** 939
[3] Yamabe J, Yoshikawa M, Matsunaga H and Matsuoka S 2016 *Procedia Structural Integrity* **2** 525
[4] Murakami Y, Kanezaki T, Mine Y and Matsuoka S 2008 *Metal. Mater. Trans.* A **39** 1327
[5] Macadre A, Artamonov M, Matsuoka S and Furtado J 2011 *Eng. Frac. Mech.* **78** 3196
[6] Vergani L, Colombo C, Gobbi G, Bolzoni F M and Fumagalli G 2014 *Procedia Engineering* **74** 468
[7] Tanaka H, Homma N, Matsuoka S and Murakami Y 2007 *Trans. Jpn. Soc. Mech. Eng.* **73** 1358
[8] Somerday B P, Sofronis P, Nibur K A, San Marchi C and Kirchheim R 2013 *Acta Mater.* **61** 6153
[9] Sun Z, Benoit G, Moriconi C, Hamon F, Halm D, Hamon F and Henaff G 2011 *Int. J. Hydrogen Energy* **36** 8641
[10] Takeuchi E, Furuya Y, Hirukawa S, Matsuo T and Matsuoka S 2013 *Trans.Jpn. Soc. Mech. Eng.* **79** 1030

Failure analysis of blots for diesel engine intercooler

Ping Ren [1], Zongquan Li [1], Jiangfei Wu [1],Yibin Guo [1], Wanyou Li [1]
1 College of Power and Energy Engineering, Harbin Engineering University, Harbin, Heilongjiang Province, China

E-mail: guoyibin@hrbeu.edu.cn

Abstract. In diesel generating sets, it will lead to the abominable working condition if the fault couldn't be recovered when the bolt of intercooler cracks. This paper aims at the fault of the blots of diesel generator intercooler and completes the analysis of the static strength and fatigue strength. Static intensity is checked considering blot preload and thermal stress. In order to obtain the thermal stress of the blot, thermodynamic of intercooler is calculated according to the measured temperature. Based on the measured vibration response and the finite element model, using dynamic load identification technique, equivalent excitation force of unit was solved. In order to obtain the force of bolt, the excitation force is loaded into the finite element model. By considering the thermal stress and preload as the average stress while the mechanical stress as the wave stress, fatigue strength analysis has been accomplished. Procedure of diagnosis is proposed in this paper. Finally, according to the result of intensity verification the fatigue failure is validation. Thereby, further studies are necessary to verification the result of the intensity analysis and put forward some improvement suggestion.

1. Introduction

Turbocharged diesel engine is characterized of high ventilation efficiency and strong power. Before sending the high temperature air into cylinders of diesel engine, inter-cooler is required to cool the hot air ,which is heated by supercharging. If inter-cooler fails to work, the compressed high temperature air will go directly into the cylinders, which resulting in knocking or even damages to the engine.

This article takes the inter-cooler of certain diesel engine as research object, which is composed of encloser, tank body and inter-cooler bed. Encloser and tank body are connected by 48 pieces of 10.9 grade M12 high strength bolts while tank body and inter-cooler bed are connected by 46 pieces of 10.9 grade M12 high strength bolts. In respect that high temperature air is cooled by inter-cooler, temperature of inter-cooler is very high. The highest temperature of the inter-cooler which is taken as research object can be 150 degrees Celsius, hence, bolts on the inter-cooler not only bear the mechanical load but are also under large thermal load, and the working condition of blots are very bad.

Faulty parts of diesel engine are bolts on inter-cooler, after running for less than 24 hours, fracture occurred to the bolts. After replacing the faulted bolts for many times, failure still remains unsolved. Analysing the microstructure and fracture morphology by electron microscope determined that fracture was caused by fatigue, but the cause of fatigue fracture is unclear. Therefore, this article carried out research on bolts failure of inter-cooler and find out the causes of fatigue fracture.

2. Analytic train of thought

Static analysis and fatigue analysis of bolts on inter-cooler were based on accurate thermal analysis and fluctuating load calculation. For fluctuating load calculation, the force source and load path should be determined. Reliable transfer function can be gotten through by precise modelling of main structure and appropriate simplification of complex boundary. However, the vibration excitation source of diesel engine is relatively complex and should be equivalent simplified, fluctuating stress calculation flow chart are as shown in figure 1.

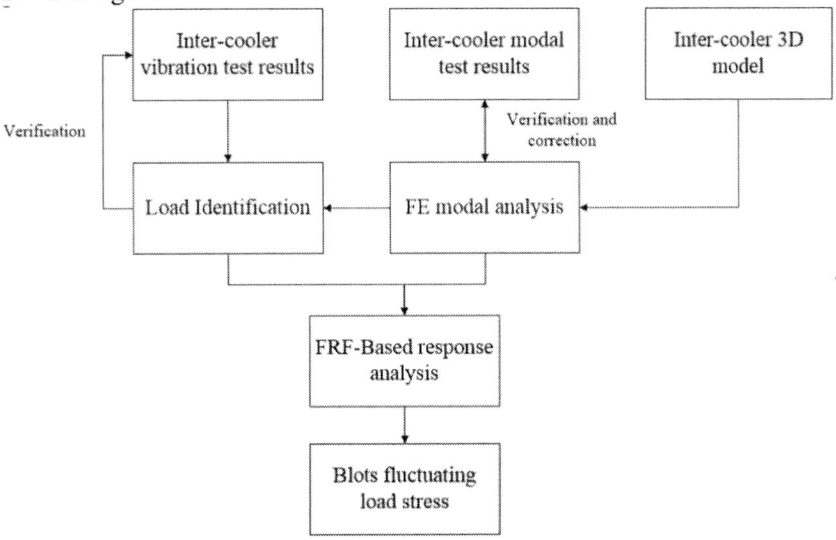

Figure 1. Flow chart of blots fluctuating load analysis

3. Forced vibration response calculation of the inter-cooler

3.1 modal analysis of the inter-cooler

The hypothesis of diesel engine vibrostand was introduced here. Due to the complexity of diesel engine structure, in respect that diesel engine is not taken as main research object, it was equivalently simplified and taken as vibrostand.

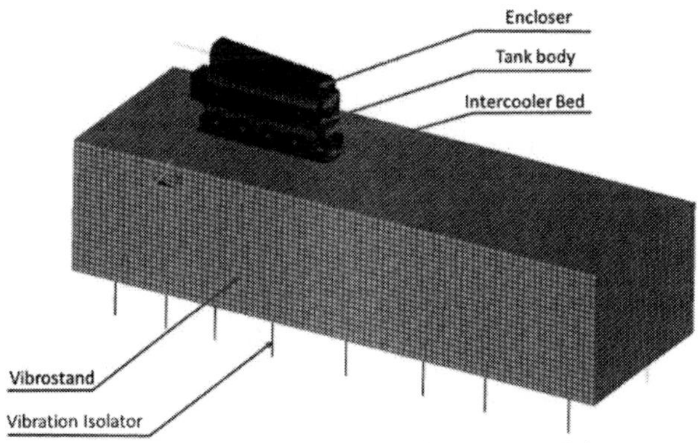

Figure 2. Diesel engine FE model

The diesel engine is equivalent to a cuboid model, the mass, centre-of-gravity position and position of vibration isolator are consistent with the actual diesel engine. The accuracy of modal calculation results is directly affected by the accuracy of finite element model. Finite element model should reflect the actual status of equipment as far as possible [1], thus establishment of finite element model is correspondingly corrected. The finite element model is continuously revised by comparing results of test modal and simulation to making it close to reality [2]. Figure 2 is the finite element model of diesel engine.

Modal of inter-cooler is calculated by finite element software ANSYS. Comparing with test modal simulation result is revised. According to the differences between the results, stiffness of blots and expansion joint is adjusted to make the calculating modal results consistent with the test modal results [3-4].Due to the variable combinations, multiple correction maybe required to obtain the ideal effect. By means of model adjustment comparison between modal analysis and test is shown in table 1.

Table 1. Results comparison of modal analysis and modal test

Modal Order	Modal Analysis (Hz)	Modal Test (Hz)	Error（Hz）	Percentage Error（%）
1	76.1	77.7	1.6	2.1
2	107.9	100.7	7.2	7.1
3	116.0	116.5	0.5	0.4
4	178.4	178.2	0.2	0.1
5	195.4	205.4	10.0	4.9
6	248.5	251.5	3.0	1.2
7	262.2	265.3	3.1	1.2
8	339.2	344.2	5.0	1.5
9	349.2	349.4	0.2	0.1

Seen from the comparison results between the natural frequencies that maximum frequency error is 11.5 Hz, maximum error percentage is 7.1%. The remaining errors are below 5%, meeting the engineering requirements.

3.2 Identification and verification of equivalent load

With the continuous development of dynamic load identification technology, it has been used in more fields. Application of frequency domain method in dynamic load identification field is long and relatively mature [5], so frequency domain method was applied in this article to identify the force source of the machine. Diesel engine vibration test result and finite element modal is taken as important parameter to identify and verify the excitation force.

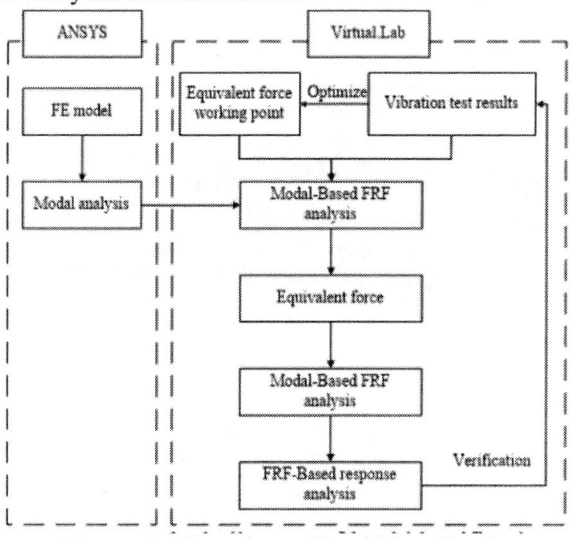

Figure 3. Flow chart of load identification

The most important factor for identification of force source is to provide reasonable location of the equivalent source points. Vibration of equipment will not be correctly obtained if setting of force source point is unreasonable, so the location of force source point is modify which should be revised according to the calculation results.[5] Overall idea of load identification is as shown in figure 3.

The research object in this article is the bolts on inter-cooler, which are mainly affected by excitation force of diesel engine, thus, the force source point should be set within the simplified diesel engine vibrostand. In this paper, the combinations of different locations were repeatedly calculated. It found when there are three equivalent source points in diesel engine, the result is consistent with the actual situation. One force source point is in the centre-of-gravity position, the other two are at both sides of the centre-of-gravity position. The force sources location are as shown in figure 4(a), the three yellow dots are the location of equivalent force points.

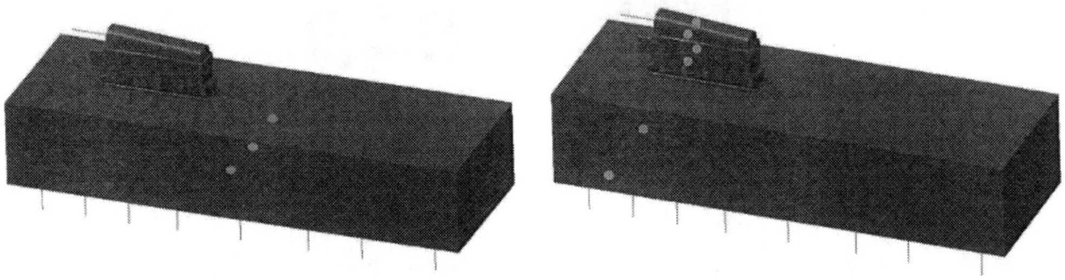

(a) (b)

Figure 4. Location of equivalent force and vibration test

3.3 Equivalent exciting force verification

The identified force source is taken as excitation force to solve the vibration response. Comparing with experimental data the error is analysed to determine whether the identified force source is accurate or not. The test point is as shown in figure 4(b), one of comparison between the actual measured responses and simulation results are shown in figure 5.

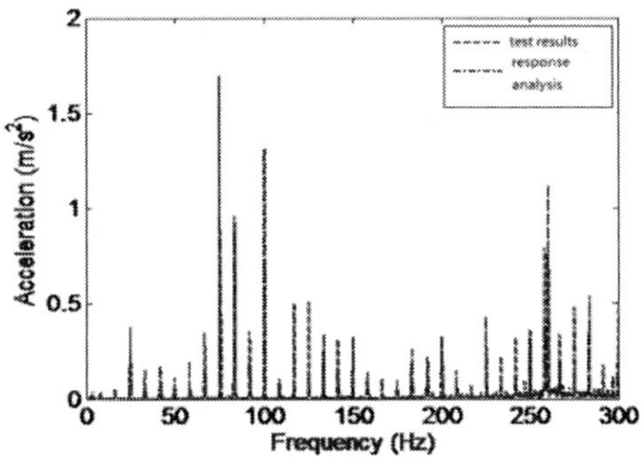

Figure 5. Comparison of test result and response analysis

Comparison between actual measured response and simulation results shows that the force source identified is relatively accurate, so the identified force source can be used to solve the bolts stress.

3.4 forced vibration response analysis of the inter-cooler

The identified force source is load on the finite element model to analyse the bolts stress, bolts position is as shown in figure 6. Yellow dots are the bolts, totally 94 pieces.

Blots stress result of FRF-Based response analysis is shown in figure 7. Seen from the data that the maximum tensile stress of vast majority of bolts appears at 75 Hz. 75 Hz is consistent with 4.5 times shaft frequency under the rated conditions of diesel engine, at the same time, test modal results of encloser found an one-order model of inter-cooler in 77.7 Hz. Therefore, this article concluded that, under the rated conditions, 4.5 times shaft frequency excitation force aroused the resonance of inter-cooler and the resonance is the main cause of the bigger bolts dynamic force.

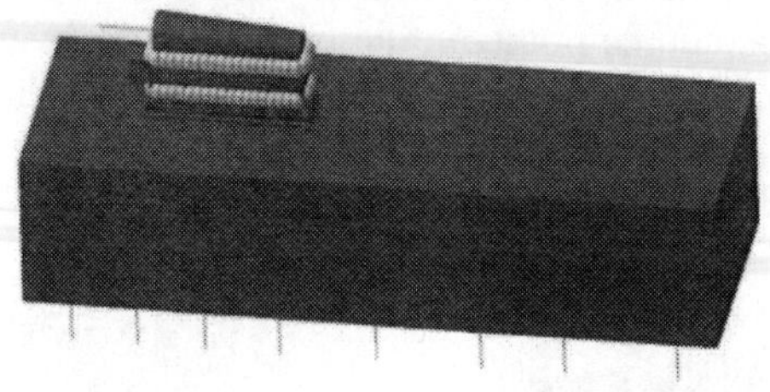

Figure 6. Location of blots

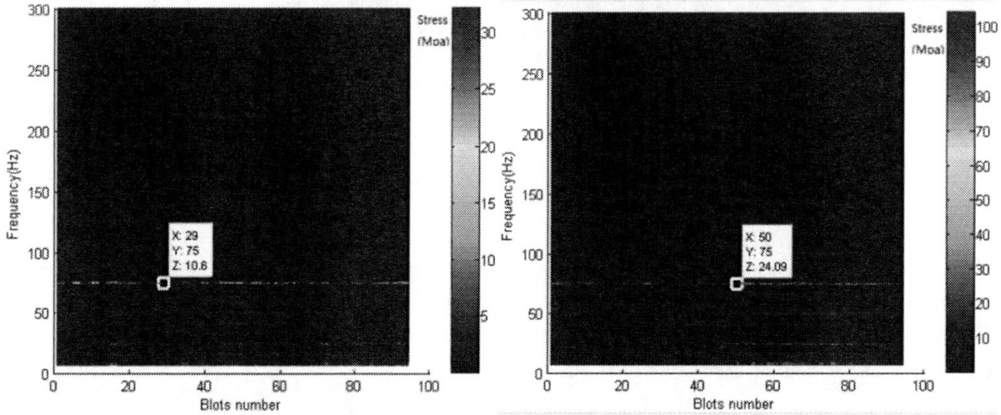

(a) Shear stress (b) Tensile stress

Figure 7. Blots stress result of FRF-Based response analysis

4. Thermodynamic analysis of the inter-cooler

High temperature compressed air can heat inter-cooler and makes its temperature higher, also, the temperature differences of each part are big, different levels of heat deformations occur to encloser, tank body and inter-cooler bed. Thus, connecting bolts on inter-cooler also suffer from different levels of heat stress. 3d model of Inter-cooler was established, initial temperature of all parts were set at 22 °C according to the environment temperature. Temperature distribution of inter-cooler took the form of decreasing from top to bottom while diesel engine was stable under the rated conditions. Thermodynamic analysis was carried out on the inter-cooler in accordance with the temperature distribution of diesel engine actually measured by factory staff. As shown in figure 8(a), the maximum temperature is 150 degrees Celsius, appeared at the connecting position between tank body and encloser, while the minimum temperature is 50 degrees Celsius in the inter-cooler bed position.

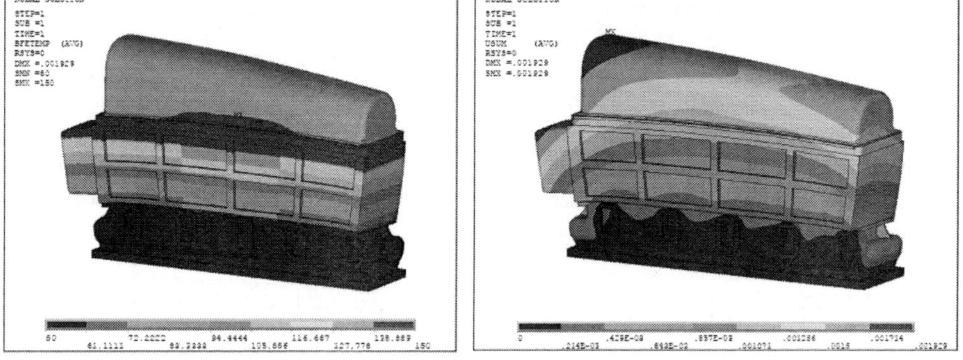

(a) (b)

Figure 8. Thermodynamic analysis result of the inter-cooler

Based on the temperature distribution of inter-cooler, hot-structure coupling analysis was carried out by applying the boundary conditions of inter-cooler (namely, displacement constraints of inter-cooler bed and pre-tightening force of bolts). Displacement contour of inter-cooler are as shown in figure 8(b), seen from the displacement contour, thermal expansion amount is small where the temperature is lower, such as inter-cooler bed and lower body of inter-cooler. Thermal expansion amount is big where the temperature is high, such as encloser and upper body of inter-cooler.

5. Bolt strength verification of the inter-cooler

5.1 static force verification

In order to prevent the air leak, some certain pre-tightening force should be exerted on bolts of inter-cooler to meet the requirement of sealing conditions. According to the information provided by the manufacturer, the pre-tightening torque of bolts is T=70 nm, calculate the pre-tightening force of bolts according to the pre-tightening force formula (1):

$$T \approx 0.2 \times F_0 \times D \tag{1}$$

F_0 is pre-tightening force of bolts (N), D are the bolts nominal diameter (m).

Therefore, pre-tightening force of M12 bolts is F_0=29167 N. The sum of pulling force that bolts receive in pre-tightening force and thermal stress is the actual tension. Due to the NBR (nitrile butadiene rubber) was applied in connection joint surface of inter-cooler, the relative stiffness coefficient of bolt is $\lambda = 0.25$ [6], and the working load of connection joint is F', then the total tensile stress can be obtained by formula (2):

$$s = \frac{F_0 + \lambda F'}{A} \tag{2}$$

Tangential working load that bolts receive is F_r, then the shearing stress that bolts receive can be obtained by formula (3):

$$\tau = \frac{4(F_r - fF_0)}{3A} \tag{3}$$

Thermodynamic analysis results shows that changes of bolts in axial direction is big while in horizontal direction is small. The horizontal displacement amount is less than spacing between the bolt and bolt hole, therefore, this article concludes that the tangential force that bolts receive is small and the bolts mainly bear tension.

Carry out static strength verification of bolts on inter-cooler, the specific verification formula is as shown in formula (4) [7]:

$$\frac{s^2}{Ft^2} + \frac{\tau^2}{Fv^2} \le 1 \tag{4}$$

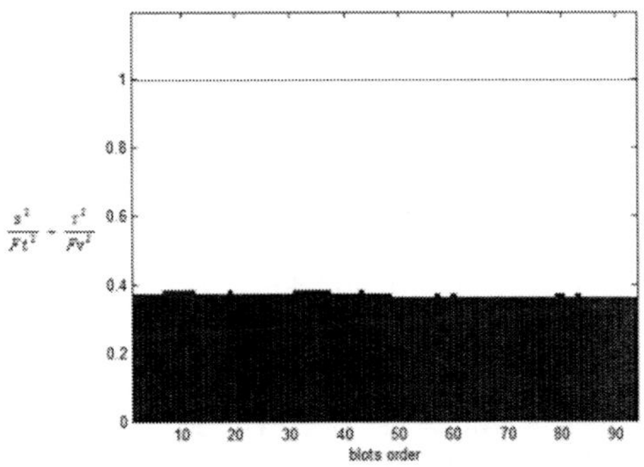

Figure 9. Results of static force verification

According to the ferritic steel standard [6], allowable tensile stress is $Ft = 0.58Su$, allowable shearing stress is $Fv = 0.24Su$. Su is tensile strength of the bolts, considering all the above-mentioned parameter tensile strength of 10.9 grade M12 bolts is 1000MPa. Verification result is shown in figure 9.

Seen from static strength verification results that the bolts of inter-cooler can satisfy the requirement of static strength, under the premise of meeting the requirement of this article, therefore, fracture failure arising from static load will not occur to bolts on inter-cooler.

5.2 fatigue verification

According to the S-N curve provided by the manufacturer, the bolts is verified through the fatigue safety coefficient method and verification formula of bolt fatigue strength is as shown in formula (5) [7]:

$$n = \frac{2\sigma_{-1} + (K_\sigma - \psi_\sigma)\sigma_{\min}}{(K_\sigma + \psi_\sigma)(2\sigma_a + \sigma_{\min})} \geq [n] \tag{5}$$

n is calculating safety coefficient, $[n]$ is allowable safety coefficient, $[n]$=1.21

σ_{-1} is symmetrical cyclic fatigue limit, according to the formula (6):

$$\sigma_{-1} = 0.23(\sigma_s + \sigma_b) \tag{6}$$

σ_s is the material yield strength limit, σ_b is the material tensile strength limit, ψ_σ is the reduction factor of stress. On the basis of mechanical design manual [6-7], ψ_σ =0.25, K_σ is fatigue limit comprehensive influence coefficient.

$$K_\sigma = \left(\frac{k_\sigma}{\varepsilon_\sigma} + \frac{1}{\beta_\sigma} - 1\right)\frac{1}{\beta_q} \tag{7}$$

k_σ is effective stress concentration factor, k_σ=5.2; ε_σ is size factor, ε_σ=1; β_σ is surface quality coefficient, β_σ=0.9; β_q is strengthening factor, $\beta_q = 2$, σ_a is stress amplitude, $\sigma_{\min}$ is minimum stress. Fatigue strength verification results of bolts are shown in figure 10.

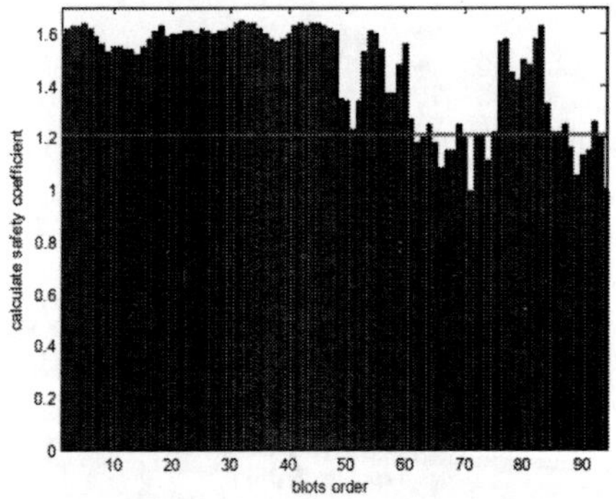

Figure 10. Results of fatigue verification

Figure 10 illustrates that fatigue ratio of bolts between tank body and inter-cooler bed is not enough for safety requirements. The possibility of fracture failure arising from fatigue is big.

6. Conclusion

This article carried out related researches on fracture failure of bolts on inter-cooler, the results show that the maximum tensile stress of the vast majority of bolts appears at 75 Hz, 75 Hz is consistent with 4.5 times shaft frequency under the rated conditions of diesel engine, at the same time, actual measured modal results of encloser found an one-order modal of inter-cooler at 77.7 Hz.

Therefore, this article concluded that, under the rated conditions of diesel engine, 4.5 times shaft frequency excitation force aroused the resonance of inter-cooler and the resonance is the main cause of the bigger bolt dynamic force. Also, bolts strength verification found that static verification of bolts reach the standard, however, fatigue ratio of bolts between body and inter-cooler bed of inter-cooler is not enough for safety requirements, which is the main reason of fracture failure of bolts.

References

[1] Toshimichi Fukuoka and Masataka Nomura 2008 Proposition of Helical Thread Modeling With Accurate Geometry and Finite Element Analysis *Journal of Pressure Vessel Technology* 011204-1---011204-6.

[2] Desanghere G and Snoeys R 1985 Indirect Identification of Excitation Forces by Modal Coordinate Transformation *Proceedings of the 3nd MAC* p685-690

[3] Bartlett F.D and Flannelly W.D 1979 Modal verification of force determination for measuring vibration loads *Journal of the American Helicopter society* 19(4): pp10-18.

[4] Hansen M and Starkey J M 1990 On Predicting and Fin-proving the condition of Modal -based indirect Force Measurement Algorithms *Proceedings of the 8th MAC* pp115-120.

[5] Kreitinger T; Luo H L 1987 Force Identification from Structural Responses *Proceedings of the SEM spring conference* June 1987： pp851- 855.

[6] Liang-Gui Pu and Ming-Gan Ji 2004 *Mechanical Design (Seventh Edition)* (China: Higher Education Press) p67

[7] Bang-Chun Wen 2010 *Machinerys Handbook (Fifth Edition)* (China Machine Press) pp5-31

Application of Risk-Based Inspection method for gas compressor station

Meng Zhang, Wei Liang*, Zeyang Qiu, Yang Lin

College of Mechanical and Transportation Engineering, China University of Petroleum-Beijing

lw@cup.edu.cn

Abstract. According to the complex process and lots of equipment, there are risks in gas compressor station. At present, research on integrity management of gas compressor station is insufficient. In this paper, the basic principle of Risk Based Inspection (RBI) and the RBI methodology are studied; the process of RBI in the gas compressor station is developed. The corrosion loop and logistics loop of the gas compressor station are determined through the study of corrosion mechanism and process of the gas compressor station. The probability of failure is calculated by using the modified coefficient, and the consequence of failure is calculated by the quantitative method. In particular, we addressed the application of a RBI methodology in a gas compressor station. The risk ranking is helpful to find the best preventive plan for inspection in the case study.
Keywords. RBI; Risk Analysis; Gas compressor station;

1. Introduction

The gas compressor station is the pressurizing part of the pipeline, once the explosion damage is great, and the shock waves generated great, huge shock waves also endanger the people and things surrounding. The integrity management in the gas compressor station is to identify and evaluate the risk of the gas compressor station by using scientific methods. It is a kind of advanced risk prevention and risk management method. Effective measures are taken to eliminate or reduce the impact of risk so that the risk can be controlled in the acceptable range. Finally, accidents are reduced to guarantee safety operation of the gas compressor station in an economic and reasonable way.

RBI is an effective way to solve pressure vessel and pipeline integrity problems in the gas compressor station. By Risk analysis of pressure vessels or pipes, the failure mechanism and testing technology of the pressure vessel or pipeline can be determined. Further, inspection plan and spare parts plan can be optimized. Besides, the scientific decision can be provided to support for prolonging the service life and shortening the maintenance period [1].

The RBI technology originated in the nuclear power industry, in the 1970s. In 1985, the American society of mechanical engineering (ASME) set up a research group of risk analysis. From 1991 to 1999, the ASME have published the RBI guidance documents in different industries [2]. Because of the Det norske veritas (DNV) has the ability in risk management and the experience in performing integrity around the world for a long time [3], American Petroleum Institute (API) seeks cooperation with DNV in the Early 1990s. They have successfully transplanted the RBI technology to the integrity management of petrochemical equipment, and have promulgated the RBI technology in the field of two landmark standards: API RP 580[4] and API RP 581[5]. With the support of the API and approved by the competent department of the United States, the RBI technology has been more and more trust and public acceptance.

2. Analysis of corrosion mechanism and corrosion risk

2.1. *Corrosion mechanism*

Stress and strain damage and corrosion are both common damage of pressure equipment. The types of the damage to pressure equipment are shown in figure 1. As an important component of the gas transportation system, the gas compressor station maintains the pressure and flow of natural gas transportation. Due to the natural gas contains H_2S, CO_2, H_2O and solid impurities, with the extension of service time, corrosion will lead to different degrees of damage to equipment and pipelines. In particular, the corrosion will be more serious in which there is a condensate.

In order to improve the reliability of risk inspection, to avoid the traditional inspection method based on the visual inspection, the mechanism of corrosion damage of equipment and pressure piping should be studied. Through the process and corrosion analysis of the gas compressor station equipment and pipeline, also refer to failure analysis of similar equipment at home and abroad, the mechanism of corrosion damage of the compressor station equipment and the pipeline is obtained [6].

According to the gas analysis data of a gas station: the CO_2 content in natural gas is 0.8878mol%. Therefore, corrosion of CO_2+ H_2O may occur in areas where water may be condensed. CO_2+ H_2O corrosion in carbon steel is a kind of electrochemical corrosion:

$$CO_2+H_2O+Fe \rightarrow FeCO_3+H_2$$

The factors that influence the CO_2+H_2O corrosion include water content, temperature, CO_2 partial pressure, pH, etc. In addition, according to the gas analysis data, the H_2S content is 0.85~2.20ppm, which has an impact on CO_2 corrosion. When the temperature is low, H_2S will accelerate the corrosion. Considering there may be there may be liquid water condensation leading to a wet H_2S environment where the temperature is low or the gas does not flow, especially, in the equipment and pipeline which are upstream of the gas-liquid junction and residual liquid tank. The equipment and pipelines may occur sulfide stress corrosion cracking (SSC) in the wet H_2S environment. Further, some other kinds of corrosion are threatening, such as atmospheric corrosion to the equipment exposure to the air, corrosion under insulation (CUI) to thermal insulation pipe, soil corrosion to buried pipelines [7-10].

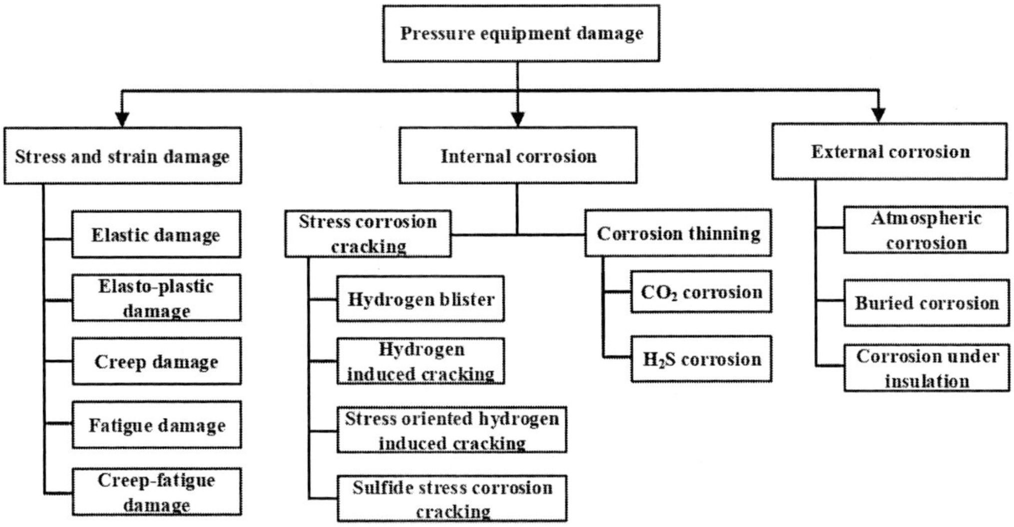

Figure 1. Types of damage to pressure equipment.

2.2. Risk analysis of the gas compressor station

The compressor gas station mostly has the following types of accidents: pipe and valve rupture; high voltage electric shock; leakage; fire and explosion [11].

Corrosion will cause the pipeline and valve failure, and thus lead to leakage. If Measures are taken at the early stage of leakage, malignant accidents can be avoided, such as human suffocation, fire, explosion, and others. For people, when the gas content in the air is above 15%, it may cause the human body hypoxia, which can lead to nerve damage, coma and even death in severe cases. The high-pressure air caused by compressor station leakage may cause serious injury even death to people near or in the spray.

It is significantly important to ensure the stable and efficient operation of gas compressor station by evaluating the risk of the gas compressor station.

3. RBI

3.1. RBI process

At present, RBI technology is most applied in the petrochemical enterprises. The data mainly come from API 581 standard data and the historical data of refinery [12].

To carry out the risk analysis of the gas compressor station based on RBI technology, it is necessary to improve the analysis model, add basic information, check maintenance records and hazard analysis, establish the RBI database for the gas compressor station based on API 581 standard. In this paper, the RBI process of gas compressor station is shown in Figure 2 [13].

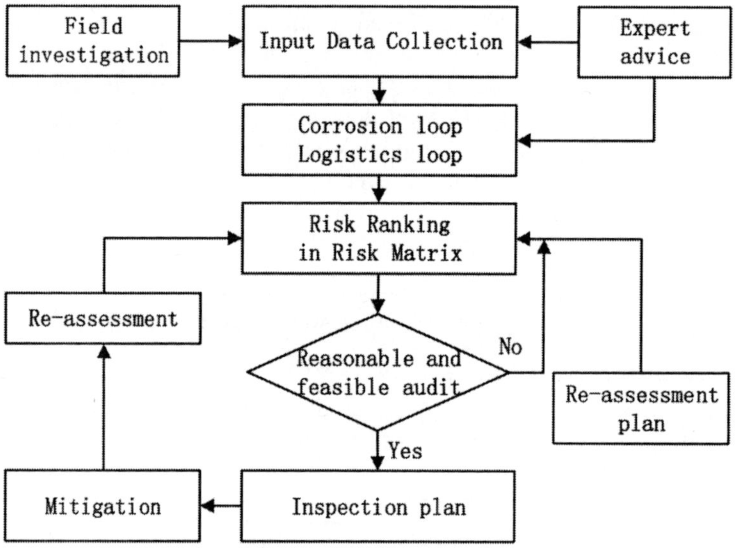

Figure 2. RBI process

3.2. Data collection

The data required for RBI includes the design and completion of the equipment, process data, inspection data, management data and financial data.

1) The design and completion of the equipment includes the installation, pipeline and valve list, general layout, pipeline diagram, completion and acceptance data.

2) Process data includes the operating pressure and temperature, flow chart of process, medium velocity of outlet and inlet, chemical analysis report of sampling point.

3) Inspection data includes the inspection plan, inspection report, history report.

4) Management data includes operating procedures and operating records.

5) Financial data includes the cost of environmental damage and business interruption, the average cost of equipment in the plant area.

3.3. Quantification of RBI
Risk Based Inspection (RBI) is the technique combining the risk of equipment and online detection to risk assessment and management of equipment based on risk analysis. In API RBI methodology the failure is defined as loss of containment, and the risk of failure is calculated using Equation (1).

$$Risk(t)=POF(t)\times COF \quad (1)$$

Where POF stands for Probability of Failure and is a function of time, t, and COF is Consequence of Failure. As shown in the Equation (1) the risk is also a function of time. The Probability is calculated based on the Equation (2):

$$POF(t)=GFF\times FMS\times EF(t) \quad (2)$$

GFF stands for Generic Failure Frequency and is a probability of failure developed for specific component types based on a large population of component data that does not include the effects of specific damage mechanisms. FMS is a factor which adjusts the generic failure frequencies for differences in process safety management systems. The factor is derived from the results of an evaluation of a facility or operating unit's management systems that affect plant risk. EF is an adjustment factor applied to the generic failure frequency to account for equipment status which is active in a component [14,15].

3.4. Equipment factor
Equipment factor is a collection of many modificatory factors which are shown in figure 3. Damage Factor (DF) is the key factor that affects the Equipment Factor. Damage Factors should be related to the failure mode of the equipment, including six modes of failure: thinning corrosion, stress corrosion cracking, external damage, brittle fracture, mechanical fatigue damage and high-temperature hydrogen damage. Therein, the stress corrosion cracking and external damage factors can be further classified [16].

In addition, other factors affect the value of the equipment factor. As the general factor, which is relate to the gas compressor station conditions, the local environment, the seismic activity and other aspects; Mechanical factors, mainly considering the initial situation of the equipment design and manufacture, is the sum of four factors: the complexity, the construction specifications, the life cycle and the safety factor; Process factors, is based on the analysis of the process and operation mode of the separator, including the continuity of the process and the safety of valve, etc.

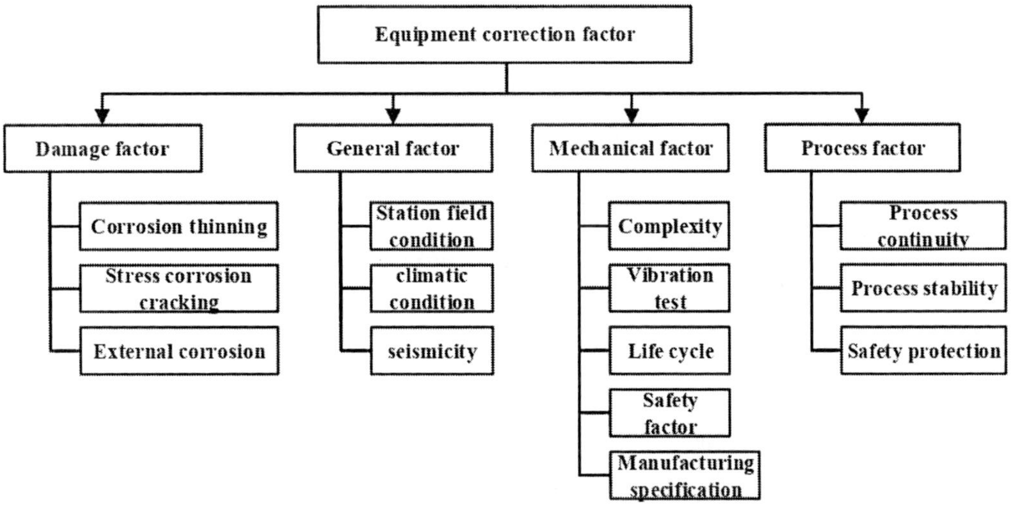

Figure 3. Overview of modificatory factors

3.5. Calculation of consequences

Risk-based inspection (RBI) technology quantifies the consequence of failure from two aspects: the consequences of failure area and the consequences of economic losses.

In the consequences of failure area analysis, there are three kinds of consequences: the consequence of combustion, the toxic consequence and the leakage of non-toxic and non-combustible medium.

Not only the consequences of failure area but also and the consequences of economic losses can be used to reflect economic losses directly. The consequences of economic losses include six categories: the cost of maintenance or replacement, the cost of other equipment damage, downtime loss, the cost of personnel Injury, the cost of environment clean-up and other expenses [17,18].

The risk is actually the product of failure probability and failure consequence. The equation is $R = P_f \bullet C$. Due to the consequences of the failure is divided into two categories: the consequences of failure area and the consequences of economic losses, so the equation is further changed into two: $R = P_f \bullet CA$ and $R = P_f \bullet FC$.

We can determine the risk level by the calculation of risk, and sort out the various types of damage to the gas compressor station, so as to optimize the arrangement of the test. The risk matrix is the most direct way to show the risk distribution of different equipment [19-22]. The recommended values for the probability level and consequence categories are shown in the following table 1.

Table 1. Quantitative risk matrix.

POF Ranking		COF				
5	>0.1	Med-high	Med-high	Med-high	High	High
4	≤0.1	Medium	Medium	Med-high	Med-high	High
3	≤0.01	Low	Low	Medium	Med-high	High
2	≤0.001	Low	Low	Medium	Medium	Med-high
1	≤0.0001	Low	Low	Medium	Medium	Med-high
COF Ranking		A	B	C	D	E
COF Types	Personal safety	No injury	Minor injury	Major injury	Single fatality	Multiple fatality
	Environment	No pollution	Slight effect	Minor local effect	Major local effect	Significant environment al effect
	Economic loss	0~100K	100K~1M	1M~10M	10M~100M	>100M
	Impact area (m^2)	0~10	10~100	100~1K	1K~10K	>10K

4. RBI application

4.1. Information of the gas compressor station

A pipeline compressor station design inlet pressure 6.5Mpa ~ 7.52MPa, the design pressure is 10Mpa. First of all, it is necessary to collect information of the gas compressor station to determine the corrosion loop and logistics loop.

The principle of corrosion loop is to classify a section of pipeline and process equipment in the same corrosion loop, which has the same damage mechanism and Continuous in the process flow. According to this principle, five corrosion loops have been decided in table 2.

Table 2. Corrosion loop table.

corrosion loop	medium	corrosion mechanism	Corrosion form	Stress corrosion cracking
C-001	CH_4	CO_2+ H_2O	Local	SSC

C-002	CH_4	—	Uniform	—
C-003	CH_4	—	Uniform	—
C-004	CH_4、H_2O	CO_2+ H_2O	Local	SSC
C-005	CH_4	CO_2+ H_2O	Local	—

The principle of Logistics loop is to define the equipment and the pipeline between the two fast cut-off points as a logistics loop. The device as a breakpoint can be an ESD Valve, a Failure Close, Valves can be manually closed within three minutes, Pumps, compressors, and Valves which is closed in normal operation. According to the situation, 20 logistics loops have been determined.

4.2. Result

19 risks of the Equipment, 443 risks of the pressure pipeline, a total of 462 risks were assessed, the risk analysis results are as follows. The total risks distribution map of compressor station is shown in Figure 4.

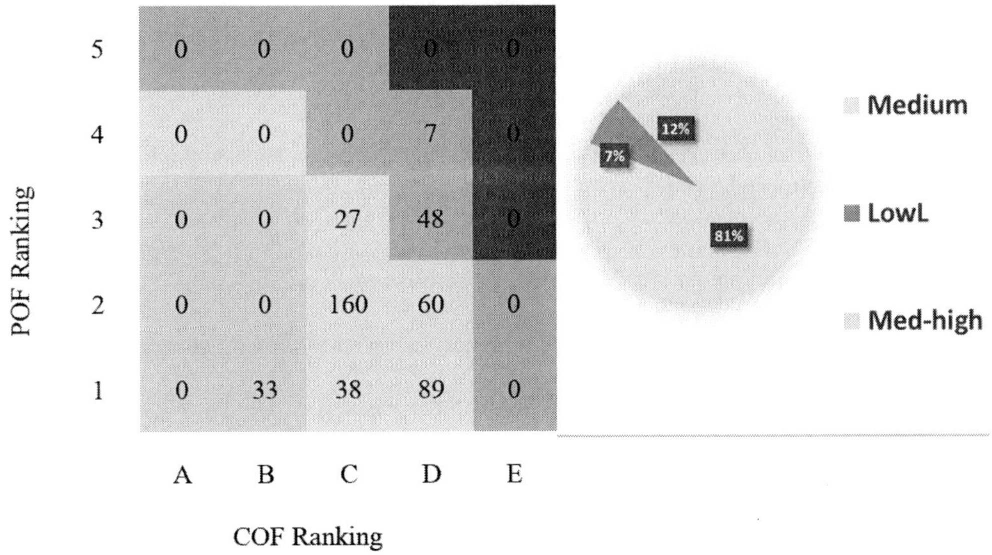

Figure 4. The total risks distribution map of compressor station

According to the risk matrix diagram, 7 risks are rank 4 of the possibility of failure, of which the probability of failure is higher.75 risks are rank 3 of the possibility of failure, 220 risks are grade 2 of the possibility of failure. And other 160 risks are in the low possibility of failure.

There are 56 of all risks that are in the med-high possibility of failure. 7 risks with rank 4 of the possibility of failure which is for the reason of the potential risk of sulfide stress corrosion cracking. The inspection should be taken immediately to reduce the likelihood of failure, so as to reduce its risk. 48 risks with rank 3 of the possibility of failure which has the internal medium of natural gas. The consequence of failure rank is D, which leads to higher total risk.

5. Conclusions

1. The results show that the RBI is effectively used in the gas compressor station. By using Risk ranking, enough attention can be paid to high-risk equipment and pipelines obviously. The cumulative risk and average risk are much higher, which is mainly due to the potential of piping and equipment for sulfide stress corrosion cracking.
2. According to the analysis results: about 10% of the equipment account for about 80% of all the risk in the plant. Developing a targeted inspection plan can effectively control the risk and reduce costs.

References

[1] Shuai J, Han K, Xu X. Risk-based inspection for large-scale crude oil tanks [J]. *Journal of Loss Prevention in the Process Industries*, 2012, 25(1):166-175.
[2] American Society of Mechanical Engineers. Research Task Force on Risk-Based Inspection Guidelines. Fossil fuel-fired electric power generating station applications [M]. *American Society of Mechanical Engineers,* 1994
[3] Det Norske Veritas. DNV-OSS-300 Offshore service specification Risk-based verification[S]. *Oslo: DNV*, 2004.
[4] American Petroleum Institute. Recommended practice 580 Risk-based inspection[S]. *Washington D C: API*, 2009.
[5] American Petroleum Institute. Recommended practice 581 Risk-based inspection technology[S]. *Washington D C: API*, 2008.
[6] Sarafraz M M, Hormozi F. Convective boiling and particulate fouling of stabilized CuO-ethylene glycol nanofluids inside the annular heat exchanger [J]. *International Communications in Heat & Mass Transfer*, 2014, 53(4):116-123.
[7] Yang yinchen. Detection of hidden danger of buried steel pipe corrosion and the formulation of the rectification scheme [J]. *Total Corrosion Control*, 2006, 20(2):20-23..
[8] Perez N. Electrochemistry and Corrosion Science [M]. *Springer US*, 2004.
[9] Medvedeva M L, Gur'Yanov V V. On the Corrosion State of Plants for Purifying Natural Gas from Acid Components [J]. *Protection of Metals and Physical Chemistry of Surfaces*, 2002, 38(3):284-288.
[10] Niu ren. Selection of piping design materials in wet hydrogen sulfide environment [J]. *Petrochemical corrosion and protection*, 2003, 20(6):6-9.
[11] Guoyan. Study on the risk assessment and control measures of fuzzy natural gas compressor station[D]. *Southwest Petroleum University*, 2014.
[12] Reynolds J. T. The Application of Risk Based Inspection Methodology in the Petroleum and Petrochemical Industry. *ASME PVP*,1996,336:125-134.
[13] CHANG Guangzhong, SHI Zhexiong, JIANG Xiaodong. Study on risk assessment technology and soft-ware development based on RBI method for storage rank [J]. *Corro-sion Science and Protection Technology*,2009,21(3) : 343-346.
[14] Latcovich J, Michalopoulos E, Selig B. Risk-based analysis tools [J]. *Mechanical Engineering*, 1998,120(11):72-75.
[15] Shishesaz M R, Bajestani M N, Hashemi S J, et al. Comparison of API 510 pressure vessels inspection planning with API 581 risk-based inspection planning approaches [J]. *International Journal of Pressure Vessels & Piping*, 2013, 111-112(6):202-208.
[16] Liu wanpeng. Research on the application of risk based inspection and maintenance strategy for pressure vessel system [D]. *East China University of Science and Technology*, 2013.
[17] Guo xiaolu. Research on risk assessment method of pressure equipment design stage [D]. *Lanzhou University of Technology*, 2012.
[18] Ma siliang. Application Analysis and research of risk based inspection method for petrochemical device [D]. *Beijing University of Chemical Technology*, 2013.

[19] Nilsson F. Risk Based approach to plant life assessment [J]. *Nuclear Engineering and Design*, 2003,221:293-300.

[20] Lee S. M., Chang Y. S., Choi J. B. et al. Application of an Enhanced RBI Method for Petrochemical Equipments [J]. *Journal of Pressure Vessel Technology*, 2006, 128: 445-453.

[21] Shim S. H., Song J. S., Kim, J. Y. et al. A Study on the Application of RBI Procedure for Petrochemical Plant 2003,*ASME Int. J.*, 27(3):416-423.

[22] Bareib J, Buck P, Matschecko B. et al. RIMAP Demonstration Project:Risk-Based Life Management of Piping System in Power Plant Heilbronn [J]. *Pressure Vessels Piping*, 2004, 81: 807-813.

A new testing method for investigating the shear-mode fatigue crack growth behavior in hydrogen environment

Y Akaki[1], T Matsuo[2, 3], Y Nishimura[4], S Miyakawa[4] and M Endo[2, 3]

[1]Graduate School of Engineering, Fukuoka University, 8-19-1 Nanakuma, Jonan-ku, Fukuoka 814-0180, Japan
[2]Development of Mechanical Engineering, Fukuoka University, 8-19-1 Nanakuma, Jonan-ku, Fukuoka 814-0180, Japan
[3]Institute of Materials Science and Technology, Fukuoka University, 8-19-1 Nanakuma, Jonan-ku, Fukuoka 814-0180, Japan
[4]Materials Engineering R&D Division, DENSO CORPORATION, Japan, 1-1 Syowa-cho, Kariya, Aichi 448-8661, Japan

E-mail: tmatsuo@fukuoka-u.ac.jp

Abstract. Ball bearing is widely used in a variety of machines including the transportation equipments of the automobiles and airplanes. Flaking failure is a common problem for ball bearing and it is caused by shear-mode fatigue crack growth under cyclic shear stress. Further, it is known that the premature flaking is attributed to the combined effect of hydrogen penetration into the material and cyclic shear stress during the operation. Therefore, in order to ensure the integrity of ball bearing, it is necessary to clarify the effect of hydrogen on shear-mode fatigue crack growth behavior, in particular, the threshold behavior. The evaluation of the shear-mode crack growth behavior is not easy because mode I crack branching occurs easily. Our previous studies revealed that it is required to apply static compression in the direction of specimen axis to attain a stable shear-mode fatigue crack growth. In addition, successive hydrogen supply to the specimen is essential for the evaluation of hydrogen effect on the fatigue threshold because hydrogen emits from the specimen during the fatigue test. In other words, the hydrogen-precharging method, commonly used for the research on hydrogen embrittlement, is not appropriate for the evaluation of fatigue threshold. In this study, to solve these problems, we have developed a novel, easy-to-use experimental method to evaluate the threshold behavior of shear-mode fatigue crack in the presence of hydrogen. The fundamental principle of the method is introduced in this paper.

1. Introduction

A premature flaking called white structure flaking (WSF) is a crucial problem in roller ball bearings especially for automobile components, wind turbine gearboxes, etc. This WSF is attributed to rolling contact fatigue, i.e. the crack growth from a surface defect or a subsurface inclusion under the combination of cyclic shear stress and compressive stress. Some previous studies [1-10] reported that invasion of hydrogen, derived from decomposition of lubricants in operation, into the bearing steel plays a key role in WSF. Most of them conducted a rolling contact fatigue test to investigate the WSF [1-7]. On the other hand, to the authors' best knowledge, there is only a few papers [8-10] that investigate the essential shear-mode fatigue crack growth behavior in the presence of hydrogen by utilizing torsional fatigue test. In order to understand the essential mechanism of WSF, it is necessary to clarify the effect of hydrogen on the shear-mode fatigue crack growth behavior.

Shear-mode fatigue crack initiation and growth behavior has often been investigated in torsional fatigue tests with round bar specimens. However, it is required to apply a static compression in the direction of specimen axis to attain the stable shear-mode fatigue crack growth since the Mode I branching occurs easily in pure torsional fatigue test. Fujita et al.[10] investigated the shear-mode fatigue characteristics of hydrogen-precharged bearing steel by cyclic torsion fatigue test with a static compression, and they figured out that Mode II (i.e. in-plane shear mode) crack growth rate was higher in hydrogen-precharged specimens than in uncharged specimens, involving the local microstructural change. In the fatigue test using hydrogen-prechaged specimen, however, hydrogen outgassing from the specimen occurs in parallel with crack initiation and growth, and consequently, hydrogen effect cannot be evaluated precisely in a long-term fatigue test. To avoid this problem, a few institutes employ the testing in high pressure hydrogen environment. However, this method is technically difficult and expensive for most others because high-pressure gas equipment such as high-pressure vessel, piping and so on needs to be installed in a laboratory.

The objective of the present paper is to establish an easy-to-use yet reasonable method for evaluation of hydrogen effect on the shear-mode fatigue crack growth behavior. A new testing system was designed to continuously circulate hydrogen charging solution in a specimen with a through-hole in the axial direction during torsional fatigue testing and its capability was discussed.

2. Experimental set-up

2.1 Developed testing method

In order to realize continuous hydrogen-charging for the round bar specimen with a through-hole during torsional fatigue test, hydrogen-charging solution was circulated in a specimen with a through-hole using a pump, flexible tubes and special jigs, which are shown in Figure 1 and Figure 2. Hydrogen charging solution was pumped into the specimen not directly but via a L-joint, and owing to this, screw holes do not need to be introduced on the specimen surface. This is an important consideration because the screw hole on the side surface of specimen would cause combined effect of stress concentration and direct exposure to hydrogen-charging solution, and would lead to failure from itself. In addition, the through-hole was on the neutral axis where the shear stress by torque is zero, and thus the effect of the through-hole on the stress field near the test part (i.e. the side surface of specimen) can be regarded as negligible. The L-joints which are exposed directly to hydrogen-charging solution were made of type 316L austenitic stainless steel (JIS-SUS316L), a material being non-susceptible to hydrogen embrittlement.

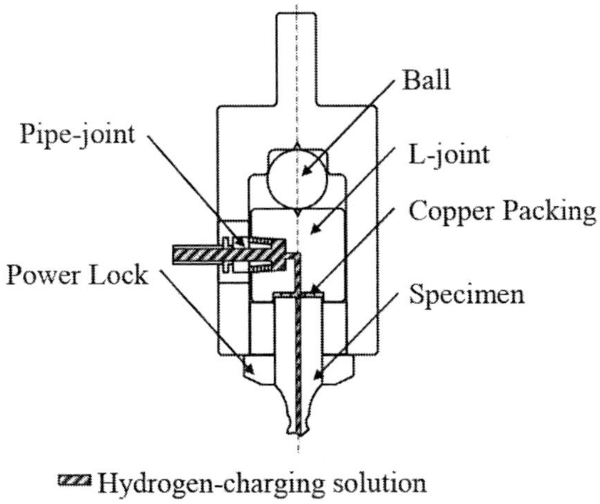

Figure 1. Cross-section view of assembled jigs.

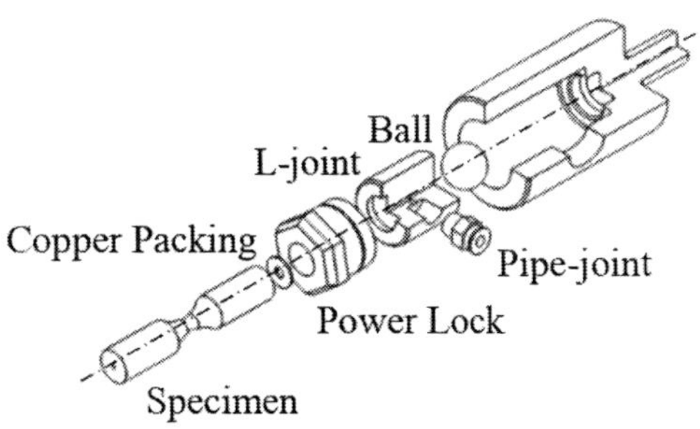

Figure 2. 3D image of jigs.

IOP Conf. Series: Journal of Physics: Conf. Series **842** (2017) 012065 doi:10.1088/1742-6596/842/1/012065

The friction-type fastener, called Power-Lock, was used to fasten the specimen to the jig. The specimen and the jigs except the Power-Lock were attached on the test machine, and a static compression was applied to them. Then, the specimen was fastened to the jig by Power-Lock. Consequently, the Power-Lock which is vulnerable to axial thrust load transmits only the torque to the specimen. A copper packing was set to prevent the leakage of solution from the space between the L-joint and the specimen by utilizing the test compression load.

2.2 Material and specimen

The material used in this study is a bearing steel, JIS-SUJ2, which is basically the same material as the ball used in a ball bearing. The shape and the dimensions of specimen are shown in Figure 3. As seen in this figure, to circulate a hydrogen-charging solution, a through-hole of 2 mm in diameter was introduced in the direction of specimen axis. Also, a shallow notch with the radius of 6.5 mm and the depth of 0.75 mm was introduced in order to localize the crack initiation. The stress concentration factor of the notch was K_t = 1.48 for axial loading and K_t = 1.06 for torsion. The notched region of the specimen was finished by buffing and electro-polishing following polishing with emery papers up to #2000. Residual stresses were measured using the X-ray diffraction method. Measurements were performed at the root of the notch before fatigue testing. The residual stresses were smaller than 5.0 MPa.

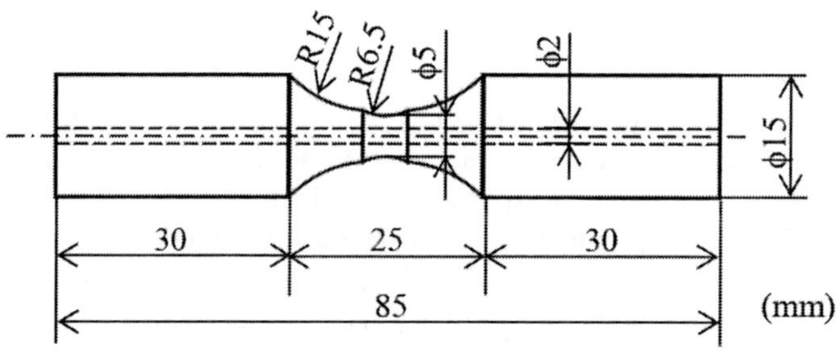

Figure 3. Fatigue test specimen.

2.3 Hydrogen-charging method

Hydrogen-charging was performed by circulation of an aqueous solution of 20 mass% NH_4SCN in a specimen. In order to achieve stable hydrogen distribution inside the specimen, hydrogen charging was started 48 h prior to fatigue testing.

2.4 Experimental condition

Fatigue tests were carried out in air at room temperature by a servo-hydraulic combined axial/torsional loading fatigue testing machine. A single specimen was used for each test under the conditions in presence and absence of hydrogen. The nominal shear stress amplitudes in cyclic torsion was τ_a = 1000 MPa. A nominal static compressive axial stress of σ_s = -1200 MPa was simultaneously applied to attain a stable shear-mode fatigue crack growth. The test frequency was f = 2 Hz and the stress ratio was R = -1. Surface cracks were observed using the replica method at every scheduled number of cycles to measure the crack length.

3. Results and discussion

Several cracks were observed on the root of notches regardless of hydrogen-charging, and one of them continued propagation as a main crack. The propagation of main crack was observed successively by the replica method, which is shown in Figures 4 and 5. For both the uncharged specimen and the specimen subjected to continuous hydrogen-charging, Mode II cracks propagated axially up to a several hundred micrometers long (see Figures 4(b)-(d) and 5(b)-(d)), and then a branching from

Mode II to Mode I occurred (see Figures 4(e) and 5(e)). As shown in Figure 4(e) and 5(e), macroscopically, Mode I crack propagated in the about 45° direction with respect to the axial direction after branching and the Mode I crack eventually propagated until the final failure of specimens. Figure 6 shows the relationship between the crack length of Mode II fatigue crack, $2a$, and the number of cycles, N, for both the uncharged specimen and the continuously hydrogen-charged specimen. The number of cycles to crack initiation of continuously hydrogen-charged specimen was smaller than that of uncharged specimen. In other words, hydrogen facilitated crack initiation. In addition, the final length of Mode II crack, i.e. the crack length at the moment of Mode I branching was 256 μm for the

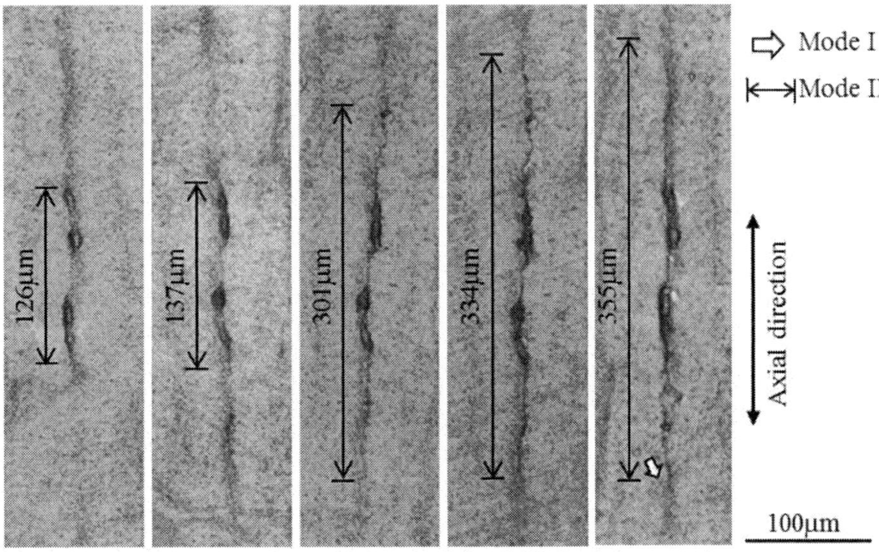

(a) 1.0×10⁴ (b) 2.0×10⁴ (c) 4.0×10⁴ (d) 4.5×10⁴ (e) 5.0×10⁴

Figure 4. Mode II crack growth behavior in the continuously hydrogen-charged specimen. (σ_s = -1200 MPa, τ_a =1000 MPa, f = 2 Hz)

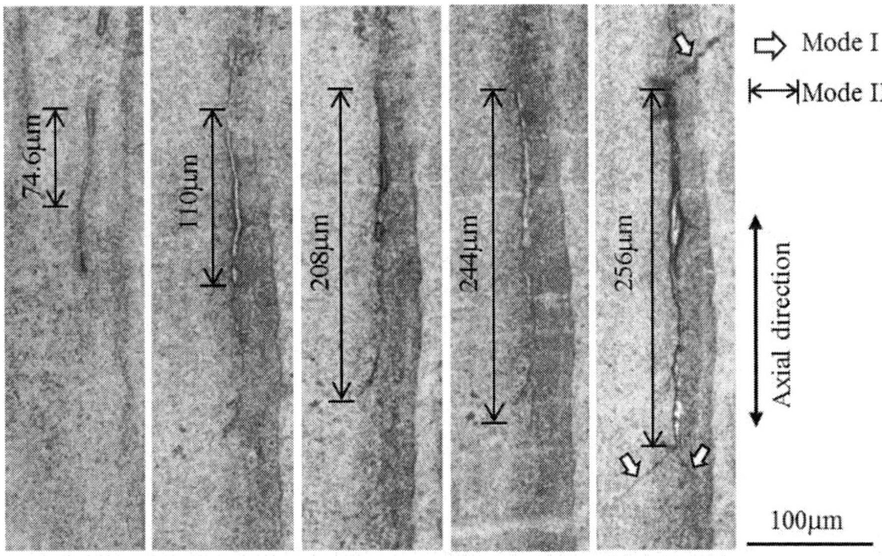

(a) 2.0×10⁴ (b) 4.0×10⁴ (c) 5.0×10⁴ (d) 6.0×10⁴ (e) 9.0×10⁴

Figure 5. Mode II crack growth behavior in uncharged specimen. (σ_s = -1200 MPa, τ_a=1000 MPa, f = 2 Hz)

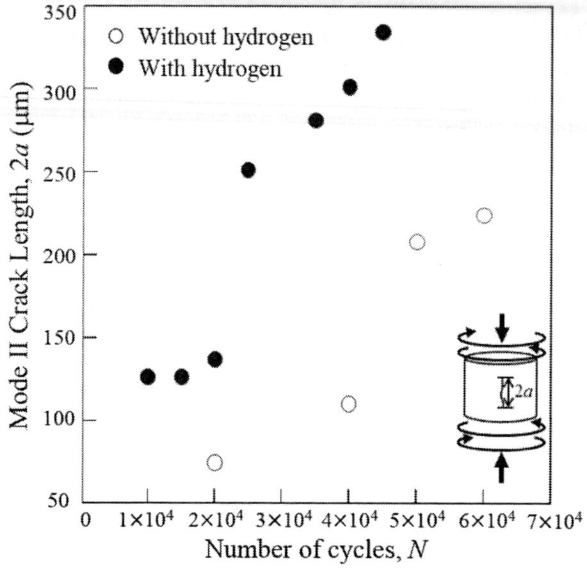

Figure 6. A comparison of Mode II crack growth behaviors between continuously hydrogen-charged and uncharged specimen. (σ_s = -1200 MPa, τ_a = 1000 MPa, f = 2 Hz)

uncharged specimen and 355 μm for the continuously hydrogen-charged specimen as shown in Figures 4(e) and 5(e) .This result suggests that hydrogen would inhibit the Mode I branching and/or enhance the stable Mode II crack growth. Regarding the Mode II crack growth rate, however, there is no remarkable hydrogen effect. As seen in Figure 6, it appears that the Mode II fatigue crack growth rate was slightly faster for the continuously hydrogen-charged specimen than for the uncharged specimen, but it is hard to say that there is the marked difference in crack growth rate between these two specimens because of the short crack growth distance, i.e. a small number of data. In contrast, regarding Mode I crack growth after Mode I branching, the crack growth rate was markedly accelerated by hydrogen. Unfortunately, no crack growth data was obtained because the extremely fast crack growth caused the unexpected failure of the hydrogen-charged specimen.

Fujita et al. [10] conducted torsional tests with a static axial compression using hydrogen-precharged specimens and they pointed out the crack initiation was earlier and the Mode II and Mode I fatigue crack growth rate was higher for the hydrogen-precharged specimen than for the uncharged specimen. According to their results, additionally, it can be seen that the final length of Mode II fatigue crack in the hydrogen-precharged specimen was slightly longer than that in the uncharged specimen, though they did not remark it. This result obtained using hydrogen-precharged specimen by Fujita el al. is similar to that obtained in this study using continuously hydrogen-charged specimen. Therefore, it can be concluded that the novel testing method developed in this study has a capability comparable to a conventional precharging method. In case of a long-term test such as a fatigue test to evaluate a fatigue threshold and a test at extreme low frequency, the new testing method may have an advantage over the conventional precharging method, because the effect of hydrogen outgassing from a specimen during a test is not negligible in the conventional precharging method.

4. Conclusions

A new testing method of torsional fatigue test with a superposed static compression using continuously hydrogen-charged specimen was developed. The following conclusions are drawn up:

1. The number of cycles to crack initiation for continuously hydrogen-charged specimen was smaller than that for uncharged specimen.
2. It appears that the Mode II fatigue crack growth rate was accelerated slightly by continuous hydrogen-charging.
3. The final length of Mode II crack, i.e. the crack length at the moment of Mode I branching, was 256 μm for uncharged specimen and 355 μm for hydrogen-charged specimen. Hydrogen would inhibit the Mode I branching and/or enhance the stable Mode II crack growth. However, Mode I crack growth after mode I branching was markedly accelerated by hydrogen.
4. The newly developed testing method in evaluating hydrogen embrittlement has a capability comparable to or would rather be better than the conventional precharging method.

References

[1] Tamada K and Tanaka H 1996 *Wear* **15** 245
[2] Evans M-H. 2012 *Mater. Sci. Technol.* **28** 22
[3] Matsumoto Y, Murakami Y and Oohori M 2002 *ASTM Bearing Steel Technology* **7** 226
[4] Evans M-H, Richardson A.D, L. Wang and Wood R.J.K 2013 *Wear* **302** 1573
[5] Shravan J, Ole W, Peder K and Niels S.J 2015 *Inter. J. Fatigue* **77** 128
[6] Iso K, Yokouhci A, and Takemura H 2005 *SAE Technical Paper* **114** 1868
[7] Kino N and Otani K 2003 *JSAE Review* **24** 289
[8] Matsubara Y and Hamada, H 2006 *J. ASTM. Inter.* **3** 1
[9] Hamada H and Matsubara Y *NTN Technical Review* **74** 54
[10] Fujita S, Matuoka S, Murakami Y and Marquis G 2010 *Inter. J. Fatigue* **32** 943

IOP Conf. Series: Journal of Physics: Conf. Series **842** (2017) 012066 doi:10.1088/1742-6596/842/1/012066

Improvement of torsional fatigue limit and rendering surface defect harmless by shot peening for spring steel

K Takahashi[1], M Nakagawa[1], H Koike[1] and H Okada[2]

[1] Yokohama National University, 79-5 Tokiwadai, Hodogaya-ku, Yokohama, 240-8501 Japan
[2] NHK Spring co. ltd., 3-10 Fukuura, Kanazawa-ku, Yokohama, 236-0004, Japan

E-mail: takahashi-koji-ph@ynu.ac.jp

Abstract. The effects of shot peening (SP) on the torsional fatigue limit of spring steel (SUP7) were investigated for specimens with Vickers hardness values of 460, 540, and 670 HV containing a semicircular surface slit. SP was conducted on smooth specimens and specimens containing a semicircular surface slit with a depth of 0.15 or 0.3 mm. Compressive residual stress was introduced into the specimens by SP. Torsional fatigue tests were carried out under a stress ratio of $R = -1$. The torsional fatigue limits of the shot peened specimens with Vickers hardness values of 460, 540, and 670 HV increased by 8%–67%, 33%–143%, and 36%–127%, respectively, in comparison with the non-shot peened specimens. The maximum depth of the slit that could be rendered harmless by SP was 0.15 mm for the 460 and 670 HV specimens. However, even a slit with a depth of less than 0.15 mm could not be rendered harmless by SP for the 540 HV specimen. Considering the improvement in the torsional fatigue limit and the size of the surface defect that could be rendered harmless by SP, the 670 HV specimen is optimal for practical use.

Keywords: Torsion, Fatigue limit, Shot peening, Spring steel, Surface defect, Residual stress.

1. Introduction

The improvement of fatigue limit of machines and components has been increasing from the viewpoints of fuel efficiency and life span. Shot peening (SP) is a surface treatment method applied to increase the fatigue strength of metals. Compressive residual stress induced by SP prevents fatigue crack propagation. Mechanical components including coil springs, drive shafts and crank shafts, are subjected to cyclic torsion. Since the maximum stress occurs on the surface under torsion and bending, the fatigue crack often initiates from surface defects. The effects of a small surface defect on the torsional fatigue limit have been studied in various metals using specimens containing a small drilled hole [1, 2], a semielliptical surface crack [3], a shallow micronotch [4], and artificial pits [5]. These studies have demonstrated that the presence of surface defects reduces the torsional fatigue limit.

If the surface defects can be rendered harmless from the viewpoint of torsional fatigue limits thorough SP, the reliability of components can be improved. Previous studies have shown that the bending fatigue limit of spring steel (JIS-SUP9A) containing a small drilled hole [6–8] or a semicircular slit [9–11] can be rendered harmless from the viewpoint of the bending fatigue limit. However, little research has been conducted on the SP-induced torsional fatigue limit improvement for high-strength steel specimens containing a crack-like surface defect [12]. In addition, the actual hardness of automobile parts used in practical situations varies over a wide range. Kuno et al. noted that the improvement ratio of the torsional fatigue limits of spring steel specimens by SP depends on the specimen hardness [13]. It is important to know the size of defects rendered harmless by SP to maintain the component reliability. However, how the hardness affects the size of the defect that can be rendered harmless by SP is not clear. Therefore, in this study, the effect of SP on the torsional fatigue limit of spring steel (JIS SUP7) was investigated for specimens with Vickers hardness values of 460, 540, and 670 HV containing a semicircular surface slit. SP was conducted on smooth specimens and specimens containing a semicircular surface slit with depths of 0.15 and 0.3 mm. Torsional fatigue tests were then carried out under a stress ratio of $R = -1$.

2. Experimental procedure

2.1 Test materials and specimens

The material used in this study was spring steel (JIS SUP7). Table 1 shows the chemical composition of JIS SUP7. Figure 1 shows the shape and dimensions of a fatigue test specimen. Smooth specimens and specimens with a semicircular slit of depth 0.15 or 0.3 mm were subjected to SP. Figure 2 shows the shape of the semicircular slit. The semicircular slit was made by electric discharge machining and was parallel to the axial direction. The width of the slit (0.05 mm) was very small to simulate an initial crack-like surface defect. Next, the specimens were oil quenched at 860 °C and tempered at different temperatures. Table 2 shows the relationship between the tempering temperatures and the Vickers hardness of the test specimens after heat treatment.

Table 1. Chemical composition of SUP7 (wt%).

C	Si	Mn	P	S
0.61	1.91	0.83	0.018	0.014

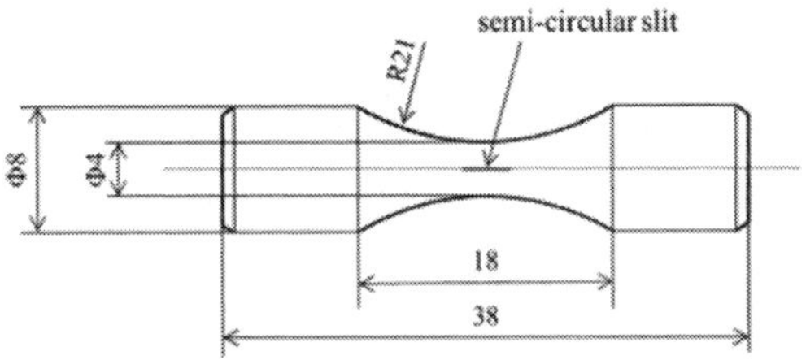

Figure 1. Shape and dimension of a specimen with a semicircular slit.

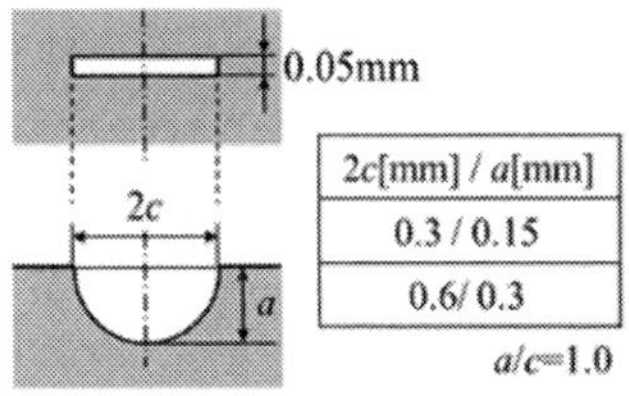

Figure 2. Shape of the semicircular slit.

Table 2. Vickers hardness and tempering temperature.

Vickers hardness (HV)	460	540	670
Tempering temperature (°C)	500	430	300

2.2 Shot peening conditions and residual stress

After heat treatment, the specimens were subjected to SP. Table 3 shows the SP conditions adopted in this study. The residual stress was measured using the X-ray diffraction method with a Cr Kα beam X-ray spectrum and an X-ray beam injection diameter of 1.0 mm. The distributions of residual stress in the thickness direction were investigated by alternately measuring the residual stress on the surface and removing the surface layer by chemical etching. The residual stress distributions 45° from the axial directions, i.e., the principal stress directions, were measured.

Table 3. Shot peening conditions.

Peening machine	Direct pressure peening
Air pressure (MPa)	0.42
Shot diameter (mm)	0.7
Shot hardness (HV)	680–750
Nozzle diameter (mm)	5
Shot time (one side) (s)	40
Shot distance (mm)	100
Coverage	300%
Arc height (mm A)	0.49

2.3 Surface roughness and hardness

The surface roughness and strain hardening after SP affect the fatigue strength of metals. Thus, we measured these values for non-shot peened (Non-SP) and shot peened (SP) specimens. Table 4 gives the arithmetic average roughness (R_a) and the maximum height of the profile (R_y). The surface roughness values increased after SP; however, they were much smaller than the depth of the semicircular slit (0.15 and 0.3 mm). Thus, the effects of surface roughness on the fatigue strength were considered to be small. The surface roughness of SP specimens decreased as increasing the Vickers hardness.

Table 4. Surface roughness before and after shot peening.

	Non-SP		SP	
	R_a (μm)	R_y (μm)	R_a (μm)	R_y (μm)
460 HV	0.53	2.79	4.57	26.27
540 HV	0.68	4.16	3.64	19.49
670 HV	0.69	3.56	1.58	10.34

Figures 3(a) and (b) show the distributions of the Vickers hardness of the non-shot peened (Non-SP) and shot peened (SP) specimens, respectively. In the Non-SP specimens, the Vickers hardness was almost constant for each specimen regardless of the distance from the surface, as shown in figure 3(a).

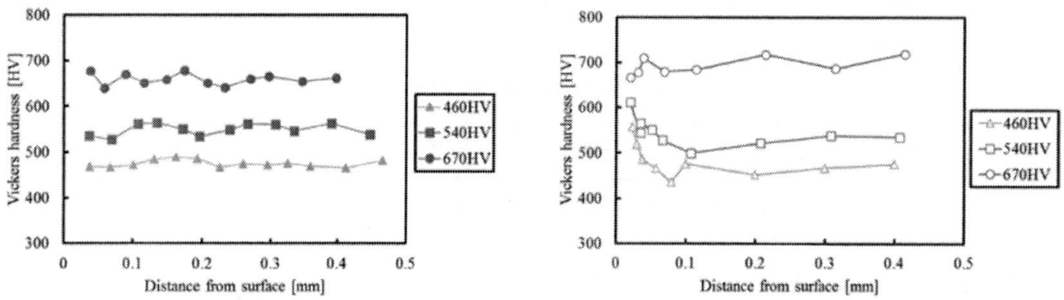

(a) Non-SP (b) SP

Figure 3. Distributions of Vickers hardness.

In the SP specimens, the Vickers hardness of the 460 and 540 HV specimens increased near the surface because of work hardening caused by the shot impact. However, the 670 HV specimen did not show such an increase in Vickers hardness near the surface. This is because the hardness of the 670 HV specimen is close to that of the shot media (see Table 3).

2.4 Fatigue tests

Torsional fatigue tests were performed on the specimens using a torsional fatigue testing machine (PDF-30, Tokyo Koki Ltd.) at a stress ratio of $R = -1$ and a cyclic frequency of 20 Hz. The torsional fatigue limit was defined as the maximum torsional stress amplitude under which the specimen endured 10^7 cycles. The fracture surfaces of all tested specimens were observed using an optical microscope and a scanning electron microscope (SEM).

3. Experimental results and discussion

3.1 Fatigue test results

Figures 4(a), (b), and (c) show *S–N* diagrams that demonstrate the relationship between the stress amplitude τ_a and the number of cycles to failure N_f for the 460, 540, and 670 HV specimens, respectively. The solid and open symbols indicate Non-SP and SP specimens, respectively. The arrows indicate tests in which no fracture occurred in 10^7 cycles. The torsional fatigue limits are given beside the arrows. The figure demonstrates that torsional fatigue limit and fatigue life were increased by SP. The asterisk symbol (*) indicates that the specimen fractured at the surface but outside the semicircular slit. The pound symbol (#) indicates that an internal fracture occurred. The specimens without symbols fractures from the semicircular slit. The differences of fracture surfaces are discussed in section 3.4.

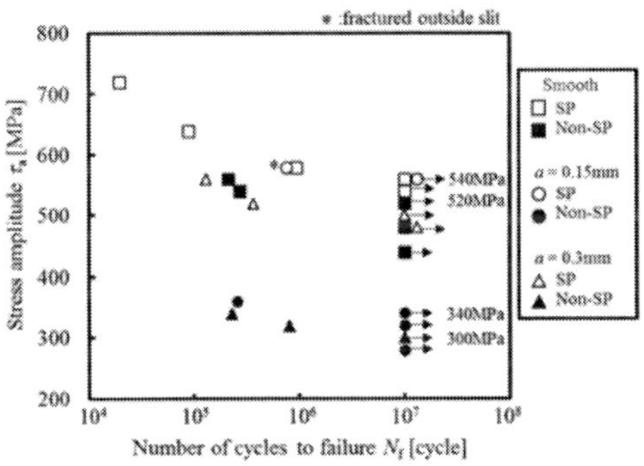

(a)

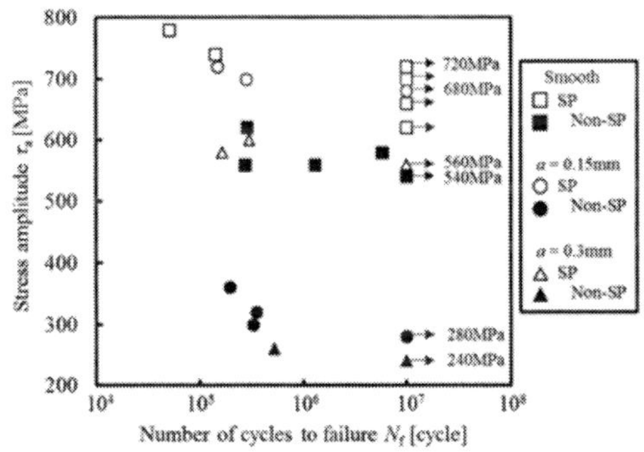

(b)

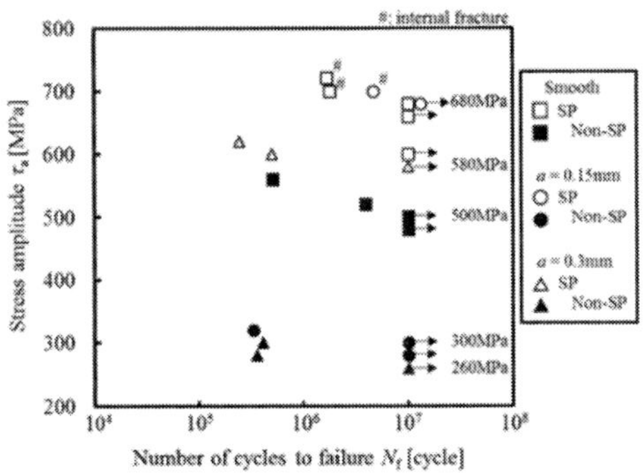

(c)

Figure 4. S–N diagrams for torsional fatigue test (SUP7, $R = -1$); (a) 460, (b) 540, and (c) 670 HV.

3.2 Residual stress before and after torsional fatigue tests

The stability of residual stress under cyclic loading is important. Thus, we measured the residual stresses before and after fatigue tests. Figure 5 shows the residual stress distributions before and after the torsional fatigue tests. The residual stress of the Non-SP specimens was much smaller than that of the SP specimens. A small compressive residual stress was induced by machining in all Non-SP specimens. In SP specimens, a compressive residual stress was induced by SP. The solid circles indicate the residual stress before the torsional fatigue tests. The values of the residual stress and crossing point (distance from the surface where the residual stress becomes zero) depend on the specimen hardness. As the specimen hardness increased, the compressive residual stress increased and the crossing point decreased. Okada *et al.* have reported that the compressive residual stress induced by SP increased in proportion with the Vickers hardness in spring steel (SUP9) from 308 to 746 HV [14]. The open circles indicate the residual stress after the torsional fatigue tests. The fatigue tests caused relaxation of the compressive residual stress in the 460 and 540 HV specimens. In contrast, in the 670 HV specimen, the fatigue tests did not change the compressive residual stress appreciably. Relaxation of the compressive residual stress occurs if the sum of the compressive residual stress and the applied stress exceeds the yield stress of the material [13]. Among the three cases, the yield stress is the highest in the 670 HV specimen, which is greater than the sum of the compressive residual stress and the applied stress. Thus, relaxation of the residual stress did not occur in the 670 HV specimen.

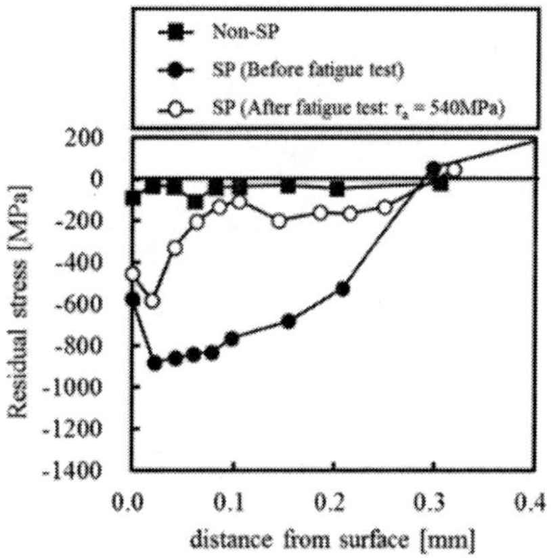

(a)

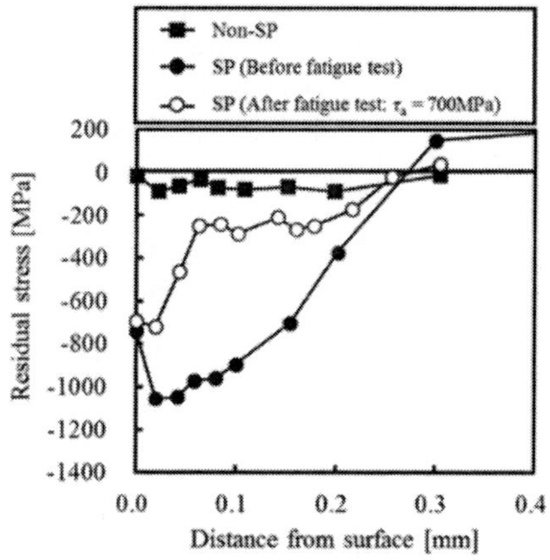

(b)

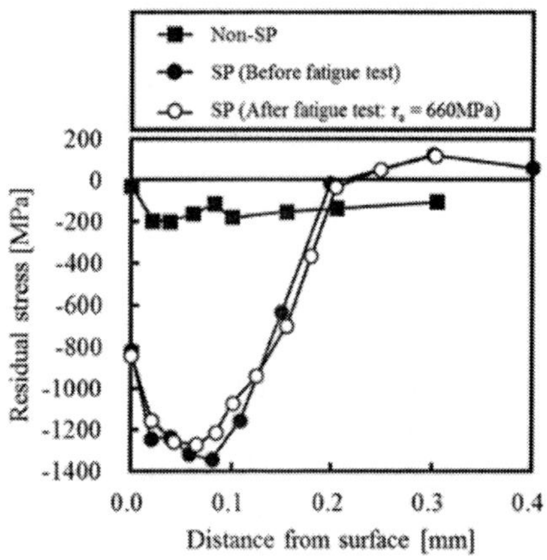

(c)

Figure 5. Residual stress distributions before and after torsional fatigue tests; (a) 460, (b) 540, and (c) 670 HV.

3.3 Effects of shot peening on the torsional fatigue limit

Figures 6(a), (b), and (c) show the relationships between the stress amplitude τ_a and the depth a of the semicircular slit for the 460, 540, and 670 HV specimens, respectively. The solid symbols represent the specimens that failed during the torsional fatigue tests. The open symbols represent the specimens that did not fail in 10^7 cycles, at which point the maximum stress amplitude corresponds to the torsional fatigue limit. The figures demonstrate that the torsional fatigue limit of the Non-SP specimens decreased with increasing slit depth. It was found that SP increased the torsional fatigue limits of the 460, 540, and 670 HV specimens by 8%–67%, 33%–143%, and 36%–127%, respectively. Compressive residual stresses are the main reason for these increases in the torsional fatigue limit. It is noted that the torsional fatigue limit of the 460 and 670 HV SP specimens with a slit of 0.15 mm depth was equivalent to that of SP specimens without a slit, as shown in figure 6(a) and (c).

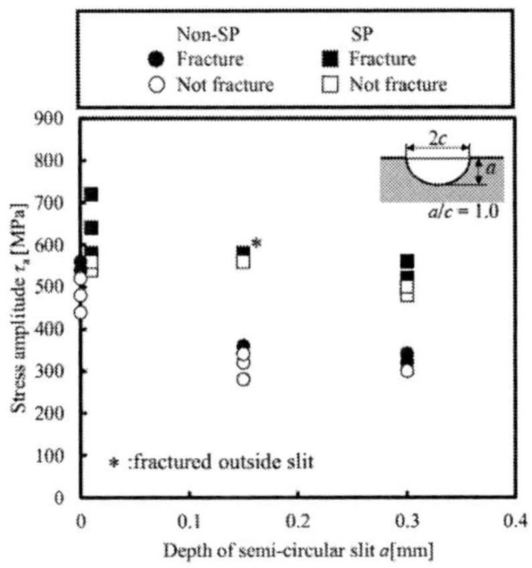

(a)

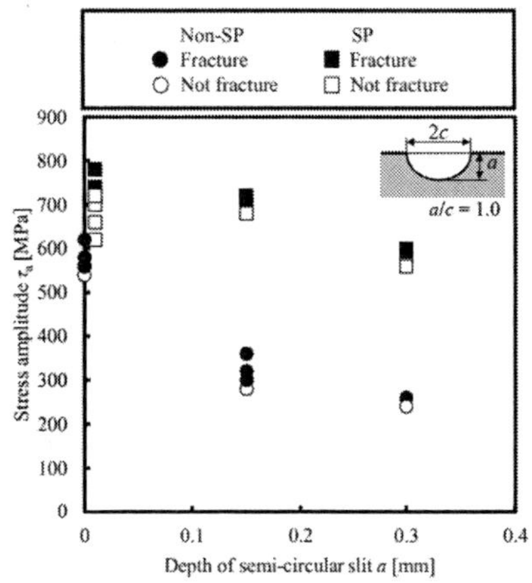

(b)

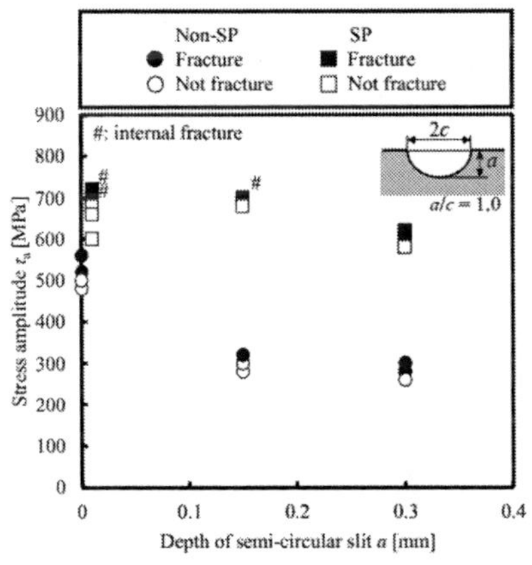

(c)

Figure 6. Relationships between stress amplitude and depth of the semicircular slit (SUP7, torsional, $R = -1$); (a) 460, (b) 540, and (c) 670 HV. In the figure, a denotes the slit depth and c the slit radius.

3.4 Fracture origin

The appearances and origins of the fractures in the tested specimens were examined using an SEM. Figures 7 and 8 show SEM images of the fracture surfaces of specimens with a semicircular slit of depth 0.15 and 0.3 mm, respectively. The mode I fatigue cracks propagated in a direction perpendicular to the principal stress direction, i.e., ±45° from the axial direction. The Non-SP specimens fractured at the semicircular slit regardless of the slit depth. The semicircular slits can be clearly observed in the Non-

IOP Conf. Series: Journal of Physics: Conf. Series **842** (2017) 012066 doi:10.1088/1742-6596/842/1/012066

SP specimens. The SP specimens with a slit of 0.3 mm depth fractured at the semicircular slit, as shown in figure 8. The semicircular slits were deformed by SP. The crack initiation site depended on the specimen hardness in the SP specimens with a slit of 0.15 mm depth. The 540 HV specimens fractured at the slit, whereas the 460 and 670 HV specimens fractured outside the slit. The crack initiation sites in all 670 HV SP specimens were matrix cracking at 0.2–0.3 mm depth where the tensile residual stress observed (see figure 5(c)). Inclusions were not observed at the crack initiation sites.

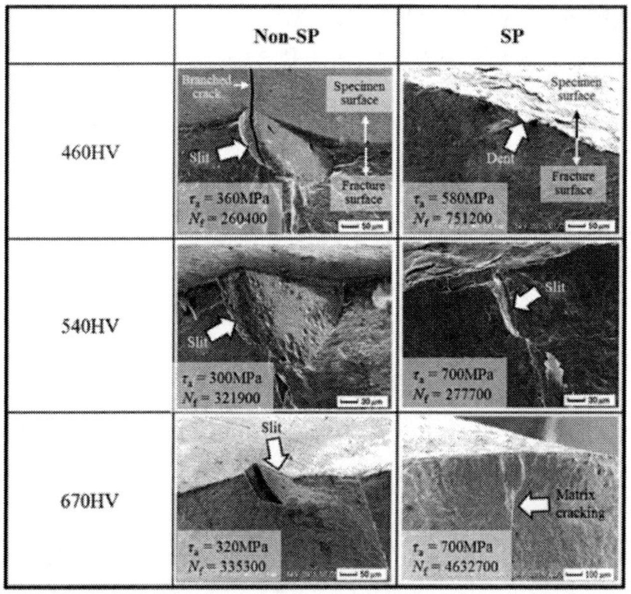

Figure 7. Fatigue fracture surface after test; slit depth = 0.15 mm.

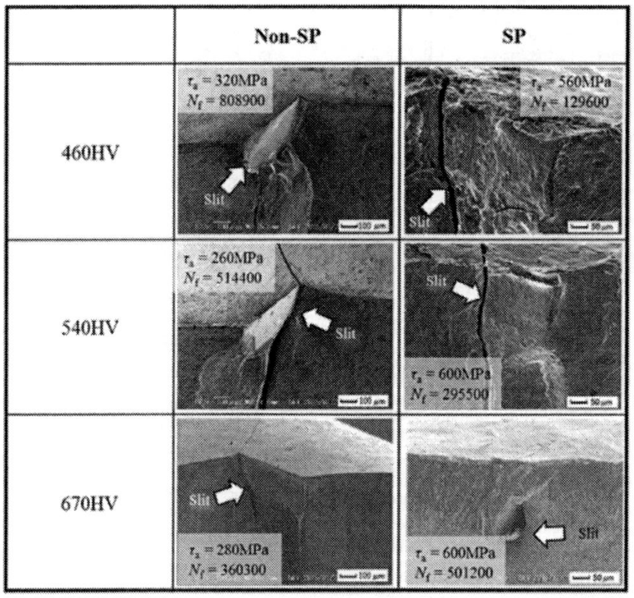

Figure 8. Fatigue fracture surface after test; slit depth = 0.3 mm.

It is important to observe non-propagating cracks of the specimens that did not fracture during fatigue test because the non-propagating crack sizes were related to the acceptable crack sizes [9, 11]. We tried to observe non-propagating fatigue cracks of the specimens tested under the torsional fatigue limit. However, non-propagating cracks could not be observed probably due to the wear of the mating crack surfaces under cyclic shear loads and compressive residual stress. In the case of bending fatigue, non-propagating cracks were clearly observed in the spring steel specimens tested under the bending fatigue limit [6-11]. In this study, the fatigue cracks propagated under mode I (see figures 7 and 8). Thus, it is thought that the torsional fatigue limit was determined by the threshold condition for non-propagation of fatigue cracks similar to the case of bending fatigue tests.

3.5 Size of defect that can be rendered harmless by shot peening

In this section, we discuss how the specimen hardness affects the sizes of semicircular slits that can be rendered harmless by SP. If the fatigue test results of an SP specimen with a semicircular slit meet either of the following two conditions, the slit is considered to have been rendered harmless.

Condition (a): The fatigue limit increased up to more than 95% of that of a SP specimen without a semicircular slit.

Condition (b): The specimen fractured outside of the slit.

These conditions were proposed by the "Acceptable Defect Size Research Committee" in Japan Society of Spring Engineers [7]. For the 460 and 670 HV specimens, the torsional fatigue limit of the SP specimens with a slit of 0.15 mm depth was equivalent to that of SP specimens without a slit (see figures 6(a) and (c)). Moreover, these specimens fractured outside of the semicircular slit (figure 7). However, the SP specimens with a slit of 0.3 mm depth fractured at the semicircular slit (figure 8). Therefore, the depth of a surface slit that can be rendered harmless by SP was 0.15 mm for the 460 and 670 HV specimens according to conditions (a) and (b). It is reported that the surface slit with a 0.15 mm depth was rendered harmless by SP in 500 HV spring steel SUP9A [12]. Thus, SUP7 tested in our experiments shows similar results. The surface slit of 0.15 mm depth reduced the torsional fatigue strength by 35% and 36%, respectively, in the 460 and 670 HV specimens. Thus, SP is a useful surface treatment technique for increasing the structural integrity of components subjected to cyclic torsion, even if the components have a detrimental surface defect.

In the 540 HV specimens, the torsional fatigue limit of SP specimens with a slit of 0.15 mm depth was lower than that of SP specimens without defects (figure 6(b)). Moreover, the SP specimens with a slit of 0.15 mm depth fractured at the semicircular slit (figure 7). Therefore, in the 540 HV specimen, the surface slit of 0.15 mm depth was not rendered harmless by SP according to conditions (a) and (b). However, the torsional fatigue limit was increased by 143% in this specimen.

From the above results, it was found that the defect size that can be rendered harmless depends on the specimen hardness. Considering the torsional fatigue strength and defect size that can be rendered harmless by SP, 670 HV is a suitable Vickers hardness value for components.

4. Conclusions

(1) SP increased the torsional fatigue limits of the 460, 540, and 670 HV specimens by 8%–67%, 33%–143%, and 36%–127%, respectively. Compressive residual stresses are the main reason for this increase in the torsional fatigue limit.

(2) In the 460 and 670 HV specimens, the torsional fatigue limit of the SP specimens with a slit of 0.15 mm depth was equivalent to that of the SP specimens without defects. Moreover, these specimens fractured outside of the slit. Therefore, the size of the surface slit that can be rendered harmless by SP was 0.15 mm.

(3) In the 540 HV specimens, the torsional fatigue limit of the SP specimens with a slit of 0.15 mm depth was lower than that of the SP specimens without defects. Moreover, the SP specimens with a slit of 0.15 mm depth fractured at the slit. Therefore, the depth of the surface slit that can be rendered harmless was less than 0.15 mm for the 540 HV specimens.

(4) SP is a useful surface treatment technique for increasing the structural integrity of components subjected to cyclic torsion, even if these components have a detrimental surface defect.

Acknowledgment

A part of this research has been conducted as a collaborative research project in a research center for Green Material Innovation at Yokohama National University. The authors gratefully acknowledge members of the collaborative research project.

References

[1] Endo M and Murakami Y 1987 Effects of an artificial small defect on torsional fatigue strength of steels *J. Eng. Mater. Technol.* **109** 124–9.

[2] Billaudeau T, Nadot Y and Bezine G 2004 Multiaxial fatigue limit for defective materials: mechanisms and experiments *Acta Mater.* **52** 3911–20.

[3] Murakami Y and Takahashi K 1998 Torsional fatigue of a medium carbon steel containing an initial small surface crack introduced by tension-compression fatigue: Crack branching, non-propagation and fatigue limit *Fatigue Fracture Eng. Mater. Struct* **21** 1473–84.

[4] Beretta S, Foletti S and Valiullin K 2011 Fatigue strength for small shallow defects/cracks in torsion *Int. J. Fatigue* **33** 287–99.

[5] Wakita M, Kuno T, Amano A, Nemoto A, Saruki K and Tanaka K 2007 Study on predicting formula for torsional fatigue strength of spring steel—effect of environment, notch and hardness *Transactions of Japan Society of Spring Engineers* **2007** 1–7.

[6] Takahashi K, Amano T, Miyamoto T, Ando K, Takahashi F, Tange A, Okada H and Ono Y 2007 Improvement of fatigue strength by shot peening for spring steel specimens containing an artificial surface defect *Transactions of Japan Society of Spring Engineers* **2007** 9–13.

[7] Acceptable Defect Size Research Committee 2008 The research committee report of the acceptable defect size for spring steel, *Transactions of Japan Society of Spring Engineers* **2008** 57–66.

[8] Takahashi K, Hayashi T, Ando K and Takahashi F 2010 Evaluation of acceptable defect size by shot peening based on fracture mechanics *Transactions of Japan Society of Spring Engineers* **2010** 25–30.

[9] Takahashi K, Amano T, Ando K and Takahashi F 2011 Improvement of fatigue limit by shot peening for high-strength steel containing a crack-like surface defect *Int. J. Struct. Integr.* **2** 281–92.

[10] Nakagawa M, Takahashi K, Osada T, Okada H and Koike H 2014 Improvement in fatigue limit by shot peening for high-strength steel containing crack-like surface defect (influence of surface crack aspect ratio) *Transactions of Japan Society of Spring Engineers* **2014** 13–18.

[11] Yasuda J, Takahashi K and Okada H 2014 Improvement of fatigue limit by shot peening for high-strength steel containing a crack-like surface defect: Influence of stress ratio *Int. J. Struct. Integr.* **5** 45–59.

[12] Takahashi K, Okada H and Ando K 2012 Effects of shot peening on the torsional fatigue limit of high strength steel containing an artificial surface defect *Int. J. Struct. Integr.* **3** 274–84.

[13] Kuno T, Wakita M, Hasegawa T, Saruki K and Tanaka K 2010 Effect of hardness and shot peening on torsional fatigue strength of high strength spring steel *Transactions of Japan Society of Spring Engineers* **2010** 19–24.

[14] Okada H, Tange A and Ando K 2003 Relationship among specimen's hardness, residual stress distribution and yield stress on the difference of shot peening methods *J. High Pressure Inst. Japan* **41** 233–42.

IOP Conf. Series: Journal of Physics: Conf. Series **842** (2017) 012067 doi:10.1088/1742-6596/842/1/012067

Fatigue limit prediction of ferritic-pearlitic ductile cast iron considering stress ratio and notch size

T Deguchi[1], H J Kim[2,3] and T Ikeda[4]

[1] Graduate School of Engineering, Fukuoka University, 8-19-1, Nanakuma, Jonan-ku, Fukuoka, 814-0180, Japan
[2] Department of Mechanical Engineering, Fukuoka University, 8-19-1, Nanakuma, Jonan-ku, Fukuoka, 814-0180, Japan
[3] Institute of Materials Science and Technology, Fukuoka University, 8-19-1, Nanakuma, Jonan-ku, Fukuoka, 814-0180, Japan
[4] R&D Center HINODE, Ltd., Iwasaki, Harakoga, Miyaki-cho, Miyaki-gun, Saga, 849-0101, Japan

E-mail: td160004@cis.fukuoka-u.ac.jp

Abstract. The mechanical behavior of ductile cast iron is governed by graphite particles and casting defects in the microstructures, which can significantly decrease the fatigue strength. In our previous study, the fatigue limit of ferritic-pearlitic ductile cast iron specimens with small defects ($\sqrt{area}$ = 80 ~1500 μm) could successfully be predicted based on the $\sqrt{area}$ parameter model by using $\sqrt{area}$ as a geometrical parameter of defect as well as the tensile strength as a material parameter. In addition, the fatigue limit for larger defects could be predicted based on the conventional fracture mechanics approach. In this study, rotating bending and tension-compression fatigue tests with ferritic-pearlitic ductile cast iron containing circumferential sharp notches as well as smooth specimens were performed to investigate quantitatively the effects of defect. The notch depths ranged 10 ~ 2500 μm and the notch root radii were 5 and 50 μm. The stress ratios were $R = -1$ and 0.1. The microscopic observation of crack propagation near fatigue limit revealed that the fatigue limit was determined by the threshold condition for propagation of a small crack emanating from graphite particles. The fatigue limit could be successfully predicted as a function of R using a method proposed in this study.

1. Introduction

Because of their low cost as well as excellent workability, ductile cast irons have been used for various machine parts and structural components such as crankshafts, railways and gas pipes. Ductile cast irons have good monotonic strength and high ductility compared to gray cast irons and malleable cast irons. However, the fatigue strength of ductile cast irons is relatively lower than those of the steels and alloys with the same level of monotonic strength because of their peculiar microstructure containing graphite particles and casting defects.

The fatigue strength of ductile cast iron was affected by the size and distribution of casting defects and graphite particles and by the phase distribution of matrix materials. Many researchers investigated the effect of casting defects on the fatigue strength and assessed the fatigue limit of ductile cast iron using matrix hardness such as Vickers hardness, *HV*. Nodat et al. [1] and Shiota et al. [2] reported that fatigue strength was decreased with the increase of the size of graphite particles and casting defects. Further, the distribution of graphite particles and casting defects affected the fatigue strength of ductile cast iron [3-5]. Noguchi et al. [6] assessed the effects of defects in ductile cast iron based on the fracture mechanics and the nondestructive method. It was shown that the matrix hardness had a remarkable effect on the fatigue strength of ductile cast iron. Sofue [7] investigated the effect of matrix hardness on the fatigue strength of ductile cast iron with single-phase matrix. It was reported that the fatigue limit was increased with the increase of matrix hardness. Several researchers [3, 8-11] proposed the fatigue limit predictions using *HV* for ductile cast iron with single-phase matrix. The prediction using *HV* of the matrix showed a good agreement with the experimental data. Therefore, the fatigue limit prediction using *HV* of ductile cast iron with single-phase matrix is possible. However, when a ductile cast iron contains two-phase matrix, the measurement of *HV* of the matrix is impossible. This is because ferrite and pearlite phases are evenly distributed within the matrix structure, in addition to the influence of interference by graphite particles.

In our previous study [12, 13], we proposed a fatigue limit prediction for ferritic-pearlitic ductile cast iron based on the $\sqrt{area}$ parameter model by using the tensile strength, σ_B, as material parameter, instead of *HV*, for stress ratio $R = -1$. This was because, in the case of ductile cast iron with two-phase matrix, the measurement of correct value of *HV* was quite difficult, as mentioned above. Alternatively, the prediction using σ_B as a material parameter shows a reasonably good agreement with the experimental data. However, ductile cast iron is frequently used under the condition where the mean stress is applied (i.e., $R \neq -1$). Accordingly, for safety designing of the products with ductile cast iron, it is necessary to understand the effect of stress ratio, *R*, on the fatigue limit.

In this study, we investigated the fatigue limit of a ductile cast iron with a two-phase matrix of almost evenly distributed ferrite and pearlite phases by using the specimens containing circumferential notches with a wide range of sizes at $R = 0.1$ in addition to $R = -1$. The purpose of this study is to present a simple yet useful method for prediction of the fatigue limit as a function of *R*.

2. Experimental procedures

The material investigated was an as-cast ductile cast iron. The chemical composition is listed in Table 1. The microstructure is shown in Figure 1. The area fractions in the microstructure were 10.5% for graphite, 45.3% for ferrite and 44.2% for pearlite, respectively. The ultimate tensile strength, σ_B, was 552 MPa. The shapes and dimensions of test specimens are shown in Figure 2(a) and 2(b). After lathe turning of specimen, the surface was finished with an emery paper up to #1000 and then by buffing with an alumina paste. Thereafter, a circumferential notch shown in Figure 2(c) was introduced into a specimen. Before the fatigue test, the surface layer of about 10 μm in thickness was removed from the specimens by electro-polishing.

The rotating bending fatigue tests were carried out using a machine of uniform moment type, and its capacity was 100 Nm and operating frequency was 50 - 67 Hz. The tension-compression fatigue tests

Table 1. Chemical compositions (wt.%.).

C	Si	Mn	P	S	Cu	Mg
3.84	2.5	0.66	0.017	0.009	0.21	0.043

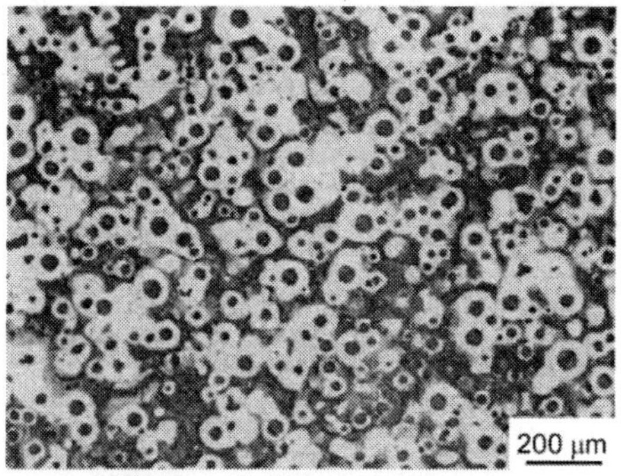

Figure 1. Microstructure.

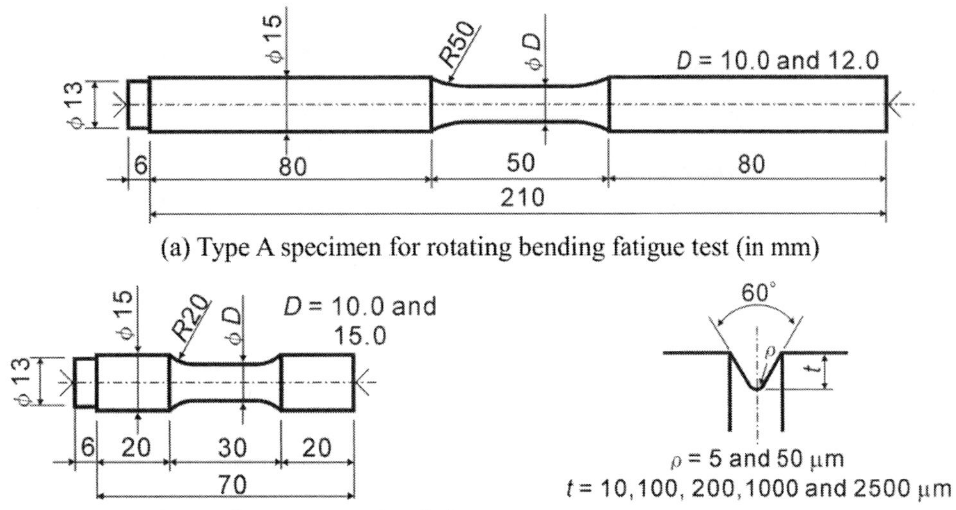

(a) Type A specimen for rotating bending fatigue test (in mm)

(b) Type B specimen for tension-compression fatigue test (in mm)

(c) Circumferential notches

Figure 2. Shapes and dimensions of test specimens and circumferential notches.

were carried out using a uniaxial hydraulic fatigue test machine at a frequency of 35 Hz. The stress ratios, R, were -1 and 0.1. Concerning the shallow-notched specimens (t = 10, 100 and 200 μm), the nominal stress was calculated with the average diameter of specimen, D. On the other hand, concerning the deep-notched specimens (t = 1000 and 2500 μm), the nominal stress was calculated with the minimum diameter of specimen, $D - 2t$. The fatigue limit was defined by the maximum stress amplitude, σ_a, for which a specimen endured $N = 10^7$ cycles without failure.

3. Results and discussion

3.1. *Crack growth and non-propagation behavior*

Figure 3 shows the *S-N* diagram of smooth and notched specimens at the stress ratios of $R = -1$ and 0.1. The results for notched specimens for t = 2500 μm at $R = -1$ and those for t = 100 μm and 2500 μm at R = 0.1 were obtained by the tension-compression fatigue tests. The results of other specimens were

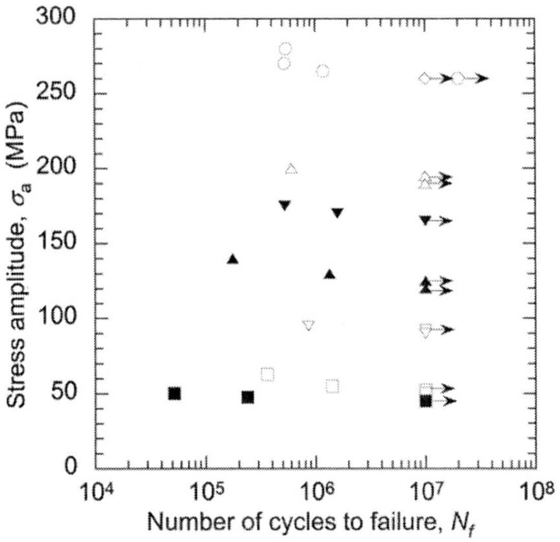

Specimen		Type	ρ (μm)	t (μm)	Fatigue limit (MPa)	Loading Type	Symbol
Smooth	R = -1	A	-		260	RB	○
Notched	R = -1	A	5	10	260	RB	◇
		A	50	100	195	RB	△
		A	50	200	165	RB	▼
		A	50	1000	92.5	RB	▽
		B	50	2500	52.5	TC	□
	R = 0.1	B	50	100	125	TC	▲
		B	50	2500	45	TC	■

RB : Rotating bending, TC : Tension-compression

Figure 3. *S-N* data.

IOP Conf. Series: Journal of Physics: Conf. Series **842** (2017) 012067 doi:10.1088/1742-6596/842/1/012067

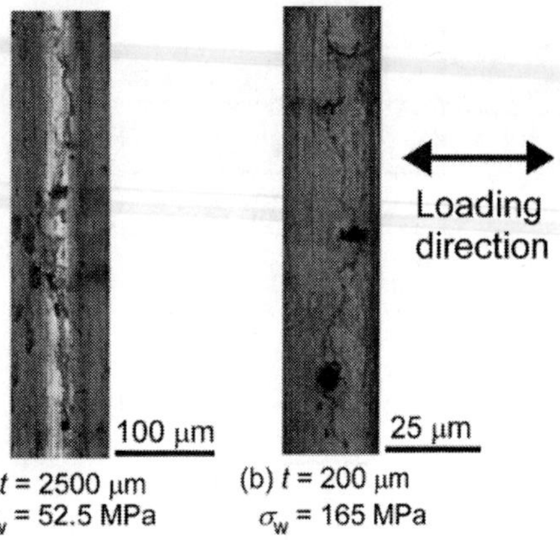

Figure 4. Non-propagating cracks emanating from the notch at the fatigue limit for R = -1 [13].

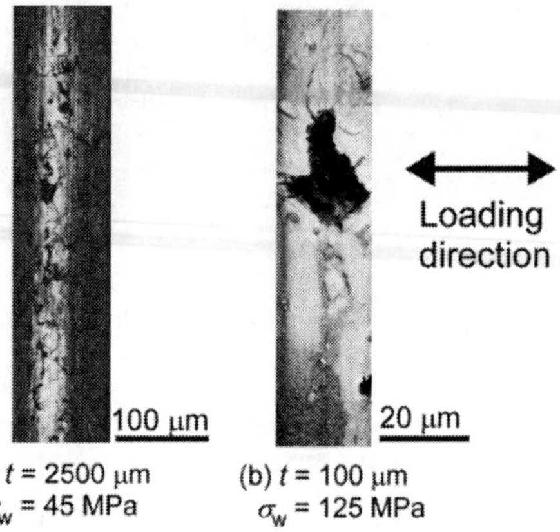

Figure 5. Non-propagating cracks emanating from the notch at the fatigue limit for R = 0.1.

obtained by the rotating bending fatigue tests. The fatigue strength decreases significantly with an increase in notch size at $R = -1$ as well as 0.1. Especially, the fatigue strength of notched specimen with t = 100 μm at R = 0.1 (σ_w = 125 MPa) decreased significantly compared to that of the identical specimen tested at $R = -1$ (σ_w = 195 MPa). On the other hand, the fatigue strength of notched specimens with t = 2500 μm at R = 0.1 (σ_w = 45 MPa) was slightly lower than that of the identical specimen tested at R = −1 (σ_w = 52.5 MPa).

The non-propagation of cracks was observed at the root of shallow- and deep-notched specimens at fatigue limit, irrespective of notch sizes and stress ratios, as shown in Figures 4 and 5. In the case of notched specimens, a lot of fatigue cracks initiated from graphite particles at the notch root, and they stopped the propagation before $N = 10^7$ cycles. In the case of smooth specimens, several fatigue cracks emanated from different graphite particles at the specimen surface and they coalesced, but they ceased propagation before $N = 10^7$ cycles [13]. Consequently, all those cracks can be regarded as a non-propagating crack and it is concluded that the fatigue limit is determined by the threshold condition for crack propagation, regardless of notch sizes and stress ratios.

3.2. Prediction of fatigue limit of ductile cast iron

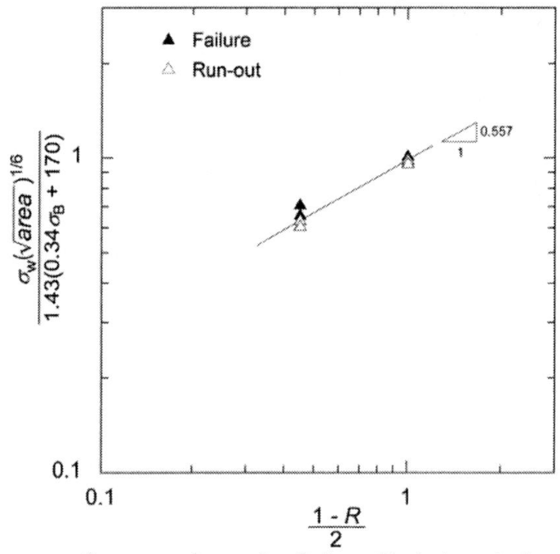

Figure 6. Influence of stress ratio on the fatigue limit (notch depth t = 100 μm).

In our previous study [12, 13], a fatigue limit prediction for ferritic-pearlitic ductile cast iron based on the $\sqrt{area}$ parameter model by using ultimate tensile strength, σ_B, instead of the Vickers hardness, HV, was proposed. Let us recall that a predictive equation of fatigue limit for smooth specimens, σ_{w0}, was rendered as:

$$\sigma_{w0} = 0.25\sigma_B + 110 \tag{1}$$

where σ_{w0} and σ_B are in MPa. Further, the following form of a predictive equation for the fatigue limit in the presence of defect, σ_w, for ductile cast irons was proposed by taking advantage of the $\sqrt{area}$ parameter model:

$$\sigma_w = \frac{F_{loc}(0.34\sigma_B + 170)}{(\sqrt{area})^{1/6}} \tag{2}$$

where F_{loc} is the correction coefficient for the location of small defect being 1.43 for surface defects, 1.56 for internal defects and 1.41 for defects just in contact with the surface [14]. Here, σ_w is in MPa and $\sqrt{area}$ is in μm. Similarly, based on the $\sqrt{area}$ parameter model, it is expected that the effect of stress ratio, R, is estimated by using a term of $\{(1\text{-}R)/2\}^{\alpha}$ as follows:

$$\sigma_w = \frac{F_{loc}(0.34\sigma_B + 170)}{(\sqrt{area})^{1/6}} \cdot \left(\frac{1-R}{2}\right)^{\alpha} \tag{3}$$

where α is a correction coefficient that depends on the material. To determine the value of α in Eq. (3), the value of $\sigma_w(\sqrt{area})^{1/6}/\{F_{loc}(0.34\sigma_B + 170)\}$ is shown as a function of $(1-R)/2$ on a logarithmic graph in Figure 6. A value of $\alpha = 0.557$ was determined from the slope of the line. The present value of $\alpha = 0.557$ is significantly higher than the value of $\alpha = 0.402$ for ductile cast iron reported in the literature [15]. The reason of higher value of α is unclear at present, and it is necessary to conduct fatigue tests with other stress ratios.

Figure 7 shows a comparison of prediction by Eqs. (1) and (3) with the experimental results for smooth specimen as well as shallow-notched specimen. The prediction of fatigue limit is in reasonable

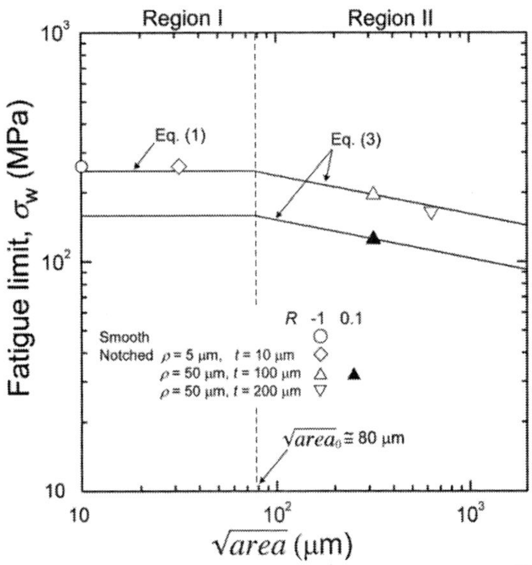

Figure 7. Relationship between fatigue limit and $\sqrt{area}$.

agreement with the experimental data. Irrespective of the stress ratio, there is a critical size of defect that has no effect on the fatigue strength of ferritic-pearlitic ductile cast iron. Defects with the $\sqrt{area}$ smaller than about 80 μm (region I), designated herein as $\sqrt{area}_0$, are not harmful to the fatigue limit. Prediction with Eq. (3) shows a good agreement with the experimental results for the artificial defects with 80 μm$\leq \sqrt{area} \leq$1000 μm (region II).

In the case of deep-notched specimen with t = 1000 μm and 2500 μm (cf. Figure 2 (c)), the fatigue limit can no longer be predicted by Eq. (3). This is because the notch size is sufficiently large beyond the region II. In this circumstance, the threshold stress intensity factor (SIF) range, ΔK_{th}, become a material constant, which is ΔK_{th} for a long crack, $\Delta K_{th,lc}$. The relationship between the fatigue limit range, $\Delta\sigma_w = 2\sigma_w$, and the threshold range, ΔK_{th}, can be derived by using the following equation based on the linear elastic fracture mechanics analysis [15, 16]:

$$K_{\mathrm{Imax}} = 0.65\sigma_0\sqrt{\pi\sqrt{area}} \tag{4}$$

where K_{Imax} is the maximum value of SIF along the front of a three-dimensional surface crack with the arbitrary shape existing in a semi-infinite body with a Poisson's ratio of 0.3 under remote tensile stress σ_0. By setting $K_{\mathrm{Imax}} = \Delta K_{th}/2$, $\sigma_0 = \sigma_w$ and F_{loc} = 1.43, Eqs. (2) and (4) render the following equation:

$$\Delta K_{th} = 3.3\times10^{-3}(0.34\sigma_B + 170)(\sqrt{area})^{1/3} \tag{5}$$

where ΔK_{th} is in MPa$\sqrt{\mathrm{m}}$, σ_B is in MPa and $\sqrt{area}$ is in μm. The effect of stress ratio can be estimated by using the term of $\{(1-R)/2\}^{\alpha}$, thus Eq. (5) becomes:

$$\Delta K_{th} = 3.3\times10^{-3}(0.34\sigma_B + 170)(\sqrt{area})^{1/3}\cdot\left(\frac{1-R}{2}\right)^{\alpha} \tag{6}$$

Figure 8 shows relationship between ΔK_{th} and $\sqrt{area}$ for R = −1 and 0.1. In the region II, the solid lines calculated by Eq. (6) shows the variations of ΔK_{th} for small defect. Prediction line shows a reasonably good agreement with the experimental results for the small defects. In the region III, where ΔK_{th} is constant, the fatigue limit can be estimated by the conventional fracture mechanics approach established for long cracks. The experiment result of the $\Delta K_{th,lc}$ with the large defects with t = 1000 and

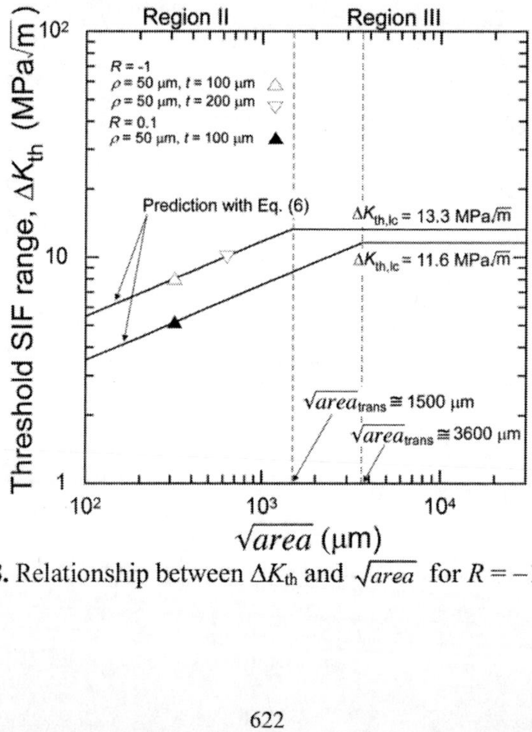

Figure 8. Relationship between ΔK_{th} and $\sqrt{area}$ for R = −1 and 0.1.

2500 μm was 13.3 MPa$\sqrt{\text{m}}$ at $R = -1$ by using a formula [17]. The translation value of $\sqrt{area}$, termed as $\sqrt{area}_{\text{trans}}$, represents the boundary between region II and III. By substituting ΔK_{th} = 13.3 MPa$\sqrt{\text{m}}$ σ_{B} = 552 MPa and $R = -1$ into Eq. (6), $\sqrt{area}_{\text{trans}} \cong 1500$ μm for $R = -1$ was obtained. Similarly, by substituting ΔK_{th} = 11.6 MPa$\sqrt{\text{m}}$, σ_{B} = 552 MPa and $R = 0.1$ into Eq. (6), $\sqrt{area}_{\text{trans}} \cong 3600$ μm for R = 0.1 was obtained.

For the region III, we express the relationship between $\Delta K_{\text{th,lc}}$ and $\Delta K_{\text{th,lc}|R=-1}$ by using the term of $\{(1-R)/2\}^{\alpha^*}$ as follows:

$$\Delta K_{\text{th,lc}} = \Delta K_{\text{th,lc}|R=-1} \cdot \left(\frac{1-R}{2}\right)^{\alpha^*} \tag{7}$$

where $\Delta K_{\text{th,lc}}$ and $\Delta K_{\text{th,lc}|R=-1}$ are in MPa$\sqrt{\text{m}}$. By substituting $\Delta K_{\text{th,lc}}$ = 11.6 MPa$\sqrt{\text{m}}$, $\Delta K_{\text{th,lc}|R=-1}$ = 13.3 MPa$\sqrt{\text{m}}$ and $R = 0.1$ into Eq. (7), $\alpha^* = 0.171$ was obtained. Accordingly, the different mean stress effect is expected for the region II ($\alpha = 0.557$) and the region III ($\alpha^* = 0.171$).

4. Conclusions

The effects of notch size and stress ratio on the fatigue limit of ferrite-pearlite ductile cast iron were investigated. The principal results and conclusions are obtained as follows:

(1) From the microscopic observation of crack growth, it is revealed that the fatigue limit is determined by the threshold condition for propagation of a crack emanating from graphite particles, regardless of notch size and stress ratio.

(2) The effect of defects on the fatigue limit can be classified into three regions depending on the defect size. In the region I where $\sqrt{area}$ of defect is smaller than 80 μm, the defect can be regarded as a non-damaging defect and the fatigue limit is the same as that of smooth specimen.

(3) In the region II, the fatigue limit can be quantitatively predicted based on the $\sqrt{area}$ parameter model in terms of a geometrical parameter of defects, $\sqrt{area}$, and a material parameter, σ_{B}. Furthermore, the fatigue limit that depends on the effect of stress ratio can be described by the term of $\{(1-R)/2\}^{\alpha}$. The value of $\alpha = 0.557$ was obtained for the investigated ferritic-pearlitic ductile cast iron.

(4) In the region III in which $\sqrt{area}$ >1500 μm for $R = -1$ and $\sqrt{area}$ >3600 μm for $R = 0.1$, the threshold SIF range, ΔK_{th}, becomes a material constant ($\Delta K_{\text{th,lc}}$ = 13.3 MPa$\sqrt{\text{m}}$ for $R = -1$ and $\Delta K_{\text{th,lc}}$ = 11.6 MPa$\sqrt{\text{m}}$ for $R = 0.1$) and the fatigue limit can be estimated by the conventional fracture mechanics approach.

References

[1] Nadot Y, Mendez J and Ranganathan N 2004 *Int. J. Fatigue* **26** 311
[2] Shiota T and Komatsu S 1982 *Imono* **54** 434
[3] Yamabe J, Kobayashi M and Nakajima N 2005 *Trans. Jpn. Soc. Mech. Eng.* **71** 116
[4] Nadot Y, Mendez J, Ranganathan N and Beranger S 1999 *Fatigue Fract. Engng. Mater. Struct.* **22** 289
[5] Horikawa N, Sameshima D, Nakamura T and Oguma H 2010 *J. Soc. Mater. Sci., Japan* **76** 425
[6] Noguchi T and Horikawa N 2013 *J. FES.* **85** 852
[7] Sofue M 1976 *Imono* **48** 441
[8] Endo M 1989 *Trans. Jpn. Soc. Mech. Eng.* **38** 1139
[9] Tanaka Y, Yang Z and Miyamoto K 1995 *Mater. Trans.* **36** 749

[10] Endo M and Yanase K 2014 *Theor. Appl. Fract. Mec.* **69** 34
[11] Costa N, Machado N and Silva F S 2010 *Int. J. Fatigue* **32** 988
[12] Deguchi T, Matsuo T, Takemoto S, Ikeda T and Endo M 2016 *International Journal of Fracture Fatigue and Wear* **4** 93
[13] Deguchi T, Matsuo T, Kim H J, Ikeda T and Endo M 2016 *11th Int. Symp. on Advanced Scienc and Technology in Experimental Mechanics* Ho Chi Minh
[14] Murakami Y 2002 *Metal Fatigue: Effects of Small Defects and Nonmetallic Inclusions*, (Elsevier)
[15] Endo M and Iseda K 2006 *Int. J. Mod. Phys. B* **20** 3817
[16] Murakami Y and Endo M 1983 *Eng. Fract. Mech.* **17** 1
[17] Benthem J P and Koiter W T 1973 *Asymptotic approximations to crack problems In: Mechanics of Fracture :Methods of Analysis and Solutions of Crack Problems* Vol. 1 (Springer Netherlands) pp 131-178

Development of testing machine for tunnel inspection using multi-rotor UAV

Tatsuya IWAMOTO, Tomoya ENAKA, and Keijirou TADA

National Institute of Technology, Ariake College, 150 Higashihagio, Omuta, Fukuoka, Japan

Email: tiwamoto@ariake-nct.ac.jp

Abstract. Many concrete structures are deteriorating to dangerous levels throughout Japan. These concrete structures need to be inspected regularly to be sure that they are safe enough to be used. The inspection method for these concrete structures is typically the impact acoustic method. In the impact acoustic method, the worker taps the surface of the concrete with a hammer. Thus, it is necessary to set up scaffolding to access tunnel walls for inspection. Alternatively, aerial work platforms can be used. However, setting up scaffolding and aerial work platforms is not economical with regard to time or money. Therefore, we developed a testing machine using a multirotor UAV for tunnel inspection. This test machine flies by a plurality of rotors, and it is pushed along a concrete wall and moved by using rubber crawlers. The impact acoustic method is used in this testing machine. This testing machine has a hammer to make an impact, and a microphone to acquire the impact sound. The impact sound is converted into an electrical signal and is wirelessly transmitted to the computer. At the same time, the position of the testing machine is measured by image processing using a camera. The weight and dimensions of the testing machine are approximately 1.25 kg and 500 mm by 500 mm by 250 mm, respectively.

1. Introduction

Concrete structures, including buildings and bridges, were built in quantity during a high-economic-growth period in Japan. The percentages of these structures are approximately 40% of all bridges and approximately 25 % of all tunnels. Many concrete structures are now at least 40 years old. The deterioration of old concrete structures is a serious problem. Accidents caused by falling concrete pieces have often occurred in recent years. For example, a concrete piece fell from Kuyamakawa bridge on the Nagasaki Expressway in January 2009. The concrete piece hit a moving vehicle. Fortunately, no one was injured in this accident. However, this is a problem that must be solved immediately because falling concrete pieces have a high probability of causing third-party damage.

A reconstruction of the structure is one solution to this problem. However, rebuilding an entire structure is difficult given recent economic conditions. In addition, with regard to CO_2 reduction, repairing only the problem parts is good for the environment. Because reconstruction of the entire structure generates a great deal more CO_2 than the repair of a problem part, it have a bad influence on the environment. Therefore, a way to correctly detect a problem point is required.

The typical inspection method for these concrete structures is the impact acoustic method carried out by workers. In the impact acoustic method, the worker taps the surface of the concrete with a hammer. Thus, it is necessary to set up scaffolding to access tunnel walls for inspection. Alternatively, aerial work platforms can be used. However, setting up scaffolding and aerial work platforms is not economical with regard to time or money. Many structures need to be checked, and the inspection area is wide, so inspection methods that excel in high working efficiency is requested. As a solution to such a problem, the method has been proposed. For instance, health monitoring of structures, damage assessment using vibration data and inspection methods using a robot. Zhou et al. proposed an approach for detecting structural damage using transmissibility with hierarchical clustering and similarity analysis [1–3]. Gillich et al.

IOP Conf. Series: Journal of Physics: Conf. Series **842** (2017) 012068 doi:10.1088/1742-6596/842/1/012068

The use of robots has attracted attention as method to check large civil-engineering structures. Takada *et al.* developed a bridge inspection robot using a permanent magnet [5]. Tahara *et al.* developed a multirotor helicopter and used it for high-altitude inspections and surveys of disaster sites [6]. Some wall-climbing robots were developed that test concrete non-destructively than the above [7-8].

In this study, we developed a testing machine using a multirotor UAV for tunnel inspection as an efficient inspection method.

2. Testing machine using multirotor UAV

Figure 1 shows a schematic diagram of a testing machine that uses a multirotor UAV. The appearance of the developed testing machine is shown in Figure 2. F330 (DJI) was used as the base of the testing machine. The weight and dimensions of the testing machine are approximately 1.25 kg and 500 mm by 500 mm by 250 mm, respectively. The testing machine consists of a thrust part (to generate thrust to press the machine against the ceiling), a traversing part (to move across the ceiling), a control section such as a microcomputer, a flight controller, a wireless communication device, a measurement part such as a hammering device, a microphone, and a battery. The position of the testing machine is measured by image processing using a camera. The constitution of all parts is as follows:

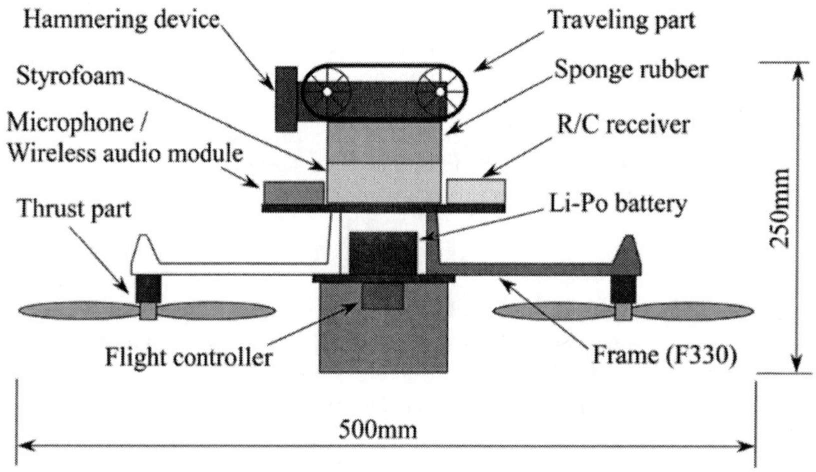

Figure 1. Schematic diagram of testing machine.

Figure 2. Appearance of testing machine.

IOP Conf. Series: Journal of Physics: Conf. Series **842** (2017) 012068 doi:10.1088/1742-6596/842/1/012068

2.1. Thrust part

The thrust part consists of four sets of propellers and brushless motors. The brushless motors are controlled by the speed controller. An electronic scale (with a smallest indication of 1 g) was used to measure the thrust. Figure 3 shows a schematic diagram of the thrust measurement. Electric power was supplied by a DC switching power supply (12 V, 1 kW), and the electric current was measured using a clamp-type ammeter. Testing machine was inversely fixed on an electronic scale. The throttle of the controller was adjusted, and the thrust and electric current were measured. The relationship between the electrical current and thrust are shown in Figure 4. From Figure 4, the maximum thrust of propeller 1 (8 × 4. 5) was 16.9 N, and that of propeller 2 (9 × 4. 5) was 22.45 N. Therefore, propeller 2 was adopted for use.

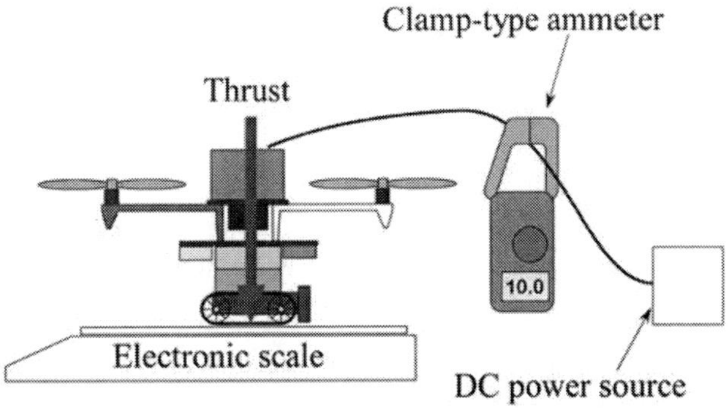

Figure 3. Schematic diagram of thrust measurement.

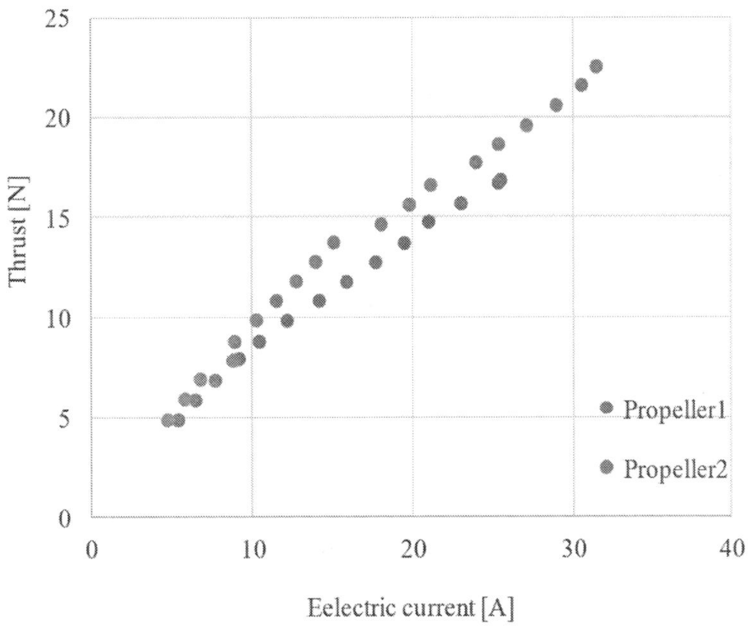

Figure 4. Relationship between electric current and thrust.

2.2. Traveling part
A tracked robot platform (Zumo robot by Arduino, Popolu) was used for the traveling part. The traveling part was forced into the ceiling by thrust, and it drives a crawler and moves the robot. Turning is also possible.

2.3. Control part
A flight controller (NAZA-M Lie, DJI) controls the brushless motors to produce stable flight. In addition, a microcomputer controls the tracked robot platform. A radio control system was used for the remote control of each controller.

2.4. Measurement part
The measurement part consists of a hammer to make an impact, and a microphone to acquire the impact sound. A push solenoid (CB0730, Takaha Kiko Co., Ltd.) was used for the hammer. The measurement value is converted into an electrical signal, and is transmitted to a computer using the wireless audio module (CPI-WAM001, CPI Technologies, Inc.).

2.5. Positioning system
The positioning system consists of a camera and a personal computer. First, a picture of the testing machine with a motion marker is acquired using the camera. Next, the location of the marker is estimated by image analysis. NI LabVIEW was used for image analyses.

3. Verification of traveling performance
To verify the traveling performance, a movement test was carried out using inclined composite panels. Figure 5 shows a schematic diagram of a moving test, and a situation for a moving test is shown in Figure 6. A movement test includes the following three patterns: (1) movement in horizontal direction (x-direction), (2) movement in the inclination direction (y-direction), and (3) turning movement. The results are as follows. When the steerability is sufficient, the results are good. The case in which the robot can move barely is poor. Control impossibility is N/A. The results of a verification test for traveling performance are listed in Table 1.

Up to an angle of inclination of 15°, all tests were good. At more than 20°, the y-direction movement and turning movement were poor. Because the crawler only partly contacts a composite panel in this case, contact areas necessary for movement are not sufficient, as shown in Figure 7. Furthermore, if the angle is greater than 45°, only movement in the x-direction is possible. Finally, when the angle was at least 55°, steering was impossible.

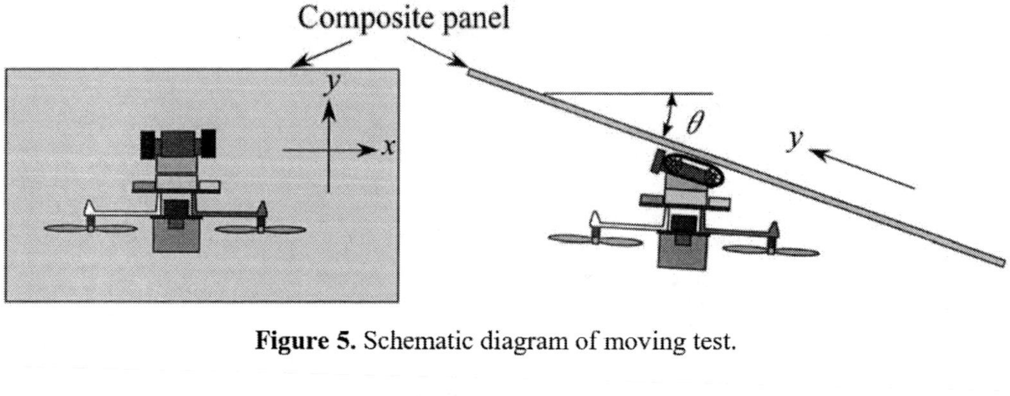

Figure 5. Schematic diagram of moving test.

Figure 6. Photograph of moving test.

Table 1. Results of verification test of traveling performance.

Angle of inclination θ [°]	x-direction movement	y-direction movement	Turning movement
0	Good	Good	Good
10	Good	Good	Good
15	Good	Good	Good
20	Good	Poor	Poor
25	Good	Poor	Poor
30	Good	Poor	Poor
35	Good	Poor	Poor
40	Good	Poor	Poor
45	Good	N/A	N/A
50	Good	N/A	N/A
55	N/A	N/A	N/A

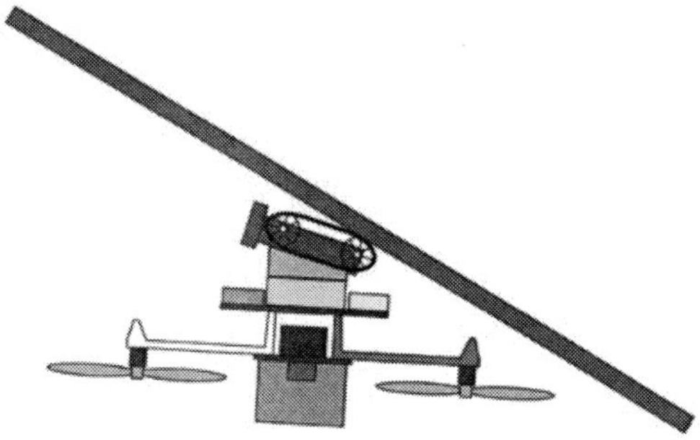

Figure 7. Situation in which part of crawler contacts the ceiling.

4. Verification of inspection performance

To verify the inspection performance, the propeller noise and affect sound were measured using a microphone for sound measurement. The experimental results are shown in Figure 8. Figure 8 (a) shows the frequency distribution of the propeller noise. An experimental result shows that the frequency mainly included in the propeller noise is less than 1 kHz. Therefore, if the frequency of the impact sound is higher than 1 kHz, it is possible to detect it. Figure 8 (b) shows the frequency distribution of the impact sound. It seems that this peak is a frequency of the impact sound. In addition, the measurement results of hitting concrete pieces during flight are shown in the Figure 9. Figure 8 (a) shows measurement sound, and Figure 9 (b) shows the frequency distribution of the measurement sound. To reduce the influence of propeller noise, measurement results passes a high-pass filter (cut-off frequency is 2kHz). Both figures can recognize the impact sound. Thus, the validity of the inspection performance has been confirmed.

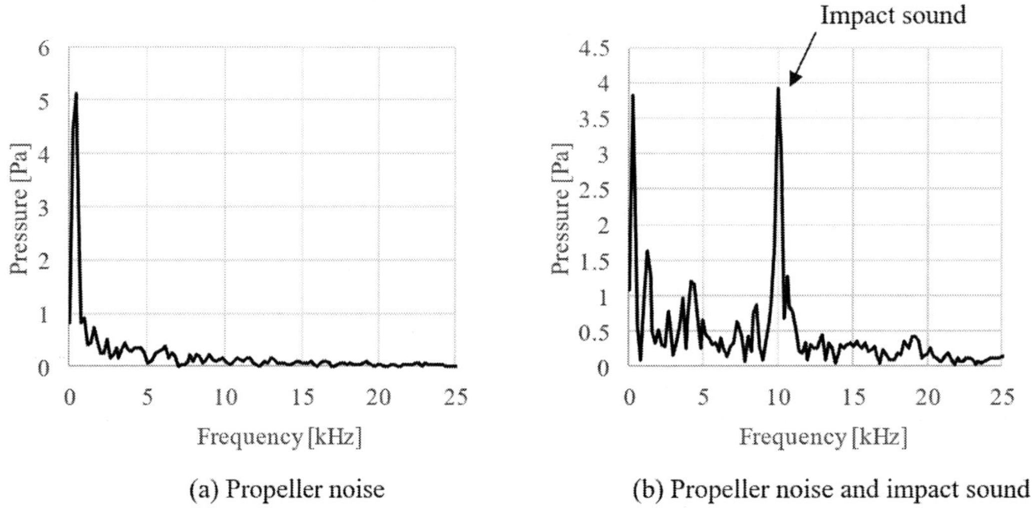

Figure 8. Results of verification of impact acoustic test.

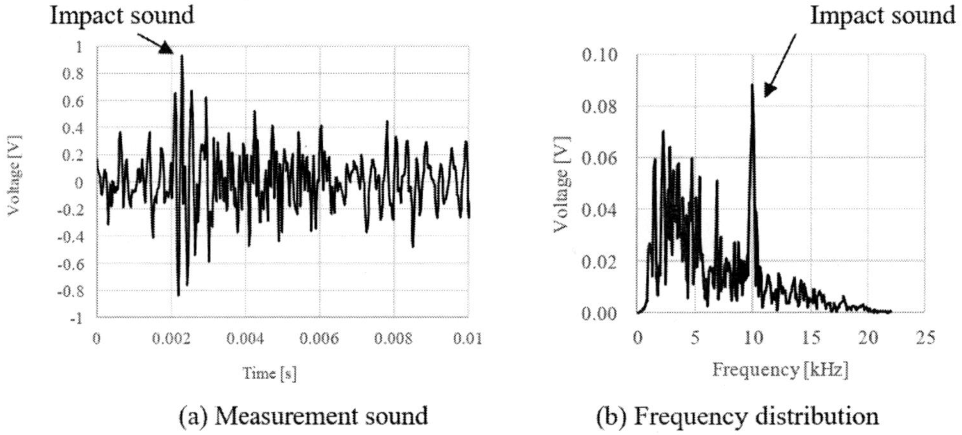

Figure 9. Measurement results during flight.

5. Conclusions

In this study, we developed a testing machine that can move along the walls at any angle, and mounted the inspection device for the tunnel ceiling. We obtained the following results:

(1) Using propeller thrust, the testing machine was developed to move across a ceiling. To verify the traveling performance, a movement test was carried out using inclined composite panels. As a result, up to an angle of inclination of 15°, the testing machine could move in all directions and turn.

(2) Impact acoustic test equipment that can be mounted was developed. To verify the inspection performance, the propeller noise and impact sound were measured using a microphone for sound measurement. The frequency mainly included in propeller noise is less than 1 kHz. Therefore, if the frequency of the impact sound is higher than 1 kHz, it is possible to detect it.

References

[1] Zhou Y-L, Maia NMM, Sampaio R, and Wahab MA 2016 *Struct. Health Monit.* doi: 10.1177/1475921716680849
[2] Zhou Y-L, Maia N, and Abdel Wahab M 2016 *J. Vib. Control* doi: 10.1177/1077546316674544
[3] Zhou Y-L and Abdel Wahab M 2016 *J. Vibroeng.* **18** 4491–4499[1] Schmidt D., Hillenbrand C. and Berns K.: Robotics, 11, pp. 1-13, April 2011
[4] Gillich G-R, Praisach Z-I, Abdel Wahab M, Gillich N, Mituletu IC and Nitescu C 2016 Shock and Vibration 2016(Article ID 2086274) 10 pages; http://dx.doi.org/10.1155/2016/2086274
[5] Takada Y, Kirimoto K, Tajiri T, and Kawai T 2013 *Trans. Jpn. Soc. Mech. Eng.* **79–805** 3135–3146
[6] Tawara M and Nonami K 2012 *Trans. Jpn. Soc. Mech. Eng.* **78–787** 872–882
[7] Elkmann N, Lucke M, Kruger T, Kunst D and Sturze T 2007 Proceedings of the 10th International Conference on Climbing and Walking Robots, pp. 169-176,
[8] Unver O, Uneri A, Aydemir A and Sitti M 2006 *Proceedings of the IEEE International Conference on Robotics and Autmation*, pp. 2329-2335
[9] Shang J, Bridge B, Sttar T P, Monal S and Brenner A 2007 *Proceedings of the 10th International Conference on Climbing and Walking Robots*, pp. 105-114
[10] Penko M-J, Haynes G-C, Saunders J-A, Cutkosky M-R, Rizzi A-A, Full R-J and Koditschek D-E 2008 *J. Field Rob.* **25 4-5**, pp. 223-242

Acknowledgments

This work was supported by JSPS KAKENHI Grant Number JP25820198.

IOP Conf. Series: Journal of Physics: Conf. Series **842** (2017) 012069 doi:10.1088/1742-6596/842/1/012069

Estimation of dispersion curves by combining Effective Elastic Constants and SAFE Method: A case study in a plate under stress

J E Quiroga[1], L Mujica[3], R Villamizar[2], M Ruiz[3] and J Camacho[3]

[1]Escuela de Ingeniería Mecánica, [2]Escuela de Ingenierías Eléctrica, Electrónica y de Telecomunicaciones (E3T), Universidad Industrial de Santander (UIS),Cra 27, Calle 9, C. Universitaria, Bucaramanga, Colombia.
3Universitat Politécnica de Catalunya (UPC) BARCELONATECH, Department of Mathematics, Escola d'Enginyeria de Barcelona Est. (EEBE), Campus Diagonal Besòs, Edifici A, C. Eduard Maristany, 10-14, 08019, Barcelona, Barcelona, Spain

Corresponding author: jabib@uis.edu.co

Abstract. This paper presents an approach to calculate dispersion curves for homogeneous and isotropic plates subject to stress, via Semi-Analytical Finite Element and the Effective Elastic Constants, since stresses in the waveguide modify the phase and group velocities of the lamb waves. In the proposed methodology an isotropic specimen subjected to anisotropic loading is emulated by proposing an equivalent stress-free anisotropic specimen. This approximation facilitates determining the dispersion curves by using the well-studied numerical solution for the stress-free cases. The lamb wave in anisotropic materials can be studied by means of the Effective Elastic Constants, which reduces the complexity of the numerical implementation. Finally, numerical data available in literature were used to validate the proposed methodology, where it could be demonstrated its effectiveness as approximated method.

1. Introduction

In last years, application of lamb waves in non-destructive tests is of a growing interest [1][2][3][4][5][6]. While bulk waves refers to waves propagated in an infinite media, guided waves require a boundary for their existence. Furthermore, when a guided wave is produced at least one dimension is smaller than the other ones. On the other hand, if the guided wave is propagating in a plate-like structure, the wave is referred as a Lamb wave. Propagation of lamb waves is a complex phenomenon, where phase and group velocity are frequency dependent and it can be graphically observed by means of called dispersion curves. The exact trace of the dispersion curves is obtained by using transcendent equations assembled by either the transfer matrix or the global matrix methods [7]. Recently, the Semi-Analytical Finite Element (SAFE) method has been proposed to compute rapidly and efficiently the dispersion curves for any constant cross section specimen. SAFE has been used to analyze wave modes in cylindrical waveguides [8][9][10][11], to calculate leaky lamb waves [12], to obtain the dispersion curves in a pipe elbow [13] and in materials with viscoelastic properties [14].

In SAFE methodology the waveguide is discretized over the cross section, while in an analytical solution is in the wave propagation direction. Based on a variational scheme by inserting the kinetic and potential energies into Hamilton´s equation, a system of linear equations can be constructed with the circular frequency and wave number as unknowns under stress-free condition. The unknowns can be solved using standard eigenvalue routines. Therefore, SAFE approach allows computing dispersion curves in waveguides with complex cross sections, such as: multilayered laminates [15] and rails [16][17][18][19], where it is often computationally complex to solve analytical solutions. For semi-infinite waveguides, SAFE has a better performance compared with the traditional FEM due to only the constant cross-section is considered, which reduces the computational cost. Additionally, SAFE does not have missing roots problem, found in matrix methods, when dispersion curves are computed [20].

The dispersion curves in stress-free isotropic and anisotropic specimens have been widely studied and can be adequately described based on the theory of elasticity. However, this formulation is insufficient to describe wave propagation in specimens under stress since small non-linearities in the stress-strain relationships become significant, which can be described by using the acoustoelasticity approach. It dictates the stress dependence of acoustic bulk wave velocities propagating in elastic media under stress. Few studies on dispersion of waves in a waveguide subjected to axial load using the acoustoelasticity have been reported [21][22][23]. On the other hand, the use of SAFE methodology for this problem is even more reduced. In [16][24][25] is demonstrated a proportionality between the stiffness matrix, required to describe the effect of axial load and the mass matrix, which makes the use of existing software (stress-free) trivial to obtain the dispersion curves for waveguides under stress.

Thus, this paper proposes the use of EEC to describe the anisotropy effect presents in the lamb wave dispersion curves when a load is applied along one direction in an isotropic plate, based on the SAFE scheme. Then, the isotropic specimen subject to loading in one direction is studied by proposing an equivalent stress-free anisotropic specimen.

2. Plate's SAFE Analytical Model

The SAFE methodology is in this case, implemented for an isotropic, homogenous stress-free plate (infinitely wide plate), where the wave propagates along direction x with wavenumber ξ_x and circular frequency ω (see fig. 1). Since, the cross-section lies in the y–z plane it is formulated as a two-dimensional problem (x,z) in the plane strain formulation. Thus, the harmonic displacement, stress and strain field components in cartesian coordinate at each point of the waveguide are expressed by equation (1)

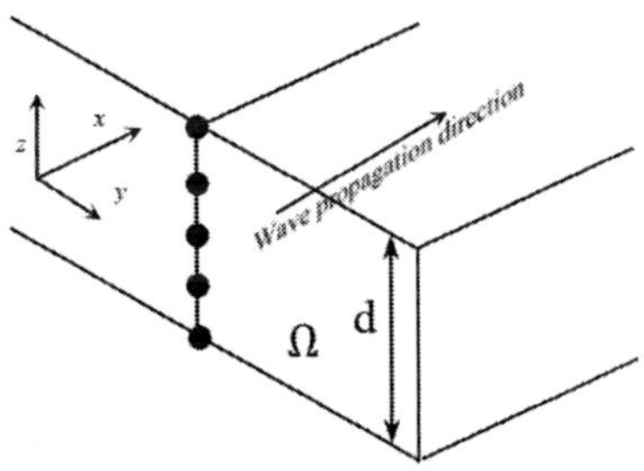

Figure 1. Schematic representation of the plate.

$$u = [u_x \; u_z]^T, \qquad \sigma = [\sigma_x \; \sigma_z \; \sigma_{xz}]^T, \qquad \varepsilon = [\varepsilon_x \; \varepsilon_z \; \gamma_{xz}]^T \tag{1}$$

The constitutive equation is given by equation (2)

$$\sigma = E\varepsilon \tag{2}$$

where, E is the elasticity matrix (real symmetric matrix for isotropic specimen) defined by equation (3)

$$E = \frac{2G}{(1-2\nu)}\begin{bmatrix} 1-\nu & \nu & 0 \\ \nu & 1-\nu & 0 \\ 0 & 0 & (1-2\nu)/2 \end{bmatrix} \quad (3)$$

where G is the shear modulus and ν the Poisson's ratio. The strain-displacement relationship can be compactly written as follows.

$$\varepsilon = Lu \quad (4)$$

where L is a three-dimensional differential operator defined as equation (5)

$$\mathrm{L} = \begin{bmatrix} \partial_x & 0 & \partial_z \\ 0 & \partial_z & \partial_x \end{bmatrix}^T \quad (5)$$

Thus, the compatibility equations can be written as equation (6)

$$\varepsilon = \left[L_x \frac{\partial}{\partial x} + L_z \frac{\partial}{\partial z}\right]^T u \quad (6)$$

where

$$\mathrm{L}_x = \begin{bmatrix} 1 & 0 & 0 \\ 0 & 0 & 1 \end{bmatrix}^T, \quad \mathrm{L}_z = \begin{bmatrix} 0 & 0 & 1 \\ 0 & 1 & 0 \end{bmatrix}^T \quad (7)$$

The cross-sectional domain of the plate, Ω, can be represented by a system of mono-dimensional finite elements with domain Ω_e, due to the simplification attributed to the waveguide symmetry. The displacement expressions, discretized over the element domain, can be written in terms of the shape functions $N_i(z)$ and the nodal unknown displacements, (U_{xi}, U_{yi}, U_{zi}), in cartesian coordinates. Such shape functions are considered in this formulation as unidimensional, isoparametrics and quadratics.

$$u^e_{(x,z)}(x,z,t) = \left[\sum_i^k N_i(z)U_{(x,z)i}\right]^e exp(-i(\xi x - \omega t)) = \mathbf{N}(z)f^e exp(-i(\xi x - \omega t))$$
(8)

where

$$\mathbf{N}(z) = \begin{bmatrix} N_1 & 0 & N_2 & 0 & N_3 & 0 \\ 0 & N_1 & 0 & N_2 & 0 & N_3 \end{bmatrix}, \quad f^e = [U_{x1}\ U_{z1}\ U_{x2}\ U_{z2}\ U_{x3}\ U_{z3}]^T \quad (9)$$

With $N_1 = \frac{\eta^2}{2} - (\frac{\eta}{2})$, $N_2 = 1 - \eta^2$ and $N_3 = \frac{\eta^2}{2} + (\frac{\eta}{2})$, represented in the local coordinates $\eta\ \epsilon[-1,1]$. Consequently, the Jacobian $|J|$ is equal to $l/2$, where l is the length of the element. For a maximum frequency of interest, the mesh criterion of [26] indicates the maximum element length, $L < 2\pi c_T/\beta\omega_{max}$, where c_T is the shear bulk wave velocity and β is 4 for the case of quadratic elements.

The strain vector in the element can be represented as a function of the nodal displacements:

$$\varepsilon^e = (\mathbf{B}_1 + i\xi\mathbf{B}_2)f^e exp(-i(\xi x - \omega t))$$
(10)

Where $\mathbf{B}_1 = L_z(\partial N/\partial z)$, $\mathbf{B}_2 = L_x N$

By considering adiabatic wave propagation, the weak form of the governing balance equation can be obtained by applying Hamilton's principle and adding up the contributions of every element as in the standard FE method. The SAFE governing equation for the stress-free plate is obtained as equation (11)

$$[\xi^2\mathbf{K}_3 + i\xi\mathbf{K}_2 + \mathbf{K}_1 - \omega^2\mathbf{M}]U = 0 \tag{11}$$

With

$$\begin{aligned}
\mathbf{K}_1 &= \sum_{e=1}^{N_{elem}} \int_{-1}^{1} \mathbf{B}_1^T\mathbf{C}\mathbf{B}_1 \, |J| d\eta, \\
\mathbf{K}_2 &= \sum_{e=1}^{N_{elem}} \mathbf{T}^T \left[\int_{-1}^{1} \mathbf{B}_1^T\mathbf{C}\mathbf{B}_2 \, |J| d\eta - \int_{-1}^{1} \mathbf{B}_2^T\mathbf{C}\mathbf{B}_1 \, |J| d\eta\right] \mathbf{T} \\
\mathbf{K}_3 &= \sum_{e=1}^{N_{elem}} \int_{-1}^{1} \mathbf{B}_2^T\mathbf{C}\mathbf{B}_2 \, |J| d\eta, \\
\mathbf{M} &= \sum_{e=1}^{N_{elem}} \int_{-1}^{1} \mathbf{N}^T \rho \mathbf{N} \quad |J| d\eta,
\end{aligned} \tag{12}$$

where, $\mathbf{K}_1, \mathbf{K}_2, \mathbf{K}_3$ are the global stiffness matrices and $\mathbf{M}$ is the global mass matrix, superscript T denotes the matrix transpose, N_{elem} is the total number discretized elements across d and U is the vector of global displacements (and also the eigenvectors) at particular circular frequency ω. Transformation matrix $\mathbf{T}$ is an orthogonal diagonal matrix introduced to eliminate the imaginary term in the eigenfunction. The u_z displacement components in $\mathbf{T}$ are equal to 1, while those of u_x are equal to the imaginary unit.

The estimation of dispersion curves is tackled by sweeping the wave number ξ in a set of real values. Thus, the eigenfunction can be solved as a standard eigenvalue problem in $\omega(\xi)$. Real, purely imaginary and complex eigenvalues are obtained, however only the real one values (that correspond to the propagating waves) are considered. In this case, the number of eigenvalues $\omega(\xi)$ obtained is the number of total Degrees Of Freedom (DOF) of the system, N. Thus, N propagating modes (ξ_m, ω_m) are determined for each $\mathbf{U}_N$ cross sectional wave structure. The phase velocity of the propagating N mode can be computed by $C_p^m = \omega/\xi^m$, while the group velocity C_g^m can be calculated using modal properties for conservatives materials, equation (13)[27].

$$C_g^m = \frac{\psi_L^m \mathbf{K}_{,\xi} \psi_R^m}{2\omega_m \psi_L^m \mathbf{M} \psi_R^m} \tag{13}$$

where ψ_L^M and ψ_R^M are the mth left and right eigenvectors of the eigenfunction, $\mathbf{K}_{,\xi}$ is $2\xi\mathbf{K}_3 + \mathbf{K}_2$ and M is the mass matrix.

3. Dispersion curves using EEC for a uniaxial stressed plate

Acoustoelasticity is a nonlinear phenomenon that explains speed changes in bulk waves (longitudinal and shear) as a function of applied stress, based on the continuum theory for small disturbances. The linearization of the relation between stress and strain is no longer valid for the case of ultrasound propagation in a media subject to stress and with finite deformations [22].

For the linear case, only the second order elastic constants λ and μ are needed as parameters to describe the linear constitutive relation stress-strain in an isotropic medium. However, for nonlinear characterization, additional third order elastic constants (TOECs) have to be added to the constitutive relation. Thus, the deformation energy, strain energy function U, has to be expressed in a third order form. On the other hand, the relation between strain energy function and stress is used to express the nonlinear relation between stress-strain, where this function is expressed as a power series in strains as equation (14)

$$T_{ij} = C_{ijkl}^{(2)} E_{kl} + C_{ijklmn}^{(3)} E_{kl} E_{mn} + \cdots \tag{14}$$

where T_{ij} is the second Piola-Kirchhoff stress tensor, E is the Lagrangian strain tensor and $C^{(2)}, C^{(3)}$,.. are increasing order tensors and correspond to the coefficient of the power series expansion. $C^{(2)}_{ijkl}$ is the second order constants for the linear case, and $C^{(3)}_{ijklmn}$ represents the third order elastic constants, for the case of isotropic materials. This tensor can be represented in terms of the Murnaghan constants l, m and n.[28]. Some papers have shown the use of EECs to tackle the influence of the acoustoelasticity in the propagation of guided waves [29][30][31].

By using a Cartesian frame to describe the incremental stresses, strains and displacements in stressed media, three deformation states are defined. Following the notation used by [30], the unstressed frame is called the "natural state." The position of a material point is given by the position vector **x** whose "natural coordinates" are $\boldsymbol{x}_1(x_1, x_2, x_3,)$. "The initial state" is a finite deformation (applied or residual) in static equilibrium is then given by the position vector $\bar{x}$, whose initial coordinates are $\overline{x_1}(\overline{x_1}, \overline{x_2}, \overline{x_3})$. Finally, when a dynamic perturbation (wave propagation) is applied at the initial state, the point material reach the third state called the "final state." The position of the material point is then defined by the position vector $\tilde{x}$ whose "final coordinates" are $\tilde{x}_1(\tilde{x}_1, \tilde{x}_2, \tilde{x}_3)$. A common Cartesian frame (ξ_1, ξ_2, ξ_3) is used to refer to the position of material points of any of the three states. The physical quantities, which refer to the natural state, are denoted by the superscript O and those that refer to the initial state by the superscript I. The initial deformation and the final deformation are denoted by $\bar{u}$ and $\tilde{u}$, respectively. Equations (15) and (16) are the equations of motion with initial coordinates and natural coordinates, respectively [31],

$$\frac{\partial}{\partial \overline{x_j}}\left[\left(A_{ijkl} + \delta_{ik} t^I_{jl}\right)\frac{\partial \tilde{u}_k}{\partial \overline{x_l}}\right] = \rho^I \frac{\partial^2 \tilde{u}_I}{\partial t^2} \tag{15}$$

$$\frac{\partial}{\partial x_j}\left[\left(\Gamma_{ijkl} + \delta_{ik} T^O_{jl}\right)\frac{\partial \tilde{u}_k}{\partial \overline{x_l}}\right] = \rho^O \frac{\partial^2 \tilde{u}_I}{\partial t^2} \tag{16}$$

where A_{ijkl} and Γ_{ijkl} are tensors that depend on the symmetry of the material and describe the relation between second and third order elastic constants and the displacements $\bar{u}$ with respect to x and $\bar{x}$ positons, ρ is the density, t^I is the Cauchy stress tensor, T^O is the Kirchhoff stress tensor or second Piola–Kirchhoff stress tensor and δ is the Kronecker delta function. In [31], a more detailed analysis of A_{ijkl} and Γ_{ijkl} is found, where it can be observed that variations of the wavespeed depend of the initial stress value and the initial displacement gradient $\partial u_i/\partial \xi$. Motion equation (16) in natural coordinates is equivalent to the equation of motion relative to a stress-free medium. In addition, the use of natural coordinates or material's specific frame avoids problems associated to the splitting terms in the EEC when initial coordinates are used. The similarity between these equations allows considering a stressed material as an unstressed material with EEC, which takes into account the disturbances linked to the presence of stress[30]. This description considers the following formalism:

$$C^a_{ijkl} = C_{ijkl} + \delta C_{ijkl} \tag{17}$$

where, C^a_{ijkl} is the tensor of the EEC, C_{ijkl} is the tensor of the second order elastic constants for a stress-free material and δC_{ijkl} is the disturbance related to the presence of applied or residual stress. In this sense C^a_{ijkl} is equal to $\Gamma_{ijkl} + \delta_{ik} T^O_{jl}$, which corresponds to the tensor of the EEC in natural coordinates. Since, EEC allows the use of the second order approach in the case of stressed materials, they can be considered as a stress-free material presenting second-order elastic constants different from those of an unstressed material. Under these condition, the eigenfunction that satisfy the guided wave modal propagation can be written in the same form as shown in equation (18):

$$\left|K\left(\xi, C_{ijkl}\eta_j\eta_l\right) - M\omega^2\right| = 0 \qquad \left|K'\left(\xi', C^a_{ijkl}\eta_j\eta_l\right) - M\omega'^2\right| = 0, \tag{18}$$

where ω and ω' correspond to the ultrasonic wave circular frequency in unstressed and stressed media respectively, the tensors K and K' is the elasticity relation for unstressed and stressed materials and $\eta_j\eta_l$ are the direction cosines of the normal to the wavefront, i.e., $\xi_p = |\xi|\eta_p$. Thus, the symmetry of a material will change in the presence of static stress with respect to ultrasonic wave propagation. The terms of C^a_{ijkl} are listed in appendix A.

4. Numerical Validation

In order to validate the proposed approach, dispersion curves obtained by [31] for an aluminum isotropic plate ($\rho = 2800\ kg/m^3$) of thickness d=6.35 mm with material constants in GPa, $l = -252.2$; $m = -324.9$; $n = -351.2$; $\lambda = 54.9$; $\mu = 26.5$, were used. It is a benchmark data from an analytical model based on the acoustoelasticity principle. Figure 2 illustrates the used plate Cartesian frames, while figure 3 shows the dependence of phase velocity on the direction of propagation, by comparing the S1 modes about 600 kHz, propagating at varying angles with respect to the x_1'. Herein, analytical solution by [31] is represented by solid lines while the proposed approach by markers. Additionally, figure 4 presents the deviation of S1 mode between the proposed scheme and the previous analytical solution

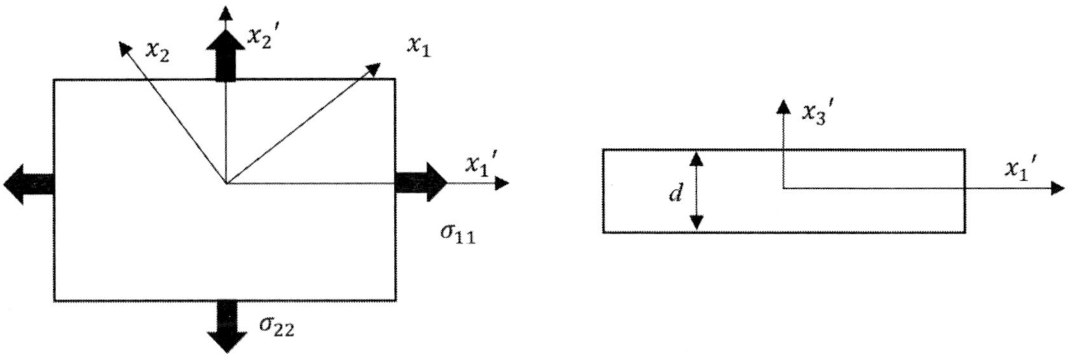

Figure 2. Plate's Cartesian frames.

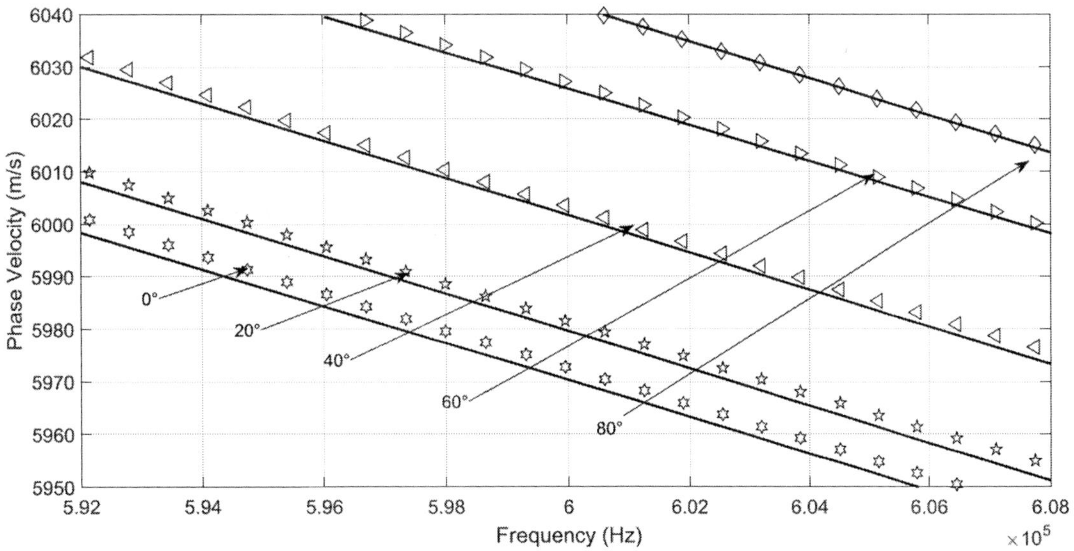

Figure 3. Comparison of angle dependence of S1 mode for a uniaxial load of 120 MPa.

Figure 5 presents the dispersion curves for the S1 mode about 600 kHz by applying different stresses values in order to observe the variations of phase velocity respect to the applied load, where all curves are at an angle of $\phi = 45°$. Moreover, figure 6 presents the deviation of S1 mode between the proposed scheme and the previous analytical solution in the case of different applied stresses.

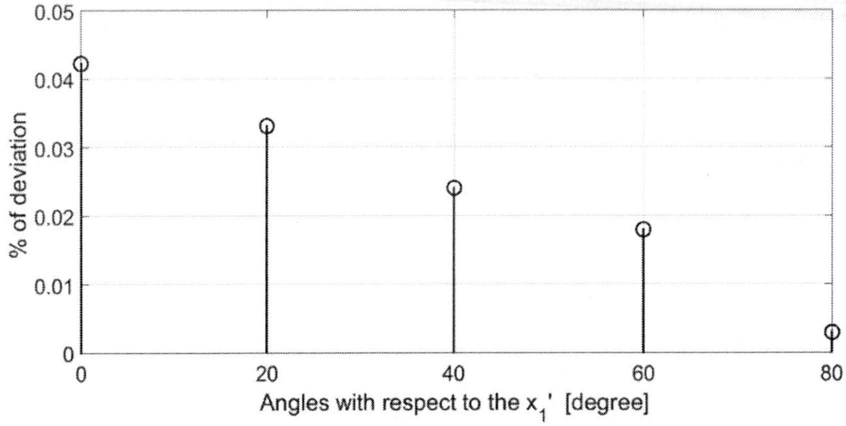

Figure 4. Deviations between proposed and analytical solution by [31] for S1 mode in [592-608] Khz

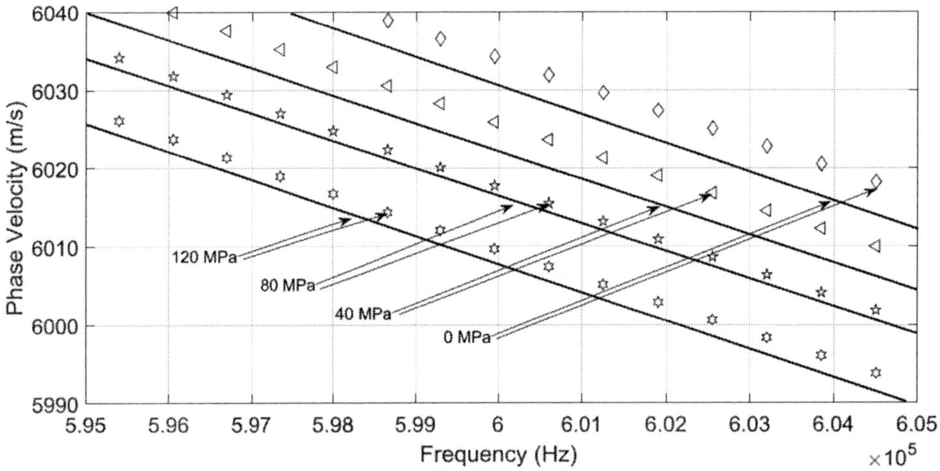

Figure 5. Comparison of stress dependence of S1 mode for a uniaxial load at φ = 45°. Analytical solution by [31] is represented by solid lines while the proposed approach by markers.

5. Results Analysis

According to figure 3, the dispersion curves of S1 mode (represented by markers) computed by means of EEC approach for a load of $\sigma_{11} = 120$ MPa are close to the analytical solution (based on the acoustoelasticity theory). However, this should not be generalized since the deviation of EECs from the theoretical model depends on the mode and frequency range under consideration [31].

In the case of the stress dependence (figure 4), it can be observed a good agreement in the range of analyzed dispersion curves between analytical model based on acoustoelasticity, and combined EEC's and SAFE proposed model.

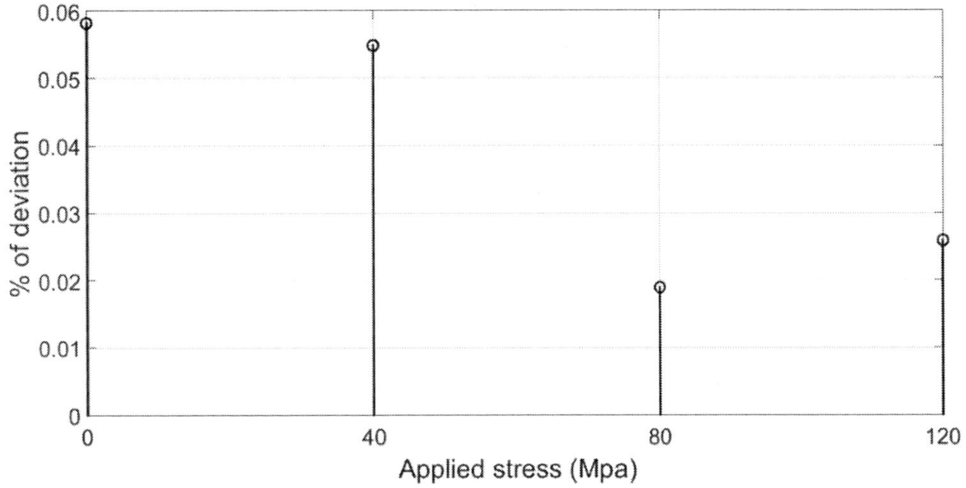

Figure 6. Deviations between proposed and analytical solution by [31] for S1 mode in [592-608] Khz

6. Conclusions

An approximate method for computing dispersion curves of lamb waves, phase and group velocity, in homogenous and isotropic plates under stress based on FEM and EEC is proposed in this work. The main contribution of the proposed approximate approach for estimating the dispersion curves of an isotropic and homogenous plate is accounting for the stress in the waveguide by using EEC combined with an SAFE scheme. These dispersion curves are a relevant tool for NDE/SHM applications, especially for identifying propagating modes and locating defects.

The proposed methodology was validated on several propagation angles and load conditions, which were previously examined by an analytical method based on the acoustoelasticity theory. It was found slight deviations by comparing the dispersion curves obtained by both solutions (proposed and previous method). It is remarked that a sufficient number of elements for discretizing the cross- section of the waveguide was ensured, thus, the observed variations can be attributed to limitations of the EEC representing the stressed state.

The main advantages of the proposed approximate approach on the available analytical methods are a low coding complexity and fast execution, with a decent accuracy compared to the previous analytical acoustoelasticity based model.

7. Acknowledges

This work has been partially funded by the Spanish Ministry of Economy and Competitiveness through the research projects DPI2014-58427-C2-1-R.

8. Appendix A

Expressions for the EEC proposed by [30]

$$C_{11}^{a} = k_2 + k_1[4\lambda + 10\mu + 4v_2 + 8v_3 + k_3(-2\mu + v_1 + 2v_2)] \quad (A.1)$$
$$C_{22}^{a} = C_{33}^{a} = k_2 + k_1[k_4 + k_3(\lambda + 2\mu + v_1 + 4v_2 + 4v_3)] \quad (A.2)$$
$$C_{44}^{a} = \mu + k_1[-2\mu - 2v_3 + k_3(\mu + v_2 + 2v_3)] \quad (A.3)$$
$$C_{55}^{a} = C_{66}^{a} = \mu + k_1[2\mu + v_3 + (1 - 2v)(v_2 + v_3)] \quad (A.4)$$
$$C_{12}^{a} = C_{13}^{a} = \lambda + k_1[\lambda + v_2 + k_3(v_2 + v_1)] \quad (A.5)$$
$$C_{23}^{a} = \lambda + k_1[-2\lambda - 2v_2 + k_3(\lambda + v_1 + 2v_2)] \quad (A.6)$$

Where λ and μ are the Lamé constants, ν_1, ν_2 and ν_3 are the Toupin and Berstein constants which are equivalent to $\nu_1 = n$, $\nu_2 = m - 0.5n$ and $\nu_3 = 1 - \nu_2$ in terms of Mourghanan (l, m, n) constants.

9. References

[1] Li W, Cho Y and Achenbach J D 2012 Detection of thermal fatigue in composites by second harmonic Lamb waves *Smart Mater. Struct.* **21** 1–8
[2] Nagy P B, Simonetti F and Instanes G 2014 Corrosion and erosion monitoring in plates and pipes using constant group velocity Lamb wave inspection *Ultrasonics* **54** 1832–41
[3] Zhu X and Rizzo P 2011 Guided waves for the health monitoring of sign support structures under varying environmental conditions *Struct. Control Heal. Monit.*
[4] Shen Y 2014 *Structural Health Monitoring Using Linear and Nonlinear (Doctoral dissertation)* (University of South Caroline)
[5] Dziendzikowski M, Dragan K, Kurnyta A, Klysz S and Leski A 2014 Damage size estimation of the aircraft structure with use of embedded sensor network generating elastic waves *Key Eng. Mater.* **598** 57–62
[6] Dai D and He Q 2014 Structure damage localization with ultrasonic guided waves based on a time-frequency method *Signal Processing* **96** 21–8
[7] Lowe M J S 1995 Matrix Techniques for Modeling Ultrasonic-Waves in Multilayered Media *Ieee Trans. Ultrason. Ferroelectr. Freq. Control* **42** 525–42
[8] Zhuang W, Shah A H and Dong S B 1999 Elastodynamic Green's Function for Laminated Anisotropic Circular Cylinders *J. Appl. Mech.* **66** 665–74
[9] Hayashi T 2004 *Guided wave animation using Semi-Analytical Finite Element method*
[10] Gravenkamp H 2016 A remark on the computation of shear-horizontal and torsional modes in elastic waveguides *Ultrasonics* **69** 25–8
[11] Marzani A 2008 Time-transient response for ultrasonic guided waves propagating in damped cylinders *Int. J. Solids Struct.* **45** 6347–68
[12] Hayashi T and Inoue D 2014 Calculation of leaky Lamb waves with a semi-analytical finite element method *Ultrasonics* **54** 1460–9
[13] Hayashi T, Kawashima K, Sun Z and Rose J L 2005 Guided Wave Propagation Mechanics Across a Pipe Elbow *J. Press. Vessel Technol.* **127** 322
[14] Bartoli I, Marzani A, Lanza di Scalea F and Viola E 2006 Modeling wave propagation in damped waveguides of arbitrary cross-section *J. Sound Vib.* **295** 685–707
[15] Gao H 2007 *Ultrasonic guided wave mechanics for composite material structural health monitoring (Doctoral dissertation)* (The Pennsylvania State University)
[16] Loveday P W and Wilcox P D 2010 Guided wave propagation as a measure of axial loads in rails *Proc. SPIE* **7650** 765023–8
[17] Predoi M V, Castaings M, Hosten B and Bacon C 2007 Wave propagation along transversely periodic structures *J. Acoust. Soc. Am.* **121** 1935–44
[18] Hayashi T, Song W J and Rose J L 2003 Guided wave dispersion curves for a bar with an arbitrary cross-section, a rod and rail example *Ultrasonics* **41** 175–83
[19] Andhavarapu E, Loveday P W, Long C S and Heyns P S 2010 Accuracy of semi-analytical finite elements for modelling wave propagation in rails *Proc. Seventh South African Conf. Comput. Appl. Mech.* 1–11
[20] Kim Y G, Moon H S, Park K J and Lee J K 2011 Generating and detecting torsional guided waves using magnetostrictive sensors of crossed coils *NDT E Int.* **44** 145–51
[21] Chen F and Wilcox P D 2007 The effect of load on guided wave propagation *Ultrasonics* **47** 111–22
[22] Gandhi N, Michaels J E and Lee S J 2012 Acoustoelastic Lamb wave propagation in biaxially

stressed plates. *J. Acoust. Soc. Am.* **132** 1284–93

[23] Bartoli I, Phillips R R, Coccia S, Srivastava A, Lanza di Scalea F, Fateh M and Carr G 2010 Stress dependence of ultrasonic guided waves in rails *J. Transp. Res. Board* **2159** 91–7

[24] Loveday P W, Long C S and Wilcox P D 2012 Semi-Analytical Finite Element Analysis of the Influence of Axial Loads on Elastic Waveguides *Finite Element Analysis - From Biomedical Applications to Industrial Developments* ed D Moratal (InTech) pp 439–54

[25] Loveday P W 2009 Semi-analytical finite element analysis of elastic waveguides subjected to axial loads *Ultrasonics* **49** 298–300

[26] Galán J M and Abascal R 2002 Numerical simulation of Lamb wave scattering in semi-infinite plates *Int. J. Numer. Methods Eng.* **53** 1145–73

[27] Han X, Liu G R, Xi Z C and Lam K Y 2002 Characteristics of waves in a functionally graded cylinder *Int. J. Numer. Methods Eng.* **53** 653–76

[28] Murnaghan F D 1937 Finite deformations of an elastic solid *Am. J. Math.* 235–60

[29] Bosher S H B and Dunstan D J 2005 Effective elastic constants in nonlinear elasticity *J. Appl. Phys.* **97** 1–7

[30] Duquennoy M, Ouaftouh M, Devos D, Jenot F and Ourak M 2008 Effective elastic constants in acoustoelasticity *Appl. Phys. Lett.* **92** 1–3

[31] Gandhi N 2010 *Determination of dispersion curves for acoustoelastic lamb wave propagation (Master dissertation)* (Georgia Institute of Technology)

Impacting load control of floating supported friction plate and its experimental verification

Keyan Ning[1,3], Yu Wang[2], Dingchuan Huang[2], Lei Yin[2]

[1] Science and Technology on Vehicle Transmission Laboratory, China North Vehicle Research Institute. P.R. CHINA
[2] The State Key Laboratory of Mechanical, Chongqing University, P.R. CHINA
[3] Corresponding author, email address: kyning@noveri.com.cn

Abstract. Friction plates are key components in automobile transmission system. Unfortunately, due to the tough working condition i.e. high impact, high temperature, fracture and plastic deformation are easily observed in friction plates. In order to reduce the impact load and increase the impact resistance and life span of the friction plate. This paper presents a variable damping design method and structure, by punching holes in the key position of the friction plate and filling it with damping materials, the impact load of the floating support friction plate can be controlled. Simulation is applied to study the effect of the position and number of damping holes on tooth root stress. Furthermore, physic test was designed and conducted to validate the correctness and effectiveness of the proposed method. Test result shows that the impact load of the new structure is reduced by 40% and its fatigue life is 4.7 times larger. The new structure provides a new way for floating supported friction plates design.
Keywords: friction plate; impact load; fatigue life; damping holes

1. Introduction

Researchers all over the world have done a lot of researches on the friction plates, most of them focus on the research of thermal stress and wear. A lot of study about wear and contact damage were conducted: Yue [1] used finite element analysis to study fretting wear under variable friction coefficient and different contact regimes, the result turns out that when considering partial slip or running-in stage of gross sliding conditions, predictions of FE models with variable friction coefficient are closer to experimental results than the models with constant friction coefficient. Further, Pereira [2] conducted multiscale analysis of the influence of roughness on fretting wear. Kumar D [3] studied the fretting fatigue stress of heterogeneous material in solid mechanics using direct numerical simulations. Bhatti [4] analysised fretting fatigue under out of phase loading conditions using finite element method, it shows that the damage initiation location is greatly influenced by the phase difference. Yue [5] studied the effect of different debris layer factors on fretting wear. Resende [6] studied the convergence of stresses in fretting fatigue analysis. Ferjaoui [7] proposed a method to predict fretting fatigue crack initiation in double lap bolted joint. Yue [8] use finite element model to study stress singularity in fretting wear. Aleksendric′ [9] predicted the wear of friction plate by neural network method considering temperature and pressure, the prediction results of neural network method were validated by bench test. Marklund [10] obtained the variation of friction coefficient, relative velocity and temperature under the condition of oil film lubrication and boundary lubrication based on the friction characteristics test of pin and plate. Zagrodzki [11] analyzes the influence of different pressure distribution on thermal stress, generation of hot spot on friction plate during clutch bonding process and influence of friction material and deformation on hot spots based on thermo elastic theory. Li [12] established a theoretical calculation model for the wear amount calculation of clutch friction plate, and a program was set up to conduct the calculation. Zhang [13] proposed a friction plate life estimation method as the life of friction plate is the ratio between the wear volume allowed to loss and the wear volume lost for one joint.

However, the research on the impact damage of friction plate during the working process is hardly involved, with the study of the wet multi friction plate brake and the wet multi friction plate clutch; it is found that the damage caused by the impact cannot be ignored. The greater the tooth stress is, the more serious the impact damage is. Therefore, the chief work of this paper is to study the impact stress control and prolong the life of friction plate by reducing the impact stress.

The tooth stress of friction plate is affected by the parameters of the friction plate, the pitch error, the installation method and so on. As the parameter of friction plate, the pitch error and the installation way are decided by the design condition and cannot be changed easily. Therefore, in this paper, variable damping is installed to reduce the tooth stress and prolong the life of friction plate.

2. Influence of damping hole on the stress of friction plate

Damping hole is arranged on friction plate. The position and the shape of the damping hole are shown in Figure. 1. Two parameters are needed to decide the position and size of damping hole, named c, the distance from the tooth tip to the centre of the damping hole, and d, the radius of the damping hole.

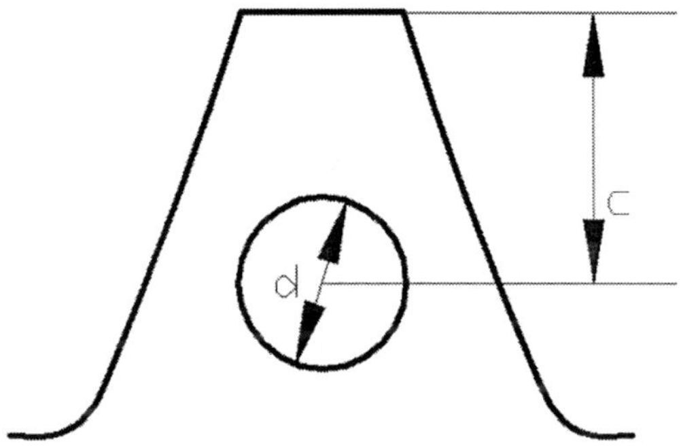

Figure. 1 Schematic diagram of position of damping hole

To study the influence of number and position of damping holes on the maximum root stress. A model of friction plate - inner hub is set up; the parameter of the model is shown in Table 1, here eccentric error exists between the friction plate and the inner hub, the amount of eccentric error is 0.025mm.

Table 1. Parameter of friction plate - inner hub model

Modulus (mm)	Tooth number	Thickness of the friction plate(mm)	Thickness of the inner hub(mm)	Eccentric error(mm)
3	122	3.5	6	0.025

ANSYS element SOLID185 was used to establish friction plate and the inner hub model, as shown in Figure. 2. Each node has 3 degrees of translational freedom. The friction plate and the inner hub are connected through point - surface contact element TARGE170 and CONTA175. The contact algorithm is Augmented Lagrange algorithm, the convergence criterion is set to 0.001. The node on both of the two side surface of the friction plate is fully constrained; torque is applied on the inner surface of the inner hub. The picture of finite element model is shown in Figure. 2, there are 4 layers of elements for the friction plate and 6 layers of elements for the inner hub in the thickness direction.

IOP Conf. Series: Journal of Physics: Conf. Series **842** (2017) 012070 doi:10.1088/1742-6596/842/1/012070

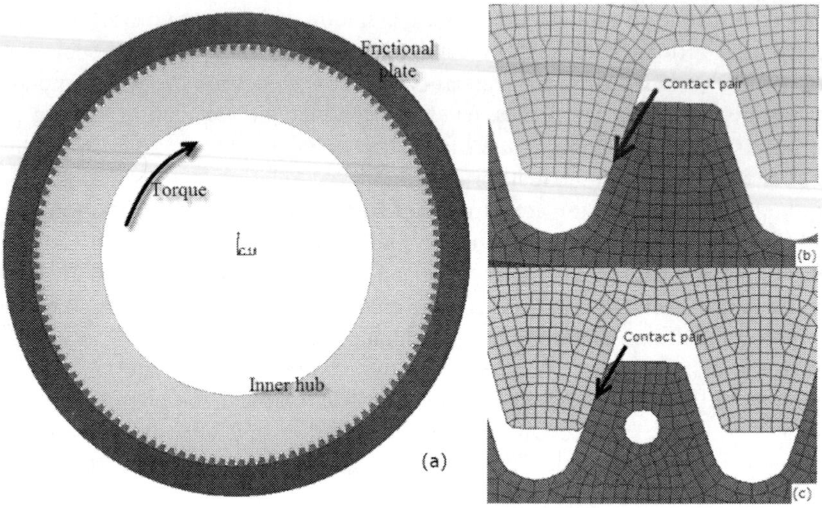

Figure. 2 Finite element model ((a) is the entire model, (b) is the contact area of

the model without holes (c) is the contact area of the model without holes)

The stress distribution obtained from simulation analysis of friction plate-inner hub model without holes is shown in Figure. 3. As eccentric error exists, the stress is not uniformly distributed among the teeth of friction plate. The position where the maximum tooth stress exists is set to 0 degree position, 120 degree, 240 degree position are set in a clockwise direction as shown in Figure. 3. The damping hole is opened at a 0 degree position or both at 120 degree and 240 degrees position.

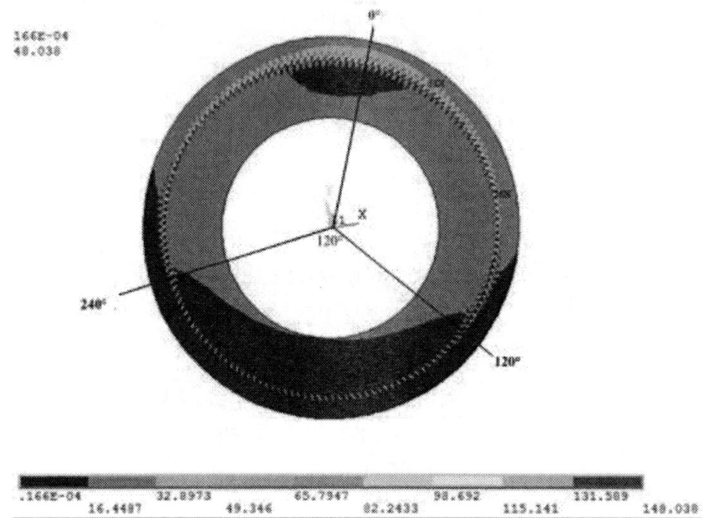

Figure. 3 Schematic diagram of the distribution of damping holes.

2.1.Stress simulation of friction plate without holes

Simulation results of complete friction plate are shown in Figure. 4 . The maximum Von Mises stress of tooth root in different position are shown in Table 2.

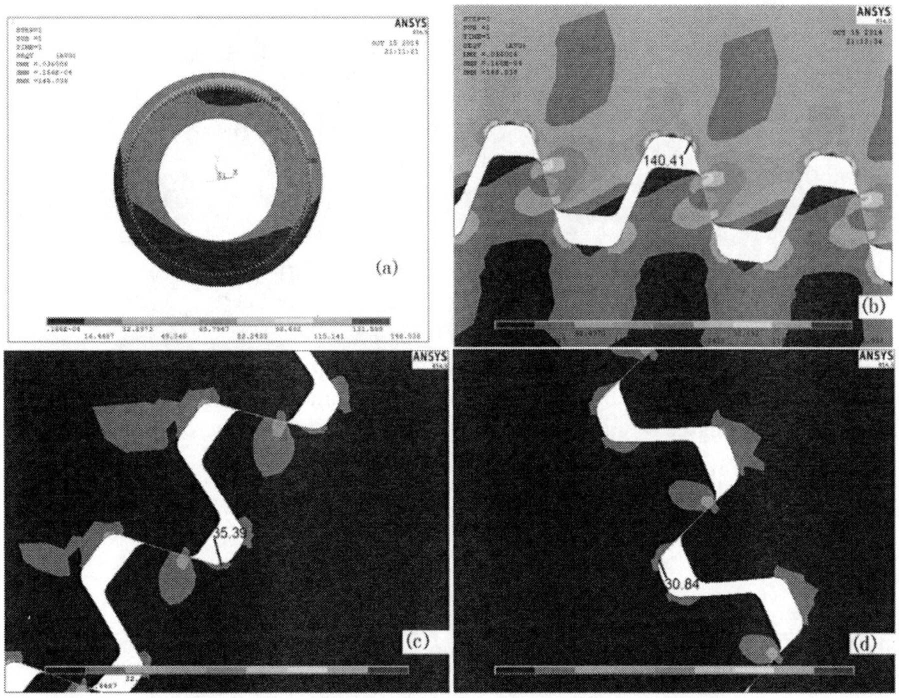

Figure. 4 Stress distribution figure of friction plate without holes ((a) is the overall stressdistribution, (b) is stress distribution of 0 degree position, (c) is stress distribution of 120 degree position, (d) is stress distribution of 240 degree position)

Table 2. Stress distribution of friction plate without holes

Position	The maximum Von Mises stress of tooth root（Mpa）
0°	140.41
120°	35.39
240°	30.84

2.2. Stress simulation of friction plate with damping holes

Stress simulation of the case where 1 damping hole is opened at 0 degree position is shown in Figure. 5. The parameters of damping hole are: d=1.6mm， c=2.87 mm. The maximum Von Mises stress of tooth root in different position are shown in Table 3.

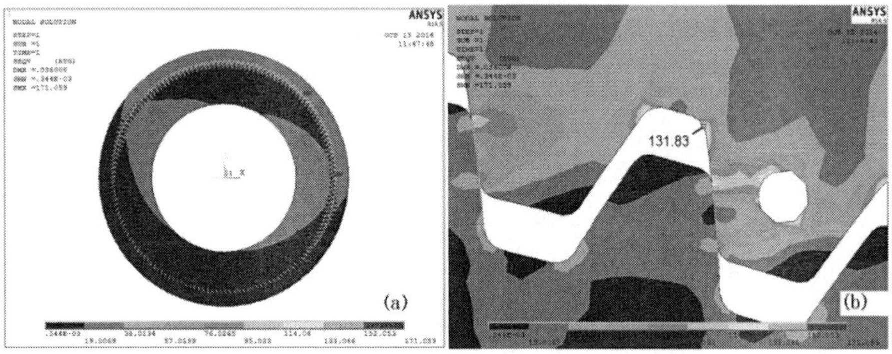

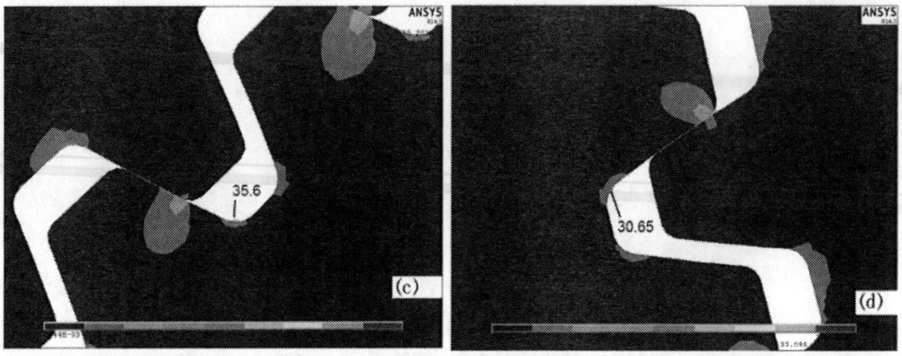

Figure. 5 Stress distribution of the case where 1 damping hole is opened at 0 degree position ((a) is the overall stress distribution, (b) is stress distribution of 0 degree position,(c) is stress distribution of 120 degree position, (d) is stress distribution of 240 degree position)

Table 3. Stress distribution of the case where 1 damping hole is opened on 0 degree position

Position	The maximum Von Mises stress of tooth root（Mpa）
0°	131.83
120°	35.6
240°	30.65

Similarly, Simulation of cases where 3, 9 damping holes opened at the 0 degree position, 1, 3, 9 damping holes opened at the 120 and 240 degrees of the position respectively are conducted. The fitting curves of simulation results of the maximum Von Mises stress of tooth root and the number of holes are shown in Figure. 6. When the damping holes are opened at the 0 degree position, the maximum Von Mises stress of tooth root decreased firstly and then increased as the number of damping holes increases. When the damping holes are opened at the 120 and 240 degrees, the maximum Von Mises stress of tooth root increases as the number of damping holes increases.

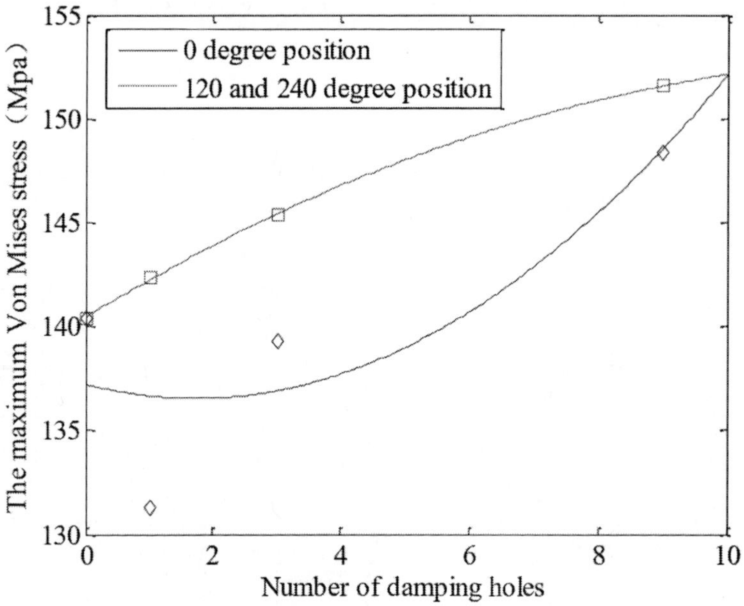

Figure. 6 The relationship between the maximum Von Mises stress of tooth root and the number, position of damping holes

3. Experimental verification

3.1. Physical experiment
In order to verify the simulation results, physical experiment was carried out. The experiment was carried out on the impact test rig. The friction plate is floating supported and the inner hub, driven by motor, will hit the friction plate. The parameters of test piece used in the experiment is the same with parameter shown in Table 1, but test piece have only 3 friction plate teeth and 2 inner hub teeth as shown in Figure 7. Experiments of friction plates with variable damping and without variable damping were carried out. Strain gauge is pasted on the tooth root of the middle tooth of the friction plate to measure the strain of friction plate, The damping hole is circular, filled with NBR material and its parameters are d=1.6mm， c=2.87mm.

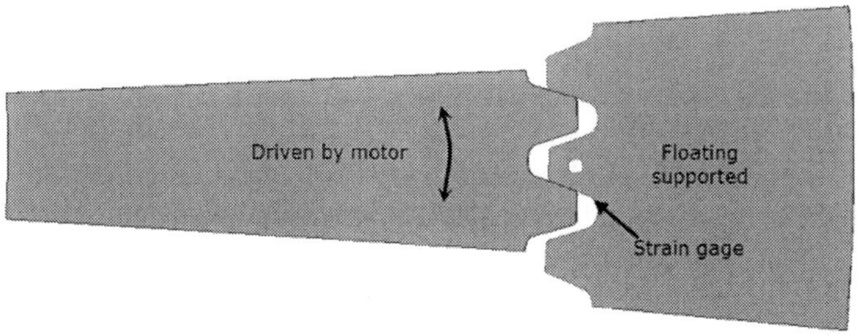

Figure. 7 Position of measuring point and damping hole

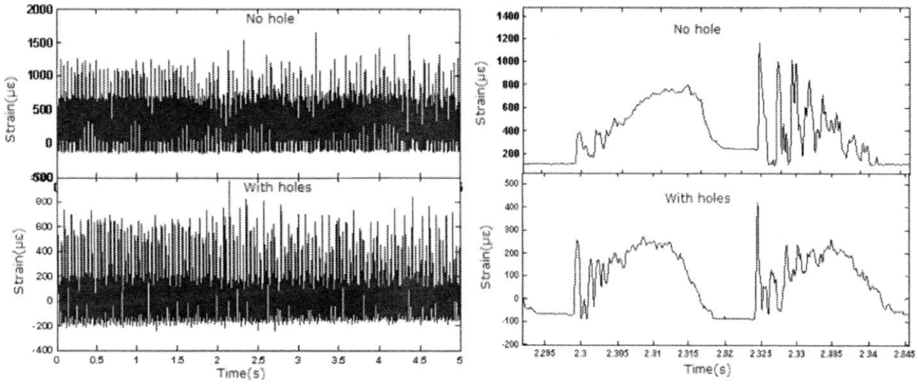

Figure. 8 Tested strain curve and its partial enlargement with and without variable damping

The tested tooth impact strain and its partial enlargement are shown in Figure. 8, the upper part shows the measured strain under the condition without variable damping and the lower part shows the measured strain under the condition with variable damping. Without variable damping, the maximum stress measured at C2 point is: 347.38Mpa, with variable damping, the maximum stress measured at C2 point is: 207.27Mpa. The maximum stress was reduced by 40% with variable damping.

3.2. *Life prediction of friction plate with and without variable damping*
According to the S-N curve of friction plate, life prediction under the condition with and without variable damping as in section 3.1 were conducted, When there is no variable damping, the life predicted is $1.46\text{x}10^6$ seconds, when there is variable damping, the life predicted is $8.31\text{x}10^6$ seconds, fatigue life increased 4.7 times with variable damping.

4. Conclusion
1) In order to reduce the impact load of floating supported friction plate, the method of damping adding was proposed. To study the influence of the position and number of damping holes to the impact load, finite element analysis was conducted; the result shows that damping holes opened at the 0 degree position can reduce the impact load. Later, experiment was carried out; the experimental results show that the proposed method is effective.

2) According to the S-N curve of friction plate, as variable damping reduces the impact load, the life of friction plate become longer.
3) The proposed method prolongs the life of friction plate by changing the body of friction plate, thus the proposed method provides a new way for the design of floating support friction plate.

References

[1] Yue T, Abdel Wahab M. Finite element analysis of fretting wear under variable coefficient of friction and different contact regimes. Tribology International, 2017,107:274-282.
[2] Pereira K, Yue T, Wahab MA. Multiscale analysis of the effect of roughness on fretting wear. Tribology International, in press(http://dx.doi.org/10.1016/j.triboint.2017.02.024).
[3] Kumar D, Biswas R, Poh LH, Abdel Wahab M. Fretting fatigue stress analysis in heterogeneous material using direct numerical simulations in solid mechanics, Tribology International, 2017,109:124-132.
[4] Bhatti NA, Wahab MA. Finite element analysis of fretting fatigue under out of phase loading conditions, Tribology International, 2017, 109: 552-562.
[5] Yue T, Abdel Wahab M. A numerical study on the effect of debris layer on frettingwear. Materials, 2016,9:597.
[6] Resende Pereira KdF, Bordas S, Tomar S, Trobec R, Depolli M, Kosec G, Abdel Wahab M. On the Convergence of Stresses in Fretting Fatigue. Materials, 2016, 9:639.
[7] Ferjaoui A, Yue T, Abdel Wahab M, Hojjati-Talemi R. Prediction of fretting fatigue crack initiation in double lap bolted joint. International Journal of Fatigue, 2015,73: 66-76 .
[8] Yue T, Abdel Wahab M. Finite element analysis of stress singularity in partial slip and gross sliding regimes in fretting wear. Wear, 2014,321: 53-63.
[9] Dragan Aleksendric′, David C Barton. Neural network prediction of disc brake performance. Tribology International, 2009, 42:1074–1080.
[10] Par Marklund, Roland Larsson. Wet clutch friction characteristics obtained from simplified pinon disc test. Tribology International, 2008, 41:824 – 830.
[11] Patankar S V, Rzemyslaw Zagrodzki. Analysis of Temperatures and Stresses in Wet Friction Disks Involving Thermally Induced Changes of Contact Pressure. SAE Technical Paper, doi:10.4271/98203.
[12] Jun Li. Theoretical estimation of wear amount of clutch friction plate of sxz1032 car. Journal of East China Jiaotong University, 1999,4:48-51.
[13] Guanxuan Xie, Siben Zhang. Llife estimation of the friction block (plate) of clutch and brake block. Forging machinery, 2000, 6:15-17.

Research on quality evaluation and forecast control in assembly process for complex products

Jihong Pang*, Xiaobo Xue, Yini Jin, Yongchao Li

College of Mechanical and Electrical Engineering, Wenzhou University, Wenzhou, P.R.China

E-mail: pangjihong@163.com

Abstract. Assembly quality affects complex product's performance in high degree because the assembly is the end of manufacture process. In this paper, a comprehensive evaluation and forecast control of various factors affecting assembly failure is presented based on theory of gray system. Firstly, the absolute correlation degree and relative degree of correlation are calculated by using data from assembly process of complex products. Then, the comprehensive correlation degree and rank analysis are obtained with the gray method. Finally, the gray system theory is used in quality analysis and forecast control of a valve manufacturer. The final result indicates that excess amount of leakage is the main factor affecting product quality in accord with the actual situation.

Keyword: quality engineering, quality control, monitoring method, complex products

1. Introduction

The product quality of complex mechanical products is mainly revealed in assembly quality. And assembly process is the most widely spread and complex production process, which has enormous impact on product quality [1]. Assembly process as the entire production process in the final stage, the highest cost is the quality of the entire product the greatest impact [2]. In the assembly process, the abnormal assembly process will affect the quality of the whole product assembly. Product assembly process involves many factors, and each factor is interrelated and mutually restricting. A number of factors to determine the decisive factors affecting the product are particularly difficult in the manufacturing industry [3]. Because many people are engaged in manufacturing personnel, the technical level and assembly site is different. But for all this, a lot of problems exist in experience of the degree of understanding of assembly process and accuracy of assembly tools [4]. These factors together determine the overall quality of product assembly, so quality of product assembly process evaluation becomes particularly important.

There is plenty of evidence to show that advanced assembly technology could improve the quality and efficiency of product assembly modeling rapidly. An assembly sequence of different assembly quality has establishment of assembly sequence based on the degree of deviation of the transfer model. So, quality evaluation of complex products were studied in many papers [5]. The assembly product quality is determined by many feature quantities. Because of the importance of each feature, assembly product quality is very different [6]. Based on gray theory, a multi-parameter diagnosis method of rolling assembly process for complex products has been developed. Therefore, gray theory is applied to the assembly of the valve quality of the various factors to evaluate and predict the impact of final quality assembly to determine the most significant factors, and to improve production efficiency.

2. Gray correlation analysis model of product assembly quality

Aimed at the product assembly quality, multiple targets and multiple solutions, according to the gray theory, the analysis model of the gray correlation analysis model is established. Based on quality model and programming and optimizing theory, the criterion of assembly quality judgment is solved in stochastic variable for online inspection. At present, three system research methods of uncertainty are probability statistics, fuzzy mathematics and gray system theory [7]. They all have one thing in common, that is, the research object has certain uncertainty [8]. The main focus of probabilistic statistics is on uncertainty of random events. Fuzzy mathematics focuses on the problem of less certainty. And the gray system theory mainly studies uncertainty of small sample and less information [9]. These areas are difficult to solve the problem of probability statistics and fuzzy mathematics [10].

In general, the study of product assembly quality often cannot provide a large number of assembly data. These data is often difficult to make a reasonable assessment of decision-making, the emergence of gray theory can solve such problems. In the valve assembly quality assessment and forecast, the overall quality of valve assembly and leakage, links, body material quality, valve core and valve seat sealing with the provisions of actuator torque, product storage and transportation are closely related to a number comprehensive enterprise valve assembly fault component composition of the reference sequence. Based on the gray system theory, the influence degree of different variables on product quality was analyzed. So, the calculation procedures of gray system theory are shown as follows:

Firstly, the length of the series n, referred to as:

$$\varphi_0 = \varphi_0(0), \varphi_0(1), \varphi_0(2), \varphi_0(3), \ldots\ldots, \varphi_0(n) \tag{1}$$

The number of failure factors by the composition of data comparison sequence. The sequence length is q, denoted as:

$$\varphi_\tau = (\varphi_\tau(1), \varphi_\tau(2), \varphi_\tau(3), \ldots\ldots, \varphi_\tau(n)) \tag{2}$$

Then, the number of product fault and the comprehensive correlation of the various factors were solved in the following.

(1) Establishing evaluation factor set

The relevant data of assembly quality of product classification, classification indicators, and independent of each indicator for the back of quality assessment were collected.

(2) Dimensional processing of data

In the evaluation, the relevance of indicators of significance, dimensions vary, and difference in the number of poor, cannot be directly calculated, usually on the data dimension of a treatment. In this paper, the actual measured values are dimensioned by using the method of initializing and initializing zero.

The numerical initialization and numerical start zero were shown by the following formula, respectively.

$$\varphi_\tau' = \left(\varphi_\tau'(1), \varphi_\tau'(2), \ldots\ldots \varphi_\tau'(n)\right) = \left(\frac{\varphi_\tau(1)}{\varphi_\tau(1)}, \frac{\varphi_\tau(2)}{\varphi_\tau(1)} \ldots\ldots \frac{\varphi_\tau(n)}{\varphi_\tau(1)}\right) \tag{3}$$

$$\varphi_\tau^0 = \left(\varphi_\tau(1) - \varphi_\tau(1), \varphi_\tau(2) - \varphi_\tau(1), \varphi_\tau(3) - \varphi_\tau(1) \ldots\ldots \varphi_\tau(n) - \varphi_\tau(1)\right)$$

$$= (\varphi_\tau^0(1), \varphi_\tau^0(2), \varphi_\tau^0(3) \ldots\ldots \varphi_\tau^0(n)) \tag{4}$$

(3) Correlation coefficient between the influencing factors

The numbers of correlation coefficient between the influencing factors can be calculated in the following:

$$|\sigma_i| = \left|\prod_{k=2}^{n-1} \varphi_i^0(k) - \frac{1}{2}\varphi_i^0(n)\right| \tag{5}$$

$$|\sigma_j| = \left|\prod_{k=2}^{n-1} \varphi_j^0(k) - \frac{1}{2}\varphi_j^0(n)\right| \tag{6}$$

$$|\sigma_i - \sigma_j| = \left|\prod_{k=2}^{n-1} (\varphi_i^0(k) - \varphi_j^0(k)) - \frac{1}{2}(\varphi_i^0(n) - \varphi_j^0(n))\right| \tag{7}$$

(4) Calculating the absolute correlation

The absolute degree of correlation reflects the degree of geometric, and two sequences are not absolutely independent. When the degree of geometric similarity is larger, the absolute degree of correlation is calculated as:

$$\varepsilon_{0\tau} = \frac{1 + |\sigma_0| + |\sigma_\tau|}{1 + |\sigma_0| + |\sigma_\tau| + |\sigma_\tau - \sigma_0|} \tag{8}$$

Next, the gray comprehensive correlation degree not only reflects the degree of similarity, but also reflects the close degree relative to the starting point. This is a more comprehensive characterization of the relationship between sequence is closely linked to a number of indicators, which can be computed as follows:

$$\rho_{0\tau} = \theta\varepsilon_{0\tau} + (1 - \theta) r_{0\tau} \tag{9}$$

In general, it is preferable that $\theta = 0.5$, θ is a little larger if the relationship between the absolute quantities, and θ may be smaller if the rate of change is heavier.

Thus, correlation analysis of the results was carried out based on above results. So the comprehensive correlation degree of each factor was compared, and the more significant factors were found. With the actual situation, the reasons were analyzed and the improvement measures were proposed in this paper.

3. A case study

Based on the gray system theory, a kind of gray modeling method for valve quality evaluation and forecast control in assembly process was proposed is this paper. These analyzing and calculation processes were characterized by high accuracy and simple algorithm. For a multivariate, nonlinear, disturbed and uncertain mathematic model of

IOP Conf. Series: Journal of Physics: Conf. Series **842** (2017) 012071 doi:10.1088/1742-6596/842/1/012071

manufacturing system, a simulation model by gray system theory was presented. The main reason of valve core and the valve seat gap is too large, and the problem usually occurs in the valve core or seat manufacturing accuracy is unqualified because assembly workers cannot properly ensure the valve seat and valve core close contact.

Firstly, the data of evaluation factor set for valve assembly process was collected and analyzed. The set of valve assembly quality assessment factors was shown in Table 1.

Table 1 Set of valve assembly quality assessment factors

Cause number of failures	First day	Second day	Third day	Fourth day
φ_1	205	197	186	214
φ_2	203	214	197	211
φ_3	220	203	189	201
φ_4	195	188	210	207
φ_5	213	220	180	195
φ_6	188	202	219	198
φ_0	1224	1224	1181	1226

From the above table it can be seen that the cause number of failures represent total number of failures, valve body leaks, torque failure, leakage exceeded, valve body produces a large number of leakage, connection leak failure, appearance of damage, respectively. And the comprehensive correlation degree is determined by the relative degree of correlation and the absolute degree of correlation. When calculating the absolute degree of correlation, a zero point of the original data is obtained. In the calculation of relative degree of correlation, the original data should be initialized. The quality of the valve assembly quality factor relative to the initial point of zero correlation can be computed by using body assembly quality factor data relative value of the initial correlation.

Then, the valve assembly quality of various factors related to the data point of zero relative to the beginning image is shown in Table 2.

Table 2 Valve assembly quality of the various factors

Zero point indicator	Object of zero point at start point						
	φ'^{0}_{0}	φ'^{0}_{1}	φ'^{0}_{2}	φ'^{0}_{3}	φ'^{0}_{4}	φ'^{0}_{5}	φ'^{0}_{6}
$\varphi_i'(1)-\varphi_i'(1)$	0.0000	0.0000	0.0000	0.0000	0.0000	0.0000	0.0000
$\varphi_i'(2)-\varphi_i'(1)$	0.0000	-0.0390	0.0542	-0.0773	-0.0359	0.0329	0.0745
$\varphi_i'(3)-\varphi_i'(1)$	-0.0351	-0.0927	-0.0296	-0.1409	-0.0769	-0.1549	0.1649
$\varphi_i'(4)-\varphi_i'(1)$	0.0016	0.0439	-0.0394	-0.0864	0.0615	-0.0845	0.0532

So, the correlation coefficient between each influencing factor and the valve assembly quality can be determined from the data in Table 1 and Table 2. The correlation coefficient between the absolute and relative degrees of each influencing factor can be presented.

Thus, the absolute association degree correlation coefficient is:

$|\sigma_0| = 42, |\sigma_1| = 22.5, |\sigma_2| = 9, |\sigma_3| = 57.5, |\sigma_4| = 14, |\sigma_5| = 35, |\sigma_6| = 50, |\sigma_1 - \sigma_0| = 19.5,$
$|\sigma_2 - \sigma_0| = 33, |\sigma_3 - \sigma_0| = 15.5, |\sigma_4 - \sigma_0| = 28, |\sigma_5 - \sigma_0| = 7, |\sigma_6 - \sigma_0| = 8$

And, the relative association degree correlation coefficient is:

$|\sigma'_0| = 0.0343, |\sigma'_1| = 0.10975, |\sigma'_2| = 0.0443, |\sigma'_3| = 0.2614, |\sigma'_4| = 0.0717, |\sigma'_5| = 0.1643, |\sigma'_6| = 0.266,$

$|\sigma'_1-\sigma'_0|$=0.07545, $|\sigma'_2-\sigma'_0|$=0.01, $|\sigma'_3-\sigma'_0|$=0.2217,

$|\sigma'_4-\sigma'_0|$=0.0375, $|\sigma'_5-\sigma'_0|$=0.13, $|\sigma'_6-\sigma'_0|$=0.2317.

So we can calculate absolute correlation degree and relative degree of correlation by using comprehensive correlation degree among the influencing factors of valve assembly quality, the absolute correlation degree is as follow:

$\varepsilon_{01} = 0.7706, \varepsilon_{02} = 0.6118, \varepsilon_{03} = 0.8664, \varepsilon_{04} = 0.6706, \varepsilon_{05} = 0.9176, \varepsilon_{06} = 0.9208$

$r_{01} = 0.9381, r_{02} = 0.9908, r_{03} = 0.8509, r_{04} = 0.9672, r_{05} = 0.9022, r_{06} = 0.8488,$

The comprehensive correlation degree ($\theta = 0.5$) is as follow:

$\rho_{01} = 0.8544, \rho_{02} = 0.8013, \rho_{03} = 0.8586, \rho_{04} = 0.8189, \rho_{05} = 0.9099, \rho_{06} = 0.8848,$

Next, the results of the above calculations were sorted as:

$$\rho_{04} > \rho_{05} > \rho_{06} > \rho_{03} > \rho_{01} > \rho_{02}.$$

Thus, having analyzed the measured statistics the result is shown below:

$$\varphi_4 > \varphi_5 > \varphi_6 > \varphi_3 > \varphi_1 > \varphi_2.$$

This shows that the importance of factors affecting the failure in turn decreased. And the result is very consistent with the actual business. From the perspective of improving product qualification rate, the valve core and valve seat machining accuracy were followed by ensuring the rationality of its installation methods and assembly skills of skilled workers. The quality evaluation model of the valve assembly demonstrates the guidance of gray system theory to the mechanical product quality evaluation and forecast control.

4. Conclusion

There are many methods for evaluating the quality of complex electromechanical products. It is necessary to ensure that the evaluation and forecast method was simple and easy to apply. In this paper, an application of gray system theory for the evaluation of product assembly quality was presented based on the gray of assembly quality with all kinds of influencing factors. The result is more accurate and the gray correlation degree can be compared. The approved method of calculating gray relational grade was somewhat complex, but it can be calculated by computer auto-made software.

In order to improve product assembly quality, the gray system theory was used in evaluation and forecast control for complex mechanical products. With the effective utilization of this machine and other equipments of the assembly line, the reducer assembly quality and the product performance were improved greatly. This assembly machine employs electric drive torque controlled wrench in quality evaluation and forecast control in assembly process for complex products, thus the physical strength of workers is largely decreased and the assembly quality was assured. For the product quality evaluation, this method was also applicable to other complex mechanical products.

Acknowledgement

This work was supported by the National Natural Science Foundation, China (No.71671130, No.71301120), and the college students' innovative projects of Wenzhou University (No. 201610351030).

References

[1] Okabe M. Mechanical device assembly: a sense of ultimate precision is indispensable for the achieving the delicate skill[J]. Journal of the Japan Society for Precision Engineering, 2014, 80:345-349.

[2] Rajamalli P, Atta S, Maity S, et al. Tolerance analysis of flexible sheet metal structures including effects of contact interaction and surface continuity of components[J]. Modares Mechanical Engineering, 2015, 2011(9):8.

[3] Kumar L R, Padmanaban K P. Design and optimization of concurrent tolerance in mechanical assemblies using bat algorithm[J]. Journal of Mechanical Science and Technology, 2016, 30(6):2601-2614.

[4] Saleem M, Zitoune R, Sawi I E, et al. Role of the surface quality on the mechanical behavior of CFRP bolted composite joints[J]. International Journal of Fatigue, 2015, 80:246-256.

[5] Laverne F, Segonds F, Anwer N, et al. Assembly based methods to support product innovation in design for additive manufacturing: an exploratory case study[J]. Journal of Mechanical Design, 2015, 137(12).

[6] Panat R, Dattaguru S, Balkan H, et al. Mechanical reliability of embedded array capacitors in ultrahigh-performance microprocessors[J]. IEEE Transactions on Device & Materials Reliability, 2014, 14(3):857-863.

[7] Camelia D. Grey systems theory in economics – a historical applications review[J]. Grey Systems, 2015, 5(2):263-276.

[8] Memon M S, Lee Y H, Mari S I. Group multi-criteria supplier selection using combined grey systems theory and uncertainty theory[J]. Expert Systems with Applications, 2015, 42(21):7951-7959.

[9] Mierzwiak R, Więcek-Janka E. The analysis of successors' competencies in family enterprises with the use of grey system theory[J]. Grey Systems, 2015, 5(3):302-312.

[10] Dejamkhooy A, Dastfan A, Ahmadyfard A. Modeling and forecasting non-stationary voltage fluctuation based on grey system theory[J]. IEEE Transactions on Power Delivery, 2015:1-1.

Damage evolution in plates subjected to fatigue loading

J Toribio, J C Matos and B González

Fracture & Structural Integrity Research Group
University of Salamanca, E.P.S. Zamora, Spain

E-mail: toribio@usal.es

Abstract. The aim of this paper is the numerical prediction of the cracking path followed by a surface crack front in plates constituted of different materials (determined by the exponent m of the Paris law), subjected to cyclic tension or cyclic bending loading. To this end, a numerical modelling was developed on the basis of the discretization of the crack front (characterized with elliptical shape) and the crack advance at each point perpendicular to such a front, according to a Paris law, using the stress intensity factor (SIF) calculated by Newman and Raju. Results show that the crack leads to a preferential propagation path that corresponds to a very shallow initial crack with a quasi-circular crack front. The increase of the Paris exponent produces a quicker convergence during fatigue crack propagation from the different initial crack shapes.

1. Introduction

The plate, a simple geometry of structural application, often presents cracks on its surface that can easily propagate due to fatigue, a reason why studying them is of great interest in fracture mechanics. The stress intensity factor (SIF) in a superficially cracked plate, subjected to tension or bending loads, has been calculated by several methods [1-11]. For tension loading the SIF increases with crack depth, showing its maximum value at the centre of the front for semi-elliptical cracks (with a ratio between semiaxes of 0.2) and at the point of the surface for semicircular cracks [1,2]. For bending moment the SIF is highest at the crack surface from relative depths greater than or equal to 0.6, regardless of the geometry of the crack front [1].

The fatigue advance of the crack front in plates subjected to tension loading or bending moment has been analyzed by experimental [1,12-16] and numerical [1,7-9,13-18] methods, applying the Paris-Erdogan law [19]. Cracks try to grow so that the SIF remains constant along the front (*iso-K condition of propagation*), but the free surfaces of the plate and the bending load prevent this behaviour [20]. Cracks asymptotically tend to a preferential propagation path [1,7-9,13-18], so that the larger the ratio of the bending loading to tension loading, the more intense the asymptotic tendency [15]. If the Paris exponent is greater, the change in the crack aspect ratio is more pronounced [15].

2. Numerical modelling

A computer application in Java programming language was developed to study the propagation path of a symmetric surface crack on the cross section of a plate (figure 1), subjected to fatigue (cyclic) loading in the form of remote uniform tension or pure bending moment (under Mode-I loading conditions).

The SIFs used were those obtained by Newman and Raju [1-3] by a 3D finite element analysis and the nodal-force method. They adjusted their results to an equation [3] that calculates the dimensionless SIF with parameters a/b, a/t, b/w and ϕ characterizing each point of the crack front (figure 2).

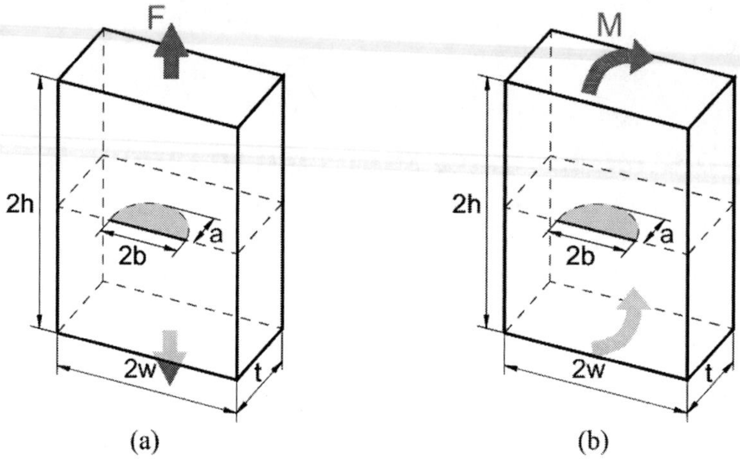

Figure 1. Surface crack in a plate: (a) tension loading; (b) bending moment.

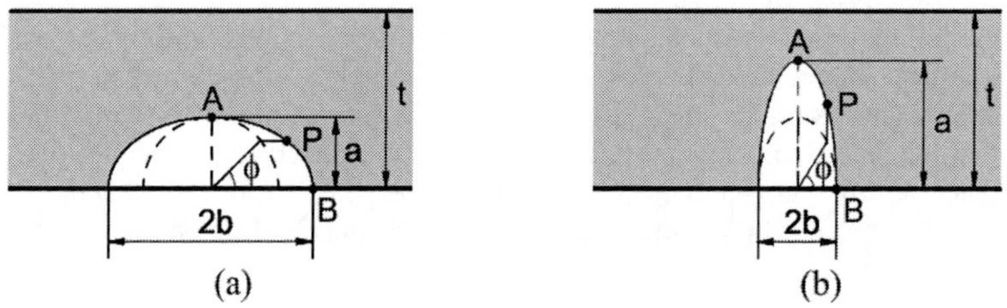

Figure 2. Parameters characterizing a point P at the crack front (a/t, a/b, ϕ): (a) $a/b \leq 1$; (b) $a/b > 1$.

Although other equations do exist, this one has been the most widely used in the bibliography due to its excellent results. This equation (valid for $0 \leq \phi \leq \pi$, $b/w < 0.5$, $0 \leq a/b \leq 2.0$ and $a/t < 1.25(a/b + 0.6)$ if $0 \leq a/b \leq 0.2$ or $a/t < 1$ if $a/b > 0.20$, but for $a/t > 0.8$ the accuracy has not been well established) has been tested experimentally and numerically by various methods [1,5,8,9,12-14,16,20]: measurements of the SIF K (photoelastic and numerical), the geometrical evolution of the front, the crack growth rate, the failure stress, etc. The SIF expression has the following form:

$$K = (S_t + H_s S_b)\sqrt{\frac{\pi a}{Q}} F_s \tag{1}$$

wherein S_t is the remote tensile stress,

$$S_t = \frac{F}{2wt} \tag{2}$$

and S_b the stress caused by the moment (maximum remote bending stress),

$$S_b = \frac{3M}{wt^2} \tag{3}$$

The parameter F_s is obtained by the following equation,

$$F_s = \left(M_1 + M_2 \left(\frac{a}{t}\right)^2 + M_3 \left(\frac{a}{t}\right)^4 \right) g f_\phi f_w \tag{4}$$

The bending multiplier H_s for bending loading is as follows:

$$H_s = H_1 + (H_2 - H_1)\sin^p\phi \tag{5}$$

The parameters needed to obtain F_s and H_s (M_1, M_2, M_3, g, f_ϕ, f_w, H_1, H_2, p and ϕ), as well as the parameter Q, are calculated by means of the mathematical expressions given in the classical paper by Newman and Raju [3]. Such parameters are, in general terms, functions of the geometry of the cracked plate, of the kind of loading (tensile remote stress or bending moment) and of the specific point at the crack front P (cf. figure 2).

The basic hypothesis of this modelling is the assumption that the crack front can be characterized as a semi-ellipse with its centre on the plate's surface and that fatigue propagation takes place in a direction perpendicular to such a front, following the Paris-Erdogan law [19],

$$\frac{da}{dN} = C\Delta K^m \tag{6}$$

where C and m are the Paris parameters.

To discretize the crack front, modelled in a semi-elliptical way, the front was divided into z segments of equal length, using Simpson's rule. The point on the surface of the plate, where a plane stress state does exist, was not taken into account. Subsequently, each of the points (i) was moved perpendicularly to the crack front, in accordance with the Paris-Erdogan law, so that the maximum crack increase, $\Delta a(\max)$, associated with the point of the maximum SIF range, $\Delta K(\max)$, was kept constant.

From this maximum increase, and using the ratio of the SIF range at a given point of the crack to the maximum SIF range over the crack front, the advance of each one of the front points (Δa_i) can be obtained as follows,

$$\Delta a_i = \Delta a(\max)\left[\frac{\Delta K_i}{\Delta K(\max)}\right]^m \tag{7}$$

Keeping in mind Newman and Raju's equations [3], the advancement at each point of the crack front can be calculated using the expression,

$$\Delta a_i = \Delta a(\max)\left[\frac{(H_s g f_\phi)_i}{(H_s g f_\phi)(\max)}\right]^m \tag{8}$$

considering that H_s is equal to 1 in the case of tension loading.

The new points, fitted by the method of least squares, form a new semi-ellipse, so that the crack advancement progress is repeated iteratively, according to the flowchart shown in figure 3. Due to the existing symmetry, only half of the problem was used in the calculations.

3. Numerical results

A convergence study was carried out in order to determine the number of parts in which each semi-ellipse is divided (crack front) and the maximum crack depth in the iterations. The parameters selected for the calculations were $z = 24$ and $\Delta a(\max) = 0.00001t$.

Figure 4 plots the evolution of the aspect ratio a/b with relative crack depth a/t. It shows how cracks propagate from distinctive initial geometry —relative crack depth $(a/t)_0 = \{0.02, 0.1, 0.2, 0.3, 0.4, 0.5\}$ and aspect ratio $(a/b)_0 = \{0.2, 0.5, 1, 1.5, 2\}$— in plates of different materials ($m = 2$, 3 and 4), subjected to tension loading or bending moment.

The crack advance from different initial geometries of the surface defect (a/b-a/t curves) tends towards a *preferential propagation path*, which corresponds to that of an initially very shallow crack with a quasi-circular front ($a/b \sim 1$).

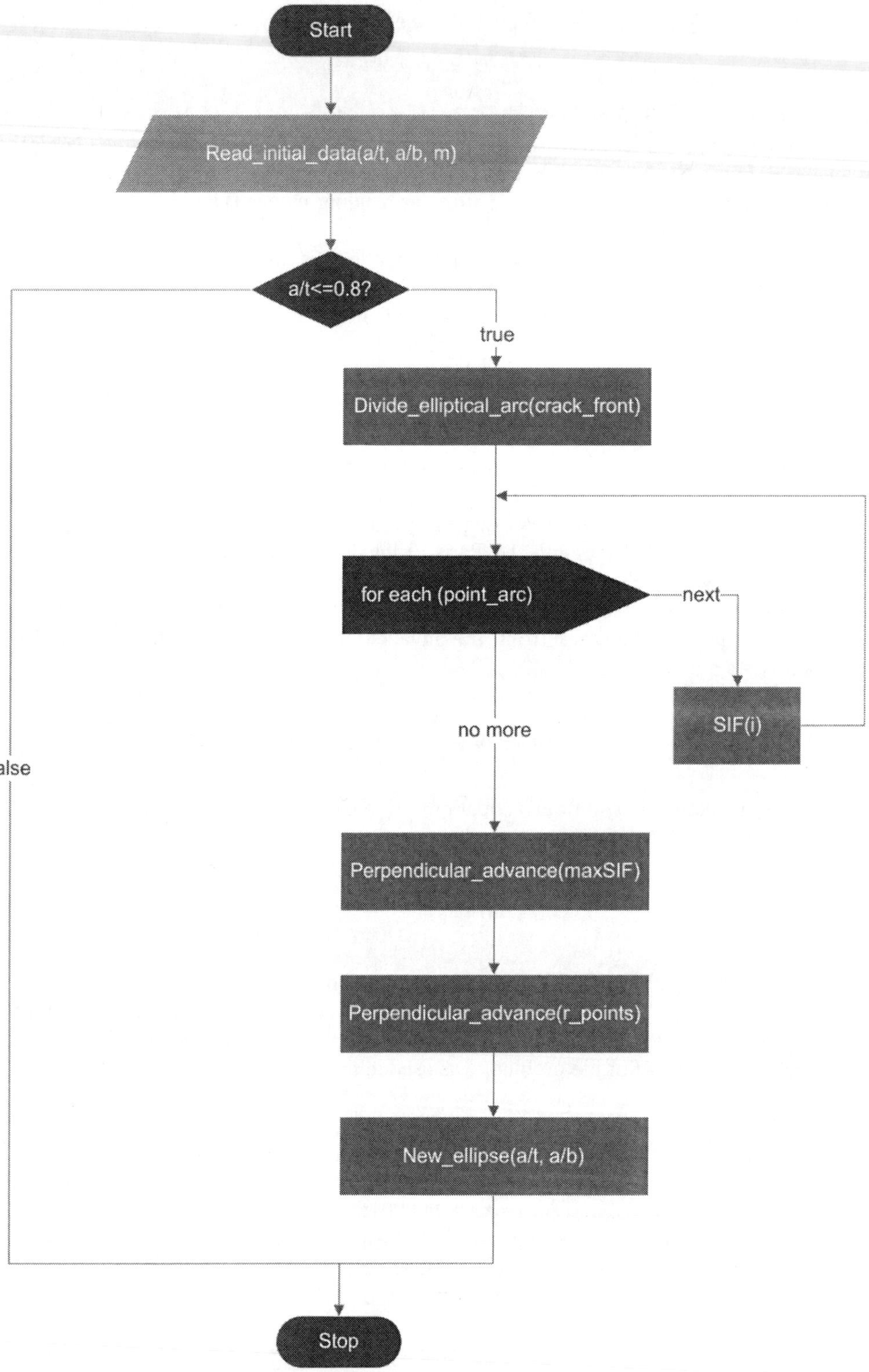

Figure 3. Flowchart of the code used for the simulations.

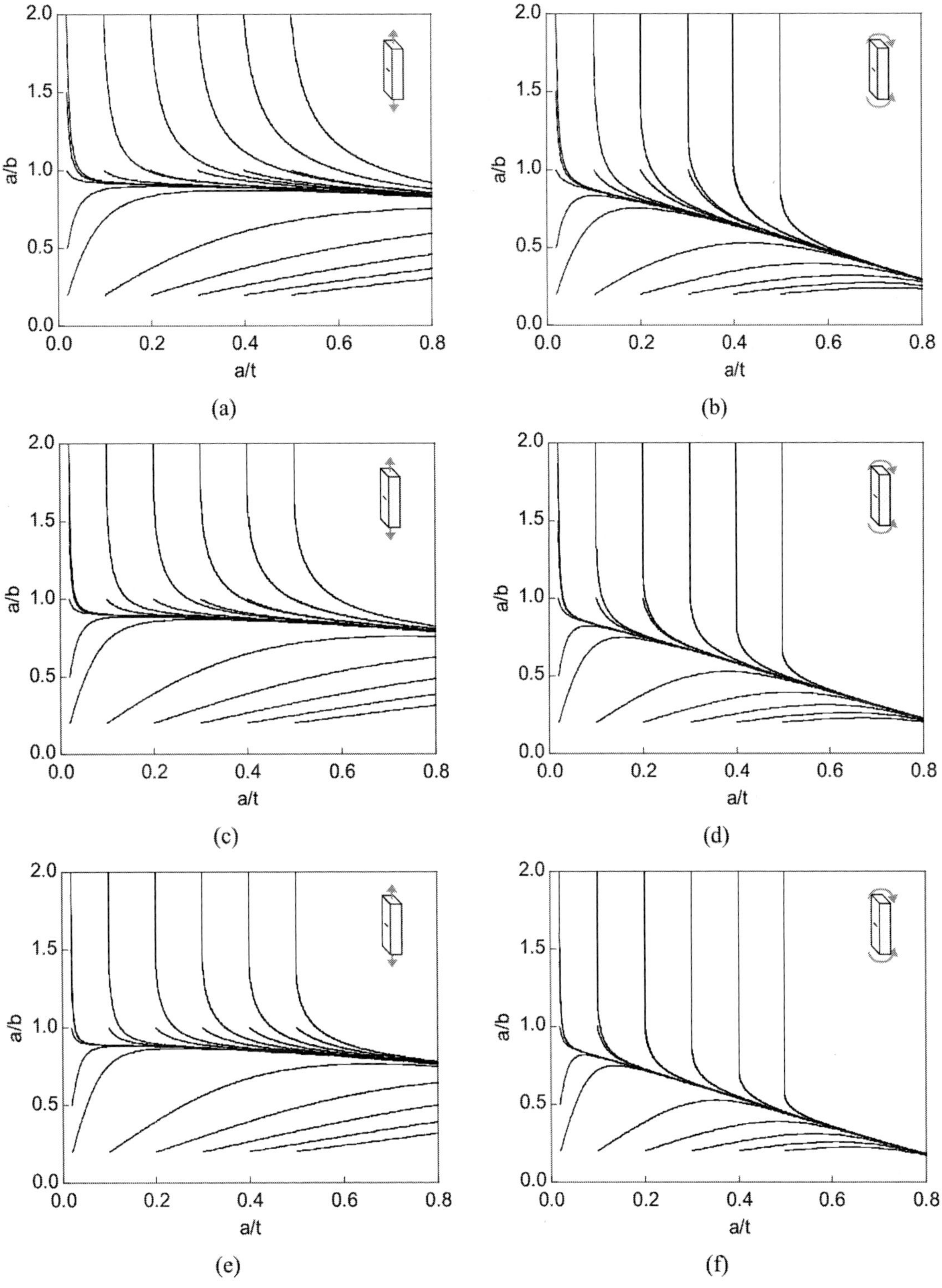

Figure 4. Crack aspect ratio evolution for: (a) $m = 2$ and tension loading; (b) $m = 2$ and bending moment; (c) $m = 3$ and tension loading; (d) $m = 3$ and bending moment; (e) $m = 4$ and tension loading; (f) $m = 4$ and bending moment.

The convergence of the solution (approach between the curves corresponding to the propagation from the different initial geometries) is higher in the case of bending than in the situation of tension, and for a larger Paris exponent *m* (although this parameter has a weaker effect on the growth pattern). The convergence is greater as the initial crack gets closer to the geometry corresponding to the path of preferential propagation. The quasi-straight cracks with elevated relative crack depth for tension loading are those showing the slowest convergence.

4. Conclusions

The following conclusions can be drawn from this research on the propagation of surface cracks in plates subjected to tension or bending cyclic loading:

(i) The cracks during fatigue growth tend towards a preferential propagation path, it corresponding to that of an initially very shallow crack with a quasi-circular front.

(ii) The convergence, approaching between the propagation curves with different initial geometries, is higher for cyclic bending than for cyclic tension and for greater Paris exponents (showing a weak effect on this parameter).

Acknowledgments

The authors wish to acknowledge the financial support provided by the following Spanish Institutions: Ministry for Science and Technology (MICYT; Grant MAT2002-01831), Ministry for Education and Science (MEC; Grant BIA2005-08965), Ministry for Science and Innovation (MICINN; Grant BIA2008-06810), Ministry for Economy and Competitiveness (MINECO; Grant BIA2011-27870) and *Junta de Castilla y León* (JCyL; Grants SA067A05, SA111A07 and SA039A08).

References

[1] Newman J C Jr and Raju I S 1979 *Analyses of Surface Cracks in Finite Plates under Tension or Bending Loads, NASA TP-1578* (Hampton: NASA)
[2] Raju I S and Newman J C Jr 1979 *Eng. Fract. Mech.* **11** 817
[3] Newman J C Jr and Raju I S 1986 Stress-intensity factor equations for cracks in three-dimensional finite bodies subjected to tension and bending loads *Computational Methods in the Mechanics of Fracture* ed S N Atluri (Amsterdam: Elsevier) chapter 9 pp. 311–334
[4] Holdbrook S J and Dover W D 1979 *Eng. Fract. Mech.* **12** 347
[5] Isida M, Noguchi H and Yoshida T 1984 *Int. J. Fract.* **26** 157
[6] Isida M, Noguchi H and Yoshida T 1990 *Eng. Fract. Mech.* **36** 889
[7] Lin X B and Smith R A 1999 *Eng. Fract. Mech.* **63** 503
[8] Wu Z 2006 *Int. J. Pres. Ves. & Piping* **83** 168
[9] Peng Y, Tong L, Zhao X-L and Xiao Z 2011 *Procedia Eng.* **14** 2601
[10] Ghajar R and Googarchin H S 2013 *Eng. Fract. Mech.* **109** 33
[11] Uslu M, Demir O and Ayhan A O 2014 *Eng. Fract. Mech.* **115** 255
[12] Vosikovsky O and Rivard A 1981 *Int. J. Fatigue* **3** 111
[13] Hosseini A and Mahmoud M A 1985 *Eng. Fract. Mech.* **22** 957
[14] Mahmoud M A and Hosseini A 1986 *Eng. Fract. Mech.* **24** 207
[15] Lin X B and Smith R A 1999 *Eng. Fract. Mech.* **63** 523
[16] Newman J C Jr and Raju I S 1981 *Eng. Fract. Mech.* **15** 185
[17] Lin X 2013 *Proc. 13th Int. Conf. Fract., ICF-13* (Beijing: ICF)
[18] Uslu M, Demir O and Ayhan A O 2014 *Eng. Fract. Mech.* **115** 284
[19] Paris P C and Erdogan F 1963 *J. Basic Eng.* **85D** 528
[20] Lin X B and Smith R A 1999 *Eng. Fract. Mech.* **63** 541

IOP Conf. Series: Journal of Physics: Conf. Series **842** (2017) 012073 doi:10.1088/1742-6596/842/1/012073

Adhesive properties and adhesive joints strength of graphite/epoxy composites

Anna Rudawska[1], Dana Stančeková[2], Nadezda Cubonova[2], Tetiana Vitenko[3], Miroslav Müller[4], Petr Valášek[4]

[1]Lublin University of Technology, Faculty of Mechanical Engineering, Nadbystrzycka 36 Str, 20-618 Lublin, Poland, e-mail: a.rudawska@pollub.pl
[2]University of Žilina, Faculty of Mechanical Engineering, Univerzitna 1, 010 26, Žilina, Slovak Republic
[3]Ternopil Ivan Pul'uj National Technical University, Faculty of Mechanical Engineering and Food Technologies, Ruska Street 56, 46001, Ternopil, Ukraine
[4]Czech University of Life Science Prague, Faculty of Engineering, 165 21, Prague, Czech Republic

Abstract. The article presents the results of experimental research of the adhesive joints strength of graphite/epoxy composites and the results of the surface free energy of the composite surfaces. Two types of graphite/epoxy composites with different thickness were tested which are used to aircraft structure. The single-lap adhesive joints of epoxy composites were considered. Adhesive properties were described by surface free energy. Owens-Wendt method was used to determine surface free energy. The epoxy two-component adhesive was used to preparing the adhesive joints. Zwick/Roell 100 strength device were used to determination the shear strength of adhesive joints of epoxy composites. The strength test results showed that the highest value was obtained for adhesive joints of graphite-epoxy composite of smaller material thickness (0.48 mm). Statistical analysis of the results obtained, the study showed statistically significant differences between the values of the strength of the confidence level of 0.95. The statistical analysis of the results also showed that there are no statistical significant differences in average values of surface free energy (0.95 confidence level). It was noted that in each of the results the dispersion component of surface free energy was much greater than polar component of surface free energy.

1. Introduction

Adhesive joints are used in many fields of engineering where surface processes, particularly adhesion, wettability and adsorption, play a significant role [1-4]. The knowledge about adhesion and adhesive properties, particularly information about methods for shaping adhesive properties of surface layer of specified materials, enables forming strong adhesive bonds which will lead to producing high-strength adhesive joints [1,3,5]. Adhesive properties can be assessed by surface free energy which describes specific properties of surface layer of a material [7]; surface free energy is affected by surface treatment methods applied [7-13]. One example of adhesive joints is adhesive joints used in a variety of structures in the machine design, automotive, aircraft and building industries [13-17].

Composite materials are nowadays widely used constructional materials due to their properties, including high strength and rigidity as well as resistance to brittle cracking [1,3,15, 18-23]. Different

forms and types of composite reinforcements and matrixes with varying properties enable formation of composites with desired properties [15, 25-27]. One of the ways of forming polymer composites is adhesive bonding [1,2, 17, 28-30]. The popularity of this technique stems from numerous advantages of adhesive-bonded joints [17], including their high strength if specified conditions for producing such joints are observed. Among others, the adhesive bonding technique does not change the structure of a material in the place of a joint.
This paper investigates the strength of adhesive joints of graphite/epoxy composites with varying thicknesses produced at constant composite production and adhesive-bonding conditions. Experimental tests involved determining surface roughness parameters of the adherends and their mechanical properties, shear strength of graphite/epoxy composites, and surface free energy of the composite adhesive joints.

2. Experimental test

2.1. Characteristics of the tested materials

The tests were conducted on specimens made of two types of graphite/epoxy composite: GR-EP 199-45-005 with a thickness of 0.48 mm and GR-EP 199-45-003 with a thickness of 0.66 mm (figure 1 and figure 2). Each composite had two layers of fabric with 90° orientation subjected to curing (in compliance with a relevant technology).

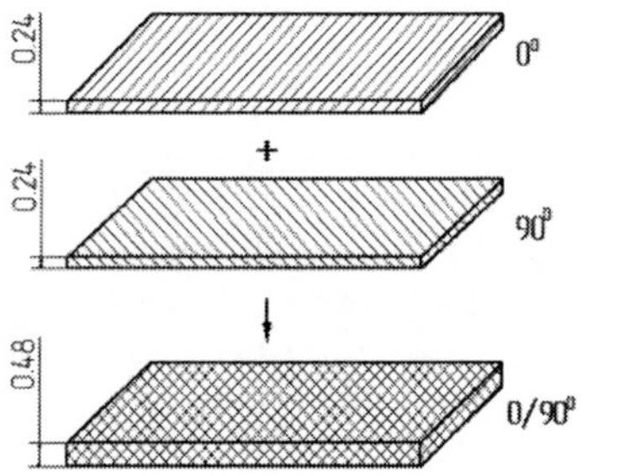

Figure 1. Thickness of fabrics and produced graphite/epoxy composites (0.48 mm).

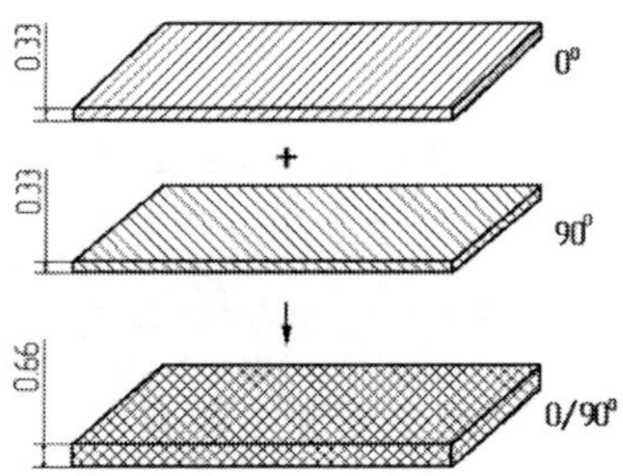

Figure 2. Thickness of fabrics and produced graphite/epoxy composites (0.66 mm).

In the first case, the fabric thickness was 0.24 mm, and in the other - it was 0.33 mm. The above composites are manufactured by an aircraft production plant and used to make different aircraft components.

2.2. Microscopic analysis of the investigated composites surfaces

The tests involved performing a microscopic analysis of the graphite/epoxy composites to determine the structure of the adherends and chemical composition of the tested composite materials. Surface morphology and chemical composition of the study was performed using scanning electron microscopy (SEM) - TESLA BS-340 analyzer [31] equipped with an element EDX (Energy Dispersive X-ray Spectroscopy) Thermo-NORAN. Scanning electron microscope with EDX detector can identify the elemental composition of the material for all elements with atomic number greater than boron.

2.3. Surface roughness

In order to determine parameters of geometrical structure of the adhesive-bonded composites, a selection of surface roughness parameters were measured. The measurements were performed on two batches of 5 specimens each. In each batch the specimens had their surface cleaned by degreasing.

IOP Conf. Series: Journal of Physics: Conf. Series **842** (2017) 012073 doi:10.1088/1742-6596/842/1/012073

Next, each specimen underwent three measurements, in total 15 measurements per each tested type of composite material. The measurements were made in compliance with the standard PN-EN ISO 4287 [32]. The measurements of the selected surface roughness parameters were made using a Perthometer 2 surface finish measurement instrument from Mahr.

2.4. Surface free energy

Surface free energy of the adhesive bonded materials was measured by the Owens-Wendt method described in numerous publications [6, 8, 11, 33, 34]. In the tests, two measurement liquids were used: distilled water and diiodomethane, the properties of which are listed in Table 1 [1].

Table 1. The surface free energy γ_L and its components of the measuring liquids [7].

Measuring liquids	The surface free energy and its components		
	γ_L (mJ/m^2)	$\gamma_L{}^d$ (mJ/m^2)	$\gamma_L{}^p$ (mJ/m^2)
Distilled water	72.8	21.8	51.0
Diidomethane	50.8	48.5	2.3

The measurements of the contact angle were performed at a temperature of 20±1°C and air humidity of 30±2%. The volume of drops of the measuring liquids ranged from 1.6 to 2.0 μl. To every sample surface (there were 7 samples per each batch) 6 drops of the measuring liquids were applied. After that, 8-12 measurements were made and the mean per each sample batch was calculated. The contact angle was measured after a few seconds following the application of a drop of the measuring liquid. The measurements were made using a PGX goniometer manufactured by Fibro System (Sweden) and a PG programme for computer image analysis Surfaceware. The relations between SFE and its components as well as a detailed description of contact angle measurements are given in the literature [7,8,11].

2.5. Adhesive joints

The strength tests were performed on shear-loaded single-lap polymer composite adhesive joints. The shape and dimensions of the tested graphite/epoxy composite adhesive joints are illustrated in figure 3 and figure 4.

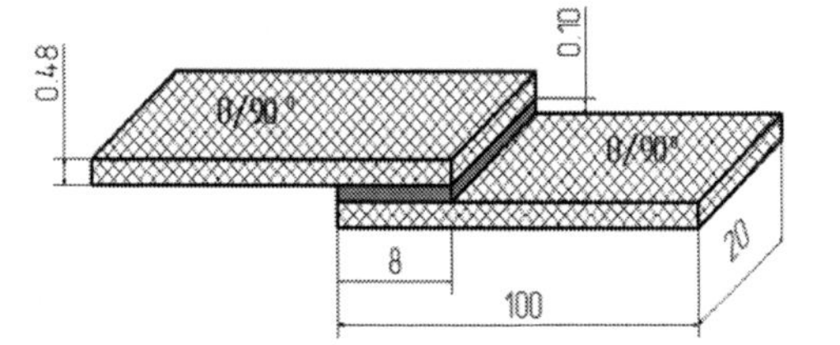

Figure 3. Shape and dimensions of the graphite/epoxy composite adhesive joint with a thickness of 0.48 mm. All dimensions are in mm.

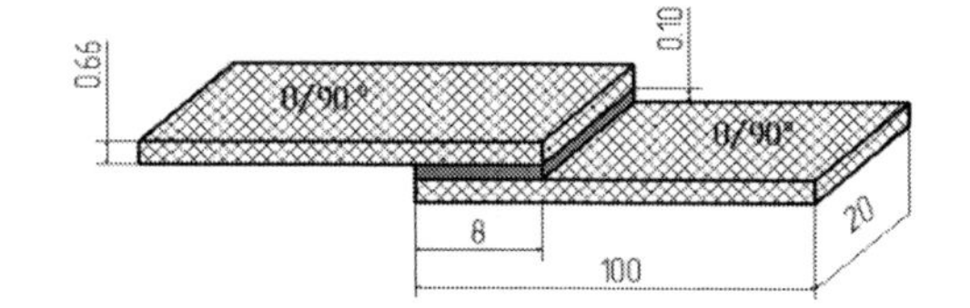

Figure 4. Shape and dimensions of the graphite/epoxy composite adhesive joint with a thickness of 0.66 mm. All dimensions are in mm.

In the tests, the surface finish treatment of the composites involved degreasing with the Loctite 7063 degreasing agent. This degreasing method is described in [8]. The adhesive joints were made using Loctite 3430 epoxy adhesive which was subjected to single-stage curing for 48 hours in an ambient temperature of 20±2°C, under a load of 0.02 MPa and a humidity of 39±1%.

2.6. Adhesive joints

The strength tests were performed on:

- adherends, i.e. graphite/epoxy composite materials,
- single-lap adhesive joints of graphite/epoxy composites.

The tests were performed on 10-11 test pieces per each batch. The tests were performed using the Zwick/Roell testing machine in compliance with the standard DIN EN 10002-1. The tests were conducted at the following parameters: modulus of elasticity: 60 MPa/s, yield point: 60 MPa/s, flow rate: 0.00025 1/s, test speed: 0.008 1/s. The following strength test results of the graphite/epoxy composite specimens were examined: E – modulus of elasticity in MPa, $R_{p0,2}$ – yield strength in MPa, R_m – tensile strength in MPa. The adhesive joints of graphite/epoxy composites were also subjected to strength testing using the Zwick/Roell 150 testing machine, in compliance with the standard DIN EN 1465. The test speed was set to 5 mm/min. Then, the fixed adhesive joints were subjected to strength testing until specimen failure. The tests helped determine the failure force for the examined adhesive joints and hence the shear strength of these joints. The strength tests were performed at an ambient temperature of 21±2°C and humidity ranging between 36% and 38%.

3. Experimental results

3.1. Description of the tested adherends

Representative images of surfaces of the tested composites obtained with the scanning electron microscope (SEM) and stereoscopic microscope are shown in figure 5. Surface morphology and chemical composition of the study was performed using scanning electron microscopy (SEM) with EDX detector (point 2.2) - figure 6 and figure 7.

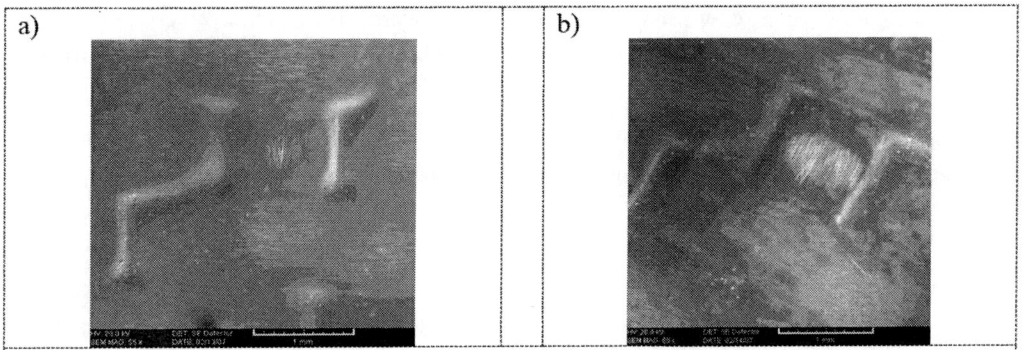

Figure 5. SEM images of the graphite/epoxy composite surface: a) 0.48 mm thickness, b) 0.66 mm thickness, magnified by 55 times.

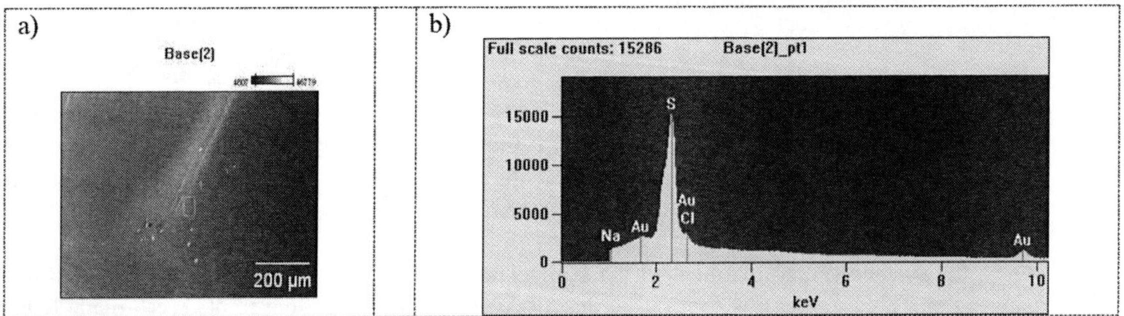

Figure 6. The chemical composition of graphite-epoxy composite of 0.48 mm thickness: a) SEM micrograph, b) chemical composition in area 1 from SEM micrograph.

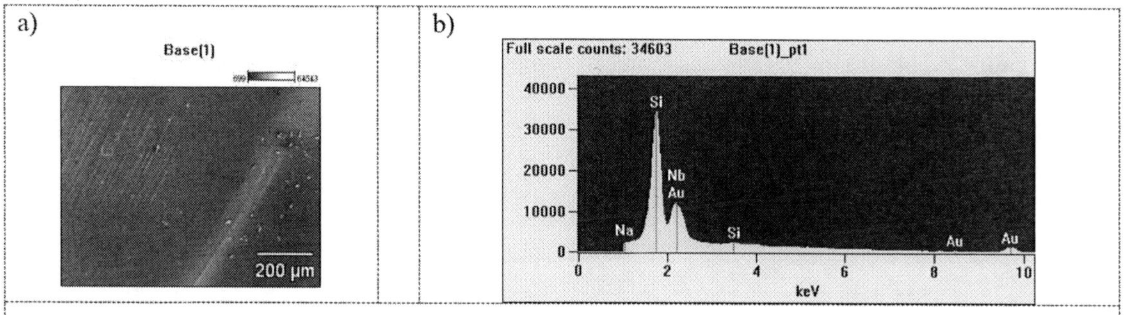

Figure 7. The chemical composition of graphite-epoxy composite of 0.66 mm thickness: a) SEM micrograph, b) chemical composition in area 1 from SEM micrograph.

In figure 6a the Base [2] it means the second research samples among 10 tested samples and pt1 (figure 6 b) it means the first point during analysis. In figure 7a the Base [1] it means the first research samples among 10 tested samples and pt1 (figure 7 b) it means the first point during analysis. The tested composites contain the following elements:

- graphite/epoxy composite with g=0.48 mm: S (highest concentration), Na, Au, Cl,
- graphite/epoxy composite with g=0.66 mm: Si (highest concentration), Na, Nb, Au.

Some differences in their compositions probably result from a slightly different chemical concentration of the fabrics that the two composites are made of.

3.2. *Surface roughness*

The surfaces of the tested graphite/epoxy composites significantly differ with respect to their geometrical structure due to the properties and design of the composite materials (and the fabrics they are made of). Table 2 lists the mean values of selected amplitude parameters of the composites measured in three measuring points (10 measurements on one point). One can immediately notice a high scatter of the results.

Table 2. Selected surface roughness parameters of the graphite/epoxy composites with g=0.48 mm.

Graphite-epoxy composite, thickness, mm	Ra (µm)	Rz (µm)	Rk (µm)	Rpk (µm)	Rvk (µm)
0.48 mm	1,30 (± 0.78)	5.45 (± 2.58)	2.83 (± 0.74)	2.13 (± 1.78)	7.68 (± 6.87)
0.66 mm	1.09 (± 0.20)	5.88 (± 1.01)	2.88 (± 0.55)	1.13 (± 0.52)	3.71 (± 1.03)

Examining the surface roughness parameters and surface profilograms we can observe significant variations in the composites' surface geometry. The numerous cavities and surface irregularities are predominantly caused by the type of the applied components (fabrics) for making the composites (figure 5). Probably this is the reason for high scatter in surface roughness. Such a structure is probably advantageous for the making of adhesive joints because the adhesive can penetrate deep into these irregularities and form mechanical fixings, which leads to higher adhesive strength of these joints. It is important that the adhesive has suitable viscosity to fill in all surface irregularities, so the real surface of adhesive joint will be higher than the theoretical one, which will also lead to higher strength.

3.3. *Surface free energy*

Table 3 lists the values of surface free energy γ_S, and its components (γ_S^d and γ_S^p) determined based on the measurements of the angle of contact of surfaces of the tested polymer composites according to relevant dependencies. The contents of individual components of surface free energy are illustrated in a graphic form in figure 8.

Table 3. The surface free energy γ_S and its components of the graphite-epoxy composites

Measuring liquids	The surface free energy and its components		
	γ_S (mJ/m^2)	γ_S^d (mJ/m^2)	γ_S^p (mJ/m^2)
Graphite-epoxy composite, 0.48 mm	45.4	39.7	5.7
Graphite-epoxy composite, 0.66 mm	45.0	41.4	3.6

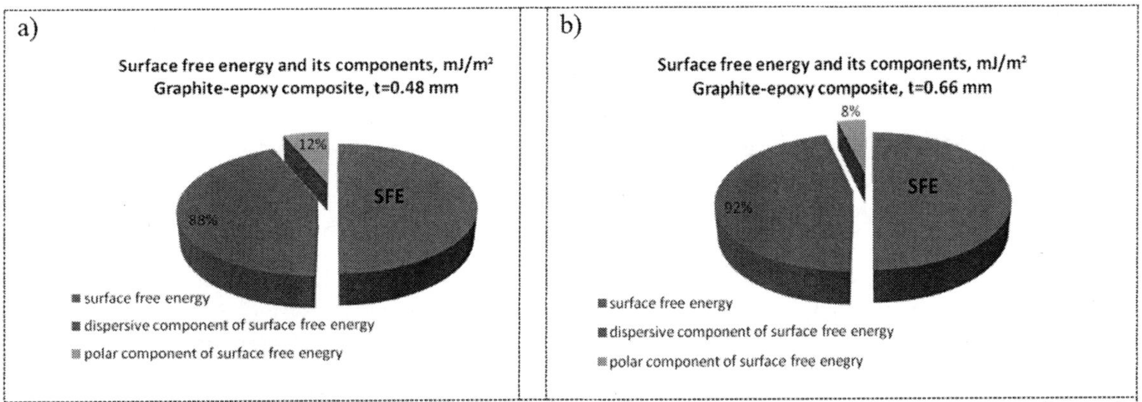

Figure 8. Graphic illustration of the contents of the dispersive and polar SFE components in total surface free energy for the graphite/epoxy composite with a thickness of: a) 0.48 mm, b) 0.66 mm.

The results of the statistical tests demonstrate that there are no significant differences in the mean SFE values (confidence level: 0.95). It can be observed that in each investigated case the dispersive component is much higher than the polar component.

3.4. Strength tests of the graphite/epoxy composites

The strength results of the tested polymeric graphite/epoxy composites are given in figures 9 and 10.

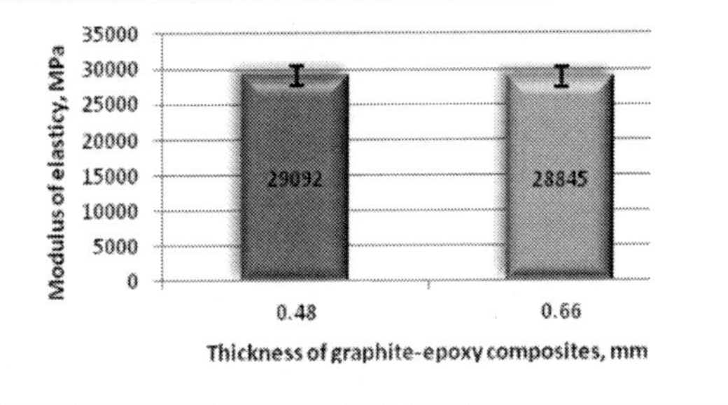

Figure 9. Modulus of elasticy of tested graphite-epoxy composites.

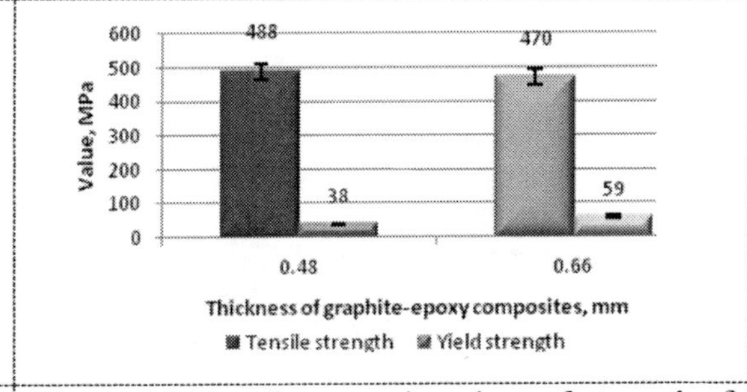

Figure 10. Tensile strength and proof strength of tested graphite-epoxy composites.

The results reveal that the tested graphite/epoxy composites have a similar modulus of elasticity E (difference is 1%) and tensile strength R_m (difference is about 3%). On the other hand, they differ significantly with respect to yield strength ($R_{p0,2}$). The specimens of the graphite/epoxy composite with a higher thickness exhibit a higher yield strength (137 MPa), while the yield strength $R_{p0,2}$ of the composite with a lower thickness (0.48 mm) is 65% of the value of $R_{p0,2}$ of the composite with a thickness of 0.66 mm.

3.5. Strength tests of the graphite/epoxy composite adhesive joints

The strength results of the examined polymer composite adhesive joints are given in figure 11. The results demonstrate that the graphite/epoxy composite with a smaller thickness of the joined materials has a higher strength of the shear-loaded adhesive joints amounting to 19.09 MPa. The shear strength of the adhesive joints of the graphite/epoxy composite with a higher thickness is 94% of the strength of the adhesive joints made of the 0.48 mm thick graphite/epoxy composite.

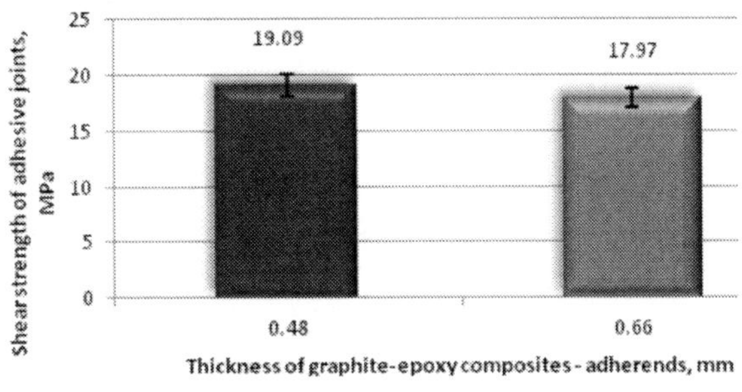

Figure 11. Shear strength of adhesive joints of the tested polymer composites: 1) graphite/epoxy composite (0.48 mm), 2) graphite/epoxy composite (0.66 mm).

The statistical analysis results reveal that there are significant differences regarding strength at a confidence level of 0.95. This can be attributed to the effect of the thickness of the joined materials on the strength of composite adhesive joints. This factor could leads to a higher bending moment resulting from a higher load eccentricity due to joining elements with higher thicknesses.

3.6. Results and discussion

Examining the geometrical structure (microscopic analysis results and surface roughness parameters) and mechanical properties (strength results) of the joined materials, the following was observed:

- graphite/epoxy composites have a similar geometrical structure, and the differences in their thickness do not lead to producing differences in the geometrical structure,
- surface roughness significantly differs over the entire area of the examined specimen in both composites, which is reflected in the microscopic images (figure 2 and figure 3) and surface roughness parameters of these composites,
- most mechanical properties of both composites are similar,
- thicker graphite/epoxy composite specimens have a higher yield strength.

The geometrical structure results suggest that such a structure is advantageous for making adhesive joints since combined with other technological factors, e.g. an adhesive with suitable properties, it can result in strong mechanical fixing of the adhesive, which will in turn lead to a higher adhesive strength of the adhesive joints due to mechanical adhesion.

Comparing the results of surface free energy (table 4) and shear strength of adhesive joints of the tested composites (figure 11), the following observations can be made:

- there is a relation between the surface free energy γ_S of the joined composites and the shear strength of adhesive joints of the tested graphite/epoxy composites. The coefficient of correlation between these quantities equals 1. This means that higher adhesive properties (defined by surface free energy) of surface of the joined materials lead to a higher strength of the adhesive joints;
- a similar relation was observed for the polar component γ_S^p of surface free energy;
- comparing the dispersive component γ_S^d of surface free energy and the shear strength of adhesive joints of the tested composites, it was possible to determine the coefficient of correlation which is equal -1. This means that there is a correlation between the examined

quantities yet the relation is inverse, i.e. with an increase in γ_S^d the shear strength of the adhesive joints decreases. The differences in the values of surface free energy components of the tested composites can results from a slightly different chemical composition of the fabrics used for making these composites, as shown in figures 4 and 5. Although the statistical analysis results demonstrate that the mean values of γ_S of surfaces of both composites are statistically equal, there are difference in the values of surface free energy components, and they can lead to obtaining a statistically significant difference in strength of adhesive joints of the tested composites. It must however be noted that the strength is also affected by the structural factors.

4. Summary

Comparing the strength test results of graphite/epoxy composite adhesive joints and the surface free energy results for these composites, it can be observed that the strength of the adhesive joints of these composites significantly differ in spite of lack of statistically significant differences in the obtained SFE results. Moreover, it was observed that the thickness of the joined elements has an effect on the strength of the adhesive joints, which also results from the occurrence of bending moment due to the fact that in the tested adhesive joints the tensile forces (in strength testing) are not located in one plane while, additionally, the adhesive layer is loaded by the bending moment. This moment increases with an increase in thickness of the adhesive-bonded elements. The adhesive joints strength is also affected by the geometrical structure of the joined materials; it was observed that the structure of the tested composites enables mechanical adhesion to a significant extent, thereby increasing the adhesive strength of the joints. The tests have therefore confirmed the synergic impact of material, technological as well as structural factors. Given the peculiar nature of the adhesive bonding technique, these problems require further research and examination.

References

[1] Magalhães AG, de Moura M F S F and Gonçalves J P M 2005 *Int. J. Adhes. Adhes.* **25** 313
[2] Moghadamzadeh H, Rahimi H, Asadollahzadeh M and Hemmati A R 2011 *Int. J. Adhes. Adhes* **31** 816
[3] Liljedahl C D M, Crocomble A D, Wahab M A and Ashcroft I A 2007 *Int. J. Adhes. Adhes.* **27** 505
[4] Camilho R D S G, de Moura M F S F, Ramantani D A, Morais J J L and Domingues J J M S 2009 *Int. J. Adhes. Adhes.* **29** 678
[5] Rudawska A 2016 Joint Strength *Joining Composites with Adhesives* ed M A Wahab ((Lancaster: DEStech Publications, Inc) chapter 3 pp 39-58
[6] Baldan A 2012 *Int. J. Adhes. Adhes.* **38** 95
[7] Zaleski R, Gorgol M, Zaleski K 2012 *Physics Proced.* **35** 92
[8] Rudawska A 2012 *J. Adhes. Sci. Technol.* **26** 1233
[9] Rudawska A, Bociąga E and Olewnik-Kruszyńska E 2017 *J. Adhes. Sci. Technol.* **31** 327
[10] Molitor P, Barron V and Young T 2001 *Int. J. Adhes. Adhes.* **21** 129
[11] Rudawska A, Danczak I, Muller M and Valasek P 2016 *Int. J. Adhes. Adhes.* **70** 176
[12] Rudawska A 2015 *Advan. Sci. Technol. Res. J.* **9** 35
[13] Grant LDR, Adams RD and da Silva LFM 2009 *Int. J. Adhes. Adhes.* **29** 535
[14] Müller M, Herák D and Valášek P 2013 *Techn. Gazet.* **20** 571
[15] de Moura M F S F, Daniaud R and Magalhaes A G 2006 *Int. J. Adhes. Adhes.* **26** 464
[16] Golio L, Rossetto M and Dragoni E 2008 *Int. J. Adhes. Adhes.* **28** 427
[17] Wahab MA 2016 *Joining composites with adhesives* (Lancaster: DEStech Publications, Inc)
[18] Ashcroft IA, Hughes DJ, Shaw SJ, Abdel Wahab M and Crocombe A 2001 *J. Adhes.***75** 61
[19] Ashcroft IA, Abdel Wahab M, Crocombe AD, Hughes DT and Shaw SJ 2001 *Comp. A* **32** 45
[20] Al-Ghamdi AH, Ashcroft IA, Crocombe AD and Abdel Wahab M. 2003 *J. Adhes.* **79** 1161
[21] Ashcroft IA, Abdel Wahab M and Crocombe AD 2003 *Mech. Adv. Mat. Struc.* **10** 227
[22] Erpolat S, Ashcroft IA, Crocombe AD and Abdel Wahab M 2004 *Comp. A* **35** 1175
[23] Erpolat S, Ashcroft IA, Crocombe AD and Abdel Wahab M 2004 *Int. J. Fatig.* **26** 1189
[24] Loh WK, Crocombe AD, Abdel Wahab M and Ashcroft IA 2005 *Int. J. Adhes. Adhes.* **25** 1
[25] Crocombe AD, Hua YX, Loh WK, Abdel Wahab M and Ashcroft IA 2006 *Int. J. Adhes. Adhes.* **26** 325
[26] Hua Y, Crocombe AD, Abdel Wahab M and Ashcroft IA 2006 *J. Adhes.* **82** 135
[27] Solana AG, Crocombe AD, Abdel Wahab M and Ashcroft IA 2007 *J. Adhes. Sci. Technol.* **21** 1343

IOP Conf. Series: Journal of Physics: Conf. Series **842** (2017) 012073 doi:10.1088/1742-6596/842/1/012073

[28] Rudawska A 2010 *Int. J. Adhes. Adhes.* **29** 451
[29] Ageorges C and Ye L 2002 *Fusion Bonding of Polymer Composites (London:* Springer)
[30] Wahab M A, Hilmy I, Ashcroft I A and Crocombe A D 2010 *J. Adhes. Sci. Technol.* **24** 325
[31] http://www.semlab.us.edu.pl/en/instruments.php (access on: 16th September 2016)
[32] PN-EN ISO 4287:1999/A1:2010 Geometrical Product Specifications
[33] Shang J, Flury M, Harsh J B and Zollars R L 2008 *J. Colloid Interface Sci.* **328** 299
[34] Della Volpe C and Siboni S 1997 *J. Colloid Interface Sci.* **195** 121

Experimental research and statistic analysis of polymer composite adhesive joints strength

Anna Rudawska[1], Izabela Miturska[1], Jakub Szabelski[1], Agnieszka Skoczylas[1], Paweł Droździel[1], Elżbieta Bociąga[2], Radovan Madleňák[3], Dariusz Kasperek[1]

[1]Lublin University of Technology, Faculty of Mechanical Engineering, Nadbystrzycka 36 Str, 20-618 Lublin, Poland, e-mail: a.rudawska@pollub.pl
[2]Czestochowa University of Technology, Department of Polymer Processing, Institute of Mechanical Technologies, Armii Krajowej Str. 19c, 42-201 Czestochowa, Poland
[3]University of Žilina, Faculty of Operation and Economics of Transport and Communications, Univerzitna 1, 010 26, Žilina, Slovak Republic

Abstract. The aim of this paper is to determine the effect of arrangement of fibreglass fabric plies in a polymer composite on adhesive joint strength. Based on the experimental results, the real effect of plies arrangement and their most favourable configuration with respect to strength is determined. The experiments were performed on 3 types of composites which had different fibre orientations. The composites had three plies of fabric. The plies arrangement in Composite I was unchanged, in Composite II the central ply had the 45° orientation, while in Composite III the outside ply (tangential to the adhesive layer) was oriented at 45°. Composite plates were first cut into smaller specimens and then adhesive-bonded in different combinations with Epidian 61/Z1/100:10 epoxy adhesive. After stabilizing, the single-lap adhesive joints were subjected to shear strength tests. It was noted that plies arrangement in composite materials affects the strength of adhesive joints made of these composites between the values of the strength of the confidence level of 0.95. The statistical analysis of the results also showed that there are no statistical significant differences in average values of surface free energy (0.95 confidence level).

1. Introduction

Currently, composite materials constitute a wide and rapidly developing area of polymers. Composites are mainly used in the production of aircraft, sports equipment, and different types of boats and tanks [1,2] as well as pipelines [3,4] and their fittings and also bearings [5] and tubular structures [6,7]. Products made of materials such as fibreglass-reinforced polymers play a significant role in industry [4]. These components include fillers and polymer matrix. The filler material can be made fibres, fabric, paper, and others [8-11]. The polymer matrix can be made of resins, such as epoxy resins, polyester resins, phenol-formaldehyde resins or maleic resins [12-14]. One of the methods of composite forming is laminating [3,15]. Laminates of fabrics exhibit the warp and weft maximum mechanical properties, while at 45° to the directions of these characteristics are much worse [12]. The textile fiber roving strands are interleaved in parallel. This allows obtaining a particularly high strength bidirectional, with a total use of considerable strength roving bands of parallel fibers. In order to achieve a high strength uni-directional in the plane is used as reinforcement fabric unidirectional (targeted), arranged in parallel layers [16]. The various laminates, which have various the

arrangements of plies and amount of plies, were tested. Katman et al. [17] tested the laminate thickness of 2 mm which was targeted using 16 plies (each ply is approximately 0.125 mm thick) with $(01/901)_{4s}$ lay-up. Bénard et al. [12] investigated the composite samples which are produced, respectively, with 8 and 10 pre-impregnate plies of carbon and glass fibres in the same orientation. There are generally two kinds of joining methods for composite structures [4,18,19]: (1) mechanical fastening and (2) adhesive bonding. Adhesive bonding is an alternative to conventional joining methods in structural applications, and it provides many advantages such as lower structural weight, improved damage tolerance and design flexibility over mechanical fasteners [6,8,17,20].
This paper investigates the effect of arrangement of individual plies of fibreglass fabric in a composite material on the strength of adhesive joints. On the basis of the analysis of the studies specified the actual impact of the arrangement of layers and the best of their arrangement affect strength of adhesive joints.

2. Experimental test

2.1. Characteristics of the tested composites

Three polymer composites made of two constituent materials were fabricated. The reinforcement material was plain weave fibreglass fabric 200g/m^2 (figure 1), while the matrix material was POLIMAL 1094 polyester resin. The schematic of plain weave fibreglass fabric was presented in figure. 2.

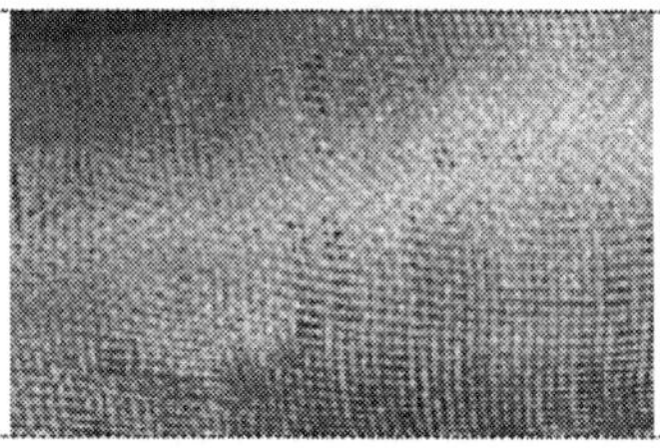

Figure 1. View of fibreglass fabric 200g/m^2.

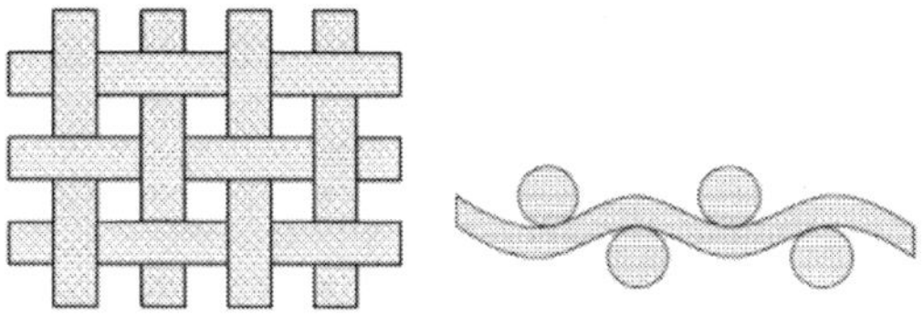

Figure 2. View of plain weave scheme of a rare woven.

Table 1. Characteristics of the tested polymer composites.

Name of material	Plies arrangement
Composite (I): plain weave fibreglass fabric 200g/m^2, Matrix - polyester, 3 plies	0°/0°/0°
Composite (II): plain weave fibreglass fabric 200g/m^2, Matrix - polyester, 3 plies	0°/45°/0°
Composite (III): plain weave fibreglass fabric 200g/m^2, Matrix - polyester, 3 plies	45°/0°/0°

The schematic of plain weave fibreglass fabric was presented in figure. 2. In plain weave, each warp thread is interwoven and intersects with each weft thread and thus the fabric is tightly bound and rigid. The composites had three plies of fabric. The plies arrangement in Composite I was unchanged, in Composite II the central ply had the 45° orientation, while in Composite III the outside ply (tangential to the adhesive layer) was oriented at 45° (figures 3 – 6 and table 1).
Prior to fabricating each composite material, fibreglass fabric and polyester resin (POLIMAL 1094, produced by Organika Sarzyna, Poland) were mixed with Metox-50 curing agent. The resin and curing agent were thoroughly mixed in a ratio of 100:10. Polimal 1094 is orthophthalic structural resin, medium-flexible, accelerated and low emission of styrene. Metox-50 is a peroxide initiator for copolymerization of unsaturated polyester resins and vinyl ester. This hardener contains methylethylketone peroxide and it is 20% solution in dimethylphthalate.

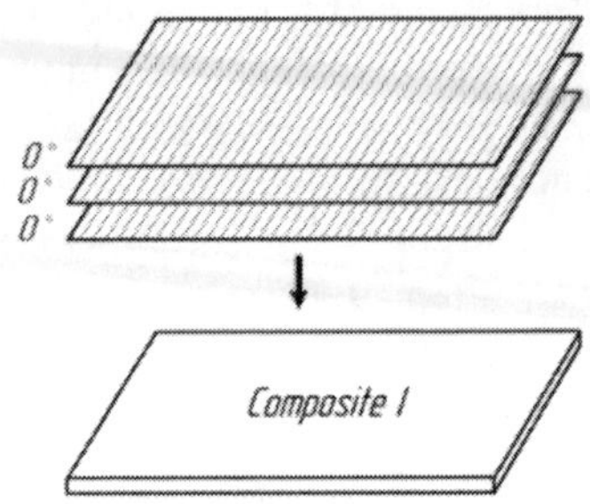

Figure 3. Schematic of an I composite.

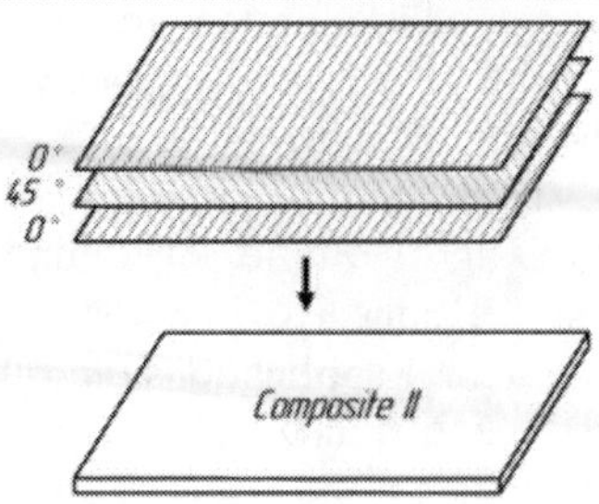

Figure 4. Schematic of an II composite.

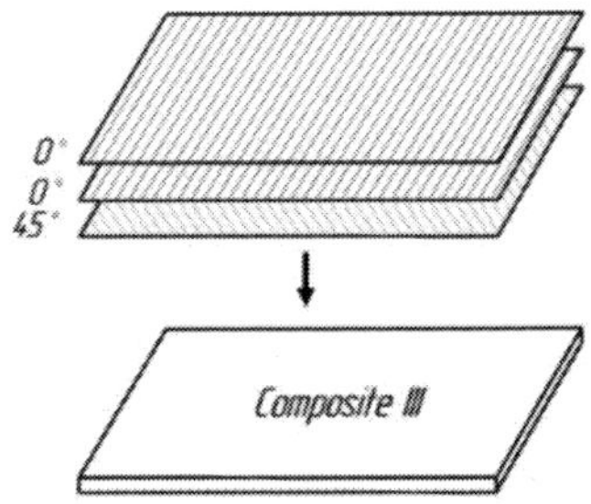

Figure 5. Schematic of an III composite.

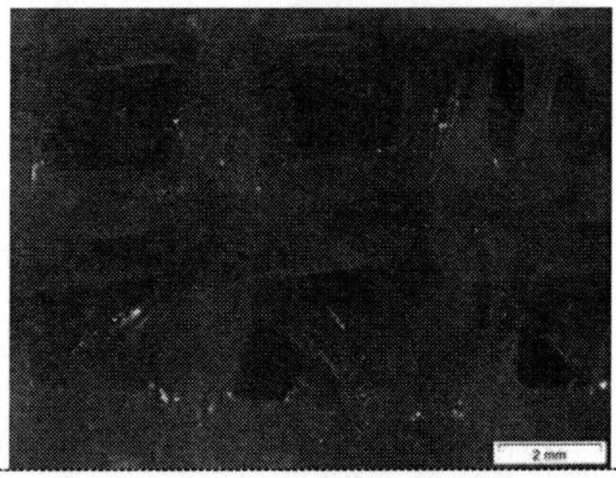

Figure 6. Surface view of composite II: 0°/45°/0° plies arrangement

The three prepared composite materials were left to cure for 4 days. After that period, each composite was removed from the mould, cut and numbered. The samples were cut by a FNU-50E milling machine. The cut composites were used four blade cutter mill with skew cuts in the working part of the CVD diamond coating for the assay JRT0600 a diameter of 6 mm. The cut was implemented with the following technological parameters: RPM 9000 min -1 and a feed of 1000 mm/min. Individual samples were numbered was to facilitate experimental tests and detection of defects.

2.2. Adhesive joints

The experiments were performed on samples produced in compliance with the above procedure. Each sample had a length of 100 mm and a width of 25 mm and a thickness of 2 mm. Based on the combinations with repetitions (1), it was calculated that the 3 types of composites combined using two constituent materials can produce samples with 6 variants of adhesive joints combinations, as listed in table 2. Calculated from the formula ([21]) (2):

$$C_k^n = \frac{(k+n-1)!}{k!\,(n-1)!} \tag{1}$$

k = 2 – k, are the combinations of elements with repetitions,
n = 3 – n, is the number of elements in the set.

Adhesive joints of polymer composite were made using an epoxy adhesive made from Epidian 6 epoxy resin and triethylenotetraamine (produced by Organika Sarzyna, Poland) curing agent which were mechanically mixed in a weight ratio of 100:10, taking into account different combinations of types of composite materials listed in table 2. The lap joint length was 15 mm and the adhesive thickness (tad) is equal 0.14±0.08 mm. The some configuration of single lap adhesive joints of polymer composites is presented in figures 7 – 9 (red plie had the 45° orientation).

Table 2. Combinations of composite adhesive joints.

Type of composite	Composite (I)	Composite (II)	Composite (III)
Composite (I)	I-I		
Composite (II)	I-II	II-II	
Composite (III)	I-III	II-III	III-III

Prior to adhesive bonding, the samples were degreased with Loctite 7063, an acetone-based degreasing agent. The procedure of degreasing was presented in [22]. The adhesive joints were made using Epidian 6/Z1/100:10 epoxy adhesive which was subjected to single-stage curing for 168 hours in an ambient temperature of 23-25°C, under a load of 0.0016 MPa and a humidity of 22-31%.
The strength tests were performed on shear-loaded single-lap polymer composite adhesive joints (figure 10) in compliance with the standard DIN EN 1465 and the test speed was 5 mm/min. The tests were performed on 10-12 test pieces per each batch.

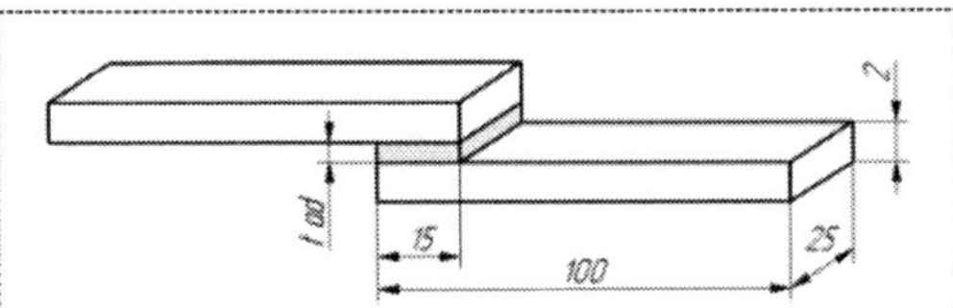

Figure 7. Schematic of an I-I composites adhesive joint.

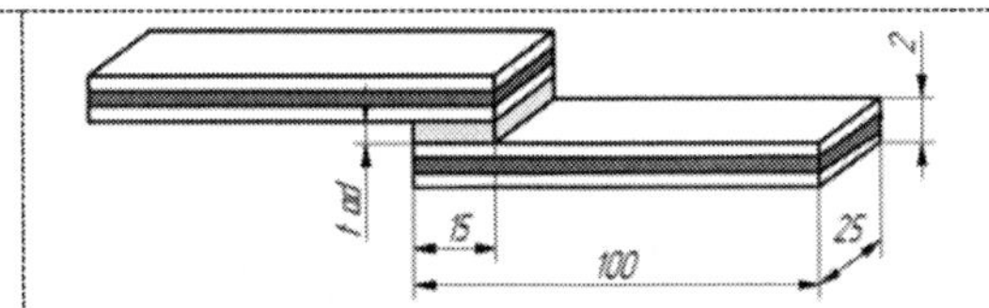

Figure 8. Schematic of an II-II composites adhesive joint.

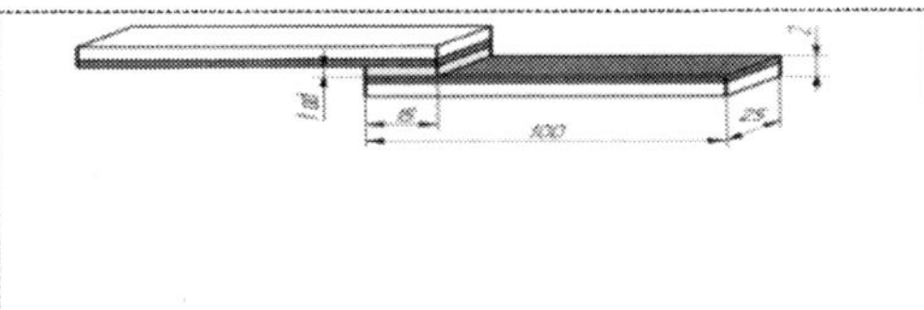

Figure 9. Schematic of an III-III composites adhesive joint.

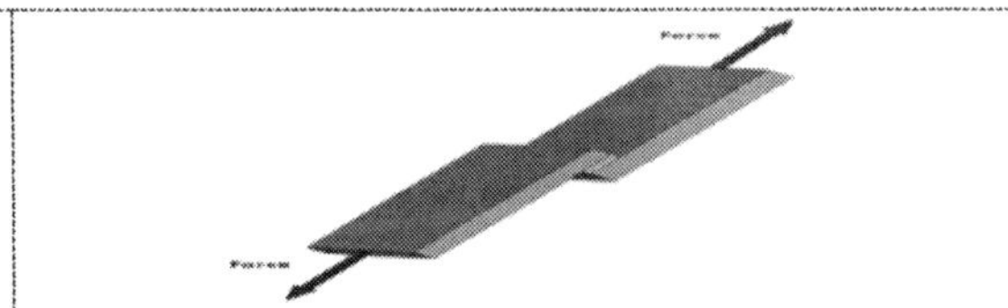

Figure 10. View of single lap shear specimen

2.3. Statistical analysis method

In order to proceed to the analysis of the test results, some parameters of descriptive statistics should be determined, which sets certain characteristics describing the properties of the tested characteristics [21]. Evaluation of distribution by means of these measures may include the position of characteristics, their diversity, symmetry of distribution and concentration. The normality of the distribution can be determined by analysing the above mentioned parameters and using the Shapiro-Wilk test, which is considered the best test to check the normality of distribution of a random variable of a small sample. The main advantage of this test is its high power, which means that for a fixed confidence level α, the probability of rejecting the null hypothesis H0 when it is false, is greater than for the other tests of this type. In order to check the veracity of the hypotheses posed in relation to the test results, the F test (Fisher Snedecor test) should be used, through which it is possible to verify the hypothesis of equality of two population's variances. If the hypothesis is not rejected, then the t-Student test should be used, and if the hypothesis is rejected, then the Cochran-Cox test should be used [21].The development of the data involved tests of significance to compare the mean values of the tested trait in two general populations. The test included the first type error, i.e. rejection of the tested

IOP Conf. Series: Journal of Physics: Conf. Series **842** (2017) 012074 doi:10.1088/1742-6596/842/1/012074

hypothesis if it is true. The first type error is called the level of significance and was assumed at $\alpha = 0.05$. The samples are to be treated as small due to the fact that the set included up to 30 elements.

2.4. Preparation for experiments: criteria for comparison
The adhesive joints of polymer composites were divided into 4 groups. In each of these groups the results were investigated with respect to a different set of comparative criteria.
Group 1 – the joints in Group 1 were compared with regard to the effect of change (shifting the turned ply) relative to the ply in Composite I that was constant in terms of its arrangement (table 3).
Group 2 – in this group, it was assumed that the plies arrangement on one of the sides was constant and had the form of 0°/45°/0° (Composite II), while it was changed on the other side of the adhesive joint (table 4).
Group 3 – the results in Group 3 were examined with respect to the effect of shifting the turned ply on the one side of the adhesive layer, while the plies arrangement was constant on the other side and had the form of 0°/0°/45°, the turned ply was located the closest to the adhesive layer (table 5).
Group 4 – the results in Group 6 were examined with respect to the effect of symmetry change in plies arrangement on the strength of adhesive joints (table 6).

Table 3. Adhesive joints – Group 1.

Composite plies arrangement		Composite plies arrangement	Type of composite adhesive joint
0°/0°/0°	Adhesive layer	0°/0°/0°	I-I
0°/0°/0°		0°/45° /0°	I-II
0°/0°/0°		45° /0°/0°	I-III

Table 4. Adhesive joints – Group 2.

Composite plies arrangement		Composite plies arrangement	Type of composite adhesive joint
0°/45°/0°	Adhesive layer	0°/0°/0°	II-I
0°/45°/0°		0°/45° /0°	II-II
0°/45°/0°		45° /0°/0°	II-III

Table 5. Adhesive joints – Group 3.

Composite plies arrangement		Composite plies arrangement	Type of composite adhesive joint
0°/0°/45°	Adhesive layer	0°/0°/0°	III-I
0°/0°/45°		0°/45° /0°	III-II
0°/0°/45°		45° /0°/0°	III-III

Table 6. Adhesive joints – Group 4.

Composite plies arrangement		Composite plies arrangement	Type of composite adhesive joint
0°/0°/0°	Adhesive layer	0°/0°/0°	I-I
0°/45°/0°		0°/45° /0°	II-II
0°/0°/45°		45° /0°/0°	III-III

3. Experimental results

3.1. Results

Tables 7 – 10 present the shear strength of adhesive joints and statistical analysis results of the samples from four groups of adhesive joints. The designation in tables 7-10 is the same like in the legend which was presented in table 7.

Table 7. Statistical analysis results of samples in Group 1.

Adhesive joints	S_s[a], MPa	Fα[b]	F[c]	Result	Tα[d]	T[e]	Conclusion
I-I/I-II	**4.66**	9.28	3.33	$\sigma_{11}^2 = \sigma_{12}^2$	1.943	-1.585	$\mu_{11}^2 = \mu_{12}^2$
I-II/I-III	**4.99**	9.28	1.68	$\sigma_{12}^2 = \sigma_{13}^2$	1.943	1.710	$\mu_{12}^2 = \mu_{13}^2$
I-I/I-III	**4.71**	9.28	2.00	$\sigma_{11}^2 = \sigma_{13}^2$	1.943	-0.223	$\mu_{11}^2 = \mu_{13}^2$

[a] Value of shear strength of adhesive joints.
[b] Value of statistic – Fα (Fisher Snedecor test).
[c] Obtained value of statistic – F.
[d] Value of statistic – Tα (t-Student test).
[e] Obtained value of statistic – T.
σ^2 – variance.
μ – mean value.

Table 8. Statistical analysis results of samples in Group 2.

Adhesive joints	S_s, MPa	Fα	F	Result	Tα	T	Conclusion
I-II/II-III	**4.66**	6.59	1.25	$\sigma_{12}^2 = \sigma_{23}^2$	1.895	-0.558	$\mu_{12}^2 = \mu_{23}^2$
Adhesive joints	S_s[a], MPa	Fα[b]	F[c]	Result	Cα	C	Conclusion
I-II/II-II	**4.99**	6.59	8.75	$\sigma_{12}^2 \neq \sigma_{22}^2$	2.827	0.567	$\mu_{12}^2 = \mu_{22}^2$
II-II/II-III	**4.71**	6.39	7.00	$\sigma_{22}^2 \neq \sigma_{23}^2$	2.776	-0.849	$\mu_{22}^2 = \mu_{23}^2$

[a] Value of statistic – Cα (Cochran-Cox test).
[c] Obtained value of statistic – C.

The statistical analysis results of the samples from Group 1 (table 3) demonstrate that the hypothesis about equality of variances of two populations is met in all cases. The hypothesis about equality of means is also confirmed in all cases, so the means in these cases are statistically equal (4.66 = 4.99, 4.99 = 4.71, 4.66 = 4.71) – table 7.
The statistical analysis results of the samples from Group 2 (table 4) demonstrate that the hypothesis about equality of variances of two populations is met in one case (I-II and II-III), while it is rejected in the remaining two cases (I-II and II-II, II-II and II-III). The hypothesis about equality of means is also confirmed in all cases, so the means in these cases are statistically equal (4.99 = 4.83, 4.99 = 5.07, 4.83 = 5.07) – table 8.
The statistical analysis results of the samples from Group 3 (table 5) demonstrate that the hypothesis about equality of variances of two populations is met in all cases. However, the hypothesis about equality of means is confirmed only in one case (I-III and III-III), so the mean in this case is statistically equal (4.71 = 4.43); however, in the remaining cases (I-III and II-III, II-III and III-III), the hypothesis is rejected, so the means in these cases are statistically different (4.71 ≠ 5.0, 5.07 ≠ 4.43) – table 9.

Table 9. Statistical analysis results of samples in Group 3.

Adhesive joints	S_s, MPa	Fα	F	Result	Tα	T	Conclusion
I-III/II-III	**4.71**	6.95	1.33	$\sigma_{13}^2 = \sigma_{23}^2$	1.895	-2.245	$\mu_{13}^2 \neq \mu_{23}^2$
II-III/III-III	**5.07**	6.39	1.50	$\sigma_{23}^2 = \sigma_{33}^2$	1.859	4.040	$\mu_{23}^2 \neq \mu_{33}^2$
I-III/III-III	**4.43**	6.59	1.13	$\sigma_{13}^2 = \sigma_{33}^2$	1.895	1.561	$\mu_{13}^2 = \mu_{33}^2$

Table 10. Statistical analysis results of samples in Group 4.

Adhesive joints	S_s, MPa	Fα	F	Result	Tα	T	Conclusion
I-I/II-II	**4.66**	6.59	2.63	$\sigma_{11}^2 = \sigma_{22}^2$	1.859	-0.499	$\mu_{11}^2 = \mu_{22}^2$
II-II/III-III	**4.83**	6.39	2.67	$\sigma_{22}^2 = \sigma_{33}^2$	1.859	1.371	$\mu_{22}^2 = \mu_{33}^2$
III-III/I-I	**4.43**	6.59	1.78	$\sigma_{33}^2 = \sigma_{11}^2$	1.895	1.084	$\mu_{33}^2 = \mu_{11}^2$

The statistical analysis results of the samples from Group 4 (table 6) demonstrate that the hypothesis about equality of variances of two populations is met in all cases. The hypothesis about equality of means is confirmed in all cases, too, so the means in these cases are statistically equal (4.66 = 4.83, 4.83 = 4.43, 4.66 = 4.43) – table 10.

3.2. Discussion

The results of adhesive joint strength demonstrate that the arrangement of composite plies affects the strength of the investigated adhesive joints. In addition, it can be observed that in the adhesive bonding of composite materials the strength of adhesive joints is also significantly affected by the number of fiberglass fabric plies.

It must be stressed that special care was taken to ensure that all composite materials are made under the same conditions, the only variables being the arrangement and number of plies. Based on the statistical analysis of the strength test results, it is found that in most cases there are no statistically significant differences in the yielded means of strength (Groups 1, 2, 4). However, the results for other tested samples (Groups 3,) reveal that such differences do occur. Group 3 involved analysis of the effect of shifting the turned ply on the one side of the adhesive layer, while the plies arrangement on other side was unchanged and had the form of 0°/0°/45° (the turned ply was located closest to the adhesive layer, table 5). Supported by a statistical analysis, these results demonstrate a clear effect of the fibreglass fabric plies arrangement on failure force. The best results in this group are yielded for the configuration II-III (plies arrangement: 0°/45°/0°/adhesive/45°/0°/0°). On the other hand, the worst results are obtained for the configuration III-III, where the plies arrangement is changed only at the adhesive layer (plies arrangement: 0°/0°/45°/adhesive/45°/0°/0°). The results in Group 4 demonstrate that the symmetrical plies arrangement has no significant effect on the strength of the examined adhesive joints. Nonetheless, it is observed that changing the ply arrangement in the vicinity of the adhesive layer (joint III-III) has a negative effect on the strength of this layer, as opposed to the case when the ply arrangement is not changed (joint I-I). However, the combination of these two plies arrangement types into one (joint II-II) reveals a certain synergy and yields better results than in the case of the two previous adhesive joints. Summing up, plies arrangement in composite materials affects the strength of adhesive joints made of these composites.

4. Summary

The paper investigated the effect of plies arrangement in a hand-made composite material on the strength of adhesive joints made from this composite. The significance of this problem is particularly true these days owing to a wide use of composite materials to design structures with required specific strength. To investigate the effect of plies arrangement on the strength of fibreglass fabric, 3 composite materials were fabricated, each having a different plies arrangement. Special care was taken to maintain the same manufacturing conditions throughout the entire process as well as to ensure that the composite material be produced with utmost accuracy so as to yield the most reliable results. Individual composite plates were cut into smaller samples and then adhesive bonded using an epoxy adhesive made from Epidian 61 epoxy resin combined with triethylenotetraamine hardener. The stabilizing period of all samples was set to 7 days. After that, each sample was measured and subjected to failure tests. Some of the procedures had to be repeated because of a significant scatter of the results. The final results were then divided into groups to examine the effect of different variants of plies arrangement. Summing up, it was found that in some cases (joint II-III) plies arrangement has a significant effect on increasing the shear strength of samples. Two types of adhesive joints samples exhibit higher shear strength: those consisting of two samples of adhesive joints made of uniform thickness composites as well as those with alternate plies arrangements. It should be noted that the both tangential plies in the plies arrangement 0°/adhesive layer/0° should not be turned in the vicinity of the adhesive layer. But Composite II (0°/45°/0°) samples exhibit the lowest tensile strength. In this composite central ply has the 45° orientation. However, the combination of two plies arrangement types into one (joint II-II) reveals a certain synergy and yields better results than in the case of the joint III-III and joint I-I. Concluding, plies arrangement is of vital importance and should be taken into account when designing composite materials for adhesive bonding. Nonetheless, this problem has not been exhaustively investigated in this paper. It can be studied further using samples which differ from those used in this study (e.g. samples with 45°/0°/45° plies arrangement).

References

[1] de Ulzurrun I D, López F, Herreros M A and Suárez J C 2007 *Eng. Fail. Analys.* **14** 310
[2] Hashim S A 1999 *Mar. Struc.* **12** 405
[3] Mouritz A P, Bannister M K, Falzon P J and Leong K H 1999 *Comp. A* **30** 1445
[4] Adams R D (2010) *Adhesive bonding. Science, technology and applications* (Cambridge: Woodhead Publishing)
[5] Kim S S and Lee D G 2006 *Comp. Struct.* **75** 222
[6] Minnetyan L and Chamis C C 1999 *Theor. Appl. Fract. Mech.* **31** 75
[7] Parashar A and Mertiny P 2012 *Int. J. Adhes. Adhes.* **38** 58
[8] Rudawska A 2015 *J. Phys. Conf. Series* **628** 012077
[9] Hang J, Luo R and Yang C 2012 *Carbon* **50** 4922
[10] Suberlyak O, Krasinskyi V, Moravskyi V, Gerlach H and Jachowicz T 2014 *Mat. Sci.* 2014 **50** 296
[11] Müller M, Valášek P and Rudawska A 2016 *6th Int. Conf. on Trends in Agricultural Engineering, 7 - 9 September 2016,* Prague, Czech Republic, 412
[12] Bénard Q, Fois M and Grisel M 2005 *Int. J. Adhes. Adhes.* **25** 404
[13] Prolongo S G, del Rosario G and Ureña A 2006 *Int. J. Adhes. Adhes.* **26** 125
[14] Rudawska A and Czarnota M 2013 *J. Adhes. Sci. Technol.* **27** 1933
[15] Kishore A N and Prasad N S 2012 *Int. J. Adhes. Adhes.* **35** 55
[16] Kang M - H, Choi J - H and Kreon J - H 2014 *Comp. Struct.* **108** 417
[17] Katnam K B, Comer A J, Stanley W F, Buggy M, Ellingboe A R, and Young T M 2011 *Int. J. Adhes. Adhes.* **31** 679
[18] Kumar P, Singh R K and Kumar R 2007 *Int. J. Adhes. Adhes.* **27** 68
[19] Wahab M A 2016 *Joining composites with adhesives* (Lancaster: DEStech Publications, Inc)
[20] Balkova R, Holcnerova S and Cech V 2002 *Int. J. Adhes. Adhes.* **22** 291
[21] Dowdy S and Wearden S 1983 *Statistics for Research* (New York: Willey)
[22] Rudawska A 2012 *J. Adhes. Sci. Technol.* **26** 1233

Impact localization on composite structures using time difference and MUSIC approach

Yongteng Zhong[1] **and Jiawei Xiang**[1]

[1] College of Mechanical and Electrical Engineering, Wenzhou University, Wenzhou, P.R.China

E-mail: zhongyongteng@wzu.edu.cn, wxw8627@163.com;

Abstract. 1-D uniform linear array (ULA) has the shortcoming of the half-plane mirror effect, which does not allow discriminating between a target placed above the array and a target placed below the array. This paper presents time difference (TD) and multiple signal classification (MUSIC) based omni-directional impact localization on a large stiffened composite structure using improved linear array, which is able to perform omni-directional 360° localization. This array contains $2M+3$ PZT sensors, where $2M+1$ PZT sensors are arranged as a uniform linear array, and the other two PZT sensors are placed above and below the array. Firstly, the arrival times of impact signals observed by the other two sensors are determined using the wavelet transform. Compared with each other, the direction range of impact source can be decided in general, 0° to 180° or 180° to 360°. And then, two dimensional multiple signal classification (2D-MUSIC) based spatial spectrum formula using the uniform linear array is applied for impact localization by the general direction range. When the arrival times of impact signals observed by upper PZT is equal to that of lower PZT, the direction can be located in x axis (0°or 180°). And time difference based MUSIC method is present to locate impact position. To verify the proposed approach, the proposed approach is applied to a composite structure. The localization results are in good agreement with the actual impact occurring positions.

1. Introduction

Structural health monitoring (SHM) is an important requirement for detecting, estimating, classifying, and predicting damage in engineering structures [1-2]. Composite materials have a number of potential advantages over conventional materials since they can be designed to increase performance in a number of applications including specific stiffness, specific strength. However, these components are sensitive to low-velocity impact damages that can considerably degrade the structural integrity [3]. Therefore, impact localization has become an important tool for SHM systems based on ultrasonic guided waves.

Several methods have been reported using compact sensor array arrangement. Linear ultrasonic phased arrays, as the simplest and most widely used ones, have been well explored and developed. To deal with the near-field monitoring problem, the authors in previous work proposed the near-field 2D-MUSIC algorithm based impact localization method for composite plate using the uniform linear array [4-7]. However, linear array based algorithm suffers from the half-plane mirror effect when it cover the direction 0° and 360°, which does not allow discriminating between a target placed above the array and a target placed below the array [8]. An improved linear array is designed for omni-directional 360° impact localization. This array contains 2M+3 PZT sensors, where 2M+1 PZT sensors are arranged as a uniform linear array, and the other two PZT sensors are placed above and below the array. Firstly, the arrival times of impact signals observed by the other two sensors are determined using the wavelet transform. Compared with each other, the direction range of impact source can be decided in general, 0°to 180° or 180°to 360°. And then, 2D-MUSIC based spatial spectrum formula using uniform linear array is applied for impact localization by the general direction range. When the arrival time of impact signal observed by upper PZT is equal to that of lower PZT, the direction can be located in x axis (0°or 180°). And time difference based method is present to locate impact position.

2. Observed signal model from uniform linear array

As seen in Figure 1, a uniform linear array (ULA) consists of 2M+3 piezoelectric (PZT) sensors on the structure, which are arranged uniformly along the x axis and asymmetric with y axis. The distance between two sensors is d. To ensure the approximation condition in the analysis, d has to be small enough to meet the condition of $d \le \lambda/2$, where λ is the wavelength of wave signal. Seen as Figure.1, θ denotes the wave propagating direction with respect to the coordinate x axis. r is defined as the distance between the impact source and the ULA which is the distance from the source to the reference sensor labeled as 0 in the sensor array. Besides the ULA arrangement, two PZT sensors are added above and below the reference sensor, labeled as PZT U and PZT D, whose function description will been given in section 3.3.

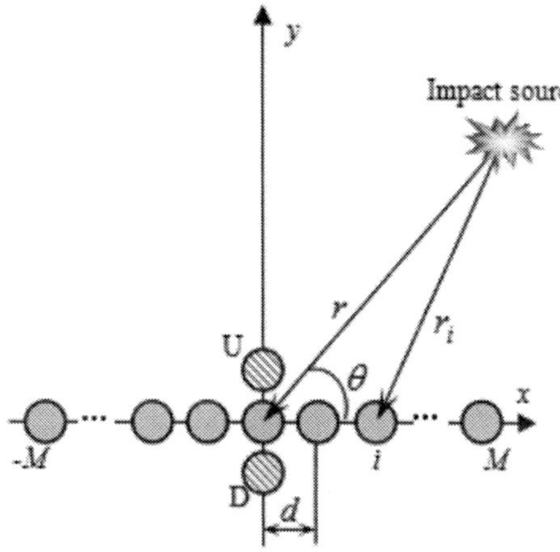

Figure 1. Impact signal model observed using improved uniform linear array

Let $\mathbf{x}_i(t)$ denote the output from PZT i of the sensor array observed at time t, it can be expressed as

$$\mathbf{x}_i(t) = \mathbf{s}_i(t) + \mathbf{n}_i(t),\ i = -M, \cdots, M \tag{1}$$

where $\mathbf{s}_i(t)$ is the impact source signal corresponding output from PZT i, $\mathbf{n}_i(t)$ is the output corresponding to the background noise.

Considering the impact waves with a certain frequency component of ω_0 arriving at the sensor array, the corresponding output $\mathbf{s}_1(t)$ from PZT 1 can be represented by Eq. (2):

$$\mathbf{s}_1(t) = u(t)e^{j(\omega_0 t - kr)} \tag{2}$$

where $u(t)$ is the signal output amplitude, $k = \omega_0 / c$ is the wavenumber, where, c is wave velocity..

According to the formula deduced by the authors, the observed data output from PZT i can be represented as Eq. (3)

$$\mathbf{x}_i(t) = \frac{r}{r_i}\mathbf{s}(t)e^{j\omega_0\tau_i} + \mathbf{n}_i(t),\ i = -M,\cdots,M \tag{3}$$

τ_i is the arriving time difference between PZT i and PZT 0. And the array steering vector $a_i(r,\theta)$ for the impact signal is

$$a_i(r,\theta) = \frac{r}{r_i}\exp(\ j\omega_0\tau_i) \tag{4}$$

The difference of propagation distance from the impact source to the reference element PZT 0 and PZT i is

$$\Delta r_i = |r - r_i| \tag{5}$$

A triangle in this figure is composed of the distance r between impact source and PZT 1, the distance r_i between impact source and PZT i, the distance $(i-1)d$ between PZT 1 and PZT i. According to this triangle, r_i can be obtained by cosine theorem

$$r_i = \sqrt{r^2 + (i-1)^2 d^2 - 2r(i-1)d\cos\theta} \tag{6}$$

Thus, Eq. (5) can be rewritten as Eq. (7).

$$\Delta r_i = r - \sqrt{r^2 + (i-1)^2 d^2 - 2r(i-1)d\cos\theta} \tag{7}$$

Thus, the arriving time difference between PZT i and PZT 0 can be obtained as

$$\tau_i = \frac{\sqrt{r^2 + (i-1)^2 d^2 - 2r(i-1)d\cos\theta} - r}{c} \tag{8}$$

To simplify the calculation, τ_i is represented by its second-order Taylor expansion about τ_1 as shown [34] by Eq. (9)

$$\tau_i = \frac{\tau'(i)|_{i=1}}{1!}(i-1) + \frac{\tau''(i)|_{i=1}}{2!}(i-1)^2 + O((i-1)^2) \tag{9}$$

By calculating the first derivative $\tau'(i)|_{i=1}$ and second derivative $\tau''(i)|_{i=1}$, Eq. (10) can be obtained.

$$\tau_i = \frac{(-d\cos\theta)}{c}(i-1) + (-\frac{d^2}{cr}\sin^2\theta)(i-1)^2 + O(\frac{d^2}{r^2}) \tag{10}$$

For the whole sensor array, the observed input signal vector can be represented as

$$\mathbf{X}(t) = \mathbf{A}(r,\theta)\mathbf{s}(t) + \mathbf{N}(t) \tag{11}$$

where

$$\mathbf{X}(t) = [\mathbf{x}_{-M}(t), \mathbf{x}_{-M+1}(t), \cdots, \mathbf{x}_M(t)]^{\mathrm{T}}$$

$$\mathbf{A}(r,\theta) = [a_{-M}(r,\theta), a_{-M+1}(r,\theta), \cdots, a_M(r,\theta)]^{\mathrm{T}}$$

$$\mathbf{N}(t) = [\mathbf{n}_{-M}(t), \mathbf{n}_{-M+1}(t), \cdots, \mathbf{n}_M(t)]^{\mathrm{T}}$$

3. TD-MUSIC based omni-directional impact localization

3.1. MUSIC based spatial spectrum formula

The covariance matrix of the narrow-band signal extracted from observed signal vector of the ULA is

$$\mathbf{R} = \mathrm{E}[\mathbf{X}\mathbf{X}^{\mathrm{H}}] = \mathbf{A}\mathrm{E}[\mathbf{s}\mathbf{s}^{\mathrm{H}}]\mathbf{A}^{\mathrm{H}} + \mathbf{A}\mathrm{E}[\mathbf{s}\mathbf{N}^{\mathrm{H}}] + \mathrm{E}[\mathbf{s}\mathbf{N}^{\mathrm{H}}]\mathbf{A}^{\mathrm{H}} + \mathrm{E}[\mathbf{N}\mathbf{N}^{\mathrm{H}}] \tag{12}$$

$\mathrm{E}[\ ]$ denotes covariance computation, and the superscript H denotes the complex conjugate transpose. Assuming that signal and noise are independent and the background noise is Gaussian white, Eq. (12) can be simplified as

$$\mathbf{R} = \mathbf{A}\mathbf{R}_S\mathbf{A}^{\mathrm{H}} + \sigma^2\mathbf{I} \tag{13}$$

where $\mathbf{R}_S$ is the covariance matrix of signal, and σ^2 is noise power. $\mathbf{I}$ denotes the covariance matrix of noise. Because noise arriving at the sensor array is of equal power and uncorrelated, $\mathbf{I}$ stands for $M \times M$ identity matrix.

The eigenvalue decomposition of $\mathbf{R}$ is:

$$\mathbf{R} = \mathbf{U}_{\mathrm{S}}\sum\nolimits_{\mathrm{S}}\mathbf{U}_{\mathrm{S}}^{\mathrm{H}} + \mathbf{U}_{\mathrm{N}}\sum\nolimits_{\mathrm{N}}\mathbf{U}_{\mathrm{N}}^{\mathrm{H}} \tag{14}$$

Here, $\mathbf{U}_{\mathrm{S}}$ denotes the signal subspace spanned by the eigenvector matrix corresponding to the largest eigenvalue. $\mathbf{U}_{\mathrm{N}}$ denotes the noise subspace spanned by the eigenvector matrix corresponding to those small eigenvalues.

Since that ideal covariance matrices $\mathbf{R}$ is unknown, it is estimated by using a finite number of data vectors. $\hat{\mathbf{R}}$ is the estimation of $\mathbf{R}$ which can be calculated by

$$\hat{\mathbf{R}} = \frac{1}{N}\mathbf{X}\mathbf{X}^{\mathrm{H}} \tag{15}$$

where N is the number of the direct wave fronts of observed signal vector.

To describe the orthogonal properties described above, the spatial spectrum is used which can be calculated by.

$$\mathbf{P}_{\mathrm{MUSIC}}(r,\theta) = \frac{1}{\mathbf{A}^{\mathrm{H}}(r,\theta)\mathbf{U}_{\mathrm{N}}\mathbf{U}_{\mathrm{N}}^{\mathrm{H}}\mathbf{A}(r,\theta)} \tag{16}$$

By varying r and θ to realize a scanning process, $\mathbf{A}(r,\theta)$ is steered to scan the whole structure area. The peak point on the spatial spectrum corresponds to the impact source point. Both the distance and direction of the source can be obtained as

$$(\hat{r},\hat{\theta}) = \arg\max \mathbf{P}_{\mathrm{MUSIC}}(r,\theta) \tag{17}$$

3.2. Omni-directional impact localization

As mentioned in section 2, two PZT sensors are added above and below the reference sensor of the sensors array shown in Figure2. When the impact source occurs above the sensors array, the signal induced by impact must arrive earlier at the upper PZT sensors than the lower PZT sensors, that is $t_{\mathrm{U}} < t_{\mathrm{D}}$. Therefore, we can compare the arrival time of two PZT sensors before scanning the whole structure using Eq. (16), and Eq. (16) could be rewritten as

$$\begin{cases} \mathbf{P}_{\mathrm{MUSIC}}(r,\theta_1) = \dfrac{1}{\mathbf{A}^{\mathrm{H}}(r,\theta_1)\mathbf{U}_{\mathrm{N}}\mathbf{U}_{\mathrm{N}}^{\mathrm{H}}\mathbf{A}(r,\theta_1)}, 0° < \theta < 180° \quad if \quad t_{\mathrm{U}} > t_{\mathrm{D}} \\ \mathbf{P}_{\mathrm{MUSIC}}(r,\theta_2) = \dfrac{1}{\mathbf{A}^{\mathrm{H}}(r,\theta_2)\mathbf{U}_{\mathrm{N}}\mathbf{U}_{\mathrm{N}}^{\mathrm{H}}\mathbf{A}(r,\theta_2)}, 180° < \theta < 360° \quad if \quad t_{\mathrm{U}} < t_{D} \end{cases} \tag{18}$$

And the distance and direction of the source can be obtained as

$$\begin{cases} (\hat{r},\hat{\theta}_1) = \arg\max \mathbf{P}_{\mathrm{MUSIC}}(r,\theta_1), t_{\mathrm{U}} > t_{\mathrm{D}} \\ (\hat{r},\hat{\theta}_2) = \arg\max \mathbf{P}_{\mathrm{MUSIC}}(r,\theta_2), t_{\mathrm{U}} < t_{\mathrm{D}} \end{cases} \tag{19}$$

However, the time of impact signal arrivals at the PZT U will be equal to the arrival time at PZT D when the impact source occurs along the x axis, shown in Fig.5. In this situation, the direction of impact can be obtained as

$$\begin{cases} \hat{\theta} = 0°, if \quad t_M < t_{-M} \\ \hat{\theta} = 180°, if \quad t_M > t_{-M} \end{cases} \tag{20}$$

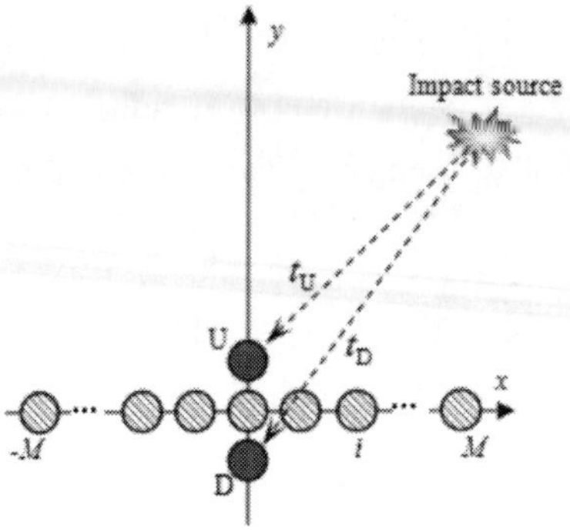

Figure 2. The time of impact signal arrival at the PZT U and PZT D ($t_U \neq t_D$)

A triangle in this figure is composed of the distance r between impact source and PZT 0, the distance r_D between impact source and PZT D, the distance d between PZT 0 and PZT D. According to this triangle, r_D can be calculated as

$$r_D = \sqrt{r^2 + d^2} \tag{21}$$

And

$$r_D - r = c\Delta t_{0D} \tag{22}$$

Thus, the distance of impact can be obtained as

$$\hat{r} = \frac{d^2 + (c_x \Delta t_{0D})^2}{c_x \Delta t_{0D}} \tag{23}$$

where, Δt_{0D} is the time difference between the impact signal arrivals at the PZT 0 and PZT D, and c_x is the velocity of impact signal propagating along the x axis. This velocity can be represented by the time difference Δt_{-MM} between the impact signal arrivals at the PZT -M and PZT M.

$$c_x = \frac{2Md}{\Delta t_{-MM}} \tag{24}$$

And Eq. (23) can be rewritten as

$$\hat{r} = \frac{(\Delta t_{-MM})^2 + 4M^2 (\Delta t_{0D})^2}{2M(\Delta t_{-MM})(\Delta t_{0D})} d \tag{25}$$

4. Experiment investigation

The experiment setup is shown in Figure 3, including the integrated structural health monitoring scanning system (ISHMS) developed by the authors is adopted as the monitoring system. This system is developed to control the excitation and sensing of the PZT sensor array.

The epoxy laminate plate with a dimension of 600 mm×600 mm×2 mm. The thickness of each ply is 0.125mm and the ply sequence is $[0_2/90_4/0_2]_S$. The array used in the experiment is a ULA bonded on the structure surface of opposing panel with 7 PZT sensors. The diameter of the PZT sensor is 8 mm. These sensors are arranged with a space of 13 mm and are labeled as PZT -3, PZT -2,..., PZT 3 respectively from the left to the right. Besides the ULA arrangement, two PZT sensors are added above and below the reference sensor, labeled as PZT U and PZT D. The sampling rate is set to be 10 MHz, and set the PZT 0 as the trigger channel, and the trigger voltage is set to 3V in the experiments. The sampling length is 30000 including 6000 pre-trigger samples.

Take the impact at (100 mm, 90°) as an example. Using the 2D-MUSIC algorithm, the impact occurring direction and distance can be simultaneously found from the spatial spectrum shown in Figure 4.

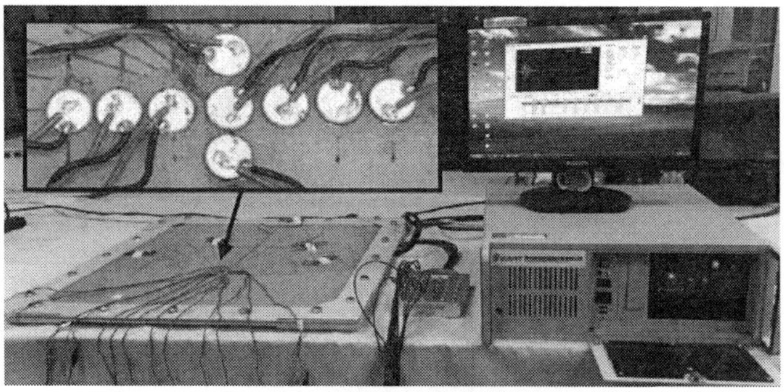

Figure 3. Experimental setup and sensor laylout

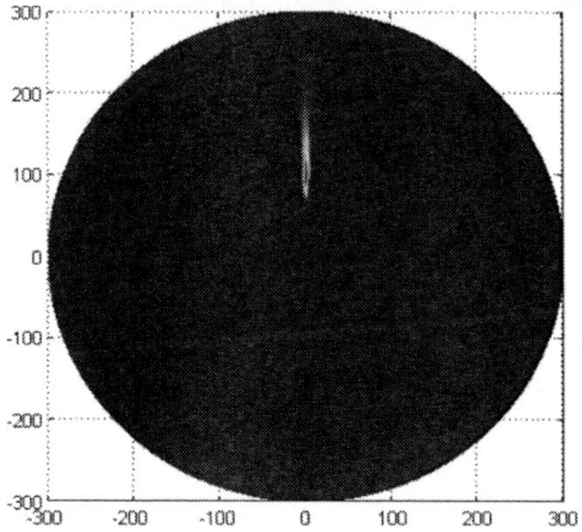

Figure 4. The estimated result by 2D-MUSIC at impact point (100 mm, 90°)

Acknowledgments

This work is supported by the National Science Foundation of China (No. 51505339, No. 51575400), the Zhejiang Provincial Natural Science Foundation of China (No. LQ16E050005, No. LR13E050002), and the Zhejiang Provincial education department scientific research project (Y201533105).

5. References

[1] Diamanti K, Soutis C S 2010 *Prog. Aerosp. Sci.* **46** 342
[2] Maillet E, Godin N, R'Mili M, Reynaud P, Fantozzi G and Lamon J 2014 *Compos Part A-Appl. S* **57** 8
[3] Yildirim U, Oguz O and Bogdanovic N 2013 *J. Vib. Control.* **19** 345
[4] Yang H, Lee Y and Lee S 2013 *J.Mech. Eng. Sci.* **227** 703
[5] Zhong Y, Yuan S and Qiu L 2015 *Struct. Infrastruct. E.* **11** 310
[6] Zhong Y, Yuan S and Qiu L 2014 *Actuat. A-Phy.* **214** 234
[7] Zhong Y, Xiang J, Gao H and Zhou Y 2016 *Struct. Control Health Monit.* **23**: 825
[8] Purekar A, Pines D, Sundararaman S and Adams D E 2004 *Smart. Mater. Struct.* **13** 838

Analysis of cracked RC beams under vibration

R Capozucca[1] and E Magagnini[2]

[1] Confirmed Associate Prof. of Structural Engineering, DICEA, University Polytechnic of Marche, Ancona, ITALY
[2] PhD Student, DICEA, University Polytechnic of Marche, Ancona, ITALY

E-mail: r.capozucca@univpm.it, e.magagnini@pm.univpm.it

Abstract. Among the methods of monitoring of integrity, vibration analysis is more convenient as non-destructive testing (NDT) method. Many aspects regarding the vibration monitoring of the structural integrity of damaged RC elements have not been completely analysed in literature. The correlation between the development of the crack pattern on concrete surface under bending loadings, as well as the width and depth of cracks, and the variation of dynamic parameters on a structural element is an important aspects that has to be more investigated. This paper deals with cracked RC beams controlled by NDT based on natural vibration, which may be correlated to damage degree due to cracking of concrete under severe state of loading. An experimental investigation on the assessment of RC beams in different scale under loading has been done through dynamic tests in different constraint conditions of edges measuring frequency values and frequency variation. Envelope of Frequency Response Functions (FRFs) are shown and the changes of natural frequency values are related to the damage degree of RC beams subjected to static tests. Finally, a comparison between data obtained by finite element analysis and experimental results is shown.

1. Introduction

Damage of reinforced structures may be the result of many problems, some of them grown during their life as insufficient reinforcement, large deflections, poor concrete quality, corrosion of steel reinforcement or insufficient capacity. All these issues cause the cracking of structural elements; cracking and depth of cracks in reinforced concrete (RC) beams depict damage degree. Among the methods of monitoring of integrity and availability of beams there are obviously static tests; on the other hand, vibration analysis is more convenient as non-destructive testing (NDT) method [1-7].

Tests of vibration of undamaged RC beams and damaged permit the evaluation of variation of frequency values correlated to the damage of RC beams due to increased bending moment [8]. Although damage is generally a local phenomenon, whereas vibration response is a global characteristic, the dynamic characteristics of damaged structures have been associated to the location and size of damage [2]. The basic concept behind vibration monitoring is that dynamic characteristics are functions of structures' physical properties, therefore, any change caused by damage results as a change in dynamic response [3]. There are several scientific reports in literature regarding the use of vibration data for locating damage in controlled conditions or in numerical simulations of elements having mainly homogeneous material. Though available results may be obtained with NDT based on the variation of dynamic parameters of RC beams [8-10], there are few results in scientific literature so that extensive investigation is yet necessary to develop convenient method of analysis. In this paper it is described an investigation of RC vibration with different constraints at the ends and with different scale. Envelope of Frequency Response Functions (FRFs) obtained through dynamic testing are shown and the changes of natural frequency values are related to the damage degree of RC beams represented by the cracking on the concrete due to static bending tests of loading. At the end, the results are discussed comparing experimental and numerical results.

2. Free vibration of beam

Experimental values of frequency are compared below with theoretical data obtained considering as first approximation the behavior of undamaged RC beam as uniform homogeneous Euler-Bernoulli (EB) uniform slender beam. The natural frequencies of EB beam are considered neglecting gravity forces, the effects of rotary inertia, shear deformation and damping. For a beam in flexure only the component of displacement v is significant for us and v is function of x and t. In the case of natural vibration of a uniform beam with homogeneous material, the expression of circular natural frequency ω_r^f for generic r mode of vibration in the case of both free ends, is obtained [11]:

$$\omega_r^f = \left(\xi_r \cdot \frac{r\pi}{L}\right)^2 \cdot \sqrt{\frac{EI}{\rho A}} \tag{1}$$

being the eigenvalue $\lambda_r = r \cdot \pi/L$ for a simply supported beam. The eigenvalue for a free end beam at the r mode $\lambda_r = r \cdot \pi/L$ may be correlated to the value λ_r for a simply supported beam: $\lambda_r^f = \xi_r \cdot \lambda_r$ with ξ_r = coefficient that depends on the different r mode of vibration, equal to 1.506, 1.25, 1.167 and 1.125, respectively, for the first four modes. Circular frequency values of uniform beam with free-free edge expressed by Eq. (1) may be as it follows:

$$\omega_r^f = (\xi_r \cdot r \cdot \pi)^2 \cdot \frac{\psi}{L \cdot \kappa} \tag{2}$$

where: $\psi = \sqrt{E/\rho}$ is the dimensional length on time coefficient of beam; $\kappa = L/i$ the slenderness and i the radius of inertia of section. The frequency ratio values for beam in different scale, B0 full scale, and BM0 (or BM1) in redacted scale, may be expressed by:

$$\frac{f_{r,B0}}{f_{r,BM0}} = \frac{\omega_{r,B0}^f}{\omega_{r,BM0}^f} = \frac{\kappa_{BM0} \cdot L_{BM0}}{\kappa_{B0} \cdot L_{B0}} \tag{3}$$

considering beams characterized by the same uniform material. The experimental concrete beams tested by static and dynamic analysis are represented by beams: B0 and BM0-BM1with different scale. The ratio of slenderness between the experimental beams BM0-BM1 and B0 adopted in the experimental vibration analysis is: $\mu = \frac{i_{B0} \cdot L_{BM0}}{i_{BM0} \cdot L_{B0}} = \frac{\kappa_{BM0}}{\kappa_{B0}} \cong 0.9$.

3. Experimental tests

Static and dynamic tests were planned for three RC beams strengthened with steel bars, beam B0 in full scale, and BM0 and BM1 beams, in reduced scale.

3.1. Static and dynamic tests on full scale RC beam B0

The dimensions of beam sections are 150mm·220mm and 1700mm in length. The steel reinforcement used was 4 bars measuring 10mm and stirrups at interval of 60mm having a diameter of 6mm. The beam was characterized by concrete having average cylinder strength equal to $f_{c,av.}$~ 44.3N/mm^2 and steel bars with an average yielding stress equal to $f_{y,av.}$~500 N/mm^2. Set-up of test for full scale beam B0 is shown in Figure 1; bending tests were carried out on beam, increasing the load P applied in two points measuring 300mm (Figure 1). The instruments were vertical jacks for load P/2 in two points; strain gauges on steel bars at midspan section; LVDTs used to record deformation on the compressed edge concrete; LVDTs to measure the deflection at midspan section and close to the support. Beam was subjected to cyclic loading paths and, successively, increasing load till failure. Every cyclic loading corresponds to a damage degree D_i with i=1,...,5. The main experimental results for beam B0 are summarized in Table 1. The failure of beam B0 was reached at P≈47.7kN with crash of compressive concrete. The dynamic test was carried out with a mobile accelerometer measures the acceleration due to impact of hammer in a fixed point at 45mm from edge of beam (Figure 2a).

Figure 1. Exp. static tests - set up of test for full scale RC beam B0.

Table 1. Exp. results from static cycle tests for beam B0.

Damage degrees	Load (kN)	Moment (kNm)	δ deflection (mm)	$\varepsilon_{c,sup}$ concrete strain (10^{-3})	ε_s steel strain (10^{-3})	χ curvature (10^{-5}·1/mm)
D_1	7.20	2.16	0.55	0.20	0.66	0.53
D_2	14.28	4.28	1.79	0.49	1.09	0.84
D_3 (yield of steel bars)	31.54	9.46	5.07	1.01	2.39	1.84
D_4	38.68	11.60	8.61	1.79	4.67	3.22
D_5 (failure)	47.76	14.32	28.02	-	-	-

The frequencies provided by 14 measuring points placed at regular distances had to be mediated so as not to coincide with the structure's nodal or null points (Figure 2a) with an average of 10 beats per location. Given the impossibility of achieving a true free-free condition, testing proceeds using a suspension system with springs with constant k~0.01 N/m. A FFT two-channel analyzer, and PULSE Labshop software were used for the dynamic tests and for experimental data acquisition. In order to evaluate the reliability of the measurements, a function referred to as coherence was considered. A comparison between the theoretical and experimental dynamic characterization of the beams without damage, D_0, is shown in Table 3 assuming parameters for RC beam B0 in Table 2.

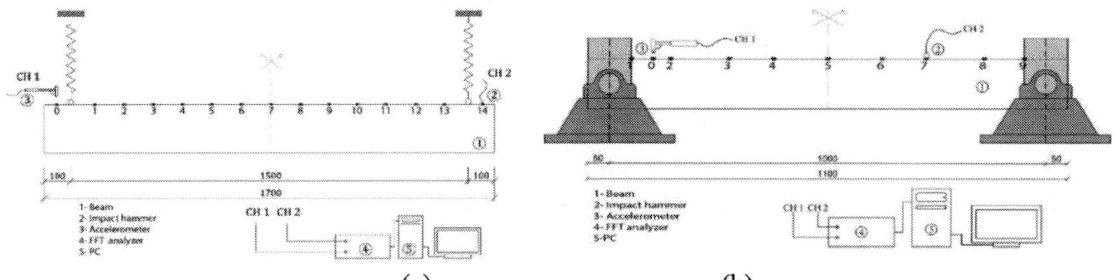

(a) (b)

Figure 2. Set-up for (a) dynamic test on free-free edge - B0 and (b) hinge-hinge edge – BM1.

In Table 3 the frequency values in damaged condition, D_i with i=1,...,4 are indicated with values determined for beam B0 in un-damaged, D_0 and the frequency variations in percent with reference to the different damage degree are shown. The FRF envelopes, for the different damage degree, for every position point of the accelerometer, are shown in Figure 3 for B0. In Figure 3 it is possible to observe the shift of the diagrams at various steps D_i with reduction of the frequency values due to increasing concrete cracking damage.

Table 2. Parameters of RC beam B0 to evaluate theoretical frequency values.

Concrete Beam	Length of beam L [m]	Section of beam A [m^2]	Young's Modulus E_c [$kN/m^2 \cdot 10^5$]	Moment of inertia I [$m^4 \cdot 10^{-3}$]	Density ρ [kg/m^3]
B0	1.7	0.03483	359.60	0.14357	2376.12

Table 3. Average frequency values for vibration of beam B0 with free-free edges.

Damage degrees	f_1 (Hz)	$\Delta f_1/f_{D0}$ (%)	f_2 (Hz)	$\Delta f_2/f_{D0}$ (%)	f_3 (Hz)	$\Delta f_3/f_{D0}$ (%)	f_4 (Hz)	$\Delta f_4/f_{D0}$ (%)
Theor. EB beam	307.73	-	848.24	-	1663.10	-	2749.01	-
D_0	275.07	-	706	-	1274	-	1878	-
D_1	249.93	9.14	685	3.04	1234	3.16	1822	2.99
D_2	227.36	17.35	635	10.06	1190	6.59	1753	6.65
D_3	215.79	21.55	580	17.80	1078	15.38	1661	11.54
D_4	180	34.56	520	26.28	1000	21.54	-	-

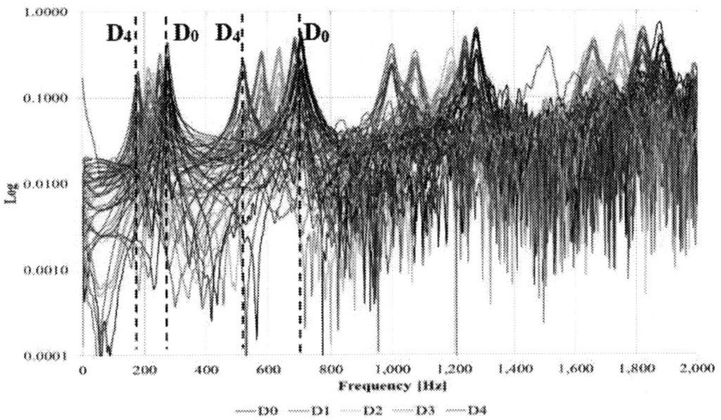

Figure 3. Envelope exp. diagrams of FRFs for B0 at damage degrees D_i i=1,...,4.

Comparison between experimental and theoretical diagrams moment versus curvature and variation in percent of frequency values is shown in Figure 4. The comparison between static and dynamic values is a convenient method to correlate dynamic response to static damage degree. In particular, we can see the influence of the RC beam' nonlinear behavior due to concrete cracking under loading on the beams in the different vibration modes considered. As shown in Figure 4, the variation percent of the frequency values as compared to the undamaged state, D_0, is always increasing from the elastic-linear uncracked phase (I-II), to the elastic-linear cracked phase (I-II), to the inelastic and plastic phase (I-III). In any case, the natural frequency values tend to describe the beam's global response relative to loss of bending stiffness and they are less sensitive to local stiffness variations. In general, the percent of frequency reduction increases more markedly in the elastic–linear cracked phase (I-II) as compared to phase (0-I) non-cracked; the frequencies' percent of variations increases considerably in phase (II-III), beyond flexing and, hence, when damage is permanent.

3.2. Static and dynamic tests of beams in scale BM0 – BM1

Main parameters of RC beam in scale, BM0 and BM1, are shown in Table 4. BM0-BM1 were reinforced with 4 steel bars and stirrups with diameter of 6mm, with average yielding stress $f_{y,av}$~500N/mm^2 and E_s ~2.1·105N/mm^2.

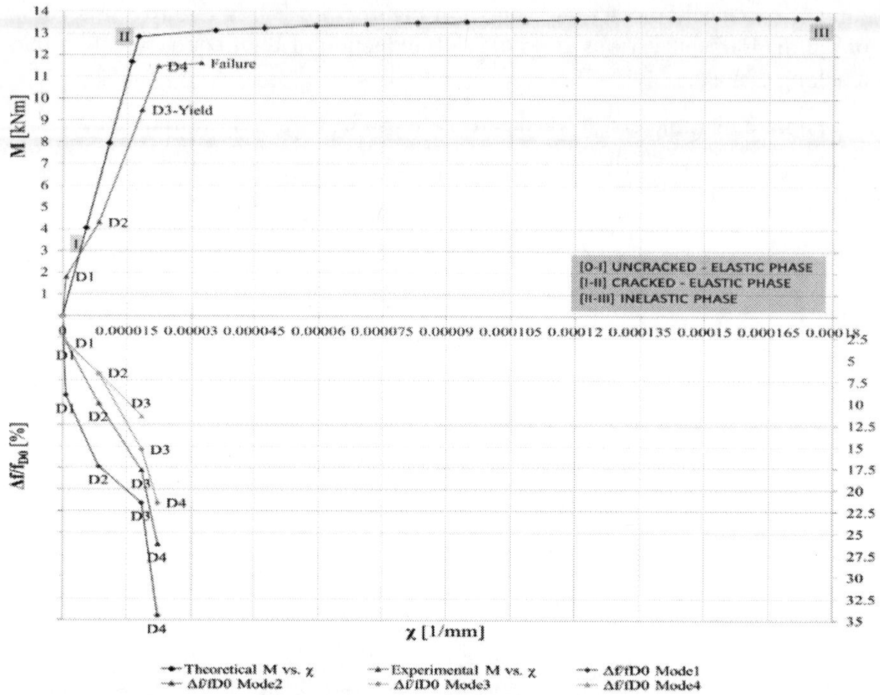

Figure 4. Comparison of exp. and theor. diagrams of moment-curvature vs variation frequency - B0.

Table 4. Experimental geometric and mechanical parameters of RC beams in scale BM0-BM1.

Concrete Beams	Width b [mm]	Height h [mm]	Length L [m]	Young's modulus E_c [N/mm^2]	Density ρ [Ns^2/mm^4]	Moment of inertia I_g [mm^4]
BM0 – BM1	80	120	1.1	35.310	$2.547\cdot10^{-9}$	$11.5\cdot10^{6}$

Beam BM1 (similar to BM0) was also subjected to the same cycle bending path with damage degrees $D_1,\ldots,D_5$. In Table 5, main experimental results recorded for beam BM1 subjected to bending tests are shown. In Figures 5 view of cracking damage on beam BM1 are shown for damage degree D_5 being P~16kN. Beam BM0 and BM1 were tested dynamically after every cycle of loading: the beam BM0 was analysed by a suspension system while the BM1 with hinge-hinge end constraints (Figure 2(b)).

Table 5. Exp. results from static cycle tests for beam in scale BM1.

Damage degrees	Load (kN)	Moment (kNm)	δ deflection (mm)	$\varepsilon_{c,sup}$ concrete strain (10^{-3})	χ curvature ($10^{-5}\cdot1$/mm)
D_1	4.00	0.70	0.67	0.15	0.53
D_2	8.00	1.40	2.08	0.56	2.30
D_3	12.42	2.17	3.61	0.85	1.90
D_4	14.00	2.45	4.41	0.91	3.20
D_5 (yield of steel bars)	16.13	2.82	6.54	0.95	3.74

In Figure 6, variation in percent of experimental frequency values $\Delta f/f_{D0}$ (%) in comparing with condition D_0 for beam BM0 is shown at different damage degree D_i for first three modes. As in the case of full scale beam B0, it is possible to see how the cracking of the concrete influenced the dynamic behavior of in scale beam. The variation in percent of frequency values presents main variation in two phases: the first, between the undamaged state and cracking phase; the second, at the beginning of plastic phase for steel bars. BM1 was tested under vibration with hinge-hinge end condition made by a specific steel support system (Figure 2(b)).

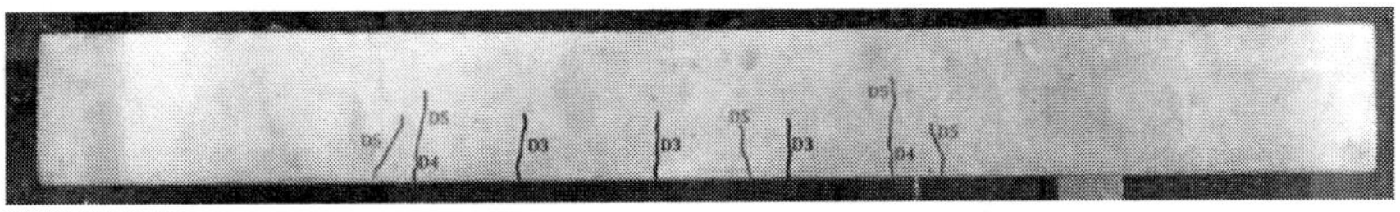

Figure 5. Exp. cracking of BM1 under static test at damage degree D_5.

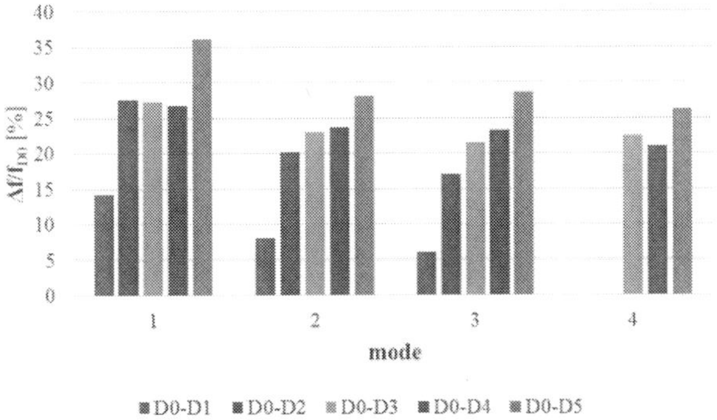

Figure 6. Variation of exp. frequency values BM0 at different damage degree at first three r modes.

Experimental and theoretical frequency values respect to D_0 for beam BM1 are shown in Tables 6 together with recorded frequency values at different damage degree D_i with i = 1,...,5 for first four r-modes 1,...,4 considering position of accelerometer no. 1.
The frequency variations in percent with reference to the undamaged condition was evaluated for each damage degree. The envelope diagrams of FRFs for BM1 are shown in Figure 7; it is possible to observe the shift of the diagrams of FRFs at various steps D_i with reduction of the frequency values due to increasing concrete cracking damage.

Table 6. Frequency values for beam in scale BM1 with hinge-hinge ends – Acc. No. 1.

Damage degrees	f_1 (Hz)	$\Delta f_1/f_{D0}$ (%)	f_2 (Hz)	$\Delta f_2/f_{D0}$ (%)	f_3 (Hz)	$\Delta f_3/f_{D0}$ (%)	f_4 (Hz)	$\Delta f_4/f_{D0}$ (%)
Theor. EB beam	175.75	-	703.00	-	1581.75	-	2812.00	-
D_0	180	-	672	-	1234	-	1794	-
D_1	164	8.89	686	-2.08	1244	-0.81	1774	1.11
D_2	178	1.11	658	2.08	1218	1.29	1746	2.67
D_3	176	2.22	650	3.27	1194	3.24	1734	3.34
D_4	176	2.22	644	4.17	1180	4.37	1722	4.01
D_5	166	7.78	492	26.78	1110	10.04	1634	8.91

By experimental tests we can conclude that the influence of cracking on the natural frequency values for increment of bending may be summarized as following: the variation in percent of frequency values is always growing respect the values at undamaged D_0 degree although it presents main variation in two phases: the first, between the undamaged state and cracking phase; the second, at the beginning of plastic phase for steel bars.

Referring to the frequency variations values in comparing to the undamaged condition, it can be noted that, in the case of measurements on beams in free-free end condition, the variation is mainly higher. For hinge end constraint conditions at beams' extremity monitoring the damage state through the vibration method is significant if the damage degree registered is high.

4. Finite element modelling

The FE modeling of RC beam B0 and in scale BM0 and BM1 was carried out with ANSYS code. Three types of models were created to simulate beams with all materials including concrete and longitudinal steel reinforcements and to reproduce the experimental constraint conditions.

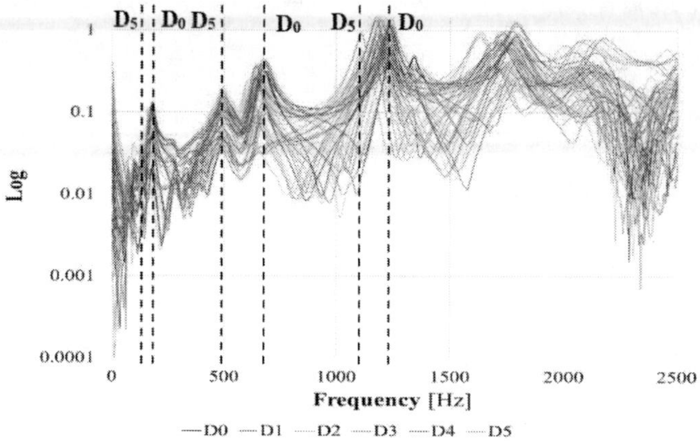

Figure 7. Envelope exp. diagrams of FRFs for beam in scale BM1 with hinge-hinge ends at different damage degrees D_i i=1,…,4.

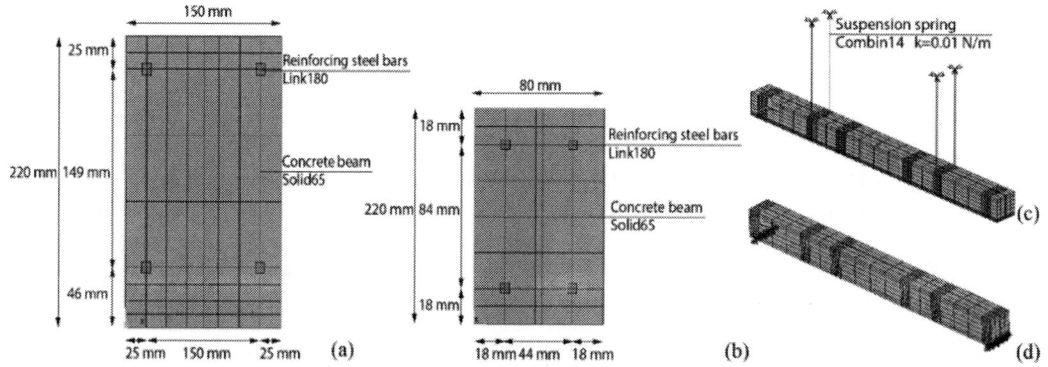

Figure 8. Typical mesh for FE analysis: (a) item of B0 beam section; (b) item of in scale BM0 and BM1 beam section; (c) beams hang by springs; (d) BM1 beam with hinge-hinge end.

The eight-node solid brick elements - Solid65 - were used to model the concrete. A Link180 element was used to model steel reinforcement. Element Combin14 was inserted in the numerical modelling in addition to the aforementioned elements in order to model the beams' suspension springs simulating the same free beam conditions in vibration. The element is defined by two nodes and a spring constant k equal to 0.01 N/m. The typical FE modelling of the reinforced beams is shown in Figure 9. A variable Young's modulus E of concrete was used to take into account damage degree due to cracking; the moment of inertia I and density ρ have been assumed constant. The natural frequency values determined for each damage degree for the free-free in scale beam B0 and for the hinge-hinge in scale beam BM1 are contained in Table 7 with the variation of frequency values respect to D_0.

Table 7. FE frequency values at damage degree D_i for BM0 and BM1.

BM0 (free-free end condition)								
Damage degree	f_1 (Hz)	$\Delta f_1/f_{D0}$ (%)	f_2 (Hz)	$\Delta f_2/f_{D0}$ (%)	f_3 (Hz)	$\Delta f_3/f_{D0}$ (%)	f_4 (Hz)	$\Delta f_4/f_{D0}$ (%)
D_0	355.72	-	924.84	-	1696.2	-	2605.3	-
D_1	310.59	12.69	806.17	12.83	1475.76	13.00	2262.2	13.17
D_2	268.63	24.48	695.47	24.80	1434.2	15.45	1939.9	25.54
D_3	269.22	24.32	697.03	24.63	1437.7	15.24	1944.5	25.36
D_4	270.85	23.86	701.36	24.16	1447.4	14.67	1957.1	24.88
D_5	242.4	31.86	625.95	32.32	1139.1	32.84	1735.6	33.38
BM1 (hinge-hinge end condition)								

D_0	183.73	-	675	-	1381.28	-	2220.12	-
D_1	183.94	-0.11	686.75	-0.12	1382.9	-0.12	2222.8	-0.12
D_2	182.23	0.82	661.25	0.82	1369.7	0.84	2201.2	0.85
D_3	181.41	1.26	648.75	1.27	1363.4	1.29	2190.9	1.32
D_4	179.98	2.04	642.5	2.06	1352.4	2.09	2172.9	2.13
D_5	170.73	7.08	561	7.14	1281	7.26	2056.2	7.38

Considering the variations of frequency values in particular for the first mode, we may note that the theoretical results obtained with FE modelling for free-free and hinge-hinge conditions are confrontable with experimental results by dynamic tests. The comparison between experimental and theoretical data confirms the suitability of experimental apparatus for dynamic tests both in the case of hinge-hinge end and free-free ends of beams.

5. Conclusions

The monitoring of RC beams during its life may be done through the dynamic response as NDT method based on the measure of frequency values of vibration. This paper describes experimental and theoretical results obtained analyzing the response of RC beams under vibration both in different undamaged and damaged degrees and in different constraints. Dynamic analysis of full and in scale RC beams allows the evaluation of the drop in frequency measure values due to damage. Variation of frequency values are influenced both damage degree due to cracking and different constraint condition.

Acknowledgement

This research was supported by research funds provided by Università Politecnica delle Marche.

References

[1] Cawley P and Adams R D 1979 The location of defects in structures from measurements of natural frequencies *J. Strain Anal. Eng. Des.* **14** 49-57

[2] Salawu O S 1997 Detection of structural damage through changes in frequency: a review *Eng. Struct.* **19** 718-23

[3] Zhou Y-L, Maia N and Abdel Wahab M 2016 Damage detection using transmissibility compressed by principal component analysis enhanced with distance measure *J. Vib. Control* doi: 10.1177/1077546316674544

[4] Zhou Y-L and Abdel Wahab M 2016 Rapid early damage detection using transmissibility with distance measure analysis under unknown excitation in long-term health monitoring *J. Vibroeng.* **18**(7) 4491-99

[5] Khatir S, Belaidi I, Serra R, Abdel Wahab M and Khatir T 2016 Numerical study for single and multiple damage detection and localization in beam-like structures using BAT algorithm *J. Vibroeng.* **18**(1) 202-13

[6] Khatir A, Tehami M, Khatir S and Abdel Wahab M 2016 Multiple damage detection and localization in beam- like and complex structures using co-ordinate modal assurance criterion combined with firefly and genetic algorithms *J. Vibroeng.* **18**(8) 5063 - 73

[7] Khatir S, Belaidi I, Serra R, Abdel Wahab M and Khatir T 2015 Damage detection and localization in composite beam structures based on vibration analysis *Mechanika* **21**(6) 472-79

[8] Capozucca R 2013 A reflection on the application of vibration tests for the assessment of cracking in PRC/RC beams Eng. Struct. **48** 508-18

[9] Casas J R. and Aparicio A C 1994 Structural damage identification from dynamic–test data *J. Struct. Eng.* **120** 2437-50

[10] Capozucca R 2013 Analysis of bond-slip effects in RC beams strengthened with NSM CFRP rods *Compos. Struct.* **102** 110-23

[11] Warburton G B 1964 *The Dynamical Behaviour of Structures* (Oxford: Pergamon Press)

Micro CT Analysis of Dynamic Damage in Laminates: Impact *vs*. blast loading

Laurence A Coles, Anish Roy, Vadim V Silberschmidt

Wolfson School of Mechanical, Electrical and Manufacturing Engineering, Loughborough University, Ashby Road, Leicestershire, LE11 3TU, UK

Abstract. Dynamic loading is often an unavoidable condition in various applications of carbon-fibre-reinforced polymers and can cause various modes of damage. Realisation of dynamic damage in composites can differ significantly from that under quasi-static loading conditions. A comprehensive study of damage in composites caused by a wide variety of impact and blast loading is currently lacking. The work presents a detailed analysis of damage in specimens of a 2×2 twill weave T300 carbon-fibre/epoxy composite subjected to ballistic loading with both steel and ice projectiles (with energies from 95 J to 865 J at 70-90 m/s and 300-500 m/s, respectively) and air blast (with incident pressures of 0.4 MPa, 0.6 MPa and 0.8 MPa and wave speeds between 650 m/s and 950 m/s). The resultant damage was analysed in-depth based on detailed volumetric data obtained with high-resolution X-ray micro computed tomography.

1. Introduction

Over the last few decades the use of fibre-reinforced composites (FRCs) has risen considerably across many areas of application including automotive, aerospace, naval, defence, energy and sport. Service conditions of all these areas can include dynamic loading events, resulting in minor or major damage to components or structures. Some examples of typical dynamic loading may include: low- and high-velocity impacts (1 to 1000 m/s) ranging from flight at Mach 1-2 (300-600 m/s) of fast jets or intercontinental ballistic missiles through clouds of airborne debris or hailstone, to lower-velocity events caused by debris on runways, railway lines, automotive race tracks etc., or even air-blast loading such as interaction with close-proximity explosions or sudden increases in local pressure. As a result, it becomes increasingly important to understand fully the effect of various dynamic loading conditions on a FRC's response, in terms of both local and global deformation as well as visible and hidden damage.

The last decade has seen a vast amount of research effort towards characterising the response of FRCs to various dynamic loading conditions [1], but drawing comparisons between these studies is challenging and often impossible due to their major differences such as type of specimens and, more importantly, experimental and analytical methodologies. In the past, some papers tried to surmise an overview of, for example, impact behaviour [2, 3] and [4, 5], but given a breadth of experimental data drew mostly qualitative conclusions regarding the effects of differences in methodology, specimen as well as projectile mass and velocity on such features as delamination behaviour.

An important aspect of this dynamic study is understanding the types of loading conditions and their effect on the specimen over time. For instance, a large part of both ballistic experimental and numerical studies was focused on the resulting deformation and damage caused by solid (typically steel) projectiles. In recent years, efforts were made to understand the impact process of ice projectiles [6-8], the blast response of composites [9-11] and the effect on composite specimens as well as distribution of load and dispersion of kinetic energy. But, typically, the analysis of the induced damage is limited to a visual inspection of external surfaces, or a use of invasive techniques to study internal damage; the latter could introduce unwanted damage complicating the analysis [12-16]. Still, some studies employed X-ray tomography to investigate damage [17].

This paper presents the experimental case studies of both ballistic impact with steel and ice projectiles and air blast loading of carbon fibre/epoxy specimens, using digital image correlation and X-ray computed tomography (CT) to investigate and compare the levels of deformation and damage.

2. Experimental Setups

2.1. Materials and Specimens

The tested carbon-fibre-reinforced/epoxy specimens, measuring approximately 195 mm x 195 mm with a thickness of 5.6 mm (SD = 0.031mm) were fabricated from 10 plies of carbon-fibre fabric, pre-impregnated with a toughened epoxy matrix (IMP530R). The 10 plies were formed to a laminate consisting of two surface plies (T300 3K fibres) and eight central bulking plies (T300 12K fibres), with a 2 × 2 twill weave 0°/90° layup configuration and with a theoretical density of 1600 kg/m^3. All specimens were manufacture using the autoclave process as a flat sheet, cured at 120°C with a 1.5°C per minute ramp rate and a soak time of 160 minutes at a pressure of 90 psi under full vacuum.

2.2. Pneumatic Gun Setup

The investigation was performed on a specially developed ballistic experimental apparatus utilising a pneumatic gun. The CFRP specimens were installed into a 4-axis positioning system, and aligned with the barrel resulting in a cantilever clamping scheme, with all specimens being subjected to a orthogonal impact (as shown in Figure 1). The projectile was accelerated to the required speed in the barrel using compressed air, and muzzle velocity measurements were collected using a light gate device installed between the specimen and the end of barrel. Projectile velocities were then determined by the time difference between the signal peaks from the light gates; the experiment was controlled by a PC via a PXI system (National Instruments). The impact process was captured using two high-speed cameras (Photron Fastcam SA5) configured in two arrangements, the first of which captured the front and top views at 25,000 - 50,000 frames per second while the second recorded the rear surface for digital image correlation (DIC) using the VIC-3D (Correlated Solutions) system at 60,000 frames per second.

2.3. Shock Tube Setup

In this setup, the specimens were positioned vertically on a three-point-bending style fixture, which consisted of 2 knife edges (slightly rounded) located 152.4 mm apart, the rest of the specimen was unsupported (as shown in Figure 2). A rubber band was used to keep the specimen firmly against the knife edges, while positioned vertically on the fixture. The shock tube apparatus (8 m in length) consisting of a driver, diaphragm and driven section, produced the air blast shockwave by pressurising the driver section until a critical pressure, at which the diaphragm ruptured creating a dynamic pressure-wave profile. The muzzle of the shock tube, with its inner diameter of 76.2 mm, was moved towards the specimen until there was only a paper thin gap between the specimen and the muzzle (approximately 0.1-0.2 mm). Pressure sensors located towards the end of the muzzle recorded the shockwave profile.

The deformation process was captured using three cameras (Photron SA1 Photron USA, Inc., CA,

USA), two cameras recording at 28,800 fps viewed the rear surface of the specimen for DIC using the VIC-3D (Correlated Solutions) system. A third camera, also recording at 28,800 fps, was placed perpendicular to the edge of the specimen to acquire side-view images and observe the mechanisms of failure for each specimen.

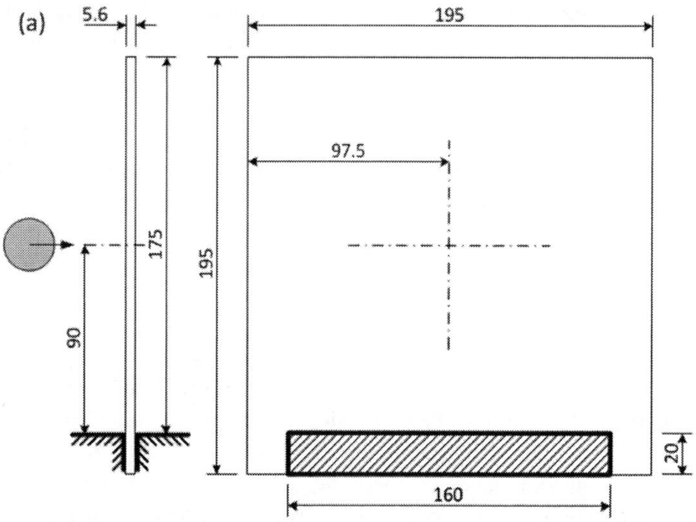

Figure 1. Ballistic impact regime and cantilever clamping fixture

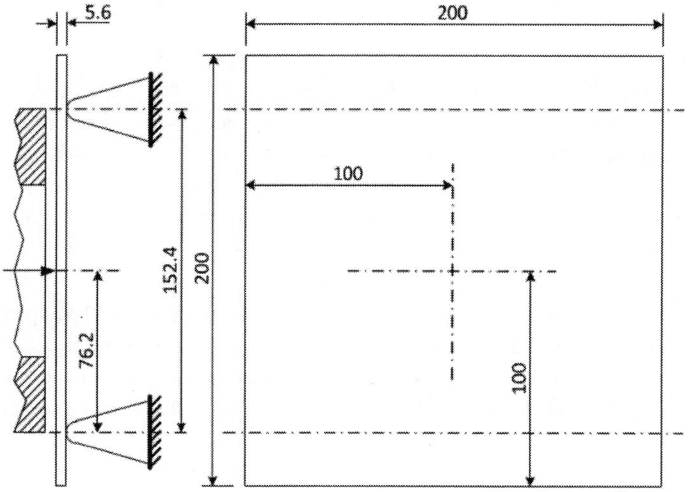

Figure 2. Air blast regime and three-point-bending fixture

2.4. Experimental Parameters
All the solid (steel) projectiles were 23.8 mm diameter with a weight of 54.7 g, and the fragmenting ice (hailstone imitation) projectiles had a diameter 25 mm with a weight of 7.5 g at a temperature of -20°C. The impact velocities for the ballistic and air blast cases were chosen to produce three levels of damage in the specimens – minor, medium and major with penetration, and were determined during calibration experiments to correspond to approximately 60, 75 and 90 m/s for the steel projectiles, 300, 400 and 500 m/s for the ice impacts and incident pressures of 0.4 MPa, 0.6 MPa and 0.8 MPa for the air blasts with respective reflected pressures of 1.35, 2.50 and 3.4 MPa, and had wave speeds of between 650 m/s and 950 m/s. A total of 28 samples were subjected to various loading conditions, 9 to impacts with the solid steel projectiles, 9 to impacts with the ice projectiles and 10 to the air blasts.

2.5. Configuration of X-Ray Tomography Scans

All the specimens were inspected using a Metris 160 H-XT X-ray micro CT system to investigate the extent of the internal damage and its spatial distribution. Each scan was conducted at 140 kV and 130 μA using a tungsten target, with 2650 radiography projections taken over the 360° rotation for each specimen at an exposure of 500 ms. In order to reduce granular noise, 8 images where taken and averaged per projection. The total volume scanned for each specimen was 180 mm x 140 mm x 20 mm at a resolution of 97 μm.

3. Experimental results and discussion

Typical impacts with both the steel and ice projectiles in the experimental studies can be seen in Figure 3, with a characteristic response of the major damage air blast condition shown in figure 4. The ice projectile initially indented the specimen before fragmenting within it, causing major wide-spread delamination of the first few plies. The impact with the steel projectile resulted in major indentation before full penetration, as expected, inducing more localised damage when compared to that of the ice projectiles. And, finally, the specimen subjected to the air blast undergoes global flexural bending between the supports with damage initiating at the centre of the rear surface.

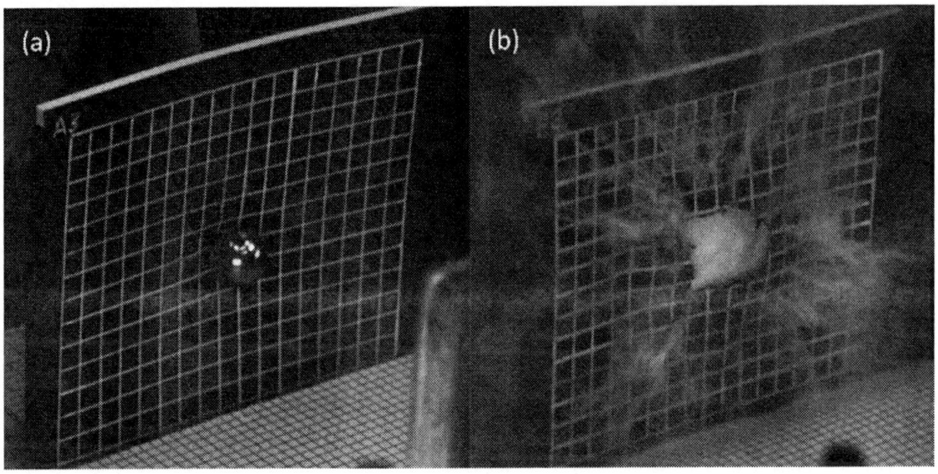

Figure 3. Typical impact behaviour of solid steel projectile at 0.40 ms (a) and fragmenting ice projectile at 0.19 ms (b)

Figure 4. Typical damage in dynamic response to air blast at 0.13 ms

3.1. Deformation Analysis

3.1.1. Ballistic Impact

Analysing impacts with the steel projectile in terms of global displacement, it is clear that due to the more rigid nature of the material, the composite demonstrated more defined indentation before the transition into global flexural bending. For the ice projectiles, this initial indentation was still present, but upon fragmentation, this local indentation was more gradually transitioned due to the distributed load, leading to global bending. This difference in the distribution of dynamic loads led to varying curvature of the specimen during loading. For the solid projectiles, the curvature was small, with transitions between concave and convex forms happening gradually, whereas for the ice projectile this transition was more abrupt and resulted in multiple damage modes. As the ice fragmented on impact, the local indentation of the specimen became more widespread causing a more distributed loading and, hence, reduced failure at the rear surface. In the case of the solid projectile, this damage occurred much sooner.

From localised deformation behaviour and the normalised response of each specimen, it became evident that the maximum out-of-plane displacement increased with the growing impact velocity (energy). The time until maximum displacement reduced with the increased velocities (energies), although this trend was not as clear due to the limitation related to time resolution (directly linked to the frame rate of the used camera).

3.1.2. Air Blast

Analysis of the centre-point displacement of each specimen demonstrated clearly that they all show an initial oscillation, achieving maximum displacement before the oscillations decayed gradually due to the contact surface behind the shockwave and resultant back pressure. Double and triple peaks were observed on the centre-point displacement oscillations, the former representing the transition of lateral oscillations to the free edges and back again. The triple peaks show similar behaviour, but the centre peak was related to delamination of a given number of plies that were then free to move with higher deflection due to reduced local stiffness. As a result, delamination was initiated at different times for each pressure magnitude, after which both the amplitude and frequency of the oscillations changed due to the expected reduction in specimen stiffness.

Observations of vertical and horizontal out-of-plane displacement line sections demonstrated that the deformation and transitions in curvature of each specimen resulting from the different air-blast magnitudes were similar, with no obvious differences, except from increasing out-of-plane displacement as excepted. Given the widely distributed loading in this case, there is minimal localised indentation, and all specimens undergo global flexural bending. As the air-blast magnitude increased, the specimens see greater deformation of the rear surface at the centre of the free edges. This was caused by the increasing damage and delamination of the rear surface because of increased bending stress. Although this is more apparent for the major damage case, the progression of increased deformation at the free edge increased at each damage level before resulting in a clear failure.

3.2. Damage Analysis

Micro CT inspections of the specimens after ballistic impact tests with the solid (steel) and fragmenting (ice) projectiles demonstrated two very distinct types of visual damage: (i) highly localised and, therefore, more penetrating damage from the steel projectiles and (ii) more widespread damage to the front surface of the specimen with some signs leading to the early stages of penetration in the case of the ice projectiles. Figure 5 shows the damage clouds for two major damage ballistic loading conditions. For the air blast loading, the minor damage case showed no signs of external damage, whereas the medium and major cases resulted in clearly increasing levels of damage at the rear surface as the air blast magnitude increased. The damage across all cases can be compared to that of standard static three-point bending. For this type of composite specimen, the first signs of damage appear along a central line, which is the line of symmetry between the supporting locations, as tensile failure of the individual plies leading to substantial delamination. At the front surface for each specimen there were no signs of any damage until complete failure, suggesting that the tensile failure and resultant damage propagates from the rear surface to the front surface. Figure 6 shows the damage clouds for the major-damage air-blast loading conditions. From the CT scans it is clear that for all cases, the main damage modes were delamination and tensile fibre damage at the rear surface, but at the employed resolution it was not possible to accurately determine any other damage mechanisms.

This was confirmed for all the three damage cases across all loading conditions by means of 3D analysis: for the steel projectile, the damage cloud remained highly localised even under increasing energy whereas for the ice projectile it increased more clearly. For the air blast cases with increasing intensity, the levels of tensile failure and delamination at the centre and free edges of the specimen grew, transforming the circular loading area into a central symmetrical band of damage.

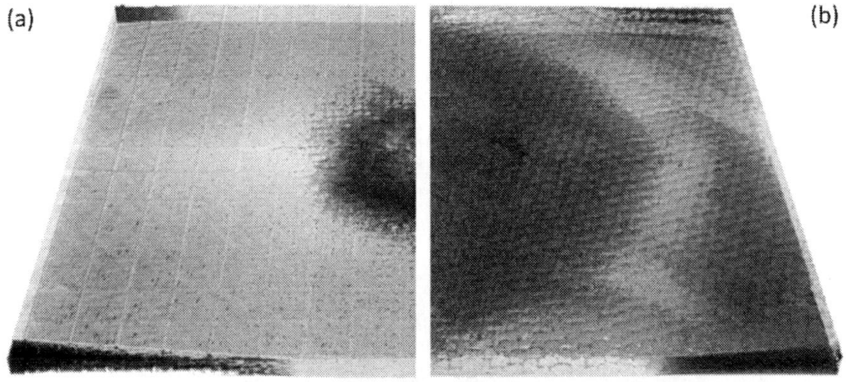

Figure 5. Typical examples of internal damage cloud in composite specimens caused by solid steel (a) and fragmenting ice (b) projectiles (half specimens are shown)

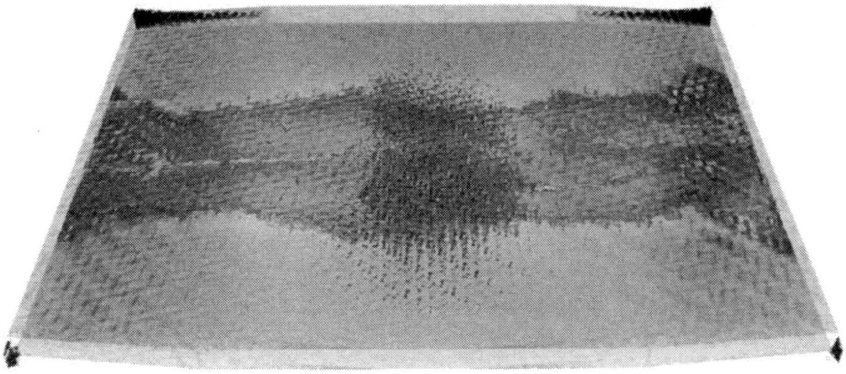

Figure 6. Typical example of internal damage cloud for major damage case

The transparent 3D tomograms shown in Figures 5 and 6 allowed assessment of the damage areas and greyscale intensity, with the latter related to the extent of damage accumulation through the specimen's thickness. For instance, the specimen impacted with the solid projectile demonstrated a dark-grey zone toward the centre, indicating greater damage (like penetration), whereas the fragmenting projectiles caused catastrophic damage across the whole front surface and first few plies of the composite target, causing a more distributed gradient of the intensity. The same can be said for the air-blast loading, where the major-damage case showed the higher intensity at the central area and free edges, i.e. a larger amount of through-thickness damage in these locations.

Observing the 2D horizontal and vertical cross sections of each specimen, it was possible to approximately measure this spread of damage with resolution of 97 μm. Figure 7 shows the final averaged damage areas against the averaged maximum out-of-plane displacement for various loading conditions.

4. Conclusions

For impacts with projectiles of the same size, a case with the solid (steel) projectiles demonstrated more defined indentation of the CFRP specimens before the transition to global flexural bending. In contrast, the fragmenting (ice) projectiles produced initial indentation that, upon fragmentation, more gradually transitioned to distributed loading leading to global flexural bending. Comparison of dynamic loading conditions showed that projectiles of the same material produced similar responses of the impacted specimen, even at different energy levels. Oppositely, two projectiles of different materials will show different levels of indentation within the specimen, even at lower energy levels. The character of transition from indentation to global flexural bending also changed with the projectile material, affecting the evolution of specimen's curvature during its deformation.

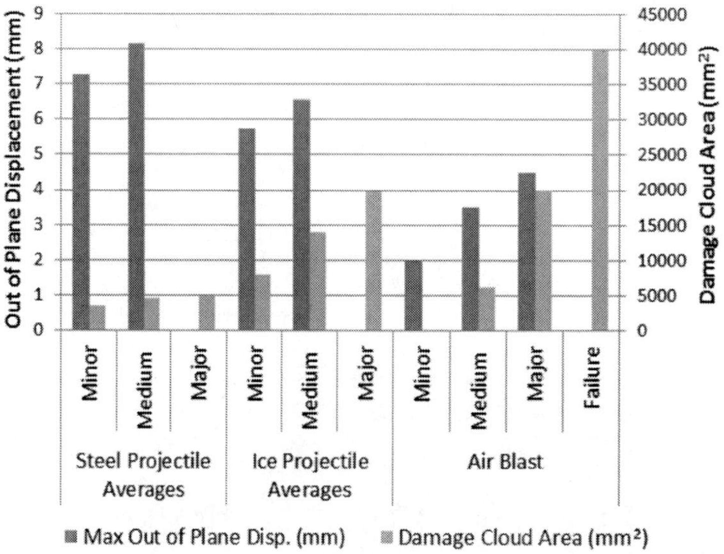

Figure 7. Maximum out-of-plane displacements and damage areas for both types of projectiles and air blast

For the case of air blast loading, the observed centre-point displacement of each specimen demonstrated changes in the oscillation period after the appearance of delamination damage (triple peaks) caused by a reduction of specimen's stiffness, not observed in the impact cases. The global deformation and transitions in curvature of each specimen remained similar as the air blast magnitude increased, whereas the out-of-plane displacement increased showing that the magnitude had no effect on the curvature and modes of oscillation during deformation for the range of air-blast magnitudes investigated. The behaviour follows a typical deformation pattern for this type of specimen fixture and size of the loading area, showing that the distributed loading led to almost instant global flexural bending, without any obvious significant localised indentation and associated damage. For the major and failure damage cases, greater deformation of the rear surface was observed at the free edges, resulting from the initiation of the tensile failure and delamination of the first few plies along the central band of damage observed.

Damage at the rear surface of the impact specimens occurred much sooner in the case of the solid projectiles; this is likely due to the almost instantaneous fragmentation of the ice projectile on impact causing a broader distribution of impact load and, therefore, a longer impact duration even at clearly higher impact energies. Although the fragmenting projectile mitigated the penetration of the specimen, the partial indentation followed by the projectile's destruction, causing delamination and removal of the first 2-3 plies from the front surface of the impacted composite plate. The resultant air-blast damage initiated and propagated from the rear surface of the specimens towards its front surface as the air blast magnitude increased, similar to that of normal three-point bending. For the minor-damage case, no significant damage was observed, but at the major- and failure-damage cases the specimens experienced severe damage in the form of tensile failure and wide-spread delamination propagating from the central line. Given the similar global flexural bending behaviour of all the specimens across all damage cases, the initiation points of damage were consistent to their centres and free edges.

Acknowledgements
The authors are grateful to groups of Prof. Arun Shukla (University of Rhode Island, Kingston, RI, USA) and Prof. Mikhail Nikhamkin (Perm National Research Polytechnical University, Russian Federation) for implementation of experimental studies.

References
[1] Silberschmidt VV (ed.), *Dynamic Deformation, Damage and Fracture in Composite Materials and Structures*. Elsevier, Amsterdam e.a., 2016, 616 pp.
[2] Abrate S. Impact on laminated composites. *ASME: Applied Mechanics Reviews*, 44:155-190, 1991.
[3] Abrate S. Impact on laminated composites: Recent advances. Composites. *ASME: Applied Mechanics Reviews*, 47:517-544, 1994.
[4] Langdon GS, Lee WC, Louca LA. The infuence of material type on the response of plates to air-blast loading. *International Journal of Impact Engineering*, 78: 150-160, 2014.
[5] Langdon GS, Cantwel WJ, Guan ZW, Nurick GN The response of polymeric composite structures to air-blast loading: A state-of-the-art. *International Materials Reviews*, 59: 159-177, 2014.
[6] Pernas-Sánchez,. Artero-Guerrero J, Varas JA, López-Puente J. Analysis of ice impact process at high velocity. *Experimental Mechanics*, 55:1669-1679, 2015.
[7] Kim H, Welch DA, Kedward KT Experimental investigation of high velocity ice impacts on woven carbon/epoxy composite panels. *Composites Part A: Applied Science and Manufacturing*, 34:25-41, 2003.
[8] Appleby-Thomas GJ, Hazell P J, Dahini G. On the response of two commercially-important CFRP structures to multiple ice impacts. *Composite Structures*, 93:2619-2627, 2011.
[9] LeBlanc J, Shukla A, Rousseau C, Bogdanovich A. Shock loading of three-dimensional woven composite materials. *Composite Structures*, 79: 344-355, 2007.
[10] Tekalur S, Shivakumar K, Shukla A. Mechanical behavior and damage evolution in E-glass vinyl ester and carbon composites subjected to static and blast loads. *Composites Part B: Engineering*, 39: 57-65, 2008.
[11] Kumar P, Stargel DS, Shukla A. Effect of plate curvature on blast response of carbon composite panels. *Composite Structures*, 99: 19-30, 2013.
[12] Nunes LM, Paciornik S, d'Almeida JRM Evaluation of the damaged area of glass-fiber-reinforced epoxy-matrix composite materials submitted to ballistic impacts. *Composites Science and Technology*, 64: 945-954, 2004.
[13] Sevkat E. Experimental and numerical approaches for estimating ballistic limit velocities of woven composite beams. *International Journal of Impact Engineering*. 45: 16-27, 2012.
[14] Shaktivesh NS, Nair ChV, Kumar S, Naik NK. Ballistic impact performance of composite targets. *Materials & Design*. 51: 833-846, 2013.
[15] Pandya KS, Pothnis JR, Ravikumar G, Naik NK. Ballistic impact behavior of hybrid composites. *Materials & Design*. 44: 128-135, 2013.
[16] Yahaya R, Sapuan SM, Jawaid M, Leman Z, Zainudin ES. Quasi-static penetration and ballistic properties of kenaf–aramid hybrid composites. *Materials & Design*. 63: 775-782, 2014.
[17] Karthikeyan K, Russell BP, Fleck NA, Wadley HNG, Deshpande VS. The effect of shear strength on the ballistic response of laminated composite plates. *European Journal of Mechanics - A/Solids*. 42: 35-53, 2013.

Evaluation of debonding strength of single lap joint by the intensity of singular stress field

Tatsujiro Miyazaki[1] **and Nao-Aki Noda**[2]

[1]University of the Ryukyus, 1 Senbaru, Nishihara-cho, Nakagami-gun, Okinawa 903-0213, Japan

[2]Kyushu Institute of Technology, 1-1 Sensui-cho, Tobata-ku, Kitakyushu-shi, Fukuoka 804-8550, Japan

E-mail: t-miya@tec.u-ryukyu.ac.jp

Abstract. In this paper, the similarity of the singular stress field of the single lap joint (SLJ) is discussed to evaluate the debonding fracture by the intensity of the singular stress field (ISSF). The practical method is proposed for analyzing the ISSF for the SLJ. The analysis method focuses on the FEM stress at the interface end by applying the same mesh pattern to the unknown and reference models. It is found that the independent technique useful for the bonded plate and butt joint cannot be applied to the SLJ because the singular stress field of the SLJ consists of two singular stress terms. The FEM stress is divided to two FEM stresses by applying the unknown and reference models to different minimum element sizes. Then, the practicality of the present method is examined by applying to the previous tensile test results of the SLJ composed of the aluminum alloy and the epoxy resin. The ISSFs for the SLJ were calculated by changing the adhesive thickness t_2 and the overlap length l_2. In the case of the SLJ with 225 mm in total length and 7 mm in adherend thickness, it was found that the similar singular stress fields are formed in the range of $0.15\,\text{mm} \leq t_2 \leq 0.9\,\text{mm}$ and $15\,\text{mm} \leq l_2 \leq 50\,\text{mm}$. It is shown that the critical ISSFs at the fracture are constant in the range.

1. Introduction

The intensity of singular stress field (ISSF) is useful for evaluating the debonding strength [1–4]. Generally, the ISSF cannot be calculated directly by the finite element method (FEM) [5–8]. The authors proposed the method for calculating the ISSF easily and accurately by the FEM [3, 4]. The method does not require the complex calculation and can be applied to various bonded structures [9–12]. In the previous studies, the butt joint was analyzed under all material combination by using the bonded plate as the reference solution [3, 4]. The singular stress field of the butt joint is expressed with a singular stress term. On the other hand, for many material combinations, the singular stress field of the single lap joint (SLJ) consists of two singular stress terms and is not discussed sufficiently. The similarity of the singular stress field needs be discussed to evaluate the debonding strength by the ISSF [10, 13]. The method for analyzing two ISSFs easily and conveniently is required.

In this paper, the practical method for calculating two ISSFs for SLJ from the stress at the interface end by FEM is proposed. When the FE analyses are performed on the reference and unknown models under the same mesh patter and the same material combination, the ratio of the FEM stresses at the interface end of the unknown model to that of the reference model

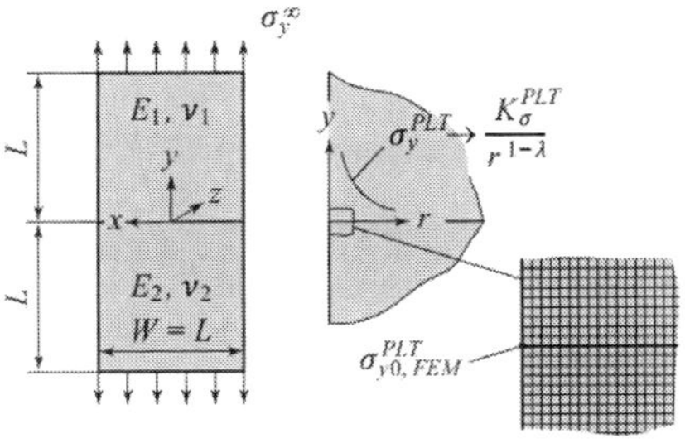

Figure 1. Bonded plate used as the reference model.

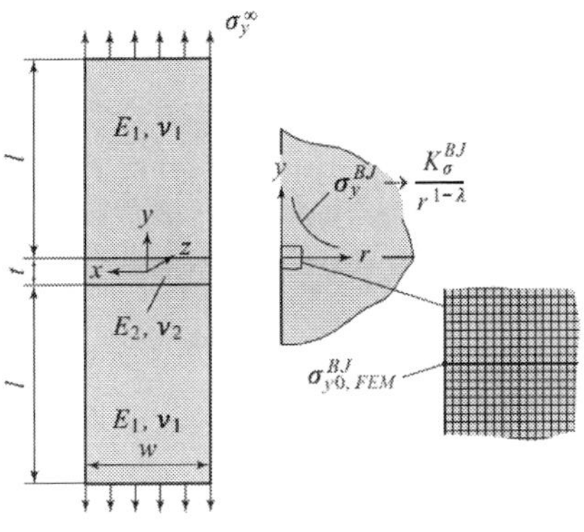

Figure 2. Butt joint used as the unknown model.

corresponds to the ratio of the ISSF of the unknown model to that of the reference model. Since the singular stress field of the SLJ consists on two singular terms, the sum of two FEM stresses is output as the nodal solution. Therefore, the FEM stress is divided to two FEM stresses by applying the unknown and reference models to different minimum element sizes. Then, two ISSFs are calculated by the divided FEM stresses. Then, the present method is applied to the previous experimental results of the SLJ. The similarity of the singular stress field and the debonding fracture criterion are discussed.

2. Mesh-independent technique useful for evaluating the ISSF for butt joint

The authors proposed the method for calculating the ISSF for the butt joint (Fig. 1) accurately by using the ISSF for the bonded plate (Fig. 2) as the reference solution [3, 4]. The real singular stresses of the bonded plate and the butt joint, σ_{ij}^{PLT} and σ_{ij}^{BJ}, are given by the following equations, respectively.

$$\sigma_{ij}^{PLT} = K_{\sigma_{ij}}^{PLT}/r^{1-\lambda} \tag{1}$$

$$\sigma_{ij}^{BJ} = K_{\sigma_{ij}}^{BJ}/r^{1-\lambda} \tag{2}$$

Here, r is the distance on the interface from the corner edge, λ is the singular index, $K_{\sigma_{ij}}^{PLT}$ and $K_{\sigma_{ij}}^{BJ}$ are ISSFs for the bonded plate and the butt joint, respectively. When the FE analyses are performed on the bonded plate and the butt joint under the same mesh pattern and the same material combination, the ratio of the FEM stresses, $\sigma_{ij0,FEM}^{BJ}/\sigma_{ij0,FEM}^{PLT}$, corresponds to the ratio of the ISSFs, $K_{\sigma_{ij}}^{BJ}/K_{\sigma_{ij}}^{PLT}$, as follows [3, 4].

$$\frac{K_{\sigma_{ij}}^{BJ}}{K_{\sigma_{ij}}^{PLT}} = \frac{\lim\limits_{r\to 0} r^{1-\lambda}\sigma_{ij}^{BJ}}{\lim\limits_{r\to 0} r^{1-\lambda}\sigma_{ij}^{PLT}} = \lim_{r\to 0}\frac{r^{1-\lambda}\sigma_{ij}^{BJ}}{r^{1-\lambda}\sigma_{ij}^{PLT}} = \lim_{r\to 0}\frac{\sigma_{ij}^{BJ}}{\sigma_{ij}^{PLT}} \simeq \frac{\sigma_{ij0,FEM}^{BJ}}{\sigma_{ij0,FEM}^{PLT}} \tag{3}$$

The real singular stress of the SLJ is given by the following equation under many material combinations [10, 13].

$$\sigma_{ij}^{SLJ}(r) = \frac{K_{\sigma_{ij},\lambda_1}^{SLJ}}{r^{1-\lambda_1}} + \frac{K_{\sigma_{ij},\lambda_2}^{SLJ}}{r^{1-\lambda_2}} = \frac{K_{\sigma_{ij},\lambda_1}^{SLJ}}{r^{1-\lambda_1}}\left(1 + \frac{C_{\sigma_{ij}}^{SLJ}}{r^{\lambda_1-\lambda_2}}\right),\ C_{\sigma_{ij}}^{SLJ} = \frac{K_{\sigma_{ij},\lambda_2}^{SLJ}}{K_{\sigma_{ij},\lambda_1}^{SLJ}} \tag{4}$$

Here, λ_1 and λ_2 ($\lambda_1 < \lambda_2$) are singular indexes, $K_{\sigma_{ij},\lambda_1}^{SLJ}$ and $K_{\sigma_{ij},\lambda_2}$ are the ISSFs. The FEM stresses which correspond to $K_{\sigma_{ij},\lambda_1}^{SLJ}/r^{1-\lambda_1}$ and $K_{\sigma_{ij},\lambda_2}^{SLJ}/r^{1-\lambda_2}$ are denoted with $\sigma_{ij0,FEM,\lambda_1}^{SLJ}$ and

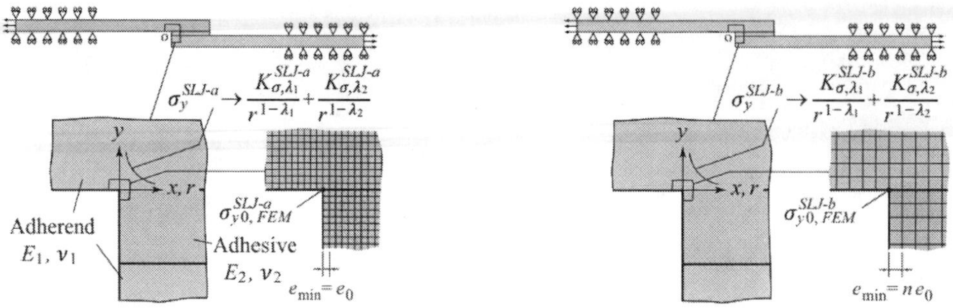

(a) Fine mesh model with minimum element size $e_{\min} = e_0$

(b) Coarse mesh model with minimum element size $e_{\min} = n\,e_0$

Figure 3. Schematic illustration of SLJ models

$\sigma^{SLJ}_{ij0,FEM,\lambda_2}$, respectively. The $\sigma^{SLJ}_{ij0,FEM}$ is expressed with $(\sigma^{SLJ}_{ij0,FEM,\lambda_1}+\sigma^{SLJ}_{ij0,FEM,\lambda_2})$ and is governed by the $\sigma^{SLJ}_{ij0,FEM,\lambda_1}$ because of $\lambda_1 < \lambda_2$. Therefore, only ISSF ratio $K^{SLJ}_{\sigma_{ij},\lambda_1}/K^{SLJ*}_{\sigma_{ij},\lambda_1}$ is determined by the FEM stress ratio $\sigma^{SLJ}_{ij0,FEM}/\sigma^{SLJ*}_{ij0,FEM}$ as follows [10, 13].

$$\frac{K^{SLJ}_{\sigma_{ij},\lambda_1}}{K^{SLJ*}_{\sigma_{ij},\lambda_1}} = \frac{\lim\limits_{r\to 0} r^{1-\lambda_1}\sigma^{SLJ}_{ij}}{\lim\limits_{r\to 0} r^{1-\lambda_1}\sigma^{SLJ*}_{ij}} = \lim_{r\to 0}\frac{r^{1-\lambda_1}\sigma^{SLJ}_{ij}}{r^{1-\lambda_1}\sigma^{SLJ*}_{ij}} = \lim_{r\to 0}\frac{\sigma^{SLJ}_{ij}}{\sigma^{SLJ*}_{ij}} \simeq \frac{\sigma^{SLJ}_{ij0,FEM,\lambda_1}}{\sigma^{SLJ*}_{ij0,FEM,\lambda_1}} \simeq \frac{\sigma^{SLJ}_{ij0,FEM}}{\sigma^{SLJ*}_{ij0,FEM}} \tag{5}$$

The $K^{SLJ}_{\sigma_{ij},\lambda_2}/K^{SLJ}_{\sigma_{ij},\lambda_1}$ is necessary to discuss the similarity of the singular stress field. However, the $K^{SLJ}_{\sigma_{ij},\lambda_2}/K^{SLJ*}_{\sigma_{ij},\lambda_2}$ cannot be calculated from the FEM stress ratio.

3. Mesh-independent technique useful for evaluating the ISSF for SLJ

3.1. Division of the FEM stress

Figure 3 shows the schematic illustrations of the single lap joint models. The model (a) is subdivided by the minimum element size $e_{\min} = e_0$. The FEM stress at the interface end and the ISSF are denoted with $\sigma^{SLJ\text{-}a}_{ij0,FEM} = \sigma^{SLJ}_{ij0,FEM}$ and $K^{SLJ\text{-}a}_{\sigma_{ij},\lambda_k} = K^{SLJ}_{\sigma_{ij},\lambda_k}$, respectively. The model (b) is as large as the model (a) and subdivided by $e_{\min} = ne_0$. The FEM stress at the interface end and the ISSF are denoted with $\sigma^{SLJ\text{-}b}_{ij0,FEM} = \sigma^{SLJ}_{ij0,FEM}|_{e_{\min}=n\,e_0}$ and $K^{SLJ\text{-}b}_{\sigma_{ij},\lambda_k}$, respectively.

The FEM stress of the model (a), $\sigma^{SLJ}_{ij0,FEM}$, is expressed as follows.

$$\sigma^{SLJ}_{ij0,FEM} = \sigma^{SLJ}_{ij0,FEM,\lambda_1} + \sigma^{SLJ}_{ij0,FEM,\lambda_2} \tag{6}$$

The $\sigma^{SLJ}_{ij0,FEM}$ has to be divided into $\sigma^{SLJ}_{ij0,FEM,\lambda_1}$ and $\sigma^{SLJ}_{ij0,FEM,\lambda_2}$ in order to calculate the $K^{SLJ}_{\sigma_{ij},\lambda_k}$.

Since the minimum element size of the model (b) is n times as large as that of the model (a), the FEM stress of the model (b), $\sigma^{SLJ}_{ij0,FEM}\big|_{e_{\min}=n\,e_0}$, is also expressed as follows [14, 15].

$$\begin{aligned}\sigma^{SLJ}_{ij0,FEM}\big|_{e_{\min}=n\,e_0} &= \sigma^{SLJ}_{ij0,FEM,\lambda_1}\big|_{e_{\min}=n\,e_0} + \sigma^{SLJ}_{ij0,FEM,\lambda_2}\big|_{e_{\min}=n\,e_0} \\ &= \frac{\sigma^{SLJ}_{ij0,FEM,\lambda_1}}{n^{1-\lambda_1}} + \frac{\sigma^{SLJ}_{ij0,FEM,\lambda_2}}{n^{1-\lambda_2}}\end{aligned} \tag{7}$$

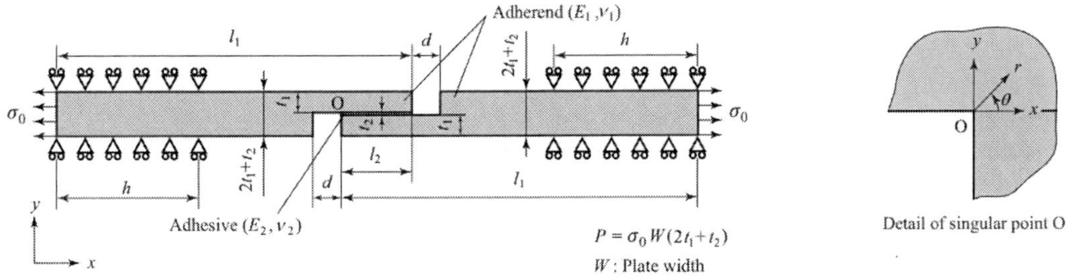

Figure 4. Schematic illustration of the thick adherend single lap joint

When the simultaneous equations (6) and (7) are solved on the $\sigma^{SLJ}_{ij0,FEM,\lambda_1}$ and $\sigma^{SLJ}_{ij0,FEM,\lambda_2}$, the following equations are obtained.

$$\sigma^{SLJ}_{ij0,FEM,\lambda_1} = \frac{\sigma^{SLJ}_{ij0,FEM}}{1-n^{\lambda_1-\lambda_2}} - \frac{\sigma^{SLJ}_{ij0,FEM}\big|_{e_{\min}=n\,e_0}}{n^{\lambda_2-1}-n^{\lambda_1-1}} \quad (8)$$

$$\sigma^{SLJ}_{ij0,FEM,\lambda_2} = -\frac{\sigma^{SLJ}_{ij0,FEM}}{1-n^{\lambda_2-\lambda_1}} + \frac{\sigma^{SLJ}_{ij0,FEM}\big|_{e_{\min}=n\,e_0}}{n^{\lambda_2-1}-n^{\lambda_1-1}} \quad (9)$$

3.2. Mesh-independent technique
The ratio of the ISSFs can be obtained from the ratios of the FEM stresses divided by Eqs. (8) and (9) as follows.

$$\frac{K^{SLJ}_{\sigma_{y0},\lambda_1}}{K^{SLJ*}_{\sigma_{y0},\lambda_1}} = \frac{\sigma^{SLJ}_{y0,FEM,\lambda_1}}{\sigma^{SLJ*}_{y0,FEM,\lambda_1}}, \quad \frac{K^{SLJ}_{\sigma_{y0},\lambda_2}}{K^{SLJ*}_{\sigma_{y0},\lambda_2}} = \frac{\sigma^{SLJ}_{y0,FEM,\lambda_2}}{\sigma^{SLJ*}_{y0,FEM,\lambda_2}} \quad (10)$$

As shown in Eq. (10), the ISSFs for the unknown model can be determined by those for the only one reference model. That is the utmost advantage obtained by dividing the FEM stresses.

4. Application to the experimental result

4.1. Experimental results used in the analysis
The experimental result of the thick adherend SLJ as shown in Fig. 4 by Park et al [16] is used. In the experiment, the adherend and adhesive are aluminum alloy 6061-T6 (Young's modulus $E_1 = 68.9\,\mathrm{GPa}$, Poisson's ratio $\nu_1 = 0.3$) and epoxy resin ($E_2 = 4.2\,\mathrm{GPa}$, $\nu_2 = 0.45$), respectively. $(2l_1 - l_2) = 225\,\mathrm{mm}$, $t_1 = 7\,\mathrm{mm}$ and $h = 37.5\,\mathrm{mm}$ are set. The adhesive thickness t_2 is varied from $0.15\,\mathrm{mm}$ to $0.9\,\mathrm{mm}$. The overlap length l_2 is vaired from $15\,\mathrm{mm}$ to $50\,\mathrm{mm}$.

Figure 5 shows the fracture load P_{af} under (a) t_2 constant condition and (b) l_2 constant condition. The P_{af} increases with increasing the l_2 as shown in Fig. 5(a). Then, the P_{af} is almost independent of the t_2 under l_2 constant condition. Figure 6 shows the average shear stress at the fracture, $\tau_c = P_{af}/(l_2 W)$, obtained from Fig. 5(a). When $l_2 < 15\,\mathrm{mm}$, the τ_c becomes constant at about $28.7\,\mathrm{MPa}$. When the overlap length is short, the cohesive fracture occurs and the τ_c becomes constant. In this study, it is supposed that debonding fracture occurs when $l_2 > 15\,\mathrm{mm}$.

4.2. Similarity of the singular stress field and debonding fracture criterion
Figure 4 shows the schematic illustration of the analysis model. Dundurs' parameters are $\alpha = -0.8699$ and $\beta = -0.06642$ [10, 13]. The SLJ has two different real singular indexes $\lambda_1 = 0.6062$ and $\lambda_2 = 0.9989$ at point O. In this analysis, all models were subdivided by the same mesh pattern (Fig. 7). The minimum element size $e_{\min}$ is changed to confirme the mesh

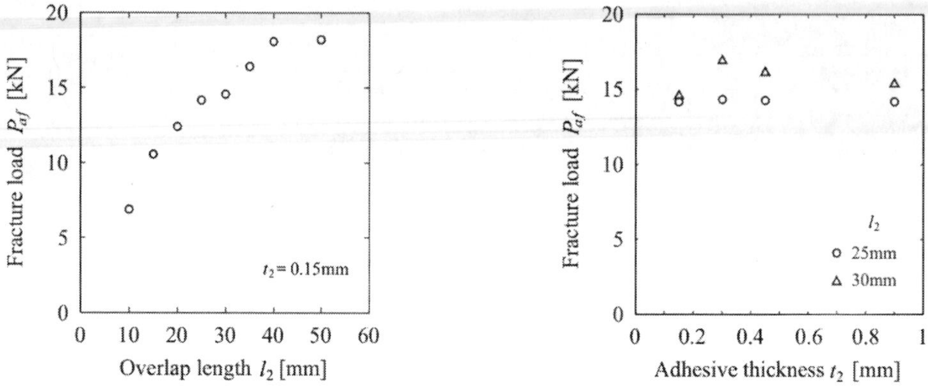

(a) Relationship between P_{af} and l_2 under $t_2 = 0.15$ mm

(b) Relationship between P_{af} and t_2 under $l_2 = 25$ mm and 30 mm

Figure 5. Experimental fracture load P_{af} of the specimens in Fig. 4 by Park et al [16]

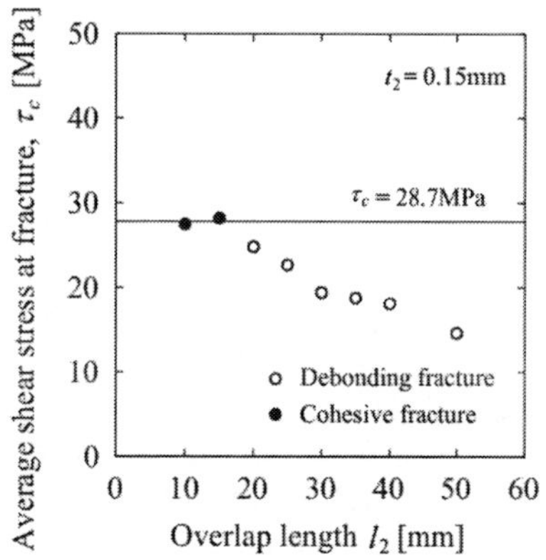

Figure 6. Avarage shear stress at fracture, $\tau_c = P_{af}/(l_2 W)$, obtained from Fig. 5(a) [10, 13]

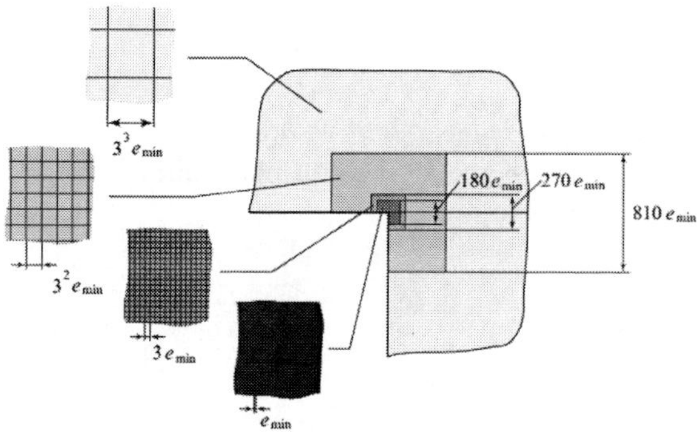

Figure 7. Mesh pattern near the interface end.

independency. $(e_{\min}, ne_{\min}) = (3^{-14}, 3^{-13})$ and $(3^{-13}, 3^{-12})$ are used.

Table 1 shows the FEM stresses of the models with $(l_2, t_2) = (25, 0.15)$, $(50, 0.15)$ and $(25, 0.90)$. The FEM stresses are quite different depending on the mesh size $e_{\min}$. Table 2 shows the $K^{SLJ}_{\sigma_{ij0},\lambda_1}/K^{SLJ*}_{\sigma_{ij0},\lambda_1}$ and the $K^{SLJ}_{\sigma_{ij0},\lambda_2}/K^{SLJ*}_{\sigma_{ij0},\lambda_2}$ obtained from the FEM stress in Table 1, where the specimen A25 model with $(l_2, t_2) = (25, 0.15)$ is used as the reference solution and $*$ is added in the superscript. The $K^{SLJ}_{\sigma_{ij0},\lambda_1}/K^{SLJ*}_{\sigma_{ij0},\lambda_1}$ by the present method is independent of the mesh size $e_{\min}$ and has the same value as the $K^{SLJ}_{\sigma_{ij0},\lambda_1}/K^{SLJ*}_{\sigma_{ij0},\lambda_1}$ by th RWCIM [10]. The $K^{SLJ}_{\sigma_{ij0},\lambda_2}/K^{SLJ*}_{\sigma_{ij0},\lambda_2}$ are little different depending on the $e_{\min}$. That is because the $|\sigma^{SLJ}_{ij0,FEM,\lambda_2}|$ is much smaller than the $|\sigma^{SLJ}_{ij0,FEM,\lambda_1}|$. Since the $K^{SLJ}_{\sigma_{x0},\lambda_2}/K^{SLJ*}_{\sigma_{x0},\lambda_2}$ by the present method has the same value as the $K^{SLJ}_{\sigma_{ij0},\lambda_2}/K^{SLJ*}_{\sigma_{ij0},\lambda_2}$ by th RWCIM, it is found that the FEM stress in the x direction on the material 1 is the most suitable for the present method in this material combination.

Figure 8 shows the $K^{SLJ}_{\sigma_{ij0},\lambda_1}/K^{SLJ*}_{\sigma_{ij0},\lambda_1}$ and the $K^{SLJ}_{\sigma_{ij0},\lambda_2}/K^{SLJ*}_{\sigma_{ij0},\lambda_2}$ obtained by changing the l_2 and the t_2 variously. When $0.15\,\text{mm} \le t_2 \le 0.9\,\text{mm}$ and $15\,\text{mm} \le l_2 \le 50\,\text{mm}$, the $K^{SLJ}_{\sigma_{ij0},\lambda_1}/K^{SLJ*}_{\sigma_{ij0},\lambda_1}$ and the $K^{SLJ}_{\sigma_{ij0},\lambda_2}/K^{SLJ*}_{\sigma_{ij0},\lambda_2}$ decrease linearly with increasing the l_2. Figure 9 shows the $C^{SLJ}_{\sigma_{ij0}}/C^{SLJ*}_{\sigma_{ij0}}$ obtained from the $K^{SLJ}_{\sigma_{ij0},\lambda_1}/K^{SLJ*}_{\sigma_{ij0},\lambda_1}$ and the $K^{SLJ}_{\sigma_{ij0},\lambda_2}/K^{SLJ*}_{\sigma_{ij0},\lambda_2}$ in Fig. 8.

Table 1. Mesh-dependent singular FEM stress at the interface end.

Model (l_2, t_2)	e_0	$\sigma^{SLJ}_{x0,FEM}$ $[=\sigma^{SLJ}_{x0,FEM,\lambda_1}+\sigma^{SLJ}_{x0,FEM,\lambda_2}]$ Mat. 1	Mat. 2	$\sigma^{SLJ}_{x0,FEM}\vert_{e_{min}=n\,e_0}$ Mat. 1	Mat. 2	$\sigma^{SLJ}_{y0,FEM}$ $[=\sigma^{SLJ}_{y0,FEM,\lambda_1}+\sigma^{SLJ}_{y0,FEM,\lambda_2}]$	$\sigma^{SLJ}_{y0,FEM}\vert_{e_{min}=n\,e_0}$	$\tau^{SLJ}_{xy0,FEM}$ $[=\tau^{SLJ}_{xy0,FEM,\lambda_1}+\tau^{SLJ}_{xy0,FEM,\lambda_2}]$	$\tau^{SLJ}_{xy0,FEM}\vert_{e_{min}=n\,e_0}$
A25 (25, 0.15)	3^{-14}	1219.634 [= 1212.633 +7.001]	2018.765 [= 2018.660 +0.105]	793.7081	1309.743	1453.725 [= 1454.046 −0.321]	943.0149	−461.4383 [= −461.4974 +0.0591]	−299.3449
	3^{-13}	793.7081 [= 786.7254 +6.9827]	1309.743 [= 1309.683 +0.060]	517.3754	405.0707	943.0149 [= 943.3633 −0.3485]	611.6741	−299.3449 [= −299.4065 +0.0616]	−194.1832
A50 (50, 0.15)	3^{-14}	927.7130 [= 922.2484 +5.4646]	1535.343 [= 1535.272 +0.071]	603.7818	996.1034	1105.601 [= 1105.858 −0.257]	717.1869	−350.9387 [= −350.9837 +0.0450]	−227.6614
	3^{-13}	603.7818 [= 598.3338 +5.4480]	996.1034 [= 996.0530 +0.0504]	393.6204	646.2558	717.1869 [= 717.4571 −0.2702]	465.1920	−227.6614 [= −227.7103 +0.0489]	−147.6819
A25-90 (50, 0.9)	3^{-14}	1223.239 [= 1216.955 +6.284]	2025.962 [= 2025.872 +0.090]	795.7958	1314.407	1458.949 [= 1459.238 −0.289]	946.4155	−463.0916 [= −463.1415 +0.0498]	−300.4207
	3^{-13}	795.7958 [= 789.5330 +6.2628]	1314.407 [= 1314.341 +0.066]	518.4778	852.7658	946.4155 [= 946.7212 −0.3057]	613.8953	−300.4207 [= −300.4757 +0.0549]	−194.8836

$n = 3$ is used in all analyses. The smallest element size of the coarse model, ne_0, is three times as large as that of the fine model.

Table 2. Mesh-independent ISSF ratio $K^{SLJ}_{\sigma_{ij},\lambda_1}/K^{SLJ*}_{\sigma_{ij},\lambda_1}$ and $K^{SLJ}_{\sigma_{ij},\lambda_2}/K^{SLJ*}_{\sigma_{ij},\lambda_2}$ obtained from the FEM stress in Table 1.

$SLJ/SLJ*$	e_0	Present method: $K^{SLJ}_{\sigma_x,\lambda_1}/K^{SLJ*}_{\sigma_x,\lambda_1}$, $K^{SLJ}_{\sigma_x,\lambda_2}/K^{SLJ*}_{\sigma_x,\lambda_2}$ Mat. 1	Mat. 2	$K^{SLJ}_{\sigma_y,\lambda_1}/K^{SLJ*}_{\sigma_y,\lambda_1}$, $K^{SLJ}_{\sigma_y,\lambda_2}/K^{SLJ*}_{\sigma_y,\lambda_2}$	$K^{SLJ}_{\tau_{xy},\lambda_1}/K^{SLJ*}_{\tau_{xy},\lambda_1}$, $K^{SLJ}_{\tau_{xy},\lambda_2}/K^{SLJ*}_{\tau_{xy},\lambda_2}$	RWCIM: $K^{SLJ}_{\sigma_{ij},\lambda_1}/K^{SLJ*}_{\sigma_{ij},\lambda_1}$, $K^{SLJ}_{\sigma_{ij},\lambda_2}/K^{SLJ*}_{\sigma_{ij},\lambda_2}$
A50 / A25	3^{-14}	0.761	0.761	0.761	0.761	
		0.781	0.678	0.800	0.761	0.761
	3^{-13}	0.761	0.761	0.761	0.761	0.780
		0.780	0.837	0.775	0.793	
A25-90 / A25	3^{-14}	1.004	1.004	1.004	1.004	
		0.898	0.859	0.899	0.843	1.003
	3^{-13}	1.004	1.004	1.004	1.004	0.891
		0.897	1.093	0.877	0.891	

SLJ: unknown model, $SLJ*$: reference model

When $0.15\,\text{mm} \le t_2 \le 0.9\,\text{mm}$ and $15\,\text{mm} \le l_2 \le 50\,\text{mm}$, the $C^{SLJ}_{\sigma_{ij0}}/C^{SLJ*}_{\sigma_{ij0}}$ is almost constant and varies from 0.9 to 1.1. It can be confirmed that the similar singular stress fields are formed in the range.

Figure 10 shows the critical ISSFs at the fracture, $K^{SLJ}_{\sigma c}/K^{SLJ*}_{\sigma c}$, in the range of $0.15\,\text{mm} \le t_2 \le 0.9\,\text{mm}$ and $10\,\text{mm} \le l_2 \le 50\,\text{mm}$. The solid line is the average $K^{SLJ}_{\sigma c}/K^{SLJ*}_{\sigma c}$. The $K^{SLJ}_{\sigma c}/K^{SLJ*}_{\sigma c}$ values are constant within about 10 % error.

5. Conclusion

In this paper, the ISSFs for the SLJ were calculated by changing the adhesive thickness t_2 and the overlap length l_2 and the similarity of the singular stress field of the SLJ was discussed. Then, it was shown that the debonding strength can be expressed as the constant value of the ISSF. The following conclusion can be drawn.

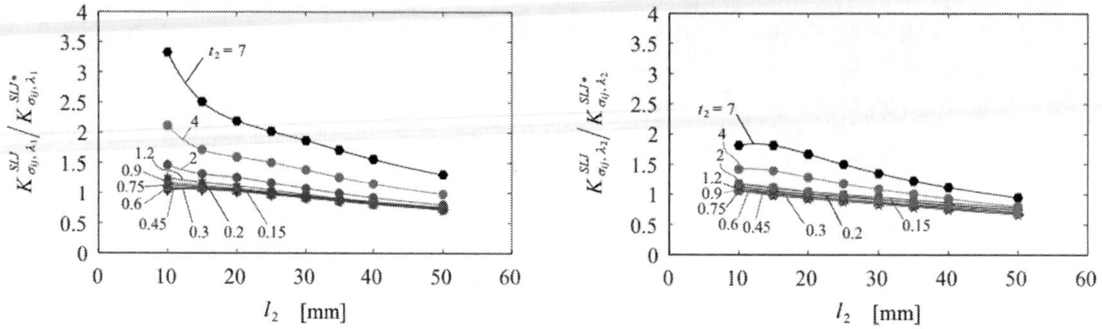

(a) $K^{SLJ}_{\sigma_{ij},\lambda_1}/K^{SLJ*}_{\sigma_{ij},\lambda_1}$ vs l_2.

(b) $K^{SLJ}_{\sigma_{ij},\lambda_2}/K^{SLJ*}_{\sigma_{ij},\lambda_2}$ vs l_2.

Figure 8. Relation between $K^{SLJ}_{\sigma_{ij},\lambda_1}/K^{SLJ*}_{\sigma_{ij},\lambda_1}$, $K^{SLJ}_{\sigma_{ij},\lambda_2}/K^{SLJ*}_{\sigma_{ij},\lambda_2}$ and l_2.

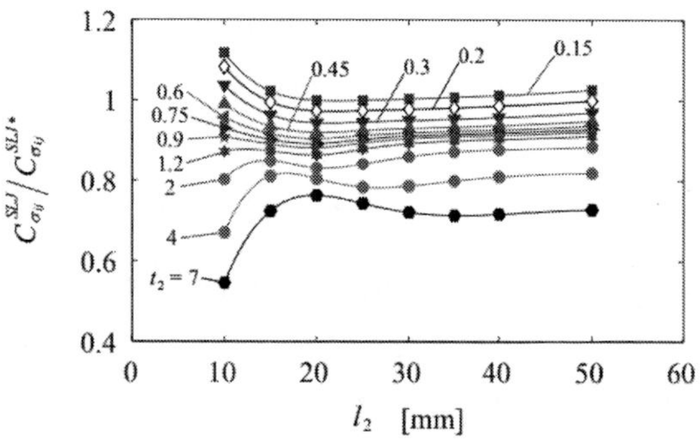

Figure 9. Relation between $C^{SLJ}_{\sigma_{ij}}/C^{SLJ*}_{\sigma_{ij}}$ and l_2.

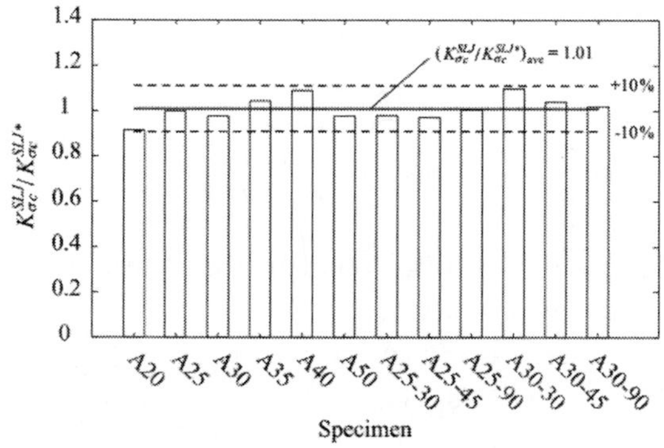

Figure 10. Debonding fracture criterion.

(i) The analysis method for calculating the ISSF is applied to the previous tensile test results of the SLJ composed of the aluminum alloy and the epoxy resin. It was found that the similar singular stress fields are formed in the range of $0.15\,\text{mm} \leq t_2 \leq 0.9\,\text{mm}$ and $15\,\text{mm} \leq l_2 \leq 50\,\text{mm}$ in the case of the SLJ with 225 mm in total length and 7 mm in adherend thickness.

(ii) When the specimens are satisfied with $0.15\,\text{mm} \leq t_2 \leq 0.9\,\text{mm}$ and $15\,\text{mm} \leq l_2 \leq 50\,\text{mm}$, the critical ISSFs at the fracture were constant within 10 % error.

(iii) It was found that the FEM stress can be divided to two FEM stresses by applying the unknown and reference models to different minimum element sizes. Two ISSFs for the SLJ can be obtained by using the divided FEM stresses.

References

[1] Qian, Z. and Akisanya, A. R., An experimental investigation of failure initiation in bonded joints, Acta Materialia, Vol. 46, No. 14 (1998), pp. 4895 - 4904.

[2] Mintzas, A. and Nowell, D., Validation of an H_{cr}-based fracture initiation criterion for adhesively, Engineering Fracture Mechanics, Vol. 80 (2012), pp. 13 - 27.

[3] Zhang, Y., Noda, N. -A., Wu, P. and Duan, M., A mesh-independent technique to evaluate stress singularities in adhesive joints, International Journal of Adhesion and adhesives, Vol. 57 (2015a), pp. 105 - 117.

[4] Zhang, Y., Noda, N. -A., Wu, P. and Duan, M., Corrigendum to "A mesh-independent technique to evaluate stress singularities in adhesive joints" [International Journal of Adhesion and adhesives, Vol. 57 (2015), pp. 105 - 117], International Journal of Adhesion and adhesives, Vol. 60 (2015b), pp. 130.

[5] Chen, D. H. and Nisitani, H., Intensity of singular stress field near the interface edge point of a bonded strip, Transactions of the Japan Society of Mechanical Engineers, Series A, Vol. 59,No. 567 (1993),pp. 2682 - 2686 (in Japanese).

[6] Li, Y. L., Hu, S. Y., Munz, D. and Yang, Y. Y., Asymptotic description of the stress field around the bond edge of a cylindrical joint, Archive of Applied Mechanics, Vol. 68, No. 7-8 (1998), pp. 552 - 565.

[7] Koguchi, H., Distribution of displacements and stresses near stress singularity field at a vertex in three-dimensional bonded structures, Transactions of the Japan Society of Mechanical Engineers, Series A, Vol. 66, No. 648 (2000), pp. 1597 - 1605 (in Japanese).

[8] Noda, N. -A., Shirao, R., Li, J. and Sugimoto, J., Intensity of singular stress at the end of a fiber under pull-out force, Transactions of the Japan Society of Mechanical Engineers, Series A, Vol. 72, No. 721, (2006), pp. 1397 - 1404 (in Japanese).

[9] Noda, N. -A., Miyazaki, T., Uchikoba, T., Li, R., Sano, Y. and Takase, Y., Convenient debonding strength evaluation based on the intensity of singular stress for adhesive joints, Journal of the Japan Institute of Electronics Packaging, Vol. 17, No. 2 (2014), pp. 132 - 142 (in Japanese).

[10] Miyazaki, T., Noda, N. -A., Uchikoba, T., Li, R. and Sano, Y., Proposal of a Convenient and Accurate Method for Evaluation of Debonding Strength, Transactions of the Society of Automotive Engineers of Japan, Vol. 45, No. 5 (2014), pp. 895 - 901 (in Japanese).

[11] Noda, N. -A., Miyazaki, T., Li, R., Uchikoba, T. and Sano, Y., Debonding strength evaluation in terms of the intensity of singular stress at the interface coner with and without fictitious crack, International Journal of Adhesion and adhesives, Vol. 61 (2015), pp. 46 - 64.

[12] Miyazaki, T., Noda, N. -A., Wang, Z. and Sano, Y., Analysis of intensity of singular stress field for bonded cylinder in comparison with bonded plate, Transactions of the JSME (in Japanese), Vol. 81, No. 829 (2015), DOI: 10.1299/transjsme.15-00210.

[13] Miyazaki, T., Noda, N. -A., Li, R., Uchikoba, T. and Sano, Y., Debonding criterion for single lap joints from the intensity of singular stress field, Journal of the Japan Institute of Electronics Packaging, Vol. 16, No. 2 (2013), pp. 143 - 151 (in Japanese).

[14] Nisitani, H., Teranishi, T., Highly accurate values of K_I and K_{II} of axially symmetrical cracked body subjected to tension obtained by FEM, Damage and fracture mechanics VI, (2000), pp. 461 - 469, WIT Press.

[15] Nisitani, H., Teranishi, T.. K_I of a circumferential crack emanating from an ellipsoidal cavity obtained by the crack tip stress method in FEM, Engineering Fracture Mechanics, Vol. 70 (2002), pp. 579 - 585.

[16] Park, J. -H., Choi, J. -H. and Kweon, J. -H., Evaluating the strength of thick aluminum-to-aluminum joints with different adhesive lengths and thicknesses, Composite Structures, Vol. 92 (2010), pp. 2226 - 2235.

IOP Conf. Series: Journal of Physics: Conf. Series **842** (2017) 012079 doi:10.1088/1742-6596/842/1/012079

Detection of Fatigue Crack in Basalt FRP Laminate Composite Pipe using Electrical Potential Change Method

Wael A. Altabey[*1,2] **and Mohammed Noori**[1,3]

[1] International Institute for Urban Systems Engineering, Southeast University, Nanjing (210096), China
[2] Department of Mechanical Engineering, Faculty of Engineering, Alexandria University, Alexandria (21544), Egypt
[3] Professor of Mechanical Engineering and ASME Fellow, California Polytechnic State University, San Luis Obispo, California, USA

Email: wael.altabey@gmail.com

Abstract. Novel modulation electrical potential change (EPC) method for fatigue crack detection in a basalt fibre reinforced polymer (FRP) laminate composite pipe is carried out in this paper. The technique is applied to a laminate pipe with an embedded crack in three layers $[0°/90°/0°]_s$. EPC is applied for evaluating the dielectric properties of basalt FRP pipe by using an electrical capacitance sensor (ECS) to discern damages in the pipe. Twelve electrodes are mounted on the outer surface of the pipe and the changes in the modulation dielectric properties of the piping system are analyzed to detect damages in the pipe. An embedded crack is created by a fatigue internal pressure test. The capacitance values, capacitance change and node potential distribution of ECS electrodes are calculated before and after crack initiates using a finite element method (FEM) by ANSYS and MATLAB, which are combined to simulate sensor characteristics and fatigue behaviour. The crack lengths of the basalt FRP are investigated for various number of cycles to failure for determining crack growth rate. Response surfaces are adopted as a tool for solving inverse problems to estimate crack lengths from the measured electric potential differences of all segments between electrodes to validate the FEM results. The results show that, the good convergence between the FEM and estimated results. Also the results of this study show that the electrical potential difference of the basalt FRP laminate increases during cyclic loading, caused by matrix cracking. The results indicate that the proposed method successfully provides fatigue crack detection for basalt FRP laminate composite pipes.

1. Introduction

A review of recent literature indicates that there is an increasing interest in the aircraft industry in monitoring the actual health condition of a structure in real time and while the structure is in service. It is anticipated that a real-time Structural Health Monitoring (SHM) could make some of the regular inspections unnecessary and allow maintenance only when required. This is particularly important for basalt FRP structures as a new composite material in which aircraft manufacturers are increasingly interested. For these new composite structures where the incidence of fatigue is low, SHM technique could potentially be economical and improve the safety of the structures. While there might be numerous benefits, there are some challenges and difficulties in operational conditions especially in early stage that need to be dealt with. This challenges may require large computational efforts in the SHM and potential maintenance. The challenges feed by SHM system has been discussed by many researchers [1-7].

Non-destructive testing (NDT) methods have been found to be useful for in-situ evaluation of composite structures, where the structural integrity of laminate composite structures can be assessed effectively. There is a diverse range of NDE techniques for detecting fatigue crack in composite structures, the capabilities and limitations of each method are different. Each technique has its specific field of applicability although there is a level of overlap based on the size of defect and accuracy of detection and the ability to detect certain types of damage. For instance, it may be necessary to combine information obtained from acoustic emission and X-ray radiography to achieve a three-dimensional map of the complex array of fatigue cracks in a composite, however, no single method is capable of easily detecting, or identifying, all fatigue cracks with high level of accuracy, and at a low-cost [8].

ECS is one of the most mature and promising methods, which measures the capacitance change of multi-electrode sensors due to the change in dielectric permittivity being imaged. It can then reconstruct the cross-section images using the measured raw data utilizing a suitable algorithm. It has the characteristics such as being a low cost, fast response, non-intrusive method with a broad range of applications and with a high level of safety [9-13]. The need for a more accurate measurement of ECS has led to the study of the factors which influence and affect ECS sensitivity and the sensitive domain of ECS electrodes. There are three factors that have been studied and found that they affect ECS measurements, e.g. pipe's material, inner dielectric permittivity [14-20] and the ratio of pipeline thickness and diameter [21-22]. Altabey [23] found that the environmental temperature also affects ECS sensitivity and sensitive domain of ECS electrodes with high percentage. Therefore, it was concluded that the environmental temperature should be considered as the fourth factor which influences the ECS measurement sensitivity.

In the present study, detection of fatigue cracks in a basalt FRP laminate composite pipe is attempted numerically using a new non-destructive evaluation (NDE) technique via utilizing ECS to detect the changes in the dielectric properties. An embedded crack due to fatigue internal pressure is created across the pipe thickness on which multiple electrodes are mounted. A finite element model (FEM) is developed to simulate the fatigue life behaviour before and after matrix cracking. Through measuring the capacitance values, capacitance change and electric potential difference of ECS electrodes are analyzed in order to obtain relationships between the electric potential differences and the fatigue life before and after cracking. The crack lengths in the basalt FRP pipe are investigated for various number of cycles to failure to determine crack growth rate. Applicability of the method for the detection of fatigue cracking is numerically investigated in detail using ANSYS and MATLAB software. The Response surfaces are adopted as a tool for solving inverse problems to estimate crack lengths from the measured electric potential differences of all segments between electrodes to validate the FEM results.

2. Principle of Electrical Capacitance sensor (ECS)

ECS was first introduced in the 1980s by a group of researchers from the US Department of Energy, at Morgantown Energy Technology Centre (METC), to measure fluidized bed systems [24-26]. The technique further developed and advanced rapidly during the past 10 years. It has gained attention and found important applications in monitoring industrial processes, due to its low cost and its operability under harsh environmental conditions.

ECS consists of an insulating pipe, measurement electrode, radial screen and earthed screen [27]. The measurement electrode is mounted symmetrically around the circumference of a pipeline. Radial screen is fitted between the electrodes to cut the electro-line external to the sensor pipeline and reduce the inter-electrode capacitance. The earthed screen surrounds the measurement electrodes to shield external electromagnetic noise. In most applications, ECS electrode is mounted outside the pipeline which is called external electrode ECS [28]. Electrical capacitance system includes a sensor.

ECS converts the permittivity of the piping system to inter-electrode capacitance, which is the ECS forward problem. Capacitance measuring circuit takes the capacitance data and transfers that to imaging

IOP Conf. Series: Journal of Physics: Conf. Series **842** (2017) 012079 doi:10.1088/1742-6596/842/1/012079

computer. Imaging computer reconstructs the distribution image with a suitable algorithm, which is called ECS inverse problem.

2.1. ECS System geometrical model

The model section comprises of an ECS column with 0.1 m inner diameter and 0.3 m height. The ECS is made up of the basalt FRP laminate composite pipe structure and having a ring of 12 electrodes (which are separated from each other by a small gap) on its outer periphery. Figure 1 shows the cross section of a 12-electrode ECS system, in which R_1 is the inner pipe radius; R_2 is the outer pipe radius; R_3 is the earthed screen radius and the radial screen is connected to the outer composite pipe. The ECS also includes radial guard electrodes to constrain the field lines from the excited electrode and to reduce the dependence of spacing between the electrodes and the screen as shown in the Figure. The function of the sensor includes measuring the capacitance between all possible combination pairs of the electrodes and converting the measured capacitance values in the voltage signals. The sensors physical specifications and the permittivity values of GFRE composite pipe are shown in Table 1.

Table 1. Sensor physical specification

ECS system	Specification
No. of electrodes	12
Space between electrodes	2 mm
Pipe diameter (d_i)	94 mm
Pipe thickness (h)	6 mm
Earth Screen diameter	110 mm
Thickness of electrodes	1mm
Permittivity Basalt fibre/Polymer	$\varepsilon_b = 2.2\ Fm^{-1}$
Permittivity of Water	$\varepsilon_w = 80\ Fm^{-1}$
Permittivity of Air	$\varepsilon_a = 1.0\ Fm^{-1}$
Excitation voltage	$\varphi = 15$ Volts

2.2. The ECS composition and working principle

For the 12-electrode system the electrodes are numbered as shown in Figure 1. These electrodes are excited with an electric potential, one at a time, in increasing order. When one electrode is excited, the other electrodes are kept at ground potential, as shown in Figure 1, and act as detector electrodes. When electrode No. 1 is excited with a potential, the change $Q_{1,j}$ induced on the electrodes, $j = 2, \cdots\cdots, N$, can be measured. Next, electrode No. 2 is excited whereas, rest of the electrodes are kept at ground potential, and the induced charges $Q_{23}, Q_{24}, \cdots, Q_{2N}$ $(N = 12)$ are measured. The measurement protocol continues until electrode $N - 1$ is excited. Using these charge measurements, the inter electrode capacitance C_{ij} can be computed using the definition of capacitance (Eq. 1) i.e.:

$$C_{ij} = \frac{Q_{ij}}{\Delta V_{ij}} \tag{1}$$

Where Q_{ij} is the charge induced on electrode j when electrode *i* is excited with a known potential, V_{ij} is the potential difference between electrodes *i* and *j* $(\Delta V_{ij} = V_i - V_j)$. So the number of independent capacitance measurements $M = 66$ using Eq. 2 is:

$$M = \frac{N(N-1)}{2} \tag{2}$$

It is important to note that these capacitances are dependent on the geometry of electrodes and are determined once the size and the location of the electrodes and the permittivity distribution $\varepsilon(x, y)$ are known. A change in the permittivity distribution $\varepsilon(x, y)$ is naturally reflected in the capacitance measurements. The actual capacitance changes measured will be very small, in the order of Pico or Femto Farad (10^{-12}F or 10^{-15}F). Sequential electrodes are referred to as adjacent electrodes; have the largest standing capacitance, while diagonally or opposing electrodes will have the smallest capacitances.

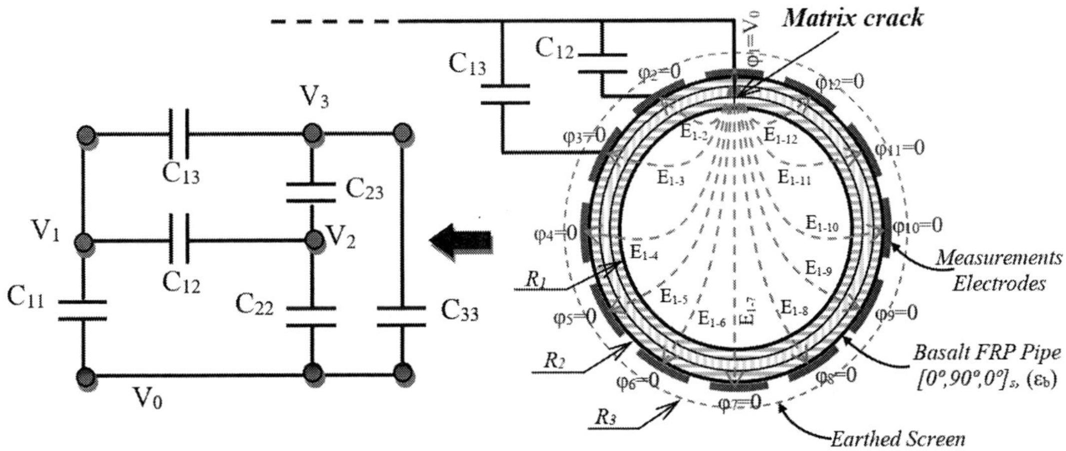

Figure 1. Schematic representation of the measurement principle of 12-electrode ECS.

3. Finite Element Simulation Model

3.1. Materials and fatigue testing Model

The fatigue model for carrying out the analysis on a basalt FRP laminate composite pipe was manufactured using three layers of basalt fibre with orientation $[0°/90°/0°]_s$ and a polymer resin matrix with inner diameter R_1=94mm and outer diameter R_2=100mm. The thickness of all plies were 2mm. The corresponding elastic modulus values were E_1=96.74GPa, E_2=E_3=22.55GPa, and the Shear modulus values were G_1=G_3=10.64GPa, G_2=8.73GPa. Poisson coefficients were υ_1=υ_3=0.3, υ_2=0.6 and the density was 2700 kg/m^3.

The basalt FRP pipe was modelled at room temperature and was subjected to internal fatigue pressure (P), using a sinusoidal wave load at a frequency of 10 Hz. One series of fatigue model was performed using the pressure ratio (P_r) between the applied (P) to the burst pressure (P_{max}) equal to 0.5, and the fatigue stress S_f and the number of cycles to failure (N) were measured. Utilization of the power formula $S_f = aN_f{}^b$ proved its suitability by giving acceptable values for the correlation factor (C.F) equal to 0.9677. The value of fatigue constants a and b for present pressure ratio were 222.6 MPa, -0.0731 respectively, and the results were plotted in Figure 2 in terms of fatigue stress versus the number of cycles to failure.

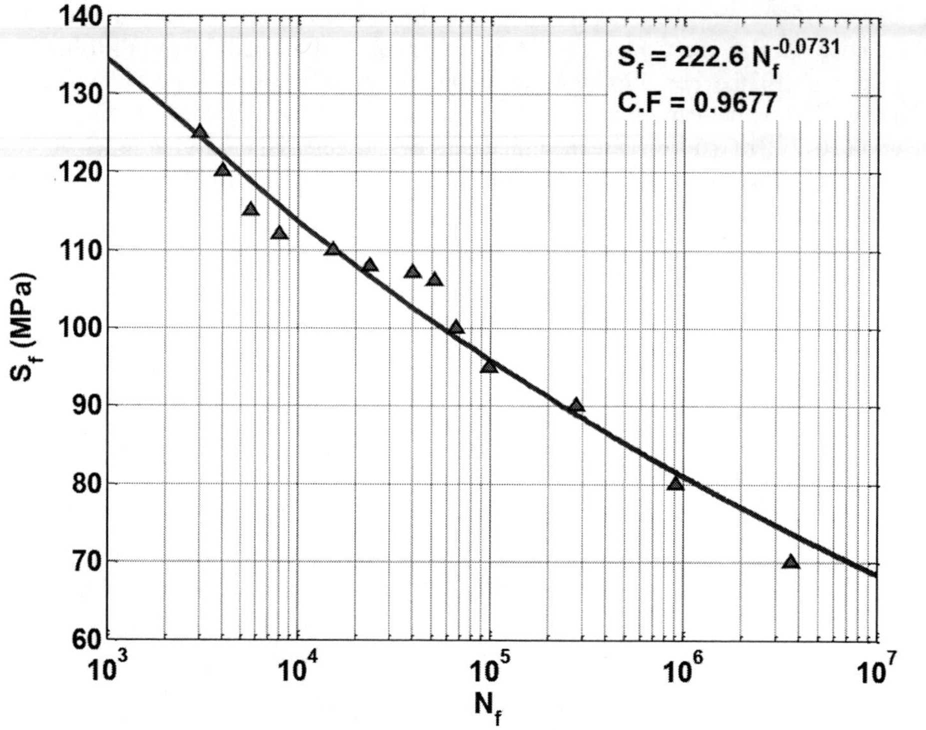

Figure 2. S–N curve for pressure ratio P_r=0.5.

3.2. Dielectric Properties Change of ECS Model
In terms of Electrical Capacitance sensor (ECS), the forward problem is the problem of calculating the capacitance matrix C from a given set of sensor design parameters and a given cross-sectional permittivity distribution (ε). The forward model proposed for ECS [29] was based on finite element simulations. It was assumed that both the flow distribution and the electrical field during the measurement set were 2D and static. Changes in axial direction were neglected within the axial electrode length. Furthermore, free charges in the flow were also neglected. Thus, the system was governed by the following Poisson equation:

$$\nabla.\varepsilon(x,y)\nabla\varphi(x,y) = 0 \tag{3}$$

The 2D models considered in this study were constructed using commercially available finite element software, ANSYS (The Electrostatic Module in the Electromagnetic subsection of ANSYS (2014)) [30, 31]. The problem space was divided into triangular elements. In a region of ideal dielectrics and space charges, the potential distribution $\varphi(x,y)$ inside the ECS was determined by solving the Poisson's equation. For the boundary condition imposed on the ECS head by the measurement system, the potential distribution $\varphi(x,y)$ could be found. The electric field vector $E(x,y)$, the electric flux density $D(x,y)$ and the potential function $\varphi(x,y)$ are related as follows:

$$E(x,y) = -\nabla\varphi(x,y) \tag{4}$$

$$D = \varepsilon(x,y)E(x,y) \tag{5}$$

The change on the electrodes, and hence the inter electrode capacitances could be found using the definition of the capacitance and Gauss's law based on the following surface integral:

$$Q_{ij} = \oint_{S_j} (\varepsilon(x,y)\nabla\varphi(x,y).\hat{n})\, ds \tag{6}$$

Where: $\varepsilon(x,y)$ is the permittivity distribution, $\nabla.\varepsilon(x,y)$ is the divergence of permittivity distribution, $\nabla\varphi(x,y)$ is the gradient of potential distribution, S_j is a surface enclosing electrode j, ds is an infinitesimal area on electrode j, $\hat{n}$ is the unit vector normal to S_j and ds is an infinitesimal area on that.

4. Response surface for the electric potential change method

The response surface is a widely adopted tool for quality engineering fields [32]. The response surface methodology comprises curve fitting with regression to obtain approximate responses, design of numerical model to obtain minimum variances of the responses and optimizations using the approximated responses. In the present study, the response surface methodology is adopted as a solver for inverse problems. For the present study, predictions of fatigue crack length (L_c) from measured electric potential differences are one of the inverse problems. The response surface methodology brings two advantages; the inverse problems can be approximately solved without consideration of modeling, and the approximated response surfaces can be evaluated using powerful statistical tools. The response surface is described as follow:

$$L_c = \beta_0 + \sum_{j=1}^{k} \beta_j E_j + \sum_{j=1}^{k} \beta_{jj} E_j^2 + \sum_{i=1}^{k-1} \sum_{j=i+1}^{k} \beta_{ij} E_i E_j \tag{7}$$

Where: k is the number of variables. In the case of 12 electrode type specimens, there are 11 electric potential difference variables; $E_{1\text{-}2}$, $E_{1\text{-}3}$,…, $E_{1\text{-}12}$.

To improve the estimation performance of the crack length, the normalizations of the measured electric potential differences (E) are performed by means of the norm of the electric potential differences vector. Each element is divided by the square root sum, and replacement the vector ($E_{1\text{-}2}$, $E_{1\text{-}3}$,..., $E_{1\text{-}12}$) in Eq. 7 with norm vector ($e_{1\text{-}2}$, $e_{1\text{-}3}$,…, $e_{1\text{-}12}$). Each element is divided by the square root sum of all results as follows:

$$(e_{1-2}, \cdots\cdots\cdots, e_{1-12}) = \left(\frac{E_{1-2}}{\sqrt{{E_{1-2}}^2 + \cdots\cdots + {E_{1-12}}^2}}, \cdots\cdots\cdots, \frac{E_{1-12}}{\sqrt{{E_{1-2}}^2 + \cdots\cdots + {E_{1-12}}^2}} \right) \tag{8}$$

In the case that varies crack length (L_c), the total number of location is n, the response surface can be expressed as follows using matrix expression.

$$L_c = E\beta + \lambda \tag{9}$$

Where: $L_c = \begin{Bmatrix} L_{c1} \\ L_{c2} \\ \vdots \\ L_{cn} \end{Bmatrix}$, $E = \begin{bmatrix} 1 & E_{11} & E_{12} & \cdots & E_{1k} \\ 1 & E_{21} & E_{2k} & \cdots & E_{2k} \\ \vdots & \vdots & \vdots & \ddots & \vdots \\ 1 & E_{n1} & E_{n2} & \cdots & E_{nk} \end{bmatrix}$, $\beta = \begin{Bmatrix} \beta_0 \\ \beta_1 \\ \vdots \\ \beta_k \end{Bmatrix}$, $\lambda = \begin{Bmatrix} \lambda_1 \\ \lambda_2 \\ \vdots \\ \lambda_n \end{Bmatrix}$

Where: λ is an error vector, the unbiased estimator b of the coefficient vector β is obtained using the well-known least square error method as follows:

$$b = (E^T E)^{-1} + E^T L_c \tag{10}$$

The variance-covariance matrix of the b is obtained as follows:

$$Cov(b_i, b_j) = \sigma^2\, (E^T E)^{-1} \tag{11}$$

Where the σ is the error of L_c. The estimated value of σ is obtained as follows:

$$\sigma^2 = \frac{SS_E}{n-k-1} \tag{12}$$

SS_E is a square sum of errors, and expressed as follows:

$$SS_E = {L_c}^T L_c - b^T E^T L_c \tag{13}$$

The lack of fit is evaluated with the adjusted coefficient of the multiple determination $R^2{}_{adj}$ [33, 34]; $R^2{}_{adj}$ is defined as:

$$R^2{}_{adj} = 1 - \frac{SS_E/(n-k-1)}{S_{yy}/(n-1)} \tag{14}$$

Where: S_{yy} is the total sum of squares:

$$S_{yy} = {L_c}^T L_c - \frac{\left(\sum_{i=1}^{n} L_{ci}\right)^2}{n} \tag{15}$$

The value of $R^2{}_{adj}$ is equal to or lower than 1.0. A higher value of $R^2{}_{adj}$ implies a better fit. When the response surface shows a very good fit, $R^2{}_{adj}$ approaches 1.0. A good fit of the response surface means that the response surface gives good estimations for the DPC method used for the regression. Lower $R^2{}_{adj}$ values means poorer estimations and the error band of the estimated result is wider.

5. Results and Discussions

5.1. Electrical potential change method for fatigue crack detection

To investigate the effect of the matrix crack on the electrical potential changes, FEM analysis of the electric field intensity of basalt FRP piping system were performed, using commercially available FEM code ANSYS ver.14. 2D FEM software.

The approach taken by ANSYS 2D is to divide the different materials and geometries into triangular elements as shown in Figure 3, because many pipes are round under the circumstances of a smaller number of pixels, we can achieve higher accuracy to use the triangular mesh instead of rectangular grids, and then to represent the electric field (see Eq. (4)) within each element with a separate polynomial at six integration points location. For FEM simulating of pipe, electrostatic module (PLANE121), triangular 6-node, and the element has one degree of freedom, voltage, at each node. The 6-node elements have compatible voltage shapes and are well suited to model curved boundaries. The total number of elements used for the analyses was 4845. To improve the accuracy of mesh, we divided the meshed region again into 9825 elements. The applied electric potential ($V=V_0$) was 15V (RMS) and another electrode was kept at ground (V=0) potential.

The software only computes the potential (φ) and the electric field (E) values during each cycle (N_f) under fatigue test at the element nodes by using looping method in ANSYS through creating an analysis file for use during looping and then interpolates between these nodes to obtain the values for other points within the elements. The Simulations, and node potential distribution of basalt FRP laminated composite pipe under fatigue internal pressure before and after matrix crack initiates, due to fatigue stress for the ANSYS 2D simulation when electrode (1) is excited, are illustrated in Figure 4 right and left respectively. The blue area represents the region of the pipe system without potential i.e. $\varphi = 0$ and the colored areas represent the region of the pipe that have the different potential (different node potential), i.e. the domain of electrode can be sensitive or detection domain.

From the Simulation Figure 4, we can make a conclusion that there is a significant difference before and after matrix crack initiates in the node potential and electric field intensity.

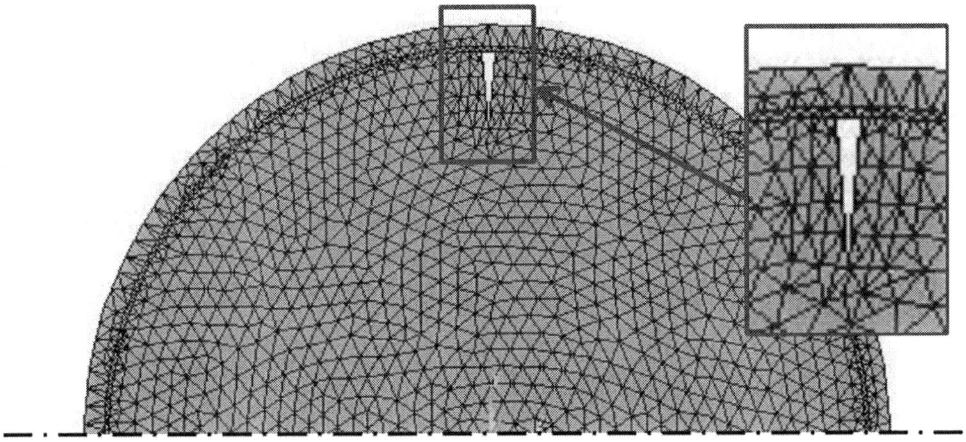

Figure 3. (9825) Element map of finite element mesh.

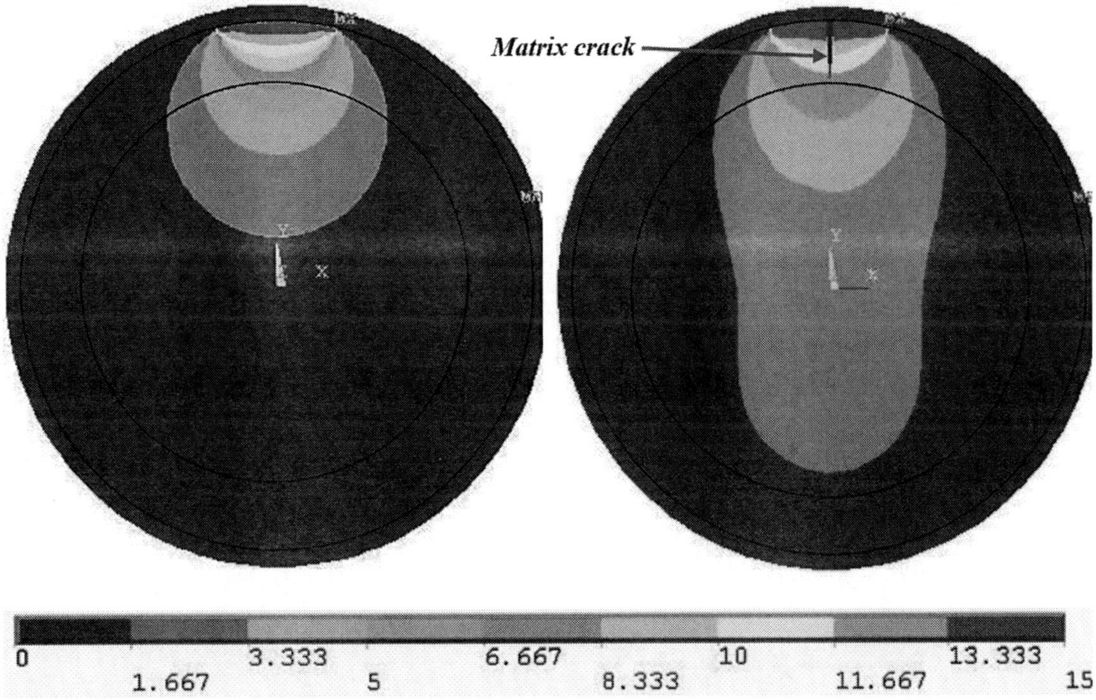

Figure 4. The node potential distribution of basalt FRP laminated composite pipe before matrix crack starts right and after matrix crack left.

Using the scripting capabilities in ANSYS we can simulate the electrical potential change during cyclic loading. Figure 5 shows the finite element results obtained for the electric potential change, during cyclic loading of the basalt FRP laminate pipe, when electrode (1) is excited. The electric potential is almost constant with the increase of the number of cycles, before 25,000 cycles. After 25,000 cycles, the electric potential starts to increase.

The main objective of the FE modelling was to investigate the crack lengths of the basalt FRP for various number of cycles to failure. Figure 6 shows the relation between crack length and the number of cycles. Crack length curve shows a sharply-increasing relationship between the crack lengths versus the number of cycles once initial cracking was detected. At first, the results show no signs of crack for an initial large set of cycles, 3000–25,000 cycles, at 31% of fatigue stress level degradation.

Using the power formula $L_c = cN_f{}^d + e$ has proved its suitability by giving acceptable values for the correlation factor (C.F) equal to 0.986, in order to fit these plots to determine crack growth rate. The value of crack length constants c, d and e are -60.04, -0.2065 and 8.077 respectively, and the results are plotted in Figure 6 in terms of crack lengths versus the number of cycles to failure.

The crack length (L_c) as a function of cycles (N_f) and applied fatigue stress level (S_f) can be determined by combining Eqs. ($L_c = cN_f{}^d + e$) and ($S_f = aN_f{}^b$) and integrating from a set of starting crack length and cycle values (L_0 and N_0) to final values (L and N):

$$\int_{L_0}^{L_f} \frac{cS_f}{a(L_c - e)}\, dL_c = \int_{N_0}^{N_f} N_f{}^{b-d}\, dN_f \tag{16}$$

Where a, b, c, d and e are the same as those determined in Eqs. ($L_c = cN_f{}^d + e$) and ($S_f = aN_f{}^b$).

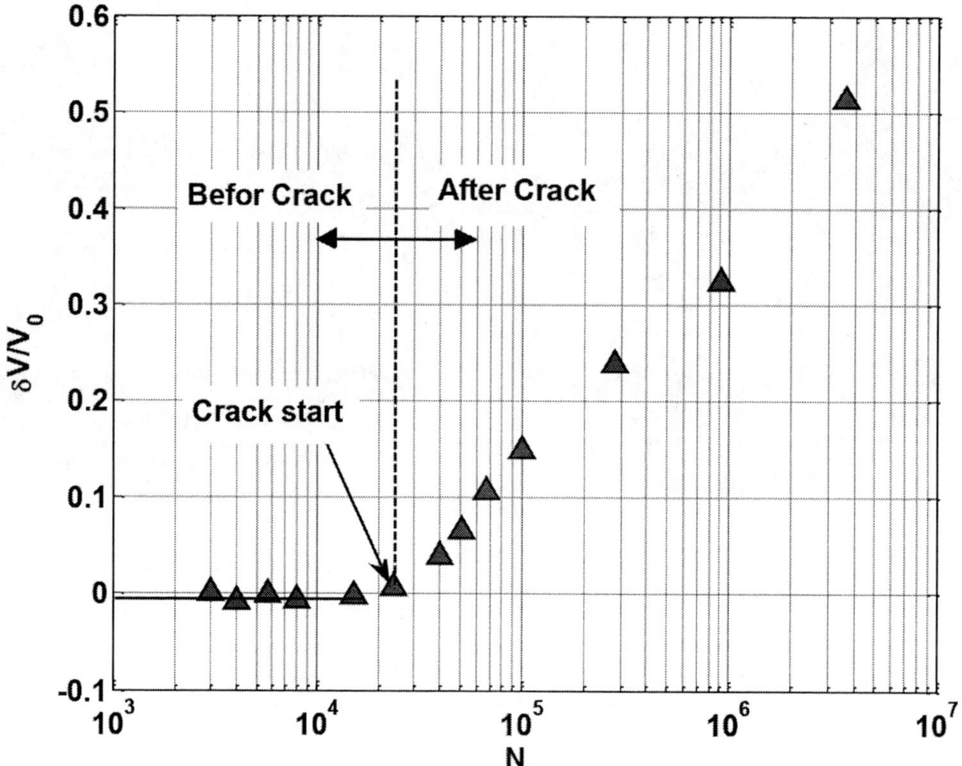

Figure 5. Change of the contact electrode potential during cyclic loading of basalt FRP laminate.

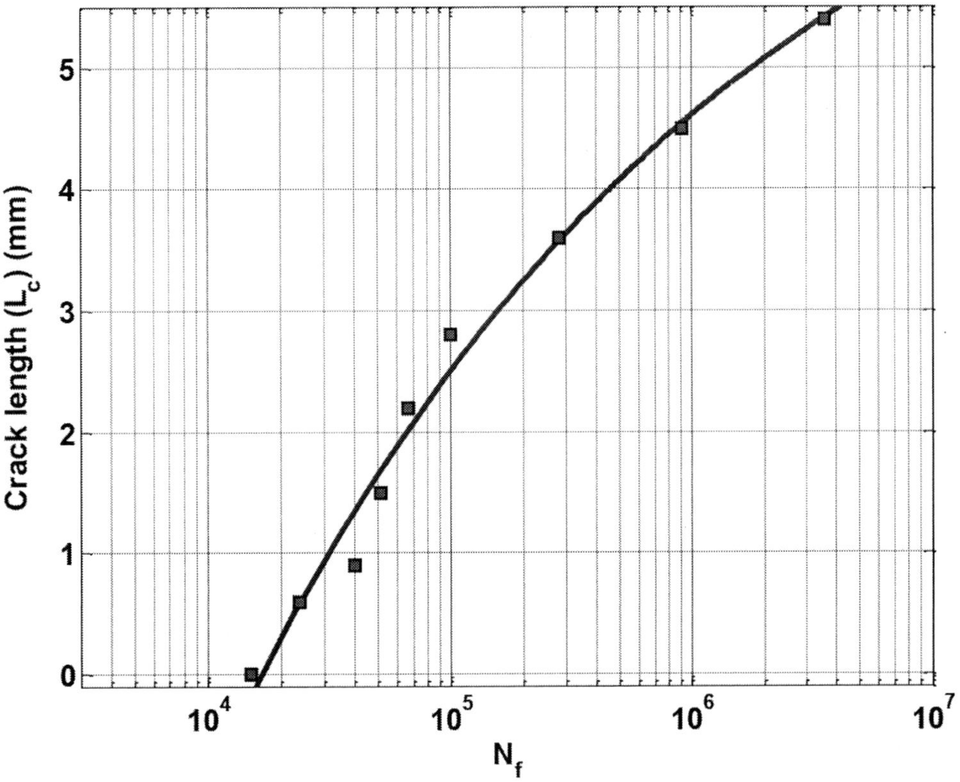

Figure 6. Crack length versus number of cycles for basalt FRP laminate.

5.2. Estimation of fatigue crack length

Table 2 show the comparison between the FEM and estimated results of the crack length (L_c) for basalt FRP laminate composite pipe of response surface method. The R^2_{adj} of this result of crack length is 0.9728. The error band is defined as the maximum error of the crack length. The error band is less than 0.0477 mm. As a result, the response surfaces gave good estimations for FE data even for extrapolations crack length for basalt FRP laminate composite pipe.

Table 2. Estimations and Errors of response surfaces method Data (Unit mm)

N_f	L_c (FEM)	Estimated L_c	Error
15132	0	0	0
23810	0.615	0.651	0.036
39998	0.9203	0.968	0.0477
51254	1.5111	1.5497	0.0386
66825	2.236	2.253	0.017
99549	2.8011	2.8462	0.0451
280670	3.6137	3.6431	0.0294

6. Conclusions

In the present work, detection of the fatigue crack in a Basalt fibre reinforced polymer (FRP) laminated composite pipe is performed using the electric potential change (EPC) method by evaluating the dielectric properties of basalt FRP pipe by using electric capacitance sensor (ECS). The following conclusions can be drawn:

1. For the Basalt fiber reinforced polymer (FRP) laminated composite pipe, matrix cracking caused a significant increase in the electric potential. The electric potential is almost constant with the increase of the number of cycles, before 25,000 cycles. After 25,000 cycles, the electric potential starts to increase.
2. The crack length is sharply increases with the number of cycles increasing, once initial cracking was detected at 25,000. No signs of crack for an initially large set of cycles, 3000–25,000 cycles, at 31% of fatigue stress level degradation.
3. The crack lengths versus the number of cycle to failure curve was fitted using the power formula $L_c = cN_f{}^d + e$, in order to determine the crack growth rate formula by combining this formula with the fatigue life formula $S_f = aN_f{}^b$.
4. These response surfaces data gave good convergence for FEM data with R^2_{adj} of this result of crack length is 0.9728 within the error band of less than 0.0477 mm.
5. Finally, the present method successfully provides fatigue crack detection for a basalt FRP laminate composite pipe including the crack lengths investigation during cyclic loading and crack growth rate estimation.

7. References

[1] Zhou Y L, Maia N, Abdel Wahab M 2016 J. Vibration and Control (doi: 10.1177/1077546316674544).
[2] Gillich G R, Praisach Z I, Abdel Wahab M, Gillich N, Mituletu I C, Nitescu C 2016 *J. Shock and Vibration* (http://dx.doi.org/10.1155/2016/2086274).
[3] Khatir S, Belaidi I, Serra R, Abdel Wahab M, Khatir T 2015 *Mechanika* **21**(6) 472-9.
[4] Gillich G R, Praisach Z I, Abdel Wahab M, Vasile O 2014 *J. Shock and Vibration* (http://dx.doi.org/10.1155/2014/607125).
[5] Zhou Y L, Abdel Wahab M 2016 *J. Vibroengineering* **18**(7) 4491-9.
[6] Khatir S, Belaidi I, Serra R, Abdel Wahab M, Khatir T 2016 *J. Vibroengineering* **18**(1) 202-13.
[7] Zhou Y L, Maia N M M, Sampaio R, Wahab M A 2016 *J. Structural health monitoring* (doi: 10.1177/1475921716680849).
[8] Mouritz A P 2003 *Fatigue in Composites* (Composites Science and Engineering) Chapter 8.
[9] Yang W Q, Stott A L, Beck M S and Xie C G 1995a *Rev. of Sci. Ins.* 66(8) 4326.
[10] Yang W Q, Beck M S and Byars M 1995b *J. Measurement & Control* **28**(9) 261.
[11] Li H and Huang Z 2000 *Special measurement technology and application* (Zhejiang University Press, Hangzhou).
[12] Mohamad E J Rahim R A Leow P L Fazalul Rahiman M H Marwah O M F Nor A N M Rahim H A and Mohd Yunus F R 2012 *J. Sensors and Actuators* A **180** 1.
[13] Zhang W Wang C Yang W Wang C 2014 *J. Advanced Powder Technology* **25** 174.
[14] Jaworski A J and Bolton G T 2000 *J. Measurement Science and Technology* **11**(6) 743.
[15] Pei T and Wang W 2009 *9th Int. Conference on Electronic Measurement & Instruments.*
[16] Al-Tabey W A 2010 *Int. Postgraduate Conference on Engineering Proceedings* (Perlis, Malaysia, October 16-17).
[17] Asencio K Bramer-Escamilla W Gutiérrez G and Sánchez I 2015 *J. Powder Technology* **270** 10.
[18] Sardeshpande M V, Harinarayan S and Ranade V V 2015 *J. chem. eng. Rese. & des.* **9**(4) 1.
[19] Mohamad E J, Rahim R A, Rahiman M H F, Ameran H L M, Muji S Z M and Marwah O M F 2016 *J. Flow Measurement and Instrumentation* **47** 62.
[20] Altabey W A 2016 *J. Structural Monitoring and Maintenance* **3**(4) 377.

[21] Daoye Y Bin Z Chuanlong X Guanghua T and Shimin W 2009 *6th International Symposium on Measurement Techniques for Multiphase Flows Proceedings J. of Physics* **147** 1.
[22] Altabey W A 2016 *J. Structural Monitoring and Maintenance* **3**(3) 297.
[23] Altabey W A 2016 *J. Structural Monitoring and Maintenance* **3**(4) 335.
[24] Fasching G E and Smith N S 1988 *US Dept. Energy* **37** DOE/METC-88/4083.
[25] Fasching G E and Smith N S 1991 *Rev. Sci. Instr.* **62** 2243.
[26] Huang S M Plaskowski A B Xie C G and Beck M S 1989 *J. Phys.* **22** 173.
[27] Yang W Q and York T A 1999 *IEEE proceedings: Meas. Sci. & Tech.* **146**(1) 47.
[28] Yang W Q 1997 *IEEE proceedings: Meas. Sci. & Tech.* **144**(5) 203.
[29] Xie C G Huang S M Hoyle B S Thorn R Lenn C Snowden D and Beck M S 1992 *IEEE Proceedings-G* **139**(1) 89.
[30] 2014 *ANSYS Low-Frequency Electromagnetic analysis Guide* PA 15317 published in the USA.
[31] Al-Tabey W A 2012 *Hand Book For Mechanical Engineers with ANSYS Tutorials*, LAP Lambert Academic Publishing, Germany, **ISBN 978-3-8454-0479-0**.
[32] Myers R and Montgomery D C 1995 *Response surface methodology process and product optimization using designed experiments* New York: Wiley Inter-science Publication.
[33] Myers R and Montgomery D C 2002 *Response surface methodology process and product optimization using designed experiments* 2nd ed. New York: Wiley Inter-science Publication.
[34] Jiang S, Li D, Zhou C and Zhang L 2014 *J. Structural Engineering and Mechanics*, **49** (1) 111.

IOP Conf. Series: Journal of Physics: Conf. Series **842** (2017) 012080 doi:10.1088/1742-6596/842/1/012080

A symplectic analytical singular element for the inclined crack terminating at the material interface of composite structures

XF Hu, QS Shen, WA Yao*

State Key Laboratory of Structural Analysis for Industrial Equipment,
International Research Center for Computational Mechanics,
Dalian University of Technology, Dalian 116024, China

*Email: ywa@dlut.edu.cn

Abstract. The inclined bimaterial interface crack problem is studied from theoretical and computational aspects. The original problem is transformed into an eigen value problem by using the symplectic approach, and the eigen solution are obtained. A Symplectic Analytical Singular Element (SASE), of which the interior fields are described by the obtained eigen solutions, is developed. The stress intensity factors (SIFs) can be calculated directly without any post-processing. Numerical example on a square bimaterial plate containing an inclined crack is provided to validate the proposed method. The comparison of the predicted results with the existing result indicates that the proposed method is accurate for the modelling of the inclined bimaterial interface crack.

1. Introduction

With the increasing of the demand from modern manufacturing, more and more composite materials are used in engineering applications. However, the strength of composite material is still facing with challenges where extensive material interfaces exist. Cracks may initiate from, propagate along or terminate at these interfaces[1, 2, 3, 4]. This study investigates a typical case where crack is terminating at the material interface, to provide an effective tool for numerical modelling. Finite element method (FEM) is popular in engineering to calculate the stress intensity factors (SIFs) to estimate the state of cracked structure. However, extensive mesh refinement around crack tip is required to ensure the accuracy of simulation if conventional elements are used[5]. Bouhala *et al.*[6] studied the inclined interface crack problem by using the extended finite element method (XFEM), the stress and displacement enrichment functions were determined by considering plane elasticity solution based on Airy functions. Natarajan *et al.* applied the extended scaled boundary finite element method (XSBFEM) to the inclined interface crack problem, and both SIFs and T stresses were solved numerically without any necessary of path independent integral[7]. Nasri *et al.* investigated the behaviour of the crack terminating at the zinc/steel bimaterial interface with different orientations by using the commercial FE software package ABAQUS[8]. Recently, cohesive crack model was applied to the modeling of the bimaterial interface cracks. Adams modelled the crack which was perpendicular to the bimaterial interface and found that the critical value for crack propagation depended upon the maximum stress of the cohesive crack model, as well as the Dundures parameters[9]. Mehidi *et al.* investigated the crack normal to the metal/alumina interface by determining the J integral and the plastic zone at the crack tip using the three-dimensional finite element methods[10].

In this study, a new symplectic analytical singular element (SASE) is developed for numerical study of inclined interface crack. By using the proposed method, mesh refinement around the crack tip is not required, and hence the solving efficiency is improved. The SIFs can be calculated directly without any post-processing. Verification of the proposed method by comparing with the benchmark has indicated that the developed method is very accurate and efficient.

2. Theoretical study

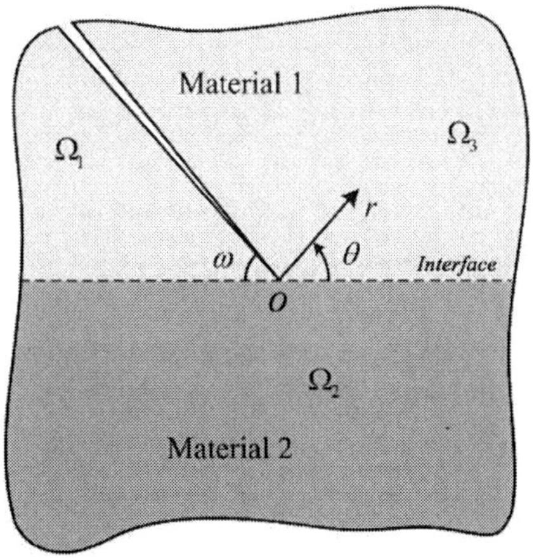

Figure 1, The configuration of inclined crack terminating at bimaterial interface

Considering an inclined interface crack terminating at the bimaterial interface as shown in Figure 1, the origin of the polar coordinate system is located at the crack tip. In the figure, $\theta=\pi-\omega$ is used to denote the crack orientation while $\pi\geq\omega\geq 0$. The discussed domain around the crack tip is divided into three regions by the crack and the material interface. The subscript "i" ($i=1,2,3$) is employed to represent each region. The Young's modulus and Poisson's ratio of the material in the ith region are E_i, ν_i. We have $E_1=E_3$ and $\nu_1=\nu_3$ since the regions #1 and #3 are occupied by the material 1. The fundamental equations can be derived from the follow variational principle:

$$\delta\sum_{i=1}^{3}\int_{\theta_i}^{\theta_{i+1}}\int_0^{\infty}\left\{\sigma_{r,i}\frac{\partial u_{r,i}}{\partial r}+\frac{\sigma_{\theta,i}}{r}\left(\frac{\partial u_{\theta,i}}{\partial\theta}+u_{r,i}\right)+\tau_{r\theta,i}\left(\frac{\partial u_{\theta,i}}{\partial r}-\frac{u_{\theta,i}}{r}+\frac{1}{r}\frac{\partial u_{r,i}}{\partial\theta}\right)\right.$$
$$\left.-\frac{1}{2E_i}\left[\sigma_{\theta,i}^2+\sigma_{r,i}^2-2\nu_i\sigma_{r,i}\sigma_{\theta,i}+2\left(1+\nu_i\right)\tau_{r\theta,i}^2\right]\right\}r\mathrm{d}r\mathrm{d}\theta=0 \tag{1}$$

where $\theta_1=-\pi-\omega$, $\theta_2=-\pi$, $\theta_3=0$ and $\theta_4=\pi-\omega$. Assume that the two materials are perfectly bounded and the crack surfaces are free of external traction, so the compatibility condition as well as the boundary condition can be obtained easily. In the following discussion, the subscript "i" ($i=1,2,3$) is omitted in the equations except where it may cause confusion. In order to solve this problem, the symplectic approach is used. By introducing the generalized coordinate

$$\xi=\ln r \tag{2}$$

and the generalized stress components

$$S_r=r\sigma_r,\quad S_\theta=r\sigma_\theta,\quad S_{r\theta}=r\tau_{r\theta} \tag{3}$$

the variational principle expressed in Eq.(1) can be further simplified into

$$\delta\sum_{i=1}^{3}\int_{\theta_i}^{\theta_{i+1}}\int_{0}^{\infty}\left\{S_{r,i}\frac{\partial u_{r,i}}{\partial\xi}+v_iS_{r,i}\left(u_{r,i}+\frac{\partial u_{\theta,i}}{\partial\theta}\right)+S_{r\theta,i}\left(\frac{\partial u_{\theta,i}}{\partial\xi}+\frac{\partial u_{r,i}}{\partial\theta}-u_{\theta,i}\right)\right.$$
$$\left.\frac{E_i}{2}\left(u_{r,i}+\frac{\partial u_{\theta,i}}{\partial\theta}\right)^2-\frac{1}{2E_i}\left[(1-v_i^2)S_{r,i}^2+2(1+v_i)S_{r\theta,i}^2\right]\right\}\mathrm{d}\xi\mathrm{d}\theta=0 \tag{4}$$

Making variation with respect to the independent variables gives the symplectic dual equation,

$$\frac{\partial \boldsymbol{Z}}{\partial\xi}=\boldsymbol{HZ} \tag{5}$$

where $\boldsymbol{Z}=[u_r,u_\theta,S_r,S_{r\theta}]^{\mathrm{T}}$ is the vector of displacements and generalized stresses. The Hamiltonian operator matrix $\boldsymbol{H}$ is presented as follows,

$$\boldsymbol{H}=\begin{bmatrix} -v & -v\frac{\partial}{\partial\theta} & \frac{1-v^2}{E} & 0 \\ -\frac{\partial}{\partial\theta} & 1 & 0 & \frac{2(1+v)}{E} \\ E & E\frac{\partial}{\partial\theta} & v & -\frac{\partial}{\partial\theta} \\ -E\frac{\partial}{\partial\theta} & -E\frac{\partial^2}{\partial^2\theta} & -v\frac{\partial}{\partial\theta} & -1 \end{bmatrix} \tag{6}$$

The symplectic equation can be solved by using the method of separation of variables, let

$$\boldsymbol{Z}(\xi,\theta)=e^{\xi\mu}\boldsymbol{\psi}(\theta) \tag{7}$$

where μ is the symplectic eigenvalue and $\boldsymbol{\psi}(\theta)=[\psi_u \quad \psi_v \quad \psi_r \quad \psi_{r\theta}]^{\mathrm{T}}$ is the corresponding symplectic eigenvector. ψ_u, ψ_v, ψ_r and $\psi_{r\theta}$ are variables which are, respectively, separated from the original variables u_r, u_θ, S_r and $S_{r\theta}$. The eigenvalue and the corresponding eigenvector can be solved according to the standard solving procedure of symplectic approach which has been discussed in Refs.[11, 12]. It is found that $\mu^{(1)}=\mu^{(2)}=0$ and generally $\mu^{(3)}$ is the first eigenvalue which brings stress singularity. After obtaining the eigen solution, the solution of the original problem can be expressed in the form of

$$\boldsymbol{Z}=\sum_{i=1}^{\infty}a^{(i)}e^{\xi\mu^{(i)}}\boldsymbol{\psi}^{(i)} \tag{8}$$

where $a^{(i)}$ is the unknown eigen expanding coefficient.

3. Symplectic analytical singular element (SASE)

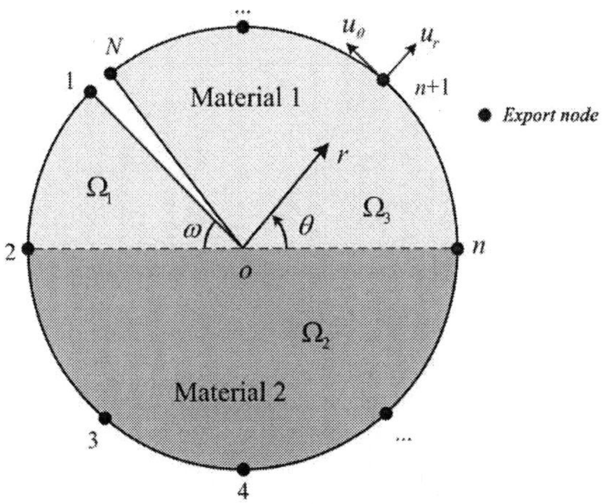

Figure 2, The geometric configuration of the SASE

Solving SIFs by using the conventional finite elements is still unsatisfactory even though refined mesh around the crack tip is used. The eigen solution obtained above should be used in the numerical method. In this study, the eigen solution is used to define a novel SASE which occupies the area around the crack tip while the other area of the structure is meshed by using conventional elements. The radius of the SASE is denoted by ρ. The node indexes are arranged from 1 to N. Assume that the unknown fields of the SASE is expressed by Eq.(8) with the first $2N$ expanding terms. Substituting Eq.(8) back into the variational principle (4) gives the deformation energy of the SASE which is specified by

$$\boldsymbol{\Pi} = \boldsymbol{a}^{\mathrm{T}}\boldsymbol{R}\boldsymbol{a} \tag{9}$$

Meanwhile, the explicit form of $\boldsymbol{R}$ can be obtained. Noted that the nodal displacements of the SASE can be expressed as

$$\boldsymbol{d} = [u_r^{(1)}, u_\theta^{(1)}, \ldots u_\theta^{(2N)}]^{\mathrm{T}} = \boldsymbol{T}_u\boldsymbol{a} \tag{10}$$

where $\boldsymbol{d} = [u_r^{(1)}, u_\theta^{(1)}, \ldots u_\theta^{(2N)}]^{\mathrm{T}}$ is the vector of the nodal displacement and $\boldsymbol{a} = [a^{(1)}, a^{(2)}, \ldots a^{(2N)}]^{\mathrm{T}}$. So the deformation energy of the SASE is transformed into

$$\boldsymbol{\Pi} = \boldsymbol{d}^{\mathrm{T}}\boldsymbol{K}\boldsymbol{d}, \text{ where } \boldsymbol{K} = \boldsymbol{T}_u^{-\mathrm{T}}\boldsymbol{R}\boldsymbol{T}_u^{-1} \tag{11}$$

It is clear that $\boldsymbol{K}$ is the stiffness matrix of the SASE. Assemble the stiffness matrices of the SASE and conventional elements, the FE system can be solved directly. Then, the expanding coefficients can be solved directly by Eq.(10).

The definition of the SIFs for inclined bimaterial interface crack in Ref.[5] is employed in this study, and the relationship between the eigen expanding coefficient and the SIF can be determined and specified as follows

$$K_{\mathrm{I}} = 2\sqrt{2\pi}\,\mathrm{Re}[a^{(3)}(L)^{-\mathrm{i}\cdot\mathrm{Im}(\mu^{(3)})}\psi_\theta^{(3)}(-\omega)],\quad K_{\mathrm{II}} = 2\sqrt{2\pi}\,\mathrm{Re}[a^{(3)}(L)^{-\mathrm{i}\cdot\mathrm{Im}(\mu^{(3)})}\psi_{r\theta}^{(3)}(-\omega)] \tag{12}$$

So, once the expanding coefficients are solved, the SIFs can be solved directly without any post-processing.

4. Verification

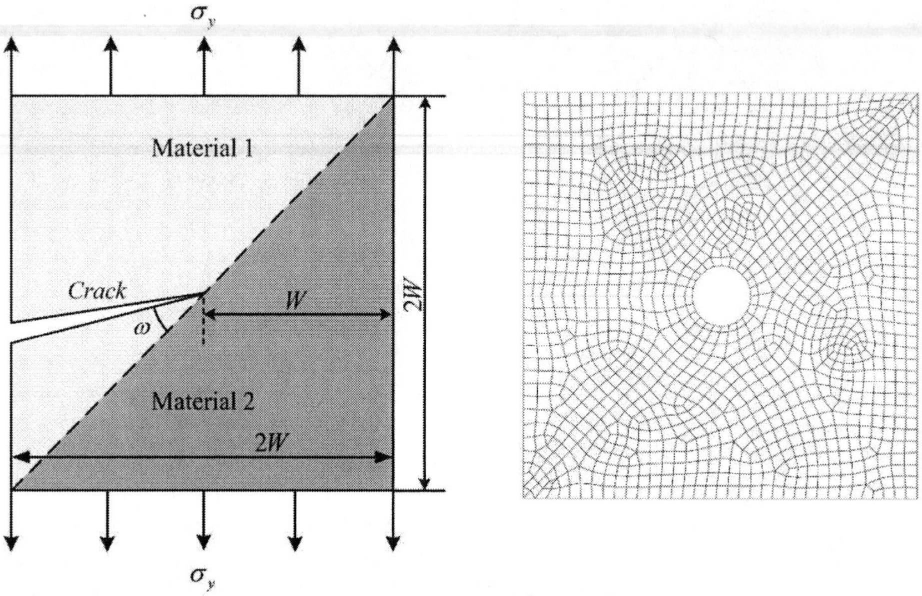

Figure 3, The configuration of a bimaterial plate containing an interior crack terminating at the interface and the FE mesh

Consider the $4mm \times 4mm$ bimaterial square plate containing a crack terminating at the interface as shown in Figure 3, the material properties are $E_1 = 73GPa$, $\nu_1 = 0.17$, $E_2 = 206GPa$ and $\nu_1 = 0.3$. The tensile loading is $\sigma_y = 1MPa$. The FE mesh is also shown in Figure 3. The numerical results are listed in Table 1. The results obtained in Ref.[5] by using extrapolation method on the FE results with extremely refined mesh are listed as a reference. It is shown that the present results agree very well with the reference results. Moreover, the computing cost of the present method is much more cheaper because mesh refinement is not required.

Table. 1 Numerical results of the non-dimensional SIFs of the inclined bimaterial interface crack

	$\omega = 15^\circ$	$\omega = 30^\circ$	$\omega = 45^\circ$	$\omega = 45^\circ$ [5]
μ_s	0.5303±0.0746i	0.5648±0.0667i	0.5996±0.0448i	0.5996±0.0448i
K_I	6.3740	7.5944	8.9794	8.9890
K_II	2.7036	2.6679	2.2099	2.0978

5. Conclusion

The inclined interface crack problem in composite has been investigated systematically in this study. The eigen solution for the inclined interface crack were derived by using the symplectic approach. Based on the derived eigen solution, a new Symplectic Analytical Singular Element (SASE) has been developed to solve the SIFs. The comparison between the present results and the benchmarks has indicated that the developed model is very accurate in capturing the stress state around the inclined interface crack and mesh refinement around crack tip is not required.

6. Acknowledgments

The supports of the National Natural Science Foundation of China (No.11502045, and No. 11372065).

7. References

[1] Ma P, Su RKL, Feng WJ. Fracture analysis of an electrically conductive interface crack with a contact zone in a magnetoelectroelastic bimaterial system. International Journal of Solids and Structures, 2015, 53: 48-57.

[2] Sharma K, Bui TQ, Zhang CZ, Bhargava RR. Analysis of a subinterface crack in piezoelectric bimaterials with the extended finite element method. Engineering Fracture Mechanics, 2013, 104: 114-139.

[3] Hemanth D, Aradhya KSS, Murthy TSR, Govinda Raju N. Strain energy release rates for an interface crack in orthotropic media--a finite element investigation. Engineering Fracture Mechanics, 2005, 72: 759-772.

[4] Agrawal A, Karlsson AM. Obtaining mode mixity for a bimaterial interface crack using the virtual crack closure technique. International Journal of Fracture, 2006, 141: 75-98.

[5]. Chang J, Xu JQ. The singular stress field and stress intensity factors of a crack terminating at a bimaterial interface. International Journal of Mechanical Sciences, 2007, 49: 888-897.

[6]. Bouhala L, Shao Q, Koutsawa Y, Younes A, Nunez P, Makradi A, Belouettar S. An XFEM crack-tip enrichment for a crack terminating at a bi-material interface. Engineering Fracture Mechanics, 2013, 102: 51-64.

[7]. Natarajan S, Song CM, Belouettar S. Numerical evaluation of stress intensity factors and T-stress for interfacial cracks and cracks terminating at the interface without asymptotic enrichment. Computer Methods in Applied Mechanics and Engineering, 2014, 279: 86-112.

[8]. Nasri K, Abbadi M, Zenasni M, Azari Z. Numerical and experimental study of crack behaviour at the zinc/TRIP steel 800 interface. Computational Materials Science, 2014, 82: 172-177.

[9]. Adams GG. Critical value of the generalized stress intensity factor for a crack perpendicular to an interface. Proc. R. Soc. A. The Royal Society, 2015, 471: 20150571.

[10]. Mehidi A, Kaddouri K, Belhouari M, Amiri A, Bachir Bouiadjra B. Three-dimensional finite element analysis of a crack normal to and terminating at a bi-material interface. Journal of the Brazilian Society of Mechanical Sciences and Engineering, 2015, 37: 1785-1792.

[11] Hu XF, Yao WA. A novel singular finite element on mixed-mode bimaterial interfacial cracks with arbitrary crack surface tractions. International Journal of Fracture, 2011, 172: 41-52.

[12] Yao WA, Hu XF. A novel singular finite element of mixed-mode crack problems with arbitrary crack tractions. Mechanics Research Communications, 2011, 38: 170-175.

IOP Conf. Series: Journal of Physics: Conf. Series **842** (2017) 012081 doi:10.1088/1742-6596/842/1/012081

Acoustic emission based damage localization in composites structures using Bayesian identification

A Kundu[1], M J Eaton[1], S Al-Jumali[2], S Sikdar[3] and R Pullin[1]

[1] Cardiff School of Engineering, Cardiff University, The Parade, Queen's Building, Cardiff, CF24 3AA, UK.
[2] Materials department, College of Engineering, University of Basrah, Basrah, Iraq.
[3] Institute of Fluid-Flow Machinery, Polish Academy of Sciences, 14, Fiszera Street, Gdansk 80-231, Poland.

E-mail: KunduA2@cardiff.ac.uk

Abstract. Acoustic emission based damage detection in composite structures is based on detection of ultra high frequency packets of acoustic waves emitted from damage sources (such as fibre breakage, fatigue fracture, amongst others) with a network of distributed sensors. This non-destructive monitoring scheme requires solving an inverse problem where the measured signals are linked back to the location of the source. This in turn enables rapid deployment of mitigative measures. The presence of significant amount of uncertainty associated with the operating conditions and measurements makes the problem of damage identification quite challenging. The uncertainties stem from the fact that the measured signals are affected by the irregular geometries, manufacturing imprecision, imperfect boundary conditions, existing damages/structural degradation, amongst others. This work aims to tackle these uncertainties within a framework of automated probabilistic damage detection. The method trains a probabilistic model of the parametrized input and output model of the acoustic emission system with experimental data to give probabilistic descriptors of damage locations. A response surface modelling the acoustic emission as a function of parametrized damage signals collected from sensors would be calibrated with a training dataset using Bayesian inference. This is used to deduce damage locations in the online monitoring phase. During online monitoring, the spatially correlated time data is utilized in conjunction with the calibrated acoustic emissions model to infer the probabilistic description of the acoustic emission source within a hierarchical Bayesian inference framework. The methodology is tested on a composite structure consisting of carbon fibre panel with stiffeners and damage source behaviour has been experimentally simulated using standard H-N sources. The methodology presented in this study would be applicable in the current form to structural damage detection under varying operational loads and would be investigated in future studies.

1. Introduction

Maintenance costs for high value assets, such as aircraft and wind turbines, make a significant contribution to operating costs and the successful implementation of structural health monitoring (SHM) systems has the potential reduce these costs by a considerable amount (up to 30M p.a., Airbus sources). Some SHM technologies that have shown good potential in lab scale demonstrations have struggled to translate this performance in to industrial applications because the increseasd structural complexity, noise levels and vatiation in operating conditions leads to increasing uncertainty and lower accuracy. The acoustic emission (AE) technique is one such approach that has been shown to perform poorly in industrial environments [1]. Despite this poor performance the AE technique has a number of desirable attributes that make it appealling for SHM applications, in particular: it is able to monitor in real-time (making it suitable for in-service monitoring), it is able to globally monitor a structure using a distributed array of sensors and it is possible to both locate and characterise damage. This list of desirable attributes means that addressing the poor performance observed in industrial environments is seen as a worthwhile activity.

Acoustic emission is the phenomenon whereby small amounts of elastic energy are released by changes whithin a material or structure. These changes result from mechanical mechanisms such as crack growth and plastic deformation as well as friction and rubbing (which are often considered noise sources). The energy is released from it point of origin (often referred to as the source) and propagates out in all directions as an elastic stress wave, akin to an ultrasond wave. As with ultrasound waves the elastic stress waves produce minute surface displacements when they arrive at or travel along a surface. These surface displacements can then be converted to a voltage response using a piezoelectric transducer. These transient signals are analysed in real time by most modern AE systems but can also be digitised and stored to allow further signal processing and analysis. If a distributed array of sensors is used to detect the energy released from a single source, or energy release, then triangulation techniques can be utilised to predict the location of the source (i.e. the damage location), in a similar fashion to determining the epicentre of an earthquake in seismology. Figure 1 presents a schematic of the AE measurement process

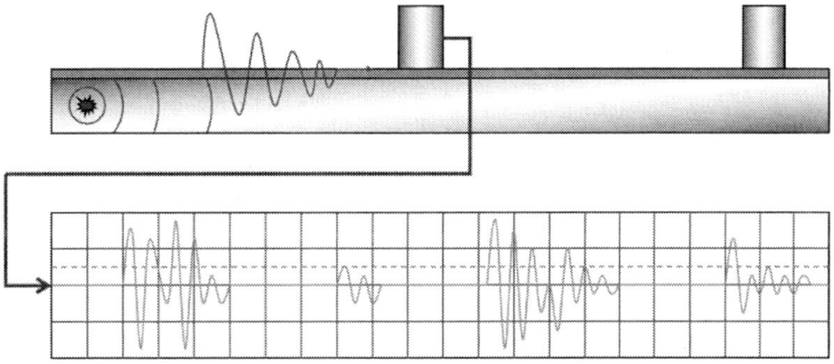

Figure 1. Schematic of AE measurement process.

and it is clear that any changes in the material and/or geometry will affect the resultant recorded signal. Accounting for the boundary uncertainty explicitly in the mathematical model of the physical process is complex and associated with significant computational overhead [2]. Thus accurate damage location using AE in complex structures presents significant challenge. The loss in accuracy observed in complex structures occurs for two specific reasons: firstly a failure to accurately determine the signal arrival time and secondly an overly simplified representation of the AE wave propagation path and velocity in the location calculations [3, 4].

This study aims to overcome the above limitations using a novel approach which trains and calibrates the damage model using a compound correlation metric between the signals recorded with the distributed sensory network. When only using the arrival time of the signal at the individual sensors (and the difference in arrival time) as the basis of training the model, a large portion of the the signal data is not being utilized fully. Additionally error is incurred in the adhoc definitions of signal threshold values which are used to calculate arrival time. Moreover, it is strongly based on the assumption of the existence of a travel path between the damage and the sensor location (this can be considered as an underlying regularization) which might often not be the case. The proposed methodology, in contrast, constructs a compressed representation of the full signal characteristics using a projected correlation metric (as discussed in section 3) which utilizes the full signal characteristics to infer the source of the incoming waves. The gaussian process based surrogate regression approach explicitly accounts for the uncertainty in lack of training data and/or the error incurred using the process described in section 4.

2. Acoustic emission source location

The traditional approach to determination of the arrival time of an AE signal, and that used in all commercially available systems, is a simple threshold approach whereby a user defined threshold level is set and data is recorded when the transducer voltage response exeeds this level with the first threshold crossing time taken as the signals arrival time. It is, however, possible for signal to be present prior to the first threshold crossing and attenuation of signal amplitude can mean variation in signal arrival time determination. A wide range of approaches have been developed in an attempt to improve the arrival time determination compared with the traditional threshold crossing technique. A range of frequency based techniques including filtering [5], cross-correlation [?] and wavelet transforms [6, 7] have been investigated, however, statistical approaches based on 6th order statistical moments [8] and the Akaike Infomation Criteria (AIC) [9, 10] have been shown to be more reliable. The AIC approach in particular has been demonstrates to be very robust across a range of materials and structures [11].

Standard AE location algorithms assume a single, constant, wave propagation speed, however, in composite materials the wave speed is seen to vary with propagation direction and is dependent on fibre orientations within the layup used. Several researchers have tried to address this challenge and some success has been by extending the traditional time of arrival optimization scheme to include a variable wave speed dependent on propagation direction [12, 13]. Ciampa and Meo [14] adopted a novel approach whereby closely spaced sensor pairs were used to reduce the number of unknown propagation velocities in a set of simultaneous non-linear equations that describe the source position. An iterative Newton appraoch was adopted to solve the unknowns in the equations and therefore yield an estimate of the source position without prior knowledge of the wave speeds in the material. Despite these advances achieved in anisotropic monolithic materials, none of the approaches are capable of accounting for geometric complixities such as access holes, curvatures and thickness changes that commonly occur in industrial structures. Alternatively a mapping approach has been proposed and rigorously validated [15, 16, 11, 17, 18] in which a structure is mapped using artificial AE sources (such as a H-N source [19, 20]) to derive an emprical relationship between the known source position and the resultant arrival times at an array of sensors. This relationship can then be used to determine the source origin for a set of measured arrival times. The empirical nature of the approach inherantly accounts for all material and structural complexity such as anisotropy and geometric features.

The previously discussed approaches are deterministic and do not consider the uncertainty in the measurements and calculations performed. This is particularly relevant to their application to industrial environments where uncertainty is seen to increase, with varying operating conditions such as temperature affecting wave propagation and therefore reducing reliability. To account for the uncertainty that can be expereinced in an industrial environment researchers have begun to adopt probabilistic approaches. Schumacher et al [21] developed an approach based on Bayesian statistics for AE source location in a reinforced concrete beams that accounts for uncertainties and errors that exist within the measurement and calculation process. A simplified model for the concrete beam was developed, in which the mean of the wave slowness, the standard deviation of the wave slowness, the event time and the standard deviation of the observed arrival times are represented as prior probability density functions (PDF). The initial PDF of each parameter was then refined using experimental data collected from H-N sources at known positions on the beam surface. The refined model could then be used to predict the

IOP Conf. Series: Journal of Physics: Conf. Series **842** (2017) 012081 doi:10.1088/1742-6596/842/1/012081

most likely position of any subsequent AE sources. The approach reduced the mean error of 22 arbitrarily located H-N sources from 40mm down to 30mm. Further work by Zarate et al [22] developed a Bayesian framework based on a ray tracing model of AE wave propagation in liquid filled storage tanks. The approach allowed structure bourne and water bourne wave paths to be considered. Using a Markov Chain Monte Carlo (MCMC) method to sample the posterior distribution of the source position in x and y coordinates the most probably source position could be determined. Both of these probabilistic approaches are limited to homogeneous materials and simple geometries, i.e. direct and uninterrupted wave paths.

This main aim to the present work is to combine the Bayesian probabilistic mapping of the source of AE to the correlation characteristics of the signal collected at the distributed sensor network. This would not only capture the essential information pertaining to the phase difference and attenuation of elastic waves travelling different distances over the surface of the composite structures to reach individual sensors but also include the complex effects of boundary reflection within the signal correlation characteristics. The uncertainty due to the measurement noise and experimental errors would be explicitly accounted for in the probabilistic model and conditioned on the training data generated with H-N sources.

3. Important signal characteristics for damage identification

The notion of correlation between the signals observed at different locations in a distributed sensory network is of prime importance in the inference of underlying parameters representative of the physical process. If we consider n_s sensors distributed on the structure under study, the collected signal is represented as

$$\mathbf{X} = \{\mathbf{x}_i : \mathbf{x}_i = [x_i[1], x_i[2], \ldots, x_i[n]]^\intercal \quad \forall i = 1, \ldots, n_s\} \qquad 1$$

where each sample $x_i[j]$ has been collected at sensor i at n discrete time points $j = 1, ldots, n$. An initiation of crack or a growth in crack size is accompanied by a packet of ultrasonic wave which propagates radially outwards from its point of origin along the plane of the composite structure. Each sensor captures signal $x_i[n]$ as show in figure 2. Additionally reflection from

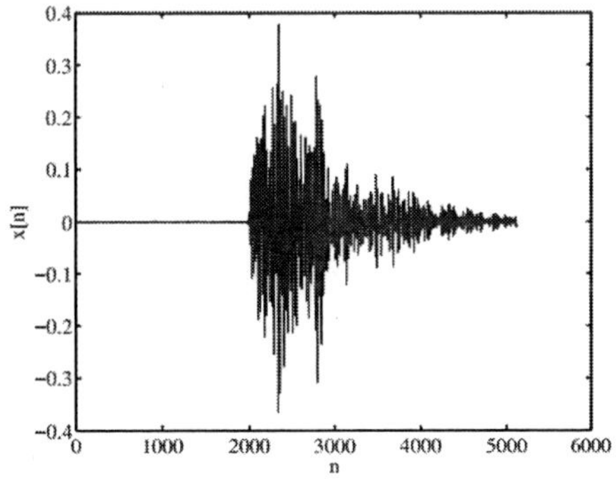

Figure 2. Sample AE signal captured at a particular sensor location.

boundaries and/or other geometrical discontinuities (such as holes) result in a complicated propagation characteristics. It is expected that the accumulated data from the distributed sensory network would contain distinguishing features (statistics) which would map the detected waveforms uniquely to the source location.

IOP Conf. Series: Journal of Physics: Conf. Series **842** (2017) 012081 doi:10.1088/1742-6596/842/1/012081

The correlation matrix $\bar{\mathbf{R}}$ between the signals collected at each of the n_s sensors is defined as

$$\bar{\mathbf{R}}_\tau = \begin{bmatrix} \mathbf{R}_{11}(\tau) & \mathbf{R}_{12}(\tau) & \cdots & \mathbf{R}_{1n_s}(\tau) \\ \mathbf{R}_{21}(\tau) & \mathbf{R}_{22}(\tau) & & \mathbf{R}_{2n_s}(\tau) \\ \vdots & & \ddots & \vdots \\ \mathbf{R}_{n_s1}(\tau) & \mathbf{R}_{n_s2}(\tau) & \cdots & \mathbf{R}_{n_sn_s}(\tau) \end{bmatrix} \quad 2$$

where each element $\mathbf{R}_{ij}(\tau)$ of the correlation matrix is defined in the continuous time domain as

$$R_{ij}(\tau) = \int_{-\infty}^{\infty} x_i(t)x_j(t+\tau)dt \quad 3$$

for real-valued signal components. For L^1 integrable signals, i.e. $\int_{-\infty}^{\infty} |x(t)|\, dt < \infty$ the Fourier transform of the signal exists and is related to the terms in $\bar{\mathbf{R}}$ matrix. In equation 3, when $i = j$, the terms $\mathbf{R}_{ii}$, i.e. the diagonal terms in the matrix $\bar{\mathbf{R}}_\tau$, give the autocorrelation measure of the signals at each sensor.

For discrete time signals the cross-correlation for real-valued signals is defined as

$$\mathbf{R}_{ij}(\tau) = \mathbb{E}\left[(\mathbf{x}_i[n] - \boldsymbol{\mu}_i)(\mathbf{x}_j[n+\tau] - \boldsymbol{\mu}_j)\right] \quad \forall i, j = 1, \ldots, n_s \quad 4$$

The cross-correlation is often normalized by the respective auto-correlation functions at zero lag such that

$$\mathbf{r}_{ij}(\tau) = \frac{\mathbf{R}_{ij}(\tau)}{\sqrt{\mathbf{R}_{ii}(0)\mathbf{R}_{jj}(0)}} \quad \text{where} \quad |\mathbf{r}_{ij}(\tau)| \leq 1 \quad \forall i, j, \tau \quad 5$$

and we denote the normalized version of $\bar{\mathbf{R}}_\tau$ as $\bar{\mathbf{r}}_\tau$ where each element of the matrix in equation 2 is normalized as per equation 5. It is important to note that $\mathbf{R}_{ij}(\tau) = \mathbf{R}_{ji}(-\tau)$ and $\mathbf{R}_{ij}(\tau) = \mathbf{R}_{ij}(-\tau)$, hence $\mathbf{r}_\tau$ is a symmetric matrix for all τ. The normalized correlation coefficient is utilized here to balancethe varying intensity of the AE source. The mapping which links training dataset to the source location assumes that the source intensity is normalized across the entire training dataset which is achieved with the normalized correlation coefficient.

The power spectral density $\mathcal{S}$ which gives the frequency content of the signal is related to the cross-correlation function by the Fourier transform dual, following Wiener-Khintchin theorem [23], as

$$\mathcal{S}_{ij}(\omega) = \int_{-\infty}^{\infty} R_{ij}(\tau)e^{-i\omega\tau}d\tau \quad \text{and} \quad R_{ij}(\tau) = \frac{1}{2\pi}\int_{-\infty}^{\infty} \mathcal{S}_{ij}(\omega)e^{-i\omega\tau}d\omega \quad 6$$

and when considering windowed signals over finite duration T, the and for discrete time signals it is of the form

$$\mathcal{S}_{ij}(\omega) = \lim_{T\to\infty} \mathbb{E}\left[\frac{1}{T}\ \widetilde{\mathbf{x}}_i^*(\omega)\widetilde{\mathbf{x}}_j(\omega)\right] \quad 7$$

where $\widetilde{\mathbf{x}}_i^*(\omega)$ is the Fourier transform of the time signal $\mathbf{x}_i[n]$. The power spectral density $\mathcal{S}_{ii}(\omega) = \lim_{T\to\infty} \mathbb{E}\left[\frac{1}{T}\ |\widetilde{\mathbf{x}}_i^*(\omega)|^2\right]$ is a measure of the signal's energy distribution over the frequency range. Thus each term of the correlation matrix $\bar{\mathbf{R}}_\tau$ has a spectral decomposition in terms of the power spectral density and are sometimes referred to as integrated spectrum.

The correlation matrix captures the essential information regarding the correlation of the signals captured using the distributed sensor network which contains essential information regarding the delay or phase of the arriving signals. The matrix $\bar{\mathbf{R}}_\tau$ is constructed over the interval $-\tau_s \leq \tau \leq \tau_s$ such that each of its elements $\mathbf{R}_{ij}(\tau)$ or the normalized cross-correlation $\mathbf{r}_{ij}(\tau)$ is a vector of dimension $2\tau_s + 1$. Since the correlation matrix is symmetric, only the upper triangular part is considered in a matrix representation

$$\bar{\mathbf{r}}_{\Delta_k,\tau_s} = \left[\mathbf{r}_{\tau_s,i} : \mathbf{r}_{\tau_s,i} \in \mathbb{R}^{2\tau_s+1} \ \text{and} \ -\tau_s \leq \tau \leq \tau_s \quad \forall i \in \mathcal{I}\right] \quad 8$$

where $\mathcal{I}$ is the set of indices associated with the ordering of the elements of the upper triangle of $\mathbf{r}_\tau$ at some instance τ and $\mathbf{r}_{\tau_s,i}$ is a vector which contains the cross-correlation measure over the interval $\pm\tau_s$. It is important to note that the subscript Δ_k of the matrix $\bar{\mathbf{r}}_{\Delta_k,\tau_s}$ $k = 1,\dots,n_t$ denotes normalized correlation data evaluated for the k-th test point out of a total of n_t training sets. Thus the full set of training correlation data is given as

$$\bar{\mathbf{r}}_{\Delta,\tau_s} = \left[\bar{\mathbf{r}}_{\Delta_1,\tau_s}, \bar{\mathbf{r}}_{\Delta_2,\tau_s}, \dots, \bar{\mathbf{r}}_{\Delta_{n_t},\tau_s}\right] \quad \text{where} \quad \bar{\mathbf{r}}_{\Delta_k,\tau_s} \in \mathbb{R}^{2\tau_s+1 \times n_\mathcal{I}}, \quad k = 1,\dots,n_t \qquad 9$$

where $n_\mathcal{I}$ is the cardinality of the set $\mathcal{I}$.

The objective here is to obtain a map between the normalized correlation matrix representing the signal data for each of the n_t training sets and the corresponding coordinates of the acoustic source. However, the computational demands of modelling the large predictor matrix $\bar{\mathbf{r}}_{\Delta,\tau_s}$ is substantial and would make the subsequent real-time model prediction infeasible. Thus we seek an optimal reduced basis which can be used to represent the data for each training set such that

$$\begin{aligned} &\text{find } \mathcal{T} : \mathbb{R}^{2\tau_s+1 \times n_\mathcal{I}n_t} \rightarrow \mathbb{R}^{n_r \times n_\mathcal{I}n_t} \text{ where } \mathcal{T}(\bar{\mathbf{r}}_{\Delta_i,\tau_s}) = \boldsymbol{\alpha}_{i,\tau_s} \;\; \forall i = 1,\dots,n_\mathcal{I}n_t \\ &\text{such that } \Phi_{\Delta,\tau_s} = \underset{\phi_{\Delta_j,\tau_s},\forall j}{\arg\inf} \; \|\Phi_{\Delta,\tau_s}[\boldsymbol{\alpha}_{1,\tau_s},\dots,\boldsymbol{\alpha}_{n_\mathcal{I}n_t,\tau_s}] - \bar{\mathbf{r}}_{\Delta_i,\tau_s}\|_F \end{aligned} \qquad 10$$

where $n_r << 2\tau_s + 1$ and $\boldsymbol{\alpha}_{i,\tau_s} \in \mathbb{R}^{n_r}$. Thus the transformation $\mathcal{T}$ expresses the normalized correlation terms using an optimal basis on which the solution would be projected and an additional orthonormality constraint of $\Phi^\intercal_{\Delta,\tau_s}\Phi_{\Delta,\tau_s} = \mathbf{I}$ can be imposed on the basis. This optimal basis can be informed by the left singular eigenvectors of the training correlation data contained in $\bar{\mathbf{r}}_{\Delta,\tau_s}$ or the principal eigenvectors associated with $\bar{\mathbf{r}}^\intercal_{\Delta,\tau_s}\bar{\mathbf{r}}_{\Delta,\tau_s}$. The compressed correlation data is be expressed in terms of this optimal basis as

$$\underline{\boldsymbol{\alpha}}_{\tau_s} = \{\underline{\boldsymbol{\alpha}}_{1,\tau_s},\dots,\underline{\boldsymbol{\alpha}}_{nt,\tau_s}\} = \Phi_{\Delta,\tau_s} \quad \text{where} \quad \bar{\mathbf{r}}^\intercal_{\Delta,\tau_s} \quad \text{where} \quad \underline{\boldsymbol{\alpha}}_{\tau_s} \in \mathbb{R}^{n_r \times n_\mathcal{I}n_t} \qquad 11$$

The matrix $\underline{\boldsymbol{\alpha}}_{\tau_s}$ consists of a collection of $n_\mathcal{I}n_t$ vectors each of dimension n_r, thus for each of the n_t training points we get a set of $n_\mathcal{I}$ vectors which contains the compressed information.

Next we introduce the linear map $vec(\cdot) : \mathbb{R}^{n\times k} \rightarrow \mathbb{R}^{nk}$ which transforms a collection of k vectors each of dimension n into a single vectors of dimension nk (where each of the k vectors are stacked end to end). The matrix $\bar{\mathbf{r}}_{\Delta,\tau_s}$ is transformed as

$$\underline{\boldsymbol{\alpha}}_v = \left\{vec(\underline{\boldsymbol{\alpha}}_{1,\tau_s}),\dots,vec(\underline{\boldsymbol{\alpha}}_{nt,\tau_s})\right\} \quad \text{where} \quad \underline{\boldsymbol{\alpha}}_v \in \mathbb{R}^{n_rn_\mathcal{I}\times n_t} \qquad 12$$

such that each of the $n_rn_\mathcal{I}$ vector in $\underline{\boldsymbol{\alpha}}_v$ corresponds to the compressed data collected at the distributed sensor network on the structural system.

The problem then defined defined in terms of finding the map $\mathcal{G} : \boldsymbol{\alpha}_{v,xy} \rightarrow r_{xy}$ where r_{xy} is a vector of the Euclidean distance of AE source from each sensor in the distributed sensory network and the subscript (xy) denotes the Cartesian coordinates of the AE soruce. The problem is posed as the following multivariate least square regression problem

$$\boldsymbol{\gamma} = \underset{\boldsymbol{\gamma}\in\mathbb{R}^{n_rn_\mathcal{I}}}{\arg\inf} \; \|\boldsymbol{\alpha}_{v,xy}\boldsymbol{\gamma} - r_{xy}\|_2 \qquad 13$$

However, for each training run, there is an uncertainty around the measured signal. We construct a scalar error indicator following from equation 13 as

$$\varepsilon_r = \|\hat{r}_{xy} - r_{xy}\| \quad \text{where} \quad \hat{r}_{xy} = \boldsymbol{\alpha}_{v,xy}\boldsymbol{\gamma} \qquad 14$$

where $\hat{r}_{xy}$ is the approximate identified location of the acoustic source associated with the training data where the known true source is at r_xy distance from the sensors. Equation 14 gives the error associated with the regression map $\mathcal{G} : \boldsymbol{\alpha}_{v,xy} \rightarrow r_{xy}$. Thus the error indicator ε_r for the training dataset is mapped on the spatial domain using a Gaussian process surrogate, discussed in the following section, based on which the probabilistic prediction of the source location is performed based on the observed signal data at the distributed sensory network.

4. Bayesian emulation of the error surface

In the previous section, the error indicator ε_r is a function of (x, y) coordinates of the AE source such that $\varepsilon_r : (x, y) \to \mathbb{R}$. In this section, we introduce a Gaussian process model [24] compute a surrogate response surface of the error indicator over the spatial domain of the structural system.

The coordinates (x, y) of the AE source over the domain $[l_x, l_y]$ serves as the parameter space or the input space over which the error indicator ε_r is trained. Here it is assumed that the observed error output vector $[\varepsilon_{r1}(x, y), \ldots, \varepsilon_{r n_t}(x, y)]$ are realisations of a Gaussian stochastic process with parametrized mean and covariance functions. The model structure is thus expressed as

$$\varepsilon_r(x, y) = \mathbf{h}(x, y)^{\mathsf{T}} \boldsymbol{\beta} + Z(x, y) \tag{15}$$

where $\mathbf{h}(x, y)$ is a vector of known functions and $\boldsymbol{\beta}$ is an unknown hyperparameter to be estimated from the data. The choice of $\mathbf{h}(\boldsymbol{\xi})$ is an active research area [25, 26]. The function $Z(\cdot)$ is a stochastic process with mean zero and covariance function

$$Cov(Z((x, y)), Z(x', y')) = \sigma_z^2 C((x, y), (x', y')) \tag{16}$$

where $C(\cdot, \cdot)$ is a correlation function and σ_z^2 is the process variance, a hyperparameter that can also be estimated from the data. In order to choose a valid positive-definite correlation function, we consider products of one-dimensional correlations[27] with functional forms

$$C((x, y), (x', y')) = \exp\Big\{ -b_i \Big|x - x'\Big|^2 - b_i \Big|y - y'\Big|^2 \Big\} \tag{17}$$

where $b_i > 0$ for all i with $x, x' \in [0, l_x]$ and $y, y' \in [0, l_y]$. The above correlation function is infinitely differentiable which is convenient when Gaussian processes are used to model not only the code output, but also its derivatives [28]. The vector of smoothness hyperparameters $\mathbf{b} = (b_1, b_2)^{\mathsf{T}}$ quantifies the rate at which the output varies over the spatial domain.

Let $\mathcal{D} = \big\{ ((x, y)_i, \varepsilon_r(x, y)_i) \big| i = 1, \ldots, n_t \big\}$ be a set of n_t training runs. Given this observed dataset, Bayes' theorem is used to estimate the hyperparameters as follows

$$\mathbb{P}(\boldsymbol{\beta}, \sigma_{\mathbf{z}}^2, \mathbf{b} | \mathcal{D}) = \frac{\mathbb{P}(\mathcal{D} | \boldsymbol{\beta}, \sigma_{\mathbf{z}}^2, \mathbf{b}) \mathbb{P}(\boldsymbol{\beta}, \sigma_{\mathbf{z}}^2, \mathbf{b})}{\mathbb{P}(\mathcal{D})} \tag{18}$$

where $\mathbb{P}(\boldsymbol{\beta}, \sigma_{\mathbf{z}}^2, \mathbf{b} | \mathcal{D})$ is the posterior probability of the hyperparameters, $\mathbb{P}(\mathcal{D} | \boldsymbol{\beta}, \sigma_{\mathbf{z}}^2, \mathbf{b})$ is the likelihood, $\mathbb{P}(\boldsymbol{\beta}, \sigma_{\mathbf{z}}^2, \mathbf{b})$ is the prior of the hyperparameters, and $\mathbb{P}(\mathcal{D})$ is the marginal likelihood. A detailed derivation of prior-to-posterior analysis along with the hyperparameter estimation is given in [29, 30].

The assumed Gaussian process prior on the code's output implies that the posterior distribution is also a Gaussian process. Once the hyperparameters are estimated, the mean of the posterior distribution approximates the output of ε_r^* at any point (x^*, y^*) on the physical domain. The variance of the posterior distribution quantifies the uncertainty that arises from having only a limited number of observations [31]. It can be shown that the posterior distribution is of the form

$$\varepsilon_r(x, y) \big| \varepsilon^{\text{obs}}, \sigma_z^2 \sim \mathcal{N}\Big(m^*(x, y), \sigma_z^2 C^*((x, y), (x', y')) \Big) \tag{19}$$

where the posterior mean and posterior variance are respectively

$$m^*(x, y) = \widehat{\boldsymbol{\beta}} + \mathbf{r}^{\mathsf{T}} \mathbf{C}^{-1} (\mathbf{y} - \mathbf{1} \widehat{\boldsymbol{\beta}}) \tag{20}$$

$$C^*((x, y), (x', y')) = \mathbf{C}((x, y), (x', y')) - \mathbf{r}^{\mathsf{T}}(x, y) \mathbf{C}^{-1} \mathbf{r}(x, y) \tag{21}$$

In the above expressions, $\mathbf{C} \in \mathbb{R}^{n_t \times n_t}$ such that $[\mathbf{C}]_{ij} = C((x_i, y_i), (x_j, y_j)),\ \forall i, j = 1, \dots, n_t$, $\mathbf{r}(x, y) \in \mathbb{R}^{n_t}$ such that $\mathbf{r}(x, y) = \left(C((x, y), (x, y)^{(1)}), \dots, C((x, y), (x, y)^{(n_t)})\right)^{\mathsf{T}}$, and $\mathbf{1} \in \mathbb{R}^n$ such that $\mathbf{1} = (1, \dots, 1)^{\mathsf{T}}$.

Finally, it has been shown that if σ_z^2 is integrated out of the posterior distribution, then

$$\frac{\varepsilon(x, y) - m^*(x, y)}{\widehat{\sigma}_z \sqrt{C^*}} \sim t_{n-1} \tag{22}$$

which is a Student's t-distribution with $n-1$ degrees of freedom. Thus, the error indicator can be evaluated at any (x, y) (location predicted by the multivariate least square regression problem in equation 13) and $\varepsilon(x, y)$ gives the uncertainty in the acoustic source location. Once the error response surface surrogate has been trained, the prediction of the error is straightforward and computationally efficient which allows for real-time identification of damage location in the structure under active operating conditions.

The error surface $\varepsilon(x, y)$ trained above is used in conjunction with the least square estimator using correlation characteristics (presented in section 3) to give a probabilistic prediction of the AE source conditional on the training data. At the prediction stage the reduced correlation matrix, constructed using the optimal basis Φ_{Δ,τ_s}, is mapped to the AE source and the associated error is evaluated from the trained Gaussian process surrogate in equation 22. Thus a robust probabilistic estimate of the AE source location is obtained using the correlation characteristics of the signal collected from the distributed sensor network. The above procedure is summarized

Algorithm 1 Source localization with Bayesian identification using sensor cross-correlation

Input: Sensor signal data: $\mathbf{X} = \{\mathbf{x}_i : \mathbf{x}_i = [x_i[1], x_i[2], \dots, x_i[n]]^{\mathsf{T}} \quad \forall i = 1, \dots, n_s\}$
Input: Equidistant test grid on plate surface: $\{\vec{r}_i : i = 1, \dots, n_d\}$
Output: Source location $\vec{r}_s$ and the uncertainty around it $\varepsilon(\vec{r}_s) \sim \pi(\theta|\vec{r}_s)$

```
for k = 1 to n_d do
    Accumulate sensor data for grid point k: {x_1, ..., x_{n_s}} where x_i ∈ R^{n_t} ∀i
    for i = 1 to n_s do
        for j = i to n_s do
            Calculate normalized cross-correlation (r̄_{Δ_k,τ_s})_{ij} ∈ R^{2τ_s+1}
        end for
    end for
    Assemble the upper triangular cross-correlation matrix r̄_{Δ_k,τ_s} ∈ R^{2τ_s+1 × n_I}
end for
Compressed representation of r̄_{Δ_k,τ_s} as T(r̄_{Δ_i,τ_s}) = α_{i,τ_s} using optimal basis Φ_{Δ,τ_s}.
Solve γ = arg inf_γ ||α_{v,τ_s} γ − r⃗||_2 which maps α_{i,τ_s} to r⃗
Approximation error ε_{r_i} = α_{v,τ_s} γ − r⃗,  ∀i ∈ n_d for each test point on grid
A prior for Gaussian process surrogate: ε_{r_i} ~ N(μ, Z(r⃗)|θ) with hyper-parameter set θ.
Obtain posterior: ε_r(r⃗)|ε^obs, θ* ~ N(μ*_θ(r⃗), Z*_θ(r⃗)) conditioned on training data ε^obs
globals  Φ_{Δ,τ_s},  μ*,  Z*,  θ*
Compute r̂_{Δ,τ_s}, the normalized cross-correlation with test data
Evaluate: α_{v,τ_s} = r̂^T_{Δ,τ_s} Φ_{Δ,τ_s} and r⃗_0 = α_{v,τ_s} γ
Prediction: r⃗_c = r⃗_0 + N(μ*_θ(r⃗_0), Z*_θ(r⃗_0))
```

in Algorithm 1 whereby the steps to obtain posterior prediction from a probabilistic surrogate has been listed.

5. Experimental setup

In order to develop and validate the above approach for damage localization, data was collected from a stiffened carbon fibre composite panel, representative of an aerospace structure.

5.1. Sample details

The manufactured stiffened panel is presented in Figure figure 3. The stiffeners were purchased from Easy Composites Ltd. (Staffordshire, UK) they consist of a 90° L-shaped cross-section with laminate thickness of 3mm and cross-section dimensions of 25×25mm and a length of 600mm. The stiffeners are made from 2×2 twill high strength carbon fibres in an epxoy matrix

(a) Stiffener side (b) Grid side

Figure 3. Manufactured panel a) from stiffener side showing sensor positions and b) from skin side showing grid used for data collection.

with fibres aligned in the 0° and 90° directions. The skin was manufactured from 12 plies of Cytec MTM28/T800HB/200/42% 2×2 twill weave carbon fibre composite material with a $(0)_{12}$ layup and was cured in an autoclave in line with the manufacturers recommended cure cycle. Following curing the thickness was 2.85mm and the skin panel was cut to 550×600mm using a water cooled diamond tipped cutting wheel. The skin and stiffeners were lightly abraded and degreased in preparation for bonding using permabond ET5429 adhesive. The final overall panel dimensions were 550×600mm with the stiffeners running vertically. Two aluminium dumby ribs (representative of attachment to a wing rib for a composite wing skin) were attached at 1/4 and 3/4 hight of the panel by drilling and bolting using 12 M4 bolts for each, as seen in 3(a). The top and bottom edges of the panel were potted into 20mm deep aluminium frames using Airtech TMR2001 high temperature laminating resin to allow application of a compressive load (not considered in this paper). A 500×500mm grid with 50mm resolution was applied to the skin side of the panel (3(b)) and was used to aid the collection of training data.

5.2. Panel instrumentation

The panel was instrumented with eight AE sensors as seen in 3(a), five McWade NS-3303 (300 kHz), the larger gold coloured sensors, and three Mistral Group Ltd. Nano30 sensors (300kHz), the smaller silver sensors. Four McWade sensors are arranged in a 275×175mm square with the fifth placed centrally. The Nano30 sensors are arranged in a 75mm spaced triangular array. The sensor outputs are amplified by 40dB using a McWade PA3303 pre-amplifier for the McWade sensors and a Mistral Group Ltd. 2/4/6 (20-1200kHz) pre-amplifier for the Nano30 sensors. A silicon adhesive (Loctite 595) was used to attatch the sensors and provide a suitable acoustic couplant, the adhsive was allowed to cure for 24 hrs before any data acquisition or testing was undertaken and the correct coupling of the sensors was assessed using a H-N source [19, 20]. The AE data was recorded with a Mistras Group Ltd. PCI-2 acquisition system using a 45dB threshold level. The detected signals were sampled at 5MHz for a duration of 1000μs: 600μs after the threshold crossing point and with a pre-trigger of 400μs. The time when the threshold crossing occurs for each test point has been recorded and shown in figure 4 and has been discussed in section 6.

5.3. Data acquisition
AE data was collected from the manufactured composite panel using a H-N source [19, 20] to excite artificial AE waves. The H-N source is recognised as a standard reference source (ASTM E976) for AE testing and requires the fracture of a 0.5mm diameter 2H pencil lead agains the sample surface at an angle of 30°. This is facilitated using a propelling pencil fitted with a plastic rocker the rocker is placed on the sample and the pencil rotated until the lead contacts the surface and then fractures. This results in minute elastic deformation of the surface under the tip of the pencil lead and when the lead fractures the elastic energy stored in the surface is rapidly released and excites a broadband elastic stress wave. This is highly representative of the rapid release of elastic energy that occurs when cracks and fractures grow in materials and hence is why it has been adopted as a standardised reference source for AE testing. The H-N source commonly excites a larger amplitdue signal than a real fracture, however, the source mechanism is still representative and it has been shown to be a suitable artificial source for training a system for the detection and analysis of AE signals from real fracture events as seen in the Delta T Mapping techniques discussed above.

The data used in this work were collected using H-N sources performed at the nodes of the grid applied to the front of the stiffened panel. Ten H-N sources were conducted at each of the grid nodes within the 500 $\times$ 500mm grid shown in 3(b) and for each H-N source eight AE signals were recorded and stored (one from each sensor).

6. Results and discussion

A sample of the collected sensor data is shown in figure 4 where each red dot signify the time (x-axis) of arrival of the wave at a sensor (8 sensors denoted by Ch 1" to "Ch 8") and its maximum amplitude (y-axis). A "hit" is recorded when the signal received at a particular sensor exceeds a preset threshold value. Figure 4 shows the *hits* recorded simultaneously at all the eight sensor locations. All the hits between the vertical blue lines on each channel indicate a test performed with the H-N source at a grid point (figure 3(b)) on the panel. Around 10 tests have been performed at each grid point due to which there are approximately 10 hits (red dots) between two consecutive blue lines. The red dots are aligned in time across all channels except for the small time difference associated with the waves travelling different distances from the source to reach each sensor. All the tests represented in figure 4 have been performed along the 11 test grid points at $x = 0$ (the bottommost line in the grid). This figure essentially represents only the time of arrival for the signal at the individual sensors, but the correlation measure between the sensors is shown in the next figure.

Figure 5 shows the elements of the normalized correlation matrix $\bar{\mathbf{r}}_\tau$ (described in equation 5) where the dimension of $\bar{\mathbf{r}}_\tau$ for a particular τ is 8×8 since we have 8 sensors distributed over the physical domain of the composite structure. Each sub-window in the figure denotes the correlation $\mathbf{r}_{ij}(\tau)$ over $-100 \leq \tau \leq 100$. The figure shows that the autocorrelation function is symmetric, i.e. $\mathbf{r}_{ij}(\tau) = \mathbf{r}_{ij}(-\tau)$ and $\mathbf{r}_{ij}(\tau) = \mathbf{r}_{ji}(tau)$. The correlation function is treated as an energy indicator of the waveforms recorded by each sensor as well as containing essential information about the phase of the waves arriving at each sensor.

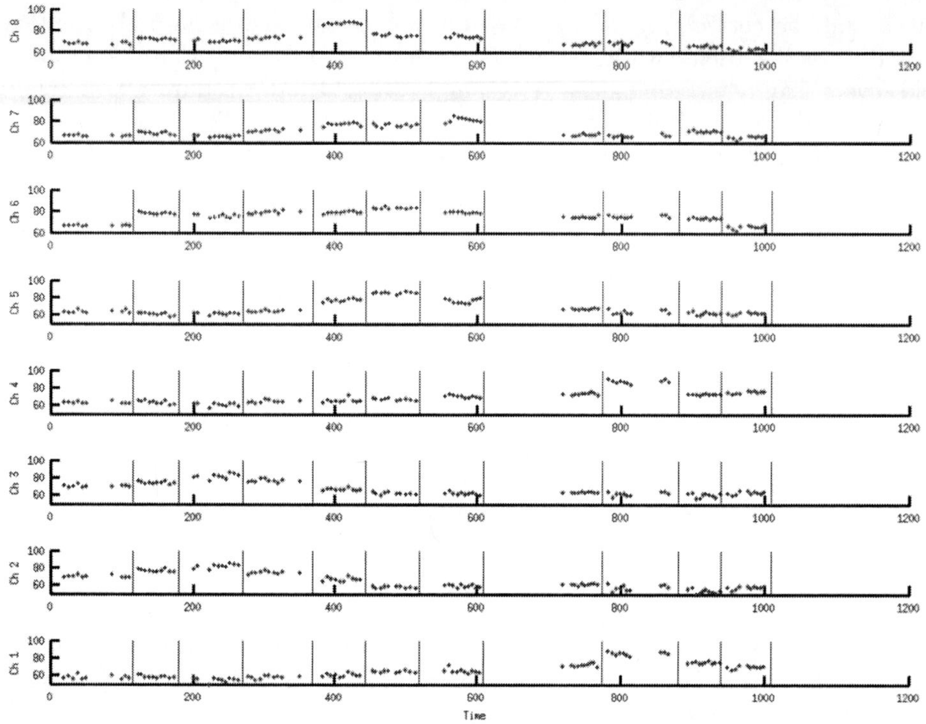

Figure 4. Normalized correlation coefficients of the signal collected at the distributed sensor network (only the first 4 out of the total of 8 sensor signals shown) over discrete time steps $-100 \leq \tau_s \leq 100$

The optimal basis on which the correlation matrix is projected to obtain a compressed representation of the information collected at the sensors is shown in figure 6 A total of 30 basis functions have been used to approximate the terms of the correlation matrix. The basis functions are orthonormalized and have been calculated using the left eigenvectors of the correlation matrix using the singular value decomposition. The rapidly decaying singular value spectrum of the ensures that a good approximation is obtained with the reduced basis. The correlation matrix projected on these 30 optimal bases is termed as the reduced correlation matrix.

The accuracy of the identified AE source using the least square estimate is shown in figure 8. The black dots indicate the training points on the 10×10 grid of the composite panel while the red dots indicate the approximate identified acoustic source locations obtained using the least square mapping of the reduced correlation matrix to the vector of the distance of the acoustic source from the sensor network (as discussed in equation 13). The error $\varepsilon_r(x, y)$ is given by the distance of the identified source locations (red dots) to the known position of AE (black squares).

This error indicator $\varepsilon_r(x, y)$ varies as per the accuracy of the fit and a Gaussian process surrogate is used to build a probabilistic error surface over the spatial domain as discussed in section 4. Figure 8 gives the mean and standard deviation of the posterior error surface conditional on the observations. This gives a distribution of the accuracy of the least sqaures

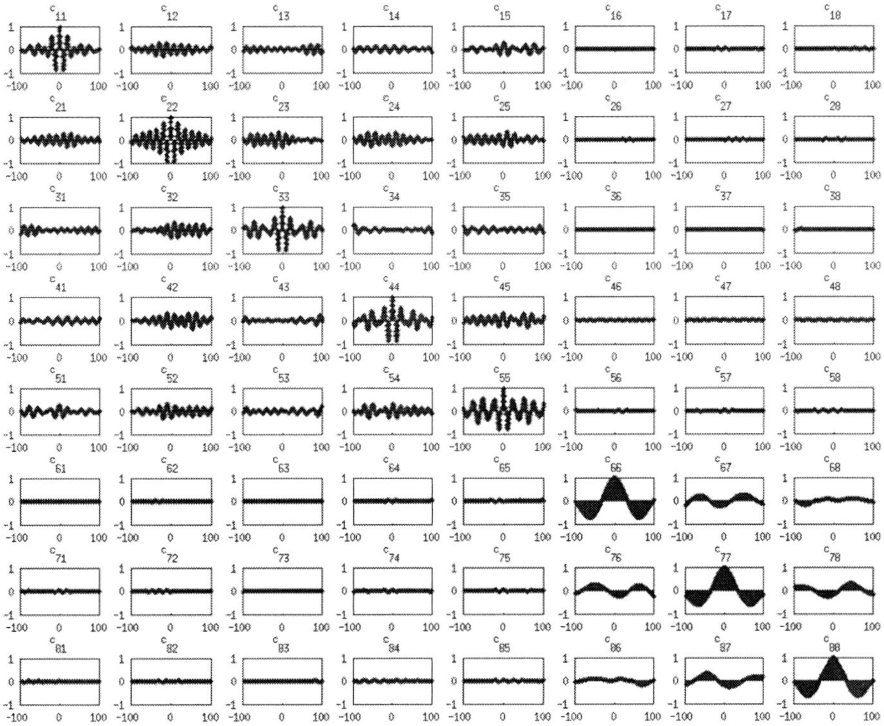

Figure 5. Normalized correlation coefficients of the signal collected at the distributed sensor network over discrete time steps $-100 \leq \tau_s \leq 100$

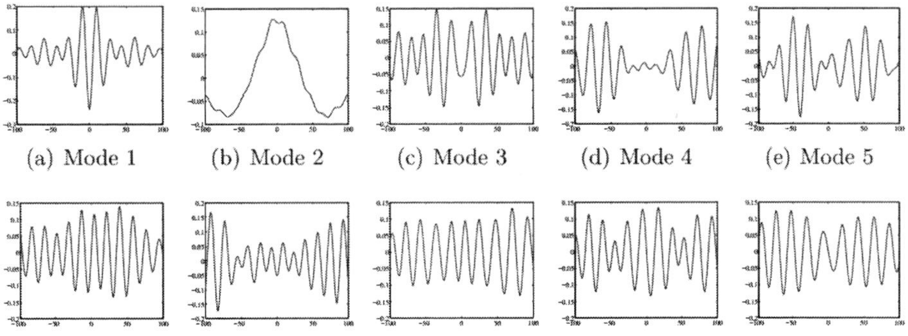

(a) Mode 1 (b) Mode 2 (c) Mode 3 (d) Mode 4 (e) Mode 5

(f) Mode 6 (g) Mode 7 (h) Mode 8 (i) Mode 9 (j) Mode 10

Figure 6. The first 10 modes of the reduced basis for the normalized correlation matrix

estimate as a function of the spatial coordinates. It can be verified that the mean error is almost zero at those training locations where the accuracy of the least square method is maximum. The locations where there are no training points show a high value of mean error and correspondingly high standard deviation of error which is an expected behavior.

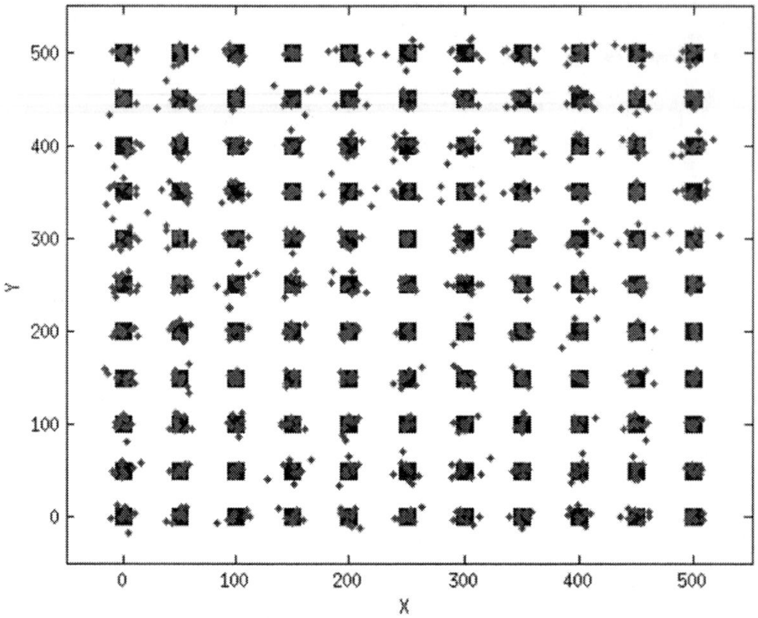

Figure 7. The accuracy of fit of the training dataset based on the observed correlation matrix over the physical domain of the composite panel. The square blocks shown in black constitute the actual test grid while the red dots are least square estimates of the AE source.

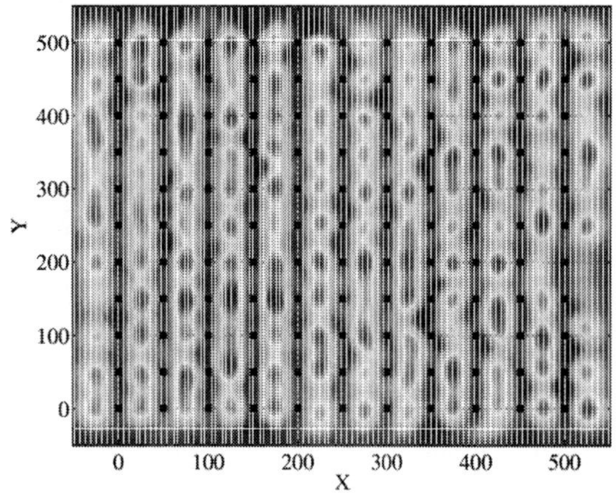

(a) Mean of the error surface

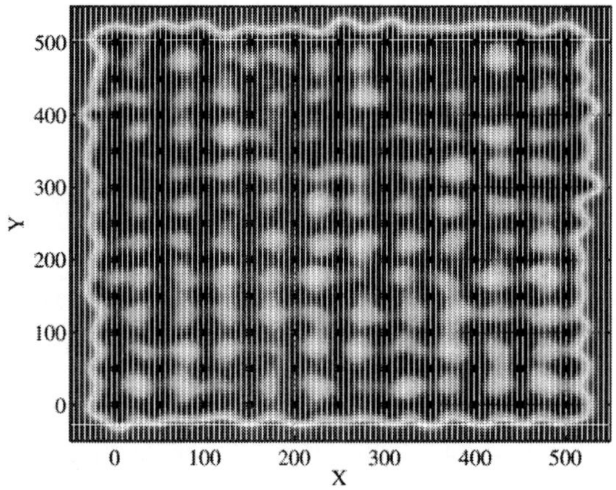

(b) Standard deviation of the error surface

Figure 8. The mean an standard deviation of the error response surface over the domain of the plate. The colormap of (a) varies between $0 - 45$mm while for the standard deviation is between $0 - 20$mm

Figure 9 shows the probability distribution associated with an identified test point. The actual location of the acoustic source is shown as the green dot (although this information has not been used to train the model). The assumed gaussian process surrogate predicts the probability distribution of the identified acoustic source location conditional on the training data

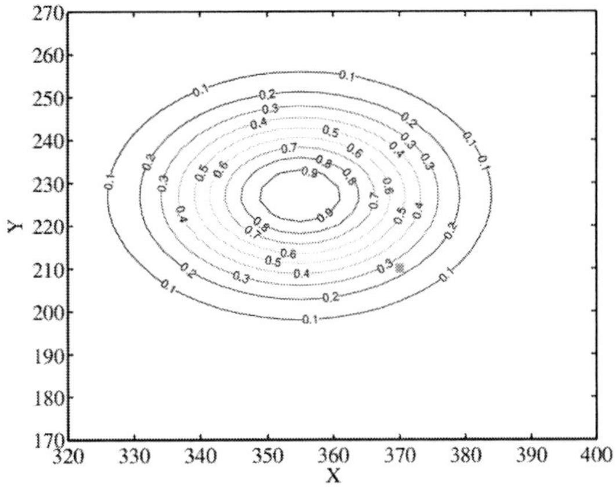

Figure 9. The probability distribution associated with a source identification. The true source is shown as a green dot.

and though the mean true location does not match the exact source location (which is expected due to experimental errors and measurement noise), the true source location is contained within the 95% probability envelope as shown in figure 9.

Thus effectiveness of the methodology is demonstrated in terms of the accuracy of identifying the source of the AE. The cost of producing the gaussian process surrogate is incurred mostly in an offline training stage while in the online identification stage the effective location of the AE source is derived efficiently from the surrogate. This allows for implementation of the technique in real-time identification procedure within the performance constraints of the signal processing platforms.

7. Conclusion and future work

The study demonstrates the effectiveness of the proposed correlation matrix based identification of AE sources in applications of non-destructive testing. The effectiveness of the method stems from using direct evaluations of normalized autocorrelation functions which effectively reduces the impact of signal noise and captures the essential correlation information between the signals recorded with the distributed sensor network. A reduced set of basis vectors which compress the correlation information reduces the memory and computational overhead associated with the method. A Gaussian process surrogate is fitted on the error surface over the spatial domain to explicitly consider the error associated with deterministic evaluators. The results demonstrate the effective of the method in predicting the source location with test data on the same panel.

The study promises a number of interesting future investigations to extend and improve the proposed methodology to address additional challenges. Some of these are

- The application of the proposed methodology to identify the generation cracks on the composite panel under fatigue loading cycles.
- The portability of the trained surrogate i.e. the ability of the trained model to predict AE sources in nominally similar test panels.
- A hierarchical probabilistic model of the AE source mapped to the correlation matrix, where identification stages can be compartmentalized to correspond to various levels of refinement. This would involve the use of a combination of probabilistic classification and regression methodology in an automated sequence to improve the accuracy and minimize the computational overhead associated with the method.

IOP Conf. Series: Journal of Physics: Conf. Series **842** (2017) 012081 doi:10.1088/1742-6596/842/1/012081

Subsequent work would also focus on using physics based model of ultrasonic wave propagation in composite panels which would provide a means of identifying the nature of damage induced in the structure which has not been addressed in this study. This would allow us to go beyond the black-box input-output mapping techniques which is expected to significantly improve the performance of the identification algorithm and provide additional information pertaining to the degradation of health of operational structures.

Acknowledgments
AK acknowledges the support from Sêr Cymru NRN, Wales through the award of the Industrial Fellowship.

References

[1] Holford K, Eaton M, Hensman J, Pullin R, Evans S, Dervilis N and Worden K 2017 **In press**
[2] Kundu A, Adhikari S and Friswell M I 2014 *International Journal for Numerical Methods in Engineering* **100** 183–221
[3] Miller R, Carlos M, Findlay R, Godinez-Azcuaga V, Rhodes M, Shu F and Wang W 2005 *Acoustic Emission Source Location* (ASNT) pp 121–146 3rd ed
[4] Eaton M, Pullin R and Holford K 2012 *Journal of Mechanical Engineering Science, Proceedings of the Institution of Mechanical Engineers Part C* **226** 2141–2153
[5] Miller R, Anastasopoulos A, Carlos M, Demeski R, Vallen H and Walker J 2005 *Acoustic Emission Signal Processing* (ASNT) pp 147–180 3rd ed
[6] Aljets D, Chong A, Wilcox S and Holford K 2010 *Journal of Acoustic Emission* **28** 85–98
[7] Hamstad M, O'Gallagher A and Gary J 2002 *Journal of Acoustic Emission* **20** 62–82
[8] Lokajicek T and Klima K 2006 *Measurement Science and Technology* **17** 2461–2466
[9] Akaike H 1974 *Annals of the Institute of Statistical Mathematics* **26** 363–387
[10] Kurz J H, Grosse C and Reinhardt H W 2005 *Ultrasonics* **43** 538–546
[11] Pearson M, Eaton M, Featherston C, Pullin R and Holford K 2016 1–18
[12] Kundu T, Das S and Jata K 2007 *Journal of the Acoustical Society of America* **122** 2057–2066
[13] Kundu T, Das S, Martin S and Jata K 2008 *Ultrasonics* **48** 193–201
[14] Ciampa F and Meo M 2010 *Composite Part A* **41** 1777–1786
[15] Baxter M, Pullin R, Holford K and Evans S 2007 *Mechanical Systems and Signal Processing* **21** 1512–1520
[16] Eaton M J, Pullin R and Holford K 2012 *Composites Part A* **43** 856–863
[17] Al-Jumaili S, Pearson M, Holford K, Eaton M and Pullin R 2016 **72-73** 513–524
[18] Scholey J, Wilcox P, Wisnom M and Friswell M 2009 *Ultrasonics* **49** 538–548
[19] Hsu N N and Breckenridge F R 1981 *Materials Evaluation* **39** 60–68
[20] ASTM 2010 *American Society for Testing and Materials,* **E976**
[21] Schumacher T, Straub D and Higgins C 2012 **331**(19)
[22] Zarate B, Pollock A, Momeni S and Ley O 2014 **24**(1)
[23] Chatfield C 2016 *The Analysis of Time Series: An Introduction, Sixth Edition* (CRC Press)
[24] Kundu A, DiazDelaO F, Adhikari S and Friswell M 2014 *Computer Methods in Applied Mechanics and Engineering* **270** 201 – 219
[25] Oakley J E and O'Hagan A 2004 *Journal of the Royal Statistical Society B* **66** 751–769
[26] Keane A and Nair P 2005 *Computational Approaches for Aerospace Design* (Chichester, UK: John Wiley & Sons)
[27] Sacks J, Welch W, Mitchell T and Wynn H 1989 *Statistical Science* **4** 409–435
[28] O'Hagan A 1992 *Bayesian Statistics 4* (Cambridge, UK: Oxford University Press) chap Some Bayesian numerical analysis, pp 345–363
[29] Haylock R and O'Hagan A 1996 *Bayesian Statistics 5* (Oxford, UK: Oxford University Press) chap On inference for outputs of computationally expensive algorithms with uncertainty on the inputs
[30] Oakley J 2002 *The Statistician* **51** 81–97
[31] Rougier J 2007 *Climatic Change* **81** 247–264

IOP Conf. Series: Journal of Physics: Conf. Series **842** (2017) 012082 doi:10.1088/1742-6596/842/1/012082

A Two-step Approach for Damage Identification in Axially Functionally Graded Beams

Z.R. Lu[1], R.Z. Yao, J.K. Liu, M. Huang

Department of Applied Mechanics, Sun Yat-sen University,
Guangzhou, Guangdong Province, 510006, P.R. China

Email: lvzhr@mail.sysu.edu.cn(Z.R. Lu), huangm7@mail.sysu.edu.cn(M. Huang)

Abstract. This study presents a two-stage approach based on residual force vector and response sensitivity analysis for structural damage identification in axially functionally graded (AFG) beams. The local damage is simulated by a reduction in the elemental Young's modulus of the beam. The residual force vector is used to find the suspicious damaged elements in the beam at first. Then, a hybrid objective function is established and a sensitivity-based model updating method is adopted to identify the perturbation of the stiffness parameter from the measured acceleration responses. Two numerical examples are investigated to illustrate the correctness and efficiency of the proposed method. The effects of measurement noise on the identification results are investigated. Studies in this paper indicate that the proposed method is efficient and robust for identifying damages in the axially functionally graded beams. The advantage of the present approach lies in that only a few number of acceleration measurements and the first several natural frequencies of the beam are needed in the identification.

1. Introduction

A plenty of non-destructive techniques have been developed for health monitoring and structural damage identification in the last few decades. Doebling et al. [1] suggest a comprehensive review of the damage detection methods by testing changes in the dynamic parameters of a structure. Housner et al. [2] provided an extensive summary on the state-of-the-art in control and health monitoring in civil engineering structures. Zou et al. [3] reviewed the development on structural condition monitoring and damage identification for composite structures.

Usually, damage identification requires a finite element model of a structure in conjunction with experimental modal parameters of the structure. There are two categories of methods for damage identification, namely, the frequency domain methods and time domain methods. The frequency domain approaches are mainly based on the changes in the natural frequencies [4], mode shapes and their derivatives [5], measured dynamic flexibility [6] or frequency response function [7]. The natural frequency is easy to be measured with a high level of accuracy, and it is the most common method using dynamic parameter for damage detection. However, problems may occur in the symmetric structures if only natural frequencies are used, since the structural symmetry would lead to non-uniqueness in the solution of damage identification. This problem can be solved by incorporating the mode shapes in the inverse analysis.

The residual force vectors, derived from the natural frequencies and mode shapes of the intact and damaged structure, have been utilized to identify structural damages. Zimmerman and Kaouk [8] proposed a theory based on an original finite element model and a subset of measured eigenvalues and eigenvectors to identify structural damages. Damage locations are determined firstly and the damage extents are assessed by a minimum rank update theory. Li and Smith [9] used residual force vectors directly to solve for damaged element parameters, but again use experimentally determined undamped mode shapes and natural frequencies in the damage identification. Liu and Yang [10] made use of the mode shape matrix and the residual force matrix to determine the number of damaged elements and localized the damage elements by the damage localization matrix. And the damage extents can be easily obtained.

Recently, because of its excellent thermal and mechanical performance, FGM has great promising applications in many fields, such as space plane frames, spacecraft heat shield, etc. An intensive study on the vibration analysis of structures made of FGM has been conducted. Accurate condition assessment of the FGM structures is vital as most of them operate in severe conditions. However, the reported study on damage identification of structures made of FGM was very limited in the literature [11-17].

In this study, a two-step approach based on residual force vector and response sensitivity analysis is proposed to identify structural damage in functionally graded beams. A hybrid sensitivity matrix is constructed and used in the finite element model updating. The advantage of the present method lies in only the first several natural frequencies and a few number of acceleration responses are needed in the identification. A single span simply supported beam and a multi-span beam are studied as two numerical examples to illustrate the correctness and efficiency of the proposed method. Good identified results can be obtained from the short time histories of a few number of measurement points and the first several natural frequencies.

2. Theory

2.1. Finite element model of intact and damaged AFG beam

As shown in Figure 1, a straight uniform AFG beam has length L, width b, thickness h, with a Cartesian coordinate system O-xyz, where x axis is taken along the central axis, the y axis is in the width direction and the z axis is in the depth direction.

In this study, it is assumed that the material properties of beam such as, Young's modulus E, mass density ρ vary continuously along the transversal direction. Both E and ρ are assumed to vary according to the following power-law functions[18]

$$E(x) = (E_L - E_R)(1 - x/L)^{\vartheta} + E_R \tag{1}$$

$$\rho(x) = (\rho_L - \rho_R)(1 - x/L)^{\vartheta} + \rho_R \tag{2}$$

where E_L and E_R are the Young's modulus of the left and the right side of the beam, ρ_L and ρ_R are the mass densities of the left and the right side of the beam, ϑ is the non-negative power-law component which dictates the material variation profile through the axis of the beam.

The equation of motion of the intact AFG beam structure under the external force can be expressed as

$$\mathbf{M}\ddot{\mathbf{d}} + \mathbf{C}\dot{\mathbf{d}} + \mathbf{K}\mathbf{d} = \mathbf{P}(t), \tag{3}$$

where $\mathbf{M}$, $\mathbf{K}$ and $\mathbf{C}$ are the system mass, stiffness and damping matrices respectively, $\ddot{\mathbf{d}}$, $\dot{\mathbf{d}}$ and $\mathbf{d}$ are the acceleration, velocity and displacement vectors of the structure, $\mathbf{P}(t)$ is the vector of external force. The well-known Rayleigh damping model is assumed in the study, i.e., $\mathbf{C} = \eta_1\mathbf{M} + \eta_2\mathbf{K}$, where η_1 and η_2 are constants to be determined from two given damping ratios that corresponding to two unequal modal frequencies of vibration. The forced vibration response of the beam under external force can be calculated using Newmark direct integration method.

When the beam is damaged, it is assumed that the local damage only leads to the change in the stiffness parameter (for instance, Young's modulus E) and the mass property remains unchanged

[19]. Instead of the absolute value of elemental Young's modulus E, its relative perturbation to the initial value E_0 is chosen as dimensionless updating parameter α in this study, it is expressed as

$$E = E_0(1+\alpha), \quad with \quad -1 \le \alpha \le 0 \tag{4}$$

Supposing the ith element is damaged, the elemental stiffness matrix $\mathbf{k}_d^i$ is stated as

$$\mathbf{k}_d^i = \mathbf{k}_0^i(1+\alpha^i) \tag{5}$$

And the global stiffness matrix $\mathbf{K}_d$ for the damaged structure can be expressed as

$$\mathbf{K}_d = \sum_i^{NE} \mathbf{k}_0^i(1+\alpha^i) \tag{6}$$

The equation of motion for free and forced vibration of the damaged beam can be written as

$$\mathbf{M}\ddot{\mathbf{r}}_s + \mathbf{K}_d\mathbf{r}_s = 0 \tag{7}$$

$$\mathbf{M}\ddot{\mathbf{d}} + \mathbf{C}_d\dot{\mathbf{d}} + \mathbf{K}_d\mathbf{d} = \mathbf{P}(t) \tag{8}$$

The natural frequencies and mode shapes of the damaged beam can be calculated from Eq. (9) .

$$(\mathbf{K}_d - \omega_{di}^2\mathbf{M})\mathbf{V}_{di} = 0 \tag{9}$$

where ω_{di} is the ith natural frequency of the damaged beam and $\mathbf{V}_{di}$ is the associated ith modeshape.

2.2. Localization of damage from residual force vector method

The change of structural stiffness matrix $\Delta\mathbf{K}$ introduced by the structural damage can be modeled by a perturbation in the stiffness parameter of the structure in this study. So the global stiffness matrix $\mathbf{K}_\text{d}$ can be expressed as

$$\mathbf{K}_\text{d} = \mathbf{K} - \Delta\mathbf{K}. \tag{10}$$

Substituting equation (10) into equation (9), we have

$$\left(\mathbf{K} - \omega_{dj}^2\mathbf{M}\right)V_{dj} = \Delta\mathbf{K}V_{dj}. \tag{11}$$

The residual force vector (RFV) (Zimmerman and Kaouk [9], Liu and Yang [10]) $\mathbf{f}_j$ from the jth mode of the damaged structure is defined as

$$\mathbf{f}_j = \Delta\mathbf{K}\boldsymbol{\varphi}_{dj} = \left(\mathbf{K} - \omega_{dj}^2\mathbf{M}\right)V_{dj}. \tag{12}$$

The residual force vector $\mathbf{f}_j$ from equation (12) will have nonzero values only in the damaged elements. Furthermore, we can identify the damaged elements according to the relation between the element number and the DOF number. To avoid miss some damage information when a single mode is used, the normalized average value of the first pth RFV is adopted, i.e.

$$\bar{\text{f}} = \sum_{j=1}^{p} \frac{f_j}{\max\left|\text{f}_j\right|} \cdot p^{-1} \tag{13}$$

where p denotes the number of mode used to calculate.

2.3. The objective function for model updating problem

In finite element model updating, an optimization problem is usually set-up in which the differences between the experimental and numerical modal parameters or dynamic responses have to be minimized by modifying some parameters of the analytical model. During the practice, the natural frequencies of the structures can be measured easily and accurately but the mode shapes cannot, especially the higher ones. On the other hand, the acceleration responses of the structure can be measured easily. In this study, we establish the objective function making use of the natural frequency and acceleration measurements. The objective functions are expressed as follows:

$$g1(\alpha)=\frac{1}{2}\sum_{p=1}^{nf}(\hat{\lambda}_p-\lambda_p(\boldsymbol{\alpha}))^{\mathrm{T}}\mathbf{W}_1(\hat{\lambda}_p-\lambda_p(\boldsymbol{\alpha}))\text{, with } \lambda_p=\omega_p^2 \quad (14a)$$

$$g2(\alpha)=\frac{1}{2}\sum_{j=1}^{l}\sum_{i=1}^{nt}(\hat{\mathbf{R}}_{ij}-\mathbf{R}_{ij}(\boldsymbol{\alpha}))^{\mathrm{T}}\mathbf{W}_2(\hat{\mathbf{R}}_{ij}-\mathbf{R}_{ij}(\boldsymbol{\alpha})) \quad (14b)$$

where l is the number of measurement locations, nt is the number of time instances of the measured data, nf is the number of eigenvalue. $\mathbf{R}(\boldsymbol{\alpha})$ is the vector of calculated response of the structure from a known set of $\boldsymbol{\alpha}$ and $\hat{\mathbf{R}}$ is the vector of measured response. $\mathbf{W}_1$ and $\mathbf{W}_2$ are the weighting matrices for the eigenvalues and dynamic responses respectively, which are taken as unit matrices in this study. $\boldsymbol{\alpha}$ is the vector of unknown damage parameters $(\alpha^1,\alpha^2,...,\alpha^{Ne})^{\mathrm{T}}$ to be identified.

2.4. Hybrid sensitivity for model updating

Based on Eq. (14a) and Eq.(14b), the increment vector of the damage parameter in the jth iteration can be calculated from the following equation using the penalty function method [20]

$$\mathbf{S}_\lambda^j\Delta\boldsymbol{\alpha}^j=\Delta\boldsymbol{\lambda}^j \quad (15a)$$

$$\mathbf{S}_R^j\Delta\boldsymbol{\alpha}^j=\Delta\mathbf{R}^j \quad (15b)$$

where

$$\Delta\boldsymbol{\lambda}^j=\hat{\boldsymbol{\lambda}}-\boldsymbol{\lambda}^j(\boldsymbol{\alpha}), \quad (16a)$$

$$\Delta\mathbf{R}^j=\hat{\mathbf{R}}^j-\mathbf{R}^j(\boldsymbol{\alpha}) \quad (16b)$$

is the difference between the measured and calculated responses at the jth iteration. $\mathbf{S}_\lambda^j$ is the sensitivity matrix of natural frequency with respective to damage parameter, which can be calculated using formulas of Fox and Kapoor [21]. $\mathbf{S}_R^j$ is the dynamic response sensitivity matrix, which can be obtained using formulas of Lu and Law [22].

Combining Eq. (15a) and Eq. (15b), we have

$$\bar{\mathbf{S}}^j\Delta\boldsymbol{\alpha}^j=\Delta\mathbf{H}^j \quad (17)$$

where $\bar{\mathbf{S}}^j=\begin{bmatrix}\mathbf{S}_\lambda^j\\ \mathbf{S}_R^j\end{bmatrix}$ is the hybrid sensitivity matrix, $\Delta\mathbf{H}^j=\begin{bmatrix}\Delta\boldsymbol{\lambda}^j\\ \Delta\mathbf{R}^j\end{bmatrix}$

Eq. (12) can be solved by the damped least-squares method (DLS) [23] and the incremental in the damage parameter is identified as

$$\Delta\boldsymbol{\alpha}^j=(\bar{\mathbf{S}}^{j\mathrm{T}}\bar{\mathbf{S}}^j+\gamma\mathbf{I})\bar{\mathbf{S}}^{j\mathrm{T}}\Delta\mathbf{H}^j \quad (18)$$

where γ is the non-negative damping (regularization) coefficient governing the participation of least-squares error in the solution. The solution of Eq. (17) equals to the minimizing the function,

$$\mathbf{J}\left(\Delta\boldsymbol{\alpha}^j,\gamma\right)=\left\|\bar{\mathbf{S}}^j\Delta\boldsymbol{\alpha}^j-\Delta\mathbf{H}^j\right\|^2+\gamma\left\|\Delta\boldsymbol{\alpha}^j\right\|^2 \quad (19)$$

with the second term in Eq. (18) providing bounds to the solution. In this paper, Tikhonov regularization approach is used to obtain the optimal regularization parameter γ [24].

The updated vector of the $(j+1)$th iteration of damage parameter $\boldsymbol{\alpha}^{j+1}$, can be obtained in the next iteration as follows:

$$\boldsymbol{\alpha}^{j+1}=\boldsymbol{\alpha}^j+\Delta\boldsymbol{\alpha}^j. \quad (20)$$

The convergence is recognized as accomplished when the following criteria is met

$$\max(\Delta\boldsymbol{\alpha}^j)\le \mathrm{Tol}. \quad (21)$$

IOP Conf. Series: Journal of Physics: Conf. Series **842** (2017) 012082 doi:10.1088/1742-6596/842/1/012082

In this study, the tolerance Tol is taken as 1×10^{-3}.

Starting with an initial damage parameter vector $\boldsymbol{\alpha}^0$, which is taken as a null vector, the flowchart of iteration is shown in Figure 1.

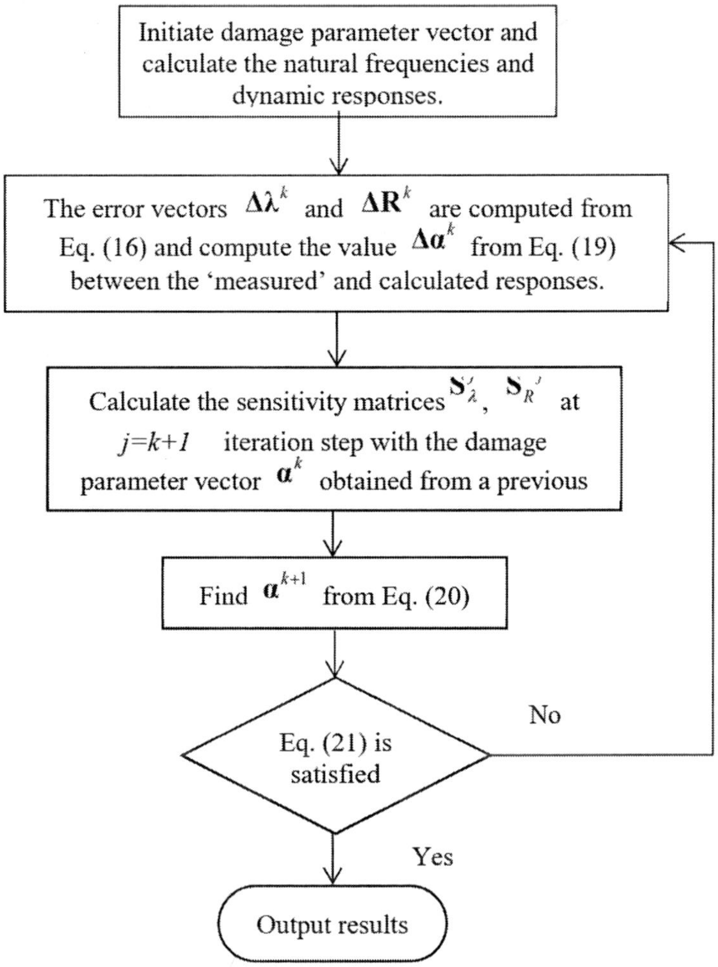

Figure 1. Flowchart of iteration

3. Numerical simulation

3.1. A simply supported AFG beam

As the first the example, an AFG beam made of steel and aluminum is examined using the proposed method as shown Figure. 2. The length of the beam under study is 2 m, the width b 0.1m and height h 0.05m. Table 1 shows the material properties of the beam. The right side of beam is pure steel and left side is pure aluminum. The non-negative power-law ϑ is taken as 1.5 in this study.

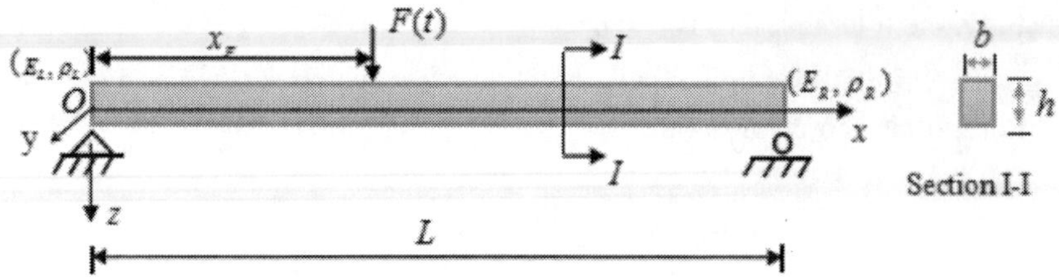

Figure 2. An axially functionally beam under study

Table 1 Properties of beam in the numerical study

Properties	Steel	Aluminum
E	210GPa	69GPa
ρ	$7800\,\text{kg/m}^3$	$2700\,\text{kg/m}^3$

MATLAB software package was employed to build the finite element model of the beam, in which the beam was discretized into twelve 2-node Euler-Bernoulli beam elements. The first five natural frequencies of the intact beam are 37.20, 151.75, 342.69, 609.54, 804.93 and 952.06 Hz, respectively. To obtain the forced vibration response of the beam, an impulsive force is supposed to act at the 7th node of the beam in the global z direction with $\begin{cases} F(t)=10^5(t-0.02)\text{ N} & (0.02\text{s}<t\le 0.04\text{s}) \\ F(t)=10^5(0.06-t)\text{ N} & (0.04\text{s}<t\le 0.06\text{s}) \end{cases}$. In calculating the dynamic response, the time increment is 0.0002 second and the time duration for the response calculation is 2.0 seconds unless otherwise specified. The two damping coefficients used for calculating Rayleigh damping matrix are both assumed to be 0.01. In this numerical example the first 6 natural frequencies and five acceleration measurements locating at nodes 3, 5, 7, 9, 10 are used in damage identification.

In this case, it is assumed that the natural frequency and the acceleration are contaminated with measurement noise. The noise polluted eigenvalues and mode shapes are expressed as

$$\hat{\lambda}_j = \lambda_j(1+\chi_i\varsigma) \tag{22}$$

$$\hat{\phi}_{ij} = \phi_{ij}(1+\chi_i\varsigma|\phi_j|_{\max}) \tag{23}$$

where $\hat{\lambda}_j$ and $\hat{\phi}_{ij}$ are the jth simulated measured noisy eigenvalue and modeshape, λ_j and ϕ_{ij} are the calculated ones, χ_i is random number with a mean equal to zero and a variance equal to one, ς is noise level in percentage. Messina et al. [25] suggested a standard error of $\pm 0.15\%$ as a benchmark figure for natural frequencies in the laboratory with the impulse hammer technique.

Two different noise levels are utilized to study the robustness of the proposed method: 1) low level noise: 0.15% noise in the natural frequencies, 5% noise in the mode shapes and 2% noise in the acceleration responses; 2) high level noise 1% noise in the natural frequencies, 15% noise in the mode shapes and 15% noise in the acceleration responses. Figure. 3 shows the identified results with noisy measurements. One can find that all the four damages have been identified successfully. The largest identification error is only 0.15% at the 3rd element for low noise level. In this case the number of iteration required is 22 and the optimal regularization parameter λ_{opt} is found to be 143.48 at the last iteration. The largest identification error is only 1.73% at the 1st element for high level noise, the number of iteration required is 26 and the optimal regularization parameter λ_{opt} is found to be 443.14 at the last iteration.

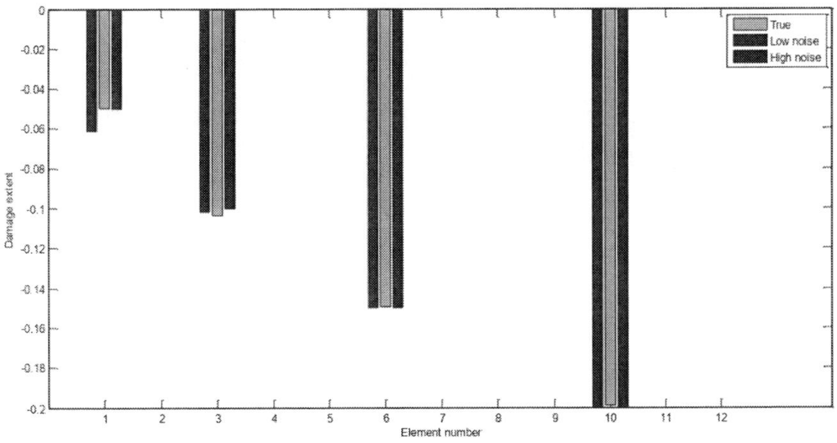

Figure 3. Comparison of damage identification results under different noise level

3.2. *A four span continuous AFG beam*

An AFG beam made of Zirconia(ZrO2)/Titanium(Ti-6Al-4V) is studied as the second numerical example as shown in Figure 4. The length of the beam is 12 m, the width b 0.2m and height h 0.2 m. Table 2 shows the material properties of the beam. The right side of beam is pure ZrO2 and left side is pure Ti-6Al-4V.

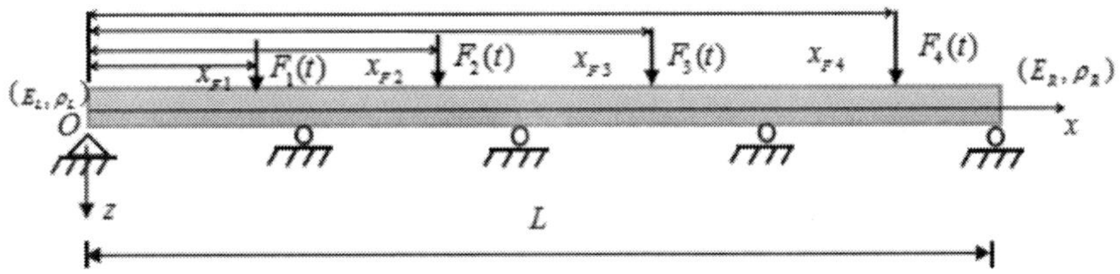

Figure 4. A four span functionally graded beam under study

Table 2. Properties of beam in the numerical study

Properties	ZrO2	Ti-6Al-4V
E	168GPa	105GPa
ρ	3000 kg/m^3	4429 kg/m^3

In the finite element, the beam was discretized into sixty 2-node Euler-Bernoulli beam elements. The first six natural frequencies of the beam are 60.29, 76.11, 99.23, 129.42, 236.44 and 282.64 Hz, respectively. To obtain the forced vibration response of the beam, four impulsive forces are supposed to act at the 8th and 17th 34th, 46th nodes of the beam in the global z direction with the same expression $\begin{cases} F(t)=10^6(t-0.02)\text{ N} & (0.02\text{s}<t\le 0.04\text{s}) \\ F(t)=10^6(0.06-t)\text{ N} & (0.04\text{s}<t\le 0.06\text{s}) \end{cases}$. In computing the dynamic response, the time increment is 0.0005 second and the time duration for the response calculation is 2.0 seconds unless otherwise specified. The two damping coefficients used for calculating Rayleigh damping matrix are both assumed to be 0.01.

In this study case, we try to use the displacement measurements instead of acceleration for damage identification. The last study case is re-examined. The same number of natural frequencies and the displacement measurements at the same nodes as the last case is used. The noise level is assumed to be 0.5% in the natural frequency and 10% in the displacement time histories respectively. After 32 iterations, the identification results converge to the seven assumed damages in the beam as shown in Figure 5. The max identified error is only 1.5% at element 8. The optimal regularization parameter λ_{opt} is found to be 14.73 at the last iteration. This case further illustrates the correctness and robustness of the proposed method.

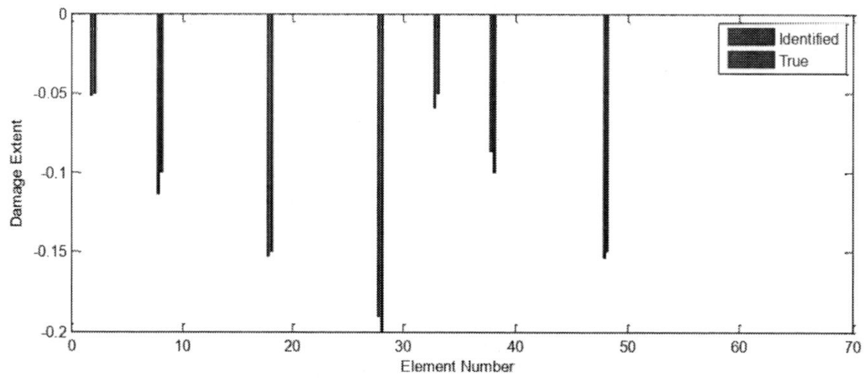

Figure 5. Damage identification using displacement measurements

4. Conclusion

In this study, a hybrid sensitivity matrix is constructed for finite element model updating. It is found to be effective and robust for structural damage identification in AFG beam structures. The proposed method only needs several natural frequencies and a few number of dynamic responses of the structure. Both acceleration and displacement measurements under impulsive or sinusoidal excitation force can be used in the identification. Damage identification results are obtained iteratively with the penalty function method with Tikhonov regularization from the measured natural frequencies and structural dynamic responses. Two numerical examples studied in this contribution show the proposed method effective in identifying multiple damages in the AFG beams. Study shows that the proposed method own excellent robustness to the artificial measurement noise.

5. Acknowledgements

The work described in this paper is supported by the National Natural Science Foundation of China (11572356), Doctoral Program Foundation of Ministry of Education of China (20130171110039), Guangdong Province Natural Science Foundation (2015A030313126), and the Guangdong Province Science and Technology Program (2016A020223006). Such financial aids are gratefully acknowledged.

References

[1] S.W. Doebling, C.R. Farrar, M.B. Prime, A review of damage identification methods that examine changes in dynamic properties, Shock Vibr. Dig. 30(1998) 91-105.

[2] G.W. Housner, L.A. Bergman, T.K. Caughey, et al., Structural control: Past, present, and future, J. Eng. Mech. 123(1997) 897-971.

[3] T. Zou, L. Tong, G.P. Steve, Vibration based model-dependent damage (delamination) identification and health monitoring for composite structures-a review, J. Sound Vib. 230(2000) 357-378.

[4] P. Cawley, R.D. Adams, The location of defects in structures from measurements of natural frequencies, J. Strain Anal. 14(1979) 49-57.

[5] A. K. Pandey, M. Biswas, M. M. Samman, Damage detection from change in curvature mode shapes, J. Sound Vib. 145(1991) 321-332.

[6] A. K. Pandey, M. Biswas, Damage detection in structures using change in flexibility, J. Sound Vib. 169(1994) 3-17.
[7] X. Liu, N.A.J. Lieven, P.J. Escamilla-Ambrosio, Frequency response function shape-based methods for structural damage localization, Mech. Syst. Signal Proc., 23(2009) 1243-1259.
[8] D.C. Zimmerman and M. Kaouk, Structural damage detection using a minimum rank update theory, J. Vib. Acoust. 116(2) (1994) 222-231.
[9] C. Li and W. Smith, Hybrid approach for damage detection in flexible structures, J. Guid. Control Dynam. 18(3) (1995) 419-425.
[10] J.K. Liu and Q.W.Yang, A new structural damage identification method, J. Sound Vib. 297(3)(2006)694-703.
[11] A. Khatir, M. Tehami, S. Khatir and M. Abdel Wahab, Multiple damage detection and localization in beam- like and complex structures using co-ordinate modal assurance criterion combined with firefly and genetic algorithms Journal of Vibroengineering 18(8) (2016) 5063-5073
[12] Y.L. Zhou, NMM. Maia, R. Sampaio and MA. Wahab, Structural damage detection using transmissibility together with hierarchical clustering analysis and similarity measure Structural health monitoring, 2016, doi: 10.1177/1475921716680849
[13] Y.L. Zhou, N. Maia and M. Abdel Wahab, Damage detection using transmissibility compressed by principal component analysis enhanced with distance measure Journal of Vibration and Control, 2016, doi: 10.1177/1077546316674544
[14] Y.L. Zhou and M. Abdel Wahab, Rapid early damage detection using transmissibility with distance measure analysis under unknown excitation in long-term health monitoring Journal of Vibroengineering 18(7) (2016) 4491-4499
[15] S. Khatir, I. Belaidi, R. Serra, M. Abdel Wahab and T. Khatir, Numerical study for single and multiple damage detection and localization in beam-like structures using BAT algorithm Journal of Vibroengineering 18(1)(2016) 202-213
[16] S. Khatir, I. Belaidi, R. Serra, M. Abdel Wahab and T. Khatir, Damage detection and localization in composite beam structures based on vibration analysis Mechanika 21(6)(2015) 472-479
[17] G.R. Gillich, Z.I. Praisach, M. Abdel Wahab, N. Gillich, IC. Mituletu and C. Nitescu, Free vibration of a perfectly clamped-free beam with stepwise eccentric distributed masses Shock and Vibration 2016(Article ID 2086274) 10 pages
[18] K. Wakashima, T. Hirano, M. Niino. Space applications of advanced structural materials. ESA 1990. SP 303-97.
[19] A. Messina, E. J. Williams, T. Contursi, Structural damage detection by a sensitivity and statistical-based method, J. Sound Vib. 216(1998) 791-808.
[20] M.I. Friswell, J.E. Mottershead, Finite Element Model Updating in Structural Dynamics, Dordrecht: Kluwer Academic Publisher, 1995.
[21] R. Fox, M. Kapoor, Rate of change of eigenvalues and eigenvectors, AIAA J. 6(1968) 2426-2429.
[22] Z.R. Lu and S.S. Law, Features of dynamic response sensitivity and its application in damage detection, J. Sound Vib. 303(1-2) (2007) 305-329.
[23] A.M. Tikhonov, On the solution of ill-posed problems and the method of regularization, Soviet Math. 4(1963)1035-1038.
[24] P.C. Hansen, D.P.O. Leary, The use of the L-curve in the regularisation of discrete ill-pose problem. SIAM J. Sci. Comput. 14(1993) 1487-503.
[25] A. Messina, I.A. Jones, E.J. Williams, Damage detection and localization using natural frequency changes, in: Proceedings of Conference on Identification in Engineering Systems, Swansea, U.K., 1996, pp. 67-76.

New damage-sensitive feature for structures with bolted joints

W F Luo[1] and L Yu[1, 2 *]

[1] School of Mechanics and Construction Engineering, Jinan University, Guangzhou 510632, China.
[2] MOE Key Lab of Disaster Forecast and Control in Engineering, Jinan University, Guangzhou 510632, China.

[*] E-mail: lyu1997@163.com

Abstract. Compared with riveted and welded joints, bolted joints have advantages of easily dismantled, high load carrying and load-transferring capacity. However, bolted joints are also weaker components of assembled structures. Structural damage detection (SDD) on bolted joints is much required in the field of structural health monitoring (SHM). A new SDD method is proposed for damage identification of structures with bolted joints based on residual error of AR model in time series analysis. Firstly, a new data standardization process is defined to maintain the information of damage location. Then, a new structural damage feature sensitive to structural damage is developed based on the standard deviation of AR model residual errors. To verify the proposed method, a bolted joint structure is designed and fabricated in laboratory, connection damages of structures are simulated by loosening the bolted joints. The acceleration responses of structures with bolted joints under healthy and damage cases are acquired. Finally, the SDD is performed by traditional DSF and the new DSF. The illustrated results show that the proposed method is a hybrid tool for the bolted joint damage detection with the new damage-sensitive feature. In addition, some related issues will be discussed as well.

1. Introduction

The vibration-based SDD technique has become an effective way in SDD [1, 2]. Normally, SDD can be achieved by comparing the structural characteristics extracted from structural reference state and damage state respectively. Zhou et al [3] used the hierarchical clustering analysis and similarity measure to distinguish structural damage state from health state. Most of these methods can be divided into two categories: model based and feature based. The model based methods need to construct the structural dynamic model. Khatir et al [4] used the co-ordinate modal assurance criterion for SDD. Khatir et al [5] used the modal scale factor and natural frequencies for structural damage detection and localization. To the feature based methods, especially for those based on time series analysis, are found to be less complicated and more sensitive to local damage [6].

Most of time series analysis based methods need to construct a time series model that fit for vibration response data and achieves SDD via comparing the extracted features from structural reference (healthy) and damage states. Chen and Yu [7] used the ARMA and GARCH model to analyze the vibration response data. There are kinds of time series models for different type of vibration data. Generally, the AR (Auto-Regressive) model, MA (Moving Average) model and ARMA (Auto-Regressive with Moving Average) model are suitable for modeling stationary time series while ARIMA (Auto-Regressive Integrated Moving Average) model is suitable for non-stationary time series. One of the critical factors to successfully apply time series analysis based methods in SDD is to choose features that are sensitive to structural damage and have good robustness. Chen et al [8] used linear model to analyze the vibration response data from a nonlinear damage structure and extracted the damage features. Nair et al [9] used the first three autoregressive (AR) components to define damage features. Carden and Brownjohn [10] used ARMA coefficients as damage features and adopted classifier to distinguish normal data and damage data. Other researchers used residual errors to construct damage features. Yao and Pakzad [11] defined the Ljung-Box statistic of AR model residual error series as the damage feature. Yu and Lin [12] defined ratio of standard deviation as damage sensitive features. Yu and Zhu [13] used skewness and kurtosis to define a feature for structural non-linear damage diagnosis. Unfortunately, most of time series analysis based methods aim at determining existence of damages but they do not have good effectiveness when considering damage localization or extents, sometimes they cannot effectively locate the damage.

In this study, bolted joints damage detection is achieved by locating the damage region. If the bolted joints are included in the identified damage region, the bolted joint damage can be successfully identified. The acceleration time histories of structures are recorded and fitted by the AR model. A new damage-sensitive feature (DSF) is developed based on the standard deviation of residual errors. In order to verify the validity of the proposed method, a series of experiments are designed and some measured data are analyzed by traditional DSF and new one, respectively. The results will be compared and discussed, and some reasonable conclusions are made finally.

2. Theoretical bbackground

The bolted joint damage will affect dynamic characteristics of structures, resulting in change of statistical feature in vibration signals. As a result, damage detection can be achieved by comparing the developed DSF under reference (healthy) and test states. In this study, AR model is adopted to model the acceleration time histories recorded from accelerometers. It is necessary to note that the AR model is limited to analyze linear stationary time series processes, which first and second moments are time invariant.

2.1. *Data standardization*

A stationary acceleration time history from reference state is shown in Figure 1. Supposing data x_i denotes the value of measured data with N samples at all l measurement points. In order to eliminate the effects caused by different loads, data standardization is necessary. In general, the data standardization process is performed by zero mean and unit variance as follows.

$$\hat{x}_i = \frac{x_i - \bar{x}}{\sigma_x}, \text{ where } \sigma_x^2 = \frac{1}{N-1}\sum_{i=1}^{N}\left(x_i - \bar{x}\right)^2 \tag{1}$$

Where $\bar{x}$, $\hat{x}_i$ and σ_x are mean, standardized value and standard deviation of x_i, respectively. However, to a certain state of structure, data variances from different measurement points are related to their positions. The standardization of unit variance ensures that the features extracted from normal and damage structures can be compared but making the variance from different measurement points are the same to be unit and thus, losing the information of damage location. As the bolted joint damage detection is achieved by locating the damage region in this study, it is necessary to maintain the information of damage location and ensure that the same channel features extracted from normal and damage structures can be compared. A new data standardization is defined as follows:

$$\hat{x}'_i = \frac{x_i - \bar{x}}{\sigma_f^2} \tag{2}$$

$$\bar{f} = \frac{1}{N}\sum_{i=1}^{N} f_i, \sigma_f^2 = \frac{1}{N-1}\sum_{i=1}^{N}\left(f_i - \bar{f}\right)^2 \tag{3}$$

Where $\hat{x}'_i$, $\bar{f}$ and σ_f^2 are new standardized value of x_i, mean and variance of force f_i, respectively. It can be found that using variance of force time histories as normalized parameter, the features extracted from different structural states can be compared, and then the feature relationship between two different measurement points can be maintained as well.

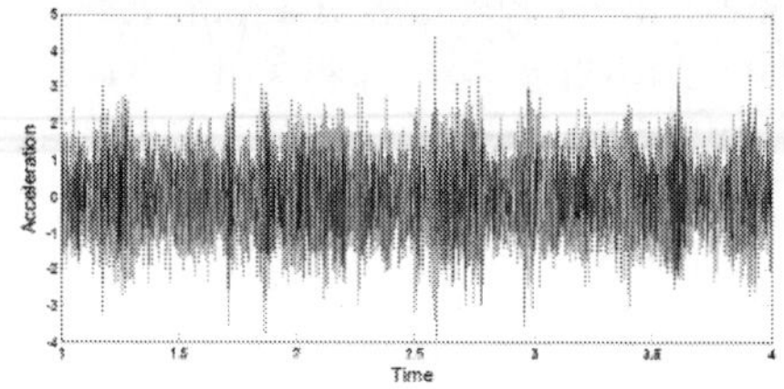

Figure 1. Acceleration time history of structures in reference state

2.2. *Time series analysis based on AR model*

The basic idea of time series analysis methods is that the development of things always keeping its inertia, which means that the future values in time series are dependent to its predecessors. In mathematical language, there are certain correlation relations in time series data and such correlation can be describe statistically. AR model is one of time series models that attempt to describe this regular pattern of time series by using the linear combination of its predecessors. The basic steps of AR model based time series analysis method includes model significance test, coefficients estimation and order determination.

As AR model is used to analysis stationary time series whose auto-correlation function (ACF) decays should gradually, but the partial auto-correlation function (PACF) would cut off after a few lags [14], a model significance test can be done by observing the changing trends of both sample ACF and PACF. Both sample ACF and PACF results from reference state are shown in Figure 2. It can be found that the behaviors of measured data are consistent with the behaviors of AR model process. Thus, AR model is adopted in this study. The coefficients $\phi_1, \phi_2, \cdots \phi_{t-p}$ are estimating by the Yule-Walker equations [14]. An AR model of order p at *j-th* measured acceleration time histories, or AR (p) , can be written as:

$$x_t = \phi_1 x_{t-1} + \phi_2 x_{t-2} + \cdots \phi_p x_{t-p} + \varepsilon_t \tag{4}$$

Where $x_t, x_{t-1}, \cdots x_p$ denote the current and previous values in time series and ε_t is the residual error.

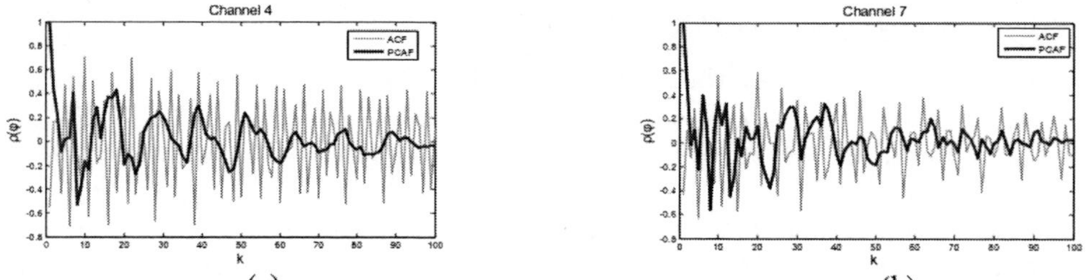

Figure 2. (a). Sample ACF and PACF at channel 4, (b). Sample ACF and PACF at channel 7

2.3. *Order determination*

The order of AR model is an unknown parameter. The high-order model has more variables making it can change more flexibly. As a result, a high-order model has better fitting effects to data than low-order model. However, the huge amount number of variables means there are more unknown factors in model resulting in the estimation precision decline and increasing the complexity of model. In contrast, a low-order model may not effectively describe the regular pattern of data. The determination of optimum order must consider fitting degree and variable quantity together. There are several criterion used in order determination, such as Akaike's information criterion (AIC), Bayesian information criterion (BIC), final prediction error (FPE) and partial auto-correlation function (PACF). Finally, the BIC is selected to assess the optimum order. The BIC is a weighted function of fitting precision and variable quantity as follow:

$$BIC = n\ln\sigma_\varepsilon^2 + (\ln n)p \tag{5}$$

Where n = number of time series data; σ_ε^2 = variance of residual error series; p = number of adjustable variables. The first term indicates the fitting precision. If the model is too simple, the variance of residual errors increases. The second term is a penalty factor which increase as the variable quantity grows. The optimum model is obtained when the responding BIC function reaches minimum and the model order is determined by the optimum model.

2.4. *Traditional damage-sensitive feature*

For structural reference (health) state, the measured acceleration time histories can be modeled by AR model, the corresponding AR coefficients Φ_k^{ref} and residual errors series ε^{ref} can be obtained as well. Similarly, for the structural test state, the acceleration time histories at the same structural position can be measured. And then, the residual error series ε^{test} can be obtained by fitting the AR model data in reference state to measured data in test state. As we know, the constructed AR model can effectively predict the data of structural health state. When the data comes from structural damage state, its regular pattern is different with the reference data. As a result, its residual error series will be different with the reference one as well. Therefore, the damage detection is performed through comparing the standard deviation (STD) of residual error series measured from same channel [13]. This method does not need to compare the features extracted from different channels and therefore the data standardization is performed by zero mean and unit variance in Equation (1). The traditional damage-sensitive index is defined as follows:

$$\varepsilon_t^{ref} = x_t^{ref} - \phi_1^{ref} x_{t-1}^{ref} - \phi_2^{ref} x_{t-2}^{ref} - \cdots - \phi_p^{ref} x_{t-p}^{ref} \tag{6}$$

$$\varepsilon_t^{test} = x_t^{test} - \phi_1^{ref} x_{t-1}^{test} - \phi_2^{ref} x_{t-2}^{test} - \cdots - \phi_p^{ref} x_{t-p}^{test} \tag{7}$$

$$\gamma^{std} = \sigma\left(\varepsilon^{test}\right) / \sigma\left(\varepsilon^{ref}\right) \tag{8}$$

The damage detection is judged by the value of STD ratio in Equation (9). When the test samples come from the structural health state, the STD ratio γ^{std} is approximately equal to one, when the test sample come from the structural damage state, γ^{std} will larger than one.

2.5. *New damage-sensitive features*

The new DSF is designed for SDD and bolted joint damage detection is achieved by locating the damage location. The STD of residual error series is still being used to describe the structural dynamic characteristics. The new DSF is constructed through residual errors relationship between measured data from two different measurement points installed on a structure, where the area within two consecutive measured points can be regarded as a region. Therefore, the whole structure is divided into several regions. The new DSF is developed through comparing regular patterns of STDs from two consecutive measured points.

Assuming that there are totally l measured points on a structure and P_k and P_{k+l} are two consecutive measured points, the data standardization is performed by newly defined data standardization process in Equations (2) and (3), and the new DSF is developed as follows:

$$\varepsilon_t^{\prime ref} = x_t^{\prime ref} - \phi_1^{ref} x_{t-1}^{\prime ref} - \phi_2^{ref} x_{t-2}^{\prime ref} - \cdots - \phi_p^{ref} x_{t-p}^{\prime ref} \tag{9}$$

$$\varepsilon_t^{\prime test} = x_t^{\prime test} - \phi_1^{test} x_{t-1}^{\prime test} - \phi_2^{test} x_{t-2}^{\prime test} - \cdots - \phi_p^{test} x_{t-p}^{\prime test} \tag{10}$$

$$DSF(k) = \sum_{n=k}^{k+1} \left| \frac{\sigma_n\left(\varepsilon^{\prime test}\right)}{\sigma_k\left(\varepsilon^{\prime test}\right) + \sigma_{k+1}\left(\varepsilon^{\prime test}\right)} - \frac{\sigma_n\left(\varepsilon^{\prime ref}\right)}{\sigma_k\left(\varepsilon^{\prime ref}\right) + \sigma_{k+1}\left(\varepsilon^{\prime ref}\right)} \right| \quad (k = 1, 2 \cdots l-1) \tag{11}$$

DSF(k) is used for detecting the region k between points k to k+1. When the damage occurs in a certain region, especially for bolted joints damage, the connection strength will be reduced, which result in the changes in load transfer characteristics or connection characteristics in the region. As a result, the relationship of standard deviation between two consecutive measured points may change as well. In other regions, the connection strength maintains as structural health state and the DSF changes a little. It is reasonable to consider that the new DSF is a sensitive to damage location as the value of DSF(k) is mostly determined by damage in region k and slightly affected by damage in other regions.

If the test sample comes from the structural health state, the connection strength in region k remains unchanged and there are no any damage occurs in other regions. Therefore, the DSF(k) should close to zero. When the test samples come from structural damage state but there is no location in region k, the DSF(k) should be a small value. When the damages are located in region k, the DSF(k) would be significant nonzero. If the identified damage location regions include the bolted joints, the new DSF can effectively detect structural damages caused by the bolted joints. Therefore, the DSF is used for detecting bolted joints damages by locating the damage region.

3. Experimental verification

In order to verify the validity of the proposed DSF for bolted joint damage detection, a steel cantilever beam, which is connected with a steel free-free beam via a bolted joint, is adopted in experiments, as shown in Figure 3. The spans of the test cantilever beam and the free-free beam are 800 mm and 400 mm respectively. The cross section is a rectangular section with 50 mm width and 10 mm height. The contact length is 50 mm.

As shown in Figure 4, a vibrator (HEV-200) together with a power amplifier and a force sensor (PCB, 208C02) are hanged on the beam at 0.7 m from the fixed end. A random excitation is generated by LMS Test.Lab and sent to the vibrator. Seven acceleration sensors (PCB, ICP 333B30) are mounted on points 1 to 7 and their positions are marked in Figure 4 by P_1 to P_7. The force and acceleration time histories from the reference state and damage state are recorded. The sampling frequency is 2048 Hz and the sampling duration is 4s for each data block and thus 8192 consecutive data samples are recorded. There are two structural states considered in experiments, i.e., the reference (healthy) and damage states, respectively. In structural reference state, 20 N·m torques are loaded by torque spanner. The damage state is simulated through loosening the bolted joints where the residual torque is reduced to 2.5 N·m. Three sets of acceleration data are measured, two of them are from structural health states, in which the first set is used as reference data while the second one as test sample named as Sample 1. The third set of data comes from structural damage state named as Sample 2. All the measured data are recorded by LMS Test.Lab and stored into the personal computer.

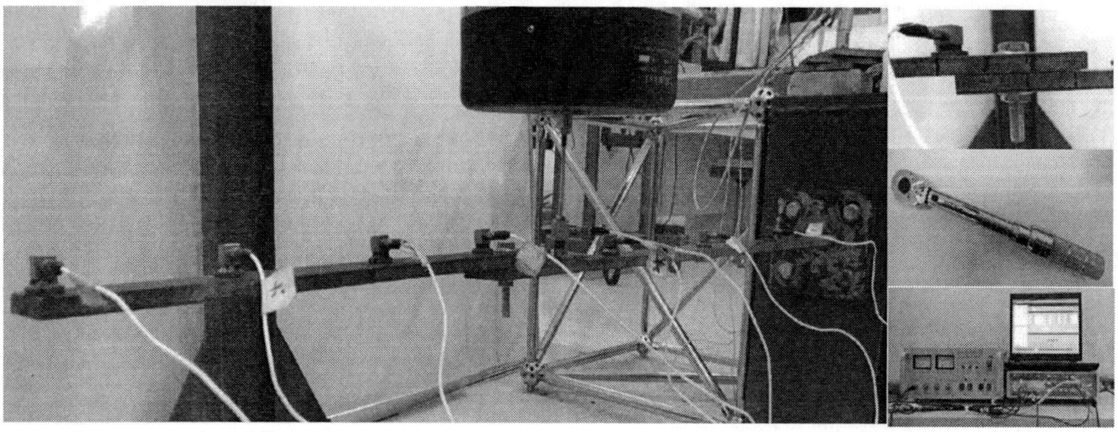

Figure 3. Experiment setup

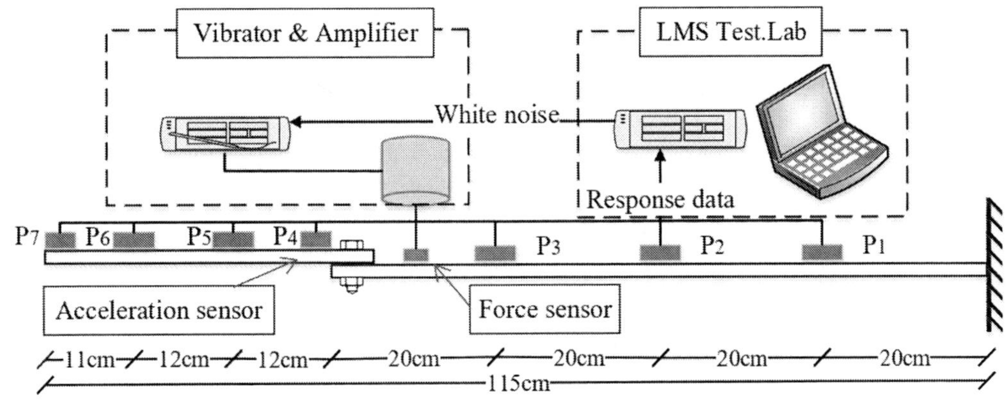

Figure 4. Simple diagram of experiment

3.1. *Order determination for AR model*

The effects of AR model orders on BIC of all measured channels in structural health state and damage state are show in Figure 5. The low order model cannot describe the regular pattern of measured data, so the BIC value is large and monotonically decrease with the increase of order at beginning. When the BIC curves reach minimum, they begins to increase gradually. It is because that the fitting precision changes a little with the increase of model order but the model becomes more and more complex. As a result, when the BIC curve reach minimum, the corresponding model is the optimum model and the corresponding order is obtained as shown in Table 1.

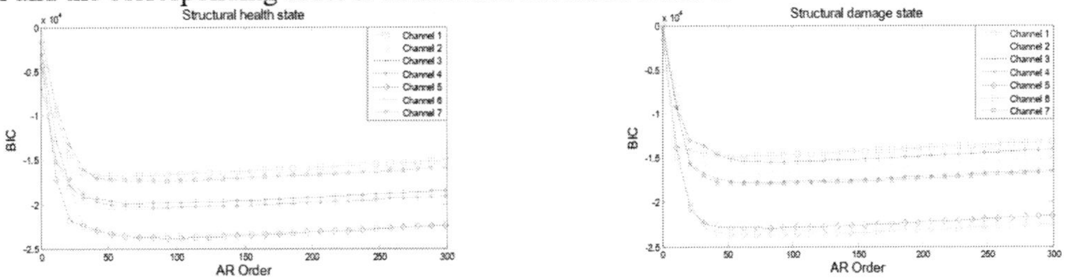

(a) (b)

Figure 5. Determination of AR model order

Table 1. AR model orders of structure

Structural states	Measuring point number and corresponding AR model order						
	1	2	3	4	5	6	7
Reference state (Torque=20 N·m)	46	50	56	88	90	46	66
Sample 1(Torque=20 N·m)	46	50	70	94	54	46	50
Sample 2(Torque=2.5 N·m)	63	55	63	85	69	67	69

3.2. *Bolted joint damage detection*

In order to perform the proposed method and assess its validity, the traditional DSF and newly defined DSF are used to detect bolted joint damage and the results will be compared and discussed.

IOP Conf. Series: Journal of Physics: Conf. Series **842** (2017) 012083 doi:10.1088/1742-6596/842/1/012083

To calculate the traditional DSF γ^{std} , the data standardization is performed using Equation (1) firstly. After obtaining the AR orders, the residual errors of test samples and reference state are calculated by reference state's AR coefficients as shown in Equations (6) and (7). The values of traditional DSF γ^{std} are calculated by Equation (8) and the results are shown in Figure 6 and Table 2. It can be found when the test sample comes from structural health state, the γ^{std} values at all measurement points are closer to one. Despite of the small errors that maybe caused by measured noises or environmental influences, the regular pattern of measured data from Sample 1 is the same as structural reference state. As a result, it can be seen that Sample 1 comes from structural health state, which is in agreement with the real situation. When the test sample comes from structural damage state, γ^{std} values at all measurement points obviously away from one, which indicates that the Sample 2 comes from structural damage state. As a result, the traditional DSF γ^{std} is successful in SDD. However, it is unable to perform damage localization by combining the γ^{std} values at different measurement points. It is not sure that the changes in structural dynamic characteristics are caused by bolted joint damage or cantilever beam damage or free-free beam damage.

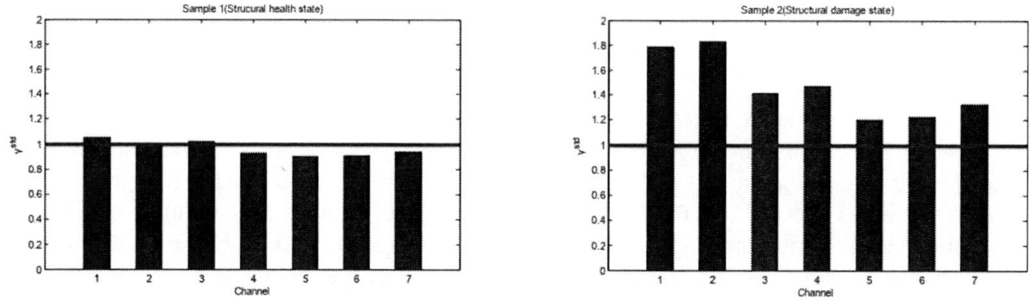

(a) (b)

Figure 6. Traditional DSF values at seven measurement points under health and damage states

Table 2. Values of traditional DSF γ^{std}

Structural states	Measuring point number and corresponding DSF values						
	1	2	3	4	5	6	7
Sample 1(Torque=20 N·m)	1.050	0.9995	1.0220	0.9370	0.9060	0.9180	0.9397
Sample 2(Torque=2.5 N·m)	1.7832	1.8266	1.4170	1.4694	1.2028	1.2268	1.3271

For the newly defined DSF, the whole structure is divided into six regions by seven measurement points. First, the data is standardized as equation (2) and equation (3). The AR coefficients of three data sets are calculated and the AR orders are shown in table 1. The residual errors of test samples and reference state are calculated by its corresponding AR coefficients as shown in equation (9) and equation (10) and then the new DSF is calculated by equation (11). DSF(k) is used for *k*-th region damage detection. It should be mentioned that the bolted joint locates in the fourth region. The values for new DSF from two samples are shown in Figure 7 and table 3. For Sample 1, it can be found that DSF values in all regions are closer to zero, which indicates that the relationship is similar between Sample 1 and reference state in each region. As a result, state 1 is identified to be structural health state.

For Sample 2, the DSF(3) is obviously away from zero and the others are small values compared with DSF(3). The DSF values in regions 1, 2, 3, 5 and 6 are pretty small. According to the definition of new DSF, it can be seen that the damage occurs in structure and it is located in region (3), which is corresponding to the actual damage situation. Therefore, bolted joint damage detection is achieved by successfully locating the damage region in the proposed method.

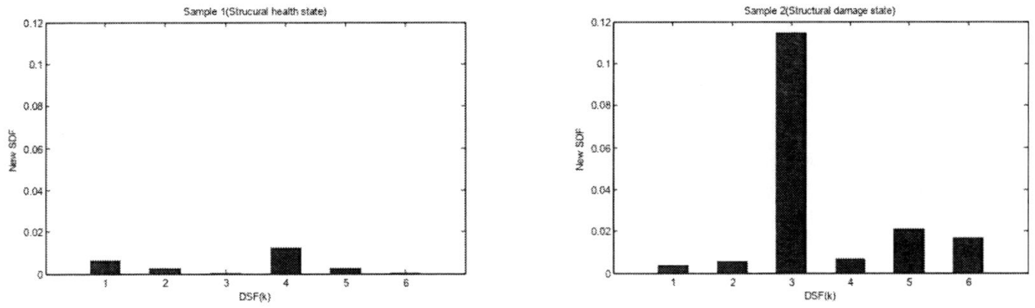

(a) (b)

Figure 7. New DFS values for six region under structural health and damage states

Table 3. Values of new DSF

Structural states	Region number and corresponding DSF values					
	1	2	3	4	5	6
Sample 1(Torque=20 N·m)	0.0063	0.0023	0.0001	0.0123	0.0025	0.0001
Sample 2(Torque=2.5 N·m)	0.0034	0.0055	0.1146	0.0066	0.0208	0.0169

3.3. Discussion and analysis

Comparing the results due to traditional DSF with ones due to new DSF, it can be found that both of them can achieve damage detection but only new SDF can achieve damage localization and ensures that the proposed method can be used in bolted joint damage detection.

Firstly, it is owed to the data standardization. When the structure with seven measurement points is excited under the same excitation force, a set of data can be measured at seven points. Their different regular patterns are related to themselves positions, their variance will also be different as well. So when data standardization of unit variance adopted, the different normalized parameters have affected the relationships of different measurement points and thus lost the location information. Secondly, when the residual error series ε^{test} are obtained by fitting the reference state AR model to the measured data in health state, the residual error series ε^{test} can describe actual regular pattern of measured data. When ε^{test} were obtained by fitting the reference state AR model to the measured data in damage state, the residual error series ε^{test} cannot describe actual regular pattern of measured data.

For the traditional DSF, the information of damage location is lost in data standardization. Even if the data at all measurement points were standardized by the same normalized parameter, the relationship data at different measurement points is influenced by measured position and inappropriate AR model together. Therefore, damage localization cannot be achieved by the traditional DSF.

For the new DSF, the corresponding AR models are constructed for each sample and the ε^{test} are obtained by fitting the corresponding AR model to measured data. Therefore, ε^{test} can describe the actual test states of structures. In addition, the relationship data at different measurement points is determined by its position via using the variance of force time histories to standardize the data. The experiment also verified that if there are any damage occurred in the region *k*, the DSF(*k*) would be obviously away from zero. It is noticed that when there are no damage occurs in region *k*, (*k*=2, 3, 5, 6), DSF(*k*) from Sample 2 are larger than those from Sample 1. It is because the damage of bolted joints has affected the structural vibration characteristics. The whole structure is depart from the reference state and the DSF(*k*) would be depart from the reference state as well, so the regular pattern at each measurement point would be changed, but the connection strength in undamaged region maintain the same. Fortunately, the new DSF has been verified to be sensitive to damage location. Therefore, the bolted joint damage detection can be achieved by the proposed method.

IOP Conf. Series: Journal of Physics: Conf. Series **842** (2017) 012083 doi:10.1088/1742-6596/842/1/012083

4. Conclusions

In this study, a new SDD method is proposed for bolted joint damage detection based on time series analysis. The measured data is standardized by variance of force time histories and modeled by AR model. The standard deviation of residual errors is used to construct the new structural DSF. The bolted joint damage detection is achieved by identifying the damage location. A series of experiments on bolted joint structure is designed to verify the proposed method. The illustrated results show that: (1) The proposed method can effectively identify the bolted joint damage. (2) Data standardization by variance of force can maintain the information of damage location, the relationship between measured data at different points is sensitive to damage location. (3) The newly defined DSF is sensitive to local damage, it can be used to identify structural damage location as well. (4) The methodology should be verified with more measured data from different types of structures.

Acknowledgments

The project is jointly supported by the National Natural Science Foundation of China with Grant Numbers 51678278 and 51278226.

References

[1] Yu L, Zhu J H and Yu L L 2013 Structural damage detection in a truss bridge model using fuzzy clustering and measured FRF data reduced by principal component projection *Adv. Struct. Eng.* **16**(1) 207-217

[2] Zhou Y L and Wahab M A 2016 Rapid early damage detection using transmissibility with distance measure analysis under unknown excitation in long-term health monitoring *J. Vibroeng.* **18**(7) 4491-99

[3] Zhou Y L, Maia N M M, Sampaio R and Wahab M A 2016 Structural damage detection using transmissibility together with hierarchical clustering analysis and similarity measure *Strut. Health monit.* doi: 10.1177/1475921716680849

[4] Khatir A, Tehami M, Khatir S and Wahab M A 2016 Multiple damage detection and localization in beam-like and complex structures using co-ordinate modal assurance criterion combined with firefly and genetic algorithms *J. Vibroeng.* **18**(8) 5063-73

[5] Khatir S, Belaidi I, Serra R, Wahab M A and Khatir T 2016 Numerical study for single and multiple damage detection and localization in beam-like structures using BAT algorithm *J. Vibroeng.* **18**(1) 202-213

[6] Yao R and Pakzad S N 2012 Autoregressive statistical pattern recognition algorithms for damage detection in civil structures *Mech. Syst. Sig. Proc.* **31**(8) 355-368

[7] Chen L J and Yu L 2013 Structural nonlinear damage identification based on time series ARMA/GARCH model *Adv. Struct. Eng.* **16**(9) 1597-1609

[8] Chen L J, Yu L and Xiao T 2015 The structural nonlinear damage detection based on linear time series algorithm *Appl. Mech. Mater.* **(744-746)** 345-350

[9] Nair K K, Kiremidjian A S and Law K H 2006 Time series-based damage detection and localization algorithm with application to the ASCE benchmark structure *J. Sound Vib.* **291**(1-2) 349-368

[10] Carden E P and Brownjohn J M 2008 ARMA modeled time-series classification for structural health monitoring of civil infrastructure *Mech. Syst. Sig. Proc.* **22**(2) 295-314

[11] Yao R and Pakzad S N 2014 Multisensor aggregation algorithms for structural damage diagnosis based on a substructure concept *J. Eng. Mech.* **141**(6) 04014164

[12] Yu L and Lin J C 2017 Cloud computing-based time series analysis for structural damage detection *J. Eng. Mech.* 134(1), C4015002

[13] Yu L and Zhu J H 2015 Nonlinear damage detection using higher statistical moments of structural responses *Struct. Eng. Mech.* **54**(2) 221-237

[14] Box G E P, Jenkins G M and Reinsel G C 1976 Time series analysis: Forecasting and control, 3rd ed. *J. Oper. Res. Soc.* **22**(2) 199-201

Delamination detection in laminated composite using Virtual crack closure technique (VCCT) and modal flexibility based on dynamic analysis

S Khatir[1], A Behtani[2], S Tiachacht[2], A Bouazouni[2], M Abdel Wahab[3], Y-L Zhou[4]

[1] Department of Electrical Energy, Systems and Automation, Faculty of Engineering and Architecture, Ghent University, Belgium
[2] Laboratory of Mechanics, Structure and Energetics (LMSE), Mouloud Mammeri University of Tizi-Ouzou, B.P.N°17 RP, 15000, Algeria.
[3] Soete Laboratory, Faculty of Engineering and Architecture, Ghent University, Technologiepark Zwijnaarde 903, B-9052 Zwijnaarde, Belgium
[4] Department of Civil and Environmental Engineering, National University of Singapore, 2 Engineering Drive 2, Singapore 117576

E-mail: Magd.AbdelWahab@UGent.be

Abstract. The delamination problem is very important failure mechanism in certain types of composite structures. Detecting this type of damage using vibration data is currently a problem of interest to the structural health monitoring community. In this paper, we used finite element method with embedded interface for analysing damaged laminated composite structures. The flexibly modal method, in which analysis data is related to finite element modelling, is used to detect and localize delamination. Several numerical examples are presented in order to evaluate the accuracy this approach.

1. Introduction

The ability to monitor a composite structure and detect, locate damage at the earliest possible stage is pervasive throughout the civil, mechanical, and aerospace engineering communities. The mechanics of delamination in composite structures has been largely studied in the recent years by procedures that employ fracture mechanics parameters. A finite element model for damage identification in such structures is presented using finite element SI12 [1]. The proposition of virtual crack closure technique is possibly the most adopted fracture mechanic's theory to simulate the propagation of delamination structure composite [2]. In recent years, a growing interest has emerged in the application of cohesive zone models to study delamination damages in laminated composite structures [3-5]. The ultrasonic assessment of a damaged CFRP after impact proved that the FBG sensor was better than the PZT sensor to identify damage [6]. Different damage detection diagnosis approaches, vibration based damage detection (VBDD) techniques [7-11] have received increased attention recently. This is particularly due to their relative simplicity and the moderate cost of dynamic analysis measurements with different structures, such as beam , trusses, and simple structures in reinforced concrete and steel [12]. A method for damage detection and localization in beam like and complex structure using co-ordinate modal assurance criterion combined with firefly and genetic algorithms was proposed in [13]. In [14], the authors presented an new approach of inverse damage detection and localization in composite beam structure graphite-epoxy based on model reduction. This problem was formulated as an inverse problem where an optimization algorithm was used to minimize the cost function expressed as the normalized difference between a frequency vector of the tested structure and its numerical model. A new damage detection and localization technique based on the changes in vibration parameters using BAT and Particle Swarm Optimization algorithm for detecting and locating damage in composite structure was presented in [15]. The authors in [16], used residual modal force vectors for damage localization because the residual force method is effective in damage localization using modal data. Furthermore, the authors in [17] demonstrated that the cyclic behaviour was related to the phase velocity of the converted wave modes in the two subsystems after delamination. The damage indicator based on vibration data was used to detect and locate defects in stratified beam structures in [18].

In [19], experimental free vibration of CFRP cantilever beam elements were investigated by dynamic tests. The damage introduced to the CFRP specimens were due to double notches close to the fixed end with different reduction of section and bending stiffness. The experimental and theoretical analysis of damaged and undamaged Carbon Fibre Reinforced Polymer (CFRP) laminate elements under free vibration of CFRP laminate specimens were experimentally investigated using a mechanical apparatus capable of simulating hinge conditions at the edges of simply supported laminate beams [20]. The analysis of vibration of undamaged and damaged Carbon Fibre Reinforced Polymer (CFRP) is carried out in this work using beam finite elements. Finite element method is a well established numerical technique, is widely used in academia and industries and have wide range of applications [21-29]. Recently, FEA was with Computer Added Design (CAD) and extended to Isogeometric analysis (IGA), which was originally introduced by Hughes et al. [30, 31]. IGA makes use of non-uniform rational B-spline (NURBS) basis functions to describe the exact geometry of a structure and approximate its field variables and has many applications including composite structures [32-43].

In the present paper, we used virtual crack closure technique (VCCT) and modal flexibility based on dynamic analysis in clamped-free and simply supported beam composite laminate to detect delamination between layers using embedded interface.

2. Damage detection method using flexibility matrix

In this paper the structural damage detection does not cause mass variation, but only a reduction in the structural stiffness. The flexibility matrix is defined as the inverse of the stiffness matrix. The reduced stiffness caused by damage results in increased flexibility and is written:

$$F = K^{-1} \tag{1}$$

Damage is detected by comparing the matrix of the flexibility of the damaged structure with undamaged structure. The sensitivity matrices F_s and F_e of undamaged and damaged structures are written, respectively, as:

$$F_s = \sum_{i=1}^{n} \frac{1}{\omega_{si}^2} \phi_{si} . \phi_{si}^T \tag{2}$$

$$F_e = \sum_{i=1}^{n} \frac{1}{\omega_{ei}^2} \phi_{ei} . \phi_{ei}^T \tag{3}$$

where:

ω_{si} and ω_{ei} : Natural frequencies of undamaged and damaged structures.

ϕ_{si} and ϕ_{ei} : Eigenvectors of undamaged and damaged structure.

n : mode number measured.

Having the eigenvectors of the undamaged and damaged structure with the introduction of some damage, the variation of the flexibility ΔF caused by the damage is:

$$\Delta F = Fs - Fe = [\delta_{ij}] \tag{4}$$

where:

δ_{ij} : The elements of the matrix of variation of flexibility ΔF .

In the following, the quantity of $\overline{\delta}_j$ represents the maximum of the absolute values of the elements j^{th} column of the variation in flexibility ΔF :

$$\overline{\delta}_j = \max\left|\Delta F_{ij}\right| \tag{5}$$

Then $\overline{\delta}_j$ is taken as a measure of the variation of flexibility at each location. The column of the flexibility matrix corresponding to the greatest change is indicative of the element where the damage is found. The presentation of this method is shown in Figure 1.

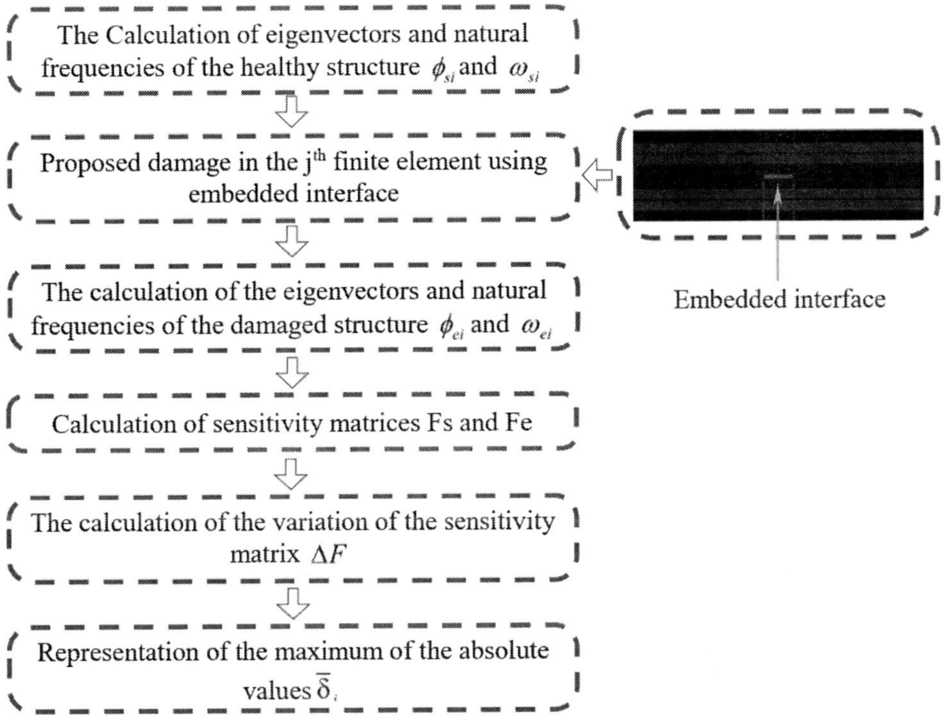

Figure 1. The Flowchart of the Flexibility Matrix Method

3. Numerical simulations

The suggested cracked element is very important for modelling the laminated composite beams structure. A triangle element with an interface crack in the two-dimension space and used it for analysis of cracked beam [44]. In [45, 46], the authors developed the formulation of a triangular finite element with an embedded interface, which can properly simulate the crack propagation phase. In our study, we used this technique in dynamic analysis of composite structure with 8 layers modelled using FEM software (PATRAN). The mechanical and geometric properties of the composite beam are given in Table 1.

IOP Conf. Series: Journal of Physics: Conf. Series **842** (2017) 012084 doi:10.1088/1742-6596/842/1/012084

Table 1. Mechanical and geometrical properties

Mechanical and geometrical properties	Values
Length L	0.5 m
Width B	0.03 m
Thickness E	0.016 m
Young's Modules	$E_1 = 144.7\,GPa$ and $E_2 = E_3 = 9.65\,GPa$
Shear modulus	$G_{12} = G_{13} = 4.14\,GPa$ and $G_{23} = 3.45\,GPa$
Poisson's coefficients	$\nu_{12} = 0.3$ and $\nu_{13} = \nu_{23} = 0.011$
Density	$\rho = 138.23\,Kg/m^3$

4. Results and discussion
In order to determine the dynamic properties of our structure, we performed some tests on three types of structures, namely a) undamaged, b) damaged with a single delamination and c) damaged with two delamination with two different types of boundary conditions, clamped-free and simply-supported, as shown in Figure 2.

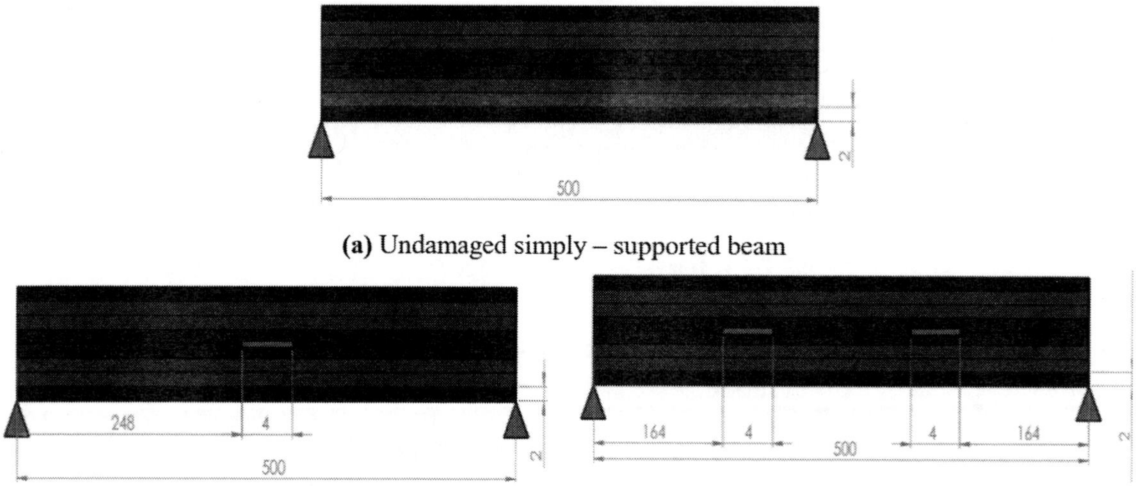

(a) Undamaged simply – supported beam

(b) Simply – supported beam with 1 damage **(c)** Simply – supported beam with 2 damage

Figure 2. Different structures studied with position of delamination

4.1. Simply – supported beam laminate
In this case, we consider the structure delamination using the same data and changing only the boundary conditions, i.e. simply-supported beam. The natural frequencies of simply-supported beam structure of undamaged and damaged with one and two delamination are listed in Table 2.

Table 2. The natural frequencies of simply-supported beam structure undamaged, damaged.

Frequencies	undamaged	Damaged with one delamination	Damaged with two delamination
Mode 1	175.927	175.202	174.439
Mode 2	688.848	677.088	679.459
Mode 3	1498.769	1493.571	1403.227
Mode 4	2551.43	2420.205	2455.505
Mode 5	3789.749	3768.298	3645.006
Mode 6	5161.819	4788.613	4500.532

The mode shapes with one and two delamination of simply supported beam are presented in Figure 3.

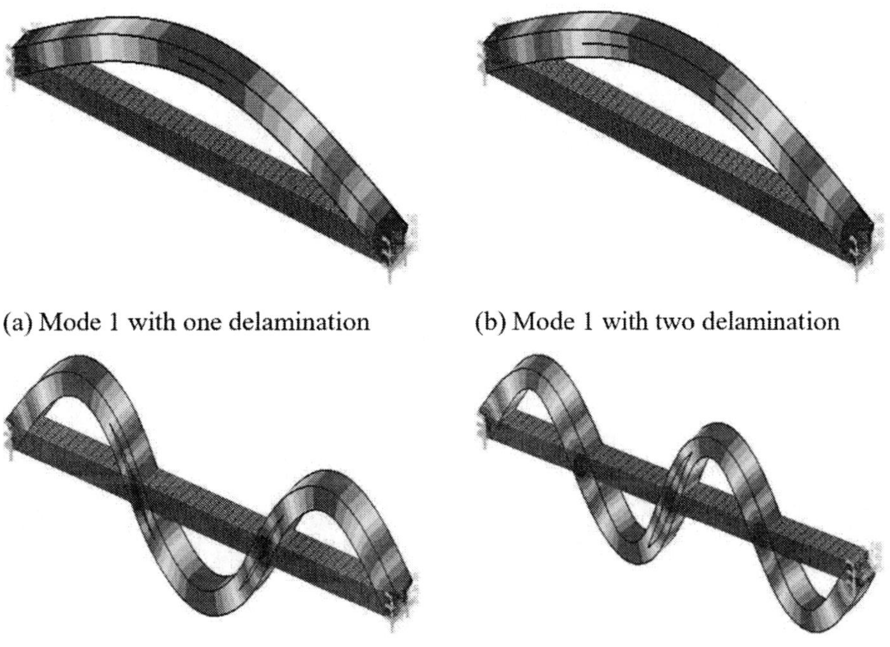

(a) Mode 1 with one delamination (b) Mode 1 with two delamination

(c) Mode 3 with two delamination (d) Mode 4 with one delamination

Figure 3. Presentations of modes with one and two delamination of simply-supported beam

In the first section, we were able to model our structure with the software PATRAN, while introducing the mechanical, physical characteristics of the beam as well as the boundary conditions. All the results obtained with eigenvectors will be used to add data in flexibility modal method in second section. In all tests, the delamination is located between layer 4 and 5 with same demensions.

4.2. Scenario 1: Simply – supported beam with one delamination

For this case, details of scenario 1 is shown in Figure 4 and the results found by the variation of flexibility are given in Figure 5. Figures 5.a to 5.c show that the variation of the flexibility is important on all nodes of the surfaces. In figure 5.d, we note that the variation in flexibility is greater in the zone where the defect is located. Furthermore, in figure 5.e, we note that the variation in flexibility is important in the zone where the defect is located is zero in the rest of the surfaces. The geometrical symmetry is reproduced due to the symmetry of the results with respect to the neutral axis of the beam.

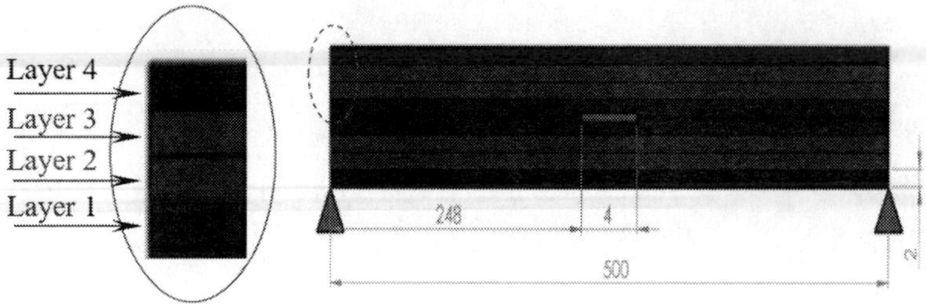

Figure 4. Presentation of Simply – supported beam with one delamination

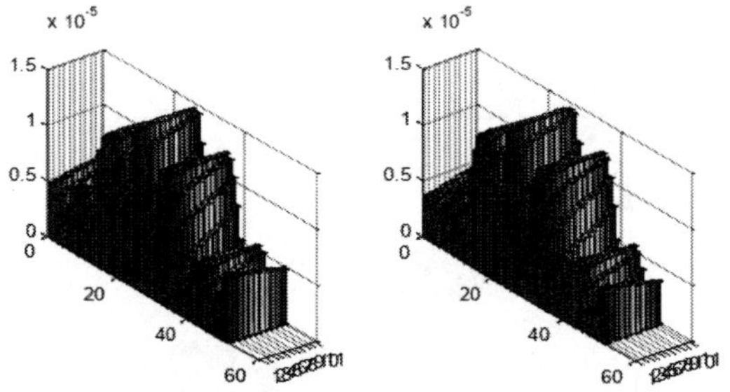

(a) The variation of the flexibility $\overline{\delta}_j$, 'lower layer 1 and Higher layer 8'.

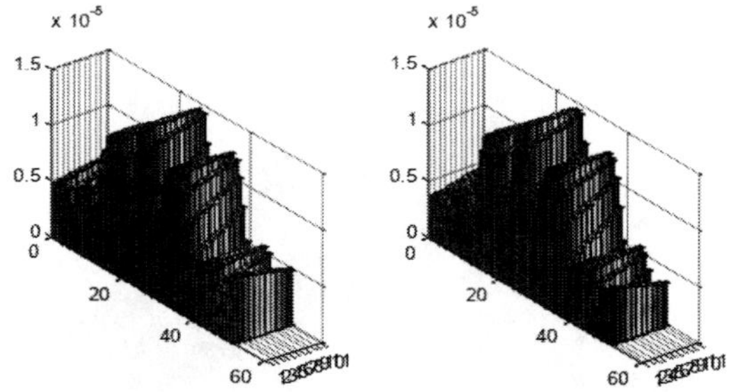

(b) The variation of the flexibility $\overline{\delta}_j$, 'Higher layer 1 and lower layer 8'.

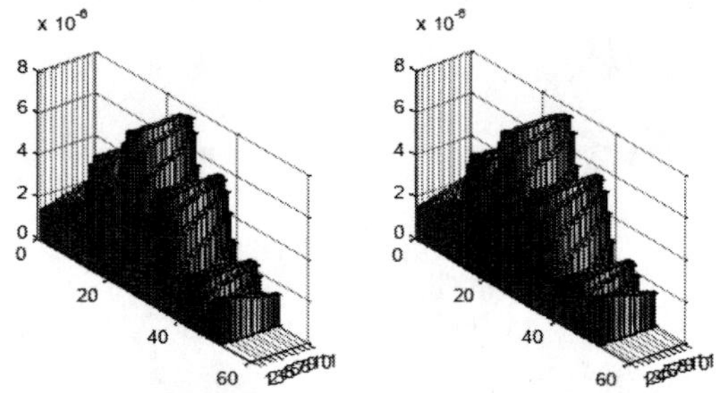

(c) The variation of the flexibility $\overline{\delta}_j$, 'lower layer 2 and Higher layer 7'.

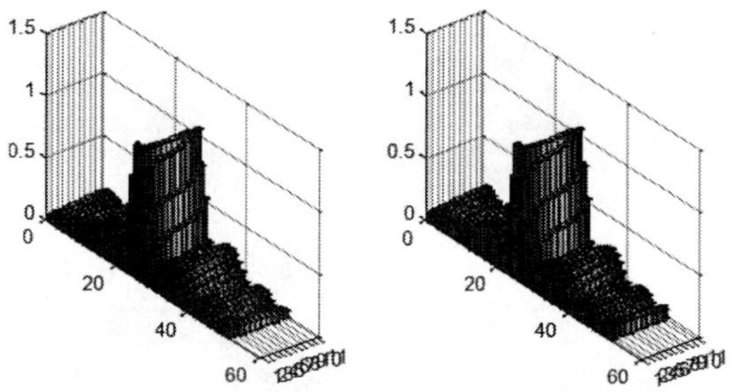

(d) The variation of the flexibility $\overline{\delta}_j$, 'lower layer 3 and Higher layer 6'.

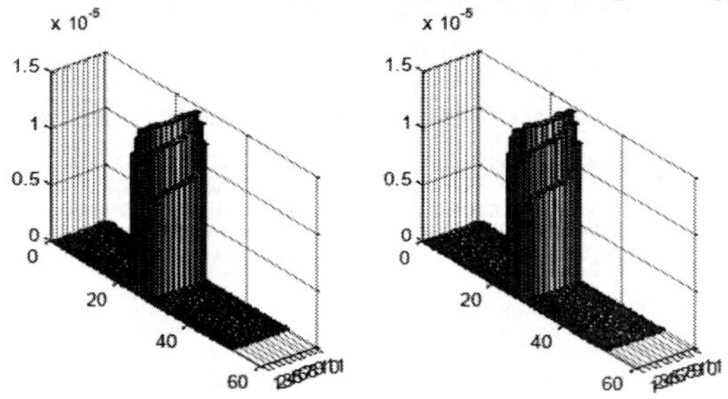

(e) The variation of the flexibility $\overline{\delta}_j$, 'lower layer 4 and Higher layer 5'.

Figure 5. The variation of flexibility of Simply-supported beam with one delamination

4.3. Scenario 2: Simply – supported beam with two delamination figure 6.

For this case, details of scenario 2 is shown in Figure 6 and the results found by the variation of flexibility are given in Figure 7. The presence of delamination reduces the rigidity of the composite structure studied. As the flexibility is the inverse of the rigidity, then the reduction of rigidity generates the increase in the flexibility of the structure, hence the change in flexibility can be used to detect and locate delamination. In this study, we found that the results obtained from this method can localize the delamination failure by the variation of the flexibility.

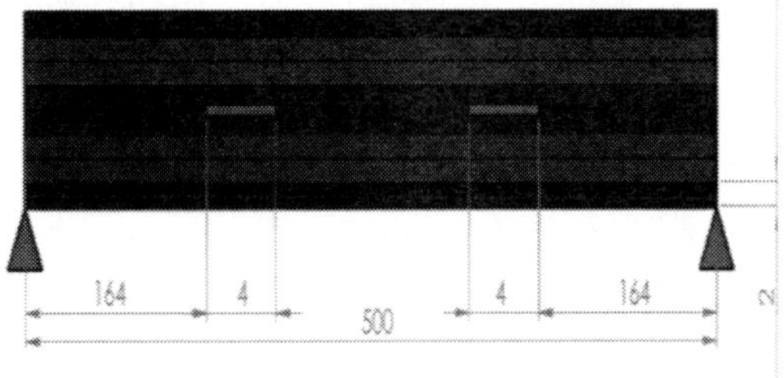

Figure 6. Simply – supported beam with two delamination

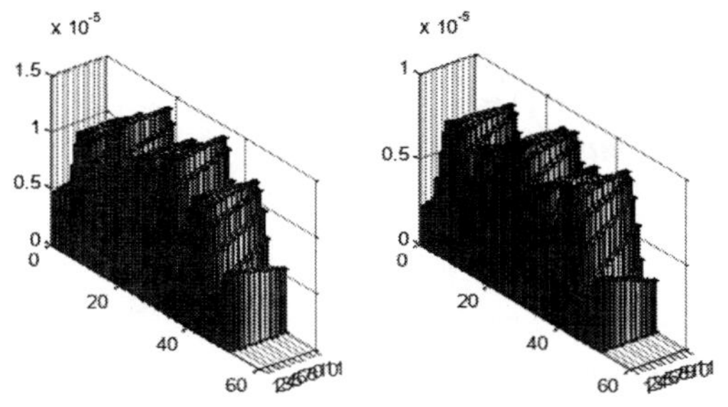

(a) The variation of the flexibility $\overline{\delta}_j$, 'lower layer 1 and 2'.

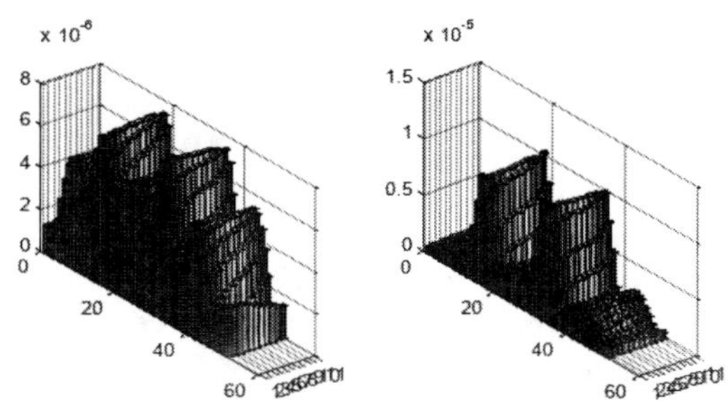

(b) The variation of the flexibility $\overline{\delta}_j$, 'lower layer 3 and 4'.

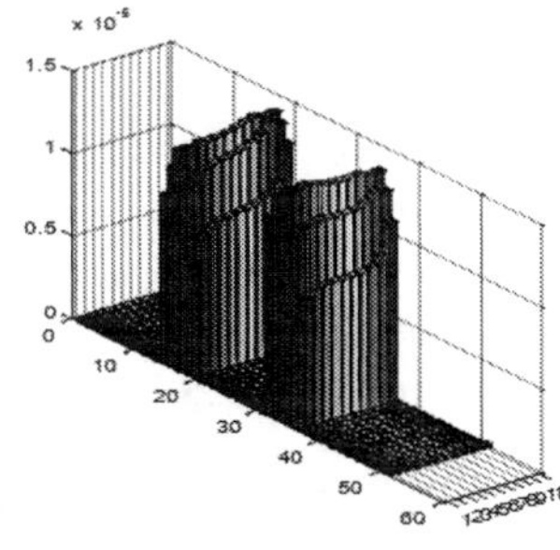

(c) The variation of the flexibility $\overline{\delta}_j$, 'lower layer 5'.

Figure 7. The variation of flexibility of Simply – supported with two delamination.

5. Conclusion

In this paper, we used a flexibly modal method analysis based on finite element modelling to localize and detect delamination in composite beam structures having 8 layers. For detecting this type of damage, we used virtual crack closure technique (VCCT). The Finite Element Method with embedded interface crack for analysing damaged laminated composite structures was proposed. The results found by flexibly modal method, show that this approach can detected the delamination with higher accuracy.

IOP Conf. Series: Journal of Physics: Conf. Series **842** (2017) 012084 doi:10.1088/1742-6596/842/1/012084

6. References

[1] Rikards R 1991 *Riga: Riga Technical University*
[2] Krueger R 2004 *Applied Mechanics Reviews* **57**(2) 109-143
[3] Tay T 2003 *Applied Mechanics Reviews* **56**(1) 1-32
[4] Mi Y, Crisfield M, Davies G and Hellweg H 1998 *Journal of composite materials* **32**(14) 1246-1272
[5] Camanho PP, Davila C and De Moura M 2003 *Journal of composite materials* **37**(16) 1415-1438
[6] Tsuda H 2006 *Composites science and technology* **66**(5) 676-683
[7] Zhou Y-L and Abdel Wahab M 2017 *Engineering Structures* **141** 175-183
[8] Zhou Y-L, Maia NMM, Sampaio R and Wahab MA 2016 *Structural health monitoring* **DOI: https://doi.org/10.1177/1475921716680849**
[9] Zhou Y-L, Maia N and Abdel Wahab M 2016 *Journal of Vibration and Control* doi: 10.1177/1077546316674544
[10] Zhou Y-L and Abdel Wahab M 2016 *Journal of Vibroengineering* **18**(7) 4491-4499
[11] Gillich G-R, Praisach Z-I, Abdel Wahab M, Gillich N, Mituletu IC and Nitescu C 2016 *Shock and Vibration* **2016**(Article ID 2086274) 10 pages; http://dx.doi.org/10.1155/2016/2086274
[12] Materazzi AL and Ubertini F 2011 *Journal of sound and vibration* **330**(26) 6420-6434
[13] Khatir A, Tehami M, Khatir S and Abdel Wahab M 2016 23rd International Congress on Sound and Vibration (ICSV23) ed editors) International Institute of Acoustics and Vibration
[14] Khatir S, Belaidi I, Serra R, Wahab MA and Khatir T 2016 *Mechanics* **21**(6) 472-479
[15] Khatir S, Belaidi I, Serra R, Wahab MA and Khatir T 2016 *Journal of Vibroengineering* **18**(1)
[16] Ricles J and Kosmatka J 1992 *AIAA journal* **30**(9) 2310-2316
[17] Gao H, Ali S and Lopez B 2010 *NDT & E International* **43**(4) 316-322
[18] BEHTANI A and BOUAZZOUNI A 2011 *20ème Congrès Français de Mécanique, 28 août/2 sept 2011-25044 Besançon, France (FR)*
[19] Capozucca R 2014 *Composite Structures* **116** 211-222
[20] Capozucca R and Bonci B 2015 *Composite Structures* **122** 367-375
[21] Yue T and Abdel Wahab M 2017 *Tribology International* **107** 274-282
[22] Pereira K, Yue T and Abdel Wahab M 2017 *Tribology International* **110** 222-231
[23] Martínez JC, Vanegas Useche LV and Wahab MA 2017 *International Journal of Fatigue* **100, Part 1** 32-49
[24] Kumar D, Biswas R, Poh LH and Abdel Wahab M 2017 *Tribology International* **109** 124-132
[25] Noda N-A, Chen X, Sano Y, Wahab MA, Maruyama H, Fujisawa R and Takase Y 2016 *MATERIALS & DESIGN* **96** 476-489
[26] Gadala I, Abdel Wahab M and Alfantazi A 2016 *MATERIALS & DESIGN* **97** 287-299
[27] Wang C, Sun Q, Abdel Wahab M, Zhang X and Xu L 2015 *Waste Management* **43** 19-27
[28] Vanegas-Useche L, Abdel Wahab M and Parker G 2015 *Waste Management* **43** 28-36
[29] Junyan Ni and Wahab MA 2017 *Computers & Structures* **186** 35–49
[30] Cottrell JA, Reali A, Bazilevs Y and Hughes TJ 2006 *Computer Methods in Applied Mechanics and Engineering* **195**(41) 5257-5296
[31] Hughes TJ, Cottrell JA and Bazilevs Y 2005 *Computer Methods in Applied Mechanics and Engineering* **194**(39) 4135-4195
[32] X. Nguyen H, N. Nguyen T, Abdel Wahab M, Bordas SPA, Nguyen-Xuan H and P. Voa T 2017 *Computer Methods in Applied Mechanics and Engineering* **313** 904-940
[33] Phung-Van P, Qui LX, Nguyen-Xuan H and Wahab MA 2017 *Composite Structures* **166** 120–135
[34] Phung-Van P, Ferreira AJM, Nguyen-Xuan H and Abdel Wahab M 2017 *Composites Part B: Engineering* **118** 125-134
[35] Tran LV, Phung-Van P, Lee J, Wahab MA and Nguyen-Xuan H 2016 *Composite Structures* **140** 655-667
[36] Thai CH, Ferreira AJM, Abdel Wahab M and Nguyen-Xuan H 2016 *Acta Mechanica* **227**(5) 1225-1250

[37] Thai C, Zenkour AM, Abdel Wahab M and Nguyen-Xuan H 2016 *Composite Structures* **139** 77-95

[38] Phung-Van P, Tran LV, Ferreira AJM, Nguyen-Xuan H and Abdel-Wahab M 2016 *Nonlinear Dynamics* 1-16; doi:10.1007/s11071-11016-13085-11076

[39] Tran Vinh L, Lee J, Nguyen-Van H, Nguyen-Xuan H and Abdel Wahab M 2015 *International Journal of Non-Linear Mechanics* **72** 42-52

[40] Tran Vinh L, Lee J, Ly HA, Abdel Wahab M and Nguyen-Xuan H 2015 *International Journal of Mechanical Sciences* **96-97** 65-78

[41] Phung Van P, Nguyen LB, Tran Vinh L, Dinh TD, Thai CH, Bordas SPA, Abdel Wahab M and Nguyen-Xuan H 2015 *International Journal of Non-Linear Mechanics* **76** 190-202

[42] Phung Van P, De Lorenzis L, Thai CH, Abdel Wahab M and Nguyen-Xuan H 2015 *Computational Materials Science* **96** 495-505

[43] Phung Van P, Abdel Wahab M, Liew KM, Bordas SPA and Nguyen-Xuan H 2015 *Composite Structures* **123** 137-149

[44] Löblein J and Schröder J 2005 *Computational materials science* **32**(3) 435-445

[45] Bolzon G and Corigliano A 2000 *International Journal for Numerical Methods in Engineering* **49**(10) 1227-1266

[46] Bolzon G 2001 *Computational Mechanics* **27**(6) 463-473

IOP Conf. Series: Journal of Physics: Conf. Series **842** (2017) 012085 doi:10.1088/1742-6596/842/1/012085

An isogeometric approach for size-dependent buckling analysis of FGM nanoplates

P Phung-Van[1], M Abdel-Wahab[1], H Nguyen-Xuan[2]

1 Department of Electrical Energy, Metals, Mechanical Constructions and Systems, Faculty of Engineering and Architecture, Ghent University, Belgium
2 Center for Interdisciplinary Research in Technology, Hutech University, Ho Chi Minh City, Vietnam

Abstract. In this paper, a suitable and simple computational formulation based on Isogeometric Analysis (IGA) integrated with higher-order shear deformation theory (HSDT) is introduced for size-dependent buckling analysis of functionally graded material (FGM) nanoplates. The material properties of FGM based on the Mori–Tanaka schemes and the rule of mixture are used. The differential nonlocal equations are utilized to take into account size effects. The nonlocal governing equations are approximated according to IGA based on HSDT, which satisfies naturally the higher-order derivatives continuity requirement in weak form of FGM nanoplates. The effect of nonlocal approach on the behaviors of the FGM nanoplates with several volume fraction exponents is investigated to show the reliability of the proposed method.

1. Introduction

There has recently been a fast growth in applications of nanoscale structures, which are primarily concerned with fabrication of FGMs. FGM nano-plates, as specific nanostructures, have been applied in the engineering and technology sectors. Therefore, the understanding of the behaviours of the FGM nano-plates is essential for the development of nanostructures because of their huge potential applications in the real life. In continuum mechanics, one of the well-known theories, which includes small scale effects with a good agreement and accuracy with molecular dynamics (MD) simulations, is the nonlocal continuum theory of Eringen [1]. Finite Element method (FEM) is a very well know numerical techniques and has been used for a wide range of applications [2-15]. Using FEM, Demir and Civalek [16] investigated small scale effects on vibration response of microtubules. The free vibration and buckling analyses of FG nano-plate using Navier solution subjected to thermal load were reported in Ref. [17]. Effect the beam thickness reduction on bending behaviour of micro-beams was observed in Ref. [18]. The mechanical behaviour for homogenous nonlinear micro beams [19], FG nonlinear micro beams [20]. A nonlinear micro beams model based on the strain gradient elasticity is introduced in Refs. [21, 22]. Size-dependent free analysis of FGM square plate based on FSDT was reported by Natarajan et al. [23]. Size-dependent analysis of FG nano-plates using IGA based on quasi-3D theory is recently examined in Ref [24]. Recently, Phung-Van et al. studied static and free vibration analyses of FG-CNTRC nano-plates [25] and nonlinear transient analysis of FGM nano-plates [26]. Furthermore, mechanical behaviours of FGM composite plates based on the local continuum theory have recently been published in Refs. [27-43]. This paper hence aims to develop a size-dependent buckling analysis of FGM nano-plates by a combination of IGA and the nonlocal continuum theory based on HSDT. In particular, we show that IGA based on HSDT fulfilling C2-continuity requirements can easily achieve the higher-order derivatives in the framework of the nonlocal continuum theory, which is of interest in this study. Size effects based on Eringen [1] in the differential nonlocal equations are performed. The effect of nonlocal approach on the behaviors of the FGM nanoplates with various volume fraction exponents are discussed in details.

2. Functional graded materials

A functionally graded material nanoplate made of ceramic and metal is considered in this research. The properties materials based on the rule of mixture can be given as:

$$P(z) = (P_c - P_m)V_c + P_m \;\; ; \;\; V_c = \left(\frac{1}{2} + \frac{z}{t}\right)^n \;\; (n \ge 0) \;\; ; \;\; V_m = 1 - V_c \tag{1}$$

where m and c represent the metal and ceramic constituents, respectively, P refers to the effective material properties including the thermal conductivity k, ν Poisson's ratio, ρ density, E Young's modulus and α thermal expansion. V_m and V_c are the volume fraction of the metal and ceramic, respectively, z is the thickness coordinate of plate and varies from $-t/2$ to $t/2$ and n is the volume fraction exponent.

3. Theoretical formulation

3.1 Nonlocal elasticity theory

Eringen [1] exhibited an equivalent form of differential equations of nonlocal stress at any points **x** as follows:

$$\left(1 - \mu\nabla^2\right)\sigma_{ij} = t_{ij} \tag{2}$$

where $\mu = e_0 l,\ 0 \le \mu \le 4$ is the small-scale effect; l is an internal characteristic length; e_0 is material constant and $\nabla^2 = \partial^2/\partial x^2 + \partial^2/\partial y^2$ is the Laplace operator.

3.2 Displacement field

The displacement field can be expressed as follows [44]:

$$\begin{aligned} u(x,y,z) &= u_0 + z\beta_x + cz^3\left(\beta_x + w_{,x}\right) \\ v(x,y,z) &= v_0 + z\beta_y + cz^3\left(\beta_y + w_{,y}\right), \quad \left(-h/2 \le z \le h/2\right) \\ w(x,y,z) &= w_0 \end{aligned} \tag{3}$$

where $c = 4/3h^2$.

The Green strain-displacement relations are now given as

$$\boldsymbol{\varepsilon} = [\varepsilon_{xx}\ \varepsilon_{yy}\ \gamma_{xy}]^T = \boldsymbol{\varepsilon}_m + z\boldsymbol{\kappa}_1 + z^3\boldsymbol{\kappa}_2 \;\; ; \;\; \boldsymbol{\gamma} = \left[\gamma_{xz}\ \gamma_{yz}\right]^T = \boldsymbol{\varepsilon}_s + z^2\boldsymbol{\kappa}_s \tag{4}$$

where

$$\boldsymbol{\varepsilon}_m = \begin{bmatrix} u_{0,x} \\ v_{0,y} \\ u_{0,y} + v_{0,x} \end{bmatrix};\ \boldsymbol{\kappa}_1 = \begin{bmatrix} \beta_{x,x} \\ \beta_{y,y} \\ \beta_{x,y} + \beta_{y,x} \end{bmatrix};\ \boldsymbol{\kappa}_2 = c\begin{bmatrix} \beta_{x,x} + w_{0,xx} \\ \beta_{y,y} + w_{0,yy} \\ \beta_{x,y} + \beta_{y,x} + 2w_{0,xy} \end{bmatrix};\ \boldsymbol{\varepsilon}_s = \begin{bmatrix} \beta_x + w_{0,x} \\ \beta_y + w_{0,y} \end{bmatrix};\ \boldsymbol{\kappa}_s = 3c\begin{bmatrix} \beta_x + w_{0,x} \\ \beta_y + w_{0,y} \end{bmatrix} \tag{5}$$

3.3 Isogeometric analysis

The displacement field of the plate can be expressed as:

$$\mathbf{u}^h(\xi,\eta) = \sum_{I=1}^{m\times n} R_I(\xi,\eta)\mathbf{d}_I \tag{6}$$

where $\mathbf{d}_I = \left[u_{0I}\ v_{0I}\ \beta_{xI}\ \beta_{yI}\ w_I\right]^T$ is the vector of degrees of freedom associated with the control point I, and R_I is the shape function as defined in Ref [45].

The in-plane and shear strains can be expressed as:

$$\left[\boldsymbol{\varepsilon}_0\ \boldsymbol{\kappa}_1\ \boldsymbol{\kappa}_2\ \boldsymbol{\kappa}_s\right]^T = \sum_{I=1}^{m\times n}\left[\mathbf{B}_I^m\ \mathbf{B}_I^{b1}\ \mathbf{B}_I^{b2}\ \mathbf{B}_I^s\right]^T \mathbf{d}_I = \sum_{I=1}^{m\times n}\left[\mathbf{B}_I^{mb}\ \mathbf{B}_I^s\right]^T \mathbf{d}_I \tag{7}$$

where

$$\mathbf{B}_I^m = \begin{bmatrix} R_{I,x} & 0 & 0 & 0 & 0 \\ 0 & R_{I,y} & 0 & 0 & 0 \\ R_{I,y} & R_{I,x} & 0 & 0 & 0 \end{bmatrix}, \quad \mathbf{B}_I^{b1} = -\begin{bmatrix} 0 & 0 & R_{I,xx} & 0 & 0 \\ 0 & 0 & R_{I,yy} & 0 & 0 \\ 0 & 0 & 2R_{I,xy} & 0 & 0 \end{bmatrix},$$
$$\mathbf{B}_I^{b2} = \begin{bmatrix} 0 & 0 & 0 & R_{I,x} & 0 \\ 0 & 0 & 0 & 0 & R_{I,y} \\ 0 & 0 & 0 & R_{I,y} & R_{I,x} \end{bmatrix}, \quad \mathbf{B}_I^s = \begin{bmatrix} 0 & 0 & 0 & R_I & 0 \\ 0 & 0 & 0 & 0 & R_I \end{bmatrix} \tag{8}$$

The governing algebraic equations for buckling analysis can be obtained as follows

$$\left(\mathbf{K} - \lambda \mathbf{K}_g\right)\mathbf{d} = \mathbf{0} \tag{9}$$

where

$$\mathbf{K} = \int_\Omega \left[\left(\mathbf{B}^{mb}\right)^T \mathbf{D}_{mb} \mathbf{B}_I^{mb} + \left(\mathbf{B}^s\right) \mathbf{D}_s \mathbf{B}^s \right] d\Omega$$
$$\mathbf{K}_g = \int_\Omega \left[\left(\mathbf{B}^g\right)^T - \lambda \nabla^2 \left(\mathbf{B}^g\right)^T \right] \mathbf{N}_0 \mathbf{B}^g d\Omega \tag{10}$$

in which

$$\mathbf{D}_{mb} = \begin{bmatrix} \mathbf{A} & \mathbf{B} & \mathbf{E} \\ \mathbf{B} & \mathbf{D} & \mathbf{F} \\ \mathbf{E} & \mathbf{F} & \mathbf{H} \end{bmatrix} \; ; \; \mathbf{B}_I^s = \begin{bmatrix} 0 & 0 & R_{I,x} & 0 & 0 \\ 0 & 0 & R_{I,y} & 0 & 0 \end{bmatrix}$$
$$A_{ij}, B_{ij}, D_{ij}, E_{ij}, F_{ij}, H_{ij} = \int_{-h/2}^{h/2} \left(1, z, z^2, f(z), zf(z), f^2(z)\right) C_{ij} \mathrm{d}z \;, \quad i,j = 1,2,6 \tag{11}$$
$$D_{sij} = \int_{-h/2}^{h/2} \left[f'(z)\right]^2 C_{ij} \mathrm{d}z \;, \quad i,j = 4,5$$

It can be seen that second and third order derivatives are included in stiffness and buckling matrixes, respectively. The order of the basic functions is at least three. Hence, it is clear that NURBS basic function is the most suitable for calculating size-dependent analysis of nanoplate structures.

4. Numerical results

In this study, the critical buckling loads $\bar{P}_{cr} = P_{cr} R^2 / D_m$ of Al/Al2O3 circular plate (radius R = 10, h = 0.34) are performed. Table 1 shows Non-dimensional buckling load of simply supported and clamped circle nanoplate. It can be observed that the present results match well with the reference solution [24].

Table 1. Material properties of FGM plates.

BCs	μ	Model	n				
			0	1	2	5	10
SSSS	0	Quasi-3D [24]	23.1059	10.0048	8.5094	7.0907	6.1583
		HSDT-Present	22.7565	9.6263	8.1811	6.8955	6.0381
	1	Quasi-3D [24]	22.1473	9.5755	8.1448	6.7891	5.8991
		HSDT-Present	21.8371	9.2172	7.8328	6.6045	5.7871
CCCC	0	Quasi-3D [24]	82.3402	33.5902	28.5574	24.4165	21.6023
		HSDT-Present	79.3204	31.5156	26.7567	23.1807	20.7108
	1	Quasi-3D [24]	71.7287	29.2705	24.8837	21.2721	18.8188
		HSDT-Present	69.1652	27.4808	23.3311	20.2129	18.0593

5. Conclusions

In this paper, a simple and effective approach using IGA based on HSDT for size-dependent buckling analysis of FGM nanoplates. The material properties of FGM based on the Mori–Tanaka schemes and the rule of mixture are considered. The differential nonlocal equations are utilized to take into account effect of the size-dependent. The buckling nonlocal governing equations of motion is approximated according to IGA based on HSDT, which satisfies naturally the 3rd derivatives of displacement field. The effects of volume fraction exponent and small scale parameter on nonlinear transint of FGM nanoplates are also performed. Numerical results proved high accuracy and reliability of the present method in comparison with other available numerical approaches.

Acknowledgements

P. Phung-Van would like to acknowledge the financial support of the Special Research Fund (BOF) of Ghent University (BOF16/PDO/88)

References

[1] Eringen AC 1983 *Journal of Applied Physics* **54**(9) 4703-4710
[2] Yue T and Abdel Wahab M 2017 *Tribology International* **107** 274-282
[3] Pereira K, Yue T and Abdel Wahab M 2017 *Tribology International* **110** 222-231
[4] Martínez JC, Vanegas Useche LV and Wahab MA 2017 *International Journal of Fatigue* **100, Part 1** 32-49
[5] Kumar D, Biswas R, Poh LH and Abdel Wahab M 2017 *Tribology International* **109** 124-132
[6] Bhatti NA and Abdel Wahab M 2017 *Tribology International* **109** 552-562
[7] Yue T and Abdel Wahab M 2016 *Materials* **9** 597; doi:510.3390/ma9070597
[8] Resende Pereira KdF, Bordas S, Tomar S, Trobec R, Depolli M, Kosec G and Abdel Wahab M 2016 *Materials* **9** 639; doi:610.3390/ma9080639
[9] Ferjaoui A, Yue T, Abdel Wahab M and Hojjati-Talemi R 2015 *International Journal of Fatigue* **73** 66-76
[10] Yue T and Abdel Wahab M 2014 *Wear* **321** 53-63
[11] Noda N-A, Chen X, Sano Y, Wahab MA, Maruyama H, Fujisawa R and Takase Y 2016 *MATERIALS & DESIGN* **96** 476-489
[12] Gadala I, Abdel Wahab M and Alfantazi A 2016 *MATERIALS & DESIGN* **97** 287-299
[13] Wang C, Sun Q, Abdel Wahab M, Zhang X and Xu L 2015 *Waste Management* **43** 19-27
[14] Vanegas-Useche L, Abdel Wahab M and Parker G 2015 *Waste Management* **43** 28-36
[15] Junyan Ni and Wahab MA 2017 *Computers & Structures* **186** 35–49
[16] Demir C and Civalek O 2013 *Applied Mathematical Modelling* **37**(22) 9355-9367
[17] Ansari R, Ashrafi M, Pourashraf T and Sahmani S 2015 *Acta Astronaut* **109** 42-51
[18] Lam DCC, Yang F, Chong ACM, Wang J and Tong P 2003 *Journal of the Mechanics and Physics of Solids* **51**(8) 1477-1508
[19] Asghari M, Kahrobaiyan MH and Ahmadian MT 2010 *International Journal of Engineering Science* **48**(12) 1749-1761
[20] Ke LL, Wang YS, Yang J and Kitipornchai S 2012 *International Journal of Engineering Science* **50**(1) 256-267
[21] Rajabi F and Ramezani S 2012 *Archive of Applied Mechanics* **82** 363-376
[22] Ghayesh MH, Amabili M and Farokhi H 2013 *International Journal of Engineering Science* **63** 52-60
[23] Natarajan S, Chakraboty S, Thangavel M, Bordas SPA and Rabczuk T 2012 *Computational Materials Science* **65** 74-80
[24] Nguyen NT, Hui D, Lee J and Nguyen-Xuan H 2015 *Computer Methods in Applied Mechanics and Engineering* **297** 191-218
[25] Phung-Van P, Lieu X Q, Nguyen-Xuan H and Abdel-Wahab M 2017 *Composite Structures* **166** 120-135
[26] Phung-Van P, Ferreira AJM, Nguyen-Xuan H and Abdel-Wahab M 2017 *Composites Part B Engineering* **118** 125-134
[27] Phung-Van P, Nguyen-Thoi T, Luong-Van H and Lieu-Xuan Q 2014 *Computer Methods in Applied Mechanics and Engineering* **270** 15-36
[28] Phung-Van P, Nguyen-Thoi T, Dang-Trung H and Nguyen-Minh N 2014 *Composite Structures* **111** 553-565

[29] Nguyen-Thoi T, Bui-Xuan T, Phung-Van P, Nguyen-Hoang S and Nguyen-Xuan H 2014 *KSCE Journal of Civil Engineering* **18**(4) 1072-1082
[30] Phung-Van P, Nguyen-Thoi T, Luong-Van H, Thai-Hoang C and Nguyen-Xuan H 2014 *Computer Methods in Applied Mechanics and Engineering* **272** 138-159
[31] Phung-Van P, Thai CH, Nguyen-Thoi T and Nguyen-Xuan H 2014 *Composites Part B-Engineering* **60** 227-238
[32] Luong-Van H, Nguyen-Thoi T, Liu GR and Phung-Van P 2014 *Engineering Analysis with Boundary Elements* **42** 8-19
[33] Nguyen-Thoi T, Luong-Van H, Phung-Van P, Rabczuk T and Tran-Trung D 2013 *International Journal for Composite Materials* **3(A)** 19-27
[34] Phung-Van P, Luong-Van H, Nguyen-Thoi T and Nguyen-Xuan H 2014 *International Journal for Numerical Methods in Engineering* **98**(13) 988-1014
[35] Phung-Van P, Nguyen-Thoi T, Le-Dinh T and Nguyen-Xuan H 2013 *Smart Materials and Structures* **22**(9)
[36] Nguyen-Thoi T, Rabczuk T, Lam-Phat T, Ho-Huu V and Phung-Van P 2014 *Theoretical and Applied Fracture Mechanics* **72** 150-163
[37] Phung-Van P, Nguyen-Thoi T, Tran LV and Nguyen-Xuan H 2013 *Computational Materials Science* **79** 857-872
[38] Phung-Van P, Tran LV, Ferreira AJM, Nguyen-Xuan H and Abdel-Wahab M 2017 *Nonlinear Dynamics* **87**(2) 879-894
[39] Phung-Van P, Nguyen LB, Tran LV, Dinh TD, Thai CH, Bordas SPA, Abdel-Wahab M and Nguyen-Xuan H 2015 *International Journal of Non-Linear Mechanics* **76** 190-202
[40] Phung-Van P, De Lorenzis L, Thai CH, Abdel-Wahab M and Nguyen-Xuan H 2015 *Computational Materials Science* **96** 495-505
[41] Nguyen-Thoi T, Luong-Van H, Phung-Van P, Rabczuk T and Tran-Trung D 2013 *International Journal of Composite Materials* **3**(A) 19-27
[42] Tran LV, Phung-Van P, Lee J, Wahab MA and Nguyen-Xuan H 2016 *Composite Structures* **140** 655-667
[43] Tran Vinh L, Lee J, Ly HA, Abdel Wahab M and Nguyen-Xuan H 2015 *International Journal of Mechanical Sciences* **96-97** 65-78
[44] Reddy JN 1984 *Journal of Applied Mechanics-Transactions of the Asme* **51**(4) 745-752
[45] Cottrell JA, Hughes TJR and Bazilevs Y 2009 Isogeometric analysis, towards integration of CAD and FEA ed editors. Wiley